Comprehensive Toxicology

Comprehensive Toxicology

Editors-in-Chief

I. Glenn Sipes
University of Arizona, Tucson, AZ, USA

Charlene A. McQueen
University of Arizona, Tucson, AZ, USA

A. Jay Gandolfi
University of Arizona, Tucson, AZ, USA

Volume 7

RENAL TOXICOLOGY

Volume Editor

Robin S. Goldstein
Sandoz Pharmaceuticals, East Hanover, NJ, USA

PERGAMON

UK	Elsevier Science Ltd., The Boulevard, Langford Lane, Kidlington, Oxford, OX5 1GB, UK
USA	Elsevier Science Inc., 655 Avenue of the Americas, New York, NY 10010, USA
JAPAN	Elsevier Science Japan, 9-15 Higashi-Azabu 1-chome, Minato-ku, Tokyo 106, Japan

First edition 1997

Library of Congress Cataloging in Publication Data
Comprehensive toxicology / editors-in-chief, I. Glenn Sipes, Charlene A. McQueen, A. Jay Gandolfi.
  p. cm.
  Includes index.
  1. Toxicology. I. Sipes, I. Glenn. II. McQueen, Charlene A., 1947– .
III. Gandolfi, A. Jay.
  [DNLM: 1. Poisons. 2. Poisoning. 3. Toxicity Tests. 4. Toxicology—methods. QV 600 C737 1997]
RA1199.C648 1997
615.9—dc20
DNLM/DLC
for Library of Congress   96-33052
                                         CIP

British Library Cataloguing in Publication Data
A catalogue record for this book is available from the British Library.

ISBN 0-08-042301-9 (set : alk. paper)
ISBN 0-08-042972-6 (Volume 7)

Typeset by Alden Bookset, Didcot, UK.
Printed and bound in Great Britain by Cambridge University Press, Cambridge, UK.

Contents

Preface

Toxicology is the study of the nature and actions of chemicals on biological systems. In more primitive times, it really was the study of poisons. Evidence of the use of venoms and poisonous herbs exists in both the Old World and the New World. Early recorded history documents the use of natural products as remedies for certain diseases, and as agents for suicides, execution, and murders. In those days, the "science of toxicology" certainly operated at the upper limit of the dose response curve. However, in the early 1500s, it was apparent to Paracelsus, that "the dose differentiates a poison and a remedy." Stated more directly, any chemical can be toxic if the dose is high enough. In addition, with respect to metals, he recognized that the effects they produce may differ depending upon the duration of exposure (i.e., acute vs. chronic). Clearly, the two most important tenets of toxicology were established during that time. The level of exposure (dose) and the duration of exposure (time) will determine the degree and nature of a toxicological response.

Since that time the discipline of toxicology has made major advances in identifying and characterizing toxicants. It is the scientific discipline that combines the elements of chemistry and biology. It not only addresses how chemicals affect biological systems, but how biological systems affect chemicals. Due to this broad definition, toxicology has always been an interdisciplinary science. It uses advances made in other disciplines to better describe a toxic event or to elucidate the mechanism of a toxic event. In addition, toxicology contributes to the advancement of other disciplines by providing chemicals with selective or specific mechanisms of action. These can be utilized to modify biochemical and physiological processes that advance our understanding of biology and the treatment of disease.

During the last quarter century, toxicology has made its greatest advances. Reasons for this include a demand by the public for protection of human and environmental health, an emphasis on understanding cellular and molecular mechanisms of toxicity, and use of such data in safety evaluation/risk assessment. Clearly, the explosion of information as a result of the new biology (molecular and cellular) provided the conceptual framework as well as tools to better elucidate mechanisms of toxicity and to understand interindividual differences in susceptibility to toxicants. Due to this new information, more sensitive end points of toxicity will be identified. This is critical because the real need in toxicology is to understand if/how exposure to trace levels of chemicals (i.e., low end of the dose–response curve) for long periods of time cause toxicity.

Thus, the time is appropriate for the development and publication of an extensive and authoritative work on toxicology. The concept of *Comprehensive Toxicology* was developed at Pergamon Press in 1989–1990. After market research documented a need for such a work, discussions took place to identify an Editor-in-Chief. The project was too large for one individual and the current team of co-Editors-in-Chief was identified. These individuals along with scientific and support staff from Pergamon Press (now an imprint of Elsevier Science, Ltd.), developed a tentative series of volumes to be edited by experts. These Volume Editors then solicited the Authors, who provided the chapters that appear in these several thousand pages. It is to the Authors that we are indebted, because without them, this project would still be a concept, not a product.

Although the title of this work is *Comprehensive Toxicology*, we certainly realize that it does not meet the literal definition of comprehensive. Due to space limitations, certain critical tissues are not included (skin, eye). Also, the time it takes to receive, edit, and print the various chapters limits inclusion of the latest information. Our goal was to provide a strong foundation. Similarly, our understanding of toxic mechanisms is much greater in certain tissues (liver, kidney, lung) than in others (cardiovascular, nervous system, hematopoietic). Thus, the scope of these chapters will differ. They reflect the state of the science and the specializations of the Editors/Authors. Not only will the reader find information on toxicology, but superb chapters on key physiological and biochemical processes in a variety of tissues. In all cases, a wealth of scientific material is presented that will be of interest to a wide audience.

As co-Editors-in-Chief we have appreciated the opportunity to be involved in this project. We are grateful to the very supportive staff at Elsevier and at the University of Arizona who have assisted us with the project. We trust that you will consider *Comprehensive Toxicology* a milestone in the literature of this very old, but very dynamic discipline.

I. Glenn Sipes
Tucson

Charlene A. McQueen
Tucson

A. Jay Gandolfi
Tucson

Contributors to Volume 7

Dr. M. Abbate
Laboratori Negri, Istituto de Ricerche Farmacologiche Mario Negri, Via Garrazzeni, 11, Bergamo
I-24125, Italy

Dr. P. H. Bach
Interdisciplinary Research Centre for Cell Modulation Studies, Faculty of Science and Health,
University of East London, Romford Road, London E15 4LZ, UK

Dr. K. F. Badr
Department of Medicine, Center for Glomerulonephritis, Emory University School of Medicine,
1840 Southern Lane, Decatur, GA 30033, USA

Dr. G. L. Bakris
Rush Presbyterian St. Luke's Medical Center, Department of Preventive and Internal Medicine,
Rush University, 1725 West Harrison Street, Suite 117, Chicago, IL 60612, USA

Dr. W. M. Bennet
Department of Medicine, Division of Nephrology, Hypertension and Clinical Pharmacology, PP Suite
262, Oregon Health Sciences University, 3314 South West US Veterans Hospital Road, Portland,
OR 97201-2940, USA

Dr. J. F. Bernardo
Center for Clinical Pharmacology, University of Pittsburgh, 623 Scaife Hall, Pittsburgh,
PA 15261, USA

Dr. J. V. Bonventre
Massachusetts General Hospital East, Harvard Medical School, 149–4002 13th Street, Charlestown,
MA 02129-2060, USA

Professor H. R. Brady
Mater Misericordiae Hospital, University College Dublin, 41 Eccles Street, Dublin 7, Ireland

Dr. R. A. Branch
Center for Clinical Pharmacology, University of Pittsburgh, 623 Scaife Hall, Pittsburgh,
PA 15261, USA

Dr. M. Brezis
Department of Medicine, Hadassah University Hospital, Mount Scopus, PO Box 24035, Jerusalem,
91240, Israel

Professor C. Cojocel
Hoechst Marion Roussel, Hoechst AG, Pharmaceutical Division (K-607), D-65926 Frankfurt am
Main, Germany

Dr. W. H. Dantzler
Department of Physiology, University of Arizona Health Sciences Center, Room 4130, Tucson,
AZ 85724-0001, USA

Mr. B. B. Davis
Department of Medicine and Biochemistry, St. Louis University and VA Medical Center,
GRECC (11G-JB), 1 Jefferson Barracks Drive, St. Louis, MO 63125-4199, USA

Dr. J. R. Diamond
Department of Medicine and Cellular and Molecular Physiology, Division of Nephrology, Milton
South Hershey Medical Center, Pennsylvania State College of Medicine, PO Box 850, Hershey,
PA 17003-0850, USA

Dr. D. Dinour
Institute of Nephrology, Shiba Hospital, Tel-Hashomez, Tel Aviv, Israel
Division of Nephrology, University of California, San Francisco, CA, USA

Dr. A. A. Elfarra
Department of Comparative Biosciences, School of Veterinary Medicine, University of
Wisconsin–Madison, 2015 Linden Drive, West Madison, WI 53706-1102, USA

Dr. S. L. Eustis
US SmithKline Beecham Pharmaceuticals, 709 Swedeland Road, Mail Code UE 0360, King of
Prussia, PA 19406, USA

Dr. A. Fogo
Department of Pathology, Vanderbilt University School of Medicine, MCN C-3310, Nashville,
TN 37232, USA

Dr. S. M. Ford
Department of Pharmaceutical Sciences, St. John's University, 8000 Utopia Parkway, Jamaica,
NY 11439, USA

Dr. B. Fowler
Program in Toxicology, University of Maryland Baltimore County, 5202 Westland Boulevard,
TEC II Building, Baltimore, MD 21227, USA

Dr. J. Gailit
Department of Dermatology, School of Medicine, Health Sciences Center, T-16, 060, State
University of New York at Stony Brook, Stony Brook, NY 11794-8165, USA

Dr. M. S. Goligorsky
Division of Nephrology and Hypertension, Department of Medicine, State University of New York
at Stony Brook, Stony Brook, NY 11794-8152, USA

Dr. G. C. Groggel
Department of Internal Medicine, University of Nebraska Medical Center, 600 South 42nd Street,
Box 983040, Omaha, NE 68198-3040, USA

Dr. L. B. Kinter
Nycomed Inc., 460 Devon Park Drive, PO Box 6630, Wayne, PA 19087-6630, USA

Dr. V. M. Lakshmi
Department of Medicine and Biochemistry, St. Louis University and VA Medical Center,
GRECC (11G-JB), 1 Jefferson Barracks Drive, St. Louis, MO 63125-4199, USA

Dr. L. H. Lash
Department of Pharmacology, Wayne State University School of Medicine, 540 East Canfield Avenue,
Detroit, MI 48201-1908, USA

Dr. S. S. Lau
Division of Pharmacology and Toxicology, School of Pharmacy, University of Texas at Austin,
Austin, TX 78712-1074, USA

Dr. L. D. Lehman-McKeeman
Miami Valley Laboratories, Proctor & Gamble, PO Box 538707, Cincinnati, OH 45253-8707, USA

Dr. M. E. I. Leibbrandt
Chiron Corporation, 4560 Horton Street, Emeryville, CA 94608-2916, USA

Dr. E. A. Lock
Central Toxicology Laboratories, Zeneca, Alderley Park, Macclesfield, Cheshire SK10 4TJ, UK

Dr. J. Mason
Sandoz Pharmaceuticals, 59 Route 10, East Hanover, NJ 07936-1080, USA

Dr. G. W. Miller
Howard Hughes Medical Institute, Department of Cell Biology, Duke University, PO Box 3709, Durham, NC 27710, USA

Dr. B. A. Molitoris
Department of Medicine, Indiana University Medical Center, 1120 South Drive, Indianapolis, IN 46202, USA

Professor T. J. Monks
Division of Pharmacology and Toxicology, School of Pharmacy, University of Texas at Austin, Austin, TX 78712-1074, USA

Dr. L. C. Moore
Department of Physiology and Biophysics, State University of New York at Stony Brook, Stony Brook, NY 11794-8661, USA

Dr. E. Noiri
Division of Nephrology and Hypertension, Department of Medicine, State University of New York at Stony Brook, Stony Brook, NY 11794-8152, USA

Dr. G. O. Rankin
Department of Pharmacology, Marshall University School of Medicine, 1542 Spring Valley Drive, Huntington, WV 25704-9388, USA

Dr. C. J. Reed
School of Biomolecular Sciences, Liverpool John Moores University, Byrom Street, Liverpool L3 3AF, UK

Dr. G. Remuzzi
Division of Nephrology and Dialysis, Ospedali Riuniti di Bergamo, Bergamo I-24128, Italy
Laboratori Negri, Istituto de Ricerche Farmacologiche Mario Negri, Via Garrazzeni, 11, Bergamo I-24125, Italy

Dr. S. D. Ricardo
Department of Medicine and Cellular and Molecular Physiology, Division of Nephrology, Milton South Hershey Medical Center, Pennsylvania State College of Medicine, PO Box 850, Hershey, PA 17003-0850, USA

Dr. V. Romanov
Membrane Biology Section, Laboratory of Mathematical Biology, NCI-Frederick Cancer Research and Development Center, Building 538, Room 124, Frederick, MD 21702-1201, USA

Dr. C. E. Ruegg
In Vitro Technologies, 1450 South Rolling Road, Baltimore, MD 21227, USA

Dr. R. L. Safirstein
Division of Nephrology, Department of Internal Medicine, University of Texas Medical Branch at Galveston, 301 University Boulevard, Room 4200 OJS, Galveston, TX 77555-0562, USA

Dr. J. M. Sands
Renal Division, Emory University School of Medicine, Department of Medicine, 1364 Clifton Road, North East, Atlanta, GA 30322, USA

Dr. R. G. Schnellmann
Division of Toxicology, University of Arkansas for Medical Sciences, 4301 Markham Street, Mail Slot 638, Little Rock, AR 72205-7199, USA

Dr. K. S. Squibb
Program in Toxicology, University of Maryland Baltimore County, 5202 Westland Boulevard, TEC II Building, Baltimore, MD 21227, USA

Dr. J. B. Tarloff
Department of Pharmacology and Toxicology, Philadelphia College of Pharmacy and Sciences, 600 South 43rd Street, Philadelphia, PA 19104-4495, USA

Dr. B. M. Tune
Laboratory of Renal Pharmacology, Division of Nephrology, Department of Pediatrics, Stanford University School of Medicine, Stanford, CA 94305, USA

Dr. J. W. Verlander
Renal Division, University of Florida School of Medicine, PO Box 100224, JHMHC, Health Science Center Room CG98, Gainesville, FL 32610, USA

Dr. R. Witzgall
Institute for Anatomy and Cell Biology, Universität Heidelberg, Im Neuenheimer Feld 307, D-69120 Heidelberg, Germany

Dr. G. H. I. Wolfgang
Chiron Corporation, 4560 Horton Street, Emeryville, CA 94608-2916, USA

Dr. S. H. Wright
Department of Physiology, University of Arizona Health Sciences Center, Room 4130, Tucson, AZ 85724-0001, USA

Dr. R. K. Zalups
Division of Basic Medical Sciences, Mercer University School of Medicine, 1550 College Street, Macon, GA 31207, USA

Dr. M. L. Zeidel
Laboratory of Epithelial Cell Biology, Renal-Electrolyte Division, University of Pittsburgh Medical Center, Room 937, Scaife Hall, 3550 Terrace Street, Pittsburgh, PA 15213, USA

Dr. T. V. Zenser
Department of Medicine and Biochemistry, St. Louis University and VA Medical Center, GRECC (11G-JB), 1 Jefferson Barracks Drive, St. Louis, MO 63125-4199, USA

Introduction

OBJECTIVES, SCOPE, AND COVERAGE

Comprehensive Toxicology is an in-depth, state-of-the-art review of toxicology with an emphasis on human systems. This series of volumes has been designed to encompass investigation from the molecular level to the intact organism. The goal is to provide a balanced presentation that integrates specific biological effects of pertinent toxicants across the various disciplines of toxicology. To accomplish this the individual chapters were written by leaders in their specific areas of toxicology or related disciplines.

These reference volumes begin with basic principles of toxicology (general aspects and biotransformation) and their utilization in toxicological testing. This is followed by a thorough systems approach to the key organs or tissues susceptible to toxic chemicals. Within the biological systems volumes the structure/function of the tissue, its response to toxic insult, approaches for evaluation of injury, and examples of specific toxicants are profiled. Particular attention will be paid to understanding the cellular and molecular mechanisms of toxicity. A portion of this treatise is dedicated to state-of-the-art *in vitro* approaches for examining the mechanisms and endpoints of toxicity. Finally, important advances in chemical carcinogenesis, including oncogenes, tumor suppressor genes, and anticarcinogens complete this major work.

In planning this series, the needs of a wide range of potential users were considered. Consequently, each area includes a thorough review as well as the latest scientific data and interpretation. Scientists and students in academic, industrial, and governmental settings will be major users of this series. University scientists will utilize this source as a basis for their research studies and as a reference text for teaching. Individuals in need of toxicological data in the chemical, pharmaceutical, agricultural, petroleum, biotechnical, mining, and semiconductor industries will find this series to be a valuable tool as will governmental scientists, regulators, and administrators. Other potential users include those with interests in environmental and medical law as well as scientists associated with research institutes and consulting firms.

The work consists of 13 volumes as described below.

Volume 1: General Principles (edited by J. Bond)

Volume 1 provides the reader with a basic overview of the field of toxicology. The volume is divided into four sections: (1) introduction, (2) toxicokinetics, (3) mechanisms of toxicity, and (4) risk assessment. There is an in-depth coverage of the field of toxicokinetics including exposure assessment, routes of exposure of toxicants, and distribution, biotransformation, and excretion of toxicants. Dosimetry modeling, including physiologically based pharmacokinetic models, and extrapolation modeling are covered as well as key mechanisms associated with the toxicity/carcinogenicity of chemicals. Concepts of risk assessment are also discussed in this volume. Volume 1 lays the groundwork for subsequent volumes of the *Comprehensive Toxicology* series where different aspects of the field of toxicology are extensively discussed.

Volume 2: Toxicological Testing and Evaluation (edited by P. D. Williams and G. H. Hottendorf)

Volume 2 provides the reader with a comprehensive overview of the routine and special toxicologic assessments performed on pharmaceutical agents, chemicals, pesticides, and consumer products. Descriptions of toxicity testing procedures emphasize experimental design considerations (dose and species selection, toxicokinetic criteria) as well as regulatory guidelines and impacts of global harmonization efforts. The risk assessment process is considered in terms of evaluating and interpreting toxicity data relative to human safety.

Volume 3: Biotransformation (edited by F. P. Guengerich)

Volume 3 deals with the enzymatic processes involved in the biotransformation of toxic and potentially toxic chemicals. The first section provides introductory material on general aspects of history, regulation, mechanism, inhibition, and stimulation of the enzymes in this group. The bulk of the volume is comprised of chapters dealing with the current status of enzymes involved in oxidation, reduction, hydrolysis, and conjugation of xenobiotic chemicals. The current status of enzyme multiplicity and nomenclature, gene and protein structure, regulation, catalytic mechanism, substrate specificity, human studies, and relevance to issues in toxicology are discussed.

Volume 4: Toxicology of the Hematopoietic System (edited by J. C. Bloom)

Volume 4, which is divided into five sections, reviews the current understanding of the hematopoietic system and toxic effects on its components. The work is arranged to provide a background on hematotoxicology, describe the components of blood and blood-forming organs, and review the complex process of hematopoiesis. Additional chapters describe toxic effects on erythrocytes, leukocytes, and hemostasis, respectively, and include discussions on putative agents and mechanisms, as well as the diagnosis and treatment of these disorders. Special attention is given to human hematopoietic stem cells, which represent an important target tissue that is amenable to *in vitro* testing. The volume concludes with risk assessment in both clinical and preclinical settings.

Volume 5: Toxicology of the Immune System (edited by D. A. Lawrence)

Volume 5 provides current knowledge regarding the intricacies of the immune system and the manner by which environmental agents can disrupt immune homeostasis and induce pathologies. The volume begins with an introduction to the field of immunotoxicology and an overview of the immune system, followed by a description of the architecture and cellular components of the immune system and how their development and reactivities can be modified. Immune functions specifically delineated include cell trafficking, processing of antigens, the manner by which cells communicate and are signaled into activation from plasma membrane events through to transcriptional events. The immunopathologies resultant from hypersensitivities are presented with emphasis on the mechanistic means by which chemicals can modify health. Animal and human examples of autoimmune diseases are included as well as the immune changes that occur with aging. Immunosuppression induced by cancers and drugs used for treatment of cancers and transplant patients are reviewed along with the immunotoxicities associated with biological response modifiers. The ability of stressors to modify immune functions help describe the importance of neuroimmunological investigations. The immunoregulatory properties of stress response proteins are also reviewed. The volume concludes with methods and applications utilized for risk assessment and the consequences of immunotoxicological analysis of humans.

Volume 6: Cardiovascular Toxicology (edited by S. P. Bishop and W. D. Kerns)

Volume 6 contains a comprehensive review of basic cardiovascular anatomy, physiology, and pharmacology as well as an analysis of the various classes of compounds affecting the heart and blood vessels. Cardiovascular biology includes embryologic development, physiology, and molecular pharmacology of the system. Methods for evaluation of the heart and blood vessels are reviewed, including both *in vivo* and *in vitro* methods of study. Separate sections deal with the response of the heart and blood vessels to injury, with emphasis on mechanisms of toxic injury. Specific compounds are classified according to their major mechanism of toxicity, and are reviewed with prime examples of those compounds for which the cellular and molecular mechanisms are best known. Emphasis is on toxicologic studies in experimental animals where mechanisms have been most thoroughly investigated, but the relationship of results from animal studies is made to toxicologic lesions in man. Finally, there is an assessment of the problems related to the use of toxicologic studies in animals for clinical application in man.

Volume 7: Renal Toxicology (edited by R. S. Goldstein)

Current concepts on mechanisms mediating chemically induced nephrotoxicity are rapidly evolving. Volume 7 is focused on capturing this rapidly growing field by providing a comprehensive, state-of-the-art review of renal pathophysiology and toxicology, written by internationally recognized experts in the field. The first section of this volume is designed to provide the reader with the required background information, including an overview of clinical nephrotoxicity, the anatomy and physiology of the kidney and urinary bladder, renal transport mechanisms, xenobiotic metabolism, and *in vivo* and *in vitro* methods used to assess renal toxicity. Current knowledge on the pathophysiology and biochemistry of acute renal failure, renal and urinary bladder carcinogenesis, and immune-mediated renal injury are covered in detail. The role of vasoactive substances, cell adhesion molecules, oxidative stress/antioxidants and membrane changes in composition/fluidity in mediating nephrotoxic and ischemic renal injury are reviewed and provide the reader with the conceptual framework for understanding key mechanisms of renal injury. In addition, the roles of gene expression and growth factors in renal injury and repair are discussed. The response of each of the major segments of the nephron to a toxic insult is discussed in detail, with chapters devoted to mechanisms mediating injury to the glomerulus, proximal tubule, collecting duct and papilla, and the tubulointerstitium. Various nephrotoxicants are discussed in detail with an emphasis on pathophysiologic and morphologic effects, and mechanisms of toxicity. Wherever possible, the relevance of experimental findings to human exposure is covered.

Volume 8: Toxicology of the Respiratory System (edited by R. A. Roth)

Volume 8 begins with a detailed description of the functional anatomy of the various regions and critical cells of the respiratory tract, from the nasal cavity to the gas exchange region. This section conveys the structural heterogeneity of the respiratory system and how it determines tissue and cellular targets for toxic agents. The next section describes functional and biochemical responses of the lungs to toxic insult and includes chapters on carcinogenic and developmental responses to chemical exposure. The final section comprises over 20 chapters describing injurious effects of selected chemicals of toxicologic interest. Human health concerns as well as mechanisms of toxicity are emphasized. Unlike other volumes that have treated this subject, attention is devoted to effects of toxic agents on both the airways and the pulmonary vasculature.

Volume 9: Hepatic and Gastrointestinal Toxicology (edited by R. S. McCuskey and D. L. Earnest)

Volume 9, which is divided into two sections, presents important new information concerning mechanisms of effect and consequences of exposure of the liver and gastrointestinal tract to a wide variety of toxicants. Sections on the liver and gastrointestinal tract provide the reader with concise introductory descriptions of relevant cellular and organ anatomy, physiology, and mechanisms of toxic injury. Specific consideration is given to methods of toxicant exposure and to processes that provide defense against injury. The discussions provide a broad focus but also contain details about molecular, cellular, and biochemical mechanisms. The section on hepatic toxicology is a comprehensive review of the toxic effects of a wide variety of specific compounds including volatile hydrocarbons, anesthetic agents, drugs of therapeutic value as well as abuse, heavy metals, natural compounds, and so on. The section on gastrointestinal tract toxicology, presenting toxicology of the esophagus, stomach, small intestine, and colon, represents the first major reference work available with a broad overview of the effect of toxicants on the mucosal, motility, and immune functions of the gastrointestinal tract as well as information about specific toxicants. The final chapter presents an overview of clinical toxicology in animals with a focus on the liver and gastrointestinal systems. This volume will be of particular interest and use to those concerned with environmental and industrial toxicology, as well as with mechanisms of toxicant injury and clinical medicine.

Volume 10: Reproductive and Endocrine Toxicology (edited by K. Boekelheide, R. Chapin, P. Hoyer, and C. Harris)

Section 1, on male reproductive toxicology, starts with a basic overview of the anatomy and physiology of the system. This is followed by reviews on the way molecules are evaluated for male reproductive toxicity in both the pharmaceutical and chemical industries, along with a strategy for evaluating the reproductive function of transgenic animals. The bulk of this section is devoted to reviewing mechanisms and manifestations of toxicants to particular targets within the male reproduction system (Sertoli cell, Leydig cell, etc.). Finally, there is an evaluation of areas of interest not often considered in the toxicologic context (fluid flow, the immune system, paracrine factors, etc.), as well as a chapter highlighting technical advances and what these mean for the field. Overall, Section 1 reviews the state of the science in a number of areas, explicitly providing investigational strategies, and identifies promising areas of future research.

Section 2 is a comprehensive overview of the field of female reproductive toxicology. Much interest has recently been focused on this area of toxicology, due to the increasing number of women in the workplace and the impact of female fertility on reproduction issues. This section begins with an overview of female reproductive physiology, with an emphasis on the complexities of its hormonal regulation. Next, the various components of the female reproductive system are described in detail, along with known and potential sites of disruption by xenobiotics. Finally, assessment of human risk is discussed from the standpoint of classical methods of evaluation, as well as recently developed, novel experimental approaches.

Section 3 provides an overview of mammalian development, from fertilization to parturition and early postnatal maturation, in terms of the stage-selective anatomical and functional characteristics of each developmental phase that may underpin the ultimate manifestations of toxicity. Several possible mechanisms of developmental toxicity are discussed in terms of the roles of biotransformation, pharmacokinetics, altered gene expression, neurobehavioral development, physiological conditions, and nutritional status. A selected list of chemical agents and environmental extremes known to produce persistent developmental abnormalities is reviewed by class and are discussed in terms of their known toxic effects and possible mechanisms of action. Strategies for the study of developmental toxicants *in vivo* and *in vitro* are described, as well as current screening and testing systems for the detection of additional potential disrupters of normal development. This section concludes with a summary and discussion of the critical need for additional understanding and the future prospects for accurate prediction, prevention, and assessment of risk in developmental toxicity.

Section 4 provides a review of the effect of toxicants on endocrine tissues. This is an emerging area of concern in toxicology. The adrenals, thyroid, parathyroid, and pancreas will be covered in this section.

Volume 11: Nervous System and Behavioral Toxicology (edited by H. E. Lowndes and K. Reuhl)

Neurotoxicology is among the fastest growing of the toxicological disciplines. Recent refinements of neuropathologic techniques, meticulous validation of behavioral tests, and introduction of the tools of molecular biology have led to rapid progress in understanding the mechanisms and consequences of neurotoxic injury. This volume is intended to provide the reader with an appreciation for the scope of approaches taken in the mechanistic study of neurotoxic agents. The various chapters, each contributed by recognized experts, are divided into three sections. Section 1 addresses the basic anatomical and biochemical components of the nervous system, with special attention placed on those areas where existing or emerging data have demonstrated toxic actions. This section is directed towards the nonspecialists, offering overviews of topics in neuroscience, as well as contemporary reviews of selected topics for researchers in the field. Section 2 focuses on neurobehavioral toxicology and alterations of the special issues. Specific behavioral domains and the functional consequences of their perturbation are discussed in detail. Particular emphasis has been given to research methodologies in neurobehavioral toxicology, especially the appropriate use, limitations, and interpretations. Section 3 provides a survey of selected classes of compounds demonstrated to have adverse effects on nervous system structure and function. It is intended to provide current reviews of chemical-specific neurotoxicity for practitioners in neurotoxicity.

Volume 12: Chemical Carcinogens and Anticarcinogens (edited by G. T. Bowden and S. M. Fischer)

Volume 12 provides a comprehensive overview of the field of chemical and radiation carcinogenesis with particular emphasis on molecular mechanisms. The actions of various chemical carcinogens as well as ultraviolet and ionizing radiation are discussed in the context of the multistage process of both experimental and human carcinogenesis. Background information concerning the target genes for genetic and epigenetic modulation by carcinogens are covered. These target genes include proto-oncogenes, tumor suppressor genes, and effector genes. Various classes of carcinogens including tumor initiating, promoting, and progressing agents are discussed in terms of how they bring about critical alterations in the target genes. The functional role of these gene alterations are covered in the context of the various phenotypic changes associated with each stage of carcinogenesis. Finally, strategies for intervening with the multistep process of carcinogenesis through chemoprevention are discussed.

Wherever possible the relevancy of more basic findings with experiment model systems to human carcinogenesis is emphasized.

Volume 13: Author and Subject Indexes

The Author Index contains a complete list of authors who are listed in the references at the end of each chapter and those cited in the text.

The Subject Index contains a comprehensive list of terms used throughout the other 12 volumes. The EMTREE Thesaurus has been used as a guide for the selection of preferred terms, along with the IUPAC Recommendations for the nomenclature of chemical terms.

References

References cited in text are numbered for easy identification with the complete reference (authors, title, year, volume, inclusive pages) in the Reference list at the end of each chapter. Each reference has been validated. While exhaustive referencing was not performed, sufficient references were used to support the statements and provide the reader with adequate sources for finding additional information.

Volume Editor's Preface

In the past 10–15 years the field of renal toxicology has matured and evolved into a discipline in which elucidation of molecular and biochemical mechanisms has clearly been at the forefront. Such mechanistic-based research has not only expanded our thinking considerably but has impacted regulatory decisions and public policies relating to human safety. The most notable example of the latter is the application of mechanistic research on the role of a male-rat specific protein, α2u-globulin, in chemically induced renal carcinogenicity in rats. This line of research provided compelling evidence for the US Environmental Protection Agency to conclude that renal toxicity–tumors in male rats associated with α2u-globulin nephropathy should not be used to estimate cancer risk in humans. Such an example clearly highlights how mechanistic-based research in renal toxicology can be incorporated into the human risk assessment process.

As will be apparent in this volume, the field of renal toxicology also has contributed to, and has been influenced by, the rapidly evolving concepts in molecular and cell biology in both health and disease. New insights from molecular biology have changed the way we view the pathogenesis of chemically induced renal toxicity and the sequelae of events leading to renal cell death, repair, regeneration, and carcinogenesis. It has become increasingly apparent that while subpopulations of renal cells appear superficially to be nonresponsive to a nephrotoxicant, they in fact may be mounting molecular–biochemical responses to the nephrotoxic insult. Continued research in this field will hopefully provide insights into the biological and toxicological significance of these molecular responses, with a greater understanding of those responses leading to cellular adaptation and survival vs. those associated with cell injury, death, and/or regeneration. Such advances ultimately should guide innovative approaches to the detection and clinical management of chemically induced renal toxicity.

This volume is focused on providing a comprehensive, state-of-the-art review of renal toxicology and capturing the remarkable progress in mechanistic-based research in the field. Our current state of knowledge of the field is reviewed at a variety of levels of anatomic–biochemical complexity, i.e., from the whole organ level to nephron segment, cell, organelle and finally, to the biochemical–molecular level. The first section of this volume is designed to provide the reader with an overview of clinical nephrotoxicity, the basic principles of anatomy, biochemistry, and physiology of the healthy kidney and appropriate *in vivo* and *in vitro* methods to evaluate renal toxicity. This section provides the foundation for an understanding of subsequent chapters which focus on the molecular, biochemical, and pathophysiological events underlying various forms of renal failure and the response of individual segments of the nephron to a nephrotoxic insult. An entire section is devoted to providing the conceptual framework for understanding key mechanisms and mediators of nephrotoxicity, with in-depth reviews on the role of vasoactive substances, cell adhesion molecules, molecular events, oxidative stress, and membrane changes in nephrotoxicity. Selected nephrotoxicants are then discussed in detail with an emphasis on morphologic, pathophysiologic, biochemical, and molecular effects as they relate to mechanisms of toxicity. Where possible, experimental findings are extrapolated to human safety.

Each chapter is written by recognized authorities in the area. The chapters are written in a manner that is intended to provide the reader with a balanced and critical review of each area of interest. Every effort was made to ensure completeness yet minimize redundancy within the volume. In instances where topics could not be comprehensively reviewed, an adequate number of references were supplied for further reading. The volume will be of value to students, researchers, industrial and regulatory scientists, and clinicians interested in renal injury. It is my hope that this volume will serve as an educational tool and will stimulate innovative research in the field, with the ultimate aim of applying our knowledge to human risk assessment and to the detection and clinical management of renal failure.

I would like to express my sincere appreciation to all of the contributing authors for their outstanding and scholarly reviews. I would also like to thank Drs. Gandolfi, McQueen, and Sipes for the opportunity to have served as editor of this volume. Their vision for a Comprehensive Series in Toxicology, their enthusiasm for this work, and their assistance and encouragement, were invaluable. Finally, on a more personal note, I would like to acknowledge Dr. Jerry Hook whose contributions to the field of nephrotoxicity are evident throughout this volume and who first introduced me to the fascinating world of renal toxicology.

Robin S. Goldstein
East Hanover

7.01
Overview: Clinical Aspects of Nephrotoxicity

WILLIAM M. BENNETT

Oregon Health Sciences University, Portland, OR, USA

7.01.1 INTRODUCTION

Toxic nephropathy may reveal itself to the physician in various ways, corresponding to the presenting symptoms and signs of renal disease in general. The clinician can categorize nephrotoxic renal disease in syndromes that correspond to how the patient presents to the doctor.[1] Nephrotoxic drugs and chemicals should be in the differential diagnosis of any renal syndrome. It is important to note that a single therapeutic agent or nephrotoxicant may cause more than one clinical picture, depending on the nature of the toxicant, dose, and duration of exposure. There are also a variety of host factors, such as age, sex, hepatic and renal drug metabolism phenotype, and so on, which individualize responses. In this chapter the framework for the specific information on individual nephrotoxicants, which make up the rest of this book, will be provided. Major clinical syndromes produced by nephrotoxic drugs and chemicals are tabulated in Table 1.

Table 1 Clinical presentations produced by drugs and nephrotoxicants.

Syndrome	Clinical examples
Direct renal tubular injury producing tubular necrosis and acute renal failure	Aminoglycosides, *cis*-platinum, amphotericin B, pentamidine, heavy metals
Prerenal azotemia due to impaired renal perfusion	Nonsteroidal anti-inflammatory drugs (NSAIDs), angiotensin-converting enzyme (ACE) inhibitors, cyclosporine, tacrolimas, diuretics
Acute interstitial nephritis	Penicillins, sulfonamides, phenytoin, diuretics, allopurinol, NSAIDs, many miscellaneous drugs
Acute obstructive uropathy	High-dose methotrexate, intravenous acyclovir, methysergide
Chronic renal failure	Analgesic-associated nephropathy, lead nephropathy, cyclosporine nephrotoxicity, intravenous drug abuse
Nephrotic syndrome	Gold, penicillamine, NSAIDs, heroin
Hyperkalemia	Beta blockers, NSAIDs, ACE inhibitors, potassium-sparing diuretics, cyclosporine
Hyponatremia	NSAIDs, chlorpropamide, thiazide diuretics, narcotics
Nephrogenic diabetes insipidus	Lithium, demeclocycline, clofibrate

7.01.2 ACUTE RENAL FAILURE

Nephrotoxic renal damage is most frequently manifested clinically by sudden deterioration of renal function over a matter of days to weeks. The severity of this renal failure depends on the nature and dose of the toxicant as well as a wide variety of individual host factors. It should be emphasized that acute renal failure can be present in a nonoliguric form with relative preservation of urine flow (>500 mL/24 h). In fact, most clinically relevant nephrotoxicants, such as aminoglycoside antibiotics, *cis*-platinum, cyclosporine, and radiocontrast media produce a nonoliguric type of acute renal failure. The precise pathophysiology of the decline in renal function differs depending on the particular mechanism of toxic injury.

7.01.2.1 Direct Toxic Tubular Injury

Aminoglycoside antibiotics are still widely used in the treatment of serious gram-negative infections because of retained efficacy and relative lack of bacterial resistance. Some 10% to 15% of therapeutic courses are complicated by acute renal dysfunction due to proximal tubular injury. This injury may be observed despite keeping peak and trough drug levels in the desired therapeutic range.[2]

The aminoglycosides are cationic drugs with molecular weights of approximately 500. They are excreted by the kidney quantitatively with a clearance only slightly less than the glomerular filtration rate (GFR). Trough serum levels are a reflection of renal excretory capacity. A small fraction of filtered aminoglycosides binds to anionic phospholipid receptors on the proximal tubular epithelium, where a high-capacity transport system internalizes the drug into tubular cells. By interfering with critical intracellular biochemical processes, aminoglycosides ultimately produce cell necrosis.[3] The drug is concentrated in the renal cortex to values 10–20 times concomitant serum levels although renal cortical levels can be dissociated from toxicity.[4] The elimination half-life of aminoglycosides, which is normally two hours, is markedly prolonged as renal function falls. Thus, maintenance-dose intervals should be extended in patients with re-existing renal dysfunction when aminoglycosides are required. Because the therapeutic efficacy of aminoglycosides is strongly correlated with peak serum levels, it is important to give an adequate initial loading dose independent of renal function. Among the clinically available aminoglycosides, the rank order of nephrotoxicity is gentamicin > tobra-mycin > amikacin > netilmicin.

Nonoliguric acute renal insufficiency is the most common manifestation of aminoglycoside nephrotoxicity. Less common are a variety of isolated tubular syndromes, for example, nephrogenic diabetes insipidus and Fanconi syndrome. Severe oliguric renal failure requiring dialysis is rare from aminoglycosides alone. A drug-induced urinary concentration defect characterized by polyuria and secondary thirst stimulation may precede the detectable rise in blood urea nitrogen and serum creatinine in patients given more that five to seven days of aminoglycoside treatment.[5] Granular casts and mild proteinuria occur frequently, but are not diagnostically helpful. Enzymuria and proteinuria often develop between the second and fourth day of administration of aminoglycosides to both infected patients and human volunteers. In particular, the quantity of β_2-microglobulin and proximal tubular brush border enzymes, namely alanine aminopeptidase and N-acetylglucosaminidase, increase significantly.

Aminoglycoside nephrotoxicity most often produces patchy proximal tubular necrosis. In many patients who satisfy the clinical criteria for aminoglycoside nephrotoxicity, cellular autophagocytosis has been observed using electron microscopy. Prominent cytosegresomes, containing whorled material known as "myeloid bodies," can be observed within the proximal tubular lumen or urine.[6]

A variety of clinical risk factors that predispose to the development of aminoglycoside

Table 2 Clinical risk factors for aminoglycoside nephrotoxicity.

Inadequate peak and elevated trough blood levels
Duration of drug treatment more than 7 days
Recent courses of aminoglycoside therapy
Age greater than 60 years
Extracellular fluid volume depletion
Potassium depletion
Magnesium depletion
Pre-existing renal dysfunction
Liver disease

---

nephrotoxicity have been identified. They are summarized in Table 2.

Since aminoglycosides are often necessary for the best antibiotic management of a critically ill patient, therapy needs to be optimally prescribed. Loading doses should be sufficient to achieve high peak levels to maximize bacterial killing, and maintenance doses should be based on calculated or measured renal function.[7] Extending the interval between doses is safer and more effective than reducing the size of individual doses in patients with renal insufficiency, since peak levels correlate with bacterial killing. Risk factors should be corrected if possible. In very high-risk patients, the least nephrotoxic congeners should be considered. Elevation of the trough level, indicating drug accumulation, usually precedes a rise in the less-sensitive serum creatinine measurement.

Other common examples of drugs which cause direct tubular injury are amphotericin B and *cis*-platinum.

Amphotericin B is a polyene antibiotic with excellent activity against a broad spectrum of fungi responsible for human infection. It has been estimated that renal function becomes impaired in more than 80% of patients given this drug.[8] This nephrotoxicity is dose related and probably inevitable when the cumulative dose is greater than 5 g in adults. Patients at high risk include elderly patients, particularly those with extracellular fluid volume depletion.

The usual clinical presentation of amphotericin B nephrotoxicity is characterized by defects in urinary concentration ability and distal-type renal tubular acidosis. If therapy is continued, nonoliguric renal failure with frank azotemia will follow. Modest proteinuria associated with a relatively normal urinary sediment is the initial finding. In addition, the presence of a magnesium-wasting syndrome has now been recognized to be a prominent feature of amphotericin nephrotoxicity. Amphotericin B may cause permanent impairment of renal function if repeated courses of therapy are given.

Sodium loading has been reported to be useful in the prevention of amphotericin nephrotoxicity. Using a sodium supplementation approach, Branch[9] has reported that the incidence of amphotericin-associated renal impairment is reduced from 67% to 12%. Newer liposomal preparations of amphotericin B have been reported to ameliorate nephrotoxicity by delaying and minimizing renal exposure.

Nephrotoxicity is a well-recognized problem with the use of platinum derivatives. Cisplatin is a useful agent that improves tumor response rates in patients with a variety of malignancies.[10] Dose-limiting nephrotoxicity is manifested by increases in serum creatinine and drops in creatinine clearance. Renal dysfunction tends to be cumulative, even if acute falls in GFR are prevented by hydration.[11]

After cisplatin administration, there is early reduction of both GFR and effective renal plasma flow. Extracellular fluid volume depletion, prior renal impairment, and increased age tend to be special risk factors for renal dysfunction. Renal magnesium wasting is present in a high percentage of patients treated with cisplatin. Magnesium depletion is thought to be important in the associated hypokalemia and hypocalcemia.[12] Renal sodium wasting may occur with polyuria, hyponatremia, and volume depletion. In as many as 10% of patients treated with cisplatin, the tubular dysfunction is severe enough to produce orthostatic hypotension. The renin–angiotensin system in these patients does not respond appropriately to the drug-induced, decreased, effective circulating volume.[13] Carboplatin, allegedly less nephrotoxic, has also caused renal sodium wasting and acute renal dysfunction.[14] Repeated cycles of platinum therapy, particularly with high doses, can also lead to chronic structural and functional renal lesions.

Good hydration and forced diuresis will minimize the acute changes in renal hemodynamics attributable to cisplatin, although tubular uptake of the drug is not prevented. Magnesium and potassium losses should be anticipated with vigorous replacement. Concomitant nephrotoxicants, such as aminoglycosides, amphotericin, and nonsteroidal anti-inflammatory drugs should be avoided. Carboplatin may be indicated in patients with pre-existing reduction of renal function and those whose cardiovascular reserve is unable to tolerate high-volume hydration regimens.

7.01.2.2 Prerenal Azotemia

Renal hemodynamics may be compromised in certain predisposed patients by a variety of drugs. A common example would be the use of nonsteroidal anti-inflammatory drugs (NSAIDs) in patients in whom renal perfusion is dependent on vasodilatory prostaglandins such as PGE_2 and PGI_2. In such circumstances, inhibition of cyclooxygenase by the NSAIDs results in intrarenal vasoconstriction and decreased renal perfusion.

Certain subsets of patients are particularly prone to the development of oliguric acute renal failure when treated with NSAIDs. These include patients with congestive heart failure, chronic renal insufficiency, cirrhosis with ascites, systemic lupus erythematosus, intravascular volume depletion, elderly patients with significant atherosclerotic disease, and those treated with diuretics.[15,16] All of these conditions may be associated with a decrease in renal perfusion and activation of the sympathetic nervous and renin–angiotensin systems. Constriction of the efferent arteriole by angiotensin II increases intraglomerular capillary hydrostatic pressure, enhancing glomerular filtration rate during periods of decreased renal perfusion. Angiotensin II is also a potent stimulator for the renal synthesis of prostaglandin E_2, which, through its vasodilatory effect, augments renal perfusion. Thus, blockade of prostaglandin synthesis can result in an acute decline in renal blood flow and glomerular filtration rate. The hemodynamic nature of NSAID-induced acute renal failure is supported by the abrupt onset of oliguria, avid sodium and water retention, and rapid reversibility after discontinuation of the prostaglandin inhibitor. Compared with other forms of acute renal dysfunction, these patients recover baseline renal function within two to three days after the NSAID is discontinued. All available NSAIDs, including parenteral forms, have been implicated as causes of acute renal failure.

Likewise, angiotensin-converting enzyme inhibitors themselves may cause acute, usually reversible renal failure in patients whose GFR in dependent on the presence of angiotensin's vasoconstrictor effect on the postglomerular efferent arteriole. Examples of such patients include those with bilateral renal artery stenosis, renal artery stenosis in a solitary kidney, transplant artery stenosis, severe intrarenal vascular disease, or patients with congestive heart failure. If such patients have volume depletion due to concurrent diuretics, they are at particular risk. Renal function rapidly improves with volume repletion. Potent diuretics per se may also produce a prerenal type of acute renal failure.

7.01.2.3 Interstitial Nephritis

Many drugs can cause rises in blood urea nitrogen (BUN) and serum creatinine in the context of evidence of systemic hypersensitivity such as fever, eosinophilia, and a diffuse skin rash. However, a renal hypersensitivity response can be present without any of these systemic manifestations. The diagnosis depends on clinical suspicion and a confirmatory renal biopsy showing acute interstitial inflammation with lymphocytes and eosinophils. The urinary sediment shows microhematuria, pyuria, and urinary eosinophils. Withdrawal of the offending drug usually results in renal functional improvement, which can be hastened by a 10- to 14-day course of steroid therapy. Acute renal failure is usually nonoliguric but can progress enough to require dialysis. Common offending drugs are allopurinol, cimetidine, sulfonamides, penicillins, rifampicin, and phenytoin. NSAIDs may rarely produce a distinctive clinical picture of progressive nonoliguric acute renal failure, plus nephrotic syndrome. In these cases, glomerular pathology is most consistent with minimal-change nephropathy, since there are a few electron-dense deposits and negative immunofluorescence. Interestingly, this unusual complication can occur after many months of therapy with NSAIDs. The renal dysfunction and proteinuria are most often reversible following drug discontinuation and steroid therapy.

7.01.2.4 Obstructive Nephropathy

Drug-induced intrarenal and extrarenal obstruction can cause acute renal failure. Intrarenal obstruction may occur when insoluble crystalline material deposits within tubular lumens. Examples include uric acid crystal deposition from tumor lysis, particularly in acid urine. Alkaline diuresis, adequate hydration, and allopurinol can be given prophylactically with good results. Intrarenal obstruction may also be caused by high-dose methotrexate; in some cases the resulting retroperitoneal fibrosis may require surgical relief. Obviously, the alert clinician should consider drug-induced obstructive nephropathy when evaluating patients with acute renal failure.

7.01.3 CHRONIC RENAL FAILURE

This clinical syndrome usually presents with nonspecific symptoms such as nocturia, weakness, fatigue, nausea, or vomiting. Since renal functional reserve is large and disease processes proceed slowly, patients may be asymptomatic until 85% to 90% of renal function is lost. With primary involvement of glomeruli or blood vessels, hypertension and proteinuria (e.g., albuminuria) are usually noted, whereas tubulo-interstitial disease is often characterized by normal blood pressure and absence of proteinuria. Obviously, these distinctions are simply rules of thumb since most chronic renal diseases involve all parenchymal elements at end stage. It is not unusual to have advanced renal disease discovered during the workup of unexplained anemia or even on a routine chemistry screen.

7.01.3.1 Chronic Tubulointerstitial Nephritis

The association between chronic interstitial nephritis and ingestion of large amounts of analgesic drugs was first noted in 1953.[17] The number of cases reported is generally thought to be a gross underestimate of the true incidence of this preventable form of renal disease. Furthermore, discontinuance of ingestion after renal insufficiency develops often results in stabilization or improvement in renal function.[18] It is widely believed that a combination of drugs causes the renal disease. Difficulties in relating the disease to a single drug result from the fact that combination analgesic therapy is the rule of most patients. The most distinctive pathologic feature of analgesic-associated nephropathy is renal papillary necrosis. Although prolonged use of aspirin alone has been reported as a cause of analgesic-associated nephropathy,[19] it is likely that a combination of drugs is usually necessary to produce nephropathy. Acetaminophen is a major metabolite of phenacetin in man. Reports of association of acetaminophen, both alone and with NSAIDs, with papillary necrosis are now known.[20,21] Since NSAIDs have become available over the counter without any consumer warnings, more patients with and without pre-existing renal disease will be exposed.

Clinically, the patient usually presents with the chronic complaints that require analgesics, that is, headache, back pain, or with some manifestation of chronic renal failure. Depression and gastrointestinal symptoms are prominent. The typical patient is a woman who is middle-aged and somewhat defensive about consumption of analgesics. To obtain accurate histories, repeated questioning and family interviews may be necessary. Anemia and pyuria occur commonly, and acute infection or obstruction may be present. In view of the potential scope of this problem physician and patient education must be given high priority if this common form of chronic renal failure is to be eliminated. With NSAIDs, the drugs may be prescribed or taken for legitimate medical reasons.[21] Periodic monitoring of such patients is recommended.

7.01.3.2 Arteriolar Nephrosclerosis

This pathologic picture is most often seen in patients with long-standing hypertension. Clinical manifestations include slowly progressive renal failure with proteinuria of moderate proportions $(1-2\,g\,d^{-1})$. Hyperuricemia is common either as a cause or result of the vascular disease. The glomeruli and interstitium show the effects of chronic ischemia. Lead nephropathy is known to produce similar clinicopathologic features.[22] Patients can be exposed to lead in homemade whiskey made in stills coated with lead. Lead can also be ingested by children from paint or toys or in rain water from porches painted with lead-containing paints. Cable jointers and battery workers are often exposed industrially. Pathologic descriptions in patients dying form chronic lead nephropathy are nonspecific, but severe nephrosclerosis predominates. Treatment is directed at removal of the source of lead and, in some cases, chelation therapy.

7.01.3.3 Chronic Glomerulonephritis

This slowly progressive syndrome of proteinuria, hypertension, and declining renal function is not commonly due to medicinal drugs, since most agents are not ingested over the long periods of time required. Immunologic renal injury induced by the intravenous injection of heroin may progress to chronic renal failure with progressive glomerular obliteration. The typical patient is an intravenous drug abuser in the second to fourth decade who has injected himself with heroin for 1 to 10 years. Some 70% to 75% of patients with proteinuria progress to end-stage renal insufficiency within four years despite steroid or cytotoxic drug therapy. Heroin or its injection vehicle may be directly nephrotoxic, or more likely, may sensitize the addict who subsequently develops immune complex nephropathy.[22] Newer agents such as methamphetamines, and cocaine also have been associated with chronic renal failure.

Progressive renal insufficiency requiring management for end-stage renal disease has been reported in cardiac and liver allograft recipients and autoimmune disease patients treated with cyclosporine.[23] Because these patients have excellent function of their allografts and have no other causes for renal disease, cyclosporine is almost certainly implicated. Although early reports emphasized large initial doses, it is disturbing that this disease continues despite dose reduction and strict maintenance of "therapeutic" blood levels. When biopsies have been performed, chronic tubulointerstitial fibrosis in a striped pattern, afferent arteriolopathy, and glomerulosclerosis have been observed. The striped pattern of fibrosis has been attributed to the arteriolar lesion, the arterioles being presumed to be a primary target for the drug in these patients. This type of change is less common in patients given doses less than $5\,mg\,kg^{-1}$, but it may

still occur. In some patients treated for autoimmune uveitis and psoriasis, these same chronic histopathologic findings have been observed.[24]

7.01.4 NEPHROTIC SYNDROME

Nephrotic syndrome is the combination of massive proteinuria, edema, and hyperlipoproteinemia. The most common causes of this syndrome are disease processes that affect glomerular permeability. Some common drugs produce nephrotic syndrome, albeit rarely, including Tridione, tolbutamide, and probenecid, as well as intravenous heroin abuse. In addition, environmental substances known to produce allergy or used to desensitize allergic patients are sometimes associated with nephrotic syndrome. Pathologic correlates of these clinical entities are sparse.

The nephrotic syndrome produced by gold-containing therapeutics is characterized pathologically by membranous nephropathy.[25] Although gold can be seen in the proximal tubules on ultrastructural examination, electron-dense deposits are noted within the glomerular basement membrane. In addition, IgG and other immunoglobulins are found on immunofluorescent examination. Gold-induced injury to tubular cells likely triggers an antigen–antibody complex reaction with some tubular component as the antigen. The nephrotic syndrome usually recedes when gold treatment is stopped. Penicillamine and captopril also have been associated with membranous nephropathy.

7.01.5 CLINICAL MANIFESTATIONS OF RENAL FAILURE

7.01.5.1 Disturbances of Body Fluids: Osmolality and Volume

Fluid and electrolyte disorders are among the most common conditions evaluated by a consulting nephrologist or internist.

7.01.5.1.1 *Hyponatremia and hypo-osmolality of body fluids*

This disturbance is common in hospitalized patients and can result from many drugs that have an ability to impair renal water excretion. These effects may be independent of the primary pharmacologic action of the drug. The ability of some drugs, such as chlorpropamide, to impair water excretion has even been used therapeutically to treat diabetes insipidus.[26] Prostaglandins also antagonize the effects of antidiuretic hormone (ADH). Consequently, prostaglandin blockade by NSAIDs can enhance ADH action resulting in a decrease in free-water excretion leading to the development of hyponatremia. A full review of the drugs that impair water excretion and their mechanisms of action is beyond the scope of this chapter.

7.01.5.1.2 *Diabetes insipidus and hyperosmolality*

This is most often nephrogenic when produced by drugs. However, central diabetes insipidus can be simulated under appropriate circumstances by the effects of ethanol and phenytoin. Drug-induced states of nephrogenic diabetes insipidus have been reviewed.[27] Therapeutic advantage may be gained by using these drugs to treat severe states of hypo-osmolality. Lithium, fluoride from halogenated anesthetic metabolism, and demeclocycline are known causes.

7.01.5.2 Acid–Base Disturbances

Virtually any disturbance of hydrogen ion balance, metabolic or respiratory, can be produced by, or related to, a drug or toxicant. The following section will catalogue such disorders, using as examples common entities that clinicians frequently see.

7.01.5.2.1 *Metabolic acidosis*

Metabolic acidosis may be conveniently divided into two types. When there is a difference between serum cations and anions of greater than $12–14\,mEq\,L^{-1}$, an "anion gap" is said to exist. Retention of unmeasured anions implies an excess of nonvolatile acids and is most commonly observed in diabetic ketoacidosis and acute renal failure. Overdoses of such toxicants as methanol, thenylene glycol, and salicylate may produce metabolic acidosis in which large anion gaps are measured.

Lactic acidosis is a cause of metabolic acidosis produced by accumulations of serum lactate ($>7\,mm\,L^{-1}$). Phenformin, an oral hypoglycemic agent, increases anaerobic metabolism of glucose by peripheral tissues with excess production of lactate. Clinical lactic acidosis most often occurs in diabetic patients with impaired renal function.

Renal tubular acidosis is a form of metabolic acidosis, usually without an increase in unmeasured anions. The kidney is unable to acidify the urine in response to systemic acidosis. The proximal type or type 2 renal tubular acidosis is associated with a decrease capacity of the proximal tubule to reabsorb filtered bicarbonate, leading to "bicarbonate leak." Often, proximal reabsorptive defects of amino acids, glucose, and phosphate also occur (Fanconi syndrome). The full-blown picture may be produced by exposure to outdated tetracycline. Bicarbonate wasting without the multiple dysfunction of the proximal tubule is a common finding in patients treated with acetazolamide, a potent carbonic anhydrase inhibitor. This may lead to metabolic acidosis, hypercalciuria, nephrolithiasis, and hypokalemia. In this type of acidosis, the urinary pH is inappropriately high. Bicarbonaturia, however, occurs only when acidosis is mild to moderate. At this level, the serum bicarbonate is above the reduced nephron reabsorptive capacity. The amount of bicarbonate necessary to raise serum bicarbonate to normal and sustain it is generally large.

Distal or type 1 renal tubular acidosis is characterized by alkaline urine and bicarbonaturia persisting despite severe degrees of systemic acidosis. The dysfunction is thought to be due to an inability of the distal nephron segments to generate or maintain steep luminal-to-peritubular fluid hydrogen ion gradients. Amphotericin B is known to produce this type of physiological defect in man. A fall in hydrogen ion excretion and a rise in urine pH usually precede overt acute renal failure. Maintenance of adequate diuresis and bicarbonate during therapy may minimize toxicity. If therapy with amphotericin B can be stopped before azotemia occurs, recovery of renal function is the rule. Hypokalemia, despite systemic acidosis, is a prominent feature of amphotericin B distal renal tubular acidosis, and early $KHCO_3$ replacement is warranted.

7.01.5.2.2 *Metabolic alkalosis*

Metabolic alkalosis may be produced by mineralocorticoid excess. Such patients characteristically have expansion of the extracellular fluid volume, hypokalemia, and alkalosis. The volume expansion causes "escape" from sodium-retaining effects of these steroids, making clinical edema uncommon. Habitual heavy users of licorice can reproduce such a syndrome because licorice has mineralocorticoid-like properties.

7.01.5.2.3 *Hyperkalemia*

Elevation of serum potassium may be produced by drugs, particularly in patients with pre-existing renal disease. Among the commonly used therapeutic agents, angiotensin converting enzyme inhibitors frequently produce hyperkalemia via inhibition of angiotensin formation with secondary impairment of aldosterone secretion.

In addition to their effects on glomerular filtration rate, NSAIDs can have prominent effects on tubular function, especially with regard to renal handling of sodium, water, and potassium. Prostaglandins stimulate the production of renin, which subsequently leads to enhanced secretion of aldosterone. Decreased production of renal prostaglandins has been implicated in the pathogenesis of hyperkalemia in patients with hyporeninemic hypoaldosteronism.[28] It is not surprising, then, that predisposed patients may develop hyperkalemia when treated with NSAIDs. This has been most commonly reported with indomethacin, but may occur with all NSAIDs. High-risk patients are characterized by pre-existent renal insufficiency, vascular disease, diabetes mellitus, and mild hyperkalemia. Other drug-associated causes of hyperkalemia are potassium-sparing diuretics, beta blockers, and large doses of heparin.

Drug-induced renal disease is increasingly common, and it is vital for the clinician to recognize these nephrotoxic syndromes, since they are usually reversible. The spectrum of drug-related nephrotoxicity is broad, and virtually any primary renal disorder can be mimicked. Drugs may lead to acute and chronic renal failure, interstitial nephritis, nephrotic syndrome, obstructive uropathy, and altered tubular function. Multiple pathogenetic mechanisms are involved including direct renal toxicity, altered hemodynamics, hypersensitivity reactions, and altered hormone metabolism. The key to the diagnosis and treatment of drug-related renal syndrome is generally dependent on early recognition of the action of the offending agent and discontinuing the agent.[29]

7.01.6 INCIDENCE/OUTCOME

The incidence of in-hospital acute renal failure (ARF) attributed to drug nephrotoxicity is currently estimated at one in five cases.[30] The interpretation of this estimate in contemporary terms is complicated since it was derived during a period of significant change in the clinical pattern of ARF. With improved technology for both the diagnosis and treatment of ARF, plus the extended longevity of the American population, today's hospitalized patient with ARF is older (>59), if oliguric, having a multifactorial etiology and an expected death rate of >50%.[31]

The presence of nonoliguria imparts a lower mortality and is more likely to be due to a single etiology, that is, nephrotoxic drugs.

The Société de Nephrologie[32] conducted a one-year survey of all patients with ARF to determine the frequency and causes of drug-induced acute renal injury. This survey has yielded the most comprehensive data by far regarding the epidemiology of drug-induced nephrotoxicity. Of the 2175 cases of ARF, 398 (18.3%) were considered to be drug induced. Table 3 provides a summary of the distribution of the cases of ARF according to offending agent. As with previous reports concerning specific drug etiologies, antibiotics were most frequently encountered, with analgesics and NSAIDs sharing the second spot. Glafenin, an analgesic that is available in Europe but not in the United States, distorted the data when compared to reports from North American (Table 4). More importantly, two-thirds of Glafenin cases represented acute ingestions associated with suicidal attempts. If these patients ($n = 79$) are excluded for purposes of calculating an incidence to compare with recent reports in the United States literature, the contribution of either antibiotics (7.8%) or contrast agents (2.9%) is still less than reported by either Rasmussen and Ibels[30] or Hou *et al.*[31]

The significant contribution that NSAIDs provided to the overall incidence of drug-induced nephrotoxicity, summarized in Table 3, is in keeping with reports concerning the nephrotoxic potential of this group of compounds.[33,34] Over half of the patients presented

Table 3   Distribution of drug-induced ARF.

Drug	No. of patients	%
Antibiotics	136	34.2
Glafenin	79	19.8
NSAIDs	62	15.6
Contrast media	50	12.6
Diuretics	18	4.5
Chemotherapy	14	3.5
Others	39	9.8
Clinical presentation		
Oliguric	175	44.8
Nonoliguric	216	55.2
Outcome		
Recovery complete	190	47.7
Residual renal impairment	153	38.4
Death	50	12.6
Lost to follow-up	5	1.3

Source: Kleinknecht *et al.*[32]

Table 4   Incidence of ARF due to nephrotoxins.

Authors	Date	n	% ARF due to		
			Antibiotics	Contrast agents	Total
Anderson *et al.*[a]	1977	92	16%	4%	20%
Galpin *et al.*[a]	1978	43	33%	9%	42%
Rasmussen and Ibels[30]	1982	143	11%	11%	22%
Hou *et al.*[31]	1983	129	7%	12%	19%
Société de Nephrologie[32]	1985	2175	16%	2%	18%

[a] Cited in Bennett *et al.*[1]

with nonoliguric ARF (NOARF). Of particular interest was the outcome of patients who survived the episode of ARF (mortality was 12.7%–50 patients); only 28.6% of all patients required acute dialysis support during the episode of ARF. Table 3 summarizes the outcome of the 398 patients. Most impressive is the unexpectedly high percentage of patients who were left with permanent renal damage, that is, 38.4%. This latter observation led the authors to speculate that "drug-induced acute renal failure" is an underestimated cause of chronic renal failure.[34] In this pivotal study nearly 15% of the 398 patients were classified as having prerenal azotemia, a category of renal presentation that is often excluded in reports concerning ARF.

Thus, in a one-year experience from practicing clinical nephrologists, nearly one in five patients who were diagnosed as having ARF were categorized as drug induced. Over 50% presented as NOARF, but almost one in four suffered significant renal sequelae.

The estimated incidence of 20% drug-induced ARF in hospitalized patients is in marked contrast to the extremely low incidence of outpatient drug-induced renal disease requiring hospitalization reported by Beard *et al.*,[35] that is, 1 : 300 000 persons/year. Part of the explanation for this low incidence may rest in the authors' exclusion of chronic pre-existing renal disease. In the study from France,[34] 74 of 398 or nearly 19% of their cases occurred in patients with pre-existing renal disease. Clearly, the difference is not traceable to a low offending drug exposure, since an accompanying summary of outpatient medication provided ample exposure to antibiotics, chemotherapeutics, and NSAIDs, but not contrast agents.

As noted from the study conducted by the Société de Nephrologie, a surprisingly large percentage of patients who suffered an episode of ARF were left with permanent renal impairment. Earlier reports from the European Dialysis and Transplant Association[36] supports a similar conclusion. In evaluating its 1982 data, the Association reported a direct correlation between the number of ARF patients per million and new ESRD patients per million. This correlation is compatible with the proposition that ARF, successfully treated, leaves a significant portion of people with residual renal damage, leading to ESRD.

A myriad of risk factors, along with certain acute insults, have been suggested as contributing to the incidence of ARF. The study by Rasmussen and Ibels[30] is particularly relevant since they were one of the first groups to apply multivariate analysis to characterizing significant contributions to clinical deterioration, with the hope being that by identifying those factors that significantly contributed to ARF, a strategy to modify or prevent the occurrence of ARF would result. Of the six risk factors evaluated, only pre-existing hypertension and renal disease were found to contribute; age>59 years, gout/hypouricemia, diabetes, and diuretics could not be implicated. Of the six acute insults evaluated, aminoglycosides, pigmenturia, and dehydration all had a significant contribution, whereas no weighted contribution was identified for sepsis, radiocontrast media, or liver disease.

Shusterman *et al.*[37] have reported on the influence of risk factors on outcome of hospitalized ARF. These authors used a case-control study design and calculated the odds ratio. Based upon logistic regression, volume depletion, aminoglycoside use, congestive heart failure, radiocontrast exposure, and septic shock contributed to the development of ARF in the hospital. Thus, although several of these factors are beyond the physician's control, for example, sepsis, volume depletion, and congestive heart failure, the decision to give aminoglycosides and to administer radiocontrast media rest with the physician and, thus, may be modified or eliminated.

7.01.7 STRUCTURE/FUNCTION

7.01.7.1 Vulnerability of the Kidney to Toxic Injury

As a vital organ, the kidney performs several unique functions that can be impaired by drug-induced renal injury including: (i) regulation of the body's fluid volume, a major contributor to the control of blood pressure; (ii) regulating the pH of the body, in concert with the lungs, through the excretion of fixed, nonvolatile acids and the conservation of body base; (iii) excretion of waste products and, equally important, the conservation of critical body constituents, for example, electrolytes and other body nutrients; (iv) detoxification of certain drugs, and (v) synthesis and release of hormones, such as renin and erythropoietin, and the conversion of vitamin D_3 to the 1,25-dihydroxy form.

To perform these functions requires the integration of many physiological actions of the kidney. Analysis of these actions provide insights as to the vulnerability of the kidney to drug-induced injuries.

One-fifth of the resting cardiac output goes to the kidneys, of which 10% undergoes glomerular filtration. Electrolyte reabsorption accounts for the principal oxygen consuming work performed by the kidneys. Regulation of plasma volume and the generation of vasoactive humoral agents, for example, renin and prostaglandins, play a key role in blood pressure regulation, which is signaled by changes in blood flow resistance.

7.01.7.2 Mechanism of Nephron Injury

In the glomerulus, renal blood flow, in concert with net glomerular hydrostatic pressure, acts on the endothelial surface area of the glomerulus to create an ultrafiltrate of plasma. Two aspects of the endothelial surface warrant comment. The filtration area has been assigned a prominent role in a novel concept proposed to explain sustained, progressive loss of renal function following acute injury or insult. The concept, known as "hyperfiltration-injury," is experimentally anchored in the work of Hostetter et al.,[38] who demonstrated that loss of a nephron unit causes the remaining nephrons to hypertrophy, with a rise in single-nephron filtration rate due to an elevated glomerular hydrostatic pressure. The latter results from preferential vasoconstriction of efferent compared to afferent arteriolar resistance. Progressive nephron loss induces glomerular hypertension in the remaining nephrons eventuating in the pathologic state of glomerulosclerosis characterized by faulty blood pressure

regulation and the diminution of renal excretory capacity.

The hyperfiltration-injury concept is attractive as it provides an explanation for the well-recognized phenomena of progressive deterioration of renal function after the acute insult has been eliminated. Animals with experimentally induced hyperfiltration have a predictable rate of deterioration of renal function that can be delayed by reducing glomerular hypertension by converting enzyme inhibition or restricting dietary protein.[39] By understanding the vulnerability of the endothelial surface, new principles of management with potential maneuvers, such as angiotensin converting enzyme inhibitors to prevent chronic renal failure, are emerging.

The negative charges of endothelial filtration surface can be modified sufficiently to cause renal injury. These fixed negative charges are responsible for variations in the sieving coefficient of neutral versus charged particles.[40] In particular, electrostatic attachment of positively charged ligands can either alter the permeability coefficient of the glomerulus (K_f), or act as "planted" antigens. Thus, cationic proteins may sequestrate in either the vascular or mesangial cells of the glomerulus to become "planted" antigens. Subsequently, a circulation antibody may attach to the "planted" antigens causing *in situ* immune complex formation. An example of this form of injury is the linear immunophoretic deposits noted in anti-glomerular basement membrane (GBM) disease. A group of environmental toxicants implicated as causing an anti-GBM lesion in the kidney are short-chain hydrocarbons resulting from the refinement of crude oil.[41]

When the complex biochemical composition of the glomerulus[42] is exposed to the wide spectrum of antigenic compounds to which the body is exposed over a lifetime, it is not surprising that a variety of glomerular injury patterns have been identified. This is not the only form of immunologically mediated glomerular injury recognized. Additional mechanisms include "circulating antigen–antibody complexes," which are formed elsewhere in the body but undergo filtration arrest in the glomerulus. The glomerulonephritis associated with systemic lupus erythematosus is an example of this; heavy metals also induce this lesion by forming circulating complexes.[43] Once bound to tissue, the complexes fix complement with subsequent activation of the complement cascade triggering the *in situ* inflammatory response.

Another mechanism of glomerular damage is cell-mediated injury. Macrophages, either leukocytes or monocytes, attach to glomerular cells resulting in a local inflammatory response due to release of cytokines, thromboxanes, and leukotrienes. This tissue damage may be augmented by a contribution from activated platelets or intrinsic glomerular cells. The therapeutic implications include both techniques, which modify the inflammatory response as well as eliminating drugs that potentiate inflammation. For example, nonsteroidal anti-inflammatory agents, because of the selective inhibition on cyclooxygenase, create a circumstance in which activation of endogenous prostaglandin production would now favor proinflammatory factors.

In the tubular cell, the renal tubules, like the glomerulus, are also vulnerable to drug-induced kidney disease. There are several tubular processes for which drug exposure can cause a unique loss of function. The medullary countercurrent multiplier system, which probably evolved as an adaption to terrestrial existence,[44] provides a mechanism of eliminating waste products while minimizing body water losses. The reabsorption and recycling of small molecular weight compounds, in particular urea, are critical to the process. Since this process is dependent on the permeability of collecting and medullary duct cells, plus the luminal to medullary interstitial concentration gradients, it follows that drugs and/or their metabolites can accumulate in the medullary interstitium. Furthermore, depending upon the chemical properties of the accumulated drugs, an inflammatory response can occur through activation of mediators similar to those involved in glomerular injury. Chronic analgesic nephropathy is an example of such injury. Mudge[45] has proposed that the analgesic-induced chronic interstitial nephritis results from the accumulation of salicylate metabolic intermediate *N*-acetyl-*p*-aminophenol (APAP). Conversely, an opposite effect during osmotic diuresis when the capacity of the distal concentration–dilution mechanism is severely curtailed results in an increase risk of either volume depletion or selective electrolyte losses. Carbenicillin is an example of a drug capable of such an "osmotic" effect since it induces substantial electrolyte depletion. Other examples of drugs that interfere with the regulation of body fluid volume are included in Table 5.

The organic acid and base transport systems of the proximal tubule are an important route of elimination (not necessarily for non-filterable molecular species). Several commonly used drugs undergo such transport (Table 6). Some drugs' mechanism of action is dependent upon the transport, for example, loop diuretics, whereas for others proximal tubular transport represents a major route of

Table 5   Drug-induced alterations of body water balance.

I. Polyuria
    A. Centrally mediated
        Alcohol
        Diphenylhydantoin
        Norepinephrine
        Lithium
    B. Renally mediated
        Lithium
        Gentamicin
        Tetracycline
        Colchicine
        Fluoride
        Amphotericin
        Diuretics

II. Water retention
    A. Centrally mediated
        Nicotine
        Carbamazepine
        Narcotics
        Vincristine
        Clofibrate
        Opiates
        Histamine
        Isoproterenol
        Thioridazine
    B. Renally mediated
        Cyclophosphamide
        Vasopressin
        Sulfonylureas
        Biquanides
        Nonsteroidal anti-inflammatory agents

Table 6   Drugs eliminated through organic transport.

Organic acid system	Organic base system
Para-aminohippurate	Isoproterenol
Phenylbutazone	Tetraethylammonium
Salicylate	Quinidine
Cephalothin	Morphine
Sulfonamides	Procaine
Salyrgan	Tolazoline
Chlorothiazide	Macamylamine
Furosemide	Piperidine
Penicillin	
Methotrexate	
Probenecid	

drug elimination from the body, for example, penicillin. Drugs transported by the organic ion system can cause renal injury either directly, for example, cellular accumulation and/or high urinary concentrations, or indirectly, for example, as competitive inhibitors to block the elimination of endogenously produced toxic end metabolites.

The process of urinary acidification can either prevent or contribute to toxic renal injury. Since urinary acidification involves sequential steps located along the course of the nephron, it is not surprising that abnormalities of acidification are found in a variety of drug-induced renal disease. Chronic lithium nephropathy is an example of a drug-induced acidification defect.[46] Other examples of drug-induced injury in which renal tubular acidosis is a prominent feature are summarized in Table 7.

Peptides or proteins that enter the urinary space are recaptured by the proximal tubule by the process of pinocytosis. Pinocytotic vesicles, once formed, will fuse with proximal tubular lysosomes, the combination allowing the digestive enzymes of the lysosomes to exert their proteolytic action on the material contained within the pinocytotic vesicle, thus insuring reclamation and recycling within the body. Aminoglycoside antibiotics undergo pinocytotic reabsorption in the proximal tubule. However, unlike proteins, the aminoglycoside arrests proteolytic digestion, creating a pathologic appearance that resembles lysosomal storage disease.[47] From experimental data, it is speculated that this interruption of digestion accounts for the proximal tubular necrosis that characterizes the more severe renal injury associated with aminoglycoside nephrotoxicity.[48]

Table 7   Drug-induced acid–base disturbances.

I. Metabolic acidosis
    Anion-gap acidosis
    Nonanion-gap acidosis
    Phenformin
    Acetazolamide
    Methanol
    Tetracyclines (old)
    Salicylates
    Amphotericin B
    Ethylene glycol
    Toluene
    Paraldehyde
    Analgesic
    Ethanol abuse
    Lithium

II. Metabolic alkalosis
    Chloride dependent
    Nonchloride dependent
    Licorice
    Bicarbonate
    Carbenoxolone
    Carbenicillin
    Diuretics
    Mannitol

Table 8   Drug-induced electrolyte disturbances.

I. Sodium
    Hyponatremia—drugs that impair water
      excretion
    Hypernatremia—saline

II. Potassium
    Hypokalemia
    Diuretics
    Antibiotics
    Tocolytic agents
    Licorice
    Hyperkalemia
    Potassium supplements
    Potassium-sparing diuretics
    Selected antihypertensive drugs

III. Calcium
    Hypocalcemia
    Aminoglycoside antibiotics
    Hypercalcemia
    Thiazide diuretics
    Vitamin D supplements

IV. Phosphorus
    Hypophosphatemia
    Parenterol nutrition
    Hyperphosphatemia
    Cytotoxic drugs

Finally, toxin-induced renal disease can cause excessive urinary loss of vital anions or cations. Examples of clinical conditions are summarized in Table 8. Often such losses coexist with either defects in the urinary concentrating–diluting mechanism or renal acidosis.

7.01.8   SITES OF NEPHRON INJURY AND DETECTION

7.01.8.1   Classification

Nephrotoxicity can be categorized according to offending agents, by clinical syndrome,[1] or by proposed site of nephron injury.[49] Each approach has its advocates and unique utility. From the standpoint of the emergency room physician, toxicologists, and clinical pharmacologists, since the suspect drug is often known but the effects are in question, a tabulation of the various groups of drugs and their renal effects is very appropriate. Using such a classification one would have categories such as analgesics; anesthetics; anti-infectives, including antibiotics; antifungal and antiviral compounds; biologic agents; cardiovascular drugs, including antihypertensives; antiarrhythmics and diuretics; cytotoxic chemotherapeutic agents; image enhancing agents; and immuno-suppressive drugs. Conversely, for the physician seeing a patient with a particular constellation of signs and symptoms suggesting renal disease, classification using clinical syndromes is more useful. Identifying drug-related syndromes could be a useful technique form identifying patients with nephrotoxicity.[1] Over the years, this approach has been modified and expanded. Table 1 summarizes Cooper and Bennett's classification of drug-related clinical syndromes.[29] In addition to defining the clinical syndrome, the common causative agents, based upon the frequency of occurrence, are included.

Defining the site of drug-induced nephron damage has depended in large part on recognition of distinctive clinical patterns of presentation. In some cases, this can be supplemented by animal experiments.[50] Since there are several suitable methods for estimating GFR and RBF, drugs inducing a dominantly glomerular injury are easier to characterize. However, since there is no satisfactory technique for assessing global renal tubular function, more indirect and inferential approaches are required. The first involves matching changes in plasma electrolytes as they reflect site-specific alteration in tubular function.[51] An example would be the renal concentrating effect that characterizes chronic analgesic nephropathy, since the pathologic lesion involves the medullopapillary portion of the kidney. Conversely, the type II renal tubular acidosis (RTA), bicarbonate wasting, that has been reported with exposure to heavy metals is consistent with the proximal tubular pathology characteristic of such exposure. If a drug-induced injury were limited to the proximal tubule, the altered tubular function would consist of a reduced reabsorption of a wide variety of filtered items, for example, glucose, bicarbonate, phosphate, uric acid, and amino acids. These alterations would be detected by glucosuria (with normal blood glucose concentration), increased fractional excretion[1] of phosphorus, a fractional excretion of uric acid >10%, and aminoaciduria. For a drug whose injury involves the distal nephron changes in both reabsorption, for example, sodium, and secretion, for example, potassium and hydrogen, the alterations would be detected by an increased FE of sodium and elevated plasma potassium and distal (type I) RTA. Finally, for drug-induced damage of the medullopapillary portion of the nephron, a decreased reabsorption of both sodium and water would result in a concentrating defect combined with an increased FE of sodium. The specificity of these patterns of findings decrease from the proximal tubule to the medullopapillary site. Unfortunately, the sensitivity of these

observations is sufficiently low that alternate techniques for assessing tubular injury have been pursued.

Another approach for evaluating tubular injury is by determining the presence of "tubular proteinuria."[52] Basically, this involves the precise measurement of low molecular weight proteins that undergo filtration but fail to be reabsorbed due to tubular cell dysfunction.[53] An example of this approach is the use of β_2-microglobulin (β_2-MG) excretion with increasing body burden of cadmium.[54,55] The clearance of β_2-MG is a sensitive measure of tubular function provided one is careful to prevent urinary proteolytic digestion by maintaining a pH > 5.5 in the sample. The excretion of Tamm-Horsfall protein (THP) has also been proposed as a measure of tubular injury;[56] however, the technical aspects of measuring THP preclude its widespread use at present.

Urinary enzymes have attracted attention as a possible biomarker of renal tubular injury because of the known differences in enzyme distribution in various segments of the nephron.[57] Unfortunately, up to 1996, few studies have been reported correlating enzymuria with histological confirmed tubular damage.[58] In part, this may be traced to the variety of nonpathologic factors that are known to increase the urinary excretion of certain enzymes.[59] The basis for linking enzymuria with tubular disease rests on the premise that the bulk of large molecular weight enzymes (>70 kDa) reach the nephron lumen by cellular release rather than filtration. On a practical basis, only a limited number of urinary enzymes have proven useful in detecting renal injury. *N*-acetyl-beta-glucosaminidase (NAG) is an example of a urinary enzyme that has been used both for screening for chemical exposure in the workplace[60] and the detection of nephrotoxic injury associated with the injection of contrast media.[61] Other enzymes that have been reported to correlate with various drugs that have effects on the kidney are alanine aminopep (AAP), intestinal alkaline phosphatase (IAP), and γ-glutamyl transpeptidase (GGT). Although enzymuria is much more sensitive than proteinuria as a biomarker of tubular disease, there still remain several unresolved questions as to site specificity and quantitation that need to be answered before universal acceptance will be possible. Urinary monoclonal antibodies to renal cell membrane have been developed and potentially offer a sensitive, specific, biomarker for assessing tubular injury.[62] Confirmation and extension of the clinical conditions studied will improve application and acceptance of this technology.

7.01.8.2 Examination of Urine and its Constituents

Although specificity is often lacking, a carefully performed urinalysis can provide vital information at little cost to the clinician evaluating a patient for drug-induced disease. Furthermore, disease activity can easily be followed over time. Specific gravity of the urine in the absence of glucose, protein, and radiographic contrast dyes gives an acceptable estimate of a directly measured urinary osmolarity. If the patient's state of hydration is known, the urinary specific gravity or osmolality can be a general guide to the renal tubular capacity for normal concentration and dilution. Formal concentration and dilution tests are seldom utilized in clinical practice because of difficult standardization and inconvenience. Inability to concentrate the urine in a setting of extracellular volume depletion or dehydration may be an early clue to toxic tubular injury. Conversely, preservation of concentrating ability despite a low measured or estimated GFR may suggest that the cause of renal dysfunction is "prerenal." Similarly, the urinary pH can be used if the patient's physiological circumstances are known, to ascertain the kidneys' ability to excrete the body's daily acid load. In the face of systemic acidosis, a urinary pH above 6 may indicate renal tubular dysfunction.

The qualitative presence of proteinuria, blood pigments, leukocytes, and glucosuria may be useful in documenting the renal effects of some drugs and toxins. Positive qualitative and/or dipstick tests should be followed up with more specific and quantitative studies to eliminate the substantial number of false positives when these sensitive studies are used for screening. Quantification of urinary protein is useful for separating glomerular diseases (>2 g per 24 h) from tubulointerstitial processes (<1.5 g per 24 h). There is a close correlation between 24 h urinary protein excretion and the ratio of protein to creatinine.[63] A normal ratio is 0.1, whereas a ratio of 2 equates with 2 g of protein per 24 h. Dipstick concentrations of protein in a single urine sample are inadequate because of the wide inter- and intrapatient variation in urine volume. Also, the presence of nonalbumin proteins in the urine may be missed by dipstick methodology. This applies to immunoglobulin light chains in multiple myeloma. Sensitive assays are now available to detect small amounts of urinary albumin excretion. So-called microalbuminuria is a predictor of subsequent renal disease in diabetic nephropathy. Its use as an indicator of toxic nephropathy has not been fully explored.

Microscopic examination of the urine sediment may provide the clinician with useful information about possible nephrotoxic renal disease. In addition to red blood cells and white blood cells, increased numbers of tubular cells may be noted in tubular necrosis, due to drugs or toxins. Eosinophilia is quite specific for acute interstitial nephritis, which is most often due to drugs and chemicals. Casts containing a matrix of Tamm-Horsfall mucoprotein form in the acidic environment of the distal nephron. Granular casts may contain aggregates of serum proteins, including darkly pigmented casts containing hemoglobin or myoglobin. Patients with tubular necrosis due to hemoglobinuria or myoglobinuria will often excrete large numbers of these pigmented casts. Red cells, white cells, and tubular cells may appear in the casts of patients with glomerulonephritis, interstitial nephritis, and tubular cell injury, respectively. Calcium oxalate crystalluria in patients with acute renal failure may suggest ethylene glycol of methoxyflurane-induced acute renal failure.

Urinary electrolyte concentrations vary greatly with intake and thus are of limited value in the diagnosis and management of toxic renal disease. When intake and volume status are known, inappropriately high urinary sodium concentrations may indicate renal tubular dysfunction. Likewise, in hypokalemic states, a high urinary potassium may indicate a renal source of potassium loss.

The presence in the urine of low molecular weight proteins, such as β_2-microglobulin and retinol-binding protein, which are filtered at the glomerulus and catabolized by the proximal tubule, indicates tubular disease or damage. Serial evaluations in patients receiving a drug or chemical may provide an early sensitive indicator of tubular cell damage. Correlation of these and other sensitive indicators of tubular cell integrity, such as enzymuria with functional and structural evidence of renal damage, are not available in patients exposed to drugs and chemicals. Until these types of studies are done, nephrotoxicity will continue to be defined by GFR estimations. Thus, it is imperative that new drugs with nephrotoxic potential in preclinical testing not be considered safe until long-term experience provides evidence that cumulative renal damage is absent with long-term, low-dose or repeated exposures.

7.01.8.3 Structural Assessment of the Kidney

The kidney may be imaged by a wide variety of techniques, such as ultrasound, computed tomography, and magnetic resonance imaging. Although these techniques all provide useful information for the clinician, they have no specific use in nephrotoxic renal disease. Definition of overall kidney size is a rough guide as to functional reserve. Small, contracted kidneys are associated, in general, with more chronic disease processes. The end-stage kidney, due to toxic nephropathy, cannot be distinguished from other causes of this process by imaging procedures. Cysts and renal masses are best delineated by computed tomography or magnetic resonance imaging. The prognosis of many acute renal insults may be gauged by computed tomography scanning to exclude renal cortical mass loss; loss of corticomedullary differentiation on magnetic resonance imaging is a sensitive indicator of a moderate to severe insult to the kidney, but provides no information about the cause of damage. Thus, renal imaging can provide useful general information, but specific details are only ascertained by renal biopsy.

Percutaneous renal biopsy needs to be performed by an experienced operator. The overall complication rate is 2% to 3%, mainly as bleeding, perirenal hematoma, or rarely, arteriovenous fistula. The primary indication for the procedure is renal failure, acute or chronic, of unknown cause. Biopsy is also helpful in defining the pathological processes involved in patients with proteinuria, hematuria, or nephrotic syndrome. Valuable prognostic information, particularly about the amount and type of interstitial involvement in a disease process, may help determine the aggressiveness or type of therapeutic approach. There are few data on a systematic morphological approach to patients with toxic renal injury.

7.01.9 RISK FACTORS—DRUG-INDUCED RENAL FAILURE

Based upon experimental findings, our understanding of drug-induced renal disease has expanded rapidly; despite this, it is still important to identify the patient groups that are at significant risk from toxin-induced renal injury. The clinical manifestations of drug-induced renal injury may be a sudden, acute deterioration of renal function or a chronic insidious loss. The different time course may simply represent dosage exposure to toxins, as it has been reported that the rapidity with which experimental renal failure occurs is dependent on the rate at which the nephrotoxins are administered.[64] Although this has not been proven for human renal disease, if true it provides an explanation for why it has been so difficult to confirm the role of suspected environmental toxicants.

Because of the frequency and well-defined clinical course of acute renal failure, much of our knowledge regarding risk factors in renal failure comes from analysis of these patients. In particular, when multivariant analysis was applied to a group of patients with acute renal failure, a defined frequency for both predisposing insults (Table 9) and clinical risk factors (Table 10) was achieved.[30]

Hypotension was the most frequent single predisposing insult, with pigmenturia second. However, for the majority of patients acute renal failure occurs in the setting of multiple, simultaneous insults. Hypotension, dehydration, and sepsis cause abnormal body fluid volume regulation, which can be monitored as changes in blood pressure. The other predisposing insults of significance are drugs or chemicals used to establish a diagnosis or treat a coexisting disease. As alluded to above, antibiotics continue to be the most frequent cause of drug-induced renal failure in hospitalized patients.[31]

The patient risk factors summarized in Table 9 indicate that although it is more common for a single risk factor to be recognized than a single acute insult, the pattern of multiple risk

Table 9 Frequency of combined acute insults in patients without previous renal disease ($n = 121$).

Acute insult	↓BP	S	AG	RM	P	LD	↓ECV
Sole insult	33	0	1	1	5	0	1
Multiple insults	78	30	10	9	11	1	1
Sepsis	29	0					
Aminoglycoside[a]	4	10	1				
Radiocontrast media	12	1	0	1			
Pigmentoria[a]	12	2	0	1	5		
Liver disease	2	3	1	0	0	0	
Dehydration[a]	20	14	9	8	11	1	

Source: Rasmussen and Ibels.[30]
[a]Significant insult in causing ARF based on discriminant multiple linear regression analysis.

Table 10   Frequency of combined risk factors in patients with ARF ($n = 143$).

Suspected risk factors	A	↑BP	↑UA	DM	RD	D
Age > 59 years	30					
Hypertension[a]	29	4				
Gout/Hyperuricemia	21	18	4			
Diabetes	11	6	4	1		
Renal disease[a]	18	12	12	6	4	
Diuretics	29	27	21	8	13	
Multiple risks	108	63	37	14	13	0

Source: Rasmussen and Ibels.[30]
[a]Significant risk contributing to ARF based on discriminant multiple linear regression analysis.

factors coexisting in the same patient occurs in the majority of patients who develop acute renal failure.

Age above 60 has been a consistent risk factor in several studies. It is reasonable to view age as biomarker of increased renal vulnerability due to the measurable decline in GFR that occurs with each decade beyond the age of 50. In searching for an explanation for this decline in renal function, it has been suggested that progressive atherosclerotic lesions, which accompany aging, could induce the progressive decline through a diminished renal blood flow.[65] An alternate explanation is that mature kidneys fail to respond to hypertrophic growth factors. In patients who act as live donors of a kidney, it has been observed that following unilateral nephrectomy, the retained kidney undergoes both functional and structural hypertrophy. However, for live donors 50 years or older, little or no functional increase occurs, which suggests that their stimulus of growth is blunted and ineffective. Whether or not renal cell regeneration is a lifelong physiological process that deteriorates with aging remains an interesting speculation.

Pre-existing renal disease is a very significant patient risk factor. Conceptually, the added burden of a hyperfiltering nephron would intensify the vulnerability of the tubular processes previously outlined. In addition, for drugs that depend upon intact renal function for elimination,[66] abnormal accumulation and excess blood levels will add to the nephrotoxic insult.

The sex of a patient has not been verified as a significant risk factor for humans. This is not the case for experimental nephrotoxicants where the animal's sex clearly influences susceptibility to drug or chemical insults.

Disease states that predispose to drug-induced renal injury include systemic hypertension, diabetes mellitus, liver disease vasculitis, and systemic lupus erythematosus (SLE). These conditions may share the existence of glomerular hypertension, an observation that has been confirmed in experimental studies where glomerular dynamics are determined using micropuncture techniques.

Aminoglycoside antibiotics are an example of a drug whose prior exposure increases the risk of renal injury. A critical accumulation of the drug, which has been reported to occur in those patients who eventually develop aminoglycoside nephrotoxicity, is consistent with this observed risk. Exactly how this critical accumulation induces risk is unclear but may be due to unusual tissue binding and/or delayed drug elimination.

To evaluate risk factors in states of chronic drug exposure is problematic. One group of patients of interest are those with chronic lithium administration. Observations on these patients indicate that 10 to 20 years of low-dose treatment are probably required for chronic lithium nephropathy to develop. Lokkegaard et al.[67] correlated a small decline in glomerular filtration rate with the duration of lithium treatment. The derived regression line intersected the 95% confidence limits for age-matched control glomerular filtration rates with the duration of lithium administration. This is important since the report by Bendz,[68] which found that no correlation existed between lithium-induced nephropathy and sex, diagnosis, duration of treatment, serum lithium levels, type of lithium preparation used, or combined lithium-neuroleptic treatment, was derived form a patient population whose average duration of treatment was only 5 years. A defect in maximum urinary concentrating capacity occurs in 30% to 100% of patients receiving chronic lithium treatment and is inversely correlated with the duration of lithium treatment.[69] Importantly, approximately 10% of patients receiving long-term lithium therapy will demonstrate measurable reductions in glomerular filtration with a larger fraction showing mild renal tubular acidosis. Chronic interstitial nephritis has been reported[70,71] in patients receiving long-term lithium therapy without acute lithium intoxication. To assign these changes to lithium seems premature since similar pathologic changes have been noted in renal biopsies from patients with affective disorders not receiving lithium. Additional risk factors for chronic lithium nephropathy include dose frequency, but not accumulative dose and volume depletion, irrespective of its cause.

Intrinsic to determining relative risk from the various factors discussed is the lack of any verifiable cause and effect. For chronic renal impairment, the long duration of environmental exposure that is probably required before clinically evident renal dysfunction is manifested further complicates the interpretation. Adding to this uncertainty is the absence of a convenient clinical measure of total tubular function as a companion to the measurement of glomerular filtration rate. The lack of a universally accepted measure of tubular function severely curtails our ability to detect early nephron changes.

Table 9 contains a list of commonly identified patient risk factors that seem to be consistent between patient populations developing acute renal failure. Furthermore, in certain cases the toxicant possesses unique features that will assist in the identification and the prevention of drug-induced renal injury.

7.01.10 PREVENTION/MODIFICATION

Experimentally, many approaches to modifying or preventing drug-induced nephrotoxicity have been tried.[72] Such techniques may vary from concomitant administration of specific drugs or chemicals through modifying drug dosage amounts or schedules to changing dietary content. To date, 1996 only a few such strategies have undergone clinical trial. The use of risk stratification is very important when developing a clinical management program directed at prevention. Obviously, patient characteristics that cannot be changed include age, sex, and coexisting disease; on the other hand, volume or electrolyte depletion, hypotension, and concomitant nephrotoxic drugs can be actively treated or eliminated prior to administering a potentially injurious drug or chemical.

Clinically tested methods for prevention or modifying drug-induced ARF include (i) volume expansion/osmotic diuresis, (ii) loop diuretics, and (iii) calcium channel blockers. Of the three approaches, volume expansion has had the greatest accumulated experience. The basis by which mannitol might prevent drug-induced ARF probably involves its action to reduce proximal tubular reabsorption of sodium and water, thus inducing an osmotic diuresis. The net effect of such limited proximal reabsorption would be to decrease the concentration of drug available for tubular cell uptake. Furthermore, the increased flow in the distal nephron and collecting duct system would limit reabsorption in this region. Much of the early clinical experience with mannitol involved its use in patients undergoing aortic surgery. We demonstrated that by introducing mannitol into the perfusate, a substantial reduction in the rate of hemolysis occurred that could also account for the reduced incidence of ARF.

The use of loop diuretics is based on the concept that converting oliguric ARF to nonoliguric ARF improves patient survival by reducing the problem of fluid management. Calcium channel blockers have been used to prevent drug-induced ARF. Although there are much experimental data to support such an approach,[73] few clinical studies have been reported. Russo and co-workers[74] found that three days of pretreatment with the calcium channel blocker nitrendipine, followed by the injection of contrast media, reduced the magnitude of the decline in GFR in patients preselected to be at high risk. As with other proposed techniques for preventing ARF, this one will have to await confirmation. Non ionic contrast media are beneficial in high-risk patients.[75]

In summary, the best prevention is avoidance of potentially nephrotoxic drugs or chemicals, especially in patients at high risk. However, in many circumstances that is not practical, since often patients are critically ill and no alternative approaches are available. Under such circumstances, it is crucial to eliminate the situations known to predispose to nephrotoxicity, for example, volume repair, restoration of blood pressure, and electrolyte replacement. It must also be emphasized that careful monitoring of renal function, usually by serial serum creatinines, is essential since most cases of drug-induced renal injury present as nonoliguric ARF.

7.01.11 REFERENCES

1. W. M. Bennett, L. W. Elzinga and G. A. Porter, in 'The Kidney,' eds. B. Brenner and F. Rector, Saunders, Philadelphia, PA, 1991, pp. 1430–1496.
2. G. R. Matzke, R. L. Lucarotti and H. S. Shapiro, 'Controlled comparison of gentamicin and tobramycin nephrotoxicity.' *Am. J. Nephrol.*, 1983, **3**, 11–17.
3. H. D. Humes, 'Arrinoglycoside nephrotoxicity.' *Kidney Int.*, 1988, **33**, 900–911.

4. D. N. Gilbert, C. A. Wood, S. J. Kohlhepp *et al.*, 'Polyasportic acid prevents experimental aminoglycoside nephrotoxicity.' *J. Infect. Dis.*, 1989, **159**, 945–953.

5. W. M. Bennett, 'Mechanisms of aminoglyroside nephrotoxicity.' *Clin. Exp. Pharmacol. Physiol.*, 1989, **16**, 1–6.

6. E. A. Burdmann, T. F. Andoh, J. Lindsley *et al.*, 'Urinary enzymes as biomarkers of renal injury in experimental nephrotoxicity of immunosuppressive drugs.' *Ren. Fail.*, 1994, **16**, 161–168.

7. S. K. Swan and W. M. Bennett, 'Drug dosing guidelines in patients with renal failure.' *West. J. Med.*, 1992, **156**, 633–638.

8. J. E. Bennett, in 'Current Clinical Topics in Infectious Disease,' eds. R. S. Remington and M. N. Swartz, McGraw-Hill, New York, 1981, pp. 54–67.

9. R. A. Branch, 'Prevention of arrphotericin B-induced renal impairment. A review on the use of sodium supplementation.' *Arch. Intern. Med.*, 1988, **148**, 2389–2394.

10. H. Calvert, I. Judson and W. J. van der Vijgh, 'Platinum complexes in cancer medicine: pharmacokinetics and pharmacodynamics in relation to toxicity and therapeutic activity.' *Cancer Surv.*, 1993, **17**, 189–217.

11. K. B. Meyer and N. E. Madias, 'Cisplatin nephrotoxicity.' *Miner. Electrolyte. Metab.*, 1994, **20**, 201–213.

12. J. D. Blachley and J. R. Hill, 'Renal and electrolyte disturbance associated with cisplatin.' *Ann. Intern. Med.*, 1981, **95**, 628–632.

13. F. N. Hutchison, E. A. Perez, D. R. Gandara *et al.*, 'Renal salt wasting in patients treated with cisplatin.' *Ann. Intern. Med.*, 1988, **108**, 21–25.

14. W. J. van der Vijgh, 'Clinical pharmacokinetics of carloplatin.' *Clin. Pharmacokinet.*, 1991, **21**, 242–261.

15. J. L. Blackshear, M. Davidman and M. T. Stillman, 'Identification of risk for renal insufficiency from nonsteroidal anti-inflammatory drugs.' *Arch. Intern. Med.*, 1993, **143**, 1130–1134.

16. D. M. Clive and J. S. Stoff, 'Renal syndromes associated with nonsteroidal anti-inflammatory drugs.' *N. Engl. J. Med.*, 1984, **310**, 563–572.

17. O. Spuler and H. U. Zollinger, 'Die chronische-interstitielle nephritis.' *Z. Klin. Med.*, 1953, **151**, 1–50.

18. T. Murray and M. Goldberg, 'Analgesic abuse and renal diseases.' *Annu. Rev. Med.*, 1975, **26**, 537–550.

19. R. H. Murray, D. H. Lawson and A. C. Linton, 'Analgesic nephropathy: clinical syndrome and prognosis.' *Br. Med. J.*, 1971, **1**, 479–482.

20. W. M. Bennett and M. E. DeBroe, 'Analgesic nephropathy—a preventable renal disease.' *N. Engl. J. Med.*, 1989, **320**, 1269–1271.

21. M. Segasothy, S. A. Samad, A. Zulfigar *et al.*, 'Chronic renal disease and popillary necrosis associated with the long-term use of nonsteroidal anti-inflammatory drugs as the role as predominant analgesic.' *Am. J. Kidney Dis.*, 1994, **24**, 17–24.

22. W. M. Bennett, 'Lead nephropathy.' *Kidney Int.*, 1985, **28**, 212–220.

23. T. K. S. Rao, A. D. Nicastri and E. A. Friedman, 'Natural history of heroin-associated nephropathy.' *N. Engl. J. Med.*, 1974, **290**, 19–23.

24. G. Feutren and M. J. Mihatsch, 'Risk factors for cyclosporine-induced nephropathy in patients with autoimmune diseases. International Kidney Biopsy Registry of Cyclosporine in Autoimmune Diseases.' *N. Engl. J. Med.*, 1992, **326**, 1654–1660.

25. J. C. Lee, M. Dushkin, E. J. Eyring *et al.*, 'Renal lesions associated with gold therapy. Light and electron microscopic studies.' *Arthritis Rheum.*, 1965, **8**, 1–13.

26. M. Miller and A. M. Moses, 'Mechanism of chlorpropamide action in diabetes insipiclus.' *J. Clin. Endocrinol. Metab.*, 1970, **30**, 488–496.

27. I. Singer and J. N. Forrest, Jr., 'Drug-induced states of nephrogenic diabetes insipidus.' *Kidney Int.*, 1976, **10**, 82–95.

28. L. H. Norby, J. Weidig and P. Ranwell, 'Possible role for impaired renal prostaglandin production in pathogenesis of hyporeninaemic hypoaldosteronism.' *Lancet*, 1979, **2**, 1118.

29. K. Cooper and W. M. Bennett, 'Nephrotoxicity of common drugs used in clinical practice.' *Arch. Intern. Med.*, 1987, **147**, 1213–1218.

30. H. H. Rasmussen and L. S. Ibels, 'Acute renal failure. Multivariate analysis of causes and risk factors.' *Am. J. Med.*, 1982, **73**, 211–218.

31. S. H. Hou, D. A. Bushindky, J. B. Wish *et al.*, 'Hospital-acquired renal insufficiency: a prospective study.' *Am. J. Med.*, 1983, **74**, 243–248.

32. D. Kleinknecht, P. Laudais and B. Goldfarb, 'Drug-associated acute renal failure: a prospective multicentre report.' *Proc. EDTA-ERA*, 1985, **22**, 1002–1007.

33. Anonymous, 'Statement on the release of ibuprofen as an over-the-counter medicine. Ad Hoc Committee for the National Kidney Foundation.' *Am. J. Kidney Dis.*, 1985, **6**, 4–6.

34. Y. Pirson and C. Van Ypersele de Strihou, 'Renal side effects of nonsteroidal anti-inflammatory drugs: clinical relevance.' *Am. J. Kidney Dis.*, 1986, **8**, 338–344.

35. K. Beard, D. R. Perera and H. Jick, 'Drug-induced parenchymal renal disease in outpatients.' *J. Clin. Pharmacol.*, 1988, **28**, 431–435.

36. Anonymous, 'Report on acute (reversible) renal failure.' *Proc. EDTA-ERA*, 1983, **20**, 64–67.

37. N. Shusterman, B. L. Strom, T. G. Murray *et al.*, 'Risk factors and outcome of hospital-acquired acute renal failure. Clinical epidemiologic study.' *Am. J. Med.*, 1987, **83**, 65–71.

38. T. H. Hostetter, H. G. Rennke and B. M. Brenner, 'The case for intrarenal hypertension in the initiation and progression of diabetic and other glomerulopathies.' *Am. J. Med.*, 1982, **72**, 375–380.

39. R. C. Blantz, K. S. Konnen and B. J. Tucker, 'Angiotentenein II effects upon the glomerular microcirculation and ultrafiltration coefficient of the rat.' *J. Clin. Invest.*, 1976, **57**, 419–434.

40. B. M. Brenner, T. H. Hostetter and H. D. Humes, 'Glomerular permselectivity: laiver function based on discrimination of molecular size changes.' *Am. J. Physiol.*, 1978, **234**, F455–F460.

41. U. Ravnskov, 'Possible mechanisms of hydrocarbon-associated glomerulonephutio.' *Clin. Nephrol.*, 1985, **23**, 294–298.

42. R. J. Glassock, in 'Renal and Electrolyte Disorders,' ed. R. W. Schrier, Little Brown, Boston, MA, 1986, pp. 591–657.

43. B. Albini, I. Glurich and G. A. Andres, in 'Nephrotoxic Mechanisms of Drug and Environmental Toxins,' ed. G. A. Porter. Plenum, New York, 1982, pp. 413–423.

44. H. W. Smith, in 'From Fish to Philosopher,' Little, Brown, MA, Boston, 1953, pp. 136–152.

45. G. H. Mudge, in 'Nephrotoxic Mechanisms of Drugs and Environmental Toxins,' ed. G. A. Porter, Plenum, New York, 1982, pp. 209–225.

46. D. Batlle, M. Gaviria, M. Grupp *et al.*, 'Distal nephron function in patients receiving chronic lithium therapy.' *Kidney Int.*, 1982, **21**, 477–486.

47. P. M. Tulkens, 'Experimental studies on nephrotoxicity of aminoglycosides at low doses.' *Am. J. Med.*, 1986, **80**, 105–114.

48. D. C. Houghton, M. V. Campbell-Boswell and W. M. Bennett, 'Myeloid bodies in the renal tubules of human relationship to gentamicin therapy.' *Clin. Nephrol.*, 1978, **10**, 140–145.

49. S. Mujais and D. Batlle, 'Functional correlates of

tubulo-interstitial damage.' *Semin. Nephrol.*, 1988, **8**, 94–99.

50. G. A. Porter, in 'Nephrotoxicity in the Experimental and Clinical Situation,' eds. P. H. Bach and E. A. Lock, Naiturus Nijhoff, Boston, MA, 1987, pp. 613–642.

51. M. G. Cogan, 'Medical staff conference. Tubulo-interstitial nephropathies—a pathophysiologic approach.' *West J. Med.*, 1980, **132**, 134–141.

52. P. A. Peterson, P. E. Evrin and I. Berggard, 'Differentiation of glomerular, tubular, and normal proteinuria: determinations of urinary excretion of beta-2-macroglobulin, albumin, and total protein.' *J. Clin. Invest.*, 1969, **48**, 1189.

53. V. Dennis and R. R. Robinson, in 'The Kidney: Physiology and Pathology,' eds. D. W. Seldin and G. Giebisch, River Press, New York, 1985, pp. 1805–1815.

54. L. Friberg, 'Cadmium in the Environment,' 2nd edn., CRC Press, Cleveland, OH, 1974.

55. R. Goyer, in 'Nephrotoxic Mechanism of Drugs and Environmental Toxins,' ed. G. A. Porter. Plenum, New York/London, 1982, pp. 305–313.

56. C. Thornley, A. Dawnay and W. R. Cattell, 'Human Tamm-Horsfall glycoprotein: urinary and plasma levels in normal subjects and patients with renal disease determined by a fully validated radioimmunoassay.' *Clin. Sci.*, 1985, **68**(5), 529–535.

57. D. J. Harrison, R. K. Kharbanda, D. S. Cunningham *et al.*, 'Distribution of glutathione *S*-transferase isoenzymes in human kidney: basis for possible markers of renal injury.' *J. Clin. Pathol.*, 1989, **42**, 624–628.

58. R. J. Portman, J. M. Kissane and A. M. Robeson, 'Use of beta 2 microglobulin to diagnose tubulo-interstitial renal lesions in children.' *Kidney Int.*, 1986, **30**, 91–98.

59. D. T. Plummer, S. Naarazar, D. K. Obatomi *et al.*, 'Assessment of renal injury by urinary enzymes.' *Uremia Invest.*, 1985, **9**, 97–102.

60. B. R. Meyer, A. Fischbein, K. Rosenman *et al.*, 'Increased urinary enzyme excretion in workers exposed to nephrotoxic chemicals.' *Am. J. Med.*, 1984, **76**, 989.

61. A. Rahimi, R. P. Edmondson and N. F. Jones, 'Effect of radiocontrast media on kidneys of patients with renal disease.' *Br. Med. J. (Clin. Res. Edn.)*, 1981, **282**, 1194–1195.

62. A. Mutti, S. Lucertini, P. P. Valcavi *et al.*, 'Urinary excretion of brush-border antigen revealed by monoclonal antibody: early indicator of toxic nephropathy.' *Lancet*, 1985, **2**, 914.

63. S. J. Schwab, R. L. Christensen, K. Dougherty *et al.*, 'Quantitation of proteinuria by the use of protein-to-creatinine ratios in single urine samples.' *Arch. Intern. Med.*, 1987, **147**, 943–944.

64. A. Aviv, E. John, J. Bernstein *et al.*, 'Lead intoxication during development: its late effects on kidney function and blood pressure.' *Kidney Int.*, 1980, **17**, 430–437.

65. L. H. Avendano and J. M. Lopez-Novoa, in 'Renal Function and Diseases in the Elderly,' eds. I. F. Macias-Nunez and J. S. Cameron, Butterworth, London, 1987, pp. 27–48.

66. W. M. Bennett, G. R. Aronoff, T. A. Golper *et al.*, in 'Drug Prescribing in Renal Failure,' American College of Physicians, Philadelphia, PA, 1994.

67. H. Lokkegaard, N. F. Andersen, E. Henricksen *et al.*, 'Renal function in 153 manic-depressive patients treated with lithium for more than five years.' *Acta. Psychiatr. Scand.*, 1985, **71**, 347–355.

68. H. D. Bendz, 'Kidney function in lithium-treated patients. A literative survey.' *Acta. Psychiatr. Scand.*, 1983, **68**, 303–324.

69. R. G. Walker, W. M. Bennett, B. M. Davis *et al.*, 'Structural and functional effects of long-term lithium therapy.' *Kidney Int. Suppl.*, 1982, **11**, S13–S19.

70. J. Hestbeck, H. E. Hansen, A. Andersen *et al.*, 'Chronic renal lesions following long-term treatment with lithium.' *Kidney Int.*, 1977, **12**, 203–205.

71. P. Plenge, E. T. Mellerup, T. G. Bolury *et al.*, 'Lithium treatment: does the kidney prefer one daily dose instead of two?' *Acta. Psychiatr. Scand.*, 1982, **66**, 121–128.

72. W. M. Bennett, C. A. Wood, D. C. Houghton *et al.*, 'Modification of experimental aminoglycoside nephrotoxicity.' *Am. J. Kidney Dis.*, 1986, **8**, 292–296.

73. S. N. Heyman, M. Brezis, Z. Greenfeld *et al.*, 'Protective role of furosemide and saline in radiocontrast-induced acute renal failure in the rat.' *Am. J. Kidney Dis.*, 1989, **14**, 377–385.

74. D. Russo, A. Testa, L. Della Volpe *et al.*, 'Randomised prospective study on renal effects of two different contrast media in humans: protective role of a calcium channel blocker.' *Nephron*, 1990, **55**, 254–257.

75. M. R. Rudnick, S. Goldfarb, L. Wexler *et al.*, 'Nephrotoxicity of ionic and nonionic contrast media in 1196 patients: a randomized trial. The Iohexol Cooperative Study.' *Kidney. Int.*, 1995, **47**, 254–261.

7.02
Functional Anatomy of the Kidney

JEFF M. SANDS

Emory University School of Medicine, Atlanta, GA, USA

and

JILL W. VERLANDER

University of Florida School of Medicine, Gainesville, FL, USA

7.02.1 INTRODUCTION

The kidney is one of the two major excretory organs in the body, the other being the liver and GI tract. The kidney's major function is to excrete waste products which the organism does not need while maintaining total body salt, water, potassium, and acid–base balance. Thus, the kidney is the primary organ responsible for preserving the constancy of the internal environment. The kidney accomplishes these tasks through a combination of three general mechanisms: (i) glomerular filtration; (ii) tubular reabsorption; and (iii) tubular secretion. Glomerular filtration refers to the movement of water and solute to form an ultrafiltrate of plasma across the glomerular membrane. Tubular reabsorption refers to the movement of a substance from tubular fluid back into the plasma. Tubular secretion refers to the movement of a substance from plasma into the tubular fluid.

The kidney is composed of a variety of highly specialized nephron segments (Figure 1).[1] These nephron segments are composed of distinct epithelial cell types which engender specific transport properties to specific nephron segments. However, the functional anatomy of the kidney results not only from the flow of tubular fluid from one nephron segment to the next, but also from the three-dimensional organization of the kidney. This is especially true in the kidney medulla where both the loop of Henle and the vasa recta are arranged in a hair-pin or U-shaped configuration.

The kidney is uniquely sensitive to toxins since they receive and filter an enormous quantity of blood. On average, the two kidneys receive 25% of the cardiac output and the glomeruli filter about 180 L of blood per day. Thus, the glomeruli are frequently exposed to toxins in the blood in excess of the exposure of other tissues. In addition, the processes of tubular reabsorption, tubular secretion, and countercurrent arrangement of structures in the kidney medulla often result in toxins being concentrated in renal tissue, plasma, or tubular fluid to levels vastly exceeding systemic plasma levels, thus exacerbating the potential for nephrotoxicity.

In addition to the kidney's role in maintaining the internal environment by reabsorbing necessary components of plasma and excreting waste products, the kidney is involved in the production and regulation of several important hormones, including erythropoietin, renin, and $1,25\text{-}(OH)_2\text{-}$cholecalciferol (vitamin D). Erythropoietin is an essential growth factor for hematopoiesis and patients with renal failure are commonly anemic unless they are treated with erythropoietin. Renin is a critical hormone produced by cells in the juxtaglomerular apparatus in response to a decrease in intrarenal vascular perfusion. Renin is released into the systemic circulation where it is converted to angiotensin II and subsequently stimulates aldosterone production by the adrenal gland. Both angiotensin II and aldosterone act to stimulate renal sodium reabsorption so as to restore intravascular volume and maintain adequate blood pressure. In pathologic states, stimulation of the renin–angiotensin II–aldosterone system can lead to hypertension or exacerbate congestive heart failure. Finally, the kidney hydroxylates 25-OH-cholecalciferol to its active form, $1,25\text{-}(OH)_2\text{-}$cholecalciferol, a critical hormone involved in regulating plasma calcium and phosphorus. Patients with renal failure often develop secondary hyperparathyroidism due to a deficiency in $1,25\text{-}(OH)_2\text{-}$cholecalciferol synthesis.

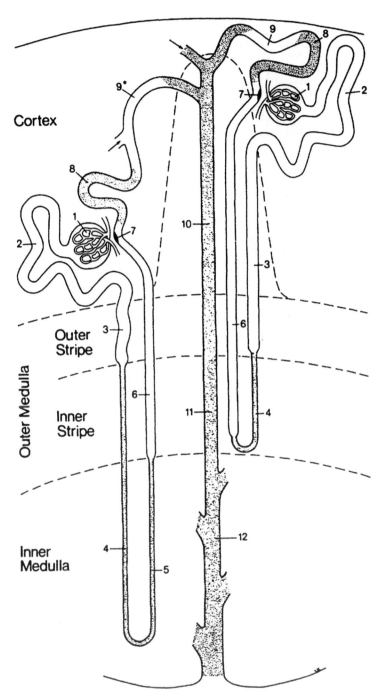

Figure 1 Diagram showing a short- and a long-looped nephron together with the collecting system (not drawn to scale). Within the cortex, a medullary ray is delineated by a dashed line. 1, Renal corpuscle including Bowman's capsule and the glomerulus (glomerular tuft); 2, proximal convoluted tubule; 3, proximal straight tubule; 4, thin descending limb; 5, thin ascending limb; 6, medullary thick ascending limb; 7, macula densa located within the final portion of the cortical thick ascending limb; 8, distal convoluted tubule; 9, connecting tubule; 10, cortical collecting duct; 11, outer medullary collecting duct; 12, inner medullary collecting duct (reproduced by permission of the American Physiology Society from *Am. J. Physiol.*, 1988, **254**, F1–F8).

In summary, the kidney is involved in a plethora of vital bodily functions, ranging from regulation of acid–base balance, maintenance of blood pressure via regulation of sodium excretion, maintenance of plasma osmolality via regulation of water excretion, regulation of plasma potassium, calcium, and phosphorus, and elaboration of erythropoietin to maintain

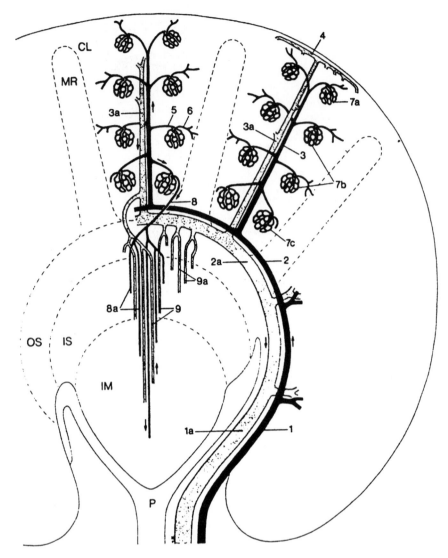

Figure 2 Diagram showing the course and distribution of the intrarenal blood vessels; peritubular capillaries are not shown (not drawn to scale). Within the cortex, the medullary rays (MR) are delineated from the cortical labyrinth (CL) by a dashed line. OS, outer stripe of the outer medulla; IS, inner stripe of the outer medulla; IM, inner medulla; P, renal pelvis (which is lined by the papillary surface epithelium). 1, 1a, Interlobar artery and vein; 2, 2a, arcuate artery and vein; 3, 3a, cortical radial artery and vein; 4, stellate vein; 5, afferent arteriole; 6, efferent arteriole; 7a, 7b, 7c, superficial, midcortical, and juxtamedullary glomerulus; 8, 8a, juxtamedullary efferent arteriole, descending vasa recta; 9, 9a, ascending vasa recta within a vascular bundle, ascending vasa recta independent of a vascular bundle (reproduced by permission of the American Physiology Society from *Am. J. Physiol.*, 1988, **254**, F1–F8).

hematocrit. In this chapter, we will provide an overview of the functional anatomy of the kidney, with an emphasis upon the major ultrastructural and functional features of each nephron segment. Our goal is to provide the reader with an anatomic and functional framework for understanding renal function and toxicology. Clearly, this chapter cannot provide an in-depth review of every aspect of renal functional anatomy and the reader is referred to any of the major nephrology texts for more a detailed discussion of a particular aspect of renal function.

7.02.2 OVERALL STRUCTURAL ORGANIZATION OF THE MAMMALIAN KIDNEY

Mammalian kidneys are paired. They are located posteriorly near the lower ribs, allowing the kidneys to be biopsied percutaneously by inserting a biopsy needle under radiologic guidance through the back in humans. When viewed anteriorly, the right kidney is situated behind the liver and the left kidney is below the spleen. Each kidney normally receives its blood supply from a single renal artery and renal vein.

The renal arteries originate from the aorta and the renal veins drain into the inferior vena cava. The urine formed in each kidney is drained via a single ureter into the bladder.

The kidney is subdivided into cortex, outer medulla, and inner medulla or papilla (Figures 1 and 2). Small mammals, such as rat and rabbit, are unipapillate, that is each kidney has a single papilla. The papilla is the pyramid-shaped bottom portion of the inner medulla and descends into the renal pelvis. Humans and larger mammals have multipapillate kidneys. Each papilla descends into a renal fornix. These fornices merge to form the renal pelvis. In all mammals, urine exits from the tip of the papilla(e) via the ducts of Bellini into the renal pelvis, which is an expanded upward extension of the ureter.

7.02.2.1 Cortex

The renal cortex consists of two regions: the cortical labyrinth and the medullary rays (Figures 1 and 2). The cortical labyrinth contains the glomeruli, proximal convoluted tubule, macula densa, distal convoluted tubule, connecting tubule, initial collecting duct, interlobular arteries, and the afferent and efferent capillary networks. The medullary rays contain the proximal straight tubule or pars recta, cortical thick ascending limb, and cortical collecting duct.

7.02.2.2 Outer Medulla

The outer medulla consists of two regions: the outer stripe and the inner stripe (Figures 1 and 2). The outer stripe is adjacent to the cortex and consists of the proximal straight tubule (also called the pars recta or thick descending limb), medullary thick ascending limb, and outer medullary collecting duct. The outer stripe is distinguished by the absence of any thin limbs. The inner stripe consists of the thin descending limb, thick ascending limb, and outer medullary collecting duct.

Descending and ascending vasa recta are located in both outer medullary regions (Figure 2). However, their is a difference in their organization between outer and inner stripe. Cone-shaped aggregations of both descending and ascending vasa recta called vascular bundles form in the outer stripe. The vascular bundles become much more prominent in the inner stripe.

7.02.2.3 Inner Medulla

The inner medulla consists of thin descending limbs, thin ascending limbs, inner medullary collecting ducts, and descending and ascending vasa recta (Figures 1 and 2). The collecting ducts merge as they descend through the inner medulla, ultimately merging to form the ducts of Bellini from which urine exits at the papillary tip. The deepest portion of the inner medulla is called the papilla. The papilla is located in the renal pelvis and is covered by the papillary surface epithelium (Figure 2).

7.02.3 NEPHRON HETEROGENEITY

Nephrons are described as superficial, mid-cortical, or juxtamedullary based upon the location of the glomerulus which gives rise to that nephron (Figures 1 and 2). In general, superficial nephrons have glomeruli located near the surface of the kidney and give rise to short-loop nephrons. These nephrons lack a thin ascending limb and do not penetrate into the inner medulla. Instead, their loop of Henle turns at the inner–outer medullary border and their thin descending limb connects to a medullary thick ascending limb. In addition, the proximal straight tubule of these nephrons is more centrally located within the medullary rays. In animals with the ability to produce concentrated urine, such as the rat, the thin descending limb descends within the vascular bundles. This organization allows counter-current exchange between the ascending vasa recta and descending thin limb of short-loop nephrons within the vascular bundles, especially within the inner stripe.[2,3]

Juxtamedullary nephrons have glomeruli which are located deep within the cortex, near the cortical–medullary border, and give rise to long-loop nephrons (Figures 1 and 2). These nephrons enter the inner medulla and have a thin ascending limb. In general, mammals with the ability to produce a highly concentrated urine have longer papillae and a higher percentage of long-loop nephrons.

Mid-cortical nephrons have glomeruli which are located between the superficial and juxtamedullary glomeruli (Figure 2). These nephrons may be either long-loop or short-loop. In general, the deeper the location of the glomerulus, the greater is the likelihood that its nephron will descend into the inner medulla.

7.02.4 RENAL VASCULATURE

Each kidney has a single renal artery which originates from the aorta. As it enters the renal sinus, the renal artery divides into several interlobar arteries (Figure 2). These enter the kidney tissue at the cortico–medullary border and become the arcuate arteries. Interlobular arteries arise from the arcuate arteries and ascend into the cortex, where they divide into afferent arterioles which supply blood to the glomerular capillaries. After traversing the renal corpuscle, the glomerular capillaries reform into a second arteriolar network, the efferent arterioles.

The fate of the efferent arteriole depends upon the location of its glomerulus (Figure 2). Efferent arterioles from superficial and mid-cortical nephrons form peritubular capillary networks which surround the proximal and distal convoluted tubules located in the cortex. However, an efferent arteriole which forms from a specific glomerulus does not necessarily provide blood to the peritubular capillary of the proximal tubule which originates from that glomerulus. The actual physiology and organization of the postglomerular vasculature is complex and poorly understood.

Efferent arterioles from the juxtamedullary nephrons form the descending vasa recta (Figure 2). The vasa recta are hair-pin or U-shaped vascular structures which supply blood to the medulla. The descending vasa recta form capillary networks at various levels within the medulla to supply it with oxygenated blood. As the vasa recta descends deeper into the medulla, the oxygen tension decreases and blood flow slows, such that the inner medulla is effectively anaerobic. This has an important clinical significance since the low oxygen tension, along with the high osmolality, produces a strong stimulus for sickling of SS (sickle cell anemia) or even SA (sickle trait) erythrocytes. Thus, papillary necrosis often occurs in patients with sickle cell anemia or sickle trait.

The ascending vasa recta provide venous drainage for the medulla (Figure 2). The hairpin configuration of the descending and ascending vasa recta allow for countercurrent exchange which permits rapid osmotic equilibration between arterial and venous capillary blood. Any increase in the osmolality of the venous outflow from the ascending vasa recta dissipates the effectiveness of the countercurrent multiplier and reduces urine concentrating ability. The ascending vasa recta merge with the venules draining the cortex to form interlobular veins. These merge to form arcuate veins, interlobar veins, and ultimately form the renal vein within the renal sinus. The renal vein drains into the inferior vena cava.

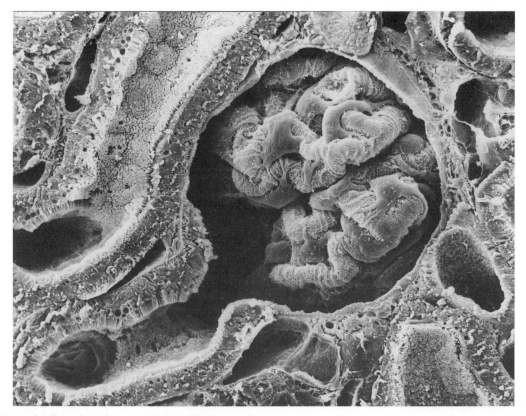

Figure 3 Scanning electron micrograph of a rat glomerulus. Bowman's capsule is cut away revealing the capillary tuft. The proximal tubule emerges from the urinary pole of Bowman's capsule (magnification: 900×).

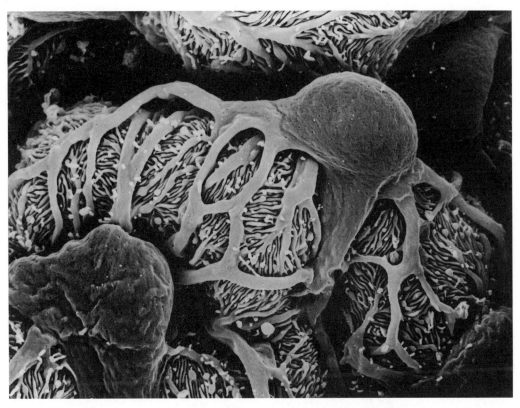

Figure 4 Scanning electron micrograph of a visceral epithelial cell. The cell body and primary and secondary foot precesses are illustrated (magnification: 5500×).

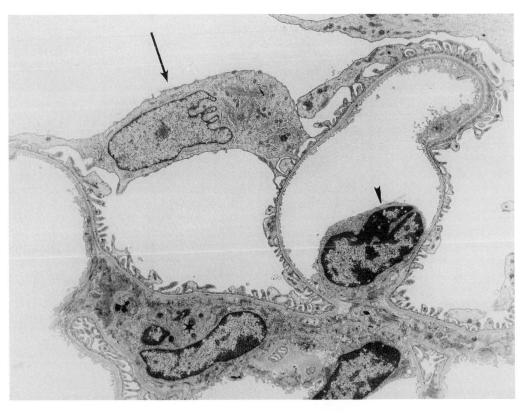

Figure 5 Transmission electron micrograph of a rat glomerulus. A visceral epithelial cell (arrow), endothelial cell (arrowhead), and mesangial cell (star) are illustrated (magnification: 4500×).

7.02.5 GLOMERULUS

7.02.5.1 Structure

The glomerulus is the filtering unit of the kidney. Each glomerulus consists of several capillary loops (Figure 3). Blood enters through the afferent arterioles and exits via the efferent arterioles. The glomerular capillaries are located between these two arteriolar systems. These capillaries are formed by large flat endothelial cells which have rounded fenestrae. The fenestrae are connected by thin diaphragms. The endothelial cells are covered by negatively charged, polyanionic

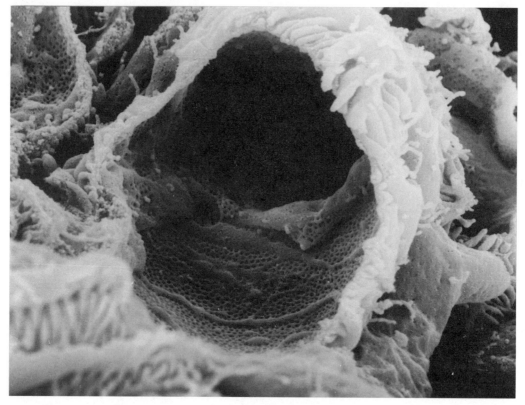

Figure 6 Scanning electron micrograph of a rat glomerular capillary cut to reveal the fenestrated endothelium that lines the lumen of the capillary. The foot processes of the visceral epithelial cells are visible on the urinary surface of the capillary (magnification: 10 800×).

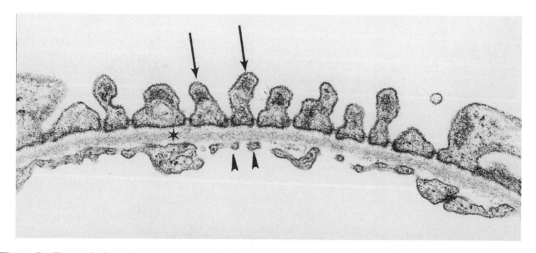

Figure 7 Transmission electron micrograph of a rat glomerular capillary wall. The foot processes of the visceral epithelial cells (arrows), the fenestrated capillary endothelium (arrowheads), and the basement membrane are illustrated (magnification: 36 900×).

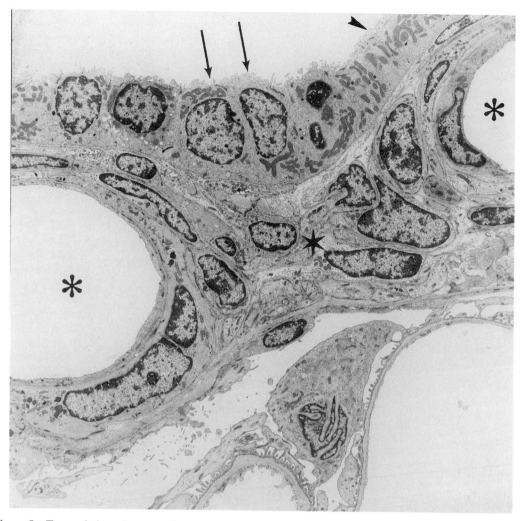

Figure 8 Transmission electron micrograph of the juxtaglomerular apparatus in rat kidney. The macula densa cells (arrows) are taller than the adjacent cortical thick ascending limb cell (arrowhead). The extraglomerular mesangium (star) is situated between the afferent and efferent arterioles (asterisks in lumens) and the macula densa. A portion of the glomerulus is visible at the bottom of the illustration (magnification: 3200×).

surface glycoproteins, of which the major protein is podocalyxin.[4]

Underneath the endothelial cells is the glomerular basement membrane, and on the opposite side are located the glomerular epithelial cells (Figure 4). Thus, the filtration barrier consists of these three structures: the endothelial cells, the glomerular basement membrane, and the visceral epithelial cells (Figure 5). Capillary blood is filtered across this barrier and an ultrafiltrate forms in Bowman's space which is the space located between the visceral and parietal epithelial cell layers (Figure 6). The proximal convoluted tubule originates from the parietal epithelial cell layer (Figure 7).

Extraglomerular mesangial cells are located in the hilum of the glomerulus and, along with the mesangial matrix, surround the afferent and efferent arterioles as they enter and exit the glomerulus (Figure 8). The glomerular mesangial cells are similar to vascular smooth muscle cells and cultured mesangial cells contract or relax in response to several vasoactive compounds when tested *in vitro*. Thus, it is hypothesized that mesangial cells may be involved in regulating glomerular hemodynamics. However, this has not been established *in vivo*. Another role proposed for mesangial cells, based upon their resemblance to smooth muscle cells, is that they proliferate during glomerular injury and are thus involved in proliferative glomerular lesions. Again, *in vivo* confirmation of this hypothesis is not available.

7.02.5.2 Function

The glomerulus is responsible for filtering the blood arriving via the afferent arteriole. The portion of plasma which crosses the glomerular membrane and enters Bowman's space is called an ultrafiltrate and is composed of water, solute, and other small molecules. The glomerular membrane forms both a size barrier and a charge barrier, and is generally less permeable to negatively charged molecules than neutral or positively charged molecules of the same size. Thus, the normal glomerular membrane prevents ultrafiltration of cells and of large anionic proteins. In disease states, both the size barrier and charge barrier are compromised and result in the appearance of proteins and cells in the ultrafiltrate and urine.[5,6]

Glomerular filtration is driven by the same Starling forces which drive fluid movement across the systemic capillary bed. The glomerular capillary is surrounded by two arterioles, the afferent and the efferent arteriole. Thus, the hydrostatic pressure in the glomerular capillaries is affected by both the afferent and efferent arteriolar tone. The glomerular capillary hydrostatic pressure normally exceeds the hydrostatic pressure in Bowman's space, promoting the production of ultrafiltrate. As ultrafiltration proceeds, the oncotic pressure within the glomerular capillary bed increases. In rat, the oncotic pressure can increase so that it equals the hydrostatic pressure, at which point ultrafiltration ceases, a condition known as filtration equilibrium. Whether filtration equilibrium occurs in humans is unknown.[5]

Glomerular filtration rate (GFR) can be calculated as follows

$$GFR$$
$$= K_f S(\Delta \text{hydraulic pressure} - \Delta \text{oncotic pressure})$$
$$= K_f S[(P_{gc} - P_{bs}) - s(\pi_p - \pi_{bs})]$$

where K_f is the ultrafiltration coefficient (or unit permeability) of the glomerular capillary wall, S is the surface area available for filtration, P is the hydraulic pressure, π is the oncotic pressure, gc is glomerular capillary, bs is Bowman's space, p is plasma, and s is the reflection coefficient of proteins across the capillary wall. In people or animals, GFR is measured using a substance which is filtered by the glomerulus but is neither reabsorbed nor secreted by the tubules such as inulin or iothalamate. These molecules are infused i.v. to achieve a stable plasma level and then their clearance is measured. $GRF = (U_i V)/P_i$, where U is the urine concentration, V is the urine flow rate, P is the plasma concentration, and i is the infused substance. In clinical practice, creati-

nine clearance is generally used rather than infusing a radioisotope. Creatinine has the advantage of being safe and inexpensive, but it undergoes tubular secretion. Thus, it is an imperfect measure of GFR. GFR can also be estimated using the following formulas which assume a stable rate of creatinine production.

For men:

$$GFR = [28 - (\text{age in years}/6)] \times \text{body weight (in kg)}$$

For women:

$$GFR = [22 - (\text{age in years}/9)] \times \text{body weight (in kg)}$$

The glomeruli filter approximately 180 L of plasma per day. This enormous rate of glomerular filtration results in the glomeruli being exposed to toxins in amounts which vastly exceed their plasma concentration. Thus, glomeruli are often damaged as innocent bystanders. For example, immune complexes can be trapped within the glomerular membrane and subsequent complement activation can result in damage to the glomerulus. Another example is aminoglycoside antibiotics which neutralize the anionic charges in the glomerular membrane and eliminate the charge barrier, ultimately leading to ultrastructural changes, reduced glomerular filtration, and acute renal failure.[7] One of the earliest manifestations of aminoglycoside-induced acute renal failure is proteinuria, presumably due to loss of the charge barrier which is normally responsible, in part, for preventing proteinuria.

The kidney is able to maintain renal blood flow and glomerular filtration over a wide range of systemic blood pressure due to autoregulatory mechanisms. While the precise mechanism(s) are not known, angiotensin II and prostaglandins are key components of renal autoregulation. Drugs which interfere with either angiotensin II or prostaglandin production can prevent autoregulation and predispose the kidney to toxic damage.

In addition to autoregulation of renal blood flow, two other mechanisms are present to preserve glomerular filtration: tubular–glomerular feedback and glomerular–tubular balance. Tubular–glomerular feedback is a process which is controlled by the macula densa cells of the juxtaglomerular apparatus. These cells are a specialized portion of the cortical thick ascending limb which connects with the hilum of the glomerulus. These cells respond to a decrease in tubular fluid, and presumably a decrease in intravascular volume, by secreting renin. Renin is converted to angiotensin II and aldosterone. These hormones cause vasoconstriction and stimulate sodium reabsorption to restore blood pressure. The increase in blood pressure increases renal perfusion, thereby increasing the hydrostatic pressure within the glomerular capillaries and restoring glomerular filtration. Thus, tubular–glomerular feedback is a mechanism by which the kidney senses decreased intravascular volume and activates the renin–angiotensin system to restore glomerular filtration. The exact signal which is sensed by the macula densa cells is controversial but is currently thought to involve some combination of tubular fluid sodium or chloride concentration and/or osmolality.

Glomerular–tubular balance is a mechanism by which proximal tubule reabsorption varies with glomerular filtration. If glomerular filtration increases, the oncotic pressure in the efferent arterioles increases, and this promotes reabsorption from the proximal convoluted tubule. Conversely, if glomerular filtration decreases, oncotic pressure does not increase as much, and proximal tubule reabsorption decreases. Thus, glomerular–tubular balance tends to decease the variation in the amount of tubular fluid delivered beyond the end of the proximal convoluted tubule.

7.02.6 PROXIMAL CONVOLUTED TUBULE

7.02.6.1 Structure

The proximal tubule originates from the parietal epithelial cell layer (Figure 9). It is commonly divided into two portions, the proximal convoluted tubule (PCT) and the proximal straight tubule (PST) or pars recta (PR). Morphologically, it is composed of three distinct segments: S_1, S_2, and S_3 (or P_1, P_2, and P_3).[8,9] S_1 consists of the first 1 mm of the proximal tubule (Figure 10). Thus, the S_1 segment is located entirely within the PCT. S_1 cells have a tall brush border (apical membrane) and an extensively interdigitated basolateral membrane. These cells have a well-developed vacuolar lysosomal system and many mitochondria. These mitochondria are predominantly located near the basolateral membrane and are typical of cells performing extensive active transport. These cells' high metabolic rate make them particularly susceptible to ischemia damage. These cells' role in

Figure 9 Scanning electron micrograph of rat proximal tubules illustrating the lush brush border on the luminal surface of the epithelium and the lateral interdigitations (magnification: 3400×).

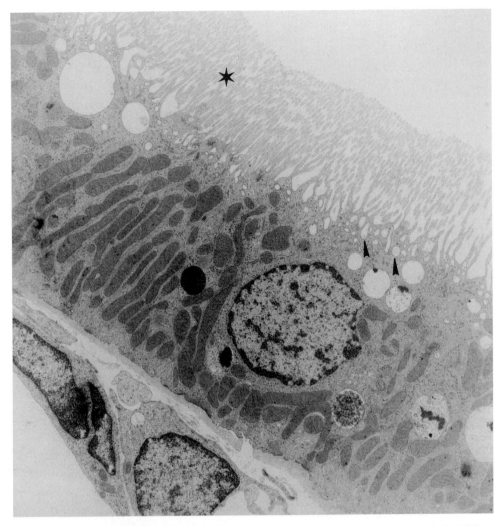

Figure 10 Transmission electron micrograph of the S_1 segment of the rat proximal tubule. This segment is characterized by a tall brush border (star), numerous apical endocytic vesicles (arrowheads), an array of vertically oriented mitochondria, and extensive infoldings of the basolateral plasma membrane (magnification: $7900\times$).

metabolizing small proteins results in their taking up many substances which appear in the ultrafiltrate. Thus, the proximal tubule cells are especially vulnerable to injury by toxins such as heavy metals and aminoglycoside antibiotics.

S_2 cells are found in both the PCT and PST (Figure 11). The transition from S_1 to S_2 is gradual, but begins around 1 mm from the glomerulus. Thus, the remainder of the PCT is composed of S_2 cells. These cells are not as tall as S_1 cells, have a shorter brush border, and have fewer mitochondria. The quantity and size of the lysosomal vacuoles in S_1 compared with S_2 vary with species and gender. For example, the lysosomes in S_2 cells are larger than in S_1 cells in male rats.[8,9]

7.02.6.2 Function

The proximal convoluted tubule (PCT) is responsible for reabsorbing 50–60% of the glomerular ultrafiltrate. Thus, it is the site for high-volume reabsorption, but it is not the site for regulation of the final composition of the urine. The latter task is the responsibility of the most distal portion of the nephron, the collecting duct.

The PCT reabsorbs solute isosmotically. Water is reabsorbed through the aquaporin-1 water channel[10,11] located in both the apical and basolateral membranes. Water reabsorption is driven by reabsorption of NaCl and $NaHCO_3$. NaCl and $NaHCO_3$ are transported via a variety of transcellular and paracellular mechanisms,

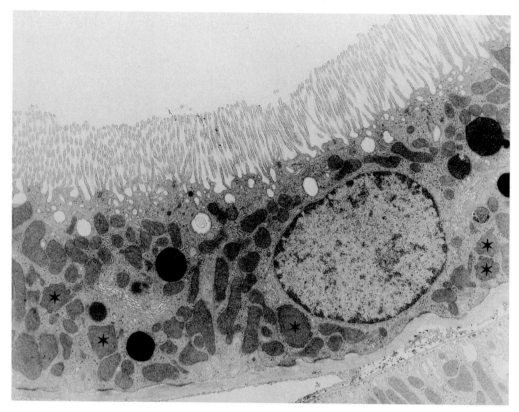

Figure 11 Transmission electron micrograph of the S_2 segment of the rat proximal tubule. The epithelium is lower than the S_1 and the S_3 segments and it has a relatively short brush border and less prominent mitochondria. Numerous peroxisomes are present in this cell which suggests that it is from the proximal straight tubule (pars recta) (magnification: $7900\times$).

including sodium–proton antiporters, sodium–bicarbonate cotransporters, and a chloride–formate antiporter. Under normal conditions, most of the filtered bicarbonate is reabsorbed.

The PCT is responsible for reabsorbing most of the glucose, amino acids, and small peptides which enter the ultrafiltrate.[12] The PCT contains numerous secondary active, sodium-coupled transporters for glucose and amino acids. Many of these transport proteins have been cloned and several isoforms of each transporter have been identified. These sodium-coupled cotransporters all take advantage of the low intracellular sodium concentration, generated by the Na/K-ATPase pump, to drive transport. Thus, the PCT has a high metabolic demand and is particularly sensitive to anoxic injury. Damage to the PCT results in Fanconi's syndrome with glucosuria, aminoaciduria, and a proximal renal tubular acidosis.

The PCT also transports numerous organic acids and has a multi-drug resistance transporter.[12,13] Secretion of organic anions, such as p-aminohippurate (PAH), occurs by sodium-dependent secondary active transport from peritubular capillaries into the PCT cell across the basolateral plasma membrane. Examples of basolateral organic anion transporters include a sodium–α-ketoglutarate cotransporter coupled to an α-ketoglutarate-organic anion countertransporter. The net result is sodium–organic anion uptake across the basolateral membrane.[14] Transport into the lumen generally does not require active transport. Examples of apical membrane secretory pathways include an OH^-/urate$^-$ exchanger and a facilitated transporter for organic anions.[14]

The PCT is the major source of cytochrome P450.[15] Organic anions such as p-aminohippurate increase arachidonic acid metabolism by cytochrome P450.[15] However, a causal link between organic anion secretion and cytochrome P450-induced metabolism of arachidonic acid has not been established.[15]

Organic cations are also secreted by sodium-dependent secondary active transport. In general, organic cations enter the cell across the basolateral plasma membrane by a facilitated transport pathway. They are secreted into the tubule lumen by an amiloride-inhibitable sodium–proton exchanger coupled to a protein–organic cation exchanger with the net result being sodium–organic cation exchange. A second apical secretory mechanism is an organic cation ATPase.[14] Many pharmaceuticals and toxins are organic cations or anions and able to be transported on these transporters and make the PCT very sensitive to damage by drugs and toxins.

Proteins and peptides which are smaller than albumin are able to enter the ultrafiltrate across the glomerulus.[16] Small proteins can be reabsorbed by nonspecific endocytotic reabsorption. Peptides are reabsorbed by sodium-coupled secondary active transport pathways. Many small peptides are taken up by PCT cells, including insulin, glucagon, and other peptide hormones. When GFR is reduced, as in renal insufficiency, the half-lives of these peptide hormones are increased since their filtration and subsequent degradation by the PCT are reduced. The PCT cells degrade these and other small peptides such as albumin in lysosomal vacuoles. Cathepsins B and L are found in highest abundance in the PCT, but are also found in the proximal straight tubule.[17]

Finally, the PCT reabsorbs divalent cations. Calcium is reabsorbed isosmotically in parallel with sodium. Magnesium is also reabsorbed passively but is less permeable than calcium. Magnesium needs to achieve a tubular fluid to plasma ratio of 1.5 before it is reabsorbed.[18]

7.02.7 PROXIMAL STRAIGHT TUBULE

7.02.7.1 Structure

The length of any proximal straight tubule (PST) is a function of the cortical location of its glomerulus (Figure 1) since all PST undergo gradual transition to thin descending limbs of Henle at the junction of the outer and inner stripes of the outer medulla.[19] PST of superficial and midcortical nephrons contain two cell types, S_2 and S_3, whereas PST of juxtamedullary nephrons are made up predominantly of S_3 cells (Figure 12). As the PST descend from the cortex into the outer stripe of the outer medulla, the cell type changes from S_2 to S_3. Thus, the outer stripe of the outer medulla contains only S_3 proximal tubule cells.[19] S_3 cells are cuboidal and have the longest brush border, most numerous peroxisomes, fewest mitochondria, least basolateral invaginations, and least developed endocytic apparatus of all three proximal

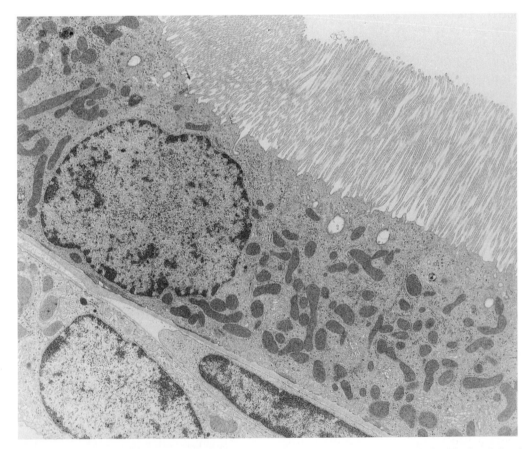

Figure 12 Transmission electron micrograph of the S_3 segment of the rat proximal tubule. The brush border is very long and relatively few endocytic vesicles and mitochondria are present (magnification: 7900×).

tubule cell types.[8] Peroxisomes are also very important in the S_2 portion of the PST.

7.02.7.2 Function

Generally, absorption rates for all solutes are lower in PST than in PCT. Sodium-coupled reabsorption of glucose, amino acids, ketone bodies, pyruvate, lactate, and short-chain fatty acids occurs by secondary active transport in PST.[12] Osmotic water permeability is extremely high in PST and occurs through aquaporin-1 water channels.[10,11]

The passive urea permeability of rabbit PST is low.[20] One study found evidence for active urea secretion using an isotopic assay,[21] while a second study using a chemical assay did not.[20] However, the second study would not have been able to detect the low rate of active urea secretion reported in the first study.[20]

Potassium transport varies along the length of the superficial PST and between superficial and juxtamedullary PST.[22] Passive K^+ permeability is highest in juxtamedullary PST, favoring significant K^+ secretion if the medullary interstitial K^+ concentration exceeds that of tubular fluid.[22] Net secretion of K^+ occurs in S_2 and S_3 segments of superficial PST and in juxtamedullary PST (which are almost exclusively made up of S_3 cells).[22]

Bicarbonate absorption in PST occurs at 20–25% of the rate seen in PCT and is greater in juxtamedullary than in superficial PST.[23] The PST does not normally contribute to net HCO_3^- absorption, because the HCO_3^- concentration in fluid leaving the PCT is already at its low steady-state level. However, a spontaneous luminal disequilibrium pH is generated in rabbit S_3, but not in rabbit S_2 or in rat PST.[24,25] The disequilibrium pH indicates lack of functional luminal carbonic anhydrase and will favor NH_3 diffusion into the lumen.[25]

7.02.8 THIN DESCENDING LIMB OF LOOP OF HENLE

7.02.8.1 Structure

Short-loop nephrons contain thin descending limbs (tDL), but no thin ascending limbs (tAL)

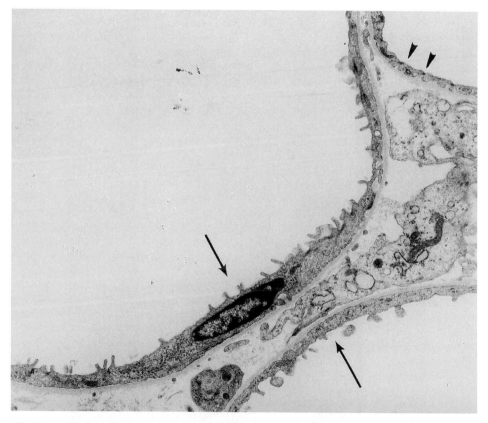

Figure 13 Transmission electron micrograph of a descending limb in the initial inner medulla of a rat. A type III epithelium is illustrated (arrows) which exhibits a relatively thin basolateral plasma membrane and prominent apical plasma membrane microprojections. A portion of a type IV epithelial cell from the thin ascending limb is also visible (arrowheads) (magnification: 9800×).

or only very short tAL (Figure 1). In short-loop nephrons, tDL are made up of one cell type. The epithelium is squamous, with little cellular interdigitation and with deep tight junctions,[26,27] and has been termed tDL type I. The tDL of long-loop nephrons have two morphologic components. The early portion is made up of cells with well-developed lateral interdigitation and shallow tight junctions. These are termed type II tDL cells. As the tDL of long-loop nephrons enter the inner medulla, the cell type changes to type III tDL cells (Figure 13), which are similar to the type I tDL cells seen in the tDL of short-loop nephrons.[26,27] This classification holds true for the rat and hamster, which have been the best studied species.

In rat, mouse, and desert sand rat,[2] the tDL of short-loop nephrons travel through the outer medulla within the vascular bundles made up of arterial and venous vasa recta. However, tDL of long-loop nephrons descend through the outer medulla in the interbundle region.[2] In contrast, the tDL of both short- and long-loop nephrons from species with lower concentrating ability (rabbit, dog, pig, human) travel through the outer medulla in the interbundle region.[2] In most species there is a relatively random distribution of tDL in the inner medulla.[2]

7.02.8.2 Function

The thin descending limb is characterized by a very high osmotic water permeability which occurs through aquaporin-1 water channels.[10,11] This high osmotic water permeability permits complete osmotic equilibration of the luminal fluid with the medullary interstitium by the process of water abstraction.

The rabbit tDL is not permeable to any solute measured thus far.[28] The impermeability of the tDL to Na^+ and urea is critical to the operation of the passive mechanism for concentrating the urine.[29,30] However, recent studies have demonstrated functional heterogeneity in tDL subsegments from hamsters and rats, unlike rabbits. The hamster tDL type II has higher Na^+, K^+, and Cl^- permeabilities and a lower urea permeability than the type I or

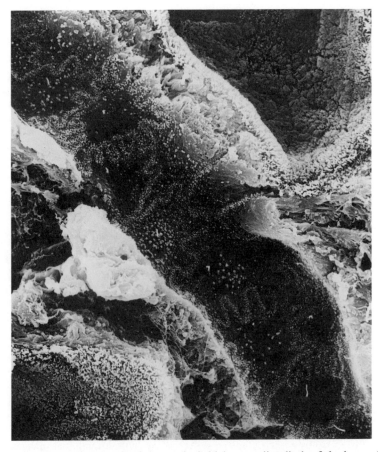

Figure 14 Scanning electron micrograph of the cortical thick ascending limb of the loop of Henle from rat. The apical surface exhibits a central cilium and short microprojections which are most prevalent near the cell margins. The cell margins exhibit a characteristic undulating path (magnification: 3000×).

type III cells.[28] The implications of this heterogeneity for the function of the concentrating mechanism remains to be determined. In all species studied, the tDL does not perform active transport,[28] consistent with the sparse mitochondria in this subsegment.

7.02.9 THIN ASCENDING LIMB OF LOOP OF HENLE

7.02.9.1 Structure

The tAL begins at or slightly before the bend of Henle's loop in most species. At the inner–outer medullary junction it abruptly becomes the medullary thick ascending limb of Henle. In all species examined, the tAL (type IV epithelial cell, Figure 14) is made up of flat, moderately interdigitated cells connected by shallow tight junctions.[26]

7.02.9.2 Function

The osmotic water permeability measured in rabbit, rat, and hamster tAL is very low.[31,32] Diffusional water permeability is also extremely low,[32] suggesting that tALs are essentially water impermeable. However, with appropriate concentration gradients (NaCl from lumen-to-bath and urea from bath-to-lumen), the tAL can dilute its luminal fluid by purely passive means.

The tAL has very different permeability characteristics from the tDL. In rabbit, rat, and hamster tAL, the Na^+ and Cl^- permeabilities are 5–20 times higher than those found in the proximal tubule.[31,32] Even the urea permeability is moderately high.[31] These permeabilities make the tAL well suited for the passive mechanism for concentrating the urine in the inner medulla.[29,30] The tAL was thought to be the only nephron segment with undetectable Na/K-ATPase activity. However, low levels of Na/K-ATPase activity are present in rat tAL.[33]

7.02.10 THICK ASCENDING LIMB OF LOOP OF HENLE

7.02.10.1 Structure

The thick ascending limb is divided into two portions: the medullary thick ascending limb (mTAL), which is located in the outer medulla, and the cortical thick ascending limb (cTAL), which is located in the medullary rays of the cortex (Figure 15). The mTAL originates from the tAL at the inner–outer medullary border. By definition, but not necessarily correlated with a transition in cell type, it becomes the cTAL at the junction of the outer medulla and cortex. Juxtamedullary nephrons have a longer mTAL than superficial or midcortical nephrons

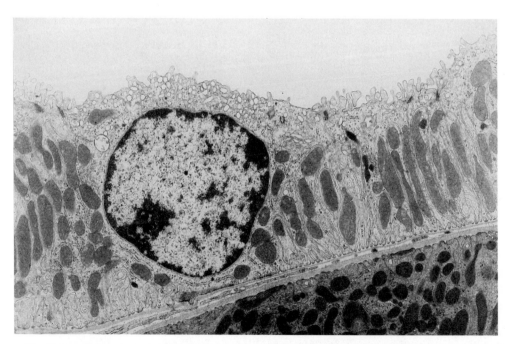

Figure 15 Transmission electron micrograph of the cortical thick ascending limb of the loop of Henle from rat. The epithelial cells are characterized by prominent vertically arranged mitochondria and extensive infoldings of the basolateral plasma membrane. Short microprojections are present on the apical surface and subapical cytoplasmic vesicles are evident over the nucleus (magnification: 8800×).

which complete their hairpin turn within the outer medulla (Figure 1). Adjacent cells of mTAL are extensively interdigitated, contain a large number of elongated rod-shaped mitochondria, are joined to each other by shallow tight junctions, and contain one or two cilia on their apical surfaces.[19] Two cell types are present in mTALs, based on whether their luminal surfaces are smooth or rough, with a predominance of the smooth type.[19]

The cTAL begins at the junction of outer medulla and cortex. This transition is not discrete and is not accompanied by an abrupt change in cellular morphology.[19] The epithelium is much thinner than that of the mTAL. Individual cTAL cells contain fewer mitochondria than those of mTAL and rough-surface cells predominate.[19] Depending on the species, the cTAL can extend a variable distance beyond the macula densa.

7.02.10.2 Function

The major reabsorbed solute in the mTAL and cTAL is NaCl. The Na^+ permeability exceeds that of Cl^- severalfold in both rabbit and mouse,[34] resulting in a lumen-positive voltage. NaCl is actively reabsorbed by an Na–K–2Cl cotransporter in the apical membrane linked to the Na/K-ATPase in the basolateral membrane. NaCl reabsorption and the lumen-positive transepithelial voltage are inhibited by loop diuretics such as furosemide and bumetanide, which inhibit the Na–K–2Cl cotransporter. The mTAL can lower the luminal NaCl concentration from 140 to 117 mM,[35,36] while the cTAL can lower the luminal NaCl concentration to 65 mM.[37] Thus, the mTAL is a high-capacity system but can generate only modest transepithelial NaCl gradients whereas the cTAL is a low-capacity system but can generate significant transepithelial NaCl gradients. In mouse, vasopressin (AVP, ADH) increases NaCl reabsorption in the mTAL.[38] Thus, vasopressin increases the reabsorption of NaCl necessary for increasing or maintaining urine concentrating ability. However, significant species differences exist: human and canine mTAL do not respond and rabbit mTAL is a weak responder.[38]

The mTAL and cTAL are the major sites for regulation of reabsorption of the divalent cations Ca^{2+} and Mg^{2+}. Reabsorption of 20% of the filtered Ca^{2+} and 50–60% of the filtered Mg^{2+} [18,39] occur in these nephron segments primarily by paracellular transport driven by the lumen-positive voltage. Thus, loop diuretics also inhibit Ca^{2+} and Mg^{2+} reabsorption.

Net reabsorption of total ammonia and HCO_3^- occur in both rat mTAL and cTAL, which is inhibitable by furosemide.[40,41] These studies suggest that active NH_4^+ absorption occurs by substitution for K^+ on both the Na–K–2Cl cotransporter and the Na/K-ATPase, providing the single effect for countercurrent multiplication of ammonia in the medulla.

The urea permeability is low in rabbit mTALs[42] and in rat mTALs from the inner stripe.[43] Rat mTALs from the outer stripe have a significantly higher urea permeability, comparable to rat cTALs.[43] This higher urea permeability allows for dilution of these segments by passive urea absorption. The reabsorbed urea may be recycled by secretion into the PST.

The osmotic water permeability of the mTAL and cTAL is essentially zero in rabbit[37] and mouse[44] and is unaffected by vasopressin. Thus, the mTAL and cTAL lower luminal fluid osmolality by net reabsorption of solute.

The thick ascending limb contains cytochrome P450.[15] This enzyme metabolizes arachidonic acid to a variety of biologically active products.[15] In spontaneously hypertensive rats, inhibition of cytochrome P450 by stannous chloride reduces blood pressure to normal values.[15]

7.02.11 DISTAL CONVOLUTED TUBULE, CONNECTING SEGMENT, AND INITIAL COLLECTING TUBULE

7.02.11.1 Structure

The nephron segment between the macula densa and the first junction of two tubules is generally referred to as the distal convoluted tubule (DCT). However, the "DCT" is actually composed of several structurally and functionally distinct segments. The first portion of the "DCT" (usually 100–150 μm) beyond the macula densa contains cells of the cTAL type.[45] The second portion of the "DCT" is short and constitutes the true DCT. This segment appears bright under the dissection microscope is made up of one unique cell type called the DCT cell (Figure 16). The DCT cell has deep tight junctions, basolateral invaginations, many long mitochondria extending apically from the basolateral cell membrane, and short blunt apical microvilli.[45] The cTAL abruptly transitions to the DCT.

The third portion of the "DCT" is called the connecting tubule or connecting segment (CNT, Figure 17). It is somewhat wider than the preceding portion of the DCT and appears granular. Thus, it is occasionally called the

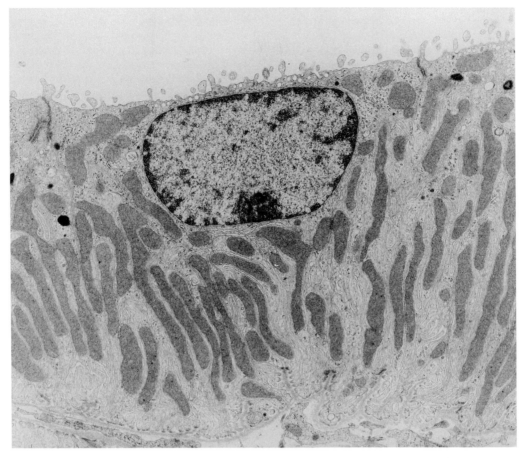

Figure 16 Transmission electron micrograph of a rat distal convoluted tubule cell (early or bright portion). The epithelium is composed of tall, cuboidal cells which contain a prominent mitochondrial compartment and extensive basolateral plasma membrane infoldings (magnification: 9500×).

DCTg. It contains two cell types: connecting tubule cells and intercalated (dark) cells.[45] The latter cells are also found in the collecting duct. The fourth and final portion of the "DCT" is the initial collecting duct (ICT). It contains intercalated cells and principal (light) cells.

7.02.11.2 Function

The osmotic water permeability of both DCT and CNT is very low and is unaffected by vasopressin.[46] Conversely, high rates of sodium reabsorption occur in both DCT and CNT,[47] resulting in a lumen-negative voltage. NaCl is reabsorbed by a thiazide-inhibitable Na–Cl cotransporter in rat DCT[47,48] and rabbit CNT.[49] Potassium transport occurs via a K–Cl cotransport pathway in the apical membrane of rat DCT and ICT [50] and a conductive K^+ channel in rabbit DCT.[51] Calcium is reabsorbed in the DCT, but at lower rates than in cTAL.[47]

7.02.12 CORTICAL COLLECTING DUCT

7.02.12.1 Structure

The CCD extends from its origin (at the convergence of ICTs) to the junction of the cortex and outer medulla (Figure 1). Because there is no distinct or abrupt change in its appearance at that point, definition of the distal end of the CCD is based on location, not cell type. Two cell types are present in CCD: the principal or light cell and the intercalated or dark cell (Figure 18). The former is the typical collecting duct cell, whereas the latter can also be found in the CNT. The ratio of principal cells to intercalated cells is approximately 2:1 in CCD.[52]

Principal cells have less electron-dense cytoplasm, fewer mitochondria, and fewer cellular organelles than do intercalated cells. The apical membrane of the principal cell has short, sparse microprojections, whereas that of the intercalated cell has long microvilli or

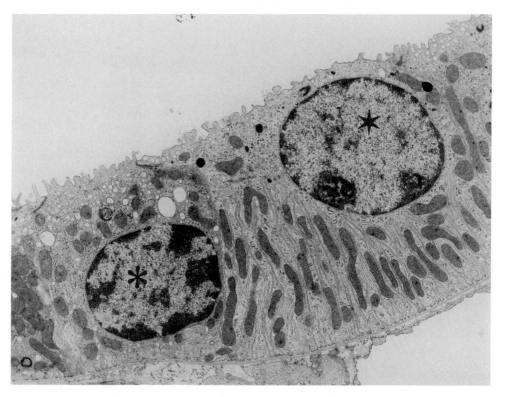

Figure 17 Transmission electron micrograph of a rat connecting segment (late or granular portion of the distal convoluted tubule). Both connecting segment cells (star) and intercalated cells (asterisk) are present in this segment (magnification: 8800×).

microplicae. Intercalated cells contain cytoplasmic vesicles that reside near the apical membrane, whereas such vesicles are rarely found in principal cells.[53] Carbonic anhydrase is present in intercalated cells, but not principal cells.[54]

Some evidence indicates that the CCD contains two populations of intercalated cells, type A or H^+-secreting cells (Figure 19) and type B or HCO_3^--secreting cells, whereas the OMCD contains only type A cells.[55] All intercalated cells in the OMCD stain for anion exchanger 1 (AE1) protein on their basolateral membrane, whereas only type A intercalated cells in the CCD stain for AE1.[56,57] Type A cells contain H^+-ATPase in their apical membrane, whereas type B cells contain H^+-ATPase in their basolateral membrane.[56] Thus, axial heterogeneity of intercalated cells exists along the collecting duct.

7.02.12.2 Function

The collecting duct is the location for fine control of urine composition. Unlike the proximal tubule, it is a low-capacity reabsorption site. However, the collecting duct regulates reabsorption of solute, water, and protons in response to hormonal stimuli.

The osmotic water permeability of CCD is low in the absence of vasopressin.[58-60] Vasopressin significantly increases the osmotic water permeability in rabbit[58] and rat.[59,60] Vasopressin induces a sustained increase in osmotic water permeability in rat CCD but only a transient increase in rabbit CCD.[59,60] Metabolites of arachidonic acid, produced by cytochrome P450, inhibit vasopressin-stimulated osmotic water permeability by a postcyclic AMP mechanism.[15]

The apical membrane is the rate-limiting barrier to water movement, and vasopressin increases apical membrane osmotic water permeability by inducing the insertion of endosomes containing Aquaporin-2 water channels;[61] vasopressin removal stimulates retrieval of these water channels. Urea permeability in CCD is very low and is not increased by vasopressin.[43,62] Thus, vasopressin increases both luminal fluid osmolality and urea concentration in CCD.

The major ionic transport processes in CCD are Na^+ reabsorption and K^+ secretion,[63] both of which are stimulated by mineralocorticoids. Cl^- absorption,[64] HCO_3^- absorption,[65] and HCO_3^- secretion[65] also occur in CCD. Sodium

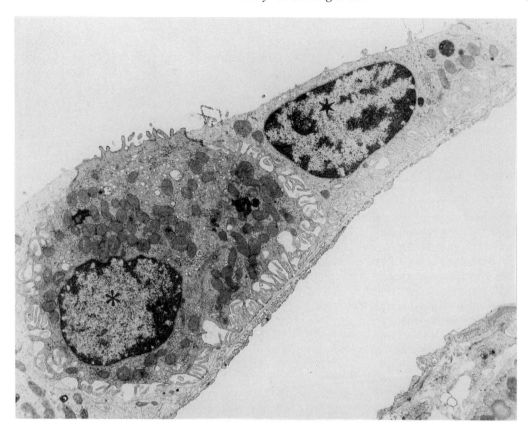

Figure 18 Transmission electron micrograph of the cortical collecting duct from rat kidney. The principal cell (star) is characterized by a relatively smooth apical surface, prominent basolateral plasma membrane infoldings, and few mitochondria. The type B intercalated cell (asterisk) is characterized by a dark cytoplasm, an eccentric nucleus, cytoplasmic vesicles throughout the cell, and numerous, clustered mitochondria (magnification: 7900×).

is actively reabsorbed in CCD and is responsible for the lumen-negative voltage.[63] Na^+ enters across the apical cell membrane via the amiloride-sensitive sodium channel.[66] Vasopressin and mineralocorticoids stimulate Na^+ reabsorption, while bradykinin inhibits it.[67] Na^+ exits the cell via Na/K-ATPase in the basolateral membrane.

Potassium secretion in CCD has both active and passive components. Both apical and basolateral cell membranes exhibit major K^+ conductances, with the apical exceeding the basolateral and thus partially accounting for net K^+ secretion.[68] The apical membrane K^+ conductance is composed of high conductance Ca^{2+}-activated (maxi) K^+ channels and low conductance K^+ channels.[69] The apical membrane also contains a K–Cl cotransport process.[70] Aldosterone increases K^+ permeability of rabbit CCD[71] and vasopressin stimulates K^+ secretion in rat CCD.[67]

Chloride is transported both paracellularly and transcellularly in CCD. In rabbit, Cl^- reabsorption is predominantly passive, although some evidence for Cl^- reabsorption against an electrochemical gradient exists.[64] In rat, net Cl^- reabsorption occurs which is stimulated by vasopressin and inhibited by bradykinin.[72]

Bicarbonate transport in CCD has been extensively studied in recent years. CCD from normal rabbits show no net HCO_3^- transport while CCD from normal rats secrete HCO_3^-.[72] However, CCD from NH_4Cl-treated rabbits and rats reabsorb HCO_3, whereas CCD from $NaHCO_3$-treated and DOCA-treated rabbits and rats secrete HCO_3^-.[65,73] Vasopressin converts HCO_3^- secretion to reabsorption in rat.[72] Ammonia secretion is dependent upon an acidic luminal disequilibrium pH in CCD due to the absence of luminal carbonic anhydrase.[74]

7.02.13 OUTER MEDULLARY COLLECTING DUCT

7.02.13.1 Structure

The outer medullary collecting duct (OMCD) begins at the junction of the cortex and medulla and ends at the inner–outer

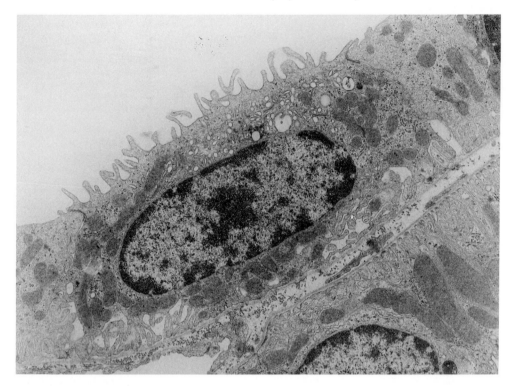

Figure 19 Transmission electron micrograph of a type A intercalated cell in the cortical collecting duct of a rat. The type A intercalated cell contains a prominent apical tubulovesicular compartment, a centrally located nucleus, and evenly distributed mitochondria. When compared with the type B intercalated cell (Figure 18), the cytoplasm is less dense and the apical plasma membrane microprojections are longer and more numerous (magnification: 9800×).

medullary border (Figure 1). However, its beginning and end are not well defined. Both morphologically and functionally the OMCD from the outer stripe resembles the CCD with the exception of the absence of type B intercalated cells and HCO_3^- secretion in the OMCD.[75] The number of intercalated cells decreases along the length of the OMCD from outer stripe to inner stripe.

7.02.13.2 Function

Osmotic water permeability is low in the absence of vasopressin, and increases in the presence of vasopressin.[42] Urea permeability, as in CCD, is quite low, and vasopressin does not increase it.[42,59] There is no net Na^+ or K^+ transport in the OMCD.[75]

The OMCD plays an important role in acidifying the urine. Both rabbit and rat OMCD reabsorb HCO_3^-.[65,73] HCO_3^- reabsorption is stimulated by aldosterone and is independent of Na^+ in both outer stripe and inner stripe OMCDs.[74] However, only outer stripe OMCDs generate a disequilibrium pH, indicating functional luminal carbonic anhydrase in the inner stripe, but not in the outer

stripe.[74] Carbonic anhydrase is present in the intercalated cells of the inner stripe.[76]

Ammonia secretion, predominantly by NH_3 diffusion, occurs three times faster in the outer stripe than the inner stripe.[74] The inner stripe OMCD possesses an H/K-ATPase which can be inhibited by omeprazole.[77] The apical membrane also contains an H^+-ATPase.[78] The basolateral membrane contains a Cl^-–HCO_3^- exchanger (AE1) and an Na^+–H^+ exchanger.[79]

7.02.14 INNER MEDULLARY COLLECTING DUCT

7.02.14.1 Structure

The inner medullary collecting duct (IMCD) consists of two morphologically distinct sub-segments.[80,81] The initial IMCD (or $IMCD_1$) is located in the third of the inner medulla closest to the inner–outer medullary junction (Figure 20).[59] This segment contains 90% principal cells and 10% intercalated cells.[80] Morphologically, these cells are similar to the cell types found in the inner stripe of the outer medulla.[80] The terminal IMCD (or $IMCD_2$ and $IMCD_3$, or papillary collecting duct) is located

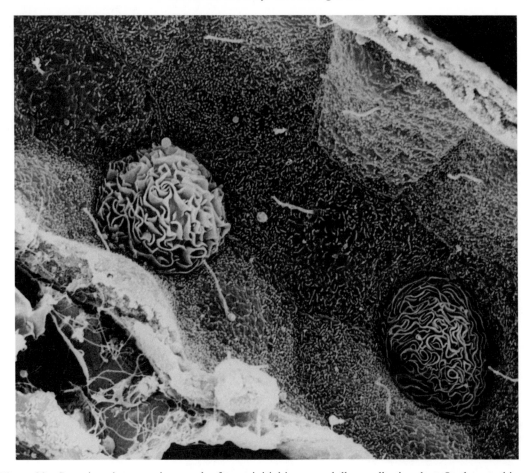

Figure 20 Scanning electron micrograph of a rat initial inner medullary collecting duct. In the rat, this segment of the collecting duct contains both principal cells and intercalated cells. The intercalated cells (arrow) are covered with numerous microplicae on the luminal surface, similar to type A intercalated cells in the cortical collecting duct. The principal cells have relatively few short apical microprojections and a central cilium (magnification: 4000×).

in the two thirds of the inner medulla closest to the papillary tip (Figure 21).[59] This segment contains a unique cell type, the IMCD cell (Figure 22).[81] IMCD cells, as compared with principal cells, are taller and more columnar with prominent microvilli.[81] They have fewer basal infoldings and no cilia.[81] Their cytoplasm is lighter staining with fewer organelles or mitochondria, but with abundant free ribosomes, small coated vesicles, and lysosomes.[81] There are no intercalated cells in the terminal IMCD.[81]

7.02.14.2 Function

The initial IMCD has a low urea permeability, either in the absence or presence of vasopressin, similar to CCD and OMCD.[59,62] In contrast, the terminal IMCD has a high basal urea permeability in rat, rabbit, and hamster.[59,62] Urea permeability is stimulated by vasopressin[59] and hypertonic peritubular

NaCl,[82] and inhibited by phloretin,[83] urea analogues,[83] and hypertonic peritubular urea.[62] This pattern of urea permeabilities allows for significant urea reabsorption into the deep inner medullary interstitium where it is needed to maximally concentrate the urine.[59,62]

Mammals fed a low-protein diet are unable to concentrate their urine. This concentrating defect is due, at least in part, to a reversal of the normal inner medullary urea concentration gradient. Normally, the highest inner medullary urea concentration is found at the papillary tip. However, in rats and sheep fed a low-protein diet, the highest urea concentration is found at the inner–outer medullary border.[84] Studies show that feeding rats a low-protein diet induces the expression of two urea transporters in the initial IMCD, a segment which normally does not transport urea: a vasopressin-stimulated, facilitated urea transporter which is normally expressed in the terminal IMCD[85,86] and a secondary active, sodium–urea cotran-

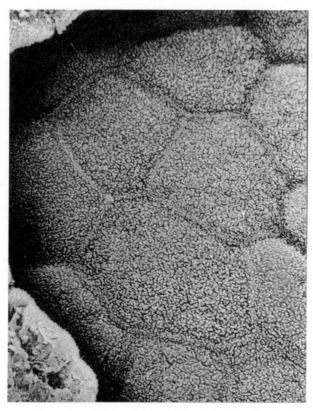

Figure 21 Scanning electron micrograph of the terminal inner medullary collecting duct in rat kidney. The epithelial cells are covered with short microprojections, lateral cell margins are prominent, and cilia are absent (magnification: 4000×).

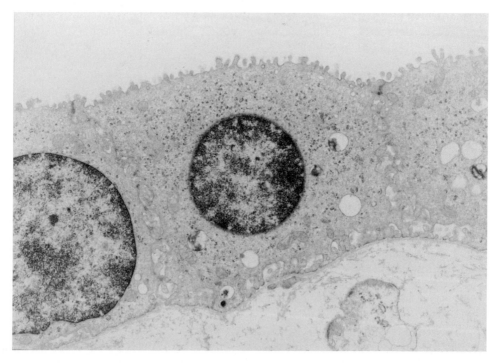

Figure 22 Transmission electron micrograph of the IMCD$_2$ segment of the rat terminal inner medullary collecting duct. The IMCD cells are cuboidal and contain relatively few mitochondria and cytoplasmic vesicles. This epithelium becomes progressively taller as the collecting duct approaches the papillary tip (magnification: 8800×).

sporter.[87] Expression of these two urea transporters in the initial IMCD will cause urea to be reabsorbed more proximally and prevent the delivery of urea to the deep inner medulla where it is needed to concentrate the urine.

The osmotic water permeability of initial IMCDs is low in the absence of vasopressin, and increases in response to it.[59] The terminal IMCD has a high osmotic water permeability even in the absence of vasopressin and it increases further in response to it.[59] Water transport occurs through Aquaporin-2 water channels in the apical membrane[61] and Aquaporin-3 and Aquaporin-4 water channels in the basolateral membrane.[88–93]

The IMCD has the highest Na^+, K^+, and Cl^- permeabilities found in the entire collecting duct system.[94] Thus, the IMCD permits significant passive movement of these ions, depending on their respective concentration gradients.

Net HCO_3^- reabsorption and ammonia secretion occur in rat terminal IMCDs.[95] HCO_3^- reabsorption is stimulated by vasopressin, mineralocorticoids, and NH_4Cl loading.[95] Rat terminal IMCDs are able to lower luminal pH, whereas rabbit IMCDs are not.[95] An acidic disequilibrium pH is present in IMCDs consistent with a lack of luminal carbonic anhydrase.

7.02.15 PAPILLARY SURFACE EPITHELIUM

7.02.15.1 Structure

The papillary surface epithelium (PSE) covers the surface of the true papilla (Figure 2). It is bathed apically by urine and basolaterally by inner medullary interstitial fluid. In hamster, the PSE is morphologically similar to IMCD cells.[96] However, in rat and rabbit, the PSE is a simple cuboidal epithelium which is distinct from cells of the IMCD, except at the papillary tip.[97–99] Near the openings of the ducts of Bellini, the papillary surface epithelium becomes columnar, similar to IMCD cells.[97,98] The apical cell membrane has scattered short microvilli surrounded by an electron-dense fuzz.[99] The cytoplasm contains relatively few mitochondria.[99] The lateral cell membrane is tortuous and contains some dilatations of the intercellular space which increase when interstitial osmolality increases.[98,100]

7.02.15.2 Function

Relatively few permeability coefficients have been measured across the papillary surface epithelium. Both the urea permeability[62] and the osmotic water permeability[101] are low and unaffected by vasopressin. Thus, it seems unlikely that significant transport of water or urea occurs across this epithelium.

The Cl^- permeability is higher than that of the terminal IMCD.[101] The apical membrane contains an Na–K–Cl cotransporter which is stimulated by vasopressin and inhibited by bumetanide.[98] The basolateral membrane contains a K^+ conductive pathway.[102,103] The PSE can acidify its apical surface.[104]

ACKNOWLEDGMENTS

The authors thank Ms. Wendy Wilber for her expert assistance in the darkroom. This work was supported by National Institutes of Health grants R01-DK41707 and R01-DK45688. J.M.S. performed part of this work during the tenure of an Established Investigatorship from the American Heart Association.

7.02.16 REFERENCES

1. W. Kriz and L. Bankir, 'A standard nomenclature for structures of the kidney.' *Am. J. Physiol.*, 1988, **254**, F1–F8.
2. K. V. Lemley and W. Kriz, 'Cycles and separations: the histotopography of the urinary concentrating process.' *Kidney Int.*, 1987, **31**, 538–548.
3. W. Kriz, 'Structural organization of the renal medullary counterflow system.' *Fed. Proc.*, 1983, **42**, 2379.
4. C. C. Tisher and B. M. Brenner, in 'Renal Pathology with Clinical and Functional Correlations,' eds. C. C. Tisher and B. M. Brenner, J. B. Lippincott, Philadelphia, PA, 1989, pp. 92–110.
5. H. Valtin, 'Renal Function: Mechanisms Preserving Fluid and Solute Balance in Health,' Little, Brown, Boston, MA, 1973, p. 127.
6. D. A. Maddox and B. M. Brenner, in 'The Kidney,' 4th edn., eds. B. M. Brenner and F. C. Rector, Jr., Saunders, Philadelphia, PA, 1991, pp. 205–244.
7. J. B. Hook and W. R. Hewitt, in 'Casarett and Doull's Toxicology: The Basic Science of Poisons,' 3rd edn., eds. C. D. Klaassen, M. O. Amdur and J. Doul, Macmillan, New York, 1986, pp. 310–329.
8. A. B. Maunsbach, 'Observations on the segmentation of the proximal tubule in the rat kidney. Comparison of results from phase contrast, fluorescence and electron microscopy.' *J. Ultrastruct. Res.*, 1966, **16**, 239–258.
9. A. B. Maunsbach, in 'Handbook of Physiology, Section 8: Renal Physiology,' ed. J. Orloff, American Physiological Society, Washington, DC, 1973, pp. 31–81.
10. G. M. Preston, T. P. Carroll, W. B. Guggino *et al.*, 'Appearance of water channels in *Xenopus* oocytes expressing red cell CHIP28 protein.' *Science*, 1992, **256**, 385–387.
11. S. Nielsen, B. L. Smith, E. I. Christensen *et al.*, 'CHIP28 water channels are localized in constitutively water-permeable segments of the nephron.' *J. Cell Biol.*, 1993, **120**, 371–383.
12. C. A. Berry and F. C. Rector, Jr., in 'The Kidney,' 4th

edn., eds. B. M. Brenner and F. C. Rector, Jr., Saunders, Philadelphia, PA, 1991, pp. 245–282.

13. N. Kartner and V. Ling, 'Multidrug resistance in cancer.' *Sci. Am.*, 1989, **260**, 44–51.

14. J. B. Pritchard, in 'Textbook of Nephrology,' 3rd edn., eds. S. G. Massry and R. J. Glassock, Williams and Wilkins, Baltimore, MD, 1995, pp. 96–101.

15. K. F. Badr, in 'Textbook of Nephrology,' 3rd edn., eds. S. G. Massry and R. J. Glassock, Williams and Wilkins, Baltimore, MD, 1995, pp. 182–91.

16. V. Johnson and T. Maack, in 'Textbook of Nephrology,' 3rd edn., eds. S. G. Massry and R. J. Glassock, Williams and Wilkins, Baltimore, MD, 1995, pp. 101–107.

17. C. J. Olbricht, J. K. Cannon, L. C. Garg *et al.*, 'Activities of cathepsins B and L in isolated nephron segments from proteinuric and nonproteinuric rats.' *Am. J. Physiol.*, 1986, **250**, F1055–F1062.

18. G. A. Quamme and J. H. Dirks, 'Magnesium transport in the nephron.' *Am. J. Physiol.*, 1980, **239**, F393–F401.

19. P. B. Woodhall, C. C. Tisher, C. A. Simonton *et al.*, 'Relationship between *para*-aminohippurate secretion and cellular morphology in rabbit proximol tubules.' *J. Clin. Invest.*, 1978, **61**, 1320–1329.

20. M. A. Knepper, 'Urea transport in nephron segments from medullary rays of rabbits.' *Am. J. Physiol.*, 1983, **244**, F622–F627.

21. S. Kawamura and J. P. Kokko, 'Urea secretion by the straight segment of the proximal tubule.' *J. Clin. Invest.*, 1976, **58**, 604–612.

22. J. Work, S. L. Troutman and J. A. Schafer, 'Transport of potassium in the rabbit pars recta.' *Am. J. Physiol.*, 1982, **242**, F226–F237.

23. D. G. Warnock and M. B. Burg, 'Urinary acidification: CO_2 transport by the rabbit proximal straight tubule.' *Am. J. Physiol.*, 1977, **232**, F20–F25.

24. J. L. Garvin and M. A. Knepper, 'Bicarbonate and ammonia transport in isolated perfused rat proximal straight tubules.' *Am. J. Physiol.*, 1987, **253**, F277–F281.

25. I. Kurtz, R. Star, R. S. Balaban *et al.*, 'Spontaneous luminal disequilibrium pH in S3 proximal tubules. Role in ammonia and bicarbonate transport.' *J. Clin. Invest.*, 1986, **78**, 989–996.

26. M. M. Schwartz, M. J. Karnovsky, J. L. Garvin *et al.*, 'Regional membrane specialization in the thin limbs of Henle's loops as seen by freeze-fracture electron microscopy.' *Kidney Int.*, 1979, **16**, 577–589.

27. M. M. Schwartz and M. A. Venkatachalam, 'Structural differences in thin limbs of Henle: physiological implications.' *Kidney Int.*, 1974, **6**, 193–208.

28. M. Imai, J. Taniguchi and K. Tabei, 'Function of thin loops of Henle.' *Kidney Int.*, 1987, **31**, 565–579.

29. J. P. Kokko and F. C. Rector, Jr., 'Countercurrent multiplication system without active transport in inner medula.' *Kidney Int.*, 1972, **2**, 214–223.

30. J. L. Stephenson, 'Concentration of urine in a central core model of the renal counterflow system.' *Kidney Int.*, 1972, **2**, 85–94.

31. M. Imai, 'Function of the thin ascending limb of Henle of rats and hamsters perfused *in vitro*.' *Am. J. Physiol.*, 1977, **232**, F201–F209.

32. M. Imai and J. P. Kokko, 'Sodium, chloride, urea, and water transport in the thin ascending limb of Henle.' *J. Clin. Invest.*, 1974, **53**, 393–402.

33. Y. Terada and M. A. Knepper, 'Na$^+$–K$^+$-ATPase activities in renal tubule segments of rat inner medulla.' *Am. J. Physiol.*, 1989, **256**, F218–F223.

34. S. C. Hebert, R. M. Culpepper and T. E. Andreoli, 'NaCl transport in mouse medullory thick ascending limbs. I. Functional nephron heterogeneity and ADH-stimulated NaCl transport.' *Am. J. Physiol.*, 1981, **241**, F412–F431.

35. M. B. Burg and N. Green, 'Function of the thick ascending limb of Henle's loop.' *Am. J. Physiol.*, 1973, **224**, 659–668.

36. M. Horster, 'Loop of Henle functional differentiation: *in vitro* perfusion of the isolated thick ascending segment.' *Pfluegers Arch.*, 1978, **378**, 15–24.

37. A. S. Rocha and J. P. Kokko, 'Sodium chloride and water transport in the medullary thick ascending limb of Henle. Evidence for active chloride transport.' *J. Clin. Invest.*, 1973, **52**, 612–623.

38. F. Morel, M. Imbert-Teboul and D. Chabardes, 'Distribution of hormone-dependent adenylate cyclase in the nephron and its physiological significance.' *Annu. Rev. Physiol.*, 1981, **43**, 569–581.

39. W. N. Suki, 'Calcium transport in the nephron.' *Am. J. Physiol.*, 1979, **237**, F1–F6.

40. D. W. Good, 'Active absorption of NH$_4^+$ by rat medullary thick ascending limb: inhibition by potassium.' *Am. J. Physiol.*, 1988, **255**, F78–F87.

41. J. L. Garvin, M. B. Burg and M. A. Knepper, 'Active NH$_4^+$ absorption by the thick ascending limb.' *Am. J. Physiol.*, 1988, **255**, F57–F65.

42. A. S. Rocha and J. P. Kokko, 'Permeability of medullary nephron segments to urea and water: effect of vasopressin.' *Kidney Int.*, 1974, **6**, 379–387.

43. M. A. Knepper, 'Urea trasnport in isolated thick ascending limbs and collecting ducts from rats.' *Am. J. Physiol.*, 1983, **245**, F634–F639.

44. D. A. Hall and D. M. Varney, 'Effect of vasopressin on electrical potential difference and chloride transport in mouse medullary thick ascending limb of Henle's loop.' *J. Clin. Invest.*, 1980, **66**, 792–802.

45. M. Crayen and W. Thoenes, 'Structure and cytological characterization of the distal tubule of the rat kidney.' *Fortschr. Zool.*, 1975, **23**, 279–288.

46. M. Imai, 'The connecting tubule: a functional subdivision of the rabbit distal nephron segments.' *Kidney Int.*, 1979, **15**, 346–356.

47. L. S. Costanzo, 'Localization of diuretic action in microperfused rat distal tubules: Ca and Na transport.' *Am. J. Physiol.*, 1985, **248**, F527–F535.

48. D. H. Ellison, H. Velazquez and F. S. Wright, 'Thiazide-sensitive sodium chloride cotransport in early distal tubule.' *Pfluegers Arch.*, 1987, **409**, 182–187.

49. T. Shimizu, K. Yoshitomi, M. Nakamura *et al.*, 'Site and mechanism of action of trichlormethiazide in rabbit distal nephron segments perfused *in vitro*.' *J. Clin. Invest.*, 1988, **82**, 721–730.

50. H. Velazquez, D. H. Ellison and F. S. Wright, 'Chloride-dependent potassium secretion in early and late renal distal tubules.' *Am. J. Physiol.*, 1987, **253**, F555–F562.

51. J. Taniguchi, K. Yoshitomi and M. Imai, 'K$^+$ channel currents in basolateral membrane of distal convoluted tubule of rabbit kidney.' *Am. J. Physiol.*, 1989, **256**, F246–F254.

52. J. B. Stokes, C. C. Tisher and J. P. Kokko, 'Structural–functional heterogeneity along the rabbit collecting tubule.' *Kidney Int.*, 1975, **14**, 585–593.

53. B. Kaissling and W. Kriz, in 'Advances in Anatomy: Embryology and Cell Biology,' eds. A. Brodal, W. Hild, J. Van Limborgh *et al.* Springer, Berlin, 1979, vol. 56, p. 1–123.

54. D. Brown, J. Roth, T. Kumpulainen *et al.*, 'Ultrastructural immunocytochemical localization of carbonic anhydrase. Presence in intercalated cells of the rat collecting tubule.' *Histochemistry*, 1982, **75**, 209–213.

55. J. W. Verlander, K. M. Madsen and C. C. Tisher,

'Effect of acute respiratory acidosis on two populations of intercalated cells in rat cortical collecting duct.' *Am. J. Physiol.*, 1987, **253**, F1142–F1156.

56. S. L. Alper, J. Natale, S. Gluck *et al.*, 'Subtypes of intercalated cells in rat kidney collecting duct defined by antibodies against erythroid band 3 and renal vacuolar H⁺-ATPase.' *Proc. Natl. Acad. Sci. USA*, 1989, **86**, 5429–5433.

57. J. W. Verlander, K. M. Madsen, P. S. Low *et al.*, 'Immunocytochemical localization of band 3 protein in the rat collecting duct.' *Am. J. Physiol.*, 1988, **255**, F115–F125.

58. J. J. Grantham and M. B. Burg, *Am. J. Physiol.*, 1965, **211**, 255–259.

59. J. M. Sands, H. Nonoguchi and M. A. Knepper, 'Vasopressin effects on urea and H₂O transport in inner medullary collecting duct subsegments.' *Am. J. Physiol.*, 1987, **253**, F823–F832.

60. M. C. Reif, S. L. Troutman and J. A. Schafer, 'Sustained response to vasopressin in isolated rat cortical collecting tubule.' *Kidney Int.*, 1984, **26**, 725–732.

61. S. Nielsen, S. R. DiGiovanni, E. I. Christensen *et al.*, 'Cellular and subcellular immunolocalization of vasopressin-regulated water channel in rat kidney.' *Proc. Natl. Acad. Sci. USA*, 1993, **90**, 11663–11667.

62. J. M. Sands and M. A. Knepper, 'Urea permeability of mammalian inner medullary collecting duct system and papillary surface epithelium.' *J. Clin. Invest.*, 1987, **79**, 138–147.

63. J. J. Grantham, M. B. Burg and J. Orloff, 'The nature of transtubular Na and K transport in isolated rabbit renal collecting tubules.' *J. Clin. Invest.*, 1970, **49**, 1815–1826.

64. M. J. Hanley and J. P. Kokko, 'Study of chloride transport across the rabbit cortical collecting tubule.' *J. Clin. Invest.*, 1978, **62**, 39–44.

65. W. E. Lombard, J. P. Kokko and H. R. Jacobson, 'Bicarbonate transport in cortical and outer medullary collecting tubules.' *Am. J. Physiol.*, 1983, **244**, F289–F296.

66. L. G. Palmer and G. Frindt, 'Amiloride-sensitive Na channels from the apical membrane of the rat cortical collecting tubule.' *Proc. Natl. Acad. Sci. USA*, 1986, **83**, 2767–2770.

67. K. Tomita, J. J. Pisano and M. A. Knepper, 'Control of sodium and potassium transport in the cortical collecting duct of the rat. Effects of bradykinin, vasopressin, and deoxycorticosterone.' *J. Clin. Invest.*, 1985, **76**, 132–136.

68. B. M. Koeppen, B. A. Biagi and G. Giebisch, 'Intracellular microelectrode characterization of the rabbit cortical collecting duct.' *Am. J. Physiol.*, 1983, **244**, F35–F47.

69. G. Frindt and L. G. Palmer, 'Low conductance K channels in apical membrane of rat cortical collecting tubule.' *Am. J. Physiol.*, 1989, **256**, F143–F151.

70. C. S. Wingo, 'Reversible chloride-dependent potassium flux across the rabbit cortical collecting tubule.' *Am. J. Physiol.*, 1989, **256**, F697–F704.

71. J. B. Stokes, 'Mineralocorticoid effect on K⁺ permeability of the rabbit cortical collecting tubule.' *Kidney Int.*, 1985, **28**, 640–645.

72. K. Tomita, J. J. Pisano, M. B. Burg *et al.*, 'Effects of vasopressin and bradykinin on anion transport by the rat cortical collecting duct. Evidence for an electroneutral sodium chloride transport pathway.' *J. Clin. Invest.*, 1986, **77**, 136–141.

73. J. L. Atkins and M. B. Burg, 'Bicarbonate transport by isolated perfused rat collecting ducts.' *Am. J. Physiol.*, 1985, **249**, F485–F489.

74. R. A. Star, M. B. Burg and M. A. Knepper, 'Luminal disequilibrium pH and ammonia transport in outer medullary collecting duct.' *Am. J. Physiol.*, 1987, **252**, F1148–F1157.

75. J. B. Stokes, M. J. Ingram, A. D. Williams *et al.*, 'Heterogeneity of the rabbit collecting tubule: localization of mineralocorticoid hormone action to the cortical portion.' *Kidney Int.*, 1981, **20**, 340–347.

76. D. Brown and T. Kumpulainen, 'Immunocytochemical localization of carbonic anhydrase on ultrathin frozen sections with protein A-gold.' *Histochemistry*, 1985, **83**, 153–158.

77. C. S. Wingo, 'Active proton secretion and potassium absorption in the rabbit outer medullary collecting duct. Functional evidence for proton–potassium-activated adenosine triphosphatase.' *J. Clin. Invest.*, 1989, **84**, 361–365.

78. M. L. Zeidel, P. Silva and J. L. Seifter, 'Intracellular pH regulation and proton transport by rabbit renal medullary collecting duct cells. Role of plasma membrane proton adenosine triphosphatase.' *J. Clin. Invest.*, 1986, **77**, 113–120.

79. M. D. Breyer and H. R. Jacobson, 'Regulation of rabbit medullary collecting duct cell pH by basolateral Na⁺/H⁺ and Cl⁻/base exchange.' *J. Clin. Invest.*, 1989, **84**, 996–1004.

80. W. L. Clapp, K. M. Madsen, J. W. Verlander *et al.*, 'Intercalated cells of the rat inner medullary collecting duct.' *Kidney Int.*, 1987, **31**, 1080–1087.

81. W. L. Clapp, K. M. Madsen, J. W. Verlander *et al.*, 'Morphologic heterogeneity along the rat inner medullary collecting duct.' *Lab. Invest.*, 1989, **60**, 219–230.

82. J. M. Sands and D. C. Schrader, 'An independent effect of osmolality on urea transport in rat terminal inner medullary collecting ducts.' *J. Clin. Invest.*, 1991, **88**, 137–142.

83. C. Chou and M. A. Knepper, 'Inhibition of urea transport in inner medullary collecting duct by phloretin and urea analogues.' *Am. J. Physiol.*, 1989, **257**, F359.

84. B. Truniger and B. Schmidt-Nielsen, in 'Urea and the Kidney,' eds. B. Schmidt-Nielsen and D. W. S. Kerr, Excerpta Medica, Amsterdam, 1970, pp. 314–322.

85. T. Isozaki, J. W. Verlander and J. M. Sands, 'Low protein diet alters urea transport and cell structure in rat initial inner medullary collecting duct.' *J. Clin. Invest.*, 1993, **92**, 2448–2457.

86. T. Isozaki, A. G. Gillin, C. E. Swanson *et al.*, 'Protein restriction sequentially induces new urea transport processes in rat initial IMCD.' *Am. J. Physiol.*, 1994, **266**, F756–F761.

87. T. Isozaki, J. P. Lea, J. A. Tumlin *et al.*, 'Sodium-dependent net urea transport in rat initial inner medullary collecting ducts.' *J. Clin. Invest.*, 1994, **94**, 1513–1517.

88. K. Ishibashi, S. Sasaki, K. Fushimi *et al.*, 'Molecular cloning and expression of a member of the aquaporin family with permeability to glycerol and urea in addition to water expressed at the basolaterol membrane of kidney collecting duct cells.' *Proc. Natl. Acad. Sci. USA*, 1994, **91**, 6269–6273.

89. T. Ma, A. Frigeri, H. Hasegawa *et al.*, 'Cloning of a water channel homolog expressed in brain meningeal cells and kidney collecting duct that functions as a stilbene-sensitive glycerol transporter.' *J. Biol. Chem.*, 1994, **269**, 21845–21849.

90. M. Echevarria, E. E. Windhager, S. S. Tate *et al.*, 'Cloning and expression of AQP3, a water channel from the medullary collecting duct of rat kidney.' *Proc. Natl. Acad. Sci. USA*, 1994, 91, 10997–11001.

91. C. A. Ecelbarger, J. Terris, G. Frindt *et al.*, 'Aquaporin-3 water channel localization and regulation in rat kidney.' *Am. J. Physiol.*, 1995, **269**, F663–F672.

92. H. Hasegawa, T. Ma, W. Skach *et al.*, 'Molecular

cloning of a mercurial-insensitive water channel expressed in selected water-transporting tissues.' *J. Biol. Chem.*, 1994, **269**, 5497–5500.

93. J. S. Jung, R. V. Bhat, G. M. Preston *et al.*, 'Molecular characterization of an aquaporin cDNA from brain: candidate osmoreceptor and regulator of water balance.' *Proc. Natl. Acad. Sci. USA*, 1994, **91**, 13052–13056.

94. J. M. Sands, H. R. Jacobson and J. P. Kokko, in 'The Kidney: Physiology and Pathophysiology,' 2nd edn., eds. D. W. Seldin and G. Giebisch, Raven Press, New York, 1992, pp. 1087–1155.

95. S. M. Wall, J. M. Sands, M. F. Flessner *et al.*, 'Net acid transport by isolated perfused inner medullary collecting ducts.' *Am. J. Physiol.*, 1990, **258**, F75–F84.

96. E. R. Lacy and B. Schmidt-Nielsen, 'Anatomy of the renal pelvis in the hamster.' *Am. J. Anat.*, 1979, **154**, 291–320.

97. F. J. Silverblatt, 'Ultrastructure of the renal pelvic epithelium of the rat.' *Kidney Int.*, 1974, **5**, 214–220.

98. J. M. Sands, M. A. Knepper and K. R. Spring, 'Na–K–Cl cotransport in apical membrane of rabbit renal

papillary surface epithelium.' *Am. J. Physiol.*, 1986, **251**, F475–F484.

99. M. R. Khorshid and D. B. Moffat, 'The epithelia lining the renal pelvis in the rat.' *J. Anat.*, 1974, **118**, 561–569.

100. J. V. Bonventre, M. J. Karnovsky and C. P. Lechene, 'Renal papillary epithelial morphology in anti-diuresis and water diuresis.' *Am. J. Physiol.*, 1978, **235**, F69–F76.

101. R. K. Packer, J. M. Sands and M. A. Knepper, 'Chloride and osmotic water permeabilities of isolated rabbit renal papillary surface epithelium.' *Am. J. Physiol.*, 1989, **257**, F218–F224.

102. J. M. Sands, E. J. Ivy and R. Beeuwkes III, 'Transmembrane potential of renal papillary epithelial cells: effect of urea and DDAVP.' *Am. J. Physiol.*, 1985, **248**, F762–F766.

103. W. B. Reeves and T. E. Andreoli, 'Apical acidification by rabbit papillary surface epithelium.' *Kidney Int.*, 1988, **33**, F893–F899.

104. P. S. Chandhoke, R. K. Packer and M. A. Knepper, 'Apical acidification by rabbit papillary surface epi-thelium.' *Am. J. Physiol.*, 1990, **258**, F893–F899.

7.03

Functional Anatomy of the Mammalian Urinary Bladder

MARK L. ZEIDEL

University of Pittsburgh Medical Center, PA, USA

7.03.1 INTRODUCTION

The mammalian bladder stores urine of strikingly different composition from that of the blood, often for prolonged periods, and then releases the urine as needed by the organism. The bladder's storage function requires appropriate responses in the epithelium, muscle, and innervation components (see Figures 1 and 2). The epithelium must prevent noxious substances in the urine from reaching the underlying bladder layers and the blood. This barrier to diffusion must expand its surface area as the bladder fills (see Figure 3). In addition, the bladder wall, composed of smooth muscle, must expand to hold the increasing load of urine, and both the intrinsic and extrinsic sphincters, composed of smooth and striated muscle, respectively, must block leakage of urine. The expansion of the bladder with filling must occur at low pressures, to prevent retrograde pressure up the ureters from

interfering with normal renal tubular function. Finally, the bladder's complex neural control system must suppress excitatory input stimuli to the bladder smooth muscle and maintain sphincter tone during urinary storage.

The release of urine requires a no less complex set of responses. The epithelium must remove apical surface area as the bladder empties. The muscle must contract in a coordinated fashion as the sphincter relaxes. The innervation must orchestrate coordinated detrusor muscular contraction and sphincter relaxation to permit emptying.

This chapter will review our current understanding of the anatomy and physiology of the bladder epithelium, muscular wall, and innervation, emphasizing how each component functions in the normal filling and emptying of the bladder. As appropriate, various bladder disease states will also be discussed, primarily to illustrate the elements of normal bladder function.

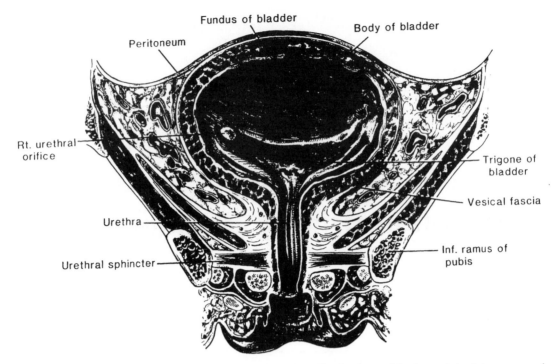

Figure 1 Anatomy of the normal human female bladder. The fundus and body contain detrusor muscle, while the trigone or base of the bladder (also called the bladder neck) participates in sphincter tone. The urethra is made up of smooth muscle. Surrounding the urethra is striated muscle which acts as the external urethral sphincter.

7.03.2 THE BLADDER EPITHELIUM: SITE OF THE PERMEABILITY BARRIER

Since the urine osmolality may vary from 50 to 1000 mOsm kg^{-1} in humans, while blood osmolality is tightly controlled between 275 and 290 mOsm kg^{-1}, it is clear that the bladder must be relatively impermeable to water. Similarly, the urine often contains high concentrations of urea, ammonium/ammonia (NH_4^+/NH_3), protons, and toxic substances. Since many of these substances are toxic to cells, it is critical that the bladder remains relatively impermeable to them as well. In addition to high concentrations of acid, ammonia, and urea, as well as widely varying osmolalities, the urine may contain exceptionally high concentrations of other toxic substances which may be secreted in the proximal tubule and concentrated by subsequent removal of salt and water from the final urine by the distal nephron. Thus, the bladder must store urine which differs markedly in composition from the blood for prolonged periods. In its barrier function, the bladder epithelium resembles other barrier epithelia, such as those of the gastric mucosa, the thick ascending limb of Henle, and the collecting duct, which each maintain large chemical gradients between the substances in their lumina and the blood.[1-3]

The permeability barrier of the mammalian bladder is localized in the apical membrane of the superficial epithelial cells.[4-8] This membrane possesses a unique ultrastructure, with 70–90% of the apical surface area occupied by plaques, and the balance of the surface occupied by normal appearing membrane.[4-7] The plaques appear to be rigid, proteinaceous structures, which exhibit a hexagonal array in freeze fracture and image-enhanced negative staining ultrastructural studies.[6,7] Interestingly, the outer (exoplasmic) membrane leaflet of the plaques is thicker (8 nm) than the usual unit membrane leaflet thickness of 4 nm; this latter value is the thickness of the inner (cytoplasmic) leaflet of the plaques and of both leaflets of the membrane in the nonplaque or "hinge" regions. Detailed electrophysiological studies of rabbit bladder epithelial cells have been performed and reveal an extremely high resistance of 20 000 Ω cm^{-2} and apical amiloride-sensitive Na^+ channels.[8] Impalements of the distinct epithelial cell layers with microelectrodes have demonstrated that the cells of different layers are not in electrical communication and that the apical membrane of the most superficial layer is the site of the electrical resistance of the epithelium.[9,10] Semiquantitative examination of the water permeability of bladder sacs has revealed that mechanical or chemical trauma to the superficial cell layer dramatically increases the water permeability of the bladder.[5,6] Application of molecules which introduce artificial pores into the apical membrane result in striking decreases in transepithelial resistance to values near zero.[11,12] In addition, these agents markedly increase water permeability (H. Negrete and M. L. Zeidel, unpublished observations). Importantly, purified apical membrane endosomes from mammalian bladder exhibit strikingly low permeabilities to water, small nonelectrolytes such as urea, and to ammmonia.[13] Taken together, these results indicate that the permeability barrier resides in the apical membrane.

The molecular nature of the bladder permeability barrier is unknown. Since varying the lipid composition of artificial membranes leads to values of membrane fluidity and water permeability which span the range of those observed in nature,[14-17] it has been assumed that the bulk lipid composition of the apical membrane was responsible for the bladder permeability barrier. However, it has been shown that the lipid structure of the cytoplasmic leaflets of biological membranes differ from those of the exoplasmic leaflet.[18-21] Moreover, in model epithelia the lipid molecules of the cytoplasmic leaflet communicate freely between the apical and basolateral domains, while the tight junctions prevent passage of lipids of the exoplasmic leaflets from apical to basolateral domains.[19,20,22] Since the apical domain of epithelial cells generally exhibits lower permeabilities, the lipids responsible for the low permeabilities likely reside in the outer membrane leaflet.[21,22] Finally, a class of enzymes which transport phospholipids from cytoplasmic to exoplasmic leaflets (phospholipid "flippases") has been described.[23-25] These results suggest that the bilayer structure of biological membranes as well as their lipid composition determines their permeability properties. Studies in isolated apical membranes have provided further confirmation of the importance of bilayer asymmetry in barrier apical membrane function.[13]

Partially purified preparations of mammalian bladder apical membranes have been shown to contain high levels of cholesterol, consistent with low permeability.[26] Interestingly, these membranes also contain high levels of cerebrosides, which are unusual in plasma membranes except for those containing myelin.[6,26] Levels of gangliosides, which may be abundant in the apical membranes of epithelia, and which may contribute to permeability properties, were not quantitated.[6,26]

The unique protein structure and the glycosaminoglycans of the apical membrane have also been proposed as critical elements in the permeability barrier.[27] The plaque areas of the apical membrane of bovine bladders were purified by density gradient centrifugation and detergent dissolution of contaminants.[28] Three proteins of molecular weight 27 kDa, 15 kDa, and 47 kDa, termed Uroplakins I, II, and III, respectively, were isolated.[28] These proteins are tightly associated *in situ* and likely form the protein matrix observed in ultrastructural studies.[28,29] Uroplakins I and II were immunolocalized to the exoplasmic membrane leaflet, while Uroplakin III was localized to both leaflets.[28] All uroplakins are also localized to endosomes which are removed from the apical membrane via apical membrane endosomes following voiding and reinsert as the bladder fills. The function of these proteins in the permeability barrier is unknown, but they may stabilize the apical membrane to withstand mechanical disruption as the bladder expands and contracts. They may also serve as attachment sites for cellular microfilaments to control insertion and removal of apical membrane (see below), or they may help reduce apical membrane permeabilities.[28,29]

It has been proposed that the glycosaminoglycans of the apical membrane might also serve as a permeability barrier, by trapping water within the sulfated sugar moeities.[27] There is already evidence that the glycosaminoglycans help prevent attatchment of bacteria to transitional epithelial cells.[30] When protamine is added to disrupt the bonds between water and the glycosaminoglycans, water, urea, and calcium permeabilities increase, both in rabbit bladders *in vitro* and in human bladders *in situ*.[27,31] These increases are reversed with heparin and artificial aminoglycans (pentosan polysulfate), which are thought to restore the activity of the native glycosaminoglycan layer.[27,31] However, studies by Lewis and coworkers have shown that the protamine treatment results in global damage to epithelial integrity, resulting in loss of resistance and marked increases in amiloride-insensitive Na^+ conductance.[32] Moreover, exposure of the apical membrane to molecules which induce pores results in striking increases in water permeability (H. Negrete and M. L. Zeidel, unpublished observations). If the glycosaminoglycan served as the permeability barrier then the introduction of pores in the membrane should not have altered water permeability. It is possible that the glycosaminoglycan impedes the movement of some molecules such as protons, but not others. Gastric mucin was shown to form a critical barrier to proton flux, helping maintain the enormous proton gradient (pH 2 in the lumen to pH 7.4 inside the cell) across the apical membrane of the gastric epithelial cell.[33]

The bladder maintains the permeability barrier when it is filled with urine and when it empties (see Figure 3). These changes in volume lead to profound changes in cell shape and the area of apical membrane exposed on the surface. Thus, when the bladder is filled, the cells are thinner and have more surface area exposed. By contrast, when the bladder empties, the apical cells are more round and have far less surface area exposed. The superficial epithelial cell adjusts its apical membrane area by endocytosis and reinsertion of apical membrane endosomes (AME). Ultrastructural studies have shown large numbers of AME (referred to as discoid vesicles and fusiform vesicles in these studies) within the cytoplasm of superficial epithelial cells of all mammalian species (including human) studied up to the mid-1990s.[34–36] These AME possess plaques which appear identical to those on the apical membrane and which are connected by normal appearing membranes.[34,35] Instillation of luminal ferritin results in incorporation into these AME within minutes. Morphometry of superficial epithelial cells of filled and empty bladders from four different rodent types demonstrated large increases in cytoplasmic AME when the bladder is empty and far fewer of them when the bladder is filled.[34] These AME have been

isolated, and their permeability properties defined. They exhibit strikingly low permeabilities to water, urea, and ammonia, but not to protons.[13] Similar trafficking of apical membrane containing specific apical surface proteins occurs in numerous other epithelial cells including those of the renal collecting duct (water channels),[3] distal convoluted tubule (calcium channels),[37] and gastric epithelial cells (H^+/K^+-ATPase).[38]

In addition to maintaining a low permeability to water, urea, ammonia, and toxins, the bladder epithelium transports Na^+ from the apical to the basolateral surface via apical amiloride-sensitive Na^+ channels coupled to basolateral Na/K-ATPase.[10,39,40] This transport has been thoroughly characterized in rabbit bladders *in vitro* and is altered by proteases which may appear in the urine.[41,42] It appears that Na^+ channels traffic into and out of the apical membrane in conjunction with AME.[42] Whether Na^+ channels are in AME remains to be determined. In addition, the physiological role of transepithelial Na^+ transport in the mammalian bladder is unclear.

7.03.3 THE BLADDER MUSCULATURE: ITS ROLE IN FILLING AND EMPTYING

The bladder musculature consists of the smooth muscle comprising the detrusor and

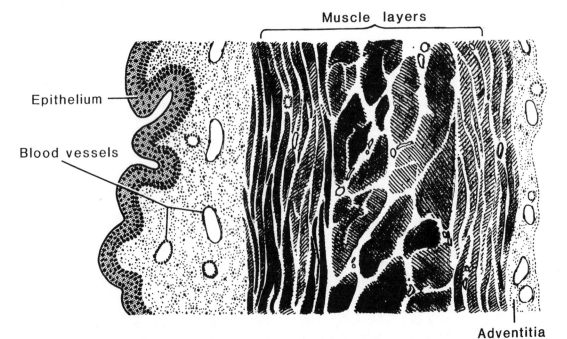

Figure 2 Normal elements of mammalian urinary bladder. The epithelium rests on a connective tissue base that contains blood vessels and afferent fibers. Underlying this are the smooth muscle layers and adventitia.

bladder neck and the striated muscle comprising the external urethral sphincter (see Figures 2 and 3).[43] During bladder filling, the smooth muscle of the detrusor relaxes, allowing filling at low intravesicular pressures, while the smooth muscle of the bladder neck and along the urethra as well as the striated muscle of the external urethral sphincter constrict, preventing release of urine. During bladder emptying, the detrusor smooth muscle contracts, increasing intravesicular pressure, while the smooth muscle of the bladder neck and the striated muscle of the external urethral sphincter relaxes, permitting urine outflow.

The three major physiological stimuli inducing smooth muscle contraction are stretch and exposure to the parasympathetic neurotransmitter, acetylcholine, as well as purinergic neurotransmitters such as ATP. Pharmacologic stimuli of contraction include the acetylcholine analogue, carbachol, caffeine, and membrane depolarization, either by excess extracellular potassium or by voltage clamp. As the bladder fills, the myocytes are stretched, leading to activation of nonselective cation channels, which permit rapid entry of sodium and some calcium.[44,45] Entry of cations depolarizes the smooth muscle cell membrane potential. If the extent of stretch is mild, there is less activation, and the membrane potential rests at a more depolarized level, predisposing the cell to activation by lower levels of muscarinic agonists. If the extent of stretch is more significant, the activation of cation channels may be sufficient to depolarize the cell sufficiently for initiation of an action potential. Although individual cells may contract, contraction of the bladder as a whole generally requires stimulation by efferent nervous activity.[46] When the membrane is sufficiently depolarized, L-type calcium channels open and calcium channels in the sarcoplasmic reticulum open, flooding the cell with calcium and resulting in an action potential.[47] The plasma membrane L-type calcium channels are gated by membrane potential and blocked by calcium channel blockers such as nifedipine. The channels which release calcium from the sarcoplasmic reticulum are also gated by calcium.[46,48] Between contractions the sarcoplasmic reticulum accumulates calcium to levels far above those of the cytosol by means of a calcium-ATPase;[48] calcium stores in the sarcoplasmic reticulum can be released in the absence of action potentials by exposure to caffeine, which renders the calcium channels sensitive to normal ambient cytoplasmic calcium levels.[48] Following the depolarization induced by an action potential, the membrane potential is repolarized by the effect of high levels of intracellular calcium to block L-type calcium channels, and the opening of potassium channels (both maxi-K^+ and K_{ATP} channels);[49,50] the latter repolarizes membrane potential by the efflux of positively charged K^+ from the cell.[47]

The rise in intracytoplasmic calcium brought on by the action potential results in the binding of calcium to calmodulin. Calcium-bound calmodulin is then capable of activating myosin light chain kinase, permitting it to phosphorylate the myosin type II light chain. Phosphorylation of the light chain allows the myosin to interact with actin, leading to force generation.[51,52] The actual geometrical arrangement of the actin–myosin complex

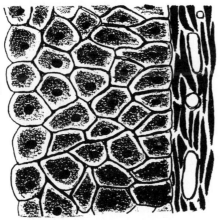

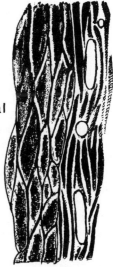

Transitional Epithelium in Empty Bladder

Transitional Epithelium in Full Bladder

Figure 3 Schematic diagram showing the bladder epithelium when the bladder is empty and filled. Note how the apical membrane surface area of each apically exposed cell increases as the bladder fills.

which permits force generation is unclear. Different isoforms of the myosin heavy chain have been defined,[51] as well as the components of the regulatory system linking cytoplasmic calcium levels to contraction.[52]

Exposure of the plasma membrane of the myocyte to acetylcholine or its analogue carbachol results in activation of the receptor, which, through an attached GTP (guanosine triphosphate) binding or G-protein, activates phospholipase C, releasing diacylglycerol and inositol trisphosphate. Diacylglycerol activates protein kinase C.[50,53] This cascade stimulates the nonselective cation channel and inhibits K_{ATP} channels, resulting in depolarization of myocyte membrane potential and activation of L-type calcium channels. The release of inositol trisphosphate stimulates release of calcium from the sarcoplasmic reticulum, further increasing cytoplasmic calcium levels.[47] As evidence for this pathway, it has been shown that ATP-sensitive K^+ channels (K_{ATP} channels) play a major role in determining the resting potential of the guinea pig detrusor smooth muscle cell.[54–56] When these channels are activated by specific activators such as cromakalim, the membrane potential is hyperpolarized and the muscle relaxes.[54–57] By contrast, inhibition of these channels causes depolarization of the membrane potential, aiding in activation of the calcium channels necessary for contraction.[50,55,56]

Activation of purinergic receptors by release of ATP from nerve endings can also result in detrusor contraction. It appears that purinergic stimulation opens nonselective cation channels, resulting in depolarization of membrane potential and the generation of action potentials.[46,53] Interestingly, purinergic mechanisms appear to be more prominent in animals which micturate partially to mark territory, than in humans, where cholinergic mechanisms predominate.[53]

An additional site of regulation concerns the relationship between intracellular calcium levels and contraction. When the smooth muscle is activated by muscarinic agonists or by stretch so that it depolarizes, an action potential occurs which results in rapid increases in intracellular calcium. This rise results in phasic contraction of the muscle. Variation in the rate of calcium entry effected by maneuvers such as changing the external pH of the cell result in closely parallel changes in phasic tension, indicating that the rate of initial calcium entry determines the extent of phasic contraction.[58] Following the phasic component of contraction, a degree of tonic contraction may persist following many stimuli after intracellular calcium levels have returned

to normal. These results suggest that other mediators may regulate the response of the contractile apparatus to changes in the levels of intracellular calcium.

In contrast to the smooth muscle of the detrusor, urethral smooth muscle is under adrenergic control, with alpha receptors responsive to norepinephrine predominating. The mechanisms underlying contraction of urethral smooth muscle have not been defined, but likely resemble those of the detrusor;[46] the steps between activation of adrenergic receptors and contraction of urethral smooth muscle are also unclear.

Striated muscle may be composed of fast or slow twitch fibers.[43,59] The former can be recruited rapidly, tend to fatigue rapidly, and perform predominantly anaerobic metabolism, while the latter are recruited slowly, fatigue slowly, and can perform high rates of oxidative metabolism.[43,59] The striated muscle of the external urethral sphincter comprises two parts, the periurethral striated muscle of the pelvic floor, which contains both fast and slow twitch fibers, and the striated muscle of the distal urethra, which contains predominantly slow twitch fibers.[60,61] Slow twitch fibers seem ideally suited to maintaining sphincter tone for prolonged periods, while fast twitch fibers may be needed to add to sphincter tone rapidly to maintain continence when intra-abdominal pressure is abruptly increased. Like that of smooth muscle, contraction of striated muscle fibers is governed by intracellular calcium, via interactions with troponin, as described in recent reviews.[43] The striated muscle of the external urethral sphincter is under somatic control (see below).

7.03.4 BLADDER INNERVATION AND CONTROL OF FILLING AND EMPTYING

Control of bladder filling and emptying requires complex reflex arcs involving several types of afferent fibers, sympathetic, parasympathetic, and somatic efferent fibers, as well as centers in the brainstem and more rostral portions of the brain. The complex innervation of the bladder is depicted in Figure 4. To summarize the function of bladder innervation, we will focus on the pathways regulating filling and micturition, and leave much of the anatomical detail to several excellent reviews.[62–67]

The pathways regulating urinary storage are depicted in Figure 5. Afferent fibers which are sensitive to low levels of tension in the bladder wall (below 5–15 mmHg, the level at which human subjects begin to be aware of bladder

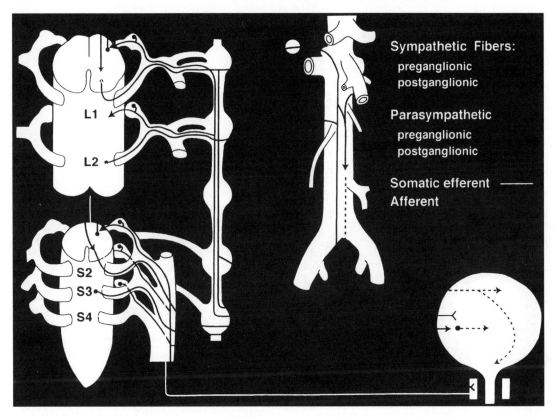

Figure 4 Innervation of the mammalian bladder. Sympathetic efferents exit the lumbar segments, synapse in the ganglia, and travel to the detrusor and base of the bladder smooth muscle. Parasympathetic fibers arising in the sacral segments synapse adjacent to the bladder and innervate the detrusor muscle. Myelinated and unmyelinated afferents travel from the bladder to the sacral segments, where they synapse and ascend to more rostral segments and to the brainstem.

filling) travel via the pelvic nerve to the sacral spinal segments, where they synapse on propriospinal neurons that send axons rostrally in the cord to the lumbar spinal segments, where they synapse on sympathetic neurons.[62–64] Low-level tension in the bladder stimulates a low level of afferent activity, which reflexly activates sacral somatic efferent fibers and lumbar sympathetic efferent fibers.[68,69] Sacral somatic fibers travel via the pudendal nerve to the striated muscle of the external sphincter and increase its tone, promoting urine retention. Lumbar sympathetic fibers travel via the hypogastric nerve to the detrusor muscle, where they inhibit the contraction that would normally result from stretch activation, and to the base of the bladder, where they increase tone and enhance sphincter function.[63,65] Sympathetic fibers may also inhibit parasympathetic nerve activity, resulting in indirect inhibition of detrusor function.[63] Nerve fibers from the pontine micturition center inhibit the activity of this reflex during the stimulation of micturition.[63,64] More rostral inputs stimulate the activity of this reflex when urinary retention is necessary.

The pathways regulating micturition are shown in Figure 6. δ-Myelinated afferent fibers responding to bladder stretch synapse on interneurons in the sacral parasympathetic nucleus as well as on spinal tract neurons which send ascending axons to the pontine micturition center.[43,62,63,68–70] C type unmyelinated fibers, which respond to noxious stimuli and not to tension, also reach the parasympathetic nucleus and project more rostrally.[43,62,63] C type fibers likely detect bladder inflammation and leakage of urinary substituents through the bladder permeability barrier. The increased stimulation of the pontine micturition center leads to enhancement of reflex stimulation of parasympathetic fibers in the sacral cord, leading to stimulation of detrusor contraction. The micturition center also inhibits sympathetic outflow (which would promote detrusor relaxation) and somatic outflow to the external sphincter, resulting in relaxation of the sphincter as the detrusor contracts.[62,63,67] Immediately after interruption of the spinal cord above the lumbosacral cord, the bladder becomes flaccid and exhibits urinary retention. However, over the course of weeks to months, lower

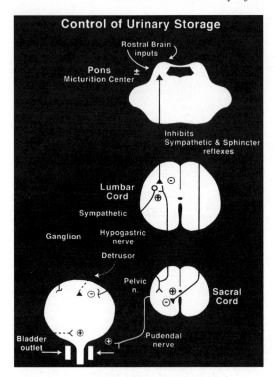

Figure 5 Control of urinary storage. Schematic view of reflex arcs controlling urinary storage. See text for details.

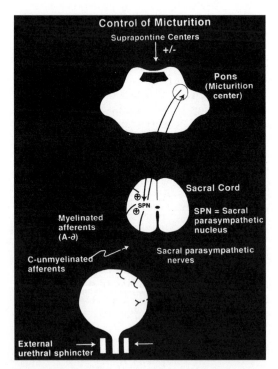

Figure 6 Control of micturition. Schematic view of reflex arcs controlling micturition. See text for details.

cord reflexes return, and the bladder develops a form of automatic micturition, so that when the bladder is sufficiently full, micturition begins.[43,62] Similar behavior is observed in newborn humans, in whom the appropriate rostral control of the lumbosacral reflex pathways has not yet developed.[62,64] It appears that activation of distinct afferent fibers results in stimulation of bladder emptying in the intact animal, as opposed to the animal with an interrupted or not fully developed spinal cord.[62,64]

Numerous neurotransmitters participate in the control of micturition. Many amino acids and peptides such as γ-aminobutyric acid (GABA), enkephalins, serotonin, vasoactive intestinal peptide, and substance P can modulate the activities of the micturition reflex when applied centrally or in the spinal cord.[46,62,67] The wide array of neurotransmitters which may regulate bladder function attests to the complexity of control mechanisms which can govern the relatively simple reflex pathways outlined above. Interestingly, release of some neuropeptides such as substance P from afferent terminals in the bladder wall by agents such as capsaicin can directly stimulate bladder contraction or stimulate bladder afferents to provoke bladder contractions via reflex pathways.[62,71,72]

7.03.5 INTEGRATIVE FUNCTION OF THE BLADDER IN HEALTH AND DISEASE

Since the bladder stores the urine, the ability of the kidneys to excrete a urine which differs markedly in composition from that of the blood depends on the integrity of the bladder permeability barrier. Failure of this barrier threatens the ability of the body to excrete a concentrated or a dilute urine, a urine high in ammonium/ammonia and of low pH, and a urine in which uremic toxins, K^+, or potentially toxic drugs may be highly concentrated. If this barrier is disrupted by mechanical, chemical, or inflammatory damage to the epithelial layer, noxious substances from the urine penetrate the connective tissue beneath the epithelium where they can provoke an inflammatory response. The noxious substances and the inflammation can stimulate C-type afferent fibers, leading to the sensation of discomfort and augmentation of micturition reflexes. The noxious substances and inflammatory mediators may directly stimulate bladder smooth muscle, increasing responsiveness to parasympathetic stimulation. For example, if high concentrations of potassium in the urine leak through the epithelial barrier to the muscle layer below, they could depolarize smooth muscle cells, leading to generation of action potentials. The increased afferent traffic as well as the direct stimulation of bladder muscle contraction cause micturition reflexes to predominate over storage reflexes at lower levels of bladder filling. Together these responses lead to urinary frequency and urgency, and can result in incontinence. In addition, many individuals with these symptoms also drink large quantities of water, increasing the rate of urine flow. These responses are adaptive in several respects. Increased fluid intake leads to dilution of many of the noxious substances in the urine, such as urea, ammonium/ammonia, and acid, reducing further irritation. Increased frequency of urination also reduces the time available for substances in the urine to diffuse down their concentration gradients into the blood through the damaged permeability barrier.

Elderly men with enlarged prostates exhibit urinary frequency and urgency, as well as incontinence. In animal models as well as patients, gradual onset of bladder outflow obstruction leads initially to incomplete bladder emptying, and to gradual chronic increase in the bladder volume.[43,62] Over time, these changes lead to marked smooth muscle hypertrophy and hyperplasia, as well as evidence for increased ramification and/or size of both efferent autonomic and afferent fibers innervating the bladder.[62] These adaptive responses plus the postvoiding residual volume of urine reduce the effective compliance of the bladder for further urine storage and increase the predominance of the micturition reflexes over the storage reflex, leading to urinary urgency and frequency.

7.03.6 SUMMARY

The mammalian bladder maintains large chemical and osmotic gradients between the urine and the blood. During storage of urine the surface area of the apical barrier membranes of the epithelial cells expands to accommodate the increasing volume. In parallel with this, the bladder smooth muscle also expands at low tension and the sphincter muscles of the bladder base and external sphincter tighten due to the predominance of storage reflexes. When the bladder has filled, increased afferent nerve activity stimulates the pontine micturition center as well as lumbosacral reflexes, leading to stimulation of the detrusor muscle, relaxation of the sphincter, and release of urine. As the bladder empties, the epithelial cells lining the cavity reduce the apical membrane surface area via a process of endocytosis. Proper functioning of the epithelial barrier, the musculature, and the innervation are essential to bladder function which, in turn permits the kidney to excrete a urine of composition which differs markedly from that of the blood.

ACKNOWLEDGMENT

I thank Dr. W. C. de Groat for his careful reading of the manuscript and helpful comments.

7.03.7 REFERENCES

1. D. Kikeri, A. Sun, M. L. Zeidel et al., 'Cell membranes impermeable to NH_3.' *Nature*, 1989, **339**, 478–480.
2. N. A. Priver, E. C. Rabon and M. L. Zeidel, 'Apical membrane of the gastric parietal cell: water, proton, and nonelectrolyte permeabilites.' *Biochemistry*, 1993, **32**, 2459–2468.
3. M. L. Zeidel, K. Strange, F. Emma et al., 'Mechanisms and regulation of water transport in the kidney.' *Semin. Nephrol.*, 1993, **13**, 155–167.
4. R. M. Hicks, 'The mammalian urinary bladder: an accommodating organ.' *Biol. Rev. Camb. Philos. Soc.*, 1975, **50**, 215–246.
5. R. M. Hicks, 'The permeability of rat transitional epithelium: keratinization and the barrier to water.' *J. Cell Biol.*, 1966, **28**, 21–31.
6. R. M. Hicks, B. Ketterer and R. C. Warren, 'The ultrastructure and chemistry of the luminal plasma membrane of the mammalian urinary bladder: a structure with low permeability to water and ions.' *Philos. Trans. R. Soc. Lond. B Biol. Soc.*, 1974, **268**, 23–38.
7. L. A. Staehelin, F. J. Chlapowski and M. A. Bonneville, 'Lumenal plasma membrane of the urinary bladder, I: three-dimensional reconstruction from freeze-etch images.' *J. Cell Biol.*, 1972, **53**, 73–91.
8. S. A. Lewis, 'A reinvestigation of the function of the mammalian urinary bladder.' *Am. J. Physiol.*, 1977, **232**, F187–F195.
9. S. A. Lewis, N. K. Wills and D. C. Eaton, 'Basolateral membrane potential of a tight epithelium: ionic diffusion and electrogenic pumps.' *J. Membrane Biol.*, 1978, **41**, 117–148.
10. S. A. Lewis and J. M. Diamond, 'Na^+ transport by rabbit urinary bladder, a tight epithelium.' *J. Membrane Biol.*, 1976, **28**, 1–40.
11. S. A. Lewis, D. C. Eaton, C. Clausen et al., 'Nystatin as a probe for investigating the electrical properties of a tight epithelium.' *J. Gen. Physiol.*, 1977, **70**, 427–440.
12. S. A. Lewis and N. K. Wills, 'Electrical properties of the rabbit urinary bladder assessed using Gramicidin D.' *J. Membrane Biol.*, 1982, **67**, 45–53.
13. A. Chang, T. G. Hammond, T. T. Sun et al., 'Permeability properties of the mammalian bladder apical membrane.' *Am. J. Physiol.*, 1994, **267**, C1483–C1492.
14. A. Finkelstein, 'Water Movement Through Lipid Bilayers, Pores and Plasma Membranes, Theory and Reality,' Wiley, New York, 1986.
15. A. Finkelstein, 'Water and nonelectrolyte permeability of lipid bilayer membranes.' *J. Gen. Physiol.*, 1976, **68**, 127–135.

16. R. Fettiplace and D. A. Haydon, 'Water permeability of lipid membranes.' *Physiol. Rev.*, 1980, **60**, 510–550.

17. M. B. Lande, J. M. Donovan and M. L. Zeidel, 'The relationship between membrane fluidity and permeabilities to water, solutes, ammonia, and protons.' *J. Gen. Physiol.*, 1995, **106**, 67–84.

18. Y. Lange, H. Matthies and T. L. Steck, 'Cholesterol oxidase susceptibility of the red cell membrane.' *Biochim. Biophys. Acta*, 1984, **769**, 551–562.

19. K. Simons and G. van Meer, 'Lipid sorting in epithelial cells.' *Biochemistry*, 1988, **27**, 6197–6202.

20. G. van Meer, 'Plasma membrane cholesterol pools.' *TIBS*, 1987, **12**, 375–376.

21. G. van Meer, 'Lipid traffic in animal cells.' *Annu. Rev. Cell Biol.*, 1989, **5**, 247–275.

22. K. Simons and S. D. Fuller, 'Cell surface polarity in epithelia.' *Annu. Rev. Cell Biol.*, 1985, **1**, 243–288.

23. A. Zachowski, E. Favre, S. Cribier *et al.*, 'Outside–inside translocation of aminophospholipids in the human erythrocyte membrane is mediated by a specific enzyme.' *Biochemistry*, 1986, **25**, 2585–2590.

24. A. Zachowski and P. F. Devaux, 'Bilayer asymmetry and lipid transport across biomembranes.' *Comments Mol. Cell Biophys.*, 1989, **6**, 63–90.

25. A. Zachowski, J. P. Henry and P. F. Devaux, 'Control of transmembrane lipid asymmetry in chromaffin granules by an ATP-dependent protein.' *Nature*, 1989, **340**, 75–76.

26. J. S. Caruthers and M. A. Bonneville, 'Isolation and characterization of the urothelial lumenal plasma membrane.' *J. Cell Biol.*, 1977, **73**, 382–399.

27. C. L. Parsons, D. Boychuk, S. Jones *et al.*, 'Bladder surface glycosaminoglycans: an epithelial permeability barrier.' *J. Urol.*, 1990, **143**, 139–142.

28. X. R. Wu, M. Manabe, J. Yu *et al.*, 'Large scale purification and immunolocalization of bovine uroplakins I, II, and III: molecular markers of urothelial differentiation.' *J. Biol. Chem.*, 1990, **265**, 19170–19179.

29. J. Yu, M. Manabe, X. R. Wu *et al.*, 'Uroplakin I: a 27 kDa protein associated with the asymmetric unit membrane of mammalian urothelium.' *J. Cell Biol.*, 1990, **111**, 1207–1216.

30. R. E. Hurst, S. W. Rhodes, P. B. Adamson *et al.*, 'Functional and structural characteristics of the glycosaminoglycans of the bladder luminal surface.' *J. Urol.*, 1987, **138**, 433–437.

31. J. D. Lilly and C. L. Parsons, 'Bladder surface glycosaminoglycans is a human epithelial permeability barrier.' *Surg. Gynecol. Obstet.*, 1990, **171**, 493–496.

32. C. J. Tzan, J. Berg and S. A. Lewis, 'Effect of protamine sulfate on the permeability properties of the mammalian urinary bladder.' *J. Membrane Biol.*, 1995, **133**, 227–242.

33. M. R. Bhaskar, P. Garik, B. S. Turner *et al.*, 'Viscous fingering of HCl through gastric mucus.' *Nature*, 1992, **360**, 458–461.

34. B. D. Minsky and F. J. Chlapowski, 'Morphometric analysis of the translocation of lumenal membranes between cytoplasm and cell surface of transitional epithelial cells during the expansion–contraction cycles of mammalian urinary bladder.' *J. Cell Biol.*, 1978, **77**, 685–697.

35. R. M. Hicks, 'The function of the Golgi complex in transitional epithelium. Synthesis of the thick cell membrane.' *J. Cell Biol.*, 1966, **30**, 623–647.

36. S. N. Sarikas and F. J. Chlapowski, 'Effect of ATP inhibitors on the translocation of luminal membrane between cytoplasm and cell surface of transitional epithelial cells during the expansion–contraction cycle

of the rat urinary bladder.' *Cell Tissue Res.*, 1986, **246**, 109–117.

37. B. J. Bacskai and P. A. Friedman, 'Activation of latent Ca^{2+} channels in renal epithelial cells by parathyroid hormone.' *Nature*, 1990, **347**, 388–391.

38. J. G. Forte and A. Soll, in 'Handbook of Physiology. The Gastrointestinal System: Gastrointestinal Secretion,' American Physiological Society, Bethesda, MD, 1990, pp. 207–228.

39. S. A. Lewis and J W. Hanrahan, 'Physiological approaches for studying mammalian urinary bladder epithelium.' *Methods Enzymol.*, 1990, **192**, 632–650.

40. S. A. Lewis and J. M. Diamond, 'Active sodium transport by mammalian urinary bladder.' *Nature*, 1975, **253**, 747–748.

41. S. A. Lewis and C. Clausen, 'Urinary proteases degrade epithelial sodium channels.' *J. Membrane Biol.*, 1991, **122**, 77–88.

42. D. D. Loo, S. A. Lewis, M. S. Ifshin *et al.*, 'Turnover, membrane insertion and degradation of sodium channels in rabbit urinary bladder.' *Science*, 1983, **221**, 1288–1290.

43. W. D. Steers, in 'Campbell's Urology,' 6th edn., eds. P. C. Walsh, A. B. Retik, T. A. Stamey *et al.*, Harcout, Brace Jovanovich, Philadelphia, PA, 1992, pp. 142–178.

44. M. C. Wellner and G. Isenberg, in 'Nonselective Cation Channels: Pharmacology, Physiology and Biophysics,' eds. D. Siemen and J. Hescheler, Birkhauser Verlag, Basel, 1993, pp. 93–99.

45. M. C. Wellner and G. Isenberg, 'Properties of stretch-activated channels in myocytes from the guinea-pig urinary bladder.' *J. Physiol. (Lond.)*, 1993, **466**, 213–227.

46. K. E. Anderson, 'Pharmacology of lower urinary tract smooth muscles and penile erectile tissues.' *Pharmacol. Rev.*, 1993, **45**, 253–308.

47. J. L. Mostwin, 'The action potential of guinea pig bladder smooth muscle.' *J. Urol.*, 1986, **135**, 1299–1303.

48. S. M. Wall, M. F. Flessner and M. A. Knepper, 'Distribution of luminal carbonic anhydrase activity along rat inner medullary collecting duct.' *J. Physiol.*, 1991, **260**, F738–F748.

49. F. Markwardt and G. Isenberg, 'Gating of maxi K^+ channels studies by Ca^+ concentration jumps in excised inside-out multi-channel patches (myocytes from guinea-pig urinary bladder).' *J. Gen. Physiol.*, 1992, **99**, 841–862.

50. A. D. Bonev and M. T. Nelson, 'Muscarinic inhibition of ATP-sensitive K^+ channels by protein kinase C in urinary bladder smooth muscle.' *Am. J. Physiol.*, 1993, **265**, C1723–C1728.

51. S. Chacko, S. S. Jacob and K. Y. Horiuchi, 'Myosin I from mammalian smooth muscle is regulated by caldesmon-calmodulin.' *J. Biol. Chem.*, 1994, **269**, 15803–15807.

52. S. White, A. F. Martin and M. Periasamy, 'Identification of a novel smooth muscle myosin heavy chain cDNA: isoform diversity in the S1 head region.' *Am. J. Physiol.*, 1993, **264**, C1252–C1258.

53. A. F. Brading, 'Ion channels and control of contractile activity in urinary bladder smooth muscle.' *Jpn J. Pharmacol.*, 1992, **58** (Suppl. 2), 120P–127P.

54. A. D. Bonev and M. T. Nelson, 'ATP-sensitive potassium channels in smooth muscle cells from guinea pig urinary bladder.' *Am. J. Physiol.*, 1993, **264**, C1190–C1200.

55. K. Fujii, C. D. Foster, A. F. Brading *et al.*, 'Potassium channel blockers and the effects of cromakalim on the smooth muscle of the guinea-pig bladder.' *Br. J. Pharmacol.*, 1990, **99**, 779–785.

56. T. L. Grant and J. S. Zuzack, 'Effects of K^+ channel

blockers and cromakalim (BRL 34915) on the mechanical activity of guinea pig detrusor smooth muscle.' *J. Pharmacol. Exp. Ther.*, 1991, **259**, 1158–1164.

57. C. D. Foster, K. Fujii, J. Kingdon *et al.*, 'The effect of cromakalim on the smooth muscle of guinea-pig urinary bladder.' *Br. J. Pharmacol.*, 1989, **97**, 281–291.

58. C. H. Fry, C. R. Gallegos and B. S. Montgomery, 'The actions of extracellular H^+ on the electrophysiological properties of isolated human detrusor smooth muscle cells.' *J. Physiol. (Lond.)*, 1994, **480**, 71–80.

59. H. A. Padykula and G. F. Gauthier, 'The ultrastructure of the neuromuscular junctions of mammalian red, white and intermediate skeletal muscle fibers.' *J. Cell Biol.*, 1970, **46**, 27–41.

60. J. A. Gosling, J. S. Dixon, O. D. Critchley *et al.*, 'A comparative study of the human external sphincter and periurethral levator ani muscles.' *Br. J. Urol.*, 1981, **53**, 35–41.

61. A. Elbadawi and M. A. Atta, 'Ultrastructural analysis of vesicourethral innervation, IV: evidence for somatomotor plus autonomic innervation of the male feline rhabdosphincter.' *Neurol. Urodynam.*, 1985, **4**, 23–29.

62. W. C. de Groat, A. M. Booth and N. Yoshimura, in 'The Autonomic Nervous System,' ed. C. A. Maggi, Harwood Academic, London, 1993, vol. 6, pp. 227–290.

63. W. C. de Groat and A. M. Booth, in 'The Autonomic Nervous System, vol. 2, Nervous Control of the Urogenital System,' ed. C. A. Maggi, Harwood Academic, London, 1992.

64. W. C. de Groat and A. M. Booth, in 'Peripheral Neuropathy,' 3rd edn., eds. P. J. Dyck, P. K. Thomas, E. Lambert *et al.*, W. B. Saunders, Philadelphia, PA, 1992.

65. Y. Wakabayashi, Y. Makiura, T. Tomoyoshi *et al.*, 'Adrenergic innervation of the urinary bladder body in the cat, with special reference to structure of the detrusor muscle: an immunohistochemical study of noradrenaline and its synthesizing enzymes.' *Arch. Histol. Cytol.*, 1994, **57**, 277–289.

66. B. S. Mitchell, 'Morphology and neurochemistry of the pelvic, and paracervical ganglia.' *Histol. Histopathol.*, 1993, **8**, 761–773.

67. J. D. Stephenson, 'Pharmacology of the central control of micturition.' *Funct. Neurol.*, 1991, **6**, 211–217.

68. P. Satchall and C. Vaughan, 'Bladder wall tension and mechanoreceptor discharge.' *Pflugers Arch.*, 1994, **426**, 304–309.

69. H. J. Habler, W. Janig and M. Koltzenburg, 'Myelinated primary afferents of the sacral spinal cord responding to slow filling and distension of the cat urinary bladder.' *J. Physiol. (Lond.)*, 1993, **463**, 449–460.

70. K. E. Creed, Y. Ito and H. Katsuyama, 'Neurotransmission in the urinary bladder of rabbits and guinea pigs.' *Am. J. Physiol.*, 1991, **261**, C271–C277.

71. C. A. Maggi and A. J. Meli, 'The role of neuropeptides in the regulation of the micturition reflex.' *Auton. Pharmacol.*, 1986, **6**, 133–162.

72. C. A. Maggi, P. Santicioli and A. Meli, 'The effects of topical capsaicin on rat urinary bladder motility *in vivo.*' *Eur. J. Pharmacol.*, 1984, **103**, 41–50.

7.04
Renal Tubular Transport of Organic Anions and Cations

WILLIAM H. DANTZLER and STEPHEN H. WRIGHT

University of Arizona Health Sciences Center, Tucson, AZ, USA

7.04.1 INTRODUCTION

Two general renal transport systems—one for organic anions and one for organic cations—play a major role in regulating the concentrations in the body of various endogenous and exogenous compounds, some of them toxic. In addition, renal toxicity produced by some compounds may depend on transport by these systems. Because there have been several comprehensive reviews on these transport pathways,[1–3] only those mechanistic aspects likely to be most significant for toxicologists are reviewed. For those who wish more insight into earlier *in vivo* and intact tissue studies, the reviews by Weiner[4,5] and by Roch-Ramel *et al.*[1] are particularly comprehensive.

7.04.2 ORGANIC ANION TRANSPORT

7.04.2.1 General Characteristics

A major system for the secretion of a wide range of hydrophobic organic anions (or weak organic acids that exist as anions at physiological pH; collectively OAs) apparently exists in the renal tubules of almost all vertebrates studied.[2,6] This system, although capable of secreting a number of endogenous compounds, is particularly effective in secreting numerous exogenous compounds, including many pharmacologically active substances, environmental toxins, and plant and animal toxins. It also secretes toxic metabolic breakdown products of exogenous and endogenous compounds. Indeed, its primary function appears to be the effective removal of such substances from the body. Some examples of compounds removed from the body by this general secretory system are shown in Table 1. Although there are a number of other systems for transporting specific types of OAs in either the secretory or reabsorptive directions,[7,8] these do not appear to be particularly significant for renal toxicity and this chapter focuses only on the major transport pathway for secretion of OAs.

Studies with numerous vertebrate species and a wide range of techniques and tissue preparations (e.g., renal clearance, renal micropuncture, renal stop-flow, isolated tubule transport) made over many years have demonstrated that this secretory process resides in the proximal tubules.[3] Additional studies with isolated mammalian renal tubules have shown that the primary site of transport in most species, probably including humans, is in the S2 segment of the proximal tubule.[9] However, net secretion also occurs in both the S1 and S3 segments. Most studies with single tubules suggest that the difference in transport between these segments reflects a difference in the number of transporters per unit length (difference in V_{max}) rather than a difference in affinity (K_t) that would suggest different transporters.[10–12] However, some studies with transport inhibitors suggest that there might yet be some difference in the characteristics of the transporters between the S2 and S3 segments.[11]

7.04.2.2 Transepithelial Transport

Net transepithelial secretion of OAs, of which *para*-aminohippurate (PAH) is the prototype, occurs by an overall saturable active process in all vertebrate species studied (see Refs. 1 and 2). During the net secretory process in isolated perfused mammalian and nonmammalian renal tubules, the steady-state concentration of OA in the tubule lumen is greater than that in the peritubular bathing medium and the concentration in the cells is greater than that in either the bathing medium or the lumen.[13,14] Because the inside of the cells is negative compared with the lumen and peritubular fluid and because these compounds are transported as anions at physiological pH, these data are compatible with transport into the cells against an electrochemical gradient at the basolateral membrane and movement from the cells into the lumen down an electrochemical gradient. Indeed, this is the accepted process for transepithelial secretion.

In mammalian renal tubules, there is no carrier-mediated reabsorptive process for compounds secreted by this general OA transport process.[1] Moreover, for a great many compounds like PAH, which share this secretory pathway, lipid solubility is not high and there is virtually no passive backdiffusion from lumen to basolateral fluid across the epithelium.[1] For some compounds, however, with relatively high lipid solubilities, the degree to which they exist in the nonionized form, as determined by the pK_a of the compound and the pH of the tubule fluid, determines the amount of passive backdiffusion that will occur.[1,3]

7.04.2.3 Basolateral Transport

7.04.2.3.1 Mechanism of basolateral transport

Transport of OAs into the tubule cells against an electrochemical gradient at the basolateral membrane is the energy requiring step in the transepithelial secretory process. Moreover, in isolated mammalian renal tubules, kinetic analyses of PAH transport based on initial rates of uptake at the basolateral membrane[12] and steady-steady rates of net transepithelial secretion[10] give almost identical values for K_t and J_{max}. These data indicate that transport of OAs into the cells at the basolateral membrane also is rate-limiting for transepithelial secretion.

The mechanism involved in this basolateral transport step has been defined. Although it could easily be shown that this step required energy,[15,16] it could never be shown that it was directly linked to the hydrolysis of ATP.[17,18] However, it became apparent early that this transport step was inhibited, not only by a reduction in energy production, but also by the removal of Na^+ or K^+ from the medium bathing isolated tubules or slices or by inhibition of Na^+-K^+-ATPase. These observations suggested that Na^+ might be important in the transport process.[16,19,20] This concept was further strengthened by evidence that the inwardly directed Na^+ gradient, not the simple presence of Na^+, was the important factor,[21,22] but direct coupling of OA entry to the Na^+ gradient could not be demonstrated either *in vivo* or *in vitro*.[23–27] However, because basolateral PAH/PAH exchange could easily be demonstrated;[23,25,26] because PAH uptake could be inhibited by the anion exchange inhibitor SITS;[28] because a number of anionic metabolites, whose entry across the basolateral membrane was directly coupled to the Na^+ gradient, could stimulate or inhibit basolateral PAH uptake;[27,29,30] and because the combination of an outwardly directed PAH or OH^- gradient and an inwardly directed Na^+ gradient could produce a brief uptake of PAH above equilibrium,[25,31] a number of investigators postulated that PAH entry across the basolateral membrane might involve exchange for an anionic metabolite.[23,27,32] In 1987, Pritchard[33] and Burckhardt and co-workers[34] extended this idea by proposing that PAH transport into the cells against an electrochemical gradient at the basolateral membrane involved countertransport for a dicarboxylate that was in turn taken up into the cells by a Na^+ cotransport process. Using rat basolateral membrane vesicles (BLMV), they were able to demonstrate that this apparently was the case, as illustrated in the model shown in Figure 1.[34,35] In this model,

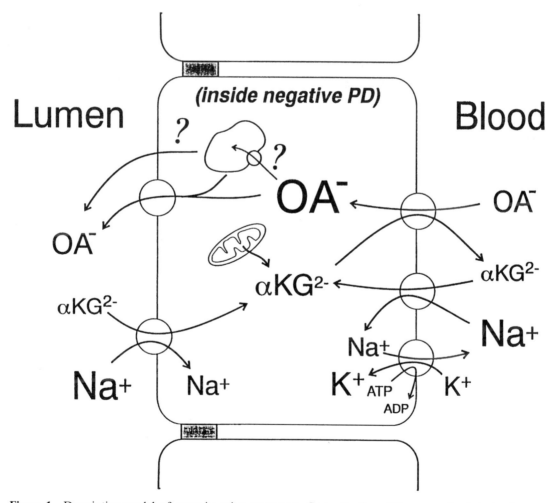

Figure 1 Descriptive model of organic anion transport. Concentration differences of substrates between inside and outside of cells are shown by the size of the symbols for the substrates. Circles represent carrier-mediated transport processes. Transport requiring direct input of energy is indicated by breakdown of ATP to ADP. Arrows with question marks indicate speculative pathways (see text).

transport into the cells of PAH (and other OAs that share this system) is a tertiary active process, in which the final step is the counter-transport of PAH against its electrochemical gradient for an intracellular dicarboxylate (physiologically, α-ketoglutarate, αKG) moving out of the cells down its electrochemical gradient. The outwardly-directed gradient for αKG is maintained by metabolism and by transport into the cells across the basolateral membrane via the Na^+-dicarboxylate cotransport system (Figure 1). The inwardly-directed Na^+ gradient driving dicarboxylate entry is maintained in turn by the primary energy-requiring transport step in this tertiary process, the transport of Na^+ out of the cells via Na^+-K^+-ATPase at the basolateral membrane (Figure 1).

This model, originally proposed on the basis of studies with BLMV,[34,35] has now been confirmed in studies with intact renal tubules.[36] Loading S2 segments of isolated, perfused rabbit proximal tubules with αKG via the basolateral Na^+-dicarboxylate cotransporter increases both the rate of uptake of PAH at the basolateral membrane and the rate of net transepithelial secretion to an approximately equal extent.[36] Kinetic analysis of the effect of basolateral αKG loading on the rate of PAH uptake indicates that stimulation results from an increase in the maximum rate of PAH transport (J_{max}) with little change in K_t.[12] However, the metabolic state of the tubules and thus their production of intracellular αKG appear to be important for maintenance of normal transport of PAH. The control rate of PAH transport is about four to five times higher in tubules perfused and bathed in bicarbonate-buffered medium than in tubules perfused and bathed in HEPES (*N*-2-hydro-xyethylpiperazine-*N'*-2-ethanesulfonic acid)-buffered medium.[37] In accordance with this observation, basolateral loading with αKG stimulates PAH transport about twofold in bicarbonate-buffered medium and about four-fold in HEPES-buffered medium.[12,36] Similarly, in isolated S3 segments, although αKG can apparently be countertransported for PAH, the metabolic state of the tubule appears more important than basolateral αKG uptake in determining the outwardly-directed gradient for αKG.[11] Indeed, in all proximal tubule segments in the intact animal, the outwardly-directed gradient for αKG may be maintained primarily by metabolic production within the cells (Figure 1), and the basolateral uptake of αKG via the Na^+-dicarboxylate cotransporter may serve primarily to reaccumulate αKG that has been exchanged for PAH or other OAs transported by this system.

Additional studies with isolated, perfused S2 segments indicate that αKG reabsorbed from the perfusate (or, physiologically, from the glomerular filtrate) via the luminal Na^+-dicarboxylate cotransporter also can function in the countertransport step for PAH at the basolateral membrane.[37] However, *in vitro* αKG reabsorbed from the lumen apparently is important for PAH transport only in the absence of the bicarbonate/CO_2 buffer system, suggesting that *in vivo* it plays no role in PAH transport unless cellular production of αKG is reduced.

7.04.2.3.2 *Specificity of basolateral transport*

Extensive studies involving *in vivo et situ* microperfusions of basolateral capillaries in rats by Ullrich and co-workers[8,38,39] have helped to characterize the substrate structure for which the basolateral OA transporter has the highest affinity. According to these studies, the affinity of the transporter for various substrates is determined by their hydrophobi-city and charge, charge strength, and charge placement. The transporter apparently accepts monovalent hydrophobic OAs with a negative or partially negative charge. In these compounds, the minimum length of the hydropho-bic domain for effective binding to the transporter is 4 Å. The transporter also appar-ently accepts divalent organic compounds with two negative or partially negative charges. The optimal distance between charges for effective binding is 6–7 Å. The divalent compounds accepted can apparently include some zwitter-ions. Because the hydrophobic domain of the divalent compounds accepted by the transpor-ter can have a maximum length of 10 Å, this domain may lie at least partially outside a direct line between the charges. Finally, the affinity of the transporter for compounds within a given series tends to increase with the strength of their charge. However, despite numerous attempts, the basolateral transporter for OAs has yet to be isolated or its molecular structure character-ized.

7.04.2.4 Luminal Transport

For OAs, the transport step down an electrochemical gradient from the cells into the tubule lumen is much less well understood than the transport step against an electroche-mical gradient into the cells at the basolateral membrane (Figure 1). In isolated renal tubules, the apparent permeability of the luminal membrane to OAs is several times greater than the apparent passive permeability of the basolateral membrane.[13,40] This favors movement of OAs that have entered the tubule cells at the basolateral side into the tubule lumen. Although this movement from the cells into the lumen could involve simple passive diffusion, the apparent permeability of the luminal membrane is much greater than would be expected for simple passive diffusion of a charged molecule across a biological membrane. Moreover, transport can be inhibited by probenecid and other inhibitors in isolated tubules and brush border membrane vesicles (BBMV) from a number of species.[23,26,28,41] Although an OA exchanger, which can be shared by PAH, urate, and hydroxyl ions has been identified in the luminal membrane of some mammalian species (e.g., dog, rat), it is not present in all species (e.g., rabbit[42,43]). Moreover, in those species in which such a luminal OA exchanger exists it appears poised to reabsorb urate, not to secrete PAH.[43] Although transport out of the cells across the luminal membrane may well involve anion exchange, it may also involve a potential driven conductive pathway (Figure 1).[3]

7.04.2.5 Intracellular Steps

It has generally been assumed that OAs, such as PAH, transported into the cells at the basolateral membrane diffuse through the cytoplasm from the basolateral side to the luminal side, despite evidence for some intracellular binding.[44,45] However, studies involving epifluorescence video-imaging and confocal microscopy techniques have indicated that fluorescein, a fluorescent anion transported by the general OA (PAH) transport system,[46] is compartmentalized within the renal tubule cells of nonmammalian vertebrates.[47] The compartmentalization appears to be within vesicular structures, to be specific for OAs that share the OA transport system, and to be dependent on metabolism.[47] Even more recent studies with isolated single S2 segments of rabbit proximal renal tubules have shown that in the absence of the bicarbonate/CO_2 buffer system, fluorescein accumulates in vesicular structures in the cytoplasm and fails to enter the tubule lumen. However, when the buffer system is changed to bicarbonate/CO_2, the physiological buffer system and one that is well-known to be optimal for tubule metabolism and OA transport (Dantzler, unpublished observations),[37,48] all evidence of vesicular sequestration disappears and fluorescein accumulates in the tubule lumen (Shpun *et al.*, unpublished observations).[11] It has been suggested that vesicular accumulation could be involved in intracellular

translocation while simultaneously protecting cytoplasmic functions from toxic OAs.[3] However, the failure of such vesicular sequestration to occur in mammalian tubules, except when metabolism is compromised, leaves its function and the control of its occurrence far from clear (Figure 1). The process by which OAs move within the cells from the basolateral to the apical pole during transepithelial transport certainly merits further study.

7.04.2.6 Interaction of OA Transporters with Nephrotoxicants

The transport process for OAs at the basolateral membrane may be significant, not only for the removal of toxic substances from the body, but also possibly for the entry of nephrotoxicants into the renal tubule cells. Indeed, for some agents, this pathway may be the most important or even sole means by which they enter the cells.

A group of compounds for which the OA transport system may be important for the production of toxicity are the nephrotoxic cysteine conjugates and their *N*-acetyl derivatives. They can enter the renal tubule cells across the basolateral membrane to produce their nephrotoxicity.[49] Although all possible entry pathways for these compounds are not yet known, evidence indicates that they can be transported into the cells via the general OA pathway.[12] Initially, studies of uptake by proximal tubule suspensions from rat kidneys suggested that the *N*-acetyl cysteine conjugates, but not the cysteine conjugates themselves could be taken up by the OA pathway.[50] However, this initial distinction was based only on the failure of high concentrations of probenecid, a compound with a very high affinity for the OA transporter,[12] to inhibit uptake of the radiolabeled cysteine conjugate. The possibility that the effect of probenecid might have been obscured by entry of the cysteine conjugate by another pathway was not considered.[50] Moreover, inhibition studies involving *in vivo* microperfusions of basolateral capillaries in rats suggested that both cysteine conjugates and their *N*-acetyl derivatives could interact with the OA transporter, although the transporter appeared to have a greater affinity for the *N*-acetylated form than for the cysteine conjugate alone.[51]

Other studies have considered the kinetics of the interactions of both toxic (DCVC, *S*-(1,2-dichlorovinyl)-L-cysteine) and nontoxic (BTC, *S*-(2-benzothiazole)-L-cysteine) cysteine conjugates and their *N*-acetyl derivatives with the basolateral OA transporter in single isolated S2 segments of rabbit proximal tubules.[12] These studies indicate that both the cysteine conjugates and their *N*-acetyl derivatives are effective competitive inhibitors of basolateral PAH transport. Moreover, the cysteine conjugates interact at least as effectively with the transporter as their *N*-acetyl derivatives. In addition, *trans*-stimulation of PAH efflux across the basolateral membrane of these isolated proximal S2 tubule segments by both the cysteine conjugates and their *N*-acetyl derivatives strongly indicates that both forms are actually transported by the PAH transporter.[12]

However, it must be kept in mind that the general OA transport system may be only one pathway by which these nephrotoxicants enter renal tubule cells. Moreover, the nephrotoxic cysteine conjugates (e.g., DCVC) apparently cause their initial toxic effects on cells of the S3 segment, not the S2 segment, of the proximal tubule. Although, OA transporters are present in this segment, no information is available about whether cysteine conjugates enter these cells by this pathway or, if they do, whether differences in the initial toxicity in the two segments reflect differences in metabolic function rather than in rates of entry.

7.04.3 ORGANIC CATION TRANSPORT

7.04.3.1 General Characteristics

Organic cations and bases (collectively, OCs) are a structurally diverse array of molecules of broad physiological, pharmacological, and toxicological importance. As in the case of OAs, the renal proximal tubule is a major site of excretion of these compounds from the body. Although capable of secreting many endogenous OCs, proximal tubule cells are particularly capable of extremely efficient, active secretion of many xenobiotic OCs (Table 1), and tubular transport plays a significant role in clearing from the body these potentially toxic molecules. Clearance studies indicate that renal secretion of OCs is typically at least as efficient as clearance of OAs, with as much as 90% of the tetraethylammonium (TEA) in the renal artery being cleared during a single circuit through the kidney.[52] Secretion of OCs arises from the concerted activity of a number of distinct transport processes found in the luminal or basolateral membrane of the proximal tubule epithelium. Although it is likely that the transepithelial secretion of OCs involves predominantly the activity of a single pair of transport processes (in the basolateral and luminal membranes), it has become apparent that several other transport pathways exist in

Table 1 Examples of organic anions and cations secreted by the renal tubules.

Organic anions	Organic cations
Benzoates	Acetylcholine
Bile acids	Amiloride
Cephalosporins	Choline
Chlorothiazide	Cimetadine
Diodrast	Dopamine
Ester glucoronides	Epinephrine
Ether glucoronides	Histamine
Furosemide	Isoproterenol
Para-aminohippurate (PAH)	*N'*-methylnicotinamide (NMN)
Penicillins	Morphine
Prostaglandins	Procainamide
Pyrazinoate	Procaine
Saccharin	Quinidine
Salicylates	Quinine
Sulfonamides	Tetraethylammonium (TEA)

these membranes which, in at least one case, may facilitate reabsorption vs. secretion of some OCs, and, in other cases, may add to the options available for secreting potentially toxic compounds. Following is a discussion of the characteristics of the several renal OC transport processes with a particular emphasis on the general mechanism of action, kinetics, specificity, and distribution within the kidney of each process. This discussion also briefly covers what is known about species differences in the expression of these transporters.

7.04.3.2 Basolateral Transport of Monovalent OCs

7.04.3.2.1 Mechanism of basolateral transport

Transepithelial secretion of OCs begins with movement of these compounds from the blood into proximal tubule cells across the basolateral membrane. For monovalent OCs, this transport step appears to involve interaction with a single transport process that displays a broad specificity for this class of compound.[39] The "basolateral OC transporter" is not coupled to gradients of either Na^+ or H^+ and, instead, behaves as an electrogenic facilitated diffusion process (Figure 2). Indeed, concentrative transport of OCs in BLMV can be driven by an inside negative membrane potential,[53] consistent with the suggestion that the inside negative membrane potential of proximal renal cells serves as the driving force supporting the accumulation of OCs in these cells.[53,54] Studies

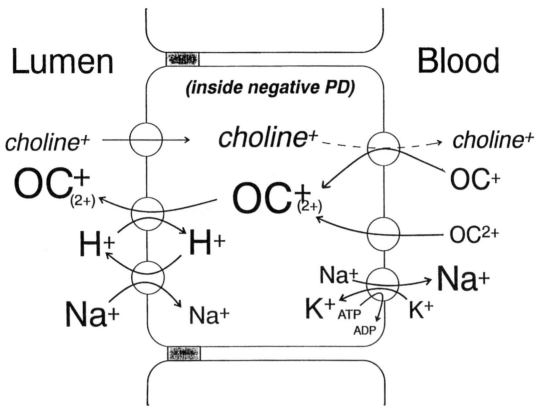

Figure 2 Descriptive model of organic cation transport. Symbols have the same meanings as in Figure 1.

with both intact tubules[55,56] and BLMV[57] also indicate that the basolateral OC transporter can operate as an electroneutral "OC/OC exchanger." The most effective intracellular substrate for supporting this operational mode of the basolateral OC transporter appears to be choline[57,58] leading to suggestions that OC/choline exchange may play a role in the reabsorptive flux of choline.[59] This suggestion is expanded upon below.

Although the basolateral OC transporter, as described above, could use the inside negative PD of renal cells to support the accumulation of OCs (e.g., up to $10\times$ the concentration of the extracellular fluid for a PD of $-60\,\text{mV}$), this operational mode of the transporter cannot move an OC against its electrochemical gradient. Thus, observations made with intact proximal tubules in which basolateral uptake results in concentrations of OCs in the cells 30 times or more of that in the bath[55,60–64] appear to challenge the contention that the basolateral OC transport does not include an "active" (i.e., energy requiring) process. However, intracellular "concentrations" of OCs need not be indicative of cytoplasmic activities of these compounds. Indeed, tubular homogenates have been shown to bind at least some OCs,[65] leading to suggestions that the high

intracellular accumulation of OCs noted in intact tubules are a consequence of purely passive properties of the cytoplasm.[3] Alternatively, OCs may be actively sequestered in intracellular compartments. Adding support to the latter hypothesis is the observation of ATP-dependent accumulation of tetraethylammonium (TEA) by endosomes isolated from renal cortical tissue.[66] In either event, it appears unlikely that the entry of OCs across the basolateral membrane is the active step in tubular secretion.

7.04.3.2.2 Kinetics of basolateral OC transport

Studies with intact perfused renal proximal tubules suggest that the rate-limiting transport step in secretion of OCs is at the luminal membrane, not the basolateral membrane. This conclusion is based on the observation[62] that, whereas single, nonperfused S1, S2, and S3 segments of proximal renal tubules display equivalent steady-state accumulations of TEA, the rate of net transepithelial secretion of TEA decreases with distance along the tubule (i.e., S1 secretion $>$ S2 $>$ S3). The rate-limiting status of the luminal membrane is, however, at odds with the results of studies on the kinetics of OC transport in isolated renal membranes. Consistently, transport measured in BLMV has been found to have a lower maximal capacity (i.e., lower J_{max}) and apparent affinity (i.e., higher K_t) than transport measured in BBMV.[56] However, using intact renal proximal tubules, Groves et al.[56] determined that basolateral TEA transport displays both high capacity and high affinity, compared to the transport of TEA in isolated BBMV. It is likely that the relatively low rates of OC transport measured in studies with isolated basolateral membranes reflect the nature of the preparation, rather than the characteristics of basolateral transport. Indeed, the kinetics of basolateral transport in intact tubules support the conclusion that luminal transport is rate limiting.

7.04.3.2.3 Specificity of basolateral OC transport

The studies of Ullrich and his colleagues[39] indicate that the basolateral OC transporter can interact with an extremely diverse array of chemical structures, thereby facilitating its presumed role as the principle avenue for renal secretion of cationic, xenobiotic agents. Indeed, it is frequently assumed that interaction of a molecule with the basolateral OC transporter simply requires that the compound possesses a net positive charge.[1,3] However, Ullrich has found that interaction of the transporter with organic bases does not display a pH sensitivity consistent with the binding of substrate being limited to the protonated species.[67] This led to the suggestion that the OC transporter senses the substrate's dipole structure and/or its ability to form hydrogen bonds, rather than the degree of ionization. Nevertheless, it is evident that organic bases and cations do display a preferential interaction with the basolateral OC transporter and that this transporter effectively excludes anionic substrates.[3] In addition to the effective influence of "charge status," studies in vivo and in vitro indicate that the apparent affinity of the basolateral transporter for OCs increases as their hydrophobicity and charge (or pK_a) increase.[3,8,56] However, whereas this general rule holds within a homologous series of compounds, a number of individual compounds (e.g., cimetidine) do not fit this pattern. Thus, a number of other factors may influence the binding of various molecules to the carrier. In addition, an increasing body of evidence suggests that the rate of transport via this basolateral carrier varies inversely with its affinity for the substrate.[56,68]

7.04.3.2.4 Axial heterogeneity of basolateral OC transport

As noted above, the secretory flux of OC decreases along the length of the proximal tubule and this has been attributed to a decrease in luminal flux of OC.[62] However, because luminal transport is believed to be rate-limiting, it is not clear whether there is a decrease of basolateral OC transport along the nephron. Although the similar degree of steady-state accumulation of OCs noted in isolated, nonperfused S1, S2, and S3 proximal tubule segments could reflect similar rates of basolateral transport, OC accumulation is also sensitive to the rate of basolateral efflux of OC from each segment, as well as to the degree of sequestration of these substrates that may exist in proximal cells of different segments. There have been no systematic studies of the rate of basolateral OC transport in different regions of the proximal tubule. It is, therefore, premature to draw a conclusion concerning whether the rate of basolateral OC transport varies along the length of the tubule.

7.04.3.2.5 Species differences in basolateral OC transport

There have been no systematic efforts to examine either quantitative or qualitative species differences in the basolateral transport of OCs. Nevertheless, studies employing membrane preparations from rabbit and rat indicate that the electrogenic facilitated diffusion mode of OC transport is operative in both species.[53,54,69] However, comparisons of transport kinetics between species have little meaning because of the differences noted between laboratories for determinations within a single species.[54,70,71] In addition, several studies employing intact flounder renal tubules have shown a basolateral OC transporter qualitatively similar to that observed in mammalian tubules.[2,72] Koepsell and co-workers succeeded in cloning the basolateral OC transporter from the rat,[73] leading to the prospect of studies on comparative aspects of the molecular physiology of this transport process.

7.04.3.3 Luminal OC Transport

7.04.3.3.1 Mechanism of luminal transport

Studies with both isolated membrane vesicles and intact, perfused proximal tubules support the conclusion that the exit of OCs across the luminal membrane is the active step in transepithelial secretion. Holohan and Ross first showed that an outwardly-directed H^+ gradient could support the active, concentrative accumulation of the endogenous OC, N'-methylnicotinamide (NMN) in dog renal BBMV[74] (see Figure 2). They concluded that luminal OC transport involves the carrier-mediated exchange of NMN for H^+. Transoriented proton gradients have subsequently been shown to support the active transport of a wide variety of OCs in BBMV from several mammalian species,[54,69] including humans.[75] Studies with intact, perfused single proximal tubules also support the contention that OC flux across the luminal membrane involves a mediated exchange with H^+.[55] The mediated exchange of monovalent OCs for H^+ is an electroneutral process,[54,76] suggesting that OC/H^+ exchange has a stoichiometry of 1:1. This was confirmed in studies of the thermodynamic coupling of OC/H^+ exchange showing that a 10-fold H^+ gradient is sufficient to support a 10-fold gradient of monovalent OC.[54] This observation also argues against the exchanger's operating in any mode other than that of a "fully occupied" exchanger complex. Therefore, electrogenic facilitated diffusion is not a mode supported by the luminal OC transporter. Nevertheless, as with the basolateral processes, the luminal OC transporter can support electroneutral OC/OC exchange,[77–79] although it is not clear whether this mode of activity represents one of physiological or pharmacological significance.

In the intact proximal tubule cell, H^+ is normally distributed well away from electrochemical equilibrium across the luminal membrane, a condition that is maintained by the activity of a Na^+/H^+ exchanger in this membrane. The resulting electrochemical H^+ gradient across the luminal membrane acts as the immediate source of energy to support active transepithelial secretion of OC (Figure 2). The functional linkage of parallel OC/H^+ and Na^+/H^+ exchangers has been confirmed in experiments in which an imposed Na^+ gradient has been shown to result in the active transport of OC arising from the secondary development of a H^+ gradient.[74,80] In the intact cell, the ultimate source of energy to support sustained OC secretion at the luminal membrane is the basolateral Na^+-K^+-ATPase which maintains the inwardly-directed Na^+ gradient and therefore the activity of the Na^+/H^+ exchanger. Moreover, activity of the Na^+-K^+-ATPase serves to maintain the outwardly directed K^+ gradient which, in turn, results in the cell's inside-negative membrane potential. This electrical gradient serves as a source of energy supporting the influx of OC across the basolateral membrane via the electrogenic OC transporter.

The luminal OC/H^+ exchanger has been found to support the mediated exchange of H^+ and the divalent OC, paraquat^{2+}.[81] Although the exchange of species with different cationic valances implies either a difference in stoichiometry of coupling or operation in an electrogenic mode, evidence concerning these issues is equivocal. Nevertheless, the demonstration that H^+ gradients stimulate mediated paraquat transport and that gradients of TEA support uphill transport of paraquat, indicate that this pathway can serve as a means for transmembrane flux of an even wider array of substrates than previously suspected.

7.04.3.3.2 Kinetics of luminal transport

Luminal OC transport is adequately described by Michaelis–Menten kinetics. The interaction of H^+ and OC at the external (and, presumably, the internal) aspect of the luminal membrane is competitive, suggesting that these substrates compete for a common site or a set of mutually exclusive binding sites.[78] The kinetic basis for the effect of H^+ at the *trans*-aspect of the transport site is less clear. In rabbit renal BBMV, Wright and Wunz[78] found that *trans*-gradients of H^+ increased the maximal rate of TEA transport without influencing the K_t for TEA uptake. However, Sokal et al.[82] found that, in dog BBMV, a *trans*-H^+ gradient increased the J_{max} and decreased the K_t for NMN uptake. While the basis for these differences in kinetic behavior is not clear, it is evident that the exchanger has a high apparent affinity for H^+ at both aspects of the membrane with effective "saturation" of the exchanger occurring by pH 6.0.[70,78,82]

7.04.3.3.3 Specificity of the luminal OC/H^+ exchange

The luminal OC/H^+ exchanger accepts a diverse array of monovalent OCs and at least some divalent OCs. OAs appear to be effectively excluded,[80] although probenecid, at high concentrations, has been found to interact competitively with the exchanger.[83,84] Beyond the issue of valence, the most important single criterion for interaction of the exchanger with substrate appears to be lipophilicity, with apparent affinity increasing in parallel with lipophilicity.[85] Molecular structure does, however, appear to exert some influence on the binding of substrate to the exchanger. Several studies have reported modest differences in inhibitory interaction of stereoisomers of several OCs.[86–91] Moreover, alignment of lipophilic "bulk" along the longitudinal rather than horizontal axis of a substrate (e.g., 4-phenyl-pyridinium vs. quinolinium compounds) appears to reduce substrate interaction with the exchanger.[85]

There has also been speculation that there is an inverse relationship between apparent affinity of the exchanger for substrate and the turnover of the substrate/exchanger complex. For example, the rabbit OC/H$^+$ exchanger displays the following sequence of apparent affinity ($1/K_t$ in μM$^{-1} \times 10^3$) for the following compounds: cimetidine (278) > amiloride (133) $\cong$ MPP$^+$ (128) $\gg$ TEA (10) $\gg$ choline (0.1). The transport capacity (J_{max} in nmol mg^{-1} min^{-1}) of the exchanger for these same compounds is reversed: choline (38) > TEA (12.4) > MPP$^+$ (3.3) = amiloride (3.3) > cimetidine (1.7) (from Ref. 92 and Wright, unpublished results). Nevertheless, the "carrier-mediated permeability" of the luminal membrane for this series of compounds (expressed as K_t/J_{max}) suggests that the substrates for which the carrier has the highest apparent affinity are those that are probably most efficiently transported at the very low concentrations to which the transporter is likely to be exposed.

7.04.3.3.4　Axial heterogeneity of luminal OC/H$^+$ exchange

As noted above, OC secretion decreases along the length of the proximal tubule (S1 > S2 > S3). Studies with intact tubules suggest that this is a consequence of a decrease in luminal transport activity along the length of the tubule.[62] This was confirmed in a study by Montrose-Rafizadeh *et al.*[93] Whereas the K_t for TEA/H$^+$ exchange was the same in rabbit renal BBMV isolated from the outer cortex (predominantly S1) and from the outer medulla (predominantly S3), the J_{max} was significantly greater in those from the outer cortex.

7.04.3.3.5　Species differences in luminal OC/H$^+$ exchange

The activity of a luminal OC/H$^+$ exchanger has been described in studies with renal BBMV isolated from rabbit, dog , rat, and human.[3] In addition, similar transport activity has been observed in membranes isolated from the kidneys of birds[94] and reptiles.[95] It is unclear whether differences in rates of transport activity or in relative affinity of these processes for substrates represent variations in preparation or real species differences. Even taken at face value, the differences noted between species

have been modest. However, studies of luminal OC transport in both reptilian (snake) BBMV and intact tubules indicate the presence of at least two OC transporters. TEA transport in the snake luminal membrane appears to involve the "standard" OC/H$^+$ exchanger. Transport of NMN, however, involves a different process that is independent of H$^+$ gradients.[95–97]

7.04.3.4　Other Pathways for Renal OC Transport

7.04.3.4.1　Choline transport

Although choline can be actively secreted by the proximal tubule,[98,99] at physiological concentrations ($<10\,\mu$M) it is typically reabsorbed.[100] Because choline was known to interact with the luminal and basolateral OC transporters responsible for secretion of these compounds, the reabsorption of this molecule represented a paradox. Indeed, efforts to explain the reabsorption of this compound in the face of the net secretion of others led to suggestions of a "kinetic asymmetry" in the activity of the luminal OC/H$^+$ exchanger.[77] However, a separate transporter for choline in the luminal membrane was described.[59] Choline uptake into proximal tubule cells across the luminal membrane appears to involve a high affinity, electrogenic process that uses the inside-negative electrical potential of the cells to accumulate this cation. Once inside the cells, choline can leave across the basolateral membrane in exchange for other OCs, as shown in studies with BLMV[57]and with intact proximal tubules.[58]

7.04.3.4.2　Guanidine transport

Guanidine has been shown to be transported by renal BBMV by a H$^+$-exchanger that is distinct from the OC/H$^+$ exchanger described above.[101] A similar process has also been found in BBMV from small intestine[102] and placenta,[103] neither of which display TEA/H$^+$ exchange activity. The role of this process in the renal handling of OCs is not clear, although it does display a modest inhibitory interaction with several polyamines.[102]

7.04.3.4.3　Polyvalent OCs

As noted earlier, the luminal OC/H$^+$ exchanger does transport the divalent OC, paraquat.[81] Paraquat does not, however, interact with the basolateral OC transporter. Rather, paraquat uptake from the blood appears to involve a separate transport pathway that shows little interaction with monovalent OCs (e.g., TEA and 4-phenyl-1-methylpyridinium; MPP^+), but which is blocked by the synthetic polyamine, methylglyoxal bis(guanyl-hydrazone)dihydrochloride (MGBG).[104] The specificity of this pathway appears, however, to be comparatively narrow: endogenous polyamines (e.g., putrescine and spermine) and structural analogues of paraquat (diquat) have no interaction with this transporter, whose role in overall renal handling of OCs is unclear.

7.04.3.5 Interaction of OC Transporters with Nephrotoxicants

The broad specificity of the several mediated pathways for the renal secretion of organic cations provides ample opportunity for cationic nephrotoxicants to use these pathways both as avenues for excretion from the body and as means for entering cells that subsequently become targets for toxicity. For example, celiptium, a chemotherapeutic agent that exerts a marked nephrotoxicity, interacts with the peritubular OC transport pathway and probably uses this process to enter proximal cells;[105] and paraquat, which enters proximal cells across the peritubular membrane using a pathway specific for divalent OCs, has also been shown to exert a marked nephrotoxicity.[106,107] However, the literature is not replete with examples of toxicity mediated by cationic agents (in contrast to the situation that exists for anionic toxicants). Nevertheless, it is likely that the principal substrates for these pathways are, in fact, potentially toxic basic plant alkaloids, for the excretion of which this extensive battery of renal transport processes has evolved. With this in mind it is significant that one of the normal sites of expression of the multidrug resistance transporter (MDR) is the brush border membrane of renal proximal cells where it may well play a routine role in the excretion of xenobiotic agents.[108]

7.04.4 SUMMARY

In summary, two general renal transport systems—one for organic anions (or weak organic acids) and one for organic cations (or weak organic bases) exist in the proximal renal tubules of almost all vertebrate species. These systems are capable of secreting a wide array of molecules of broad physiological, pharmacological, and toxicological importance. Both transport systems play a significant role in eliminating exogenous compounds and regulating the concentrations of endogenous compounds. They may also be significant for the entry of nephrotoxicants into the renal tubule cells. Both secretory processes involve concerted transport activities at the basolateral and luminal membranes of renal tubule cells. For organic anions (OAs), the initial step in the secretory process is transport into the cells against an electrochemical gradient at the basolateral membrane via a tertiary active transport system, the final step in which involves countertransport for a dicarboxylate (physiologically, αKG) moving out of the cells down its electrochemical gradient (Figure 1). The outwardly directed gradient for αKG is maintained by metabolism and by Na^+-dicarboxylate cotransport into the cells at the basolateral membrane and, possibly, at the luminal membrane (Figure 1). The relative importance of each of these pathways for maintaining the intracellular αKG concentration is not yet clear. The inwardly-directed Na^+ gradient responsible for Na^+-dicarboxylate cotransport is maintained in turn by Na^+ transport out of the cells via basolateral Na^+-K-ATPase, the primary energy-requiring step in the tertiary process (Figure 1). The affinity of the basolateral transporter for OAs is determined by their hydrophobicity and charge, charge strength, and charge placement. OAs transported into the cells across the basolateral membrane then move out of the cells across the luminal membrane down an electrochemical gradient via a carrier-mediated process, the nature of which is poorly understood (Figure 1). Movement of OAs through the cytoplasm from the basolateral to the luminal membrane may involve diffusion, vesicular accumulation, or a combination of these, but this process too is poorly understood.

For organic cations (OCs), the initial step in transepithelial secretion involves movement into the cells across the basolateral membrane down an electrochemical gradient via a transporter that has now been cloned from rat renal tissue (Figure 2). Although the inside-negative membrane potential appears to drive this carrier, it may also operate as an electroneutral OC/OC exchanger (Figure 2). Interaction of OCs with the basolateral transporter is influenced by the charge and hydrophobicity of the OC. OCs transported into the cells by this basolateral process are transported out of the cells against an electrochemical gradient at the luminal membrane via a tertiary active process, the final step in which OC/H^+ countertransport (Figure 2). The inwardly directed H^+ gradient responsible for driving this countertransport process is maintained by the luminal Na^+/H^+ exchanger (Figure 2). The inwardly-directed

Na^+ gradient responsible for driving the Na^+/H^+ exchanger is maintained in turn by Na^+ transport out of the cells via basolateral Na^+-K^+-ATPase, the primary energy-requiring step in the tertiary process (Figure 2). The apparent affinity of this luminal transporter for OCs appears to increase with the lipophilicity of the OC, although other structural factors may also be important. As in the case of OAs, OCs may be accumulated in intracellular compartments (e.g., endosomes) during transcellular movement, but this has not been determined with certainty.

7.04.5 REFERENCES

1. F. Roch-Ramel, K. Besseghir and H. Murer, in 'Handbook of Physiology. Section 8. Renal Physiology,' ed. E. E. Windhager, Oxford University Press, New York, 1992, vol. II, pp. 2189–2262.
2. J. B. Pritchard and D. S. Miller, 'Comparative insights into the mechanisms of renal organic anion and cation secretion.' *Am. J. Physiol.*, 1991, **261**, R1329–R1340.
3. J. B. Pritchard and D. S. Miller, 'Mechanisms mediating renal secretion of organic anions and cations.' *Physiol. Rev.*, 1993, **73**, 765–796.
4. I. M. Weiner. in 'Handbook of Physiology. Renal Physiology,' eds. J. Orloff and R. W. Berliner, Am. Physiol. Soc., Washington, DC, 1973, pp. 521–554.
5. I. M. Weiner. in 'The Kidney: Physiology and Pathophysiology,' eds. D. W. Seldin and G. Giebisch, Raven Press, New York, 1985, pp. 1703–1724.
6. W. H. Dantzler, 'Organic acid (or anion) and organic base (or cation) transport by renal tubules of nonmammalian vertebrates.' *J. Exp. Zool.*, 1989, **249**, 247–257.
7. H. Murer, M. Manganel and F. Roch-Ramel, in 'Handbook of Physiology, Section 8. Renal Physiology,' ed. E. E. Windhager, Oxford, New York, 1992, vol. II, pp. 2165–2188.
8. K. J. Ullrich, G. Rumrich and G. Fritzsch. in 'Progress in Cell Research,' eds. E. Bamberg and H. Passow, Elsevier, Amsterdam, 1992, pp. 315–321.
9. P. B. Woodhall, C. C. Tisher, C. A. Simonton *et al.*, 'Relationship between *para*-aminohippurate secretion and cellular morphology in rabbit proximal tubules.' *J. Clin. Invest.*, 1978, **61**, 1320–1329.
10. A. Shimomura, A. M. Chonko and J. J. Grantham, 'Basis for heterogeneity of *para*-aminohippurate secretion in rabbit proximal tubules.' *Am. J. Physiol.*, 1981, **240**, F430–F436.
11. S. Shpun, K. K. Evans and W. H. Dantzler, 'Interaction of α-KG with basolateral organic anion transporter in isolated rabbit renal S3 proximal tubules.' *Am. J. Physiol.*, 1995, **268**, F1109–F1116.
12. W. H. Dantzler, K. K. Evans and S. H. Wright, 'Kinetics of interactions of *para*-aminohippurate, probenecid, cysteine conjugates, and *N*-acetyl cysteine conjugates with basolateral organic anion transporter in isolated rabbit proximal renal tubules.' *J. Pharmacol. Exp. Ther.*, 1995, **272**, 663–672.
13. B. M. Tune, M. B. Burg and C. S. Patlak, 'Characteristics of *p*-aminohippurate transport in proximal renal tubules.' *Am. J. Physiol.*, 1969, **217**, 1057–1063.
14. W. H. Dantzler, 'PAH transport by snake proximal tubules: differences from urate transport.' *Am. J. Physiol.*, 1974, **226**, 634–641.
15. R. L. Chambers, V. Beck and M. Belkin, 'Secretion in tissue cultures. I. Inhibition of phenol red accumulation in the chick kidney.' *J. Cell. Comp. Physiol.*, 1935, **6**, 425–439.
16. R. J. Cross and J. V. Taggart, 'Renal tubular transport: accumulation of *p*-aminohippurate by rabbit kidney slices.' *Am. J. Physiol.*, 1950, **161**, 181–190.
17. J. Maxild, 'Effect of externally added ATP and related compounds on active transport of *p*-aminohippurate and metabolism in cortical slices of the rabbit kidney.' *Arch. Int. Physiol. Biochim.*, 1978, **86**, 509–530.
18. C. R. Ross and I. M. Weiner, 'Adenine nucleotides and PAH transport in slices of renal cortex: effects of DNP and CN.' *Am. J. Physiol.*, 1972, **222**, 356–359.
19. M. B. Burg and J. Orloff, 'Effect of strophantidin in electrolyte content and PAH accumulation of rabbit kidney slices.' *Am. J. Physiol.*, 1962, **202**, 565–571.
20. R. A. Podevin and E. F. Boumendil-Podevin, 'Monovalent cation and ouabain effects on PAH uptake by rabbit kidney slices.' *Am. J. Physiol.*, 1977, **232**, F239–F247.
21. R. A. Podevin, E. F. Boumendil-Podevin and C. Priol, 'Concentrative PAH transport by rabbit kidney slices in the absence of metabolic energy.' *Am. J. Physiol.*, 1978, **235**, F278–F285.
22. M. I. Sheikh and J. V. Moller, 'Na^+-gradient dependent stimulation of renal transport of *p*-aminohippurate.' *Biochem. J.*, 1982, **208**, 243–246.
23. W. Berner and R. Kinne, 'Transport of *p*-aminohippurate acid by plasma membrane vesicles isolated from rat kidney cortex.' *Pflugers Arch.*, 1976, **361**, 269–277.
24. J. Eveloff, R. Kinne and W. B. Kinter, '*p*-Aminohippuric acid transport into brush border vesicles isolated from flounder kidney.' *Am. J. Physiol.*, 1979, **237**, F291–F298.
25. J. S. Kasher, P. D. Holohan and C. R. Ross, 'Na^+ gradient-dependent *p*-aminohippurate (PAH) transport in rat basolateral membrane vesicles.' *J. Pharmacol. Exp. Ther.*, 1983, **227**, 122–129.
26. J. L. Kinsella, P. D. Holohan, N. I. Pessah *et al.*, 'Transport of organic ions in renal cortical luminal and antiluminal membrane vesicles.' *J. Pharmacol. Exp. Ther.*, 1979, **209**, 443–450.
27. K. J. Ullrich, G. Rumrich, G. Fritzsch *et al.*, 'Contraluminal *para*-aminohippurate (PAH) transport in the proximal tubule of the rat kidney. I. Kinetics, influence of cations, anions, and capillary preperfusion.' *Pflugers Arch.*, 1987, **409**, 229–235.
28. W. H. Dantzler and S. K. Bentley, 'Bath and lumen effects of SITS on PAH transport by isolated perfused renal tubules.' *Am. J. Physiol.*, 1980, **238**, F16–F25.
29. I. Kippen and J. R. Klinenberg, 'Effects of renal fuels on uptake of PAH and uric acid by separated renal tubules of the rabbit.' *Am. J. Physiol.*, 1978, **235**, F137–F141.
30. K. J. Ullrich, G. Rumrich, G. Fritzsch *et al.*, 'Contraluminal *para*-aminohippurate transport in the proximal tubule of rat kidney. II. Specificity: aliphatic dicarboxylic acids.' *Pflugers Arch.*, 1987, **408**, 38–45.
31. J. Eveloff, '*p*-Aminohippurate transport in basal-lateral membrane vesicles from rabbit renal cortex: stimulation by pH and sodium gradients.' *Biochim. Biophys. Acta*, 1987, **897**, 474–480.
32. M. I. Sheikh and J. V. Moller, 'Nature of Na^+-independent stimulation of renal transport of *p*-aminohippurate by exogenous metabolites.' *Biochem. Pharmacol.*, 1983, **32**, 2745–2749.
33. J. B. Pritchard, 'Luminal and peritubular steps in the renal transport of *p*-aminohippurate.' *Biochim. Biophys. Acta*, 1987, **906**, 295–308.

34. H. Shimada, B. Moewes and G. Burckhardt, 'Indirect coupling to Na+ of *p*-aminohippuric acid uptake into rat renal basolateral membrane vesicles.' *Am. J. Physiol.*, 1987, **253**, F795–F801.

35. J. B. Pritchard, 'Coupled transport of *p*-aminohippurate by rat kidney basolateral membrane vesicles.' *Am. J. Physiol.*, 1988, **255**, F597–F604.

36. V. Chatsudthipong and W. H. Dantzler, 'PAH/α-KG countertransport stimulates PAH uptake and net secretion in isolated rabbit renal tubules.' *Am. J. Physiol.*, 1992, **263**, F384–F391.

37. W. H. Dantzler and K. K. Evans, 'Effect of αKG in lumen on PAH transport by isolated perfused proximal renal tubules.' *FASEB J.*, 1995, **9**, A565

38. K. J. Ullrich and G. Rumrich, 'Contraluminal transport systems in the proximal renal tubule involved in the secretion of organic anions.' *Am. J. Physiol.*, 1988, **254**, F453–F462.

39. K. J. Ullrich, 'Specificity of transporters for 'organic anions' and 'organic cations' in the kidney.' *Biochim. Biophys. Acta*, 1994, **1197**, 45–62.

40. W. H. Dantzler, in 'Amino Acid Transport and Uric Acid Transport', eds. S. Silbernagl, F. Lang and R. Greger, Georg Thieme, Stuttgart, 1976, pp. 169–179.

41. W. H. Dantzler and S. K. Bentley, 'Effects of inhibitors in lumen on PAH and urate transport by isolated renal tubules.' *Am. J. Physiol.*, 1979, **236**, F379–F386.

42. J. W. Blomstedt and P. S. Aronson, 'pH-gradient-stimulated transport of urate and *p*-aminohippurate in dog renal microvillous membrane vesicles.' *J. Clin. Invest.*, 1980, **65**, 931–934.

43. P. S. Aronson, 'The renal proximal tubule: a model for diversity of anion exchangers and stilbene-sensitive anion transporters.' *Annu. Rev. Physiol.*, 1989, **51**, 419–441.

44. W. O. Berndt, 'Probenecid binding by renal cortical slices and homogenates.' *Proc. Soc. Exp. Biol. Med.*, 1967, **126**, 123–126.

45. J. Eveloff, W. K. Morishige and S. K. Hong, 'The binding of phenol red to rabbit renal cortex.' *Biochim. Biophys. Acta*, 1976, **448**, 167–180.

46. L. P. Sullivan, J. A. Grantham, L. Rome *et al.*, 'Fluorescein transport in isolated proximal tubules *in vitro*: epifluorometric analysis.' *Am. J. Physiol.*, 1990, **258**, F46–F51.

47. D. S. Miller, D. E. Stewart and J. B. Pritchard, 'Intracellular compartmentation of organic anions within renal cells.' *Am. J. Physiol.*, 1993, **264**, R882–R890.

48. K. G. Dickman and L. J. Mandel, 'Relationship between HCO3-transport and oxidative metabolism in rabbit proximal tubule.' *Am. J. Physiol.*, 1992, **263**, F342–F351.

49. J. M. Weinberg, in 'Diseases of the Kidney,' eds. R. W. Schrier and C. W. Gottschalk, 5th edn., Little, Brown, Boston, MA, 1993, pp. 1031–1097.

50. G. H. Zhang and J. L. Stevens, 'Transport and activation of *S*-(1,2,-dichlorovinyl)-L-cysteine and *N*-acetyl-*S*-(1,2,-dichlorovinyl)-L-cysteine in rat kidney proximal tubules.' *Toxicol. Appl. Pharmacol.*, 1989, **100**, 51–61.

51. K. J. Ullrich, G. Rumrich, M. Gemborys *et al.*, 'Tranformation and transport: how does metabolic transformation change the affinity of substrates for the renal contraluminal anion and cation transporters?' *Toxicol. Lett.*, 1990, **53**, 19–27.

52. J. S. Petersen and S. Christensen, 'Superiority of tetraethylammonium to *p*-aminohippurate as a marker for renal plasma flow during furosemide diuresis.' *Ren. Physiol.*, 1987, **10**, 102–109.

53. C. Montrose-Rafizadeh, F. Mingard, H. Murer *et al.*, 'Carrier-mediated transport of tetraethylammonium across rabbit renal basolateral membrane.' *Am. J. Physiol.*, 1989, **257**, F243–F251.

54. S. H. Wright and T. M. Wunz, 'Transport of tetraethylammonium by rabbit renal brush-border and basolateral membrane vesicles.' *Am. J. Physiol.*, 1987, **253**, F1040–F1050.

55. W. H. Dantzler, O. Brokl and S. H. Wright, 'Brush-border TEA transport in intact proximal tubules and isolated membrane vesicles.' *Am. J. Physiol.*, 1989, **256**, F290–F297.

56. C. E. Groves, K. K. Evans, W. H. Dantzler *et al.*, 'Peritubular organic cation transport in isolated rabbit proximal tubules.' *Am. J. Physiol.*, 1994, **266**, F450–F458.

57. P. P. Sokol and T. D. McKinney, 'Mechanism of organic cation transport in rabbit renal basolateral membrane vesicles.' *Am. J. Physiol.*, 1990, **258**, F1599–F1607.

58. W. H. Dantzler, S. H. Wright, V. Chatsudthipong *et al.*, 'Basolateral tetraethylammonium transport in intact tubules: specificity and *trans*-stimulation.' *Am. J. Physiol.*, 1991, **261**, F386–F392.

59. S. H. Wright, T. M. Wunz and T. P. Wunz, 'A choline transporter in renal brush-border membrane vesicles: energetics and structural specificity.' *J. Membr. Biol.*, 1992, **126**, 51–65.

60. E. Brändle and J. Greven, 'Transport of cimetidine across the basolateral membrane of rabbit kidney proximal tubules: characterization of transport mechanisms.' *J. Pharmacol. Exp. Ther.*, 1991, **258**, 1038–1045.

61. C. T. Hawk and W. H. Dantzler, 'Tetraethylammonium transport by isolated perfused snake renal tubules.' *Am. J. Physiol.*, 1984, **246**, F476–F487.

62. C. Schäli, L. Schild, J. Overney *et al.*, 'Secretion of tetraethylammonium by proximal tubules of rabbit kidneys.' *Am. J. Physiol.*, 1983, **245**, F238–F246.

63. H. Hohage, D. M. Mörth, I. U. Querl *et al.*, 'Regulation by protein kinase C of the contraluminal transport system for organic cations in rabbit kidney S2 proximal tubules.' *J. Pharmacol. Exp. Ther.*, 1994, **268**, 897–901.

64. J. B. Tarloff and P. H. Brand, 'Active tetraethylammonium uptake across the basolateral membrane of rabbit proximal tubule.' *Am. J. Physiol.*, 1986, **251**, F141–F149.

65. W. O. Berndt, 'Organic base transport: a comparative study.' *Pharmacology*, 1981, **22**, 251–262.

66. J. B. Pritchard, D. B. Sykes, R. Walden *et al.*, 'ATP-dependent transport of tetraethylammonium by endosomes isolated from rat renal cortex.' *Am. J. Physiol.*, 1994, **266**, F966–F976.

67. K. J. Ullrich and G. Rumrich, 'Renal contraluminal transport systems for organic anions (*para*-amino-hippurate, PAH) and organic cations (*N'*-methyl-nicotinanide, NMeN) do not see the degree of substrate ionization.' *Pflugers Arch.*, 1992, **421**, 286–288.

68. K. J. Ullrich, F. Papavassiliou, C. David *et al.*, 'Contraluminal transport of organic cations in the proximal tubule of the rat kidney. I. Kinetics of *N'*-methylnicotinamide and tetraethylammonium, influence of K+, HCO3, pH; inhibition by aliphatic primary, secondary and tertiary amines, and mono- and bisquaternary compounds.' *Pflugers Arch.*, 1991, **419**, 84–92.

69. M. Takano, K. I. Inui, T. Okano *et al.*, 'Carrier-mediated transport systems of tetraethylammonium in rat renal brush-border and basolateral membrane vesicles.' *Biochim. Biophys. Acta*, 1984, **773**, 113–124.

70. J. S. Jung, Y. K. Kim and S. H. Lee, 'Characteristics

of tetraethylammonium transport in rabbit renal plasma-membrane vesicles.' *Biochem. J.*, 1989, **259**, 377–383.

71. C. Rafizadeh, F. Roch-Ramel and C. Schäli, 'Tetraethylammonium transport in renal brush border membrane vesicles of the rabbit.' *J. Pharmacol. Exp. Ther.*, 1987, **240**, 308–313.

72. P. M. Smith, J. B. Pritchard and D. S. Miller, 'Membrane potential drives organic cation transport into teleost renal proximal tubules.' *Am. J. Physiol.*, 1988, **255**, R492–R499.

73. D. Gründemann, V. Gorboulev, S. Gambarian *et al.*, 'Drug excretion mediated by a new prototype of poly-specific transporter.' *Nature*, 1994, **372**, 549–552.

74. P. D. Holohan and C. R. Ross, 'Mechanisms of organic cation transport in kidney plasma membrane vesicles: 2. ΔpH studies.' *J. Pharmacol. Exp. Ther.*, 1981, **216**, 294–298.

75. R. J. Ott, A. C. Hui, G. Yuan *et al.*, 'Organic cation transport in human renal brush-border membrane vesicles.' *Am. J. Physiol.*, 1991, **261**, F443–F451.

76. P. P. Sokol, P. D. Holohan and C. R. Ross, 'Electroneutral transport of organic cations in canine renal brush border membrane vesicles (BBMV).' *J. Pharmacol. Exp. Ther.*, 1985, **233**, 694–699.

77. P. D. Holohan and C. R. Ross, 'Mechanisms of organic cation transport in kidney plasma membrane vesicles: 1. Countertransport studies.' *J. Pharmacol. Exp. Ther.*, 1980, **215**, 191–197.

78. S. H. Wright and T. M. Wunz, 'Mechanism of *cis*- and *trans*-substrate interactions at the tetraethylammonium/H^+ exchanger of rabbit renal brush-border membrane vesicles.' *J. Biol. Chem.*, 1988, **263**, 19494–19497.

79. K. D. Lazaruk and S. H. Wright, 'MPP^+ is transported by the TEA^+-H^+ exchanger of renal brush-border membrane vesicles.' *Am. J. Physiol.*, 1990, **258**, F597–F605.

80. S. H. Wright, 'Transport of N'-methylnicotinamide across brush border membrane vesicles from rabbit kidney.' *Am. J. Physiol.*, 1985, **249**, F903–F911.

81. S. H. Wright and T. M. Wunz, 'Paraquat^{2+}/H^+ exchange in isolated renal brush-border membrane vesicles.' *Biochim. Biophys. Acta*, 1995, **1240**, 18–24.

82. P. P. Sokol, P. D. Holohan, S. M. Grassl *et al.*, 'Proton-coupled organic cation transport in renal brush-border membrane vesicles.' *Biochim. Biophys. Acta*, 1988, **940**, 209–218.

83. R. J. Ott, A. C. Hui and K. M. Giacomini, 'Mechanisms of interactions between organic anions and the organic cation transporter in renal brush border membrane vesicles.' *Biochem. Pharmacol.*, 1990, **40**, 659–661.

84. P. H. Hsyu, L. G. Gisclon, A. C. Hui *et al.*, 'Interactions of organic anions with the organic cation transporter in renal BBMV.' *Am. J. Physiol.*, 1988, **254**, F56–F61.

85. S. H. Wright, T. M. Wunz and T. P. Wunz, 'Structure and interaction of inhibitors with the TEA/H^+ exchanger of rabbit renal brush border membranes.' *Pflugers Arch.*, 1995, **429**, 313–324.

86. R. J. Ott, A. C. Hui, P. H. Hsyu *et al.*, 'Interactions of quinidine and quinine and (+)- and (−)-pindolol with the organic cation/proton antiporter in renal brush membrane vesicles.' *Biochem. Pharmacol.*, 1991, **41**, 142–145.

87. P. H. Hsyu and K. M. Giacomini, 'Stereoselective renal clearance of pindolol in humans.' *J. Clin. Invest.*, 1985, **76**, 1720–1726.

88. L. T. Wong, D. D. Smyth and D. S. Sitar, 'Stereoselective inhibition of renal organic cation transport in human kidney.' *Br. J. Clin. Pharmacol.*, 1992, **34**, 438–440.

89. R. Bendayan, E. M. Sellers and M. Silverman, 'Inhibition kinetics of cationic drugs on N'-methylnicotinamide uptake by brush border membrane vesicles from the dog kidney cortex.' *Can. J. Physiol. Pharmacol.*, 1990, **68**, 467–475.

90. L. T. Wong, D. D. Smyth and D. S. Sitar, 'Interference with renal organic cation transport by (−)- and (+)-nicotine at concentrations documented in plasma of habitual tobacco smokers.' *J. Pharmacol. Exp. Ther.*, 1992, **261**, 21–25.

91. A. S. Gross and A. A. Somogyi, 'Interaction of the stereoisomers of basic drugs with the uptake of tetraethylammonium by rat renal brush-border membrane vesicles.' *J. Pharmacol. Exp. Ther.*, 1994, **268**, 1073–1080.

92. S. H. Wright and T. M. Wunz, 'Correlation between J_{max} and K_t of organic cation/H^+ exchange in renal brush border membranes.' *FASEB J.*, 1990, **4**, A728

93. C. Montrose-Rafizadeh, F. Roch-Ramel and C. Schäli, 'Axial heterogeneity of organic cation transport along the rabbit renal proximal tubule: studies with brush-border membrane vesicles.' *Biochim. Biophys. Acta*, 1987, **904**, 175–177.

94. A. R. Villalobos and E. J. Braun, 'Characterization of organic cation transport by avian renal brush-border membrane vesicles.' *Am. J. Physiol.*, 1995, **269**, R1050–R1059.

95. W. H. Dantzler, S. H. Wright and O. H. Brokl, 'Tetraethylammonium transport by snake renal brush-border membrane vesicles.' *Pflugers Arch.*, 1991, **418**, 325–332.

96. W. H. Dantzler and O. H. Brokl, 'N'-methylnicotinamide transport by isolated perfused snake proximal renal tubules.' *Am. J. Physiol.*, 1986, **250**, F407–F418.

97. W. H. Dantzler and O. H. Brokl, 'NMN transport by snake renal tubules: choline effects, countertransport, H^+-NMN exchange.' *Am. J. Physiol.*, 1987, **253**, F656–F663.

98. K. Besseghir, L. B. Pearce and B. Rennick, 'Renal tubular transport and metabolism of organic cations by the rabbit.' *Am. J. Physiol.*, 1981, **241**, F308–F314.

99. M. Acara, F. Roch-Ramel and B. Rennick, 'Bidirectional renal tubular transport of free choline: a micropuncture study.' *Am. J. Physiol.*, 1979, **236**, F112–F118.

100. M. Acara and B. Rennick, 'Regulation of plasma choline by the renal tubule: bidirectional transport of choline.' *Am. J. Physiol.*, 1973, **225**, 1123–1128.

101. Y. Miyamoto, C. Tiruppathi, V. Ganapathy *et al.*, 'Multiple transport systems for organic cations in renal brush-border membrane vesicles.' *Am. J. Physiol.*, 1989, **256**, F540–F548.

102. Y. Miyamoto, V. Ganapathy and F. H. Leibach, 'Transport of guanidine in rabbit intestinal brush-border membrane vesicles.' *Am. J. Physiol.*, 1988, **255**, G85–G92.

103. V. Ganapathy, M. E. Ganapathy, C. N. Nair *et al.*, 'Evidence for an organic cation–proton antiport system in brush-border membranes isolated from the human term placenta.' *J. Biol. Chem.*, 1988, **263**, 4561–4568.

104. C. E. Groves, M. N. Morales, A. J. Gandolfi *et al.*, 'Peritubular paraquat transport in isolated renal proximal tubules.' *J. Pharmacol. Exp. Ther.*, 1995, **275**, 926–932.

105. S. J. McGuiness, D. R. Rankin, P. Bezrutczyk *et al.*,

'Nephrotoxicity and transport of celiptium in rabbit renal proximal tubule suspensions.' *In Vitro Tox.*, 1995, **8**, 169–175.

106. J. L. Ecker, J. B. Hook and J. E. Gibson, 'Nephrotoxicity of paraquat in mice.' *Toxicol. Appl. Pharmacol.*, 1975, **34**, 178–186.

107. D. G. Clark, T. F. McElligot and E. W. Hurst, 'The toxicity of paraquat.' *Br. J. Ind. Med.*, 1966, **23**, 126–132.

108. M. M. Gottesman and I. Pastan, 'Biochemistry of multidrug resistance mediated by the multidrug transporter.' *Annu. Rev. Biochem.*, 1993, **62**, 385–427.

7.05
Renal Xenobiotic Metabolism

EDWARD A. LOCK
Zeneca, Macclesfield, UK

and

CELIA J. REED
Liverpool John Moores University, UK

7.05.1 INTRODUCTION

In mammals, the primary route of xenobiotic excretion is the kidney, and it is becoming increasingly recognized that elucidation of renal excretory mechanism(s) is important in understanding the pharmacological efficacy and duration of a drug and, with certain compounds, the mechanism of nephrotoxicity.[1-4]

It is now well established that the kidney contains xenobiotic biotransformation enzymes,[5-10] but there are still two important questions that need to be answered. First, what role do these enzymes play in the normal physiological function of the kidney, and second, what is their role in drug metabolism *in vivo*? It has been known for some time that the kidney plays an important role in the hydroxylation of fatty acids, arachidonic acid (AA) and 25-hydroxyvitamin D_3.[11,12] Renal metabolism of AA by cytochrome P450 epoxygenase and ω-1 hydroxylases results in the formation of metabolites which possess a broad spectrum of biological activity. Those of significance to renal physiology are involved in regulation of salt and water balance, and include inhibition of sodium reabsorption and potassium excretion in the cortical collecting tubule, inhibition of sodium transport by the proximal tubule, inhibition of vasopressin-stimulated water reabsorption in the collecting tubule, intrarenal vasoconstriction and modulation of sodium potassium ATPases.[13–16] Thus it is becoming increasingly clear that renal drug metabolizing enzymes play a crucial role in the normal physiology of the kidney.

Extrahepatic metabolism of a xenobiotic can be assumed when its systemic blood clearance is greater than hepatic blood flow or when there is no significant change in its metabolic clearance in a patient with severe hepatic dysfunction, such as cirrhosis, or an anhepatic condition, such as occurs during liver transplantation. For example, it is known that renal metabolism of morphine contributes significantly to the total biotransformation of this drug in the dog,[17] sheep[18] and humans.[19] The kidney also contributes to the metabolism of this opiate in patients undergoing liver transplantation during the anhepatic phase.[20] Reevaluation of the renal toxicity of the anaesthetic methoxyflurane, which is metabolized to liberate fluoride ion,[21] indicates that *in situ* metabolism within the kidney may be an important factor in determining its nephrotoxicity.[22] Thus there is also an increasing awareness that renal xenobiotic biotransformation has an important role not only in the overall *in vivo* metabolism of certain chemicals, but also in the development of nephrotoxicity.

Early studies in the field of renal drug metabolism used microsomal or cytosolic preparations to determine enzyme activity with various substrates. Such methodologies have limited usefulness in a tissue with extensive cellular heterogeneity, low levels, and marked regional distribution of drug metabolizing enzymes. The advent of molecular biological techniques (for instance, *in situ* hybridization and Northern blotting using oligonucleotide probes, cloning, sequencing and expression of renal enzymes) has resulted in major advances

in our knowledge of renal drug metabolism. This chapter reviews advances in renal biotransformation from 1990 to 1995 and builds on an earlier review.[10] For discussion of the relevance of renal xenobiotic metabolism to nephrotoxicity see Refs. 23–25 and other chapters in this volume.

7.05.2 RELEVANCE OF RENAL STRUCTURE AND FUNCTION TO XENOBIOTIC BIOTRANSFORMATION

A detailed discussion of renal structure and function lies outside the scope of this review, and the reader is referred to Chapter 2, this volume. However, several aspects of the physiology of the kidney are of particular relevance to renal drug metabolism and toxicity, and these will be mentioned briefly.

First, the renal cortex is exposed to relatively high concentrations of foreign compounds for a number of reasons: (i) it receives about 80% of total renal blood flow (which is 25% of cardiac output) and thus foreign compounds in the blood stream are readily delivered to this region; (ii) xenobiotics which are filtered at the glomerulus enter the tubular lumen where they are concentrated, thus exposing the tubular cells to high concentrations; and (iii) high concentrations of xenobiotics may accumulate in proximal tubular cells following either passive or active reabsorption from the tubular lumen, or active secretion from the blood stream into the tubular lumen. Thus those cells containing the highest levels of drug metabolizing enzymes (see later) are also exposed to the highest concentrations of xenobiotics, and this has important toxicological consequences. For example, the nephrotoxic chemicals paracetamol and chloroform concentrate in male murine proximal tubular cells where they are believed to be activated by cytochrome P450 2E1.[26–28] Similarly the mercapturic acid of hexachloro-1,3-butadiene concentrates in proximal tubular cells via the organic anion transport system.[29,30] It is then deacetylated,[31] prior to activation by the enzyme cysteine conjugate β-lyase which is located in that part of the nephron.[32,33] Furthermore, since the cortex has a high oxygen consumption it is particularly susceptible, especially within the S3 segment, to compounds which produce anoxia.[34]

The renal medulla receives a much lower blood supply than the cortex, in general contains lower levels of drug metabolizing enzymes, and is more anaerobic in its metabolism. Thus it is generally less important than the cortex, both as a site for renal xenobiotic biotransformation and as a site for nephrotoxicity. However, it is the site of the countercurrent mechanism which is responsible for concentrating the urine and this mechanism can lead to xenobiotics becoming concentrated in the medulla and papilla to concentrations many times those found in the plasma.[35,36] Thus for some specific chemicals the medulla is the site of activation and toxicity.[37]

7.05.3 PHASE 1 XENOBIOTIC BIOTRANSFORMATION ENZYMES

7.05.3.1 Cytochromes P450

7.05.3.1.1 *Introduction*

The cytochromes P450 (P450s) are a superfamily of haemoproteins which play a central role in the metabolism of xenobiotics and are also involved in both the synthesis and catabolism of endogenous compounds. Mammalian P450s have been divided into two major classes based on their intracellular location and the enzyme from which they receive electrons. The majority of P450s are located in the endoplasmic reticulum membrane, and electrons are donated via the flavoprotein NADPH-P450 oxidoreductase and, in some cases, cytochrome b5. These enzymes are the ones most frequently implicated in xenobiotic metabolism, and they catalyze a wide variety of chemical transformations including oxidative reactions such as aliphatic and aromatic hydroxylation, N- and S-oxidation, epoxidation, N-, S-, and O-dealkylation, desulfuration, and deamination, as well as reductive reactions such as dehalogenation and reduction of azo and nitro groups. The other class are the mitochondrial cytochromes P450, side chain cleavage enzymes, and in the kidney these are involved in catalyzing the hydroxylation of 25-hydroxy vitamin D_3.[38] These enzymes receive electrons from the iron–sulfur protein adrenodoxin via NADPH-adrenodoxin oxidoreductase.

Multiple forms of cytochrome P450 have been identified in many tissues including the kidney, and their nomenclature is based on amino acid sequence homology.[39] Members of families (designated by arabic numerals) and subfamilies (designated by letters, capitals in all species except the mouse) share greater than 39% and 55% sequence homology, respectively. Individual subfamily members are designated by arabic numerals after the subfamily designation (for example CYP1A1 and CYP1A2). Four gene families (CYP1, CYP2, CYP3, and CYP4) contain those P450s primarily responsible for xenobiotic metabolism. Cytochromes P450 are, in general, inducible and this may have important consequences with regard to the balance of detoxification and activation, either within a particular tissue or in the whole animal. P450 subfamilies fall into the following categories on the basis of their inducibility: CYP1A, polycyclic aromatic compound inducible; CYP2B, phenobarbital inducible; CYP2E, ethanol and solvent inducible; CYP3A, steroid inducible and CYP4A, peroxisome proliferator inducible. For further reading on cytochromes P450 see the following reviews.[40–43]

In vitro studies with renal cytochromes P450 have demonstrated metabolism of a wide range of substrates, including both xenobiotics and endogenous compounds. The renal concentration of total, spectrally determined P450 is low compared to the liver, $0.1–0.2 \, nmol \, mg^{-1}$ protein compared to $1.0–1.5 \, nmol \, mg^{-1}$ protein. However, there are marked regional differences in P450 content, with the S2 and S3 segments of the proximal tubule generally having the highest concentrations (Figure 1).[44] Thus the use of whole kidney, as opposed to renal cortex or cell preparations enriched in proximal tubular cells, can grossly underestimate the metabolic capacity of certain regions in the kidney. Immunochemical analysis[27,45,46] and studies using microdissection of the nephron and specific substrates,[13] have confirmed the preferential localization of certain specific P450s within the proximal tubule (Figure 2). However, this does not mean that other parts of the nephron lack cytochromes P450, for example AA epoxygenase is also present in high concentrations within cells of the thick ascending limb of the loop of Henle of the rabbit[47] and rat[13] (Figure 2).

Multiple forms of P450 have been identified in renal tissue from several species, and this review will now focus on each P450 family in turn.

7.05.3.1.2 *CYP1 gene family*

The CYP1 gene family consists of three members, namely CYP1A1, CYP1A2, and CYP1B1. Substrates for these enzymes include aromatic hydrocarbons such as benzo(*a*)pyrene and aflatoxin B_1, and in the liver CYP1A1/2 activity is important in the generation of mutagenic and carcinogenic metabolites of these compounds. This gene family is inducible by 3-methylcholanthrene, 2,3,7,8-tetrachlorodibenzo-*p*-dioxin (TCDD) and has high catalytic activity towards polycyclic aromatic hydrocarbons and arylamines. In the kidney,

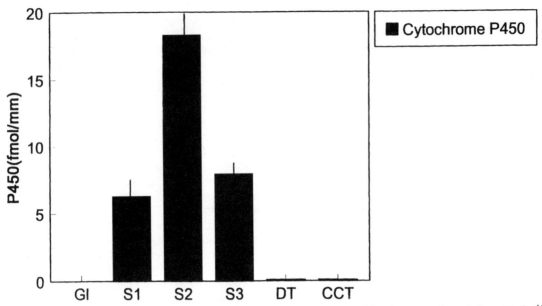

Figure 1 Localization of cytochrome P450 along the rabbit nephron. The data are adapted from Endo.[44] Abbreviations of the nephron segments are: glomerulus (Gl), segments of the proximal tubule (S1–S3), distal tubule (DT) and cortical collecting duct (CCT). Data are shown as mean ± SE.

early studies reported the induction of the CYP1 subfamily by 3-methylcholanthrene, benzo(*a*)pyrene, certain polychlorinated and polybrominated biphenyl compounds, TCDD and β-naphthoflavone in rat, mouse, guinea pig and hamster, and human renal cortical microsomes have been shown to be capable of some arylhydrocarbon metabolism.[10] Renal CYP1A1 is detected only after treatment with inducers in the rat[45,48] and rabbit[46,49] and is located in high concentration in the S2 and S3 segments of the proximal tubules. In contrast, renal CYP1A2 is constitutively expressed in rabbit and can be induced with TCDD.[46] The member of this subfamily, CYP1B1,[50] appears to be involved in hormonal regulation in steroidogenic tissues such as the adrenal cortex, ovary and testis, and is inducible by polycyclic aromatic hydrocarbons. In the kidney CYP1B1 is primarily located in the glomerulus. The mechanism of regulation of the CYP1 subfamily has been extensively studied and reviewed.[51,52]

7.05.3.1.3 CYP2 gene family

The CYP2 gene family is by far the largest of the P450 families, consisting of at least eight subfamilies. Members of this family are involved in the hydroxylation of endogenous substrates such as testosterone, oestradiol, progesterone and AA, and a wide range of xenobiotic compounds from small molecules such as ethanol to drugs such as debrisoquine. Certain subfamilies are also readily inducible by chemicals such as phenobarbital and ethanol. For further reading see Henderson and Wolf.[53]

P450s of the CYP2A subfamily are expressed in mouse,[54–58] but not rat,[59] kidney. In the mouse, two Cyp 2a-4 proteins which are under androgenic control and which are responsible for the 15α-hydroxylation of testosterone are expressed, namely type I in male kidney and type II in female kidney. In the hamster kidney three enzyme activities characteristic of the CYP2A subfamily, coumarin 7-hydroxylase, testosterone 15α-hydroxylase and testosterone 7α-hydroxylase have been characterized.[60] These activities are induced by both 3-methylcholanthrene and phenobarbital, and in the case of testosterone 7α-hydroxylase, there is a marked sex difference with female hamsters having higher activity than males.[60]

The renal CYP2B genes are interesting because they show a species-specific constitutive expression and induction by phenobarbital. Two enzymes, CYP2B1 and CYP2B2, have been extensively studied in rat liver and shown to have similar broad and overlapping substrate specificities. Neither CYP2B1 or CYP2B2 are expressed constitutively in rat or mouse kidney, nor can they be detected following administration of phenobarbital or a nonplanar polychlorinated biphenyl.[10,61] Similarly CYP2B3 is absent from rat kidney.[62] Cyp 2b-1 has been detected in the kidneys of female

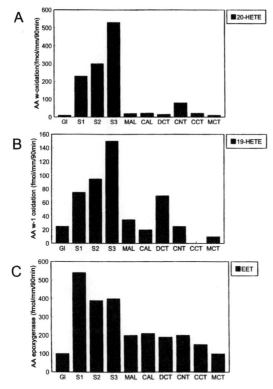

Figure 2 Localization of cytochrome P450-dependent arachidonic acid ω, ω-1 and epoxygenase activities along the rat nephron. The data are adapted from Laniado-Schwartzman and Abraham.[13] (A) ω-oxidation to form 20-hydroxyeicosatetraenoic acid (20-HETE); (B) ω-1-oxidation to form 19-hydroxyeicosatetraenoic acid (19-HETE); and (C) epoxygenase activity to form epoxyeicosatetraenoic acid (EET). Abbreviations for the nephron segments are: glomerulus (Gl), segments of the proximal tubule (S1–S3), medullary ascending limb (MAL), cortical ascending limb (CAL), distal convoluted tubule (DCT), connecting tubule (CNT), cortical collecting tubule (CCT) and medullary collecting tubule (MCT).

been purified from rabbit renal cortex and shown to catalyze the epoxidation of AA, resulting in the formation of 11,12- and 14,15-epoxyeicosatrienoic acids (EETs)[65,66] (see Figure 3). cDNAs coding for rat renal AA epoxygenases[67,68] have been cloned and the proteins expressed and shown to metabolize AA to 8,9-, 11,12- and 14,15-EETs.[69] Sequence analysis showed that the 3′- and 5′-terminal regions of the cDNA were identical to those of CYP2C23, suggesting that CYP2C23 is responsible, in part, for the production of EETs in rat kidney.[68] Determination of the activity of AA epoxygenase along the rat nephron has shown that the proximal tubules contain the highest activity, although EETs were also produced to a significant extent in distal nephron segments[13] (Figure 2). Human kidney also contains AA epoxygenase activity, the major metabolites being 11,12-EET and 11,12-dihydroxyeicosatrienoic acid (DHET).[70] It is now becoming clear that the majority of renal enzymes metabolizing AA by epoxidation are members of the 2C gene family, with preferential metabolism occurring at the 11,12 double bond. This is of physiological interest since rat renal AA epoxygenase is induced by salt loading[69] which can protect against hypotension.[14] For additional information on P450 mediated AA metabolism (see Refs. 12, 13, and 71).

Members of the CYP2D subfamily have been purified from rat, mouse, and human liver and cloning studies have shown at least five genes designated CYP2D1–CYP2D5. The CYP2D proteins carry out oxidation of the drugs debrisoquine and bufuralol. Little is known about their presence in the kidney. Members of the CYP2D subfamily were not detected in mouse[56,57] or human renal microsomes,[72] but there is some evidence suggesting that they are present in rat kidney.[40]

The CYP2E subfamily contains an important form of P450, CYP2E1, which metabolizes small molecular weight chemicals and is implicated in the hepatotoxicity induced by a variety of xenobiotics. In the kidney, this enzyme is induced by ethanol, acetone, pyridine, pyrazole, and 4-methylpyrazole.[73–76] In the mouse, renal CYP2E1 is also under testosterone regulation, being present in males to a much greater extent than in females, although treatment of female mice with testosterone will markedly induce the protein.[27,77,78] The mechanism by which CYP2E1 is regulated is quite distinct from that of other P450s, treatment with inducers such as testosterone or fasting leads to acetone production which increases CYP2E1 mRNA levels[75,77,78] due to posttranscriptional mRNA stabilization. The presence of CYP2E1 in the

mice treated with testosterone,[56,57] however this observation requires further confirmation. In contrast, in the rabbit, hamster and minipig, renal CYP2B1/2 is present at low levels in untreated animals and is inducible with phenobarbital.[10,61]

The CYP2C subfamily is generally thought to represent a class of constitutively expressed proteins, and these enzymes are noted for their sex-specificity and developmentally regulated expression in rats. Several have been purified or cloned from the liver and the reader is referred to reviews for further details,[40,53] CYP2C2, but not CYP2C1 or CYP2C3, have been detected in rat and rabbit kidney, and in the latter species its level is induced threefold following treatment with phenobarbital.[10,63,64] A member of the CYP2C2 subfamily (AA epoxygenase) has kidney is believed to account for the nephro-toxicity or nephrocarcinogenicity of a number of chemicals including chloroform,[26] 4-ipomeanol,[79] 1,1-dichloroethene,[80] paracetamol,[28] and N-nitrosodimethylamine.[81] Thus males are markedly more sensitive to the toxic effects of these xenobiotics, and the proximal tubule, the site of necrosis, is where CYP2E1 is primarily localized.[27]

7.05.3.1.4 CYP3 gene family

The CYP3 family consists of a single subfamily with multiple hepatic members in rats and humans. In the liver these P450s are inducible by glucocorticoids, such as pregnenolone-16α-carbonitrile, and are involved in steroid hydroxylation and metabolism of macrolide antibiotics such as cyclosporine. This family has been little studied in the kidney. mRNAs coding for members of the CYP3A family were detected in rat kidney at a very low level during development,[82] and in human embryonic and adult tissue CYP3A4 was just detectable or absent.[72,83] Murray et al.[84] showed some cross reactivity of an antibody to human liver cytochrome P450hA7 (CYP3A4) in human renal proximal tubules,

and human renal cortical slices have been shown to hydroxylate cyclosporine.[85] Clearly further studies with this family which, in human liver, is the predominant form of P450 are needed.

7.05.3.1.5 CYP4 gene family

Members of the CYP4 gene family catalyze the metabolism of fatty acids and AA, and are induced by hypolipidaemic drugs and chemicals that cause peroxisome proliferation. In the rat kidney the main constitutive member of this family is CYP4A2 which is not inducible by hypolipidaemic drugs, while CYP4A1 is expressed at a low level, but is highly inducible.[86–88] The renal expression of CYP4A2 is higher in male than female rats and appears to be regulated by testosterone, growth hormone and thyroid hormone.[88–90] CYP4A2 has been isolated and purified from rat kidney,[91–93] along with two other isoforms of this subfamily which have been partially characterized. Form K4, like CYP4A2, has high ω and ω-1 hydroxylase activity towards lauric acid, and also catalyzes ω and ω-1 hydroxylation of AA (see Figure 3). Form K2 has low ω and ω-1 hydroxylase activities with AA but efficiently

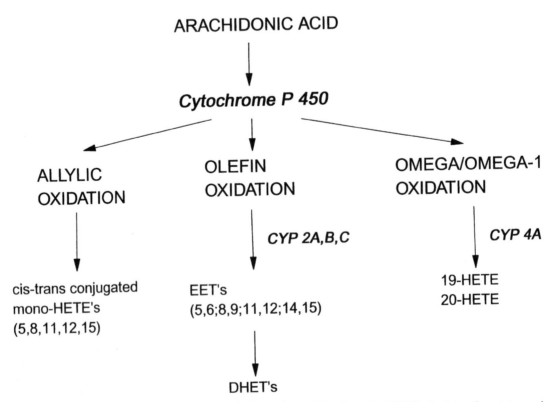

Figure 3 Cytochrome P450-mediated metabolism of arachidonic acid. HETE, hydroxyeicosatetraenoic acid; EET, epoxyeicosatetraenoic acid; DHET, dihydroxyeicosatetraenoic acid.

catalyses ω-1, but not ω, hydroxylation of lauric acid.[91,92] Studies with dissected nephron segments have shown that ω-oxidation of AA to 20-HETEs occurs primarily in the proximal tubule (Figure 2), while ω-1 oxidation of AA to 19-HETEs is highest in the proximal tubule, but considerable activity was also found in the distal convoluted tubule (DCT) and connecting tubule (CNT)[13] (Figure 2). Rabbit kidney contains not only CYP4A1 and CYP4A2, but also CYP4A5, CYP4A6, and CYP4A7 which catalyze the ω and ω-1 hydroxylation of fatty acids and, in the case of CYP4A6 and CYP4A7, the ω hydroxylation of prostaglandin A1 and A2. These forms have all been cloned and heterologously expressed.[94–98] In the mouse, a member of this family, designated Cyp4a-10, has been cloned and its developmental and hormonal regulation studied in the kidney. The enzyme was expressed gradually from birth, there being more in males than females, and testosterone was an important regulatory factor.[99]

A cytochrome P450 (CYP4A11) that catalyzes ω and ω-1 hydroxylation of fatty acids but has no activity towards prostaglandin A1 has been isolated and purified from human renal microsomes,[93] and also independently cloned from a human kidney cDNA library.[100,101] CYP4A11 is a major human renal P450 and may be the orthologous counterpart of rat CYP4A2.

7.05.3.1.6 Summary

Most major families of P450 are present in the kidney. The majority of renal P450s are localized in high concentrations in the proximal tubule, although some members are concentrated at other sites along the nephron such as in the distal tubules. In general, renal P450s appear to favour the hydroxylation of substrates such as fatty acids, prostaglandins, and arachidonic acid rather than xenobiotics. However, some xenobiotics have been shown to undergo activation by renal P450s in the proximal tubule and to produce nephrotoxicity at this site. Renal P450s in the rat, mouse, and guinea pig show little or no induction following *in vivo* administration of phenobarbital, consistent with the absence of the CYP2B subfamily in the kidneys of these species.

7.05.3.2 Epoxide Hydrolases

Epoxide hydrolases are enzymes that catalyze the conversion of epoxides to *trans*-dihydrodiols, thus preventing or reducing their reaction with cellular macromolecules such as protein or DNA. Three forms of epoxide hydrolase have been identified in the liver, two membrane bound forms and one in the cytosolic fraction. Of the two known membrane bound epoxide hydrolases, one catalyzes the conversion of cholesterol 5,6-epoxide to the corresponding diol and displays no activity towards xenobiotic epoxides.[102] The second and major microsomal form, which is both immunologically and enzymatically distinct from the cholesterol 5,6-epoxide hydrolase, is responsible for the hydration of a wide variety of xenobiotic arene and alkene oxides, including benzo(*a*)pyrene-4,5-oxide and styrene oxide.[103] The primary sequence of this enzyme isolated from human liver has been determined, and shown to exhibit high amino acid sequence identity with the rat and rabbit enzymes.[104,105] Cytosolic epoxide hydrolase catalyzes the hydration of several alkene oxides (e.g., *trans*-stilbene oxide and styrene oxide), and is inducible with hypolipidaemic agents such as clofibrate and nafenopin.[106,107]

The kidney expresses both microsomal and cytosolic epoxide hydrolases, albeit at levels lower than those found in the liver, and the renal enzymes have been shown to metabolize a number of xenobiotic epoxides. Despite the prominent role of these enzymes in xenobiotic detoxification, only very limited information exists on their intrarenal distribution, ontogeny, and induction. Microsomal and cytosolic epoxide hydrolase activities have been detected in adult human renal cortex and medulla,[108] and in the rat, the microsomal enzyme has been immunolocalized to the proximal tubular cells.[109] In rats, mRNA for microsomal epoxide hydrolase can be detected at low levels in fetal kidney, and remains at these levels for at least 10 days after birth.[82] Both mRNA and protein levels change little with age over a 4–56 week period.[82,110] In humans mRNA for microsomal epoxide hydrolase has been detected in fetal kidney, but at much lower levels than in fetal liver or adrenal glands.[111] Renal epoxide hydrolases can be induced by chemicals such as *trans*-stilbene oxide, agents that cause peroxisome proliferation, and heavy metals such as lead (see earlier review[10]).

7.05.3.3 Microsomal Flavin Containing Monooxygenases

The microsomal flavin containing monooxygenases (FMOs) catalyze the NADPH-dependent oxidative metabolism of a wide variety of nitrogen-, sulfur- and phosphorus-containing compounds in the liver and extrahepatic tissues, thereby contributing to xenobiotic detoxification and bioactivation.[112,113] Since 1990 significant advances have been made in the study of these enzymes using molecular techniques. Five distinct FMO genes (FMO1–5) have been identified, each belonging to a separate subfamily,[114] and molecular characterization of these subfamilies has been performed in a variety of species including humans, although the rabbit has been most extensively studied (see Shehin-Johnson *et al.*[115] for references).

With the exception of FMO3, all of the five subfamilies have been detected in rabbit kidney.[115–117] FMO1 mRNA was detectable in female, but not male kidney,[115] while FMO4 mRNA was found in the kidney but not in the liver or lung.[117] In other species, including humans, evidence for multiple forms of renal FMOs has been reported.[118–122] The distribution of FMO was studied in rat kidney using an antibody raised against rabbit lung FMO2, and the immunoreactivity shown to be localized in the proximal and distal tubules, and in the collecting ducts of the medulla.[123] Thus various forms of FMO are present in the kidney, where they may play an important role in the detoxification, and possibly also the activation,[23,124] of xenobiotics.

7.05.3.4 Prostaglandin H-Synthase

Prostaglandin *H*-synthase (PHS) is a bifunctional enzyme containing both cyclooxygenase and peroxidase activities. PHS first converts AA to prostaglandin G2 (cyclooxygenase), and then reduces this metabolite to the corresponding alcohol, prostaglandin H2 (peroxidase). PHS-dependent bioactivation of xenobiotics can occur by three principle mechanisms. First, and arguably most important, the peroxidase can directly oxidize chemicals. Second, peroxy radicals generated during prostaglandin biosynthesis can be potent oxidizing agents, while the third mechanism involves secondary oxidant species formed by the peroxidase. For reviews on this area see Eling *et al.*[125] and Smith *et al.*[126]

The kidney contains relatively high levels of PHS,[125] which is not distributed uniformly along the nephron (see Chapter 30, this volume for an in-depth review). In the rabbit, PHS activity is highest in the inner medulla, intermediate in the outer medulla and barely detectable in the cortex.[127] This gradation is the converse of that seen with rabbit renal cytochrome P450.[44,128] Immunocytochemical studies using rabbit and guinea pig kidney have shown that PHS is primarily localized in

the collecting ducts and interstitial cells of the medulla and papilla.[129,130] Several xenobiotics are substrates for renal PHS, including acetaminophen[131] and other nonsteroidal anti-inflammatory drugs, antipyretics and mild analgesics, some of which can cause papillary necrosis.[37] Furthermore, diethylstilbestrol, and certain steroid estrogens which are renal carcinogens in hamsters, can undergo metabolism by rabbit and hamster renal PHS.[132,133] Thus renal PHS may contribute to the papillary toxicity induced by a number of xenobiotics.

7.05.3.5 Carboxylesterases

The carboxylesterases are a family of primarily microsomal enzymes which hydrolyze xenobiotics containing an ester, thioester or amide group.[134–136] This is generally a detoxification process as the resulting metabolites are more hydrophilic and hence more readily excreted, and the carboxylesterases play an important role in the detoxification of organophosphorus pesticides such as malathion.[137] There are multiple forms of carboxylesterase present in numerous mammalian tissues, although the exact number and a well defined nomenclature has yet to be determined. In general activity is highest in the liver, where it is moderately inducible by a variety of chemicals including phenobarbital.[135,138,139] and clofibrate.[140,141]

In the rodent kidney carboxylesterase-dependent hydrolysis of a number of substrates including malathion,[137,140,142] palmitoyl CoA,[143] *p*-nitrophenyl acetate[140,144] and several acetophenetidines and acetamidophenols[145] has been detected. There are however, species differences in renal activity, for example, with *p*-nitrophenyl acetate as substrate the activity is about fourfold higher in the mouse compared to the rat, whereas with malathion as substrate the activity is about sevenfold higher in the rat compared to the mouse.[140] A member of the carboxylesterase family, designated hydrolase B, has been cloned from a rat kidney cDNA library, and its mRNA shown to be present at higher levels in renal compared to hepatic tissue.[146] Immunocytochemical studies with an antibody to hydrolase B has shown that the enzyme is localized in the proximal tubules.[146] Other studies have investigated hydrolases A and C within rat kidney. mRNA to hydrolases A and C could be detected, albeit at low levels,[147,148] but immunohistochemical studies with an antibody to hydrolase A could not demonstrate the presence of this protein.[147]

Rat renal carboxylesterase activity with *p*-nitrophenyl acetate as substrate is inducible to a small extent by prior treatment of animals with clofibrate,[140] benzo(a)pyrene or 3-methylcholanthrene but not benz(a)anthracene.[144] In contrast, in the mouse kidney carboxylesterase activity with p-nitrophenyl acetate is not induced by treatment with clofibrate, whereas with malathion and diethylsuccinate as substrates there is a small induction.[140] Organophosphorus compounds such as di-isopropylfluorophosphate (DFP),[143] tri-ortho-tolyl phosphate (TOCP)[149] and a series of substituted trifluoroketones[140] are potent inhibitors of carboxylesterases. Two organophosphorus compounds, isomalathion and O,S,S,-trimethylphosphorodithioate, which have been detected in technical grade malathion, are potent inhibitors of rat plasma, hepatic, and renal carboxylesterases, thereby potentiating the toxicity of malathion.[142] The presence of these carboxylesterase inhibitors as impurities in malathion is believed to have been responsible for an epidemic of malathion poisoning among workers in a malaria control programme in Pakistan.[150]

Further studies are needed to determine the number and type of carboxylesterases present in the kidney, and in particular in human kidney.

7.05.4 CONJUGATION REACTIONS (PHASE 2 METABOLISM)

Conjugation of the parent chemical or a phase 1 metabolite with either glutathione, glucuronic acid or sulfate frequently occurs, making the xenobiotic more hydrophilic and hence more readily excretable by the kidneys. The key renal enzymes involved in these reactions will be discussed, and advances since 1990 reviewed. For information prior to 1990 the reader is referred to Lock[10] and references therein.

7.05.4.1 Glutathione S-transferases

Glutathione (GSH) conjugation is the formation of a thioether link between the tripeptide GSH (γ-glutamylcysteinylglycine) and a compound with an electrophilic centre. This process is catalyzed by the glutathione S-transferases (GSTs), which are a multifunctional group of enzymes involved in metabolic detoxification,[151,152] intracellular binding and transport of hydrophobic compounds,[153] and catalysis of key steps in the synthesis of prostaglandins and leukotrienes.[154] The majority of GSTs are cytosolic, dimeric proteins consisting of two identical or closely related subunits, which have been separated into four classes, alpha, mu, pi and theta.[151,152,155] Rat tissues have been shown to contain at least 18 dimers made up of 13 different subunits. There is also a microsomal GST which is trimeric and immunologically distinct from the cytosolic forms.[156] Table 1 shows the current nomenclature used for the classification of the cytosolic enzymes in rats and humans. The kidney contains all of the four classes of cytosolic GSTs, while the microsomal enzyme is present but at a low level and is not activated by N-ethylmaleimide.[157] Studies since 1990 have primarily focused on the intrarenal distribution of these enzymes, their ontogeny, and their induction with various xenobiotics.

In rat kidney, biochemical studies have demonstrated that the major GST subunits are 1, 2, 4, 7, and 8.[158–161] Immunochemical localization has demonstrated that the alpha subunits (1, 2, and 8) are localized to the proximal and distal tubules, while the mu (3, 4, and 6) and pi (7) subunits are located primarily in the distal tubules.[162–164] However, there was not complete concordance in these studies, Rozell et al.[162] reporting the absence of alpha subunits in the distal tubules. The immunoreactivity to the different forms of GSTs was predominantly cytoplasmic, although nuclear

Table 1 Nomenclature for rat and human cytosolic glutathione S-transferases.

	Glutathione S-transferase class												
Species	*Alpha (basic)*					*Mu (neutral)*					*Pi (acid)*	*Theta*		
Rat	1	1′	2	8	10	3	4	6	9	11	7	5	12	13
GST subunit	Ya$_1$	Ya$_2$	Yc	Yk	Yl	Yb$_1$	Yb$_2$	Yb$_3$ Yn$_1$	Yn$_2$	Yo	Yf Yp			
Human	α	β	GSTA 2	GSTA 3		GSTM 1a	GSTM 1b	GSTM 4	GSTM 5		GSTP 1	GSTT 1	GSTT 2	

localization was reported with subunits 1 and 2 by Rozell et al.,[162] and with subunits 3 and 4 by Davies et al.[163] In the hamster, renal GSTs are predominantly alpha class,[165] and are localized primarily in the proximal tubules.[166] In the mouse kidney, activity towards a variety of GST substrates has been demonstrated, and members of the alpha, mu and pi classes detected, the latter at low levels only.[167] There is a marked sex difference, with female mice expressing more alpha class than male mice.[167] Murine GSTs are also localized primarily in the proximal tubules.[167] Members of the alpha, mu, and pi classes of GST have been localized in rabbit kidney.[168] Mu and pi class GSTs were widely distributed throughout the kidney, whereas one member of the alpha class showed a localization restricted to the proximal tubules and another was found primarily within the collecting ducts of the inner medulla.[168]

Differences in the rat renal GST subunit profile between 1 and 7 weeks of age have been reported by Moser et al.[169] They showed that the mu and pi classes did not alter with time, whilst subunit 2 increased throughout the seven week period, subunits 1 and 1' peaked at four weeks, and subunit 10 declined during development and was not detected at seven weeks of age (Figure 4).

Early studies reported that rat renal GST activity towards various substrates was induced following treatment of rats with 3,4-benzo(a)-pyrene, phenobarbital and 3-methyl-cholanthrene.[170,171] However, work using either reverse phase HPLC to separate the various subunits or immunoblotting with antisera to

GST classes have shown that phenobarbital and 3-methylcholanthrene do not alter the GST profile in rat kidney.[172] In contrast, administration of the antioxidant ethoxyquin results in a fourfold increase in subunit 7 (pi class), and a doubling in subunit 4 (mu class).[172] Similarly, the antioxidant butylated hydroxyanisole (BHA) induces GST subunits in mouse kidney.[167] Moser et al.[169] and Oberley et al.[164] have shown that, in the rat, chronic lead acetate exposure combined with a low calcium diet results in a marked increase in the expression of various renal GST subunits, particularly 4 and 7.

Little new information is available on human kidney since the earlier review,[10] with the exception of some additional data on alpha and pi class GSTs during renal development and their immunolocalization in both normal and polycystic kidneys.[173] There are some discrepancies in the literature regarding the presence of theta class GSTs in the kidney (see earlier review[10]). Some preliminary data suggests that 2-chloropropionic acid (a theta class specific substrate) is metabolized, albeit with low activity, in both rat and human renal cytosol.[174]

7.05.4.1.1 *Processing of GSH conjugates*

The kidney plays a very important role in the degradation of GSH conjugates to their corresponding mercapturic acids.[175–178] First the isopeptide bond between the α-carbonyl of glutamate and the amino group of cysteine is cleaved by γ-glutamyl transferase, and then

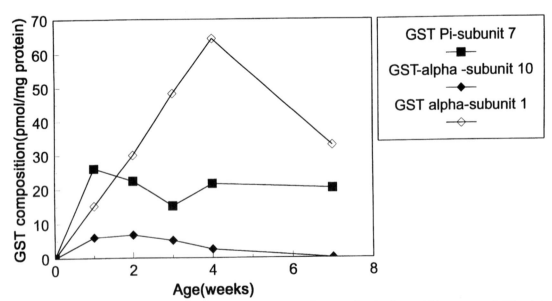

Figure 4 Postnatal expression of three glutathione *S*-transferase subunits in rat kidney cytosol. The data are adapted from Moser *et al.*[169]

cysteinyl glycine dipeptidase and aminopeptidase M cleave the resulting *S*-cysteinylglycine conjugate to yield the cysteine conjugate. In the rat these enzymes are localized in high concentrations on the brush border membrane of the proximal tubule. The *S*-cysteine conjugate is then usually reabsorbed, and mercapturic acid biosynthesis completed by *N*-acetylation which is catalyzed by a microsomal *N*-acetyltransferase. The formation, processing and transport of GSH conjugates occurs in a number of organs including the liver, biliary tract, gastrointestinal tract, and kidney, and involves considerable inter- and intraorgan cooperativity.[175–178]

7.05.4.1.2 *Cysteine* **S**-*conjugate* β-*lyase*

Cysteine conjugate β-lyase (β-lyase) is a pyridoxal phosphate dependent enzyme which is able to metabolize *S*-cysteine conjugates via both transamination and β-elimination. It has been shown to be identical to glutamine transaminase K, and catalyzes the transamination of hydrophobic amino acids such as phenylalanine.[179,180] In addition, other studies have shown that β-lyase also possessed kynurenine aminotransferase (KAT) activity.[181–183] In the rat kidney this enzyme is present mainly in the cytosol, but is also found to a lesser extent in the mitochondria.[179,180,184,185]

β-Lyase has been shown to play a critical role in the nephrotoxicity of certain *S*-cysteine conjugates such as 1,2-dichlorovinyl-L-cysteine (DCVC) or 1,2,3,4,4-pentachloro-1,3-butadienyl-L-cysteine (PCBC).[178,186] Immunocytochemical studies have localized β-lyase to the proximal tubules of rat kidney,[180,187] the primary site of the morphological damage produced by nephrotoxic cysteine conjugates. The enzyme is present in the kidney of all species examined to date (see earlier review[10] and references therein) and has been isolated and purified from human renal cytosol.[188] The β-lyase activity of human renal cytosol with several haloalkane and haloalkene *S*-cysteine conjugates is much lower than that found in rat renal cytosol (Figure 5).[189,190]

The gene coding for rat renal β-lyase/glutamine transaminase K has been cloned, sequenced and expressed in COS cells.[191] The sequence showed a pyridoxal phosphate binding site, and the deduced amino acid sequence was identical to that reported for the purified enzyme.[185] The expressed enzyme possessed catalytic activity with both *S*-cysteine conjugates (β-lyase activity) and phenylalanine (transaminase activity). KAT has also been cloned and characterized from the cytosol of rat brain, and shown to be identical to rat renal β-lyase.[192] Thus renal β-lyase, glutamine transaminase K, and KAT appear to be the same

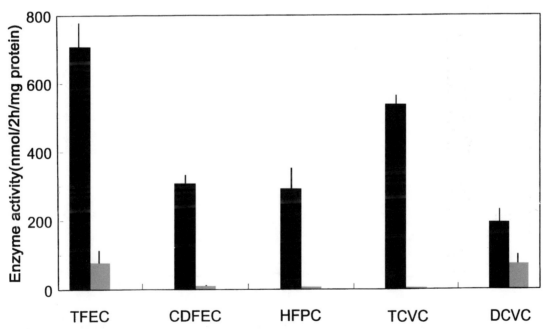

Figure 5 Cysteine conjugate β-lyase activity in rat and human renal cytosol with several haloalkene *S*-conjugates as substrates. The data are adapted from McCarthy *et al.*[190] Rat (■) and human (▨). TFEC, *S*-(1,1,2,2-tetrafluoroethyl)-L-cysteine; CDFEC, *S*-(2-chloro-1,1-difluoroethyl)-L-cysteine; HFPC, *S*-(1,1,2,3,3,3-hexafluoropropyl)-L-cysteine; TCVC, *S*-(1,2,2-trichlorovinyl)-L-cysteine; DCVC, *S*-(1,2-dichlorovinyl)-L-cysteine. Data are shown as mean ± SE.

enzyme, which can either catalyse β-elimination reactions with certain *S*-cysteine conjugates leading to nephrotoxicity, or can catalyse the transamination of hydrophobic molecules such as phenylalanine and kynurenine. Human cysteine conjugate β-lyase has also been cloned, sequenced and expressed in COS cells. It possesses an overall 82% deduced amino acid sequence homology with the rat enzyme, with 90% agreement around the pyridoxal phosphate binding site.[193] Expression of human renal β-lyase confirmed β-elimination activity with DCVC as substrate and transaminase activity with phenylalanine as substrate[193] (Figure 6).

7.05.4.1.3 Cysteine S-conjugate N-acetyltransferase

The final step in the mercapturic acid pathway is *N*-acetylation by cysteine *S*-conjugate *N*-acetyltransferase.[194,195] This enzyme has been isolated and characterized from rat kidney microsomes,[196] and shown to be localized in cells of the proximal tubule.[197] In rodents, the capacity for *N*-acetylation and/or active secretion of the mercapturate into the tubular lumen is greater in the straight portion of the proximal tubule than in the convoluted portion.[198] *N*-acetyl cysteine *S*-conjugates are not substrates for the enzyme β-lyase, therefore this pathway is considered to be a detoxification one.

7.05.4.1.4 N-Deacetylases

The kidney also possesses enzymes that can deacetylate mercapturic acids, and this activity has been demonstrated with several *S*-alkyl, *S*-aralkyl, and *S*-aryl mercapturic acids in the rat, rabbit, and guinea pig kidney.[199] In rat kidney there are multiple enzymes which can catalyze deacetylation reactions; acylase I and III can deacetylate mercapturic acids and are more active in the kidney than in other tissues.[200] Deacetylation of haloalkene *S*-conjugates can make them substrates for β-lyase, and hence can be considered a bioactivation pathway. Indeed, renal deacetylation of several nephrotoxic haloalkene *S*-conjugates has been reported.[201–203] Thus the balance between *N*-acetylation and *N*-deacetylation of haloalkane and haloalkene *S*-conjugates plays a key role in the nephrotoxicity of these chemicals.[204]

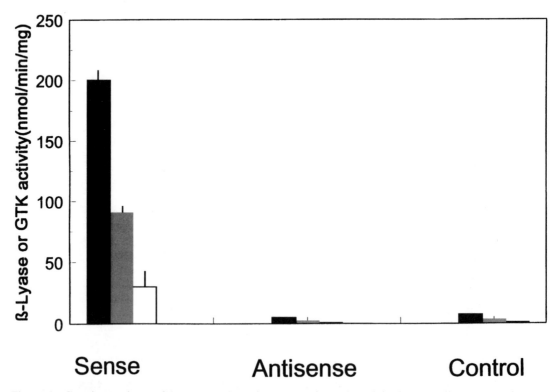

Figure 6 Cysteine conjugate β-lyase and glutamine transaminase K activity in cytosolic extracts of COS-1 cells transfected with human β-lyase cDNA. The data are adapted from Perry *et al.*[193] Glutamine transaminase K activity (×0.1)(■), β-lyase activity with TFEC as substrate (▨) and β-lyase activity with DCVC as substrate (□). COS cells were either transfected with sense or antisense cDNA constructs, or were not transfected.

7.05.4.2 Uridine Diphosphate Glucuronosyltransferases

Conjugation with glucuronic acid is a major pathway in the biotransformation and elimination of a wide variety of lipophilic compounds of both endogenous and exogenous origin.[6,204,205] The glucuronides formed are, in general, less biologically active and more polar than their aglycone substrates and are hence more readily excreted in the urine. Glucuronidation reactions are catalyzed by two families of closely related uridine diphosphate glucuronosyltransferases (UDPGTs).[206] Cloning studies using cDNA libraries from human liver have identified a number of novel UDPGT isoforms, resulting in a suggested nomenclature for this gene superfamily which is similar to that for the cytochromes P450.[207] Each UDPGT has a root symbol, UGT, followed by an arabic number denoting the family, a letter designating the subfamily and an arabic numeral representing the individual gene, for example, UGT2B3. The UGT1 subfamily contains four isoforms, two of which will glucuronidate bilirubin while the other two glucuronidate small planar and bulky phenols. The UGT2 subfamily isoforms have been shown to glucuronidate steroids, including bile acids and catechol estrogens.[207]

Renal glucuronidation of a wide range of substrates has been demonstrated, including both phenols and steroids such as bilirubin and testosterone, suggesting that both UDPGT families are represented in this organ (see earlier review for details[10]). Sex and species differences in renal glucuronidation have been identified, for example, rat kidney is unable to conjugate morphine[208] while human kidney is deficient in bilirubin conjugation.[209] This latter observation has been confirmed by Sutherland et al.[210] who reported conjugation of phenol but not bilirubin. Renal UDPGTs can be induced by many classical inducing agents, including TCDD, Arochlor 1254, *trans*-stilbene oxide, β-naphthoflavone, and clofibrate.[10] In general the levels of induction are not very high, for example, TCDD treatment has been shown to induce renal 6-hydroxychrysene conjugation by UGT1A1 only twofold.[211]

UDPGTs are not uniformly distributed throughout renal tissue, but follow a pattern similar to that found for cytochromes P450. Thus studies with a number of different substrates have demonstrated activities which are highest in the cortex and lowest in the medulla.[108,212–216] These results have been confirmed immunohistochemically in both adult rats and humans. Using an antibody that cross-reacted with all rat UDPGT isoforms, intense staining was demonstrated in rat proximal tubules, with less immunoreactivity in the distal tubules.[217] In humans, a monoclonal antibody which inhibits human liver UDPGT activities with bilirubin, 4-nitrophenol and 4-methylumbelliferone, was localized exclusively to proximal tubules.[218] A study has examined the ontogeny of UDPGTs in human embryonic and fetal kidney using an antibody raised against a rat liver UDPGT which catalyzes the glucuronidation of testosterone and 4-nitrophenol.[219] During early gestation UDPGT immunoreactivity was found to be widespread throughout the collecting duct and tubular systems, but as the nephrons develop reactivity became predominantly localized to the proximal tubules.[219] In the rat, no age differences between 3 and 30 months were observed in the renal UDPGT activity with acetaminophen as substrate.[220]

In summary, at present it appears that the activity of most UDPGTs is lower in the kidney than the liver. However, renal UDPGTs appear to be concentrated primarily, although not exclusively, in the proximal tubules, and thus cells in this region of the nephron may have activities comparable to those found in hepatocytes. The marked species differences that exist for certain substrates require that caution be exercised in extrapolating findings in experimental animals to humans. As more UDPGTs are cloned and expressed the classification, substrate specificity and inducibility of these enzymes in the kidney will become clearer.

7.05.4.3 Sulfotransferases

Sulfation is an important pathway of metabolism for a range of endogenous compounds including steroid hormones, bile acids, neurotransmitters and small peptides. Xenobiotics also undergo conjugation by sulfation which generally, but not exclusively, leads to reduced biological activity.[221] Sulfation reactions, which involve the transfer of the sulfate group from 5′-phosphoadenosine 3′-phosphosulfate to an appropriate functional group, for example, hydroxyl or amino, are catalyzed by a family of sulfotransferase isoenzymes present in the cytosol of most tissues including the kidney. In rats, early studies using a number of different substrates suggested the existence of several sulfotransferase subfamilies, and cDNA sequence analysis has confirmed the presence of at least three classes in rat liver, namely hydroxysteroid, phenol and oestrone sulfotransferases.[222–225] Renal tissue from a number of species has been shown to sulfate a variety of substrates (see references in earlier review[10]), but renal sulfotransferases are relatively uncharacterized.

In the rabbit kidney sulfotransferase activity is predominantly localized to the proximal tubule.[215] Studies by Hume and Coughtrie[226] using an antibody to rat liver paracetamol sulfotransferase have localized this enzyme in human embryonic and fetal kidney. As the nephrons evolve and develop they become increasingly more sulfotransferase immunoreactive, and in the mature kidney the proximal tubules show the most intense staining.[226] The localization and ontogeny of paracetamol sulfotransferase in rat and human kidney closely resemble that reported for the UDPGTs.[217,219] Several groups have examined the gene expression and immunohistochemical localization of rat liver hydroxysteroid sulfotransferases in rat and human kidney, and found little or no expression of this subfamily in renal tissue.[227–229] Any detectable immunoreactivity in human fetal kidney was localized to the collecting ducts.[229]

Thus, although there is some evidence to suggest expression of multiple forms of sulfotransferase in the kidney, many questions regarding these enzymes remain to be answered. The application of molecular biological techniques should provide information regarding which forms are expressed in the kidney, species and sex differences and whether or not renal sulfotransferases are inducible. These questions are of particular importance in light of the known activation of certain procarcinogens, for instance *N*-hydroxy-2-acetylaminofluorene and 5-hydroxymethylchrysene, by sulfation.[230,231]

7.05.5 SUMMARY

The kidney possesses most of the common drug metabolizing enzymes, and thus is able to make an important contribution to the body's metabolism of drugs and other xenobiotics. In general, catalytic activity of the various enzymes in the whole kidney is lower than in the liver, although there are notable exceptions such as those enzymes involved in the processing of GSH conjugates to their mercapturates. Xenobiotic metabolizing enzymes are not evenly distributed along the nephron, cytochromes P450 and those enzymes involved in conjugation with GSH, glucuronic acid or sulfate are present in much higher concentrations in cells of the proximal tubules than in other parts of the nephron. However, some isoenzymes of cytochrome P450 and the glutathione *S*-transferases are selectively localized in the cells of the thick ascending limb and distal tubules, while prostaglandin H-synthase appears to be concentrated in the collecting

ducts in the medulla. Thus the proximal tubule, the principal site of xenobiotic biotransformation, is particularly susceptible to chemical insult, while the localization of prostaglandin H-synthase to the inner medulla and papilla may be a contributory factor to the toxicity produced by chemicals in this part of the nephron.

Molecular biological techniques have enabled many xenobiotic biotransformation enzymes to be cloned and sequenced, and their presence in the kidney detected. Thus, most of the major families of cytochrome P450 involved in xenobiotic metabolism have been detected in the kidneys of experimental animals and humans. Members of the CYP2B subfamily appear to be absent from rat and mouse kidney but present in the rabbit and hamster. Epoxidation of AA appears to be catalyzed by the CYP2C subfamily, these isoenzymes playing an important role in the renal regulation of salt and water balance. Members of the CYP4A subfamily involved in ω and ω-1 oxidation of fatty acids and AA are expressed at particularly high levels in the kidney. The glutathione *S*-transferases have been most extensively characterized in rat and human kidney, this tissue having a different subunit profile to that of the liver, with a predominance of alpha and pi class subunits. The renal enzyme cysteine conjugate β-lyase involved in the bioactivation of haloalkene *S*-cysteine conjugates has been cloned and sequenced from both rat and human kidney, and shown to be identical to glutamine transaminase K and kynurenine aminotransferase. Cloning approaches have also demonstrated the presence in the kidney of multiple forms of UDP-glucuronosyltransferase, flavin monooxygenase, carboxylesterase, and sulfotransferase.

Studies on the presence and localization of xenobiotic biotransformation enzymes in human kidney are still limited, although information on embryonic and fetal tissue is beginning to emerge. Many of the enzymes discussed have, in addition to their ability to metabolize drugs or foreign compounds, an endogenous function in the kidney, for instance, in the regulation of salt and water balance and the synthesis of vitamin D.

7.05.6 REFERENCES

1. J. J. Grantham and A. M. Chonko, in 'The Kidney,' 4th edn, eds. B. M. Brenner and F. C. Rector, Saunders, Philadelphia, PA, 1991, pp. 483–509.
2. T. B. Vree, E. W. J. Van Ewijk-Beneken Kolmer, E. W. Wuis *et al.*, 'Interindividual variation in the capacity-limited renal glucuronidation of probenecid by humans.' *Pharm. World Sci.*, 1993, **15**, 197–202.

3. D. Grundemann, V. Gorboulev, S. Gambaryam *et al.*, 'Drug excretion mediated by a new prototype of polyspecific transporter.' *Nature*, 1994, **372**, 549–552.

4. D. R. Krishna and U. Klotz, 'Extrahepatic metabolism of drugs in humans.' *Clin. Pharmacokinet.*, 1994, **26**, 144–160.

5. C. L. Litterst, E. G. Mimnaugh, R. L. Reagan *et al.*, 'Comparison of *in vitro* drug metabolism by lung, liver and kidney of several common laboratory species.' *Drug Metab. Dispos.*, 1975, **3**, 259–265.

6. M. W. Anders, 'Metabolism of drugs by the kidney.' *Kidney Int.*, 1980, **18**, 636–647.

7. M. D. Burke and S. Orrenius, 'Isolation and comparison of endoplasmic reticulum membranes and their mixed function oxidase activities from mammalian extrahepatic tissues.' *Pharmacol. Ther.*, 1979, **7**, 549–599.

8. J. B. Tarloff, R. S. Goldstein and J. B. Hook, 'Xenobiotic biotransformation by the kidney: pharmacological and toxicological aspects.' *Prog. Drug Metab.*, 1990, **12**, 1–39.

9. G. J. Kaloyanides, 'Metabolic interactions between drugs and renal tubulointestitial cells: role in nephrotoxicity.' *Kidney Int.*, 1991, **39**, 531–540.

10. E. A. Lock, in 'Mechanisms of Injury in Renal Disease and Toxicity,' ed. R. S. Goldstein, CRC Press, Boca Raton, FL, 1994, pp. 173–206.

11. R. Kumar, J. Schaefer, J. P. Grande *et al.*, 'Immunolocalization of calcitriol receptor, 24-hydroxylase cytochrome P-450, and calbindin D$_{28k}$ in human kidney.' *Am. J. Physiol.*, 1994, **266**, F477–485.

12. E. H. Oliw, 'Oxygenation of polyunsaturated fatty acids by cytochrome P450 monooxygenases.' *Prog. Lipid Res.*, 1994, **33**, 329–354.

13. M. Laniado-Schwartzman and N. G. Abraham, 'The renal cytochrome P-450 arachidonic acid system.' *Pediatr. Nephrol.*, 1992, **6**, 490–498.

14. K. Makita, K. Takahashi, A. Karara *et al.*, 'Experimental and/or genetically controlled alterations of the renal microsomal cytochrome P450 epoxygenase induce hypertension in rats fed a high salt diet.' *J. Clin. Invest.*, 1994, **94**, 2414–2420.

15. Y. H. Ma, M. L. Schwartzman and R. J. Roman, 'Altered renal P-450 metabolism of arachidonic acid in Dahl salt-sensitive rats.' *Am. J. Physiol.*, 1994, **267**, R579–589.

16. B. Escalante, D. Erlig, J. R. Falck *et al.*, 'Cytochrome P-450 arachidonate metabolites affect iron fluxes in rabbit medullary thick ascending limb.' *Am. J. Physiol.*, 1994, **266**, C1775–C1782.

17. E. Jacqz, S. Ward, R. Johnson *et al.*, 'Extrahepatic glucuronidation of morphine in the dog.' *Drug Metab. Dispos.*, 1986, **14**, 627–630.

18. P. A. Sloan, L. E. Mather, C. F. McLean *et al.*, 'Physiological disposition of i.v. morphine in sheep.' *Brit. J. Anaesth.*, 1991, **67**, 378–386.

19. J. X. Mazoit, P. Sandouk, P. Zetlaoui *et al.*, 'Pharmacokinetics of unchanged morphine in normal and cirrhotic subjects.' *Anaesth. Analg.*, 1987, **66**, 293–298.

20. A. Bodenham, K. Quinn and G. R. Park, 'Extrahepatic morphine metabolism in man during the anhepatic phase of orthotopic liver transplantation.' *Br. J. Anaesth.*, 1989, **63**, 380–384.

21. M. J. Cousins and R. I. Mazze, 'Methoxyflurane nephrotoxicity. A study of dose response in man.' *JAMA*, 1973, **225**, 1611–1616.

22. E. D. Kharasch, D. J. Hankins and K. E. Thummel, 'Human kidney methoxyflurane and sevoflurane metabolism. Intrarenal fluoride production as a possible mechanism of methoxyflurane nephrotoxicity.' *Anesthesiology*, 1995, **82**, 689–699.

23. L. H. Lash, in 'Mechanisms of Injury in Renal Disease and Toxicity,' ed. R. S. Goldstein, CRC Press, Boca Raton, FL, 1994, pp. 207–234.

24. J. B. Hook and R. S. Goldstein (eds.), 'Toxicology of the Kidney,' 2nd edn., Raven, New York, 1993.

25. M. W. Anders, W. Dekant, D. Henschler *et al.* (eds.), 'Renal Disposition and Nephrotoxicity of Xenobiotics,' Academic Press, New York, 1993.

26. J. H. Smith, 'Role of renal metabolism in chloroform nephrotoxicity.' *Comments Toxicol.*, 1986, **1**, 125–144.

27. J. J. Hu, W. B. Rhoten and C. S. Yang, 'Mouse renal cytochrome P450IIE1: immunocytochemical localization, sex-related differences and regulation by testosterone.' *Biochem. Pharmacol.*, 1990, **40**, 2597–2602.

28. J. J. Hu, M. J. Lee, M. Vapiwala *et al.*, 'Sex-related differences in mouse renal metabolism and toxicity of acetaminophen.' *Toxicol. Appl. Pharmacol.*, 1993, **122**, 16–26.

29. E. A. Lock and J. Ishmael, 'Effect of the organic acid transport inhibitor probenecid on renal cortical uptake and proximal tubular toxicity of hexachloro-1,3-butadiene and its conjugates.' *Toxicol. Appl. Pharmacol.*, 1985, **81**, 32–42.

30. E. A. Lock, J. Odum and P. Ormond, 'Transport of N-acetyl-S-pentachloro-1,3-butadienylcysteine by rat renal cortex.' *Arch. Toxicol.*, 1986, **59**, 12–15.

31. I. S. Pratt and E. A. Lock, 'Deacetylation and further metabolism of the mercapturic acid of hexachloro-1,3-butadiene by rat kidney cytosol *in vitro*.' *Arch. Toxicol.*, 1988, **62**, 341–345.

32. M. MacFarlane, J. R. Foster, G. G. Gibson *et al.*, 'Cysteine conjugate β-lyase of rat kidney cytosol: characterization, immunocytochemical localization and correlation with hexachlorobutadiene nephrotoxicity.' *Toxicol. Appl. Pharmacol.*, 1989, **98**, 185–197.

33. T. W. Jones, C. Qiu, V. H. Schaeffer *et al.*, 'Immunohistochemical localization of glutamine transaminase K, a rat kidney cysteine conjugate β-lyase, and the relationship to the segment specificity of cysteine conjugate nephrotoxicity.' *Mol. Pharmacol.*, 1988, **34**, 621–627.

34. M. A. Venkatchalam, D. B. Bernard, J. F. Donohoe *et al.*, 'Ischemic damage and repair in the rat proximal tubule: differences along the S1, S2 and S3 segments.' *Kidney Int.*, 1978, **14**, 31–49.

35. G. G. Duggin and G. H. Mudge, 'Analgesic nephropathy: renal distribution of acetaminophen and its conjugates.' *J. Pharmacol. Exp. Ther.*, 1976, **199**, 1–9.

36. G. H. Mudge, M. W. Gemborys and G. G. Duggin, 'Covalent binding of metabolites of acetaminophen to kidney protein and depletion of renal glutathione.' *J. Pharmacol. Exp. Ther.*, 1978, **206**, 218–226.

37. P. H. Bach and J. W. Bridges, 'Chemically induced renal papillary necrosis and upper urothelial carcinoma. Part 1.' *Crit. Rev. Toxicol.*, 1985, **15**, 217–329.

38. H. F. DeLuca, 'The vitamin D story: a collaborative effort of basic science, and clinical medicine.' *FASEB J.*, 1988, **2**, 224–236.

39. D. R. Nelson, T. Kamataki, D. J. Waxman *et al.*, 'The P450 superfamily: update on new sequences, gene mapping, accession numbers, early trivial names of enzymes, and nomenclature.' *DNA Cell Biol.*, 1993, **12**, 1–51.

40. F. J. Gonzalez, 'Molecular genetics of the P-450 superfamily.' *Pharmac. Ther.*, 1990, **45**, 1–38.

41. A. J. Paine, 'The cytochrome P450 gene superfamily.' *Int. J. Exp. Pathol.*, 1991, **72**, 349–363.

42. S. A. Wrighton and J. C. Stevens, 'The human hepatic cytochromes P450 involved in drug metabolism.' *Crit. Rev. Toxicol.*, 1992, **22**, 1–21.

43. B. Testa, 'The Metabolism of Drugs and Other Xenobiotics: Biochemistry of Redox Reactions', Academic Press, New York, 1995.

44. H. Endou, 'Cytochrome P-450 monooxygenase system in the rabbit kidney: its intranephron localization and its induction.' *Jpn. J. Pharmacol.*, 1983, **33**, 423–433.

45. J. R. Foster, C. R. Elcombe, A. R. Boobis *et al.*, 'Immunocytochemical localization of cytochrome P-450 in hepatic and extra-hepatic tissues of the rat with a monoclonal antibody against cytochrome P-450c.' *Biochem. Pharmacol.*, 1986, **35**, 4543–4554.

46. J. H. Dees, B. S. Masters, U. Muller-Eberhard *et al.*, 'Effect of 2,3,7,8-tetrachlorodibenzo-*p*-dioxin and phenobarbital on the occurrence and distribution of four cytochrome P-450 isozymes in rabbit kidney, lung and liver.' *Cancer. Res.*, 1982, **42**, 1423–1432.

47. N. R. Ferreri, M. L. Schwartzman, N. G. Ibraham *et al.*, 'Arachidonic acid metabolism in cell suspension isolated from rabbit renal outer medulla.' *J. Pharmacol. Exp. Ther.*, 1984, **231**, 441–448.

48. J. D. Wilson, H. Miller, H. V. Gelboin *et al.*, 'Variation in inducibility of P-450c and aryl hydrocarbon hydroxylase in rat liver, lung, kidney, pancreas and nasopharynx.' *Pharmacology*, 1990, **41**, 256–262.

49. D. R. Koop, R. M. Laethem, A. L. Goldner *et al.*, 'Cytochrome P450 expression and metabolism in isolated rabbit renal epithelium.' *Methods Enzymol.*, 1991, **206**, 364–371.

50. K. K. Bhattacharyya, P. B. Brake, S. E. Eltom *et al.*, 'Identification of a rat adrenal cytochrome P450 active in polycyclic hydrocarbon metabolism as rat CYP1B1. Demonstration of a unique tissue-specific pattern of hormonal and aryl hydrocarbon receptor-linked regulation.' *J. Biol. Chem.*, 1995, **270**, 11595–11602.

51. D. W. Nebert, A. Puga and V. Vasiliou, 'Role of the Ah receptor and the dioxin-inducible [Ah] gene battery in toxicity, cancer, and signal transduction.' *Ann. NY Acad. Sci.*, 1993, **685**, 624–640.

52. O. Hankinson, 'The aryl hydrocarbon receptor complex.' *Annu. Rev. Pharmacol. Toxicol.*, 1995, **35**, 307–340.

53. C. J. Henderson and C. R. Wolf, 'Molecular analysis of cytochrome P450s in the CYP2 gene family.' *Prog. Drug Metab.*, 1992, **13**, 73–139.

54. E. J. Squires and M. Negishi, 'Reciprocal regulation of sex-dependent expression of testosterone 15α-hydroxylase (P-450$_{15\alpha}$) in liver and kidney of male mice by androgen. Evidence for a single gene.' *J. Biol. Chem.*, 1988, **263**, 4166–4171.

55. E. J. Squires and M. Negishi, 'Tissue-specific regulation of cytochrome P-450 dependent testosterone 15α-hydroxylase.' *Can. J. Physiol. Pharmacol.*, 1990, **68**, 769–776.

56. C. J. Henderson, A. R. Scott, C. S. Yang *et al.*, 'Testosterone-mediated regulation of mouse renal cytochrome P-450 isoenzymes.' *Biochem. J.*, 1990, **266**, 675–681.

57. C. J. Henderson and C. R. Wolf, 'Evidence that the androgen receptor mediates sexual differentiation of mouse renal cytochrome P450 expression.' *Biochem. J.*, 1991, **278**, 499–503.

58. P. Salonpaa, M. Iscan, M. Pasanen, *et al.*, 'Cerium-induced strain-dependent increase in Cyp2a-4/5 (cytochrome P4502a-4/5) expression in the liver and kidneys of inbred mice.' *Biochem. Pharmacol.*, 1992, **44**, 1269–1274.

59. D. J. Waxman, G. A. Dannan and F. P. Guengerich, 'Regulation of rat hepatic cytochrome P-450: age-dependent expression, hormonal imprinting and xenobiotic inducibility of sex-specific isoenzymes.' *Biochemistry*, 1985, **24**, 4409–4417.

60. P. Pelkonen, M. Lang and M. Pasanen, 'Tissue and sex-dependent differences in CYP2A activities in hamsters.' *Arch. Toxicol.*, 1994, **68**, 416–422.

61. R. Ryan, S. W. Grimm, K. M. Kedzie *et al.*, 'Cloning, sequencing, and functional studies of phenobarbital-inducible forms of cytochrome P450 2B and 4B expressed in rabbit kidney.' *Arch. Biochem. Biophys.*, 1993, **304**, 454–463.

62. D. Labbe, A. Jean and A. Anderson, 'A constitutive member of the rat cytochrome P450IIB subfamily: full-length coding sequence of the P40IIB3 cDNA.' *DNA*, 1988, **7**, 253–260.

63. J. K. Leighton and B. Kemper, 'Differential induction and tissue-specific expression of closely related members of the phenobarbital-inducible rabbit cytochrome P-450 gene family.' *J. Biol. Chem.*, 1984, **259**, 11165–11168.

64. M. J. Finlayson, J. H. Dees, B. S. Masters *et al.*, 'Differential expression of cytochrome P-450 1 and related forms in rabbit liver and kidney.' *Arch. Biochem. Biophys.*, 1987, **252**, 113–120.

65. R. M. Laethem, C. L. Laethem and D. R. Koop, 'Purification and properties of a cytochrome P450 arachidonic acid epoxygenase from rabbit renal cortex.' *J. Biol. Chem.*, 1992, **267**, 5552–5559.

66. R. M. Laethem and D. R. Koop, 'Identification of rabbit cytochromes P450 2C1 and 2C2 as arachidonic acid epoxygenases.' *Mol. Pharmacol.*, 1992, **42**, 958–963.

67. A. Karara, K. Makita, H. R. Jacobson *et al.*, 'Molecular cloning, expression, and enzymatic characterization of the rat kidney cytochrome P-450 arachidonic acid epoxygenase.' *J. Biol. Chem.*, 1993, **268**, 13565–13570.

68. S. Imaoka, P. J. Wedlund, H. Ogawa *et al.*, 'Identification of CYP2C23 expressed in rat kidney as an arachidonic acid epoxygenase.' *J. Pharmacol. Exp. Ther.*, 1993, **267**, 1012–1016.

69. J. H. Capdevila, S. Wei, J. Yan *et al.*, 'Cytochrome P-450 arachidonic acid epoxygenase. Regulatory control of the renal epoxygenase by dietary salt loading.' *J. Biol. Chem.*, 1992, **267**, 21720–21726.

70. M. L. Schwartzman, P. Martasek, A. M. Rios *et al.*, 'Cytochrome P450-dependent arachidonic acid metabolism in human kidney.' *Kidney Int.*, 1990, **37**, 94–99.

71. J. C. McGiff, 'Cytochrome P-450 metabolism of arachidonic acid.' *Annu. Rev. Pharmacol. Toxicol.*, 1991, **31**, 339–369.

72. I. de Waziers, P. H. Cugnenc, C. S. Yang *et al.*, 'Cytochrome P450 isoenzymes, epoxide hydrolase and glutathione transferases in rat and human hepatic and extrahepatic tissues.' *J. Pharmacol. Exp. Ther.*, 1990, **253**, 387–394.

73. X. X. Ding, D. R. Koop, B. L. Crump *et al.*, 'Immunocytochemical identification of cytochrome P-450 isozyme 3a (P-450 ALC) in rabbit nasal and kidney microsomes and evidence for differential induction by alcohol.' *Mol. Pharmacol.*, 1986, **30**, 370–378.

74. T. H. Ueng, F. K. Friedman, H. Miller *et al.*, 'Studies on ethanol-inducible cytochrome P-450 in rabbit liver, lungs and kidneys.' *Biochem. Pharmacol.*, 1987, **36**, 2689–2691.

75. H. Kim, S. G. Kim, M. Y Lee *et al.*, 'Evidence for elevation of cytochrome P4502E1 (alcohol-inducible form) mRNA levels in rat kidney following pyridine administration.' *Biochem. Biophys. Res. Commun.*, 1992, **186**, 846–853.

76. D. Wu and A. I. Cederbaum, 'Characterization of pyrazole and 4-methyl-pyrazole induction of cytochrome P4502E1 in rat kidney.' *J. Pharmacol. Exp. Ther.*, 1994, **270**, 407–413.

77. J. Pan, J. Y Hong and C. S. Yang, 'Post-transcriptional regulation of mouse renal cytochrome P4502E1 by testosterone.' *Arch. Biochem. Biophys.*, 1992, **299**, 110–115.

78. J. F. Davis and M. R. Felder, 'Mouse ethanol-inducible cytochrome P-450 (P-450IIE1). Characterization of cDNA clones and testosterone induction in kidney tissue.' *J. Biol. Chem.*, 1993, **268**, 16584–16589.

79. M. R. Boyd and J. S. Dutcher, 'Renal toxicity due to reactive metabolites formed *in situ* in the kidney, investigations with 4-ipomeanol in the mouse.' *J. Pharmacol. Exp. Ther.*, 1981, **216**, 640–646.

80. P. Speerschneider and W. Dekant, 'Renal tumorigenicity of 1,1-dichloroethene in mice: the role of male-specific expression of cytochrome P450 2E1 in the renal bioactivation of 1,1-dichloroethene.' *Toxicol. Appl. Pharmacol.*, 1995, **130**, 48–56.

81. J. Y. Hong, J. M. Pan, S. M. Ning *et al.*, 'Molecular basis for the sex-related difference in renal *N*-nitrosodimethylamine demethylase in C3H/HeJ mice.' *Cancer Res.*, 1989, **49**, 2973–2979.

82. D. L. Simmons and C. B. Kasper, 'Quantitation of mRNAs specific for the mixed-function oxidase system in rat liver and extrahepatic tissues during development.' *Arch. Biochem. Biophys.*, 1989, **271**, 10–20.

83. H. Y. Yang, Q. P. Lee, A. E. Rettie *et al.*, 'Functional cytochrome P4503A isoforms in human embryonic tissues: expression during organogenesis.' *Mol. Pharmacol.*, 1994, **46**, 922–928.

84. G. I. Murray, T. S. Barnes, H. F. Sewell *et al.*, 'The immunocytochemical localization and distribution of cytochrome P-450 in normal hepatic and extrahepatic tissues with a monoclonal antibody to human cytochrome P-450.' *Brit. J. Clin. Pharmacol.*, 1988, **25**, 465–475.

85. A. E. Vickers, 'Use of human organ slices to evaluate the biotransformation and drug-induced side-effects of pharmaceuticals.' *Cell Biol. Toxicol.*, 1994, **10**, 407–414.

86. J. P. Hardwick, B. J. Song, E. Huberman *et al.*, 'Isolation, complementary DNA sequence, and regulation of rat hepatic lauric acid ω-hydroxylase (cytochrome P-450AL omega). Identification of a new cytochrome P-450 gene family.' *J. Biol. Chem.*, 1987, **262**, 801–810.

87. D. R. Bell, R. G. Bars and C. R. Elcombe, 'Differential tissue-specific expression and induction of cytochrome P450IVA1 and acyl-CoA oxidase.' *Eur. J. Biochem.*, 1992, **206**, 979–986.

88. S. S. Sundseth and D. J. Waxman, 'Sex-dependent expression and clofibrate inducibility of cytochrome P450 4A fatty acid ω-hydroxylases. Male specificity of liver and kidney CYP4A2 MRNA and tissue-specific regulation by growth hormone and testosterone.' *J. Biol. Chem.*, 1992, **267**, 3915–3921.

89. S. Imaoka, Y. Yamazoe, R. Kato *et al.*, 'Hormonal regulation of rat renal cytochrome P450s by androgen and the pituitary.' *Arch. Biochem. Biophys.*, 1992, **299**, 179–184.

90. S. Imaoka, M. Nakamura, T. Ishizaki *et al.*, 'Regulation of renal cytochrome P450s by thyroid hormone in diabetic rats.' *Biochem. Pharmacol.*, 1993, **46**, 2197–2200.

91. S. Imaoka, S. Tanaka and Y. Funae, 'ω and ω-1 hydroxylation of lauric acid and arachidonic acid by rat renal cytochrome P-450.' *Biochem. Int.*, 1989, **18**, 731–740.

92. S. Imaoka, K. Nagashima and Y. Funae, 'Characterization of three cytochrome P450s purified from renal microsomes of untreated male rats and comparison with human renal cytochrome P450.' *Arch. Biochem. Biophys.*, 1990, **276**, 473–480.

93. H. Kawashima, E. Kusunose, I. Kubota *et al.*, 'Purification and NH₂-terminal amino acid sequences of human and rat kidney fatty acid ω-hydroxylases.' *Biochim. Biophys. Acta*, 1992, **1123**, 156–162.

94. E. Kusunose, A. Sawamura, H. Kawashima *et al.*, 'Isolation of a new form of cytochrome P-450 with prostaglandin A and fatty acid ω-hydroxylase activities from rabbit kidney cortex microsomes.' *J. Biochem. (Tokyo)*, 1989, **106**, 194–196.

95. R. Yoshimura, E. Kusunose, N. Yokotani *et al.*, 'Purification and characterization of two forms of fatty acid ω-hydroxylase cytochrome P-450 from rabbit kidney cortex microsomes.' *J. Biochem. (Tokyo)*, 1990, **108**, 544–548.

96. E. F. Johnson, D. L. Walker, K. J. Griffin *et al.*, 'Cloning and expression of three rabbit kidney cDNAs encoding lauric acid ω-hydroxylases.' *Biochemistry*, 1990, **29**, 873–879.

97. A. Sawamura, E. Kusunose, K. Satouchi *et al.*, 'Catalytic properties of rabbit kidney fatty acid ω-hydroxylase cytochrome P-450ka2 (CYP4A7).' *Biochim. Biophys. Acta*, 1993, **1168**, 30–36.

98. L. J. Roman, C. N. Palmer, J. E. Clark *et al.*, 'Expression of rabbit cytochromes P4504A which catalyse the ω-hydroxylation of arachidonic acid, fatty acids, and prostaglandins.' *Arch. Biochem. Biophys.*, 1993, **307**, 57–65.

99. C. J. Henderson, T. Bammler and C. R. Wolf, 'Deduced amino acid sequence of a murine cytochrome P-450 Cyp4a protein: developmental and hormonal regulation in liver and kidney.' *Biochim. Biophys. Acta*, 1994, **1200**, 182–190.

100. C. N. Palmer, T. H. Richardson, K. J. Griffin *et al.*, 'Characterization of a cDNA encoding a human kidney, cytochrome P-450 4A fatty acid ω-hydroxylase and the cognate enzyme expressed in *Escherichia coli*.' *Biochim. Biophys. Acta.*, 1993, **1172**, 161–166.

101. S. Imaoka, H. Ogawa, S. Kimura *et al.*, 'Complete cDNA sequence and cDNA-directed expression of CYP4A11, a fatty acid ω-hydroxylase expressed in human kidney.' *DNA Cell Biol.*, 1993, **12**, 893–899.

102. F. Oesch, C. W. Timms, C. H. Walker *et al.*, 'Existence of multiple forms of microsomal epoxide hydrolases with radically different substrate specificities.' *Carcinogenesis*, 1984, **5**, 7–9.

103. F. Oesch, N. Kaubisch, D. M. Jerina *et al.*, 'Hepatic epoxide hydrase. Structure activity relationships of substrates and inhibitors.' *Biochemistry*, 1971, **10**, 4858–4866.

104. C. Hassett, S. T. Turnblom, A. DeAngeles *et al.*, 'Rabbit microsomal epoxide hydrolase: isolation and characterization of the xenobiotic metabolizing enzyme cDNA.' *Arch. Biochem. Biophys.*, 1989, **271**, 380–389.

105. R. C. Skoda, A. Demierre, O. W. McBride *et al.*, 'Human microsomal xenobiotic epoxide hydrolase. Complementary DNA sequence, complementary DNA-directed expression in COS-1 cells, and chromosomal localization.' *J. Biol. Chem.*, 1988, **263**, 1549–1554.

106. B. D. Hammock and K. Ota, 'Differential induction of cytosolic epoxide hydrolase, microsomal epoxide hydrolase and glutathione S-transferase activities.' *Toxicol. Appl. Pharmacol.*, 1983, **71**, 254–265.

107. F. Waechter, P. Bentley, F. Bieri *et al.*, 'Organ distribution of epoxide hydrolases in cytosolic and microsomal fractions of normal and nafenopin-treated male DBA/2 mice.' *Biochem. Pharmacol.*, 1988, **37**, 3897–3903.

108. G. M. Pacifici, A. Viani, M. Franchi *et al.*, 'Profile of drug-metabolizing enzymes in the cortex and medulla of the human kidney.' *Pharmacology*, 1989, **39**, 299–308.

109. J. E. Sheehan, H. C. Pitot and C. B. Kasper, 'Transcriptional regulation and localization of the tissue-specific induction of epoxide hydrolase by lead acetate

in rat kidney.' *J. Biol. Chem.*, 1991, **266**, 5122–5127.

110. S. G. Kim and Y. H. Kim, 'Gender-related expression of rat microsomal epoxide hydrolase during maturation: post-transcriptional regulation.' *Mol. Pharmacol.*, 1992, **42**, 75–81.

111. C. J. Omiecinski, L. Aicher and L. Swenson, 'Developmental expression of human microsomal epoxide hydrolase.' *J. Pharmacol. Exp. Ther.*, 1994, **269**, 417–423.

112. D. M. Ziegler, in '*N*-Oxidation of Drugs: Biochemistry, Pharmacology, and Toxicology,' 1st edn., eds. P. Hlavica and L. A. Damani, Chapman and Hall, London, 1991, pp. 60–70.

113. D. M. Ziegler, 'Recent studies on the structure and function of multisubstrate flavin-containing monooxygenases.' *Annu. Rev. Pharmacol. Toxicol.*, 1993, **33**, 179–199.

114. M. P. Lawton, J. R. Cashman, T. Cresteil *et al.*, 'A nomenclature for the mammalian flavin-containing monooxygenase gene family based on amino acid sequence identities.' *Arch. Biochem. Biophys.*, 1994, **308**, 254–257.

115. S. E. Shehin-Johnson, E. D. Williams, S. Larsen-Su *et al.*, 'Tissue-specific expression of flavin-containing monooxygenase (FMO) forms 1 and 2 in the rabbit.' *J. Pharmacol. Exp. Ther.*, 1995, **272**, 1293–1299.

116. E. Atta-Asafo-Adjei, M. P. Lawton and R. M. Philpot, 'Cloning, sequencing, distribution, and expression in *Escherichia coli* of flavin-containing monooxygenase 1C1. Evidence for a third gene subfamily in rabbits.' *J. Biol. Chem.*, 1993, **268**, 9681–9689.

117. V. L. Burnett, M. P. Lawton and R. M. Philpot, 'Cloning and sequencing of flavin-containing monooxygenases FMO3 and FMO4 from rabbit and characterization of FMO3.' *J. Biol. Chem.*, 1994, **269**, 14314–14322.

118. G. A. Dannan and F. P. Guengerich, 'Immunochemical comparison and quantitation of microsomal flavin-containing monooxygenase in various hog, mouse, rat, rabbit, dog and human tissues.' *Mol. Pharmacol.*, 1982, **22**, 787–794.

119. R. E. Tynes and E. Hodgson, 'Catalytic activity and substrate specificity of the flavin-containing monooxygenase in microsomal systems: characterization of the hepatic, pulmonary and renal enzymes of the mouse, rabbit and rat.' *Arch. Biochem. Biophys.*, 1985, **240**, 77–93.

120. R. E. Tynes and R. M. Philpot, 'Tissue- and species-dependent expression of multiple forms of mammalian microsomal flavin-containing monooxygenase.' *Mol. Pharmacol.*, 1987, **31**, 569–574.

121. A. Lemoine, M. Johann and T. Cresteil, 'Evidence for the presence of distinct flavin-containing monooxygenases in human tissues.' *Arch. Biochem. Biophys.*, 1990, **276**, 336–342.

122. A. J. M. Sadeque, A. C. Eddy, G. P. Meier *et al.*, 'Stereoselective sulfoxidation by human flavin-containing monooxygenase. Evidence for catalytic diversity between hepatic, renal, and fetal forms.' *Drug Metab. Dispos.*, 1992, **20**, 832–839.

123. S. Bhamre, S. K. Shankar, S. V. Bhagwat *et al.*, 'Catalytic activity and immunohistochemical localization of flavin-containing monooxygenase in rat kidney.' *Life Sci.*, 1993, **52**, 1601–1607.

124. P. J. Sausen and A. A. Elfarra, 'Cysteine conjugate S-oxidase. Characterization of a novel enzymatic activity in rat hepatic and renal microsomes.' *J. Biol. Chem.*, 1990, **265**, 6139–6145.

125. T. E. Eling, D. C. Thompson, G. L. Foureman, *et al.*, 'Prostaglandin H Synthase and xenobiotic oxidation.' *Annu. Rev. Pharmacol. Toxicol.*, 1990, **30**,

126. B. J. Smith, J. F. Curtis and T. E. Eling, 'Bioactivation of xenobiotics by prostaglandin H synthase.' *Chem. Biol. Interact.*, 1992, **79**, 245–264.

127. T. V. Zenser, M. B. Mattammal and B. B. Davis, 'Demonstration of separate pathways for the metabolism of organic compounds in the rabbit kidney.' *J. Pharmacol. Exp. Ther.*, 1979, **208**, 418–421.

128. T. V. Zenser, M. B. Mattammal and B. B. Davis, 'Differential distribution of the mixed-function oxidase activities in rabbit kidney.' *J. Pharmacol. Exp. Ther.*, 1978, **207**, 719–725.

129. W. L. Smith and G. P. Wilkin, 'Immunochemistry of prostaglandin endoperoxide-forming cyclooxygenases: the detection of the cyclooxygenases in rat, rabbit and guinea-pig kidneys by immunofluorescence.' *Prostaglandins*, 1977, **13**, 873–892.

130. H. B. Mikkelsen, I. T. Rumessen and L. Thuneberg, 'Prostaglandin H synthase immunoreactivity localized by immunoperoxidase technique (PAP) in the small intestine and kidney of rabbit and guinea-pig.' *Histochemistry*, 1990, **93**, 363–367.

131. J. Mohandas, G. G. Duggin, J. S. Horvath *et al.*, 'Metabolic oxidation of acetaminophen (paracetamol) mediated by cytochrome P-450 mixed function oxidase and prostaglandin endoperoxide synthetase in rabbit kidney.' *Toxicol. Appl. Pharmacol.*, 1981, **61**, 252–259.

132. A. Freyberger and G. H. Degen, 'Prostaglandin-H synthase catalysed formation of reactive intermediates from stilbene and from steroid-estrogens: covalent binding to proteins.' *Arch. Toxicol. Suppl.*, 1989, **13**, 206–210.

133. G. H. Degen, G. Blaich and M. Metzler, 'Multiple pathways for the oxidative metabolism of estrogens in Syrian hamster and rabbit kidney.' *J. Biochem. Toxicol.*, 1990, **5**, 91–97.

134. W. Junge and K Kirsch, 'The carboxylesterases/amidases of mammalian liver and their possible significance,' *Crit. Rev. Toxicol.*, 1975, **3**, 371–435.

135. E. Heymann, in 'Enzymatic Basis of Detoxification,' ed. W. B. Jakoby, Academic Press, New York, 1980, pp. 291–323.

136. T. Satoh, 'The role of carboxylesterases in xenobiotic metabolism' *Rev. Biochem. Toxicol.*, 1987, **8**, 155–181.

137. R. E. Talcott, 'Hepatic and extrahepatic malathion carboxylesterases. Assay and localization in the rat.' *Toxicol. Appl. Pharmacol.*, 1979, **47**, 145–150.

138. S. Kaur and B. Ali, 'The effects of phenobarbital, 3-methylcholanthrene and benzo(*a* on the hydrolysis of xenobiotics in the rat.' *Biochem. Pharmacol.*, 1983, **32**, 3479–3480.

139. M. Raftell, K. Berzins and F.Blomberg, 'Immunochemical studies of a phenobarbital-inducible esterase in rat liver microsomes.' *Arch. Biochem. Biophys.*, 1977, **181**, 534–541.

140. M. B. A. Ashour, D. E. Moody and B. D. Hammock,' Apparent induction of microsomal carboxylesterase activities in tissues of clofibrate-fed mice and rats.' *Toxicol. Appl. Pharmacol.*, 1987, **89**, 361–369.

141. E. W. Morgan, B.Yan, D. Greenway *et al.*, 'Regulation of two rat liver microsomal carboxylesterase isoenzymes: species differences, tissue distribution, and the effects of age, sex and xeniobiotic treatments of rats.' *Arch. Biochem. Biophys.*, 1994, **315**, 513–526.

142. D. L. Ryan and T. R. Fukuto, 'The effect of isomalathion and *O,S,S*-trimethyl phosphorodithioate on the *in vivo* metabolism of malathion in rats.' *Pest. Biochem. Physiol.*, 1984, **21**, 349–357.

143. T. Tsujita and H. Okuda, 'Palmitoyl-coenzyme A hydrolyzing activity in rat kidney and its relationship of carboxylesterase.' *J. Lipid Res.* 1993, **34**,

1773–1781.

144. U. Nousianinen, R. Torronen and O. Hanninen, 'Differential induction of various carboxylesterases by certain polycyclic aromatic hydrocarbons in the rat.' *Toxicology*, 1984, **32**, 243–251.

145. J. Baumann, F. von Bruchhausen and G. Wurm, '*In vitro* deacetylation studies of acetamidophenolic compounds in rat brain, liver and kidney.' *Arzneimittelforschung*, 1984, **34**, 1278–1282.

146. B. Yan, D. Yang, M. Brady *et al.*, 'Rat kidney carboxylesterase. Cloning, sequencing, cellular localization, and relationships to rat liver hydrolase.' *J. Biol. Chem.*, 1994, **269**, 29688–29696.

147. B. Yan, D. Yang, M. Brady *et al.*, 'Rat testicular carboxylesterase: cloning, cellular localization, and relationship to liver hydrolase A.' *Arch. Biochem. Biophys.*, 1995, **316**, 899–908.

148. B. Yan, D. Yang and A Parkinson, 'Cloning and expression of hydrolase C, a member of the rat carboxylesterase family.' *Arch. Biochem. Biophys.*, 1995, **317**, 222–234.

149. E. H. Silver and S. D. Murphy, 'Potentiation of acryl-ate ester toxicity by prior treatment with the carboxyl-esterase inhibitor triorthotolyl phosphate (TOTP).' *Toxicol. Appl. Pharmacol.*, 1981, **57**, 208–219.

150. W. N. Aldridge, J. W. Miles, D. L. Mount *et al.*, ' The toxicological properties of impurities in malathion.' *Arch. Toxicol.*, 1979, **42**, 95–106.

151. B. Mannervik and U. H. Danielson, 'Glutathione transferases—structure and catalytic activity.' *Crit. Rev. Biochem.*, 1988, **23**, 283–337.

152. B. Ketterer, D. J. Meyer and A. G. Clark, in 'Glutathione Conjugation: Mechanisms and Biological Significance,' eds. H. Sies and B. Ketterer, Academic Press, London, 1988, pp. 73–135.

153. I. Listowsky, M. Abramovitz, H. Homma *et al.*, 'Intracellular binding and transport of hormones and xenobiotics by glutathione S-transferase.' *Drug Metab. Rev.*, 1988, **19**, 305–318.

154. M. Chang, J. R. Burgess, R. W. Scholz *et al.*, 'The induction of specific rat liver glutathione S-transferase subunits under inadequate selenium nutrition causes an increase in prostaglandin $F_{2\alpha}$ formation.' *J. Biol. Chem.*, 1990, **265**, 5418–5423.

155. D. J. Meyer, B. Coles, S. E. Pemble *et al.*, 'Theta, a new class of glutathione transferases purified from rat and man', *Biochem. J.*, 1991, **274**, 409–414.

156. R. Morgenstern and J. W. DePierre, in 'Glutathione Conjugation: Mechanisms and Biological Significance,' eds. H. Sies and B. Ketterer, Academic Press, London, 1988, pp. 157–174.

157. R. Morgenstern, G. Lundqvist, G. Andersson *et al.*, 'The distribution of microsomal glutathione transferase among different organelles, different organs and different organisms.' *Biochem. Pharmacol.*, 1984, **33**, 3609–3614.

158. C. Guthenberg, H. Jensson, L. Nystrom *et al.*, 'Isoenzymes of glutathione transferase in rat kidney cytosol.' *Biochem. J.*, 1985, **230**, 609–615.

159. G. M. Trakshel and M. D. Maines, 'Characterization of glutathione S-transferases in rat kidney. Alteration of composition by *cis*-platinum.' *Biochem. J.*, 1988, **252**, 127–136.

160. D. J. Meyer, E. Lador, B. Coles *et al.*, 'Single-step purification and h.p.l.c. analysis of glutathione transferase 8–8 in rat tissues.' *Biochem. J.*, 1989, **260**, 785–788.

161. J. A. Johnson, K. A. Finn and F. L. Siegel, 'Tissue distribution of enzymic methylation of glutathione S-transferase and its effects on catalytic activity. Methylation of glutathione S-transferase 11–11 inhibit conjugating activity towards 1-chloro-

162. 2,4-dinitrobenzene.' *Biochem. J.*, 1992, **282**, 279–289.

162. B. Rozell, H. A. Hansson, C. Guthenberg *et al.*, 'Glutathione transferases of classes α, μ and π show selective expression in different regions of rat kidney.' *Xenobiotica*, 1993, **23**, 835–849.

163. S. J. Davies, R. D'Sousa, H. Philips *et al.*, 'Localization of α, μ and π class glutathiones S-transferases in kidney: comparison with CuZn superoxide dismutase.' *Biochim. Biophys. Acta*, 1993, **1157**, 204–208.

164. T. D. Oberley, A. L. Friedman, R. Moser *et al.*, 'Effects of lead administration on developing rat kidney. II. Functional, morphologic, and immunohistochemical studies.' *Toxicol. Appl. Pharmacol.*, 1995, **131**, 94–107.

165. J. J. Boogaards, B. van Ommen and P. J. van Bladeren, 'Purification and characterization of eight glutathione S-transferase isoenzymes of hamster. Comparison of subunit composition of enzymes from liver, kidney, testis, pancreas and trachea.' *Biochem. J.*, 1992, **286**, 383–388.

166. T. D. Oberley, L. W. Oberley, A. F. Slattery *et al.*, 'Immunohistochemical localization of glutathione-S-transferase and glutathione peroxidase in adult Syrian hamster tissues and during kidney development.' *Am. J. Pathol.*, 1991, **139**, 355–369.

167. L. I. McLellan, D. J. Harrison and J. D. Hayes, 'Modulation of glutathione S-transferases and glutathione peroxidase by the anticarcinogen butylated hydroxyanisole in murine extrahepatic organs.' *Carcinogenesis*, 1992, **13**, 2255–2261.

168. L. H. Overby, S. Gardlik, R. M. Philpot *et al.*, 'Unique distribution profiles of glutathione S-transferases in regions of kidney, ureter, and bladder of rabbit.' *Lab. Invest.*, 1994, **70**, 468–478.

169. R. Moser, T. D. Oberley, D. A. Daggett *et al.*, 'Effects of lead administration on developing rat kidney. I. Glutathione S-transferase isoenzymes.' *Toxicol. Appl. Pharmacol.*, 1995, **131**, 85–93.

170. G. Clifton, N. Kaplowitz, J. D. Wallin *et al.*, 'Drug induction and sex differences of renal glutathione S-transferases in the rat.' *Biochem. J.*, 1975, **150**, 259–262.

171. G. Clifton and N. Kaplowitz, 'Effect of dietary phenobarbital, 3,4-benzo(α)pyrene and 3-methylcholanthrene on hepatic, intestinal and renal glutathione S-transferase activities in the rat.' *Biochem. Pharmacol.*, 1978, **27**, 1284–1287.

172. M. Derbel, T. Igarashi and T. Satoh, 'Differential induction of glutathione S-transferase subunits by phenobarbital, 3-methycholanthrene and ethoxyquin in rat liver and kidney.' *Biochim. Biophys. Acta*, 1993, **1158**, 175–180.

173. C. G. Hiley, M. Otter, J. Bell *et al.*, 'Immunocytochemical studies of the distribution of alpha and pi isoforms of glutathione S-transferase in cystic renal diseases.' *Pediatr. Pathol.*, 1994, **14**, 497–504.

174. E. A. Lock, V. Rodilla and G. M. Hawksworth, unpublished observations.

175. J. L. Stevens and D. P. Jones, in 'Glutathione: Chemical, Biochemical and Medical Aspects,' eds. D. Dolphin, R. Poulson and O. Avramovic, Wiley, New York, 1989, pp. 46–84.

176. L. H. Lash, in 'Mechanisms of Injury in Renal Disease and Toxicity', ed. R. S. Goldstein, CRC Press, Boca Raton, FL, 1994, pp. 207–234.

177. C. A. Hinchman and N. Ballatori, 'Glutathione conjugation and conversion to mercapturic acids can occur as an intrahepatic process.' *J. Toxicol. Environ. Health*, 1994, **41**, 387–409.

178. J. N. Commandeur, G. J. Stijntjes and N. P. Vermeulen, 'Enzymes and transport systems involved in the formation and disposition of glutathione S-conjugates. Role of bioactivation and detoxication mechanisms of

xenobiotics.' *Pharmacol. Rev.*, 1995, **47**, 271–330.

179. J. L. Stevens, J. D. Robbins and R. A. Byrd, 'A purified cysteine conjugate β-lyase from rat kidney cytosol. Requirement for an alpha-keto acid or an amino acidoxidase for activity and identity with soluble glutamine transaminase K.' *J. Biol. Chem.*, 1986, **261**, 15529–15537.

180. M. MacFarlane, J. R. Foster, G. G. Gibson *et al.*, 'Cysteine conjugate β-lyase of rat kidney cytosol: characterization, immunocytochemical localization and correlation with hexachlorobutadiene nephrotoxicity.' *Toxicol. Appl. Pharmacol.*, 1989, **98**, 185–197.

181. J. L. Stevens, 'Isolation and characterization of a rat liver enzyme with both cysteine conjugate β-lyase and kynureninase activity.' *J. Biol. Chem.*, 1985, **260**, 7945–7950.

182. L. D. Buckberry, I. S. Blagborough, B. W. Bycroft *et al.*, 'Kynurenine aminotransferase activity in human liver: identity with human hepatic C-S lyase activity and a physiological role for this enzyme.' *Toxicol. Lett.*, 1992, **60**, 241–246.

183. M. Mosca, L. Cozzi, J. Breton *et al.*, 'Molecular cloning of rat kynurenine aminotransferase: identity with glutamine transaminase K.' *FEBS Lett.*, 1994, **353**, 21–24.

184. J. L. Stevens, N. Ayoubi and J. D. Robbins, 'The role of mitochondrial matrix enzymes in the metabolism and toxicity of cysteine conjugates.' *J. Biol. Chem.*, 1987, **263**, 3395–3401.

185. A. Yamauchi, G. J. Stijntnes, J. N. Commandeur *et al.*, 'Purification of glutamine transaminase K cysteine conjugate β-lyase from rat renal cytosol based on hydrophobic interaction HPLC and gel permeation FPLC.' *Protein Exp. Purif.*, 1993, **4**, 552–562.

186. W. Dekant, M. W. Anders and T. J. Monks, in 'Renal Disposition and Nephrotoxicity of Xenobiotics,' eds. M. W. Anders, W. Dekant, D. Henschler *et al.*, Academic Press, San Diego, CA, 1993, pp.187–215.

187. T. W. Jones, C. Qiu, V. H. Schaeffer *et al.*, 'Immunohistochemical localization of glutamine transaminase K, a rat kidney cysteine conjugate β-lyase, and the relationship to the segment specificity of cysteine conjugate nephrotoxicity.' *Mol. Pharmacol.*, 1989, **34**, 621–627.

188. L. H. Lash, R. M. Nelson, R. A. Van Dyke *et al.*, 'Purification and characterization of human kidney cytosolic cysteine conjugate β-lyase activity.' *Drug Metab. Dispos.*, 1990, **18**, 50–54.

189. T. Green, J. Odum, J. A. Nash *et al.*, 'Perchloroethylene-induced rat kidney tumors: an investigation of the mechanisms involved and their relevance to humans.' *Toxicol. Appl. Pharmacol.*, 1990, **103**, 77–89.

190. R. I. McCarthy, E. A. Lock and G. M. Hawksworth, 'Cytosolic C-S lyase activity in human kidney samples—relevance for the nephrotoxicity of halogenated alkenes in man.' *Toxicol. Ind. Health*, 1994, **10**, 103–112.

191. S. J. Perry, M. A. Schofield, M. MacFarlane *et al.*, 'Isolation and expression of a cDNA coding for rat kidney cytosolic cysteine conjugate β-lyase.' *Mol. Pharmacol.*, 1993, **43**, 660–665.

192. D. Alberati-Giani, P. Malherbe, C. Kohler *et al.*, 'Cloning and characterization of a soluble kynurenine aminotransferase from rat brain: identity with kidney cysteine conjugate β-lyase.' *J. Neurochem.*, 1995, **64**, 1448–1455.

193. S. Perry, H. Harries, C. Scholfield *et al.*, 'Molecular cloning and expression of a cDNA for human kidney cysteine conjugate β-lyase.' *FEBS Lett.*, 1995, **360**, 277–280.

194. E. A. Barnsley, N. A. Eskin, S. P. James *et al.*, 'The

195. R. M. Green and J. C. Elce, 'Acetylation of S-substituted cysteines by a rat liver and kidney microsomal N-acetyltransferase.' *Biochem. J.*, 1975, **147**, 283–289.

196. M. W. Duffel and W. B. Jakoby, 'Cysteine S-conjugate N-acetyltransferase from rat kidney microsomes.' *Mol. Pharmacol.*, 1982, **21**, 444–448.

197. R. P. Hughey, B. B. Rankin, J. S. Elce *et al.*, 'Specificity of a particulate rat renal peptidase and its localization along with other enzymes of mercapturic acid synthesis.' *Arch. Biochem. Biophys.*, 1978, **186**, 211–217.

198. A. Heuner, W. Dekant, J. S. Schwegler *et al.*, 'Localization and capacity of the last step of mercapturic acid biosynthesis and the reabsorption and acetylation of cysteine S-conjugates in the rat kidney.' *Arch. Pflugers*, 1991, **417**, 523–527.

199. H. G. Bray and S. P. James, 'The formation of mercapturic acids. Deacetylation of mercapturic acids by the rabbit, rat and guinea pig.' *Biochem. J.*, 1960, **74**, 394–397.

200. Y. Endo, 'N-Acyl-L-aromatic amino acid deacylase in animal tissues.' *Biochim. Biophys. Acta*, 1978, **523**, 207–214.

201. S. Vamvakas, W. Dekant, K. Berthold *et al.*, 'Enzymatic transformation of mercapturic acids derived from halogenated alkenes to reactive and mutagenic intermediates.' *Biochem. Pharmacol.*, 1987, **36**, 2741–2748.

202. I. S. Pratt and E. A. Lock, 'Deacetylation and further metabolism of the mercapturic acid of hexachloro-1,3-butadiene by rat kidney cytosol *in vitro*.' *Arch. Toxicol.*, 1988, **62**, 341–345.

203. J. N. Commandeur, F. J. DeKanter and N. P. Vermeulen, 'Bioactivation of the cysteine S-conjugate and mercapturic acid of tetrafluoroethylene to acylating reactive intermediates in the rat: dependence of the activation and deactivation activities on acetyl coenzyme A availability.' *Mol. Pharmacol.*, 1989, **36**, 654–663.

204. G. J. Dutton, in 'Glucuronidation of Drugs and Other Compounds,' ed. G. J. Dutton, CRC Press, Boca Raton, FL, 1980, pp.149–158.

205. M. W. H. Coughtrie, 'Role of molecular biology in the structural and functional characterization of the UDP-glucuronosyltransferases.' *Prog. Drug Metab.*, 1992, **13**, 35–71.

206. B. Burchell and M. W. Coughtrie, 'UDP-glucuronosyltransferases.' *Pharmacol. Ther.*, 1989, **43**, 261–289.

207. B. Burchell, D. W. Nebert, D. R. Nelson *et al.*, 'The UDP-glucuronosyltransferase gene superfamily: suggested nomenclature based on evolutionary divergence.' *DNA Cell Biol.*, 1991, **10**, 487–494.

208. G. F. Rush and J. B. Hook. 'Characteristics of renal UDP-glucuronyltransferase.' *Life Sci.*, 1984, **35**, 145–153.

209. W. H. Peters and P. L. Jansen, 'Immunocharacterization of UDP-glucuronyltransferase isoenzymes in human liver, intestine and kidney.' *Biochem. Pharmacol*, 1988, **37**, 564–567.

210. L. Sutherland, T. Ebner and B. Burchell, 'The expression of UDP-glucuronosyltransferases of the UGT1 family in human liver and kidney, and in response to drugs.' *Biochem. Pharmacol.*, 1993, **45**, 295–301.

211. P. A. Munzel, M. Bruck and K. W. Bock, 'Tissue-specific constitutive and inducible expression of rat phenol UDP-glucuronosyltransferase.' *Biochem. Pharmacol.*, 1994, **47**, 1445–1448.

212. I. H. Stevenson and G. J. Dutton, 'Glucuronide

synthesis in kidney and gastrointestinal tract.' *Biochem. J.*, 1962, **82**, 330–340.

213. R. Hobkirk, R. N. Green, M. Nilsen *et al.*, 'Formation of estrogen glucosiduronates by human kidney homogenates.' *Can. J. Biochem.*, 1974, **52**, 9–14.

214. B. A. Fowler, G. E. Hook and G. W. Lucier, 'Tetrachlorodibenzo-*p*-dioxin induction of renal microsomal enzyme systems; ultrastructural effects on pars recta (S3) proximal tubular cells of the rat kidney.' *J. Pharmacol. Exp. Ther.*, 1977, **203**, 712–721.

215. J. T. Hjelle, G. A. Hazelton, C. D. Klaassen *et al.*, 'Glucuronidation and sulphation in rabbit kidney.' *J. Pharmacol. Exp. Ther.*, 1986, **236**, 150–156.

216. Q. Y. Yue, I. Odar-Cedarlof, J. O. Svensson *et al.*, 'Glucuronidation of morphine in human kidney microsomes.' *Pharmacol. Toxicol.*, 1988, **63**, 337–341.

217. J. R. Chowdhury, P. M. Novikoff, N. R. Chowdhury *et al.*, 'Distribution of UDP-glucuronosyltransferase in rat tissue.' *Proc. Natl. Acad. Sci.*, 1985, **82**, 2990–2994.

218. W. H. Peters, W. A. Allebes, P. L. Jansen *et al.*, 'Characterization and tissue specificity of a monoclonal antibody against human uridine 5'-diphosphate-glucuronosyltransferase.' *Gastroenterology*, 1987, **93**, 162–169.

219. R. Hume, M. W. Coughtrie and B. Burchell, 'Differential localization of UDP-glucuronosyltransferase in kidney during human embryonic and fetal development.' *Arch. Toxicol.*, 1995, **69**, 242–247.

220. J. B. Tarloff, R. S. Goldstein, R. S. Sozio *et al.*, 'Hepatic and renal conjugation (Phase II) enzyme activities in young adult, middle-aged, and senescent male Sprague-Dawley rats.' *Proc. Soc. Exp. Biol. Med.*, 1991, **197**, 297–303.

221. G. J. Mulder and W. B. Jakoby, in 'Conjugation Reactions in Drug Metabolism: An Integrated Approach: Substrates, Co-Substrates, Enzymes and their Interaction *In Vivo* and *In Vitro*,' ed. G. J. Mulder, Taylor and Francis, London, 1990, pp. 107–162.

222. K. Ogura, J. Kajita, H. Narhita *et al.*, 'Cloning and sequence analysis of a rat liver cDNA encoding hydroxysteroid sulfotransferase.' *Biochem. Biophys. Res. Commun.*, 1989, **165**, 168–174.

223. K. Ogura, J. Kajita, H. Narhita *et al.*, 'cDNA cloning of the hydroxysteroid sulfotransferase STa sharing a strong homology in amino acid sequence with the senescence marker protein SMP-2 in rat livers.' *Biochem. Biophys. Res. Commun.*, 1989, **166**, 1494–1500.

224. S. Ozawa, K. Nagata, D. W. Gong *et al.*, 'Nucleotide sequence of a full-length cDNA (PST-1) for aryl sulfotransferase from rat liver.' *Nucleic Acids Res.*, 1990, **18**, 4001.

225. W. F. Demyan, C. S. Song, D. S. Kim *et al.*, 'Estrogen sulfotransferase of the rat liver: complementary DNA cloning and age- and sex-specific regulation of messenger RNA.' *Mol. Endocrinol.*, 1992, **6**, 589–597.

226. R. Hume and M. W. Coughtrie, 'Phenolsulphotransferase: localization in kidney during human embryonic and fetal development.' *Histochem. J.*, 1994, **26**, 850–855.

227. S. Sharp, E. V. Barker, M. W. Coughtrie *et al.*, 'Immunochemical characterisation of a dehydroepiandrosterone sulfotransferase in rats and humans.' *Eur. J. Biochem.*, 1993, **211**, 539–548.

228. M. A. Runge-Morris, 'Sulfotransferase gene expression in rat hepatic and extrahepatic tissues.' *Chem. Biol. Interact.*, 1994, **92**, 67–76.

229. C. R. Parker, Jr., C. N. Falany, C. R. Stockard *et al.*, 'Immunohistochemical localization of dehydroepiandrosterone sulfotransferase in human fetal tissues.' *J. Clin. Endocrinol. Metab.*, 1994, **78**, 234–236.

230. G. J. Mulder, E. D. Kroese and J. H. N. Meerman, in 'Metabolism of Xenobiotics,' eds. J. W. Gorrod, H. Oelschlager and J. Caldwell, Taylor & Francis, London, 1988, pp. 243–250.

231. H. Okuda, H. Nojima, N. Watanabe *et al.*, 'Sulphotransferase-mediated activation of the carcinogen 5-hydroxymethyl-chrysene. Species and sex differences in tissue 5-hydroxymethyl-chrysene. Species and sex differences in tissue distribution of the enzyme activity and a possible participation of hydroxysteroid sulphotransferases.' *Biochem. Pharmacol.*, 1989, **38**, 3003–3009.

7.06
In Vivo Methodologies Used to Assess Renal Function

JOAN B. TARLOFF

Philadelphia College of Pharmacy and Science, PA, USA

and

LEWIS B. KINTER

Nycomed Inc., Wayne, PA, USA

7.06.1 INTRODUCTION

In the late nineteenth century, Claude Bernard, the French physician and physiologist, noted that higher animals "have really two environments: an external environment (milieu exterieur) in which the organism is situated, and an internal environment (milieu interieur) in which the tissue elements live."[1] Bernard recognized that while the external environment was variable and beyond the control of vital mechanisms, "all the vital mechanisms, however varied they may be, have only one object, that of preserving the conditions of life in the internal environment." Since the 1940s our knowledge of the pivotal role of the kidneys in the regulation of the volume and composition of Bernard's "milieu interieur" has increased to include molecular descriptions of the genetic control of the mechanisms involved. Coincident with elucidation of the mechanisms of renal functions since the mid-1970s, an increasingly detailed understanding and appreciation of the mechanisms of chemical-induced renal toxicity has emerged.

The susceptibility of the mammalian kidney to the adverse effects of toxicants can be attributed largely to the anatomic, physiologic, and biochemical features of the kidney. The kidneys receive a disproportionately large blood flow to support glomerular filtration and sustain renal metabolism. However, high blood flow to the kidneys results in greater exposure to blood-borne chemicals than occurs in other less-well perfused organs. During the process of solute and water reabsorption along the nephron, the tubular fluid becomes enriched in those chemicals that remain in the tubular fluid, exposing the cells of the nephron to potentially high concentrations of toxicants. Chemicals with low aqueous solubilities may precipitate or crystallize in the tubular fluid, causing obstruction of the nephron. Chemicals that are substrates for organic solute transport systems may accumulate within and damage the cells of the proximal tubule. Intrarenal metabolism of nontoxic chemicals can produce toxic metabolites that disrupt essential cellular functions, resulting in damage to renal structures and functions.

The results of toxic injury to the kidneys may be varied, depending on the particular nephron segment or function altered by the injury. The kidneys serve a major role in a number of mechanisms to maintain the integrity of Bernard's "milieu interieur." For example, the kidneys participate in the regulation of extracellular fluid volume and osmolarity through the action of antidiuretic hormone on the collecting duct. By this mechanism, the kidneys reabsorb water in the absence of solutes, allowing excretion of urine that is more concentrated than plasma. Water is retained in extracellular fluid, thereby compensating for dehydration and decreasing the osmolarity of body fluids. A toxicant may alter the ability of the kidneys to respond to antidiuretic hormone, causing the excretion of large amounts of dilute urine and interfering with the ability to maintain sodium and water balance. The kidneys participate in the maintenance of the acid–base balance by reabsorbing filtered sodium bicarbonate and secreting hydrogen ions formed during intermediary metabolism, and a toxicant may adversely affect the ability of the kidneys to maintain pH control. The kidneys are the primary mechanism by which excess salt and water are eliminated, preventing persistent

elevations of arterial blood pressure. A consequence of toxicant-induced renal injury may be loss of the ability to excrete excess salt and water, thereby interfering with normal blood pressure control.

Finally, one consequence of the kidneys' role in the maintenance of Bernard's "milieu interieur" is that the microenvironments of the renal cells, especially those of the medulla, are highly variable and can change rapidly with regard to both composition and tonicity, depending upon changes in the external environment and intake of solutes and water. It is notable that most of the cells within the "milieu interieur" could not survive for long in the variable milieu of the kidney. Erythrocytes are observed to alternatively shrink (crenate) and swell as they pass through the renal medulla, surviving only by the flexibility of their structure and rapidity with which they equilibrate their internal volume. The stresses of this environment are seen in the extreme in small desert rodents whose urine osmolality may exceed 9 M, including concentrations of salts and urea approaching 5 M that can denature proteins. It follows that the mechanisms by which the cells of the renal medulla regulate their internal environment are different from those of cells bathed in the "milieu interieur" maintain their homeostasis. The significance of these observations extends beyond their implications for cellular biochemistry. Any technique that removes renal tissues from their as yet poorly defined *in vivo* environment results in the study of those tissues in a necessarily arbitrary environment whose composition is only tenuously related to reality, at best. The implications can be profound. In a study of peptide antagonists of adenylate-cyclase coupled vasopressin receptors (V2 receptors), evaluations of V2 receptor functions in kidney homogenates, isolated perfused renal tubules, and semipurified renal membranes (including human tissues) failed to detect V2 agonist activity of these compounds, which caused these compounds to fail in clinical trials.[2] Perhaps Homer Smith foresaw this concept when he commented to colleagues that in his opinion (paraphrased), "when one utilized a preparation that allowed the investigator to see the kidney or renal tissues, it was likely that the kidney/kidney tissue was not functioning normally."

Clearly, the kidney participates in a variety of regulatory mechanisms and toxic injury may manifest as changes in one or more of those functions. Alternately, changes in renal function are not always indicative of toxicity. For example, prolonged ingestion of nonsteroidal anti-inflammatory drugs (NSAIDs) may lead to destruction of the renal papilla, the site of maximal urinary concentrating ability. One manifestation of NSAID-induced renal injury is persistent diuresis and an inability to form concentrated urine. However, the same effects, that of diuresis and an inability to form concentrated urine, are the desired consequences of diuretic therapy. Thus, not all changes in renal function are associated with toxicity. In addition, the clinical manifestation of renal injury may yield little or no information concerning the mechanism of that injury. With diuretic therapy, the change in renal function is due to an alteration in nephron function so that less fluid and solutes are reabsorbed. We might be tempted to predict that NSAID-induced renal injury is related to interference with solute and water reabsorption since the clinical symptoms resemble those seen with diuretic therapy. However, in NSAID-induced renal disease, the observed changes in renal function are due to destruction of the portion of the kidney responsible for forming concentrated urine and are distinctly different from the mechanism of action of most diuretic agents.

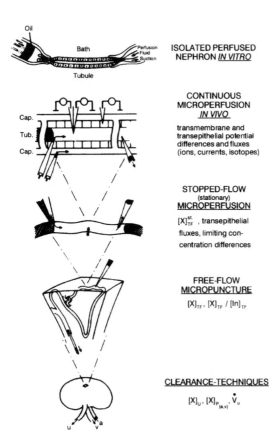

Figure 1 An overview of methodologies used to assess renal function—from clearance studies to isolated tubules and cells (K. Hierholzer, unpublished, courtesy of G. Giebisch).

Techniques used to investigate the biology of the kidney span simple diagnostic procedures used in clinical and veterinary medicine to the cloning, expression, and translation of nucleic acid sequences harvested from renal cells (Figure 1). In this chapter, we briefly review the techniques used to evaluate kidney function in intact organisms and give several examples of studies in which those techniques were employed.

7.06.2 OVERVIEW OF *IN VIVO* KIDNEY METHODOLOGIES

In vivo methodologies range from collection of blood and urine to analysis of the contents of a single tubular segment. In general, *in vivo* techniques permit the collection of large amounts of data but are limited in the ability to define biochemical mechanisms of injury. However, *in vivo* techniques are indispensable to demonstrate that the kidney is a site of injury for a particular drug or chemical and to begin to identify the site of the injury. In addition, *in vivo* techniques have the advantage of yielding tissue for morphological analysis following administration of the drug or chemical to an experimental animal.

7.06.2.1 Clinical Biochemistry of Blood and Urine

In toxicology studies, urinalysis may be conducted on spontaneously voided urine specimens collected from test animals. Rodents often spontaneously void urine when handled, and can be induced to void into a funnel or metabolism cage by gentle abdominal massage. Rabbits, dogs, and primates may be housed in metabolism cages for collection of spontaneously voided urine. Dogs and nonhuman primates may also be briefly catheterized to collect bladder urine under sterile conditions. Quantitative urine collection methods are used to more precisely document treatment-related changes in renal excretory functions and to define renal clearances. Quantitative urine collection for renal excretory function differs from urinalysis of spontaneously voided urine only in that quantities of urine are collected over specific intervals. To assist in complete collection of voided urine, animals are often housed in metabolism cages.

A metabolism cage consists of an animal chamber mounted above an excrement collection system. The animal chamber must be equipped with feeder and waterer units if an animal is to be housed in the metabolism cage for more than a few hours. The collection system usually consists of a funnel and a urine/feces separator. Many design variations of metabolism cages are available and one example is shown in Figure 2. Important considerations in the selection of a metabolism cage include:

(i) *Animal chamber*. The chamber size must meet floor space and size as specified by the USDA (US Department of Agriculture) (Animal Welfare Act). Feeder and waterer systems must be easy for animals to learn to use and must have provisions to prevent spills or leakage from contaminating urine and fecal matter and vice versa.

(ii) *Urine/feces collection system*. The collection system is generally a funnel. Systems with steep funnels with minimal surface area enhance collection efficiency and reduce residue formation.

(iii) *Urine/feces separator*. The greatest variations in metabolism cages are in these systems. The simplest urine/feces separators are screens placed in the funnel that retain the feces while allowing the urine to pass through. The principal drawback of these simple systems is that urine may contact or "percolate" through feces. These systems are often sufficient for renal excretory studies, but not for studies using radioactive tracers. More complicated systems act to divert urine and feces into separate collection vessels. These systems rely on urine adhering to and running along the surface of the funnel, while (dry) feces "ricochet" off the funnel into a collection pan or vessel. These systems are sufficient for both renal excretory and pharmacokinetic/biodistribution studies. However, the urine/feces separator efficiency may be defeated by high rates of urine flow or watery or mucoid stools.

Whatever type of metabolism cage is used, the following general precautions are offered:

(i) *Animal husbandry*. For studies of >24 h duration, test animals should be acclimatized to the metabolism cage for several days prior to study initiation. During this period, animals should be monitored frequently to insure that they learn to use the feeder and waterer systems properly. Test animals should be maintaining or gaining weight prior to study initiation. Feeder and waterer systems should provide sufficient food and water to meet the animal's needs and all separator systems should function properly. For chronic studies (>5–7 d duration) it is useful to have a complete exchange of feeder, waterer, and urine/feces collector and separator systems so that soiled units may be rapidly exchanged with clean, dry/filled units at regular intervals.

(ii) *Cages*. Cages should be decontaminated, cleaned, rinsed with distilled/deionized water, and thoroughly dried prior to use. Surfaces used to collect urine may be siliconized or sprayed with a suitable hydrophobic material (PAM, American Home Foods) to facilitate urine collection.

(iii) *Urine and feces collection*. Urine may be collected under mineral oil to prevent evaporative losses. For very small or antidiuretic animals (e.g., hamsters, gerbils, desert mice) placing the cage over a shallow pan of oil and skimming feces from the surface may be necessary while collecting urine with a pipette from under the oil. All surfaces contacting urine should be rinsed with distilled/deionized water or appropriate solvents to collect any residuals at appropriate intervals.

Figure 2 A NALGENE metabolic cage (650 series). These cages are useful for quantitative urine collection and are easily assembled. The unit is constructed to allow the animal to reach food and water. Units may be used singly or placed on a rack that holds about a dozen units. An attractive feature of the NALGENE unit is the ability to collect both urine and feces.

In urinalysis, some parameters of principal interest are shown in Table 1. Urine osmolality (or specific gravity), and pH (hydrogen ion concentration) are basic parameters that provide an overview of renal function. Urine volume may be included if urine has been collected over a specified interval. The principal

Table 1 Parameters of major interest in urinalysis.

Urine osmolarity or specific gravity
Urinary pH
Urine volume
Urinary electrolyte concentrations
  sodium
  potassium
  chloride
  urea
Creatinine excretion
Glucose excretion
Aminoaciduria
Proteinuria
  low molecular weight proteinuria
    β_2-microglobulin
    α_1-microglobulin
    retinol-binding protein (α_2-microglobulin)
    albuminuria

urinary solutes are the electrolytes, sodium, potassium, and chloride, and the nonelectrolyte, urea. Excretion of other monovalent electrolytes (ammonium, bicarbonate) and divalent electrolytes (calcium, magnesium, phosphate, sulfate) generally accounts for only a small fraction of the total osmotically active solute. Creatinine, a by-product of protein metabolism, is a useful urinary solute as its excretion in urine is almost entirely by glomerular filtration. Carbohydrates and amino acids are normally completely removed from the urine by reabsorptive processes along the nephron. Similarly, plasma proteins are nearly completely excluded from urine by the filtration process, and the small (but not insignificant) amounts that do enter the nephron are generally scavenged by reabsorptive processes in the proximal tubule. Additional proteins (e.g., Tamm–Horsfall glycoprotein, γ-glutamyl transpeptidase) are added to the urine from renal sources. Finally, hormones, second messengers, cytokines, autocoids, nucleotides, drugs, and metabolites may all be excreted in varying quantities in urine.

The principal feature of steady-state urinary solute and water excretion is that it represents the balance between solute and water intake and metabolism. In short, whatever is left over is excreted in the urine. Hence, interpretation of urinary excretory data must include consideration of intake and the metabolic status of the animal. In addition, the quality and consistency of the environment must be known and considered, as environmental temperature and humidity directly influence insensible water and solute losses (respiration, fecal, evaporation, sweat). Finally, stress factors, including indivi-

dual or gang housing, appropriateness of lighting cycle, and quality of caging and husbandry practices must be considered. The kidney can neither excrete distilled water nor precipitate solids; water is required to excrete excess solutes, and additional solute is required to excrete water. Water and solute excretion are linked through the function of the urine concentrating mechanism.

From a practical perspective, it is useful to apply a tiered strategy in analyses of urine to gain different levels of perspective on kidney function (Figure 3). Urine osmolality, urine volume, and urinary pH provide basic insight into overall fluid homeostatic and acid–base state. These parameters are highly dependent upon intakes and metabolic state and treatment-related changes are most easily detected by comparisons with untreated controls or appropriate historical controls. Urine volume reflects the excess between ingested water and water produced as a by-product of metabolism and insensible water losses. Urine osmolality is the ratio of excreted osmotically active solute (mosmoles) to excreted water (kg). This parameter reflects the urine concentrating ability, and is susceptible to changes in either water or solute excretion. Urine pH is an indicator of

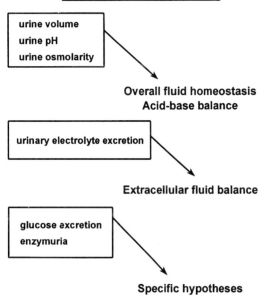

Figure 3 Schematic representation of steps involved in urinalysis. As an initial approach, urine volume, pH, and osmolarity provide an overview of kidney function. Second, electrolyte excretion provides insight into extracellular fluid balance. Third, excretion of specific markers such as glucose, amino acids, or site-specific proteins, may assist in identification of sites of injury.

relative total body acid (or base) load. Animals maintained (or studied) in a controlled environment will exhibit urine osmolality and pH values within characteristic ranges, unless a disturbance has taken place.

Second, renal excretion of sodium, potassium, chloride, and urea accounts for most of the total urinary solutes. The sum of sodium and potassium concentrations ($mEq\,L^{-1}$) or excretions ($mEq\,time^{-1}$) should almost equal chloride ($mEq\,L^{-1}$ or $mEq\,time^{-1}$):

$$sodium + potassium \approx chloride \qquad (1)$$

Additionally, the sum of the electrolytes plus urea should approximately equal urine osmolality:

$$sodium + potassium + chloride + urea$$
$$\approx urine\ osmolality \qquad (2)$$

Deviations represent electrolyte and nonelectrolyte "gaps" and suggest the presence of unusual solutes in urine. Endogenous creatinine is a useful index of the glomerular filtration rate and can be used to calculate the fractional excretion of other solutes with respect to glomerular and tubular components. Expressing urinary solutes (e.g., Na^+, K^+, Cl^-) per mg (or mmol) creatinine accounts for changes related to increases or decreases in filtered loads of individual solutes.

Third, renal excretion of divalent cations (such as Mg^{2+}, PO_4^{2-}, and Ca^{2+}), other electrolytes and nonelectrolytes, and specific sugars, amino acids, proteins, hormones, vitamins, and other chemicals may be determined. In general, these analyses are only conducted to test specific hypotheses when discrepancies have been established based upon the prior tiers of testing or when testing for stone formation (urolithiasis).

Several investigators have used high resolution ${}^1H$ NMR spectroscopy to analyze low molecular weight compounds present in urine.[3–5] Both quantitative and qualitative analyses are possible using NMR spectroscopy. For example, Bales *et al.* compared concentrations of urinary creatinine determined by NMR and the traditional Jaffe reaction.[4] There was excellent agreement between the two methods (creatinine concentration of $9.98 \pm 0.36\,mmol\,L^{-1}$ determined by NMR vs. $10.2 \pm 0.1\,mmol\,L^{-1}$ determined colorimetrically) in samples of human urine, suggesting that ${}^1H$ NMR spectroscopy can be used for quantitative urinalysis.[4] Gartland *et al.* used qualitative ${}^1H$ NMR spectroscopy to assess urinary excretion of low molecular weight

compounds following administration of chemicals that selectively injure the proximal tubule or collecting duct.[3] Sodium chromate, mercuric chloride, hexachlorobutadiene, and cisplatin injure the proximal tubule.[6–9] These compounds increased urinary excretion of glucose, acetate, amino acids (such as alanine, lysine, and valine), lactic acid, and acetate.[3] In contrast, compounds that selectively injure the renal papilla, such as propyleneimine[10] and 2-bromoethylamine hydrobromide,[11] increased urinary excretion of trimethylamine *N*-oxide, dimethylamine, acetate, succinate, lactate, and alanine.[3] Although more research is needed to validate the hypothesis, it is possible that site-specific injury to the nephron may be identified by ${}^1H$ NMR spectroscopic urinalysis.

Urinalysis is an invaluable tool for obtaining an initial assessment of the effect of a drug or chemical on kidney function. However, relatively little specific information concerning a site of injury or mechanism of toxicity can be gained through urinalysis. Increased excretion of certain substances, such as glucose or amino acids, may suggest that the proximal tubule is the site of injury since these substances are reabsorbed by the proximal tubule.[12] Excretion of specific proteins in urine can be measured by enzymatic analysis or by subjecting urine to polyacrylamide gel electrophoresis. Proteinuria may be due to glomerular injury, allowing high molecular weight proteins, such as albumin, to be filtered, or to tissue injury, causing cells to release proteins for excretion in urine.[13] By analyzing urine for specific protein content, a site of injury may be postulated. For example, γ-glutamyl transpeptidase and alanine aminopeptidase are proteins localized in the brush border of the proximal tubule.[14,15] Identification of site-specific proteins, such as those listed in Table 2, in urine may suggest that a chemical is producing tubular injury. For example, hydroquinone causes increased excretion of protein originating from the proximal tubule as well as glucose and creatinine, suggesting that the proximal tubule is the site of injury.[16] Identification of albuminuria suggests that glomerular damage may have occurred. For example, puromycin aminonucleoside causes proteinuria and increases glomerular permeability. Electrophoretic analysis of urinary protein following administration of puromycin aminonucleoside to rats indicates that the excreted proteins are largely albumin.[17,18]

Blood samples for analysis may be collected from animals at appropriate intervals for calculation of renal clearances. Blood may also be analyzed to provide information on kidney function. Plasma creatinine and blood urea nitrogen (BUN) concentrations reflect

Table 2　Localization of specific enzyme markers in the kidney.

Enzyme	Localization	Ref.
Glutathione *S*-transferase-α	proximal tubule	19
Glutathione *S*-transferase-π	distal convoluted tubule	
	loop of Henle (thin limb)	
	collecting duct	19
Alkaline phosphatase	proximal tubule > distal tubule	20
N-Acetyl-β-glucosaminidase	papillary collecting duct	21
	proximal tubule > distal tubule	
Alanine aminopeptidase	proximal tubule	20
γ-Glutamyl transpeptidase	proximal tubule (brush border)	14
β-Galactosidase	proximal tubule > distal tubule	20
Tamm–Horsfall glycoprotein	thick ascending limb	22,23
Cathepsin B	proximal tubule (cytosol)	24
Lactate dehydrogenase	distal tubule > proximal tubule	20

directly normal and abnormal kidney functions. Creatinine and urea are both by-products of nitrogen metabolism and are excreted primarily by renal filtration. In the absence of changes in protein intake or metabolism, increases in plasma creatinine and BUN concentrations reflect decreases in glomerular filtration rate. However, plasma creatinine and BUN concentrations are relatively insensitive indices of changes in renal function. Approximately 50% of renal function must be destroyed before serum creatinine or BUN concentrations rise (Figure 4).

7.06.2.2　Renal Hemodynamics and Blood Flow Probes

The kidneys receive about 20% of the cardiac output at rest.[25] This high renal blood flow is essential in allowing glomerular filtration and

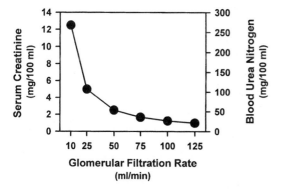

Figure 4　Relationship between glomerular filtration rate, serum creatinine, and blood urea nitrogen concentrations. In general, about 50% of renal function must be lost before serum creatinine or blood urea nitrogen concentrations rise appreciably.

subsequent reabsorption of the bulk of the filtered fluid. When renal perfusion is decreased, as by a drug or chemical, glomerular filtration is also reduced, triggering a series of events to allow the kidneys to conserve sodium and water. Therefore, the kidneys excrete small amounts of highly concentrated urine containing little sodium. Drugs or chemicals may alter renal blood flow either directly, by causing vasoconstriction of the renal artery or arterioles, or indirectly, by releasing one or more of the many regulators of renal blood flow. For example, prostaglandins, such as PGE_2, are vasodilators while thromboxanes are vasoconstrictors. Drugs that inhibit prostaglandin synthesis, such as nonsteroidal antiinflammatory agents, have the potential to decrease renal blood flow by removing one vasodilatory influence.

Prolonged reductions in renal blood flow may have deleterious effects on the kidneys by causing hypoxia. Since the kidneys are highly metabolic organs, hypoxia or anoxia causes tissue injury, particularly of the proximal tubule and thick ascending limb of the loop of Henle. To determine if tissue injury is a direct effect of a drug or chemical or an indirect effect due to reductions of renal blood flow, measuring renal blood flow is necessary. A variety of techniques are available for analysis of renal blood flow.

Perhaps the simplest technique to measure renal blood flow is to measure the renal clearance of *para*-aminohippurate (PAH), as described below. There are several drawbacks to using PAH clearance as a measure of renal blood flow. First, only 90% of PAH is actually cleared by the kidneys, so that PAH clearance underestimates renal blood flow by about 10%. Second, determination of PAH clearance requires continuous infusion of PAH and timed collections of blood and urine, so that experimental animals must be surgically prepared and the risk of infection or blockage of a cannula is a possibility. Finally, PAH clearance approaches renal blood flow because PAH is avidly secreted by the proximal tubules. Therefore, proximal tubule damage will probably decrease PAH clearance in the absence of any changes in renal blood flow, making interpretation of data difficult or ambiguous.

Several other methods are available for determination of renal blood flow, although these alternative methods are more invasive than is determination of PAH clearance. However, the alternative methods have the advantage of allowing unambiguous interpretation of renal blood flow in the presence of a toxicant that might reduce organic anion or cation secretion. Thus, flow probes (electromagnetic, Doppler) may be positioned around a renal artery to allow direct measurement of renal blood flow.[26,27] Detection of blood flow using Doppler systems is based on changes in the emitted ultrasonic frequency, a Doppler shift, caused by reflection of the signal off moving blood cells. The Doppler shift is proportional to the velocity of blood flow. Another method involves injection of radiolabeled microspheres with subsequent determination of the amount of radioactivity present in kidney tissue.[28] The theory used with microspheres is that the size of the sphere determines where along the vascular tree the sphere will impact. Spheres are milled to a diameter that allows impaction in the glomerular capillaries, thereby enabling measurement of effective renal blood flow, that is, the amount of blood that actually reaches the glomeruli and is therefore available for filtration.[28]

When renal blood flow was measured in dogs using either electromagnetic flow probes or radioactive microspheres, excellent correlation ($r = 0.96$) was obtained between the two methods, as shown in Figure 5.[29] Similarly, renal blood flow determined by electromagnetic flowmeter was highly correlated with flow determined by chromium-51 labeled ethylene diaminetetraacetic acid (^{51}Cr-EDTA) clearance over a wide range of renal blood flows in dogs.[29] In contrast to the excellent correlation observed with electromagnetic flow probes, radioactive microspheres, or ^{51}Cr-EDTA clearance, renal blood flow measured by PAH clearance was not as well-correlated. For example, in control dogs the difference between renal blood flow determined by flowmeter vs. PAH clearance averaged 19%. Following cisplatin administration, the difference between the two methods was even greater (57%).[29] Following cisplatin administration, PAH clearance decreased from 112 ±

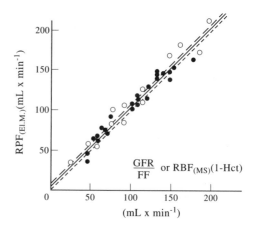

Figure 5 Comparison between the values for renal plasma flow measured by electromagnetic flowmeter (●) and values calculated from the clearance and extraction fraction of ^{51}Cr-EDTA (○——○), or values calculated from microsphere distribution and hematocrit (○ – – – ○). The dotted line is the line of identity. Abbreviations: RPF = renal plasma flow; El.M. = electromagnetic flowmeter; GFR = glomerular filtration rate; FF = filtration fraction (GFR/RPF); Hct = hematocrit (reproduced by permission of the Danish Medical Bulletin from *Dan. Med. Bull.*, 1990, **37**, 1–12).

4 mL min^{-1} in control dogs to 37 ± 5 mL min^{-1} within 72 h, while the extraction of PAH fell from 0.73 ± 0.03 to 0.42 ± 0.04 mL min^{-1}.[29] Since cisplatin damages the proximal tubule, changes in PAH clearance in these experiments might be due to cisplatin-induced alterations in PAH secretion rather than changes in renal blood flow.

7.06.2.3   Renal Clearance Techniques

Clearance methods may be useful in determining the effects of toxicants on glomerular filtration rate and renal blood flow. In addition, clearance methods are used to determine the mechanism(s) involved in the renal excretion of a toxicant. Simply defined, the renal clearance of a compound is the volume of plasma from which that compound is completely cleared by the kidneys per unit time.[25]

Renal clearance (*Cl*) of any compound (X) can be determined by comparing the urinary excretion rate of compound X to the plasma concentration of compound X. The urinary excretion rate is calculated as:

urinary excretion rate (mg min^{-1})

$$= U_X \, (\text{mg mL}^{-1}) \times V \, (\text{mL min}^{-1}) \qquad (3)$$

where U_X represents the concentration of substance X in urine (in mg mL^{-1}) and V

represents the volume of urine collected per unit time (in mL min^{-1}). Thus, the clearance equation may be constructed:

$$Cl_X \text{ (mL min}^{-1})$$
$$= \frac{U_X \text{ (mg mL}^{-1}) \times V \text{ (mL min}^{-1})}{P_X \text{ (mg mL}^{-1})} \quad (4)$$

where P_X is the concentration of compound X in plasma (in mg mL^{-1}). More comprehensive reviews of the theories underlying clearance methods and assumptions involved in calculating renal clearances may be found in Vander[30] and Valtin.[31]

The preferred indicator for the estimation of glomerular filtration rate (GFR) is inulin, a fructose polysaccharide (MW ~ 5200). Since inulin is freely filtered and then is neither secreted nor reabsorbed, all of the inulin that appears in urine must have originated from plasma through the process of glomerular filtration. Measurement of inulin clearance serves as the "gold standard" method for estimating GFR.[32] Other indicators suitable for measuring GFR include creatinine, isotopes of vitamin B$_{12}$, sodium iodothalamate, and metal chelates of EDTA and diethylaminotriaminepentaacetic acid (DPTA).[33] Of these only creatinine requires no infusion of an exogenous substance and is particularly useful when one wishes to monitor GFR over a protracted period of time (days to decades) in the same individual. Creatinine is reabsorbed and its clearance tends to underestimate the inulin clearance in some species by 10% or more.

The indicator most commonly used for estimation of renal plasma flow (RPF) is PAH, which is both freely filtered by the glomeruli and actively secreted by the organic acid transport pathway of the proximal tubule. PAH extraction by the kidneys varies from about 70% to 90% in rats, dogs, and humans.[34] One reason that PAH extraction is incomplete is that blood flow to the inner cortical glomeruli is directed to the medulla and does not perfuse proximal tubule segments. Heterogeneity in the distributions of organic acid transporters in the cortex may also contribute to less than complete extraction of PAH. Therefore, RPF estimated using PAH clearance is often designated effective renal plasma flow (ERPF). As renal venous plasma is difficult to obtain, PAH extraction is often assumed to be complete. Using these assumptions, RPF is slightly underestimated. RPF is converted to renal blood flow (RBF) by dividing RPF by the plasma fraction of whole blood, as estimated from the hematocrit (Hct):

$$RBF = RPF/(1 - Hct) \quad (5)$$

The clearances of other compounds can be compared with inulin clearances to determine how the kidney functions in the elimination of the test compound. Typically, the renal clearances of a compound of interest (X) and inulin are determined. From these data, a clearance ratio is constructed by dividing the renal clearance of the test compound (X) by the renal clearance of inulin:

$$\text{Clearance ratio}$$
$$= \frac{\text{renal clearance of test compound}}{(Cl_X, \text{ mL min}^{-1})}{\text{renal clearance of inulin } (Cl_{\text{inulin}}, \text{ mL min}^{-1})}$$
$$(6)$$

For example, if a substance has a clearance less than that of simultaneously administered inulin, then some test substance must be reabsorbed following filtration to account for the lower clearance relative to inulin. Glucose is an example of a compound that is reabsorbed and the clearance ratio for glucose ($Cl_{\text{glucose}}/Cl_{\text{inulin}}$) is always less than unity (Figure 6). Alternately, if a test substance has a clearance greater than simultaneously administered inulin, then the test compound must be filtered and secreted by the renal tubules to account for the higher clearance relative to inulin. PAH is

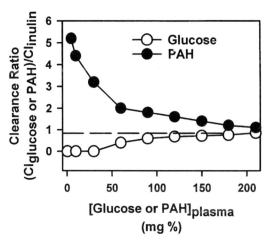

Figure 6 Interpretation of renal clearance data. Clearances of compounds such as glucose (○) or PAH (●) are determined in comparison with inulin clearance. The dashed line indicates unity, the value expected when a substance is freely filtered and neither reabsorbed nor secreted. In contrast, glucose is extensively reabsorbed, as indicated by clearance ratios less than unity, while PAH is extensively secreted, as indicated by clearance ratios of greater than unity. The clearance of both glucose and PAH approaches that of inulin when plasma concentrations of each compound are elevated, indicating saturation of transport pathways.

an example of a compound that is secreted by the renal tubules and the clearance ratio for PAH (Cl_{PAH}/Cl_{inulin}) is always greater than unity (Figure 6).

7.06.2.4   *In Vivo* Micropuncture

Micropuncture, while used infrequently in toxicology, is an important method for determining the effects of drugs or chemicals on single nephron function in the intact kidney. In addition, micropuncture can provide information concerning tubular handling of a drug or chemical. Much of our knowledge of glomerular filtration comes from micropuncture studies done in the early 1920s by Richards and co-workers.[35] In the 1940s, micropuncture techniques were extended to mammalian kidneys.[36,37]

Micropuncture is a highly specialized technique that requires considerable equipment and experience to be utilized properly. Studies may be done in small rodents and dogs, and studies in various strains of rats are most common. Munich-Wistar rats are frequently used for micropuncture studies because the glomeruli are visible and accessible at the surface of the kidney.[38] The basic experimental design in micropuncture is to identify the specific nephron segment of interest and insert a micropipette into the lumen of that segment. A column of castor oil is initially injected into the tubular lumen to prevent collection of fluid downstream from the micropuncture site and to allow complete collection of tubular fluid that reaches the micropipette.[38] Nephron segments are identified by using a dye such as lissamine green.[38] Experiments are done under microscopic control to visualize the micropipettes and tubule segments of interest.

A particularly useful aspect of micropuncture relates to the ability to collect glomerular filtrate from a single nephron by positioning the micropipette in a glomerulus rather than in a tubular lumen. Using glomerular micropuncture, the ultrafiltration coefficient of a substance may be calculated based on comparison of the tubular fluid/plasma concentration ratio of the substance in question with the tubular fluid/plasma concentration ratio of a standard, such as inulin. For a substance that is freely filtered, this ratio should be approximately unity. Substances that are protein-bound or in some other manner restricted from filtration will have a ratio less than unity. Using this technique, Burnatowska-Hledin *et al.* determined that, at plasma aluminum concentrations ranging from 2–10 mg L^{-1}, 2–10% of plasma aluminum was filtered at the glomerulus in Munich-Wistar rats.[39] In contrast, at plasma concentrations ranging from 0.01 to 23.5 mg L^{-1}, ultrafilterable aluminum *in vitro* ranged from 2 to 108%, regardless of the type of filtration cone used.[39] The high variability observed *in vitro* suggested that the membranes used to determine ultrafiltration retained aluminum and that *in vitro* methods were unreliable for determination of ultrafilterable aluminum concentrations.[39] Using their more reliable estimates of 2–10%, these workers determined that the fractional excretion of aluminum averaged $\sim$40%, suggesting that $\sim$60% of filtered aluminum was reabsorbed prior to excretion.

Micropuncture techniques were used to characterize the renal handling of cadmium and the cadmium chelate, cadmium-pentaacetic acid (Cd-DTPA).[40] The kidneys accumulate cadmium, as well as other divalent cations, but the mechanism of metal accumulation is unclear and could involve luminal reabsorption, basolateral uptake, or a combination of both pathways. Cadmium uptake across the luminal membrane of the proximal tubules was assessed directly using micropuncture. Cadmium acetate was infused in Munich-Wistar rats to an average plasma concentration of $\sim$10 μM. Under these conditions, the ultrafilterable fraction of plasma cadmium (calculated as the glomerular filtrate/plasma ratio, UF/P) was 0.20 ± 0.01 while the tubular fluid/plasma concentration ratio (TF/P) for cadmium at the late proximal tubule was 0.10 ± 0.01.[40] In comparison, the UF/P ratio for inulin was 1.07 ± 0.01 while the TF/P ratio for inulin was 1.57 ± 0.28 at the end of the proximal tubule. Using these values, the fraction of filtered cadmium reaching the end of the proximal tubule was 0.32, meaning that about 68% of filtered cadmium was reabsorbed by proximal tubular cells.[40] In contrast, the cadmium chelate Cd-DTPA was freely filtered (UF/P ratio = 1.08 ± 0.07) and was neither reabsorbed nor secreted by the proximal tubule (TF/P ratio for Cd-DPTA = 1.32 ± 0.11 vs. TF/P ratio of inulin = 1.32 ± 0.11).[40] These studies enabled a precise determination of cadmium reabsorption by the proximal tubule that could not have been obtained by other methods except, perhaps, by using isolated perfused tubules. Further, micropuncture techniques allowed these investigators to determine that cadmium acetate was reabsorbed to a substantial extent, whereas chelated cadmium (Cd-DTPA) was filtered and excreted without accumulating in the kidneys.

In some cases, obtaining tubular fluid samples from thin loops of Henle or collecting ducts is possible. For these experiments, tubule segments in the papilla are subjected to micropuncture. Micropuncture of papillary structures requires the use of animals in which the papilla can be exposed, such as golden hamsters and young Wistar rats.[38] Alternately, loop function can be inferred by comparing tubular fluid obtained from the late proximal convoluted tubule and distal tubule. Using this method, Senekjian *et al.* determined that in male Sprague-Dawley rats, the fractional delivery rate of gentamicin to the late portion of the proximal convoluted tubule was $65.5 \pm 5.3\%$ compared with a fractional delivery rate of $30.7 \pm 3.7\%$ to the distal tubule.[41] They interpreted these data to suggest that some gentamicin was reabsorbed during transit through the straight portion of the proximal tubule and/or the loop of Henle. However, a more precise localization of the site of gentamicin reabsorption could not be inferred from micropuncture studies since the loop of Henle is inaccessible in the strain of rats used.

Micropuncture can be useful in defining the effects of drugs or chemicals on tubular function. For example, when administered *in vivo*, gentamicin produces increases in urine output and sodium chloride excretion within 15 min of injection.[42] These effects could be due to diminished proximal tubular reabsorption, since gentamicin has been shown to damage the renal proximal tubule.[43] Alternatively, gentamicin-induced increases in fluid and electrolyte excretion could be due to damage along the loop of Henle.[41] Kidwell *et al.* used micropuncture to determine that proximal tubular function was not altered by gentamicin administration, suggesting that the early effects of gentamicin on urine composition were unrelated to any effects along the accessible portion of the proximal tubule.[42] These investigators determined that chloride transport in the isolated, perfused thick ascending limb of the loop of Henle was markedly inhibited when gentamicin (1 mM) was included in the luminal perfusate but not the bathing medium. They postulated that the loop of Henle was an early target of gentamicin, accounting for the early increases in fluid and electrolyte excretion observed following gentamicin administration.[42]

7.06.2.5 *In Vivo* Microperfusion

Microperfusion is a useful technique to determine the function of a specific segment of the nephron. The segment is isolated from the glomerular filtrate between a wax and oil block, allowing the segment to be perfused with a solution of controlled composition (Figure 7). The solution is then quantitatively collected

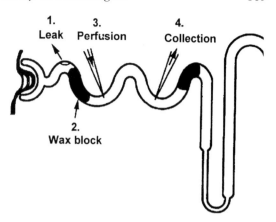

Figure 7 Schematic representation of microperfusion. A tubule with several surface loops is identified by infusing stained fluid in an early portion of the tubule segment (1). The initial pipette is withdrawn, leaving a hole for glomerular filtrate to leak out of the tubule. A paraffin wax block is inserted into the next visible loop (2) and the tubule is perfused with a pipette attached to a microperfusion pump (3). An additional pipette is inserted further down the length of the proximal tubule to be perfused so that the perfusate may be collected following placement of a distal mineral oil block (4) (reproduced by permission of CRC Press from "Renal Methods in Toxicology," 1996).

from the same nephron segment following perfusion. Essentially, microperfusion is analogous to *in vitro* perfusion of isolated nephron segments except that a single nephron is perfused *in situ*. With microperfusion, it is possible to perfuse the loop of Henle by inserting the perfusion pipette into the last accessible loop of the proximal tubule and the collection pipette into the first accessible loop of the distal tubule. Similarly, the distal tubule may be perfused as long as multiple loops are visible at the surface of the kidney.[38]

In addition to micropuncture, microperfusion was used to define the renal handling of cadmium. In these experiments, radiolabeled cadmium (1 mM as $^{109}CdCl_2$) was microinjected into the lumens of proximal or distal tubules and the amount of cadmium excreted in the ureteral or pelvic urine was determined.[44] When injected into the proximal tubule, fractional recovery of cadmium in ureteral urine was $30 \pm 5\%$. In contrast, fractional recovery of cadmium injected into the distal tubule lumen was $92 \pm 4\%$ of the injected amount.[44] Luminal uptake of cadmium–metallothionein (Cd–Mt) was significantly lower than luminal uptake of inorganic cadmium. Whereas fractional recovery of inorganic cadmium was 30% of the amount injected, fractional recovery of Cd–Mt was 80% of the amount injected.[44] It is likely that Cd–Mt as well as inorganic cadmium contributes to the nephrotoxicity observed following chronic exposure to this metal.[45]

Capillary microperfusion is a form of microperfusion experiment in which the peritubular capillaries are perfused with solutions of known composition, allowing examination of basolateral events in the intact kidney. This technique has been used extensively to investigate interactions among organic anions that undergo tubular secretion, as well as to identify substrates that interact with both the organic anion and cation transport systems.[46–48]

7.06.2.6 Electron Probe Microanalysis

Three major techniques have been applied to the study of the renal medulla. Micropuncture techniques permit direct sampling of fluid from late distal tubule and from the terminal collecting duct at the papillary tip. This type of "end-to-end" measurement is of limited value since the intervening segment is composed of at least three segments with differing transport properties and responses to hormones, and passes through several interstitial regions, including the cortical medullary ray, the outer medulla, and inner medulla. Isolated perfused tubule studies have provided much detailed data from defined segments of collecting ducts, but in an artificial and necessarily arbitrary bathing medium. Thus, while isolated perfused tubule studies do provide segment-specific transport and permeability data missing from the micropuncture studies, they do so only under conditions that may have little resemblance to the natural tubular environment. The microcatheterization studies of Hilgar *et al.*[49] and Sonnenberg,[50] while apparently solving both the problems of segmental localization and natural environment, add the complication of almost complete obstruction of the duct. Thus, the procedure, as practiced, must alter tubular flow rates and pressures. Damage to the tubule wall is frequent and tubular fluid specimens are sometimes collected containing erythrocytes.

The 'micropunch' is an instrument designed and constructed to permit sampling of frozen collecting duct fluid (or tissue) from anatomically defined locations.[51] This unique device includes an anvil provided with a hollow tip bored to a diameter of 25 μm and a precisely aligned punch having a square end that exactly fits into the tip of the anvil. These components have been made to allow material within the bore of the anvil to be removed through an opening ground in its top (Figure 8). The tissue specimen is collected and frozen as rapidly as possible to preserve tissue architecture. The

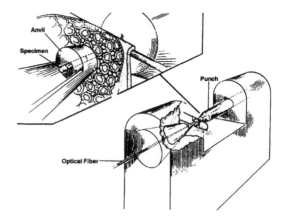

Figure 8 Schematic representation of the principle of the micropunch. A frozen section on a wire loop is positioned in front of the anvil. By means of a punch 25 μm in diameter, a small cylinder of ice may be removed from the lumen of the collecting duct.

frozen tissue is trimmed and sectioned in a cryostat microtome ($\sim -40\,^{\circ}$C at the cutting surface). Serial sections (10–20 μm) are cut to reconstruct the exact anatomical position of micropunch specimens. To maintain the selected tissue section and the specimens to be collected from it in a frozen condition, the micropunch assembly is operated in an insulated box partially filled with liquid nitrogen. The tissue section is adjusted on the surface of the anvil with the desired tubule lumen or tissue specimen positioned over the bore. An optical fiber is used to project a spot of light through the bore of the anvil, illuminating the area to be punched from the section. The punch is driven through 'the tissue section, depositing the selected specimen at the hollowed out part of the anvil. The specimen is picked off the punch tip with micromanipulators and the specimen transferred to the surface of a pure silicon wafer. The silicon wafer is always kept slightly warmer than the surrounding environment to prevent condensation of moisture and to slowly freeze-dry the punched specimens, whose residue adheres firmly. Each residue spot consists of the entire solute content of a known volume (punch diameter × tissue section thickness) of duct fluid or tissue. When all punched specimens have been collected, the punch apparatus and silicon wafer are slowly brought up to room temperature, and the wafer removed and stored desiccated until analyzed using electron probe microelemental analyses techniques. A micropunch specimen of collecting duct fluid will have total solute concentrations ranging from 10–2000 mOsm, or from 0.5–10 pmol of typical urinary solutes. The electron probe has been used for analysis of sodium, potassium, and chlorine in this range. In addition, a microanalytical technique has been reported for the analysis of urea in punched specimens by electron probe analysis.[52] Although not widely used at this time, due primarily to the lack of or difficulty of microanalytical techniques, microsampling of anatomically defined fluids and tissues coupled with modern specific microanalytical techniques permits precise evaluation of anatomically defined regions within the normal or damaged kidney, *in situ*.

7.06.2.7 Imaging Modalities

7.06.2.7.1 *Renal imaging with radionuclides*

The present era of radionuclide imaging was initiated with the development of the scintillation camera by Hal Anger in 1957. The Anger camera enabled rapid image formation, allowing acquisition of sequential images taken in rapid succession. The addition of this dynamic capability initiated the imaging of physiological and pharmacological processes in the kidney and other organ systems.

Washout of radioactive inert gases (krypton, xenon) or radioactive indicators (^{99c}Tc-DPTA, ^{99c}Tc-thioacetyltriglycine) may be used to measure renal blood or plasma flows and glomerular filtration rate. These techniques require intraarterial or intravenous administration of the tracer followed by monitoring of the amount of tracer in the kidneys with an external detector.[53] These techniques do not yield absolute flow, but rather flow per unit volume of tissue mass. Uptake of various radioactive microspheres will yield absolute blood flows but requires removal of the kidney at the end of the experiment. Finally, indicator/thermal dilution methods or electromagnetic, ultrasonic, and Doppler flow probe techniques require access to the renal artery and/or vein for injection and sampling of indicator, or placement of the probes. Progress with nuclear magnetic resonance imaging techniques has shown that the same concepts underpinning washout of inert gases to measure renal blood flow may be used with the appropriate paramagnetic tracers to estimate glomerular filtration rate.[33] The ideal radiopharmaceutical for measurement and imaging of glomerular filtration rate should be inert, freely filtered by the glomeruli, not reabsorbed or secreted by the renal tubule, not bound to the parenchyma, not accumulated in other tissues, not toxic, not metabolized, and freely excreted in urine. Radionuclides used for the measurement and imaging of glomerular filtration rate are listed in Table 3.

Imaging renal blood flow can contribute to assessments of renal morphology and renal artery stenosis, and integrity of the renal cortex. With radiopharmaceuticals, renal blood flow is estimated using the Fick Principle, taking advantage of the strong kinetics of the weak

Table 3 Radiopharmaceutical products available for determining GFR, RBF, and for imaging renal paremchyma.

Radiopharmaceuticals for estimating GFR

^{3}H-inulin	^{125}I-diatrizoate
^{14}C-inulin	^{125,131}I-iothalamate (Conray-60)
^{14}C-carboxy inulin	^{131}I-diatrizoate (Hypaque, renografin)
^{14}C-hydroxy-methyl inulin	^{51}Cr-, ^{99m}Tc-, ^{111,113m}In-, ^{140}La-, ^{169}Yb-EDTA
^{131}I-chloroiodopropyl inulin	^{51}Cr-, ^{99m}Tc-, ^{111,113m}In-, ^{140}La-, ^{169}Yb-DTPA
^{131}I-propargyl inulin	^{57,58}Co-hydroxycobalamin
	^{57,58}Co-cyanocobalamin

Radiopharmaceuticals for estimating RBF

^{3}H-*para*-aminohippuric acid (PAH)	$^{125,\,131}$I-iodopyracet (Diodrast)
^{14}C-PAH	^{131}I-*ortho*-iodohippurate (Hippuran, OIH)
^{99m}Tc-hippuran analogues	^{67,68}Ga-*N*-succinyl desferioxamine
^{99m}Tc-iminodiacetic PAH (PAHIDA)	^{97}Ru-ruthenocenyl-glycine (Ruppuran)
^{99m}Tc-mercaptoacetyltriglycine (^{99m}Tc-MAG3)	^{99m}Tc-thiodiglycolic acid
^{99m}Tc-mercaptosuccinyltriglycine (^{99m}Tc-MSG3)	
^{99m}Tc-*N*,*N*'-bis(mercaptoacetyl)-2,3-diaminopropanoate (CO2-DADS-A)	

Radiopharmaceuticals for imaging renal parenchyma

^{197,203}Hg-chlormerodrin	^{197}Hg-bichloride
^{99m}Tc-penicillamine (TPEN)	^{99m}Tc-penicillamine-acetazolamide (TPAC)
^{99m}Tc-2,3-dimercaptosuccinic acid (^{99m}Tc-DMSA)	^{99m}Tc-RGD (a ligand for integrins)
^{99m}Tc-gluconate and ^{99m}Tc-glucoheptonate	
^{203}Hg-3-chloromercuri-2-methoxypropyl urea (neohydrin)	

organic acid secretory system located in the pars recta segment of the proximal nephron. The ideal radiopharmaceutical for measurement and imaging of renal blood flow should be inert, 100% extracted by the kidney in a single pass, not reabsorbed by the renal tubule, not bound to the parenchyma, not accumulated in other tissues, not toxic, not metabolized, and freely excreted in urine. Radionuclides used for the measurement of renal blood flow are listed in Table 3.

In addition to their use in estimating glomerular filtration rate and renal blood flow, imaging techniques are extremely useful for viewing renal tissue. In designing radiopharmaceuticals to assist in imaging renal tissue, one attempts to design an agent that will be rapidly and specifically extracted from blood by the kidney and retained for a sufficient period in the renal tissue to permit imaging. The ideal radiopharmaceutical for imaging the renal parenchyma should be inert, 100% extracted in a single pass, irreversibly bound to the parenchyma, not accumulated by other tissues and not excreted in the urine. Radionuclides used to image the renal parenchyma are listed in Table 3.

A variety of radionuclides may be used in scintigraphic imaging of the kidney, as listed in Table 4. ^{99m}Tc represents an almost ideal radionuclide for scintigraphic imaging. The gamma-photon of 140 keV energy is well within the range of optimally imaged energies and there are no particle emissions. The half-life is 6 h and the nuclide is produced from the longer lived ^{99}Mo parent (67 h half-life). These characteristics have led to the development of many ^{99m}Tc-labeled radiopharmaceuticals, several of which have replaced older less-desirable agents. Dominance of ^{99m}Tc has resulted in modifications of the gamma-camera that optimize imaging with this radionuclide, even to the detriment of other higher energy radionuclides.

7.06.2.8 Ultrasonic and Magnetic Resonance Imaging

Ultrasonic imaging (UI) and magnetic resonance imaging (MRI) technologies offer substantial increases in spatial resolution and image quality above that provided by scintigraphy. The quality and utility of these imaging modalities have improved dramatically and further improvements are to be anticipated.

Ultrasonic imaging relies upon processing and display of ultrasonic radiation reflected from body tissues. Animals are usually tranquilized or anesthetized to reduce body movements. A probe is positioned manually on the abdomen or flank to collect images of the kidney (Figure 9). Pulsed Doppler and phase shift signal processing techniques can be applied to collect additional information on regional blood flow in the kidney. Ultrasonic imaging will detect relatively gross abnormalities in renal structure, including large infarcts and tumor masses. Ultrasonic imaging is relatively inexpensive and easy to perform. However, use of ultrasonic techniques to quantitate renal functions is limited by manual (e.g., variable) use of the probe, and the current lack of contrast media that interact with renal functions in a predictable fashion.

MRI produces the highest spatial resolution and image quality. Images are based upon radio frequency information of proton spin shifts induced by changes in an imposed magnetic field. Spatial resolution is a function of differences in proton microenvironments and magnetic field strength. For highest quality images, animals are usually anesthetized prior to being placed within the imaging magnet. Serial images of the area of interest may be collected permitting a three-dimensional reconstruction of the kidney. MRI will detect much smaller abnormalities in renal structure than ultrasonic imaging. For example, MRI imaging

Table 4   Isotopes used for imaging.

Isotope	Gamma energies (keV)	Abundance (%)	Beta-emission	Half-life (h)
^{67}Ga	93	38	no	78.3
	185	21		
	300	17		
^{111}In	171	91	no	67.9
	245	94		
^{123}I	159	83	no	13.2
^{131}I	364	81	yes	193
^{99m}Tc	140	89	no	6.02

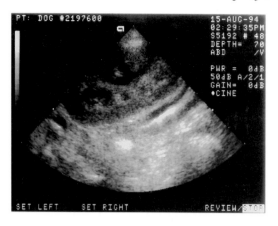

Figure 9 An ultrasonic image of an *in situ* dog kidney acquired using Acuson XP10 and a mildly tranquilized beagle dog. The image, located in the upper portion of the sweep, is a sagittal projection with the renal pelvis at the top. The alternating dark and light areas correspond to the individual renal pyramids and columns.

at 1.5 T will detect swelling of the renal cortex and dissipation of cortical–medullary intensity gradients associated with treatment with a loop diuretic.[54] Using a combination of MRI at 4.7 T and histopathology, Bosch *et al.*[55] observed marked cortical hypertrophy and increased glomerular size in kidneys of obese Zucker rats in comparison with kidneys from lean littermates. In addition, paramagnetic-based MRI contrast agents can be used to image glomerular filtration rate and regional renal blood flow.[33]

7.06.3 SPECIFIC EXAMPLES

Much of our understanding of renal function originated from *in vivo* studies in amphibians and mammals. In addition, selective kidney injury caused by xenobiotics can only be assessed when such compounds are administered to intact animals and simple urinalysis, histopathology, and possibly a few clearance studies are done. Reviewing the literature of *in vivo* studies examining toxic interactions in the kidney is impossible, and we will not attempt to do so. Rather, several studies are presented here in which *in vivo* techniques were invaluable in addressing some specific hypotheses or establishing some specific technique for further investigation.

7.06.3.1 *In Vivo* Techniques to Assess Cooperation Between Liver and Kidney

The kidneys are involved not only in the excretion of xenobiotics, but in some instances may play a role in the metabolism of compounds. The role of intrarenal metabolism is important for some toxicants, particularly glutathione conjugates of some compounds. The role of the kidney in the bioactivation of nephrotoxic glutathione conjugates has been intensively investigated and a complex pattern of interorgan cooperativity has emerged. Compounds, such as bromobenzene, hexachlorobutadiene, dichlorovinyl-L-cysteine, and possibly *para*-aminophenol are metabolized in the liver to form glutathione conjugates.[56–60] These conjugates are released from hepatocytes into bile or plasma. In either case, the conjugates are eventually filtered at the glomerulus of the kidney and enter the luminal fluid. In the tubular lumen, glutathione conjugates are sequentially metabolized to cysteinylglycine conjugates by the action of γ-glutamyl transpeptidase and to cysteine conjugates by the action of cysteinylglycine dipeptidase (Figure 10). The cysteine conjugates enter tubular epithelial cells where they may be further metabolized to *N*-acetyl cysteine conjugates by the action of *N*-acetyltransferase or to reactive intermediates by the action of cysteine conjugate β-lyase (Figure 10).

Styrene is a nephrotoxic compound that is extensively metabolized in rats. Several styrene metabolites are glutathione conjugates that may be responsible for the nephrotoxicity of styrene. When administered to male Fischer-344 rats, *S*-[(1 and 2)-phenyl-2-hydroxyethyl] glutathione (PHEG), a mixture of styrene–glutathione conjugates, caused significant increases in the excretion of glucose, γ-glutamyl transpeptidase, glutamate dehydrogenase (a mitochondrial enzyme), *N*-acetyl-β-glucosaminidase, and lactate dehydrogenase (a cytosolic enzyme),[61] suggesting proximal tubular damage. Histopathology was consistent with this interpretation and renal proximal tubules contained pyknotic nuclei and numerous vacuoles. Acivicin is an inhibitor of γ-glutamyl transpeptidase, and pretreatment of rats with acivicin prevented PHEG-induced increases in glucose and enzyme excretion, confirming that γ-glutamyl transpeptidase is involved in metabolism and/or cellular uptake of PHEG. Similarly, rats pretreated with phenylalanylglycine, an inhibitor of cysteinylglycine dipeptidase, were protected from PHEG toxicity, indicating that metabolism and/or cellular uptake of PHEG involved cysteinylglycine dipeptidase. In contrast, pretreatment of rats with aminooxyacetic acid, an inhibitor of cysteine conjugate β-lyase, failed to protect rats from PHEG-induced glucosuria and enzymuria, indicating that cysteine conjugate β-lyase is not involved in PHEG toxicity.[61] From studies such as these,

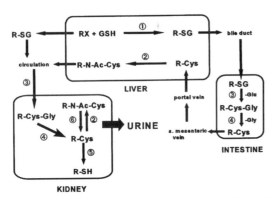

Figure 10 A schematic diagram of interactions of glutathione conjugates in liver, intestine, and kidney. Glutathione conjugates are initially formed by enzymatic or nonenzymatic reactions in the liver (1). These conjugates (R–SG) may be exported into bile or plasma. In bile, the conjugates enter the intestinal tract where they are sequentially metabolized to a cysteinylglycine conjugate (R–Cys–Gly) by γ-glutamyl transpeptidase (2) and a cysteine conjugate (R–Cys) by cysteinylglycine dipeptidase (3). The cysteine conjugate may undergo enterohepatic recirculation and reenter the liver for *N*-acetylation to a mercapturic acid (R–N–Ac–Cys) by *N*-acetyltransferases (4). Both the mercapturic acid and glutathione conjugates may be released into plasma and be filtered at the glomerulus of the kidney. In the tubular fluid, glutathione conjugates are accumulated in tubular epithelial cells by the action of γ-glutamyl transpeptidase and cysteinylglycine dipeptidase, forming the cysteine conjugate. The cysteine conjugate may undergo *N*-acetylation, catalyzed by *N*-acetyl transferases, or may be converted to a thiol-containing intermediate (R–SH) by the action of cysteine conjugate β-lyase (5). Drawing courtesy of Dr. P. Harvison, Philadelphia College of Pharmacy and Science.

the current understanding of the complex handling of glutathione conjugates has emerged. While we previously considered glutathione conjugation to be a detoxification reaction, for some compounds, glutathione conjugates may clearly represent a bioactivation pathway.

p-Aminophenol (PAP), a metabolite of acetaminophen and phenacetin, is a potent nephrotoxicant. When administered to Fischer-344 rats, PAP causes depletion of renal and hepatic glutathione, increases in blood urea nitrogen concentration and kidney weight, and severe proximal tubular necrosis.[62–66] Radioactivity from [³H]-PAP becomes covalently bound to proteins and DNA in both liver and kidney, with covalent binding in kidney two to three times greater than in liver.[62,67] Since inducers and inhibitors of cytochrome P450 failed to consistently alter PAP-induced nephrotoxicity, it was considered that PAP underwent intrarenal metabolism, possibly by autoxidation, to

produce nephrotoxic intermediates.[68] However, subsequent studies have suggested that PAP undergoes oxidation in the liver to produce glutathione conjugates.[69] These glutathione conjugates are more potent nephrotoxicants that PAP itself.[67] In support of the hypothesis that PAP–GSH (glutathione) conjugates formed in liver are responsible for nephrotoxicity, biliary cannulation partially protected Fischer-344 from PAP-induced renal functional and histologic changes.[60] Pretreating rats with buthionine sulfoximine to deplete hepatic and renal glutathione completely prevented PAP-induced nephrotoxicity,[60] confirming the importance of glutathione conjugation in PAP bioactivation. If PAP–GSH conjugates are accumulated in renal tissue similarly to other glutathione conjugates, such as styrene and bromobenzene, inhibition of γ-glutamyl transpeptidase should prevent PAP nephrotoxicity. Indeed, in isolated renal tubular cells, coincubation with acivicin significantly decreased PAP-induced cell lethality, measured as trypan blue exclusion.[69] However, protection by acivicin could not be duplicated when PAP or PAP–GSH conjugates were administered to rats *in vivo* and, in fact, a modest potentiation of PAP-induced elevations of blood urea nitrogen concentration and glucosuria were observed when PAP and acivicin were coadministered.[70,71] Thus, PAP appears to undergo bioactivation either enzymatically or nonenzymatically in the liver, forming PAP–GSH conjugates. These conjugates, gain entry into proximal tubular cells, possibly by a mechanism independent of γ-glutamyl transpeptidase, and exert nephrotoxicity.

7.06.3.2 *In Vivo* Techniques to Assess Renal Handling of Compounds

In vivo techniques are particularly useful to investigate the mechanisms by which compounds are eliminated by the kidneys. In addition to filtration, substances may be removed from plasma by tubular secretion. Alternately, compounds may be reabsorbed by the renal tubules following glomerular filtration. Various *in vitro* techniques are available to determine tubular uptake of compounds, presumably the first step in tubular secretion. However, examining tubular reabsorption is difficult unless an investigator uses microperfusion or isolated perfused tubule segments. These techniques are of limited value for some compounds because of the extremely small quantities of tubular fluid that may be obtained. Thus, clearance methods are invaluable aids in determining how compounds are handled by the kidneys.

The renal handling of cisplatin was investigated in anesthetized dogs by simultaneously determining cisplatin and inulin clearances. Cisplatin clearance was compared with inulin clearances to determine the mechanisms involved in cisplatin elimination.[72] In these studies, anesthetized dogs received cisplatin at a dosage of 5 mg kg^{-1} i.v. infused over 10 min. In addition, the dogs received a continuous infusion containing inulin to maintain stable plasma concentration for the duration of the experiment. Urine was collected for 20 min intervals for 2 h following cisplatin administration with blood collected at the midpoint of the collection interval. To elucidate the mechanisms involved in the tubular secretion of cisplatin, dogs received probenecid (100 mg kg^{-1}), an inhibitor of organic anion transport; or cimetidine (20 mg kg^{-1}), ranitidine (5 mg kg^{-1}), or quinidine (2.2 mg kg^{-1}), inhibitors of organic cation transport. In the absence of transport inhibitors, cisplatin clearance was 3.78 ± 0.4 mL min^{-1} kg^{-1} and inulin clearance was 2.85 ± 0.47 mL min^{-1} kg^{-1}, suggesting that cisplatin was secreted by the kidneys. During infusion of probenecid, cisplatin clearance was 2.52 ± 0.68 mL min^{-1} kg^{-1}, while inulin clearance was 2.9 ± 0.36 mL min^{-1} kg^{-1}, confirming that cisplatin was secreted by a mechanism inhibited by probenecid. In addition, since the fractional clearance of cisplatin ($Cl_{cisplatin}/Cl_{inulin}$) was less than one (0.83 ± 0.09), the authors suggested that cisplatin may be reabsorbed. Compounds that inhibit organic cation transport, such as cimetidine, ranitidine, and quinidine, also significantly decreased cisplatin clearance. However, fractional clearance of cisplatin in the presence of organic cation transport inhibitors was never less than one, failing to confirm the observation of tubular reabsorption of cisplatin.[72] These studies examining the renal handling of cisplatin could not have been accomplished using microtechniques since the detection limit of cisplatin using atomic absorption spectroscopy is 10 ng mL^{-1},[66] a concentration not likely to be reached using microperfusion or isolated perfused tubules.

7.06.3.3 *In Vivo* Techniques to Assess Renal Blood Flow

Radiocontrast agents are extremely useful for imaging the kidney, as has been previously discussed. However, these agents have been associated with acute renal failure that is believed to involve intense renal vasoconstriction.[73] The mechanisms underlying radiocontrast-induced vasoconstriction are not entirely clear, but may involve endothelin, a potent vasoconstrictor.[74,75] Brooks and DePalma[76] used ultrasonic flow probes in dogs to determine the role of endothelin in radiocontrast-induced renal vasoconstriction.

In these studies, dogs were anesthetized and surgically prepared for administration of drugs, collection of blood and urine, and measurement of systemic blood pressure. Renal blood flow was determined using an ultrasonic flow probe positioned around the left renal artery. The radiocontrast agent used was Hypaque (diatrizoate meglumine, Winthrop Pharmaceuticals) administered at a dose of 5 mg kg^{-1} over a 10 min infusion period. Following Hypaque administration, urine flow and sodium excretion were significantly increased and the diuresis and natriuresis were unaltered by infusion of a specific endothelin receptor antagonist, SB 209670. Renal blood flow was not significantly altered following Hypaque administration but was significantly elevated when dogs received both Hypaque and SB 209670.[76] These data suggested that blockade of endothelin receptors increased renal blood flow in anesthetized dogs but did not address the role of endothelin in Hypaque-induced vasoconstriction. Therefore, the experiments were repeated in dogs pretreated with indomethacin to block any prostaglandin-dependent vasodilation that might prevent Hypaque-induced vasoconstriction. In indomethacin-pretreated dogs, Hypaque produced a sustained and significant decrease in renal blood flow that was reversed

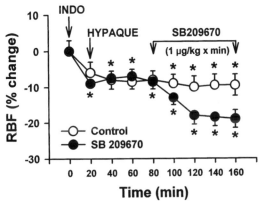

Figure 11 Effect of renal arterial infusion of vehicle ($n = 8$) or (±)-SB 209670 (1 µg kg^{-1} min^{-1}, $n = 8$) on renal blood flow (RBF) in indomethacin-treated anesthetized dogs. Baseline RBF was 117 ± 7 mL min^{-1} and 106 ± 5 mL min^{-1} in vehicle and (±)-SB 209670-treated dogs, respectively. Asterisks indicate points that are significantly different from time = zero ($p < 0.05$) (reproduced by permission of Karger from *Nephron*, **72**, 629–636).

when SB 209670 was infused (Figure 11). Thus, these data suggest that radiocontrast-induced vasoconstriction does involve activation of endothelin receptors, presumably through activation of endothelin production.[76] Monitoring renal blood flow using a flow probe was critical to these experiments since measurement of renal blood flow using PAH clearance would undoubtedly be complicated by competition for secretion between PAH and SB 209670, a weak organic acid.[76]

Compounds that promote peripheral vasodilation would be therapeutically useful in the treatment of various disorders, including hypertension, heart failure, and acute renal failure. For maximal patient compliance, these therapeutic agents should be orally active with a long duration of action. Most peripheral vasodilators are dopamine agonists that are chemically related to catecholamines. The catechol moiety is a prime site for metabolism and elimination, giving these agents poor oral availability and short durations of action. One strategy to avoid these potential problems is to provide the agonist in a prodrug form, as has been done with ibopamine, the diisobutyryl ester of *N*-methyldopamine[76] and docarpamine, the diethyloxycarbonate ester of dopamine.[78] Using conscious animals to test the oral efficacy and duration of action of these agents is desirable, and the chronically instrumented dog is frequently used for this purpose.

In this type of study, a flow probe is placed around the renal artery using aseptic surgical techniques and the lead wires are tunneled to the back of the neck. The dog is allowed to recover and studies are done in conscious animals. Using these techniques, Christie and Smith[79] compared 5,6-dihydroxy-3-phenyl-1-aminomethyl-isochroman (A-68930) with fenoldopam and dopamine for efficacy and duration of action in promoting an increase in renal blood flow. With intravenous infusion, the pED_{40} values were 9.7 ± 0.2 for A-68930, 8.6 ± 0.4 for fenoldopam, and 7.8 ± 0.1 for dopamine, corresponding to infusion rates of 0.2, 2.4, and $16.8 \, nmol \, kg^{-1} \, min^{-1}$, respectively.[79] When the infusion was stopped, the half-time for restoration of normal renal blood flow was $14.8 \pm 2.5 \, min$ for A-68930, $2.9 \pm 0.5 \, min$ for fenoldopam, and $1.4 \pm 0.4 \, min$ for dopamine. Thus, the rank order of potency and duration of action was A68930 > fenoldopam > dopamine.[79] When administered orally to conscious dogs, the pED_{40} for A-68930 was 7.5 ± 0.2 vs. 5.6 ± 0.1 for fenoldopam, corresponding to oral doses of $32 \, nmol \, kg^{-1}$ and $2.7 \, \mu mol \, kg^{-1}$. When given orally at doses that produced similar maximal changes in renal blood flow ($0.1 \, \mu mol \, kg^{-1}$ A-68930 and

$3 \, \mu mol \, kg^{-1}$ fenoldopam to increase renal blood flow by $47 \pm 7\%$ and $42 \pm 7\%$, respectively), the onset and duration of action of the two compounds were dissimilar. A-68930 required 60 min to achieve maximum response, whereas fenoldopam produced a maximal response within 10 min of dosing. The duration of action of A-68930 was about 3 h compared with 1 h for fenoldopam.[79] By using conscious dogs, these investigators could assess oral activity as well as any central nervous system (CNS) stimulation that might occur upon stimulation of central dopamine receptors. Such CNS stimulation would obviously be absent in anesthetized animals.

7.06.3.4   Magnetic Resonance Imaging in Determination of Glomerular Filtration Rate

A problem with imaging glomerular filtration or renal blood flow is exposure to radioactive chemicals. To minimize internal radioactivity, several paramagnetic agents have been tested for utility in magnetic resonance imaging. Heavy metal chelates of diethylenetriamine pentaacetic acid (DTPA) are useful paramagnetic agents for imaging glomerular filtration rate since these compounds are relatively inert, nontoxic, and are handled by the kidneys in the same manner as inulin.[80–83]

Ytterbium-DTPA (Yb-DTPA) was evaluated as a potential imaging agent in male Sprague-Dawley rats. Simultaneously determined inulin and Yb-DTPA clearances were highly correlated although Yb-DTPA clearance was approximately 15% higher than inulin clearance.[33] Using whole-body imaging at 1.5 T, the signal intensity in the renal papilla was decreased within 11 min of administration of Yb-DTPA and was restored to normal intensity by 47 min after injection.[33] In contrast, administration of gadolinium-DTPA (Gd-DTPA) produced a decrease in signal intensity within 5 min of injection and papillary signal intensity remained at low and variable levels for an additional 30 min, suggesting that Gd-DTPA persists in the renal papilla long after the compound has been eliminated from the cortex.[33] Paramagnetic agents are effective for imaging because they induce changes in the relaxation behavior of protons through dipole–dipole interactions between the spin of unpaired electrons of the paramagnetic metal ion and the nuclear spin of protons. However, contrast generated by Gd-DTPA is due to dipole–dipole relaxation and differences in susceptibility between capillaries containing contrast agent and surrounding tissue. Since gadolinium has a higher magnetic moment than ytterbium,[84] susceptibility-induced changes following Gd-DTPA would be greater than those due to Yb-DTPA and the signal intensity loss with Gd-DTPA would be larger than the signal loss with Yb-DTPA. These differences between gadolinium and ytterbium can be clearly seen in the plots of signal intensity in papilla as compared with adjacent muscle (Figure 12). Although Gd-DTPA and Yb-DTPA give similar estimates for glomerular filtration rate (determined as clearance of Gd-DTPA and Yb-DTPA), Yb-DTPA may be a better agent for imaging because of clearer resolution of signal intensity.

7.06.4 CONCLUSIONS

Historically, *in vivo* studies laid the foundation for our understanding of renal physiology. Clearance studies were indispensable for interpretation of the action of diuretic agents on the kidney. Micropuncture studies provided the first evidence for glomerular filtration in amphibians and mammals. Clearance experiments are still the only reliable method to determine the mechanisms involved in renal elimination of drugs and chemicals. The presence or absence of nephrotoxicity can be established unequivocally only using *in vivo* techniques where morphologic examination of tissue is possible. In addition, *in vivo* methods are indispensable for examining integrated renal responses, such as urinary concentrating ability or regulation of renal blood flow. Unfortunately, *in vivo* methods shed little light on the mechanism of toxicity for most chemicals and investigators generally turn to *in vitro* methods to identify cellular responses to toxic agents.

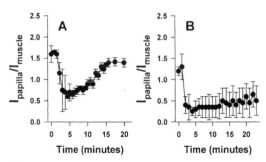

Figure 12 Plot of the ratio of MR signal intensity of rat papilla to adjacent muscle, obtained from GRASS (gradient recalled acquisition in steady state) images, as a function of time after administration of Yb-DTPA (panel A) and Gd-DPTA (Panel B) (reproduced by permission of Academic Press from *Magn. Reson. Med.*, 1991, **17**, 328–335).

7.06.5 REFERENCES

1. H. W. Smith, 'The Kidney,' Oxford University Press, New York, 1951.
2. L. B. Kinter, S. Caltabiano and W. F. Huffman, 'Anomalous antidiuretic activity of antidiuretic hormone antagonists.' *Biochem. Pharmacol.*, 1993, **45**, 1731–1737.
3. K. P. Gartland, F. W. Bonner and J. K. Nicholson, 'Investigations into the biochemical effects of region-specific nephrotoxins.' *Mol. Pharmacol.*, 1989, **35**, 242–250.
4. J. R. Bales, D. P. Higham, I. Howe *et al.*, 'Use of high-resolution proton nuclear magnetic resonance spectroscopy for rapid multi-component analysis of urine.' *Clin. Chem.*, 1984, **30**, 426–432.
5. J. K. Nicholson, M. P. O'Flynn, P. J. Sadler *et al.*, 'Proton-nuclear-magnetic-response studies of serum, plasma and urine from fasting normal and diabetic subjects.' *Biochem. J.*, 1984, **217**, 365–375.
6. A. P. Evan and W. G. Dail, Jr., 'The effects of sodium chromate on the proximal tubules of the rat kidney. Fine structural damage and lysozymuria.' *Lab. Invest.*, 1974, **30**, 704–715.
7. B. H. Haagsma and A. W. Pound, 'Mercuric chloride-induced renal tubular necrosis in the rat.' *Br. J. Exp. Pathol.*, 1979, **60**, 341–352.
8. E. A. Lock and J. Ishmael, 'The acute-toxic effects of hexachlor-1,3-butadiene on the rat kidney.' *Arch. Toxicol.*, 1979, **43**, 47–57.
9. B. J. Leonard, E. Eccleston, D. Jones, *et al.*, 'Antileukaemic and nephrotoxic properties of platinum compounds.' *Nature*, 1971, **234**, 43–45.
10. J. Halman and R. G. Price, in 'Selected Topics in Clinical Enzymology,' eds. D. M. Goldberg and M. Werner, Walter de Gruyter, Berlin, 1984, vol. 2, pp. 435–444.
11. G. Murray, R. G. Wyllie, G. S. Hill *et al.*, 'Experimental papillary necrosis of the kidney. I. Morphologic and functional data.' *Am. J. Pathol.*, 1972, **67**, 285–302.
12. P. J. S. Chiu, *Drug Devel. Res.*, 1994, **32**, 247.
13. A. Bernard and R. R. Lauwerys, 'Proteinuria: changes and mechanisms in toxic nephropathies.' *Crit. Rev. Toxicol.*, 1991, **21**, 373–405.
14. P. D. Dass and T. C. Welbourne, 'Evidence for lumenal and antilumenal localization of gamma-glutamyltranspeptidase in rat kidney.' *Life Sci.*, 1981, **28**, 355–360.
15. R. C. Cottrell, C. E. Agrelo, S. D. Gangolli *et al.*, 'Histochemical and biochemical studies of chemically induced acute kidney damage in the rat.' *Food Cosmet. Toxicol.*, 1976, **14**, 593–598.
16. R. J. Boatman, J. C. English, L. G. Perry *et al.*, 'Differences in the nephrotoxicity of hydroquinone among Fischer 344 and Sprague-Dawley rats and B6C3F1 mice.' *J. Toxicol. Environ. Health*, 1996, **47**, 159–172.
17. P. J. Chiu and A. D. Brown, 'Effect of captopril on preexisting and aminonucleoside-induced proteinuria in spontaneously hypertensive rats.' *Res. Commun. Chem. Pathol. Pharmacol.*, 1981, **31**, 419–433.
18. T. Yamada, K. Tomioka, M. Horie *et al.*, 'Effects of YM264, a novel PAF antagonist, on puromycin aminonucleoside-induced nephropathy in the rat.' *Biochem. Biophys. Res. Commun.*, 1991, **176**, 781–785.
19. A. Sundberg, E. L. Appelkvist, G. Dallner *et al.*, 'Glutathione tranferases in the urine: sensitive methods for detection of kidney damage induced by nephrotoxic agents in humans.' *Environ. Health Perspect.*, 1994, **102**, 293–296.
20. M. D. Stonard, in 'Nephrotoxicity in the Experimental and Clinical Situation,' eds. P. H. Bach and E. A. Lock, Martinus Nijhoff, Boston, MA, 1987, pp. 563–592.

21. M. D. Stonard, C. W. Gore, G. J. Oliver *et al.*, 'Urinary enzymes and protein patterns as indicators of injury to different regions of the kidney.' *Fundam. Appl. Toxicol.*, 1987, **9**, 339–351.

22. C. Thornley, A. Dawney and W. R. Cattell, 'Human Tamm–Horsfall glycoprotein: urinary and plasma levels in normal subjects and patients with renal disease determined by a fully validated radioimmunoassay.' *Clin. Sci. (Colch.)*, 1985, **68**, 529–535.

23. J. H. Hoyer, S. P. Sisson and R. L. Vernier, 'Tamm–Horsfall glycoprotein: ultrastructural immunoperoxidase localization in rat kidney.' *Lab. Invest.*, 1979, **41**, 168–173.

24. J. S. Woods, in 'Methods in Renal Toxicology,' eds. L. H. Lash and R. K. Zalups, CRC Press, Boca Raton, FL, 1996, pp. 19–33.

25. R. F. Pitts, in 'Physiology of the Kidney and Body Fluids: An Introductory Text,' 3rd edn., Year Book Medical Publishers, Chicago, IL, 1974.

26. Y. Yagil, 'Acute effect of cyclosporin on inner medullary blood flow in normal and postischemic rat kidney.' *Am. J. Physiol.*, 1990, **258**, F1139–F1144.

27. J. R. Haywood, R. A. Shaffer, C. Fastenow *et al.*, 'Regional blood flow measurement with pulsed Doppler flowmeter in conscious rat.' *Am. J. Physiol.*, 1981, **241**, H273–H278.

28. F. G. Knox, E. L. Ritman and J. C. Romero, 'Intrarenal distribution of blood flow: evolution of a new approach to measurement.' *Kidney Int.*, 1984, **25**, 473–479.

29. G. Daugaard, 'Cisplatin nephrotoxicity: experimental and clinical studies. [Review].' *Dan. Med. Bull.*, 1990, **37**, 1–12.

30. A. J. Vander, 'Renal Function,' 3rd edn., McGraw-Hill, New York, 1985.

31. H. Valtin, 'Renal Function, Mechanisms Preserving Fluid and Solute Balance in Health,' 2nd edn., Little Brown, Boston, MA, 1983.

32. D. R. Finco, H. Tabaru, S. A. Brown *et al.*, 'Endogenous creatinine clearance measurement of glomerular filtration rate in dogs.' *Am. J. Vet. Res.*, 1993, **54**, 1575–1578.

33. S. K. Sarkar, R. E. Rycyna, R. E. Lenkinski *et al.*, 'Yb-DTPA, a novel contrast agent in magnetic resonance imaging: application to rat kidney.' *Magn. Reson. Med.*, 1991, **17**, 328–335.

34. B. M. Brenner, R. Zatz, and I. Ichikawa, in 'The Kidney,' 3rd edn., eds. B. M. Brenner and F. C. Rector, Jr., Saunders, Philadelphia, PA, 1976, pp. 93–123.

35. J. T. Wearn and A. N. Richards, *Am. J. Physiol.*, 1924, **71**, 209.

36. A. M. Walker and J. Oliver, *Am. J. Physiol.*, 1941, **134**, 562.

37. A. M. Walker, P. A. Bott, J. Oliver *et al.*, *Am. J. Physiol.*, 1941, **134**, 580.

38. C. R. Ramsey and F. G. Knox, in 'Renal Methods in Toxicology,' eds. L. H. Lash and R. K. Zalups, CRC Press, Boca Raton, FL, 1996, pp. 79–96.

39. M. A. Burnatowska-Hledin, G. H. Mayor and K. Lau, 'Renal handling of aluminum in the rat: clearance and micropuncture studies.' *Am. J. Physiol.*, 1985, **249**, F192–F197.

40. E. Felley-Bosco and J. Diezi, 'Fate of cadmium in rat renal tubules: a micropuncture study.' *Toxicol. Appl. Pharmacol.*, 1989, **98**, 243–251.

41. H. O. Senekjian, T. F. Knight and E. J. Weinman, 'Micropuncture study of the handling of gentamicin by the rat kidney.' *Kidney Int.*, 1981, **19**, 416–423.

42. D. T. Kidwell, J. W. McKeown, J. S. Grider *et al.*, 'Acute effects of gentamicin on thick ascending limb function in the rat.' *Eur. J. Pharmacol.*, 1994, **270**, 97–103.

43. J. C. Kosek, R. I. Mazze and M. J. Cousins, 'Nephrotoxicity of gentamicin.' *Lab. Invest.*, 1974, **30**, 48–57.

44. E. Felley-Bosco and J. Diezi, 'Fate of cadmium in rat renal tubules: a microinjection study.' *Toxicol. Appl. Pharmacol.*, 1987, **91**, 204–211.

45. B. A. Fowler, *Environ. Health Perspect.*, 1992, **100**, 57.

46. K. J. Ullrich, G. Fritzsch, G. Rumrich *et al.*, 'Polysubstrates: substances that interact with renal contraluminal PAH, sulfate, and NMeN transport: sulfamoyl-, sulfonylurea-, thiazide- and benzeneaminocarboxylate (nicotinate) compounds.' *J. Pharmacol. Exp. Ther.*, 1994, **269**, 684–692.

47. K. J. Ullrich, G. Rumrich, F. Papavassiliou *et al.*, 'Contraluminal *p*-aminohippurate transport in the proximal tubule of the rat kidney. VII. Specificity: cyclic nucleotides, eicosanoids.' *Pflugers Arch.*, 1991, **418**, 360–370.

48. K. J. Ullrich, G. Rumrich, F. Papavassiliou *et al.*, 'Contraluminal *p*-aminohippurate transport in the proximal tubule of the rat kidney. VIII. Transport of corticosteroids.' *Pflugers Arch.*, 1991, **418**, 371–382.

49. H. H. Hilgar, J. D. Klumper and K. J. Ullrich, *Pflugers Arch.*, 1958, **267**, 218.

50. H. Sonnenberg, 'Medullary collecting-duct function in antidiuretic and in salt- or water-diuretic rats.' *Am. J. Physiol.*, 1974, **226**, 501–506.

51. L. B. Kinter, Ph.D. Thesis, Harvard University, 1978, pp. 275–292.

52. R. Beeuwkes III, J. M. Amberg and L. Essandoh, 'Urea measurement by x-ray microanalysis in 50 picoliter specimens.' *Kidney Int.*, 1977, **12**, 438–442.

53. E. Fommei and D. Volterrani, 'Renal nuclear medicine.' *Semin. Nucl. Med.*, 1995, **25**, 183–194.

54. S. K. Sarkar, G. A. Holland, R. E. Lenkinski *et al.*, 'Renal imaging studies at 1.5 and 9.4 T: effects of diuretics.' *Magn. Reson. Med.*, 1988, **7**, 117–124.

55. C. S. Bosch, J. J. H. Ackerman, R. G. Tilton *et al.*, 'In vivo NMR imaging and spectroscopic investigation of renal pathology in lean and obese rat kidneys.' *Magn. Reson. Med.*, 1993, **29**, 335–344.

56. S. S. Lau and T. J. Monks, 'The contribution of bromobenzene to our current understanding of chemically-induced toxicities. [Review].' *Life. Sci.*, 1988, **42**, 1259–1269.

57. T. J. Monks, T. W. Jones, B. A. Hill *et al.*, 'Nephrotoxicity of 2-bromo-(cystein-*S*-yl) hydroquinone and 2-bromo-(*N*-acetyl-L-cystein-*S*-yl) hydroquinone thioethers.' *Toxicol. Appl. Pharmacol.*, 1991, **111**, 279–298.

58. E. A. Lock, in 'Nephrotoxicity in the Experimental and Clinical Situation,' eds. P. H. Bach and E. A. Lock, Martinus Nijhoff, Boston, MA, 1987, pp. 429–461.

59. M. W. Anders, L. Lash, W. Dekant *et al.*, 'Biosynthesis and biotransformation of glutathione S-conjugates to toxic metabolites. [Review].' *Crit. Rev. Toxicol.*, 1988, **18**, 311–341.

60. K. P. Gartland, C. T. Eason, F. W. Bonner *et al.*, 'Effects of biliary cannulation and buthionine sulphoximine pretreatment on the nephrotoxicity of para-aminophenol in the Fischer 344 rat.' *Arch. Toxicol.*, 1990, **64**, 14–25.

61. S. Chakrabarti and M. A. Malick, 'In vivo nephrotoxic action of an isomeric mixture of *S*-(1-phenyl-2-hydroxyethyl)glutathione and *S*-(2-phenyl-2-hydroxyethyl) glutathione in Fischer-344 rats.' *Toxicology*, 1991, **67**, 15–27.

62. C. A. Crowe, A. C. Yong, I. C. Calder *et al.*, 'The nephrotoxicity of *p*-aminophenol. I. The effect on microsomal cytochromes, glutathione and covalent binding in kidney and liver.' *Chem. Biol. Interact.*, 1979, **27**, 235–243.

63. R. Shao and J. B. Tarloff, *Fundam. Appl. Toxicol.*, 1996, **31**, 268.

64. J. F. Newton, C. H. Kuo, M. W. Gemborys *et al.*, 'The nephrotoxicity of *p*-aminophenol. I. The effect on microsomal cytochromes, glutathione and covalent binding in kidney and liver.' *Toxicol. Appl. Pharmacol.*, 1983, **65**, 336–344.

65. J. B. Tarloff, R. S. Goldstein, D. G. Morgan *et al.*, 'Acetaminophen and *p*-aminophenol nephrotoxicity in aging male Sprague-Dawley and Fischer 344 rats.' *Fundam. Appl. Toxicol.*, 1989, **12**, 78–91.

66. J. F. Newton, M. Yoshimoto, J. Bernstein *et al.*, 'Acetaminophen nephrotoxicity in the rat. II. Strain differences in nephrotoxicity and metabolism of *p*-aminophenol, a metabolite of acetaminophen.' *Toxicol. Appl. Pharmacol.*, 1983, **69**, 307–318.

67. L. M. Fowler, R. B. Moore, J. R. Foster *et al.*, 'Nephrotoxicity of 4-aminophenol glutathione conjugate.' *Hum. Exp. Toxicol.*, 1991, **10**, 451–459.

68. I. C. Calder, A. C. Yong, R. A. Woods *et al.*, '*p*-Aminophenol nephrotoxicity: biosynthesis of toxic glutathione conjugates.' *Chem. Biol. Interact.*, 1979, **27**, 245–254.

69. C. Klos, M. Koob, C. Kramer *et al.*, 'Effect of ascorbic acid, acivicin and probenecid on the nephrotoxicity of 4-aminophenol in the Fischer 344 rat.' *Toxicol. Appl. Pharmacol.*, 1992, **115**, 98–106.

70. L. M. Fowler, J. R. Foster and E. A. Lock, 'The nephrotoxicity of *p*-aminophenol. II. Th effect of metabolic inhibitors and inducers.' *Arch. Toxicol.*, 1993, **67**, 613–621.

71. M. L. Anthony, C. R. Beddell, J. C. Lindon *et al.*, 'Studies on the effects of L-(alpha-*S*, 5*S*)-alpha-amino-3-chloro-4,5-dihydro-5-isoxazoleacetic acid (AT-125) on 4-aminophenol-induced nephrotoxicity in the Fischer 344 rat.' *Arch. Toxicol.*, 1993, **67**, 696–705.

72. J. Klein, Y. Bentur, D. Cheung *et al.*, 'Renal handling of cisplatin: interactions with organic anions and cations in the dog.' *Clin. Invest. Med.*, 1991, **14**, 388–394.

73. L. B. Talner and A. J. Davidson, 'Renal hemodynamic effects of contrast media 1968 [classical article].' *Invest. Radiol.*, 1990, **25**, 365–372.

74. S. N. Heyman, B. A. Clark, N. Kaiser *et al.*, 'Radiocontrast agents induce endothelin release *in vivo* and *in vitro*.' *J. Am. Soc. Nephrol.*, 1992, **3**, 58–65.

75. K. B. Margulies, F. L. Hildebrand, D. M. Heublein *et al.*, 'Radiocontrast increases plasma and urinary endothelin.' *J. Am. Soc. Nephrol.*, 1991, **2**, 1041–1045.

76. D. P. Brooks and P. D. DePalma, *Nephron*, 1996, **72**, 629–636.

77. C. Casagrande, F. Santangelo, C. Saini *et al.*, 'Synthesis and chemical properties of ibopamine and of related esters of N-substituted dopamines—synthesis of ibopamine metabolites.' *Arzneim. Forsch.*, 1986, **36**, 291–303.

78. I. Yamaguchi, S. Nishiyama, Y. Akimoto *et al.*, 'A novel orally active dopamine prodrug TA-870. I. Renal and cardiovascular effects and plasma levels of free dopamine in dogs and rats.' *J. Cardiovasc. Pharmacol.*, 1989, **13**, 879–886.

79. M. I. Christie and G. W. Smith, 'Cardiovascular and renal hemodynamic effects of A-68930 in the conscious dog: a comparison with fenoldopam.' *J. Pharmacol. Exp. Ther.*, 1994, **268**, 565–570.

80. R. R. Bailey, T. G. H. Rogers and J. J. Tait, 'Measurement of glomerular filtration rate using a single injection of ^{51}Cr-Edetic acid.' *Australas. Ann. Med.*, 1970, **19**, 255–258.

81. C. Chantler, E. S. Garnett, V. Parsons *et al.*, 'Glomerular filtration rate measurement in ma n by the single injection methods using ^{51}Cr-EDTA.' *Clin. Sci.*, 1969, **37**, 169–180.

82. J. F. Klopper, W. Hauser, H. L. Atkins *et al.*, 'Evaluation of ^{99m}Tc-DTPA for the measurement of glomerular filtration rate.' *J. Nucl. Med.*, 1972, **13**, 107.

83. R. C. Reba, F. Hosain and N. H. Wagner, Jr., 'Indium-113m diethylenetriaminepentaacetic acid (DTPA): a new radiopharmaceutical for study of the kidneys.' *Radiology*, 1968, **90**, 147–149.

84. F. A. Cotton and G. Wilkinson, 'Advanced Inorganic Chemistry: A Comprehensive Text,' 4th edn., Wiley, New York, 1980, p. 984.

7.07
In Vitro Toxicity Systems

SUE M. FORD

St. John's University, Jamaica, NY, USA

7.07.1 INTRODUCTION

Renal toxicologists have at their disposal numerous *in vitro* techniques which are well-characterized with respect to the correspondence of their function and metabolism to that in the intact animal. *In vitro* models can be useful for descriptive toxicology, that is, to characterize the structural and functional lesions resulting from xenobiotic exposure. The models have proven valuable in mechanistic toxicology, for examining the biochemical and molecular means by which xenobiotics cause damage. In the applied areas of research, *in vitro* methods hold much potential for predictive toxicology (screening), whereby test compounds and new drug candidates are compared and ranked with the intention of predicting *in vivo* results for human safety testing. If the correlation between *in vitro* models and *in vivo* models were perfect, the use of animals in testing could be greatly reduced and risk assessment could be improved. However cells *in vitro* are often different in many ways than those *in vivo* and it is necessary to understand the changes which occur in renal tissue when it is removed from the body in order to appropriately select and use *in vitro* techniques.

7.07.1.1 Advantages of *In Vitro* Methods for Renal Toxicology

In vitro methods provide humanitarian, economic, and scientific benefits for basic research and safety testing (Table 1). Reduction in the number of animals used for humane reasons is an important end in itself. Although *in vitro* models generally start with fresh tissue taken from animals, the kidneys from a single animal are often sufficient for a complete study rather than using individual animals as treatment units. Animal use can be eliminated altogether in some cases. For example kidneys taken from the control animals of nonrenal studies can be frozen intact, thawed later, and used to prepare plasma membrane vesicles for transport studies.

Many of the economic benefits of *in vitro* models are derived from isolating the tissue from the whole animal and treating only the target tissue. Considerable savings are possible when using scarce or expensive reagents because the investigator need use only enough of the compound to treat the tissue rather than the whole animal.

In vitro models also offer concrete scientific advantages for studying organ-specific toxicity. Isolated tubules and cells offer the ability to study the effects of toxicants on discrete regions of the nephron. In the intact animal, the response of a given tissue to a xenobiotic can be the result of indirect actions, direct actions, or both, and *in vitro* models can help distinguish between the possibilities. Numerous treatment units can be prepared from the same batch of tissue, which reduces genetic and other sources of biological variability among treatment units. Frozen cultured cells and membrane vesicles can be shipped conveniently from laboratory to laboratory or stored for experiments performed years apart.

Experimental conditions *in vitro* can be pushed beyond those possible in the intact animal. Inhibitors can be used *in vitro* which would be toxic to the whole animal, and inhibition experiments can be done rapidly, since most *in vitro* models can be manipulated quickly. The dose reaching the target cells can be adjusted more accurately with *in vitro* experiments and much higher concentrations can be used without concern about effects on the rest of the animal. *In vitro* models are not only suitable for conventional biochemical and radioactive assays, but also lend themselves to techniques such as scanning electron microscopy, and nuclear magnetic resonance (NMR) spectroscopy.[1,2]

Table 1 Benefits of *in vitro* models for toxicology research.

Animal Welfare	Replacement of animals
	Greater number of treatment units per animal
Economic	Reduced cost of animals and animal care
	Decreased amount of scarce or expensive test compound needed
	Decreased disposal costs for radioactive or hazardous reagents
	Decreased disposal costs for contaminated carcasses, bedding, excreta
	In vitro models can be easily adapted for automated assays
Scientific	Isolation of tissue from extrarenal pharmacological effects of test compound
	Isolation of tissue from extrarenal drug metabolism
	Ability to study specific cells or nephron segments
	Ability to manipulate media composition and physical environment
	Greater control over duration of exposure to test compounds
	Ability to change experimental conditions more quickly
	Unphysiological conditions or high concentrations of test compounds can be used
	Greater flexibility in experimental design

7.07.1.2 Limitations of *In Vitro* Methods

There are a number of shortcomings common to all of the *in vitro* methods (Table 2). Although the number of animals used is reduced, it is still necessary to use animals for the source of tissue and for reagents such as hormones and serum. Cell culture models require the use of costly disposable plasticware or the personnel to properly wash, prepare, and sterilize glassware. A significant investment in equipment and the training of technical personnel may be needed for the isolated perfused kidney (IPK) and for cell culture.

Among the most important problems with *in vitro* methods are the functional, metabolic, and structural changes that occur when the kidneys are taken from the body. For preparations that will be used within a few hours, the acute risks are related to physical damage during the isolation procedures and to the risk of cellular hypoxia. *In vitro* models suffer from limited viability, typically in the 4–10 h range, though this may be extended with specialized and costly apparatus. Cultured cells do have extended viability and can be exposed to test agents for a longer interval. Yet for such models where the viability is prolonged, the trade-off is that many differentiated functions are diminished.

There are also technical difficulties with *in vitro* technologies which need to be considered in experimental design. Problems related to *in vitro* testing involving media pH, drug solubility, osmolality, binding of test compound to materials, and microbial contamination are discussed in a monograph from the Johns Hopkins Center for Alternatives to Animal Testing.[3]

7.07.2 ISOLATED PERFUSED KIDNEY

The IPK has the most complex level of organization of the *in vitro* models. Kidneys from treated or naive animals are removed with the renal artery, renal vein, and ureter attached, and are then perfused with physiologic media.

7.07.2.1 Description of Method

The procedures involved in preparation of the rat IPK have been exhaustively described in several excellent reviews.[4–8] The animal is anesthetized and a laparotomy performed. The right ureter is cannulated to permit collection of urine for monitoring renal function and analysis of excreted drugs and metabolites. The renal artery is cannulated via the anterior mesenteric artery and the perfusion is started. At this point, the kidney can be removed from the animal and transferred to a temperature-controlled perfusion chamber (Figure 1) or left *in situ* for the experiment.

The perfusate may pass through the kidney once or it can be recirculated to conserve a scarce reagents or test compounds. Radiolabeled inulin is included so that glomerular filtration rate (GFR) may be calculated by drawing aliquots of the perfusate and urine at intervals, though newer techniques allow continuous monitoring of GFR without taking samples.[8] Prior to the experimental period, the perfused IPK is equilibrated for 10–15 min while the perfusion pressure decreases and stabilizes. During the subsequent treatment period, urine and perfusate samples are obtained at 10 min collection intervals. When the experiment is completed the kidney can be frozen for analysis of drug or metabolite residues.

The physiologic function of the organ must be maintained within a reasonable range during the experiment in order for the results to be pertinent to the intact animal. Thus, the perfusion pressure, perfusate flow rate, urine flow rate, GFR, glucose reabsorption, and fractional excretion of Na^+ and K^+ must be

Table 2 Limitations of *in vitro* models for toxicology research.

Animal Welfare	Animals are the source of material for most *in vitro* models
	Animal sources are needed for some reagents, eg. serum for cell culture
Economic	Investment in equipment and apparatus can be prohibitive for occasional use
	Reagents and disposable labware may be expensive
	Specialized training and experience of personnel may be required
Scientific	Viability of tissue is limited
	Loss of structural and functional cell interactions found in whole tissue
	Isolation from extrarenal influences that may contribute to toxic mechanisms
	Hypoxic damage
	Difficulty in reproducing the *in vivo* environment to maintain cellular functions
	Reduction of differentiated characteristics of cultured cells
	Preparations of specific cells or nephron segments may be contaminated with other types

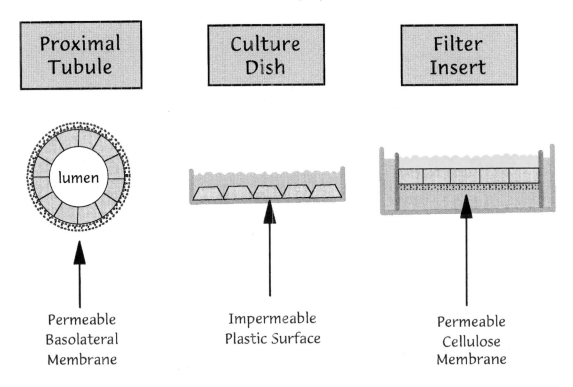

Figure 1 The basolateral side of proximal tubule epithelial cells *in vivo* (left) rests on a permeable basement membrane which allows diffusion of solutes from the blood. Cultured cells in a typical culture dish (center) grow on an impermeable plastic surface which will decrease access of solutes in the culture media to the basolateral membrane. The same cells grown on permeable membranes (right) exhibit structural and functional characteristics absent in those grown on plastic.[99]

monitored to provide information on the status of the preparation in order to facilitate interpretation of the data and comparison of results among laboratories. The renal vasculature constricts in response to many physiologic and mechanical stimuli, consequently the tissue must be handled carefully during and after surgery. As is the case *in vivo*, the isolated kidneys are sensitive to anoxia. Reduction in blood or perfusate flow due to constriction of the blood vessels or inadequate oxygenation can cause significant tissue damage within a short period of time.[8]

Bicarbonate buffers such as Krebs–Ringer or Krebs–Henseleit are typically used as perfusion media[4] and are oxygenated during use. The performance of the IPK and maintenance of tissue integrity can be affected by the perfusate composition, particularly with respect to energy substrates, amino acids, oxygen, and colloid content. The rate of sodium transport, for example, can be maintained at a higher level by increasing the available energy substrates and the oxygen content of the medium.[5] However with greater rates of transport the demand for oxygen might exceed that delivered, leading to hypoxic tissue damage in the medulla. Amino

acids have been shown to reduce hypoxic injury though the mechanism(s) responsible for such protection are uncertain. Addition of albumin to the perfusate increases the fractional reabsorption of sodium and the glomerular filtration rate is decreased but more stable.[4] Other possibilities for maintaining IPK function and integrity include addition of synthetic colloidal material, red blood cells,[9] or synthetic oxygen carriers.

7.07.2.2 Advantages

Unlike other *in vitro* systems, the IPK retains the complex morphological and anatomical relationships within the kidney. The blood vessels to and around the nephrons are intact so that circulation patterns which are involved in glomerular filtration, tubular secretion of xenobiotics, and concentration of urine are in place. The arrangements of the loop of Henle and the vasa recta which contribute to the intrarenal retention of certain xenobiotics are preserved. Such factors are important when employing the IPK to study the excretion and intrarenal distribution of a test compound.

Since the organ is left intact, the cells are not exposed to proteolytic enzymes nor are the cells damaged physically by homogenization or slicing procedures.

The IPK offers flexibility for experimental design. The perfusate can be manipulated to control the concentrations of ions and transport substrates. The investigator can sample venous outflow and urine from the IPK, which is not possible with any of the other *in vitro* methods. By simple surgical manipulation, a nonfiltering IPK can be employed to study the localization of secretion or reabsorption of xenobiotics.[10] The entire cortex and medulla are present in the IPK so that the complete renal metabolism of a xenobiotic can be studied; conversely, the renal filtration, secretion, and reabsorption of a test compound can be studied without interference by metabolites produced in other organs. The IPK enables the use of high concentrations of test compounds which may have extrarenal effects in the intact animal. For example the renal tubular transport of cimetidine, which has cardiovascular effects at high doses, was studied using the IPK.[11]

7.07.2.3 Limitations

The IPK requires a significant initial investment in equipment set-up and in the training of personnel for the surgical preparation and maintenance of the preparation. However, the technique is a very powerful one once it has been mastered. Great care must be taken in handling the organ in order to prevent vaso-constriction and subsequent ischemia.[8] The composition of the perfusate still needs to be optimized; in particular, most physiological buffers do not have the oxygen-carrying capacity of whole blood so that the function of the organ deteriorates quickly. This limits the interval in which the preparation can be used to several hours so that the IPK is unsuitable for studying the progression of chronic functional and structural toxicity *in vitro*.

Many cell types are present in the IPK and it can be difficult with this preparation to determine where or by which cell type a xenobiotic is metabolized. By isolating the kidneys from the body, critical extrarenal effects may be lost. The IPK is also impractical for screening purposes since it is a labor-intensive procedure.

7.07.2.4 Applications to Toxicology

The isolated perfused kidney is a good *in vitro* model to examine the functional effects, renal handling, and renal metabolism of a xenobiotic. It is the only *in vitro* technique appropriate for studying glomerular function.[5,8] The IPK has been useful in identifying the spectrum of metabolites resulting from renal biotransformation of xenobiotics.[7,8] The renal handling and metabolism[12,13] of acetaminophen, and the acetaminophen-induced depletion of GSH have been investigated with the IPK. It has been particularly valuable in exploring renal metabolism for acetaminophen and salicylic acid[4,14] because the liver produces metabolites similar

Table 3 Effect of ifosfamide and its metabolites in the isolated perfused rat kidney.

	Period[a]	No drug (n = 4)	Ifosfamide 470 µM (n = 4)	Acrolein 470 µM (n = 4)	Chloroacetaldehyde 210 µM (n = 4)
Urine flow rate[b]	C	152 ± 58[c]	88 ± 27	58 ± 16	99 ± 26
	E	201 ± 33	105 ± 33	64 ± 38	382 ± 90
Sodium reabsorption[d]	C	67 ± 12	52 ± 10	44 ± 11	85 ± 11
	E	75 ± 12	50 ± 9	39 ± 8	18 ± 6
Glucose reabsorption[e]	C	2.7 ± 0.6	2.0 ± 0.5	1.8 ± 0.5	3.4 ± 0.7
	E	3.1 ± 0.5	2.0 ± 0.5	1.5 ± 0.4	1.4 ± 0.2
Phosphate reabsorption[e]	C	0.9 ± 0.2	0.7 ± 0.1	0.5 ± 0.1	1.2 ± 0.2
	E	1.0 ± 0.2	0.7 ± 0.1	0.5 ± 0.1	0.4 ± 0.1
Sulfate reabsorption[e]	C	1.5 ± 0.3	1.3 ± 0.2	1.0 ± 0.2	1.7 ± 0.3
	E	1.5 ± 0.3	1.2 ± 0.2	0.9 ± 0.2	1.2 ± 0.3

[a]During each experiment the test chemical was absent from the perfusate in the control period (C) followed by an experimental period (E) during which the tested chemical was present. [b]µl min^{-1} g kidney weight^{-1}. [c]Results are mean ± S.E. [d]µEq min^{-1} g kidney weight^{-1}. [e]µM min^{-1} g kidney weight^{-1}.

to those produced by the kidneys. Additionally, the IPK would be a good choice to investigate potential drug interactions during metabolism or renal excretion.

The IPK was used to investigate the role of the nephrotoxic anticancer drug ifosfamide and its metabolites in the development of the Fanconi syndrome.[15] It was found that the metabolite chloroacetaldehyde reduced the reabsorption of sodium, glucose, phosphate, and sulfate (Table 3) without concomitant effects on the GFR or perfusion rate. Neither the parent compound nor another metabolite, acrolein, had such effects. The IPK can be helpful in distinguishing the relative contributions of transport and metabolism to toxicity, as for the case of thidione, a glutathione conjugate of menadione.[16]

Although the IPK preparation has limited viability, it is possible to use the preparation *ex vivo* to examine renal function several days after treatment of laboratory animals with test compounds. The functional effects of the anticancer drug cisplatin were assessed in untreated animals and those treated 48 h prior to the IPK experiment.[17] The glomerular and tubular dysfunctions caused by the drug could be dissociated from the changes in hemodynamics that occur *in vivo*, thus providing evidence of direct toxicity of the drug.

7.07.3　RENAL SLICES

Many important discoveries in renal physiology, pharmacology, and toxicology were initiated with slices, include seminal work in renal gluconeogenesis, Na^+ and K^+ reabsorption, amino acid reabsorption, and organic ion secretion. The functional and metabolic behavior of slices correlates well with renal function *in vivo*; cortical slices transport the same substrates and respond to the same transport inhibitors as the proximal tubule. Most toxicology studies utilizing slices have been done with those from the cortex due to the importance of proximal tubule transporters in renal toxicology.

7.07.3.1　Description of Method

The simplest method to prepare renal slices is freehand, using a razor blade and a glass microscope slide as a slicing guide. Slices should be uniform in thickness and taken only from the part of the kidney desired; that is, when collecting slices from the renal cortex, care should be taken not to cut into the medulla. Adult rats can provide a mass of 200–250 mg of

cortical slices per animal, generally enough for 2–3 beakers; using too much tissue in an incubation flask can produce invalid results due to depletion of metabolic or transport substrates from the media during the incubation period.

The incubation is typically done in a Dubnoff metabolic bath which permits control over temperature, shaking rate, and the gas environment. Typically, gluconeogenesis, amino acid transport, and other functional assays are carried out in a bicarbonate-based buffer such as Krebs–Ringer in an atmosphere of 95% O_2/5% CO_2 at 37 °C with a brisk rate of shaking. However, early reports[18–20] indicated that organic ion transport should be carried out at 25 °C and 100% O_2 for tissue integrity was prolonged at the lower temperature. The Krumdieck mechanical tissue slicer has made possible the production of precision-cut renal slices.[21] These are smaller, thinner, and more uniform than those prepared freehand, and O_2 diffusion is less of a problem.

Many toxicologic endpoints can be assayed in treated slices. Slice viability can be assessed with assays of membrane integrity such as lactate dehydrogenase (LDH) leakage. The rate of oxygen consumption (QO_2) or the ability to maintain intracellular K^+ levels[22] can also be used as indicators of general tissue status. Kidney-specific functions such as gluconeogenesis, and the activity of secretory systems for organic anions such as *p*-aminohippurate (PAH) and organic cations such as tetraethylammonium (TEA) or *N*-methylnicotinamide (NMN), have been found to be more sensitive toxicity indicators than LDH release.[22,23] When interpreting results about organic ion transport, one has to be careful to distinguish between a true untoward effect on cellular function caused by the test compound and competition with the indicator substrate for transport or binding sites. Other measurements of toxicological relevance such as glutathione concentration and lipid peroxidation can be assayed in slices.[23]

7.07.3.2　Advantages

The slice technique is a versatile tool for toxicological studies and, if appropriate endpoints and conditions are chosen, the results correspond well with toxicity *in vivo*.[23] Data obtained from slice experiments are generally very reproducible, even with the freehand method. Preparation of slices from the cortex or medulla is quick and inexpensive. Slices can be made from the kidneys of many species, from fish to mice to humans, and from the kidneys of newborn laboratory animals. All of the cell types found in the tissue *in vivo* are present in slices and the relationships between the cells are maintained. The effects of xenobiotic concentration, time of exposure, and influence of other media constituents can be conveniently studied. Data that can be difficult to obtain by other methods can be acquired quickly with slice experiments. For example, slices can be manipulated quickly so that kinetic parameters of transport and inhibition can be estimated. Slices can be used to study both the effects of test compounds applied *in vitro* as well as the functional effects in tissue taken from treated animals.

The viability and structural integrity of the tissue are maintained for several hours of incubation. If the incubation period is short enough, the cells will use endogenous substrates for energy obviating the need for added substrates. The activity of most transport and metabolic processes also remain high so that small quantities of tissue or costly reagents can be used. Newer techniques enable investigators to maintain slices for up to 24 h.[21,24] These sophisticated and complex arrangements permit exposure periods more relevant to the *in vivo* situation, as well as allowing sufficient time for the development of histological lesions.

7.07.3.3 Limitations

Much of the controversy about the use of slices has focused on two issues: first, there have been concerns about the viability of cells in slices, and second, the lumen of the tubules have been reported to be collapsed which would imply that slices should not be used to assess transport processes at the brush border. Rose *et al.*[25] and Kleinzeller[18] evaluated several of the objections about the use of slices and concluded that with proper technique and conditions, these problems could be avoided.

With regard to the first concern, thin slices are needed to permit rapid diffusion of gases and solutes to and from the interior and prevent anoxia and deterioration of function. This is especially important for slices made from the renal cortex which is a tissue with a high rate of oxygen consumption. The limiting thickness of the slice is around 0.3–0.4 mm when incubated at 25 °C under 100% oxygen.[18] The lower temperature helps to alleviate the problem of gas diffusion by both reducing the metabolic rate of the tissue and increasing the solubility of oxygen in the media. In contrast, the studies showing histological abnormalities in freehand slices after incubation were generally done at 37 °C.[18] Another problem is the presence of damaged and crushed cells and nephrons on the cut surfaces. The damaged units, while only representing 5–10% of the population of cells in slices 300 μ thick,[26] will represent a greater proportion in thinner slices. Thus, slices shouldn't be too thin, since the proportion of damaged cells would be increased[18] and the slices would be friable.

The second major concern with slice work is based on the premise that the tubules will collapse once the slices are made and the glomerular filtration process is eliminated, subsequently reducing access of reabsorbed sub-strates to the lumen. However, the lumina of tubules in rabbit cortical slices were clearly observed to be open after 90 min incubation in transport buffer (personal observation). Furthermore, slices accumulate the sugar α-methylglucoside (AMG) at the same rate as both isolated tubules and the cortex *in vivo*.[18] In fact, many investigators have used productively slices to study reabsorptive processes occurring from the lumen. Nonetheless, it is likely that solutes will have slower access to the lumen than to the basolateral side and the results will be qualitatively valid, rather than quantitatively applicable.

7.07.3.4 Applications to Toxicology

The slice technique has long been used to study the renal handling of xenobiotics postulated to be secreted by the kidneys; because of its ease and versatility it will remain important for such studies. Slices are well suited for studying the biotransformation of xenobiotics, as they contain cytochrome P450-dependent drug-metabolizing enzymes which metabolize and bioactivate some nephrotoxicants, and sections can be taken from either the cortex or medulla and the metabolic products isolated. The toxicity of metabolites from extrarenal sources can be investigated by co- or pre-incubation with a second tissue such as liver slices. The content of GSH can be manipulated to examine the role of GSH depletion and possible lipid peroxidation in injury.[23] Furthermore, slices possess the pathways of intermediary metabolism which may be targets for toxicants. Kidney slices can be used to elucidate the mechanism of action of a known toxicant by modulating the activities of transporters, drug-metabolizing enzymes, or pathways of intermediary metabolism. Many xenobiotics are toxic because of their high accumulation in the kidneys; the use of transport inhibitors can help determine the role of transport-dependent accumulation in the toxicity of a xenobiotic. Slices have been used to compare the relative importance of renal secretion, renal metabolism, and extra-renal metabolism in the toxicity of bioactivated halogenated hydrocarbons.[22,27] Slices can be used to study effects of toxicants on cellular energy metabolism and by use of the appropriate conditions, can be used to answer questions about metabolic pathways.[28]

The ease of preparation and the good correlation of injury *in vitro* with that seen *in vivo*[23,29] makes slices a suitable system in which to develop rapid screening tests for nephrotoxicity. These tests can be based on any number of quickly measured endpoint assays, such as oxygen consumption, transporter activity, leakage of cellular enzymes, ion content, and metabolic activity.[30]

7.07.4 ISOLATED GLOMERULI, TUBULE SEGMENTS, AND CELLS

The goals of the isolation procedures are to physically separate the cells or nephron segments from each other with a minimum of damage and to subsequently enrich the sample in the desired cell type. There are three basic steps involved: (i) dissociation of the tissue into nephron segments or single cell suspensions, (ii) enrichment or purification of the units of interest, and (iii) characterization of the final preparation with respect to both viability and identity. The important variables in a dissociation protocol include the type and concentration of proteolytic enzymes, the length and temperature for incubation, and the ion content of the buffer. The optimal conditions required will depend on the product; nephron segments can be dissociated with gentler treatment than isolated cells.

7.07.4.1 Description of Methods

7.07.4.1.1 *Isolation of nephron segments*

For some *in vitro* studies only a few isolated tubule segments are needed, these can be teased individually from tissue slices with a dissecting needle.[31] When nephron segments are to be used for biochemical or functional assays, methods for quickly obtaining larger quantities of homogeneous units are needed.

Several laboratories have described methods to isolate nephron segments following perfusion of the intact kidney with digestive enzymes or calcium chelators in order to expose efficiently the interior of the organ to the medium.[32–37] Mannitol may be added to the perfusate in order to keep the tubular lumina

open[38] and chelators such as EDTA (ethylenediaminetetraacetate) included to weaken intercellular junctions by removing calcium.[39] When costly reagents are used, the perfusate may be recirculated through the kidney.[37] The investigator may seek to maintain renal metabolism at the *in vivo* level during the perfusion interval, in which case the system is held at 37°C. The medium must be well-oxygenated and contain metabolic substrates. Alternatively, ischemic damage can be alleviated by perfusion with cold media.

More often, dissociation of the tissue is done after dissection of the kidney. The cortical tissue contains primarily the glomeruli and the proximal and distal tubules, and the inner portion is a source of loop of Henle, collecting ducts, and interstitial cells. Finer divisions can be made in rabbit kidneys where the outer stripe of the medulla can be distinguished.[40] Once the separation has been made, nephrons are dissociated from the tissue. For the rabbit cortex, this can be easily accomplished by employing a Dounce homogenizer with a loose clearance. Dickman and Mandel noted that the commercial source and quality of the rabbits influenced the ease of release of tubules from the cortex,[41] which agrees with our own experience. Papillae or kidneys with more fibrotic tissue will require more rigorous treatment to dissociate the tubules;[38] age is probably another factor.

The tissue is treated with proteolytic enzymes after it is sliced or minced, with the size of the minces determining the length of the tubules in the final preparation. Glomeruli and tubules released by the enzymatic action are harvested at intervals by letting the heavier tissue pieces settle down and then removing the tubule-containing supernatant. For rat tissue, it may be helpful to add bovine serum albumin (BSA) to prevent clumping of the tubules.

Comparison of isolation procedures indicated that collagenase treatment decreased oxidation of succinate and the rate of gluconeogenesis by tubules.[33] Other investigators observed no differences between collagenase-treated and untreated tubules.[41] Enzyme treatment should be as brief as possible to minimize damage, although disrupting the basement membrane with collagenase treatment is necessary to enable outgrowth of cells if the tubules are to be cultured.[42] When especially gentle treatment is necessary, lower temperature can be used to control protease activity, and soybean trypsin inhibitor can be added to collagenase digestions. There are several types[43] and grades of collagenase, some of which are recommended for renal tissue.[33,34,41] The grades are not strictly comparable among manufacturers and the potency of the enzyme and presence of contaminating proteolytic enzymes can vary among lots from the same source. Most companies which sell collagenase allow the customer to test samples of different lots and to reserve large quantities of those which give the best results.

7.07.4.1.2 *Isolation of single cell suspensions*

Isolation of single cells from epithelia requires more vigorous enzymatic treatment than for nephron segments since the cells must be separated from each other and released from the extracellular matrix and basement membrane. The initial preparation may be as simple as mincing the tissue containing the desired cells, incubating the pieces with proteolytic enzymes and harvesting the cells as described in Section 7.23.4.1.1.

Minimizing cell damage requires that the cells be exposed to the proper digestive enzymes for the shortest duration needed. Collagenase alone is sufficient for isolating male rat cortical cells by perfusion[44] and soybean trypsin inhibitor or other protease inhibitors[45,46] can be added to reduce the activity of contaminating enzymes. Trypsin, collagenase, pronase, dispase and others can be used alone or in combination.[47,48] Many commercially-available enzyme preparations are contaminated with several other activities, which explains the observation that crude trypsin is sometimes more effective for tissue dissociation than the highly purified enzyme.[49]

When the cells have been isolated, the action of the enzymes must be halted. Trypsin can be inhibited by adding soybean trypsin inhibitor or serum. Inhibitors of other enzymes are rarely used due to their cytotoxicity. Instead, the enzyme activity is stopped by rapidly cooling the solutions on ice, centrifuging the sample, and removing the enzyme-containing media.

7.07.4.1.3 *Enrichment methods*

After the cells or nephron segments have been dissociated from the matrix, the heterogenous mixture is fractionated to purify the desired populations. Reported enrichment of proximal tubules by the various methods generally runs in the 90–95% range.[38] Often, separation is based on density gradient methods such as Percoll[35,39,40,50–52] or Ficoll,[39,53] although other methods can be used. Lash and Tokarz[39] reported that separation of cells on Percoll results in a Gaussian distribution rather than bands as is found with tubule fragments.

Enrichment of tubules or glomeruli can also be accomplished by taking advantage of size differences. The various segments can be retained on nylon fabric or other precision-mesh sieves. The corresponding nephron segments of animal species vary in size;[54–56] fortunately nylon material is available with a wide spectrum of meshes. Another approach to separation of biological material is to use techniques based on surface properties, such as two-phase aqueous partitioning.[57] Nephron segments or cells which will be cultured can be purified by immunodissection, where monoclonal antibodies are used to attach populations of specific nephron segments to culture dishes.[58–60] Glomeruli can also be eliminated from a tubule suspension by perfusion of the intact kidneys with a paramagnetic iron oxide solution. The iron oxide particles become lodged in the glomeruli which can be removed later from the suspension by attracting them to magnetic stirring bars.[33]

7.07.4.1.4 *Assessment of viability and purity*

The viability of the final tubule or cell suspension should be evaluated[61] with a suitable method such as trypan blue dye exclusion,[50] NADH penetration,[37] LDH leakage,[50] neutral red dye accumulation,[62,63] ATP content,[64,65] intracellular K^+ levels,[65] or the rate of oxygen consumption.[50,64,65] The efficiency of separation of proximal from distal tubules or cells can be determined by marker enzymes, hormone responsive adenylate cyclase, segment-specific specific transport activities, or light microscopy. These procedures will serve both to assess the enrichment of the desired fraction compared to the starting suspension as well as to document the degree of contamination with other cells in the final suspension. Marker enzymes for proximal tubule cells include alkaline phosphatase, gamma glutamyl transpeptidase, and glucose-6-phosphatase, while hexokinase activity and renal kallikrein are used as markers for distal tubule cells.[35,50] Stimulation of adenylate cyclase by hormones (parathyroid hormone, calcitonin, vasopressin) is also helpful to characterize tubule and cell preparations[35,65,66] since the nephron segments have distinct patterns of hormone response. Other markers for cells of renal origin can be used.[36,44,67]

7.07.4.2 Advantages

The procedures to isolate cells or nephron segments require time and care to ensure the viability and to confirm the identity of the preparation. However, the methods are simple and do not require elaborate equipment. Many toxic endpoints can be used with isolated nephron segments[38] including the viability assays mentioned in the previous section, transport functions for the proximal tubule, and indices of cell metabolism such as intracellular ATP content. Isolated nephron segments are more suited than the IPK or renal slices for studying cell-specific reactions to toxicants because particular parts of the nephron can be treated. Proximal tubule suspensions accumulate TEA to levels threefold higher than slices, and this is thought to be due to the higher proportion of transporting tubules in the preparation.[68] The cells in nephron segments retain their polarity and relationships to one another as they are not removed from the underlying basement membrane. Thus the orientation of transport is similar to that of the nephron *in vivo*.[31]

Oxygenation of the cells and substrate access to the basolateral (and probably the luminal) side in isolated nephron segments is quicker than that of slices since there are no tissue layers to hinder diffusion. Balaban *et al.*[32] reported that rabbit cortical tubule segments maintained a higher ATP content (7.7 ± 0.32 nmol ATP mg^{-1} protein) compared to cortical slices (4.0 ± 0.26 nmol ATP mg^{-1} protein). Furthermore, the rate of oxygen consumption (QO_2) by slices dropped off much more quickly with decreasing media pO_2 compared to tubules, indicating that diffusion of oxygen limited the metabolism of slices. Tubule preparations maintain a constant QO_2 for at least 6 h[53] and the viability can be extended somewhat by alterations in experimental conditions.[41] Tyson *et al.*[69] found that inhibiting lipid peroxidation by adding deferoxamine to the collagenase perfusion of kidneys will more than double the period of viability of the resulting tubules.

7.07.4.3 Limitations

Isolated tubule fragments and single cell suspensions do not have the same problems of oxygen and substrate diffusion as do slices. However, the amount of material in an incubation flask relative to the oxygen or substrate concentration is critical, since the medium may be depleted of oxygen at high rates of utilization. The oxygen-carrying capacity of media should be calculated or preliminary studies done to assure that QO_2 or cellular metabolism is linearly related to the amount of tissue. Tissue dissociation into single cells causes significant damage to cell membranes which can alter the response of cells to toxic chemicals. Consequently, single cells are fragile and

may be damaged by vigorous shaking during incubations.[37] It has been found that tubule suspensions maintain a higher viability than single-cell suspensions and can be maintained up to 24 h.[41] Extra care may be needed when working with single cell suspensions or tubule segments of some species as some cells or nephron segments stick to glassware. This can be decreased by the use of plasticware or siliconized glassware.[37] Alternatively, BSA in the media can help prevent this as well as the clumping of tubules.

Investigators who prepare suspensions of cells or nephron segments often have to balance purity against yield in the procedure. At one extreme the identity is certain when using individually dissected segments. However microdissection is exacting and produces little material. In toxicology, it is most useful for microperfusion work[70] or culture work where contamination by other regions is to be strictly avoided.[58,71,72] More commonly, populations of tubules are needed and the goal is to separate the sample into two or more homogeneous fractions such as proximal and distal tubule suspensions. Proximal tubule cells are the predominate type in the cortex and generally can be obtained with the greatest abundance. For example, the recovery of tubules from pooled rat kidneys separated on Percoll was 77%, with 67% recovered for proximal tubules but only 10% for distal.[35] Quantitatively similar results were obtained with suspensions of proximal and distal tubule cells.[39]

7.07.4.4 Applications to Toxicology

Isolated glomeruli, tubule suspensions, and single-cell suspensions are particularly well-suited for studies examining the biochemical and functional responses to xenobiotics. For example, single tubules can be perfused in order to monitor directional transport of test compounds or the effect of xenobiotics on the transport of model substrates. Tubules can be used in the aggregate for numerous assays of toxic endpoints such as oxygen consumption, organic ion transport, or other biochemical measures of response to toxicants in mechanistic or screening studies. Rat proximal tubule suspensions were used to investigate the toxic mechanism of the styrene metabolite S-[(1 and 2)-phenyl-2-hydroxyethyl]-cysteine (PHEC).[64] Based on changes in QO_2 and intracellular ATP content in the first 30–45 min of exposure, the authors concluded that mitochondria are an early target of PHEC. They also showed that prevention of PHEC-induced lipid peroxidation did not alleviate cytotoxicity as indicated by tubule respiration and LDH release. Such assays can be used with isolated cells. Additionally, single cell suspensions allow sophisticated microscopic examination of changes in intracellular Ca^{2+} distribution[58] following exposure of the cells to xenobiotics. Cell suspensions are especially valuable to obtain information on how specific cell types react to toxicants and are more convenient than tubules for assays requiring uniform particles.

When cells are detached from the substratum or basement membrane the polarity is lost.[73,74] Transport studies with substances that are transported unidirectionally across tubular epithelia must be designed with care when single cells are used. Nonetheless, it is possible to acquire valuable information about the transport of xenobiotics. Rat proximal tubule cells accumulated indomethacin and salicylic acid by a probenecid-sensitive mechanism, similar to findings using the IPK.[75] Compared to other models, suspended rat proximal tubule cells were found to be the most sensitive to a variety of toxicants, exhibiting decreased α-methylglucoside (AMG) uptake at lower concentrations of the test compounds.[76]

7.07.5 CULTURED CELLS

Early studies which explored the use of cultured cells for toxicology studies relied on gross measures of cell viability and comparison of the ability of test compounds to kill cells. However, since the late 1970s the use of cultured cells in toxicology has undergone a revolution as the potential for use of *in vitro* models for mechanistic studies been recognized. Unlike research into basic cellular processes, toxicology studies require that the cells maintain the differentiated characteristics of the original tissue in order to accurately reflect the *in vivo* responses of target tissues to toxic chemicals. As a result, basic and applied research in toxicology is driving the development of increasingly sophisticated cell culture models and the introduction of commercial devices to make such models feasible for routine use.

7.07.5.1 Description of Methods

The culture environment is drastically different from that which the cells inhabit *in vivo* and the cells consequently undergo considerable changes. They may show little or limited proliferation; this is what happens with some types of organ culture. Otherwise, culture conditions are often intended to foster proliferation, which further changes the nature of the differentiated cells.

The types of renal material used to initiate cultures includes complex pieces such as cortical slices and tissue minces, tubules, or single cells. The methods to isolate, purify, and identify such material are similar to those described in the preceding section except that clean or aseptic technique should be used. Sterile techniques and procedures for the maintenance of cultured cells are ideally learned from a working tissue culture laboratory, although excellent manuals are available.[47,77]

7.07.5.1.1 Types of culture systems

In the case of organ culture the objective is to maintain the structural integrity of the sample during the culture period.[78] When growth and cell proliferation do occur, the tissue retains its identity as a unit. Organ cultures include cultured embryos, limb buds, embryonic kidneys,[79] and renal slices. The advantage of organ culture is that the complex interactions of the various cells in a tissue can be retained. A major problem is that the tissue cross-section in all planes must be small enough so that the interior of the section maintains adequate gas and nutrient exchange with the culture medium. Ultimately, the viability of the tissue is limited. Organ culture has been useful for teratogenicity studies since embryonic cells will not only proliferate but will develop into more differentiated structures.

Cell culture, on the other hand, involves the isolation and maintenance of populations of cells which are not organized into tissues.[78] The cells can be isolated prior to culturing, or explants (tissue sections, minces, or nephron segments) can be placed in culture and the cells will dissociate over the course of a few days. One can select an area such as the cortex to culture, or use the procedures described in the previous section to isolate specific cells or nephron segments.

When cells are taken directly from a plant or animal and cultured, the model is referred to as primary culture. The cells are used for experimentation after they have proliferated, adapted to the culture environment, and in many cases, become confluent. Numerous renal primary culture models have been developed including the rabbit proximal tubule,[42,54] medullary interstitial cells,[80,81] medullary thick ascending limb cells,[65] and glomeruli.[82,83] In contrast to primary cultures, cell lines are cultures which have been transferred at least once to another vessel (subcultured), and more often transferred multiple times. In some cases an alteration occurs in such cells which allows them to be passaged indefinitely; subsequently they are referred to as a continuous cell line.[78] This process, called *in vitro* transformation, can be spontaneous or it can be induced by chemicals or radiation. The nature of this change is unknown, but for many cultures it is distinct from malignant transformation. Continuous renal cell lines which can be considered for toxicology work include LLC-PK$_1$,[84] Madin-Darby Canine Kidney (MDCK),[85] opossum kidney (OK) cells, and LLC-RK$_1$.[86]

7.07.5.1.2 Refinement of culture methodology

In addition to altered proliferation, cultured cells often exhibit deterioration of cell-specific function and morphology, a phenomenon referred to as dedifferentiation. This is an important issue in toxicology studies because responses to xenobiotics *in vivo* may depend on the activity of enzymatic pathways and transporters which may be diminished when cells are cultured. Significant improvements in cell culture technology will be made as the stimuli which maintain differentiated characteristics *in vivo* are identified. It is expedient in the meantime for the toxicologist to empirically alter culture conditions to suit a particular need. The parameters that are generally modified include the nutrient composition of the culture media, the presence and concentrations of various hormones and chemical mediators, and the physical environment.

Formulation and preparation of cell culture media are time-consuming and painstaking procedures.[87] Laboratories which employ cultured kidney cells generally use standard formulas developed in the past to support rapidly-proliferating embryo or tumor cells. Consequently, most media are not optimized for the maintenance of nonmalignant differentiated cells and contain unphysiologic concentrations of substrates such as glucose or amino acids. These conditions may foster the glycolytic metabolism found in cultured cells,[88] rather than the oxidative metabolism upon which the renal cortex depends *in vivo*. Since the 1970s, a greater selection of commercial media have become available, though it remains a process of trial-and-error to find a suitable one.

Biological extracts such as fetal bovine serum (FBS) are added to culture media to provide unknown hormones and growth factors. However, serum is a poorly defined reagent. In addition to the identified factors which can

be assayed there are others which remain unknown. Serum fosters the overgrowth of fibroblasts in some primary renal cultures and contains proteins which can bind test agents. Serum-free, hormonally defined media have been developed for primary renal cultures,[54,89] several renal cell lines, including LLC-PK$_1$[90] and MDCK.[91,92]

The behavior and morphology of cells *in vivo* and *in vitro* are affected by the substratum on which they rest. Tissue culture plastic can be coated with substances such as collagen, laminin, fibronectin, or other biological extracts. These substances enhance attachment of the cells to the plastic as well as improve differentiated morphology and functions[93–96] and influence polarity.[97] *In vivo*, the basement membrane of the renal tubules is permeable which allows substances to diffuse from the blood to the basolateral side of the tubular epithelium. In contrast, kidney cells grown in tissue culture dishes are situated with the basolateral membrane attached to non-porous plastic and the apical side is bathed with the culture medium (Figure 2). Where the junctions are closed in a monolayer of transporting epithelium grown on plastic, fluid-filled blisters called domes are formed.[98] Diffusion of solutes from the apical to the basolateral surfaces is limited at these closed junctions[73] so that exposure of the basolateral side to nutrients and test compounds is hindered.

To facilitate access to the basolateral membrane, cells can be grown on porous membrane filters. Commercial filter inserts are available in a selection of configurations and filter materials to fit culture wells of different sizes. The filters can be coated with suitable materials to enhance cell growth and differentiation if desired. Cells cultured on porous membranes (Figure 2) exhibit structural and functional characteristics lacking in cells grown on plastic.[73,84,99] Tight junctions in LLC-PK$_1$ monolayers grown on filters open transiently when the media are disturbed,[100] an observation that has important implications for transport studies with cultured cells.

Although some tissue-specific functions may decline, cultured cells are responsive to modulating factors. When rabbit cortical collecting tubule cells were cultured, the transport behavior of the cells depended on the seeding density.[97] Cells seeded at lower concentrations became bicarbonate-secreting intercalated cells, whereas those inoculated at high density became acid-secreting intercalating cells. This phenomenon was related to the synthesis of an extracellular matrix protein by the cells seeded at higher density. Such information, when available, could be used to manipulate the

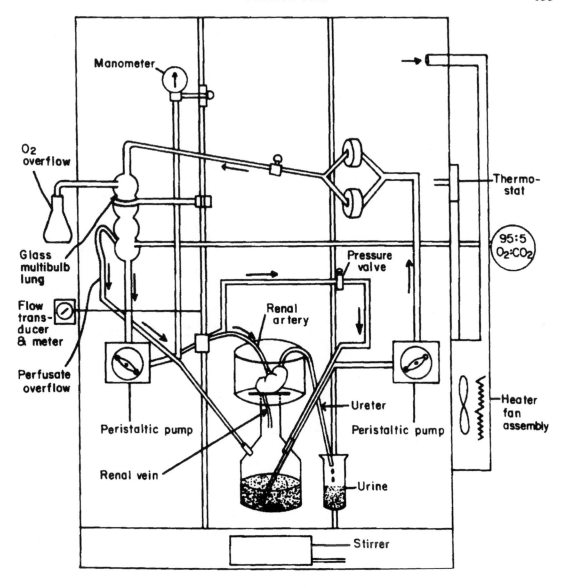

Figure 2 Isolated perfused kidney apparatus.[6]

cells and produce the desired phenotype. It is likely that the optimal conditions for expression of differentiated functions will be highly specific for each cell type. In fact, it is known that different growth factors are needed to promote the growth of specific segments of the nephron.[72]

7.07.5.2 Advantages

The extended viability of cultured cells has particular value for toxicology studies when it is necessary to treat the cells with a test compound for several days or when the events following toxic injury are to be studied. The ability of cultured cells to divide *in vitro* makes it possible to use them to examine the effects of test

compounds on developing tissues, as well as to study proliferation[101] following toxic injury. *In vitro* propagation of cells is helpful to make the most of scarce material such as human or primate kidney tissue as it becomes available.[82,94,102] Different cell types (glomeruli, proximal tubule, distal tubule, etc.) can be cultured from the same animal.

Cell lines and primary cultures each have distinct advantages. Older cell lines such as LLC-PK$_1$ were usually obtained from a cortical mince so that the initial cultures were comprised of multiple cell types. Currently the stock LLC-PK$_1$ line at ATCC appears to be mostly proximal tubular in nature, but does have some distal tubule characteristics.[84,103] Other renal cell lines, including MDCK cells, also have unclear origins within the nephron, whereas with primary cultures the desired cells can be enriched and characterized at the time of isolation. Primary cultures are often closer in function and metabolism to the corresponding cells *in vivo*, unlike cell lines which become altered in character as they are passaged in culture. On the other hand, cell lines are easier to handle and manipulate. They can be conveniently frozen[104] and used in the future, which facilitates comparisons among experiments. Cell lines can be more reliably cloned to provide a homogeneous population or to select cells with specific responses.

7.07.5.3 Limitations

While biochemical features may be maintained in primary cultures long enough for toxicology experiments to be performed, some transporters and biotransformation enzymes show dramatic reduction in activity in the first few days after the cells are removed from the animal.[105–107] One way to overcome diminished drug-metabolizing capacity for experimental purposes is to add it back in the form of freshly prepared renal S9 (9000 × g supernatant).[108]

Cell culture requires special facilities for aseptic handling of cells and technical personnel must be trained in tissue culture maintenance. Fastidious laboratory housekeeping procedures are needed, particularly when cells are grown in serum-free media. Often this factor is not fully appreciated by individuals coming from a nontissue culture background. Even skilled tissue culture laboratories may occasionally experience intervals when the cells suddenly do not grow or the cultures behave abnormally. The situation is apt to correct itself with little or no insight into the nature of the problem.

Nutrient media for cultured cells are also good environments for the growth of mold, yeast, bacteria, and mycoplasma. The use of antimicrobial agents in culture media should be discouraged because resistant microbes can survive at low levels in antibiotic-containing media.[109] Detection and identification of mycoplasma contamination requires special procedures and skill,[109–111] so that sending samples outside for routine testing can be cost-effective for many laboratories.

Contaminating cells such as fibroblasts in primary cultures may grow at a faster rate than epithelial cells, so that the cells at the start of an experiment may be different than those initially isolated. Continuous cell lines are also exposed to selection pressures in the culture environment and, in addition, may experience genetic drift. Cell lines which are passaged for long periods of time eventually exhibit changes in morphology and behavior, as do cell populations from the same stock but grown in different laboratories.[111] Proper cloning and maintenance of stock cell lines[112] can minimize these problems. It is necessary for the investigator to characterize the cultures as described in Section 7.23.4.1.4 in order to confirm that the population is the desired one as well as to measure the activity of functions that are important in the experiment planned.

Common laboratory procedures which work well with fresh tissue often need to be modified for cell culture use. Cell fractionation protocols may require modifications because the nature of the cytoskeleton of cultured cells can make it difficult to release intracellular organelles during conventional homogenization.[113,114] Cell culture work usually involves a much smaller amount of tissue than similar work with fresh tissue. Biochemical assays used might need to be reduced in scale to accommodate the small mass of available cells[51] and detection can be a problem for analytical methods lacking appropriate sensitivity. There are apparatus such as roller bottles and microcarrier beads for increasing the production of cells.[115] However, these require special equipment and the quantities of reagents and supplies needed to support scaled-up cell production can be expensive.

7.07.5.4 Applications to Toxicology

Although cultured cells have reduced activities of certain functions, this can be used advantageously. For example, since most biotransformation activity is negligible, the role of individual metabolites in the toxicity of a xenobiotic can be studied without inhibitors. Despite the loss of some activities, cell cultures do retain many differentiated structures and functions.[74,116] Organic cation transport has been detected in several culture models including LLC-PK$_1$,[117–121] MDCK,[122] and primary proximal tubule cultures.[123] The few papers reporting PAH transport in cultured cells have been with rabbit[123] and flounder[124] primary proximal tubule cultures and the OK cell line.[125] Rabbit proximal tubule cultures retain the apical membrane transporters characteristic of the tissue *in vivo*.[123,126,127] Thus, they were very suitable to use for comparing the effects of several organoplatinum compounds on glucose and amino acid transport.[128] The OK cell line possesses receptor-mediated endocytosis and was used to investigate the mechanism of proteinuria caused by ochratoxin.[129]

The relationship between uptake and toxicity of nephrotoxicants has been studied with cultured cells. The proportion of cadmium taken up from the apical and basolateral sides of LLC-PK$_1$ cells grown on filters was shown to be dependent on the concentration of the metal.[130] *In vivo*, cadmium-metallothionein (Cd-MT) is accumulated by the kidneys more avidly than cadmium, and is more toxic. Numerous studies have shown that Cd-MT is less toxic than cadmium to primary rodent proximal tubule cultures and to LLC-PK$_1$,[130–132] which was attributed to the greater uptake of cadmium than Cd-MT by the cultured cells.[132] In the mid 1990s it was reported that mercury but not Hg-MT was accumulated by the isolated perfused rat proximal tubule.[70] Similar to the case of cadmium, these mercury data are inconsistent with those obtained *in vivo*. Thus, the cultured cells respond similarly to tubules and these results with the two *in vitro* models taken together indicate that there are unknown factors which mediate the uptake of the metal-MT complexes *in vivo*.

Culture models offer opportunities to study the effects of xenobiotics on development and differentiation,[79,133] proliferation,[134] repair, and genotoxicity.[135] Cell cultures can be synchronized to allow studies of the effects of toxicants on cell cycles. For example, the hamster kidney cell line (HaK) was used to study the effect of ochratoxin A on DNA degradation in cells which were synchronized in G$_1$- and S-phase.[136] Cell culture models also have significant potential in the areas of gene fusion, cell hybridization, and molecular biology. Mutants or variants with desired characteristics can be produced and used to study basic biochemical processes.[137]

Cell cultures facilitate the ability to treat and observe living cells and are particularly suited to techniques such as NMR,[1,2] monitoring intracellular Ca^{2+} distribution,[138–142] and autoradiography. Cultured cells are convenient for ultrastructural studies using scanning or transmission electron microscopy as well as newer ion microscopy techniques.[143] Despite the tremendous differences between the *in vivo* and *in vitro* environments, cultured cells show the same distinctive histological changes following exposure to xenobiotics as found in the kidneys *in vivo*.[144,145]

Cultured cells will become increasingly important for screening and predictive toxicology as the technology for growing differentiated cells improves.[29,30,126] Cultured cells provide unparalleled convenience and economy for simultaneously assessing the relative nephrotoxicity of many drugs. The cells can be grown in multiwell dishes, with 6–96 potential treatment units on one plate. Automated devices are available to perform assays in

such plates and spectrophotometric or fluorometric plate readers can detect analytes without transferring the cells from the plate.[146,147] Potential endpoints for cell viability include vital dye exclusion (trypan blue, nigrosine), neutral red uptake,[62,63] or LDH release.[148] Sublethal indexes such as mitochondrial MTT reduction[149] or the maintenance of intracellular K$^+$ are generally more sensitive than viability assays, and measures of kidney-specific functions are even more so. For example, uptake of AMG was a more sensitive index of cadmium toxicity than was LDH release,[132] and transepithelial resistance was more sensitive than the ability to maintain intracellular K$^+$.[130] In addition to assessing conventional drug toxicity, cell culture also holds potential for evaluating the toxicity of biotechnology products[150] and the effects of xenobiotics on the renal handling of proteins.[129]

7.07.6 MEMBRANE VESICLES

Active reabsorption and secretion by cells are ultimately driven by cellular energy; however few of the plasma membrane transport systems in the proximal tubule are primarily active. Rather, there are multiple interrelationships among the transporters such that secondary and tertiary active transport systems predominate, most of which ultimately depend on the sodium gradient established by Na$^+$/K$^+$-ATPase on the basolateral membrane. Isolated plasma membrane vesicles can be used to determine the mechanism of entry of a xenobiotic into the cell and to study the actions of toxicants on membrane transporters separately from possible effects on the energy supply or on the ion gradients established by the sodium pump.

7.07.6.1 Description of Method

Isolation of membrane vesicles from kidneys of rats and rabbits is initiated by dissection of the cortex, homogenization, and centrifugation to remove cellular debris. During these steps, proteolysis should be inhibited by keeping the preparation cold and by the addition of protease inhibitors.[46] Subsequent purification and separation of the brush border membrane vesicles (BBMV) from the basolateral membrane vesicles (BLMV) can be accomplished by taking advantage of differences in membrane density, differences in surface properties, or both. One of the most frequently used methods uses the difference in surface charge between the apical and basolateral membranes to effect the separation in buffers containing Ca^{2+}, Mg^{2+}, or Mn^{2+}. Two-phase aqueous partitioning,[46,57] also uses surface properties to separate membrane vesicles. Other methods to separate and purify vesicles from the cortex or other areas include density gradient centrifugation and electrophoresis.[19,46,151,152]

It is crucial to monitor and assess the enrichment, separation, and purification of the membranes by assaying measuring marker enzymes in the starting material and the final vesicle preparations.[46] The orientation of membrane vesicles isolated by a particular procedure should be determined since vesicles may form with an inside- or right side-out orientation.[19,153] Other factors that need to be assessed, depending on the type of experiment, include vesicle volume, vesicle permeability, and nonspecific binding of the substrate to the vesicles.[46]

After the vesicles are prepared, they are used for experiments with a rapid filtration technique. The vesicles are preloaded with the appropriate solution by dialysis. A small volume of the vesicle preparation is transferred to a test tube with a significantly larger volume of the transport buffer in order to initiate the assay. After the designated interval the contents are poured on to the filter and rinsed with buffer while a vacuum is applied. The vesicles are then removed along with the filter for assay of substrate uptake. Typically, the transport substrate is radiolabeled and the filter is dissolved in a scintillation cocktail formulated for that purpose. Various protocols allow evaluation of co-transport, antiport, inwardly and outwardly directed transport, and inhibition.

7.07.6.2 Advantages

Membrane vesicles facilitate investigation of the transport of a xenobiotic without concern for toxic effects on cellular functions. The need for an intact cellular energy source to establish and maintain an ion gradient is obviated since the driving gradient is applied by the investigator. It would, however, be possible to study energy-dependent transport by using caged ATP. One can manipulate the intra- and extravesicular buffer composition to examine different modes of transport and the ion requirement for such processes. Uptake can be measured at very early stages because the samples can be rapidly manipulated and obtaining data in intervals as short as 5 s is possible. Kinetic parameters such as K_m, V_{max}, and inhibition constants can be determined quickly and easily.

Numerous experiments can be done with the vesicles prepared from a single dog or rabbit kidney. The vesicles can be frozen and aliquots thawed as needed. It is possible to freeze whole kidneys or purchase commercially frozen kidneys for preparation of vesicles at a convenient time[154] and vesicles can be prepared from cultured cells.[155] The equipment required is available in most research laboratories (i.e., ultracentrifuge, scintillation counter) or inexpensive to purchase, and the transport assays are simple to learn, quick and have excellent precision.

7.07.6.3 Limitations

Although the steps involved in making vesicles are simple, the process is time-consuming. Depending on the method employed it may take the better part of two days to prepare a batch. Each batch of vesicles must be characterized with marker enzymes[46] as they are made to estimate contamination by other membranes that may transport the test compound. Additionally, a given substrate may be transported by multiple systems on the same membrane.[19,156] Within each experiment, controls should be run to assure that the experiment will be valid. For example, when the transport of a new substrate is investigated, the researcher needs to determine the extent, if any, of nonspecific binding to the vesicle.[46] Binding to the filter membrane or to labware should also be measured. Plasma membrane vesicles are sensitive to osmolality and the uptake of a solute may vary with vesicle size or stretching of the membrane. Thus, differences in osmolality between the intracellular and extracellular media should be avoided, and possible changes in vesicle volume during an experiment should be considered.[19]

Due to the small amount of material available for transport studies with vesicles, the transport substrate must be radiolabeled, usually at a high specific activity. In some cases, the cost of radioisotopes may be prohibitive.

7.07.6.4 Applications to Toxicology

Plasma membrane vesicles are particularly valuable for the study of transepithelial transport, which involves movement of a solute into the cell at one side and movement out of the cell at the other. The membranes of the apical and basolateral sides of the proximal tubule can be isolated and studied separately. Inwardly and outwardly directed transport can be examined with respect to kinetic behavior, cotransport substrates, and inhibition. Thus, it is possible to get a complete picture of the nature of the transport of a test compound. These features have been used to study the transport of therapeutic agents, such as the antibiotic cefadroxil,[157] or environmental agents such as ochratoxin A.[158]

Vesicles can also be used to investigate the nature of toxic actions of a xenobiotic on membranes in the absence of simultaneous effects on other cellular targets. Vesicles have been used to study the toxicity of styrene metabolites,[159] maleate and 4-pentanoate,[160] and cadmium.[161–163]

7.07.7 REFERENCES

1. A. W. Jans, E. Kellenbach, B. Griewel *et al.*, in 'Physiological NMR Spectroscopy, from Isolated Cells to Man,' ed. S. M. Cohen, Academy of Sciences, New York, 1987, pp. 443–444.
2. B. D. Ross and J. C. Bellinger, in 'Renal Disposition and Nephrotoxicity of Xenobiotics,' eds. M. W. Anders, W. Dekant, D. Heenschler *et al.*, Academic Press, London, 1993, pp. 75–93.
3. J. M. Frazier and J. A. Bradlaw (eds.), 'Technical Problems Associated with *In Vitro* Toxicity Testing Systems,' 1989, Technical Report No. 1, Johns Hopkins Center for Alternatives to Animal Testing, Baltimore, MD.
4. I. Bekersky, 'Use of the isolated perfused kidney as a tool in drug disposition studies.' *Drug Metab. Rev.*, 1983, **14**, 931–960.
5. G. L. Diamond, in 'Methods in Renal Toxicology,' eds. R. K. Zalups and L. H. Lash, CRC Press, Boca Raton, FL, 1996, pp. 59–77.
6. H. M. Mehendale, in 'Principles and Methods of Toxicology,' 2nd edn., ed. A. W. Hayes, Raven Press, New York, 1989, pp. 699–740.
7. J. F. Newton, Jr. and J. B. Hook, in 'Methods in Enzymology. Vol. 77. Detoxication and Drug Metabolism: Conjugation and Related Systems,' ed. W. B. Jakoby, Academic Press, London, 1981, pp. 94–105.
8. J. B. Tarloff and R. S. Goldstein, in *'In Vitro* Assessment of Toxicology,' ed. S. C. Gad, Raven Press, New York, 1994, pp. 149–194.
9. K. Higaki, K. Kadono, S. Goto *et al.*, 'Stereoselective renal tubular secretion of an organic anion in the isolated perfused rat kidney.' *J. Pharmacol. Exp. Ther.*, 1994, **270**, 329–335.
10. P. Silva, in 'Methods in Enzymology. Vol. 191. Biomembranes. Part V. Cellular and Subcellular Transport: Epithelial Cells,' eds. S. Fleischer and B. Fleischer, Academic Press, London, 1990, pp. 31–34.
11. S. P. Boom, M. M. Moons and F. G. M. Russel, 'Renal tubular transport of cimetidine in the isolated perfused kidney of the rat.' *Drug Metab. Dispos.*, 1994, **22**, 148–153.
12. J. F. Newton, W. E. Braselton, Jr., C. H. Kuo *et al.*, 'Metabolism of acetaminophen by the isolated perfused kidney.' *J. Pharmacol. Exp. Ther.*, 1982, **221**, 76–79.
13. B. D. Ross, J. Tange, K. Emslie *et al.*, 'Paracetamol metabolism by the isolated perfused rat kidney.' *Kidney Int.*, 1980, **18**, 562–570.
14. M. Laznicek and A. Laznickova, 'Kidney and liver contributions to salicylate metabolism in rats.' *Eur. J. Drug Metab. Pharmacokinet.*, 1994, **19**, 21–26.
15. M. J. Zamlauski-Tucker, M E. Morris and J. E. Springate, 'Ifosfamide metabolite chloroacetaldehyde causes Fanconi syndrome in the perfused rat kidney.' *Toxicol. Appl. Pharmacol.*, 1994, **129**, 170–175.
16. F. A. Redegeld, G. A. Hofman, P. G. van de Loo *et al.*, 'Nephrotoxicity of the glutathione conjugate of menadione (2-methyl-1,4-naphthoquinone) in the isolated perfused rat kidney. Role of metabolism by gamma-glutamyltranspeptidase and probenecid-sensitive transport.' *J. Pharmacol. Exp. Ther.*, 1991, **256**, 665–669.
17. K. Miura, R. S. Goldstein, D. A. Pasino *et al.*, 'Cisplatin nephrotoxicity: role of filtration and tubular transport of cisplatin in isolated perfused kidneys.' *Toxicology*, 1987, **44**, 147–158.
18. A. Kleinzeller, in 'Renal Physiology. People and Ideas,' eds. C. W. Gottschalk, R. Berliner and G. Geibisch, American Physiological Society, Bethesda, MD, 1987, pp. 131–163.
19. G. H. Mudge, 'Studies on potassium accumulation by rabbit kidney slices: effect of metabolic activity.' *Am. J. Physiol.*, 1951, **165**, 113–127.
20. J. A. Schafer and T. E. Andreoli, in 'Membrane Transport in Biology. Vol. IVA. Transport Organs,' ed. G. Giebisch, Springer, New York, 1979, pp. 473–528.
21. C. E. Ruegg, A. J. Gandolfi, R. B. Nagle *et al.*, 'Preparation of positional renal slices for study of cell specific toxicity.' *J. Pharmacol. Methods*, 1987, **17**, 111–123.
22. G. H. I. Wolfgang, A. J. Gandolfi and K. Brendel, 'Evaluation of organic nephrotoxins in rabbit renal cortical slices.' *Toxicol. In Vitro*, 1989, **3**, 341–350.
23. J. H. Smith, 'The use of renal cortical slices from the Fischer 344 rat as an *in vitro* model to evaluate nephrotoxicity.' *Fundam. Appl. Toxicol.*, 1988, **11**, 132–142.
24. A. J. Gandolfi and K. Brendel, *'In vitro* systems for nephrotoxicity studies.' *Toxicol. In Vitro*, 1990, **4**, 337–345.
25. R. C. Rose, J. Bianchi and S. A. Schuette, 'Effective use of renal cortical slices in transport and metabolic studies.' *Biochim. Biophys. Acta*, 1985, **821**, 431–436.
26. J. Martel-Pelletier, D. Guerette and M. Bergeron, 'Morphologic changes during incubation of renal slices.' *Lab. Invest.*, 1977, **36**, 509–518.
27. E. A. Lock, J. Odum and P. Ormond, 'Transport of N-acetyl-S-pentachloro-1,3-butadienylcysteine by rat renal cortex.' *Arch. Toxicol.*, 1986, **59**, 12–15.
28. R. S. Goldstein, L. R. Contardi, D. A. Pasino *et al.*, 'Mechanisms mediating cephaloridine inhibition of renal gluconeogenesis.' *Toxicol. Appl. Pharmacol.*, 1987, **87**, 297–305.
29. S. M. Ford, D. A. Laska, G. H. Hottendorf *et al.*, 'Correlation between the *in vitro* and *in vivo* nephrotoxicity of parenteral antibiotics in the rabbit.' *Toxicol. Methods*, 1993, **3**, 1–17.
30. P. D. Williams and G. F. Rush, in *'In Vitro* Toxicity Testing. Applications to Safety Evaluation,' ed. J. M. Frazier, Dekker, New York, 1992, pp. 85–110.
31. A. M. Chonko and J. J. Grantham, in 'Methods in Pharmacology. Vol. 4A. Renal Pharmacology,' ed. M. Martinez-Maldonado, 1976, pp. 47–71.
32. R. S. Balaban, S. P. Soltoff, J. M. Storey *et al.*, 'Improved renal cortical tubule suspension, spectrophotometric study of O_2 delivery.' *Am. J. Physiol.*, 1980, **238**, F50–F59.
33. K. Brendel, J. T. Hjelle and E. Meezan, in 'Methods in Toxicology. Vol. IA. *In Vitro* Biological Systems,' eds. C. A. Tyson and J. M. Frazier, Academic Press, London, 1993, pp. 330–338.
34. J. E. Dabbs, C. E. Green and C. A. Tyson, in 'Methods in Toxicology. Vol. IA. *In Vitro* Biological Systems,' eds. C. A. Tyson and J. M. Frazier, Academic Press, London, 1993, pp. 348–356.

35. F. A. Gesek, D. W. Wolff and J. W. Strandhoy, 'Improved separation method for rat proximal and distal renal tubules.' *Am. J. Physiol.*, 1987, **253**, F358–F365.

36. R. K. H. Kinne, in 'Methods in Enzymology. Vol. 191. Biomembranes. Part V Cellular and Subcellular Transport: Epithelial Cells,' eds. S. Fleischer and B. Fleischer, Academic Press, London, 1990, pp. 380–409.

37. K. Ormstad, S. Orrenius and D. P. Jones, in 'Methods in Enzymology. Vol. 77. Detoxication and Drug Metabolism: Conjugation and Related Systems,' ed. W. B. Jakoby, Academic Press, London, 1981, pp. 137–146.

38. J. F. Sina and M. O. Bradley, in '*In Vitro* Methods of Toxicology,' ed. R. R. Watson, CRC Press, Boca Raton, FL, 1992, pp. 81–92.

39. L. H. Lash and J. J. Tokarz, 'Isolation of two distinct populations of cells from rat kidney cortex and their use in the study of chemical-induced toxicity.' *Anal. Biochem.*, 1989, **182**, 271–279.

40. C. E. Ruegg and L. J. Mandel, in 'Methods in Toxicology. Vol. IA. *In Vitro* Biological Systems,' eds. C. A. Tyson and J. M. Frazier, Academic Press, London, 1993, pp. 357–365.

41. K. G. Dickman and L. J. Mandel, in 'Methods in Toxicology. Vol. IA. *In Vitro* Biological Systems,' eds. C. A. Tyson and J. M. Frazier, Academic Press, London, 1993, pp. 339–348.

42. M. L. Taub, in 'Methods in Toxicology. Vol. IA. *In Vitro* Biological Systems,' eds. C. A. Tyson and J. M. Frazier, Academic Press, London, 1993, pp. 366–373.

43. M. Weber, 'Basement membrane proteins.' *Kidney Int.*, 1992, **41**, 620–628.

44. L. H. Lash, in 'Renal Disposition and Nephrotoxicity of Xenobiotics,' eds. M. W. Anders, W. Dekant, D. Heenschler *et al.*, Academic Press, London, 1993, pp. 3–26.

45. M. Davies, J. Martin, G. J. Thomas *et al.*, 'Proteinases and glomerular matrix turnover.' *Kidney Int.*, 1992, **41**, 671–678.

46. W. H. Evans, in 'Biological Membranes. A Practical Approach,' eds. J. B. C. Findlay and W. H. Evans, IRL Press, Oxford, 1987, pp. 1–36.

47. R. I. Freshney, 'Culture of Animal Cells. A Manual of Basic Technique,' 3rd edn., Wiley-Liss, New York, 1994.

48. C. Waymouth, in 'Cell Separation: Methods and Selected Applications,' eds. T. G. Pretlow, II and T. P. Pretlow, Academic Press, London, 1982, vol. 1., pp. 1–29.

49. M. M. Bashor, in 'Methods in Enzymology. Vol. LVII. Cell Culture,' eds. W. B. Jakoby and I. H. Pastan, Academic Press, London, 1979, pp. 119–131.

50. L. H. Lash, in 'Methods in Toxicology. Vol. IA. *In Vitro* Biological Systems,' eds. C. A. Tyson and J. M. Frazier, Academic Press, London, 1993, pp. 397–410.

51. P. Vinay, A. Gougoux and G. Lemieux, 'Isolation of a pure suspension of rat proximal tubules.' *Am. J. Physiol.*, 1981, **241**, F403–F411.

52. G. Wirthensohn and G. W. Guder, in 'Methods in Enzymology. Vol. 191. Biomembranes. Part V Cellular and Subcellular Transport: Epithelial Cells,' eds. S. Fleischer and B. Fleischer, Academic Press, London, 1990, pp. 325–340.

53. D. W. Scholer and I. S. Edelman, 'Isolation of rat kidney cortical tubules enriched in proximal and distal segments.' *Am. J. Physiol.*, 1979, **237**, F350–F359.

54. S. D. Chung, N. Alavi, D. Livingston *et al.*, 'Characterization of primary rabbit kidney cultures that express proximal tubule functions in a hormonally defined medium.' *J. Cell Biol.*, 1982, **95**, 118–126.

55. D. Schlondorff, in 'Methods in Enzymology. Vol. 191. Biomembranes. Part V Cellular and Subcellular Transport: Epithelial Cells,' eds. S. Fleischer and B. Fleischer, Academic Press, London, 1990, pp. 130–152.

56. D. A. Troyer and J. I. Kreisberg, in 'Methods in Enzymology. Vol. 191. Biomembranes. Part V Cellular and Subcellular Transport: Epithelial Cells,' eds. S. Fleischer and B. Fleischer, Academic Press, London, 1990, pp. 31–34.

57. S. Bamberger, D. E. Brooks, K. A. Sharp *et al.*, in 'Partitioning in Aqueous Two-Phase Systems. Theory, Methods, Uses, and Applications to Biotechnology,' eds. H. Walter, D. E. Brooks and D. Fisher, Academic Press, London, 1985, pp. 85–130.

58. M. F. Horster and M. Sone, in 'Methods in Enzymology. Vol. 191. Biomembranes. Part V Cellular and Subcellular Transport: Epithelial Cells,' eds. S. Fleischer and B. Fleischer, Academic Press, London, 1990, pp. 409–426.

59. W. L. Smith and A. Garcia-Perez, 'Immunodissection: use of monoclonal antibodies to isolate specific types of renal cells.' *Am. J. Physiol.*, 1985, **248**, F1–F7.

60. R. C. Stanton, D. L. Mendrick, H. G. Rennke *et al.*, 'Use of monoclonal antibodies to culture rat proximal tubule cells.' *Am. J. Physiol.*, 1986, **251**, C780–786.

61. L. H. Lash, in '*In Vitro* Methods of Toxicology,' ed. R. R. Watson, CRC Press, Boca Raton, FL, 1992, pp. 115–122.

62. H. Babich and E. Borenfreund, in '*In Vitro* Methods of Toxicology,' ed. R. R. Watson, CRC Press, Boca Raton, FL, 1992, pp. 237–251.

63. E. Borenfreund and J. A. Puerner, 'Toxicity determined *in vitro* by morphological alterations and neutral red absorption.' *Toxicol. Lett.*, 1985, **24**, 119–124.

64. S. K. Chakrabarti and C. Denniel, '*S*-[1 and 2)-phenyl-2-hydroxyethyl]-cysteine-induced cytotoxicity to rat renal proximal tubules.' *Toxicol. Appl. Pharmacol.*, 1996, **137**, 285–294.

65. M. E. Chamberlin, A. LeFurgey and L. J. Mandel, 'Suspension of medullary thick ascending limb tubules from the rabbit kidney.' *Am. J. Physiol.*, 1984, **247**, F955–F964.

66. C. L. Bell, H. S., Tenenhouse and C. R. Scriver, 'Initiation and characterization of primary mouse kidney epithelial cells.' *In Vitro Cell. Dev. Biol.*, 1988, **24**, 683–698.

67. P. J. Boogaard, G. J. Mulder and J. F. Nagelkerke, 'Isolated proximal tubular cells from rat kidney as an *in vitro* model for studies on nephrotoxicity. I. An improved method for preparation of proximal tubular cells and their functional characterization by α-methylglucose uptake.' *Toxicol. Appl. Pharmacol.*, 1989, **101**, 135–143.

68. C. E. Groves, H. V. Sheevers and S. J. McGuinness, in 'Methods in Toxicology. Vol. 1B. *In Vitro* Toxicity Indicators,' eds. C. A. Tyson and J. M. Frazier, Academic Press, London, 1994, pp. 108–120.

69. C. A. Tyson, J. E. Dabbs, P. M. Cohen *et al.*, 'Studies of nephrotoxic agents in an improved renal proximal tubule system.' *Toxicol. In Vitro*, 1990, **4**, 403–408.

70. R. K. Zalups, M. G. Cherian and D. W. Barfuss, 'Lack of luminal or basolateral uptake and transepithelial transport of mercury in isolated perfused proximal tubules exposed to mercury-metallothion-ein.' *J. Toxicol. Environ. Health*, 1995, **44**, 101–113.

71. M. Horster, 'Primary culture of mammalian nephron epithelia: requirements for cell outgrowth and proliferation from defined explanted nephron segments.' *Pflugers Arch.*, 1979, **382**, 209–215.

72. M. F. Horster, in 'Advance Physiology Science.

Vol. 11. Kidney and Body Fluids,' ed. L. Takacs, 1980, pp. 169–173.

73. M. Cereijido, J. Ehrenfeld, S. Fernández-Castelo *et al.*, in 'Hormonal Regulation of Epithelial Transport of Ions and Water,' eds. W. N. Scott and D. B. P. Goodman, New York Academy of Sciences, New York, 1981, pp. 422–441.

74. E. Rodriguez-Boulan and W. J. Nelson, 'Morphogenesis of the polarized epithelial cell phenotype.' *Science*, 1989, **245**, 718–725.

75. P. G. Cox, C. H. Van Os and F. G. Russel, 'Accumulation of salicylic acid and indomethacin in isolated proximal tubular cells of the rat kidney.' *Pharmacol. Res.*, 1993, **27**, 241–252.

76. P. J. Boogaard, G. J. Mulder and J. F. Nagelkerke, 'Isolated proximal tubular cells from rat kidney as an *in vitro* model for studies on nephrotoxicity. II. α-methylglucose uptake as sensitive parameter for mechanistic studies of acute toxicity by xenobiotics.' *Toxicol. Appl. Pharmacol.*, 1989, **101**, 144–157.

77. W. B. Jakoby and I. H. Pastan, 'Methods in Enzymology. Vol. 57. Cell Culture,' Academic Press, New York, 1979.

78. W. I. Schaeffer, 'Terminology associated with cell, tissue and organ culture, molecular biology and molecular genetics. Tissue Culture Association Terminology Committee,' *In Vitro Cell. Dev. Biol.*, 1990, **26**, 97–101.

79. E. D. Avner, W. E. Sweeney Jr. and D. Ellis, in 'Methods for Serum-Free Culture of Epithelial and Fibroblastic Cells,' eds. D. W. Barnes, D. A. Sirbasku and G. H. Sato, Liss, New York, 1984, pp. 33–41.

80. G. M. Hawksworth, E. M. Cockburn, J. G. Simpson *et al.*, in 'Methods in Toxicology. Vol. IA. *In Vitro* Biological Systems,' eds. C. A. Tyson and J. M. Frazier, Academic Press, London, 1993, pp. 385–396.

81. E. E. Muirhead, W. A. Rightsel, J. A. Pitcock *et al.*, in 'Methods in Enzymology. Vol. 191. Biomembranes. Part V Cellular and Subcellular Transport: Epithelial Cells,' eds. S. Fleischer and B. Fleischer, Academic Press, London, 1990, pp. 152–167.

82. T. D. Oberley, P. M. Burkholder and M. D. Mills, 'Culture of human glomerular cells.' *Am. J. Pathol.*, 1979, **96**, 101–119.

83. T. Weinstein, R. Cameron, A. Katz *et al.*, 'Rat glomerular epithelial cells in culture express characteristics of parietal, not visceral, epithelium.' *J. Am. Soc. Nephrol.*, 1992, **3**, 1279–1287.

84. G. Gstraunthaler and W. Pfaller, in '*In Vitro* Methods of Toxicology,' ed. R. R. Watson, CRC Press, Boca Raton, FL, 1992, pp. 93–114.

85. C. R. Gaush, W. L. Hard and T. F. Smith, 'Characterization of an established line of canine kidney cells (MDCK).' *Proc. Soc. Exp. Biol. Med.*, 1966, **122**, 931–935.

86. P. D. Williams and D. A. Laska, in 'Methods in Toxicology. Vol. IA. *In Vitro* Biological Systems,' eds. C. A. Tyson and J. M. Frazier, Academic Press, London, 1993, pp. 411–419.

87. R. G. Ham and W. L. Mckeehan, in 'Methods in Enzymology. Vol. 57. Cell Culture,' eds. W. B. Jakoby and I. H. Pastan, Academic Press, London, 1979, pp. 44–93.

88. G. Gstraunthaler, W. Pfaller and P. Kotanko, 'Biochemical characterization of renal epithelial cell cultures (LLC-PK$_1$ and MDCK),' *Am. J. Physiol.*, 1985, **248**, F536–F544.

89. M. Taub, in 'Methods for Serum-Free Culture of Epithelial and Fibroblastic Cells,' eds. D. W. Barnes, D. A. Sirbasku and G. H. Sato, Liss, New York, 1984, pp. 3–24.

90. M. H. Saier Jr. in 'Methods for Serum-Free Culture of Epithelial and Fibroblastic Cells,' eds. D. W. Barnes,

D. A. Sirbasku and G. H. Sato, Liss, New York, 1984, pp. 25–31.

91. M. Taub and G. H. Sato, 'Growth of kidney epithelial cells in hormone-supplemented, serum-free medium.' *J. Supramol. Struct.*, 1979, **11**, 207–216.

92. M. Taub, in 'Mammalian Cell Culture. The Use of Serum-Free Hormone-Supplemented Media,' ed. J. P. Mather, Plenum Press, New York, 1984, pp. 129–150.

93. D. Barnes, in 'Mammalian Cell Culture. The Use of Serum-Free Hormone-Supplemented Media,' ed. J. P. Mather, Plenum Press, New York, 1984, pp. 195–237.

94. C. J. Detrisac, M. A. Sens, A. J. Garvin *et al.*, 'Tissue culture of human kidney epithelial cells of proximal tubule origin.' *Kidney Int.*, 1984, **25**, 383–390.

95. K. von der Mark, H. von der Mark and S. Goodman, 'Cellular responses to extracellular matrix.' *Kidney Int.*, 1992, **41**, 632–640.

96. A. H. Yang, J. Gould-Kostka and T. D. Oberley, '*In vitro* growth and differentiation of human kidney tubular cells on a basement membrane substrate.' *In Vitro Cell. Dev. Biol.*, 1987, **23**, 34–46.

97. Q. Al-Awqati, J. Van Adelsberg and J. Takito, in 'Current Topics in Membranes. Vol. 41. Cell Biology and Membrane Transport Processes,' ed. M. Caplan, Academic Press, London, 1994, pp. 109–122.

98. J. E. Lever, 'Regulation of dome formation in differentiated epithelial cell cultures.' *J. Supramol. Struct.*, 1970, **12**, 259–272.

99. S. M. Ford, P. D. Williams, S. Grassl *et al.*, 'Transepithelial acidification by cultures of rabbit proximal tubules grown on filters.' *Am. J. Physiol.*, 1990, **259**, C103–C109.

100. K. M. Morshed and K. E. McMartin, 'Transient alterations in cellular permeability in cultured human proximal tubule cells: implications for transport studies.' *In Vitro Cell. Dev. Biol. Anim.*, 1995, **31**, 107–114.

101. F. Dolbeare and M. Vanderlaan, in 'Methods in Toxicology. Vol. 1B. *In Vitro* Toxicity Indicators,' eds. C. A. Tyson and J. M. Frazier, Academic Press, London, 1994, pp. 178–200.

102. D. F. Green, K. H. Hwang, U. S. Ryan *et al.*, 'Culture of endothelial cells from baboon and human glomeruli.' *Kidney Int.*, 1992, **41**, 1506–1516.

103. G. Gstraunthaler and W. Pfaller, in 'Renal Disposition and Nephrotoxicity of Xenobiotics,' eds. M. W. Anders, W. Dekant, D. Heenschler *et al.*, Academic Press, London, 1993, pp. 27–46.

104. B. Grout, J. Morris and M. McLellan 'Cryopreservation and the maintenance of cell lines.' *Trends Biotechnol.*, 1990, **8**, 293–297.

105. M. D. Aleo, M. L. Taub, J. R. Olson *et al.*, 'Primary cultures of rabbit renal proximal tubule cells. II. Selected phase I and phase II metabolic capacities.' *Toxicol. In Vitro*, 1990, **4**, 727–733.

106. P. Bellemann, 'Primary monolayer culture of liver parenchymal cells and kidney cortical tubules as a useful new model for biochemical pharmacology and experimental toxicology. Studies *in vitro* on hepatic membrane transport, induction of liver enzymes, and adaptive changes in renal cortical enzymes.' *Arch. Toxicol.*, 1980, **44**, 63–84.

107. I. M. Bruggemann, J. J. W. M. Mertens, J. H., M. Temmink *et al.*, 'Use of monolayers of primary rat kidney cortex cells for nephrotoxicity studies.' *Toxicol. In Vitro*, 1989, **3**, 261–269.

108. G. H. Hottendorf, D. A. Laska, P. D. Williams *et al.*, 'Role of desacetylation in the detoxification of cephalothin in renal cells in culture.' *J. Toxicol. Environ. Health*, 1987, **22**, 101–111.

109. G. J. McGarrity, in 'Methods in Enzymology. Vol. 57.

Cell Culture,' eds. W. B. Jakoby and I. H. Pastan, Academic Press, London, 1979, pp. 18–29.

110. M. G. Gabridge and D. J. Lundin, 'Cell Cultures User's Guide to Mycoplasma Detection and Control,' Bionique Laboratories, Saranac Lake, NY, 1989.

111. N. L. Simmons, in 'Methods in Enzymology. Vol. 191. Biomembranes. Part V. Cellular and Subcellular Transport: Epithelial Cells,' eds. S. Fleischer and B. Fleischer, Academic Press, London, 1990, pp. 426–436.

112. K. Wolf, in 'Methods in Enzymology. Vol. 57. Cell Culture,' eds. W. B. Jakoby and I. H. Pastan, Academic Press, London, 1979, pp. 116–119.

113. J. M. Graham, in 'Methods in Molecular Biology. Vol. 19. Biomembrane Protocols. I. Isolation and Analysis,' eds. J. Graham and J. Higgins, Humana Press, Clifton, NJ, 1993, pp. 97–108.

114. K. E. Howell, E. Devaney and J. Gruenberg, 'Subcellular fractionation of tissue culture cells.' *Trends Biochem. Sci.*, 1989, **14**, 44–47.

115. W. G. Thilly and D. W. Levine, in 'Methods in Enzymology. Vol. 57. Cell Culture,' eds. W. B. Jakoby and I. H. Pastan, Academic Press, London, 1979, pp. 184–194.

116. C. Le Grimellec, G. Friedlander and M. C. Giocondi, 'Asymmetry of plasma membrane lipid order in Madin–Darby canine kidney cells.' *Am. J. Physiol.*, 1988, **255**, F22–F32.

117. R. Bendayan, B. Lo and M. Silverman, 'Characterization of cimetidine transport in LLC-PK$_1$ cells.' *J. Am. Soc. Nephrol.*, 1994, **5**, 75–84.

118. T. D. McKinney, C. DeLeon and K. V. Speeg, Jr., 'Organic cation uptake by a cultured renal epithelium.' *J. Cell. Physiol.*, 1988, **137**, 513–520.

119. T. W. McKinney, M. B. Scheller, M. Hosford *et al.*, 'Basolateral transport of tetraethylammonium by a clone of LLC-PK$_1$ cells.' *J. Am. Soc. Nephrol.*, 1992, **1**, 1507–1515.

120. H. Saito, M. Yamamoto, K. I. Inui *et al.*, 'Transcellular transport of organic cation across monolayers of kidney epithelial cell line LLC-PK$_1$.' *Am. J. Physiol.*, 1992, **262**, C59–C66.

121. M. B. Scheller, M. Hosford and T. D. McKinney, 'Organic cation transport by cultured renal proxi-. mal tubular epithelia.' *J. Tissue Cult. Methods*, 1991, **13**, 195–198.

122. B. F. Pan, A. Dutt and J. A. Nelson, 'Enhanced trans-epithelial flux of cimetidine by Madin–Darby canine kidney cells overexpressing human *p*-glycoprotein.' *J. Pharmacol. Exp. Ther.*, 1994, **270**, 1–7.

123. M. J. Palmoski, B. A. Masters, O. P. Flint *et al.*, 'Characterization of rabbit primary proximal tubule kidney cell cultures grown on Millicell-HA membrane filters.' *Toxicol. In Vitro*, 1992, **6**, 557–567.

124. K. G. Dickman and J. L. Renfro, 'Primary culture of flounder renal tubule cells: transepithelial transport.' *Am. J. Physiol.*, 1986, **251**, F424–F432.

125. R. Hori, M. Okamura, A. Takayama *et al.*, 'Transport of organic anion in the OK kidney epithelial cell line.' *Am. J. Physiol.*, 1993, **264**, F975–F980.

126. S. M. Ford and P. D. Williams, in 'In Vitro Methods of Toxicology,' ed. R. R. Watson, CRC Press, Boca Raton, FL, 1992, pp. 123–142.

127. S. J. Scheinman, R. Reid, R. Coulson *et al.*, 'Transepithelial phosphate transport in rabbit proximal tubular cells adapted to phosphate deprivation.' *Am. J. Physiol.*, 1994, **266**, C1609–1618.

128. F. Courjault-Gautier, D. Hoet, D. Leroy *et al.*, 'Dissimilar alterations of sodium-coupled uptake by platinum-coordination complexes in renal proximal

tubular cells in primary culture.' *J. Pharmacol. Exp. Ther.*, 1994, **270**, 1097–1104.

129. M. Gekle, S. Mildenberger, R. Freudinger *et al.*, 'The mycotoxin ochratoxin-A impairs protein uptake in cells derived from the proximal tubule of the kidney (opossum kidney cells).' *J. Pharmacol. Exp. Ther.*, 1994, **271**, 1–6.

130. O. Kimura, T. Endo and M. Sakata, 'Comparison of cadmium uptakes from apical and basolateral membranes of LLC-PK$_1$ cells.' *Toxicol. Appl. Pharmacol.*, 1996, **137**, 301–306.

131. S. Blumenthal, D. Lewand, S. K. Krezoski *et al.*, 'Comparative effects of Cd^{2+} and Cd-metallothionein on cultured kidney tubule cells.' *Toxicol. Appl. Pharmacol.*, 1996, **136**, 220–228.

132. J. Liu, Y. Liu and C. D. Klaassen, 'Nephrotoxicity of CdCl$_2$ and Cd-metallothionein in cultured rat kidney proximal tubules and LLC-PK$_1$ cells.' *Toxicol. Appl. Pharmacol.*, 1994, **128**, 264–270.

133. L. Sorokin and P. Ekblom, 'Development of tubular and glomerular cells of the kidney.' *Kidney Int.*, 1992, **41**, 657–664.

134. L. K. Tay, C. L. Bregman, B. A. Masters *et al.*, 'Effects of *cis*-diamminedichloroplatinum (II) on rabbit kidney *in vivo* and on rabbit renal proximal tubule cells in culture.' *Cancer Res.*, 1988, **48**, 2538–2543.

135. M. Brezis, 'Forefronts in nephrology: summary of the newer aspects of renal cell injury.' *Kidney Int.*, 1992, **42**, 523–539.

136. J. C. Seegers, L. H. Böhmer, M. C. Kruger, 'A comparative study of ocharatoxin A-induced apoptosis in hamster kidney and HeLa cells.' *Toxicol. Appl. Pharmacol.*, 1994, **129**, 1–11.

137. J. S. Handler, F. M. Perkins and J. P. Johnson, 'Studies of renal cell function using cell culture techniques.' *Am. J. Physiol.*, 1980, **238**, F1–F9.

138. P. D. Holohan, P. P. Sokol, C. R. Ross *et al.*, 'Gentamicin-induced increases in cytosolic calcium in pig kidney cells (LLC-PK$_1$).' *J. Pharmacol. Exp. Ther.*, 1988, **247**, 349–354.

139. R. Martínez-Zaguilán, in 'In Vitro Methods of Toxicology,' ed. R. R. Watson, CRC Press, Boca Raton, FL, 1992, pp. 217–236.

140. A. P. Thomas, in 'Methods in Toxicology. Vol. 1B. In Vitro Toxicity Indicators,' eds. C. A. Tyson and J. M. Frazier, Academic Press, London, 1994, pp.

141. B. F. Trump, I. K. Berezesky, K. A. Elliget *et al.*, 'Nephrotoxicity *in vitro*: role of ion deregulation in signal transduction following injury-studies utilizing digital imaging fluorescence microscopy.' *Toxicol. In Vitro*, 1990, **4**, 409–414.

142. S. Vamvakas, V. K. Sharma, S. S. Sheu *et al.*, 'Perturbations of intracellular calcium distribution in kidney cells by nephrotoxic haloalkenyl cysteine *S*-conjugates.' *Mol. Pharmacol.*, 1990, **38**, 455–461.

143. S. Chandra and G. H. Morrison, 'Imaging elemental distribution and ion transport in cultured cells with ion microscopy.' *Science*, 1985, **228**, 1543–1544.

144. R. Hori, K. Yamamoto, H. Saito *et al.*, 'Effect of aminoglycoside antibiotics on cellular functions of kidney epithelial cell line (LLC-PK$_1$): a model system for aminoglycoside nephrotoxicity.' *J. Pharmacol. Exp. Ther.*, 1984, **230**, 742–748.

145. D. Steinmassl, W. Pfaller, G. Gstraunthaler *et al.*, 'LLC-PK$_1$ epithelia as a model for *in vitro* assessment of proximal tubular nephrotoxicity.' *In Vitro Cell. Dev. Biol. Anim.*, 1995, **31**, 94–106.

146. C. Shopsis and B. Eng, 'Rapid cytotoxicity testing using a semi-automated protein determination on cultured cells.' *Toxicol. Lett.*, 1985, **26**, 1–8.

147. P. D. Williams and G. H. Hottendorf, in 'Modern Analysis of Antibiotics,' ed. A. Aszalos, Dekker, New York, 1986, pp. 495–523.

148. M. Balls and R. H. Clothier, in '*In Vitro* Methods of Toxicology,' ed. R. R. Watson, CRC Press, Boca Raton, FL, 1992, pp. 37–52.

149. T. Mosmann, 'Rapid colorimetric assay for cellular growth and survival: application to proliferation and cytotoxicity assays.' *J. Immunol. Methods*, 1983, **65**, 55–63.

150. M. C. Diemert, V. Tricottet, L. Benel *et al.*, 'Use of a renal tubule cell line (LLC-PK$_1$) to study the nephrotoxic potential of a kappa-type Bence–Jones protein.' *In Vitro Cell. Dev. Biol. Anim.*, 1995, **31**, 716–723.

151. E. Kinne-Saffran and R. K. H. Kinne, in 'Methods in Enzymology. Vol. 191. Biomembranes. Part V. Cellular and Subcellular Transport: Epithelial Cells,' eds. S. Fleischer and B. Fleischer, Academic Press, London, 1990, pp. 450–468.

152. C. R. Ross and P. D. Holohan, 'Transport of organic anions and cations in isolated renal plasma membranes.' *Annu. Rev. Pharmacol. Toxicol.*, 1983, **23**, 65–85.

153. J. L. Kinsella, P. D. Holohan, N. I. Pessah *et al.*, 'Isolation of luminal and antiluminal membranes from dog kidney cortex.' *Biochim. Biophys. Acta*, 1979, **552**, 468–477.

154. A. G. Booth and A. J. Kenny, 'A rapid method for the preparation of microvilli from rabbit kidney.' *Biochem. J.*, 1974, **142**, 575–581.

155. K. Inui, H. Saito and R. Hori, 'H$^+$-gradient-dependent active transport of tetraethylammonium cation in apical-membrane vesicles isolated from kidney epithelial cell line LLC-PK$_1$.' *Biochem. J.*, 1985, **227**, 199–203.

156. A. Dutt, L. A. Heath and J. A. Nelson, '*p*-Glycoprotein and organic cation secretion by the mammalian kidney.' *J. Pharmacol. Exp. Ther.*, 1994, **269**, 1254–1260.

157. M. Ries, U. Wenzel and H. Daniel, 'Transport of cefadroxil in rat kidney brush-border membranes is mediated by two electrogenic H$^+$-coupled systems.' *J. Pharmacol. Exp. Ther.*, 1994, **271**, 1327–1333.

158. P. P. Sokol, G. Ripich, P. D. Holohan *et al.*, 'Mechanism of ochratoxin A transport in kidney.' *J. Pharmacol. Exp. Ther.*, 1988, **246**, 460–465.

159. S. Chakrabarti, D. D. Vu and M. G. Côté, 'Effects of cysteine derivatives of styrene on the transport of *p*-aminohippurate ion in renal plasma membrane vesicles.' *Arch. Toxicol.*, 1991, **65**, 366–372.

160. J. F. Pouliot, A. Gougoux and R. Béliveau, 'Brush border membrane proteins in experimental Fanconi's syndrome induced by 4-pentenoate and maleate.' *Can. J. Physiol. Pharmacol.*, 1992, **70**, 1247–1253.

161. D. W. Ahn and Y. S. Park, 'Transport of inorganic phosphate in renal cortical brush-border membrane vesicles of cadmium-intoxicated rats.' *Toxicol. Appl. Pharmacol.*, 1994, **133**, 239–243.

162. K. R. Kim and Y. S. Park, 'Phlorizin binding to renal outer cortical brush-border membranes of cadmium-injected rabbits.' *Toxicol. Appl. Pharmacol.*, 1994, **133**, 244–248.

163. R. K. Kinne, H. Schütz and E. Kinne-Saffran, 'The effect of cadmium chloride *in vitro* on sodium-glutamate cotransport in brush border membrane vesicles isolated from rabbit kidney.' *Toxicol. Appl. Pharmacol.*, 1995, **135**, 216–221.

7.08
Acute Renal Failure

DGANIT DINOUR

University of California, San Francisco, CA, USA

and

MAYER BREZIS

Hadassah University Hospital, Jerusalem, Israel

7.08.1 INTRODUCTION

7.08.1.1 Definitions

Acute renal failure (ARF) is a clinical syndrome characterized by rapid deterioration of renal function, resulting in the accumulation of nitrogenous wastes, such as urea and creatinine. Detectable increase in serum creatinine usually implies a 30–50% decrease in glomerular filtration rate (GFR) when the baseline creatinine is within normal limits. If ARF is superimposed on pre-existing renal insufficiency, obvious rises in serum creatinine occurs from small decrements in GFR. Therefore, different criteria are often used to define ARF according to baseline serum creatinine.[1–3] The term acute means a decline in GFR occurring within days (as opposed to weeks or months, in subacute, or years in chronic renal failure).

The above definition encompasses a variety of parenchymal renal diseases and extrarenal abnormalities, commonly classified as prerenal, renal, and postrenal diseases (Table 1). The term ARF is sometimes used in a more restrictive sense, interchangeably with acute intrinsic renal failure or acute tubular necrosis. For the purpose of this review, ARF will generally refer to intrinsic renal failure. This definition excludes prerenal and postrenal azotemia, where extrarenal causes can be rapidly identified and corrected, as well as parenchymal renal diseases that may be amenable to specific therapy. Prerenal azotemia and ARF represent two extremes in a spectrum of

Table 1 Major causes for an acute decline in GFR.

Prerenal azotemia
    Renal hypoperfusion not associated with
    structural damage to the kidney
Renal azotemia
    Acute intrinsic renal failure: ARF from toxic
        and/or ischemic injury, not immediately
        reversed upon discontinuation of the insult
        and associated with structural damage to the
        kidney (syn: acute tubular necrosis, ARF)
    Acute interstitial nephritis: acute interstitial
        inflammation
    Acute glomerulonephritis or vasculitis: acute
        glomerular or vessel inflammation
    Acute renovascular disease: acute obstruction of
        renal artery or vein in a single functioning
        kidney, or bilaterally
Postrenal azotemia
    Acute obstruction in the urinary collecting
    system (syn: obstructive uropathy)

Source: Brezis *et al.*[4]

renal responses to hypoperfusion.[4] Hemodynamic and tubular changes, similar to those operating in ARF, may be involved in the pathogenesis of obstructive uropathy, and interstitial inflammation may contribute to the impairment of renal function not only in acute interstitial nephritis but also in glomerulonephritis and obstructive uropathy.[5,6]

7.08.1.2 Incidence

ARF is a serious complication of medical and surgical illness, which, despite major advances in its prevention and treatment, still accounts for a remarkable morbidity and mortality. The incidence of ARF varies among studies depending on the definition criteria and the population. According to a community based study, the incidence of severe ARF (serum creatinine above $500 \mu mol L^{-1}$) in adults is about 170 cases per million annually.[3] Acute renal insufficiency, present in 1% of patients at the time of their admission to the hospital,[2] develops in about 5% during the course of their hospitalization[1] and is severe enough to require dialysis in 1–2% of the cases.[7–9] Certain populations are especially at risk: the incidence of ARF is 15–30% in intensive care units[10,11] or after cardiovascular surgery.

7.08.1.3 Course and Prognosis in ARF

7.08.1.3.1 Causes

A decrease in renal function may begin abruptly following surgery or hypotension, insidiously after the administration of a nephrotoxicant, or stepwise with the additive effects of multiple sequential insults, as often seen in critically ill patients. The renal failure phase usually lasts 1–3 weeks, followed by a recovery phase. The improvement in renal function is recognized by a decrease towards normal in serum creatinine and by a progressive rise in urine output in oliguric patients. Occasionally, profuse diuresis up to several L a day develops, described as the polyuric phase of ARF.[4]

7.08.1.3.2 Complications

Complications of the renal failure phase include volume overload, metabolic acidosis, hyperkalemia, hyponatremia, hyperphosphatemia, hypocalcemia, hypermagnesemia, and uremia. Infection is a common complication and a leading cause of death in patients with

ARF, followed by cardiovascular and respiratory complications.[4,12] The combination of acute stress and hemorrhagic diathesis in ARF predisposes to gastrointestinal bleeding, but with the widespread prophylactic use of antacids in critically ill patients and the early institution of dialysis, the incidence of this complication has been remarkably reduced.[11] While in many cases the recovery of renal function is incomplete, most patients who survive the acute illness recover enough renal function to discontinue dialysis, even after prolonged dialysis support.[13] The risk of permanent dialysis-dependence is higher in patients with pre-existing chronic renal insufficiency.[13]

7.08.1.3.3 Mortality

The mortality rate of patients during the acute phase of ARF has remained 40–60% despite technical advances in renal replacement therapy and supportive care.[7,11,12,14–17] The major determinant of survival is the presence of comorbid conditions. Mortality rates increases from 7% in prerenal azotemia[2] to more than 90% in critically ill patients with multiorgan failure.[18] The adverse prognostic effect of advanced age remains controversial.[7,8,14,17,19] ARF may be, in fact, a marker for a diseased organism which predisposes the kidneys to fail when challenged by toxicants or hypoperfusion.

7.08.2 CLINICAL SYNDROMES ASSOCIATED WITH ARF

7.08.2.1 Prerenal Azotemia

Prerenal azotemia, the most common cause of acute azotemia on admission,[2] is due to volume depletion, hypotension, congestive heart failure, or drug-induced renal vasoconstriction (Table 2). Enhanced tubular reabsorption of sodium and water, in association with increased urinary concentration, is the hallmark of prerenal azotemia and distinguishes this condition from other azotemic states (Table 3). Correction of the underlying cause (e.g., volume repletion, improvement of cardiac function) promptly reverses kidney function to baseline within 24–72 h.

7.08.2.2 Acute Intrinsic Renal Failure

Acute intrinsic renal failure, the major type of ARF in hospitalized patients[1,15,16] is characterized by an abrupt decline in GFR (after an

Table 2 Causes of renal hypoperfusion.

Intravascular volume depletion
  Gastrointestinal loss: diarrhea, vomiting, drainage
  Renal loss: diuretic use, osmotic diuresis, salt wasting nephropathy, hypoadrenalism
  Hemorrhage: trauma, gastrointestinal bleeding, obstetrical, surgical
  Skin, respiratory loss: sweat, burns, insensible losses
  Third space sequestration: acute pancreatitis, hypoalbuminemia, mechanical or paralytic ileus, crush injury

Decreased cardiac output
  Severe congestive heart failure, low output syndrome: ischemic heart disease, cardiomyopathies, pericardial or valvular disease
  Pulmonary hypertension, massive pulmonary embolism

Increased ratio of renal/systemic vascular resistances
  Sepsis
  Liver failure
  Systemic vasodilatation: antihypertensive drugs, anesthesia, anaphylactic shock
  Renal vasoconstriction: α-adrenergic agonists, cyclosporin A, amphotericin, hypercalcemia
  Intrarenal rheological alterations: sickle-cell anemia

Renovascular obstruction
  Renal artery: atherosclerosis, embolism, thrombosis, dissecting aneurysm
  Renal vein: thrombosis, compression

Interference with renal autoregulation
  Prostaglandin synthesis inhibitors (NSAID)
  Angiotensin converting enzyme inhibitors

Table 3 Laboratory tests to distinguish prerenal azotemia from intrinsic ARF.

Test	Prerenal azotemia	Intrinsic ARF
Urine osmolality	>500 mosm kg^{-1}	isosthenuria
Urine sodium	<20 mEq L^{-1}	>40 mEq L^{-1}
FeNa[a]	$<1\%$	$>1\%$
BUN/plasma creatinine	$>20:1$	$10–15:1$
Urinalysis	normal or near normal	granular casts, tubular epithelial cells, tubular cell casts

[a]FeNa (%) = Fractional excretion of sodium = 100 × urine Na × plasma creatinine/plasma Na × urine creatinine.

ischemic or toxic renal insult) not immediately reversible upon removal of the inciting factor. Reduced GFR is often accompanied by tubular dysfunction, manifested by loss of tubular concentrating ability, decreased sodium reabsorption, and the presence of typical granular casts in the urinary sediment (Table 3).

If severe enough, all causes of prerenal azotemia can deteriorate into ARF. This is especially true when such conditions are combined with exposure of the kidney to endogenous toxins (e.g., myoglobin) or exogenous toxicants (Table 4). It should be emphasized that in humans as opposed to most experimental models, ARF is often multifactorial, typically occurring in patients subjected to nephrotoxicants in the setting of renal hypoperfusion.[19] For instance, in the nephrotoxicity from radiological contrast agents, the incidence of ARF is directly predicted from the number of comorbid conditions (Figure 1).[20] Recognizing the multifactorial nature of human ARF is crucial for prevention, which remains the most efficient treatment modality.[21]

7.08.2.3 Differential Diagnosis of ARF

7.08.2.3.1 Acute interstitial nephritis

Acute interstitial nephritis accounts for approximately 15% of cases of ARF evaluated by kidney biopsy.[22,23] It is an allergic response to drugs[23,24] or may be secondary to infection, sarcoidosis, or antitubular antibodies.[23,25] Manifestations include eosinophilia, fever, and rash,[23,26] while the urine sediment often shows white cells, white cell casts, red cells, and eosinophils.[22,27] In general, protein excretion is only mildly increased. Less commonly, nonsteroidal anti-inflammatory agents (NSAIDs) may induce a nephrotic syndrome (edema and hypoalbuminemia associated with proteinuria above $3 g day^{-1}$, with minimal changes in glomeruli). The diagnosis, suggested by the clinical picture and a positive gallium scan[28] is confirmed by a kidney biopsy, which is indicated if renal failure is severe or persistent. A short course of steroid treatment is recommended in severe cases.[27]

7.08.2.3.2 Acute glomerulonephritis and vasculitis

Glomerular diseases and vasculitides usually present themselves with the nephritic or nephrotic syndromes and subacute or chronic renal insufficiency. Some glomerular and vascular diseases, such as the hemolytic–uremic syndrome or toxemia of pregnancy may present with ARF.

7.08.2.3.3 Obstructive uropathy

Urinary tract obstruction, due to prostatic diseases, pelvic tumors, or stones, is an important cause for ARF.[2,3,29] A renal ultrasound is indicated in every case of ARF from unclear etiology to recognize and correct obstruction.

Table 4 Agents associated with toxic ARF.

Antibiotics	**Chemotherapy and immunosuppression**	**Miscellaneous**
Aminoglycosides	Cisplatin	Dextran
Amphotericin B	Cyclosporin A	EDTA
Pentamidine	Methotrexate	Radiation
Acyclovir	Ifosfamide	Intravenous gamma-globulins
Foscarnet	Mitomycin	Mannitol overdose
Vancomycin	5-Azacitidine	ε-aminocaproic acid
	Nitrosourea	Fumaric acid esters
Anesthetic agents		
Methoxyflurane	**Poisons**	**Heavy metals**
Enflurane	Insecticides	Cadmium
	Herbicides	Mercury
Radiocontrast media	Rodenticides	Lead
Water soluble agents	Mushrooms	
	Snake bites	
Analgesics	Stings	
All NSAIDs		
Diuretics		
Ticrynafen		
Mercurials		

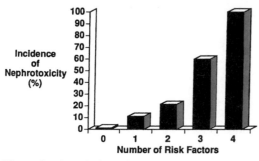

Figure 1 Association of nephrotoxicity from radiological contrast agents with presence of comorbid conditions (after Rich and Crecelius[20]).

7.08.3 CLINICAL EVALUATION OF THE PATIENT WITH ARF

ARF may present with oliguria, fluid overload, or uremia. With the widespread use of biochemical monitoring of hospitalized patients in general, and of critically ill patients in particular, the first manifestation of ARF is often a rise in plasma urea or creatinine level. To find the cause of ARF, important clues in the patient's history are the timing of renal function deterioration, fluid intake and output, and the occurrence of any insult, such as sepsis, hypotension, or administration of drugs. Physical examination is helpful in evaluating the volume status of the patient and the presence and severity of extrarenal diseases including heart failure, liver cirrhosis, or preipheral vascular disease.

It should be emphasized that a significant fall in renal perfusion can occur in circulatory insufficiency as a consequence of systemic vasoconstriction without remarkable changes in blood pressure. Hypotension is documented in 50% of patients with postoperative ARF.[1] Since quantification of intrarenal blood flow or oxygenation in humans is not readily available, clinical recognition of kidney hypoperfusion or parenchymal hypoxia will remain indirect until technologies such as positron emission tomography are refined, applied, and found suitable to this aim.[21,30]

The distinction between prerenal azotemia and ARF poses a special problem as both can occur in the same circumstances, yet vigorous fluid administration can completely reverse prerenal state but may be harmful in established ARF. As depicted in Table 3, low urinary osmolality is one of the earliest and most sensitive findings in ARF, reflecting predominant injury to renal medulla.[21] While other tests have been suggested,[31] there is a continuum between prerenal azotemia and ARF, and the proposed urinary indices often overlap.

While most cases of ARF are diagnosed on clinical grounds, a renal biopsy may alter management if nephritis or vasculitis is suspected.[32] The histologic findings of ARF are subtle, consisting mainly of brush border loss (along proximal tubules) and focal tubular cell necrosis, most prevalent in the renal medulla.[33] As discussed below, prominent proximal tubular necrosis, typical in experimental models, is rarely found in human ARF. Therefore, the histologic diagnosis of ARF is often made by exclusion.

7.08.4 PATHOLOGICAL EXPRESSION OF HYPOXIC RENAL INJURY

Anoxia induces a diversity of molecular and morphological alterations in the kidney, influenced by the intrarenal gradients of oxygenation.

7.08.4.1 Cellular, Molecular, and Functional Alterations

As in any cell, anoxia in kidney cells results in a multitude of events, that is, depletion of energy stores, collapse of electrolyte gradients, disruption of the actin cytoskeleton, activation of phospholipases, and changes in gene expression.[34] Renal hypoxia induces loss of epithelial polarity along proximal tubules,[35] and selective induction of growth response genes (e.g., *Egr*-1) with rapid DNA fragmentation (suggestive of apoptosis) along medullary thick limbs.[36,37]

Ischemic injury to renal vessels causes abnormal reno–vascular reactivity[38] and predisposes to secondary ischemic insults from hypotension during recovery from ARF. Ischemia induces the expression of histocompatibility antigens in tubules and of intercellular adhesion molecules on endothelial cells, leading to local aggregation of neutrophils and platelets.[39] Antibodies to intercellular adhesion molecules and antagonists to platelet activating factor protect the kidney from ischemic injury.[39] Intrarenal congestion after ischemia is prominent in the outer medulla, because of regional hypoxia and because the vasa recta are easily compressed by surrounding tubular edema.

7.08.4.2 Acute Morphological Lesions in Tubules

Anoxic damage along tubules is governed by the intrinsic vulnerability of the different nephron segments and by the tissue gradients of oxygenation. Glomeruli and collecting ducts are relatively resistant to lack of oxygen. By contrast, both proximal and distal tubules (especially medullary thick limbs) are intrinsically susceptible to hypoxia.[40,41] Nevertheless, the distribution of tubular damage *in vivo* appears to be largely determined by intrarenal oxygen gradients.[40]

7.08.4.2.1 Distribution of gradients of hypoxia within the kidney

Intrarenal gradients of oxygenation determine the distribution of tubular injury after kidney ischemia.[42] The outer medulla is a major target because of regional hypoxia and the presence of tubules vulnerable to hypoxia (those in the straight portion of the proximal tubules (S_3) and the thick limbs). The inner medulla is far less vulnerable because its structures have lower metabolic demands. In the cortex, the medullary rays, perfused by venous blood emerging from the medulla, are an expected target because they are located farthest away from the oxygen supply and contain a susceptible structure, the S_3 tubules.[42]

7.08.4.2.2 Structural expression of tubular hypoxic injury

After mild depletion of cell energy stores, the proximal tubules swell and lose their microvilli.[43] While visually striking, these lesions do not necessarily cause renal failure[4] and have often distracted attention from focal damage along distal tubules, marked by cytoplasmic fragmentation.[41,43] Reduction of transport work minimizes hypoxic injury in medullary thick limbs but does not prevent cell swelling in proximal tubules.[44–46] S_3 tubules respond to hypoxia by either cell fragmentation, or swelling—if transport is inhibited.[46] These disparate responses to hypoxia relate to functional differences between nephron segments: to allow bulk reabsorption of glomerular filtrate, proximal tubules are by far, 104-fold, more permeable to water than medullary thick limbs, which are designed water-tight for the best urinary concentration.[34]

The exact mechanisms by which the kidney fails in human ischemic or toxic ARF remain unclear. Vasomotor changes, tubular obstruction and backleak are probably operative to various degrees according to the severity of intrarenal damage.

7.08.5 PATHOPHYSIOLOGY

7.08.5.1 Susceptibility of the Kidney to Hypoperfusion

Why is the kidney so susceptible to compromised oxygenation? A central paradox in patients with ARF is that the most severe hypoperfusion damage occurs in the best oxygenated organ (compared with other organs, the total blood flow rate to the kidney is one of the highest in the body). The bulk of this blood supply is directed to the renal cortex, while the medulla is poorly oxygenated (Figure 2). The phenomenon of medullary hypoxemia, a property of all mammalian kidneys, may be considered a price paid for an efficient urinary concentrating mechanism, and may account for the disproportionate vulnerability of the kidney to ischemic injury.[21] Medullary oxygen balance appears to be regulated by multiple paracrine homeostatic mechanisms, which act in concert to regulate renal medullary oxygen homeostasis (Figure 3).

Cells in the outer medulla have receptors for mediators which control medullary oxygen supply by vasoconstriction or vasodilatation (e.g., prostaglandin E_2). Oxygen demand, a function of the rate of tubular reabsorption, is determined by the GFR, by the delivery of urine to the thick limbs and by the regulation of transport by a variety of local mediators. For instance, inhibition of prostaglandin synthesis causes a profound decline in medullary blood flow and oxygenation (Figure 4). Adenosine, released from ATP during hypoxia in any tissue, tends to restore regional oxygen balance: it is generally a vasodilator and inhibits cardiac contraction and neurotransmission.[47] In the kidney, adenosine induces both cortical vasoconstriction (with reduction of GFR) and medullary vasodilation with inhibition of tubular transport, suggesting an intrarenal homeostatic role to attenuate medullary hypoxia.[48,49] Adenosine A1 receptors appear to mediate cortical vasoconstriction, while A2 receptors may mediate medullary vasodilatation (Figure 5). Similarly, observations suggest that endothelin, a potent renal cortical vasoconstrictor, induces medullary vasodilatation by activation of B receptors.[50] Duality of different receptors for the same vasoactive mediator with opposing effects in cortex and medulla, may be a rule in the kidney to combine cortical vasoconstriction (and reduction of workload for reabsorption) with medullary vasodilation, as a homeostatic mechanism to attenuate medullary hypoxia.

A variety of agents and events can either exacerbate or ameliorate medullary hypoxia, in rat kidneys (Table 5). NSAIDs, radiocontrast

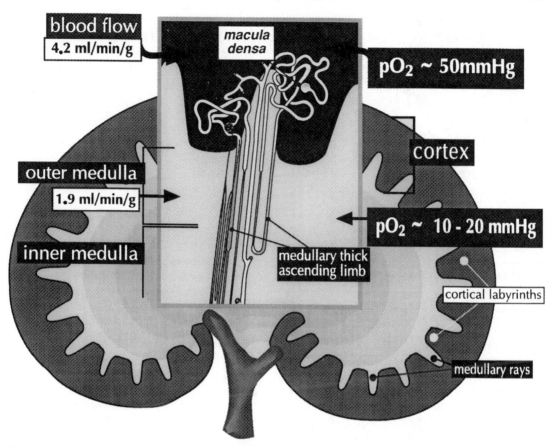

Figure 2 Anatomical and physiological features of the kidney comparing cortex and medulla. The cortex, which receives an ample blood supply to optimize glomerular filtration, is generally well oxygenated, except for the medullary rays areas devoid of glomeruli, which are supplied by venous blood ascending from the medulla. The medulla, which receives a meager blood supply to optimize concentration of the urine, is poorly oxygenated. Medullary hypoxia results both from countercurrent exchange of oxygen within the vasa recta and from oxygen consumption by the medullary thick ascending limbs. Renal medullary hypoxia is an obligatory price the mammalian kidney pays for successful urinary concentration.

media, mannitol, myoglobin, and polyene antibiotics aggravate medullary hypoxia and predispose to medullary thick limb damage. Systemic oxygenation has paradoxically little effect upon medullary pO_2, because of the isolation of medullary oxygenation by the countercurrent exchange mechanism. Reduction of tubular work (by loop-diuretics or decreased GFR) remarkably ameliorates medullary hypoxia. Prostaglandins, adenosine and nitric oxide improve medullary oxygenation, by increasing local blood flow and/or decreasing tubular reabsorption (Table 5). Accordingly, many drugs could directly or indirectly affect medullary oxygen balance, as illustrated for NSAIDs in Figure 4.

Thus, a coordinated self-defense system is present within the kidney to allow an efficient and safe urinary concentrating mechanism and reduce the risk of medullary injury, either by increasing local medullary blood flow or decreasing transport work and restoring medullary oxygen sufficiency. Failure or over activation of these mechanisms, especially in the presence of nephrotoxicants, may lead to medullary injury and/or ARF (Figure 3).

7.08.5.2 Synergism of Hypoxic and Toxic Insults in Renal Injury

For many years, research on ARF has directed attention at the prominent injury induced to proximal tubules in a variety of experimental models using a single insult inflicted to healthy animals and intensive enough to conveniently generate reproducible renal failure. Due to its excessive sensitivity to hypoxic injury, the proximal tubule readily manifests lesions which are not, in themselves, sufficient to cause renal failure and which may have distracted attention from less apparent,

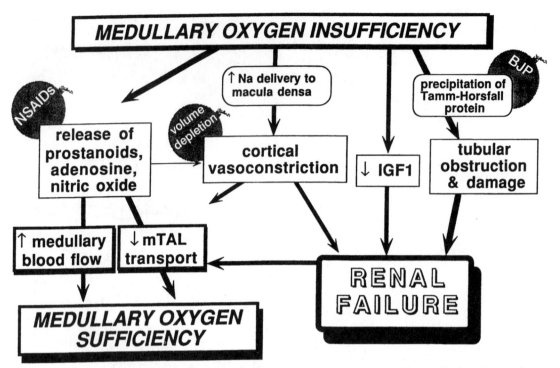

Figure 3 Putative mechanisms leading from medullary cell hypoxia to renal failure. On the left, are shown physiological homeostatic signals that improve medullary oxygenation (by increased blood flow and decreased transport) and often contribute to reduced renal function. On the right are shown some pathophysiological consequences of more advanced medullary hypoxia, such as tubular damage and reduced insulin-like growth factor 1 (IGF-1). The potential adverse effects of some nephrotoxicants and volume depletion are shown. Nonsteroidal antiinflammatory agents (NSAIDs) disable the beneficial prostanoid-mediated medullary vasodilatatory response to local hypoxia. Volume depletion enhances the tubuloglomerular feedback reflex decrease in glomerular filtration. In myeloma kidneys, Bence Jones proteins (BJP) co-precipitate with the Tamm–Horsfall protein released by damage to medullary thick ascending limbs, increasing the likelihood of tubular obstruction and renal failure from other insults (e.g., NSAIDs, volume depletion, or radiographic contrast agent). Renal failure results from tubular obstruction (by casts), back leak of glomerular filtrate from the lumen to blood (through damaged epithelium), impaired intrarenal microcirculation, such as activation of tubuloglomerular feedback (by increased distal delivery of solute to the macula densa), and failure of the local production of growth factors.

perhaps more meaningful damage in the distal nephron.[21] The relevance of "single-insult" experimental models to clinical ARF should be considered suspect, since at the bedside,

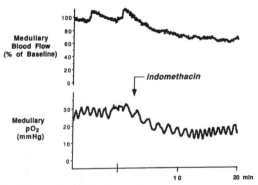

Figure 4 Effects of inhibition of prostaglandin synthesis by indomethacin upon medullary blood flow and oxygen tension in the rat kidney (after Agmon and Brezis[51]).

combinations of multiple factors rather than single insults are generally operative; chronic underlying conditions (e.g., diabetes mellitus or renal hypoperfusion) often predispose the patient to toxicity of drugs given at therapeutic dosages. Therapeutic successes described in "single-insult" animal models have not so far materialized in clinical trials, further evidence that these models may not resemble human ARF, where the only proven therapy is prevention (Table 6).

The recognition of a critical role for synergism between renal hypoperfusion and nephrotoxicity in the pathogenesis of human ARF has led to the development of "multiple-insults" models, combining factors that would not by themselves have caused significant renal injury. In these models, using for instance radiocontrast with indomethacin in salt-depleted rats[66] or gentamicin with renal hypoperfusion,[67] prominent injury is observed in the outer

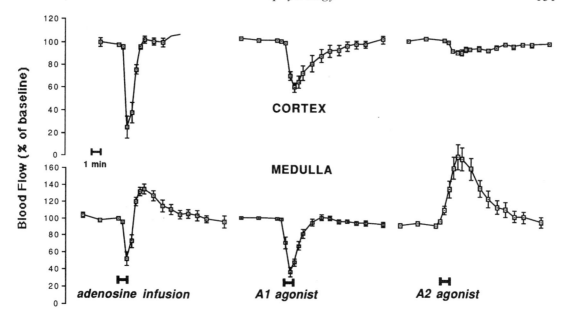

Figure 5 Effects of intrarenal adenosine infusion upon intrarenal blood flow in the rat kidney. Adenosine A1 receptors appear to mediate cortical vasoconstriction, while A2 receptors may mediate medullary vasodilatation.[48,49]

Table 5 Agents and conditions affecting medullary oxygenation.

Agent/condition	Baseline pO_2 (mm Hg)	pO_2 after the insult (mm Hg)	Ref.
Exacerbating effect			
Nonsteroidal anti-inflammatory drugs	28 ± 5	12 ± 4	52
Contrast media	26 ± 3	9 ± 2	(6)
Mannitol	20 ± 3	15 ± 3	52
Myoglobin	22 ± 4	14 ± 3	53
Amphotericin B	34 ± 4	18 ± 4	21,44,54,55
Compensatory renal hypertrophy			56
No effect			
Acetazolamide	17 ± 7	16 ± 6	57
Dopamine	16 ± 4	18 ± 4	58
Systemic oxygenation			59
Ameliorating effect			
Decreased tubular work			
Decreased GFR			44,45,60
Loop-diuretics	15 ± 4	34 ± 5	57
Prostaglandin E_2 and bradykinin			61
Adenosine	17 ± 3	40 ± 5	48,62
Nitroprusside, nitric oxide	21 ± 2	39 ± 2	63–67

medulla, along the medullary thick ascending limb or in S_3. Interference with the homeostatic mechanisms controling medullary oxygen balance may predispose to focal hypoxic injury at sites of strategic importance for overall renal function.[68,69] The synergism between renal hypoperfusion and toxic insults could derive from several factors (Figure 6), including increased intrarenal concentration of toxicants at a time when sodium reabsorption and urine concentration is increased and oxygen supply is reduced. Additional modes of interaction between intrarenal hypoxia and nephrotoxicants are, at the cellular level, increased oxygen demand to compensate for toxic membrane damage and/or mitochondrial dysfunction.[67]

Table 6   Comparison between clinical ARF and two modes of experimental ARF.

	Clinical ARF	"Single-insult" models	"Multiple-insults" models
Incidence	Variable	100%	<100%
Number of insults	Often ≥2	1	≥2
Chronic risk factor	Frequently present	Absent	Sometimes present
Renal hypoperfusion	Often present, usually moderate	Frank ischemia	Moderate
Drug dosage	Uusually close to therapeutic	Frankly toxic	Therapeutic or minimally toxic
Morphology	Focal tubular necrosis (distal > proximal) interstitial edema, vasa recta congestion and infiltration	Prominent cortical injury–proximal tubular necrosis	Prominent outer medulla injury–S_3 and/or mTAL necrosis
Mechanism	?	Toxic or hypoxic damage usually predominant in the proximal tubule	Interference with medullary O_2 balance
Response to therapy	None effective so far	ANP, T4, EGF, IGF, HGF	?ANP
Prevention	Volume expansion, avoidance of toxins	Glycine, scavengers of oxygen free radicals, Ca^{2+} channel blockers	Volume expansion

Drug interference (e.g., NSAIDs) with renal protective mechanisms, increased angiotensin II production and tubuloglomerular feedback, and increased likelihood of intraluminal precipitation of crystals or coprecipitation of a toxicant with Tamm–Horsfall protein (released from thick limbs into the urine, see below) also conspire to precipitate renal failure in combined hypoxic and toxic insults.[68] The susceptibility of the kidney to hypoperfusion is thus greatly aggravated by synergistic insults which uncover a special renal medullary vulnerability to hypoxia.

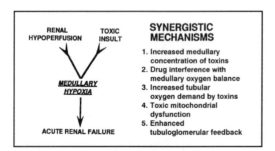

Figure 6   Synergistic mechanisms in the pathophysiology of ARF. Combinations of hypoxic and toxic insults disrupt the delicate and unstable balance of medullary oxygenation in a variety of synergistic mechanisms.

In clinical ARF, medullary injury is suggested by the early and consistent loss of urinary concentrating ability and by the presence in the urine of typical tubular casts, made from Tamm–Horsfall protein.[21] Clinicopathological studies have confirmed the predominance of distal tubular and outer medullary injury in human ARF.[33] Renal medullary hypoxia may play an important role in the susceptibility of the kidney to ARF from synergistic hypoxic and toxic insults. Therapy of ARF awaits a better understanding of "multiple-insults" models, an improved characterization of intrarenal homeostatic mechanisms, and the developments of potent and specific humoral antagonists (e.g., antiendothelin). Until then, prevention of the synergism between renal hypoperfusion and toxic insults by adequate volume repletion and avoidance of nephrotoxicants in high risk patients remains the most effective therapy for human ARF.

The example of radiocontrast nephrotoxicity is particularly instructive. In animals, as in humans, radiocontrast administration, even in large doses, causes little injury to the kidneys (Figure 1). Clinical nephrotoxicity occurs particularly in the presence of multiple risk factors, such as diabetes mellitus and preexisting kidney damage, often associated with compromised renal circulation.[4,69] To produce radiocontrast nephrotoxicity, rats were pre-treated by simultaneous inhibition of prostaglandin and nitric oxide before the administration of the contrast agent. Blocking these protective mechanisms (often referred to as endothelial-derived vasorelaxation) transformed a medullary vasodilator response to the contrast agent into profound vasoconstriction, selective necrosis of medullary thick ascending limbs and renal failure (Figure 7).[65]

Impaired endothelium-derived vasorelaxation in diabetes mellitus, hypertension, and atherosclerosis results in paradoxical vasoconstriction and regional hypoxia. Since these diseases are frequently associated with contrast agent toxicity,[4] endothelial dysfunction in chronic renal and vascular diseases may predispose patients to medullary injury from contrast agents, illustrating the vulnerability to disruption of the delicate oxygen balance within the medulla.

Interference with the homeostatic mechanisms controling medullary oxygen balance predisposes to focal hypoxic injury at sites of strategic importance for renal function. The susceptibility of the kidney to hypoperfusion is thus greatly aggravated by synergistic insults that uncover the renal medullary vulnerability to hypoxia, culminating in cell injury and organ failure. Taken together, the experimental and clinical data suggest that synergism between hypoxic and toxic insults in regions of lowest oxygen supply within the kidney may be an important mechanism for renal cell injury (Figures 3 and 6).

7.08.6 MANAGEMENT OF THE PATIENT WITH ARF

7.08.6.1 Prevention

Since the prognosis of ARF has not significantly improved over the years, the major challenge remains its prevention. Effective preventive measures include early recognition of renal hypoperfusion (by early detection and correction of hypovolemia or hypotension) and avoidance of nephrotoxicants in high risk patients (e.g., uncorrected renal hypoperfusion). Since insults are synergistic, it is imperative to avoid combination of risks, such as hypoxic and toxic injuries. Aminoglycosides can often be replaced by other broad-spectrum, less nephrotoxic antibiotics (quinolones, novel β-lactams). If used, the doses of aminoglycosides should be adjusted to renal functional status and the blood levels carefully monitored. The use of contrast agents should be limited to conditions with a clear clinical indication, where noncontrast imaging methods are insufficient and with consideration to risk/benefit ratios. NSAIDs and angiotensin converting enzyme inhibitors should be avoided

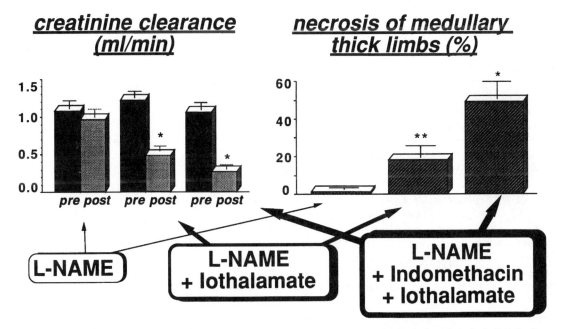

Figure 7 Predisposition to radiocontrast nephropathy by pharmacologically induced endothelial dysfunction. Combined inhibition of nitric oxide (by L-NAME) and prostaglandin synthesis (by indomethacin) was necessary and sufficient to predispose the rat to nephrotoxicity from radiological contrast media (iothalamate). Data are from Ref. 64. *$p < 0.0005$, **$p < 0.05$ (vs. baseline or controls).

in patients with reduced effective blood volume.

Fluid loading has been the cornerstone of preventive measures for many years. Hydration and salt loading reduce the work of urine concentration which predisposes the renal medulla to hypoxic and toxic damage, and are therefore a major mode of prophylaxis against ischemic and nephrotoxic renal injury.[21] This is supported by the results of several clinical trials demonstrating a protective role of fluid loading from cisplatin, amphotericin, and contrast toxicity, as well as in postoperative setting.[4,70,71]

Mannitol, proposed by animal studies to prevent ARF[72] had been used for prophylaxis in high risk surgery, rhabdomyolysis,[73–75] and prior to contrast administration.[76] Studies have shown no additional benefit of mannitol use over adequate hydration in preventing postoperative[77] or contrast induced[71] renal failure. In large doses, mannitol may be toxic[78] and can by itself induce ARF,[79] perhaps because osmotic diuresis aggravates medullary hypoxia.[52] As to loop diuretics, furosemide has increased the risk of contrast nephrotoxicity compared to the use of fluids alone, probably by causing negative fluid balance.[71,80] Low-dose dopamine was suggested to reduce the incidence of postoperative ARF in some studies.[81–83] In a randomized double-blind controlled study, low-dose dopamine offered no advantage to euvolemic patients after major vascular surgery.[84] In diabetic patients, dopamine increased the risk of contrast nephrotoxicity,[85] perhaps because of aggravated medullary hypoxia.[86] Benefits have been suggested with the prophylactic use of calcium channel blockers before the administration of radiocontrast and in renal transplantation[87] but not in high-risk surgical patients.[88] In summary, the efficacy of mannitol, loop-diuretics, dopamine, and calcium channel blockers in preventing ARF in high risk patients has not been proven. Furosemide and mannitol have been associated with an increased risk and therefore are not recommended.

A few specific measures are available to protect the kidney from exposure or injury due to specific toxins. These include the early administration of *N*-acetylcysteine in acetaminophen overdose, the use of dimercaprol in heavy metal intoxication, and the use of diuresis plus alkalinization of the urine to reduce the toxicity of uric acid and myoglobin.

7.08.6.2 Supportive Treatment

Conservative measures consist of fluid restriction; modification of diet to minimize intake of protein, sodium, potassium, and phosphate while providing enough calories; use of phosphate binders (calcium carbonate or limited course of aluminum hydroxide); avoidance of potential nephrotoxic drugs; adjustment of medications doses to the degree of renal function impairment; early detection and aggressive treatment of infections and the use of estrogens and desmopressin to reduce hemorrhagic complications.

While parenteral nutritional therapy in ARF has evoked intense interest, there is no convincing evidence for its efficacy.[89] A positive nitrogen balance cannot be achieved in most patients with multiorgan failure.[90] Early clinical studies suggested enhanced renal recovery and improved survival in patients with ARF receiving amino acids and glucose rather than glucose alone,[91] but later studies were unable to show an improved survival with the use of amino acid therapy.[90,92] The risks of aggressive nutritional therapy are of great concern. Parenteral nutrition may cause significant complications including electrolyte disturbances, acid–base disorders, hyperlipidemia, and adverse effects of central lines. Finally, increased urea formation and fluid load necessitate more aggressive dialysis treatment.

7.08.6.3 Pharmacological Interventions in Established ARF

7.08.6.3.1 Conversion of oliguric into nonoliguric ARF

Nonoliguric ARF has a better outcome than oliguric ARF,[17,93] perhaps because of less severe disease and higher GFR.[94,95] There is no convincing evidence that pharmacological measures to correct oliguria also improve the outcome in ARF.[72,96]

7.08.6.3.2 "Renal dose" dopamine

Since a renal-sparing potential for low dose dopamine has been suggested in oliguric ARF,[97] this drug became very popular in intensive care units. Low dose dopamine has been shown to increase urine output in oliguric critically ill patients,[98,99] including patients who remained oliguric despite high dose loop diuretics or mannitol,[100] but no controlled study has shown an improved outcome with the use of low dose dopamine.[84,101] No dose is clearly only a "renal dose:" even low doses are associated with tachyarrhythmias, myocardial ischemia, increase pulmonary shunt,[102] digital necrosis[103] and intestinal damage.[101] Until efficacy and safety are demonstrated in prospective controlled trials, the use of dopamine as a renal sparing agent cannot be recommended.[84,101] A small double-blind cross-over study has shown that while low dose dopamine is a better diuretic, dobutamine is more effective in improving renal function in critically ill patients, probably due to its greater hemodynamic effects.[102] A beneficial effect of dobutamine on outcome in ARF is yet to be shown.

7.08.6.3.3 Atrial natriuretic peptide (ANP)

ANP has been shown to afford some protection in a variety of animal models of ARF.[104,105] A small open clinical trial[106] showed that while mortality was not significantly reduced, creatinine clearance was improved and the need for dialysis was reduced in patients treated with ANP in comparison to controls. In 1995, a large, prospective, controlled trial suggests a beneficial effect in oliguric ARF.[107] These promising results await confirmation.

7.08.6.3.4 Growth factors

Tubular regeneration following ischemic or toxic injury is associated with increased expression of growth response genes, growth factors, and growth factors receptors.[108] In several but not all animal studies, epidermal growth factor,[109] insulin-like growth factor-1,[110,111] and hepatocyte growth factor[112–114] have been shown to accelerate recovery of renal function, to improve histologic injury and/or to reduce mortality in ARF. The mechanism of action include enhancement of tubular regeneration, vasodilatation, and stimulation of systemic anabolism.[115] While insulin-like growth factor-1 has been safely administered to patients with chronic renal failure,[116] clinical studies are needed to evaluate growth factors in the therapy of human ARF.

7.08.6.4 Renal Replacement Therapy

Many patients with ARF require dialysis, particularly those with oliguria, hypercatabolism, and failure of other organs. Indications for dialysis include volume overload, hyperkalemia or acidosis, and uremic manifestations, such as nausea and vomiting, encephalopathy, pericarditis, or bleeding diathesis. The optimal timing, frequency, and intensity of dialysis therapy are controversial.[117]

The availability of highly permeable, biocompatible membranes and the advances in continuous renal replacement therapies have expanded the armamentarium of dialysis modalities.[118] Experimental[119] and clinical[120,121] evidence suggest that dialysis membrane biocompatibility may have a significant impact on the rate of renal function recovery in ARF and, perhaps, also on patient survival.

For critically ill and highly catabolic patients with multiorgan failure, hemodynamically unstability and fluid overload, continuous hemofiltration with or without hemodialysis is the treatment of choice. Improved clearance of large molecules, including cytokines (e.g., TNF-α, IL-1, and IL-6) may be advantageous in sepsis.[118] Continuous veno–venous hemofiltration and hemodiafiltration are more expensive and technically demanding than continuous arterio–venous therapies, but avoid the risks of an arterial access, enable higher filtration volumes and provide better clearance of small molecules.[122,123] While continuous renal replacement therapies allow removal of large amounts of fluids without hemodynamic compromise and an excellent nutritional support, overall patient outcome remains as poor as with dialysis.[118]

7.08.7 REFERENCES

1. S. H. Hou, D. A. Bushinsky, J. B. Wish *et al.*, 'Hospital-acuired renal insufficiency: a prospective study.' *Am. J. Med.*, 1983, **74**, 243–248.
2. J. Kaufman, M. Dhakal, B. Patel *et al.*, 'Community-acquired acute renal failure.' *Am. J. Kidney Dis.*, 1991, **17**, 191–198.
3. T. G. Feest, A. Round and S. Hamad, 'Incidence of severe acute renal failure in adults: results of a community based study.' *BMJ*, 1993, **306**, 481–483.
4. M. Brezis, S. Rosen and F. H. Epstein, in 'The Kidney,' 4 edn., eds. J. F. Rector and B. M. Brenner, W. B. Saunders, Philadelphia, 1991, pp. 993.
5. S. Klahr, 'New insights into the consequences and mechanisms of renal impairment in obstructive nephropathy.' *Am. J. Kidney Dis.*, 1991, **18**, 689–699.
6. K. P. Harris, S. Klahr and G. Schreiner, 'Obstructive nephropathy: from mechanical disturbance to immune activation?' *Exp. Nephrol.*, 1993, **1**, 198–204.
7. H. W. Lange, D. M. Aeppli and D. C. Brown, 'Survival of patients with acute renal failure requiring dialysis after heart surgery: early prognostic indicators.' *Am. Heart J.*, 1987, **113**, 1138–1143.
8. B. K. Novis, M. F. Roizen, S. Aronson *et al.*, 'Association of preoperative risk factors with postoperative acute renal failure.' *Anesth. Analg.*, 1994, **78**, 143–149.
9. G. Zanardo, P. Michielon, A. Paccagnella *et al.*, 'Acute renal failure in the patient undergoing cardiac operation. Prevalence, mortality rate, and main risk factors.' *J. Thorac. Cardiovasc. Surg.*, 1994, **107**, 1489–1495.
10. P. I. Menashe, S. A. Ross and J. E. Gottlieb, 'Aquired renal insufficiency in critically ill patients.' *Crit. Care Med.*, 1988, **16**, 1106–1109.

11. A. B. Groeneveld, D. D. Tran, J. van der Meulen et al., 'Acute renal failure in the medical intensive care unit: predisposing, complicating factors, and outcome.' Nephron, 1991, **59**, 602–610.

12. N. Lameire, E. Matthys, R. Vanholder et al., 'Causes and prognosis of acute renal failure in elderly patients.' Nephrol. Dial. Transplant, 1987, **2**, 316–322.

13. R. F. Spurney, W. J. Fulkerson and S. J. Schwab, 'Acute renal failure in critically ill patients: prognosis for recovery of kidney function after prolonged dialysis support.' Crit. Care Med., 1991, **19**, 8–11.

14. K. Abreo, V. Moorthy and M. Osborne, 'Changing patterns and outcome of acute renal failure requiring hemodialysis.' Arch. Intern. Med., 1986, **146**, 1338–1341.

15. J. H. Turney, D. H. Marshall, A. M. Brownjohn et al., 'The evolution of acute renal failure, 1956–1988.' Q. J. Med., 1990, **74**, 83–104.

16. L. Frost, R. S. Pedersen, S. Bentzen et al., 'Short- and long-term outcome in a consecutive series of 419 patients with acute dialysis-requiring renal failure.' Scand. J. Urol. Nephrol., 1993, **27**, 453–462.

17. I. K. Barton, P. J. Hilton, N. A. Taub et al., 'Acute renal failure treated by haemofiltration: factors affecting outcome.' Q. J. Med., 1993, **86**, 81–90.

18. A. L. Beal and F. B. Cerra, 'Multiple organ failure syndrome in the 1990s. Systemic inflammatory response and organ dysfunction.' JAMA, 1994, **271**, 226–233.

19. H. H. Rasmussen and L. S. Ibels, 'Acute renal failure. Multivariant analysis of causes and risk factors.' Am. J. Med., 1982, **73**, 211–218.

20. M. W. Rich and C. A. Crecelius, 'Incidence, risk factors, and clinical course of acute renal insufficiency after cardiac catheterization in patients 70 years of age or older. A prospective study.' Arch. Intern. Med., 1990, **150**, 1237–1242.

21. M. Brezis and S. Rosen, 'Hypoxia of the renal medulla—its implications for disease.' N. Engl. J. Med., 1995, **332**, 647–653.

22. A. L. Linton, W. F. Clark, A. A. Driedger et al., 'Acute interstitial nephritis due to drugs: review of the literature with a report of nine cases.' Ann. Intern. Med., 1980, **93**, 735–741.

23. E. G. Neilson, 'Pathogenesis and therapy of interstitial nephritis.' Kidney Int., 1989, **35**, 1257–1270.

24. K. M. Murray and W. R. Keane, 'Review of drug-induced acute interstitial nephritis.' Pharmacotherapy, 1992, **12**, 462–467.

25. J. S. Cameron, 'Allergic interstitial nephritis: clinical features and pathogenesis.' Q. J. Med., 1988, **66**, 97–115.

26. M. L. Levin, 'Patterns of tubulo–interstitial damage associated with nonsteroidal antiinflammatory drugs.' Semin. Nephrol., 1988, **8**, 55–61.

27. J. E. Galpin, J. H. Shinaberger, T. M. Stanley et al., 'Acute interstitial nephritis due to methicillin.' Am. J. Med., 1978, **65**, 756–765.

28. A. L. Linton, J. M. Richmond, W. F. Clark et al., 'Gallium-67 scintigraphy in the diagnosis of acute renal disease.' Clin. Nephrol., 1985, **24**, 84–87.

29. S. H. Sachs, S. Aparicio and A. Bevan, 'Late renal failure due to prostatic outflow obstruction: a preventable disease.' BMJ, 1989, **298**, 156.

30. C. A. Rabito, F. Panico, R. Rubin et al., 'Non-invasive, real-time monitoring of renal function during critical care.' J. Am. Soc. Nephrol., 1994, **4**, 1421–1428.

31. F. Steinhauslin, M. Burnier, J. L. Magnin et al., 'Fractional excretion of trace lithium and uric acid in acute renal failure.' J. Am. Soc. Nephrol., 1994, **4**, 1429–1437.

32. N. T. Richards, S. Darby, A. J. Howie et al., 'Knowledge of renal histology alters patient management in over 40% of cases.' Nephrol. Dial. Transplant., 1994, **9**, 1255–1259.

33. T. S. Olsen and H. E. Hansen, 'Ultrastructure of medullary tubules in ischemic acute tubular necrosis and acute interstitial nephritis in man.' APMIS, 1990, **98**, 1139–1148.

34. M. Brezis and F. H. Epstein, in 'Ann. Rev. Med.,' ed. C. Coggins, 1993, pp. 44.

35. B. A. Molitoris, 'Ischemia-induced loss of epithelial polarity: potential role of the actin cytoskeleton.' Am. J. Physiol., 1991, **260**, F769–F778.

36. J. V. Bonventre, 'Mechanisms of ischemic renal failure.' Kidney Int., 1993, **43**, 1160–1178.

37. R. Beeri, Z. Symon, M. Brezis et al., 'Rapid DNA fragmentation from hypoxia along the thick ascending limb of rat kidneys.' Kidney Int., 1995, **47**, 1806–1810.

38. J. D. Conger, J. B. Robinette and R. W. Schrier, 'Smooth muscle calcium and endothelium-derived relaxing factor in the abnormal vascular responses of acute renal failure.' J. Clin. Invest., 1988, **82**, 532–537.

39. K. J. Kelly, W. W. Williams, Jr., R. B. Colvin et al., 'Antibody to intercellular adhesion molecule 1 protects the kidney against ischemic injury.' Proc. Natl. Acad. Sci. USA, 1994, **91**, 812–816.

40. C. E. Ruegg and L. J. Mandel, 'Bulk isolation of renal PCT and PST. II. Differential responses to anoxia or hypoxia.' Am. J. Physiol., 1990, **259**, F176–F185.

41. M. Brezis, S. Rosen, P. Silva et al., 'Selective vulnerability of the medullary thick ascending limb to anoxia in the isolated perfused rat kidney.' J. Clin. Invest., 1984, **73**, 182–190.

42. P. F. Shanley, M. D. Rosen, M. Brezis et al., 'Topography of focal proximal tubular necrosis after ischemia with reflow in the rat kidney.' Am. J. Pathol., 1986, **122**, 462–468.

43. M. Brezis, P. Shanley, P. Silva et al., 'Disparate mechanisms for hypoxic cell injury in different nephron segments. Studies in the isolated perfused rat kidney.' J. Clin. Invest., 1985, **76**, 1796–1806.

44. M. Brezis, S. Rosen, P. Silva et al., 'Polyene toxicity in renal medulla: injury mediated by transport activity.' Science, 1984, **224**, 66–68.

45. M. Brezis, S. Rosen, P. Silva et al., 'Transport activity modifies thick ascending limb damage in the isolated perfused kidney.' Kidney Int., 1984, **25**, 65–72.

46. P. F. Shanley, M. Brezis, K. Spokes et al., 'Transport dependent cell injury in the S_3 segment of the proximal tubule.' Kidney Int., 1986, **29**, 1033–1037.

47. R. A. Olsson and J. D. Pearson, 'Cardiovascular purinoceptors.' Physiol. Rev., 1990, **70**, 761–845.

48. D. Dinour and M. Brezis, 'Effects of adenosine upon intrarenal oxygenation.' Am. J. Physiol., 1991, **261**, F787–F791.

49. Y. Agmon, D. Dinour and M. Brezis, 'Disparate effects of adenosine A1- and A2-receptor agonists on intrarenal blood flow.' Am. J. Physiol., 1993, **265**, F802–F806.

50. K. Gurbanov, I. Rubinstein, A. Hoffman et al., 'Role of endothelin-1 in the regulation of intra-renal blood flow in the rat kidney.' Annual Meeting of the Israeli Society of Nephrology and Hypertension, Zichron Yaacov, Israel, 1995.

51. Y. Agmon and M. Brezis, 'Effects of nonsteroidal antiinflammatory drugs upon intrarenal blood flow: selective medullary hypoperfusion.' Exp. Nephrol., 1993, **1**, 357–363.

52. S. N. Heyman, M. Brezis, F. H. Epstein et al., 'Early renal medullary hypoxic injury from radiocontrast and indomethacin.' Kidney Int., 1991, **40**, 632–642.

53. S. N. Heyman, *Kidney Int.*, 1997, in press.
54. M. Brezis, S. N. Heyman and A. M. Sugar, 'Reduced amphotericin toxicity in albumin vehicle.' *J. Drug Target.*, 1993, **1**, 185–189.
55. S. N. Heyman, M. Brezis, F. H. Epstein *et al.*, 'Chronic amphotericin nephropathy: selective inner stripe and medullary ray injury.' *Clin. Res.*, 1991, **39**, 248.
56. F. H. Epstein, P. Silva, K. Spokes *et al.*, 'Renal medullary Na-K-ATPase and hypoxic injury in perfused rat kidneys.' *Kidney Int.*, 1989, **36**, 768–772.
57. M. Brezis, Y. Agmon and F. H. Epstein, 'Determinants of intrarenal oxygenation: I. Effects of diuretics.' *Am. J. Physiol.*, 1994, **267**, F1059–F1062.
58. S. N. Heyman, N. Kaminski and M. Brezis, 'Dopamine increases medullary blood flow without improving regional hypoxia.' *Exp. Nephrol.*, **3**, 331–337.
59. F. H. Epstein, Y. Agmon and M. Brezis, 'Physiology of renal hypoxia.' *Ann. NY Acad. Sci.*, 1994, **718**, 72–81, discussion 81–82.
60. M. Brezis, D. Dinour, Z. Greenfeld *et al.* in 'Advances in Nephrology,' eds. J. Grunfeld and M. H. Maxwell, Mosby Year Book, Chicago, IL, 1991, pp. 41.
61. P. Silva, S. Rosen, K. Spokes *et al.*, 'Influence of endogenous prostaglandins on mTAL injury.' *J. Am. Soc. Nephrol.*, 1990, **1**, 808–814.
62. D. Dinour, Y. Agmon and M. Brezis, 'Adenosine: an emerging role in the control of renal medullary oxygenation?' *Exp. Nephrol.*, 1993, **1**, 152–157.
63. M. Brezis, S. N. Heyman and F. H. Epstein, 'Determinants of intrarenal oxygenation. II. Hemodynamic effects.' *Am. J. Physiol.*, 1994, **267**, F1063–F1068.
64. M. Brezis, S. N. Heyman, D. Dinour *et al.*, 'Role of nitric oxide in renal medullary oxygenation. Studies in isolated and intact rat kidneys.' *J. Clin. Invest.*, 1991, **88**, 390–395.
65. Y. Agmon, H. Peleg, Z. Greenfeld *et al.*, 'Nitric oxide and prostanoids protect the renal outer medulla from radiocontrast toxicity in the rat.' *J. Clin. Invest.*, 1994, **94**, 1069–1705.
66. S. N. Heyman, M. Brezis, C. A. Reubinoff *et al.*, 'Acute renal failure with selective medullary injury in the rat.' *J. Clin. Invest.*, 1988, **82**, 401–412.
67. R. A. Zager, 'Gentamicin nephrotoxicity in the setting of acute renal hypoperfusion.' *Am. J. Physiol.*, 1988, **254**, F574–F581.
68. M. Brezis and F. H. Epstein. in 'Toxicology of the Kidney,' 2nd edn., eds. R. S. Goldstein and J. B. Hook, Raven Press, New York, 1993, pp. 129.
69. M. Brezis, S. Rosen and F. Epstein. in 'Acute Renal Failure,' 3rd edn., eds. M. Lazarus and B. M. Brenner, Churchill-Livingstone, Edinburgh, 1993, pp. 207.
70. B. J. Barrett, 'Contrast nephrotoxicity.' *J. Am. Soc. Nephrol.*, 1994, **5**, 125–137.
71. R. Solomon, C. Werner, D. Mann *et al.*, 'Effects of saline, mannitol, and furosemide on acute decreases in renal function induced by radiocontrast agents.' *N. Engl. J. Med.*, 1994, **331**, 1416–1420.
72. I. Shilliday and M. E. Allison, 'Diuretics in acute renal failure.' *Ren. Fail.*, 1994, **16**, 3–17.
73. S. R. Powers, A. Boba, W. Hostnik *et al.*, 'Prevention of postoperative acute renal failure with mannitol in 100 cases.' *Surgery*, 1964, **35**, 15–21.
74. J. L. Dawson, 'Postoperative renal function in obstructive jaundice: effect of a mannitol diuresis.' *BMJ*, 1965, **1**, 82–86.
75. D. Ron, U. Taitelman and M. Michaelson *et al.*, 'Prevention of acute renal failure in traumatic rhabdomyolysis.' *Arch. Intern. Med.*, 1984, **144**, 277–280.
76. H. R. Anto, S. Y. Chou, J. G. Porush *et al.*, 'Infusion intravenous pyelographt and renal function. Effects of hypertonic mannitol in patients with chronic renal insufficiency.' *Arch. Intern. Med.*, 1981, **141**, 1652–1656.
77. J. M. Gubern, J. J. Sancho, J. Simo *et al.*, 'A randomized trial on the effect of mannitol on postoperative renal function in patients with obstructive jaundice.' *Surgery*, 1988, **103**, 39–44.
78. H. F. Borges, J. Hocks and C. M. Kjellstrand, 'Mannitol intoxication in patients with renal failure.' *Arch. Intern. Med.*, 1982, **142**, 63–66.
79. H. R. Dorman, J. H. Sondheimer and P. Cadnapaphorenchai, 'Mannitol-induced acute renal failure.' *Medicine*, 1990, **69**, 153–159.
80. J. M. Weinstein, S. Heyman and M. Brezis, 'Potential deleterious effect of furosemide in radiocontrast nephropathy.' *Nephron.*, 1992, **62**, 413–415.
81. R. J. Polson, G. R. Park, M. J. Lindop *et al.*, 'The prevention of renal impairment in patients undergoing orthotropic liver grafting by infusion of low dose dopamine.' *Anaesthesia*, 1987, **42**, 15–19.
82. L. J. Pass, R. C. Eberhart, J. C. Brown *et al.*, 'The effect of mannitol and dopamine on the renal response to thoracic aortic cross-clamping.' *J. Thorac. Cardiovasc. Surg.*, 1988, **95**, 608–612.
83. T. H. Swygert, L. C. Roberts, T. R. Valeck *et al.*, 'Effects of intraoperative low-dose dopamine on renal function in liver transplant recipients.' *Anesthesiology*, 1991, **75**, 571–576.
84. L. Baldwin, A. Henderson and P. Hickman, 'Effects of postoperative low-dose dopamine on renal function after elective major vascular surgery.' *Ann. Intern. Med.*, 1994, **120**, 744–747.
85. L. S. Weisberg, P. B. Kurnik and B. R. Kurnik, 'Dopamine and renal blood flow in radiocontrast-induced nephropathy in humans.' *Ren. Fail.*, 1993, **15**, 61–68.
86. L. S. Weisberg, P. B. Kurnik and B. R. Kurnik, 'Risk of radiocontrast nephropathy in patients with and without diabetes mellitus.' *Kidney Int.*, 1994, **45**, 259–265.
87. H. H. Neumayer and U. Kunzendorf, 'Renal protection with the calcium antagonists.' *J. Cardiovasc. Pharmacol.*, 1991, **18**, Suppl. 1, S11–S18.
88. G. Tataranni, F. Malacarne, R. Farinelli *et al.*, 'Beneficial effects of verapamil renal-risk surgical patients.' *Ren. Fail.*, 1994, **16**, 383–390.
89. H. Sponsel and J. D. Conger, 'Is parenteral nutrition therapy of value in acute renal failure patients?' *Am. J. Kidney Dis.*, 1995, **25**, 96–102.
90. E. I. Feinstein, M. J. Blumenkrantz, M. Healy *et al.*, 'Clinical and metabolic responses to parenteral nutrition in acute renal failure. A controlled double-blind study.' *Medicine*, 1981, **60**, 124–137.
91. R. M. Abel, C. H. Beck, Jr., W. M. Abbott *et al.*, 'Improved survival from acute renal failure after treatment with intravenous essential b-amino acids and glucose. Results of a prospective double-blind study.' *N. Engl. J. Med.*, 1973, **288**, 695–699.
92. S. C. Spreiter, B. D. Myers and R. M. Swenson, 'Protein-energy requirements in subjects with acute renal failure receiving intermittent hemodialysis.' *Am. J. Clin. Nutr.*, 1980, **33**, 1433–1437.
93. R. J. Anderson, S. L. Linas, A. S. Berns *et al.*, 'Nonoliguric acute renal failure.' *N. Engl. J. Med.*, 1977, **296**, 1134–1138.
94. N. Honda and A. Hishida, 'Pathophysiology of experimental nonoliguric acute renal failure.' *Kidney Int.*, 1993, **43**, 513–521.
95. S. N. Rahman and J. D. Conger, 'Glomerular and tubular factors in urine flow rates of acute renal failure patients.' *Am. J. Kidney Dis.*, 1994, **23**, 788–793.
96. C. B. Brown, C. S. Ogg and J. S. Cameron, 'High-dose frusemide in acute renal failure: a controlled trial.' *Clin. Nephrol.*, 1981, **15**, 90–96.
97. L. I. Goldberg, 'Cardiovascular and renal actions of dopamine: potential clinical applications.' *Pharmacol. Rev.*, 1972, **24**, 1–29.

98. S. Parker, G. C. Carlon, M. Isaac *et al.*, 'Dopamine administration in oliguria and oliguric renal failure.' *Crit. Care Med.*, 1981, **9**, 630–632.

99. L. Flancbaum, P. S. Choban and J. F. Dasta, 'Quantitative effects of low-dose dopamine on urine output in oliguric surgical intensive care unit patients.' *Crit. Care Med.*, 1994, **22**, 61–68.

100. I. S. Henderson, T. J. Beattie and A. C. Kennedy, 'Dopamine hydrochloride in oliguric states.' *Lancet*, 1980, **2**, 827–828.

101. B. T. Thompson and B. A. Cockrill, 'Renal-dose dopamine: a siren song?.' *Lancet*, 1994, **344**, 7–8.

102. G. J. Duke, J. H. Briedis and R. A. Weaver, 'Renal support in critically ill patients: low-dose dopamine or low-dose dobutamine?' *Crit. Care Med.*, 1994, **22**, 1919–1925.

103. S. Greene and J. Smith, 'Letter: dopamine gangrene.' *N. Engl. J. Med.*, 1976, **294**, 114.

104. D. M. Pollock and T. J. Opgenroth, 'Atrial natriuretic peptides in the treatment of acute renal failure.' *Ren. Fail.*, 1993, **15**, 439–449.

105. J. D. Conger, S. A. Falk and W. S. Hammond, 'Atrial natriuretic peptide and dopamine in established acute renal failure in the rat.' *Kidney Int.*, 1991, **40**, 21–28.

106. S. N. Rahman, G. E. Kim, A. S. Mathew *et al.*, 'Effects of atrial natriuretic peptide in clinical acute renal failure.' *Kidney Int.*, 1994, **45**, 1731–1738.

107. S. N. Rahman, A. R. Butt, T. D. Dubose *et al.*, 'Differential clinical effects of anaritide atrial natriuretic peptide (ANP) in oliguric and nonoliguric ATN.' *J. Am. Soc. Nephrol.*, 1995, **6**, 913.

108. R. Safirstein, 'Gene expression in nephrotoxic and ischemic acute renal failure.' *J. Am. Soc. Nephrol.*, 1994, **4**, 1387–1395.

109. H. D. Humes, D. A. Cieslinski, T. M. Coimbra *et al.*, 'Epidermal growth factor enhances renal tubule cell regeneration and repair and accelerates the recovery of renal functions in postischemic acute renal failure.' *J. Clin. Invest.*, 1989, **84**, 1757–1761.

110. S. B. Miller, D. R. Martin, J. Kissane *et al.*, 'Insulin-like growth factor I accelerates recovery from ischemic acute tubular necrosis in the rat.' *Proc. Natl. Acad. Sci. USA*, 1992, **89**, 11876–11880.

111. H. Ding, J. D. Kopple, A. Cohen *et al.*, 'Recombinant human insulin-like growth factor-I accelerates recovery and reduces catabolism in rats with ischemic acute renal failure.' *J. Clin. Invest.*, 1993, **91**, 2281–2287.

112. S. B. Miller, D. R. Martin, J. Kissane *et al.*, 'Hepatocyte growth factor accelerates recovery from acute ischemic renal injury in rats.' *Am. J. Physiol.*, 1994, **266**, F129–F134.

113. T. Igawa, K. Matsumoto, S. Kanda *et al.*, 'Hepatocyte growth factor may function as a renotropic factor for regeneration in rats with acute renal failure.' *Am. J. Physiol.*, 1993, **265**, F61–F69.

114. K. Kawaida, K. Matsumoto, H. Shimazu *et al.*, 'Hepatocyte growth factor prevents acute renal failure and accelerates renal regeneration in mice.' *Proc. Natl. Acad. Sci. USA*, 1994, **91**, 4357–4361.

115. M. R. Hammerman and S. B. Miller, 'Therapeutic use of growth factors in renal failure.' *J. Am. Soc. Nephrol.*, 1994, **5**, 1–11.

116. M. H. O'Shea, S. B. Miller and M. R. Hammerman, 'Effects of IGF-I on renal function in patients with chronic renal failure.' *Am. J. Physiol.*, 1993, **264**, F917–F922.

117. D. M. Gillum, B. S. Dixon, M. J. Yanover *et al.*, 'The role of intensive dialysis in acute renal failure.' *Clin. Nephrol.*, 1986, **25**, 249–255.

118. R. L. Mehta, 'Therapeutic alternatives to renal replacement for critically ill patients with acute renal failure.' *Semin. Nephrol.*, 1994, **14**, 64–82.

119. G. Schulman, A. Fogo, A. Gung *et al.*, 'Complement activation retards resolution of ischemic renal failure in the rat.' *Kidney Int.*, 1991, **40**, 1069–1074.

120. R. M. Hakim, R. L. Wingard and R. A. Parker, 'Effect of the dialysis membrane in the treatment of patients with acute renal failure.' *N. Engl. J. Med.*, 1994, **331**, 1338–1342.

121. H. Schiffl, S. M. Lang, A. Konig *et al.*, 'Biocompatible membranes in acute renal failure: prospective case-controlled study.' *Lancet*, 1994, **344**, 570–572.

122. W. L. Macias, B. A. Mueller, K. Scarim *et al.*, 'Continuous venovenous hemofiltration: An alternative to continuous arteriovenous hemofiltration and hemodialfiltration in acute renal failure.' *Am. J. Kidney Dis.*, 1991, **18**, 451–458.

123. W. R. Clark, B. A. Mueller, K. J. Alaka *et al.*, 'A comparison of metabolic control by continuous and intermittent therapies in acute renal failure.' *J. Am. Soc. Nephrol.*, 1994, **4**, 1413–1420.

7.09
Adaptation in Progression of Renal Injury

AGNES FOGO

Vanderbilt University School of Medicine, Nashville, TN, USA

7.09.1 INTRODUCTION

Adaptation is broadly viewed as a change in cell or organ function and/or structure in response to a change in its environment. These responses may initially be protective against further injury, for example, squamous metaplasia of respiratory epithelium in a smoker. However, many forms of adaptation ultimately lead to further injury, such as, increased respiratory infections contributed to by lack of normal cilia in the smoker. Injury to the kidney induces changes in function, structure and metabolism of many of its elements, which may in the short term serve to restore kidney function towards normal by increasing, for example, filtration, renal mass, or metabolism of various nephron segments. The characteristic end stage histological pattern ultimately shared by various renal diseases has lead to the postulate of common adaptive mechanisms triggered by many different primary pathogenic insults (Figure 1). Initial pathogenic insults are presumed to lead to a loss of functioning nephrons through disease-specific pathophysiologic processes. The initial loss of nephrons then induces alterations of function and metabolism in intact glomeruli which are ultimately self-damaging in nature, and lead to further loss of nephrons. The ultimate outcome of this vicious cycle is an end stage, scarred kidney and chronic renal failure. Evidence has emerged

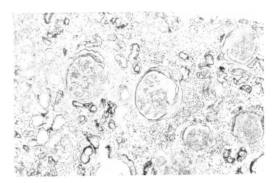

Figure 1 The common appearance of the end stage kidney is characterized by heterogeneous sclerosis even at a given time point, with focal and segmental distribution of sclerotic lesions. (Jones' silver stain, ×100).

that these adaptive processes are more diverse than initially thought. Not only do incidences and rates of progression vary amongst individuals with the same apparent initial disease process, mechanisms of progression may be varied in different experimental as well as human disease settings. This chapter will review adaptations to renal injury and impact on progression of glomerulosclerosis in experimental animal models, and will discuss relevance and implications for human renal disease.

7.09.2 POSSIBLE MECHANISMS OF PROGRESSIVE RENAL DISEASE

Ultimately, adaptive mechanisms not only fail to maintain glomerular filtration rate (GFR), but contribute to progressive scarring. The resulting end stage kidney is characterized by glomerulosclerosis in a focal and segmental pattern, and proportional tubular atrophy and interstitial fibrosis. Focal segmental glomerulosclerosis (FSGS) is defined by involvement of some, ("focal") but not all, glomeruli with sclerosis. The sclerotic process is defined by glomerular capillary collapse with increase in matrix. This scarring only affects a portion of the glomerular tuft ("segmental"), unlike global glomerulosclerosis which occurs non-specifically with aging. Thus, the very distribution of injury is heterogeneous. FSGS is a common endpoint in a variety of settings in human renal disease, including idiopathic FSGS, or scarring secondary to other glomerular disease or systemic abnormalities, such as immune complex mediated injury, obesity, HIV infection, hypertension and diabetic nephropathy. The broad spectrum of underlying diseases associated with histological lesions of

focal sclerosis in all likelihood reflects the presence of diverse pathogenic mechanisms leading to focal sclerosis. Regardless of the etiology, patients experience similar symptoms of chronic renal failure: electrolyte and fluid imbalances, hypertension, and uremia, due to loss of normal regulation of fluids and loss of renal filtration function with accumulation of nitrogenous waste products.

Glomerulosclerosis develops when the initial insult activates complex adaptations of the resident cells within the glomerulus in the remaining nephrons which lead to mesangial matrix accumulation and capillary collapse. Possible intermediary mechanisms include, but are not limited to, hemodynamic factors, shear stress, growth factors, and metabolic factors such as reactive oxygen species, hyperlipidemia or a diabetic milieu. The interplay of these factors determines the balance of cell growth and proliferation vs. cell death by necrosis or apoptosis, and the balance of matrix accumulation vs. degradation.

7.09.2.1 Systemic and Glomerular Hypertension

Systemic hypertension is both the result of, and contributor to, progression of renal damage. Systemic blood pressure not only can affect the glomerulus by transmission of pressures, but may affect other determinants of progressive injury. In addition, intrarenal factors related to the unique aspects of kidney structure and function may be important. The glomerulus has both an afferent and efferent arteriole, which permits modulation of perfusion and pressure within the glomerular capillary bed differentially from systemic blood pressure. Therefore, attention has been focused on the potential impact of local (i.e., glomerular) hemodynamic changes, rather than systemic blood pressure, on structural glomerular injury.

The remnant kidney model was initially described at the end of the nineteenth century, and has been restudied extensively in the last decades. In this model removal of a large portion of renal mass (i.e., removal of one entire kidney and two-thirds of the contralateral kidney, a total of five-sixths nephrectomy), resulted in progressive hyperperfusion, hyperfiltration and sclerosis in a focal and segmental pattern.[1-3]

The initial observations that single nephron function was increased after renal ablation led to further studies,[1] and the hypothesis that hyperfiltration was a maladaptive response to nephron loss and ultimately contributed to progression of renal injury.[3] This maladaptive change following removal of nephrons was postulated to cause scarring and further loss of glomeruli, perpetuating a cycle of hyperfiltration and glomerulosclerosis.[4] Manipulations of hyperfiltration by feeding low protein diet, or by giving angiotensin I converting enzyme inhibitors (ACEI), lipid lowering agents, or heparin, were effective in ameliorating glomerular sclerosis. However, glomerular sclerosis could be decreased without altering glomerular hyperfiltration.[5] Some experimental maneuvers which increased glomerular hyperperfusion, such as thromboxane synthetase inhibitors or exercise training, slowed the process. Finally, glomerular sclerosis occurred in some settings in the absence of intervening hyperperfusion.[6,7]

The absence of a tight link between hyperperfusion and glomerular sclerosis shifted focus from hyperperfusion/hyperfiltration to glomerular hypertension. Maneuvers which increased glomerular capillary pressure, such as therapy with erythropoietin, glucocorticoids or high protein diet, exacerbated the severity and accelerated the course of development of glomerulosclerosis. Maneuvers which induced systemic and/or glomerular hypertension, such as uninephrectomy superimposed on other forms of renal disease, including nephrotoxic serum nephritis, immune complex nephritis, or diabetic nephropathy, worsened both functional and structural injury.[6,7]

Conversely, decreased glomerular pressure, in response to dietary protein[8] restriction or antihypertensive drugs,[6,7,9,10] was associated with slower progression. As discussed below, these maneuvers also have other effects. Thus, factors other than pressure likely contribute to glomerular damage. One study in rats examined glomerular filtration and pressure repeatedly by micropuncture in the same nephrons over a six week period after renal ablation. The extent of hyperfiltration or glomerular hypertension did not correlate with the degree of sclerosis in the same glomerulus.[11] Further, ACEI ameliorated glomerulosclerosis in puromycin aminonucleoside nephropathy, although glomerular capillary pressure was not affected.[12]

In addition, several models were found where glomerulosclerosis did not develop despite glomerular hypertension. The diabetic nephropathy model in the rat, although manifesting glomerular capillary hypertension, develops only mild mesangial expansion rather than overt glomerulosclerosis.[6] Despite similar degrees of glomerular hyperperfusion, hyperfiltration and hypertension after ureter diversion or after renal ablation in rats, significant glomerulosclerosis developed only in the renal ablation group. Of note, increased glomerular

growth occurred only in the renal ablation rats, and not after ureter diversion.[13] These results suggest that glomerular hyperfiltration, hyperperfusion, or glomerular hypertension alone do not fully explain the development of glomerular hypertrophy or sclerosis.

Interaction of glomerular hypertrophy and glomerular pressures has been postulated to augment glomerulosclerosis by increasing capillary wall tension. Wall tension is determined by the product of pressure and diameter of the glomerular capillary by LaPlace's law.[14,15] However, studies have shown that at least after uninephrectomy, glomerular growth occurs primarily by increased capillary length rather than increased capillary diameter.[16] Other synergistic contributions proposed to be deleterious include systemic blood pressure, which may modulate glomerular injury either by direct transmission of altered shear stress and pressures, or by altering the elaboration of vasoactive substances that affect the microcirculation.

The mechanisms whereby increased pressure might promote sclerosis have been investigated *in vitro*. Mesangial cells in culture were grown with pulsatile mechanical stretch/relaxation. The cells changed both type and amount of collagen production to a "wound healing" phenotype, which may be pivotal in the scarring process *in vivo*.[17] Increased pressure or shear stress can also affect other cell types in the kidney. Endothelial cells grown under shear stress conditions changed phenotype, and differentiated towards the *in vivo* morphology (see below).[18] Platelet-derived growth factor (PDGF) B-chain mRNA was increased by shear stress, and media from glomerular endothelial cells subjected to shear stress induced proliferation in both mesangial and epithelial cells.[19] These responses may in turn effect matrix accumulation (see below).

7.09.2.2 Growth Factors

Adaptations to increase renal mass and renal function presumably occur in attempts to restore renal function towards normal after injury. Glomerular sclerosis often is closely associated with glomerular hypertrophy in a variety of animal and human diseases.[7] After initial loss of nephrons, many growth factors are increased. However, stimuli which induce glomerular hypertrophy often ultimately accelerate glomerular sclerosis which has focused research on possible roles of growth factors in progression of renal disease (Figure 2).

Whole kidney weight increases shortly after ablation due to adaptive growth of tubules and

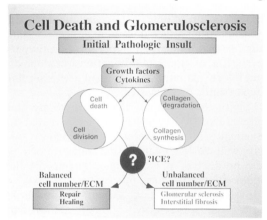

Figure 2 Glomerulosclerosis is postulated to result from the net effects of growth promoters and inhibitors on cell growth and matrix accumulation. The balance of cell proliferation vs. apoptosis and matrix production vs. degradation determine whether scarring or healing ensue in response to injury (modified by permission of Raven Press from "Immunologic Renal Diseases," 1997).

glomeruli via hyperplasia and hypertrophy. So-called glomerular hypertrophy represents both cellular hypertrophy (increase in cell size) and hyperplasia (increase in cell number). The relative contributions of cellular hyperplasia and hypertrophy to the increase in glomerular size are dependent on both amount of renal mass lost and age.[20] Hyperplasia occurred to greater extent after subtotal (5/6) nephrectomy, whereas hypertrophy was the predominant form of growth after uninephrectomy. Of note, the increased growth in the remnant kidney model in the rat occurred before increases in single nephron GFR.[21] The patterns of gene expression were also markedly different in hypertrophy vs. hyperplasia, with a rapidly, transiently increased expression of the protooncogenes c-fos, c-myc and c-Ha-ras in hyperplasia in contrast to a gradual, progressive increase when hypertrophy occurred after uninephrectomy.[22,23] With loss of renal parenchyma and in children with reflux nephropathy, glomerular growth occurred primarily by lengthening of the capillaries without significant increase in diameter.[16,24] In experimental diabetes and with toxic nephropathy due to lithium, growth was accomplished by new capillary branching.[25,26] In more extreme situations, dilatation of capillaries may, however, occur.[27] Hyperplasia was more prominent in the young and was greater after extensive nephron loss, although new nephrons do not form after term birth in humans.[28,29] Of interest, in young animals where maturational growth is occurring, injury after renal ablation was more severe than in adults.[29,30] Detailed analysis of the distribution of this glomerular injury revealed more focal and severe glomerulosclerosis in the deep vs. superficial glomeruli in young rats. Hemodynamic factors were not different among the two age groups. Although glomerular enlargement occurred to greater degree in the young compared to the adult rat after renal ablation, glomerular enlargement occurred proportionally in both superficial and deep nephron populations. The more severe injury of the deep glomeruli in the young immature rat was therefore postulated to be related to factors unique in the young growing kidney which is characterized by centripetal growth and differentiation.[30] The mechanisms that promote hyperplasia rather than hypertrophy in the accentuated growth and wound healing in response to nephron loss, may also promote scarring.

Many experimental maneuvers and therapies affect the renal growth response. Uninephrectomy, a stimulus for hypertrophy, accelerated glomerular sclerosis when superimposed in several models, including diabetes, immune complex injury and toxic injury.[6,7] In animal studies, high and low protein-content diet feedings, which modulate glomerular size in normal animals,[31] increased and decreased, respectively, the renal hypertrophy following nephrectomy or in diabetic nephropathy.[32] High protein diet increased incidence of FSGS without alteration in glomerular hemodynamics.[33] High salt diet increased glomerular volume and sclerosis and proteinuria following subtotal nephrectomy without influencing arterial or intraglomerular pressures.[15] Excess growth hormone, whether endogenously produced, that is, transgenic mice,[34] or exogenous,[35,36] is associated with marked glomerular hypertrophy and accelerated glomerulosclerosis. Even in normal young rats, recombinant human growth hormone caused glomerular hypertrophy and severe sclerosis.[36] Some human conditions associated with enlarged glomeruli include solitary kidney (renal agenesis or post nephrectomy), congenital cyanotic heart disease, sickle cell disease, diabetes, oligomeganephronia and marked obesity (all of which may, interestingly, be associated with FSGS).[7] In human FSGS, increased glomerular size precedes development of the sclerotic lesions. In pediatric patients with nephrotic syndrome and apparent minimal change disease on initial biopsies, those patients with subsequent progression to overt focal segmental glomerular sclerosis had significantly enlarged glomerular areas in those first biopsies (Figure 3).[37] These findings were confirmed in a study that included adult patients.[38] Similarly, adult patients with existing FSGS had significantly larger glomerular size than adult patients with minimal

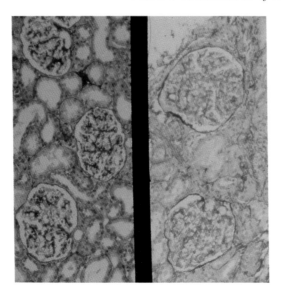

Figure 3 Glomerular size in FSGS vs. minimal change disease. The two panels show initial biopsies from patients with apparent minimal change disease. The 5 year old girl's initial biopsy on the right was indistinguishable from age-matched typical minimal change disease, left, except for marked glomerular enlargement. Her subsequent biopsy, 50 months later, showed FSGS (Jones' silver stain, ×160) (reproduced by permission of Blackwell Science Publishers Inc. from *Kidney Int.*, 1990, **38**, 115–123).

change disease. In patients with glomerular tip lesions (i.e., segmental sclerosis only at the tubular pole of the glomerulus), where prognosis is thought to be similar to that of minimal change disease, glomerular enlargement was not found.[39] Thus, in cases of nephrotic syndrome without evident segmental glomerular sclerosis, patients with normal glomerular size appear to have a good prognosis. In contrast, those patients with markedly enlarged glomerular size have a high risk of subsequent progression to overt FSGS.

Of interest, this association of increased glomerular size and early pathogenetic events of glomerular sclerosis was also seen in the setting of recurrent FSGS in the transplant. Thus, children who received an adult transplant who did not develop FSGS in their transplants did not show increased glomerular size over the initial months after transplantation. In contrast, those who developed recurrent FSGS had marked, abnormal glomerular growth, preceding overt manifestation of FSGS.[40]

Abnormal glomerular size distribution may also contribute to differences in populations in severity and incidence of progressive renal disease. Kidneys from African American and Caucasian normal adults who died suddenly were assessed morphometrically. Subjects were matched for body mass index, age, and gender. Average glomerular size in African Americans was significantly greater than in Caucasians. Of note, the distribution, while Gaussian in African Americans, showed a biphasic distribution in Caucasians. We speculate that the subpopulation of Caucasians with larger glomeruli may have increased risk for progressive renal disease.[41] In the late 1990s, studies have begun to investigate whether differences in genotype and/or glomerular number may underlie this difference in size. Abnormal glomerular growth may thus be taken as an indicator of abnormal growth stimuli which ultimately result in the maladaptive responses of augmented cell growth, matrix accumulation and then sclerosis.

While hypertrophic stimuli accelerate glomerular sclerosis, the converse is also true, that is, decreasing the hypertrophic response is associated with decreased glomerular sclerosis. Rats with diminished capacity for hypertrophy, whether due to a defect in growth hormone or an intrinsic strain characteristic, were resistant to the development of glomerulosclerosis following renal ablation.[7,35,42] Decreasing salt or phosphate prevented both glomerular hypertrophy and sclerosis in animal models, in addition to salt effects to lower blood pressure.[43,44] Alterations of hormonal status also affect glomerular injury. Treatment with androgen, a potent stimulator of renal growth, accelerated injury.[43] Conversely, castration in male rats substantially ameliorated glomerular hypertrophy, proteinuria and glomerular structural injury following uninephrectomy.[45] Anemia ameliorated glomerular sclerosis, decreased glomerular volume in addition to altering hemodynamic patterns, whereas treatment with erythropoietin worsened glomerular injury.[46] Treatment with low protein diets ameliorated sclerosis; however, it may be effective only when the diet is low enough in calories to inhibit growth.[47]

Ang II actions in addition to vasoconstriction, include effects on mesangial and vascular smooth muscle cell growth and matrix production, implicating it as a key factor promoting sclerosis. Therapies which inhibit its actions are therefore of particular interest. Animal models have shown efficacy of antihypertensive drugs which inhibit Ang II, either ACEI, or a newly developed angiotensin II receptor antagonist (AIIRA). These agents not only are effective systemic antihypertensives, but they also appear to be more efficacious than other, nonspecific antihypertensives in protecting against progressive renal injury.[48,49] This has been postulated to relate to ACEIs unique actions to decrease glomerular capillary pressure by preferential dilatation of the efferent arteriole.[6] ACEIs also act as kininase inhibitors, and activate bradykinin, which underlies its effects on the efferent arteriole.[50] Indeed, AIIRA does not have this kininase activity, and therefore did not cause the marked decrease in efferent arteriolar resistance seen after ACEI.[50] Since both modes of inhibition of the renin angiotensin system are equally effective in protecting against glomerulosclerosis, the specific effect of ACEI to decrease efferent arteriolar resistance, does not appear to be critical for amelioration of glomerulosclerosis. Rather, inhibition of Ang II actions on matrix is postulated to be involved in the effect of these drugs on sclerosis. Thus, although ACEI decreased while AIIRA had no effect on proteinuria in acute PAN nephropathy, sclerosis was equally ameliorated by both treatments.[51] Overall, these studies indicate that adaptive changes in renal hemodynamics and filtration and permeability may be mediated by different cellular mechanisms than those involved with glomerular or tubular interstitial scarring.

In vivo and *in vitro* evidence point to the capability of many cytokines to promote growth of glomerular cells and also to enhance extracellular matrix release, thus promoting sclerosis and growth in tandem. Several growth factors appear to play key roles in progression of glomerular and tubulointerstitial scarring. The specific factor(s) and the roles they play may differ at the various stages of injury. Altered gene expressions in pathophysiologic settings implicate many factors, including the following: PDGF, transforming growth factor-β (TGF-β), TGF-α, insulin-like growth factor-1 (IGF-1), growth hormone, epidermal growth factor, interleukins 1 and 6, tumor necrosis factor α (TNF-α), Ang II and endothelin.[7,52–54] To date, the relative importance of these specific growth factors in *in vivo* settings of development of FSGS is not defined. Several growth factors are now considered to play key causal roles in glomerulosclerosis, including Ang II, PDGF and TGF-β, based on results of infusion, transfection and inhibition of these factors.

In vitro studies have demonstrated that Ang II, a potent vasoconstrictor, also induces the smooth muscle cell c-fos oncogene.[55] Infusion of Ang II *in vivo* also activates the early growth response genes c-fos and Egr-1.[56] Although implications for long term structural effects of these early protooncogene changes have not yet been delineated, the induction of proto-oncogenes may be a common pathway for hypertrophic agents, allowing transition from a quiescent to an active state. Studies of cardiovascular disease have also shown key

interactions between Ang II and PDGF: Specifically, exogenous Ang II infusion, in doses which caused moderate hypertension, induced marked vascular, glomerular, and tubulointerstitial injury in association with upregulation of PDGF-B chain mRNA.[31] Of interest, increased protein and proteoglycan synthesis in the aorta in response to exogenous Ang II preceded increased systemic blood pressure.[43] Further, the mitogenic effect of Ang II on vascular smooth muscle cells under mechanical strain was largely eliminated by neutralizing antibody to PDGF-AB protein.[45] Ang II also induced TGF-β in smooth muscle cells in culture.[57] PDGF is increased in many models of glomerular injury. PDGF protein infusion or gene transfection increased mesangial cell proliferation and matrix.[58,59] Conversely, inhibition of PDGF with neutralizing antibody inhibited mesangial cell proliferation in the anti-Thy1 model.[60]

Elevated expression of TGF-β has been found in numerous animal models associated with renal scarring,[61,62] and studies show increased mesangial matrix after transfection with TGF-β.[58] Conversely, pretreatment of rats with antibody to TGF-β in the anti-Thy1 model, characterized by mesangial proliferation and increased production and activity of TGF-β, dramatically attenuated injury.[63] *In vitro* studies suggest that TGF-β effects may be dependent on cell cycle,[62] and that TGF-β acts synergistically with some, but not all, growth factors. For example, epidermal growth factor alone had no effect on mesangial cell matrix production, whereas the combination of TGF-β and epidermal growth factor doubled matrix production vs. TGF-β alone. TGF-β also is a potent inducer of PDGF-β receptor, and thus could enhance responsiveness to PDGF.[61,62]

A note of caution is in order, however, in extrapolating directly from animal models to human disease. Studies have begun to map some of these key growth factors, showing upregulation in human renal disease. It should also be noted that all growth promoters do not act equally. Moreover, even a given growth promoter may affect different cells differently, inducing proliferation, hypertrophy or even inhibiting growth, depending on the host cells' milieu and phenotype status (see stage of injury below).

7.09.2.3 Metabolic Alterations

7.09.2.3.1 *Reactive oxygen species*

Reactive oxygen species (ROS) have been shown to be a major component of many renal disease models, both acute and chronic. These include toxic models such as puromycin amino-nucleoside (PAN) or adriamycin nephropathy (see Section 7.09.3.4), remnant kidney model, and post-ischemic and tubular injury models such as the heme or glycerol acute renal failure models.[31,32,64–71] The mechanism of the epithelial cell injury induced by PAN or adriamycin has been shown to be, at least in part, due to oxidative damage on cell structure through a direct reduction to a semiquinone radical or by producing ROS via the xanthine oxidase system. *In vitro* experiments support a direct effect of H_2O_2 on glomerular epithelial cell morphology.[72] In addition to tubular injury and glomerular hypoperfusion (see below), ROS can result in proteinuria without morphologic injury. Acute infusion of H_2O_2 into the renal artery induced transient massive proteinuria associated with glomerular size-selective dysfunction without causing appreciable histological changes. Pretreatment with the antioxidant catalase or the iron chelator deferoxamine largely prevented this proteinuria, implying that while H_2O_2 is essential for the development of proteinuria, metabolism through an iron-mediated reaction is necessary to induce a glomerular permselectivity defect.[73] This may be the primary mechanism for the development of proteinuria in pathophysiological conditions with excess ROS, and may underlie the physiological so-called "functional proteinuria," associated with increased oxygen consumption (such as exercise). ROS has also been shown to affect degradation of glomerular basement membranes by activating metalloproteinases. Whether ROS-mediated proteinuria then affects glomerular sclerosis has not been directly demonstrated. Studies of a protein overload model have demonstrated direct activation of tubulointerstitial fibrosis in this experimental setting.[74] Possible mechanisms include tubular cell activation in response to specific components of the ultrafiltrate.[75]

Of interest, vitamin E, an antioxidant, ameliorates proteinuria and results in lessened glomerular sclerosis and tubular interstitial scarring in the adriamycin model. These effects are associated with reduction in glomerular hypertrophy, and reduced malondialdehyde content in the kidney.[76] Further, in the remnant kidney model, oxygen consumption is increased due to adaptive increases in cell metabolism, associated with increased oxidized/reduced glutathione ratio, an index of oxidative stress, and increased urinary excretion of lipid peroxidation products.[77] Nath *et al.* have further shown the complex interactions of various modulators of ROS in oxidative injury, including heme, ferritin, and antioxidant enzymes. Increased generation of some of these

metabolites is, at least in part, balanced by adaptive increases in the cell's antioxidant enzymes. The net sum of oxidants and antioxidants then ultimately determines the severity of oxidative injury. Augmentation of antioxidants lessened progressive functional impairment, as well as structural damage in several animal models. Pyruvate, which may serve as an antioxidant, protected against glycerol-induced acute renal failure,[78] but cross-protection was not seen, indicating injury-specific mechanisms of toxicity. Taurine, an endogenous antioxidant, ameliorated both glomerular and tubulointerstitial damage in the puromycin model of progressive renal disease[79] (see below). Cortical levels of reactive oxygen products were decreased in parallel with this protective effect, pointing to a direct effect on ROS mediated tubulointerstitial damage. Altered antioxidant enzymes also has been shown to affect human disease progression. Fish oil in the diet not only increased endogenous antioxidant genes, but also decreased cytokines and eicosanoid production, thus decreasing infiltration of inflammatory cells and ROS.[80] Treatment with fish oil in humans had benefits on renal function in stable renal transplant patients and in IgA nephropathy.[80–82]

Of note, one of the most commonly used therapeutic agents in human renal disease, glucocorticoid, upregulates glomerular antioxidant enzymes and protects glomeruli from oxidant injuries. In animals given glucocorticoid, both total and manganese superoxide dismutase (Mn-SOD), glutathione peroxidase and catalase were upregulated within glomeruli. Treatment with glucocorticoid largely prevented PAN-induced proteinuria and was associated with decreased ROS generation and decreased structural epithelial cell injury.[83] The protective effects of Mn-SOD are not equal in all nephron segments. Proximal tubules are particularly sensitive to ischemic or toxic injuries, which show site specific injury dependent on the type of initial insult. Interestingly, baseline distribution of Mn-SOD mRNA is also heterogeneous along the nephron. Thus, highest expression of its mRNA is found in the distal tubule segments, while proximal tubules show low levels.[84] Injury is correspondingly diverse following ischemia: proximal tubules show greatest structural injury following ischemia, while distal tubule segments are resistant to ischemic injury. This heterogeneous localization of expression is further accentuated after ischemia-reperfusion injury, with even greater increase in Mn-SOD mRNA expression in distal tubular segments. Even within a given cell population, the nature of the initial injury determines the cell antioxidant response. Thus, volume depletion, a known predisposing clinical factor for contrast media nephropathy, decreases glomerular antioxidant enzyme levels. In contrast, low grade transient renal ischemia results in upregulation of glomerular antioxidant enzyme levels. These adaptations to injury also have functional consequences. When volume depleted animals are exposed to contrast media, marked renal dysfunction develops in association with these lowered antioxidant enzymes. However, animals without volume depletion and intact ROS defenses show negligible injury when exposed to the same dose of contrast media.[85] When the initial insult is ischemia, rather than volume depletion, the resulting upregulated antioxidant levels confers resistance to a second ischemic episode, with remarkably maintained renal function.[86] Different injuries thus lead to divergent adaptations and are not necessarily cross-protective. These studies further imply that not only severity of local injury, but cell ability to adapt and quickly induce antioxidant defense systems, determine final structural lesions and injury pattern.

In a model of PAN nephropathy with superimposed dietary induced deficiency of antioxidants, increased renal, but not somatic, growth and increased ammoniagenesis were observed with increased size of tubules and glomeruli and severe tubulointerstitial disease. These findings, as reviewed, suggest a potential role for ammoniagenesis in the growth and injury seen with antioxidant deficiencies.[87] These studies suggest ROS may injure renal cells and contribute to glomerular dysfunction, as well as glomerular structural damage. The study also illustrates the interaction of several cell adaptations that occur after loss of renal mass, namely aberrant growth response, increased ROS, suppression of factors which inhibit matrix accumulation, and increased ammonia. Each of these adaptive responses appear to contribute to hypertrophy, sclerosis and tubulointerstitial scarring.

7.09.2.3.2 *Lipids*

Histologic evidence of abnormal renal lipid accumulation in chronic renal failure has long been noted. Lipid deposits were described in early studies of kidneys of patients with nephrotic syndrome. However, renal disease is not a common finding in the more common forms of primary hyperlipidemias. Patients with minimal change disease or membranous glomerulonephritis, characterized by hyperlipidemia as part of their nephrotic syndrome

usually do not develop glomerular scarring.[88]

Evidence of the effects of lipid metabolism on progression of renal disease currently is limited to animal studies.[88,89] Hypercholesterolemia modulates vascular tone by reducing vasodilators and increasing vasoconstrictors such as thromboxane A_2 and endothelin. Indirect effects of hypercholesterolemia include altered fatty acid metabolism. Excess cholesterol in the diet increased glomerular injury in rats with pre-existing glomerular disease, ablation nephropathy or hypertension.[88] The degree of injury was proportional to glomerular hypertrophy, and the glomerular pressures were not affected. Cholesterol also worsened injury in the PAN rat model. Conversely, lipid-lowering agents decreased glomerular injury in several models. The glomerular injury in rats with renal ablation was lessened by lovastatin without affecting systemic or glomerular pressures or glomerular hypertrophy.[90] Probucol, another lipid-lowering agent, was effective in decreasing injury in PAN nephropathy and in the remnant kidney model.[91,92] Zucker rats which have an autosomal recessive disorder associated with hyperphagia, hyperlipidemia and glomerular sclerosis, do not have glomerular hemodynamic abnormalities. The early lesion is characterized by mesangial proliferation and macrophage influx. Their lean litter mates develop neither lipid abnormalities nor glomerular lesions. Lipid lowering agents lessened both the hyperlipidemia and glomerulosclerosis in the affected rats.[93]

Macrophage accumulation resulting from high cholesterol levels has been postulated to be a key early step for glomerulosclerosis, analogous to atherosclerosis.[94,95] These cells are potential sources of numerous cytokines and eicosanoids which affect the glomerulus. Increased cholesterol intake in PAN-treated rats increased macrophage infiltration and TGF-β mRNA levels.[96] Further, the protective effects of essential fatty acid deficiency were closely associated with decreased macrophage influx.[97] Abnormal lipid metabolism present in human and animal forms of chronic renal failure may result in glomerular epithelial cell injury. An avid lipoprotein transport system present in glomerular epithelial cells may contribute to their own injury.[44] Direct effects of lipids include mesangial cell cytotoxicity mediated by oxidized LDL.[98,99] Endothelial cells may be affected directly or by increased adhesion of monocytes.[100] Elevated lipids may thus contribute to glomerulosclerosis through effects on different glomerular cell types. Some lipid-lowering therapies have additional effects, for example, alpha tocopherol decreases lipids and is an antioxidant. Its efficacy in progressive adriamycin nephropathy may result from these dual actions.[101] These results further illustrate the interaction of pathogenic mechanisms and multiple actions of therapies to target these processes.

7.09.2.3.3 *Diabetes*

Only ~40% of patients with diabetes mellitus develop diabetic nephropathy with clinical onset of renal disease on average 15 years after onset of diabetes. Therefore much interest has focused on not only risks for renal disease, but mechanisms of progression, once nephropathy is established. Early changes in diabetics in addition to the hyperglycemia and its consequences, include hyperfiltration and increased renal growth.[102] Patients with diabetic nephropathy showed significantly larger mean glomerular size and mesangial volume compared to both normal controls and diabetics without nephropathy.[103] Severity of glomerular sclerosis correlated with extent of glomerular hypertrophy. However, biopsies performed at early stages of microalbuminuria showed variable lesions.[104,105] The potential importance of the diabetic milieu in promoting the glomerular hypertrophy and mesangial matrix expansion of diabetic nephropathy was evident in a study of diabetic patients with renal and pancreas transplants. Only minimal mesangial expansion was seen in the transplant kidneys from patients with combined transplant, who had metabolic cure of their diabetes. The pancreas transplant patients also had smaller mean glomerular volume and mesangial volumes than control patients.[106] In contrast, diabetic patients receiving only kidney transplant without abolishment of their underlying disease, showed more severe diabetic lesions. Control of glucose has always been advocated to minimize complications of diabetes, and data have indicated that tight control of glucose decreases the rate of loss of GFR.[107] A tight correlation has also been found with degree of glucose control and mesangial matrix expansion in diabetic kidney transplant patients.[108] However, there are no long-term studies establishing whether the degree and/or presence of glomerular hypertrophy or mesangial expansion are reliable markers of long-term progression.[109]

Although *in vitro* experiments show augmented proliferation and/or matrix production when mesangial or endothelial cells were exposed to hyperglycemia,[110,111] effects of hyperglycemia *in vivo* are probably indirect. Consequences of diabetes include both abnormal hormonal mediators, such as insulin, insulin-like growth factor-1 (IGF-1) and growth hormone, and abnormal metabolic products, such as advanced glycosylation end products (AGE).

IGF-1 induces renal and glomerular hypertrophy *in vivo* as demonstrated in transgenic mice, and causes mesangial cell hyperplasia *in vitro*, but may require synergistic actions of other cytokines to result in a sclerotic lesion.[34] Specific receptors for insulin-like growth factor show increased expression in diabetic rat mesangial cells.[110,111] Several other growth factors including basic fibroblastic growth factor, PDGF and epidermal growth factor may be altered in diabetes and affect glomerular growth and sclerosis.[110,111] Somatostatin analogues and ACEI protected from injury in diabetic animal models.[102,112]

Hyperglycemia results in accumulation of AGEs. The glycosylated products can form stable adducts, that is, AGEs, via protein cross-linking and polymerization. AGEs have direct effects to upregulate multiple cytokines which in turn augment matrix production. Further, AGEs bind to collagens and may thereby prevent normal protease digestion. Infusion of AGEs into normal mice lead to glomerular hypertrophy and increased expression of matrix genes, while inhibition of AGE cross-linking with aminoguanidine significantly blunted these effects.[113] Chronic administration of AGEs in mice resulted in significant glomerulosclerosis and proteinuria. Aminoguanidine protected against this structural injury, and also decreased injury in diabetic animals without altering glucose levels.[98,114,115] Similarly, increased circulating AGE levels paralleled the severity of renal dysfunction in patients with diabetic nephropathy.[116]

7.09.3 CELL ADAPTATIONS

Adaptation may occur in any of the cellular components of the kidney, including glomerular resident and/or infiltrating cells, tubules and interstitial cells (for more detailed consideration of tubulointerstitial factors, see Chapter 15, this volume). Differences in the renal cell populations affected by different primary injuries may dictate the subsequent response. The glomerular resident cells are highly specialized, and include the endothelial, epithelial and mesangial cells, each with unique adaptations following injury. These alterations in response to either the initial injury or to a decrease in remaining renal mass after initial injury, contribute to maintaining renal function in the short term. However, in the long term these adaptive changes promote matrix accumulation and thereby renal malfunction.

7.09.3.1 Endothelial Cell

The endothelial cell is the first contact point for circulating substances. The endothelial cells interact with circulating cells and underlying vascular smooth muscle or mesangial cells, and can release both growth promoting and growth inhibiting factors. The glomerular endothelial cell is unique due to its fenestrae, allowing it to function as part of the glomerular capillary wall filtration barrier. Evidence suggests that the endothelium may play a role in this filtration barrier by active transport of charged molecules.[117]

Endothelial cells are heterogeneous, both phenotypically and in their response to a given injury. For example, diabetic vasculopathy occurs only in small arteries and arterioles, while atherosclerotic plaques are localized to large arteries. Both micro- and macrovascular endothelial cells grown under shear stress conditions change phenotype. However, only glomerular, and not aortic endothelial cells develop fenestrae under these conditions. Conversely, only the aortic origin endothelial cells show cholesterol clefts in response to shear stress.[18] Normal glomerular endothelial cells show heterogeneous expression of mRNAs for glomerular basement membrane components. For example, collagen type IV α-1 and α-2 mRNAs are localized to endothelial cells in small diameter capillary loops.[118] Subtle differences in baseline expression of basement membrane collagen in different portions of the kidney and even within a glomerulus may then result in differential responses at these sites to injury.

The glomerular endothelial cell has a long half-life, approximately 100 days. Therefore any toxic injury primarily affecting this cell may not fully manifest until well after the initial injury. For example, radiation injury to the kidney, which has a primary effect on endothelial cells, does not result in morphologic changes until weeks after exposure in the rat, when activation of the endothelial cell with increased organelles and filaments are detectable. Mesangiolysis, thrombotic microangiopathy, microaneurysms and finally sclerosis occur approximately 8 weeks after radiation in rats. In humans, radiation nephritis manifests after a latent period of 6–12 months after exposure. Bone marrow transplantation patients also show delayed renal injury after induction therapy. This injury may include radiation toxic effects on endothelial cells, and associated oxygen stress, in addition to toxic effects of chemotherapy. Hemolytic uremic syndrome, characterized by vascular injury, with microangiopathy and mesangiolysis in glomeruli, and associated thrombosis, can occur in patients receiving cytoreductive therapy. However, this syndrome occurs 4–12 months after drug exposure.[119] This particular insult appears to involve reactive oxygen species-mediated injury as well as apoptosis. Endothelial cell apoptosis as well as proliferation and migration also occur in several animal models of glomerular injury.[120–122]

Endothelial cells produce numerous factors which modulate mesangial or vascular smooth muscle cell growth (e.g., vascular endothelial growth factor, VEGF; endothelial derived relaxing factor, EDRF; endothelin and PDGF). VEGF is an endothelial cell specific mitogen that within the kidney originates largely from epithelial cells, and may mediate physiologic and pathologic angiogenesis. VEGF gene expression is induced by hypoxia and has been implicated in the pathogenesis of diabetic retinopathy.[123,124] Of note, new capillary growth also occurs in the glomerulus in, for example, diabetes,[26,125] suggesting an analogous role for VEGF in glomerular, as well as ocular, lesions with abnormal growth of vessels. Nitric oxide, thought to account for the activity of EDRF, results in marked inhibition of mitogen-induced mesangial cell proliferation.[126] Endothelin, a powerful vasoconstrictor, is released from endothelial cells in response to a variety of stimuli and also after injury. Endothelin results in the adaptive responses of hyperplasia, hypertrophy and increased matrix in mesangial cells in culture.[127] Endothelin also enhances PDGF release, another key factor augmenting these processes.[59,128–130] Thus, endothelial cell damage potentially may have profound consequences on collagen production, both directly via elaboration of the above substances, and by loss of synergistic inhibition of mesangial cell production of collagen. In addition, damage to endothelial cells may attenuate their ability to degrade toxic substances, such as reactive oxygen species.[131,132]

Several anticoagulant agents and platelet inhibitors have been shown to ameliorate glomerular sclerosis. In addition to direct anticoagulant effects, these agents may regulate the proliferation of glomerular cells. Inhibition of platelet aggregation prevents release of PDGF and other active substances. Heparin, which is released locally within the glomerulus, is of particular interest, since it can also attenuate the progression of glomerular sclerosis.[133–136] In addition to inhibiting coagulation and consequent platelet/endothelial cell interaction and activation, heparin suppresses the growth of, as well as extracellular matrix accumulation from, mesangial cells in a dose dependent manner.[137] Of note, a nonanticoagulant heparin fraction is equally potent in suppressing the matrix accumulation ameliorating progression of glomerular sclerosis *in vivo*.[135,136] Heparin also modulates other growth factor activities. Endothelial cell growth factor (ECGF) and heparin together, but not ECGF alone, markedly inhibit vascular smooth muscle cell collagen I and IV gene expression.[138]

7.09.3.2 Epithelial Cell

The glomerular epithelial cell is highly specialized. It serves not only as part of the capillary wall barrier, but also contributes extracellular matrix to the normal glomerular basement membrane, and is a source of increased matrix in pathological settings. Alterations of glomerular epithelial cell structure and metabolism precede glomerular sclerosis. In many human and animal forms of glomerular injury, effacement of visceral epithelial cells precedes major histological changes in the glomerular basement membrane (GBM) and mesangium. This structural change is closely associated with a loss of fixed anionic charge. Early epithelial cell attenuation and subsequent adhesions at the vascular pole are associated with early sclerosis in one animal model.[139] Both puromycin and adriamycin injection cause initial changes similar to minimal change disease followed by progressive sclerosis. Animals treated with PAN have a marked loss of glomerular heparan sulfate proteoglycan, a GBM constituent[140-142] and a defect in glycosylation of podocalyxin, a glomerular epithelial cell sialoglycoprotein.[141] Moreover, studies of glomerular epithelial cells in culture demonstrate that PAN, which induces focal glomerular sclerosis, alters the morphology, localization of anionic sites, and metabolism of epithelial cells.[143] This injury is at least in part attributable to ROS.[72] Antioxidants protect against foot process effacement *in vivo*, but do not affect the structural injury of the epithelial cell bodies or major processes.[144]

Studies have described an epithelial cell specific tyrosine phosphokinase, GLEPP-1, that shows altered distribution associated with early sclerosis or injury.[145] Specific antigens on the epithelial cells also serve as targets for specific injuries. Glomerular epithelial cells appear to be a primary site of injury in many settings, including adriamycin and puromycin aminonucleoside nephropathies. The glomerular epithelial cells and tubular cells express megalin (also called Gp330), a newly described glycoprotein localized in clathrin-coated pits. Megalin functions in uptake of many macro-molecules. Antibodies against megalin interfere with GBM and glomerular epithelial cell interaction, and result in subepithelial deposits.[146,147]

Altered epithelial cell function may be important for initial acute glomerular injury, as well as modulating progression. Integrins mediate cell–cell interactions and cell–matrix binding. For example, antibodies to β1 integrins or epithelial slit pores cause proteinuria acutely.[148,149] In early sclerotic lesions, α-3 and α-5 integrins are decreased and lose polar distribution. More severe loss of integrins is present in advanced sclerotic lesions, associated with loss of filamentous actin.[150] Together, these biochemical changes in response to injury may underlie foot process structural change and altered cell–cell and cell–matrix interactions.

Cell culture experiments suggest that epithelial cells may normally regulate mesangial cell proliferation by producing a heparinase-sensitive substance capable of inhibiting mesangial cell growth.[151] Thus, epithelial cell injury, via alteration of its capacity to regulate mesangial function, may underlie development of major changes in mesangial structure. Negatively charged heparin may protect against injury by interfering with the loss or neutralization of negative charge and distortion of epithelial cells.[152]

Glomerular visceral epithelial cells have limited capacity for hyperplasia.[153] Mesangial and endothelial cells do not show similar limitations, and can rapidly increase turnover in response to injury. Glomerular growth and sclerosis are accelerated when renal ablation is superimposed on adriamycin injury. The increased glomerular size occurs in this model without proportional increase in number of glomerular visceral epithelial cells. Consequentially, epithelial cells frequently are detached from the underlying basement membrane, especially in areas of hyalinosis. These epithelial cell defects are postulated to play an important role in progressive scarring due to injury from exudation of plasma proteins.[154] Since increased glomerular growth occurs following many injuries leading to glomerulosclerosis, this mechanism may be important for initiating glomerular injury.[155]

Human FSGS is manifested by epithelial injury and by segmental sclerosis. Epithelial injury is, of note, present even in nonsclerosed glomeruli. Circulating substances have been implicated in the pathogenesis of FSGS. Recurrence of FSGS in transplants with proteinuria within hours of transplant supports this concept. Likewise, plasmapheresis with a protein A adsorptive column decreased proteinuria, and induced remission in several patients.[156,157] The specific identity of this factor(s) is not yet known, although promising data from an intriguing *ex vivo* model are emerging.[158] These experiments point to a cytokine as a primary effector of increased glomerular permeability. The specific site of action, whether epithelial cell, endothelial cell, and/or GBM, has also not yet been determined. Possible sources for the postulated cytokine include activated T cells and/or monocytes.

7.09.3.3 Mesangial Cell

Because sclerosis ultimately is comprised of obliterated capillary loops and excess mesangial matrix, the mesangial cell presumably plays a central role in the process of glomerular sclerosis.[159] The mesangial cells share many characteristics with vascular smooth muscle cells. After initial injury, the activated mesangial cell changes phenotype, expressing fibroblast-like myosin.[160] This wound-healing phenotype is associated with increased matrix generation. Alpha-actin expression is also increased after injury, along with increased PDGF-B, recapitulating their developmental patterns of expressions in mesangial cells.[161]

Many growth factors initially may be released from infiltrating macrophages, platelets or resident glomerular cells (i.e., endothelial and epithelial cells, see above) which affect the mesangial cells. The renin angiotensin system has been studied extensively, not only because of its influences on other key factors, but because of the readily available clinically used inhibitors of its activity. Ang II induces mesangial cell hypertrophy and increases matrix production from mesangial cells *in vitro*[127] and induces the smooth muscle cell *c-fos* oncogene. TGF-β, a cytokine released from a variety of cells, appears to be most upregulated at later stages of sclerosis. TGF-β increases both collagen and proteoglycan production by mesangial cells, and depending on the cell cycle stage, acts synergistically with other growth factors.[61,62] Mesangial cells in culture release both interleukin-1 and a PDGF-like factor, and proliferate in response to these substances, suggesting both autocrine and paracrine regulation of mesangial cell proliferation. PDGF-B is increased in several models of glomerulosclerosis, especially at early time points after injury.[162,163] PDGF increases both matrix and proliferation of mesangial cells in culture and *in vivo*, and has been implicated in several settings of renal injury (see above).

7.09.3.4 Tubules

Efforts to define structural injury of glomeruli which mirrors the patient's GFR have been fraught with difficulty. This should not be surprising, since adaptation of hemodynamics and/or glomerular growth may allow increased filtration, and thus serve to normalize GFR, even with advanced glomerular injury, and because of difficulties in assessing the true extent of the typical heterogeneous, that is, focal and segmental, pattern of glomerular sclerosis on a single section. In contrast, tubulointerstitial fibrosis assessed on a single section is a relatively sensitive indicator of renal function.[164–166] Further, the extent of tubular interstitial damage has been shown to be a predictor of subsequent renal failure in one study of children with FSGS.[167] These findings also illustrate the importance of the tubular injury in determining renal function. Emerging evidence has shown that the tubulointerstitial compartment participates actively in the process of renal scarring, and does not undergo fibrosis merely as a process secondary to glomerular injury (for a detailed discussion of these factors, see Chapter 15, this volume). After injury, tubules undergo adaptations of metabolism and structural changes. For example, oxygen consumption is increased, antioxidant enzymes are induced in some, but not all segments, cell hyperplasia, and hypertrophy occur in some segments, whereas other segments respond to injury, particularly ischemia, by undergoing atrophy. Thus, various segments of the tubule show heterogeneity both in structural and biochemical responses to injuries (see Section 7.09.2.3.1).

Cell turnover is prominent after tubulointerstitial injury. In a study of acute tubular necrosis (ATN) in native kidneys due to various causes, either ischemic or toxic, most tubules with regenerating morphologic features showed distal tubule phenotype marking. In kidney transplants with ATN, particularly after cyclosporine treatment, there was an equal distribution of distal and proximal tubules with apparent regenerating change.[89] These findings suggest that the transplant kidney shows a different pattern of susceptibility to ATN than native kidneys, supporting that different mechanisms of injury effect ATN in these different clinical settings.

Likewise, atrophic tubules with different morphologic appearance arise from different nephron segments. The typical atrophic tubule segments in the end stage kidney originate largely from proximal tubules, while those tubules appearing thyroidized, that is, dilated with flattened epithelium and filled with proteinaceous material (i.e., appearing superficially similar to the thyroid histologically), are of distal origin. Specific distribution of nephron injury may then ultimately be used as an indicator of pathogenesis and for diagnosis of specific etiologies of preceding injuries.

Importantly, tubular injury may directly perpetuate fibrosis by extracellular matrix production from tubular cells. Evidence also suggests that transformation of injured tubular cells to a fibroblast-like phenotype may occur, with migration of these fibrogenic cells into the interstitial compartment. A fibroblast-specific protein, FSP1, related structurally to the S100 protein family, has been described. Although its function is not yet elucidated, its expression was highly restricted to fibroblasts, and its *de novo* expression localized to atrophic injured tubular segments.[168] This molecule may therefore serve as a marker or even target for fibrogenic processes in the tubulointerstitium. *In vitro* results suggest that altered surrounding matrix may allow this process to reverse, possibly leading to therapies directly aimed at healing interstitial injuries. Interactions of cells with matrix and altered adhesion molecules, such as β1 integrins, are possible modulators of these processes.[169,170]

Other toxic insults directly effect tubulointerstitial injuries as well as affecting the glomerulus. That different mechanisms mediate injury to different cell types is further underscored by the different response to intervention of these injuries in a rat cyclosporine model. This chronic cyclosporine nephropathy model is characterized by interstitial fibrosis and tubular injury as well as arteriolopathy and glomerular hypoperfusion. Treatment with ACEI remarkably ameliorates the tubulointerstitial injury, while exacerbating renal glomerular function.[171] In contrast, antagonizing endothelin activity has the opposite effects, improving GFR without ameliorating tubulointerstitial fibrosis. These studies point to the often divergent mechanisms underlying decreased glomerular filtration and fibrosis in renal injuries, and support a role for Ang II in fibrotic processes.

7.09.4 ADAPTATION AT DIFFERENT STAGES OF INJURY

Of interest, specific growth factors are upregulated during each stage of development of sclerosis (see below). These factors interact with each other, and modulate response to stimuli, initiating a cascade which culminates in matrix accumulation. Thus, this process appears to be effected by a complex interaction of growth factor increase, inhibition of matrix degradation, and altered receptors.

Type IV collagen mRNA is a major component of the normal glomerulus which accumulates in most instances of progressive human and experimental glomerulosclerosis. Thus, the detection of glomerular mRNA could serve as a prognostic indicator of glomerulosclerosis. Severity of sclerosis correlated with mRNA assessed in individual glomeruli by RT-PCR and tissue immunofluorescence levels for α1 and α2 type IV collagens in mice models and in human glomerulosclerosis.[172,173] Of note, this marked increase in collagen mRNAs was already evident at an early stage when sclerosis was minimal. These findings suggest that this increase in collagen mRNA may not merely mirror existing sclerosis, but may have a predictive value early in the disease. Future advances may help map the risk for progression, by molecular staging of disease in individual patients from mRNA profiles of key mediators of sclerosis.

Matrix accumulation is also affected by rate of degradation. The metalloproteinase PUMP-1, which can degrade collagen, has been found to be expressed in human proliferative glomerulonephritis, but not in normal glomeruli.[174] Tissue inhibitors of metalloproteinases (TIMP) 1 and 2, which modulate basement membrane degradation, were upregulated in patients with glomerulosclerosis.[175] Thus, degradation of collagen may be impaired in concert with upregulation of matrix producing genes, both promoting sclerosis. The balance of these processes might provide an index of possible therapeutic efficacy long before changes would be manifest by light microscopy or by functional assessment.

The increased collagen expression contributing to sclerosis appears to be promoted by varying mediators over the course of disease. Of note, glomerulosclerosis is heterogeneous in its morphological appearance. Emerging evidence also suggests that pathogenetic factors are not equal at all stages of morphologic injury (Figure 4). PDGF expression was elevated at early stages after renal ablation, but not at later stages.[162] Increased interstitial PDGF-β receptor expression was detected in glomerular, as well as interstitial areas in proliferative human diseases, providing a mechanism for autocrine activation of proliferation.[176] The increased PDGF-B chain expression correlated with extent of proliferation.[177] In contrast, TGF-β has shown enhanced expression even at late stages of sclerosis. Glomerular TGF-β expression was increased in biopsies with overt diabetic nephropathy compared to biopsies with nonprogressive renal disease (minimal

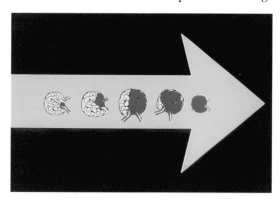

Figure 4 Glomerulosclerosis progresses over time, from early segmental lesions to complete sclerosis. This progression is postulated to be driven by growth factors, with distinct modulators predominating at each stage of injury.

change disease and thin basement membrane disease).[178] Quantity of staining correlated with the severity of glomerular sclerosis. Increased mature and/or latency-associated peptide and mRNA for TGF-β were also increased in proliferative human nephritides.[179,180] Within each biopsy, strongest staining was seen in sclerosed glomeruli, implicating TGF-β in the more advanced stages of sclerosis. Message RNA intensity did not correlate with infiltrating cells, implicating glomerular resident cells as a source of TGF-β.[179]

Cell turnover is an ongoing event which is affected by different mediators over the course of disease. Markers for cell proliferation, such as Ki-67, PCNA and bromodeoxyuridine, were upregulated in tissue repair and remodeling.[153] Interestingly, this increased turnover and increased extracellular matrix mRNA persisted at the same high levels even in late stage glomerulosclerosis, indicating that ongoing remodeling can take place.[181] Of note, growth factors affect cell cycle events other than those related to proliferation (Figure 2). TGF-β may directly cause apoptosis, an active form of cell death which does not elicit an inflammatory response. Withdrawal of some growth factors, including PDGF or IGF-1, may activate apoptosis.[182] Apoptosis as well as necrosis occurs with acute tubular injury, and growth factor therapies may diminish this injury.[183] Increased cell growth was accompanied by increased apoptosis both in proliferative and sclerotic human glomerular diseases.[184] In experimental mesangial proliferative glomerulonephritis, apoptosis of the mesangial cells was associated with healing and return of structure to normal.[122] Increased apoptosis was associated with overexpression of interleukin-1 β converting enzyme, a key initiator of apoptosis, in several animal models of glomeruloscle-

sis.[185] Studies of polycystic kidney disease have shown increased apoptosis in this setting of renal injury as well.[186]

The mechanisms which underlie the differential effect of growth factors on cell cycling may relate to altered cell–cell or cell–matrix interaction, which profoundly modulate growth responses. Multiplication of cells *in vitro* in response to any stimuli is avid until the cells reach confluency. In quiescent cells, stimuli generally augment extracellular matrix accumulation over cell growth. For example, TGF-β is growth stimulatory or inhibitory *in vitro*, depending in large part on degree of confluency of cells and cell origin. Local matrix environment also modifies the response to growth factors. Mesangial cells in routine culture readily proliferate in response to PDGF-B and express PDGF-β receptors, whereas culture under three-dimensional conditions within a matrix network renders mesangial cells of a distinctly different phenotype:[187] The PDGF-β receptor is down-regulated and the cells do not proliferate in response to PDGF. Enhanced matrix accumulation in even a segment of the glomerulus *in vivo* could similarly lead to downregulation of PDGF expression in later stages of sclerosis, when other growth factors are activated, with resultant loss of response to maneuvers which target PDGF. The possibility exists that the state of differentiation and receptor modulation may be markedly altered locally at a site of injury, with resultant augmented local response to specific growth factors.

7.09.5 INTERNEPHRON HETEROGENEITY

Sclerosis and aberrant glomerular growth are heterogeneously distributed even at any given time. Within a given remnant kidney of a subtotally nephrectomized animal, a strong correlation was found between the degree of total glomerular sclerosis and the maximum glomerular planar area, determined by serial section analysis in both animals and humans.[188] There was a positive correlation between these two indices in glomeruli with early sclerosis and a negative correlation in glomeruli with advanced sclerosis (Figure 5). Thus, it appears that the development of glomerular sclerosis takes place in a biphasic pattern, that is, initially with the hypertrophy of the glomerulus, and further advancement of sclerosis with shrinkage in glomerular size. It further appears that the marked interglomerular heterogeneity in the end-stage kidneys represents a variation of the onset of the overt sclerosing process among individual glomeruli.

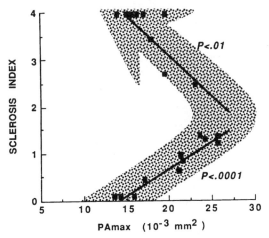

Figure 5 Relationship of glomerular growth and sclerosis in individual glomeruli within one kidney at 12 weeks after five-sixths nephrectomy. Sclerosis index (0–4 scale by serial section) and maximum planar area (PAmax) in individual glomeruli showed a significant biphasic correlation. Thus, glomeruli with advanced sclerosis showed negative correlation with glomerular size, whereas glomeruli with early stage sclerosis had positive correlation with size (reproduced with permission of W.B. Saunders Co. from *Semin. Nephrol.*, 1989, **9**, 329–342).

Figure 6 Multifocal distribution of sclerotic areas in a patient following loss of more than one kidney. The lesions were mapped by three-dimensional computer assisted analysis (reproduced with permission of Blackwell Scientific Publishers Inc. from *Kidney Int.*, 1995, **48**, 155–162).

Even at a specific time point glomerular injury is heterogeneous, that is, focal/segmental. The presence of glomeruli without overt sclerosis in human FSGS is therefore of potential great importance for therapeutic intervention, since these glomeruli and those at early stages of disease are most amenable to current therapy (see below). However, assessment on a single slide may not be accurate in detection of segmental and focal lesions. In one study, the distribution of sclerosis in adult and pediatric patients with FSGS was examined by serial section analysis. The lesions were focal, involving 23% of glomeruli in children vs. 48% in adults. A similar three-dimensional analysis in four adult patients with loss of more than one kidney revealed extensive total sclerosis, with only 8% of glomeruli remaining intact (Figure 6). Sclerosis was distributed multifocally, suggesting that pathologic processes occur simultaneously at anatomically discrete locations within the glomerulus.[189] FSGS tended to show a peripheral pattern of sclerosis in children, whereas multiple areas of the tuft, including a prominent hilar component, tended to be involved in adults.[190] It is possible that these observations reflect differences in pathogenesis between children and adults.

The heterogeneity of disease processes within the glomerulus may also affect response to therapy. In this regard, response to therapy in children with FSGS appears better than in adults. Heterogeneity of scarring is paralleled by heterogeneity in distribution of several key growth factors. Angiotensin II type I receptors are particularly highly expressed in the mesangial axial areas,[191] and ACEI is particularly effective in progressive renal disease in animal models and in humans. It is possible therefore that benefits of ACEI therapy could be linked to its potential greatest effect on mediators which are particularly activated in the mesangial axial areas. Inhibition of Ang II actions *per se*, or indirect effects of Ang II inhibition on other cytokines such as PDGF may underlie ACEI efficacy on sclerosis.[45,163]

This heterogeneity of stage of glomerular injury was reflected in the varying response of individual glomeruli to therapy initiated when FSGS was established. Early stage sclerosis was highly susceptible to therapy with an ACE inhibitor, whereas glomeruli that were more severely sclerosed at onset of therapy 8 weeks after renal ablation in the rat, showed ongoing sclerosis despite the intervention. The effect of an even higher dose of ACE inhibitor, beyond levels necessary to control systemic or local glomerular pressures was also investigated. Rats were treated with normal or high dose ACE inhibitor starting 8 weeks after renal ablation, a time where established glomerulosclerosis was present and verified by renal biopsy. At sacrifice 4 weeks later, glomerulosclerosis was ameliorated by normal dose ACE inhibitor, as described above. Of note, high dose ACE inhibitor therapy had even reversed glomerulosclerosis in some of these animals, with significantly less glomerulosclerosis present at autopsy than at biopsy in the same animals.[188] Similar responsiveness of lesions was seen in another model of progressive glomerulosclerosis. ACEI ("normal" dose) or low protein diet was started at the time of puromycin injection, and rats were sacrificed at 12, 18, or 24 weeks. At 24 weeks, sclerosis was less than at 12 weeks in rats treated with either protocol. Glomerular size was also less in the low protein group vs. control. In contrast, untreated control rats continued to show progression at 24 weeks.[192] These studies indicate that ACEI effects on sclerosis are heterogeneous and dependent, at least in part, on stage of injury.

Of interest in this regard, the distribution of PDGF-B chain varied amongst the heterogeneously affected glomeruli. Glomeruli with early sclerotic lesions had enhanced PDGF-B chain protein, whereas those glomeruli with more advanced sclerosis did not show such expression. This enhanced expression was normalized by treatment with angiotensin II receptor antagonist, but not with nonspecific antihypertensive treatment.[163] The stage of injury with upregulated PDGF may therefore be particularly sensitive to therapies which inhibit Ang II, known to affect PDGF expression (see below). PDGF expression also has its own negative feedback, in that enhanced matrix downregulates the PDGF receptor (see above).[187] Of note, inhibition of Ang II may also affect other growth factors, including TGF-β.[193] Ang II promotes conversion of TGF-β to its active form,[57] indicating that maximal TGF-β activation may follow increased Ang II. Thus, modulators at each stage of injury vary, and are under dynamic control with complex interactions.

7.09.6 REFERENCES

1. A. B. Morrison and R. M. Howard, 'The functional capacity of hypertrophied nephrons. Effect of partial nephrectomy on the clearance of inulin and PAH in the rat.' *J. Exp. Med.*, 1966, **123**, 829–844.
2. T. Shimamura and A. B. Morrison, 'A progressive glomerulosclerosis occurring in partial five-sixths nephrectomized rats.' *Am. J. Pathol.*, 1975, **79**, 95–106.
3. T. H. Hostetter, J. L. Olson, H. G. Rennke *et al.*, 'Hyperfiltration in remnant nephrons: a potentially adverse response to renal ablation.' *Am. J. Physiol.*, 1981, **241**, F85–F93.
4. B. M. Brenner, T. W. Meyer and T. H. Hostetter, 'Dietary protein intake and the progressive nature of kidney disease: the role of hemodynamically mediated glomerular injury in the pathogenesis of progressive glomerular sclerosis in aging, renal ablation, and intrinsic renal disease.' *N. Engl. J. Med.*, 1982, **307**, 652–659.
5. K. A. Nath, S. M. Kren and T. H. Hostetter, 'Dietary protein restriction in established renal injury in the rat. Selective role of glomerular capillary pressure in progressive glomerular dysfunction.' *J. Clin. Invest.*, 1986, **78**, 1199–1205.
6. J. R. Neuringer, S. Anderson and B. M. Brenner, in 'The Progressive Nature of Renal Disease,' 2nd edn., ed. W. E. Mitch, Churchill Livingstone, New York, 1992, pp. 1–21.
7. A. Fogo and I. Ichikawa, in ' Contemporary Issues in Nephrology: The Progressive Nature of Renal Disease,' 2nd edn., ed. W. E. Mitch, Churchill Livingstone, New York, 1992, pp. 23–54.
8. T. H. Hostetter, T. W. Meyer, H. G. Rennke *et al.*, 'Chronic effects of dietary protein in the rat with intact and reduced renal mass.' *Kidney Int.*, 1986, **30**, 509–517.
9. S. Anderson, T. W. Meyer, H. G. Rennke *et al.*, 'Control of glomerular hypertension limits glomerular injury in rats with reduced renal mass.' *J. Clin. Invest.*, 1985, **76**, 612–619.
10. Y. Yoshida, T. Kawamura, M. Ikoma *et al.*, 'Effects of anti-hypertensive drugs on glomerular morphology.' *Kidney Int.*, 1989, **36**, 626–635.
11. Y. Yoshida, A. Fogo, H. Shiraga *et al.*, 'Serial micropuncture analysis of single nephron function in subtotal renal ablation.' *Kidney Int.*, 1988, **33**, 855–867.
12. A. Fogo, Y. Yoshida, A. D. Glick *et al.*, 'Serial micropuncture analysis of glomerular function in two rat models of glomerular sclerosis.' *J. Clin. Invest.*, 1988, **82**, 322–330.
13. Y. Yoshida, A. Fogo and I. Ichikawa, 'Glomerular hemodynamic changes vs. hypertrophy in experimental glomerular sclerosis.; *Kidney Int.*, 1989, **35**, 654–660.
14. L. D. Dworkin, 'Effects of calcium antagonists on glomerular hemodynamics and structure in experimental hypertension.' *Am. J. Kidney Dis.*, 1991, **17**, 89–93.
15. B. S. Daniels and T. H. Hostetter, 'Adverse effects of growth in the glomerular microcirculation.' *Am. J. Physiol.*, 1990, **258**, F1409–F1416.
16. J. R. Nyengaard, 'Number and dimensions of rat glomerular capillaries in normal development and after nephrectomy.' *Kidney Int.*, 1993, **43**, 1049–1057.
17. R. C. Harris, M. A. Haralson and K. F. Badr, 'Continuous stretch–relaxation in culture alters mesangial cell morphology, growth characteristics, and metabolic activity.' *Lab. Invest.*, 1992, **66**, 548–554.
18. M. J. Ott, J. L. Olsen and B. J. Ballermann, 'Phenotypic differences between glomerular capillary (GE) and aortic (AE) endothelial cells *in vitro* (abstract).' *J. Am. Soc. Nephrol.*, 1993, **4**, 564.
19. M. J. Ott and B. J. Ballermann, 'Shear stress augments glomerular endothelial cell (GEN) PDGF mRNA expression and mitogen production (abstract).' *J. Am. Soc. Nephrol.*, 1992, **3**, 476.
20. G. Celsi, B. Jakobsson and A. Aperia, 'Influence of age on compensatory renal growth in rats.' *Pediatr. Res.*, 1986, **20**, 347–350.
21. C. A. Miskell and D. P. Simpson, 'Hyperplasia precedes increased glomerular filtration rate in rat remnant kidney.' *Kidney Int.*, 1990, **37**, 758–766.
22. J. T. Norman, R. E. Bohman, G. Fischmann *et al.*, 'Patterns of mRNA expression during early cell growth differ in kidney epithelial cells destined to undergo compensatory hypertrophy versus regenerative hyperplasia.' *Proc. Natl. Acad. Sci. USA*, 1988, **85**, 6768–6772.
23. B. D. Cowley, Jr., L. J. Chadwick, J. J. Grantham *et al.*, 'Sequential protooncogene expression in regenerating kidney following acute renal injury.' *J. Biol. Chem.*, 1989, **254**, 8389–8393.
24. K. Akaoka, R. H. R. White and F. Raafat, 'Glomerular morphometry in childhood reflux

nephropathy, emphasizing the capillary changes.' *Kidney Int.*, 1995, **47**, 1108–1114.

25. N. Marcussen, J. R. Nyengaard and S. Christensen, 'Compensatory growth of glomeruli is accomplished by an increased number of glomerular capillaries.' *Lab Invest.*, 1994, **70**, 868–874.

26. J. R. Nyengaard and R. Rasch, 'The impact of experimental diabetes mellitus in rats on glomerular capillary number and sizes.' *Diabetologia*, 1993, **36**, 189–194.

27. L. D. Dworkin, R. I. Levin, J. A. Bernstein *et al.*, 'Effects of nifedipine and enalapril on glomerular injury in rats with deoxycorticosterone-salt hypertension.' *Am. J. Physiol.*, 1990, **259**, F598-F604.

28. L. C. Aschinberg, O. Koskimies, J. Bernstein *et al.*, 'The influence of age on the response to renal parenchymal loss.' *Yale J. Biol. Med.*, 1978, **51**, 341–345.

29. M. P. O'Donnell, B. L. Kasiske, L. Raij *et al.*, 'Age is a determinant of the glomerular morphologic and functional responses to chronic nephron loss.' *J. Lab. Clin. Med.*, 1985, **106**, 308–313.

30. M. Ikoma, T. Yoshioka, I. Ichikawa *et al.*, 'Mechanism of the unique susceptibility of deep cortical glomeruli of maturing kidneys to severe focal glomerular sclerosis.' *Pediatr. Res.*, 1990, **28**, 270–276.

31. I. Ichikawa, M. L. Purkerson, S. Klahr *et al.*, 'Mechanism of reduced glomerular filtration rate in chronic malnutrition.' *J. Clin. Invest.*, 1980, **65**, 982–988.

32. J. R. Abrams, D. C. Tapp and M. A. Venkatachalam. in 'The Progressive Nature of Renal Disease,' ed. W. E. Mitch, Churchill Livingstone, New York, 1992, pp. 133–148.

33. M. P. O'Donnell, B. L. Kasiske, P. G. Schmitz *et al.*, 'High protein intake accelerates glomerulosclerosis independent of effects on glomerular hemodynamics.' *Kidney Int.*, 1990, **37**, 1263–1269.

34. T. Doi, L. J. Striker, C. C. Gibson *et al.*, 'Glomerular lesions in mice transgenic for growth hormone and insulinlike growth factor-I. I. Relationship between increased glomerular size and mesangial sclerosis.' *Am. J. Pathol.*, 1990, **137**, 541–552.

35. A. M. el Nahas, A. H. Bassett, G. H. Cope *et al.*, 'Role of growth hormone in the development of experimental renal scarring.' *Kidney Int.*, 1991, **40**, 29–34.

36. D. B. Allen, A. Fogo, R. el-Hayek *et al.*, 'Effects of prolonged growth hormone administration in rats with chronic renal insufficiency.' *Pediatr Res.*, 1992, **31**, 406–410.

37. A. Fogo, E. P. Hawkins, P. L. Berry *et al.*, 'Glomerular hypertrophy in minimal change disease predicts subsequent progression to focal glomerular sclerosis.' *Kidney Int.*, 1990, **38**, 115–123.

38. A. N. Vats, J. M. Basgen, M. W. Steffes *et al.*, 'Mean glomerular volume (GV) in minimal change nephrotic syndrome (MCNS), focal segmental glomerulosclerosis (FSGS), normal children and adults.' (abstract) *J. Am. Soc. Nephrol.*, 1994, **5**, 797.

39. J. C. Jennette, A. Marquis, R. J. Falk *et al.*, 'Glomerulomegaly in focal segmental glomerulosclerosis (FSGS) but not minimal change glomerulopathy (MCG).' (abstract) *Lab. Invest.*, 1990, **62**, 48A.

40. A. Fogo, E. P. Hawkins, R. Verani *et al.*, 'Focal segmental glomerulosclerosis (FGS) in renal transplants is associated with marked glomerular hypertrophy.' (abstract) *J. Am. Soc. Nephrol.*, 1991, **2**, 797.

41. C. Pesce, K. Schmidt, A. Fogo *et al.*, 'Glomerular size and the incidence of renal disease in African Americans and Caucasians.' *J. Nephrol.*, 1994, **7**, 355–358.

42. H. Yoshida, T. Mitarai, M. Kitamura *et al.*, 'The effect of selective growth hormone defect in the

43. D. S. Lax, J. A. Benstein, E. Tolbert *et al.*, 'Effects of salt restriction on renal growth and glomerular injury in rats with remnant kidneys.' *Kidney Int.*, 1992, **41**, 1527–1534.

44. T. Shimamura, 'Prevention of 11-deoxycorticosterone-salt-induced glomerular hypertrophy and glomerulosclerosis by dietary phosphate binder.' *Am. J. Pathol.*, 1990, **136**, 549–556.

45. U. Gafter, M. Ben-Bassat and J. Levi, 'Castration inhibits glomerular hypertrophy and proteinuria in uninephrectomized male rats.' *Eur. J. Clin. Invest.*, 1990, **20**, 360–365.

46. H. M. Lafferty, D. L. Garcia, H. G. Rennke *et al.*, 'Anemia ameliorates progressive renal injury in experiemental DOCA-Salt hypertension.' *J. Am. Soc. Nephrol.*, 1991, **1**, 1180–1185.

47. D. C. Tapp, W. G. Wortham, J. F. Addison *et al.*, 'Food restriction retards body growth and prevents end-stage renal pathology in remnant kidneys of rats regardless of protein intake.' *Lab. Invest.*, 1989, **60**, 184–195.

48. B. L. Kasiske, R. S. Kalil, J. Z. Ma *et al.*, 'The effect of antihypertensive therapy on the kidney in patients with diabetes: a meta-regression analysis.' *Ann. Intern. Med.*, 1993, **118**, 129–138.

49. Y. Kakinuma, T. Kawamura, T. Bills *et al.*, 'Blood pressure-independent effect of angiotensin inhibition on vascular lesions of chronic renal failure.' *Kidney Int.*, 1992, **42**, 46–55.

50. V. Kon, A. Fogo and I. Ichikawa, 'Bradykinin causes selective efferent arteriolar dilation during angiotensin I converting enzyme inhibition.' *Kidney Int.*, 1993, **44**, 545–550.

51. R. Tanaka, V. Kon, T. Yoshioka *et al.*, 'Angiotensin converting enzyme inhibitor modulates glomerular function and structure by distinct mechanisms.' *Kidney Int.*, 1994, **45**, 537–543.

52. L. G. Fine, M. R. Hammerman and H. E. Abboud, 'Evolving role of growth factors in the renal response to acute and chronic disease.' *J. Am. Soc. Nephrol.*, 1992, **2**, 1163–1170.

53. H. E. Abboud, 'Growth factors and the mesangium.' *J. Am. Soc. Nephrol.*, 1992, **2**, S185–S189.

54. L. J. Striker, E. P. Peten, S. J. Elliott *et al.*, 'Mesangial cell turnover: effect of heparin and peptide growth factors.' *Lab. Invest.*, 1991, **64**, 446–456.

55. M. B. Taubman, B. C. Berk, S. Izumo *et al.*, 'Angiotensin II induces c-fos mRNA in aortic smooth muscle. Role of Ca^{2+} mobilization and protein kinase C activation.' *J. Biol. Chem.*, 1989, **264**, 526–530.

56. M. E. Rosenberg and T. H. Hostetter, 'The effect of angiotensin II on early growth response genes in the rat kidney.' (abstract, update) *J. Am. Soc. Nephrol.*, 1990, **1**.

57. G. H. Gibbons, R. E. Pratt and V. J. Dzau, 'Vascular smooth muscle cell hypertrophy vs. hyperplasia. Autocrine transforming growth factor-beta if expression determines growth response to angiotensin II.' *J. Clin. Invest.*, 1992, **90**, 456–461.

58. Y. Isaka, Y. Fujiwara, N. Ueda *et al.*, 'Glomerulosclerosis induced by *in vivo* transfection of transforming growth factor-β or platelet-derived growth factor gene into the rat kidney.' *J. Clin. Invest.*, 1993, **92**, 2597–2601.

59. J. Floege, E. Eng, B. A. Young *et al.*, 'Infusion of platelet-derived growth factor or basic fibroblast growth factor induces selective glomerular mesangial cell proliferation and matrix accumulation in rats.' *J. Clin. Invest.*, 1993, **92**, 2952–2962.

60. R. J. Johnson, E. W. Raines, J. Floege *et al.*, 'Inhibition of mesangial cell proliferation and matrix

expansion in glomerulonephritis in the rat by anti-body to platelet-derived growth factor.' *J. Exp. Med.*, 1992, **175**, 1413–1416.

61. K. Sharma and F. N. Ziyadeh, 'The emerging role of transforming growth factor-β in kidney diseases.' *Am. J. Physiol.*, 1994, **266**, F829–F842.

62. W. A. Border and N. A. Noble, 'Transforming growth factor β in tissue fibrosis.' *N. Engl J. Med.*, 1994, **331**, 1286–1292.

63. W. A. Border, S. Okuda, L. R. Languino *et al.*, 'Suppression of experimental glomerulonephritis by antiserum against transforming growth factor β1.' *Nature*, 1990, **346**, 371–374.

64. S. V. Shah, 'Effect of enzymatically generated reactive oxygen metabolites on the cyclic nucleotide content in isolated rat glomeruli.' *J. Clin. Invest.*, 1984, **74**, 393–401.

65. S. V. Shah, W. H. Baricos and A. Basci, 'Degradation of human glomerular basement membrane by stimulated neutrophils. Activation of a metalloproteinase(s) by reactive oxygen metabolites.' *J. Clin. Invest.*, 1987, **79**, 25–31.

66. A. Rehan, K. J. Johnson, R. C. Wiggins *et al.*, 'Evidence for the role of oxygen radicals in acute nephrotoxic nephritis.' *Lab. Invest.*, 1984, **51**, 396–403.

67. A. Rehan, K. J. Johnson, R. G. Kunkel *et al.*, 'Role of oxygen radicals in phorbol myristate acetate-induced glomerular injury.' *Kidney Int.*, 1985, **27**, 503–511.

68. A. Rehan, R. C. Wiggins, R. G. Kunkel *et al.*, 'Glomerular injury and proteinuria in rats after intrarenal injection of cobra venom factor. Evidence for the role of neutrophil-derived oxygen free radicals.' *Am. J. Physiol.*, 1986, **123**, 57–66.

69. M. S. Paller, J. R. Hoidal and T. F. Ferris, 'Oxygen free radicals in ischemic acute renal failure in the rat.' *J. Clin. Invest.*, 1984, **74**, 1156–1164.

70. P. D. Walker and S. V. Shah, 'Evidence suggesting a role for hydroxyl radical in gentamicin-induced acute renal failure in rats.' *J. Clin. Invest.*, 1988, **81**, 334–341.

71. P. D. Walker and S. V. Shah, 'Gentamicin enhanced production of hydrogen peroxide by renal cortical mitochondria.' *Am. J. Physiol.*, 1987, **253**, C495–C499.

72. S. D. Ricardo, J. F. Bertram and G. B. Ryan, 'Reactive oxygen species in puromycin aminonucleoside nephrosis: *in vitro* studies.' *Kidney Int.*, 1994, **45**, 1057–1069.

73. T. Yoshioka, I. Ichikawa and A. Fogo, 'Reactive oxygen metabolites cause massive, reversible proteinuria and glomerular sieving defect without apparent ultrastructural abnormality.' *J. Am. Soc. Nephrol.*, 1991, **2**, 902–912.

74. A. A. Eddy, 'Experimental insights into the tubulointerstitial disease accompanying primary glomerular lesions.' *J. Am. Soc. Nephrol.*, 1994, **5**, 1273–1287.

75. T. M. Danoff, M. Y. Chiang, A. Ortiz *et al.*, 'Transcriptional regulation of murine RANTES (MuR) in proximal tubular cells (abstract).' *J. Am. Soc. Nephrol.*, 1993, **4**, 599.

76. H. Trachtman, N. Schwob, J. Maesaka *et al.*, 'Dietary vitamin E supplementation ameliorates renal injury in chronic puromycin aminonucleoside nephropathy.' *J. Am. Soc. Nephrol.*, 1995, **5**, 1811–1819.

77. K. A. Nath, M. Fischereder and T. H. Hostetter, 'The role of oxidants in progressive renal injury.' *Kidney Int. Suppl.*, 1994, **45**, S111–S115.

78. A. K. Salahudeen, E. C. Clark and K. A. Nath, 'Hydrogen peroxide-induced renal injury. A protective role for pyruvate *in vitro* and *in vivo*.' *J. Clin. Invest.*, 1991, **88**, 1886–1893.

79. H. Trachtman, R. Del Pizzo, S. Futterweit *et al.*,

80. B. Chandrrasekar and G. Fernandes, 'Decreased pro-inflammatory cytokines and increased antioxidant enzyme gene expression by omega-3 lipids in murine lupus nephritis.' *Biochem. Biophys. Res. Commun.*, 1994, **200**, 893–898.

'Taurine attenuates renal disease in chronic puromycin aminonucleoside nephropathy.' *Am. J. Physiol.*, 1992, **262**, F117–F123.

81. J. J. van der Heide, H. J. Bilo, J. M. Donker *et al.*, 'Effect of dietary fish oil on renal function and rejection in cyclosporine-treated recipients of renal transplants.' *N. Engl. J. Med.*, 1993, **329**, 769–773.

82. J. V. Donadio, Jr., E. J. Bergstrahl, K. P. Offord *et al.*, 'A controlled trial of fish oil in IgA nephropathy. Mayo Nephrology Collaborative Group.' *N. Engl. J. Med.*, 1994, **331**, 1194–1199.

83. T. Kawamura, T. Yoshioka, T. Bills *et al.*, 'Glucocorticoid activates glomerular antioxidant enzymes and protects glomeruli from oxidant injuries.' *Kidney Int.*, 1991, **40**, 291–301.

84. S. Kiyama, T. Yoshioka, I. M. Burr *et al.*, 'Strategic locus for the activation of the superoxide dismutase gene in the nephron.' *Kidney Int.*, 1995, **47**, 536–546.

85. T. Yoshioka, A. Fogo and J. K. Beckman, 'Reduced activity of antioxidant enzymes underlies contrast media-induced renal injury in volume depletion.' *Kidney Int.*, 1992, **41**, 1008–1015.

86. T. Yoshioka, T. Bills, T. Moore-Jarrett *et al.*, 'Role of intrinsic antioxidant enzymes in renal oxidant injury.' *Kidney Int.*, 1990, **38**, 282–288.

87. K. A. Nath, M. K. Hostetter and T. H. Hostetter, 'Increased ammoniagenesis as a determinant of progressive renal injury.' *Am. J. Kidney Dis.*, 1991, **17**, 654–657.

88. W. F. Keane, W. S. Mulcahy, B. L. Kasiske *et al.*, 'Hyperlipidemia and progressive renal disease.' *Kidney Int. Suppl.*, 1991, **31**, S41–S48.

89. J. F. Moorhead, 'Lipids and progressive kidney disease.' *Kidney Int. Suppl.*, 1991, **31**, S35–S40.

90. B. L. Kasiske, M. P. O'Donnell, W. J. Garvis *et al.*, 'Pharmacologic treatment of hyperlipidemia reduces glomerular injury in rat 5/6 nephrectomy model of chronic renal failure.' *Circ. Res.*, 1988, **62**, 367–374.

91. T. Hirano and T. Morchoshi, 'Treatment of hyperlipidaemia with probucol suppresses the development of focal and segmental glomerulosclerosis in chronic aminonucleoside nephrosis.' *Nephron*, 1992, **60**, 443–447.

92. K. S. Modi, G. F. Schreiner, M. L. Purkerson *et al.*, 'Effects of probucol on renal function and structure in rats with subtotal kidney ablation.' *J. Lab. Clin. Med.*, 1992, **120**, 310–317.

93. B. L. Kasiske, M. P. O'Donnell, M. P. Cleary *et al.*, 'Treatment of hyperlipidemia reduces glomerular injury in obese Zucker rats.' *Kidney Int.*, 1988, **33**, 667–672.

94. J. R. Diamond, 'Analogous pathobiologic mechanisms in glomerulosclerosis and atherosclerosis.' *Kidney Int. Suppl.*, 1991, **31**, S29–S34.

95. H. van Goor, V. Fidler, J. J. Weening *et al.*, 'Determinants of focal and segmental glomerulosclerosis in the rat after renal ablation. Evidence for involvement of macrophages and lipids.' *Lab. Invest.*, 1991, **64**, 754–765.

96. G. Ding, I. Pesek-Diamond and J. R. Diamond, 'Cholesterol, macrophages, and gene expression of TGF-β1 and fibronectin during nephrosis.' *Am. J. Physiol.*, 1993, **33**, F577–F584.

97. G. F. Schreiner, B. Rovin and J. B. Lefkowith, 'The anti inflammatory effects of essential fatty acid deficiency in experimental glomerulonephritis. The modulation of macrophage migration and eicosanoid metabolism.' *J. Immunol.*, 1989, **143**, 3192–3199.

98. M. P. Cohen, K. Sharma, Y. Jin *et al.*, 'Prevention of

diabetic nephropathy in db/db mice with glycated albumin antagonists. A novel treatment strategy.' *J. Clin. Invest.*, 1995, **95**, 2338–2345.

99. B. Pitt, '"Escape" of aldosterone production in patients with left ventricular dysfunction treated with an angiotensin converting enzyme inhibitor: implications for therapy.' *Cardiovasc. Drugs Ther.*, 1995, **9**, 145–149.

100. W. F. Keane, B. L. Kasiske and M. P. O'Donnell, 'Hyperlipidemia and the progression of renal disease.' *Am. J. Clin. Nutr.*, 1988, **47**, 157–160.

101. M. Washio, F. Nanishi, S. Okuda *et al.*, 'Alpha tocopherol improves focal glomerulosclerosis in rats with adriamycin-induced progressive renal failure.' *Nephron*, 1994, **68**, 347–352.

102. S. Anderson, H. G. Rennke, B. M. Brenner *et al.*, 'Short and long term effects of antihypertensive therapy in the diabetic rat.' *Kidney Int.*, 1989, **36**, 526–536.

103. R. W. Bilous, S. M. Mauer, D. E. Sutherland *et al.*, 'Mean glomerular volume and rate of development of diabetic nephropathy.' *Diabetes*, 1989, **38**, 1142–1147.

104. H. J. Bangstad, R. Østerby, K. Dahl-Jørgensen *et al.*, 'Early glomerulopathy is present in young, type 1 (insulin-dependent) diabetic patients with microalbuminuria.' *Diabetologia*, 1993, **36**, 523–529.

105. B. M. Chavers, R. W. Bilous, E. N. Ellis *et al.*, 'Glomerular lesions and urinary albumin excretion in type I diabetes without overt proteinuria.' *N. Engl. J. Med.*, 1989, **320**, 966–970.

106. R. W. Bilous, S. M. Mauer, D. E. Sutherland *et al.*, 'The effects of pancreas transplantation on the glomerular structure of renal allografts in patients with insulin-dependent diabetes.' *N. Engl. J. Med.*, 1989, **321**, 80–85.

107. Anonymous, 'The effect of intensive treatment of diabetes on the development and progression of long-term complications in insulin-dependent diabetes mellitus. The Diabetes Control and Complications Trial Research Group.' *N. Engl. J. Med.*, 1993, **329**, 977–986.

108. J. Barbosa, M. W. Steffes, D. E. Sutherland *et al.*, 'Effect of glycemic control on early diabetic renal lesions. A 5-year randomized controlled clinical trial of insulin-dependent diabetic kidney transplant recipients.' *JAMA*, 1994, **272**, 600–606.

109. J. A. Breyer, 'Diabetic nephropathy in insulin-dependent patients.' *Am. J. Kidney Dis.*, 1992, **20**, 533–547.

110. F. N. Ziyadeh, 'The extracellular matrix in diabetic nephropathy.' *Am. J. Kidney Dis.*, 1993, **22**, 736–744.

111. J. I. Kreisberg, 'Hyperglycemia and microangiopathy. Direct regulation by glucose of microvascular cells.' *Lab. Invest.*, 1992, **67**, 416–426.

112. A. Flyvbjerg, S. M. Marshall, J. Frystyk *et al.*, 'Octreotide administration in diabetic rats: effects on renal hypertrophy and urinary albumin excretion.' *Kidney Int.*, 1992, **41**, 805–812.

113. C. W. Yang, H. Vlassara, E. P. Peten *et al.*, 'Advanced glycation end products up-regulate gene expression found in diabetic glomerular disease.' *Proc. Natl. Acad. Sci. USA*, 1994, **91**, 9436–9440.

114. H. Vlassara, L. J. Striker, S. Teichberg *et al.*, 'Advanced glycation end products induce glomerular sclerosis and albuminuria in normal rats.' *Proc. Natl. Acad. Sci. USA*, 1994, **91**, 11704–11708.

115. C. W. Yang, H. Vlassara, G. E. Striker *et al.* 'Administration of AGEs *in vivo* induces genes implicated in diabetic glomerulosclerosis.' *Kidney Int. Suppl.*, 1995, **49**, S55–S58.

116. Z. Makita, S. Radoff, E. J. Rayfield *et al.*, 'Advanced glycosylation end products in patients with diabetic nephropathy.' *N. Engl. J. Med.*, 1991, **325**, 836–842.

117. W. D. Comper and E. F. Glasgow, 'Charge selectivity in kidney ultrafiltration.' *Kidney Int.*, 1995, **47**, 1242–1251.

118. L. K. Lee, A. S. Pollock and D. H. Lovett, 'Asymmetric origins of the mature glomerular basement membrane.' *J. Cell Physiol.*, 1993, **157**, 169–177.

119. R. A. Zager, 'Acute renal failure in the setting of bone marrow transplantation.' *Kidney Int.*, 1994, **46**, 1443–1458.

120. R. J. Johnson, 'The glomerular response to injury: progression or resolution?' *Kidney Int.*, 1994, **45**, 1769–1782.

121. A. Shimizu, H. Kitamura, Y. Masuda *et al.*, 'Apoptosis in the repair process of experimental proliferative glomerulonephritis.' *Kidney Int.*, 1995, **47**, 114–121.

122. A. J. Baker, A. Mooney, J. Hughes *et al.*, 'Mesangial cell apoptosis: the major mechanism for resolution of glomerular hypercellularity in experimental mesangial proliferative nephritis.' *J. Clin. Invest.*, 1994, **94**, 2105–2116.

123. L. P. Aiello, R. L. Avery, P. G. Arrigg *et al.*, 'Vascular endothelial growth factor in ocular fluid of patients with diabetic retinopathy and other retinal disorders.' *N. Engl. J. Med.*, 1994, **331**, 1480–1487.

124. J. Pe'er, D. Shweiki, A. Itin *et al.*, 'Hypoxia-induced expression of vascular endothelial growth factor by retinal cells is a common factor in neovascularizing ocular diseases.' *Lab. Invest.*, 1995, **72**, 638–645.

125. W. Min and N. Yamanaka, 'Three-dimensional analysis of increased vasculature around the glomerular vascular pole in diabetic nephropathy.' *Virchows Arch. A Pathol. Anat. Histopathol.*, 1993, **423**, 201–207.

126. P. Shultz, D. Ruble and L. Raij, 'S-nitro-*n*-acetyl-penicillamine (SNAP) inhibits mitogen-induced mesangial cell proliferation.' (abstract) *Kidney Int.*, 1990, **37**, 203.

127. T. Homma, R. L. Hoover, I. Ichikawa *et al.*, 'Angiotensin II (AII) induces hypertrophy and stimulates collagen production in cultured rat glomerular mesangial cell (MC).' (abstract) *Clin. Res.*, 1990, **38**, 358A.

128. J. Floege, C. E. Alpers, M. W. Burns *et al.*, 'Glomerular cells, extracellular matrix accumulation, and the development of glomerulosclerosis in the remnant kidney model.' *Lab. Invest.*, 1992, **66**, 485–497.

129. L. Stavenow, T. Kjellstrøm and J. Malmquist, 'Stimulation of collagen production in growth-arrested myocytes and fibroblasts in culture by growth factor(s) from platelets.' *Exp. Cell. Res.*, 1981, **136**, 321–325.

130. A. J. Owen 3rd, R. P. Geyer and H. N. Antoniades, 'Human platelet-derived growth factor stimulates amino acid transport and protein synthesis by human diploid fibroblasts in plasma-free media.' *Proc. Natl. Acad. Sci. USA*, 1982, **79**, 3203–3207.

131. K. A. Nath, A. C. Woolley and T. H. Hostetter, 'O$_2$ consumption (QO$_2$) and oxidant stress in the remnant nephron (abstract).' *Clin. Res.*, 1987, **35**, 553.

132. R. L. Hoover, J. M. Robinson and M. J. Karnovsky, 'Adhesion of polymorphonuclear leukocytes to endothelium enhances the efficiency of detoxification of oxygen-free radicals.' *Am. J. Pathol.*, 1987, **126**, 258–268.

133. M. L. Purkerson, P. E. Hoffsten and S. Klahr, 'Pathogenesis of the glomerulopathy associated with renal infarction in rats.' *Kidney Int.*, 1976, **9**, 407–417.

134. M. L. Purkerson, J. H. Joist, J. M. Greenberg

et al., 'Inhibition by anticoagulant drugs of the progressive hypertension and uremia associated with renal infarction in rats.' *Thromb. Res.*, 1982, **26**, 227–240.

135. M. L. Purkerson, D. M. Tollefsen and S. Klahr, '*N*-desulfated/acetylated heparin ameliorates the progression of renal disease in rats with subtotal renal ablation.' *J. Clin. Invest.*, 1988, **81**, 69–74.

136. J. R. Diamond and M. J. Karnovsky, 'Nonanticoagulant protective effect of heparin in chronic aminonucleoside nephrosis.' *Ren. Physiol.*, 1986, **9**, 366–374.

137. T. Homma, C. Broadley, R. L. Hoover *et al.*, 'Heparin reduces synthesis of collagenous proteins in rat glomerular mesangial cells (MC) (abstract).' *Kidney Int.*, 1990, **37**, 508.

138. E. M. Tan, G. R. Dodge, T. Sorger *et al.*, 'Modulation of extracellular matrix gene expression by heparin and endothelial cell growth factor in human smooth muscle cells.' *Lab. Invest.*, 1991, **64**, 474–482.

139. W. Kriz, H. Hosser, J. L. Simons *et al.*, 'In Fawn-hooded (FHH) rats focal segmental glomerulosclerosis (FSGS) preferentially develops at the vascular pole (abstract).' *J. Am. Soc. Nephrol.*, 1995, **6**, 1019.

140. L. A. Mynderse, J. R. Hassell, H. K. Kleinman *et al.*, 'Loss of heparan sulfate proteoglycan from glomerular basement membrane of nephrotic rats.' *Lab. Invest.*, 1983, **48**, 292–302.

141. D. Kerjaschki, A. T. Vernillo and M. G. Farquhar, 'Reduced sialylation of podocalyxin—the major sialoprotein of the rat kidney glomerulus—in aminonucleoside nephrosis.' *Am. J. Pathol.*, 1985, **118**, 343–349.

142. B. S. Kasinath, A. K. Singh, Y. S. Kanwar *et al.*, 'Effect of puromycin aminonucleoside on HSPG core protein content of glomerular epithelial cells.' *Am. J. Physiol.*, 1988, **255**, F590–F596.

143. J. A. Fishman and M. J. Karnovsky, 'Effects of the aminonucleoside of puromycin on glomerular epithelial cells *in vitro*.' *Am. J. Pathol.*, 1985, **118**, 398–407.

144. S. D. Ricardo, J. F. Bertram and G. B. Ryan, 'Antioxidants protect podocyte foot processes in puromycin aminonucleoside-treated rats.' *J. Am. Soc. Nephrol.*, 1994, **4**, 1974–1986.

145. P. E. Thomas, M. Goyal, B. L. Wharram *et al.*, 'Presence of multiple forms of GLEPP1 in the renal glomerulus.' *J. Am. Soc. Nephrol.*, 1994, **5**, 734.

146. J. A. Bruijn and E. de Heer, 'Adhesion molecules in renal disease.' *Lab. Invest.*, 1995, **72**, 387–394.

147. M. G. Farquhar, D. Kerjaschki, M. Lundstrom *et al.*, 'gp330 and RAP: the Heyman nephritis antigenic complex.' *Ann. NY Acad. Sci.*, 1994, **737**, 96–113.

148. S. Adler and X. Chen, 'Anti-Fx1A antibody recognizes a β1-integrin on glomerular epithelial cell and inhibits adhesion and growth.' *Am. J. Physiol.*, 1992, **242**, F770–F776.

149. D. J. Salant, 'The structural biology of glomerular epithelial cells in proteinuric diseases.' *Curr. Opin. Nephrol. Hypertens.*, 1994, **3**, 569–574.

150. E. Kemeny, M. J. Mihatsch, U. Durmuller *et al.*, 'Podocytes loose their adhesive phenotype in focal segmental glomerulosclerosis.' *Clin. Nephrol.*, 1995, **43**, 71–83.

151. J. J. Castellot, Jr., R. L. Hoover, P. A. Harper *et al.*, 'Heparin and glomerular epithelial cell-secreted heparin-like species inhibit mesangial-cell proliferation.' *Am. J. Pathol.*, 1985, **120**, 427–435.

152. F. Pugliese, A. K. Singh, B. S. Kasinath *et al.*, 'Glomerular epithelial cell polyanion neutralization is associated with enhanced prostanoid production.' *Kidney Int.*, 1987, **32**, 57–61.

153. T. Nadasdy, Z. Laszik, K. E. Blick *et al.*, 'Prolifera-

tive activity of intrinsic cell populations in the normal human kidney.' *J. Am. Soc. Nephrol.*, 1994, **4**, 2032–2039.

154. J. W. Fries, D. J. Sandstrom, T. W. Meyer *et al.*, 'Glomerular hypertrophy and epithelial cell injury modulate progressive glomerulosclerosis in the rat.' *Lab Invest.*, 1989, **60**, 205–218.

155. H. G. Rennke, 'How does glomerular epithelial cell injury contribute to progressive glomerular damage?' *Kidney Int. Suppl.*, 1994, **45**, S58–S63.

156. M. L. Artero, R. Sharma, V. J. Savin *et al.*, 'Plasmapheresis reduces proteinuria and serum capacity to injure glomeruli in patients with recurrent focal glomerulosclerosis.' *Am. J. Kidney Dis.*, 1994, **23**, 574–581.

157. J. Dantal, E. Bigot, W. Bogers *et al.*, 'Effect of plasma protein adsorption on protein excretion in kidney-transplant recipients with recurrent nephrotic syndrome.' *N. Engl. J. Med.*, 1994, **330**, 7–14.

158. M. Sharma, R. Sharma, S. Gunwar *et al.*, 'Partial purification and characterization of a circulating protein (HFSF) that increases glomerular albumin permeability in human focal segmental glomerular sclerosis (abstract).' *J. Am. Soc. Nephrol.*, 1992, **3**, 751.

159. M. Kashgarian and R. B. Sterzel, 'The pathobiology of the mesangium.' *Kidney Int.*, 1992, **41**, 524–529.

160. R. J. Johnson, J. Floege, A. Yoshimura *et al.*, 'The activated mesangial cell: a glomerular "myofibroblast"?' *J. Am. Soc. Nephrol.*, 1992, **2**, S190–S197.

161. C. E. Alpers, R. A. Seifert, K. L. Hudkins *et al.*, 'Developmental patterns of PDGF B-chain, PDGF-receptor, and alpha-actin expression in human glomerulogenesis.' *Kidney Int.*, 1992, **42**, 390–399.

162. J. Floege, M. W. Burns, C. E. Alpers *et al.*, 'Glomerular cell proliferation and PDGF expression precede glomerulosclerosis in the remnant kidney model.' *Kidney Int.*, 1992, **41**, 297–309.

163. R. Tanaka, K. Sugihara, A. Tatematsu *et al.*, 'Internephron heterogeneity of growth factors and sclerosis-modulation of platelet derived growth factor by angiotensin II.' *Kidney Int.*, 1995, **47**, 131–139.

164. G. E. Striker, L. I. Schainuck, R. E. Cutler *et al.*, 'Structural-functional correlations in renal disease. I. A method for assaying and classifying histopathological changes in renal disease.' *Hum. Pathol.*, 1970, **1**, 615–630.

165. R. A. Risdon, J. C. Sloper and H. E. De Wardener, 'Relationship between renal function and histological changes found in renal-biopsy specimens from patients with persistent glomerular nephritis.' *Lancet*, 1968, **2**, 363–366.

166. L. I. Schainuck, G. E. Striker, R. E. Cutler *et al.*, 'Structural-functional correlations in renal disease. II. The correlations.' *Hum. Pathol.*, 1970, **1**, 631–641.

167. J. G. Mongeau, P. O. Robitaille, M. J. Clermont *et al.*, 'Focal segmental glomerulosclerosis (FSG) 20 years later. From toddler to grown up.' *Clin. Nephrol.*, 1993, **40**, 1–6.

168. F. Strutz, J. Tomazewski, P. Heeger *et al.*, 'Generation of a murine fibroblast-specific antibody.' (abstract) *J. Am. Soc. Nephrol.*, 1993, **4**, 666.

169. A. Zuk and E. D. Hay, 'Expression of β1 integrins changes during transformation of avian lens epithelium to mesenchyme in collagen gels.' *Dev. Dyn.*, 1994, **201**, 378–393.

170. E. D. Hay, 'Extracellular matrix alters epithelial differentiation.' *Curr. Opin. Cell Biol.*, 1993, **5**, 1029–1035.

171. V. Kon, T. E. Hunley and A. Fogo, 'Combined antagonism of endothelin A/B receptors links endothelin to vasoconstriction whereas angiotensin II effects fibrosis. Studies in chronic cyclosporine nephrotoxicity in rats.' *Transplantation*, 1995, **60**, 89–95.

172. E. P. Peten, L. J. Striker, M. A. Carome *et al.*, 'The

contribution of increased collagen synthesis to human glomerulosclerosis: a quantitative analysis of α2IV collagen mRNA expression by competitive polymerase chain reaction.' *J. Exp. Med.*, 1992, **176**, 1571–1576.

173. E. P. Peten, L. J. Striker, A. Garcia-Perez *et al.*, 'Studies by competitive PCR of glomerulosclerosis in growth hormone transgenic mice.' *Kidney Int. Suppl.*, 1993, **39**, S55–S58.

174. H. P. Marti, L. McNeil, G. Thomas *et al.*, 'Molecular characterization of a low-molecular-mass matrix metalloproteinase secreted by glomerular mesangial cells as PUMP-1.' *Biochem. J.*, 1992, **285**, 899–905.

175. M. A. Carome, L. J. Striker, E. P. Peten *et al.*, 'Human glomeruli express TIMP-1 mRNA and TIMP-2 protein and mRNA.' *Am. J. Physiol.*, 1993, **264**, F923–F929.

176. L. Gesualdo, S. Di Paolo, S. Milani *et al.*, 'Expression of platelet-derived growth factor receptors in normal and diseased human kidney. An immunohistochemistry and *in situ* hybridization study.' *J. Clin. Invest.*, 1994, **94**, 50–58.

177. K. Nabeshima, A. Yoshimura, K. Inui *et al.*, 'Relationship of PDGF B-chain mRNA expression identified by *in situ* hybridization (ISH) and disease severity in various human glomerular diseases.' (abstract) *J. Am. Soc. Nephrol.*, 1993, **4**, 777.

178. T. Yamamoto, T. Nakamura, N. A. Noble *et al.*, 'Expression of transforming growth factor B is elevated in human and experimental diabetic nephropathy.' *Proc. Natl. Acad. Sci. USA*, 1993, **90**, 1814–1818.

179. K. Yoshioka, T. Takemura, K. Murakami *et al.*, 'Transforming growth factor-β protein and mRNA in glomeruli in normal and diseased human kidneys.' *Lab. Invest.*, 1993, **68**, 154–163.

180. M. Iwano, Y. Akai, Y. Fujii *et al.*, 'Glomerular expression of TGF-β mRNA in human glomerulonephritis.' (abstract) *J. Am. Soc. Nephrol.*, 1993, **4**, 680.

181. C. M. Pesce, L. J. Striker, E. Peten *et al.*, 'Glomerulosclerosis at both early and late stages is associated with increased cell turnover in mice transgenic for growth hormone.' *Lab. Invest.*, 1991, **65**, 601–605.

182. J. Savill, 'Apoptosis in disease.' *Eur. J. Clin. Invest.*, 1994, **24**, 715–723.

183. M. R. Hammerman and S. B. Miller, 'Growth factor gene expression in tubular epithelial injury.' *Curr. Opin. Nephrol. Hypertens.*, 1995, **4**, 258–262.

184. M. J. Szabolcs, L. Ward, R. Buttyan *et al.*, 'Apoptosis elucidated by labeling for DNA fragmentation in human renal biopsies.' (abstract) *Lab. Invest.*, 1994, **70**, 160A.

185. T. Oikawa, J. Kakuchi and A. Fogo, 'Interleukin 1β converting enzyme (ICE) is increased in settings of increased apoptosis *in vivo*.' (abstract) *Lab. Invest.*, 1995, **72**, 160A.

186. D. Woo, 'Apoptosis and loss of renal tissue in polycystic kidney diseases.' *N. Engl. J. Med.*, 1995, **333**, 18–25.

187. M. Marx, T. O. Daniel, M. Kashgarian *et al.*, 'Spatial organization of the extracellular matrix modulates the expression of PDGF-receptor subunits in mesangial cells.' *Kidney Int.*, 1993, **43**, 1027–1041.

188. M. Ikoma, T. Kawamura, Y. Kakinuma *et al.*, 'Cause of variable therapeutic efficiency of angiotensin converting enzyme inhibitor on the glomerular lesions.' *Kidney Int.*, 1991, **40**, 195–202.

189. A. Remuzzi, M. Mazarska, G. N. Gephardt *et al.*, 'Three-dimensional analysis of glomerular morphology in patients with subtotal nephrectomy.' *Kidney Int.*, 1995, **48**, 155–162.

190. A. Fogo, A. D. Glick, S. L. Horn *et al.*, 'Is focal segmental glomerulosclerosis really focal? Distribution of lesions in adults and children.' *Kidney Int.*, 1995, **47**, 1690–1696.

191. Y. Kakinuma, A. Fogo, T. Inagami *et al.*, 'Intrarenal localization of angiotensin II type 1 receptor mRNA in the rat.' *Kidney Int.*, 1993, **43**, 1229–1235.

192. G. N. Marinides, G. C. Groggel, A. H. Cohen *et al.*, 'Enalapril and low protein reverse chronic puromycin aminonucleoside nephropathy.' *Kidney Int.*, 1990, **37**, 749–757.

193. M. Ketteler, N. A. Noble and W. A. Border, Transforming growth factor-β and angiotensin II: the missing link from glomerular hyperfiltration to glomerulosclerosis?' *Annu. Rev. Physiol.*, 1995, **57**, 279–295.

7.10
Vasoactive Substances as Mediators of Renal Injury

KAMAL F. BADR

Emory University School of Medicine, Decatur, GA, USA

7.10.1   INTRODUCTION

Vasoactive hormones comprise a chemically heterogeneous group which includes proteins, peptides, lipids, nucleosides, and amino acid-derived molecules. This chapter is a synopsis of the physiologic actions of vasoactive hormones that modulate renal function and circumstances whereby these hormones mediate maladaptive or toxic responses (Tables 1 and 2). Where appropriate, advances in molecular biology that have further elucidated the physiology of these hormones are emphasized. In particular, genetically designed animals in which a hormone is overexpressed (transgenic) provide good models for its "toxic" potential and actions.[1] Cellular mechanisms of hormone action and receptor-mediated signal transduction have been reviewed in several excellent publications.[2-6]

Table 1   Circulating hormones with renal vascular actions.

Hormone	*Renal effects*[a]
Arginine vasopressin (AVP)	↓RBF, ↓GFR
Angiotensin II	↔↓RBF or GFR
Atrial natriuretic peptide (ANP)	↑RBF, ↑GFR
Catecholamines	↓RBF, ↓GFR
Endothelin	↓RBF, ↓GFR

[a]RBF = renal blood flow; GFR = glomerular filtration rate.

7.10.2   MODULATION OF RENAL FUNCTIONS BY VASOACTIVE HORMONES: AN OVERVIEW

7.10.2.1   Modulation of the Glomerular Filtration Rate (GFR)

The major physiological determinants of single nephron GFR (snGFR) are glomerular plasma flow (Q_A), glomerular transcapillary hydraulic pressure (P_{GC}), and the ultrafiltration coefficient (K_f).[7] These variables are in part determined by the contractile state of the afferent arteriole, efferent arteriole, and mesangial cells. K_f also varies with alterations in the hydraulic permeability of the capillary filtration barrier which consists of endothelial cells, visceral epithelial cells, and the glomerular basement membrane. By binding to specific receptors on cellular and structural components of the glomerulus, circulating and locally produced hormones influence one or more of the physiologic determinants of GFR. As the same substance can act at different sites in the glomerular unit, the net effect on GFR will depend on whether its actions are antagonistic or complementary. For example, atrial natriuretic peptide (ANP) decreases afferent arteriolar resistance while increasing efferent arteriolar resistance resulting in an augmented P_{GC} and a rise in GFR.[8] Under certain conditions ANP also increases K_f which in turn contributes to enhancing GFR.[9] On the other hand, angiotensin II (AII)-mediated constriction of both afferent and efferent arterioles results in opposite effects on glomerular plasma

Table 2 Locally acting substances with renal vascular actions.

Substance	Renal effects[a]
Urodilantin (RNP)	↑RBF, ↑GFR
Dopamine	↔, ↑, or ↓RBF/GFR (dose-dependent)
Kinins	↑RBF, ↔GFR
Anenosine	↔ or ↓RBF, GFR
Nitric oxide	↑RBF, ↔GFR
Prostaglandins	↑RBF, ↔GFR
Leukotrienes	↓RBF, ↓GFR

[a]RBF = renal blood flow; GFR = glomerular filtration rate.

flow and P_{GC} and, therefore, no change in snGFR.[10] Regulation of glomerular hemodynamics is further complicated by multiple interactions between hormones in the kidney. For example, infusion of AII along with a cyclooxygenase inhibitor causes a significant decrease in snGFR, suggesting that endogenous prostaglandin production antagonizes the glomerular effects of AII.[11] The net effects of renally-relevant hormones on GFR are discussed in more detail later.

7.10.2.2 Modulation of Tubule Water and Electrolyte Transport by Vasoactive Hormones

Arginine vasopressin (AVP) binds almost exclusively to cells in the collecting tubule (CT) where it influences water and urea absorption.[12] As in the glomerulus, vasoactive hormones may modulate their own actions by altering the production of counter-regulatory hormones; AVP induces local synthesis of prostaglandin E (PGE) which opposes AVP's effect on water permeability in the CT.

7.10.3 ARGININE VASOPRESSIN

The primary physiological role of AVP, or antidiuretic hormone (ADH), is maintenance of normal plasma osmolality through modulation of water excretion by the kidney. Its other physiological functions include regulation of blood pressure and the response to stress.

7.10.3.1 Synthesis and Secretion of AVP

AVP is a 9-amino acid neuropeptide synthesized by magnocellular neurons in the paraventricular and supraoptic nuclei of the hypothalamus.[13] The most potent stimulus for AVP secretion into the bloodstream is increased plasma osmolality. As little as 1% change in plasma osmolality leads to a change in AVP concentration that is sufficient to modify renal water excretion.[14] AVP secretion is almost completely suppressed when plasma osmolality decreases below an average of $280\,mOsm\,kg^{-1}\,H_2O$ in humans.

7.10.3.2 Mechanisms of Action of AVP

AVP exerts its biological actions by binding to specific receptors on the cell surface. Three types of AVP receptors have been identified: V_{1a}, V_{1b}, and V_2.[15] V_{1a} receptors are linked to G_{plc}, a phospholipase C-coupled G protein, which mediates intracellular production of 1,2-diacylglycerol (DAG) and inositol 1,4,5-triphosphate (IP_3). DAG stimulates protein kinase C (PKC), while IP_3 increases cytosolic Ca^{2+}, thus initiating the second messenger cascade responsible for AVPs cellular actions. Biological effects of AVP mediated by the V_{1a} receptor include platelet aggregation and increased glycogenolysis and glucogenesis in the liver.[15,16] In addition, high circulating levels of AVP cause arteriolar vasoconstriction by binding to V_{1a} receptors on vascular smooth muscle cells.[17] The V_{1b} receptor, also coupled to G_{plc}, is present on neurons in the anterior pituitary (adenohypophysis) and is thought to mediate AVP-induced corticotropin secretion.[15,18] V_2 receptors, however, are coupled to an adenylate cyclase stimulatory G-protein (Gs).[15] Binding of AVP to V_2 receptors on renal epithelial cells results in increased cAMP levels, activation of protein kinase A (PKA), insertion of water channels into the apical membrane, and antidiuresis.[12]

7.10.3.3 Renal Actions of AVP

The principal effect of AVP in the kidney is increased water permeability throughout the collecting tubule.[12,19] Following binding of AVP to V_2 receptors on the basolateral membrane of principal cells and terminal inner medullary collecting duct (IMCD) cells, preformed water channels are inserted into the apical membrane, thus increasing water permeability.[12,19]

In addition to its antidiuretic effect, AVP regulates urea permeability and ion transport in the collecting duct.

7.10.3.4   Transgenic Mouse Models of AVP Expression

Majzoub and colleagues have developed a line of transgenic mice which expresses AVP in a tissue-specific manner, demonstrate appropriate osmotic regulation of transgenic vasopressin mRNA, and have normal water metabolism.[20] Despite increased hypothalamic levels of AVP, both heterozygous and homozygous transgenic mice had normal AVP levels in the posterior pituitary. Plasma AVP levels, however, were increased three- to fourfold in homozygous mice, but basal water homeostasis remained normal. The latter observation suggests that renal resistance to AVP could have developed.

7.10.3.5   Maladaptive Actions of AVP

Secretion of AVP is also influenced by alterations in intravascular volume and blood pressure sensed by baroreceptors located in the heart, aortic arch, and carotid sinus.[14] Whereas a 5%–10% decrease in blood volume or systemic arterial pressure has little effect, further hemodynamic compromise leads to a steep increase in circulating AVP levels. Significant reductions (20–30%) in effective circulatory volume or blood pressure can override osmoregulation and result in markedly increased AVP levels in the face of decreased plasma osmolality.[21] Other potent stimuli for AVP secretion include fever and emesis.[22] Thus, under conditions of "inappropriately" elevated plasma levels, due either to overriding nonosmotic stimuli or ectopic excretion, this vasoactive hormone will mediate excessive water retention leading to often severe hyponatremia and plasma dilution. When severe, this condition can be fatal. These actions of AVP can be viewed as being maladaptive in nature.

7.10.4   THE RENIN–ANGIOTENSIN SYSTEM

7.10.4.1   Regulation of Renin Production and Secretion

Renin is produced and stored in granular juxtaglomerular (JG) cells, which are modified aortic smooth muscle cells found in the media of afferent arterioles.[23–25] Genomic analysis of the renin gene identified a single locus in humans and rats, but in the mouse there are two renin genes designated *Ren-1* and *Ren-2*.[23] Renin is synthesized in an inactive precursor form, preprorenin. Cleavage of the signal peptide from the carboxyl terminal of preprorenin results in prorenin, which is also biologically inactive. Subsequent glycosylation and proteolytic cleavage leads to formation of renin, a 37–40 kDa proteolytic enzyme. Stimulation of renin release by JG cells is mediated by increased intracellular cAMP, while a rise in cytosolic free calcium is inhibitory.[26] Physiological regulators of renin secretion include urinary sodium chloride (NaCl) concentration sensed by macula densa cells in the distal tubule, activity of the sympathetic nervous system, alterations in intrarenal perfusion pressure, and endocrine and paracrine hormones and growth factors. Decreased NaCl delivery to macula densa cells stimulates renin secretion, while increased urinary NaCl exerts an opposite effect.[27] Stimulation of postjunctional β-adrenergic receptors increases renin release. The role of α-adrenergic receptors, however, is controversial.[28] Ample evidence suggests that dopamine stimulates renin secretion by direct activation of DA-1 receptors on juxtaglomerular cells.[28,29] Renin release responds inversely to changes in renal perfusion pressure.[25] Elevation in intrarenal arterial pressure inhibits renin release and induces a "pressure" natriuresis.

Several endocrine and paracrine hormones regulate renin secretion by the kidney. ANP has been shown to inhibit renin release from isolated JG cells.[30] Other inhibitory hormones include AVP, endothelin, and adenosine (A$_1$-receptor agonists).[25,27] Regulation of renin secretion by angiotensin II is probably the most physiologically relevant.[31] Angiotensin II inhibits renin secretion and renin gene expression in a negative feedback loop. Arachidonic acid metabolites produced in the kidney also play an important role in renin secretion.[27] Intrarenal infusion of arachidonic acid increases, while indomethacin decreases, plasma renin activity in the rabbit.[32] Several studies have since confirmed that prostaglandins of the PGI series are potent stimulators of renin secretion.[27] On the other hand, lipoxygenase products of arachidonic acid metabolism (12-HPETE, 15-HPETE, and 12-HETE) and cytochrome P450-mediated epoxides (14,15-epoxyeicosatrienoic acid) have been shown to inhibit renin release in renal cortical slices.[33,34]

7.10.4.2   Physiologic and Pathophysiologic ("Toxic") Actions of Angiotensin II

Renin secretion into the circulation, as in situations of decreased effective intravascular volume, initiates an enzymatic cascade that leads to the production of angiotensin.[23]

Angiotensinogen, derived from hepatocytes, is a 55–65 kDa globular glycoprotein which is cleaved by renin to form the decapeptide angiotensin I. Angiotensin-converting enzyme (ACE) then converts angiotensin I to the biologically active octapeptide angiotensin II. ACE is a dipeptidyl carboxypeptidase present in nearly all mammalian tissues and body fluids.[23] Endothelial cells constitute a major source of ACE activity.

Circulating angiotensin II exerts its biological effects by binding to specific receptors on the cell surface.[23,35] Two subtypes of angiotensin II receptors have been identified. AT_1 receptors bind angiotensin II with higher affinity than angiotensin III, are selectively blocked by the biphenylimidazole compound Losartan, and are positively coupled to phospholipase C and negatively to adenylate cyclase. AT_2 receptors bind angiotensin II and III with similar affinity and are selectively blocked by tetrahydro-imidazopyridines, such as PD123177. AT_1 receptors have been shown to mediate many angiotensin II functions including cell contraction, aldosterone secretion, pressor and tachycardic responses, increased thirst, and hypertension secondary to renal artery stenosis.[36] In the adult kidney, angiotensin II receptors are present on afferent and efferent arterioles, glomerular mesangial cells, and proximal tubule cells.[35]

The predominant function of the renin–angiotensin system is regulation of vascular tone and renal salt excretion in response to changes in extracellular fluid (ECF) volume or blood pressure. Angiotensin II represents the effector limb of this hormonal system by acting on several organs including the vascular system, heart, adrenal glands, central nervous system (CNS), and the kidneys.

7.10.4.3 Vascular Actions of Angiotensin II

Through direct action on smooth muscle cells angiotensin II significantly increases arteriolar resistance in renal, mesenteric, dermal, coronary, and cerebral vascular beds.[37] Skeletal muscle and pulmonary vessels, however, are not affected, due to angiotensin II-stimulated production of vasodilatory prostaglandins by endothelial and smooth muscle cells in these vascular beds.[38,39] Angiotensin II exerts indirect pressor effects via the central and peripheral nervous systems. Its effects in the CNS include increased sympathetic discharge and decreased vagal tone.[40] Peripherally, angiotensin II augments the vasoconstrictive response to renal nerve stimulation in dogs,[41] and its inhibition

attenuates the pressor response to norepinephrine in humans.[42] Experimental data suggest the presence of a local renin–angiotensin system in the vasculature which contributes to the regulation of vascular tone.[23]

7.10.4.4 Adrenal Actions of Angiotensin II

Angiotensin II stimulates aldosterone synthesis and secretion by zona glomerulosa cells of the adrenal cortex.[43] This is an important mechanism by which the renin–angiotensin system regulates Na^+, water, and K^+ homeostasis.

7.10.4.5 Central Nervous System Actions of Angiotensin II

In addition to its CNS-mediated pressor effects, angiotensin II stimulates thirst and salt appetite.[44,45] It also increases secretion of vasopressin and oxytocin from the posterior pituitary, and adrenocorticotrophic hormone (ACTH), prolactin, and leutinizing hormone from the anterior pituitary.[46]

7.10.4.6 Growth-promoting Actions of Angiotensin II

Angiotensin II has been shown to induce hypertrophy and mitogenesis in cultured vascular smooth muscles.[47,48] This effect is at least in part mediated through autocrine production of growth factors, such as platelet-derived growth factor (PDGF) and transforming growth factor-β (TGF-β).[49] Some studies suggest that the renin–angiotensin system contributes to neointimal formation and restenosis after angioplasty.[23] Angiotensin II has direct inotropic, chronotropic, mitogenic, and hypertrophic effects on isolated atria and ventricles.[23] Amelioration of hypertensive cardiomyopathy with ACE inhibitors suggests that the renin–angiotensin system plays a role in cardiac hypertrophy.[23] Angiotensin II also modulates mesangial cell growth as discussed in the next section.

7.10.4.7 Renal Actions of Angiotensin II

Angiotensin II serves at least three important functions in the kidney: autoregulation of GFR, reduction of salt excretion through direct and indirect actions on renal tubule cells, and growth modulation in renal cells expressing AT_1 receptors.

In conditions of decreased renal blood flow (RBF), GFR is preserved at nearly constant value over a wide range of perfusion pressures. This phenomenon is known as autoregulation of GFR and is critically dependent on angiotensin II.[50] Micropuncture studies have shown that angiotensin II infusion preferentially vasoconstricts efferent arterioles while leaving afferent arteriolar resistance unaltered.[10] The disproportionate increase in postglomerular resistance results in a marked increase in P_{GC}, ultrafiltration pressure, and filtration fraction (FF), thus preserving GFR in the face of declining RBF. The selectivity of angiotensin II's vasoconstrictive action to the efferent arteriole is dose-dependent. In addition, angiotensin II increases both afferent and efferent resistance in the presence of a cyclooxygenase inhibitor.[11] Under certain pathophysiologic conditions, afferent arteriolar constriction predominates leading to a decrease in both RBF and GFR.[51] In addition to its vascular effects, angiotensin II induces mesangial cell contraction which leads to decreased K_f in vivo.[52,53] This effect, however, is attenuated by the concomitant production of prostaglandins by mesangial cells.[54]

Angiotensin II induces hypertrophy of proximal tubule epithelial cells in vitro.[55] It also exerts similar growth-promoting effects on mesangial cells.[56] These observations suggest a potential role for angiotensin II in mediating renal hypertrophic responses to nephron loss following acute ischemic/toxic injury or uninephrectomy.

7.10.4.8 Overexpression of the Renin–Angiotensin System in Transgenic Rats

Transgenic animals provide the opportunity to study effects of hormone overexpression on physiologic parameters or pathologic phenotypes. The importance of the renin–angiotensin system in systemic blood pressure control led to interest in its possible contribution to the pathogenesis of hypertension. Mullins et al. introduced the mouse Ren-2 gene into normotensive rats, thus creating a transgenic strain that expresses high levels of Ren-2 mRNA in the adrenal glands and to a much lesser extent in the kidneys.[57,58] The renin transgenic rats developed fulminant hypertension between 5 and 10 wk of age. The hypertensive phenotype was dependent on angiotensin II because treatment with low-dose ACE inhibitors or with angiotensin II antagonists normalized blood pressure.[59] Moreover, male rats had significantly higher blood pressures than females possibly due to the stimulatory effect of androgens on tissue renin–angiotensin systems.[60] This observation is consistent with epidemiologic data demonstrating higher prevalence of hypertension in men. The fact that adrenalectomy normalized blood pressure in the Ren-2 transgenic animals led to the conclusion that a stimulated local adrenal renin–angiotensin system rather than systemically elevated renin levels contributed to hypertension in these animals.[58]

As activation of the renin–angiotensin system has been implicated in the progression of chronic renal disease,[61] Ganten and colleagues studied progression of glomerular sclerosis following subtotal nephrectomy in Ren-2 transgenic rats.[58] Compared to pressure-matched spontaneously hypertensive rats, the transgenic animals had significant acceleration of glomerulosclerosis suggesting a pathogenetic role for the intrarenal renin–angiotensin system in progression of renal failure.

7.10.5 ATRIAL NATRIURETIC PEPTIDE

In 1981, de Bold and colleagues demonstrated that injection of crude rat atrial, but not ventricular, extracts into anesthetized rats induces more than 30-fold increase in urinary sodium excretion, a 10-fold rise in urine volume, and a sustained decrease in systemic blood pressure.[62] Further biochemical studies localized the natriuretic activity to atrial granules and led to the isolation and characterization of ANP.[63,64]

7.10.5.1 Molecular and Biochemical Properties of ANP

The cDNA for human ANP was isolated in 1984 and shortly after the gene was localized to the short arm of chromosome 1.[65,66] The chromosomal gene consists of three exons and two introns encoding for a mature mRNA transcript approximately 900 bases in length.[66] Translation of human ANP mRNA results in a 151-amino acid preprohormone.[67] ProANP, a 126-residue molecule, is formed following cleavage of the signal peptide sequence of preproANP and represents the major storage form of the hormone in atrial granules.[68] The circulating, biologically active form of ANP, often referred to as ANP_{99-126} or ANP_{1-28}, is a peptide comprised of the 28 carboxy-terminal amino acids of the parent molecule.[64,67]

7.10.5.2 Secretion and Physiological Regulation of ANP_{99-126}

Cardiac atria contain the highest concentrations of ANP and serve as the major source of circulating hormone.[67] ANP is present in much lower concentrations in other tissues, such as the brain, spinal cord, pituitary gland, adrenal medulla, and kidney.

ANP_{99-126} secretion from cardiomyocytes occurs largely in response to atrial stretch resulting from increased atrial transmural pressure.[67] Physiological stimuli for the release of ANP_{99-126} include acute salt and volume loading, supine posture (head-down tilt), and head out water immersion.[69-71] Increased rate of atrial contraction has also been shown to stimulate ANP_{99-126} secretion.[72,73] Angiotensin II, vasopressin, epinephrine, and phenylephrine stimulate ANP_{99-126} secretion from the heart largely due to their systemic vasopressor effects.[74] On the other hand, glucocorticoids and endothelin raise ANP_{99-126} levels possibly by acting directly on atrial myocytes.[75,76]

7.10.5.3 Physiological Actions of ANP_{99-126}

Two subtypes of ANP receptors have been identified.[77,78] ANP-R1, previously referred to as B ("biologically active") receptor, has intrinsic guanylate cyclase activity which catalyzes production of $3',5'$-cyclic GMP following ligand binding. cGMP then serves as an intracellular second messenger which mediates the biological activities of ANP_{99-126}. ANP-R2, previously known as C ("clearance") receptor, is devoid of guanylate cyclase activity and, therefore, does not confer biological activity. ANP-R2 receptors are thought to mediate clearance of circulating ANP and other related hormones, such as brain natriuretic peptide (BNP).[79] Utilizing molecular biological techniques, a third subtype of ANP receptors has been identified.[80] ANP-R3 is homologous to ANP-R1 and is biologically active. Although ANP-R3 binds ANP and stimulates production of cGMP, it appears to have a 50-fold higher affinity for a related natriuretic factor originally purified from porcine brain and known as C-type natriuretic peptide (CNP).[81]

The major sites of action of ANF_{99-126} are the kidneys, adrenal glands, and vascular smooth muscle.[82] Short-term administration of ANF_{99-126} in experimental animals and in humans induces pronounced natriuresis, diuresis, alteration in renal hemodynamics and tubular function, suppression of renin release, inhibition of aldosterone secretion by the adrenal glands, and decreased vasomotor tone

resulting in transient drop in systemic blood pressure. Based on these actions, it is postulated that ANP plays an important physiological role in protecting against extracellular volume overload.[83]

7.10.5.4 Renal Actions of ANP_{99-126}

ANP-induced increase in GFR is well established.[84-86] ANP_{99-126} decreases afferent arteriolar resistance while increasing efferent arteriolar resistance resulting in increased P_{GC} and filtration fraction.[8] In addition, ANP_{99-126} relaxes mesangial cells *in vitro* suggesting that it can increase filtration area and K_f *in vivo*.[87] Indeed, when baseline K_f is low, as in water-deprived animals, ANP_{99-126} enhances GFR mainly by increasing K_f.[9] The effect of ANP on RBF is variable. Under conditions of pre-existing vascular constriction in the isolated perfused kidney, ANP tends to vasodilate renal vessels and increase RBF.[88] In the intact animal, however, ANP_{99-126} infusion causes either a decline or no change in RBF.[89] The effects of ANP_{99-126} on RBF are influenced by its systemic actions on blood pressure and the renin–angiotensin system. Finally, ANP has been reported to induce redistribution of blood flow from the cortex to the medulla and to increase vasa recta flow leading to dissipation of the medullary solute gradient.[90,91]

Several studies provide evidence that ANP_{99-126} induces natriuresis and diuresis by mechanisms independent of increased GFR. ANP_{99-126} has both direct and indirect effects on tubular transport of sodium and water. In the proximal tubule, ANP_{99-126} antagonizes angiotensin II-induced sodium reabsorption.[92] In the inner medullary collecting duct, it directly inhibits sodium transport by binding to ANP-R1 receptors and influencing amiloride-sensitive sodium channels and the activity of basolateral Na/K/2Cl cotransporters.[93-95] Other mechanisms by which ANP_{99-126} induces natriuresis and diuresis include suppression of renin and aldosterone release,[96,97] inhibition of the tubular actions of AVP,[98] and dissipation of the medullary solute gradient.[90,91]

7.10.5.5 Overexpression of ANP in Transgenic Mice

Field and associates have generated a transgenic mouse which overexpresses proANP in hepatocytes resulting in chronically elevated plasma levels of the prohormone.[99-101] The transgenic animals exhibited a hypotensive phenotype (20–30 mmHg lower than control littermates) without compensatory tachycardia. GFR remained within normal limits despite hypotension. Moreover, significant diuresis or natriuresis during steady state was not detected. Also contrary to observations made following short-term infusion of ANP_{99-126}, plasma renin activity did not change, while aldosterone levels were elevated in proANP transgenic mice. In summary, the transgenic model suggests that compensatory mechanisms neutralize the natriuretic and diuretic effects of ANP despite continued secretion of the hormone. Vasodilatation and hypotension, however, persist.

7.10.6 URODILATIN OR RENAL NATRIURETIC PEPTIDE (RNP)

Urodilatin is best described as a paracrine RNP.[102] Its amino acid sequence is identical to ANP_{99-126} except for an additional four amino acids at the amino terminal. Despite its high degree of homology to ANP_{99-126}, specific anti-human RNP polyclonal antibody has been generated, and RNP levels can be measured using a radioimmunoassay.[103] To date, RNP has not been detected in the circulation and the kidney is presumed to be its site of synthesis and action.[102]

RNP binds to ANP receptors in the kidney and stimulates cGMP production.[104] Its renal actions parallel those of ANP_{99-126} and include hyperfiltration, diuresis, and natriuresis.[105,106] Systemic infusion of RNP results in effects similar to those of ANP_{99-126}. These include decreases in mean arterial pressure, right atrial pressure, stroke volume, and cardiac output.[105,107] The vascular effects of RNP, however, seem to be less potent than those of ANP_{99-126}.[105]

7.10.7 CATECHOLAMINES

Catecholamines play an important role in the regulation of RBF, GFR, renin secretion, and tubular transport. Elevated endogenous levels or exogenous administration, however, can lead to severe renal dysfunction. The endogenous catecholamines that act on the kidney are epinephrine, norepinephrine, and dopamine. These catecholamines derive from renal efferent nerves (norepinephrine and to a lesser extent dopamine), from the circulation (epinephrine and norepinephrine), and from renal proximal tubule cells (dopamine).

Catecholamines exert their actions by binding to specific membrane receptors. Four major adrenergic receptor classes are responsible for the functions of norepinephrine and epinephrine: α_1, α_2, β_1, and β_2.[108] α_1 receptors are expressed on postsynaptic effector sites, such as smooth muscles and mediate catecholamine-induced vasoconstriction. α_2 receptors are predominantly presynaptic and are believed to mediate feedback inhibition of norepinephrine release from nerve endings. Postsynaptic α_2 receptors have also been described and their functions include vasoconstriction, inhibition of lipolysis, and suppression of insulin release. β_1 and β_2 receptors present on cardiac tissue are responsible for the positive chronotropic and inotropic effects of catecholamines. In contrast, stimulation of β_2 receptors on extracardiac tissues results in relaxation of vascular and bronchial smooth muscles. β_1 and β_2 receptors signal through a stimulatory G protein which activates adenylate cyclase and increases intracellular cAMP levels. In contrast, stimulation of α_2 receptors inhibits adenylate cyclase activity. Activation of α_1 receptors stimulates phospholipase C-mediated generation of inositol triphosphate (IP_3) which releases ionized calcium from intracellular stores into the cytosol.

Although high concentrations of dopamine can stimulate α- and β-adrenergic receptors, specific high-affinity dopamine receptors exist.[109] Two classes of dopamine receptors have been characterized: DA_1 and DA_2. Both receptors mediate vasodilatation albeit by different mechanisms. DA_1 receptors are expressed on smooth muscle cells where ligand binding directly decreases vascular tone by increasing intracellular cAMP production. DA_2 receptors, however, are located presynaptically and their activation leads to inhibition of norepinephrine release. Postsynaptic DA_2 receptors have also been detected in the renal vasculature (see below). Using molecular cloning techniques two DA_1-like and three DA_2-like receptor cDNAs have been isolated to date.[110,111]

7.10.7.1 Effects of α-Adrenergic Stimulation in the Kidney

Both α_1- and α_2-adrenergic receptors are present in the renal cortex.[112,113] Adrenergic stimulation causes renal vasoconstriction (increased afferent and efferent arteriolar resistance) by activating α_1 receptors on vascular smooth muscle cells.[114] α_1-Mediated renal vasoconstriction results in decreased RBF and GFR. In the proximal convoluted tubule, where α_1- and α_2-adrenergic receptors are expressed in high density, norepinephrine increases sodium and water reabsorption.[113,114] Moreover, administration of α_2-adrenergic antagonists to humans or experimental animals with sodium retention results in natriuresis and diuresis.[113] Using *in vivo* microperfusion technique, DiBona and co-workers demonstrated that renal nerve excitation stimulates NaCl reabsorption in the loop of Henle.[115] α_2 receptor activation also enhances NaCl reabsorption in the thick ascending limb as shown in experiments using isolated perfused rat kidneys.[116]

7.10.7.2 Effects of β-Adrenergic Stimulation in the Kidney

β-Adrenergic receptors have been identified in the glomerulus, juxtaglomerular apparatus, thick ascending limb of loop of Henle, distal convoluted tubule, and collecting duct.[113,114] β_1-Stimulation enhances renin release from the juxtaglomerular cells of the afferent arterioles. Otherwise, there are few β-receptors in renal vessels. Although β-receptors have not been localized to the proximal tubule, physiological studies suggest that β-adrenergic stimulation increases sodium and fluid transport in this nephron segment independent of enhanced renin secretion and angiotensin II production.[113] In the thick ascending limb, β-adrenergic receptor activation stimulates cAMP production and NaCl reabsorption.[117] β-Agonists also increase Cl^-/HCO_3^- exchange and H^+/K^+-ATPase activity in the collecting duct.[113] The latter effect results in enhanced potassium reabsorption by type A intercalated cells (and apparent decrease in potassium secretion).[113]

7.10.7.3 Dopamine Synthesis and Action in the Kidney

Although dopamine-containing nerve endings are present within the kidney, their functional role remains to be determined.[118] It is well established, however, that dopamine is synthesized by proximal tubule cells and that it acts locally to modulate sodium excretion.[119] Both DA_1 and DA_2 receptors are present in large renal vessels and to a lesser extent in glomeruli.[109] DA_1 receptors are also present on all segments of the nephron with the highest concentration expressed in the proximal tubule.[109,120] Renally produced dopamine plays a central role in the regulation of sodium excretion.

In the whole kidney, dopamine increases RBF and GFR through its DA_1 receptor-mediated vasodilatory effects.[119,121] Supraphysiologic concentrations of dopamine, however, stimulate α-adrenergic receptors which leads to vasoconstriction and decreased RBF. The natriuretic and vasodilating effects of dopamine have earned it a therapeutic role in the volume expanded patient, particularly when administered in low doses that do not activate adrenergic receptors. Moreover, dysfunction of the renal dopamine system has been postulated to contribute to the pathogenesis of systemic hypertension.[119] Results from at least two studies suggest that defects in renal generation of dopamine are common in patients with essential hypertension.[122,123]

7.10.8 THE RENAL KALLIKREIN–KININ SYSTEM

Kinins are vasoactive peptides that possibly play a role in the modulation of salt and water excretion by the kidney and in the control of blood pressure.[124,125] The kallikrein–kinin system consists of four components: kallikreins, kininogens, kinins, and kininases.[126] Kallikreins are serine proteases which act on kininogens to generate the biologically active kinins. Kallikreins exist in either circulating (plasma) or tissue (glandular) form. Plasma kallikrein participates in the clotting cascade. Tissue kallikreins are present in the kidney, small intestine, pancreas, and salivary glands where they are responsible for local production of kinins. Kininases are peptidases that cleave kinins into inactive products. Due to the presence of kininases in the vascular endothelium and in epithelial cells of the proximal and distal nephron segments, circulating kinin has a very short half-life and is unlikely to affect renal function.[125]

Renal kallikrein activity is predominantly found in the cortex where it has been detected in glomeruli, proximal tubule, and distal tubule. The majority of kallikrein found in the urine is of the tissue form and is secreted by the kidney. Renal and urinary kallikrein act on low- and high-molecular weight kininogens to release the decapeptide lysyl bradykinin (kallidin). The lysyl group of kallidin is then cleaved by aminopeptidase to produce bradykinin. In addition to the presence of kininases (specifically kininase II, a peptidyl dipeptidase which is also known as angiotensin I converting enzyme) in the vascular endothelium and renal tubule cells, the proximal tubule brush border contains endopeptidase II (enkephalase) which also inactivates kallidin and bradykinin.[127] Aprotinin, a polypeptide purified from the lung, inhibits the activity of renal and other tissue kallikreins.[125]

7.10.8.1 Renal Actions of Kinins

Experimental evidence suggests that kinins regulate renal blood flow and renal excretion of sodium and water.[124,125] Acute bradykinin infusion into the renal artery dilates both preglomerular and postglomerular arterioles. The net effect is increased RBF without a change in GFR.[128] Acute infusion of bradykinin also induces significant natriuresis and diuresis in the absence of GFR alteration.[128] In experiments where bradykinin was administered for several days, the acute rise in salt and water excretion was not sustained while renal vasodilatation persisted.[128] Inhibition of endogenous bradykinin using specific antibodies or aprotinin has been shown to blunt the natriuretic and diuretic effect of saline infusion.[125,129] Siragi *et al.* have since demonstrated that infusion of a bradykinin antagonist (BKA) into the renal arteries of dogs on a low-sodium diet causes antidiuresis and significant decrease in fractional excretion of sodium.[130] There were no changes in GFR, plasma aldosterone concentration, plasma renin activity, or systemic arterial pressure during intrarenal BKA administration. Although lower doses of BKA did not affect estimated renal plasma flow, a higher dose caused a significant decrease. These results suggest that endogenous kinins can act as natriuretic substances.

7.10.9 ADENOSINE

Adenosine, a purine nucleoside, is a paracrine hormone that regulates cellular and physiological functions in many tissues.[131,132] Intracellular generation of adenosine results from the action of 5′-nucleotidase on adenosine monophosphate (AMP) during hypoxia. Adenosine, produced intracellularly, can traverse cell membranes by facilitated diffusion and function in a paracrine or autocrine fashion. Extracellular production of adenosine from AMP is also possible due to the presence of ecto 5′-nucleotidase on the surface of many cell types. In the kidney, ecto 5′-nucleotidase activity is expressed on tubular luminal membranes, fibroblasts, and mesangial cells and is believed to be the major source of renal adenosine.[133]

At least two subtypes of adenosine receptors, high-affinity A_1- and low-affinity A_2-receptors, have been identified and their respective genes isolated.[80,134] The A_1-receptor is a 36 kDa transmembrane polypeptide coupled to G_i protein. Activation of A_1-receptors leads to inhibition of adenyl cyclase activity and decreases intracellular cAMP levels. In contrast, binding of adenosine to the G_s protein-linked A_2-receptor (MW of 45 kDa) stimulates adenyl cyclase and increases cAMP production. The opposite effects of A_1- and A_2-receptor activation on cAMP-mediated intracellular signaling possibly underlies their contrasting physiological actions. Activation of other second messenger systems, such as phospholipase C, by adenosine has been observed and further subdivision of adenosine receptor classes has been suggested.[134]

7.10.9.1 The Renal Adenosine System

Both high-affinity A_1-receptors and low-affinity A_2-receptors are widely distributed throughout the renal vasculature and the nephron.[132,134] The renal effects of adenosine are diverse and include alterations in RBF, GFR, hormone production, neurotransmitter release, and tubular absorption.

Infusion of adenosine into the renal artery of experimental animals results in transient reduction of RBF secondary to A_1-receptor-mediated afferent arteriolar vasoconstriction followed by a delayed A_2-receptor-mediated postglomerular vasodilatation and return of RBF toward normal.[135,136] On the other hand, adenosine induces a sustained decrease in GFR secondary to reduced P_{GC}.[137] Adenosine infusion in humans results in an insignificant increase in RBF and a significant, moderate decrease in GFR.[138,139] It is postulated that adenosine-mediated reduction in GFR constitutes the underlying mechanism of tubuloglomerular feedback.[140,141] The hypothesis states that increased solute delivery to the macula densa stimulates sodium transport resulting in ATP hydolysis and generation of adenosine. Adenosine, in turn, completes the feedback loop by decreasing GFR and normalizing solute delivery to the distal nephron. The hypothesis is supported by experiments demonstrating that A_1-receptor blockade inhibits tubuloglomerular feedback.[142]

Several studies strongly indicate that adenosine suppresses renin release by the kidney,[134] likely mediated by binding of adenosine to high affinity A_1-receptors.[143] In contrast, recent data demonstrate that agonists selective for the low-affinity A_2-receptor stimulate rather than suppress renin release particularly when administered in higher doses.[134,144] This suggests that adenosine regulates renin release by exerting either an inhibitory or stimulatory effect depending on its local concentration. Adenosine also plays a similar regulatory role in erythropoietin production: A_1-receptor stimulation inhibits while A_2-receptor stimulation enhances erythropoietin synthesis by renal cells.[145]

7.10.9.2 The Role of Adenosine in Renal Pathophysiology

Pathophysiologic conditions associated with increased renal production of adenosine include acute renal ischemia, myoglobinuric acute renal failure, and mercuric chloride-induced acute renal failure.[146–148] It is postulated that adenosine plays a role in balancing oxygen supply and demand during renal hypoxia by regulating RBF, GFR, renin secretion, and solute.

7.10.10 EICOSANOIDS

The eicosanoids are a group of locally acting hormones or autacoids that are derived from dietary polyunsaturated fatty acids. In humans, arachidonic acid, an essential fatty acid esterified into cellular membrane phospholipids, is the most abundant and important precursor. Following deesterification by phospholipases, free arachidonic acid may either rapidly re-esterify into membrane lipids, avidly bind intracellular proteins, or undergo enzymatic oxygenation to yield the various biologically active molecules referred to as eicosanoids. The type of product formed depends on the enzymes involved in the oxygenation process.[149] Oxygenation of arachidonic acid by cyclooxygenase results in prostaglandin and thromboxane synthesis. Oxygenation by lipoxygenase generates hydroxyeicosatetraenoic acids and leukotrienes. Finally, oxygenation by cytochrome P450 generates epoxyeicosatrienoic acids, their corresponding diols, and monooxygenated arachidonic acid derivatives. These three major enzymatic pathways are all expressed in the kidney.[150,151] The specific nature of the products generated varies with both cell type and initial stimulus for arachidonic acid release. Eicosanoids have diverse biologic effects in the kidney, the significance of which will be discussed below.

7.10.11 PROSTAGLANDINS

Prostaglandins (PGs) are a unique group of cyclic fatty acids with diverse biologic effects that are produced throughout the body. The kidney is a major site of PG production, metabolism, and action.[150,151] PGs are important modulators of renal function both in physiologic and pathophysiologic settings. The spectrum of their effects in the kidney encompasses modulation of RBF, GFR, salt and water transport, and the release of renal hormones. It is within the setting of compro-

mised renal status that maintenance of renal function is most dependent on PGs. Under these circumstances, inhibition of PG synthesis with nonsteroidal anti-inflammatory drugs (NSAIDs) is likely to impair renal function.

7.10.11.1 Biosynthesis and Metabolism

Arachidonic acid (eicosatetraenoic acid) is the major substrate for the synthesis of PGs in humans. The initial step is catalyzed by the cyclooxygenase (COX) which inserts molecular oxygen onto the carbon backbone structure of arachidonic acid with the concomitant cyclization of carbons C-8 to C-12 to form a cyclic endoperoxide, PGG2 (15-hydroperoxide). In the presence of reduced glutathione-dependent peroxidase, PGG2 is converted to the 15-hydroxy derivative, PGH2. These unstable endoperoxide intermediates (PGG2 and PGH2) have a half-life of about 5 min. PGH2 is further transformed to yield the biologically active PGs and thromboxanes (TX) collectively referred to as prostanoids. In the presence of isomerase and reductase enzymes, PGH2 is converted to PGE2 and PGF2α, respectively. TX synthase converts PGH2 into TXA2, and prostacyclin synthase, abundant in vascular endothelial cells, catalyzes the biosynthesis of PGI2 (prostacyclin).[149–151] PGD2 formation in the kidney is uncertain.

The rate of PG production is dependent on the release of free arachidonic acid from tissue stores by phospholipase A2 (PLA2). Arachidonate tissue stores vary with dietary essential fatty acid intake and can be depleted when intake is deficient.[152] Fish oil diets (rich in ω-3 polyunsaturated fatty acids) will compete for the arachidonate oxidation process and inhibit formation of active products.[153] PLA2 activity is influenced by a large number of agents, such as hormones and growth factors.[150,151,154,155] COX gene expression and biologic activity are upregulated by cytokines and growth factors, such as IL-1[156] and PDGF.[157] On the other hand, corticosteroids inhibit PLA2 and COX gene expression.[158] Several renal pathophysiological states, such as glomerulonephritis and ureteral obstruction, are associated with increased prostanoid production.[159–161]

The COX pathway is the major pathway for arachidonic acid metabolism in the kidney.[150,151] In both animals and humans, two separate COX enzymes have been identified that are encoded by two separate genes: COX-1[162] and COX-2.[163] The human COX-1 enzyme is a 68.5 kDa protein that is constitutively present in arterial and arteriolar endothelial cells,[151,164] mesangial cells,[151,154,165,166] glomerular epithelial cells,[167] renal interstitial cells,[155,168] and along most segments of the tubule, although in markedly varying concentrations.[169,170] The COX-2 enzyme is a 603 amino acid protein whose expression has been demonstrated in the macula densa and renal papillae under normal conditions,[171] and in mesangial cells subjected to continuous stretch-relaxation in culture.[172] PGI2 is the major bioactive product released by renal arterial and arteriolar endothelial cells.[170] Whole glomeruli generate several prostanoids; the predominant product varies among species, but in humans it appears to be PGI2, rather than PGE2, as seen in rat and rabbit.[173] Cultured human mesangial cells are capable of generating PGE2, as well as PGF2α and PGI2 (PGE2 > PGF2α > PGI2 = TXA2).[154,165] Glomerular epithelial and endothelial cells generate PGs, but the pattern of the products remains controversial.[151,167] Intraglomerular macrophages, localized in the mesangium, are a potential source of prostanoids and other eicosanoids particularly following glomerular injury.[152] In the rest of the nephron, the collecting tubule, particularly its medullary portion, is a predominant site of PG synthesis; PGE2 being the major prostanoid synthesized.[174]

The PGs and TXA2 undergo rapid destruction and inactivation within the kidney by cytosolic degradative enzymes.[149] Elimination of PGE2, PGF2α, and PGI2 proceeds through enzymatic oxidation and nonenzymatic hydrolysis, while that of TXA2 is exclusively nonenzymatic. The initial degradative step is catalyzed by 15-hydroxyprostaglandin dehydrogenase with formation of biologically inactive 15-ketoPGs. These metabolites are further degraded by a PG reductase. PGI2 and TXA2 undergo rapid nonenzymatic degradation to 6-keto-PGF1α and TXB2, respectively. As the kidney metabolizes and also excretes circulating PGs, it is difficult to evaluate the net rate of renal production of a particular PG. Intact PGs and stable hydrolysis products are excreted in the urine and largely reflect the rate of renal PG production.[175] Excretion of PGE2 in the urine probably reflects renal production, although there are a number of variables that can render measurement erroneous, including contamination with seminal fluid.[175,176] Urinary excretion of 6-keto-PGF1α, the hydrolytic product of PGI2, may reflect both augmented systemic and PGI2 synthesis.[176] Active PGs are also detected in the renal venous effluent but, apart from PGI2, they are destroyed in the lungs and do not enter the systemic circulation.[175,177]

7.10.11.2 Biologic Actions of COX Products

PGs have diverse actions, in part related to their site of synthesis and the cells on which they act. Their principal physiologic role is mediation and/or modulation of hormone action.[151,154,155,165,175] Thus, cortical production by arterioles and glomeruli is related to regulation of RBF, GFR, and renin release. Other cortical sites of PG production affect ammoniagenesis[178] and calcium and phosphorus transport.[179] Medullary PG production is directed to regulating vasa recta blood flow, tubular sodium and chlorine transport, and the response of the collecting duct to AVP. Inhibition of COX activity in the absence of exogenous administration or endogenous release of hormones, such as angiotensin II or AVP, has little effect on renal functional parameters.[180] Once their local release is enhanced, COX products may themselves stimulate the local generation of other hormones. PG-stimulated renin release is an example of this mode of action.[173] Under pathophysiologic conditions, such as inflammatory injury, local release of prostanoids may mediate some of the functional derangements which characterize these conditions.[159–161]

Prostanoids act through specific and distinct receptors.[181] The cDNA for numerous prostanoid receptors including receptors for TXA2,[182,183] PGF2α,[184] and PGE2[185–188] have now been cloned and sequenced. All these receptors are members of the G-protein coupled family of receptors. Multiple subtypes of each of these prostanoid receptors may exist, such as the case with the PGE2 receptor (EP receptor), thus explaining the apparently contrasting effects mediated by PGE2 on smooth muscle and collecting duct water permeability.[189] Three of the four proposed PGE2 receptor subtypes (EP-1, EP-2, EP-3) have been cloned, sequenced, and expressed.[185–188] The EP-1 and EP-3 receptors mediate smooth muscle constriction, whereas the EP-2 receptor mediates smooth muscle relaxation.[190] Activation of the EP-1 receptor causes an increase in intracellular Ca^{2+} (accounting for smooth muscle constriction) and is presumably phosphatidylinositol-mediated.[185] The EP-3 receptor predominantly signals through the inhibitory G-protein, G_i, diminishing hormone stimulated cAMP generation.[191] In contrast, EP-2 receptors display dose-dependent stimulation of cAMP generation with PGE2 concentrations above 10^{-8} M.[186,187] In mesangial cells the PGF2α receptor (FP receptor) seems to be coupled to increased intracellular Ca^{2+}. At higher concentrations, PGF2α also stimulates EP receptors.[192,193] The TXA2 receptor

(TP receptor) appears to signal via phosphatidyl inositol hydrolysis leading to increased intracellular Ca^{2+}.[182] There is pharmacologic evidence for existence of TP receptors in the glomerulus.[192] The PGI2 receptor (IP receptor) signals via stimulation of cAMP generation.[194,195] PGI2 has been demonstrated to play an important vasodilator role in the glomerular microvasculature where the effects of PGI2 and PGE2 to stimulate cAMP generation were distinct and additive.[197]

7.10.11.3 Renal Vasoactive Actions of COX Products

There are some species differences in the renal actions of PGs, and this must be taken into account when extrapolating data from animals to humans. In general PGE2 and PGI2 are vasodilators in most species, while TXA2, PGF2α, and PGE2 (in certain circumstances) are vasoconstrictors.[196,197] PGE2 relaxes rat and rabbit afferent arterioles.[198] PGI2 is a potent vasodilator in both man and dog.[199,200] It is also a highly potent relaxant of rabbit afferent, efferent, and interlobular arteriolar smooth muscle,[198] but has little intrinsic vasoactive properties in the rat kidney.[201] PGF2α is without effect on any arterial segment in the rabbit kidney,[198] but is a mild vasoconstrictor in the dog.[200] TXA2 analogues exert constrictor effects on rat arteriolar smooth muscle, and result in renal vasoconstriction accompanied by a severe reduction in filtration fraction, suggesting a predominant preglomerular action.[202,203] The contribution of these vasoactive properties of COX products to the regulation of renal vascular tone under normal physiologic conditions is likely minimal.[204,205] This is best exemplified by the minimal or absent change in RBF and GFR in euvolemic rats or humans following COX inhibition[33-35] or selective antagonism of TXA2 synthesis[204] or actions.[202]

In contrast, the local release of vasodilator PGs (PGE2 and PGI2) in response to renal vasoconstrictors plays an important role in maintaining RBF and GFR. There is compelling evidence indicating that mesangial cell synthesis and release of PGE2 and PGI2 modulate the constrictor actions of angiotensin II, norepinephrine, and AVP.[151,154,206,207] Activation of the renin–angiotensin and sympathetic nervous systems leading to enhanced release of angiotensin, catecholamines, and AVP occurs in conditions such as hemorrhage, volume depletion, general anesthesia, and cardiac failure. While serving to maintain the systemic blood pressure, these hormones constrict mesangial cells and glomerular arterioles. Fortunately, their enhancement of renal PG release locally opposes their constrictor effects. The vasodilatory action of PGs on the afferent arteriole serves to maintain renal perfusion, whereas their relaxant effects on mesangial cells maintains the effective surface area for filtration.[198] Inhibition of PG generation in these circumstances is associated with a dramatic fall in RBF and GFR.[208-210] Vasodilator PGs, in particular PGI2, may also counteract the vasoconstrictor responses to calcium in man.[211]

In addition to modulating the effects of vasoconstrictors, endogenous PGs mediate the actions of some vasodilator agents. These include a role for PGI2 in mediating the vasorelaxant actions of dopamine[212] and magnesium[213] in humans, and of hydralazine[214] and EGF[215] in dogs. PGs may also mediate the renal vasodilatory response to a protein meal in humans.[216] Conversely, TXA2, synthesis of which is increased in experimental glomerular immune injury and ureteral obstruction, may cause glomerular contraction.[159,217]

7.10.11.4 Role of PGs in Renal Injury Disease States

PGs, through their vasodilator effects, play a salutary role in maintaining RBF and GFR in several prerenal conditions, such as hemorrhage, septic shock, and low cardiac output states. Studies in patients with congestive heart failure have confirmed that enhanced PG synthesis is crucial in protecting kidneys from the effects of elevated vasoconstrictor levels in these patients.[218] Renal artery stenosis is another condition associated with increased ipsilateral renal PG secretion[219] that may locally act to enhance renal perfusion. Administration of COX inhibitors in all these settings with renal hypoperfusion is associated with adverse effects on RBF and GFR.[219,220]

With regard to intrinsic renal diseases, COX products have been implicated in modulating and/or mediating renal injury. Following experimental reduction of renal mass, glomerular synthesis and urinary excretion of prostanoids per remaining nephron increases severalfold and likely contributes to the compensatory hypertrophy which follows renal ablation.[221-224] In this setting, nonselective inhibition of COX activity reduces nephron perfusion and K_f, implying a predominantly salutary role for vasodilator PGs.[222,223] In contrast, selective inhibition of TXA2 synthesis is associated with an increase in GFR, lessening of proteinuria, and preservation of renal histology[223,224] in renal ablation.

Enhanced TXA2 production has been implicated in the pathophysiology of the intense vasoconstriction which characterizes the post-obstructed kidney.[225,226] and in mediating the drop in RBF and GFR that occurs in the early phase of nephrotoxic serum nephritis.[159,160] In patients with lupus nephritis an inverse relation between TXA2 biosynthesis and GFR has been demonstrated.[227] In this setting, improvement of renal function was seen following short-term therapy with a TX receptor antagonist, but not with aspirin.[227,228] In addition, administration of TXA2 synthesis inhibitors or receptor antagonists has been associated with improved renal function in experimental animals with allograft rejection and cyclosporine toxicity.[198]

The role of COX products in mediating diabetic nephropathy remains controversial. Vasodilator PGs may contribute to the hyperfiltration that occurs in early stages of diabetic nephropathy, whereas TXA2 may play a role in the subsequent development of albuminuria and basement membrane changes.[229] A role for decreased PGI2 synthesis in type IV RTA associated with diabetes mellitus has also been suggested.[230] Diminished vasodilator renal PG synthesis has also been implicated in the pathogenesis of the severe sodium retention that occurs in patients with the hepatorenal syndrome.[231] Pregnancy is associated with increased glomerular synthesis and urinary excretion of PGE2, PGF2α, and PGI2.[232] Augmented renal vasodilator PG production does not appear to regulate GFR and RBF in normal pregnancy; however, diminished synthesis of PGI2 has been demonstrated in humans[233] and animal models[234] with pregnancy-induced hypertension. A beneficial effect of reducing TXA2 generation, while preserving PGI2 synthesis, by low-dose $(60–100\,mg\,d^{-1})$ aspirin therapy has been demonstrated in patients at risk for pregnancy-induced hypertension.[235,236] In patients with hypertension, COX inhibition with NSAIDs is associated with increased salt retention and resistance to the diuretic action of thiazides and furosemide.[237,238] Short-term use of some NSAIDs was found to increase the mean arterial pressure of hypertensive patients.[239] On the other hand, attempts to treat hypertension with PG analogues have generally been disappointing.[240]

Finally, chronic inhibition of COX by regular use of NSAIDs leads to gastrointestinal toxicity and may increase the risk for chronic renal disease especially in older patients and those with heart disease.[241,242] Selective COX-2 inhibitors have been developed and have been shown to spare gastric PG production. These nontraditional COX-2-selective anti-inflammatory agents may represent a significant advance for the treatment of acute and chronic inflammatory disorders.[243,244]

7.10.12 LIPOXYGENASE PRODUCTS

7.10.12.1 Biosynthesis and Metabolism

Enzymatic lipoxygenation of arachidonic acid leads to the generation of leukotrienes (LTs), lipoxins (LXs), and hydroxyeicosatetraenoic acids (HETEs).[245–247] Formation of these compounds is initiated by 5-, 12-, or 15-lipoxygenase (LO) whereby a hydroperoxy group is introduced onto arachidonic acid at C-5, C-12, or C-15, respectively, to yield the corresponding 5-, 12-, or 15-hydroperoxyetraenoic acid (HPETE). HPETEs are unstable compounds that get transformed into the corresponding 5-, 12-, and 15-HETE, which in turn undergo enzymatic modification leading to the generation of the various LTs and LXs. The 5-LO pathway is a major route of arachidonic acid metabolism in the polymorphonuclear (PMN) cells and macrophages leading to the formation of 5-HETE and LTs.[245–247] 5-LO requires activation by a cell membrane bound protein called the 5-lipoxygenase activating protein (FLAP).[248] LTA4 (leukotriene A_4) is an early pivotal intermediate in the 5-LO pathway whose metabolism leads to the production of the LT series of metabolites.[245,247] Formation of LTB4 requires LTA4 hydrolase activity, whereas generation of the peptidyl-leukotrienes (LTC4, -D4, and -E4) requires the enzymatic action of glutathione-*S*-transferase.[245] Unlike 5-LO, which is largely restricted to cells of myeloid lineage, these enzymes are widely distributed among different cell types.[245–247,249] The 15-LO enzyme catalyzes the production of 15-HETE, and initiates another major pathway of arachidonic acid metabolism in leukocytes.[245–247] In activated neutrophils and macrophages, sequential lipoxygenation of arachidonic acid at carbons C-15 and C-5 yields trihydroxy derivatives, the LXs.[245,246] LX synthesis can also occur in other cells, such as mesangial cells and platelets, by uptake of leukocyte-generated LTA4, and its transformation to LXs by either 15- or 12-LO.[247] The main LXs derived from 15-HETE are designated LXA4, LXB4 and 7-*cis*-11-*trans*-LXA4.[247] In the kidney, production of lipoxygenase products is largely by infiltrating leukocytes or resident cells of macrophage/monocyte origin,[6] but intrinsic renal cells are capable of generating LTs and LXs either directly or through transcellular metabolism of intermediates.[250–253]

7.10.12.2 Biologic Effects of Lipoxygenase Products

The LTs are potent proinflammatory molecules.[245] LTB4 has minimal spasmogenic properties, but is the most potent chemotactic substance yet described for polymorphonuclear cells, and promotes their activation and adhesion to the endothelium.[245] It has no significant effects on renal hemodynamics in normal animals, but amplifies glomerular inflammation and proteinuria in animals with glomerulonephritic injury.[254] The peptidyl-LTs contract vascular, pulmonary, and gastrointestinal smooth muscle, and increase vascular permeability to macromolecules.[245] LTC4 and LTD4 exert potent effects on glomerular hemodynamics. In rats, systemic administration of LTC4 leads to reduction in RBF and GFR.[255] Similarly, infusion of either LTC4 or LTD4 in the isolated perfused kidney results in dramatic increase in renal vascular resistance and reduction in GFR.[256] LTD4 mediates these effects by causing a significant increase in efferent arteriolar resistance (R_e) leading to a fall in glomerular plasma flow rate (Q_A), and a rise in glomerular capillary hydraulic pressure (P_{GC}).[257] In addition, it markedly reduces the glomerular capillary ultrafiltration coefficient (K_f) and, therefore, its overall effect is to decrease single nephron filtration rate (SNGFR).[257] LTC4 and LTD4 contract mesangial cells,[258,259] and LTD4 stimulates neutrophil adhesion to these cells.[260] In both rats and humans, specific mesangial cell LTD4 receptors have been identified.[261,262] Intracelluar signaling for LTD4 in these cells involves receptor-activated PIP2 hydrolysis, release of inositol phosphates, and increased intracellular calcium concentrations.[261,262]

15-S-HETE and LXA4 antagonize some of the actions of leukotrienes.[245,247,263] 15-S-HETE decreases LTB4 generation by leukocytes, antagonizes neutrophil chemotaxis to LTB4, and suppresses leukocyte activation in response to ionophore or other activators.[264-267] These effects likely result from the incorporation and storage of 15-S-HETE in the phosphatidyl inositol fraction of membrane lipids with subsequent release and generation of structurally altered second messenger diacylglycerol.[264,266] LXA4 attenuates LTB4-induced neutrophil chemotaxis, and inhibits natural killer cell cytotoxicity.[245,248] The effects of LXA4 are mediated primarily by functional high-affinity LXA4 receptors.[268] In rat glomerular mesangial cells, LXA4 competes with LTD4 at a common receptor whereby LXA4-mediates partial agonist/antagonist effects.[269] Different lipoxins display distinct effects on renal hemodynamics.[247,263,270] In the rat, LXA4 causes a selective decrease in afferent arteriolar resistance (R_a), thereby increasing RBF, P_{GC}, and GFR.[270] The LXA4-induced increase in GFR, however, is partially offset by its mild effect in decreasing K_f.[269,270] The vasodilator actions of LXA4 are mediated by prostaglandins.[270] LXB4 and 7-cis-11-trans-LXA4 display vasoconstrictive effects on renal hemodynamics in the rat that are independent of cyclooxygenase activity.[270]

7.10.12.3 Role of Lipoxygenase Products in Kidney Injury and Disease

LTs are increasingly recognized as major mediators of glomerular hemodynamic and structural deterioration during the early phases of experimental glomerulonephritis.[263,271,272] Increased glomerular generation of LTB4 and peptidyl-LTs have been demonstrated in several models of glomerular injury.[263,271,272] LTB4 likely worsens glomerular injury by augmenting leukocyte recruitment and activation, and the peptidyl LTs by depressing K_f and GFR.[254-260] Selective blockade of the 5-LO pathway in the course of glomerular injury is associated with significant amelioration of the deterioration of renal hemodynamic and structural parameters.[273,274] In addition, dietary deprivation of essential fatty acids, which results in arachidonic acid and eicosanoid deficiency, confers protection against the histopathologic and the functional consequences of immune-initiated injury in the glomerulus.[275] LTs are likely involved in the pathophysiology of human glomerulonephritis. In this regard, 5-LO and FLAP mRNA expression have been detected in kidney biopsy specimens of some patients with IgA nephropathy (IgAN) and mesangial proliferative glomerulonephritis (MPGN), and were associated with clinically deteriorating renal status.[276] Also, urinary LTE4 levels have been found to be elevated in patients with active systemic lupus erythematosus.[277] A pathophysiologic role for leukotrienes has also been described in experimental acute allograft rejection, cyclosporine toxicity, and acute ureteral obstruction.[272] LXA4 and 15-S-HETE are also generated during experimental glomerular injury and may exert salutary effects on glomerular function by antagonizing the proinflammatory actions of LTs.[263-267] LT-deficient mice have been created by targeted disruption of the 5-LO gene.[278] In comparison with normal controls, initial studies with these 5-LO transgenic mice show significant differences in their inflammatory responses to some injurious stimuli.[278] Further studies on these mice will help elucidate the contribution of 5-LO products to the pathophysiology of several inflammatory disease states. Additionally, the availability of safe 5-LO pathway inhibitors, as well as leukotriene antagonists, will be useful to test the potential benefit of blocking the effects of 5-LO products in the course of human glomerulonephritis and other inflammatory diseases.[249]

7.10.13 CYTOCHROME P450 PRODUCTS

The microsomal NADPH-dependent cytochrome P450 (cyt P450) enzyme system metabolizes arachidonic acid to a wide variety of oxygenated products. Three types of reactions can take place: (i) allylic oxidation leading to the formation of HETEs; (ii) epoxidation resulting in the formation of epoxyeicosatrienoic acids (EETs) or epoxides which can be hydrolyzed to their respective dihydroxyeicosatrienoic acids or vicinal diols; and (iii) monooxygenation yielding ω and ω-1 hydroxylated acids.[279–281] The kidney is a rich source of cyt P450 metabolites; cyt P450 activity resides mostly in the proximal tubule, TALH (thick ascending loop of Henle), and CCT (cortical collecting tubule).[280–284] Cyt P450 enzymes constitute a multigene super family. Evidence suggests that the major arachidonate epoxygenase in the kidney is a member of the cyt P450 2C family, whereas ω/ω-1 hydroxylation appears to be primarily mediated by members of the cyt P450 4A gene family.[285–288] Activity of the cyt P450 is modulated by hormones, such as EGF, AVP, angiotensin II, calcitonin, and corticosteroids.[282,289]

In the kidney, arachidonate cyt P450 metabolites have been assigned several biological properties that include vasoactivity,[290–294] effects on tubular water and ion transport,[281,295–297] and local modulation/mediation of the activity of the renin–angiotensin system and other peptide hormones.[296–298] The vasoactivity of the various P450 metabolites varies among different species. Both vasoconstrictor and vasodilator effects have been described.[3,22] For example, 5,6-EET and 20-HETE are vasodilatory in the rabbit kidney, but are constrictors of rat blood vessels.[291,294] The effects of the most potent vasoactive cyt P450 metabolites, 5,6-EET and 20-HETE, are COX-dependent.[223,281] In fact the cyt P450 and the COX enzyme systems appear to interact transcellularly. In this manner the COX pathway may modify the products of cyt P450 and vice versa.[281] Such transcellular metabolism of 5,6-EET and 20-HETE by the COX pathway appears to be essential for expression of their

vasoactive effects.[281,293] With regard to their effects on salt and water balance, the cyt P450 metabolites are capable of inhibiting both ion transport in the medullary TALH,[292,295] and the hydraulic conductivity of AVP in the cortical collecting duct.[291,296] Intrarenal administration of 20-HETE, one of the major cyt P450 metabolites formed in the medullary TALH cells, resulted in natriuretic and diuretic responses in rats.[297] In addition, all EETs and their corresponding DHT (dihydrotetraene) metabolites acted as inhibitors of the hydroosmotic effect of AVP in rabbit cortical collecting tubules.[28] With regard to their interaction with the renin–angiotensin system, the cyt P450 metabolites, 12(R)HETE and 14,15-EET inhibit renin release by the rat kidney.[298] On the other hand, angiotensin II stimulates the release of 5,6-EET from rat isolated proximal tubules which mediates angiotensin II-induced increases in cytosolic calcium[299] and sodium absorption.[300]

Overall, it appears likely that the cyt P450 pathway may have an important role in modulating renal function both in health and disease. In this regard dietary sodium loading has been demonstrated to selectively induce the renal isoforms of cyt P450 and increase renal excretion of cyt P450 metabolites.[301] Induction of the cyt P450 enzyme system has also been described following unilateral nephrectomy[302] and experimental diabetes mellitus.[302] The induction of cyt P450 in all these conditions may serve as a physiologic adaptation aimed at increasing salt excretion by modulating epithelial transport, as well as affecting renal hemodynamics. Cyt P450 metabolism is enhanced in women with pregnancy-induced hypertension.[303] The effects of cyt P450 products on blood pressure differ among the various metabolites. Studies show that ω/ω-1 hydroxylation products may play a key role in the genesis of hypertension in spontaneously hypertensive rats.[304,305] Conversely, arachidonate cyt P450 epoxides are described as antihypertensive because their selective inhibition induces hypertension in rats fed high-salt diets.[306]

7.10.14 ENDOTHELIN

Endothelin, originally isolated from vascular endothelial cells, is the most potent and long lasting vasoconstrictor yet found.[307] The role of endothelin, however, extends beyond being a vasoconstrictor to include effects on cell growth and proliferation, ion transport, eicosanoid synthesis, renin and ANP release, and a host of other actions.[308–310] The kidney is an important site of endothelin production and expresses a high density of endothelin receptors.[310]

Endothelin may therefore act in an autocrine and paracrine manner to influence renal hemodynamics, tubular function, and mesangial cell biology.

7.10.14.1 Biochemistry, Synthesis, and Receptor Biology

The term endothelin (ET) refers to a family of homologous 21-amino acid vasoconstrictor peptides found in three distinct isoforms: ET-1, ET-2, and ET-3. In humans each isoform is encoded by a separate gene.[311,312] The initial ET peptide translation product is a large (approximately 200 amino acids) isopeptide-specific prohormone named preproET. Post-translational processing of preproET to mature ET requires two steps: the first involves proteolytic cleavage of the preproET by dibasic-pair-specific endopeptidases on Lys–Arg and Arg–Arg pairs, which respectively flank the N- and C- terminals of the preproET molecule, to yield an intermediate 38- or 39-amino acid proET polypeptide. The subsequent step is accomplished by proteolytic cleavage of proET between Trp21–Val22 by a putative "endothelin converting enzyme."[313,314] This enzyme, which is likely to be an important target for pharmacological antagonists of ET secretion, has been difficult to characterize and has not been cloned to date. ET-1 differs from ET-2 and ET-3 by two and six amino acids, respectively, with the greatest variation occurring at amino acid residues 4–7.[312,315] All ET isopeptides have a hairpin loop configuration structure imparted by two intrachain disulfide bonds bridging amino acid residues 1 to 15 and 3 to 11, the reduction of which leads to a twofold loss of biological activity.[316] The three ET isoforms are highly homologous in amino acid sequences and tertiary structure to certain scorpion and snake venoms, the sarafotoxins, suggesting common genetic evolutionary origins.[317] While all isoforms of ET are potent vasoconstrictors, there are significant cell- and tissue-specific differences in secretion of, and biologic responses to different isoforms.[311,318]

Initial studies identified ET based on its release from large vessel endothelial cells. Since then ET immunoreactivity has been detected in the kidney, spleen, skeletal muscle, and lung.[319] ET production is not confined to vascular endothelial cells, but can be synthesized by certain nonendothelial cells.[320–323] In humans, neurons of the spinal cord and dorsal root ganglia,[321] cultured vascular smooth muscle,[322] and breast epithelial cells[323] produce ET. In the kidney, the arcuate arteries, veins, glomerular arterioles, and capillaries are a rich source of ET.[324] In the glomerulus, there is evidence for ET secretion by mesangial, endothelial, and epithelial cells.[310] In the rest of the nephron, the inner medullary collecting duct (IMCD) has been demonstrated to be a major site of ET-1 and ET-3 production.[310,325–327]

Normally, blood vessels produce very little ET, and the normal circulating level of ET is extremely low.[315] Secretion of ET by endothelial cells is controlled at the level of transcription, and these cells do not store ET for future release.[313,328] ET peptide secretion is upregulated by various humoral mediators, such as thrombin, bradykinin, insulin, angiotensin II, arginine vasopressin, endotoxin, interleukin-1, TGF-β, and tumor necrosis factor.[329,330] These mediators may be responsible for the increase in ET observed in various pathophysiologic states. Hypoxia is also an important stimulus for ET production.[331] In the kidney, increasing urine osmolarity serves as a stimulus for tubular production of ET.[332,333] On the other hand, nitric oxide, atrial natriuretic factor and prostacyclin exert inhibitory influences on ET synthesis and release.[334–336]

Endothelin acts via specific G protein-coupled receptors that have been identified in a variety of tissues.[337] Three receptor subtypes have been cloned: the ET-A receptor binds ET-1 and ET-2 with high affinity, the ET-B receptor recognizes all three ET isoforms with equal affinity, and the recently cloned ET-C receptor binds ET-3 selectively.[310,337–339] Both ET-A and ET-B receptors are expressed on vascular smooth muscle and mediate vasoconstriction. The kidney expresses abundant mRNA transcripts for ET-A and ET-B receptors.[310,337,340,341] Expression of ET receptors is especially prominent in the renal artery, glomerular arterioles, endothelium, and mesangium, vasa recta bundles, and collecting duct.[308,310,337] Both receptor subtypes are expressed in the glomerulus. Vascular smooth muscle cells of the arcuate arteries and the renal medullary interstitial cells display ET-A receptors. Epithelial cells of the cortical, inner medullary, and outer medullary collecting duct have ET-B receptors.[310]

Activation of the ET receptor on vascular smooth muscle cells leads to (i) activation of phospholiapse C with formation of inositol triphosphate and diacylglycerol, and (ii) elevation of free intracellular calcium, which is thought to mediate the contractile response to ET.[342–344] ET increases free intracellular calcium by release of Ca^{2+} from intracellular stores via phosphoinositide-mediated mechanisms, and/or by influx of extracellular calcium due to activation of cell membrane voltage-dependent dihydropyridine-sensitive calcium channels.[343,344] Calcium channel blockers inhibit ET-induced vasoconstriction in smaller blood vessels where the contribution of extracellular calcium is important, such as intramyocardial coronary arteries.[345] Endothelial cells can express ET-B receptors linked to formation of nitric oxide and prostacyclin, and mediate endothelium-dependent vasorelaxation.[346] In nephron segments and other renal structures, ET mediates its effects via a multiplicity of intracellular signal transduction pathways that involve phospholipase activation, tyrosine pophorylation of proteins, and elevation of intracellular free calcium.[310]

7.10.14.2 Biological Effects of Endothelin in the Kidney

Endothelin is a potent renal vasoconstrictor, as much as thirtyfold more potent in this regard than angiotensin II.[307–310,343–347] In the isolated perfused kidney, ET-1 administration reduces GFR and causes a dose-dependent increase in renal vascular resistance. In the intact animal, systemic ET infusion induces a decline in cortical and medullary blood flow, GFR, and urine volume.[308–310,343,347] The direct effects of ET-1 on preglomerular and postglomerular resistances are quantitatively similar at lower doses so the glomerular transcapillary hydraulic pressure and GFR are maintained.[347,348] However, at higher doses, a greater increase in preglomerular resistance occurs which, in addition to a drop in glomerular capillary K_f, leads to a decline in GFR.[343,347,348] Both ET-1 and ET-2 are equally potent in constricting microdissected glomerular arterioles, whereas ET-3 is considerably less potent.[349] Dihydropyridines inhibit ET-mediated vasoconstriction exclusively in the afferent arterioles.[349] Renal hemodynamics may also be influenced by indirect effects of ET, such as modulation of arachidonic acid metabolism and renin release.[348,350] Local generation of prostaglandins, such as prostaglandin F2α, may mediate some of the vasoconstrictor effects of ET.[350] ET-1 directly inhibits renin release, but the net renin secretory response *in vivo* varies with the ET-1 dose, as well as with the state of activation of intrarenal baroreceptors and the macula densa-mediated pathway.[348]

Despite compromise of RBF and GFR, infusion of nonpressor doses of ET in experimental animals is associated with an increase in urinary flow and sodium excretion.[308,309,343] In addition, studies in the isolated perfused kidney

have shown that ET increases sodium excretion despite a dramatic decline in GFR.[309] These effects on sodium and water balance are largely due to the ability of ET to reduce Na-K-ATPase activity and reversibly inhibit AVP-stimulated cAMP generation and water transport in the IMCD.[309,351] Current evidence supports a physiologic role for ET in regulating sodium and water transport in the IMCD in experimental animals. In this regard, increasing the osmolality of the culture medium of IMCD cells with either NaCl or mannitol, but not urea, causes a dose-dependent reduction in ET-1 release.[332] Moreover, ET-1 mRNA is lower in the medulla of dehydrated, as compared to salt-loaded rats, supporting the hypothesis that extracellular osmolality may physiologically regulate ET production by IMCD cells.[332] Since the human inner medulla contains a high concentration of ET receptors,[340] coupled with the demonstration of ET production by human IMCD cells,[326] it is conceivable that a similar physiologic role for intrarenal ET may also be operative in humans. ET also affects sodium balance indirectly by stimulating release of ANP.[352] ET-induced increases in urinary sodium excretion in experimental animals are markedly blunted following pretreatment with ANP antibodies.[353]

ET is well known to induce contraction of mesangial cells in culture, and results of micropuncture experiments demonstrate that ET-1 directly reduces the coefficient of ultrafiltration.[343] ET is also recognized as a growth factor with mitogenic effects on mesangial cells in culture, inducing changes in mesangial cell phenotype and gene expression.[308–310,337]

7.10.14.3 Pathophysiological Significance of Endothelin in Kidney Disease

Declining renal function is associated with increase in plasma ET levels.[354] Hemodialysis (HD) patients have higher ET levels in comparison with patients on peritoneal dialysis or undialyzed uremic subjects.[355] Erythropoietin administration as well as acute volume contraction during HD may contribute to the elevation of ET-1 level in hemodialysis patients.[356,357] Several disease states such as hypertension, atherosclerosis, cardiogenic shock, congestive heart failure, cyclosporine administration, and endotoxemia have also been associated, in some but not all studies,[354] with increased plasma ET levels.[308–310,358,359] While the significance of elevated plasma levels of a primarily autocrine/paracrine hormone, such as ET, remains questionable at present,

ET-secreting hemangioendotheliomas produce marked hypertension.[360]

A pathophysiologic role for ET has been reported in several conditions affecting the kidney.[309,310,359–371] Urinary ET-1 excretion increases in patients with several forms of chronic progressive glomerulopathies.[361] In experimental animals with surgical reduction of renal mass, ET-1 gene expression increases in parallel with proteinuria and glomerulosclerosis.[362] Renal ischemia is a potent stimulus for ET-1 production.[363] In models of ischemic renal injury ET-neutralizing antibodies, as well as ET-A receptor antagonists, attenuate the decline in renal functional and structural parameters.[364,365] Radiocontrast stimulates ET synthesis by endothelial cells in culture, and increases plasma and urinary ET levels in animals.[310,366,367] Cyclosporine infusion causes a transient rise in ET plasma levels in animals.[368] ET antiserum and ET receptor antagonism partially ameliorate the renal hypoperfusion and hypofiltration that follows intravenous cyclosporine administration.[369,370] Patients with the hepatorenal syndrome have significantly elevated plasma ET-1 and ET-3 levels that may in part be due to release of ET by the kidney.[371] Selective ET-A and combined ET-A and ET-B receptor antagonists have been developed, and most likely will be helpful to characterize the therapeutic potential of antagonizing ET in various disease states.[345]

In summary, ET is a potent vasoconstrictor peptide. In the kidney, it reduces RBF and GFR, contracts mesangial cells, and may function as a paracrine-autocrine factor in modulation of sodium and water balance. It is a potential mediator of growth and proliferative changes within the kidney. It is thought to play a pathophysiological role in a number of kidney diseases. ET receptor antagonists may prove to be beneficial in certain conditions.

7.10.15 NITRIC OXIDE

In the early 1980s, Furchgott et al.[372] demonstrated that the relaxation of isolated rabbit aorta and other arteries induced by acetylcholine and other agonists for muscarinic receptors depended on the presence of endothelial cells in the preparations. They also showed that endothelium-dependent relaxation by acetylcholine results from the release of a labile humoral relaxing substance/substances, later termed endothelium-derived relaxing factor (EDRF). Within a few years of the discovery of endothelium-dependent relaxation by acetylcholine, many other vasodilators were found

to produce EDRF from endothelium of various vascular beds including renal artery.

7.10.15.1 Mechanism of Action of EDRF

While it is likely that there is more than one EDRF released by vascular endothelium, several studies suggest that one of the EDRFs is nitric oxide (NO).[373] NO is a potent vasodilator as are nitrovasodilator agents such as sodium nitroprusside, glyceryl trinitrate, and other organic nitrate esters that stimulate guanylate cyclase and have a very short half-life of 3–5 s similar to EDRF. Both EDRF and NO are protected from inactivation by superoxide dismutase, and both are inhibited by hemoglobin and methylene blue.[374] Palmer et al.[373] demonstrated that bradykinin-stimulated endothelial cells release NO in amounts sufficient to account for the biological effects of EDRF on vascular smooth muscle.

NO is formed by vascular endothelial cells from the terminal guanido nitrogen atom of the amino acid, L-arginine.[375] Although the steps involved in the synthesis of NO from L-arginine are not known, there is a evidence that a soluble NADPH-dependent enzyme is capable of the coversion.[376] The effect of this enzyme is inhibited by the L-arginine analogue NG (nitro gamma)-monomethyl L-arginine (L-NMMA),[377] an effect abolished by the simultaneous administration of L-arginine.[377] In well-oxygenated systems, NO is rapidly inactivated by oxidation to inorganic nitrite and nitrate in the presence of oxyhemoglobin, which reaction is blocked by superoxide dismutase. Due to this chemical lability, it is highly unlikely that NO functions as a circulating hormone. Instead, it likely diffuses from sites of local generation to immediately adjacent smooth muscle cell targets within vessel walls. The lipophilic NO readily permeates plasma membranes and binds to the heme group of soluble guanylate cyclase to cause enzyme activation and stimulation of cyclic guanosine monophosphate (cGMP) formation.

7.10.15.2 Renal Action of EDRF

Several investigators have provided evidence that EDRF/NO might be produced in renal endothelial cells and play a physiological role in the the regulation of renal hemodynamics and glomerular function.

Shultz et al.[378] reported that cGMP levels in rat glomerular mesangial cells coincubated with bovine aortic endothelial cells increased after bradykinin stimulation. This effect was potentiated by superoxide dismutase and inhibited by hemoglobin and L-NMMA, suggesting that NO is involved in this effect of bradykinin on endothelium and mesangial cells. They also showed that NO attenuated the effect of angiotensin II on mesangial cell contraction. In another study, Marsden *et al.*[379] showed that bovine glomerular mesangial cells coincubated with glomerular endothelial cell produce cGMP in response to bradykinin. This effect of bradykinin was blocked by methylene blue, suggesting that glomerular endothelial cells may modulate the function of adjacent mesangial cells, which possess vascular smooth muscle-like contractile ability, by releasing EDRF. These findings suggest that EDRF/NO may participate in the regulation of glomerular filtration via modulating the contraction and relaxation of mesangial cells. Marsden and Ballerman have provided evidence for release of EDRF from mesangial cells in reponse to tumour necrosis factor, a macrophage-derived proinflammatory peptide, suggesting potential autocrine function.[379] King *et al.*[380] reported that L-NMMA, infused systemically into rats, caused an increase in blood pressure and decrease in renal plasma flow rate. They also demonstrated that L-NMMA attenuated the increase in glomerular filtration rate (hyperfiltration) induced by amino acid infusion. From these results, it is suggested that NO influences basal systemic and renal vascular tone, and contributes to the renal hemodynamic changes induced by amino acid infusion.

Tolins *et al.*[381] demonstrated that systemically infused acetylcholine in rats caused a significant decrease in blood pressure and an increase in urinary excretion of cGMP and renal plasma flow rate, while during L-NMMA infusion these changes caused by acetylcholine were prevented. Loutzenhiser *et al.*[382] also showed that hemoglobin, which presumably inactivates NO, attenuated the acetylcholine-induced vasodilation in both afferent and efferent arterioles of isolated hydronephrotic rat kidneys. These observations suggest that, in addition to mesangial cells, EDRF/NO may be involved in the modulation of renal vascular tone, particularly in reponse to certain vasodilator agents, such as acetylcholine.

In addition to modulating baseline vascular tone, NO may partially ameliorate the vasoconstrictor actions of other locally or systemically generated hormones/autacoids, including ET. Boulanger *et al.*[383] demonstrated that the endothelin production by porcine aorta with intact endothelium during stimulation with thrombin was enhanced by L-NMMA and by methylene blue, and inhibited by superoxide dismutase and by 8-bromo-cGMP. They con-cluded that endothelium-derived NO released during stimulation with thrombin inhibited the production of endothelin via a cGMP-dependent pathway. Vidal *et al.*[384] have provided evidence suggesting that EDRF inhibits renin release from canine renal cortical slices. These findings indicate that EDRF may regulate the local circulation not only by relaxing vascular smooth muscle cells directly, but also by inhibiting the production of vasoconstrictive substances.

7.10.15.3 Studies at the Single Nephron Level

NO has been shown to participate in the regulation of renal hemodynamics, and the administration of a competitive inhibitor of NO production, L-NMMA, to normal rats causes dramatic glomerular hemodynamic changes including reduced single nephron plasma flow, augmented afferent and efferent arteriolar resistances, decreased ultrafiltration coefficient, and increased P_{GC}.[385–388] Chronic oral supplementation with an L-arginine inhibitor in rats caused proteinuria, increased P_{GC}, and glomerular hemodynamic changes as described above.[387] These observations suggest that NO might be an important regulator of P_{GC} and that its dysregulation might be involved in the development of glomerular sclerosis through increases in P_{GC}.

Dietary supplementation with L-arginine ameliorates the progression of renal disease in rats with subtotal nephrectomy.[388] Research suggests that L-arginine supplementation prevents the progression of glomerular sclerosis in subtotally nephrectomized rats, at least in part due to its inhibitory effects on the development of glomerular hypertension.[389]

Although endothelial cells are widely viewed as the major source of NO, emerging evidence demonstrates that NO is also produced in various other tissues and cells, including macrophages,[390] hepatocytes,[391] Kupffer cells,[392] fibloblasts,[393] vascular smooth muscle cells,[394,395] and mesangial cells.[396] In contrast to the NO synthase in endothelial cells, which is Ca^{2+}/calmodulin-dependent and constitutively expressed, some of these cells have been shown to possess NO synthases which are distinct from the constitutive enzyme in endothelial cells.[397] This type of enzyme is Ca^{2+}/calmodulin-independent and not constitutively expressed but induced by several stimuli including lipopolysaccharide, tumor necrosis factor, interleukin-1, or interferon-c.[397] The inducible NO synthase has also been identified in endothelial cells.[398,399]

In view of previous studies demonstrating the release of NO by ET,[382] the potential role of

ET-mediated NO release in causing systemic vasorelaxation and hypotension in the presence of manidipine, a calcium channel blocker, was assessed.[400] The infusion of ET in manidipine pretreated rats in a separate group of animals also receiving the NO-synthase inhibitor L-NMMA was repeated. Despite all the alterations evoked by L-NMMA in systemic and renal parameters during the premanidipine-infusion period, inhibition of NO synthesis had no significant effect on the systemic and renal responses to manidipine administration, suggesting no major role for NO in mediating manidipine-induced changes in hemodynamics. However, the existence of L-NMMA totally prevented ET-induced hypotension in these animals and, furthermore, attenuated the associated deterioration of renal hemodynamics. In addition, administration of L-arginine 10 min after L-NMMA was given, partially restored the ET-induced exaggeration of the manidipine-mediated fall in MAP and the accompanying sharp declines in GFR and RBF. Taken together, these observations indicated a major role for NO in mediating the ET-induced hypotension under Ca^{2+}-blockade.

7.10.16 REFERENCES

1. V. F. Norwood and R. A. Gomez, 'Bridging the gap between physiology and molecular biology: new approaches to perpetual questions.' *Am. J. Physiol.*, 1994, **267**, R865–R878.
2. J. H. Exton, 'Phosphoinositide phospholipases and G proteins in hormone action.' *Annu. Rev. Physiol.*, 1994, **56**, 349–369.
3. O. H. Petersen, C. C. Petersen and H. Kasai, 'Calcium and hormone action.' *Annu. Rev. Physiol.*, 1994, **56**, 297–319.
4. M. E. Gnegy, 'Calmodulin in neurotransmitter and hormone action.' *Annu. Rev. Pharmacol. Toxicol.*, 1993, **33**, 45–70.
5. R. R. Reichel and S. T. Jacob, 'Control of gene expression by lipophilic hormones.' *FASEB J.*, 1993, **7**, 427–436.
6. M. S. Goligorsky, in 'Contemporary Issues in Nephrology: Hormones, Autacoids, and the Kidney,' eds. S. Goldfarb and F. N. Ziyadeh, Churchill Livingstone, New York, 1991, vol. 23, pp. 1–26.
7. D. A. Maddox and B. M. Brenner, in 'The Kidney,' 4th edn., eds. B. M. Brenner and F. C. Rector, Saunders, Philadelphia, PA, 1991, pp. 205–244.
8. B. R. Dunn, I. Ichikawa, J. M. Pfeffer *et al.*, 'Renal and systemic hemodynamic effects of synthetic atrial natriuretic peptide in the anesthetized rat.' *Circ. Res.*, 1986, **59**, 237–246.
9. T. A. Fried, R. N. McCoy, R. W. Osgood *et al.*, 'Effect of atriopeptin II on determinants of glomerular filtration rate in the *in vitro* perfused dog glomerulus.' *Am. J. Physiol.*, 1986, **250**, F1119–F1122.
10. B. D. Myers, W. M. Deen and B. M. Brenner *et al.*, 'Effects of norepinephrine and angiotensin II on the determinants of glomerular ultrafiltration and proximal fluid reabsorption in the rat.' *Circ. Res.*, 1975, **37**, 101–110.
11. C. E. Hura and R. T. Kunau, Jr., 'Angiotensin II-stimulated prostaglandin production by canine renal afferent arterioles.' *Am. J. Physiol.*, 1988, **254**, F734–F738.
12. H. W. Harris, Jr., K. Strange and M. L. Zeidel, 'Current understanding of the cellular biology and molecular structure of the antidiuretic hormone-stimulated water transport pathway.' *J. Clin. Invest.*, 1991, **88**, 1–8.
13. J. A. Majzoub, in 'Vasopressin,' ed. R. W. Schrier, Raven Press, New York, 1985, pp. 465–474.
14. G. L. Robertson, in 'The Kidney: Physiology and Pathophysiology,' 2nd edn., eds. D. W. Seldin, G. H. Giebisch, Raven Press, New York, 1992, pp. 1595–1613.
15. M. C. Carmichael and R. Kumar, 'Molecular biology of vasopressin receptors.' *Semin. Nephrol.*, 1994, **14**, 341–348.
16. J. M. Launay, D. Vittet, M. Vidaud *et al.*, 'V$_{1a}$-vasopressin specific receptors on human platelets: potentiation by ADP and epinephrine and evidence for homologous downregulation.' *Thromb. Res.*, 1987, **45**, 323–331.
17. V. Gopalakrishnan, Y. J. Xu, P. V. Sulakhe *et al.*, 'Vasopressin (V$_1$) receptor characteristics in rat aortic smooth muscle cells.' *Am. J. Physiol.*, 1991, **261**, H1927–H1936.
18. S. Jard, R. C. Gaillard, G. Guillon *et al.*, 'Vasopressin antagonists allow demonstration of a novel type of vasopressin receptor in the rat adenohypophysis.' *Mol. Pharmacol.*, 1986, **30**, 171–177.
19. M. A. Knepper, S. Nielsen, L. Chou *et al.*, 'Mechanisms of vasopressin action in the renal collecting duct.' *Semin. Nephrol.*, 1994, **14**, 302–321.
20. F. D. Grant, J. Reventos, J. W. Gordon *et al.*, 'Expression of the rat arginine vasopressin gene in transgenic mice.' *Mol. Endocrinol.*, 1993, **7**, 659–667.
21. P. Gines, W. T. Abraham and R. W. Schrier, 'Vasopressin in pathophysiological states.' *Semin. Nephrol.*, 1994, **14**, 384–397.
22. E. T. Cunningham, Jr. and P. E. Sawchenko, 'Reflex control of magnocellular vasopressin and oxytocin secretion.' *Trends Neurosci.*, 1991, **14**, 406–411.
23. K. K. Griendling, T. J. Murphy and R. W. Alexander, 'Molecular biology of the renin–angiotensin system.' *Circulation*, 1993, **87**, 1816–1828.
24. R. E. Pratt, J. E. Carleton, J. P. Richie *et al.*, 'Human renin biosynthesis and secretion in normal and ischemic kidneys.' *Proc. Natl. Acad. Sci. USA*, 1987, **84**, 7837–7840.
25. E. Hackenthal, M. Paul, D. Ganten *et al.*, 'Morphology, physiology, and molecular biology of renin secretion.' *Physiol. Rev.*, 1990, **70**, 1067–1116.
26. A. Kurtz, 'Intracellular control of renin release—an overview.' *Klin. Wochenschr.*, 1986, **64**, 838–846.
27. J. N. Lorenz, S. G. Greenberg and J. P. Briggs *et al.*, 'The macula densa mechanism for control of renin secretion.' *Semin. Nephrol.*, 1993, **13**, 531–542.
28. U. C. Kopp and G. F. DiBona, 'Neural regulation of renin secretion.' *Semin. Nephrol.*, 1993, **13**, 543–551.
29. A. Kurtz, R. Della Bruna, J. Pratz *et al.*, 'Rat juxtaglomerular cells are endowed with DA-1 dopamine receptors mediating renin release.' *J. Cardiovasc. Pharmacol.*, 1988, **12**, 658–663.
30. A. Kurtz, R. Della Bruna, J. Pfeilschifter *et al.*, 'Atrial natriuretic peptide inhibits renin release from isolated renal juxtaglomerular cells by a cGMP-mediated process.' *Proc. Natl. Acad. Sci. USA*, 1986, **83**, 4769–4773.
31. K. D. Burns, T. Homma and R. C. Harris, 'The intrarenal renin–angiotensin system.' *Semin.*

Nephrol., 1993, **13**, 13–30.

32. C. Larsson, P. Weber and E. Anggard, 'Arachidonic acid increases and indomethacin decreases plasma renin activity in the rabbit.' *Eur. J. Pharmacol.*, 1974, **28**, 391–394.

33. I. Antonipallai, J. L. Nadler, E. C. Robin *et al.*, 'The inhibitory role of 12- and 15-lipoxygenase products on renin release.' *Hypertension*, 1987, **10**, 61–66.

34. W. L. Heinrich, J. R. Falck and W. B. Campbell, 'Inhibition of renin release by 14,15-epoxyeicosatrienoic acid in renal cortical slices.' *Am. J. Physiol.*, 1990, **258**, E269–E274.

35. K. K. Griendling and R. W. Alexander, 'The angiotensin (AT1) receptor.' *Semin. Nephrol.*, 1993, **13**, 558–566.

36. K. E. Bernstein and R. W. Alexander, 'Counterpoint: molecular analysis of the angiotensin II receptor.' *Endocr. Rev.*, 1992, **13**, 381–386.

37. R. P. Forsyth, B. I. Hoffbrand and K. L. Melmon, 'Hemodynamic effects of angiotensin in normal and environmentally stressed monkeys.' *Circulation*, 1971, **44**, 119–129.

38. M. A. Gimbrone, Jr. and R. W. Alexander, 'Angiotensin II stimulation of prostaglandin production in cultured human vascular endothelium.' *Science*, 1975, **89**, 219–220.

39. A. Hassid and C. Williams, 'Vasoconstrictor-evoked prostaglandin synthesis in cultured vascular smooth muscle.' *Am. J. Physiol.*, 1983, **245**, C278–C282.

40. I. A. Reid, 'Actions of angiotensin II on the brain: mechanisms and physiologic role.' *Am. J. Physiol.*, 1984, **246**, F533–F543.

41. P. C. Wong, S. D. Hart and P. B. Timmermans, 'Effect of angiotensin II antagonism on canine renal sympathetic nerve function.' *Hypertension*, 1991, **17**, 1127–1134.

42. D. H. Smith, J. M. Neutel and M. A. Weber, 'Effects of angiotensin II on pressor responses to norepinephrine in humans.' *Life Sci.*, 1991, **48**, 2413–2421.

43. G. Aquitlera and E. T. Marusic, 'Role of the renin angiotensin system in the biosynthesis of aldosterone.' *Endocrinology*, 1971, **89**, 1524–1529.

44. J. T. Fitzsimmons, 'Angiotensin stimulation of the central nervous system.' *Rev. Physiol. Biochem. Pharmacol.*, 1980, **87**, 117–167.

45. J. P. Coghlan, P. J. Considine, D. A. Denton *et al.*, 'Sodium appetite in sheep induced by cerebral ventricular infusion of angiotensin: comparison with sodium deficiency.' *Science*, 1981, **214**, 195–197.

46. T. Unger, E. Badoer, D. Ganten *et al.*, 'Brain angiotensin: pathways and pharmacology.' *Circulation*, 1988, **75**, 140–154.

47. B. C. Berk, V. Vekshtein, H. M. Gordon *et al.*, 'Angiotensin II-stimulated protein synthesis in cultured vascular smooth muscle cells.' *Hypertension*, 1989, **13**, 305–314.

48. G. A. Stouffer and G. K. Owens, 'Angiotensin II-induced mitogenesis of spontaneously hypertensive rat-derived cultured smooth muscle cells is dependent on autocrine production of transforming growth factor-beta.' *Circ. Res.*, 1992, **70**, 820–828.

49. G. H. Gibbons, R. E. Pratt and V. J. Dzau, 'Vascular smooth muscle cell hypertrophy vs. hyperplasia. Autocrine transforming growth factor-β_1 expression determines growth response to angiotensin II.' *J. Clin. Invest.*, 1992, **90**, 456–461.

50. L. D. Dworkin, I. Ichikawa and B. M. Brenner, 'Hormonal modulation of glomerular function.' *Am. J. Physiol.*, 1983, **244**, F95–F104.

51. I. Ichikawa and R. C. Harris, 'Angiotensin actions in the kidney: renewed insight into the old hormone.' *Kidney Int.*, 1991, **40**, 583–596.

52. D. A. Ausiello, J. I. Kreisberg, C. Roy *et al.*, 'Contraction of cultured rat glomerular cells of apparent mesangial origin after stimulation with angiotensin II and arginine vasopressin.' *J. Clin. Invest.*, 1980, **65**, 754–760.

53. R. C. Blantz, K. S. Konnen and B. J. Tucker, 'Angiotensin II effects upon glomerular microcirculation and ultrafiltration coefficient of the rat.' *J. Clin. Invest.*, 1976, **57**, 419–434.

54. J. B. Foidart and P. Mahieu, 'Glomerular mesangial cell contractility *in vitro* is controlled by an angiotensin-prostaglandin balance.' *Mol. Cell. Endocrinol.*, 1986, **47**, 163–173.

55. G. Wolf and E. G. Neilson, 'Angiotensin II induces cellular hypertrophy in cultured murine proximal tubular cells.' *Am. J. Physiol.*, 1990, **259**, F768–F777.

56. A. Huwiler, S. Stabel, D. Fabbro *et al.*, 'Platelet-derived growth factor and angiotensein II stimulate the mitagen-activated protein kinase cascade in renal mesangial cells: comparison of hypertrophic and hyperplastic agonists.' *Biochem. J.*, 1995, **305**, 777–784.

57. J. J. Mullins, J. Peters and D. Ganten, 'Fulminant hypertension in transgenic rats harbouring the mouse *Ren-2* gene.' *Nature*, 1990, **344**, 351–354.

58. J. Wagner and D. Ganten, 'The renin–angiotensin system in transgenic rats: characteristics and functional studies.' *Semin. Nephrol.*, 1993, **13**, 586–592.

59. M. Bader, Y. Zhao, M. Sander *et al.*, 'Role of tissue renin in the pathophysiology of hypertension in TGR(mREN2)27.' *Hypertension*, 1992, **19**, 681–686.

60. D. Wagner, R. Metzger, M. Paul *et al.*, 'Androgen dependence and tissue specificity of renin messenger RNA expression in mice.' *J. Hypertens.*, 1990, **8**, 45–52.

61. S. Anderson, H. G. Rennke and B. M. Brenner, 'Therapeutic advantage of converting enzyme inhibitors in arresting progressive renal disease associated with systemic hypertension in the rat.' *J. Clin. Invest.*, 1986, **77**, 1993–2000.

62. A. J. de Bold, H. B. Borenstein, A. T. Veress *et al.*, 'A rapid and potent natriuretic response to intravenous injection of atrial myocardial extract in rats.' *Life Sci.*, 1981, **28**, 89–94.

63. R. Garcia, M. Cantin, G. Thibault *et al.*, 'Relationship of specific granules to the natriuretic and diuretic activity of rat atria.' *Experientia*, 1982, **38**, 1071–1073.

64. K. Kangawa and H. Matsuo, 'Purification and complete amino acid sequence of alpha-human atrial natriuretic polypeptide (alpha-hANP).' *Biochem. Biophys. Res. Commun.*, 1984, **18**, 131–139.

65. S. Oikawa, M. Imai, A. Ueno *et al.*, 'Cloning and sequence analysis of cDNA encoding a precursor for human atrial natriuretic polypeptide.' *Nature*, 1984, **309**, 724–726.

66. M. Nemer, M. Chamberland, D. Sirois *et al.*, 'Gene structure of human cardiac hormone precursor, pronatriodilatin.' *Nature*, 1984, **312**, 654–656.

67. T. G. Yandle, 'Biochemistry of natriuretic peptides.' *J. Int. Med.*, 1994, **235**, 561–576.

68. K. D. Bloch, A. J. Scott, J. B. Zisfein *et al.*, 'Biosynthesis and secretion of proatrial natriuretic factor by cultured rat cardiocytes.' *Science*, 1985, **230**, 1168–1171.

69. E. A. Espiner, 'Physiology of natriuretic peptides.' *J. Int. Med.*, 1994, **235**, 527–541.

70. Y. Shenker, R. S. Sider, E. A. Ostafin *et al.*, 'Plasma levels of immunoreactive atrial natriuretic factor in healthy subjects and in patients with edema.' *J. Clin. Invest.*, 1985, **76**, 1684–1687.

71. G. P. Hodsman, P. A. Phillips, K. Ogawa *et al.*, 'Atrial natriuretic factor in normal man: effects of tilt,

posture, exercise, and haemorrhage.' *J. Hypertens.*, 1986, Suppl. 4, S503–S505.

72. R. J. Schiebinger and J. Linden, 'Effect of atrial contraction frequency on atrial natriuretic peptide secretion.' *Am. J. Physiol.*, 1986, **251**, H1095–H1099.

73. K. Nishimura, T. Ban, Y. Saito *et al.*, 'Atrial pacing stimulates secretion of atrial natriuretic polypeptide without elevation of atrial pressure in awake dogs with experimental complete atrioventricular block.' *Circ. Res.*, 1990, **66**, 115–122.

74. P. T. Manning, D. Schwartz, N. C. Katsube *et al.*, 'Vasopressin-stimulated release of atriopeptin: endocrine antagonists in fluid homeostasis.' *Science*, 1985, **229**, 395–397.

75. D. G. Gardner, S. Hane, D. Trachewsky *et al.*, 'Atrial natriuretic peptide mRNA is regulated by glucocorticoids *in vivo*.' *Biochem. Biophys. Res. Commun.*, 1986, **139**, 1047–1054.

76. J. P. Stasch, C. Hirth-Dietrich, S. Kazda *et al.*, 'Endothelin stimulates release of atrial natriuretic peptides *in vitro* and *in vivo*.' *Life Sci.*, 1989, **45**, 869–875.

77. R. Takayanagi, R. M. Snajdar, T. Imada *et al.*, 'Purification and characterization of two types of atrial natriuretic factor receptors from bovine adrenal cortex: guanylate cyclase-linked and cyclase-free receptors.' *Biochem. Biophys. Res. Commun.*, 1987, **144**, 244–250.

78. D. C. Leitman, J. W. Andresen, R. M. Catalano *et al.*, 'Atrial natriuretic peptide binding, cross-linking and stimulation of cyclic GMP accumulation and particulate guanylate cyclase activity in cultured cells.' *J. Biol. Chem.*, 1988, **263**, 3720–3728.

79. T. Maack, M. Suzuki, F. A. Almeida *et al.*, 'Physiological role of silent receptors of atrial natriuretic factor.' *Science*, 1987, **238**, 675–678.

80. J. Linden, A. L. Tucker and K. R. Lynch, 'Molecular cloning of adenosine A1 and A2 receptors.' *Trends Pharmacol. Sci.*, 1991, **12**, 326–328.

81. K. J. Koller, D. G. Lowe, G. L. Bennet *et al.*, 'Selective activation of B natriuretic peptide receptor by C-type natriuretic peptide.' *Science*, 1991, **252**, 120–123.

82. B. M. Brenner, B. J. Ballermann, M. E. Gunning *et al.*, 'Diverse biological actions of atrial natriuretic peptide.' *Physiol. Rev.*, 1990, **70**, 665–699.

83. M. G. Nicholls, 'Minisymposium: the natriuretic peptide hormones. Introduction, editorial and historical review.' *J. Int. Med.*, 1994, **235**, 507–514.

84. C. L. Huang, J. Lewicki, L. J. Johnson *et al.*, 'Renal mechanism of action of rat atrial natriutic factor.' *J. Clin. Invest.*, 1985, **75**, 769–773.

85. M. G. Cogan, 'Atrial natriuretic factor can increase renal solute excretion primarily by raising glomerular filtration.' *Am. J. Physiol.*, 1986, **250**, F710–F714.

86. T. Yukimura, K. Ito, T. Takenaga *et al.*, 'Renal effects of synthetic alpha-human atrial-human natriuretic polypeptide (alpha-hANP) in anesthetized dogs.' *Eur. J. Pharmacol.*, 1984, **103**, 363–366.

87. P. C. Singhal, S. DeCandido, J. A. Satriano *et al.*, 'Atrial natriuretic peptide and nitroprusside causes relaxation of cultured mesangial cells.' *Am. J. Physiol.*, 1989, **257**, C86–C93.

88. M. J. F. Camargo, H. D. Kleinert, S. A. Atlas *et al.*, 'Ca-dependent hemodynamic and natriuretic effects of atrial extract in isolated rat kidney.' *Am. J. Physiol.*, 1984, **246**, F447–F456.

89. D. M. Pollock and W. J. Arendshorst, 'Effect of atrial natriuretic factor on renal hemodynamics in the rat.' *Am. J. Physiol.*, 1986, **251**, F795–F801.

90. K. Takezawa, A. W. Cowley, Jr., M. Skeleton *et al.*, 'Atriopeptin III alters renal medullary hemodynamics and the pressure-diuresis response in rats.' *Am. J. Physiol.*, 1987, **252**, F992–F1002.

91. A. van de Stolpe and R. L. Jamison, 'Micropuncture study of the effect of ANP on the papillary collecting duct in the rat.' *Am. J. Physiol.*, 1988, **254**, F477–F483.

92. P. J. Harris, D. Thomas and T. O. Morgan, 'Atrial natriuretic peptide inhibits angiotensin-stimulated proximal tubular sodium and water reabsorption.' *Nature*, 1987, **326**, 697–698.

93. D. B. Light, E. M. Schwiebert, K. H. Karlson *et al.*, 'Atrial natriuretic peptide inhibits a cation channel in renal innermedullary collecting duct cells.' *Science*, 1989, **243**, 383–385.

94. A. S. Rocha and L. H. Kudo, 'Atrial peptide and cGMP effects on NaCl transport in inner medullary collecting duct.' *Am. J. Physiol.*, 1990, **259**, F258–F268.

95. M. L. Ziedel, 'Hormonal control of inner medullary collecting duct sodium transport.' *Am. J. Physiol.*, 1993, **265**, F159–F173.

96. W. L. Henrich, E. A. McAlister, P. B. Smith *et al.*, 'Direct inhibitory effect of atriopeptin III on renin release in primate kidney.' *Life Sci.*, 1987, **41**, 259–264.

97. C. H. Metzler and D. J. Ramsay, 'Physiological doses of atrial peptide inhibit angiotensin II-stimulated aldosterone secretion.' *Am. J. Physiol.*, 1989, **256**, R1155–R1159.

98. M. A. Dillingham and R. J. Anderson, 'Inhibition of vasopressin action by atrial natriuretic factor.' *Science*, 1986, **231**, 1572–1573.

99. G. Y. Koh, M. G. Klug and L. J. Field, 'Atrial natriuretic factor and transgenic mice.' *Hypertension*, 1993, **22**, 634–639.

100. L. J. Field, 'Transgenic mice in cardiovascular research.' *Annu. Rev. Physiol.*, 1993, **55**, 97–114.

101. M. E. Steinhelper, K. L. Cochrane and L. J. Field, 'Hypotension in transgenic mice expressing atrial natriuretic factor fusion genes.' *Hypertension*, 1990, **16**, 301–307.

102. J.-P. Valentin and M. H. Humphreys, 'Urodilatin: a paracrine renal natriuretic peptide.' *Semin. Nephrol.*, 1993, **13**, 61–70.

103. C. Drummer, F. Fielder, A. Bub *et al.*, 'Development and application of a urodilatin (CDD/ANP-95-126)-specific radioimmunoassay.' *Pfluegers Arch.*, 1993, **423**, 372–377.

104. J. M. Heim, S. Kiefersauer, H. J. Fulle *et al.*, 'Urodilatin and β-ANF: binding properties and activation of particulate guanylate cyclase.' *Biochem. Biophys. Res. Commun.*, 1989, **163**, 37–41.

105. H. Saxenhofer, A. Raselli, P. Weidmann *et al.*, 'Urodilatin, a natriuretic factor from kidneys, can modify renal and cardiovascular function in men.' *Am. J. Physiol.*, 1990, **259**, F832–F838.

106. D. A. Hidebrandt, H. L. Mizelle, M. W. Brands *et al.*, 'Comparison of the renal actions of urodilatin and atrial natriuretic peptide.' *Am. J. Physiol.*, 1992, **262**, R395–R399.

107. G. A. Riegger, D. Elsner, W. G. Forssmann *et al.*, 'Effects of ANP-(95-126) in dogs before and after induction of heart failure.' *Am. J. Physiol.*, 1990, **259**, H1643–H1648.

108. D. B. Bylund, D. C. Eikenberg, J. B. Hieble *et al.*, 'International Union of Pharmacology Nomenclature of adrenoceptors.' *Pharmacol. Rev.*, 1994, **46**, 121–136.

109. P. A. Jose, J. R. Raymon, M. D. Bates *et al.*, 'The renal dopamine receptors.' *J. Am. Soc. Nephrol.*, 1992, **2**, 1265–1278.

110. O. Civelli, J. R. Bunzow and D. K. Grandy, 'Molecular diversity of the dopamine receptors.' *Annu. Rev. Pharmacol. Toxicol.*, 1993, **33**, 281–307.

111. J. A. Gingrich and M. G. Caron, 'Recent advances in the molecular biology of dopamine receptors.' *Annu. Rev. Neurosci.*, 1993, **16**, 299–321.

112. T. Calianos II and K. H. Muntz, 'Autoradiographic quantification of adrenergic receptors in the rat

kidney.' *Kidney Int.*, 1990, **38**, 39–46.

113. D. Rouse and W. N. Suki, 'Effects of neural and humoral agents on the renal tubules in congestive heart failure.' *Semin. Nephrol.*, 1994, **14**, 412–426.

114. W. B. Jeffries and W. A. Pettinger, 'Adrenergic signal transduction in the kidney.' *Miner. Electrolyte Metab.*, 1989, **15**, 5–15.

115. G. F. DiBona and L. L. Sawin, 'Effect of renal nerve stimulation on NaCl and H$_2$O transport in Henle's loop of the rat.' *Am. J. Physiol.*, 1982, **243**, F576–F580.

116. D. D. Smyth, S. Umemura and W. A. Pettinger, 'Alpha2-adrenoreceptors and sodium reabsorption in the isolated perfused rat kidney.' *Am. J. Physiol.*, 1984, **247**, F680–F685.

117. C. Bailley, M. Imbert-Teboul, N. Roinel *et al.*, 'Isoproterenol increases Ca, Mg, and NaCl reabsorption in mouse thick ascending limb.' *Am. J. Physiol.*, 1990, **258**, F1224–F1231.

118. P. Soares-da-Silva, 'Study on the neuronal and non-neuronal stores of dopamine in rat and rabbit kidney.' *Pharmacol. Res.*, 1992, **26**, 161–171.

119. A. Aperia, 'Dopamine action and metabolism in the kidney.' *Curr. Opin. Nephrol. Hypertens.*, 1994, **3**, 39–45.

120. F. Takemoto, T. Satoh, H. T. Cohen *et al.*, 'Localization of dopamine-1 receptors along the microdissected rat nephron.' *Pfluegers Arch.*, 1991, **419**, 243–248.

121. R. A. Felder, C. C. Felder, G. M. Eisner *et al.*, 'The dopamine receptor in adult and maturing kidney.' *Am. J. Physiol.*, 1989, **257**, F315–F327.

122. O. Kuchel and S. Shigetomi, 'Defective dopamine generation from dihydroxyphenylalanine in stable essential hypertensive patients.' *Hypertension*, 1992, **19**, 634–638.

123. B. A. Clark, R. M. Rosa, F. H. Epstein *et al.*, 'Altered dopaminergic responses in hypertension.' *Hypertension*, 1992, **19**, 589–594.

124. H. Margolies, 'The kallikrein–kinin system and the kidney.' *Annu. Rev. Physiol.*, 1984, **46**, 309–326.

125. A. G. Scicli and O. A. Carretero, 'Renal kallikrein-kinin system.' *Kidney Int.*, 1986, **29**, 120–130.

126. D. W. Coyne and A. R. Morrison, in 'Contemporary Issues in Nephrology: Hormones, Autacoids, and the Kidney,' eds. S. Goldfarb and F. N. Ziyadeh, Churchill Livingstone, New York, 1991, vol. 23, pp. 263–280.

127. N. Ura, O. A. Carretero and E. G. Erdos, 'Role of renal endopeptidase 24.11 in kinin metabolism *in vitro* and *in vivo*.' *Kidney Int.*, 1987, **32**, 507–513.

128. J. P. Granger and J. E. Hall, 'Acute and chronic actions of bradykinin on renal function and arterial pressure.' *Am. J. Physiol.*, 1985, **248**, F87–F92.

129. M. M. Grez, 'The influence of antibodies against bradykinin on isotonic saline diuresis in the rat. Evidence for kinin involvement in renal function.' *Pfluegers Arch.*, 1974, **350**, 231–239.

130. H. M. Siragy, 'Evidence that intrarenal bradykinin plays a role in regulation of renal function.' *Am. J. Physiol.*, 1993, **265**, E648–E654.

131. R. J. Anderson, in 'Contemporary Issues in Nephrology: Hormones, Autacoids, and the Kidney,' eds. S. Goldfarb and F. N. Ziyadeh, Churchill Livingstone, New York, 1991, vol. 23, pp. 281–296.

132. W. S. Spielman and L. J. Arend, 'Adenosine receptors and signaling in the kidney.' *Hypertension*, 1991, **17**, 117–130.

133. M. Le Hir and B. Kaissling, 'Distribution and regulation of renal ecto-5'-nucleotidase: implications for physiological functions of adenosine.' *Am. J. Physiol.*, 1993, **264**, F377–F387.

134. D. E. McCoy, S. Bhattacharya, B. A. Olson *et al.*, 'The renal adenosine system: structure, function, and regulation.' *Semin. Nephrol.*, 1993, **13**, 31–40.

135. N. F. Rossi, P. Churchill, V. Ellis *et al.*, 'Mechanism of adenosine receptor-induced renal vasoconstriction in rats.' *Am. J. Physiol.*, 1988, **255**, H885–H890.

136. N. F. Rossi, P. Churchill, K. A. Jacobson *et al.*, 'Further characterization of the renovascular effect of N6-cyclohexyladenosine in the isolated perfused rat kidney.' *J. Pharmacol. Exp. Ther.*, 1987, **240**, 911–915.

137. H. Osswald, W. S. Spielman and F. G. Knox, 'Mechanism of adenosine-mediated decreases in glomerular filtration rate in dogs.' *Circ. Res.*, 1978, **43**, 465–469.

138. A. Edlund, H. Ohlsen and A. Sollevi, 'Renal effects of local infusion of adenosine in man.' *Clin. Sci. (Colch.)*, 1994, **87**, 143–149.

139. A. Edlund and A. Sollevi, 'Renal effects of i.v. adenosine infusion in humans.' *Clin. Physiol.*, 1993, **13**, 361–371.

140. J. Schnermann and J. P. Briggs, 'The role of adenosine in cell-to-cell signaling in the juxtaglomerular apparatus.' *Semin. Nephrol.*, 1993, **13**, 236–245.

141. H. Osswald, B. Muhlbauer and F. Schenk, 'Adenosine mediates tubuloglomerular feedback response: an element of metabolic control of kidney function.' *Kidney Int.*, 1991, **32** Suppl., S128–S131.

142. J. Schnermann, H. Weihprecht and J. P. Briggs *et al.*, 'Inhibition of tubuloglomerular feedback during adenosine 1-receptor blockade.' *Am. J. Physiol.*, 1990, **258**, F553–F561.

143. H. Weihprecht, J. N. Lorenz, J. Schnermann *et al.*, 'Effect of adenosine1-receptor blockade on renin release from rabbit isolated perfused juxtaglomerular apparatus.' *J. Clin. Invest.*, 1990, **85**, 1622–1628.

144. N. Levens, M. Beil and M. Jarvis, 'Renal actions of a new adenosine agonist, CGS 21680A selective for the A2-receptor.' *J. Pharmacol. Exp. Ther.*, 1991, **257**, 1005–1012.

145. M. Ueno, J. Brookings, B. Beckman *et al.*, 'A1 and A2 adenosine receptor regulation of erythropoietin production.' *Life Sci.*, 1988, **43**, 229–237.

146. C. M. Erley, S. H. Duda, S. Schlepckow *et al.*, 'Adenosine antagonist theophylline prevents the reduction of glomerular filtration rate after contrast media application.' *Kidney Int.*, 1994, **45**, 1425–1431.

147. A. K. Bidani, P. C. Churchill and W. Packer, 'Theophylline-induced protection in myoglobinuric acute renal failure: further characterization.' *Can. J. Physiol. Pharmacol.*, 1987, **65**, 42–45.

148. N. Rossi, V. Ellis, T. Kontry *et al.*, 'The role of adenosine in HgCl$_2$-induced acute renal failure in rats.' *Am. J. Physiol.*, 1990, **258**, F1554–F1560.

149. B. Samuelsson, 'An elucidation of the arachidonic acid cascade. Discovery of prostaglandins, thromboxane, and leukotrienes.' *Drugs*, 1987, **33**, 2–9.

150. A. R. Morrison, 'Biochemistry and pharmacology of renal arachidonic acid metabolism.' *Am. J. Med.*, 1986, **80**, 3–11.

151. D. Schlondorff and R. Ardaillou, 'Prostaglandins and other arachidonic acid metabolites in the kidney.' *Kidney Int.*, 1986, **29**, 108–119.

152. J. B. Lefkowith and G. Schreiner, 'Essential fatty acid deficiency depletes rat glomeruli of resident macrophages and inhibits angiotensin II-induced eicosanoid synthesis.' *J. Clin. Invest.*, 1987, **80**, 947–956.

153. V. E. Kelley, A. Ferretti, S. Izui *et al.*, 'A fish oil diet rich in eicosapentaenoic acid reduces cyclooxygenase metabolites and suppresses lupus in MRL-lpr mice.' *J. Immunol.*, 1985, **134**, 1914–1919.

154. L. A. Scharschmidt and M. J. Dunn, 'Prostaglandin synthesis by rat glomerular mesangial cells in culture. Effects of angiotensin II and arginine vasopressin.' *J. Clin. Invest.*, 1983, **71**, 1756–1764.

155. R. M. Zusman and H. R. Keiser, 'Prostaglandin biosynthesis by rabbit renomedullary interstitial cells in tissue culture. Stimulation by angiotensin II, bradydinin, and arginine vasopresssin.' *J. Clin. Invest.*, 1977, **60**, 215–223.

156. A. H. Lin, M. J. Bienkowski and R. R. Gorman, 'Regulation of prostaglandin H synthase mRNA levels and prostaglandin biosynthesis by platelet-derived growth factor.' *J. Biol. Chem.*, 1989, **264**, 17379–17383.

157. J. A. Maier, T. Hla and T. Maciag, 'Cyclooxygenase is an immediate-early gene induced by interleukin-1 in human endothelial cells.' *J. Biol. Chem.*, 1990, **265**, 10805–10808.

158. D. A. Kujubu and H. R. Herschman, 'Dexamethasone inhibits mitogen induction of the TIS10 prostaglandin synthase/cyclooxygenase gene.' *J. Biol. Chem.*, 1992, **267**, 7991–7994.

159. E. A. Lianos, G. A. Andres and M. J. Dunn, 'Glomerular prostaglandin and thromboxane synthesis in rat nephrotoxic serum nephritis. Effects on renal hemodynamics.' *J. Clin. Invest.*, 1983, **72**, 1439–1448.

160. J. E. Stork and M. J. Dunn, 'Hemodynamic roles of thromboxane A2 and prostaglandin E2 in glomerulonephritis.' *J. Pharmacol. Exp. Ther.*, 1985, **233**, 672–678.

161. T. Okegawa, P. E. Jonas, K. DeSchryver *et al.*, 'Metabolic and cellular alterations underlying the exaggerated renal prostaglandin and thromboxane synthesis in ureter obstruction in rabbits. Inflammatory response involving fibroblasts and mononuclear cells.' *J. Clin. Invest.*, 1983, **71**, 81–90.

162. C. Yokohama and T. Tanabe, 'Cloning of human gene encoding prostaglandin endoperoxide synthase and primary structure of the enzyme.' *Biochem. Biophys. Res. Commun.*, 1989, **165**, 888–894.

163. D. A. Jones, D. P. Carlton, T. M. McIntyre *et al.*, 'Molecular cloning of human prostaglandin endoperoxide synthase type II and demonstration of expression in response to cytokines.' *J. Biol. Chem.*, 1993, **268**, 9049–9054.

164. H. Satoh and S. Satoh, 'Prostaglandin E2 and I2 production in isolated dog renal arteries in the absence of presence of vascular endothelial cells.' *Biochem. Biophys. Res. Commun.*, 1984, **118**, 873–876.

165. D. Schlondorff, J. Perez and J. A. Satriano, 'Differential stimulation of PGE2 synthesis in mesangial cells by angiotensin and A23187.' *Am J. Physiol.*, 1985, **248**, C119–C126.

166. S. Govindarajan, C. C. Nast, W. L. Smith *et al.*, 'Immunohistochemical distribution of renal prostaglandin endoperoxide synthase and prostacyclin synthase: diminished endoperoxide synthase in the hepatorenal syndrome.' *Hepatology*, 1987, **7**, 654–659.

167. W. L. Smith and T. G. Bell, 'Immunohistochemical localization of the prostaglandin-forming cyclooxygenase in renal cortex.' *Am. J. Physiol.*, 1978, **235**, F451–F457.

168. C. A. Brown, R. M. Zusman and E. Haber, 'Identification of an angiotensin receptor in rabbit renomedullary interstitial cells in culture. Correlation with prostaglandin biosynthesis.' *Circ. Res.*, 1980, **46**, 802–807.

169. N. Alavi, E. A. Lianos and C. J. Bentzel, 'Prostaglandin and thromboxane synthesis by highly enriched rabbit proximal tubular cells in culture.' *J. Lab. Clin Med.*, 1987, **110**, 338–345.

170. J. P. Bonvalet, P. Pradelles and N. Farman, 'Segmental synthesis and actions of prostaglandins along the nephron.' *Am. J. Physiol.*, 1987, **253**, F377–F387.

171. R. C. Harris, J. A. McKanna, Y. Akai *et al.*, 'Cyclooxygenase-2 is associated with the macula densa of rat kidney and increases with salt restriction.' *J. Clin. Invest.*, 1994, **94**, 2504–2510.

172. Y. Akai, T. Homma, K. F. Badr *et al.*, 'Mechanical stretch/relaxation stimulates release of arachidonic acid and induces transient expression of the mitogen sensitive PGH synthase (PGHS-2) gene in cultured rat mesangial cells.' *J. Am. Soc. Nephrol.*, 1992, **3**, 450.

173. R. A. Stahl, M. Paravicini and P. Schollmeyer, 'Angiotensin II stimulation of prostaglandin E2 and 6-keto-F1α formation by isolated human glomeruli.' *Kidney Int.*, 1984, **26**, 30–34.

174. N. Farman, P. Pradeles and Bonvalet, 'PGE2, PGF2α, 6-keto-PGE1α, and TXB2 synthesis along the rabbit nephron.' *Am. J. Physiol.*, 1987, **252**, F53–F59.

175. C. Patrono and M. J. Dunn, 'The clinical significance of inhibition of renal prostaglandin synthesis.' *Kidney Int.*, 1987, **32**, 1–12.

176. C. J. Lote and J. Haylor, in 'Prostaglandins, Leukotrienes and Essential Fatty Acids.' Churchill Livingstone, New York, 1989, pp. 203–217.

177. S. H. Ferreira and J. R. Vane, 'Prostaglandins: their disappearance from and release into the circulation.' *Nature*, 1967, **216**, 868–873.

178. E. R. Jones, T. R. Beck, S. Kapoor *et al.*, 'Prostaglandins inhibit renal ammoniagenesis in the rat.' *J. Clin. Invest.*, 1984, **74**, 992–1002.

179. W. R. Holt and C. Lechene, 'ADH-PGE2 interactions in cortical collecting tubules. II. Inhibition of Ca and P reabsorption.' *Am. J. Physiol.*, 1981, **241**, F461–F467.

180. R. D. Zisper, 'Effects of selective inhibition of thromboxane synthesis on renal function in humans.' *Am J. Physiol.*, 1985, **248**, F753–F756.

181. R. A. Coleman and P. P. A. Humphrey, in 'Therapeutic Applications of Prostaglandins,' eds. J. R. Vane and J. O'Grady, Edward Arnold, London, 1993, pp. 15–36.

182. M. Hirata, Y. Hayashi, F. Uhikubi *et al.*, 'Cloning and expression of cDNA for a human thromboxane A2 receptor.' *Nature*, 1991, **349**, 617–620.

183. R. M. Nusing, M. Hirata, A. Kakizuka *et al.*, 'Characterization and chromosomal mapping of the human thromboxane A2 receptor gene.' *J. Biol. Chem.*, 1993, **268**, 25253–25259.

184. K. Sakamoto, T. Ezashi, K. Miwa *et al.*, 'Molecular cloning and expression of a cDNA of the bovine prostaglandin F2-alpha receptor.' *J. Biol. Chem.*, 1994, **269**, 3881–3886.

185. A. Watabe, Y. Sugimoto, A. Honda *et al.*, 'Cloning and expression of cDNA for a mouse EP1 subtype of prostaglndin E receptor.' *J. Biol. Chem.*, 1993, **268**, 20175–20178.

186. A. Honda, Y. Sugimoto, T. Namba *et al.*, 'Cloning and expression of a cDNA for mouse prostaglandin E receptor EP2 subtype.' *J. Biol. Chem.*, 1993, **268**, 7759–7762.

187. S. An, J. Yang, M. Xia *et al.*, 'Cloning and expression of the EP2 subtype of human receptors for prostaglandin E2.' *Biochem. Biophys. Res. Commun.*, 1993, **197**, 263–270.

188. Y. Sugimoto, T. Namba, A. Honda *et al.*, 'Cloning and expression of a cDNA for mouse prostaglandin E receptor EP3-subtype.' *J. Biol. Chem.*, 1992, **267**, 6433–6436.

189. R. L. Hebert, H. R. Jacobson, D. Fredin *et al.*, 'Evidence that separate PGE2 receptors modulate water and sodium transport in rabbit cortical collecting duct.' *Am. J. Physiol.*, 1993, **265**, F643–F650.

190. R. A. Lawrence and R. L. Jones, 'Investigation of the prostaglandin E (EP-) receptor subtype mediating relaxation of the rabbit jugular vein.' *Br. J. Pharmacol.*, 1992, **265**, 817–824.

191. T. Namba, Y. Sugimoto, M. Negishi *et al.*, 'Alternative splicing of C-terminal tail of prostaglandin E receptor

subtype EP3 determines G-protein specificity.' *Nature*, 1993, **365**, 166–170.

192. P. Mene, M. S. Simonson and M. J. Dunn, 'Physiology of the mesangial cell.' *Physiol. Rev.*, 1989, **69**, 1347–1424.

193. J. B. Stokes, III, 'Modulation of vasopressin-induced water permeability of the cortical collecting tubule by endogenous and exogenous prostaglandins.' *Miner. Electrolyte Metab.*, 1985, **11**, 240–248.

194. J. H. Veis, M. A. Dillingham and T. Berl, 'Effects of prostacyclin on the cAMP system in cultured rat inner medullary collecting duct cells.' *Am. J. Physiol.*, 1990, **258**, F1218–F1223.

195. A. Chaudhari, S. Gupta and M. A. Kirschenbaum, 'Biochemical evidence for PGI2 and PGE2 receptors in the rabbit renal preglomerular microvasculature.' *Biochim. Biophys. Acta*, 1990, **1053**, 156–161.

196. M. D. Breyer and K. F. Badr, in 'The Kidney,' 5th edn., eds. B. M. Brenner and F. C. Rector, Jr., Saunders, Philadelphia, PA, 1995, vol. 1, pp. 754–788.

197. R. E. Garrick, in 'Contemporary Issues in Nephrology: Hormones, Autacoids, and the Kidney,' eds. S. Gold-farb and F. N. Ziyadeh, Churchill Livingstone, New York, 1991, vol. 23, pp. 231–262.

198. R. M. Edwards, 'Effects of prostaglandins on vasoconstrictor action in isolated renal arterioles.' *Am. J. Physiol.*, 1985, **248**, F779–F784.

199. P. M. Bolger, G. M. Einser, P. W. Ramwell *et al.*, 'Renal action of prostacyclin.' *Nature*, 1978, **271**, 467–469.

200. T. Hashimoto, 'Effects of prostaglandin E2, I2 and F2α on systemic and renal hemodynamics, renal function and renin secretion in anesthetized dogs.' *Jpn. J. Pharmacol.*, 1980, **30**, 173–186.

201. T. Yoshioka, A. Yared, H. Miyazawa *et al.*, '*In vivo* influence of prostaglandin I2 on systemic and renal circulation in the rat.' *Hypertension*, 1985, **7**, 867–872.

202. C. Baylis, 'Effects of administered thomboxanes on the intact, normal rat kidney.' *Ren. Physiol.*, 1987, **10**, 110–121.

203. R. Loutzenhiser, M. Epstein, C. Horton *et al.*, 'Reversal of renal and smooth muscle actions of the thromboxane mimetic U-44069 by diltiazem.' *Am. J. Physiol.*, 1986, **250**, F619–F626.

204. R. D. Zisper, 'Effects of selective inhibition of thromboxane synthesis on renal function in humans.' *Am. J. Physiol.*, 1985, **248**, F753–F756.

205. H. Gullner, J. R. Gill, Jr., F. C. Bartter *et al.*, 'The role of the prostaglandin system in the regulation of renal function in normal women.' *Am J. Med.*, 1980, **69**, 718–724.

206. K. Munger and C. Baylis, 'Sex differences in renal hemodynamics in rats.' *Am. J. Physiol.*, 1988, **254**, F223–F231.

207. L. A. Scharschmidt, J. G. Douglas and M. J. Dunn, 'Angiotensin II and eicosanoids in the control of glomerular size in the rat and human.' *Am. J. Physiol.*, 1986, **250**, F348–F356.

208. W. L. Henrich, R. J. Anderson, A. S. Berns *et al.*, 'The role of renal nerves and prostaglandins in control of renal hemodynamics and plasma renin activity during hypotensive hemorrhage in the dog.' *J. Clin. Invest.*, 1978, **61**, 744–750.

209. C. Patrono and M. J. Dunn, 'The clinical significance of inhibition of renal prostaglandin synthesis.' *Kidney Int.*, 1987, **32**, 1–12.

210. A. Yared, V. Kon and I. Ichikawa, 'Mechanism of preservation of glomerular perfusion and filtration during acute extracellular fluid volume depletion. Importance of intravenal vasopressin–prostaglandin interaction for protecting kidneys from constrictor action of vasopressin.' *J. Clin. Invest.*, 1985, **75**, 1477–1487.

211. J. L. Nadler, M. McKay, V. Campese *et al.*, 'Evidence

212. C. Manoogian, J. Nadler, L. Enrlich *et al.*, 'The renal vasodilating effect of dopamine is mediated by calcium flux and prostacyclin release in man.' *J. Clin. Endocrinol. Metab.*, 1988, **66**, 678–683.

213. J. L. Nadler, S. Goodson and R. K. Rude, 'Evidence that prostacyclin mediates the vascular action of magnesium in humans.' *Hypertension*, 1987, **9**, 379–383.

214. M. Panzenbeck, A. Baez and G. Kaley, 'Prostaglandins participate in the renal vasodialtion due to hydralazine in dogs.' *J. Pharmacol. Exp. Ther.*, 1986, **237**, 525–528.

215. B. S. Gan, K. L. MacCannell and M. D. Hollenberg, 'Epidermal growth factor-urogastrone causes vasodilatation in the anesthetized dog.' *J. Clin. Invest.*, 1987, **80**, 199–206.

216. Y. F. Vanrenterghem, R. K. Verberckmoes, L. M. Roels *et al.*, 'Role of prostaglandins in protein-induced glomerular hyperfiltration in normal humans.' *Am. J. Physiol.*, 1988, **254**, F463–F469.

217. A. R. Morrison, K. Nishikawa and P. Needleman, 'Unmasking of thromboxane A2 synthesis by ureteral obstruction in the rabbit kidney.' *Nature*, 1977, **267**, 259–260.

218. K. F. Badr and I. Ichikawa, 'Prerenal failure: a deleterious shift from renal compensation to decompensation.' *N. Engl. J. Med.*, 1988, **319**, 623–629.

219. Y. Tabuchi, T. Ogihara and Y. Kumahara, 'Renal vein prostaglandins in renovascular hypertensive patients.' *Prostaglandins Leuko. Med.*, 1985, **19**, 219–226.

220. D. Schlondorff, 'Renal complications of nonstroidal antiinflammatory drugs.' *Kidney Int.*, 1993, **44**, 643–653.

221. J. L. Logan, S. M. Lee, B. Benson *et al.*, 'Inhibition of compensatory renal growth by indomethacin.' *Prostaglandins*, 1986, **31**, 253–261.

222. J. L. Logan and B. Benson, 'Further studies on the effects of cyclooxygenase inhibitors on compensatory renal growth.' *Prostaglandins Leuko. Med.*, 1987, **30**, 9–15.

223. C. D. Mistry, C. J. Lote and W. J. Currie, 'Effects of sulindac on renal function and prostaglandin synthesis in patients with moderate chronic renal insufficiency.' *Clin. Sci. (Colch.)*, 1986, **70**, 501–505.

224. M. L. Purkerson, J. H. Joist and J. Yates, 'Inhibition of thromboxane synthesis ameliorates progressive renal disease of Dahl-S rats.' *Kidney Int.*, 1988, **33**, 77.

225. M. H. Loo, D. N. Marion, E. D. Vaughan, Jr. *et al.*, 'Effect of thromboxane inhibition on renal blood flow in dogs with complete unilateral ureteral obstruction.' *J. Urol.*, 1986, **136**, 1343–1347.

226. C. R. Albrightson, A. S. Evers, A. C. Griffin *et al.*, 'Effect of endogenously produced leukotrienes and thromboxane on renal vascular resistance in rabbit hydronephrosis.' *Circ. Res.*, 1987, **61**, 514–522.

227. C. Patrono, G. Ciabattoni, G. Remuzzi *et al.*, 'Functional significance of renal prostacyclin and thromboxane A2 production in patients with systemic lupus erythematosus.' *J. Clin. Invest.*, 1985, **76**, 1011–1018.

228. A. Pierucci, B. M. Simonetti, G. Pecci *et al.*, 'Improvement of renal function with selective thromboxane antagonism in lupus nephritis.' *N. Engl. J. Med.*, 1989, **320**, 421–425.

229. F. R. DeRubertis and P. A. Craven, 'Eicosanoids in the pathogenesis of the functional and structural alterations of the kidney in diabetes.' *Am. J. Kidney Dis.*, 1993, **22**, 727–735.

230. J. L. Nadler, F. O. Lee, W. Hseuh *et al.*, 'Evidence of prostacyclin deficiency in the syndrome of hyporeni-

nemic hypoaldosteronism.' *N. Engl. J. Med.*, 1986, **314**, 1015–1020.

231. M. Epstein and M. Lifschitz, 'Renal eicosanoids as determinants of renal function in liver disease.' *Hepatology*, 1987, **7**, 1359–1367.

232. I. Gregoire, J. P. Dupouy, P. Fievet *et al.*, 'Effect of pregnancy on plasma renin activity and glomerular synthesis of prostaglandins and thromboxane in rats.' *Agents Actions*, 1987, **22** Suppl., 147–154.

233. O. Ylikorkala, F. Pekonen and L. Viinikka, 'Renal prostacyclin and thromboxane in normotensive and preeclamptic pregnant women and their infants.' *J. Clin. Endocrinol. Metab.*, 1986, **63**, 1307–1312.

234. K. P. Conrad and M. C. Colpoys, 'Evidence against the hypothesis that prostaglandins are the vasodepressor agents of pregnancy. Serial studies in chronically instrumented, conscious rats.' *J. Clin. Invest.*, 1986, **77**, 236–245.

235. E. Schiff, E. Peleg, M. Goldenberg *et al.*, 'The use of aspirin to prevent pregnancy-induced hypertension and lower the ratio of thromboxane A2 to protacyclin in relatively high-risk pregnancies.' *N. Engl. J. Med.*, 1989, **321**, 351–356.

236. B. M. Sibai, S. N. Catitis, E. Thom *et al.*, 'Prevention of preeclampsia with low-dose aspirin in healthy, nulliparous pregnant women. The National Institute of Child Health and Human Development Network of Maternal–Fetal Medicine Units.' *N. Engl. J. Med.*, 1993, **329**, 1213–1218.

237. C. Patrono and M. J. Dunn, 'The clinical significance of inhibition of renal prostaglandin synthesis.' *Kidney Int.*, 1987, **32**, 1–12.

238. P. P. Koopmans, T. Thien and C. M. Thomas, 'The effects of sulindac and indomethacin on the antihypertensive and diuretic action of hydrochlorothiazide in patients with mild to moderate essential hypertension.' *Br. J. Clin. Pharmacol.*, 1986, **21**, 417–423.

239. J. E. Pope, J. J. Anderson and D. T. Felson, 'A metaanalysis of the effects of nonsteroidal antiinflammatory drugs on blood pressure.' *Arch. Intern. Med.*, 1993, **153**, 477–484.

240. J. M. Ritter, J. R. Ludgin and L. A. Scharschmidt, 'Effects of a stable prostaglandin analogue, L-644,122, in healthy and hypertensive men.' *Eur. J. Clin. Pharmacol.*, 1985, **28**, 685–688.

241. D. P. Sandler, F. R. Burr and C. R. Weinberg, 'Nonsteroidal antiinflammatory drugs and the risk for chronic renal disease.' *Ann. Intern. Med.*, 1991, **115**, 165–172.

242. T. V. Perneger, P. K. Whelton and M. J. Klag, 'Risk of kidney failure associated with the use of acetaminophen, aspirin, and nonsteroidal antiinflammatory drugs.' *N. Eng. J. Med.*, 1994, **331**, 1675–1679.

243. J. R. Vane and R. M. Botting, 'A better understanding of anti-inflammatory drugs based on isoforms of cyclooxygenase (COX-1 and COX-2).' *Adv. Prostaglandin Thromboxane Leukot. Res.*, 1995, **23**, 41–48.

244. P. Isakson, K. Seibert, J. Masferrer *et al.*, 'Discovery of a better aspirin.' *Adv. Prostaglandin Thromboxane Leukot. Res.*, 1995, **23**, 49–54.

245. B. Samuelsson, S. E. Dahlen, J. Lindgren *et al.*, 'Leukotrienes and lipoxins: structures, biosynthesis, and biological effects.' *Science*, 1987, **237**, 1171–1176.

246. A. A. Spector, J. A. Gordon and S. A. Moore, 'Hydroxyeicosatetraenoic acids (HETEs).' *Prog. Lipid. Res.*, 1988, **27**, 271–323.

247. C. N. Serhan, 'Lipoxin biosynthesis and its impact in inflammatory and vascular events.' *Biochim. Biophys. Acta*, 1994, **1212**, 1–25.

248. R. A. Dixon, R. E. Diehl, E. Opas *et al.*, 'Requirement of a 5-lipoxygenase-activating protein for leukotriene synthesis.' *Nature*, 1990, **343**, 282–284.

249. W. R. Henderson, Jr., 'The role of leukotrienes in inflammation.' *Ann. Intern. Med.*, 1994, **121**, 684–697.

250. J. B. Lefkowith, A. R. Morrison and G. F. Schreiner, 'Murine glomerular leukotriene B4 snthesis. Manipulation by (n-6) fatty acid deprivation and cellular origin.' *J. Clin. Invest.*, 1988, **68**, 1655–1660.

251. V. Cattell, H. T. Cook, J. Smith *et al.*, 'Leukotriene B4 produciton in normal rat glomeruli.' *Nephrol. Dial. Transplant.*, 1987, **2**, 154–157.

252. J. Sraer, M. Rigaud, M. Bens *et al.*, 'Metabolism of arachidonic acid via the lipoxygenase pathway in human and murine glomeruli.' *J. Biol. Chem.*, 1983, **258**, 4325–4330.

253. K. Jim, A. Hassid, F. Sun *et al.*, 'Lipoxygnease activity in rat kidney glomeruli, glomerular epithelial cells, and cortical tubules.' *J. Biol. Chem.*, 1982, **257**, 10294–10299.

254. A. Yared, C. Albrightson-Winslow, D. Griswold *et al.*, 'Functional significance of leukotriene B4 in normal and glomerulonephritic kidneys.' *J. Am. Soc. Nephrol.*, 1991, **2**, 45–56.

255. K. F. Badr, C. Baylis, J. M. Pfeffer *et al.*, 'Renal and systemic hemodynamic responses to intravenous infusion of leukotriene C4 in the rat.' *Circ. Res.*, 1984, **54**, 492–499.

256. A. Rosenthal and C. R. Pace-Asciak, 'Potent vasoconstriction of the isolated perfused kidney by leukotrienes C4 and D4.' *Can. J. Pharmacol.*, 1983, **61**, 325–328.

257. K. F. Badr, B. M. Brenner and I. Ichikawa, 'Effects of leukotriene D4 on glomerular dynamics in the rat.' *Am. J. Physiol.*, 1987, **253**, F239–F243.

258. R. Barnett, P. Goldwasser, L. A. Scharschimdt *et al.*, 'Effects of leukotrienes on isolated rat glomeruli and cultured mesangial cells.' *Am. J. Physiol.*, 1986, **250**, F838–F844.

259. M. S. Simonson and M. J. Dunn, 'Leukotriene C4 and D4 contract rat glomerular mesangial cells.' *Kidney Int.*, 1986, **30**, 524–531.

260. H. R. Brady, U. Persson, B. J. Ballerman *et al.*, 'Leukotrienes stimulate neutrophil adhesion to mesangial cells: modulation with lipoxins.' *Am. J. Physiol.*, 1990, **259**, F809–F815

261. K. F. Badr, S. Mong, R. L. Hoover *et al.*, 'Leukotriene D4 binding and signal transduction in rat glomerular mesangial cells.' *Am. J. Physiol.*, 1989, **257**, F280–F287.

262. M. S. Simonson, P. Mene, G. R. Dubyak *et al.*, 'Identification and transmembrane signaling of leukotriene D4 receptors in human mesangial cells.' *Am. J. Physiol.*, 1988, **255**, C771–C780.

263. K. F. Badr, '15-Lipoxygenase products as leukotriene antagonists: therapeutic potential in glomerulonephritis.' *Kidney Int.*, 1992, **38**, S101–S108.

264. M. E. Brezinski and C. N. Serhan, 'Selective incorporation of 15-S-hydroxyeicosatetraenoic acid in phosphotidylinositol of human neutrophils: agonist induced deacylation and transformation of stored hydroxyeicosanoids.' *Proc. Natl. Acad. Sci. USA*, 1990, **87**, 6248–6252.

265. D. B. Fischer, J. W. Christman and K. F. Badr, 'Fifteen-S-hydroxyeicosatetraenoic acid (15-S-HETE) specifically antagonizes the chemotactic action and glomerular synthesis of leukotriene B4 in the rat.' *Kidney Int.*, 1992, **41**, 1155–1160.

266. A. B. Legrand, J. A. Lawson, B. O. Meyrick *et al.*, 'Substitution of 15-hydroxyeicosatetraenoic acid in the phosphoinositide signaling pathway.' *J. Biol. Chem.*, 1994, **266**, 7570–7577.

267. T. H. Lee, C. E. Horton, U. Kyan-Aung *et al.*, 'Lipoxin A4 and lipoxin B4 inhibit chemotactic responses of human neutrophils stimulated by leuko-

triene B4 and *N*-formyl-L-methionyl-leucyl-L-phenyl-alanine.' *Clin. Sci. (Colch.)*, 1989, **77**, 195–203.

268. S. Fiore, J. F. Maddox, H. Perez *et al.*, 'Identification of a human cDNA encoding a functional high affinity lipoxin A4 receptor.' *J. Exp. Med.*, 1994, **180**, 253–260.

269. K. F. Badr, D. K. DeBoer, M. Schwetzberg *et al.*, 'Lipoxin A4 antagonizes cellular and in vivo actions of leukotriene D4 in rat glomerular mesangial cells: evidence for competition at a common receptor.' *Proc. Natl. Acad. Sci. USA*, 1989, **86**, 3438–3442.

270. T. Katoh, K. Takahashi, D. K. DeBoer *et al.*, 'Renal hemodynamic actions of lipoxins in rats: a comparative physiological study.' *Am. J. Physiol.*, 1992, **263**, F436–F442.

271. K. F. Badr, 'Five-lipoxygenase products in glomerular immune injury.' *J. Am. Soc. Nephrol.*, 1992, **3**, 907–915.

272. G. M. Nassar and K. F. Badr, 'Role of leukotrienes and lipoxygenases in glomerular injury.' *Miner. Electrolyte Metab.*, 1995, **21**, 262–270.

273. D. Fischer, K. Takahashi, J. Ebert *et al.*, 'Limited early therapy with a novel 5-lipoxygenase (5-LO) activating protein (FLAP) antagonist, MK 886, during heterologous rat nephrotoxic serum (NTS) nephritis totally prevents proteinuria in the autologous phase.' *J. Am. Soc. Nephrol.*, 1990, **1**, 628.

274. K. F. Badr, G. F. Schreiner, M. Wasserman *et al.*, 'Preservation of the glomerular capillary ultrafiltration coefficient during rat nephrotoxic serum nephritis by a specific leukotriene D4 receptor antagonist.' *J. Clin. Invest.*, 1988, **81**, 1702–1709.

275. K. Takahashi, T. Kato, G. F. Schreiner *et al.*, 'Essential fatty acid deficiency normalizes function and histology in rat nephrotoxic nephritis.' *Kidney Int.*, 1992, **41**, 1245–1253.

276. A. Rifai, H. Sakai and M. Yagame, 'Expression of 5-lipoxygenase and 5-lipoxygenase activation protein in glomerulonephritis.' *Kidney Int.*, 1993, **39** Suppl., S95–S99.

277. K. V. Hackshaw, N. F. Voelkel, R. B. Thomas *et al.*, 'Urine leulotriene E4 levels are elevated in patients with active systemic lupus erythematosus.' *J. Rheumatol.*, 1992, **19**, 252–258.

278. C. D. Funk, X. S. Chen, U. Kurre *et al.*, 'Leukotriene-deficient mice generated by targeted disruption of the 5-lipoxygenase gene.' *Adv. Prostaglandin Thromboxane Leuko. Res.*, 1995, **23**, 145–150.

279. F. A. Fitzpatrick and R. C. Murphy, 'Cytochrome P450 metabolism of arachidonic acid: formation and biological actions of "epoxygenase-" derived eicosanoids.' *Pharmacol. Rev.*, 1988, **40**, 229–241.

280. D. L. Hirt and H. R. Jacobson, 'Functional effects of cytochrome P450 arachidonate metabolites in the kidney.' *Semin. Nephrol.*, 1991, **11**, 148–155.

281. J. C. McGiff, C. P. Quilley and M. A. Carroll, 'The contribution of cytochrome P450-dependent arachidonate metabolites to integrated renal function.' *Steroids*, 1993, **58**, 573–579.

282. D. Schlondorf and R. Ardaillou, 'Prostaglandins and other arachidonic acid metabolites in the kidney.' *Kidney Int.*, 1986, **29**, 108–119.

283. M. L. Schwartzman, P. Martasek, A. R. Rios *et al.*, 'Cytochrome P450-dependent arachidonic acid metabolism in human kidney.' *Kidney Int.*, 1990, **37**, 94–99.

284. R. E. Garrick, in 'Contemporary Issues in Nephrology: Hormones, Autacoids, and the Kidney,' eds. S. Goldfarb and F. N. Ziyadeh, Churchill Livingstone, New York, 1991, vol. 23, pp. 231–262.

285. A. Karara, K. Makita, H. R. Jacobson *et al.*, 'Molecular cloning, expression, and enzymatic characterization of the rat kidney cytochrome P450 arachidonic acid epoxygenase.' *J. Biol. Chem.*, 1993, **268**, 13565–13570.

286. M. D. Breyer, J. Capdevila, R. Dubois *et al.*, 'Localization of rat renal cytochrome P450 isoforms.' *J. Am. Soc. Nephrol.*, 1993, **4**, 451.

287. S. Imaoka, K. Nagashima and Y. Funae, 'Characterization of three cytochrome P450s purified from renal microsomes of untreated male rats and comparison with human renal cytochrome P450.' *Arch. Biochem. Biophys.*, 1990, **276**, 473–480.

288. S. Kimura, J. P. Hardwick, C. A. Kozak *et al.*, 'The rat clofibrate-inducible CYP4A subfamily. II. cDNA sequence of IVA3, mapping of the Cyp4a locus to mouse chromosome 4, and coordinate and tissue-specific regulation of the CYP4A genes.' *DNA*, 1989, **8**, 517–525.

289. J. Quilley, C. P. Auilley and J. C. McGiff, in 'Hypertension: Pathophysiology, Diagnosis, and Management,' 2nd edn., eds. J. H. Laragh and B. M. Brenner, Raven Press, New York, pp. 829–840.

290. M. A. Carroll, M. P. Garcia, J. R. Falck *et al.*, '5,6-Epoxyeicosatrienoic acid, a novel arachidoniate metabolite. Mechnism of vasoactivity in the rat.' *Circ. Res.*, 1989, **67**, 1082–1088.

291. K. Takahashi, J. Capdevila, A. Karara *et al.*, 'Cytochrome P450 arachidonate metabolites in rat kidney: characterization and hemodynamic responses.' *Am. J. Physiol.*, 1990, **258**, F781–F789.

292. M. A. Carroll, M. Schwartzman, J. Capdevila *et al.*, 'Vasoactivity of arachidonic acid epoxides.' *Eur. J. Pharmacol.*, 1987, **38**, 281–283.

293. M. A. Carroll, M. P. Garcia, J. R. Flack *et al.*, 'Cyclo-oxygenase dependency of the renovascular actions of cytochrome P450-derived arachidonate metabolites.' *J. Pharmacol. Exp. Ther.*, 1992, **260**, 104–109.

294. B. Escalante, W. C. Sessa, J. R. Flack *et al.*, 'Vasoactivity of 20-hydroxyeicosatetraenoic acid is dependent on metabolism by cyclooxygenase.' *J. Pharmacol. Exp. Ther.*, 1989, **248**, 229–232.

295. B. Escalante, D. Erlij, J. R. Falck *et al.*, 'Effect of cytochrome P450 arachidonate metabolites on ion transport in rabbit kidney loop of Henle.' *Science*, 1991, **251**, 799–802.

296. D. L. Hirt, J. Capdevila, J. R. Falck *et al.*, 'Cytochorme P450 metabolites of arachidonic acid are potent inhibitors of vasopressin action on rabbit cortical collecting duct.' *J. Clin. Invest.*, 1989, **84**, 1805–1812.

297. M. Schwartzman, N. R. Ferreri, M. A. Carroll *et al.*, 'Renal cytochrome P450-related arachidonate metabolite inhibits ($Na^+ + K^+$) ATPase.' *Nature*, 1985, **314**, 620–622.

298. W. L. Henrich, J. R. Falck and W. B. Campbell, 'Inhibition of renin release by 14,15-epoxyeicosatrienoic acid in renal cortical slices.' *Am. J. Physiol.*, 1990, **258**, E269–E274.

299. Z. T. Madhun, D. A. Goldwait, D. McKay *et al.*, 'An epoxygenase metabolite of arachidonic acid mediates angiotensin II-induced rises in cytosolic calcium in rabbit proximal tubule epithelial cells.' *J. Clin. Invest.*, 1991, **88**, 456–461.

300. M. F. Romero, J. G. Douglas and U. Hopfer, in 'Proceedings of 7th International Conference on Prostaglandins and Related Compounds, Florence, Italy, 1990.'

301. J. H. Capdevila, S. Wei, J. Yan *et al.*, 'Cytochrome P450 arachidonic acid epoxygenase. Regulatory control of the renal epoxygenase by dietary salt loading.' *J. Biol. Chem.*, 1992, **267**, 21720–21776.

302. K. F. Badr, K. Takahashi and J. R. Falck, 'Induction of cytochrome P450-linked metabolism of arachidonic acid following unilateral nephrectomy and during experimental diabetes and midterm pregnancy in the mammalian kidney.' *Clin. Res.*, 1988, **36**, 513A.

303. F. Catella, J. A. Lawson, D. J. Fitzgerald *et al.*, 'Endogenous biosynthesis of arachidonic acid

epoxides in humans: increased formation in pregnancy-induced hypertension.' *Proc. Natl. Acad. Sci. USA*, 1990, **87**, 5893–5897.

304. D. Sacerdoti, N. G. Abraham, J. C. McGiff *et al.*, 'Renal cytochrome P450-dependent metabolism of arachidonic acid in spontaneously hypertensive rats.' *Biochem. Pharmacol.*, 1988, **37**, 521–527.

305. D. Sacerdoti, B. Escalante, N. G. Abraham *et al.*, 'Tratment with tin prevents the development of hypertension in spontaneously hypertensive rats.' *Science*, 1989, **243** (4889), 388–390.

306. J. H. Capdevila, K. Takahashi and H. R. Jacobson, 'Inhibition of the kidney P450 arachidonate epoxygenase causes hypertension in salt loaded rats.' *J. Am. Soc. Nephrol.*, 1993, **4**, 508.

307. M. Yanagisawa, H. Kurihara, S. Kimura *et al.*, 'A novel potent vasoconstrictor peptide produced by vascular endothelial cells.' *Nature*, 1988, **332**, 411–415.

308. M. S. Simonson and M. J. Dunn, 'Endothelin peptides and the kidney.' *Annu Rev. Physiol.*, 1993, **55**, 249–265.

309. V. Kon and K. F. Badr, 'Biological actions and pathophysiological significance of endothelin in the kidney.' *Kidney Int.*, 1991, **40**, 1–12.

310. E. P. Nord, 'Renal actions of endothelin.' *Kidney Int.*, 1993, **44**, 451–463.

311. G. M. Rubanyi and L. H. Botelho, 'Endothelins.' *FASEB J.*, 1991, **5**, 2713–2729.

312. A. Inoue, M. Yanagisawa, S. Kimura *et al.*, 'The human endothelin family: three structurally and pharmacologically distinct isopetides predicted by three separate genes.' *Proc. Natl. Acad. Sci. USA*, 1989, **86**, 2863–2867.

313. P. A. Marsden, M. S. Goligorsky and B. M. Brenner, 'Endothelial cell biology in relation to current concepts of vessel wall structure and function.' *J. Am. Soc. Nephrol.*, 1991, **1**, 931–948.

314. S. Kimura, Y. Kasuya, T. Sawamura *et al.*, 'Conversion of big endothelin-1 to 21-residue endothelin-1 is essential for expression of full vasoconstricor activity: structure-activity relationships of big endothelin-1.' *J. Cardiovasc. Pharmacol.*, 1989, **13**, S5–S7.

315. T. F. Luscher, C. M. Boulanger, Y. Dohi *et al.*, 'Endothelium-derived contracting factors.' *Hypertension*, 1992, **19**, 117–130.

316. S. Kimura, Y. Kasuya, T. Sawamura *et al.*, 'Structure-activity relationships of endothelin: importance of the C-terminal moiety.' *Biochem. Biophys. Res. Commun.*, 1988, **156**, 1182–1186.

317. G. Landan, A. Bdolah, Z. Wollberg *et al.*, 'Evolution of the sarafotoxin/endothelin superfamily of proteins.' *Toxicon*, 1991, **29**, 237–244.

318. E. E. Anggard, R. M. Botting and J. R. Vane, 'Endothelins.' *Blood Vessels*, 1990, **27**, 269–281.

319. J. Pernow, A. Hemsen and J. M. Lundberg, 'Tissue specific distribution, clearance and vascular effects of endothelin in the pig.' *Biochem. Biophys. Res. Commun.*, 1989, **161**, 647–653.

320. T. Yoshizawa, O. Shinmi, A. Giaid *et al.*, 'Endothelin: a novel peptide in the posterior pituitary system.' *Science*, 1990, **247**, 462–464.

321. A. Giaid, S. J. Gibson, N. B. Ibrahim *et al.*, 'Endothelin 1, an endothelium-derived peptide, is expressed in neurons of the human spinal cord and dorsal root ganglia.' *Proc. Natl. Acad. Sci. USA*, 1989, **86**, 7634–7638.

322. T. J. Resink, A. W. Hahn, T. Scott-Burden *et al.*, 'Inducible endothelin mRNA expression and peptide secretion in cultured human vascular smooth muscle cells.' *Biochem Biophys. Res. Commun.*, 1990, **168**, 1303–1310.

323. P. A. Baley, T. J. Resink, U. Eppenberger *et al.*, 'Endothelin messenger RNA and receptors are differentially expressed in cultured human breast epithelial and stromal cells.' *J. Clin. Invest.*, 1990, **85**, 1320–1323.

324. B. M. Wilkes, M. Susin, P. F. Mento *et al.*, 'Localization of endothelin-like immunoreactivity in rat kidneys.' *Am. J. Physiol.*, 1991, **260**, F913–F920.

325. D. E. Kohan, 'Endothelin synthesis by rabbit renal tubule cells.' *Am. J. Physiol.*, 1991, **261**, F221–F226.

326. D. E. Kohan, 'Endothelin production by human inner medullary collecting duct cells.' *J. Am. Soc. Nephrol.*, 1993, **3**, 1719–1721.

327. K. Ujiie, Y. Terada, H. Nonoguchi *et al.*, 'Messenger RNA expression and synthesis of endothelin-1 along rat nephron segments.' *J. Clin. Invest.*, 1992, **90**, 1043–1048

328. T. F. Luscher, H. A. Bock, Z. H. Yang *et al.*, 'Endothelium-derived relaxing and contracting factors: perspectives in nephrology.' *Kidney Int.*, 1991, **39**, 575–590.

329. M. S. Simonson, 'Endothelins: multifunctional renal peptides.' *Physiol. Rev.*, 1993, **73**, 375–411.

330. T. Masaki, 'Endothelins: homeostatic and compensatory actions in the circulatory and endocrine systems.' *Endocr. Rev.*, 1993, **14**, 256–268.

331. H. Rakugi, S. Tabuchi, M. Nakamura *et al.*, 'Evidence for endothelin-1 release from resistance vessels of rats in response to hypoxia.' *Biochem. Biophys. Res. Commun.*, 1990, **169**, 973–977.

332. D. E. Kohan and E. Pallida, 'Osmolar regulation of endothelin-1 production by rat medullary collecting duct.' *J. Clin. Invest.*, 1993, **91**, 1235–1240.

333. T. Yang, Y. Terada, H. Nonoguchi *et al.*, 'Effect of hyperosmolality on production and mRNA expression of ET-1 in inner medullary collecting duct.' *Am. J. Physiol.*, 1993, **264**, F684–F689.

334. C. M. Boulanger and T. F. Luscher, 'Hirudin and nitrates inhibit the thrombin-stimulated release of endothelin from the intact porcine aorta.' *Circ. Res.*, 1991, **68**, 1768–1772.

335. M. Kohno, T. Horio, M. Ikeda *et al.*, 'Natriueretic peptides inhibit mesangial cell production of endothelin induced by arginine vasopressin.' *Am. J. Physiol.*, 1993, **264**, F678–F683.

336. O. Saijonmaa, A. Ristirnaki and F. Fhyrquist, 'Atrial natriuretic peptide, nitroglycerin, and nitroprusside reduce basal and stimulated endothelin production from cultured endothelial cells.' *Biochem. Biophys. Res. Commun.*, 1990, **173**, 514–520.

337. M. S. Simonson, 'Endothelin peptides and compensatory growth of renal cells.' *Curr. Opin. Nephrol. Hypertens.*, 1994, **3**, 73–85.

338. T. F. Luscher, B. S. Oemar, C. M. Boulanger *et al.*, 'Molecular and cellular biology of endothelin and its receptors. Part I.' *J. Hypertens.*, 1993, **11**, 7–11.

339. T. F. Luscher, B. S. Oemar, C. M. Boulanger *et al.*, 'Molecular and cellular biology of endothelin and its receptors. Part II.' *J. Hypertens.*, 1993, **11**, 121–126.

340. Y. K. Terada, K. Tomita, H. Nonoguchi *et al.*, 'Different localization of two types of endothelin receptor mRNA in microdissected rat nephron segments using reverse transcription and polymerase chain reaction assay.' *J. Clin. Invest.*, 1992, **90**, 107–112.

341. F. E. Karet, R. E. Kuc, A. P. Davenport *et al.*, 'Novel ligands BQ123 and BQ2030 characterize endothelin receptor subtypes ETA and ETB in human kidney.' *Kidney Int.*, 1993, **44**, 36–42.

342. M. A. Simonson, S. Wann, P. Mene *et al.*, 'Endothelin stimulates phospholiase C, Na$^+$/H$^+$ exchange, c-fos expression, and mitogenesis in rat mesangial cells.' *J. Clin. Invest.*, 1989, **83**, 708–712.

343. K. F. Badr, J. J. Murray, M. D. Breyer *et al.*,

'Mesangial cell, glomerular and renal vascular responses to endothelin in the rat kidney. Elucidation of signal transolution pathways.' *J. Clin Invest.*, 1989, **83**, 336–342.

344. K. Goto, Y. Kauya, N. Matsuki *et al.*, 'Endothelin activates the dihydropyridine-sensitive, voltage-dependent Ca2+ channel in vascular smooth muscle.' *Proc. Natl. Acad. Sci. USA*, 1989, **86**, 3915–3918.

345. T. F. Luscher, 'Endothelin, endothelin receptors, and endothelin antagonists.' *Curr. Opin. Nephrol. Hypertens.*, 1993, **3**, 92–98.

346. M. Clozel, G. A. Gray, V. Breu *et al.*, 'The endothelin ETB receptor mediates both vasodilation and vasoconstriction *in vivo*.' *Biochem. Biophys. Res. Commun.*, 1992, **186**, 867–873.

347. A. J. King and B. M. Brenner, 'Endothelium-derived vasoactive factors and the renal vasculature.' *Am. J. Physiol.*, 1991, **260**, R653–R662.

348. H. Lin, M. Sangmal, M. J. Smith, Jr., *et al.*, 'Effect of endothelin-1 on glomerular hydraulic pressure and renin release in dogs.' *Hypertension*, 1993, **21**, 845–851.

349. R. M. Edwards, W. Trizna and E. H. Ohlstrein, 'Renal microvascular effects of endothelin.' *Am. J. Physiol.*, 1990, **259**, F217–F221.

350. K. A. Munger, K. Takahashi. M. Awazu *et al.*, 'Maintenance of endothelin-induced renal arteriolar constriction in rats is cyclooxygenase dependent.' *Am. J. Physiol.*, 1993, **264**, F637–F644.

351. M. L. Zeidel, H. R. Brady, B. C. Kone *et al.*, 'Endothelin, peptide inhibitor of Na+-K+-ATPase in intact renal tubular epithelial cells.' *Am. J. Physiol.*, 1989, **257**, C1101–C1107.

352. T. Horio, M. Kohno and T. Takeda, 'Cosecretion of atrial and brain natriuretic peptides stimulated by endothelin-1 from cultured rat atrial and ventricular cardiocytes.' *Metabolism*, 1993, **42**, 94–96.

353. K. A. Munger, M. Suguira and K. Takahashi, 'A role for atrial natriuretic peptide in endothelin-induced natriuresis.' *J. Am. Soc. Nephrol.*, 1991, **1**, 1278–1283.

354. F. Mallamaci, S. Parlongo and C. Zoccali, 'Influence of cardiovascular damage and residual renal function on plasma endothelin in chronic renal failure.' *Nephron*, 1993, **63**, 291–295.

355. A. N. Warrens, M. J. Cassidy, K. Takahashi *et al.*, 'Endothelin in renal failure.' *Nephrol. Dial. Transplant.*, 1990, **5**, 418–422.

356. R. D. Ross, V. Kalidindi, J. A. Vincent *et al.*, 'Acute changes in endothelin-1 after hemodialysis for chronic renal failure.' *J. Pediatr.*, 1993, **122**, S74–S76.

357. K. Takahashi, K. Totsune, Y. Imai *et al.*, 'Plasma concentrations of immunoreactive-endothelin in patients with chronic renal failure treated with recombinant human erythropoietin.' *Clin., Sci. (Colch.)*, 1993, **84**, 47–50.

358. A. L. Clavell and J. C. Burnett, Jr., 'Physiologic and pathophysiologic roles of endothelin in the kidney.' *Curr. Opin. Nephrol. Hypertens.*, 1994, **3**, 66–72.

359. G. M. Nassar and K. F. Badr, 'Endothelin in kidney disease.' *Curr. Opin. Nephrol. Hypertens.*, 1994, **3**, 86–91.

360. K. Yokokawa, H. Tahara, M. Kohno *et al.*, 'Hypertension associated with endothelin-secreting malignant hemangioendothelioma.' *Ann. Intern. Med.*, 1991, **114**, 213–215.

361. N. Perico and G. Remuzzi, 'Role of endothelin in glomerular injury.' *Kidney Int.*, 1993, **39** Suppl., S76–S80.

362. S. Oriso, A. Benigni, I. Bruzzi *et al.*, 'Renal endothelin gene expression is increased in remnant kidney and correlates with disease progression.' *Kidney Int.*, 1993, **43**, 354–358.

363. J. D. Firth and P. J. Ratcliffe, 'Organ distribution of the three rat endothelin messenger RNAs and the effects of ischemia on renal gene expression.' *J. Clin. Invest.*, 1992, **90**, 1023–1031.

364. N. Mino, M. Kobayashi, A. Nakajima *et al.*, 'Protective effect of a selective endothelin receptor antagonist, BQ-123, in ischemic acute renal failure in rats.' *Eur. J. Pharmacol.*, 1992, **221**, 77–83.

365. A. Lopez-Farre, D. Gomez-Garre, F. Bernabeu *et al.*, 'A role for endothelin in the maintenance of post-ischaemic renal failure in the rat.' *J. Physiol. (London)*, 1992, **444**, 513–522.

366. S. N. Heyman, B. A. Clark, N. Kaiser *et al.*, 'Radiocontrast agents induce endothelin release *in vivo* and *in vitro*.' *J. Am. Soc. Nephrol.*, 1992, **3**, 58–65.

367. K. B. Marguilies, F. L. Hildebrand, D. M. Heublein *et al.*, 'Radiocontrast increases plasma and urinary endothelin.' *J. Am. Soc. Nephrol.*, 1991, **2**, 1041–1045.

368. V. Kon, M. Sugiura, T. Inagami *et al.*, 'Role of endothelin in cyclosporine-induced glomerular dysfunction.' *Kidney Int.*, 1990, **37**, 1489–1491.

369. I. T. Bloom, F. R. Bentley and R.N. Garrison, 'Acute cyclosporine-induced renal vasoconstriction is medi-ated by endothelin-1.' *Surgery*, 1993, **114**, 480–487.

370. D. M. Lanese and J. D. Conger, 'Effects of endothelin receptor antagonist on cyclosporine-induced vasoconstriction in isolated rat renal arterioles.' *J. Clin. Invest.*, 1993, **91**, 2144–2149.

371. K. Moore, J. Wendon, M. Frazer *et al.*, 'Plasma endothelin immunoreactivity in liver disease and the hepatorenal syndrome.' *N. Engl. J. Med.*, 1992, **327**, 1774–1778.

372. R. F. Furchgott and J. V. Zawadzki, 'The obligatory role of endothelial cells in the relaxation of arterial smooth muscle by acetylcholine.' *Nature*, 1980, **288**, 373–376.

373. R. M. Palmer, A. G. Ferrige and S. Moncada, 'Nitric oxide release accounts for the biological activity of endothelium-derived relaxing factor.' *Nature*, 1987, **327**, 524–526.

374. L. J. Ignarro, 'Endothelium-derived nitric oxide: actions and properties.' *FASEB J.*, 1989, **3**, 31–36.

375. R. M. Palmer, D. S. Ashton and S. Moncada, 'Vascular endothelial cells synthesize nitric oxide from L-arginine.' *Nature*, 1988, **333**, 664–666.

376. R. M. Palmer and S. Moncada, 'A novel citrulline-forming enzyme implicated in the formation of nitric oxide by vascular endothelial cells.' *Biochem. Biophys. Res. Commun.*, 1989, **158**, 348–352.

377. D. D. Rees, R. M. Palmer, H. F. Hodson *et al.*, 'A specific inhibitor of nitric oxide formation from L-arginine attenuates endothelium-dependent relaxation.' *Br. J. Pharmacol.*, 1989, **96**, 418–424.

378. P. J. Shultz, A. E. Schorer and L. Raij, 'Effects of endothelium-derived relaxing factor and nitric oxide on rat mesangial cells.' *Am. J. Physiol.*, 1990, **258**, F162–F167.

379. P. A. Marsden, T. A. Brock and B. J. Ballermann, 'Glomerular endothelial cells respond to calcium mobilizing agonists with release of endothelium derived relaxing factor.' *Clin. Res.*, 1989, **37**, 496A.

380. A. J. King, J. L. Troy, S. J. Downes *et al.*, 'Effects of N-monomethyl-L-arginine (L-NMMA) on basal renal hemodynamics and the response to amino acid infusion.' *Kidney Int.*, 1990, **37**, 371.

381. J. P. Tolins, R. M. J. Palmer, S. Moncada *et al.*, 'Role of endothelium-derived relaxing factor in regulation of renal hemodynamic responses.' *Am. J. Physiol.*, 1990, **258**, H655–H662.

382. R. Loutzenhiser, K. Hayashi and M. Epstein, 'Evidence for multiple endothelial-derived relaxing factors in the renal microcirculation.' *Kidney Int.*, 1990, **37**, 373.

383. C. Boulanger and T. F. Luescher, 'Release of endothelin from the porcine aorta. Inhibition by endothelium-derived nitric acid.' *J. Clin. Invest.*, 1990, **85**, 587–590.

384. M. J. Vidal, J. C. Romero and P. M. Vanhoutte, 'Endothelium-derived relaxing factor inhibits renin release.' *Eur. J. Pharmacol.*, 1988, **149**, 401–402.

385. K. Takahashi, T. Katoh and K. F. Badr, in 'International Yearbook of Nephrology, 1991,' eds. V. E. Andreucci and L. G. Fine, Kluwer Academic, Boston, MA, 1991, pp. 3–19.

386. R. Ferrario, A. Fogo, K. Takahashi *et al.*, 'Micro-circulatory responses to inhibition of nitric oxide (NO) synthesis in normal kidneys and during acute glomerulonephritis in the rat.' *J. Am. Soc. Nephrol.*, 1991, **2**, 504.

387. C. Baylis, L. Samsell and A. Deng, 'A new model of systemic hypertension with high glomerular capillary blood pressure (PGC) and proteinuria: chronic blockade of endogenous endothelial derived relaxing factor (EDRF).' *J. Am. Soc. Nephrol.*, 1991, **2**, 471.

388. A. A. Reyes, M. L. Purkerson, I. Karl *et al.*, 'Dietary supplementation with L-arginine ameliorates the progression of renal disease in rats with subtotal nephrectomy.' *Am. J. Kidney Dis.*, 1992, **20**, 168–176.

389. T. Katoh, K. Takahashi, S. Klahr *et al.*, 'Dietary supplementation with L-arginine ameliorates glomerular hypertension in rats with subtotal nephrectomy.' *J. Am. Soc. Nephrol.*, 1994, **4**, 1690–1694.

390. D. J. Stuehr and C. F. Nathan, 'Nitric oxide. A macrophage product responsible for cytostasis and respiratory inhibition in tumor target cells.' *J. Exp. Med.*, 1989, **169**, 1543–1555.

391. R. D. Curran, T. R. Billiar, D. J. Stuehr *et al.*, 'Multiple cytokines are required to induce hepatocyte nitric oxide production and inhibit total protein synthesis.' *Ann. Surg.*, 1990, **212**, 462–469.

392. T. R. Billiar, R. D. Curran, D. J. Stuehr *et al.*, 'Evidence that activation of Kupffer cells results in production of L-arginine metabolites that release cell-associated iron and inhibit hepatocyte protein synthesis.' *Surgery*, 1989, **106**, 364–372.

393. G. Werner-Felmayer, E. R. Werner, D. Fuchs *et al.*, 'Tetrahydrobiopterin-dependent formation of nitrite and nitrate in murine fibroblasts.' *J. Exp. Med.*, 1990, **172**, 1599–1607.

394. R. Busse and A. Mulsch, 'Induction of nitric oxide synthase by cytokines in vascular smooth muscle cells.' *FEBS Lett.*, 1990, **275**, 87–90.

395. V. B. Schini, D. C. Junquero, T. Scott-Burden *et al.*, 'Interleukin-1β induces production of an L-arginine-derived relaxing factor from cultured smooth muscle cells from rat aorta.' *Biochem. Biophys. Res. Commun.*, 1991, **176**, 114–121.

396. J. Pfeilschifter and K. Vosbeck, 'Transforming growth factor 2β inhibits interleukin 1β- and tumor necrosis factor a-induction of nitric oxide synthase in rat renal mesangial cells.' *Biochem. Biophys. Res. Commun.*, 1991, **175**, 372–379.

397. U. Foerstermann, H. H. Schmidt, J. S. Pollock *et al.*, 'Isoforms of nitric oxide synthase. Characterization and purification from different cell types.' *Biochem. Pharmacol.*, 1991, **42**, 1849–1857.

398. R. G. Kilbourn and P. Belloni, 'Endothelial cell production of nitrogen oxides in response to inter-feron-γ in combination with tumor necrosis factor, interleukin-1, or endotoxin.' *J. Natl. Cancer Inst.*, 1990, **82**, 772–776.

399. M. W. Radomski, R. M. Palmer and S. Moncada, 'Glucocorticoids inhibit the expression of an inducible, but not the constitutive, nitric oxide synthase in vascular endothelial cells.' *Proc. Natl. Acad. Sci. USA*, 1990, **87**, 10043–10047.

400. K. Takahashi, T. Katoh, M. Fukunaga *et al.*, 'Studies on the glomerular microcirculatory actions of manidipine and its modulation of the systemic and renal effects of endothelin.' *Am. Heart J.*, 1993, **125**, 609–619.

7.11
The Glomerulus: Mechanisms of Injury

MAURO ABBATE

Istituto de Ricerche Farmacologiche Mario Negri, Bergamo, Italy

and

GIUSEPPE REMUZZI

Ospedali Riuniti di Bergamo and Istituto de Ricerche Farmacologiche Mario Negri, Bergamo, Italy

Table 1   Forms of glomerular injury caused by drugs and toxic substances in humans or experimental animals.

Toxic effect	Clinico-pathologic picture	Agents
Epithelial cell disease	Minimal change disease, proteinuria	Puromycin aminonucleoside Adriamycin Cimetidine Gold Lithium Cadmium Fenoprofen Nonsteroidal anti-inflammatory drugs Heroin Antibiotics Leucocyte alpha-interferon
Endothelial cell disease	Hemolytic uremic syndrome/ Thrombotic thrombocytopenic purpura	Mitomycin Cyclosporin-A
Immune complex disease Subepithelial deposits	Membranous glomerulonephritis	Mercuric chloride Gold Cadmium Penicillamin Captopril Hydrocarbons Trimethadione Chloromethiazole
Anti-glomerular basement membrane antibodies	Proliferative glomerulonephritis	Mercuric chloride Volatile hydrocarbons Solvents
Subendothelial, mesangial deposits	Systemic lupus erythematosus-like	Hydralazine Procainamide Nifedipine

Source: Hook and Goldstein, p. 154.[3]

7.11.1 INTRODUCTION

Drugs and chemicals may cause glomerular structural or functional changes both in experimental animals and humans (Table 1).[1,2] Adriamycin and puromycin for instance induce direct damage to glomerular epithelial cell as consistently documented by experimental studies showing forms of nephrotic syndrome in animals caused by the above toxics. Mitomycin is instead toxic to glomerular endothelial cells and determine microvascular thrombosis *in vivo*. However, the toxic effect of drugs on the glomerulus are not necessarily associated with structural changes. Cyclosporine (CsA), gentamicin, and amphotericin B are examples of drugs that, at least upon acute administration, alter glomerular function and lead to significant decline in glomerular filtration rate (GFR) without changes in its structural integrity.

In humans reports of glomerular disease causally related to drugs or toxic agents are less consistent than for acute tubular necrosis or tubulo-interstitial diseases, suggesting a role for additional, genetic factors, which are ill defined. Moreover many glomerular diseases are immune-mediated so that the glomerulus has been viewed as a secondary target of nephrotoxic agents. Thus, loss of glomerular permselectivity in cases of drug-induced minimal change disease or focal segmental glomerulosclerosis was suspected to result from abnormalities in cell-mediated immunity and cytokine production. In immunecomplex glomerulonephritis, drugs or toxicants are thought to interfere with the immune system with eventual deposition of pathogenic antibodies in the glomerulus. For example, rather than targeting glomerular cells, procainamide, hydralazine, and nifedipine would activate antinuclear and antihistone antibody formation with subsequent formation of immune complexes that deposit in subendothelial and mesangial space. Further examples are captopril and enalapril, angiotensin converting enzyme-inhibitors (ACEi) which induce a rare form of membranous glomerulophaty in humans by mechanisms which are not yet fully clarified. An additional problem in studying glomerular toxicity is that the effects of many of these compounds cannot be reproduced in animals. A notable exception is experimental mercuric chloride glomerulonephritis, in which polyclonal B-cell activation leads to autoantibodies production reacting with components of the glomerular basement membrane (GBM), that result in immune aggregates forming within the GCW. However, this is again the case of the glomerulus studied as a secondary (i.e., immune-mediated) target of toxicity.

In part due to such limitations, current knowledge of the mechanism(s) of glomerular toxicity of drugs and chemicals is largely descriptive, merely reflecting the associated pathology, or biochemical and functional consequences of exposing glomerular cells to a given toxic molecule. However, understanding of glomerular toxicity is expanding as is glomerular pathophysiology in other fields, like glomerular inflammation and immune reactions, orienting toward new cell surface molecules, soluble mediators, and mechanisms of injury. Furthermore, studies of drug toxicity has allowed us to better define functional and structural determinants of glomerular filtration and to identify pathways of abnormal permeability to water and macromolecules that may cause progressive renal disease. Thanks to major advances in molecular cell biology and powerful tools of investigation recently made available, distinctive molecular targets of toxicity have recently been identified, as a preliminary step to more clear understanding of glomerular pathophysiology.

In this chapter the mechanisms of proteinuria and altered glomerular filtration induced by agents that directly interfere with the glomerular capillary wall (GCW) are reviewed. The potential relevance of mechanisms and mediators of glomerular injury will be discussed, including our current state of knowledge on the mechanisms of toxicity by which drugs like CsA and amphotericin B affect GFR and impair renal function.

7.11.2 GLOMERULAR PERMSELECTIVE PROPERTIES

The GCW functions as a complex filter composed of highly differentiated endothelial and epithelial cells resting on a structurally unique basement membrane. A peculiar feature of the glomerular filter is that despite its remarkably high hydraulic conductivity, which in humans allows it to filter approximately 150 L of water daily, only some 150 mg of the 18 000 g of albumin that pass through glomeruli each day reach Bowman's space.[4] It has long been known that this is possible because the glomerular capillaries are endowed with both size and charge selective properties,[5] so that large molecular weight proteins are denied access to the urine. *In vivo* studies with macromolecular solutes have elucidated the basis of size and charge glomerular permselectivity.[6–11] Test molecules such as dextrans are neither secreted nor reabsorbed along the nephron, and the clearance of those molecules relative to inulin, or fractional clearance, reflects permselective properties intrinsic to the GCW. Both in humans and rats no restriction to filtration is found for molecules <20 Å, with nearly complete restriction for molecules >42 Å.[6–8] However, the filtration of anionic molecules such as albumin is restricted to a much greater extent than is that of neutral dextran having the same effective molecular radius of approximately 36 Å.[9,10] The fractional clearance of dextran sulfate, an anionic polymer of dextran,[10] is also lower for any given size than that of neutral dextran, indicating that native anionic residues of the GCW retard the transcapillary passage of circulating polyanions.

7.11.2.1 Structural Determinants of Size and Charge Selectivity of the Normal Glomerular Filtration Barrier

The determinants of glomerular permeability have been investigated by morphological, biochemical, and theoretical studies, in the attempt to discriminate the relative role of each individual layer of the capillary wall. A novel approach to the dissection of the overall glomerular filtration of water and macromolecules has been proposed by Daniels *et al.*[12,13] who measured the dextran permeance of filters of isolated glomerular basement membrane (GBM) or acellular glomeruli. They determined that the hydraulic conductivity of isolated GBM is nearly 10 times higher than that observed for the glomerulus *in vivo*,[12,13] implicating major roles for cellular components. By computing velocity and pressure fields within the normal GCW on the basis of dimensional data from ultrastructural studies, Drumond and Deen have estimated that approximately 60% of the overall restriction to water flow occurs in the basement membrane and 40% at the slit diaphragms in the epithelial layer.[14] Little restriction in water flux would be expected to occur at the level of the endothelium due to the very large radius (averaging 600 Å) of the glomerular fenestrae. Similarly, the resistance to glomerular albumin permeability resides in cells as well as the GBM.[12,13,15] Fenestrated endothelial cells would confer a fraction of the overall charge selectivity by negative charges in their sialoglycoprotein coat.[16–18] Small amounts of plasma percolate from the fenestrae through channels present in the mesangium and find their way into the interstitium, or into the Bowman's space across the paramesangial basement membrane and the epithelial cell interface.[19] This surface represents a very small fraction of the filtration surface of the glomerulus, and the actual contribution of the mesangial pathway to glomerular permselectivity has never been clarified.

Several studies have emphasized the role of the GBM in preventing albumin loss. This has relied in part on findings that physicochemical and permselective properties become altered even in the absence of ultrastructural changes of resident glomerular cells.[20–23] Immunohistochemical data can also be interpreted to indicate that this structure is effective both as size and charge selective barrier for endogenous proteins.[24,25] Other studies, however, focus on the visceral epithelial cell as key component of the glomerular filter with highly specialized structure and function.

7.11.2.2 The Role of the Glomerular Basement Membrane in Restricting Protein Flow Across the Capillary Wall

The GBM is composed of the three major classes of constituents of the extracellular matrix: collagens, noncollagenous structural glycoproteins, and proteoglycans. The lamina densa of the GBM with its network of fibrils containing type IV collagen[26,27] might offer a significant size selective barrier to plasma proteins.[28] Chains of type IV collagen (α1, α2, α3, and α4) have been detected in the GBM[29,30] and the data indicate that the α3(IV) chain (globular, or noncollagenous, domain) is present in the central zone[27,31] as well as the laminae rarae interna and externa,[27] whereas the α1(IV) chain (NC domain) is excluded from the lamina rara externa.[27] Thanks to such differences in distribution, the normal GBM is a polarized structure, probably reflecting a functional asymmetry that is necessary to maintain an intact filtration barrier. Other protein components of the GBM include type V[32] and type VI collagen,[27] laminin,[33–35] fibronectin, and entactin/nidogen.[33,36–39] There is disagreement over the exact location of many of these components in the GBM[37] as well as their intrinsic roles in restricting the filtration of circulating molecules. Conversely the laminae rarae interna and externa contain "fixed" negatively charged sites which appear spaced at regular intervals of approximately 60 nm and arranged in a lattice-like network, as stained with cationic ferritin or ruthenium red.[40] Kanwar and Farquhar demonstrated that such sites are enriched with heparan sulfate proteoglycan (HSPG), which is composed of heparan sulfate-rich glycosaminoglycan chains covalently bound to a core peptide.[41,42] Importantly, the removal of glycosaminoglycans by enzymatic digestion resulted in an increase in the permeability of the basement membrane to both ferritin[43] and bovine albumin.[44]

Based on the molecular composition of the extracellular matrix, a novel function of basement membranes has been suggested that is possibly relevant to the glomerular permselectivity, in particular to the charge barrier. Extracellular matrix components may act as a functional reservoir by binding circulating or locally secreted molecules in a relatively selective manner, as documented for some growth factors, cytokines, and adhesion molecules. Bound molecules might influence permeability by virtue of electric charge and perhaps more specific biologic actions on the capillary wall. Heparin-containing proteoglycans, for instance, can bind polypeptides of the family of vascular permeability factor/vascular endothelial growth factor (VPF/VEGF), molecules with potent angiogenic and permeabilizing activity.[45] VPF/VEGF is also synthesized within the glomerulus[46,47] where it probably stimulates endothelial receptors,[48] and has been reported to induce proteinuria when perfused into rat kidneys *ex vivo*.[49] Whether this binding has a regulatory role in the normal glomerulus is unknown, nor are the pathways by which VEGF/VPF disrupt permeability. However, such a mechanism is presumably perturbed when heparan sulfate proteoglycans are removed or neutralized, and may represent a novel target of toxicity. Similarly, another vasoactive molecule, AII, that caused proteinuria in isolated and perfused rat kidney preparations,[50] might bind to basement membrane components so as to mediate proteinuria through yet unclear actions on the cellular components of the wall, or the GBM itself.

7.11.2.3 The Visceral Glomerular Epithelial Cell is the Specialized Component of the Glomerular Barrier

While confirming the role of the GBM in maintaining normal permselectivity,[51] studies by Daniels *et al.*,[12,13] and Drumond and Deen,[14] have underlined the functional significance of glomerular epithelial cells.[4] The visceral glomerular epithelial cells adhere to the GBM by their specialized extensions, the foot processes, and expose most of the surface of the processes and the main cell body into the urinary space. Attention has been focused on the slit diaphragms, now considered as a unique type of tight junction with anionic charge, which connect foot processes of adjacent cells and cover 3–10% of the GBM area.[4] In fact, the function of slit diaphragms in glomerular permselectivity was first hypothesized by Rodewald and Karnowsky, who identified its regular porous structure.[52] It has been speculated that dynamic assembly and disassembly of junctions could also be involved in regulation of the hydraulic conductance across the GCW.[53] Extensive dislocation and rearrangement of slit diaphragms was observed in rat kidneys following perfusion with polycations, like protamine sulfate and poly-L-lysine. The slit diaphragms appeared dislocated toward the urinary space, and typical tight junctions were formed close to the GBM at the original location at the slit diaphragm.[53–55] Proteins that may participate in these processes are beginning to be identified, and include the tight junction protein, ZO-1, associated with the anchoring site of the slit diaphragm in the cell membrane,[56] and a 51 kDa protein on the slit diaphragms and the epithelial aspect of the GBM in the rat. The latter antigen has not been characterized yet, but injection of a monoclonal antibody which is apparently specific induced proteinuria in rats.[57,58] The surface coat of visceral glomerular epithelial cells in proximity of the slit diaphragm is likely to support the charge barrier by virtue of its high content in negatively charged sialic acid-rich glycoproteins, which form the "epithelial polyanion."[59,60] A large proportion of sialic acid in the epithelial polyanion is carried by a major glomerular glycoprotein, podocalyxin, which has an apparent molecular weight of 140 kD in the rat and forms a doublet of 160 and 170 kDa in humans.[61] Podocalyxin and sialic acid residues have been localized in the luminal domain above the slit diaphragm both on the foot processes and the main cell body.[18,62] Of interest, besides sialic acid, sulfate residues in podocalyxin also contribute to the negative charge of the epithelial polyanion.[63] Available cloning and sequencing data of rabbit podocalyxin indicate that its amino acid sequence is not homologous to those of other sequenced proteins.[64]

A specific cell associated form of heparan sulfate proteoglycan was also detected in the same membrane domain which contains podocalyxin.[65] Like for proteoglycans in the GBM, a potential role of such a component besides providing a negative charge is to indirectly influence the glomerular barrier by binding bioactive molecules.

Both digestion[43,44] and neutralization[66,67] experiments documented that anionic charges influence the permselective properties, but the question of which is the most important site of the charge dependent barrier still remains unanswered. This is in part due to the fact that sialic acid, sulfate, and carboxyl residues on structural (glyco)proteins which are considered to represent the biochemical basis for the charge selectivity are not invariably segregated in specialized structures in the GCW.[41,42,59,60,63,68] More importantly, the size and charge barriers seem to be intimately interrelated. Thus, sulfate and carboxyl groups in the GBM may play a role both for charge and size selectivity, as suggested by the observation that experimental infusion of a polycation, hexadimethrine, induces an increase in pore size.[69] A defective charge density of the GCW might determine reorientation and defective interactions of individual basement membrane components, and perhaps of critical structures of the epithelial cell layer including the slit diaphragms, leading to altered size selectivity.[70] Conversely, an enlargement of the effective radius of the theoretical "pores" responsible for the size selective barrier would prevent anionic components to restrict the passage of macromolecules across the capillary, resulting in an apparent charge-selective defect.[71] Perhaps future experiments in animals to knock-out protein components of the capillary wall, like proteins of the slit diaphragm and podocalyxin, will eventually unravel the function of such molecules in the glomerulus and their relative importance in restricting glomerular filtration.

7.11.2.4 Glomerular Basement Membrane–Epithelial Cell Interactions and Related Functional Attributes of the Glomerular Podocyte

Besides producing proteinuria, the exposure of the rat kidney to polycations, such as protamine and polylysine, causes the foot processes to "efface," that is, to be replaced by a continuous cytoplasmic layer along the lamina rara externa.[67,72] Perfusion of the rat kidney with neuraminidase, which removes sialic acid, caused detachment of both endothelial and epithelial cells from the GBM.[66] Thus, there appears to be a close relationship between the shape of podocytes, their surface negative charge and proteinuria.[73] The denudation of the GBM resulting from epithelial cell detachment is the site of increased permeability[74] and may explain anatomically the "shunt pathway" described by mathematical models of glomerular permeability.[4] However, little is known of the cell biological mechanisms that underlie the retraction of foot processes and other important changes of glomerular visceral epithelial cells associated with loss of permselectivity. Since the investigation of such mechanisms relies on the knowledge of the molecular inventory and the functional properties of the podocyte, it is necessary to briefly consider some additional features of these highly differentiated cells (Table 2).

The normal relationship between the podocyte and the GBM is mediated by the membrane domain of the foot processes which lies below the slit diaphragms and establishes direct contact with the basement membrane. The only adhesion molecule localized by immunoelectron microscopy in this domain is a β1 integrin,[75] apparently associated with an integrin chain of the α3 type.[76] This integrin matrix receptor presumably binds via the β1-chain to arginine-glycine-aspartic acid (RGD)-motifs that are contained in fibronectin and other ligands, as well as collagen and laminin within the basement membrane matrix (for review see Ref. 77). Matrix receptors and their participation in substrate adhesion have been further characterized on rat glomerular epithelial cells in culture. Like the podocyte *in vivo*, these cells have been reported to express only the α3 β1-type of integrin complexes.[78] The cells were almost completely prevented from adhering to laminin, fibronectin, and collagen by anti-FX1A antibodies that recognize β1 integrins *in vitro*.[78] Importantly, such antibodies also caused epithelial cell detachment from the substrate, raising the intriguing possibility that the disruption of epithelial cell adhesion to the GBM is an essential part of the mechanism by which anti-Fx1A antibody

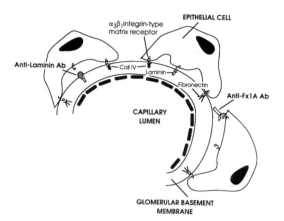

Figure 1 The normal relationship between the podocyte and the GBM is maintained by an integrin matrix receptor of the α3 β1-type. Antibodies which recognize integrin matrix receptors and perhaps other membrane proteins at the surface of the podocyte may produce detachment of epithelial cells from the basement membrane and disrupt glomerular permselectivity (after G. Remuzzi, C. Zoja and T. Bertani, *Curr. Opin. Nephrol. Hyperten.*, 1993, **2**, 465–474).

induce proteinuria *in vivo* (Figure 1).[78] Presumably antibodies that interact with matrix-binding molecules of the glomerular epithelial cells other than the β-1 integrin, possibly including VLA-3[79] and even gp330 and dipeptidyl peptidase IV,[80,81] cause proteinuria by a similar mechanism.[78] Actually, the β1-integrin does not seem to represent the sole anchoring mechanism of podocyte attachment in the normal capillary *in vivo*, since the perfusion of rat kidneys with a competing peptide, RGDS, that is known to break the interaction of cell membranes with the matrix, did not cause the shape of podocytes to change, at least in preliminary experiments.[73] Another protein at the cell surface of podocytes, aminopeptidase A, has been identified as the target of pathogenic antibodies that cause formation of immune deposits in the subepithelial space, podocyte detachment, and proteinuria in mice.[82] However, this is an enzyme with angiotensinase activity that might serve to control locally available AII. The latter molecule is indicated by pharmacological studies as a candidate mediator of altered permeability,[50,83,84] so that antibodies directed against the degrading enzyme promote proteinuria presumably by increasing local AII levels rather than affecting directly epithelial-basement membrane interactions.[82] It is likely that extensive investigation will identify the molecules and precise mechanisms which support or interfere with adhesion of the glomerular epithelial cell to the basement membrane.

One such mechanism could be related to the

Table 2 Functional properties of glomerular epithelial cells relevant in maintaining glomerular permselectivity.

- Synthesis of extracellular matrix components
- Expression of receptors for matrix proteins
- Assembly and maintenance of slit diaphragms
- Synthesis of glycocalyx and negatively charged molecules of the glomerular basement membrane
- Genesis and maintenance of specialized cytoskeletal structures

important function of the visceral glomerular epithelial cell in the synthesis of components of the basement membrane (for review see Tisher and Brenner[51]). Morphological studies and immunohistology clearly established that glomerular epithelial cells have major roles in the assembly of the GBM during development and in the insertion of type IV collagen and laminin.[37,85] In a recent study using *in situ* reverse transcription, mRNA expression of heparan sulfate proteoglycan, laminin A, B-2 and S chains, and entactin have been detected in these cells.[78] Glomerular epithelial cells also participate in the remodeling of the basement membrane with the synthesis of matrix-degrading proteins, such as metalloproteinases.[86] This emphasizes the active role of the glomerular epithelium in maintaining the integrity of the substrate to which it is anchored *in vivo*. Perturbation of this function may disrupt glomerular permselectivity by altering the composition of the GBM and the attachment of the epithelial cell.

Finally, the visceral epithelial cell cytoskeleton has been proposed to support cell adhesion to the GBM through interactions with matrix receptors at the plasma membrane.[87] The foot processes are endowed with a cytoskeletal network which contains actin, myosin, and α-actinin and is apparently linked to microtubules in the cell body.[87,88] Vinculin and talin have also been identified in the plasma membrane of the foot process base.[87] In nonmuscle cells, most likely including the podocyte,[89,90] actin filaments play key roles in determination of cell shape by associating with the plasma membrane.[87] Importantly, such organization can change rapidly, which has major significance in glomerular podocytes that are induced to undergo rapid and dramatic changes in shape without losing contact with the basement membrane. Furthermore, actin, myosin and their associated proteins together with the microtubules and the intermediate filaments, organize the cytoplasm and provide a structure to which other proteins or organelles can bind so that different proteins are localized to different regions of the cell. In particular, in epithelial cells the microtubules have been suggested to govern the genesis and maintenance of cell polarity.[91] In the complex area of the cell biology of the renal glomerulus future studies should attempt to verify whether there are molecules that confer the podocytes with their characteristic shape and to establish molecular mechanisms that in disease reverse their phenotype toward a less differentiated state.

Besides controlling cell adhesion and shape, and polarity, the differentiated cytoskeletal arrangement of the glomerular podocyte might participate in the functional regulation of the filtration of water and macromolecules,[87] perhaps by influencing dynamic assembly and disassembly of junctions. This type of regulation might involve hormones or proteins that modulate glomerular permeability and function. Atrial natriuretic peptide,[92,93] AII,[92,94] and endothelin[95,96] mediate changes in a variety of cell biological processes, signal transduction and cytoskeletal function. In particular, AII has been reported to cause both actin polymerization in epithelial cells of isolated, intact glomeruli[94] and rearrangement of F-actin with decreased actin staining at the periphery of glomerular epithelial cells in culture.[92] These effects might contribute to altered permselectivity in the presence of excess AII in the glomerulus.

7.11.2.5 Mechanisms and Mediators that Regulate Glomerular Ultrafiltration

To yield a nearly ideal ultrafiltrate of plasma the process of glomerular filtration is tightly controlled by mechanisms that act on two major determinants, the net ultrafiltrate pressure and the ultrafiltration coefficient (K_f). Such mechanisms are in turn regulated by the interaction of a variety of hormonal and vasoactive substances either circulating in the plasma or produced locally within the kidney at or near the affected sites.[97] Afferent and efferent arterioles have long been recognized as major sites for regulation of glomerular blood flow and glomerular capillary hydraulic pressure. The glomerular mesangium has been demonstrated to have an important role as both a target and a source of vasoactive molecules that regulate the glomerular capillary surface and therefore K_f.[98] Actually the glomerular mesangial cells are endowed with numerous cytoplasmic contractile myofilaments and plasma membrane specific receptors for several vasoactive substances, such as AII, TxA2, PA$_F$, and endothelin,[99–101] which stimulate the contraction of mesangial cells *in vitro*. In addition, the mesangial cell synthesizes many of these vasoactive substances,[99,102] that may act by an autocrine pattern on the glomerular filtration. Neurohormonal mediators regulate afferent and/or efferent arteriolar tone resulting in changes in renal blood flow (RBF) and glomerular plasma flow rate which translate into changes in GFR in health and disease. Thus, afferent and efferent arteriolar resistances are enhanced by direct and reflex increases in *renal nerve activity*.[103,104] The enhancement of adrenergic tone also results in a fall in K_f, likely through a reduction in available filtering surface area.[105] The *renin–angiotensin system* is well known modulator of GFR. Intravenous infusion of AII induces an increase in afferent and efferent arteriolar resistances[103] as well as a significant reduction in K_f.[106] AII is thought to mediate the *tubuloglomerular feedback* (TGF) mechanism.[107] This is a reflex that reduces both glomerular plasma flow rate and K_f in response to increased chloride uptake by the macula densa due to increase in the flow rate at the loop of Henle,[108,109] thus contributing to the control of single nephron GFR. *Arachidonate metabolites* also influence GFR. Among metabolites derived from the cyclooxygenase pathway, TxA_2 is a potent vasoconstrictor that reduces GFR in normal rats by increasing glomerular afferent and efferent arteriolar resistances.[110] TxA_2 also stimulates contraction of cultured rat mesangial cells through an elevation of cytosolic free calcium.[101] In contrast, prostaglandin E_2 (PGE_2) and prostaglandin I_2 (PGI_2) act as vasodilators in the kidney, reduce renal vascular resistance, and antagonize AII-induced and TxA_2-induced mesangial contraction.[111] Sulfidopeptide leukotrienes LTC_4 and LTD_4, arachidonate metabolites formed via 5-lipoxygenase pathway, increase renal vascular resistance and reduce GFR when infused in the intact animal[112] as well as in the isolated perfused rat kidney.[113] The infusion of *endothelin*, a constrictor peptide derived from vascular endothelial cells,[114] elicits in the rat a concentration-dependent increase in renal vascular resistance and reduction in GFR.[115] Similarly, in isolated perfused rat kidneys, proportional decreases in GFR and renal plasma flow (RPF) have been reported over a wide range of endothelin doses.[116] Further, endothelin contracts mesangial cells *in vitro* and lowers K_f through stimulation of inositol 1,4,5-triphosphate generation associated with elevation of intracellular calcium.[117–119] Nitric oxide, a potent vasorelaxing factor of endothelial origin, may contrast the actions of renal vasoconstricting factors on arteriolar tone and mesangial cell contractile activity.[120,121]

Advances on the mechanisms that regulate glomerular ultrafiltration are extensively discussed in recent reviews.[97,119,122] Importantly, the development of novel specific antagonists and inhibitors has allowed us to dissect the role of vasoactive molecules, in particular angiotensin II (AII), endothelin, and nitric oxide, in regulating glomerular function and to examine their relevance in renal pathophysiology, including toxicant-induced glomerular dysfunction.

7.11.3 THE GLOMERULAR CAPILLARY WALL AS A TARGET

7.11.3.1 Visceral Glomerular Epithelial Cell

The most well defined forms of glomerular toxicity in experimental animals are those induced by puromycin aminonucleoside (PA) and adriamycin (ADR). Such agents cause ultrastructural changes that selectively affect glomerular visceral epithelial cells. The associated pathology is the only example of toxic "epithelial cell disease." The occurrence of epithelial cell damage and nephrotic syndrome in the absence of major glomerular inflammatory changes is a typical feature of both human minimal change disease and segmental glomerulosclerosis and hyalinosis[123] Therefore, the glomerular toxicities of PA and ADR have been extensively investigated as relevant models for human disease.

An interesting model of single nephron glomerular epithelial cell injury has been developed most recently by direct manipulation of glomeruli *in vivo*[124,125] The selective injection of a detergent, saponin, into the Bowman's space of individual surface glomeruli in female Munich Wistar Fromter (MWF)/Ztm rats altered the glomerular permeability to albumin, as visualized by *in vivo* fluorescence microscopy of glomeruli.[125] Saponin apparently caused segmental lysis of podocytes without evidence of ultrastructural lesions to other intrinsic glomerular cells or to the GBM.[124,125] These data, combined with those obtained with saponin in other vascular systems, indicated the podocyte as a major glomerular target of this agent. Evidence that direct damage to visceral epithelial cells leads rapidly to albuminuria further supports the role of the podocyte as a major determinant that restricts protein traffic across the GCW.

7.11.3.1.1 *Puromycin aminonucleoside and adriamycin caused glomerular structural injury and abnormal permselectivity*

Puromycin-aminonucleoside (6-dimethyl-aminopurine, 3-amino-D-ribose) (PA) is a purine antagonist with antibiotic activity that inhibits RNA synthesis at the level of the ribosome.[126] Besides rats, monkeys, and humans[127] no other species[128,129] appear sensitive to PA nephrotoxicity. There are distinct patterns of glomerular toxicity of PA. In the "single shot" model in rats, proteinuria starts a few days after a single intravenous or intraperitoneal injection of PA, reaching peak values of up to 900 mg protein/day after one to two weeks, and disappears within 4 weeks.[130–132] Ultrastructural changes to the visceral epithelium can be seen as early as 24 hours after injection, and consist of effacement of foot processes with cytoplasmic vacuoles, protein reabsorption droplets, and villous transformation.[131] When proteinuria ensues, areas of detachment of covering epithelium are detected along the GBM.[133] Such changes are reversible with the remission of proteinuria.[130] Acute tubulointerstitial nephritis with macrophage accumulation has been also described in this single shot model.[134] A chronic form of disease develops upon direct administration of PA into the central venous circulation,[135] which induces early mesangial cell proliferation and matrix expansion and obliteration of glomerular capillary lumens.[135,136] With continuing proteinuria, glomerular lesions markedly worsen and segmental glomerulosclerosis is detectable several weeks after injection.[135] The high severity of this progressive variant by central venous injection[130] has been related to higher PA levels in the kidney. A form of chronic PA nephrosis can also be induced by repeated subcutaneous doses.[137,138]

The acute renal effects of ADR, similar to those of PA, include morphological changes of glomerular podocytes, proteinuria, and relatively preserved renal function.[139–141] Adriamycin, the trademark for doxorubicin, is an anthracyclin which inhibits RNA and DNA synthesis through intercalation into DNA double helix[142] and is one of the most employed antineoplastic drugs for clinical use. The drug also binds to the cell membrane[143] and induces the formation of free oxygen radicals.[144] Unlike PA, a single intravenous injection of ADR in the rat induces persisting proteinuria and glomerulosclerosis.[145,146] Chronic ADR nephrosis in rats is marked by sustained proteinuria, tubular cast formation and tubulointerstitial inflammation. Glomerular sclerotic lesions affect a high proportion of glomeruli several weeks after the drug administration and are associated with progressive decreases in GFR.[146–148] The rabbit kidney is also sensitive to ADR toxicity.[149] In humans the use of ADR (cycles of $1\,mg\,kg^{-1}$ daily for 2–6 days) is associated with a cumulative dose-related cardiac toxicity, but not renal toxicity. Cardiotoxicity manifests at total doses higher than $800\,mg$.[150,151]

Several studies have addressed the issue of the primary structural target of PA and ADR toxicity in the rat kidney. That proteinuria is the result of a direct insult to the glomerulus was demonstrated with transplant experiments[152] and by consistent findings of unilateral renal damage following selective unilateral exposure of the kidney to either drug *in vivo*.[138,153,154] Whether severe proteinuria is secondary to changes in GBM or podocyte still remains unclear. A reason is that, as outlined above, such components of the GCW may both serve to restrict the passage of proteins to the Bowman's space. Further, the effacement of foot processes, which is clearly associated with epithelial cell dysfunction, can be regarded as consequence of proteinuria rather than a primary toxic effect of the drug.[155,156] Despite theoretical limitations, evidence strongly suggests that toxic molecules act by targeting directly the visceral epithelial cell. Thus, glomerular epithelial cells exposed *in vitro* to PA or ADR lose their capacity to adhere to the plastic substratum and become unable to proliferate.[157,158] As discussed in Section 7.11.4, studies have demonstrated many other toxic effects *in vitro* on functions of the glomerular epithelium, which include synthesis of extracellular matrix, maintenance of intact cytoskeletal structures,[159] and generation of inflammatory mediators.[160–162]

7.11.3.1.2 Toxic agents cause abnormal glomerular permeability to proteins by altering size and charge permselective functions

In PA nephrosis, both size and the charge permselectivity are impaired as documented by studies with macromolecular tracers.[133,163] That ADR-induced proteinuria result from a size selective defect has been documented by the increased clearance of neutral dextran with a molecular radius greater than 50 Å[140,164] and by the increased glomerular sieving coefficient of albumin.[165] Studies using charged dextrans of different molecular radii indicated that the charge barrier is also impaired by ADR.[165,166]

The defect in charge-selectivity in both models has been attributed to the loss of negatively charged sites on the GCW. Early studies with quantitative ultrastructural autoradiography showed no differences between normal and PA nephrotic rats in the binding of ^{125}I cationic ferritin to the GBM,[167] and data of cationic ferritin and ruthenium red staining[74] failed to demonstrate loss of polyanions in the GBM in this model. However, negatively charged sites of the GCW appeared to be significantly reduced as reflected by reduced binding of lysozyme, alcian blue,[168,169] and protamine–heparin aggregates.[170] In addition, decreased binding of cuprolinic blue in PA treated rats indicated a decreased anionic charge, particularly of that endowed by sulfated groups and perhaps carboxyl groups present on structural proteins of the GBM.[171] Mahan and more recently Whiteside *et al.*, using a quantitative method with a cationic probe, polyethyleneimine, that presumably labels heparan sulfate proteoglycans, found significantly fewer anionic sites in the GBM in both PA[172,173] and ADR treated rats[173] than in control animals. The charge defect was observed 1 to 5 days following PA and ADR injection, respectively, and preceded foot process effacement and nephrotic proteinuria.[172,173] Immunohistochemical studies have demonstrated that reduced charge density of the GCW following PA or ADR is due to loss of heparan sulfate proteoglycans.[174,175] In the PA model, recent data with a gold-conjugated poly-L-lysine probe raised the suggestion that the decreased density of negatively charged sites can be associated with increased mobility of these charges within the GBM.[176] These and other previous studies have demonstrated that sialic acid residues at the surface of visceral glomerular epithelial cells are markedly reduced in rats injected with PA,[62,176–178] thus demonstrating that early observations of reduced colloidal iron staining by light microscopy reflected a charge loss rather than reduction of cell surface consequent to obliteration of foot processes. A study by Orci *et al.* revealed reduced binding of a gold-conjugated lectin to the plasma membrane domain at the foot process base as a possible result of loss of *N*-acetyl-D-galactosamine and *N*-acetyl-D-glucosamine residues; this was associated with freeze fracture findings of structural perturbation of the plasma membrane.[179]

The loss of size selective barrier in epithelial cell disease has been related to focal detachment of podocytes from the GBM, at sites where increased passage of macromolecular tracers such as ferritin and peroxidase is visualized.[133,180] Actually, the development of heavy proteinuria appears to be strictly associated with the detachment of the podocyte in toxic models. Early studies with peroxidase and catalase on PA nephrosis demonstrated that exogenous tracers penetrate into basal pockets and vacuoles of epithelial cytoplasm overlying the GBM, suggesting that plasma proteins which permeate the GBM could then be transferred across the epithelial layer of the GCW to the Bowman's space through a system of cytoplasmic vacuoles.[180,181] That this may represent a pathway for protein loss is indicated by ultrastructural data of serially sectioned glomeruli at different stages of acute PA nephrosis.[153] It was proposed that a slow insudation of plasma across sites of denuded GBM into basal

pockets of the epithelium may cause them to expand and form large vacuoles, which eventually rupture into the Bowman's space.[153] A similar sequence of events has been described in both PA and ADR nephrosis.[173] However, the contribution of such an intracellular pathway of proteinuria is not yet clear because the incidence of transepithelial ruptures on transmission electron microscopy appeared much lower than that of focal detachments.[173] Even more important is the concept that few areas of complete denudation would allow circulating protein to pass the barrier in large amounts, perhaps restricted by a leaky GBM only. Dislocation and rearrangement of slit diaphragms have been described in PA nephrosis,[73] changes similar to those elicited by infusion of polycations[53–55] and representing an important structural correlate of the altered permeability barrier. Even in the absence of detachment or apparent change in shape of the podocyte, the slit diaphragm is a major candidate structure for extracellular pathways of altered permeability. However, it is unknown whether processes that lead to abnormal slit diaphragms are a consequence or a cause of associated flattening of podocytes, nor whether these and even more subtle changes of such specialized structures may impair the glomerular restriction to protein flux.

7.11.3.2 The Glomerulus as a Target of Drugs that Impair Ultrafiltration

Toxic effects of drugs to the glomerulus are not invariably associated with glomerular structural changes. CsA, gentamicin, and amphotericin B are examples of drugs that can affect glomerular function without loss of structural integrity, thereby leading to a significant decline in GFR. Acute renal vasoconstriction seems to play a major role in both CsA and amphotericin B nephrotoxicity. In the latter, evidence has been provided that decreased K_f also significantly contributes to reduction of GFR. A decrease in K_f has been shown to account for the hemodynamic component of gentamicin nephrotoxicity, even in the absence of significant changes in the rate of renal perfusion. Molecular mechanisms underlying impaired ultrafiltration in different forms of toxicity will be discussed in the last section of this chapter. However, it should be underlined that the contractile activity of the mesangial cell may influence the surface area available for filtration, and this is suspected to represent a major pathway of decreased K_f in glomeruli that are targeted by toxic drugs.

7.11.3.2.1 Cyclosporine

The clinical use of (CsA) can be associated with acute, dose related decrease in renal function.[182,183] The episodes of acute renal insufficiency are usually rapidly reversible after the daily dose of CsA is reduced.[184] The acute effect of CsA on renal function in animals varies with respect to species and strain as well as dose, mode of administration, and duration of treatment.[185] Studies to identify the structural targets of acute CsA nephrotoxicity have documented damage to proximal tubule epithelial cells,[186–188] and leakage into the urine of lysosome-associated enzymes in rats.[189] However, studies provided evidence for elevated renovascular resistance, diminished RBF, and declining GFR[185,190,191] after CsA administration even before tubular damage develops.[192] Thus, a consensus has accumulated that CsA impairs renal function by primarily affecting renal vessels rather than proximal tubules. In order to divorce direct effects on renal function from indirect consequences of systemic CsA toxicity, esperiments were performed with the model of isolated perfused rat kidney.[193,194] CsA caused a dose-dependent fall in renal perfusate flow associated with a concomitant increase in renal vascular resistance.[193] Accordingly, a concentration-dependent increase in perfusion pressure was found in isolated rat kidneys perfused at constant flow and exposed to CsA.[194] The morphologic correlate to these functional changes was provided by a study with scanning electron microscopy, showing progressive narrowing in afferent arteriolar diameter which paralleled the decrease in inulin clearance.[195]

The form of renal injury that develops after long-term exposure to the drug is the main clinical problem related to CsA administration in humans.[196–198] Clinical trials have demonstrated a sustained increase in serum creatinine in CsA-treated patients but not in those on conventional immunosuppression.[199–203] Evidence is available that liver[204] or pancreas[205] allograft recipients, and patients with autoimmune diseases[206–208] have low GFR. In heart transplant recipients with healthy native kidneys both high[209,210] and low[210] doses of CsA, if given for more than 12 months, were associated with a marked decline in RPF and GFR. Chronic renal injury, once established, may not be reversible even upon withdrawal of CsA.[209,210] Progressive deterioration in renal function has been documented also in CsA-treated cadaveric renal transplant recipients.[211]

A major feature of chronic CsA nephrotoxicity is vascular damage, the so-called CsA-associated arteriolopathy, that affects almost exclusively arterioles, particularly the afferent arterioles, and small arteries.[186,188,212] Associated with vascular lesions are tubular atrophy, interstitial fibrosis, and chronic inflammation,[186,188,209,213,214] usually in narrow stripes which apparently correspond to cortical areas with afferent arteriolar lesions.[187] These lesions have all been described in CsA-treated transplant recipients[188,204,210,212] as well as in patients with autoimmune diseases,[207] which indicates that CsA therapy leads to chronic renal damage irrespective of the underlying disease. Glomerular changes associated with CsA nephrotoxicity include hyperplasia of the juxtaglomerular apparatus, thickening and wrinkling of capillary basement membrane, and focal and segmental sclerosis.[196,197,204,209,210,215,216]

Studies performed in heart transplant patients showed that the glomerular hypofiltration associated with chronic CsA therapy reflects not only reduction in RPF but also depressed K_f[196,209,210] that could not be attributed to low performance of cardiac graft. Cardiac transplant recipients treated with CsA for 1 year[210] or more[217] develop an occlusive arteriolopathy that contributes to permanently elevated renal vascular resistances. Measurements of renal artery-to-peritubular capillary pressure gradient indicated a preferential vasoconstriction at preglomerular level.[210] Notably, the estimated K_f in these patients apparently did not decline further after 12 months of CsA therapy, which is consistent with the relative constancy of GFR with time despite progressive obliteration of cortical microvessels.[210]

Myers *et al.* performed morphometric analysis of glomerular cross-sectional areas in renal biopsies from heart transplant recipients given CsA for 12 months.[210] The pattern of distribution of glomerular capillary tuft cross-sectional area, which was bell-shaped in normal kidneys, was lost in CsA-treated patients, with a shift to both smaller and larger size glomeruli, presumably reflecting collapse and sclerosis of some glomeruli and the appearance of a subset of spared remnant glomeruli which became enlarged. It is possible that remnant perfused glomeruli increase their single nephron GFR to compensate those glomeruli undergoing progressive sclerosis.[210] Thus, in these patients, the relative stability of GFR despite progressive lesions in the kidney may reflect relative hypertrophy of a population of glomeruli compensating for the loss of surface area of more damaged ones. Three-dimensional morphometric analysis of individual glomeruli in renal biopsies taken from cardiac transplant patients on CsA for more than 2 years showed that 40% glomeruli had global or segmental sclerosis. Moreover, smaller (42%) and larger (24%) than normal glomeruli were found, the former being the ones with more global or segmental sclerosis.[218] The reader is referred to Chapter 29, this volume, for a more in-depth review of CsA nephrotoxicity.

7.11.3.2.2 Gentamicin

Proximal tubule cell toxicity is the predominant cause of impaired renal function after gentamicin administration.[219] However, micropuncture studies have documented that gentamicin also affects hemodynamic determinants of glomerular filtration,[220–222] thus raising the possibility that primary disturbances in the glomerular filtration contribute to gentamicin-induced reduction of GFR. In the study of Baylis *et al.*, Munich Wistar rats treated with gentamicin for 10 d at the dose of 4 or $40\,mg\,kg^{-1}\,d^{-1}$ exhibited decreases in the mean values of both total and single nephron-GFR.[220] A small reduction in the mean value of glomerular plasma flow was found only in animals given the highest dose. The primary cause for the reduction in single nephron GFR was a marked decrease in the average value of K_f. In spite of extensive tubular damage, there was no evidence of altered tubular permeability to inulin.[220] Electron microscopy examination revealed no obvious glomerular abnormalities.[220] In contrast, dose-dependent decrements in glomerular endothelial fenestral size, density, and area were documented by others in rats given gentamicin.[223] Although the specificity of glomerular endothelial abnormalities in this as well as other forms of acute renal failure remains uncertain,[224] their presence would suggest that glomerular endothelial cell dysfunction may contribute to reduced K_f and impaired glomerular filtration. The reader is referred to Chapter 26, this volume, for a more in-depth discussion.

7.11.3.2.3 Amphotericin B

The use of amphotericin B is frequently complicated by azotemia and depressed GFR. Renal functional changes , that were reported in more than 80% of treated patients,[225] are dose-related and usually reversible after the withdrawal of the drug.[225] Pathological examination in humans shows acute tubular necrosis, with tubular dilatation and calcifications.[226] Glomerular and vascular lesions have also been reported including thickening of the GBM, glomerular hypercellularity and hyalinization,[227,228] and vacuolization of the media of small arteries and arterioles[226] Altered renal hemodynamics contribute to impair renal function in amphotericin B toxicity.[229–232] Acute administration of amphotericin B decreases RBF in dogs,[1] most likely by afferent arteriole constriction. In the rat, tubular "backleak" of tubular fluid also seems to contribute to the decreased inulin clearance.[233] Renal dysfunction after intravenous infusion in rats was reversible within 24 hours, which in the absence of obvious histological abnormalities is also in favor of a functional nature of amphotericin B toxicity.[233] That renal vasoconstriction operates in a chronic model of amphotericin B nephrotoxicity was suggested by Tolins and Raij, who found a marked increase in renal vascular resistance and a fall in RPF and GFR in rats treated with the drug for 21 days.[1] As with gentamicin, however, there is evidence that amphotericin B impairs GFR by evoking a direct glomerular response. A calcium channel blocker, verapamil, inhibited completely the renal vasoconstrictive response to infusion of amphotericin B in rats. However, the decline in GFR was delayed and blunted only,[234] and in the early postinfusion period GFR fell to a similar extent to that in rats not pretreated with verapamil. Therefore, the renal vasoconstrictive effect, preventable by calcium channel blockade, did not fully account for the decline in GFR, which was mediated at least in part by a decrease in K_f.[234] Actually, a micropuncture study from the group of Badr documented that intrarenal infusion of amphotericin B induces both significant increases in pre- and postglomerular resistances and a significant (25%) decrease in K_f.[235] These changes were associated with decreases in single nephron RPF and GFR in the absence of effects on mean arterial pressure and net hydrostatic pressure difference across the glomerular capillary, thereby indicating that amphotericin B causes reduction in single nephron GFR both through a fall in RPF due to afferent and efferent arterial vasoconstriction and through a reduction in K_f.[235] The reader is referred to Chapter 25, this volume, for a more comprehensive review of amphotericin B nephrotoxicity.

7.11.4 MECHANISMS AND MEDIATORS OF INJURY

7.11.4.1 Altered Metabolic Pathways and Biochemical Mediators of Glomerular Permeability Dysfunction in Epithelial Cell Disease and Progressive Renal Injury

7.11.4.1.1 Metabolism of heparan sulfate proteoglycan and sialoglycoproteins

Studies on glomerular toxicity have contributed important support to the theory that changes in the content of HSPG of the capillary wall represent a major determinant of the increased permeability to macromolecules. In keeping with the loss of heparan sulfate proteoglycan in PA nephrotic rats,[174] mRNA studies reported marked reduction in the glomerular HSPG expression.[236,237] These changes could be partially reversed by treatment with glucocorticosteroids, which ameliorated proteinuria in this model.[237] Of interest, dexamethasone caused dose and time dependent increases in transcription and content of HSPG core protein in cultured glomerular epithelial cells.[238] Moreover both PA and ADR inhibit the expression of a restricted set of proteins, that include HSPG, in glomerular epithelial cells.[159] Wapstra *et al.* have reported the loss of both the core protein and the side chain of HSPG in glomeruli of rats with established ADR nephrosis. Interestingly, an angiotensin-converting enzyme inhibitor (ACEi), lisinopril, reduced proteinuria and reversed these abnormalities.[175] Since AII inhibited HSPG synthesis in glomerular mesangial cells *in vitro*,[239] the possibility exists that the reduced HSPG in ADR nephrosis is related to the inhibitory action of local AII on HSPG metabolism of glomerular cells, including epithelial cells that actually bind AII and exhibit a variety of responses to this molecule *in vitro*.[92,94,240]

Both in PA[177] and ADR nephrosis[140,145] glomerular sialoproteins were markedly reduced by light microscopy examination after colloidal iron staining but unchanged when the binding of cationic molecules were examined ultrastructurally.[241] The interpretation that reduced colloidal iron staining by light microscopy reflected a reduction in cell surface area due to retraction of foot processes was challenged by Kerjaschki *et al.*[178] who found in PA nephrosis a markedly reduced sialylation of podocalyxin of the glomerular epithelial cell. Similar results have been obtained with lectin-gold techniques[62,176] and by measuring the total membrane-bound sialic acid content of isolated glomeruli obtained from PA-treated rats before the onset of nephrotic proteinuria.[173] It seems that the composition of other sugars of the N-linked side chains remains unchanged[73] and that the reduced sialylation selectively affects podocalyxin.[178] That abnormal sialylation occurs also in the ADR model is indicated by reduced labeling of sialic acid residues within the capillary wall with a biotinylation procedure.[68] Although the functional significance of this finding in ADR nephrosis needs to be explored further,[173] a loss of sialoglycoproteins may favour focal detachment of the podocyte and altered size permselectivity that is found in many types of glomerular disease.[26,44,74,178]

7.11.4.1.2 *Synthesis and degradation of laminin, collagen, and fibronectin*

Besides proteoglycans and sialoglycoproteins toxic agents interfere with the synthesis and degradation of other structural proteins and enzymes in the glomerulus that may have roles in maintaining glomerular permeability. Thus, during the acute phase of PA nephrosis mRNA levels for laminin, collagen and fibronectin are abnormal in the kidney. Actually, the renal content of mRNA for type IV collagen and laminin decreased early[242] followed by a marked increase of glomerular mRNAs for these molecules[236,242] and fibronectin[237] in the nephrotic stage of the disease. *In vitro* experiments with glomerular epithelial cells exposed to PA and ADR have suggested that these drugs may interfere directly with laminin metabolism.[159] Findings that the neutral proteinase metalloendopeptidase and tissue inhibitor of metalloproteinases can be released by cultured glomerular mesangial cells and that metalloendopeptidase extensively degrades the GBM and type IV collagen suggest that these enzymes play a role in the turnover of GBM glycoproteins *in vivo*.[243] Imbalances in the glomerular content of metalloendopeptidase and tissue inhibitor of metalloendopeptidase may result in altered production of type IV collagen and changes in the permselective properties of the GCW. In this regard, Davin *et al.*[244] in acute PA nephrosis found a significant increase in urinary excretion of neutral proteinases, which correlated both with proteinuria and urinary excretion of type IV collagen and laminin, suggesting that neutral proteinases may be involved in the altered glomerular permeability. That changes in these enzymatic pathway occur in PA nephrosis is further indicated by the finding of higher renal mRNA levels for tissue inhibitor of metalloproteinases at 1 week after PA injection.[242]

The significance of changes in extracellular matrix protein metabolism for glomerular permselectivity can be inferred from the functional characteristics of the podocyte, as described above in this chapter. Altered macromolecular assembly of the GBM may either cause abnormal permselectivity of the capillary wall directly,[70] or modify cell shape[245] and function[246] (Figure 2). In particular, one can speculate that both altered matrix composition and changes in glomerular synthesis or distribution of molecules that anchor the cell membrane to the GBM, like β−1 integrins,[79,178,247,248] contribute to podocyte detachment. This effect is likely part of the mechanism by which antibodies that interfere

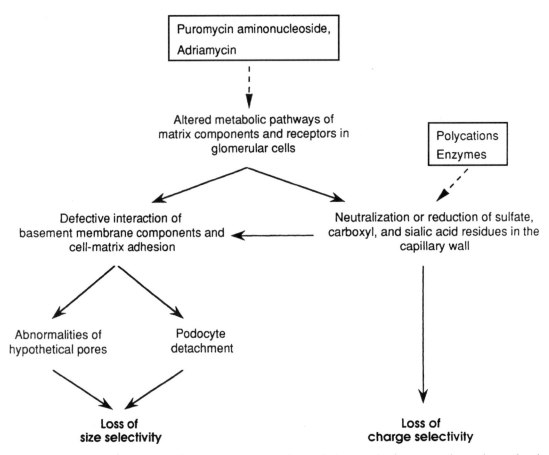

Figure 2   Potential mechanisms of altered glomerular size and charge selective properties and proteinuria induced by drugs and toxic substances (after Hook and Goldstein, p. 164[3]).

directly with the epithelial cell attachment to the basement membrane disrupt glomerular permselectivity and cause proteinuria.[78] Consistent with this possibility is the observation that PA and ADR caused changes in intensity and distribution of $\beta - 1$ integrin staining by immunofluorescence, as well as decreased expression of laminin in cultured glomerular epithelial cells, and cell detachment from the substrate.[159] Finally, effects of toxic agents on the synthesis of extracellular matrix can be mediated by molecules that have been shown to participate in the development of proteinuria or to the progression of disease in these models. They include thromboxane, oxygen free radicals, AII, and cytokines.

7.11.4.1.3   Assembly of slit diaphragms and aggregation of podocyte cyoskeleton

Freeze fracture findings that slit diaphragms[53–55] and plasma membrane domains[179] have lost their architecture in podocytes of PA treated rats have prompted investigation of possible biochemical changes underlying

such abnormalities. Evidence indicates that dislocation and rearrangement of slit diaphragms following exposure of kidneys to either protamine sulfate or PA is mediated by an active process which involves tyrosin phosphorylation of ZO-1.[249] This tight junction protein is also detected in the altered slit diaphragms in protamine treated rats and PA nephrosis.[53] Slit diaphragms normally separate two membrane domains that differ markedly in composition and function.[73,179] Slit diaphragm perturbation may be followed by disruption of cell polarity and shape. Severe podocyte cytoskeleton disorganization, already documented by early studies in both PA[250] and ADR nephrosis,[145] has been confirmed using more sophisticated immunocytochemical techniques showing that in acute PA nephrosis early aggregation of cytoskeletal actin occurs followed by disappearance of actin from foot processes.[88,251] Actin loss coincided with significant reduction in the number of attachment fibers in the lamina rara externa at the moment of onset of massive proteinuria.[88] Epithelial detachment and defective endocytosis[168] are among the potential consequences of these events, but how phenotypic changes relate to cytoskeletal dysfunction and which are signaling systems and molecules involved is not known. That the toxicity of PA and ADR are related to their effect on cytoskeleton organization and microtubules is consistent with data that both molecules caused changes in staining pattern and intensity of actin, tubulin and intermediate filaments in glomerular epithelial cells in culture.[159] However, causal relationships to concomitant cell detachment, decreased staining and redistribution of $\beta - 1$ integrin, and decreased expression of laminin and heparan sulfate have not been established.[159]

7.11.4.1.4 Oxygen free radicals

Experimental studies have convincingly documented that reactive oxygen species (ROS) damage the glomerulus and induce changes in permeability.[252–254] Diamond *et al.* provided the first evidence that ROS are important mediators of epithelial cell injury in PA nephrosis, even in the absence of significant glomerular inflammatory lesions,[136] by showing that treatment with superoxide dismutase or allopurinol, an inhibitor of xanthine oxidase, significantly reduced urinary protein excretion and limited glomerular epithelial cell structural damage. It was suggested that the nephrotoxic effects of PA are related to the formation of its metabolic byproduct, hypoxanthine, which can serve as a substrate for O_2^- production via the xanthine oxidase system.[136] PA might also provide the intracellular compartment with hypoxanthine precursor by increasing membrane permeability to adenosine in glomerular epithelial cell.[157] In contrast to many earlier reports, the results of later studies suggest that PA glomerular toxicity is mediated by ROS generated in pathways not involving xantine oxidase.[160,161,255,256] For an oxidant injury to occur, available ROS should exceed antioxidants in a given cell. Superoxide dismutases (which dismutes O_2^- to H_2O_2),[257] and the enzymes that metabolize H_2O_2 (catalase and glutathione peroxidase),[258,259] are major intracellular antioxidant systems. In PA nephrosis methylprednisolone raised renal manganese-superoxide dismutase and catalase activity while reducing renal cortical malondialdehyde, and proteinuria. These effects were associated with less epithelial cell foot process fusion and cell vacuolization.[260] Conversely, an inhibitor of superoxide dismutase and a glutathione-depleting agent both worsened the disease.[261,262] Recently, the protective role of endogenous selenium-dependent glutathione peroxidase has been documented with acute

PA nephrosis or Heymann nephritis in rats fed a selenium-deficient diet, which markedly worsened proteinuria presumably by diminishing the activity of this enzyme.[263] Beside O_2^-, OH· scavengers and an iron chelator, deferoxamine, were protective against PA-induced proteinuria, suggesting that OH·, most likely generated from the interaction of O_2^- and H_2O_2 in the presence of iron, were responsible for altered glomerular permeability to proteins in this model.[264]

Oxygen free radicals also may contribute to ADR nephrotoxicity. A free radical scavenger, dimethylthiourea, was protective against glomerular injury in this model, possibly by enhancing glutathione metabolism.[265] Moreover allopurinol, which inhibits renal xanthine oxidase, effectively reduced ADR-induced proteinuria.[266] ADR is reduced to semiquinone radical, which in turn generates O_2^-, H_2O_2, and OH· from oxygen.[144] The beneficial effect of lowering intrarenal ROS applies to other models. In passive Heymann nephritis, a complement-dependent rat model of membranous nephropathy, either scavengers of OH· (dimethylthiourea and sodium benzoate) or an iron chelator (deferoxamine) significantly reduced proteinuria.[267] OH· also is implicated in damaging glomerular permeability functions in models of immune complex glomerulonephritis.[268]

A direct role for ROS in altering glomerular permeability has been documented by Yoshioka *et al.*[269] infusing H_2O_2 into the renal artery in Munich–Wistar rats. They produced massive proteinuria with increased fractional clearance of neutral dextran of large molecular radii (42–60 Å), which was prevented by catalase and colloid-conjugated deferoxamine.[269] *In vitro*, ROS can damage extracellular matrix macromolecules. Specifically, ROS generated by xanthine oxidase and hypoxanthine or by stimulated neutrophils, degrade collagen and cause depolimerization of hyaluronic acid.[270–272] ROS also enhance susceptibility of proteins to proteolysis,[270,273] and consistent with this possibility are data by Fligiel *et al.*[274] indicating that pretreament of rat GBM or fibronectin *in vitro* with H_2O_2 increased their sensitivity to subsequent proteolytic attack.

ROS-induced damage to glomerular epithelial cells in PA and ADR nephrosis is also caused by lipid peroxidation of cell membranes.[275] Other potential effects of free radicals (Table 3) implicate the inhibition of membrane-associated enzymes. There are data that ATPase activity is decreased in kidney sections incubated either with ADR or with a superoxide generating system, and that such an effect is inhibited by superoxide dismutase.[276]

Table 3 Effects of reactive oxygen species on glomerular capillary wall structure that may disrupt permeability properties.

Glomerular basement membrane
- Direct degradation of collagen and other matrix proteins
- Increased susceptibility of matrix components to proteolytic attack

Podocyte
- Lipid peroxidation of cell membranes
- Interference with the activity of membrane-associated enzymes
- Inhibition of synthesis or processing of proteoglycans and sialoproteins
- Altered cell shape, cytoskeleton, and adhesion
- Interaction with transcription factors (NFkB) and altered regulation of specific genes

---

ROS also alter proteoglycans and sialoprotein synthesis[277] as well as intracellular and extracellular pathways of cell adhesion.[160,255]

Studies with endothelial cells in culture documented that xanthine/xanthine oxidase increased albumin transfer and induced changes in cell shape and calcium efflux.[278] These effects were reversed by a calcium antagonist and mimicked by calcium ionophore A23187. These data suggest that oxygen radicals promoted changes in cell shape and cytoskeleton even in cells other than epithelial cells and that such changes were related to altered calcium homeostasis. Findings that antioxidants prevented foot process retraction without affecting the shape of major processes in PA-treated animals[256] and in kidney sections exposed to PA *in vitro*[162] may reflect inhibiting actions of these drugs on ROS-dependent cytoskeletal changes in foot processes. Finally an emerging role for ROS is gene regulation, and experiments with mesangial cells in culture support ROS as major signals for activation of the transcription factor, NF-κB, in response to immune stimuli.[279] This raises the intriguing possibility that a similar type of regulation is responsible for the activation of specific genes in the glomerulus, perhaps even in podocytes challenged with ADR, PA, and other toxic or inflammatory stimuli.

7.11.4.1.5 *Thromboxane, platelet activating factor, and cytokines*

Altered arachidonic acid metabolism seems involved in proteinuria of epithelial cell disease. Results on the role of thromboxane (Tx) A_2 in PA nephrosis are conflicting. Some authors found that urinary excretion and glomerular production of TxA_2, measured as immunoreactive TxB_2, did not increase,[280,281] while others showed that inhibiting TxA_2 synthesis ameliorated PA nephrosis.[282] In rats with ADR nephrosis a significant increase in both glomerular synthesis and urinary excretion of TxB_2 was documented. The increase in TxB_2 paralleled onset of proteinuria, and highest levels were measured at the time of peak proteinuria. A selective TxA_2 synthase inhibitor, UK-38485, reduced renal TxB_2 production and lowered proteinuria.[283] Experiments with isolated perfused kidneys of ADR-rats documented that the exaggerated TxB_2 synthesis was of resident renal cell origin.[284] It is possible that inhibiting TxA_2 synthesis induces proteinuria by interfering with the organization of extracellular matrix network.[285] The role of TxA_2 in regulating the expression of extracellular matrix components *in vivo* was suggested by the observation that in mice with non-insulin dependent diabetes a TxA_2 synthase inhibitor, U-63557, prevented the increase in type IV collagen mRNA levels of basement membrane of kidney cortex.[286] Findings that the TxA_2 mimetic, U-46619, stimulates [^{3}H]-thymidine uptake in quiescent human mesangial cells in culture[287] also suggest that TxA_2 could contribute to the process of glomerulosclerosis by inducing mesangial cell proliferation. Further, U-46619 increases the expression of genes for type IV collagen (α1 and α2) via activation of DNA regulatory regions in mesangial cells.[288] A lipid mediator of inflammation, platelet activating factor (PAF), is released by resident glomerular cells and might cooperate with TxA_2 in inducing proteinuria in toxic models of epithelial cell disease. A specific PAF receptor antagonist ameliorated proteinuria and glomerular injury in ADR and PA nephrosis.[289,290] One of the possible sources of PAF in both models were glomerular epithelial cells, which in culture are stimulated to generate PAF by these drugs.[290] Perfused into isolated rat kidney,[291] PAF caused proteinuria without altering distribution of polyethyleneimine stainable anionic sites and increased fractional clearance of large (radius >50 Å) but not of small neutral dextrans (24–48 Å).[292]

Schreiner *et al.* have found that 10–14 days after injection of PA to the rat, glomerular macrophages increased in number in association with development of massive proteinuria.[293] Essential fatty acid deficiency prevented the increase in glomerular macrophage and reduced TxA_2 formation[280] and proteinuria.[280] In rats with acute ADR nephrosis an increase in Ia-bearing glomerular macrophages was also found to correlate with proteinuria.[294] Other studies seem to implicate the role for mesangial cell-derived cytokines in both models. Increased secretion of tumor necrosis factor (TNF) was detected from both isolated, intact glomeruli of nephrotic rats and glomeruli and mesangial cells stimulated *in vitro* with either ADR or PA.[290] Of interest, TNF might mediate glomerular effects of PAF, as indicated by findings that a PAF receptor antagonist, lowered proteinuria[289,295] and glomerular synthesis of PAF and TNF as compared to untreated, nephrotic controls.[290] The role of cytokines in PA nephrosis, specifically TNF-α and IL1-β, has received additional support from data of protective effects with specific neutralizing antibodies.[296]

That cytokines mediate toxic glomerular damage is further suggested by data of enhanced TNF-α and transforming growth factor (TGF)-β positive cells, probably macrophages, in the mesangium of PA nephrotic rats.[297,298] In this model upregulated TGF-β1 gene expression coincided temporally with macrophage infiltration in the mesangium.[299] Cholesterol feeding of rats, which worsened glomerular injury, further increased glomerular TGF-β1 as well as fibronectin mRNA levels. Upregulated glomerular TGF-β1 gene expression and infiltration macrophage were also found in ADR nephrosis.[300] It has to be established whether upregulation of TGF-β1 and macrophage infiltration are the consequence of the toxic effect of PA and ADR or follows increased traffic of proteins through the glomerular capillary as documented in nontoxic model.[301,302] TGF-β1 of macrophage origin stimulate the synthesis of fibronectin by mesangial cells[237] and perhaps other glomerular cells thus favoring the process of glomerulosclerosis invariably associated with chronic models of PA and ADR nephrosis.

7.11.4.1.6 *Roles of the renin–angiotensin system and endothelin in the abnormal protein traffic through the glomerular barrier and the progression of renal disease*

In analogy to various human diseases of the kidney and other experimental models of chronic proteinuria, ADR nephrosis and chronic variants of PA nephrosis progress to glomerulosclerosis and end stage renal failure. Some forms of progressive renal disease in humans, such as the nephropathy of acquired immune deficiency syndrome and heroin associated nephropathy, are also suspected to result from a primary insult to the glomerular visceral epithelial cell.[123] Several studies have begun to explore the precise biochemical and molecular

mechanisms of mesangial matrix expansion and glomerulosclerosis following exposure to drugs that cause epithelial cell damage and proteinuria. Strictly linked to this issue is the hypothesis that abnormal transit of plasma proteins through the GCW is a key factor in glomerulosclerosis and progressive renal injury.[156] Glomerular obsolescence induced by toxic drugs, extensive renal ablation, or aging shares a sustained abnormality in glomerular permeability to macromolecules.[156] Increased flux of macromolecules may induce dysfunction of mesangial cells, glomerular epithelial cells, and tubular epithelial cells in such a way as to favor the process of glomerulosclerosis in these different proteinuric conditions. Within the glomerulus, enhanced mesangial traffic of macromolecules may stimulate mesangial cells to proliferate and produce extracellular matrix.[303,304] Visceral glomerular epithelial cells react to filtered proteins with functional and structural abnormalities,[156] such as reduced release of heparin-like factor(s) which control the proliferation of mesangial cells,[156,157] and focal detachment from the basement membrane. The latter further favors the passage of macromolecules across the glomerular capillaries,[123] that in glomerular toxicity may superimpose on the primary insult. Thus, bioactive molecules that disrupt the glomerular permeability barrier, especially those that influence the metabolism of matrix components, may act in concert with growth factors which permeate abnormally the glomerulus to induce matrix accumulation and sclerotic lesions. Some inflammatory and vasoactive mediators, like TxA$_2$, interleukin-1, platelet derived growth factor, and transforming growth factor-*p* are theoretically implicated in such processes[156] as well as in the proliferation of mesangial cells, and their possible relevance has been discussed above.

An impressive number of studies have explored the role of the renin angiotensin system in many different experimental settings as well as in humans with proteinuria and progressive renal insufficiency (for review see Burnier and Brunner[122]). In line with the knowledge derived from most of these studies are reports of the beneficial effect of (ACEi) in retarding renal disease progression in PA nephrosis.[305,306] Anderson *et al.*[305] first found that ACEi is effective in attenuating the progressive deterioration of glomerular architecture in this model. Long term protection by ACEi has been confirmed by Tanaka *et al.*, who by using an AII-receptor antagonist provided direct evidence of the role of AII in the progressive glomerulosclerosis induced by PA.[306] A plausible explanation for the renoprotective effects of ACEi is that ACEi selectively reduced efferent arteriolar resistance and, thereby, glomerular hypertension.[307] This action of ACEi was proposed to explain the advantage of using ACEi vs. other anti-hypertensive drugs, which did not affect glomerular transcapillary hydraulic pressure, against proteinuria and renal structural abnormalities in experimental diabetes.[307] Importantly, glomerular capillary pressure in early PA nephrosis is normal[146,305] but increases later in the course of the disease, and in the latter phase ACEi lessened both abnormal urinary protein excretion and glomerulosclerosis.[305] However, beside reducing glomerular capillary pressure, ACEi has a direct effect on glomerular barrier function by ameliorating glomerular size selectivity both in experimental animals[83] and human disease.[308–310] Parallel studies, with AII blockers have suggested that the antiproteinuric effect of inhibiting ACE is most likely due to interference with an AII-dependent mechanism.[84,311] Much less information is available on the precise biological processes that mediate the protective effect of ACEi and AII receptor antagonists on the kidney. One can only speculate that these drugs interfere with AII-induced mesangial cell proliferation,[312,313] and induction of genes which encode for extracellular matrix proteins.[314] AII also affects directly the glomerular filtration barrier[94,106] and mesangial trafficking of macromolecules,[315] which may favour the action of other factors of disease progression. Early proteinuria that follows a single injection of PA in the rat was markedly attenuated by ACEi,[306] in contrast to massive proteinuria that follows repeated injections of PA.[148,305] Relevant to this issue are data that antiproteinuric and hemodynamic effects of ACEi can be dissociated both in animals with spontaneous proteinuria[316] and in human renal disease.[317] Studies in rats exposed to relatively high doses of ADR indicated that ACEi were ineffective in the face of either normal[148] or increased[139,318] glomerular capillary hydraulic pressures. In contrast, lisinopril exerted antiproteinuric effects in animals in which the disease was induced by a lower ADR dose.[319] Like in other models and human diseases, however, treatment was consistently ineffective if started late during disease.[306,320] Conflicting results with ACEi both in acute PA nephrosis and ADR nephrosis underline the concept that the severity of disease as well as the susceptibility to drug treatment are variable with the protocol used to elicit renal injury, so that different mechanisms of injury are predominant in different dosing regimens. This is especially relevant in PA and ADR nephrosis, in which the toxic insults affect a variety of complex biological processes.

Besides its potential to regulate directly growth, matrix production, and glomerular filtration, recent evidence is available to indicate that AII modulates the renal synthesis of a major vasoconstrictor peptide, endothelin-1 (ET-1) (for review see Simonson and Dunn[119]). ET-1 secretion by cultured mesangial cells was stimulated by AII in a concentration dependent manner.[321] In addition, the infusion of AII in the rat resulted in an enhanced urinary excretion of ET-1, which likely reflected the renal synthesis of the peptide.[322] Endothelin has potent biologic actions on renal microvasculature and, independent on the regulation by AII, is regarded as a mediator of renal disease progression by virtue of its vasoconstrictor effect together with the ability to stimulate mesangial cell proliferation and synthesis of extracellular matrix components.[119,323] In PA nephrosis a selective upregulation of ET-1 mRNA and ETB receptor mRNA expression was detected in glomeruli isolated during the nephrotic phase of the disease.[324] The administration of methylprednisolone to PA rats resulted in rapid disappearance of proteinuria and partially attenuated the increased ET-1 and ETB receptor gene expression in the glomeruli. ET-1 synthesis by glomerular cells may act by paracrine and autocrine pathways on mesangial cells and smooth muscle cells in the glomerulus, modulating vascular tone and capillary permselectivity.[119]

It now seems clear that, as a consequence of glomerular permeability dysfunction, the filtered protein load to the tubular compartment may translate into tubular injury and an interstitial inflammation which might contribute to renal disease progression.[325] The relationship between proteinuric glomerulopathies and tubulo-interstitial damage was highlighted by early observations on ADR nephrosis[326] and then several other animal models, such as PA nephrosis[134] and aging.[296] Modifications of proximal tubule cell functional properties in response to albumin and other proteins contribute to the process of tubulo-interstitial damage and subsequent scarring.[301] Activation of a lysosomal pathway by filtered proteins as well as peroxidative damage following loading with endogenous molecules that are potentially toxic, like transferrin-iron, are some of the mechanisms by which enhanced and protracted protein over-reabsorption may perturb tubule cell functions. Studies have begun to elucidate the molecular consequences of such perturbation. Tubular cells are thus induced to upregulate ET-1, monocyte chemoattractant protein-1 (MCP-1) as well as other vasoactive and inflammatory mediators that are secreted at the basolateral pole of the cell. The inflammatory reaction to these substances in turn determines renal over-expression of chemokines of macrophage origin, like TGF-β1, that by stimulating inter-stitial fibroblasts to upregulate extracellular matrix genes functions as a potent promoter of fibrosis. Inflammatory cells surrounding the glomeruli also release factors, many of which are implicated as mediators of proteinuria and progressive renal injury (TxA$_2$, PAF, TNF-α, TGF-β, ET-1), that might even permeate the mesangium where they stimulate cell prolifera-tion and matrix production.[156]

7.11.4.2 Mechanisms and Mediators by which Nephrotoxic Drugs Impair Glomerular Filtration

7.11.4.2.1 Cyclosporine

Several mechanisms have been proposed to mediate CsA-induced impairment of glomeru-lar filtration. The long suspected role of the renin angiotensin system[327,328] is still contro-versial, primarily because experiments with ACEi in rats given CsA have given conflicting results.[191,329,330] Likewise, little progress has been made with regard to the relevance of renal sympathetic nerves,[191,331] that was challenged by studies showing that kidney denervation[332] and pharmacological manipulation of sympa-thetic activity were both ineffective against CsA renal toxicity.[205]

In contrast, there is increasing evidence that, among other arachidonic acid metabolites implicated in CsA renal toxicity,[191,333-335] TxA$_2$ is responsible for the actions of CsA on the renal microcirculation and consequent impairment of function.[336] Urinary excretion of TxA$_2$ metabolites were increased both in experimental animals[337-342] and humans,[343,344] whereas TxA$_2$ synthase inhibition or TxA$_2$ receptor blockade prevented or reversed CsA-associated renal failure.[345,346] Many studies also indicated that the beneficial effects of inhibiting TxA$_2$ were due to hemodynamic mechanisms.[346,347,348] Importantly, TxA$_2$ receptor antagonism also prevented the dete-rioration of renal function and prolonged the survival of rats with renal isograft treated with CsA for one year,[336] evidence that TxA$_2$ is also key mediator of chronic CsA toxicity.

The mechanism by which CsA may increase renal TxA$_2$ is not yet clear, but it seems linked to stimulation of endothelial cells. Consistent with this possibility are data that incubation of cultured bovine endothelial cells with CsA caused cell injury associated with dose and time-dependent increases in TxA$_2$ production.[335] TxA$_2$ as well as other vasoactive substances might also be generated by inflammatory cells and platelets entrapped within the glomerular tuft after CsA adminis-tration.[339] Results of biochemical measure-ments and morphological evidence of intra-capillary hypercellularity *in vivo*, but not in preparations of perfused kidneys from rats previously treated with CsA, suggested that both resident renal cells and inflammatory cells contribute to the exaggerated urinary excretion of TxA$_2$ stable metabolites in CsA-treated animals.[342]

The vasoconstrictive action of TxA$_2$ may favour chronic glomerular ischemia and sub-sequent sclerosis in CsA-treated animals and humans. TxA$_2$ may also promote glomerulo-sclerosis by stimulating an excessive formation of extracellular matrix components.[288,349] In this regard, renal fibroblasts from mice given CsA for 4 weeks have a dose dependent increase in mRNA expression for protocollagen type I and IV,[350] indicating that during chronic CsA administration spatial and temporal expression of specific proteins of extracellular matrix are modified, as occurs in other conditions of excessive renal formation of TxA$_2$.[286]

The fact that inhibiting TxA$_2$ synthesis or activity ameliorates but not normalizes renal function indicates that other factors besides TxA$_2$ are involved in the acute reduction of GFR induced by CsA. Among arachidonate metabolites, LTC$_4$ and LTD$_4$ seem to contri-bute an important part, in that the combined administration of a specific TxA$_2$ receptor antagonist and a LTC$_4$/LTD$_4$ receptor antago-nist completely abolished the fall in GFR and RPF in acute CsA toxicity.[347]

Other vasoactive substances have the poten-tial to modify renal vascular resistance follow-ing CsA administration. Several recent studies have confirmed the role of endothelin. In relation to the altered arachidonate metabo-lism, it is interesting that a stimulus for the release of endothelin is TxA$_2$, as documented by the findings that both endothelial cells and glomerular mesangial cells in culture were induced by a TxA$_2$ mimetic to upregulate endothelin gene expression and to release the peptide into the medium.[102] That CsA stimu-lates endothelin synthesis and/or release was shown both with human aortic and glomerular endothelial cells in culture[351] and by acute infusion of the drug in rats, which increased serum endothelin levels.[352] Evidence for the actual role of endothelin in acute CsA nephro-toxicity was provided using specific anti-endothelin antibodies which prevented the GFR-depressing effect of CsA in isolated and perfused kidneys as well as in intact rats.[193]

Micropuncture studies by Kon et al.[352] documented that anti-endothelin antibodies were effective by inhibiting CsA-induced acute glomerular vasoconstriction. Of note, the finding that as compared to other vascular beds, renal vessels are more sensitive to the vasoconstrictory effect of endothelin[353] may explain the clinical observation that CsA has a site-selective effect on the kidney. Of further interest is the observation that even in the face of apparent increase in endothelin production, CsA-treated animals appear to have upregulation of endothelin receptor as suggested by enhanced endothelin binding in glomeruli.[354] Experiments with receptor antagonists of endothelin's actions suggest that the local endothelin activation by which CsA causes acute hypoperfusion and hypofiltration in the rat is associated with increased density of renal ET_A receptors.[218] Experiments with rat renal arterioles *in vitro* demonstrated that CsA has a constrictor effect on the afferent arteriole requiring the mediation of the ET_A receptor.[355] Other data indicate that CsA stimulates myosin light chain phosphorylation in mesangial cells, a biochemical marker of contraction.[356] No upregulation of ET_A mRNA by CsA could be detected either in glomeruli *in vivo*[357] and in cultured mesangial cells[358] even using RT-PCR, a very sensitive technique. However, mesangial cells were induced by CsA to upregulate both ETB mRNA and pre-proendothelin mRNA, and to release endothelin into the medium,[358] responses that might have significance for the regulation of glomerular tone and ultrafiltration coefficient *in vivo*.

Studies have also investigated the possible contribution of endothelin to chronic CsA nephrotoxicity. Renal synthesis of endothelin, as reflected by high urinary excretion of endothelin without altered plasma levels of the peptide, was increased in rats given CsA for 30 days compared to pretreatment values.[359] A significant correlation was found in these animals between the urinary excretion of endothelin and the serum creatinine levels.[359]

In addition to directly eliciting renal vasoconstriction, other actions of endothelin may be involved in chronic CsA nephrotoxicity. For example, endothelin stimulates 15-lipoxygenase activity leading to 15-hydroxyeicosatetraenoic acid (15-HETE) formation.[360] 15-HETE is chemotactic for inflammatory cells,[361] and may therefore represent a signal for circulating inflammatory cells and platelets to accumulate within the glomerular capillaries in rats chronically given CsA.[342,362] Relevant to this response, CsA has been shown to increase leukocyte adhesion to endothelial cells under physiological flow conditions, most likely by modulating surface expression of adhesion molecules like E-selectin and VCAM-1 that mediate endothelial–leukocyte cognate interactions.[345]

In the light of the role of nitric oxide in regulating renal hemodynamics,[363] studies have explored NO-dependent responses in CsA toxicity.[120,364,365] Actually CsA markedly impaired endothelium dependent vascular smooth muscle relaxation in aortic rings.[364] Furthermore, the administration of CsA to rats for 8 days decreased single nephron plasma flow and GFR and blunted the renal hemodynamic response to glycine, which normally causes renal vasodilation mediated by nitric oxide.[365] The administration of the substrate for NO synthesis, L-arginine, decreased preglomerular resistance and restored the NO-dependent glomerular hemodynamic response in CsA treated rats, suggesting that CsA interferes with normal NO-dependent relaxation of the afferent arteriole.[365] However, others reported that the urinary excretion of stable NO end products was unaltered in rats treated for 30 days with CsA. These findings might indicate that after CsA treatment NO production is preserved but insufficient to antagonize CsA-induced vasoconstriction and that the kidney is unable to increase NO production in response to stimuli.[120] The inhibition of NO synthesis induces efferent constriction and decrease of K_f, indicating that endothelial NO production is an important regulator of the tone of mesangial cells that determines filtration surface area. A potential target of CsA is the glomerular endothelial cell, which by producing NO may control mesangial tone and hence K_f as suggested by the observation that NO of endothelial origin inhibits *in vitro* mesangial cell contractility in response to AII.[121]

The aforementioned studies raise the possibility that CsA-induced imbalances between vasoconstrictive and relaxing factors may underlie the low ultrafiltration coefficient that has been estimated by micropuncture studies.[120,352] The contractile state of the glomerulus seems to be a major determinant of the ultrafiltration coefficient and RBF, and reports have determined that CsA induces contraction of isolated, intact glomeruli *in vitro*.[366,367] Experiments with free oxygen scavengers suggested that ROS, (O_2^-, H_2O_2, and OH·) are major mediators in this response.[367] These effects of CsA might be intimately related with those on TxA_2, endothelin, and NO metabolism. CsA was also shown to induce H_2O_2 formation in mesangial cells in culture, which may have relevance to CsA-induced mesangial contractions and the decreased ultrafiltration coefficient *in vivo*.[366] Increased cortical malondialdehyde, an index of lipid peroxidation, was detected in kidneys of CsA treated rats.[368] Interestingly, CsA-induced lipid peroxidation, presumably free-radical mediated, appears to be an acute transient phenomenon,[368] similar to the daily renal hypoperfusion and hypofiltration reported with the regular daily doses of CsA in renal transplant patients.[369]

7.11.4.2.2 Gentamicin

Relatively few studies have investigated the mechanisms underlying the defect in glomerular ultrafiltration associated with gentamicin. Schor *et al.* examined the effects of AII blockade or long-term high salt intake on gentamicin nephrotoxicity in male Munich–Wistar rats.[221] Inhibiting AII generation almost completely prevented gentamicin-induced changes in single nephron-GFR, K_f and glomerular plasma flow. That ACEi protects from the K_f-lowering action of gentamicin has been confirmed in female Munich–Wistar rats.[222] However, in other experiments, the ACEi, captopril, did not prevent the decrease in GFR and the development of tubular and glomerular morphological changes in male Sprague-Dawley rats given gentamicin at the dose of $80 \, mg \, kg^{-1} \, d^{-1}$ for 10 d.[370] Further when gentamicin was administered to rats with chronic potassium depletion, captopril worsened renal function, morphology, and glomerular ultrafiltration.[371] In addition to differences in doses and rat strains, these apparently conflicting results may be reconciled by the observation that captopril, besides inhibiting AII formation, exerts other actions including stimulation of the renal release of TxB_2 and kinins, which may have adversely influenced renal function in potassium depleted rats given gentamicin.[372] *Ex vivo* perfused kidneys from potassium depleted rats that received both gentamicin and captopril produced significantly greater amounts of TxA_2, measured as immunoreactive stable metabolite TxB_2 in the venous effluent, than did kidneys from potassium depleted rats given gentamicin alone.[371] In rats given gentamicin and captopril, RBF and the clearance of inulin increased during intrarenal infusion of imidazole, a thromboxane synthase inhibitor, or aprotinine, a kallikrein antagonist.[371] Thus, although the glomerular toxicity of gentamicin is likely mediated by mechanisms which are sensitive to converting enzyme inhibition, hemodynamic mediators other than AII, such as TxB_2 and kinins, may also be involved in the decreased GFR observed in this setting.

7.11.4.2.3 Amphotericin B

The mechanisms of the renal vascular effect of amphotericin B are unclear. It seems unlikely that amphotericin B is directly toxic to renal vessels by the same mechanisms that account for its cytotoxic effects on tubular cells. Rather, the observation that amphotericin B contracts mesangial cells *in vitro*[235] would suggest local release of vasoactive agents. Alternatively, activation of the tubuloglomerular feedback may be implicated. Salt loading, furosemide or teophylline administration, maneuvers that inhibit TGF inhibit amphotericin-induced reduction in GFR in experimental animals.[231,373] Likewise, high sodium intake improves GFR in human amphotericin B nephrotoxicity.[232] Since AII modulates TGF,[107] studies addressed the possible role of renin angiotensin system in amphotericin B-induced fall of GFR. Pretreatment with sargly AII, an AII receptor blocker, failed to prevent the decreases in GFR induced in the rat by infusing a vasoconstrictive dose of amphotericin B[1]. Similarly, adrenergic blocking agents[230] and unilateral renal denervation[234] failed to protect animals against the GFR-depressing effect of amphotericin B, indicating that the sympathetic nervous system may not be implicated. The effect of verapamil of ameliorating renal perfusion in amphotericin B nephrotoxicity[234] may be related to a direct vasodilatory action of calcium channel blocking. Changes in cytoplasmic calcium concentration may also act to attenuate the generation of vasoactive mediators, such as arachidonic acid metabolites and PAF,[374] that might contribute to impaired renal function in this setting. Further studies are needed to elucidate in more detail to which extent vasoactive hormones participate to amphotericin B nephrotoxicity.

7.11.5 REFERENCES

1. Ph. J. Hoedemaker, in 'Drugs and Kidney,' eds. T. Bertani, G. Remuzzi and S. Garattini, Raven Press, New York, 1986, pp. 75–94.
2. V. Cattell, in 'Drugs and Kidney,' eds. T. Bertani, G. Remuzzi and S. Garattini, Raven Press, New York, 1986, pp. 26–35.
3. J. B. Hook and R. S. Goldstein (eds.), 'Toxicology of the Kidney,' 2nd edn., Raven Press, New York, 1993.
4. B. S. Daniels, 'The role of the glomerular epithelial cell in the maintenance of the glomerular filtration barrier.' *Am. J. Nephrol.*, 1993, **13**, 318–323.
5. B. M. Brenner, M. P. Bohrer, C. H. Baylis *et al.*, 'Determinants of glomerular permselectivity: Insights derived from observations *in vivo.' Kidney Int.*, 1977, **12**, 229–237.
6. G. Arturson, T. Groth and G. Grotte, 'Human

glomerular membrane porosity and filtration pressure: dextran clearance data analyzed by theoretical models.' *Clin. Sci.*, 1971, **40**, 137–158.

7. J. Hardwicke, J. S. Cameron, J. F. Harrison *et al.*, in 'Proteins in Normal and Pathological Urine,' eds. Y. Manuel, J. P. Revillard and H. Betuel, University Park Press, Baltimore, 1970, pp. 111–152.

8. A. Verniory, R. Du Bois, P. Decoodt *et al.*, 'Measurement of the permeabiility of biological membranes. Application to the glomerular wall.' *J. Gen Physiol.*, 1973, **62**, 489–507.

9. G. M. Eisenbach, J. B. Liew, J. W. Boylan *et al.*, 'Effect of angiotensin on the filtration of protein in the rat kidney: a micropuncture study.' *Kidney Int.*, 1975, **8**, 80–87.

10. M. Gaizutis, A. J. Pesce and J. E. Lewy, 'Determination of nanogram amounts of albumin by radioimmunoassay.' *Microchem. J.*, 1972, **17**, 327.

11. R. L. S. Chang, W. M. Deen, C. R. Robertson *et al.*, 'Permselectivity of the glomerular capillary wall: III. Restricted transport of polyanions.' *Microchem. J.*, 1975, **8**, 212–218.

12. B. S. Daniels, E. B. Hauser, W. M. Deen *et al.*, 'Glomerular basement membrane: *in vitro* studies of water and protein permeability.' *Am. J.. Physiol.*, 1992, **262**, F919–F926.

13. B. S. Daniels, W. M. Deen, G. Mayer *et al.*, 'Glomerular permeability barrier in the rat. Functional assessment by *in vitro* methods.' *J. Clin. Invest.*, 1993, **92**, 929–936.

14. M. C. Drumond and W. M. Deen, 'Structural determinants of glomerular capillary hydraulic permeability.' *Am. J. Physiol.*, 1994, **266**, F1–F12.

15. J. P. Caulfield and M. G. Farquhar, 'The permeability of glomerular capillaries to graded dextrans. Identification of the basement membrane as the primary filtration barrier.' *J. Cell. Biol.*, 1974, **63**, 883–903.

16. D. Kerjaschki, D. J. Sharkey and M. G. Farquhar, 'Identification and characterization of podocalyxin—the major sialoprotein of the renal glomerular epithelial cell.' *J. Cell. Biol.*, 1984, **98**, 1591–1596.

17. R. Horvat, A. Hovoka, G. Dekan *et al.*, 'Endothelial cell membranes contain podocalyxin—the major sialoprotein of visceral glomerular epithelial cells.' *J. Cell. Biol.*, 1986, **102**, 484–491.

18. H. Sawada, H. Stukenbrok, D. Kerjaschki *et al.*, 'Epithelial polyanion (podocalyxin) is found on the sides but not the soles of the foot processes of the glomerular epithelium.' *Am. J. Pathol.*, 1986, **125**, 309–318.

19. H. Latta and S. Fligiel, 'Mesangial fenestrations, sieving, filtration, and flow.' *Lab. Invest.*, 1985, **52**, 591–598.

20. B. M. Brenner, T. H. Hostetter and M. D. Humes, 'Glomerular permselectivity: barrier function based on discrimination of molecular size and charge.' *Am. J. Physiol.*, 1978, **234**, F455–F460.

21. J. Bray and G. B. Robinson, 'Influence of charge on filtration across renal basement membrane films in vitro.' *Kidney Int.*, 1984, **25**, 527–533.

22. G. A. Kaysen, J. D. Myers, W. G. Couser *et al.*, 'Mechanisms and consequences of proteinuria.' *Lab. Invest.*, 1986, **54**, 479–498.

23. H. Latta and W. H. Johnston, 'The glycoprotein inner layer of glomerular capillary basement membrane as a filtration barrier.' *J. Ultrastruct. Res.*, 1976, **57**, 65–67.

24. L. Ghitescu, M. Desjardins and M. Bendayan, 'Immunocytochemical study of glomerular permeability to anionic, neutral and cationic albumins.' *Kidney Int.*, 1992, **42**, 25–32.

25. Y. Fujigaki, M. Nagase, S. Kobayashi *et al.*, 'Intra-GBM site of the functional filtration barrier for endogenous proteins in rats.' *Kidney Int.*, 1993, **43**, 567–574.

26. Y. S. Kanwar, 'Biophysiology of glomerular filtration and proteinuria.' *Lab Invest.*, 1984, **51**, 7–21.

27. D. Zhu, Y. Kim, M. W. Steffes *et al.*, 'Application of electron microscopic immunocytochemistry to the human kidney: distribution of type IV and type VI collagen in normal human kidney.' *J. Histochem. Cytochem.*, 1994, **42**, 577–584.

28. S. R. Batsford, R. Rohrbach and A. Vogt, 'Size restriction in the glomerular capillary wall: Importance of lamina densa.' *Kidney Int.*, 1987, **31**, 710–717.

29. R. J. Butkowski, J. Wieslander, M. Kleppel *et al.*, 'Basement membrane collagen in the kidney: regional localization of novel chains related to collagen IV.' *Kidney Int.*, 1989, **35**, 1195–1202.

30. Y. K. Kim, M. M. Kleppel, R. Butkowski *et al.*, 'Differential expression of basement membrane collagen chains in diabetic nephropathy.' *Am. J. Pathol.*, 1991, **138**, 413–420.

31. M. Desjardins and M. Bendayan, 'Heterogeneous distribution of type IV collagen, entactin, heparan sulfate proteoglycan and laminin among renal basement membranes as demonstrated by quantitative immunocytochemistry.' *J. Histochem. Cytochem.*, 1989, **37**, 885–897.

32. A. S. Martinez-Hernandez, S. Gay and E. J. Miller, 'Ultrastructural localization of type V collagen in rat kidney.' *J. Cell. Biol.*, 1982, **92**, 343–349.

33. J. A. Madri, F. J. Roll, H. Furthmayr *et al.*, 'Ultrastructural localization of fibronectin and laminin in basement membranes of murine kidney.' *J. Cell. Biol.*, 1980, **86**, 682–687.

34. P. J. Courtoy, R. Timpl and M. G. Farquhar, 'Comparative distribution of laminin, type IV collagen, and fibronectin in rat glomerulus.' *J. Histochem. Cytochem.*, 1982, **30**, 874–886.

35. R. Timpl, H. Rohde, P. G. Robey *et al.*, 'Laminin—a glycoprotein from basement membranes.' *J. Biol. Chem.*, 1979, **254**, 9933–9937.

36. P. J. Courtoy, Y. S. Kanwar, R. O. Hynes *et al.*, 'Fibronectin localization in the rat glomerulus.' *J. Cell. Biol.*, 1980, **87**, 691–696.

37. D. R. Abrahamson, 'Structure and development of the glomerular capillary wall and basement membrane.' *Am. J. Physiol.*, 1987, **253**, F783–F794.

38. B. Carlin, R. Jaffe, B. Bender *et al.*, 'Entactin, a novel basal lamina-associated sulfated glycoprotein.' *J. Biol. Chem.*, 1981, **256**, 5209–5214.

39. A. Katz, A. J. Fish, M. M. Kleppel *et al.*, 'Renal entactin (nidogen): isolation, characterization, and tissue distribution.' *Kidney Int.*, 1991, **40**, 643–652.

40. Y. S. Kanwar and M. G. Farquhar, 'Anionic sites in the glomerular basement membrane. *In vivo* and *in vitro* localization to the lamina rarae by cationic probes.' *J. Cell. Biol.*, 1981, **81**, 137–153.

41. Y. S. Kanwar and M. G. Farquhar, 'Presence of heparan sulfate in the glomerular basement membrane.' *Proc. Natl. Acad. Sci. USA*, 1979, **76**, 1303–1307.

42. Y. S. Kanwar and M. G. Farquhar, 'Isolation of glycosaminoglycans (heparan sulfate) from glomerular basement membranes.' *Proc. Natl. Acad. Sci. USA*, 1979, **76**, 4493–4497.

43. Y. S. Kanwar, A. Linker and M. G. Farquhar, 'Increased permeability of the glomerular basement membrane to ferritin after removal of glycosaminoglycans (heparan sulfate) by enzyme digestion.' *J. Cell. Biol.*, 1980, **86**, 688–693.

44. L. J. Rosenzweig and Y. S. Kanwar, 'Removal of sulfated (heparan sulfate) or nonsulfated (hyaluronic acid) glycosaminoglycans results in increased permeability of the glomerular basement membrane to ^{125}I-bovine serum albumin.' *Lab. Invest.*, 1982, **47**, 177–184.

45. K. A. Houck, D. W. Leung, A. M. Rowland

et al., 'Dual regulation of vascular endothelial growth factor bioavailability by genetic and proteolytic mechanisms.' *J. Biol. Chem.*, 1992, **267**, 26031–26037.

46. W. T. Monacci, M. J. Merrill and E. M. Oldfield, 'Expression of vascular permeability factor/vascular endothelial growth factor in normal rat tissues.' *Am. J. Physiol.*, 1993, **264**, C995–C1002.

47. L. F. Brown, B. Berse, K. Tognazzi, *et al.*, 'Vascular permeability factor mRNA and protein expression in human kidney.' *Kidney Int.*, 1992, **42**, 1457–1461.

48. K. Uchida, S. Uchida, K. Nitta *et al.*, 'Glomerular endothelial cells in culture express and secrete vascular endothelial growth factor.' *Am. J. Physiol.*, 1994, **266**, F81–F88.

49. B. Klanke, H. J. Grone, H. Gros *et al.*, 'Effects of vascular endothelial growth factor (VEGF) on hemodynamics and permselectivity of the isolated perfused rat kidney (IPRK).' *J. Am. Soc. Nephrol.*, 1994, **5**, 582 (abstract).

50. R. Lapinski, N. Perico, A. Remuzzi *et al.*, 'Angiotensin II modulates glomerular capillary permselectivity in rat isolated perfused kidney.' *J. Am. Soc. Nephrol.*, 1996, **7**, 653–660.

51. C. C. Tisher and B. M. Brenner, in 'Renal Pathology,' 2nd edn., eds. C. C. Tisher and B. M. Brenner, J. B. Lippincott Company, Philadelphia, PA, 1994, pp. 143–161.

52. R. Rodewald and M. J. Karnovsky, 'Porous structure of the glomerular slit diaphragm in the rat and mouse.' *J. Cell. Biol.*, 1974, **60**, 423–433.

53. H. Kurihara, J. M. Anderson, D. Kerjaschki *et al.*, 'The altered glomerular filtration slits seen in puromycin aminonucleoside nephrosis and protamine sulfate-treated rats contain the tight junction protein ZO-1.' *Am. J. Pathol.*, 1992, **141**, 805–816.

54. M. W. Seiler, H. G. Rennke, M. A. Venkatachalam *et al.*, 'Pathogenesis of polycation-induced alteration ("fusion") of glomerular epithelium.' *Lab. Invest.*, 1977, **36**, 48–61.

55. D. Kerjaschki, 'Polycation-induced dislocation of slit diaphragms and formation of cell junctions in rat kidney glomeruli: the effects of low temperature, divalent cations, colchicine, and cytochalasin B.' *Lab Invest.*, 1978, **39**, 430–440.

56. E. Schnabel, J. M. Anderson and M. G. Farquhar, 'The tight junction protein ZO-1 is connected along the slit diaphragms of the glomerular epithelium.' *J. Cell. Biol.*, 1990, **111**, 1255–1263.

57. M. Orikasa, Y. Matsui, T. Oite *et al.*, 'Massive proteinuria induced in rats by a single intravenous injection of a monoclonal antibody.' *J. Immunol.*, 1988, **141**, 807–814.

58. R. C. Blantz, F. B. Gabbai, O. Peterson *et al.*, 'Water and protein permeability is regulated by the glomerular epithelial slit diaphragm.' *J. Am. Soc. Nephrol.*, 1994, **4**, 1957–1964.

59. D. B. Jones, 'Mucosubstances of the glomerulus.' *Lab. Invest.*, 1969, **21**, 119–125.

60. H. Latta, W. H. Johnston and T. M. Stanley, 'Sialoglycoproteins and filtration barriers in the glomerular capillary wall.' *J. Ultrastruct. Res.*, 1975, **51**, 354–376.

61. D. Kerjaschki, H. Poczewski, G. Dekan *et al.*, 'Identification of a major sialoprotein in the glycocalyx of human visceral glomerular epithelial cells.' *J. Clin. Invest.*, 1986, **78**, 1142–1149.

62. P. M. Charest and J. Roth, 'Localization of sialic acid in kidney glomeruli: regionalization in the podocyte plasma membrane and loss in experimental nephrosis.' *Proc. Natl. Acad. Sci. USA*, 1985, **82**, 8508–8512.

63. G. Dekan, C. A. Gabel and M. G. Farquhar, 'Sulfate contributes to the negative charge of podocalyxin, the major sialoglycoprotein of the glomerular filtration slits.' *Proc. Natl. Acad. Sci USA*, 1991, **88**(12), 5398–5402.

64. M. Thomas, B. Goval, J. Wharren *et al.*, 'Molecular cloning of rabbit podocalyxin cDNA.' *J. Am. Soc. Nephrol.*, 1992, **3**, 728 (abstract).

65. J. L. Stow, H. Sawada and M. G. Farquhar, 'Basement membrane heparan sulfate proteoglycans are concentrated in the lamina rarae and in podocytes in the rat renal glomerulus.' *Proc. Natl. Acad. Sci. USA*, 1985, **82**, 3296–3300.

66. Y. S. Kanwar and M. G. Farquhar, 'Detachment of endothelium and epithelium from the glomerular basement membrane produced by kidney perfusion with neuraminidase.' *Lab. Invest.*, 1980, **42**, 375–382.

67. M. W. Seiler, M. A. Venkatachalam and R. S. Cotran, 'Glomerular epithelium: structural alterations induced by polycations.' *Science*, 1975, **189**, 390–393.

68. J. A. Bertolatus, 'Affinity cytochemical labeling of glomerular basement membrane anionic sites using specific biotinylation and colloidal gold probes.' *J. Histochem. Cytochem.*, 1990, **38**, 377–384.

69. J. A. Bertolatus, M. Abuyousef and L. G. Hunsicker, 'Glomerular sieving of high molecular weight proteins in proteinuric rats.' *Kidney Int.*, 1987, **31**, 1257–1266.

70. B. Lelongt, H. Makino and Y. S. Kanwar, 'Status of glomerular proteoglycans in aminonucleoside nephrosis.' *Kidney Int.*, 1987, **31**, 1299–1310.

71. R. J. Glassock, in 'The Nephrotic Syndrome,' eds. J. S. Cameron and R. J. Glassock, Marcel Dekker, New York, 1988, pp. 219–249.

72. J. L. Barnes, R. A. Radnik, E. P. Gilchrist *et al.*, 'Size and charge selective permeability defects induced in glomerular basement membrane by polycation.' *Kidney Int.*, 1984, **25**, 11–19.

73. D. Kerjaschki, 'Dysfunctions of cell biological mechanisms of visceral epithelial cell (podocytes) in glomerular diseases.' *Kidney Int.*, 1994, **45**, 300–313.

74. Y. S. Kanwar and L. J. Rosenzweig, 'Altered glomerular permeability as a result of focal detachment of the visceral epithelium.' *Kidney Int.*, 1982, **21**, 565–574.

75. D. Kerjaschki, P. P. Ojha, M. Susani *et al.*, 'A β1 integrin receptor for fibronectin in human kidney glomeruli.' *Am. J. Pathol.*, 1989, **134**, 481–489.

76. M. Korhonen, J. Ylanne, L. Laitinen *et al.*, 'Distribution of β1 and β3 integrins in human fetal and adult kidney.' *Lab Invest.*, 1990, **62**, 616–625.

77. S. Adler, 'Integrin receptors in the glomerulus: potential role in glomerular injury.' *Am. J. Physiol.*, 1992, **262**, F697–F704.

78. S. Adler and X. Chen, 'Anti-Fx1A antibody recognizes a β1 integrin on glomerular epithelial cells and inhibits adhesion and growth.' *Am. J. Physiol.*, 1992, **2622**, F770–F776.

79. F. G. Cosio, D. D. Sedmak and S. N. Nahman, Jr., 'Cellular receptors for matrix proteins in normal human kidney and human mesangial cells.' *Kidney Int.*, 1990, **38**, 886–895.

80. D. L. Mendrick, D. C. Chung and H. G. Rennke, 'Heymann antigen GP330 demonstrates affinity for fibronectin, laminin, and type I collagen and mediates rat proximal tubule epithelial cell adherence to such matrices *in vitro*.' *Exp. Cell Res.*, 1990, **188**, 23–35.

81. B. Bauvois, 'A collagen-binding glycoprotein on the surface of mouse fibroblasts is identified as dipeptidyl peptidase IV.' *Biochem J.*, 1988, **252**, 723–731.

82. K. J. Assmann, J. P. van Son, H. B. Dijkman *et al.*, 'A nephritogenic rat monoclonal antibody to mouse aminopeptidase A. Induction of massive albuminuria

after a single intravenous injection.' *J Exp. Med.*, 1992, **175**, 623–635.

83. A. Remuzzi, S. Puntorieri, C. Battaglia *et al.*, 'Angiotensin converting enzyme inhibition ameliorates glomerular filtration of macromolecules and water and lessens glomerular injury in the rat.' *J. Clin. Invest.*, 1990, **85**, 541–549.

84. G. Mayer, R. A. Lafayette, J. Oliver *et al.*, 'Effects of angiotensin II receptor blockade on remnant glomerular permselectivity.' *Kidney Int.*, 1993, **43**, 346–353.

85. H. Sarrriola, R. Timpl, K. von der Mark *et al.*, 'Dual origin of glomerular basement membrane.' *Dev. Biol.*, 1984, **101**, 86–96.

86. R. Johnson, H. Yamabe, Y. P. Chen *et al.*, 'Glomerular epithelial cells secrete a glomerular basement membrane-degrading metalloproteinase.' *J. Am. Soc. Nephrol.*, 1992, **2**, 1388–1397.

87. D. Drenckhahn and R. P. Franke, 'Ultrastructural organization of contractile and cytoskeletal proteins in glomerular podocytes of chicken, rat, and man.' *Lab Invest.*, 1988, **59**, 673–682.

88. C. I. Whiteside, R. Cameron, S. Munk *et al.*, 'Podocytic cytoskeletal disaggregation and basement-membrane detachment in puromycin aminonucleoside nephrosis.' *Am. J. Pathol.*, 1993, **142**, 1641–1653.

89. P. M. Andrews, 'Investigations of cytoplasmic contractile and cytoskeletal elements in the kidney glomerulus.' *Kidney Int.*, 1981, **20**, 549–562.

90. D. Kerjaschki, 'Polycation-induced dislocation of slit diaphragms and formation of cell junctions in rat kidney glomeruli: the effects of low temperature, divalent cations, colchicine and cytochalasin B.' *Lab. Invest.*, 1978, **39**, 430–440.

91. E. Rodriguez-Boulan and W. J. Nelson, 'Morphogenesis of the polarized epithelial cell phenotype.' *Science*, 1989, **245**, 718–725.

92. R. Sharma, H. B. Lovell, T. B. Wiegmann *et al.*, 'Vasoactive substances induce cytoskeletal changes in cultured rat glomerular epithelial cells.' *J. Am. Soc. Nephrol.*, 1992, **3**, 1131–1138.

93. R. L. Chevalier, R. J. Fern, M. Garmey *et al.*, 'Localization of cGMP after infusion of ANP or nitroprusside in the maturing rat.' *Am. J. Physiol.*, 1992, **262**, F417–F424.

94. J. G. Shake, R. C. Brandt and B. S. Daniels, 'Angiotensin II induces actin polymerisation within the glomerular filtration barrier: possible role in the local regulation of ultrafiltration.' *J. Am. Soc. Nephrol.*, 1992, **3**, 568 (abstract).

95. S. Furuya, S. Naruse, T. Nakayama *et al.*, 'Effect and distribution of intravenously injected ^{125}I-endothelin-1 in rat kidney and lung examined by electron microscopic radioautography.' *Anat. Embryol.*, 1992, **185**, 87–96.

96. J. M. Rebibou, C. J. He, F. Delarue *et al.*, 'Functional endothelin-1 receptors on human glomerular podocytes and mesangial cells.' *Nephrol. Dial. Transplant*, 1992, **7**, 288–292.

97. D. A. Maddox, W. M. Deen and B. M. Brenner, in 'Handbook of Physiology,' 2nd edn., ed. E. E. Windhager, Oxford University Press, New York, 1992, pp. 545–638.

98. E. G. Mimnaugh, M. A. Trush and T. E. Gram, 'A possible role for membrane lipid peroxidation in anthracyclin nephrotoxicity.' *Biochem. Pharmacol.*, 1986, **35**, 4327–4335.

99. D. Schlondorff, 'The glomerular mesangial cell: an expanding role for a specialized pericyte.' *FASEB J.*, 1987, **1**, 272–281.

100. K. F. Badr, K. A. Munger, M. Sugiura *et al.*, 'High and low affinity bindings sites for endothelin on cultured rat glomerular mesangial cells.' *Biochem. Biophys. Res. Commun.*, 1989, **161**, 776–781.

101. P. Mené and M. J. Dunn, 'Contractile effects of TxA_2 and endoperoxide analogues on cultured rat glomerular mesangial cells.' *Am. J. Physiol.*, 1986, **251**, F1029–F1035.

102. C. Zoja, S. Orisio, N. Perico *et al.*, 'Constitutive expression of endothelin gene in cultured human mesangial cells and its modulation by transforming growth factor-β, thrombin, and a thromboxane A_2 analogue.' *Lab Invest.*, 1991, **64**, 16–20.

103. B. D. Myers, W. M. Deen and B. M. Brenner, 'Effects of norepinephrine and angiotensin II on the determinants of glomerular ultrafiltration and proximal tubule fluid reabsorption in the rat.' *Circ. Res.*, 1975, **37**, 101–110.

104. G. DiBona, 'The function of the renal nerves.' *Rev. Physiol. Biochem. Pharmacol.*, 1982, **94**, 75.

105. V. Kon and M. J. Karnovsky, 'Morphologic demonstration of adrenergic influences on the glomerulus.' *Am. J. Pathol.*, 1989, **134**, 1039–1046.

106. R. C. Blantz, K. S. Konnen and B. J. Tucker, 'Angiotensin II effects upon the glomerular microcirculation and ultrafiltration coefficient in the rat.' *J. Clin. Invest.*, 1976, **57**, 419–434.

107. D. W. Ploth, J. Rudolph, R. LaGrange *et al.*, 'Tubuloglomerular feedback and single nephron function after converting enzyme inhibition in the rat.' *J. Clin. Invest.*, 1979, **64**, 1325–1335.

108. J. D. Dworkin and B. M. Brenner, in 'The Kidney. Physiology and Pathophysiology,' eds. D. W. Seldin and G. Giebisch, Raven Press, New York, 1985, pp. 397–426.

109. R. C. Blantz and J. C. Pelayo, 'A functional role for the tubuloglomerular feedback mechanism.' *Kidney Int.*, 1984, **25**, 739–746.

110. C. Baylis, 'Effects of administered thromboxanes on the intact, normal rat kidney.' *Renal Physiol.*, 1987, **10**, 110–121.

111. N. Schor, I. Ichikawa and B. M. Brenner, 'Mechanisms of action of various hormones and vasoactive substances on glomerular ultrafiltration in the rat.' *Kidney Int.*, 1981, **20**, 442–451.

112. K. F. Badr, C. Baylis, J. M. Pfeffer *et al.*, 'Renal and systemic hemodynamic responses to intravenous infusion of leukotriene C4 in the rat.' *Circ. Res.*, 1984, **54**, 492–499.

113. A. Rosenthal and C. R. Pace-Asciak, 'Potent vasoconstriction of the isolated perfused rat kidney by leukotrienes C4 and D4.' *Can. J. Physiol. Pharmacol.*, 1983, **61**, 325–328.

114. M. Yanagisawa, H. Kurihara, S. Kimura *et al.*, 'A novel potent vasoconstrictor peptide produced by vascular endothelial cells.' *Nature*, 1988, **332**, 411–415.

115. A. J. King, B. M. Brenner and S. Anderson, 'Endothelin: a potent renal and systemic vasoconstrictor peptide.' *Am. J. Physiol.*, 1989, **256**, F1051–F1058.

116. N. Perico, J. Dadan, M. Gabanelli *et al.*, 'Cycloxygenase products and atrial natriuretic peptide modulate renal response to endothelin.' *J. Pharmacol. Exp. Ther.*, 1990, **252**, 1213–1220.

117. M. S. Simonson, S. Wann, P. Mené *et al.*, 'Endothelin stimulates phospholipase C, Na^+/H^+ exchange, *c-fos* expression and mitogenesis in rat mesangial cells.' *J. Clin Invest.*, 1989, **83**, 708–712.

118. K. F. Badr, J. J. Murray, M. D. Breyer *et al.*, 'Mesangial cell, glomerular and renal vascular responses to endothelin in the rat kidney. Elucidation of signal transduction pathways.' *J. Clin. Invest.*, 1989, **83**, 336–342.

119. M. S. Simonson and M. J. Dunn, 'Renal actions of endothelin peptides.' *Curr. Opin. Nephrol. Hyperten.*, 1993, **2**, 51–60.

120. N. A. Bobadilla, E. Tapia, M. Franco *et al.*, 'Role of

nitric oxide in renal hemodynamic abnormalities of cyclosporine nephrotoxicity.' *Kidney Int.*, 1994, **46**, 773–779.

121. P. J. Shultz, A. E. Schorer and L. Raij, 'Effects of endothelium-derived relaxing factor and nitric oxide on rat mesangial cells.' *Am. J. Physiol.*, 1990, **258**, F162–F167.

122. M. Burnier and H. R. Brunner, 'Angiotensin II receptor antagonists and the kidney.' *Curr. Opin. Nephrol. Hyperten.*, 1994, **3**, 537–545.

123. H. G. Rennke and P. S. Klein, 'Pathogenesis and significance of nonprimary focal and segmental glomerulosclerosis.' *Am. J. Kidney Dis.*, 1989, **13**, 443–456.

124. W. E. Laurens, Y. F. Vanrenterghem, P. S. Steels *et al.*, 'A new single nephron model of focal and segmental glomerulosclerosis in the Munich-Wistar rat.' *Kidney Int.*, 1994, **45**, 143–149.

125. W. Laurens, C. Battaglia, C. Foglieni *et al.*, 'Direct podocyte damage in the single nephron leads to albuminuria *in vivo*.' *Kidney Int.*, 1995, **47**, 1078–1086.

126. S. Siegrist, S. Velitchkovitch, N. Moreau *et al.*, 'Effect of P and A site substrates on the binding of a macrolide to ribosomes. Analysis of the puromycin-induced stimulation.' *Eur. J. Biochem.*, 1984, **143**, 23–26.

127. I. Nussenzveig, C. Vilela de Faria, J. Lopez de Faria *et al.*, 'Aminonucleoside induced nephrotic syndrome in human beings: report of five cases.' Proceedings 2nd International Congress of Nephrology, Prague, 1963, **78**, 506.

128. A. F. Michael, H. D. Venters, H. G. Worthen *et al.*, 'Experimental renal disease in monkeys.' *Lab. Invest.*, 1962, **11**, 1266.

129. S. G. F. Wilson, D. B. Hackel, S. Harwood *et al.*, 'Aminonucleoside nephrosis in rats.' *Pediatrics*, 1985, **21**, 963.

130. R. Lannigan, 'The production of chronic renal disease in rats by single intravenous injection of aminonucleoside of puromycin and effect of low dosage continuous hydrocortisone.' *Br. J. Exp. Pathol.*, 1962, 326.

131. G. B. Ryan and M. J. Karnovsky, 'An ultrastructural study of the mechanisms of proteinuria in aminonucleoside nephrosis.' *Kidney Int.*, 1975, **8**, 219–232.

132. W. H. Baricos, S. Cortez-Schwartz and S. V. Shah, 'Renal neuraminidase. Characterization in normal rat kidney and measurement in experimentally induced nephrotic syndrome.' *Biochem. J.*, 1986, **239**, 705–710.

133. J. L. Olson, H. G. Rennke and M. A. Venkatachalam, 'Alterations in the charge and size selectivity barrier of the glomerular filter in aminonucleoside nephrosis in rats.' *Lab. Invest.*, 1981, **44**, 271–279.

134. A. A. Eddy and A. F. Michael, 'Acute tubulointerstitial nephritis associated with aminonucleoside nephrosis.' *Kidney Int.*, 1988, **33**, 14–23.

135. J. R. Diamond and M. J. Karnovsky, 'Focal and segmental glomerulosclerosis following a single intravenous dose of puromycin aminonucleoside.' *Am. J. Pathol.*, 1986, **122**, 481–487.

136. J. R. Diamond, J. V. Bonventre and M. J. Karnovsky, 'A role for oxygen free radicals in aminonucleoside nephrosis.' *Kidney Int.*, 1986, **29**, 478–483.

137. J. A. Velosa, R. J. Glasser, T. E. Nevins *et al.*, 'Experimental model of focal sclerosis II. Correlation with immunopathologic changes, macromolecular kinetics, and polyanion loss.' *Lab. Invest.*, 1977, **36**, 527–534.

138. J. Grond, J. J. Weening and J. D. Elema, 'Glomerular sclerosis in nephrotic rats. Comparison of the long-term effects of adriamycin and aminonucleoside.' *Lab. Invest.*, 1984, **51**, 277–285.

139. M. P. O'Donnell, L. Michels, B. Kasiske *et al.*, 'Adriamycin-induced chronic proteinuria: a structural and functional study.' *J. Lab. Clin. Med.*, 1985, **106**, 62–67.

140. J. J. Weening and H. G. Rennke, 'Glomerular permeability and polyanion in adriamycin nephrosis in the rat.' *Kidney Int.*, 1983, **24**, 152–159.

141. G. Remuzzi, C. Zoja, A. Remuzzi *et al.*, 'Low-protein diet prevents glomerular damage in adriamycin-treated rats.' *Kidney Int.*, 1985, **28**, 21–27.

142. R. B. Painter, 'Inhibition of DNA replicon initiation by 4-nitroquinoline-1-oxide, adriamycin and ethyleneimine.' *Cancer Res.*, 1978, **38**, 4445–4449.

143. R. C. Young, R. F. Ozols and C. E. Myers, 'The anthracycline antineoplastic drugs.' *N. Engl. J. Med.*, 1981, **305**, 139–153.

144. N. R. Bachur, G. L. Gordon and M. V. Gee, 'Anthracyclin antibiotic augmentation of microsomal electron transport and free radical formation.' *Mol. Pharmacol.*, 1977, **13**, 901–910.

145. T. Bertani, A. Poggi, R. Pozzoni *et al.*, 'Adriamycin-induced nephrotic syndrome in rats: sequence of pathologic events.' *Lab. Invest.*, 1982, **46**, 16–23.

146. T. Bertani, G. Rocchi, G. Sacchi *et al.*, 'Adriamycin-induced glomerulosclerosis in the rat.' *Am. J. Kidney Dis.*, 1986, **7**, 12–19.

147. S. Okuda, Y. Oh, H. Tsuruda *et al.*, 'Adriamycin-induced nephropathy as a model of chronic progressive glomerular disease.' *Kidney Int.*, 1986, **29**, 502–510.

148. A. Fogo, Y. Yoshida, A. D. Glick *et al.*, 'Serial micropuncture analysis of glomerular function in two rat models of glomerular sclerosis.' *J. Clin. Invest.*, 1988, **82**, 322–330.

149. L. F. Fajardo, J. R. Eltringham, J. R. Stewart *et al.*, 'Adriamycin nephrotoxicity.' *Lab Invest.*, 1980, **43**, 242–253.

150. P. Calabresi and R. E. Parks, Jr., in 'The Pharmacological Basis of Therapeutics,' 7th edn., eds. A. Goodman-Gilman, L. S. Goodman, T. W. Rall *et al.*, Macmillan, New York, 1985, pp. 1283–1285.

151. J. L. Speyer, M. D. Green, E. Kramer *et al.*, 'Protective effect of the bis(piperazinedione) ICRF-187 against doxorubicin-induced cardiac toxicity in women with advanced breast cancer.' *N. Engl. J. Med.*, 1988, **319**, 745–752.

152. J. R. Hoyer, J. Ratte, A. H. Potter *et al.*, 'Transfer of aminonucleoside nephrosis by renal transplantation.' *J. Clin. Invest.*, 1972, **51**, 2777–2780.

153. J. R. Hoyer, S. M. Mauer and A. F. Michael, 'Unilateral renal disease in the rat. I. Clinical morphologic and glomerular mesangial functional features of the experimental model produced by renal perfusion with the aminonucleoside of puromycin.' *J. Lab. Clin. Med.*, 1975, **85**, 756–768.

154. T. Bertani and G. Remuzzi, in 'Glomerular injury 300 years after Morgagni,' eds. T. Bertani and G. Remuzzi, Wichtig, Milano, 1983, pp. 163–177.

155. M. M. Schwartz, Z. Sharon, B. U. Pauli *et al.*, 'Inhibition of glomerular visceral epithelial cell endocytosis during nephrosis induced by puromycin aminonucleoside.' *Lab. Invest.*, 1984, **51**, 690–696.

156. G. Remuzzi and T. Bertani, 'Is glomerulosclerosis a consequence of altered glomerular permeability to macromolecules?' *Kidney Int.*, 1990, **38**, 384–394.

157. J. A. Fishman and M. J. Karnovsky, 'Effects of aminonucleoside of puromycin on glomerular epithelial cells *in vitro*.' *Am. J. Pathol.*, 1985, **118**, 398–407.

158. F. Ginevri, G. M. Ghiggeri, R. Bertelli *et al.*, 'Extracellular sites for a cytotoxic effect of adriamycin (ADR) in glomerular epithelial cells *in vitro*.' *J. Am. Soc. Nephrol.*, 1990, **1**, 611 (abstract).

159. W. Coers, S. Huitema, M. L. van der Horst et al., 'Puromycin aminonucleoside and adriamycin disturb cytoskeletal and extracellular matrix protein organization, but not protein synthesis of cultured glomerular epithelial cells.' *Exp. Nephrol.*, 1994, **2**, 40–50.

160. M. Kawaguchi, M. Yamada, H. Wada et al., 'Roles of active oxygen species in glomerular epithelial cell injury *in vitro* caused by puromycin aminonucleoside.' *Toxicology*, 1992, **72**, 329–340.

161. G. M. Ghiggeri, G. Cercignani, F. Ginevri et al., 'Puromycin aminonucleoside metabolism by glomeruli and glomerular cells *in vitro*.' *Kidney Int.*, 1991, **40**, 35–42.

162. S. D. Ricardo, J. F. Bertram and G. B. Ryan, 'Reactive oxygen species in puromycin aminonucleoside nephrosis: *in vitro* studies.' *Kidney Int.*, 1994, **45**, 1057–1069.

163. M. P. Bohrer, C. Baylis, C. R. Robertson et al., 'Mechanisms of the puromycin-induced defects in the transglomerular passage of water and macromolecules.' *J. Clin. Invest.*, 1977, **60**, 152–161.

164. A. Remuzzi, C. Battaglia, L. Rossi et al., 'Glomerular size selectivity in nephrotic rats exposed to diets with different protein content.' *Am. J. Physiol.*, 1987, **253**, F318–F327.

165. J. A. Bertolatus and L. G. Hunsicker, 'Glomerular sieving of anionic and neutral bovine albumins in proteinuria rats.' *Kidney Int.*, 1985, **28**, 467–476.

166. A. Remuzzi, C. Battaglia and G. Remuzzi, 'Effect of low protein diet on size and charge permeability in experimental nephrosis.' *Kidney Int.*, 1989, **35**, 472 (abstract).

167. Y. S. Kanwar and M. L. Jakubowski, 'Unaltered anionic sites of glomerular basement membrane in aminonucleoside nephrosis.' *Kidney Int.*, 1984, **25**, 613–618.

168. J. P. Caulfield and M. G. Farquhar, 'Loss of anionic sites from the glomerular basement membrane in aminonucleoside nephrosis.' *Lab. Invest.*, 1978, **39**, 505–512.

169. J. P. Caulfield, 'Alterations in the distribution of alcian blue-staining fibrillar anionic sites in the glomerular basement membrane in aminonucleoside nephrosis.' *Lab. Invest.*, 1979, **40**, 503–511.

170. M. W. Seiler, J. R. Hoyer and T. E. Krueger, 'Altered localization of protamine–heparin complexes in aminonucleoside nephrosis.' *Lab. Invest.*, 1980, **43**, 9–17.

171. S. Kobayashi, M. Nagase, N. Honda et al., 'Analysis of anionic charge of the glomerular basement membrane in aminonucleoside nephrosis.' *Kidney Int.*, 1989, **35**, 1405–1408.

172. J. D. Mahan, S. Sisson-Ross and R. L. Vernier, 'Glomerular basement membrane anionic charge site changes early in aminonucleoside nephrosis.' *Am. J. Pathol.*, 1986, **125**, 393–401.

173. C. Whiteside, K. Prutis, R. Cameron et al., 'Glomerular epithelial detachment, not reduced charge density, correlates with proteinuria in adriamycin and puromycin nephrosis.' *Lab Invest.*, 1989, **61**, 650–660.

174. L. A. Mynderse, J. R. Hassel, H. K. Kleinman et al., 'Loss of heparan sulfate proteoglycan from glomerular basement membrane of nephrotic rats.' *Lab. Invest.*, 1983, **48**, 292–302.

175. F. H. Wapstra, J. van den Born, J. H. M. Berden et al., 'Ace-inhibition restores loss of heparan sulfate (HS) in the glomerular basement membrane (GBM) of rats with established adriamycin (ADM) nephrosis.' *J. Am. Soc. Nephrol.*, 1994, **5**, 797.

176. P. Russo, D. Gingras and M. Bendayan, 'Poly-L-lysine-gold probe for the detection of anionic sites in normal glomeruli and in idiopathic and experimentally-induced nephrosis. A comparative ultrastructural study.' *Am. J. Pathol.*, 1993, **142**, 261–271.

177. A. F. Michael, E. Blau and R. L. Vernier, 'Glomerular polyanion. Alteration in aminonucleoside nephrosis.' *Lab. Invest.*, 1970, **23**, 649–657.

178. D. Kerjaschki, A. T. Vernillo and M. G. Farquhar, 'Reduced sialylation of podocalyxin—the major sialoprotein of the rat kidney glomerulus—in aminonucleoside nephrosis.' *Am. J. Pathol.*, 1985, **118**, 343–349.

179. L. Orci, A. Kunz, M. Amherdt et al., 'Perturbation of podocyte plasma membrane domains in experimental nephrosis. A lectin-binding and freeze-fracture study.' *Am. J. Pathol.*, 1984, **117**, 286–297.

180. M. A. Venkatachalam, M. J. Karnovsky and R. S. Cotran, 'Glomerular permeability. Ultrastructural studies in experimental nephrosis using horseradish peroxidase as a tracer.' *J. Exp. Med.*, 1969, **130**, 381–399.

181. M. A. Venkatachalam, R. S. Cotran and M. J. Karnovsky, 'An ultrastructure study of glomerular permeability in aminonucleoside nephrosis using catalase as tracer protein.' *J. Exp. Med.*, 1970, **132**, 1168–1180.

182. J. R. Salaman, 'Cyclosporine in renal trasplantation: a guide to management.' *Lancet*, 1984, **2**, 269–271.

183. P. R. Powell-Jackson, B. Young, R. Y. Calne et al., 'Nephrotoxicity of parenterally administered cyclosporine after orthotopic liver transplantation.' *Transplantation*, 1983, **36**, 505–508.

184. S. M. Flechner, C. van Buren, R. H. Kerman et al., 'The nephrotoxicity of cyclosporine in renal transplant recipients.' *Transplant. Proc.*, 1983, **15** (Suppl. 1), 2689–2694.

185. B. A. Sullivan, L. J. Hak and W. F. Finn, 'Cyclosporine nephrotoxicity: studies in laboratory animals.' *Transplant. Proc.*, 1985, **17**, 145–154.

186. M. J. Mihatsch, G. Thiel, V. Basler et al., 'Morphological patterns in cyclosporine-treated renal transplant recipients.' *Transplant. Proc.*, 1985, **17**, 101–116.

187. M. J. Mihatsch, G. Thiel and B. Ryffel, 'Histopathology of cyclosporine nephrotoxicity in the rat.' *Transplant. Proc.*, 1988, **20**, 759–771.

188. M. J. Mihatsch, G. Thiel, M. P. Spicktin et al., 'Morphological findings in kidney transplants after treatment with cyclosporine.' *Transplant. Proc.*, 1983, **15** (Suppl. 1), 2821.

189. P. H. Whiting, A. W. Thompson, J. T. Blair et al., 'Experimental cyclosporin A nephrotoxicity.' *Br. J. Exp. Pathol.*, 1982, **63**, 88–94.

190. H. D. Humes, N. M. Jackson, R. P. O'Connor et al., 'Pathogenetic mechanisms of nephrotoxicity: insights into cyclosporine nephrotoxicity.' *Transplant. Proc.*, 1985, **17**, 51–62.

191. B. M. Murray, M. S. Paller and T. F. Ferris, 'Effect of cyclosporine administration on renal hemodynamics in conscious rat.' *Kidney Int.*, 1985, **28**, 767–774.

192. H. Dieperink, H. Starklint and P. P. Leyssac, 'Nephrotoxicity of cyclosporine—an animal model: study of the nephrotoxic effect of cyclosporine on overall and renal tubular function in conscious rats.' *Transplant. Proc.*, 1983, **15**, 2763–2741.

193. N. Perico, J. Dadan and G. Remuzzi, 'Endothelin mediates the renal vasoconstriction induced by cyclosporine in the rat.' *J. Am. Soc. Nephrol.*, 1990, **1**, 76–83.

194. N. F. Rossi, P. C. Churchill, F. D. McDonald et al., 'Mechanism of cyclosporine A-induced renal vasoconstriction in the rat.' *J. Pharmacol. Exp. Ther.*, 1989, **250**, 896–901.

195. J. English, A. Evan, D. C. Houghton et al.,

'Cyclosporine-induced acute renal dysfunction in the rat. Evidence of arteriolar vasoconstriction with preservation of tubular function.' *Transplantation*, 1987, **44**, 135–141.

196. B. D. Myers, J. Ross, L. Newton *et al.*, 'Cyclosporine-associated chronic nephropathy.' *N. Engl. J. Med.*, 1984, **311**, 699–705.

197. M. Moran, S. Tomlanovich and B. D. Myers, 'Cyclosporine-induced chronic nephropathy in human recipients of cardiac allografts.' *Transplant. Proc.*, 1988, **20**, 759–790.

198. G. Remuzzi and T. Bertani, 'Renal vascular and thrombotic effects of cyclosporine.' *Am. J. Kidney Dis.*, 1989, **13**, 261–272.

199. T. E. Starzl, T. R. Hakala, J. T. Rosenthal, *et al.*, 'Colorado–Pittsburg cadaveric renal transplantation study with cyclosporine.' *Transplant. Proc.*, 1983, **15**, 2463–2468.

200. S. M. Flechner, W. D. Payne, C. Van Buren *et al.*, 'The effect of cyclosprine on early graft function in human renal transplantation.' *Transplantation*, 1983, **36**, 268–272.

201. J. S. Najarian, R. M. Ferguson, D. E. Sutherland *et al.*, 'A prospective trial of the efficacy of cyclosporine in renal transplantation at the University of Minnesota.' *Transplant. Proc.*, 1983, **15**, 438–441.

202. Anonymous, 'A randomized clinical trial of cyclosporine in cadaveric renal transplants.' *N. Engl. J. Med.*, 1983, **309**, 809–815.

203. Anonymous, 'Cyclosporine in cadaveric renal transplantation: one-year follow-up of multicentre trial.' *Lancet.*, 1983, **2**, 986–989.

204. F. E. Dische, J. Neuberger, J. Keating *et al.*, 'Kidney pathology in liver allograft recipients after long-term treatment with cyclosporin A.' *Lab. Invest.*, 1988, **58**, 395–402.

205. J. P. Bantle, M. S. Paller, R. J. Boudreau *et al.*, 'Long-term effects of cyclosporine on renal function in organ transplant recipients.' *J. Lab. Clin. Med.*, 1990, **115**, 233–240.

206. C. R. Stiller, P. A. Keown, D. Heinrichs *et al.*, 'The effect of cyclosporine on renal function in newly diagnosed diabetics.' *Transplant. Proc.*, 1985, **17**, 202–208.

207. B. von Graffenreid and W. B. Harrison, 'Renal function in patients with autoimmune diseases treated with cyclosporine.' *Transplant. Proc.*, 1985, **17**, 215–231.

208. E. W. Young, C. N. Ellis, J. M. Messana *et al.*, 'A prospective study of renal structure and function in psoriasis patients treated with cyclosporin.' *Kidney Int.*, 1994, **46**, 1216–1222.

209. B. D. Myers, R. Sibley, L. Newton *et al.*, 'The long-term course of cyclosporine-associated chronic nephropathy.' *Kidney Int.*, 1988, **33**, 590–600.

210. B. D. Myers, L. Newton, C. Boshkos *et al.*, 'Chronic injury of human renal microvessels with low-dose cyclosporine therapy.' *Transplantation*, 1988, **46**, 694–703.

211. P. Sweny, S. F. Lui, J. E. Scoble *et al.*, 'Conversion of stable renal allografts at one year from cyclosporin A to azathioprine: a randomized controlled study.' *Transplant. Int.*, 1990, **3**, 19–22.

212. H. Nizze, M. J. Mihatsch, H. U. Zollinger *et al.*, 'Cyclosporine-associated nephropathy in patients with heart and bone marrow transplants.' *Clin. Nephrol.*, 1988, **30**, 248–260.

213. S. Thiru, E. R. Maher, D. V. Hamilton *et al.*, 'Tubular changes in renal transplant recipients on cyclosporin.' *Transplant. Proc.*, 1983, **15**, 2846.

214. G. Klintmalm, S. O. Bohman, E. Sundelin *et al.*, 'Interstitial fibrosis in renal allografts after 12 to 46 months of cyclosporine treatment: beneficial effect of lower doses in early post-transplantation period.' *Lancet.*, 1984, **2**, 950–954.

215. R. Devinemi, N. McKenzie, N. Wall *et al.*, 'Renal function in patients receiving cyclosporine for orthotopic cardiac transplantation.' *Transplant. Proc.*, 1985, **17**, 1256–1257.

216. A. C. Wallace, 'Histopathology of cyclosporine.' *Transplant. Proc.*, 1985, **17**, 117–122.

217. T. Bertani, A. Schieppati and P. Ferrazzi, 'Nature and extent of glomerular injury induced by cyclosporine A (CyA) in heart transplant patients.' *Kidney Int.*, 1991, **40**, 243–250.

218. A. Fogo, S. E. Hellings, T. Inagami *et al.*, 'Endothelin receptor antagonism is protective *in vivo* acute cyclosporine toxicity.' *Kidney Int.*, 1992, **42**, 770–774.

219. D. H. Humes, 'Aminoglycoside nephrotoxicity.' *Kidney Int.*, 1988, **33**, 900–911.

220. C. Baylis, H. R. Rennke and B. M. Brenner, 'Mechanisms of the defect in glomerular ultrafiltration associated with gentamicin administration.' *Kidney Int.*, 1977, **12**, 344–353.

221. N. Schor, I. Ichikawa, H. G. Rennke *et al.*, 'Pathophysiology of altered glomerular function in aminoglycoside-treated rats.' *Kidney Int.*, 1981, **19**, 288–296.

222. C. Baylis, 'Gentamicin-induced glomerulotoxicity in the pregnant rat.' *Am. J. Kidney Dis.*, 1989, **13**, 108–113.

223. F. C. Luft and A. P. Evan, 'Glomerular filtration barrier in aminoglycoside-induced nephrotoxic acute renal failure.' *Renal Physiol.*, 1980, **3**, 265–271.

224. R. E. Bulger, G. Eknoyan, D. J. Purcell, II *et al.*, 'Endothelial characteristics of glomerular capillaries in normal, mercuric chloride-induced, and gentamicin-induced acute renal failure in the rat.' *J. Clin. Invest.*, 1983, **72**, 128–141.

225. M. A. Sande and G. L. Mandell, in 'The Pharmacological Basis of Therapeutics,' 7th edn., eds. A. Goodman-Gilman, L. S. Goodman, T. W. Rall *et al.*, Macmillan, New York, 1980, pp. 453–464.

226. P. T. Wertlake, W. T. Butler, G. J. Hill *et al.*, 'Nephrotoxic tubular damage and calcium deposition following amphotericin B therapy.' *Am. J. Pathol.*, 1963, **43**, 449.

227. D. K. McCurdy, M. Frederic and J. R. Elkinton, 'Renal tubular acidosis due to amphotericin B.' *N. Engl. J. Med.*, 1968, **278**, 124–130.

228. J. B. Douglas and J. K. Healy, 'Nephrotoxic effects of amphotericin B, including renal tubular acidosis.' *Am. J. Med.*, 1969, **46**, 154–162.

229. J. P. Utz, J. E. Bennett, M. D. Brandiss *et al.*, 'Amphotericin B toxicity.' *Ann. Intern. Med.*, 1964, **61**, 334.

230. W. T. Butler, G. J. Hill, C. F. Szwed *et al.*, 'Amphotericin B renal toxicity in the dog.' *J. Pharmacol. Exp. Ther.*, 1964, **143**, 47.

231. J. F. Gerkens and R. A. Branch, 'The influence of sodium status and furosemide on canine acute amphotericin B nephrotoxicity.' *J. Pharmacol. Exp. Ther.*, 1980, **214**, 306–311.

232. H. T. Heidemann, J. F. Gerkens, W. A. Spickard *et al.*, 'Amphotericin B nephrotoxicity in humans decreased by salt repletion.' *Am. J. Med.*, 1983, **75**, 476–481.

233. J. T. Cheng, R. T. Witty, R. R. Robison *et al.*, 'Amphotericin B nephrotoxicity: increased renal resistance and tubule permeability.' *Kidney Int.*, 1982, **22**, 626–633.

234. J. P. Tolins and L. Raij, 'Adverse effect of amphotericin B administration on renal hemodynamics in the rat. Neurohumoral mechanisms and influence of calcium channel blockade.' *J. Pharmacol. Exp. Ther.*, 1988, **245**, 594–599.

235. R. Sabra, K. Takahashi, R. A. Branch *et al.*, 'Mechanisms of amphotericin B-induced reduction of the glomerular filtration rate: a micropuncture study.' *J. Pharmacol. Exp. Ther.*, 1990, **253**, 34–37.

236. T. Nakamura, I. Ebihara, I. Shirato *et al.*, 'Modulation of basement membrane component gene expression in glomeruli of aminonucleoside nephrosis.' *Lab. Invest.*, 1991, **64**, 640–647.

237. S. Suzuki, T. Nakamura, I. Ehibara *et al.*, 'Modulation of basement membrane (BM) component gene expression in glomeruli of PAN nephrosis treated with or without methylprednisolone (MPSL).' *J. Am. Soc. Nephrol.*, 1990, **1**, 555 (abstract).

238. B. S. Kasinath, A. K. Singh, Y. S. Kanwar *et al.*, 'Dexamethasone increases heparan sulfate proteoglycan core protein content of glomerular epithelial cells.' *J. Lab. Clin. Med.*, 1990, **115**, 196–202.

239. N. F. van Det, N. A. M. Verhagen, J. T. Tamsma *et al.*, 'Angiotensin II inhibits the production of heparan sulfate proteoglycans by human adult mesangial cells *in vitro*.' *J. Am. Soc. Nephrol.*, 1994, **5**, 822 (abstract).

240. L. C. Racusen, D. H. Prozialeck and K. Solez, 'Glomerular epithelial cell changes after ischemia or dehydration. Possible role of angiotensin II.' *Am. J. Pathol.*, 1984, **114**, 157–163.

241. T. Faraggiana and E. Grishman, 'The podocyte cell coat in experimental nephrosis.' *J. Ultrastruct. Res.*, 1981, **74**, 296–306.

242. C. Jones, S. Buch, M. Post *et al.*, 'Production of extracellular matrix in purine aminonucleoside (PAN) nephrosis.' *J. Am. Soc. Nephrol.*, 1990, **1**, 550.

243. J. Martin, M. Davies, G. Thomas *et al.*, 'Human mesangial cells secrete a GBM-degrading neutral proteinase and a specific inhibitor.' *Kidney Int.*, 1989, **36**, 790–801.

244. J. C. Davin, M. Davies, J. M. Foidart, *et al.*, 'Urinary excretion of neutral proteinases in nephrotic rats with a glomerular disease.' *Kidney Int.*, 1987, **31**, 32–40.

245. F. Grinnel, J. R. Head and J. Hopepuir, 'Fibronectin and cell shape *in vivo*: studies on the endometrium during pregnancy.' *J. Cell. Biol.*, 1982, **94**, 579–606.

246. U. Hedin, B. A. Bottger, E. Forsberg *et al.*, 'Diverse effects of fibronectin and laminin on fenotypic properties of cultured arterial smooth muscle cells.' *J. Cell. Biol.*, 1988, **107**, 307–319.

247. Y. M. O'Meara, Y. Natori, A. W. Minto *et al.*, 'Nephrotoxic antiserum identifies a β1-integrin on rat glomerular epithelial cells.' *Am. J. Physiol.*, 1992, **262**, F1083–F1091.

248. A. V. Cybulsky, S. Carbonetto, Q. Huang *et al.*, 'Adhesion of rat glomerular epithelial cells to extracellular matrices: role of β1 integrins.' *Kidney Int.*, 1992, **42**, 1099–1116.

249. H. Kurihara, J. M. Anderson and M. G. Farquhar, 'Increased Tyr phosphorylation of ZO-1 during modification of tight junctions between glomerular foot processes.' *Am. J. Physiol.*, 1995, **268**, F514–F524.

250. T. Sato, A. Ito and Y. Mori, 'Interleukin-6 enhances the production of tissue inhibitor of metalloproteinases (TIMP) but not that of matrix metalloproteinases by human fibroblasts.' *Biochem. Biophys. Res. Commun.*, 1990, **170**, 824–829.

251. M. Lachapelle and M. Bendayan, 'Contractile proteins in podocytes: immunocytochemical localization of actin and alpha-actinin in normal and nephritic rat kidney.' *Virch. Arch. B Cell Pathol. Incl. Mol. Pathol.*, 1991, **60**, 105–111.

252. A. Rehan, K. J. Johnson, R. C. Wiggins *et al.*, 'Evidence for the role of oxygen radicals in acute nephrotoxic nephritis.' *Lab. Invest.*, 1984, **51**, 396–403.

253. N. W. Boyce and S. R. Holdsworth, 'Hydroxyl radical mediation of immune renal injury by desferrioxamine.' *Kidney Int.*, 1986, **30**, 813–817.

254. G. Remuzzi, C. Zoja and N. Perico, 'Proinflammatory mediators of glomerular injury and mechanisms of activation of autoreactive T cells.' *Kidney Int. (Suppl.)*, 1994, **44**, S8–S16.

255. F. Ginevri, R. Gusmano, R. Oleggini *et al.*, 'Renal purine efflux and xanthine oxidase activity during experimental nephrosis in rats: difference between puromycin aminonucleoside and adriamycin nephrosis.' *Clin. Sci. (Colch.)*, 1990, **78**, 283–293.

256. S. D. Ricardo, J. F. Bertram and G. B. Ryan, 'Antioxidants protect podocyte foot processes in puromycin aminonucleoside-treated rats.' *J. Am. Soc. Nephrol.*, 1994, **4**, 1974–1986.

257. I. Fridovich, 'Superoxide dismutases.' *Annu. Rev. Biochem.*, 1975, **44**, 147–159.

258. A. Deisseroth and A. L. Dounce, 'Catalase: physical and chemical properties, mechanism of catalysis, and physiological role.' *Physiol. Rev.*, 1970, **50**, 319–375.

259. P. B. McCay, D. D. Gibson, K. L. Fong *et al.*, 'Effect of glutathione peroxidase activity on lipid peroxidation in biological membranes.' *Biochim. Biophys. Acta*, 1976, **431**, 459–468.

260. T. Kawamura, T. Yoshioka, T. Bills *et al.*, 'Glucocorticoid activates glomerular antioxidant enzymes and protects glomeruli from oxidant injuries.' *Kidney Int.*, 1991, **40**, 291–301.

261. T. Hara, T. Miyai, T. Iida *et al.*, 'Aggravation of puromycin aminonucleoside (PAN) nephrosis by the inhibition of endogenous superoxide dismutase (SOD).' XI International Congress of Nephrology Tokyo, 15–20 July, 1990, 442 (abstract).

262. H. Miyai, T. Hara, K. Yamada *et al.*, 'Aggravation of puromycin aminonucleoside (PAN) nephrosis by glutathione-depleting agent.' XIth International Congress of Nephrology Tokyo, 15–20 July, 1990, 442.

263. R. Baliga, M. Baliga and S. V. Shah, 'Effect of selenium-deficient diet in experimental glomerular disease.' *Am. J. Physiol.*, 1992, **263**, F56–F61.

264. V. Thakur, P. D. Walker and S. V. Shah, 'Evidence suggesting a role for hydroxyl radical in puromycin aminonucleoside-induced proteinuria.' *Kidney Int.*, 1988, **34**, 494–499.

265. L. S. Milner, S. H. Wei, S. J. Stohs *et al.*, 'Role of glutathione metabolism in the reduction of proteinuria by dimethylthiourea in adriamycin nephrosis.' *Nephron*, 1992, **62**(2), 192–197.

266. F. Ginevri, G. M. Ghiggeri, R. Oleggini *et al.*, 'Renal xanthine oxidase in the pathogenesis of experimental adriamycin nephrosis in rats.' VIII Congress of the International Pediatric Nephrology Association August 27–September 1, 1989 (abstract).

267. S. V. Shah, 'Evidence suggesting a role for hydroxyl radical in passive Heymann nephritis in rats.' *Am. J. Physiol.*, 1988, **254**, F337–F344.

268. M. A. Rahman, S. S. Emancipator and J. R. Sedor, 'Hydroxyl radical scavengers ameliorate proteinuria in rat immune complex glomerulonephritis.' *J. Lab. Clin. Med.*, 1988, **112**, 619–626.

269. T. Yoshioka, T. Moore-Jarrett and A. Yared, 'Reactive oxygen species (ROS) of extrarenal origin can induce massive "functional" proteinuria.' *Kidney Int.*, 1990, **37**, 497 (abstract).

270. R. A. Greenwald and W. W. Moy, 'Inhibition of collagen gelation by action of the superoxide radical.' *Arthritis Rheum.*, 1979, **22**, 251–259.

271. S. F. Curran, M. A. Amoruso, B. D. Goldstein *et al.*,

'Degradation of soluble collagen by ozone or hydroxyl radicals.' *FEBS Lett.*, 1984, **176**, 155–160.

272. R. A. Greenwald and W. W. Moy, 'Effect of oxygen-derived free radicals on hyaluronic acid.' *Arthritis Rheum.*, 1980, **23**, 455–463.

273. S. P. Wolf and R. T. Dean, 'Fragmentation of proteins by free radicals and its effect on their susceptibility to enzymatic hydrolysis.' *Biochem. J.*, 1982, **234**, 399.

274. S. E. Fligiel, E. C. Lee, J. P. McCoy *et al.*, 'Protein degradation following treatment with hydrogen peroxide.' *Am. J. Pathol.*, 1984, **115**, 418–425.

275. B. Halliwell and J. M. C. Gutteridge, Free Radicals in Biology and Medicine. Clarendon Press, Oxford, 1985, pp. 20–64.

276. W. W. Bakker, D. Kalicharan, J. Donga *et al.*, 'Decreased ATPase activity in adriamycin nephrosis is independent of proteinuria.' *Kidney Int.*, 1987, **31**, 704–709.

277. N. Kashihara, T. Dalecki, Y. Watanabe *et al.*, 'Effect of reactive oxygen species on glomerular extracellular matrix proteoglycans.' *J. Am. Soc. Nephrol.*, 1990, **1**, 527.

278. D. M. Shasby, S. E. Lind, S. S. Shasby *et al.*, 'Reversible oxidant-induced increases in albumin transfer across cultured endothelium: alterations in cell shape and calcium homeostasis.' *Blood*, 1985, **65**, 605–614.

279. J. Satriano and D. Schlondorff, 'Activation and attenuation of transcription factor NF-κB in mouse glomerular mesangial cells in response to tumor necrosis factor-α, immunoglobulin G, and adenosine 3′,5′-cyclic monophosphate. Evidence for involvement of reactive oxygen species.' *J. Clin. Invest.*, 1994, **94**, 1629–1636.

280. K. P. G. Harris, J. B. Lefkowith, S. Klahr *et al.*, 'Essential fatty acid deficiency ameliorates acute renal dysfunction in the rat after the administration of the aminonucleoside of puromycin.' *J. Clin. Invest.*, 1990, **86**, 1115–1123.

281. J. R. Diamond, I. Pesek, S. Ruggieri *et al.*, 'Essential fatty acid deficiency during acute puromycin nephrosis ameliorates late renal injury.' *Am. J. Physiol.*, 1989, **257**, F798–F807.

282. T. Goto, M. Mune, K. Mayoba *et al.*, 'Effects of selective thromboxane A₂ synthetase inhibitor on aminonucleoside induced nephrotic rat.' Proceedings 10th International Congress of Nephrology, 1987, 227 (abstract).

283. G. Remuzzi, L. Imberti, M. Rossini *et al.*, 'Increased glomerular thromboxane synthesis as a possible cause of proteinuria in experimental nephrosis.' *J. Clin. Invest.*, 1985, **75**, 94–101.

284. A. Benigni, N. Perico, J. Dadan *et al.*, 'The increased thromboxane B₂ (TxB₂) urinary excretion in experimental nephrosis originates from resident renal cells.' *Kidney Int.*, 1990, **37**, 345 (abstract).

285. L. A. Bruggeman, E. A. Horigan, S. Horikoshi *et al.*, 'Thromboxane stimulates synthesis of extracellular matrix proteins *in vitro*.' *Am. J. Physiol.*, 1991, **261**, F488–F494.

286. S. Ledbetter, E. J. Copeland, D. Noonan *et al.*, 'Altered steady-state mRNA levels of basement membrane proteins in diabetic mouse kidneys and thromboxane synthase inhibition.' *Diabetes*, 1990, **39**, 196–203.

287. P. Mené, H. E. Abboud and M. J. Dunn, 'Regulation of human mesangial cell growth in culture by thromboxane A₂ and prostacyclin.' *Kidney Int.*, 1990, **38**, 232–239.

288. L. A. Bruggeman, P. D. Burbelo, Y. Yamada *et al.*, 'Regulation of renal transcriptional factors for type IV collagen gene expression by thromboxane.' XI

International Congress of Nephrology Tokyo, 15–20 July, 1990, 30A (abstract).

289. J. Egido, A. Robles, A. Ortiz *et al.*, 'Role of platelet-activating factor in adriamycin-induced nephropathy in rats.' *Eur. J. Pharmacol.*, 1987, **138**, 119–123.

290. M. Gomez-Chiarri, A. Ortiz, J. L. Lerma *et al.*, 'Involvement of tumor necrosis factor and platelet-activating factor in the pathogenesis of experimental nephrosis in rats.' *Lab. Invest.*, 1994, **70**, 449–459.

291. N. Perico, F. Delaini, M. Tagliaferri *et al.*, 'Effect of platelet-activating factor and its specific receptor antagonist on glomerular permeability to proteins in isolated perfused rat kidney.' *Lab. Invest.*, 1988, **58**, 163–171.

292. N. Perico, A. Remuzzi, J. Dadan *et al.*, 'Platelet-Activating Factor alters glomerular barrier size selectivity for macromolecules in the rat.' *Am. J. Physiol.*, 1991, **261**, F85–F90.

293. G. F. Schreiner, R. S. Cotran and E. R. Unnanue, 'Modulation of Ia and leukocyte common antigen expression in rat glomeruli during the course of glomerulonephritis and aminonucleoside nephrosis.' *Lab. Invest.*, 1984, **51**, 524–533.

294. T. R. Bricio, F. Mampaso, J. Egido *et al.*, 'Interleukin-1 cytokine production in adriamycin induced nephrosis.' *Cancer Chemother. Rep.*, 1990, **1**, 517 (abstract).

295. M. Gomez-Chiarri, J. L. Lerma, E. Gonzalez, *et al.*, 'Involvement of tumor necrosis factor (TNF) in the pathogenesis of experimental nephrotic syndrome.' *J. Am. Soc. Nephrol.*, 1990, **1**, 522 (abstract).

296. T. Bertani, C. Zoja, M. Abbate *et al.*, 'Age-related nephropathy and proteinuria in rats with intact kidneys exposed to diets with different protein content.' *Lab. Invest.*, 1989, **60**, 196–204.

297. J. R. Diamond and I. P. Pesek, 'Glomerular tumor necrosis factor and interleukin-1 during acute aminonucleoside nephrosis. An immunohistochemical study.' *Lab. Invest.*, 1991, **64**, 21–28.

298. J. R. Diamond and G. Ding, 'Glomerular transforming growth factor-β1 gene expression in puromycin aminonucleoside nephrosis.' *J. Am. Soc. Nephrol.*, 1992, **3**, 466 (abstract).

299. G. Ding, I. Pesek-Diamond and J. R. Diamond, 'Cholesterol, macrophages, and gene expression of TGF-β1 and fibronectin during nephrosis.' *Am. J. Physiol.*, 1993, **264**, F577–F584.

300. K. Tamaki, S. Okuda, T. Ando *et al.*, 'TGF-β1 in glomerulosclerosis and interstitial fibrosis of adriamycin nephropathy.' *Kidney Int.*, 1994, **45**, 525–536.

301. G. Remuzzi, 'Abnormal protein traffic through the glomerular barrier induces proximal tubular cell dysfunction and causes renal injury.' *Curr. Opin. Nephrol. Hyperten.*, 1995, **4**, 339–342.

302. H. R. Brady, 'Leukocyte adhesion molecules and kidney diseases.' *Kidney Int.*, 1994, **45**, 1285–1300.

303. J. Grond, M. S. Schilthuis, J. Koudstaal *et al.*, 'Mesangial function and glomerular sclerosis in rats after unilateral nephrectomy.' *Kidney Int.*, 1982, **22**, 338–343.

304. J. Grond, J. Koudstaal and J. D. Elema, 'Mesangial function and glomerular sclerosis in rats with aminonucleoside nephrosis.' *Kidney Int.*, 1985, **27**, 405–410.

305. S. Anderson, J. R. Diamond, M. J. Karnovsky *et al.*, 'Mechanisms underlying transition from acute glomerular injury to late glomerular sclerosis in a rat model of nephrotic syndrome.' *J. Clin. Invest.*, 1988, **82**, 1757–1768.

306. R. Tanaka, V. Kon, T. Yoshioka *et al.*, 'Angiotensin converting enzyme inhibitor modulates glomerular function and structure by distinct mechanisms.' *Kidney Int.*, 1994, **45**, 537–543.

307. R. Zatz, B. R. Dunn, T. W. Meyer *et al.*, 'Prevention

of diabetic glomerulopathy by pharmacological amelioration of glomerular capillary hypertension.' *J. Clin. Invest.*, 1986, **77**, 1925–1930.

308. E. Morelli, N. Loon, T. Meyer *et al.*, 'Effects of converting-enzyme inhibition on barrier function in diabetic glomerulopathy.' *Diabetes*, 1990, **39**, 76–83.

309. A. Remuzzi, P. Ruggenenti, L. Mosconi *et al.*, 'Effect of low-dose enalapril on glomerular size-selectivity in human diabetic nephropathy.' *J. Nephrol.*, 1993, **6**, 36–43.

310. A. Remuzzi, E. Perticucci, P. Ruggenenti *et al.*, 'Angiotensin converting enzyme inhibition improves glomerular size-selectivity in IgA nephropathy.' *Kidney Int.*, 1991, **39**, 1267–1273.

311. A. Remuzzi, N. Perico, C. S. Amuchastegui *et al.*, 'Short- and long-term effect of angiotensin II receptor blockade in rats with experimental diabetes.' *J. Am. Soc. Nephrol.*, 1993, **4**, 40–49.

312. R. A. Lafayette, G. Mayer, S. K. Park *et al.*, 'Angiotensin II receptor blockade limits glomerular injury in rats with reduced renal mass.' *J. Clin. Invest.*, 1992, **90**, 766–771.

313. Y. Fujiwara, T. Takama, S. Shin *et al.*, 'Angiotensin II stimulates mesangial cell growth through phosphoinositide cascade.' *Kidney Int.*, 1989, **35**, 172 (abstract).

314. S. Kagami, W. A. Border, D. Miller *et al.*, 'Angiotensin II stimulates extracellular matrix protein synthesis through induction of transforming growth factor-β expression in rat glomerular mesangial cells.' *J. Clin. Invest.*, 1994, **93**, 2431–2437.

315. W. F. Keane and L. Raij, 'Relatioship among altered glomerular barrier permselectivity, angiotensin II, and mesangial uptake of macromolecules.' *Lab Invest.*, 1985, **52**, 599–604.

316. A. Remuzzi, O. Imberti, S. Puntorieri, *et al.*, 'Dissociation between antiproteinuric and antihypertensive effect of angiotensin converting enzyme inhibitors in rats.' *Am. J. Physiol.*, 1994, **267**, F1034–F1044.

317. R. T. Gansevoort, D. de Zeeuw and P. E. de Jong, 'Dissociation between the course of the hemodynamic and antiproteinuric effects of angiotensin I converting enzyme inhibition.' *Kidney Int.*, 1993, **44**, 579–584.

318. J. W. Scholey, P. L. Miller, H. G. Rennke *et al.*, 'Effect of converting enzyme inhibition on the course of adriamycin-induced nephropathy.' *Kidney Int.*, 1989, **36**, 816–822.

319. R. T. Gansevoort, F. H. Wapstra, J. J. Weening *et al.*, 'Sodium depletion enhances the antiproteinuric effect of ACE inhibition in established experimental nephrosis.' *Nephron.*, 1992, **60**, 246–247.

320. E. Podjarny, J. L. Bernheim, A. Pomeranz *et al.*, 'Effect of timing of antihypertensive therapy on glomerular injury: comparison between captopril and diltiazem.' *Nephrol. Dial. Transplant.*, 1993, **8**, 501–506.

321. M. Kohno, T. Horio, M. Ikeda *et al.*, 'Angiotensin II stimulates endothelin-1 secretion in cultured rat mesangial cells.' *Kidney Int.*, 1992, **42**, 860–866.

322. Z. A. Abassi, H. Klein, E. Golomb *et al.*, 'Regulation of the urinary excretion of endothelin in the rat.' *Am. J. Hypertens.*, 1993, **6**, 453–457.

323. S. Orisio, A. Benigni, I. Bruzzi, *et al.*, 'Renal endothelin gene expression is increased in remnant kidney and correlates with disease progression during acute puromycin nephrosis.' *Kidney Int.*, 1993, **43**, 354–358.

324. T. Nakamura, I. Ebihara, M. Fukui *et al.*, 'Modulation of glomerular endothelin and endothelin-receptor gene expression in aminonucleoside-induced nephrosis.' *J. Am. Soc. Nephrol.*, 1995, **5**, 1585–1590.

325. A. A. Eddy, 'Interstitial nephritis induced by protein-overload proteinuria.' *Am. J. Pathol.*, 1989, **135**, 719–723.

326. T. Bertani, F. Cutillo, C. Zoja *et al.*, 'Tubulo-interstitial lesions mediate renal damage in adriamycin glomerulopathy.' *Kidney Int.*, 1986, **30**, 488–496.

327. C. R. Baxter, G. G. Duggin, B. M. Hall *et al.*, 'Stimulation of renin release from rat renal cortical slices by cyclosporin A.' *Res. Commun. Chem. Pathol. Pharmacol.*, 1984, **43**, 417–423.

328. A. Kurtz, R. Della-Bruna and K. Kuhn, 'Cyclosporine A enhances renin secretion and production in isolated juxtaglomerular cells.' *Kidney Int.*, 1988, **33**, 947–953.

329. E. J. Barros, M. A. Boim, H. Ajzen *et al.*, 'Glomerular hemodynamics and hormonal participation in cyclosporine nephrotoxicity.' *Kidney Int.*, 1987, **32**, 19–25.

330. S. Jao, W. Waltzer and L. A. Arbeit, 'Acute cyclosporine induced decrease in GFR is mediated by changes in renal blood flow and renal vascular resistance.' *Kidney Int.*, 1986, **29**, 431.

331. N. G. Moss, S. L. Powell and R. J. Falk, 'Intravenous cyclosporine activates afferent and efferent renal nerves and causes sodium retention in innervated kidneys in rats.' *Proc. Natl. Acad. Sci. USA*, 1985, **82**, 8222–8226.

332. J. J. Curtis, R. G. Luke and E. Dubovsky, 'Cyclosporin in therapeutic doses increases renal allograft vascular resistance.' *Lancet*, 1986, **2**, 477–479.

333. D. C. Lau, K. L. Wong and W. S. Hwang, 'Cyclosporine toxicity on cultured rat microvascular endothelial cells.' *Kidney Int.*, 1989, **35**, 604–613.

334. D. Adu, C. J. Lote, J. Michael *et al.*, 'Does cyclosporine inhibit renal prostaglandin synthesis?' *Proc. Eur. Dial. Transpl. Assoc. Eur. Ren. Assoc.*, 1985, **21**, 969–972.

335. C. Zoja, L. Furci, F. Ghilardi *et al.*, 'Cyclosporin-induced endothelial cell injury.' *Lab. Invest.*, 1986, **55**, 455–462.

336. N. Perico, M. Rossini O. Imberti *et al.*, 'Thromboxane receptor blockade attenuates chronic cyclosporine nephrotoxicity and improves survival in rats with renal isograt.' *J. Am. Soc. Nephrol.*, 1992, **2**, 1398–1404.

337. N. Perico, C. Zoja, A. Benigni *et al.*, 'Effect of short-term cyclosporine administration in rats on renin–angiotensin and thromboxan A_2: possible relevence to the reduction in glomerular filtration rate.' *J. Pharmacol. Exp. Ther.*, 1986, **239**, 229–235.

338. A. Kawaguchi, M. H. Goldman, R. Shapiro *et al.*, 'Increase in urinary thromboxane B_2 in rats caused by cyclosporine.' *Transplantation*, 1985, **40**, 214–216.

339. N. Perico, A. Benigni, C. Zoja, *et al.*, 'Functional significance of exaggerated renal thromboxane A_2 synthesis induced by cyclosporin A.' *Am. J. Physiol.*, 1986, **251**, F581–F587.

340. C. Smeesters, P. Chaland, L. Giroux *et al.*, 'Prevention of acute cyclosporine A nephrotoxicity by thromboxane synthetase inhibitor.' *Transplant. Proc.*, 1988, **20**, 663–669.

341. T. M. Coffman, D. R. Carr, W. E. Yarger *et al.*, 'Evidence that renal prostaglandin and thromboxane production is stimulated in chronic cyclosporine nephrotoxicity.' *Transplantation*, 1987, **43**, 282–285.

342. A. Benigni, C. Chiabrando, A. Piccinelli *et al.*, 'Increased urinary excretion of thromboxane B_2 and 2,3-dinor-TxB_2 in cyclosporin A nephrotoxicity.' *Kidney Int.*, 1988, **34**, 164–174.

343. U. Forsteemann, K. Kuhn, O. Vesterquist *et al.*, 'An increase in the ratio of thromboxane A_2 to peostacyclin in association with increased blood prsesure in patients on cyclosporine A.' *Prostaglandins*, 1989, **37**(5), 567–575.

344. S. R. Smith, E. A. Creech, A. V. Schaffer *et al.*, 'Effects of thromboxane synthase inhibition with CG513080

in human cyclosporine nephrotoxicity.' *Kidney Int.*, 1992, **41**(1), 199–205.

345. M. J. Gallego, C. Zoja, M. Morigi *et al.*, 'Cyclosporine enhances leukocyte adhesion to vascular endothelium under physiologic flow conditions.' *Am. J. Kidney Dis.*, 1996, **28**(1), 23–31.

346. M. Rossini, A. Belloni, G. Remuzzi *et al.*, 'Thromboxane receptor blockade attenuates the toxic effect of cyclosporine in experimental renal transplantation.' *Circulation*, 1990, **81**, I61–I67.

347. N. Perico, M. Pasini, F. Gaspari *et al.*, 'Co-operation of thromboxane A$_2$ and leukotriene C$_4$ and D$_9$ in mediating cyclosporine-induced acute renal failure.' *Transplantation*, 1991, **52**, 873–878.

348. L. Elzinga, V. E. Kelley, D. C. Houghton *et al.*, 'Modification of experimental nephrotoxicity with fish oil as the vehicle of cyclosporine.' *Transplantation*, 1987, **43**, 271–274.

349. P. E. Klotman, L. Bruggeman, J. Hassell *et al.*, 'Regulation of extracellular matrix by thromboxane.' *Kidney Int.*, 1989, **35**, 294.

350. G. Wolf, P. D. Killen and E. G. Neilson, 'Cyclosporine A transcription and procollagen secretion in tubulointerstitial fibroblasts and proximal tubular cells.' *J. Am. Soc. Nephrol.*, 1990, **1**(6), 918–922.

351. T. E. Bunchman and C. A. Brookshire, 'Cyclosporine-induced synthesis of endothelin by cultured human endothelial cells.' *J. Clin. Invest.*, 1991, **88**, 310–314.

352. V. Kon, M. Sugiura, T. Inagami *et al.*, 'Role of endothelin in cyclosporine-induced glomerular dysfunction.' *Kidney Int.*, 1990, **37**, 1487–1491.

353. J. Pernow, J. F. Boutier, A. Franco-Cereceda *et al.*, 'Potent selective vasoconstrictor effects of endothelin in the pig kidney *in vivo*.' *Acta Physiol. Scand.*, 1988, **134**, 573–574.

354. M. Awazu, M. Sugiura, T. Inagami *et al.*, 'Cyclosporine promotes glomerular endothelin binding *in vivo*.' *J. Am. Soc. Nephrol.*, 1991, **1**, 1253–1258.

355. D. M. Lanese and J. D. Conger, 'Effects of endothelin receptor antagonist on cyclosporine-induced vasoconstriction in isolated rat renal arterioles.' *J. Clin. Invest.*, 1993, **91**, 2144–2149.

356. M. Takeda, M. D. Breyer, T. D. Noland, *et al.*, 'Endothelin-1 receptor antagonist: effects on endothelin- and cyclosporine-treated mesangial cells.' *Kidney Int.*, 1992, **41**, 1713–1719.

357. S. Iwasaki, T. Homma and V. Kon, 'Site specific regulation in the kidney of endothelin and its receptor subtypes by cyclosporine.' *Kidney Int.*, 1994, **45**, 592–597.

358. M. Takeda, S. Iwasaki, S. E. Hellings *et al.*, 'Divergent expression of EtA and EtB receptors in response to cyclosporine in mesangial cells.' *Am. J. Pathol.*, 1994, **144**, 473–479.

359. A. Benigni, N. Perico, J. R. Ladny *et al.*, 'Increased urinary excretion of endothelin-1 and its precursor, bio-endothelin-1, in rats chronically treated with cyclosporine.' *Transplantation*, 1991, **52**(1), 175–177.

360. T. Nagase, Y. Fukuccckuchi, C. Jo *et al.*, 'Endothelin-1 stimulates arachidonate 15-lipoxygenase activity and oxygen radical formation in the rat distal lung.' *Biochem. Biophys. Res. Commun.*, 1990, **168**, 485–489.

361. E. J. Goetzl and W. C. Pickett, 'The human PMN leukocyte chemotactic activity of complex hydroxy-eicosatetraenoic acids (HETEs).' *J. Immunol.*, 1980, **125**, 1789–1791.

362. T. Bertani, N. Perico, M. Abbate *et al.*, 'Renal injury induced by long-term administration of cyclosporin A to rats.' *Am. J. Pathol.*, 1987, **127**, 569–579.

363. S. Ito, 'Nitric oxide in the kidney.' *Curr. Opin. Nephrol. Hyperten.*, 1995, **4**, 23–30.

364. D. Diederich, Z. Yang and T. F. Luscher, 'Chronic cyclosporine therapy impairs endothelium-dependent relaxation in the renal artery of the rat.' *J. Am. Soc. Nephrol.*, 1992, **2**, 1291–1297.

365. L. De Nicola, S. C. Thomson, L. M. Wead *et al.*, 'Arginine feeding modifies cyclosporine nephrotoxicity in rats.' *J. Clin. Invest.*, 1993, **92**, 1859–1865.

366. A. Wolf, N. Clemann, W. Frieauff *et al.*, 'Role of reactive oxygen formation in the cyclosporin A-mediated impairment of renal functions.' *Transplant. Proc.*, 1994, **26**, 2902–2907.

367. B. L'Azou, B. Lakhdar and J. Cambar, 'Use of two *in vitro* glomerular models to study the renal vasoreactivity of cyclosporine.' *Transplant. Proc.*, 1994, **26**, 2883–2885.

368. C. Wang and A. K. Salahudeen, 'Lipid peroxidation accompanies cyclosporine nephrotoxicity: effects of vitamin E.' *Kidney Int.*, 1995, **47**, 927–934.

369. N. Perico, P. Ruggenenti, F. Gaspari *et al.*, 'Daily renal hypoperfusion induced by cyclosporine in patients with renal transplantation.' *Transplantation*, 1992, **54**, 56–60.

370. F. C. Luft, G. R. Aronoff, A. P. Evan *et al.*, 'The renin–angiotensin system in aminoglycoside-induced acute renal failure.' *J. Pharmacol. Exp. Ther.*, 1982, **220**, 433–439.

371. P. E. Klotman, J. E. Boatman, B. D. Volpp *et al.*, 'Captopril enhances aminoglycoside nephrotoxicity in potassium-depleted rats.' *Kidney Int.*, 1985, **28**, 118–127.

372. S. Gotoh, T. Ogihara, M. Nakamaru *et al.*, 'Effect of captopril on renal vascular resistance, renin, prostaglandins and kinin in the isolated perfused kidney.' *Life Sci.*, 1983, **33**, 2409–2415.

373. J. F. Gerkens, T. H. Heidemann, E. K. Jackson *et al.*, 'Aminophylline inhibits renal vasoconstriction induced by intrarenal hypertonic saline.' *J. Pharmacol. Exp. Ther.*, 1983, **225**, 611–615.

374. K. F. Badr, D. K. DeBoer, K. Takahashi *et al.*, 'Glomerular responses to platelet-activating factor in the rat: role of thromboxane A$_2$.' *Am. J. Physiol.*, 1989, **256**, F35–F43.

7.12
The Renal Proximal Tubule: Factors Influencing Toxicity and Ischemic Injury

CHARLES E. RUEGG

In Vitro Technologies, Baltimore, MD, USA

7.12.1 INTRODUCTION

Many anatomical, physiological and biochemical factors influence the expression of toxicity in renal proximal tubular segments. As indicated by their name, renal proximal tubules are the first nephron segments of the kidney to receive and process the primary glomerular ultrafiltrate, a fluid almost identical to deproteinized plasma. The lumenal surface of proximal tubules contains an extensive brush border membrane with an enormous surface area that aids in the reabsorption of over 67% of the sodium and water filtered through the glomerulus, over 80% of the filtered chloride and bicarbonate, and over 98% of the filtered glucose, amino acids, small peptides, phosphate, and sulfate. Most of this reabsorptive capacity is accomplished through a variety of active transport processes, often coupled to the sodium gradient. Proximal tubules also contain a high capacity for secreting a variety of endogenous and exogenous organic anions and cations across the tubules from the plasma space to the urinary space.[1,2] In most mammalian species, these reabsorptive and secretory transport processes are anatomically distributed in different regions of the proximal tubule allowing chemical substances that are carried on these transporters to gain access within different regions of the proximal tubule. The heterogeneous distributions of many enzmes involved in energy metabolism, chemical bioactivation, and chemical detoxification pathways may also influence the expression of toxicity in different regions of the proximal tubule. In contrast to the other, more distal nephron segments, which are primarly involved in the process of concentrating the urine and making final adjustments to the water, acid, and base balance of the organism, renal proximal tubules tend to be the predominant target of many nephrotoxic chemicals due to the multitude of physiological functions they conduct in handling the primary glomerular ultrafiltrate. The aim of this chapter is to highlight anatomical and biochemical features which influence the expression of toxicity in different regions of the renal proximal tubule. These features will then be utilized to review the *in vivo* and *in vitro* literature on ischemic and nephrotoxicant injury in renal proximal tubules showing how these anatomical and biochemical features relate to the expression of regional toxicity to renal proximal tubules.

7.12.2 ANATOMICAL FACTORS INFLUENCING TOXICITY TO PROXIMAL TUBULES

The mammalian kidney is a highly organized tissue comprised of more than a dozen anatomically distinct nephron segments.[3,4] The anatomical relationships between the nephron and blood flow delivery patterns in this organ help to explain why local regions in the kidney are prone to injury by nephrotoxicants or ischemia. The following breif review of renal anatomy provides a foundation for interpreting mechanisms of injury to specific regions of the proximal

tubule and helps to explain some apparent discrepancies observed among various experimental models of ischemia and nephrotoxicity. Detailed anatomical descriptions of these nephron/vascular relationships are available in the monographs by Kaissling and Kriz[3] and by Beeuwkes.[5,6]

7.12.2.1 Anatomical Location of Proximal Tubules

Nephrons are classified by the cortical location of their glomerulus into superficial, midcortical, or juxtamedullary which characteristically have short, medium, or long loops of Henle, respectively (Figure 1 (top)). The first tubular portion of the nephron begins with the proximal tubule which is classified into two major divisions comprising three distinct epithelial cell types. The proximal convoluted tubule (PCT; composed of the entire segment 1 (S1) and most of segment 2 (S2) cell types) begins as an extension of the outer leaflet of Bowman's capsule and is found exclusively within the cortical labyrinth. Proximal straight tubules (PST; composed of the remainder of S2 and all of segment 3 (S3)) from midcortical and superficial nephrons are located within the medullary rays and end at the border between the outer and inner medullary stripes. The PST from juxtamedullary nephrons are not associated with medullary rays but traverse the outer stripe in association with vasa recta elements at the edge of decending vascular bundles.[4] Ultrastructural differences between S1, S2, and S3 regions of the proximal tubule are described in Table 1.

7.12.2.2 Anatomical Location of Other Nephron Segments

The nephron segment following the PST is the descending thin limb of Henle which has a variable length and makes a hairpin turn into the ascending thin limb of Henle at different levels within the lower inner stripe (for superficial nephrons) and throughout the inner medulla (for midcortical and juxtamedullary nephrons). The thick limb of Henle generally begins at the border between the inner stripe and the inner medulla and ascends adjacent to its own descending nephron elements (thin limb and PST). Hence thick limbs of Henle from superficial or midcortical nephrons ascend within the medullary rays until they reach the approximate height of their originating glomerulus before exiting the ray, while thick limbs

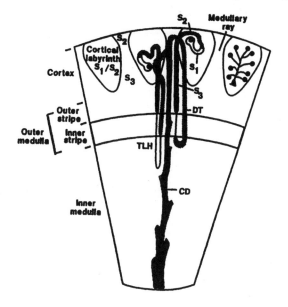

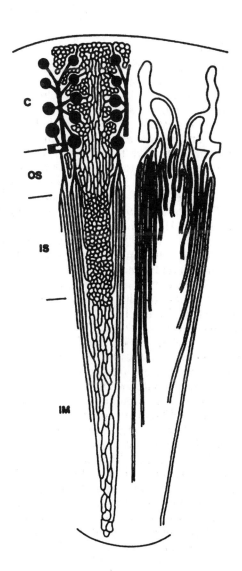

from juxtamedullary nephrons ascend adjacent to the vascular bundles some distance from the medullary rays. Once the thick limbs enter the cortical labyrinth they form a union with their own originating glomerulus at a site between the afferent and efferent arterioles called the macula densa. The distal convoluted tubule is the next short nephron segment which joins the macula densa region to the connecting tubule or arcade which leads to a common collecting duct located in the medullary ray. The collecting ducts then traverse through the outer and inner stripe before fusing into fewer and larger collecting ducts which empty into the renal pelvis at the papillary tip (Figure 1 (top)).

7.12.2.3 Renal Blood Flow Patterns

The vascular beds of the renal cortex and medulla are also highly organized forming two anatomically and functionally distinct vascular beds. Initially blood is delivered to the kidneys via the renal artery which branches as it enters the renal parenchyma forming the arcuate arteries at the border between the cortex and the outer medullary stripe[3,5] (Figure 1 (bottom)). The interlobular arteries then radially penetrate the renal cortex forming vascular poles which give rise to the afferent arterioles delivering blood into glomeruli. At this point about 20% of the blood volume is filtered through the glomeruli into the PCT with the remaining exiting the glomeruli via the efferent arterioles. Efferent arterioles exiting from midcortical and superficial glomeruli (representing 90% of all glomeruli) branch into a dense round-meshed peritubular capillary plexus within the cortical labyrinth which is continuous with a less dense long-meshed plexus present in cortical medullary rays. The blood delivered through these plexuses then

Figure 1 Anatomical localization of renal proximal tubules and microvasculature beds in the rat kidney. (Top) Note the location of S1 and S2 segments of renal proximal tubules within the cortical labyrinth, the late S2 segments within the upper regions of the medullary rays, and the S3 segments within the base of the medullary rays extending into the outer stripe of the outer medulla. (Bottom) Renal microvasculature. Left side shows arterial vessels and capillaries. An arcuate artery is shown by the arrow. The right side shows the venous vessels. C, cortex; OS, outer stripe; IS, inner stripe; IM, inner medulla; TLH, thin limb of Henle; DT, distal tubule; CD, collecting duct (reproduced by permission of Raven Press from 'Toxicology of the Kidney,' 2nd edn., 1993).

Table 1   Anatomical heterogeneity of renal proximal tubular segments.

Anatomical feature	Relative distribution	
	Proximal tubules	Other segments
Apical brush border surface area	S1 > S2 > S3	Much lower
Basolateral membrane interdigitations	S1 ≫ S2 ≫ S3	Variable[a]
Mitochondrial density	S1 > S2 > S3	Variable[a]
Mitochondrial size and elongation	S1 > S2 > S3	Variable[a]
Endocytic vacuoles (lysosomes)	S1 > S2 > S3	Much lower
Peroxisomes	S2 > S1 ≫ S3	Much lower
Tight junctions	S3 > S2 ≥ S1	Much higher
Tubular length	S2 ≫ S3 ≥ S1	Variable

[a]Higher in thick ascending limbs of Henle and distal tubules, but generally lower in other segments.

drains back through the cortex from the center of the rays into the interlobular veins located at and extending from the vascular poles into the cortical labyrinth and the cortical cortices. In contrast, the efferent arterioles of juxtamedullary glomeruli (representing only 10% of all glomeruli) descend into the outer medullary stripe directly beneath the interlobular arterial poles establishing a separate vascular bed which perfuses the outer stripe, inner stripe, and inner medulla (Figure 1 (bottom)). Only a few small arteriolar branches perfuse the sparse capillary network of the outer stripe, while the major branch of the juxtamedullary efferent arterioles traverse this region and then divide into several arterial descending vasa recta forming vascular bundles which originate at the border between the outer and inner medullary stripe. Branches located on the outside edges of these bundles then exit at various points within the inner stripe and inner medulla. The ascending venous vasa recta coming from lower regions of the inner stripe or the inner medulla rises through the inner stripe within the vascular bundles adjacent to their descending vascular elements allowing for countercurrent oxygen exchange. In contrast, ascending branches coming from the middle and upper regions of the inner stripe do not reassociate with the vascular bundles but rise within medullary rays. At the border of the inner and outer stripe these ascending venous vasa recta branch further as they traverse the outer stripe and drain into the arcuate vein or lower portions of the interlobular vein. Hence, the outer medullary stripe is mainly perfused by ascending branches of venous vasa recta which ascend either through the vascular bundles or through medullary rays. The blood flow then exits the kidney via the arcuate and renal veins which lie adjacent to their arteries.[3,5,7–9]

7.12.2.4   Renal Gradients of Oxygenation

Although renal blood flow represents approximately 25% of cardiac output, its asymmetric distribution (90% to the cortex and 10% to the medulla) and countercurrent anatomical orientation result in the formation of two separate circulatory beds each containing steep regional gradients of oxygenation.[10] Investigations using oxygen electrodes[11] or spectral analysis of aa3, an oxygen-sensitive cytochrome,[12,13] have found steep PO_2 gradients in local adjacent regions of the renal cortex and within the outer medullary stripe located just below the cortical–medullary junction. The local PO_2 measured in these regions *in situ* is below 10 mm Hg, a value far below the PO_2 of the renal vein, indicating that (i) considerable countercurrent shunting of oxygen occurs in the renal microvascular beds[9,11,14,15] and (ii) certain regions of the kidney normally function on the verge of anoxia despite the high PO_2 content measured in the renal vein.[9,12,13,16] Since the outer medullary stripe and lower portions of cortical medullary rays are perfused mainly by the ascending branches of the medullary vasa recta, countercurrent oxygen shunting within the vascular bundles would maintain a high local PO_2 in the ascending vasa recta which perfuse the tubules surrounding vascular bundles, while the ascending vasa recta perfusing the medullary ray associated tubules would be relatively oxygen poor[7,9] (Figure 1 (bottom)). Hence, nephron segments located in and directly below the lower portions of medullary rays (PST, thin and thick limbs of Henle, and collecting ducts of superficial and midcortical nephrons) are situated in the distal capillary beds of the renal cortex and medulla where oxygen tensions are the lowest. These anatomical relationships help explain many of the responses observed among the various *in vivo* and *in vitro* models of "ischemia" and nephrotoxicant exposures as discussed later in this chapter.

7.12.3 BIOCHEMICAL FACTORS INFLUENCING TOXICITY TO PROXIMAL TUBULES

Table 2 provides a summary outline of the distributions of several biochemical functions in different renal proximal tubular segments relative to other nephron segments. An excellent discussion detailing renal proximal tubular biochemical heterogeneity is available in a review by Goldstein.[17]

The distributions of enzymes involved in energy metabolism, cellular transport, and chemical metabolism (Table 2) also help to explain why different segments of the renal proximal tubule are susceptible to toxicity and ischemic injury. In the following section, the biochemical and anatomical features summarized above will be utilized to highlight how these distributions predispose specific regions of the proximal tubule to injury by toxicants and ischemia.

7.12.4 ANATOMICAL AND BIOCHEMICAL FACTORS INFLUENCING ISCHEMIC INJURY IN PROXIMAL TUBULES

Since the 1950s a variety of *in vitro* models have been developed and utilized to investigate innate cellular responses to nephrotoxicant or "ischemic" exposure under more controlled experimental conditions. Although these models help to discriminate between direct cellular mechanisms and hemodynamic mechanisms of injury, each model has its own set of attributes and pitfalls which often give the impression that the results from many *in vitro* models contradict one another and sometimes appear to contradict the results of *in vivo* studies. However, critical evaluation of these apparent discrepancies with reference to specific anatomical and biochemical attributes of these models eliminates much of the apparent ambiguity and has provided insights into the mechanism of injury for a variety of chemicals. The intent of this section is to clarify the apparent discrepancies existing between several *in vivo* and *in vitro* nephrotoxicity models aimed at elucidating pathophysiological mechanisms of injury to specific regions of the renal

Table 2 Biochemical heterogeneity in renal proximal tubular segments.

	Relative distribution		
Biochemical activity	*Proximal tubules*	*Other segments*	*Refs.*
Energy related activities			
ATP content (per millimeter)	S2 > S3 > S1	Lower	18
Glycolytic enzymes	S3 > S2 ≥ S1	Much higher	19–21
Gluconeogenic enzymes	S1 ≥ S2 > S3	Much lower	20
Fatty acids as energy sources	S2 > S1 = S3	Variable[a]	18
Glucose as energy source	S3 ≫ S2 ≥ S1	Much higher	21
Transport related activities			
Glucose, phosphate, sulfate	S1 > S2 > S3	Much lower	22–25
Organic anions (PAH)[b]	S2 > S1 = S3	Much lower	26,27
Organic cations (TEA)[b]	S1 > S2 > S3	Much lower	28,29
Sodium potassium ATPase	S1 = S2 ≫ S3	Variable[a]	30,31
Protein endocytosis	S1 > S2 ≫ S3	Much lower	32–34
Cysteine	S3 > S2 > S1	Much lower	35
Amino acids	S1 ≫ S2 ≥ S3	Much lower	36
Xenobiotic metabolism enzymes			
Cytochrome P450	S2 ≥ S3 > S1	Lower	37–40
Prostaglandin H-synthase	Very low	High	41–43
Glucuronidation and sulfation	S3 > S2 ≥ S1	Lower	40,44
Glutathione catabolism	S3 > S2 > S1	Lower	45
Glutathione synthesis	S3 = S2 ≫ S1	Much lower	46
Beta lyase	S3 > S2 = S1	Much lower	47,48

[a]Higher in thick ascending limbs of Henle and distal tubules, but generally lower in other segments. [b]PAH = *p*-aminohippuric acid; TEA = tetraethylammonium.

proximal tubule. Renal oxygen deprivation models (ischemia, anoxia, hypoxia) and selected nephrotoxic agents will be utilized to show how comparing and contrasting data from several *in vivo* and *in vitro* systems allows one to discriminate direct nephrotoxic actions from ischemic injury to specific regions of the proximal tubule.

7.12.4.1 Renal Oxygen Deprivation Models of Injury

To discriminate between direct nephrotoxicant and ischemic actions in the kidney it is appropriate to first examine and clarify the various responses observed among numerous ischemic, anoxic, and hypoxic oxygen deprivation models of injury. There are a wide variety of preparations utilized to investigate renal oxygen deprivation mechanisms of injury ranging from cells in culture to several *in vivo* models, many of which demonstrate a differential responsiveness by targeting either PCT, PST, or medullary thick limbs of Henle (MTAL; Table 3). These differential responses observed among the various oxygen deprivation models have led to much debate and confusion regarding (i) which nephron segments are susceptible to oxygen deprivation and (ii) the utility of some model systems for investigating nephrotoxic mechanisms of injury. However, critical evaluation of these differential responses with respect to the model's attributes and pitfalls helps to delineate the mechanisms responsible for selective injury *in vivo*. This section will attempt to compare

and contrast the differential responses within these various oxygen deprivation models in an effort to clarify and unify this enormous, apparently conflicting, body of literature.

7.12.4.2 Uniform and Nonuniform Oxygen Deprivation Models

Many of the pathological discrepancies observed among the various oxygen deprivation models can be explained and unified by categorizing the models into those which induce a "uniform" vs. a "nonuniform" oxygen deprivation (Table 3). Most studies done *in vivo* or in the isolated perfused kidney represent "nonuniform" models of oxygen deprivation since oxygenated blood or medium is delivered via the vasculature allowing gradients of oxygen to be produced via the countercurrent arrangements in the renal microvascular beds. Hence models which reduce renal blood flow (and/or oxygen delivery) such as partial aortic ligation, hypoxia in the isolated perfused kidney, or the reflow period following ischemia result in lesions to the PST segments located in the distal renal microvascular beds where oxygenation is limiting under normal circumstances.[16,49,56,57] Conversely, most of the *in vitro* models and those *in vivo* models which uniformly inhibit oxygen delivery or utilization, such as clamping of the renal artery or exposure of the isolated perfused kidney to mitochondrial inhibitors, induce a "uniform oxygen deprivation" which mainly targets PCT segments (Table 3). Although most of the ischemic literature has been done using the renal arterial

Table 3   Renal oxygen deprivation models of injury and their various targets.

Models	Experimental conditions	Segmental targets	Refs.
Nonuniform oxygen deprivation			
Partial aortic ligation	Reduced blood flow	PST > PCT	49
Hypovolemic shock	Reduced blood flow	PST > PCT	50,51
Norepinephrine	Reduced blood flow	PST > PCT	52
Reflow following arterial clamping	Reduced blood flow	PST > PCT	53–55
Isolated perfused kidney	Hypoxia	MTAL > PST > PCT	56
Uniform oxygen deprivation			
Arterial clamping without reflow	Anoxia	PCT > PST	53,55,57
Isolated perfused kidney	MI	PCT > PST	58
In vitro exposures of			
Organ explants	Hypoxia	PCT > PST	59
Renal slices	Hypoxia	PCT > PST	60
Isolated PCT & PST suspensions	Anoxia or MI	PCT > PST	61,62
PCT & PST cell cultures	Anoxia	PST > PCT	63

PCT = proximal convoluted tubules, PST = proximal straight tubules, MTAL = medullary thick limbs of Henle, MI = mitochondrial inhibitors.

clamping with reflow model, this model represents a hybrid beginning with a period of uniform oxygen deprivation (clamping of renal artery) followed by a period of non-uniform oxygen deprivation (reflow to a compromised organ) with most pathological lesions being observed in PST segments following reflow.[49,52-55] The hypovolemic shock and norepinephrine models of acute renal failure are also hybrid models which have only been examined following a recovery period demonstrating a similar PST susceptibility.[50,51] Very few studies have examined the responses following each independent experimental phase in these hybrid models. In three such studies, PCT segments were the most susceptible to injury during the uniform oxygen deprivation phase of the experiment but they recover during the nonuniform oxygen delivery phase while PST progress to an irreversible state of injury.[53,55,57] The PCT injury observed during uniform oxygen deprivation is time dependent since no differential PCT/PST susceptibility is observed in the clamped kidney prior to 30 min.[54] Hence, by categorizing the various model systems into those which induce uniform vs. nonuniform oxygen deprivation a clear and consistent trend emerges where PCT segments are susceptible to uniform oxygen deprivation while PST segments are susceptible under nonuniform oxygen deprivation settings[61] (Table 3). Although these trends do have exceptions as seen with cell culture models and in the MTAL lesions of the isolated perfused kidney, many of these differences can be explained by the physiologic and metabolic state of these models as described below.

7.12.4.3 ATP-dependent Mechanisms of Injury

Mechanistically, the loss of intracellular ATP content has been suggested as a critical mediator in the pathophysiological cascade of events leading to nephrotoxic and ischemic cell injury.[52,59,64-68] Several reports have shown that the ATP content drops 60–90% within minutes following the initiation of oxygen deprivation from approximately $9 \, nmol \, mg^{-1}$ protein to a maintenance level of about $2 \, nmol \, mg^{-1}$ protein.[66,68-71] The ATP content is maintained at this level for about 30 min before dropping further with extended periods of oxygen deprivation.[66,68,71] This second decline in ATP to below $2 \, nmol \, mg^{-1}$ protein coincides with the point of irreversible injury where the nephron segments do not fully recover upon reoxygenation.[66,68] Since sodium transport is critically and linearly dependent on

the cellular ATP content,[72] loss of ATP results in marked changes in intracellular and trans-epithelial ion balance leading to cell swelling in ATP-deficient segments.[51,52,59]

These ATP-dependent mechanisms can also help explain the differential susceptibility of nephron segments observed in uniform and nonuniform models of oxygen deprivation. For instance, in nonuniform oxygen deprivation models, reduced oxygen delivery to the distal capillary beds, which normally function on the verge of anoxia,[12,13] would induce a local reduction in ATP content causing the nephron segments in these regions to swell.[52] Swelling in PST segments may occur first since these segments have a much lower glycolytic capacity as compared to MTAL or collecting ducts and have a high ATP-dependent transport capacity which would tend to deplete ATP quickly in these segments.[19,66,69] Cell swelling secondary to ATP loss is thought to induce a local mechanical occlusion of the ascending vasa recta traversing the outer medullary stripe resulting in the commonly observed zonal hyperemia at the border between the outer and inner medullary stripes.[51,52,73-75] This hyperemia may further reduce the O_2 delivery to the outer stripe regions directly below medullary rays inducing a more severe local oxygen deprivation shunting more of the medullary blood flow towards the papillary tip, where an as yet unexplained increase in blood flow has been observed immediately following ischemia.[73,75,76] Since countercurrent oxygenation gradients in the renal cortex and in medullary vascular bundles would maintain a higher PO_2 delivery to the tubules located in the cortical labyrinth and adjacent to vascular bundles (Figure 1), the tubules located in and beneath the lower portions of the medullary rays are most prone to oxygen deficiency[9] (Figure 1). In the isolated perfused kidney model this is clearly demonstrated since graded hypoxia induces more injury in lower medullary ray associated PST than in vascular bundle associated PST, demonstrating that the pattern of injury strongly correlates and is probably dependent on the hemodynamic flow patterns present in the intact organ.[9,56] Likewise, following transient or sustained hypotension, similar patterns of PST injury are observed which are probably dependent on the poor perfusion of the outer medullary stripe leaving PST segments anoxic.[49-51]

It has been shown both *in vivo* and *in vitro* that procedures aimed at boosting or maintaining cellular ATP content during oxygen deprivation, and/or maneuvers aimed at limiting cell swelling, such as mannitol infusion prior to or following ischemia, provide significant protection and greatly diminish the degree of hyperemia observed in these models.[52,66,67,77–82] ATP-dependent mechanisms also play a role in determining the site of injury in uniform oxygen deprivation models where innate PCT susceptibility to oxygen deprivation is observed under conditions which prohibit hemodynamic gradients of oxygenation (Table 3). It has been shown that PST segments exposed to anoxia, hypoxia, the mitochondrial inhibitors rotenone, antimycin A, or the uncoupler FCCP, maintain a higher ATP content than PCT segments exposed under identical conditions. This indicates that PST segments may have a higher glycolytic capacity enabling them to maintain their ATP content longer under uniform anaerobic conditions.[61,62] PST protection was also found to be glucose dependent since removal of glucose during anoxia induced similar injury in both PCT and PST segments.[61] These results correlate with morphological observations in the isolated perfused kidney model.[83] In these studies, 2-deoxyglucose, a competitive inhibitor of glycolysis, was able to abolish the differential PST protection observed following rotenone or antimycin A induced anoxia. Although these results indicate that PCT segments are innately more susceptible to oxygen deprivation than PST segments due to segmental glycolytic differences, in the nonuniform oxygen deprivation setting commonly encountered *in vivo*, PST segments are more susceptible due to their anatomical location within the oxygen-limited distal regions of the renal microvascular beds.[9,61]

7.12.4.4   Oxygen Deprivation in Isolated Perfused Kidney Models

Although the isolated perfused kidney (IPK) model has provided many mechanistic details regarding the segment-specific susceptibilities following uniform and nonuniform oxygen deprivation, this preparation also results in medullary thick ascending limb of Henle (MTAL) injury which is not commonly observed in other models.[9,16,49,60,84–86] This exquisite MTAL susceptibility is mainly observed in "erythrocyte-free" IPK models and probably results from a unique physiological imbalance between oxygen delivery, glomerular filtration and sodium transport dependent work within this model.[9,84–86] MTAL injury can be blocked by: (i) increasing oxygen delivery with the addition of erythrocytes, hemoglobin,[9,85,86] or oxygen-carrying flourocarbons;[10,85] (ii) inhibiting MTAL sodium transport with furosemide or ouabain;[87] or (iii) reducing glomerular filtration rate with oncotic agents.[87] Conversely, MTAL

injury can be enhanced in the erythrocyte-free IPK model either by reducing oxygen tensions or stimulating epithelial transport with nystatin or amphotericin.[85,86,88] Under all these circumstances, injury to PCT and PST segments are still observed with the differential responsiveness dependent on the uniform vs. nonuniform mode used.[9,55,56,58,86,89] In hypovolemic shock and other hypoperfusion models where renal blood flow and oxygen delivery are greatly reduced, proximal tubular injury in PST segments is observed without injury to MTAL.[49,51] Hence, the specific MTAL injury observed under erythrocyte-free conditions of the IPK model are mainly due to a physiological imbalance between transport dependent work and oxygen delivery to MTAL regions in this model.[9,84,86]

7.12.4.5   Oxygen Deprivation in Cell Culture Models

Although several models of uniform oxygen deprivation demonstrate that PCT segments are innately more susceptible to injury than PST segments, these segments grown in culture respond quite differently to uniform oxygen deprivation exposures[63] (Table 3). These differences may be explained by comparing the metabolic differences between these experimental models. Most *in vitro* investigations of anoxia utilize freshly isolated tubular segments, slices, or perfused kidneys provided with substrates which maintain the oxidative metabolism and transport functions close to those observed under *in vivo* conditions.[21,60,86,90–92] In contrast, when proximal tubular segments are placed into primary culture, they rapidly stimulate lactate production and markedly reduce their ATP content, intracellular potassium content, and respiration rate as they induce glycolytic enzyme capacity up to eightfold between 4 and 6 days in culture.[93,94] This rapid and marked glycolytic induction observed in renal proximal tubular cultures represents a fundamental change from the oxidative dependence of these segments observed *in vivo*, and could significantly alter the *in vitro* segment-specific responses to anoxia. Indeed, the cultured segments examined by Wilson and Schrier[63] were barely affected by 45 min of anoxia (between 10 and 17% loss of viability) as compared to freshly isolated renal tubules (40–60% viability loss following 30–40 min of anoxia).[61,66,68,95] Hence, fresh metabolically active tissue isolates exposed *in vitro* to anoxia better reflect the innate cellular responses observed immediately following ischemia *in vivo*.[53,55,60,61]

Although tissue culture media and renal cell culture systems have steadily improved since the 1960s allowing for the study of nephrotoxic mechanisms of injury,[63,96,97] fundamental changes in central metabolic state need to be recognized in the mechanistic interpretation of these data. Dickman and Mandel[93] reported that the marked glycolytic alterations observed when proximal tubules are placed into dish cultures was not observed if the tubules were maintained in suspension cultures. However, suspension cultures exposed to graded hypoxia (between 1 and 3% O_2) switch to a glycolytic mode of respiration producing the same amount of lactate as proximal tubules in dish cultures. These results suggest that proximal tubules in dish cultures exposed to 95% air/5% CO_2 become hypoxic and switch from an oxidative phenotype to a glycolytic phenotype. Reshaking the dish cultures before attachment, to increase medium oxygenation, causes these segments to switch back from a glycolytic to oxidative phenotype,[93] suggesting that cell culture systems might retain their oxidative metabolic properties if the cells could recieve a higher oxygen delivery.

7.12.4.6 Postischemic Reperfusion Injury Models

Finally, the role of xanthine oxidase and oxygen free radicals in renal postischemic reperfusion injury has been a controversial issue among several different research groups. Although some *in vivo* studies report that maneuvers directed at blocking oxygen free radical generation (allopurinol and oxypurinol, inhibitors of xanthine oxidase) or pretreatment with various superoxide or hydroxy radical scavengers (superoxide dismutase, catalase, dimethylthiourea, or mannitol) protect against postischemic reperfusion injury,[98–101] other *in vivo* studies[102–105] and several *in vitro* studies[102,106,107] find little evidence of superoxide- or hydroxy-radical mediated injury. As discussed earlier, the post-ischemic reperfusion period is equivalent to a nonuniform oxygen deprivation model of injury. Hence, it is interesting that allopurinol and oxypurinol also increase cellular ATP content[105,107,108] while allopurinol, superoxide dismutase, catalase, and mannitol (an impermeable hydroxy radical scavenger) all decrease either vascular resistance or cell swelling, leading to decreased congestion and increased medullary blood flow.[52,77,109–111]

It has been suggested that oxygen free radical inhibitors and scavengers might act at the vascular level to block neutrophil adhesion in the vasa recta, which is thought to induce the vascular congestion and hyperemia observed following ischemia.[112–114] Several studies suggest that neutrophil adhesion induces the release of many vasoconstrictive prostaglandin mediators which also play an important role in the "reperfusion induced injury."[114,115] Since manuevers aimed at: (i) inhibiting vasoconstrictive prostaglandin production, (ii) inhibiting neutrophil adhesion, (iii) increasing intracellular ATP content, or (iv) increasing medullary blood flow, have been shown to protect against nonuniform oxygen deprivation injury,[52,67,73,77,81,112,114,116,117] it is reasonable to conclude that the protection from "oxygen free radical mediated injury," when observed, may in fact be due to improvements in local renal hemodynamics and/or adenine nucleotide content, independent of a direct oxygen free radical induced injury at the cellular level. Although this conclusion is reasonable, it remains controversial.

In summary, this section demonstrates that comparing and contrasting the responses from various model systems exposed to renal oxygen deprivation allows one to discriminate between hemodynamic and innate cellular mechanisms of injury. From the analysis of *in vivo* studies alone, it was difficult to determine whether PST injury following ischemia with reflow was due to an innate sensitivity of this segment to oxygen deprivation or if hemodynamic mechanisms were responsible for this injury.[50,54,57,74] By examining the response of fresh tissue isolates exposed *in vitro* to oxygen deprivation, where vascular oxygen delivery patterns are absent, it became clear that PCT segments are innately more susceptible to oxygen deprivation than PST segments and that the striking differences observed between *in vivo* and *in vitro* studies during reoxygenation are probably due to *in situ* hemodynamic delivery patterns for oxygen, which ultimately lead to the PST site of injury.[21,60–62] Similar comparative approaches allow one to discriminate direct nephrotoxicant actions from ischemic responses as discussed in the next section.

7.12.5 ANATOMICAL AND BIOCHEMICAL FACTORS INFLUENCING NEPHROTOXICITY IN PROXIMAL TUBULES

Although most nephrotoxicants express their toxicity asymmetrically within either the PCT or PST regions of the proximal tubule (Table 4; Weinberg,[118] Hook and Hewitt[119]), determining whether this differential susceptibility is primarily due to direct cellular mechanisms

Table 4   Chemicals affecting the proximal tubule following *in vivo* exposure.

Chemicals affecting PCT	*Chemicals affecting PST*
Aminoglycosides	Cyclosporin
Cadmium	Mercuric chloride
Chromate	Uranyl nitrate
Cephaloridine	*cis*-Platinum
Citrinin	Hexachloro-1,3-butadiene
Ethylene	S(1,2)-dichlorovinyl cysteine
dibromide	Acetaminophen
	Bromobenzene
	Dibromochloropropane
	Carbon tetrachloride
	Maleic acid

Source: Weinberg.[118]

(i.e., segment-specific transport and/or bioactivation to a toxic metabolite) or indirect blood flow delivery patterns and ischemic mechanisms is more difficult to sort out. This issue, however, is fundamental to the development of rational therapeutic methods aimed at circumventing the nephrotoxic side effects of these agents. This section will focus on the review of selected aspects of the nephrotoxicity of metals, cyclosporin, and cysteine conjugates as examples where anatomical and biochemical differences between PCT and PST segments help to mechanistically distinguish direct nephrotoxicant actions from indirect ischemic actions.

7.12.5.1 Nephrotoxic Metals

Potassium dichromate, cadmium chloride, mercuric chloride, and *cis*-platinum are examples of toxic metals which express their effects within different regions of the proximal tubule[120-123] (Table 4). The proposed mechanisms by which metals, and other compounds, induce segment-specific injury within PST segments include: (i) passive concentration above toxic threshold levels within the lumenal space as salt and water are progressively reabsorbed along the proximal tubule;[124-126] (ii) reductions in renal blood flow either directly or indirectly via a tubuloglomerular feedback activation of the renin–angiotensin system in response to decreased sodium reabsorption (leading to nonuniform oxygen deprivation);[127] or (iii) direct cytotoxicity due to selective accumulation of metals within targeted segments.[128,129] Similarly, metal-induced segment-specific PCT injury has been explained by: (i) blood flow patterns which initially deliver compounds directly to PCT segments

via glomerular filtration and the peritubular capillary networks of the cortical labyrinth (Figure 1 (bottom)); or (ii) direct cytotoxicity due to selective accumulation within this segment.[60,119,120] Although most of these mechanisms were proposed following extensive *in vivo* investigation, discriminating between the blood flow dependent mechanisms and the direct cellular mechanisms is demonstrated most clearly by comparing *in vivo* responses with segment-specific *in vitro* responses.

7.12.5.1.1 Mercuric chloride

In vitro exposures of renal cortical slices[60] or microdissected nephron segments[18] to mercuric chloride result in selective injury to PST segments. This *in vitro* segment-specific response matches that observed *in vivo* and demonstrates that PST segments are innately sensitive to mercuric chloride independent of passive urinary concentration[124-126] or renin–angiotensin-induced ischemia,[127,129] mechanisms which do not operate in these *in vitro* systems. Although these results cannot exclude the possibility that the above mechanisms contribute to the progressive renal injury, they do indicate that the primary mechanism responsible for selective injury is related to innate PST properties.

Studies have shown that mercuric chloride interactions with glutathione (GSH) can account for its transport-dependent renal uptake and toxicity.[130-133] Tanaka *et al.*[130] demonstrated that blockade of GSH breakdown with acivicin or selective depletion of hepatic GSH with 1,2-dichloro-4-nitrobenzene (DCNB) results in an increased fractional excretion of mercury complexed to glutathione, decreased renal uptake of the metal and decreased nephrotoxicity. Similar results were observed following acivicin or widespread GSH depletion with diethylmealate (DEM) and/or buthionine sulfoxamine (BSO),[131,134,135] except in some cases the benificial effects on renal toxicity were not observed despite the reduced renal uptake of the metal. This difference may be due to the loss of intracellular renal GSH as a cytoprotective agent since selective depletion of hepatic GSH[130] or exogenous enhancement of renal GSH[132] is protective in these models. These results suggest that mercuric chloride forms a GSH conjugate in the liver which requires further breakdown into its cysteinyl-glycine- or cysteine- conjugate before accumulating and causing nephrotoxicity. This mechanism could also explain the innate sensitivity of PST segments since these segments contain the highest gamma-glutamyl transpeptidase and cysteine transport activity in the kidney.[35,45] These results also underscore the dual role of GSH as an intracellular cytoprotective agent as well as an extracellular carrier molecule which may prove to be useful in targeting some pharmaceutical agents to the kidney.

7.12.5.1.2 Cadmium chloride

Cadmium chloride is another divalent cation similar to mercuric chloride, but its major target following *in vivo* exposure is the PCT segment.[136] Differential targeting of cadmium complexes *in vivo* suggests that this metal is also directly toxic independent of renal hemodynamic or blood delivery patterns.[121] Administration of cadmium chloride *in vivo* results in its marked liver accumulation, complexation with metallothionein (MT), and release into the blood as the cadmium–MT complex. This complex is then filtered at the glomeruli and reabsorbed by the renal endocytic pathways into lysosomes within PCT segments, where the MT is degraded releasing free cadmium locally.[136–138] Conversely, cadmium–MT complexes or cadmium coadministered *in vivo* with cysteine are not accumulated in the liver, but rather accumulate in the kidney causing PCT or PST injury, respectively.[121,137,139,140] These results highlight a transport-dependent mechanism of injury and suggest that this metal, and possibly other pharmaceutical agents, can be differentially targeted within the kidney by complexation with either metallothionein (for PCT) or cysteine (for PST).

7.12.5.1.3 Potassium dichromate

Potassium dichromate ($K_2Cr_2O_7$) at physiological pH exists almost entirely as chromate (CrO_4^{2-}), a divalent oxyanion,[141] which induces selective injury within PCT segments both *in vivo*[120,142] and *in vitro*.[60] This *in vivo/in vitro* similarity indicates that PCT segments are innately sensitive to this corrosive metal independent of *in situ* delivery patterns. Selective accumulation of chromate within PCT segments probably occurs due to its structural similarity with sulfate, which is reabsorbed primarily in PCT regions of the kidney and competes with chromate uptake in several systems.[23,143,144] Once inside the cell chromate can be converted into its carcinogenic form, chromium (VI), or interact in a futile cycle to deplete intracellular ATP as it can substitute for the terminal phosphate in this molecule when used in the process of activating sulfate for various sulfotransferases.[144]

7.12.5.1.4 cis-*Platinum*

The antineoplastic agent *cis*-diamminedichloroplatinum (II), *cis*-platinum, is a zwitterionic molecule which in rats causes a selective injury to PST segments.[122,145,146] However, *in vitro* exposures of rabbit renal cortical slices to *cis*-platinum result in injury to PCT segments.[147] This *in vivo/in vitro* targeting difference seems to be species dependent since *in vivo* *cis*-platinum exposure in rabbits was also shown to cause PCT injury.[123] Although it is not clear why *cis*-platinum targets different segments in these rat and rabbit studies, these species differences suggest that the transport-dependent mechanism which accumulates this metal is either distributed differently in these two species or *cis*-platinum may utilize different uptake mechanisms within each species. Further studies are needed to clarify this issue. Several studies and reviews indicate that hemodynamic changes occur following *cis*-platinum exposure, but usually as a secondary process which temporally follows earlier changes in mitochondrial and transport functions.[146–150] Although the mechanism(s) of *cis*-platinum nephrotoxicity remain a mystery, this class of compound is becoming safer since several structural platinum analogues are now available which retain their antineoplastic benifits with a significantly reduced nephrotoxic risk.[145–147,151] The nephrotoxicity side-effects of *cis*-platinum can also be blocked without altering its antineoplastic effects by the administration of 4-methylthiobenzoic acid.[152]

7.12.5.2 Cyclosporin

Unlike the segment-specific transport dependent accumulation of metals, the primary effects of acute cyclosporin exposure are mainly related to alterations in renal hemodynamics leading to a nonuniform oxygen deprivation (ischemia) affecting PST segments in the distal microvascular beds.[153–156] These acute effects can be reversed or blocked by: (i) removing cyclosporin; (ii) infusing verapimil and Mg-ATP; or (iii) infusing vasodilatory prostaglandins or blockers of vasoconstrictive thromboxanes, platelet activating factor (PAF), or endothelin.[155,157–160] *In vitro* exposure of cyclosporin in isolated perfused glomeruli or mesangial cell cultures results in acute dose-dependent decreases in glomerular function and mesangial cell contractility, which could directly account for the acute hemodynamic changes observed *in vivo*.[159,161,162] Calcium-dependent mesangial cell contractility has been correlated with the nephrotoxicity (independent of immunosuppresive activity) of several cyclosporin analogues and can be blocked by verapimil and PAF antagon-ists.[159,162] Although all of these responses indicate a hemodynamic mechanism of acute cyclosporin nephrotoxicity, verapimil, and Mg-ATP might also protect against direct mito-chondrial injury by blocking mitochondrial calcium accumulation and boosting intracellu-lar ATP content.[163–164] It is interesting that although the literature base for cyclosporin nephrotoxicity is enormous, very few reports are available which utilize nonperfused *in vitro* exposures in fresh renal slices, cells or tubular fragments. This void might reflect an inability to induce direct cytotoxicity with reasonable doses of cyclosporin since vascular-mediated cellular injury does not occur in these *in vitro* systems. Alternatively, cyclosporin may require longer exposure periods *in vitro* to detect any direct cytotoxicity since its toxicity is delayed in time relative to other nephrotoxicants.[118] In this regard, direct cellular toxicity has been observed in renal cell culture models exposed *in vitro* to cyclosporin,[165–167] yet these con-ditions induce mainly PCT injury.[167] This response differs from the *in vivo* PST pattern of injury discussed earlier and is difficult to interpret in light of the metabolic changes which occur in cultured cells as discussed earlier in Section 7.12.4.5. These results suggest that acute cyclosporin toxicity is mainly mediated by changes in renal hemodynamics inducing an ischemic reaction in the kidney. Chronic exposure to cyclosporin results in irreversible injury which has been associated with a marked interstitial fibrotic response localized within the subcapsular cortex, cortical medullary rays, and much of the medulla.[168–170] Although the mechanism for this chronic fibrosis is currently unknown, similar responses have been observed following ischemia, suggesting that this may be a response to long term oxygen deprivation or alterations in renal prostaglandins.[155,171] However, cyclosporin also stimulates an inter-stitial fibrotic response in other nonrenal tissues indicating that cyclosporin may stimulate inter-stitial fibrosis independent of local ischemic mechanisms.[155] Further research into the mechanisms of interstitial matrix formation and proliferation are needed to address the mechanisms underlying chronic cyclosporin exposures.

7.12.5.3   Cysteine Conjugates

Although glutathione-*S*-conjugation with xenobiotics is normally associated with cellular detoxification mechanisms, conjugation with some halogenated hydrocarbons can result in the formation of reactive episulphonium ions or reactive thiols following enzymatic cleavage of the corresponding cysteine conjugate in the kidney.[172–174] Dichlorovinylcysteine (DCVC), a metabolite of trichloroethylene, and hexa-chlorobutadiene (HCBD) are examples of nephrotoxic halogenated hydrocarbons which selectively effect PST segments following *in vivo* and *in vitro* exposure.[175–179] These results indicate that PST segments are innately sensi-tive to these compounds independent of hemo-dynamic mechanisms. Extensive *in vivo* and *in vitro* investigations have demonstrated that the glutathione conjugates of halogenated hydro-carbons are mainly formed in the liver and excreted into the bile where they are broken down further and reabsorbed from the intestine as the cysteine conjugate.[174,180] Upon second passage through the liver the cysteine conju-gates are *N*-acetylated and released into the circulation for renal excretion.[181,182] Although most cysteine conjugates are delivered to the kidney as their *N*-acetyl derivatives, some of the glutathione or cysteine conjugate might also reach the kidney.[172–174] The mechanism responsible for specific PST targeting of cysteine- or *N*-acetylcysteine- conjugates remains unclear, but probably depends on the differential distributions or specific rate con-stants for transport, *N*-deacetylation, *N*-acet-ylation, CS-beta-lyase and/or *S*-methylation activities (the enzymes responsible for generat-ing or detoxifying reactive thiols).[48,172–174,179,183] In this regard, the transport-dependent toxicity of DCVC and its *N*-acetyl derivative (NAC-DCVC) occurs via several inhibitable apically located amino acid transport systems or by basolateral probenecid-sensitive organic anion transport, respectively.[179,184] Once inside the proximal tubule NAC-DCVC is "slowly" converted by *N*-deacetylase into DCVC, which is "rapidly" metabolized by CS-beta-lyase (glutamine transaminase K) into a reactive thiol that covalently binds to cellular macromole-cules leading to toxicity.[48] This metabolism-dependent binding and toxicity can also be blocked by inhibiting *N*-deacetylase or CS-beta-lyase activities.[172–174,179,183,185] Hence, a differential distribution of CS-beta-lyase and/or *N*-deacetylase activities could also explain the PST targeting of these conjugates. Although immunocytochemical localization of CS-beta-lyase has been investigated by two independent groups, MacFarlane *et al.*[47] found this enzyme to be specifically localized within PST segments while Jones *et al.*[48] found it homogeneously distributed in all regions of the proximal tubule. The reason for this descrepancy is presently unclear but may be related to the purity or specificity of these antibodies reacting with the mitochondrial or cytosolic forms of CS-beta-lyase.[186] Since the differential distribution or activity of *N*-deacetylase has not been reported yet, the mechanisms underlying selective PST injury remain unclear. However, the transport and metabolism dependent toxicity of cysteine conjugates determined both *in vivo* and *in vitro* clearly demonstrate that these agents are directly nephrotoxic due to innate cellular mechanisms.

In summary, comparisons of specific renal proximal tubular target susceptibilities following *in vivo* and *in vitro* exposure enable one to easily discriminate direct nephrotoxicity due to transport/metabolism (metals, cysteine conjugates) from hemodynamic-induced ischemic mechanisms as seen with cyclosporin. Although this review focused on evaluating how anatomical and biochemical differences in PCT and PST are useful, several reviews are available which go into greater depth regarding the biochemical and molecular mechanisms of nephrotoxicant-induced injury.[17,119,187–191]

7.12.6 REFERENCES

1. K. J. Ullrich and G. Rumrich, 'Contraluminal transport systems in the proximal renal tubule involved in secretion of organic anions.' *Am. J. Physiol.*, 1988, **254**, F453–F462.
2. K. J. Ullrich, G. Rumrich and S. Kloss, 'Contraluminal organic anion and cation transport in the proximal renal tubule: V. Interaction with sulfamoy- and phenoxy diuretics, and with B-lactam antibiotics.' *Kidney Int.*, 1989, **36**, 78–88.
3. B. Kaissling and W. Kriz, 'Structural analysis of the rabbit kidney.' *Adv. Anat. Embryol. Cell Biol.*, 1979, **56**, 1–123.
4. W. Kriz and L. Bankir, 'A standard nomenclature for structures in the kidney.' *Am. J. Physiol.*, 1988, **254**, F1.
5. R. Beeuwkes, III and J. V. Bonventre, 'Tubular organization and vascular-tubular relations in the dog kidney.' *Am. J. Physiol.*, 1975, **229**, 695–713.
6. R. Beeuwkes, in 'Renal Pathophysiology—Recent Advances.' eds. A. Leaf, G. Giebisch, L. Bolis *et al.*, Raven Press, New York, 1980, pp. 155–163.
7. W. Kriz, 'Structural organization of renal medullary circulation.' *Nephron*, 1982, **31**, 290–295.
8. W. Kriz and A. F. Lever, 'Renal countercurrent mechanisms: structure and function.' *Am. Heart Journal*, 1969, **78**, 101–118.
9. H. J. Schurek and W. Kriz, 'Morphologic and functional evidence for oxygen deficiency in the isolated perfused rat kidney.' *Lab. Invest.*, 1985, **53**, 145–155.
10. H. Franke and G. Gronow, in 'Kidney Hormones,' ed. J. W. Fisher, Academic Press, London, 1986, vol. 3, pp. 177–216.
11. H. P. Leichtweiss, D. W. Lubbers, C. Weiss *et al.*, 'The oxygen supply of the rat kidney: measurements of intrarenal PO_2.' *Pflugers Arch.*, 1969, **309**, 328–349.
12. R. S. Balaban and A. L. Sylvia, 'Spectrophotometric monitoring of O_2 delivery to the exposed rat kidney.' *Am. J. Physiol.*, 1981, **241**, F257–F262.
13. F. H. Epstein, R. S. Balaban and B. D. Ross, 'Redox state of cytochrome aa3 in isolated perfused rat kidney.' *Am. J. Physiol.*, 1982, **243**, F356–F363.
14. M. N. Levy and G. Sauceda, 'Diffusion of oxygen from arterial to venous segments of renal capillaries.' *Am. J. Physiol.*, 1959, **196**, 1336.
15. M. N. Levy and E. S. Imperial, 'Oxygen shunting in renal cortical and medulary capillaries.' *Am. J. Physiol.*, 1961, **200**, 159.
16. M. Brezis, S. Rosen, P. Silva and F. H. Epstein, 'Renal ischemia: a new perspective.' *Kidney Int.*, 1984, **26**, 375–383.
17. R. S. Goldstein, in 'Toxicology of the Kidney,' 2nd edn., eds. J. B. Hook and R. S. Goldstein, Raven Press Ltd., New York, 1993, pp. 201–247.
18. K. Y. Jung, S. Uchida and H. Endou, 'Nephrotoxicity assessment by measuring cellular ATP content I. Substrate specificities in the maintenance of ATP content in isolated rat nephron segments.' *Toxicol. Appl. Pharmacol.*, 1989, **100**, 369–382.
19. B. D. Ross and W. G. Guder, 'in 'Metabolic Compartmentation,' ed. H. Sies, Academic Press, New York, 1982, pp. 363–409.
20. W. G. Guder and B. D. Ross, 'Enzyme distribution along the nephron.' *Kidney Int.*, 1984, **26**, 101–111.
21. C. E. Ruegg and L. J. Mandel, 'Bulk isolation of renal PCT and PST. I. Glucose-dependent metabolic differences.' *Am. J. Physiol.*, 1990, **259**, F164–F175.
22. J. P. Bonjour and J. Caverzasio, 'Phosphate transport in the kidney.' *Rev. Physiol. Biochem. Pharmacol.*, 1984, **100**, 161–214.
23. C. Lechene, 'Site of sulfate reabsorption along the rat nephron.' *Kidney Int.*, 1974, **6**, 64A.
24. J. W. McKeown, P. C. Brazy, and V. W. Dennis, 'Intrarenal heterogeneity for fluid, phosphate, and glucose absorption in the rabbit.' *Am. J. Physiol.*, 1979, **237**, F312–F318.
25. H. Valtin, 'Renal Function, Mechanisms Preserving Fluid and Solute Balance in Health,' 2nd edn., Little Brown and Co. Boston, MA, 1983.
26. P. B. Woodhall, C. C. Tisher, C. A. Simonton *et al.*, 'Relationship between para-aminohippurate secretion and cellular morphology in rabbit proximal tubules.' *J. Clin. Invest.*, 1978, **61**, 1320–1329.
27. A. Shimomura, A. M. Chonko and J. J. Grantham, 'Basis for heterogeneity of para-aminohippurate secretion in rabbit proximal tubules.' *Am. J. Physiol.*, 1981, **240**, F430–F436.
28. C. Schali, L. Schild, J. Overney *et al.*, 'Secretion of tetraethylammonium by proximal tubules of rabbit kidneys.' *Am. J. Physiol.*, 1983, 245, F238–F246.
29. T. D. McKinney, 'Heterogeneity of organic base secretion by proximal tubules.' *Am. J. Physiol.*, 1982, **243**, F404–F407.
30. S. R. Gullans and S. C. Hebert, in 'The Kidney,' 4th edn., eds. B. M. Brenner and F. C. Rector, Saunders, Philadelphia, PA, 1991, pp. 76–109.
31. A. I. Katz, A. Doucet and F. Morel, 'Na-K-ATPase activity along the rabbit, rat, and mouse nephron.' *Am. J. Physiol.*, 1979, **237**, F114–F120.
32. J. E. Bordeau, F. A. Carone and C. E. Ganote, 'Serum albumin uptake in isolated perfused renal tubules. Quantitative and electron microscope radioautographic studies in three anatomical segments of the rabbit nephron.' *J. Cell Biol.*, 1972, **54**, 382–398.
33. M. A. Cortney, L. L. Sawin and D. D. Weiss, 'Renal tubular protein absorption in the rat.' *J. Clin. Invest.*, 1970, **49**, 1–4.
34. K. M. Madsen and C. H. Park, 'Lysosome distribution and cathepsin B and L activity along the rabbit proximal tubule.' *Am. J. Physiol.*, 1987, **253**,

F1290–F1301.

35. H. Volkl, S. Silbernagl and A. Ascher, 'Mutual inhibition of L-cystine/L-cysteine and other neutral amino acids during tubular reabsorption. A microperfusion study in rat kidney.' *Pflugers Arch.*, 1982, **395**, 190–195.

36. S. Silbernagl, 'The renal handling of amino acids and oligopeptides.' *Physiol. Rev.*, 1988, **68**, 911–1007.

37. J. B. Tarloff, R. S. Goldstein and J. B. Hook, in 'Nephrotoxicity in the experimental and clinical situation,' eds. P. H. Boch and E. A. Lock, Martinus Nijhoff, Lancaster, PA, 1987, pp. 371–405.

38. C. Cojocel, K. Maita, D. A. Pasino *et al.*, 'Metabolic heterogeneity of the proximal and distal kidney tubules.' *Life Sci.*, 1983, **33**, 855–861.

39. H. Endou, 'Cytochrome P-450 monooxygenase system in the kidney: its intranephron localization and its induction.' *Jpn. J. Pharmacol.*, 1983, **33**, 423–433.

40. K. A. Moore and C. E. Ruegg, 'Phase I and phase II metabolic activity in rabbit renal proximal convoluted (PCT) and straight (PST) tubules.' *The Toxicologist*, 1995, **15**, 299.

41. M. A. Carroll, M. Schwartzman, N. G. Abraham *et al.*, 'Cytochrome P450-dependent arachidonate metabolism in renomedullary cells: formation of Na⁺-K⁺-ATPase inhibitor.' *J. Hypertens. Suppl.*, 1986, **4**, S33–S42.

42. M. A. Carroll, M. Schwartzman, M. Baba *et al.*, 'Formation of biologically active cytochrome P450-arachidonate metabolites in renomedullary cells.' *Adv. Prostaglandin Thromboxane Leukot. Res.*, 1987, **17B**, 714–718.

43. T. V. Zenser and B. B. Davis, in 'Toxic Interactions,' eds. R. S. Goldstein, W. H. Hewitt and J. B. Hook, Academic Press, San Diego, CA, 1990, pp. 62–86.

44. P. H. Brand and B. B. Taylor, 'Lactate oxidation by three segments of the rabbit proximal tubule.' *Proc. Soc. Exp. Biol. Med.*, 1986, **182**, 454–460.

45. H. Shimada, H. Endou and F. Sakai, 'Distribution of gamma-glutamyl transpepidase and glutaminase isoenzymes in the rabbit single nephron.' *Jpn. J. Pharmacol.*, 1982, **32**, 121–129.

46. L. G. Fine, E. J. Goldstein, W. Trizna *et al.*, 'Glutathione-*S*-transferase activity in the rabbit nephron: segmental localization in isolated tubules and formation of thiol adducts of ethacrynic acid.' *Proc. Soc. Exp. Biol. Med.*, 1978, **157**, 189–193.

47. M. MacFarlane, J. R. Foster, G. G. Gibson *et al.*, 'Cysteine conjugate B-lyase of rat kidney cytosol: characterization, immunocytochemical localization, and correlation with hexachlorobutadiene nephrotoxicity.' *Toxicol. Appl. Pharmacol.*, 1989, **98**, 185–197.

48. T. W. Jones, C. Qin, V. H. Schaeffer *et al.*, 'Immunohistochemical localization of glutamine transaminase K, a rat kidney cysteine conjugate B-lyase, and the relationship to the segment specificity of cysteine conjugate nephrotoxicity.' *Mol. Pharmacol.*, 1988, **34**, 621–627.

49. R. A. Zager, 'Partial aortic ligation: a hypoperfusion model of ischemic acute renal failure and a comparison with renal artery occlusion.' *J. Lab. Clin. Med.*, 1987, **110**, 396–405.

50. J. I. Kreisberg, R. E. Bulger, B. F. Trump *et al.*, 'Effects of transient hypotension on the structure and function of rat kidney.' *Virchows Arch. B: Cell Pathol.*, 1976, **22**, 121–133.

51. D. C. Dobyan, R. B. Nagle and R. E. Bulger, 'Acute tubular necrosis in the rat kidney following sustained hypotension: Physiologic and morphologic observations.' *Lab Invest.*, 1977, **37**, 411–422.

52. J. Mason, 'The pathophysiology of ischaemic acute renal failure. A new hypothesis about the initiation

phase.' *Renal Physiol.*, 1986, **9**, 129–147.

53. B. Glaumann and B. F. Trump, 'Studies on the pathogenesis of ischemic cell injury. III. Morphological changes of the proximal pars recta tubules (P3) of rat kidney made ischemic *in vivo*.' *Virchows Arch. B: Cell. Pathol.*, 1975, **19**, 303–323.

54. M. A. Venkatachalam, D. B. Bernard, J. F. Donohoe *et al.*, 'Ischemic damage and repair in the rat proximal tubule: differences among S1, S2, and S3 segments.' *Kidney Int.*, 1978, **14**, 31–49.

55. P. F. Shanley, M. D. Rosen, M. Brezis *et al.*, 'Topography of focal proximal tubular necrosis after ischemia with reflow in the rat kidney.' *Am. J. Pathol.*, 1986, **122**, 462–468.

56. P. F. Shanley, M. Brezis, K. Spokes *et al.*, 'Hypoxic injury in the proximal tubule of the isolated perfused rat kidney.' *Kidney Int.*, 1986, **29**, 1021–1032.

57. M. Kashgarian, N. J. Siegel, A. L. Ries *et al.*, 'Hemodynamic aspects in development and recovery phases of experimental postischemic acute renal failure.' *Kidney Int. Suppl.*, 1976, **10**, S160–S168.

58. M. Brezis, P. Shanley, P. Silva *et al.*, 'Disparate mechanisms for hypoxic cell injury in different nephron segments. Studies in the isolated perfused rat kidney.' *J. Clin. Invest.*, 1985, **76**, 1796–1806.

59. B. F. Trump, I. K. Berezesky and R. A. Cowley, in 'Pathophysiology of Shock, Anoxia and Ischemia.' eds. R. A. Cowley and B. F. Trump, Waverly Press, Baltimore, MD, 1982, pp. 6–46.

60. C. E. Ruegg, A. J. Gandolfi, R. B. Nagle *et al.*, 'Differential patterns of injury to the proximal tubule of renal cortical slices following *in vitro* exposure to mercuric chloride, potassium dichromate, or hypoxic conditions.' *Toxicol. Appl. Pharmacol.*, 1987, **90**, 261–273.

61. C. E. Ruegg and L. J. Mandel, 'Bulk isolation of renal PCT and PST. II. Differential responses to anoxia or hypoxia.' *Am. J. Physiol.*, 1990, **259**, F176–F185.

62. C. E. Ruegg and L. J. Mandel, 'Differential effects of anoxia or mitochondrial inhibitors in renal proximal straight (PST) and convoluted (PCT) tubules.' *Kidney Int.*, 1990, **37**, 529.

63. P. D. Wilson and R. W. Schrier, 'Nephron segment and calcium as determinants of anoxic cell death in renal cultures.' *Kidney Int.*, 1986, **29**, 1172–1179.

64. K. D. Dickman, W. R. Jacobs and L. J. Mandel, 'Renal metabolism and acute renal failure.' *Pediatr. Nephrol.*, 1987, **1**, 359–366.

65. P. W. Hochachka, 'Defense strategies against hypoxia and hypothermia.' *Science*, 1986, **231**, 234–241.

66. L. J. Mandel, T. Takano, S. P. Soltoff *et al.*, 'Mechanisms whereby exogenous adenine nucleotides improve rabbit renal proximal function during and after anoxia.' *J. Clin. Invest.*, 1988, **81**, 1255–1264.

67. N. J. Siegel, W. B. Glazier, I. H. Chaudry *et al.*, 'Enhanced recovery from acute renal failure by the postischemic infusion of adenine nucleotides and magnesium chloride in rats.' *Kidney Int.*, 1980, **17**, 338–349.

68. J. M. Weinberg, 'Oxygen deprivation-induced injury to isolated rabbit kidney tubules.' *J. Clin. Invest.*, 1985, **76**, 1193–1208.

69. J. Bastin, N. Cambon, M. Thompson *et al.*, 'Change in energy reserves in different segments of the nephron during brief ischemia.' *Kidney Int.*, 1987, **31**, 1239–1247.

70. T. D. Stinsteden, F. J. O'Neal, S. Hill *et al.*, 'The role of high-energy phosphate in norepinephrine-induced acute renal failure in the dog.' *Circ. Res.*, 1986, **59**, 93–104.

71. C. T. Warnick and H. M. Lazarus, 'Recovery of nucleotide levels after cell injury.' *Can. J. Biochem.*,

1981, **59**, 116–121.

72. S. P. Soltoff and L. J. Mandel, 'Active ion transport in the renal proximal tubule. III. The ATP dependence of the Na pump.' *J. Gen. Physiol.*, 1984, **84**, 643–662.

73. G. W. Schmid-Schonbein, 'Capillary plugging by granulocytes and the no-reflow phenomenon in the microcirculation.' *Fed. Proc.*, 1987, **46**, 2397–2401.

74. K. Yamamoto, D. R. Wilson and R. Baumal, 'Outer medullary circulatory defect in ischemic acute renal failure.' *Am. J. Pathol.*, 1984, **116**, 253–261.

75. P. O. A. Hellberg, A. Bayati, O. Kallskog *et al.*, 'Red cell trapping after ischemia and long-term kidney damage. Influence of hematocrit.' *Kidney Int.*, 1990, **37**, 1240–1247.

76. Y. Yagil, M. Miyamoto and R. L. Jamison, 'Inner medullary blood flow in postischemic acute renal failure in the rat.' *Am. J. Physiol.*, 1989, **256**, F456–F461.

77. R. A. Zager, J. Mahan and A. J. Merola, 'Effects of mannitol on the postischemic kidney. Biochemical, functional and morphologic assessments.' *Lab. Invest.*, 1985, **53**, 433–442.

78. W. M. Abbott and W. G. Austin, 'The reversal of renal cortical ischemia during aortic occlusion by mannitol.' *J. Surg. Res.*, 1974, **16**, 482–489.

79. W. A. Franklin, C. E. Ganote and R. B. Jennings, 'Blood reflow after renal ischemia. Effects of hypertonic mannitol on reflow and tubular necrosis after transient ischemia in the rat.' *Arch. Pathol.*, 1974, **98**, 106–111.

80. J. Flores, D. R. DiBona, C. H. Beck *et al.*, 'The role of cell swelling in ischemic renal damage and the protective effect of hypertonic solute.' *J. Clin. Invest.*, 1972, **51**, 118–126.

81. M. A. Venkatachalam, J. I. Kreisberg, J. H. Stein *et al.*, 'Salvage of ischemic cells by impermeant solute and adenosinetriphosphate.' *Lab. Invest.*, 1983, **49**, 1–3.

82. J. M. Weinberg and H. D. Humes, 'Increases of cell ATP produced by exogenous adenine nucleotides in isolated rabbit kidney tubules.' *Am. J. Physiol.*, 1986, **250**, F720–F733.

83. P. F. Shanley, M. Brezis, K. Spokes *et al.*, 'Differential responsiveness of proximal tubule segments to metabolic inhibitors in the isolated perfused rat kidney.' *Am. J. Kidney Dis.*, 1986, **7**, 76–83.

84. F. H. Epstein, M. Brezis and S. Rosen, in 'Acute Renal Failure in the Intensive Care Unit,' eds. D. Bihari and G. Neild, Springer-Verlag, Heidelberg, Berlin, 1990, pp. 91–102.

85. M. Brezis, S. Rosen, P. Silva *et al.*, 'Selective vulnerability of the medullary thick ascending limb to anoxia in the isolated perfused rat kidney.' *J. Clin. Invest.*, 1984, **73**, 182–190.

86. Z. H. Endre, P. J. Ratcliffe, J. D. Tange *et al.*, 'Erythrocytes alter the pattern of renal hypoxic injury: predominance of proximal tubular injury with moderate hypoxia.' *Clin. Sci.*, 1989, **76**, 19–29.

87. M. Brezis, S. Rosen, K. Spokes *et al.*, 'Transport-dependent anoxic cell injury in the isolated perfused rat kidney.' *Am. J. Pathol.*, 1984, **116**, 327–341.

88. M. Brezis, S. Rosen, P. Silva *et al.*, 'Polyene toxicity in renal medulla: injury mediated by transport activity.' *Science*, 1984, **224**, 66–68.

89. P. F. Shanley, M. Brezis, K. Spokes *et al.*, 'Transport-dependent cell injury in the S3 segment of the proximal tubule.' *Kidney Int.*, 1986, **29**, 1033–1037.

90. K. L. Klein, M. S. Wang, S. Torikai *et al.*, 'Substrate oxidation by isolated single nephron segments of the rat.' *Kidney Int.*, 1981, **20**, 29–35.

91. L. J. Mandel, 'Metabolic substrates, cellular energy production, and the regulation of proximal tubular transport.' *Annu. Rev. Physiol.*, 1985, **47**, 85–101.

92. R. S. Balaban and L. J. Mandel, 'Metabolic substrate utilization by rabbit proximal tubule. An NADH fluorescent study.' *Am. J. Physiol.*, 1988, **254**, F407–F416.

93. K. D. Dickman and L. J. Mandel, 'Glycolytic and oxidative metabolism in primary renal proximal tubule cultures.' *Am. J. Physiol.*, 1989, **257**, C333–C340.

94. M. J. Tang, K. R. Suresh and R. L. Tannen, 'Carbohydrate metabolism by primary cultures of rabbit proximal tubules.' *Am. J. Physiol.*, 1989, **256**, C532–C539.

95. T. Takano, S. P. Soltoff, S. Murdaugh *et al.*, 'Intracellular respiratory dysfunction and cell injury in short-term anoxia of rabbit renal proximal tubules.' *J. Clin. Invest.*, 1985, **76**, 2377–2384.

96. P. J. Boogaard, J. P. Zoeteweij, T. J. van Berkel *et al.*, 'Primary culture of proximal tubular cells from normal rat kidney as an *in vitro* model to study mechanisms of nephrotoxicity. Toxicity of nephrotoxicants at low concentrations during prolonged exposure.' *Biochem. Pharmacol.*, 1990, **39**, 1335–1345.

97. P. H. Bach, C. P. Ketley, S. E. Benns *et al.*, in 'Renal Heterogeneity and Target Cell Toxicity.' eds. P. H. Bach and E. A. Lock, Wiley, Chichester, 1985, pp. 505–518.

98. M. S. Paller, J. R. Hoidal and T. F. Ferris, 'Oxygen free radicals in ischemic acute renal failure in the rat.' *J. Clin. Invest.*, 1984, **74**, 1156–1164.

99. G. L. Baker, R. J. Corry and A. P. Autor, 'Oxygen free radical induced damage in kidneys subjected to warm ischemia and reperfusion. Protective effect of superoxide dismutase.' *Ann. Surg.*, 1985, **202**, 628–641.

100. S. L. Linas, D. Whittenberg and J. E. Repine, 'O₂ metabolites cause reperfusion injury after short but not prolonged renal ischemia.' *Am. J. Physiol.*, 1987, **253**, F685–F691.

101. S. L. Linas, D. Whittenberg and J. E. Repine, 'Role of xanthine oxidase in ischemia reperfusion injury.' *Am. J. Physiol.*, 1990, **258**, F711–F716.

102. L. M. Gamelin and R. A. Zager, 'Evidence against oxidant injury as a critical mediator of postischemic acute renal failure.' *Am. J. Physiol.*, 1988, **255**, F450–F460.

103. M. Joannidis, G. Gstraunthaler and W. Pfaller, 'Xanthine oxidase: evidence against a causative role in renal reperfusion injury.' *Am. J. Physiol.*, 1990, **258**, F232–F236.

104. R. A. Zager, 'Hypoperfusion-induced acute renal failure in the rat: an evaluation of oxidant tissue injury.' *Circ. Res.*, 1988, **62**, 430–435.

105. R. A. Zager and D. J. Gmur, 'Effects of xanthine oxidase inhibition on ischemic acute renal failure in the rat.' *Am. J. Physiol.*, 1989, **257**, F953–F958.

106. S. C. Borken and J. H. Schwartz, 'Role of oxygen free radical species in *in vitro* models of proximal tubular ischemia.' *Am. J. Physiol.*, 1989, **257**, F114–F125.

107. R. B. Doctor and L. J. Mandel, 'Minimal role of xanthine oxidase and oxygen free radicals in rat renal tubular reoxygenation injury.' *J. Am. Soc. Nephrol.*, 1991, **1**, 959–969.

108. R. D. Lasley, S. W. Ely, R. M. Berne *et al.*, 'Allopurinol enhanced adenine nucleotide repletion after myocardial ischemia in the isolated rat heart.' *J. Clin. Invest.*, 1988, **81**, 16–20.

109. R. Hansson, B. Gustafsson, D. Jonsson *et al.*, 'Effect of xanthine oxidase inhibition on renal circulation after ischemia.' *Transplant. Proc.*, 1982, **14**, 51.

110. R. Hansson, D. Jonsson, S. Lundstam *et al.*, 'Effects of free radical scavengers on renal circulation after ischemia in the rabbit.' *Clin. Sci.*, 1983, **65**, 605–610.

111. K. Ouriel, N. G. Smedira and J. J. Ricotta, 'Protection of the kidney after temporary ischemia: free radical

scavengers.' *J. Vasc. Surg.*, 1985, **2**, 49–53.

112. G. Ojteg, A. Bayati, O. Kallskog *et al.*, 'Renal capillary permeability and intravascular red cell aggregation after ischemia, I. Effects of xanthine oxidase activity.' *Acta Physiol. Scand.*, 1987, **129**, 295–304.

113. R. F. DelMaestro and M. Planher, 'Evidence for the participation of superoxide anion radical in altering the adhesive interaction between granulocytes and endothelium *in vivo*.' *Int. J. Microcirc. Clin. Exp.*, 1982, **1**, 105.

114. J. M. Klausner, I. S. Paterson, G. Goldman *et al.*, 'Postischemic renal injury is mediated by neutrophils and leukotrienes.' *Am. J. Physiol.*, 1989, **256**, F794–F802.

115. M. Braide, A. Blixt and U. Bagge, 'Leukocyte effects on the vascular resistance and glomerular filtration of the isolated rat kidney at normal and low flow states.' *Circ. Shock*, 1986, **20**, 71–80.

116. K. M. Mullane, J. A. Salmon and R. Kraemer, 'Leukocyte-derived metabolites of arachidonic acid in ischemia-induced myocardial injury.' *Fed. Proc.*, 1987, **46**, 2422–2433.

117. S. Lelcuk, F. Alexander, L. Kobzik *et al.*, 'Prostacyclin and thromboxane A2 moderate postischemic renal failure.' *Surgery*, 1985, **98**, 207–212.

118. J. M. Weinberg, 'Issues in the pathophysiology of nephrotoxic renal tubular cell injury pertinent to understanding cyclosporin nephrotoxicity.' *Transplant. Proc.*, 1985, **17** (Suppl.1), 81–90.

119. J. B. Hook and W. R. Hewitt, in 'Casarett and Doull's Toxicology the Basic Science of Poisons.' 3rd edn., eds. C. D. Klaassen, M. O. Amdur and J. Doull, Macmillan, New York, 1986, pp. 310–329.

120. T. U. L. Biber, M. Mylle, A. D. Baines *et al.*, 'A study by micropuncture and microdissection of acute renal damage in rats.' *Am. J. Med.*, 1968, **44**, 664–705.

121. M. Murakami and M. Webb, 'A morphological and biochemical study of the effects of L-cysteine on the renal uptake and nephrotoxicity of cadmium.' *Br. J. Exp. Pathol.*, 1981, **62**, 115–130.

122. D. C. Dobyan, J. Levi, C. Jacobs *et al.*, 'Mechanism of *cis*-platinum nephrotoxicity: II. Morphologic observations.' *J. Pharmacol. Exp. Ther.*, 1980, **213**, 551–556.

123. L. K. Tay, C. L. Bregman, B. A. Masters *et al.*, 'Effects of cis-diamminedichloroplatinum (II) on rabbit kidney *in vivo* and on rabbit renal proximal tubule cells in culture.' *Cancer Res.*, 1988, **48**, 2538–2543.

124. F. E. Cuppage and A. Tate, 'Repair of the nephron following injury with mercuric chloride.' *Am. J. Pathol.*, 1967, **51**, 405–429.

125. A. E. Rodin and C. N. Crowson, 'Mercury nephrotoxicity in the rat. I. Factors influencing the localization of the tubular lesions.' *Am. J. Pathol.*, 1962, **41**, 297.

126. F. L. Siegel and R. E. Bulger, 'Scanning and transmission electron microscopy of mercuric chloride-induced acute tubular necrosis in the rat kidney.' *Virchows Arch. B: Cell Pathol.*, 1975, **18**, 243–262.

127. B. H. Haagsma and A. W. Pound, 'Mercuric chloride-induced renal tubular necrosis in the rat.' *Br. J. Exp. Pathol.*, 1979, **60**, 341–352.

128. E. M. McDowell, R. B. Nagle, R. C. Zalme *et al.*, 'Studies on the pathophysiology of acute renal failure. I. Correlation of ultrastructure and function in the proximal tubule of the rat following administration of mercuric chloride.' *Virchows Arch. B: Cell Pathol.*, 1976, **22**, 173–196.

129. R. C. Zalme, E. M. McDowell, R. B. Nagle *et al.*, 'Studies on the pathophysiology of acute renal failure. II. A histochemical study of the proximal tubule of the rat following administration of mercuric chloride.' *Virchows Arch. B: Cell Pathol.*, 1976, **22**,

197–216.

130. T. Tanaka, A. Naganuma and N. Imura, 'Role of gamma-glutamyltranspeptidase in renal uptake and toxicity of inorganic mercury in mice.' *Toxicology*, 1990, **60**, 187–198.

131. G. Guillermina, T. M. Adriana and E. M. Monica, 'The implication of renal glutathione levels in mercuric chloride nephrotoxicity.' *Toxicology*, 1989, **58**, 187–195.

132. A. Naganuma, M. E. Anderson and A. Meister, 'Cellular glutathione as a determinant of sensitivity to mercuric chloride toxicity. Prevention of toxicity by giving glutathione monoester.' *Biochem. Pharmacol.*, 1990, **40**, 693–697.

133. R. K. Zalups, 'Organic anion transport and action of gamma-glutamyl transpeptidase in kidney linked mechanistically to renal tubular uptake of inorganic mercury.' *Toxicol. Appl. Pharmacol.*, 1995, **132**, 289–298.

134. D. R. Johnson, 'Role of renal cortical sulfhydryl groups in development of mercury-induced renal tox-icity.' *J. Toxicol. Environ. Health*, 1982, **9**, 119–126.

135. W. O. Berndt, J. M. Baggett, A. Blacker *et al.*, 'Renal glutathione and mercury uptake by kidney.' *Fundam. Appl. Toxicol.*, 1985, **5**, 832–839.

136. L. E. Sendelbach, W. M. Bracken and C. D. Klaassen, 'Comparisons of the toxicity of $CdCl_2$ and Cd-metallothionein in isolated rat hepatocytes.' *Toxicology*, 1989, **55**, 83–91.

137. K. S. Min, K. Kobayashi, S. Onosaka *et al.*, 'Tissue distribution of cadmium and nephropathy after administration of cadmium in several chemical forms.' *Toxicol. Appl. Pharmacol.*, 1986, **86**, 262–270.

138. C. Dorian and C. D. Klaassen, 'Protection by zinc-metallothionein (ZnMT) against cadmium-metallothionein-induced nephrotoxicity.' *Fundam. Appl. Toxicol.*, 1995, **26**, 99–106.

139. M. G. Cherian, R. A. Goyer and L. Delaquerriere-Richardson, 'Cadmium-metallothionein-induced nephropathy.' *Toxicol. Appl. Pharmacol.*, 1976, **38**, 399–408.

140. A. Kennedy, 'The effect of L-cysteine on the toxicity of cadmium.' *Br. J. Exp. Pathol.*, 1968, **49**, 360–364.

141. F. A. Cotton and G. Wilkinson, in 'Advanced Inorganic Chemistry,' 3rd edn., Interscience, New York, 1972.

142. J. Oliver, M. MacDowell and A. Tracy, 'The pathogenesis of acute renal failure associated with traumatic and toxic injury: Renal ischemia, nephrotoxic damage and the ischemic episode.' *J. Clin. Invest.*, 1951, **30**, 1307.

143. C. E. Ruegg, A. J. Gandolfi and K. Brendel, in 'Nephrotoxicity: *In Vitro* to *In Vivo*, Animals to Man,' eds. P. H. Bach and E. A. Lock, Plenum Press, New York, 1989, pp. 107–112.

144. K. W. Jennette, 'The role of metals in carcinogenesis: biochemistry and metabolism.' *Environ. Health. Perspect.*, 1981, **40**, 233–252.

145. J. P. Fillastre and G. Raguenez-Viotte, 'Cisplatin nephrotoxicity.' *Toxicol. Lett.*, 1989, **46**, 163–175.

146. G. Daugaard and U. Abildgaard, 'Cisplatin nephrotoxicity. A review.' *Cancer Chemother. Pharmacol.*, 1989, **25**, 1–9.

147. J. S. Phelps, A. J. Gandolfi, K. Brendel *et al.*, 'Cisplatin nephrotoxicity: *in vitro* studies with precision-cut rabbit renal cortical slices.' *Toxicol. Appl. Pharmacol.*, 1987, **90**, 501–512.

148. D. Miura, R. S. Goldstein, D. A. Pasino *et al.*, 'Cis-platinum nephrotoxicity: role of filtration and tubular transport of *cis*-platinum in isolated perfused

kidneys.' *Toxicology*, 1987, **44**, 147–158.

149. H. R. Brady, B. C. Kone, M. E. Stromski *et al.*, 'Mitochondrial injury: an early event in *cis*-platinum toxicity to renal proximal tubules.' *Am. J. Physiol.*, 1990, **258**, F1181–F1187.

150. G. Singh, 'A possible cellular mechanism of *cis*-platinum-induced nephrotoxicity.' *Toxicol.*, 1989, **58**, 71–80.

151. M. E. I. Leibbrandt and G. H. I. Wolfgang, 'Differential toxicity of *cis*-platinum, carboplatin, and CI-973 correlates with cellular platinum levels in rat renal cortical slices.' *Toxicol. Appl. Pharmacol.*, 1995, **132**, 245–252.

152. P. J. Boogaard, E. L. Lempers, G. J. Mulders *et al.*, '4-Methylthiobenzoic acid reduced *cis*-platinum nephrotoxicity in rats without compromising anti-tumour activity.' *Biochem. Pharmacol.*, 1991, **41**, 1997–2003.

153. J. English, A. Evan, D. C. Houghton *et al.*, 'Cyclosporine-induced acute renal dysfunction in the rat. Evidence of arteriolar vasoconstriction with preservation of tubular function.' *Transplantation*, 1987, **44**, 135–141.

154. N. M. Jackson, C. H. Hsu, G. E. Visscher *et al.*, 'Alterations in renal structure and function in a rat model of cyclosporin nephrotoxicity.' *J. Pharmacol. Exp. Ther.*, 1987, **242**, 749–756.

155. J. B. Kopp and P. E. Klotman, 'Cellular and molecular mechanisms of cyclosporin nephrotoxicity.' *J. Am. Soc. Nephrol.*, 1990, **1**, 162–179.

156. G. Remuzzi and T. Bertani, 'Renal vascular and thrombotic effects of cyclosporin.' *Am. J. Kidney Dis.*, 1989, **13**, 261–272.

157. A. Erman, B. Chen-Gal and J. Rosenfeld, 'The role of eicosanoids in cyclosporin nephrotoxicity in the rat.' *Biochem. Pharmacol.*, 1989, **38**, 2153–2157.

158. N. Perico, J. Dadan and G. Remuzzi, 'Endothelin mediates the renal vasoconstriction induced by cyclosporin in the rat.' *J. Am. Soc. Nephrol.*, 1990, **1**, 76–83.

159. D. Rodriguez-Puyol, S. Lamas, A. Olivera *et al.*, 'Actions of cyclosporin A on cultured rat mesangial cells.' *Kidney Int.*, 1989, **35**, 632–637.

160. B. Sumpio, A. E. Baue and I. H. Chaudry, 'Treatment with verapamil and adenosine triphosphate-MgCl$_2$ reduces cyclosporin nephrotoxicity.' *Surgery*, 1987, **101**, 315–322.

161. T. B. Wiegmann, R. Sharma, D. A. Diederich *et al.*, '*In vitro* effects of cyclosporin on glomerular function.' *Am. J. Med. Sci.*, 1990, **299**, 149–152.

162. H. J. Goldberg, P. Y. Wong, E. H. Cole *et al.*, 'Dissociation between the immunosuppressive activity of cyclosporin derivatives and their effects on intracellular calcium signaling in mesangial cells.' *Transplantation*, 1989, **47**, 731–733.

163. T. Strzelecki, S. Kumar, R. Khauli *et al.*, 'Impairment by cyclosporin of membrane-mediated functions in kidney mitochondria.' *Kidney Int.*, 1988, **34**, 234–240.

164. B. Sumpio, A. E. Baue and I. H. Chaudry, 'Alleviation of cyclosporin nephrotoxicity with verapamil and ATP-MgCl$_2$. Mitochondrial respiratory and calcium studies.' *Ann. Surg.*, 1987, **206**, 655–660.

165. L. A. Vernetti, A. J. Gandolfi and R. B. Nagle, 'Selective alteration of cytokeratin intermediate filament by cyclosporin A is a lethal toxicity in PTK2 cell cultures.' *Adv. Exp. Med. Biol.*, 1991, **283**, 847–851.

166. R. J. Walker, V. A. Lazzaro, G. G. Duggin *et al.*, 'Structure-activity relationships of cyclosporines. Toxicity in cultured renal tubular epethelial cells.' *Transplantation*, 1989, **48**, 321–327.

167. P. D. Wilson and D. Hreniuk, 'Nephrotoxicity of cyclosporin in renal tubule cultures and attenuation

by calcium restriction.' *Transplant. Proc.*, 1988, **20**, Suppl. 3, 709–711.

168. H. Dieperink, P. P. Leyssac, H. Starklint *et al.*, 'Long-term cyclosporin nephrotoxicity in the rat: effects on renal function and morphology.' *Nephrol. Dial. Transplant*, 1988, **3**, 317–326.

169. D. M. Gillum, L. Truong, J. Tasby *et al.*, 'Chronic cyclosporin nephrotoxicity. A rodent model.' *Transplantation*, 1988, **46**, 285–292.

170. P. H. Whiting and A. W. Thomson, in 'Cyclosporin: Mode of Action and Clinical Applications.' ed. A. W. Thomson, Kluwer Academic Publ., Boston, MA, 1989, pp. 303–323.

171. E. Matthys, M. K. Patton, R. W. Osgood *et al.*, 'Alterations in vascular function and morphology in acute ischemic renal failure.' *Kidney Int.*, 1983, **23**, 717–724.

172. M. W. Anders, L. Lash, W. Dekant *et al.*, 'Biosynthesis and biotransformation of glutathione-*S*-conjugates to toxic metabolites.' *Crit. Rev. Toxicol.*, 1988, **18**, 311–341.

173. E. A. Lock, in 'Selectivity and Molecular Mechanisms of Toxicity.' eds. F. DeMatteis and E. A. Lock, Macmillan, New York, 1987, pp. 59–83.

174. E. A. Lock, 'Studies on the mechanism of nephrotoxicity and nephrocarcinogenicity of halogenated alkenes.' *Crit. Rev. Toxicol.*, 1988, **19**, 23–42.

175. P. O. Darnerud, I. Brandt, V. J. Feil *et al.*, '*S*-(1,2-dichloro-[^{14}C]vinyl)-L-cysteine (DCVC) in the mouse kidney: correlation between tissue-binding and toxicity.' *Toxicol. Appl. Pharmacol.*, 1988, **95**, 423–434.

176. J. Ishmael, I. Pratt and E. A. Lock, 'Necrosis of the pars recta (S3 segment) of the rat kidney produced by hexachloro-1:3-butadiene.' *J. Pathol.*, 1982, **138**, 99–113.

177. D. R. Jaffe, A. J. Gandolfi and R. B. Nagle, 'Chronic toxicity of *S*-(trans-1,2-dichlorovinyl)-L-cysteine in mice.' *J. Appl. Toxicol.*, 1984, **4**, 315–319.

178. G. H. I. Wolfgang, A. J. Gandolfi, C. L. Krumdieck *et al.*, 'Evaluation of organic nephrotoxins in rabbit renal cortical slices.' *Toxicol. In Vitro*, 1989, **3**, 341.

179. G. H. I. Wolfgang, A. J. Gandolfi, J. L. Stevens *et al.*, '*N*-acetyl *S*-(1,2-dichlorovinyl)-L-cysteine produces a similar toxicity to *S*-(1,2-dichlorovinyl)-L-cysteine in rabbit renal slices: Differential transport and metabolism.' *Toxicol. Appl. Pharmacol.*, 1989, **101**, 205–219.

180. J. A. Nash, L. J. King, E. A. Lock *et al.*, 'The metabolism and disposition of hexachloro-1:3-butadiene in the rat and its relevant nephrotoxicity.' *Toxicol. Appl. Pharmacol.*, 1984, **73**, 124–137.

181. J. E. Bakke, J. Rafter, G. L. Larsen *et al.*, 'Enterohepatic circulation of the mercurpturic acid and cysteine conjugates of prochlor.' *Drug Metab. Dispos.*, 1981, **9**, 525–528.

182. M. Inoue, K. Okajama and Y. Morino, 'Hepato-renal cooperation in biotransformation, membrane transport and elimination of cysteine *S*-conjugates of xenobiotics.' *J. Biochem. (Tokyo)*, 1984, **95**, 247–254.

183. G. Zhang and J. L. Stevens, 'Transport and activation of *S*-(1,2-dichlorovinyl)-L-cysteine and *N*-acetyl-*S*-(1,2-dichlorovinyl)-L-cysteine in rat kidney proximal tubules.' *Toxicol. Appl. Pharmacol.*, 1989, **100**, 51–61.

184. V. H. Schaeffer and J. L. Stevens, 'Mechanism of transport for toxic cysteine conjugates in rat kidney cortex membrane vesicles.' *Mol. Pharmacol.*, 1987, **32**, 293–298.

185. E. A. Lock and R. G. Schnellmann, 'The effect of haloalkene cysteine conjugates on rat renal glutathione reductase and lipoyl dehydrogenase activities.' *Toxicol. Appl. Pharmacol.*, 1990, **104**, 180–190.

186. J. L. Stevens, 'Cysteine conjugate B-lyase activities in the rat kidney cortex: subcellular localization and

relationship to the hepatic enzyme.' *Biochem. Biophys. Res. Comm.*, 1985, **129**, 499–504.

187. J. N. Commandeur and N. P. Vermeulen, 'Molecular and biochemical mechanisms of chemically induced nephrotoxicity: a review.' *Chem. Res. Toxicol.*, 1990, **3**, 171–194.

188. J. B. Hook and J. H. Smith, 'Biochemical mechanisms of nephrotoxicity.' *Transplant. Proc.*, 1985, **17**, 41–50.

189. J. B. Hook and R. S. Goldstein (eds.), 'Toxicology of the Kidney,' 2nd edn., Raven Press, New York, 1993.

190. F. Morel, 'Sites of hormone action in the mammalian nephron.' *Am. J. Physiol.*, 1981, **240**, F159–F164.

191. D. W. Seldin and G. H. Giebisch (eds.) 'The Kidney: Physiology and Pathophysiology.' Raven Press, New York, 1985.

7.13
Biochemical Mechanisms of Proximal Tubule Cellular Death

GARY W. MILLER
Duke University, Durham, NC, USA

and

RICK G. SCHNELLMANN
University of Arkansas for Medical Sciences, Little Rock, AR, USA

7.13.1 INTRODUCTION

The kidneys maintain fluid and electrolyte homeostasis in the body through the reabsorption of filtered water, ions and nutrients, and excretion of wastes.[1] The numerous renal transport mechanisms combined with the large renal blood flow, 20% of the cardiac output, results in the kidneys being exposed to high concentrations of blood-borne toxicants. One of the most common targets in toxicant-induced renal dysfunction is the proximal tubule. The purpose of this chapter is to describe some of the biochemical mechanisms by which chemicals cause proximal tubular cell death. The focus is on the interactions of nephrotoxicants with intracellular targets and the resulting cascade of events that lead to cell damage and death.

Due to the high exposure to blood-borne chemicals and the numerous transport and metabolic enzymes in the proximal tubule, compounds such as drugs, heavy metals, and solvents can attain damaging concentrations. Since the proximal tubule is responsible for reabsorption of 60–80% of the solute and water filtered by the glomerulus, damage to the proximal tubule by toxicant exposure can severely impair renal function. Exposure to toxicants can lead to a continuum of effects from minor impairment to cell death. Nephrotoxicants generally target a particular segment of the proximal tubule. For example, the mycotoxin citrinin and the aminoglycoside antibiotics primarily damage the S1 and S2 segments, while mercuric chloride, cyclosporine, the antineoplastic agent cisplatin, bromobenzene, and cysteine conjugates of halogenated hydrocarbons affect the S3 segment.[2] The exact reason for the segmental susceptibility is unclear, but may be due to differences inherent in the different segments including transport and uptake processes, toxicant delivery, expression of biotransformation enzymes, intracellular targets and functions, and cytoprotective mechanisms within the cell.

7.13.2 CELLULAR DEATH

Cellular death is believed to occur through two distinct mechanisms, namely necrosis and apoptosis.[3–5] Apoptosis is a highly regulated process characterized by death of individual cells scattered throughout the tissue. The cells shrink and break into small membrane-bound pieces, which are phagocytized by adjoining cells (Figure 1). Apoptosis is often, but not always, accompanied by endonuclease-mediated DNA degradation at the internucleosomal linker regions, which appears as a DNA "ladder" upon agarose gel electrophoresis (see Section 7.13.4.4). Apoptosis is distinct from "programmed cell death" which refers to situations in which cells are programmed to die at a particular time, for example, during development.[3] While the morphology of apoptosis and "programmed cell death" is similar, the term apoptosis is more appropriate when referring to cellular death induced by xenobiotics or ischemia.

Necrosis or necrotic cellular death is characterized by damage to single or multiple contiguous cells. Within the kidney, necrosis is usually confined to discrete regions or cell types. Intracellular organelles swell and the integrity of the cellular membranes becomes compromised due to proteolysis and lipid hydrolysis of membrane components (Figure 1). There is a marked increase in cellular volume which is followed by rupture of the cellular membrane.

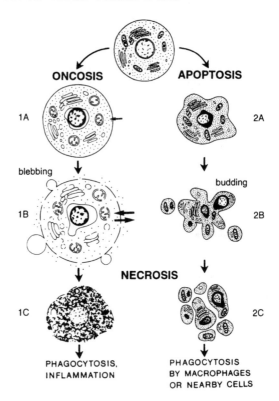

Figure 1 Comparison of morphological changes in oncosis and apoptosis. At the top is a normal cell. (1A) Cell and organelle swelling; (1B) vacuolization, blebbing, increased membrane permeability, and lysis; (1C) necrotic changes–coagulation, shrinkage, and karyolysis; (2A) cell shrinkage and pyknosis; (2B) budding and karyorrhexis; (2C) necrotic changes–breakup into cluster of apoptotic bodies (reproduced by permission of the American Society for Investigative Pathology from *Am. J. Pathol.*, 1995, **146**, 3–15).

This leads to the release of intracellular molecules and triggers an inflammatory reaction in the area of damage.[3,5] Majno and Joris[3] have noted that "necrotic cell death" is an inappropriate term to describe cellular death characterized by cellular volume increases and cell lysis. Historically, the term "necrosis" has been used to describe drastic changes to the tissue and includes karyolysis, pyknosis, condensation of the cytoplasm, intense eosinophilia, and loss of structure and fragmentation. These dramatic changes occur well after the cell has actually lost viability. The necrotic changes are independent of the mechanism of cellular death, whether it occurs from ischemia, radiation, chemicals, or apoptosis.

Majno and Joris[3] have proposed that the term "oncosis" (from *onkos*, meaning swelling) be used to describe cellular death characterized by cellular swelling, organelle swelling, blebbing, and lysis. The term "oncosis" was first coined by von Recklinghausen in 1910 to describe cellular death characterized by swelling.[6] Thus, in toxicant-induced cellular death oncosis is the counterpart to apoptosis. Within a tissue or group of cells, oncosis can occur independently of apoptosis or in the presence of apoptosis. Toxicant-induced damage in the kidney is believed to occur predominantly through oncosis.

7.13.3 TOXICANT INTERACTION AND BIOACTIVATION

The observed cellular effects of a toxicant are a function of its interaction with cellular targets. Toxicants can interact directly with a variety of targets, including membrane phospholipids, proteins, DNA, receptors, enzymes, and intracellular organelles. For example, mercuric chloride readily reacts with sulfhydryl groups on tubular proteins resulting in inhibition of a number of enzyme systems. Some compounds such as antimycin A and cyanide primarily target the electron transport system of the mitochondria and cause inhibition of mitochondrial respiration.[7,8] Other compounds do not react with cellular targets until they have been bioactivated. The kidney contains many of the xenobiotic metabolizing enzymes found in the liver, although the concentrations in the kidney are generally lower. Renal cytochrome P450 concentration is approximately 20% of that found in the liver and the distribution level varies among different renal cells.[9] In the rabbit, cytochrome P450 levels are highest in the S2 segment followed by S3 and S1.[10] Multiple forms of cytochrome P450 have been identified in rat kidney including IA1, IIC2, IIE1, IIIA, IVA1, and IVA2[9] and there are significant interspecies and gender differences in the expression of cytochrome P450 enzymes. For example, in mouse kidney, IIA, IIC, and IIE are present in males, but essentially absent in females.[11] While the liver cytochrome P450 system is known to play an important role in the bioactivation of a large number of hepatotoxic compounds there are only a few compounds known to be nephrotoxic through their bioactivation by renal cytochrome P450.

Renal cytochrome P450 plays a role in the renal toxicity of acetaminophen and chloroform.[12–15] Renal cytochrome P450 metabolizes acetaminophen to the reactive intermediate *N*-acetyl-*p*-benzoquinonimine, which has been demonstrated to bind covalently to numerous intracellular proteins.[14] Similarly, chloroform is metabolized by renal cytochrome P450 to the unstable intermediate trichloromethanol. Trichloromethanol releases HCl to form phos-

gene which can bind to cellular molecules to produce toxicity.[15]

Conjugation of compounds by glutathione is generally known as a detoxification pathway. However, numerous glutathione conjugates formed outside the kidney have been demonstrated to be nephrotoxic, including halogenated alkenes and aromatics[15–17] and mercuric chloride.[18] Indeed, the glutathione and mercapturic acid moieties of the conjugate may be the substrate for transport into the proximal tubular cell (Figure 2).

7.13.4 DEGRADATIVE PATHWAYS AND PROCESSES

After the toxic species interact with the cellular targets, the subsequent course of events will depend on the normal function of the targets, the extent of damage to the targets, and any compensatory or repair processes. If the target is a major component involved in the regulation of intracellular calcium concentrations and it is severely impaired, perturbations of calcium homeostasis and resultant alterations in calcium-dependent functions would be expected (Figure 3). Likewise, damage to enzymes crucial to the generation of ATP would disrupt any function dependent on ATP. There are several pathways that can follow the initial interaction between the toxicant and cellular target: (i) formation of reactive oxygen species, (ii) alterations in cellular ion homeostasis, and (iii) activation of degradative enzymes, such as proteinases, phospholipases, and endonucleases. It is unlikely that a toxicant activates only one of these processes. A more plausible explanation would be that multiple pathways are activated in parallel and any one of them may ultimately lead to cell death.

7.13.4.1 Reactive Oxygen Species

Oxidative stress is a condition in which there is an increased production of reactive oxygen species (ROS) (see Chapter 21, this volume).[19–21] Toxicants can indirectly cause oxidative stress by increasing the production of ROS. For example, isolated rat mitochondria treated with mercuric chloride or gentamicin have increased H_2O_2 production.[22,23] Chemicals can also produce oxidative stress by redox cycling. Certain chemicals, especially quinones, can undergo one-electron reduction to a semiquinone radical followed by a second single-electron reduction to the hydroquinone. The hydroquinone is oxidized to the quinone and the cycle begins again. During the reduction, superoxide anion is formed from oxygen and

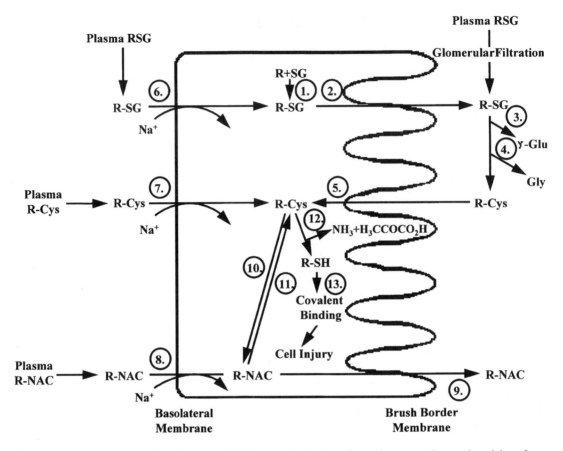

Figure 2 Diagram representing the renal tubular uptake, biotransformation, excretion and toxicity of conjugates of glutathione and cysteine, and mercapturic acid. (1) Formation of glutathione (R-SG) within the renal cell. (2) Secretion of glutathione into the tubular lumen. (3) Hydrolysis of a glutathione conjugate by gamma-glutamyl transferase. (4) Hydrolysis of cysteinylglycine conjugate by a dipeptidase. (5) Renal uptake of a cysteine conjugate (R-Cys). (6) Na^+-dependent uptake of a glutathione conjugate by the renal cell. (7) Na^+-dependent uptake of a cysteine conjugate by the renal cell. (8) Na^+-dependent uptake of a mercapturic acid (*N*-acetyl-cysteine conjugate; R-NAC) by the renal cell. (9) Secretion of a mercapturic acid into the tubular lumen. (10) Acetylation of a cysteine conjugate by a *N*-acetyl transferase. (11) Deacetylation of a mercapturic acid. (12) Conversion of the penultimate nephrotoxic species by cysteine conjugate β-lyase to a reactive thiol, pyruvate, and ammonia. (13) Binding of the reactive thiol to cellular macromolecules, such as proteins and lipids, which in turn initiates cellular injury (reproduced by permission of Williams and Wilkins from *Drug Metab. Dispos.*, 1987, **15**, 437–441).

oxidative stress follows. Brown *et al.*[24] have demonstrated that menadione (2-methyl-1,4-napthoquinone) produces toxicity in isolated rat renal epithelial cells by its ability to redox cycle and cause oxidative stress. However, the ability of quinones to undergo redox cycling varies with the quinone and some quinones produce toxicity through their ability to arylate cellular macromolecules such as protein sulfhydryls.[25,26]

ROS produce toxicity by reacting with numerous cellular constituents. Some of the major detrimental events mediated by ROS are lipid peroxidation, inactivation of cellular enzymes by oxidizing protein sulfhydryl or amino groups, and induction of chromosome

and DNA strand breaks. Many nephrotoxicants are thought to produce injury by oxidative stress. These structurally diverse compounds include mercuric chloride,[22,27] haloalkene cysteine conjugates,[28–31] cisplatin,[32,33] and cyclosporine A.[34]

7.13.4.2 Role of Altered Calcium Homeostasis

Intracellular free Ca^{2+} is important in the maintenance of normal cellular function because of its role as a second messenger and its role in many crucial intracellular processes. Under normal conditions, the intracellular

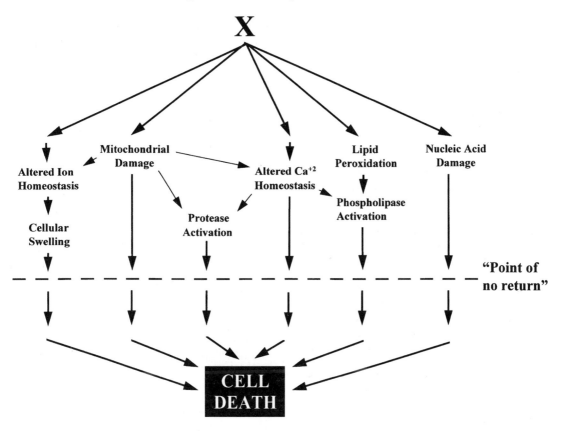

Figure 3 Schematic representation of toxicant-induced renal cell death. A toxicant may activate multiple pathways that lead to cellular injury and cell death. Any one of these representative pathways may be sufficient to cause cell death. The "point of no return" represents that point at which cell injury is not reversible and the cell will die.

concentration of cytosolic free Ca^{2+} is maintained at approximately 100 nM, despite an extracellular to intracellular ratio of 10 000 :1. Maintenance of this intracellular concentration of cytosolic free Ca^{2+} is accomplished by pumps and channels located on the endoplasmic reticulum and plasma membranes, sequestration in microsomes, and anionic binding to cellular macromolecules.[35,36] Mitochondria are not thought to contribute to the regulation of cytosolic free Ca^{2+} under normal physiological conditions. After toxicant exposure, mitochondria do accumulate Ca^{2+} but only after lethal cell injury.[1,35] Since endogenous agonists can increase cytosolic free Ca^{2+} under normal physiological conditions without any adverse effects, it is thought that a toxicant must cause uncontrolled supraphysiologic increases or sustained increases in cytosolic free Ca^{2+} in order to produce deleterious effects such as activation of calcium-dependent degradative enzymes, including proteinases, phospholipases, and endonucleases. However, the precise role of Ca^{2+} in toxicant-induced cell death is

not clear and has been the topic of considerable debate.

When rabbit renal proximal tubule suspensions are exposed to mitochondrial inhibitors or anoxia, cytosolic free Ca^{2+} does not increase until immediately prior to cell lysis[37,38] and decreasing the concentration of extracellular calcium attenuates cellular death.[38-40] These findings suggest that cytosolic free Ca^{2+} does indeed play a role in the late phase of cellular injury, although the exact mechanism by which Ca^{2+} contributes to cellular injury is not known. It has been demonstrated that inhibitors of calcium-activated neutral proteinases (calpains) inhibit cellular death produced by a variety of diverse toxicants and anoxia in renal proximal tubule suspensions, further supporting a role for Ca^{2+} in toxicant-induced cell injury.[41] Several nephrotoxicants increase cytosolic free Ca^{2+}, including oxidants,[42,43] pentachlorobutadienyl-glutathione,[44] pentachlorobutadienyl-L-cysteine,[45] tetrafluoroethyl-L-cysteine,[45] dichlorovinyl-L-cysteine,[44-49] and mercuric chloride.[50,51]

7.13.4.3   Phospholipases

Phospholipase A_2 (PLA_2) is a family of enzymes that hydrolyze the 2-acyl ester bond at the sn-2 position of phospholipids to yield arachidonic acid and lysophospholipids.[52] The enzymes in this family vary in their substrate preferences, biochemical characteristics, and calcium dependency. Phospholipase A_2 has received considerable attention due to its potential role as a mediator of renal cellular injury and death. It has been proposed that toxicant-induced activation, above the normal physiological activity, can result in the degradation of membrane phospholipids and thus impair membrane function. Further, since many of the PLA_2 enzymes are calcium-dependent, it is thought that PLA_2 activation may be the result of toxicant-induced increases in cytosolic free Ca^{2+}. The lysophospholipids and free fatty acids that are produced can also contribute to cellular injury by interfering with mitochondrial respiration and altering membrane permeability. The precise role of PLA_2 in renal cell injury is not known.[52–59] While PLA_2 activity is altered in many forms of renal cell injury, it is unclear if this activation or inhibition is causative to the cell damage or coincidental. There also appears to be considerable variability in the different experimental models and initiators of cellular death. Increased PLA_2 activity has been demonstrated in rabbit renal proximal tubule suspensions exposed to anoxia,[56] and the phospholipase inhibitors mepacrine and dibucaine decreased hypoxia-induced rat renal proximal tubule cellular death.[59] Similarly, the oxidant t-butyl-hydroperoxide increases arachidonic acid release from rabbit renal proximal tubules, and arachidonic acid increase and cellular death are both attenuated by mepacrine and dibucaine.[58] In contrast, rabbit renal proximal tubule suspensions exposed to the mitochondrial inhibitor antimycin A neither showed an increase in arachidonic acid release, nor were protected by mepacrine or dibucaine.[58] Thus, PLA_2 activation does not represent a common final pathway in proximal tubule cellular death and the role of PLA_2 in proximal tubule cellular injury is dependent on the toxic insult.

7.13.4.4   Endonucleases

Endonucleases have been suggested to be activated during apoptotic cellular injury and play a role in renal cellular death.[60–64] Much of what is known about endonuclease activation comes from studies of liver injury. In the liver, activation of Mg^{2+}-dependent endonucleases causes DNA to be degraded into fragments of 30–50 kb, 200–300 kb, and 700 kb.[65–68] Ca^{2+} potentiates the process, although it is not necessary. The subsequent internucleosomal cleavage requires both Ca^{2+} and Mg^{2+}. The DNA fragments that result from internucleosomal cleavage give rise to the DNA "ladder" pattern observed following agarose gel electrophoresis.

The evidence for a role of endonucleases in renal cellular injury is mixed. Schumer *et al.*[69] reported that DNA extracted from rat kidneys 12–48 h postischemia exhibited a "ladder" pattern upon agarose gel electrophoresis, and Ueda and Shah[70] showed that DNA from LLC-PK1 cells exposed to H_2O_2 was fragmented. In contrast, Schnellmann *et al.*[71] did not observe DNA fragmentation prior to or subsequent to the onset of cellular death in rabbit renal proximal tubule suspensions exposed to the mitochondrial inhibitors antimycin A or carbonyl cyanide p-(trifluoromethoxy)phenyl-hydrazone, the oxidant t-butylhydroperoxide, or the calcium ionophore ionomycin. Iwata *et al.*[72] demonstrated that minimal DNA fragmentation occurred in postischemic rat kidneys and posthypoxic isolated rat proximal tubule segments. Further, the DNA fragmentation was reported to be associated with oncosis, not apoptosis. Interestingly, when the human leukemic cell line U937 is subjected to osmotic shock, large DNA fragments are formed similar to those seen in apoptosis.[73] Subsequently, the large DNA fragments are degraded into a continuous spectrum of small fragments (smear pattern). Thus, DNA laddering is not exclusive to apoptotic death and may be observed in some forms and stages of oncosis. In conjunction with studies reporting the artifactual appearance of DNA fragmentation during renal DNA isolation,[74,75] these studies suggest that care needs to be taken when describing toxicant-induced endonuclease activation and DNA fragmentation as apoptosis in the absence of morphological characteristics.

7.13.4.5   Lysosomes

Lysosomes are a key intracellular target of aminoglycoside antibiotics and in α_{2u}-globulin nephropathy. The aminoglycoside antibiotics induce lysosomal dysfunction and cause acute renal failure.[76–79] The aminoglycosides are filtered by the glomerulus, bound to anionic phospholipids in the lumen, reabsorbed in the S1 and S2 segments of the proximal tubule, and accumulated in the lysosomes. With time, the number and the size of the lysosomes increase and myeloid bodies appear. The myeloid bodies consist of undegraded phospholipids, and are thought to be the result of aminoglycoside-induced inhibition of lysosomal hydrolases, such as sphingomyelinase and phospholipases. The steps between the lysosomal phospholipid overload and tubular cell death are not clear.

α_{2u}-Nephropathy is observed in male rats following exposure to such compounds as *d*-limonene, unleaded gasoline, 1,4-dichloro-benzene, lindane, and tetrachloroethylene.[80–83] α_{2u}-Globulin is synthesized in the liver of male rats and is under androgenic control. Serum α_{2u}-globulin is filtered by the glomerulus and approximately 50% is reabsorbed by endocytosis in the proximal tubule. Binding of the aforementioned agents to α_{2u}-globulin prevents its degradation and results in the accumulation of the α_{2u}-globulin in the lysosomes of the proximal tubule. The accumulation leads to a characteristic protein droplet morphology and single-cell oncosis, granular casts at the junction of the proximal tubule and the thin loop of Henle, and eventually diffuse cellular degeneration. However, it is not clear how the lysosomal overload leads to cell death. Long-term exposure of these compounds to rats can lead to chronic nephropathy and in some instances leads to an increased incidence of renal adenomas and carcinomas by nongenotoxic mechanisms.

α_{2u}-Globulin nephropathy appears to be limited to male rats. Humans, female rats, male and female mice, rabbits, and guinea pigs do not synthesize α_{2u}-globulin and do not exhibit α_{2u}-globulin nephropathy. Humans are not thought to be at risk of α_{2u}-globulin nephropathy since they do not produce α_{2u}-globulin, they secrete fewer proteins in the urine than rats, and the low molecular weight proteins found in human urine do not resemble α_{2u}-globulin or bind to those compounds which bind α_{2u}-globulin.

7.13.4.6 Proteinases

Intracellular proteolysis is a highly regulated process that is responsible for controlling protein turnover, processing of nascent polypeptide chains, and inactivating bioactive peptides. Like phospholipases and endonucleases, uncontrolled activation of proteinases can be detrimental to normal cellular function. Lysosomal proteinases degrade proteins by acid hydrolysis and disruption of the lysosomal membrane could potentially lead to uncontrolled proteolysis. However, studies have failed to demonstrate that lysosomal rupture occurs during a variety of injurious insults.[1] Further, depletion of lysosomal enzymes did

not provide any protective effect to human and rabbit renal proximal tubules in primary culture exposed to cyclosporine A.[79] Interestingly, the cysteine proteinase inhibitor E64 was demonstrated to be cytoprotective to these cells. Other cysteine and serine proteinase inhibitors were shown to be ineffective in protecting renal proximal tubular segments from tetrafluoroethyl-L-cysteine, bromohydroquinone, antimycin A, and *t*-butylhydroperoxide.[84]

The calcium-activated neutral proteinases (calpains) are found in the cytoplasm and have cytoskeletal proteins, membrane proteins, and enzymes as substrates.[85] The kidney expresses both the high- and low-calcium affinity forms of the enzyme. Schnellmann *et al.*[41] demonstrated that both calpain inhibitor I and calpain inhibitor II decreased cellular death in renal proximal tubule suspensions exposed to the diverse toxicants antimycin A, bromohydroquinone, tetrafluroethyl-L-cysteine, *t*-butylhydroperoxide, and anoxia. Edelstein[86] has reported that administration of E64 decreased calpain activity and cellular death in rat renal proximal tubules exposed to hypoxia. Further, Sarang and Schnellmann[87] showed that calpain activity increased in the late phase of cellular injury produced by antimycin A and that calpain inhibitor I inhibited calpain activity and decreased renal proximal tubular cellular death. These data suggest that calpains participate in the pathogenesis of toxicant-induced cellular death, though it is not known how toxicant exposure can lead to uncontrolled activation of calpains or precisely how calpains exert harmful effects on the cell.

7.13.4.7 Mitochondrial Damage

The ability of the renal proximal tubule to reabsorb solute and water depends upon a high rate of ATP production. In the proximal tubule approximately 95% of the ATP is formed via oxidative phosphorylation.[88,89] The level of oxidative phosphorylation that occurs within renal cells varies with the different segments of the nephron. Therefore, toxicants that impair mitochondrial function will have a greater effect in regions that have a low glycolytic capacity, such as the S1 and S2 segments of the proximal tubule.

Since numerous processes are often targeted during toxicant exposure, it is unclear if mitochondrial inhibition is a consequence of some other insult or the actual cause of cellular death.[90,91] Cell swelling is integral in oncosis and cell volume is primarily regulated by ATP-dependent ion transporters. Thus, mitochondrial dysfunction is likely to play an important role in toxicant-induced oncosis. The steps that occur following mitochondrial dysfunction are not clear. ATP production decreases as mitochondrial function is impaired and this decrease in ATP can interfere with the ability of cell protein kinases to maintain phosphorylation of proteins and ATP-dependent ion transporters. For example, incorporation of phosphate into proteins decreases in renal proximal tubule suspensions exposed to anoxia[92] and the mitochondrial inhibitor antimycin A.[93] In addition, the protein phosphatase inhibitors calyculin A and okadaic acid decrease cellular death in renal proximal tubules exposed to either anoxia or antimycin A.[92-94] Therefore, protein dephosphorylation subsequent to mitochondrial dysfunction and ATP depletion appears to play an important role in renal proximal tubule cellular death.

Nephrotoxicants can alter mitochondrial function by several different mechanisms. Toxicants can directly inhibit enzymes involved in the electron transport system, dissipate the proton gradient across the inner mitochondrial membrane, inhibit substrate transport, or uncouple oxidative phosphorylation from electron transport.[8,90] For example, in renal proximal tubules, pentachlorobutadienyl-L-cysteine initially uncouples oxidative phosphorylation by dissipating the proton gradient,[95-98] while tetrafluoroethyl-L-cysteine inhibits state 3 respiration by inhibiting sites I and II of the electron transport chain.[98] Several other nephrotoxicants have been demonstrated to alter mitochondrial function including mercuric chloride,[99,100] cisplatin,[101-103] citrinin,[104-108] ochratoxin A,[109-111] cephaloridine,[112,113] and N-(3,5-dichlorophenyl)-succinimide.[114] Since the proximal tubule relies on ATP-dependent maintenance of proper ion gradients for normal function, this can lead to inhibition of transport functions and an inability to maintain ion homeostasis. Ion homeostasis is integral in the control of cell volume with marked cell swelling resulting from ATP depletion.

7.13.4.8 Ion Homeostasis and Cell Volume Regulation

Intracellular concentrations of Na^+ and K^+ are primarily controlled by the electrogenic Na^+,K^+-ATPase, which extrudes three Na^+ for every two K^+ potassium ions transported into the cell. In the renal proximal tubule, the Na^+,K^+-ATPase consumes approximately half of the cellular ATP under normal conditions. Therefore, toxicants that block ATP production lead to inhibition of the Na^+,K^+-

ATPase, K^+ efflux, Na^+ influx, and dissipation of the normally negative membrane potential.[115,116] Loss of the negative membrane potential leads to an obligatory entry of Cl^- into the cell along with additional Na^+ influx, water influx, and cell swelling.[115,116] Treatment of rabbit renal proximal tubule suspensions with the mitochondrial inhibitor antimycin A inhibits cellular respiration within a minute followed by ATP depletion and dissipation of the Na^+ and K^+ gradients over the next 5–10 min.[8] Miller and Schnellmann have demonstrated that Cl^- influx does not occur in the initial 0–15 min, but between 15–30 min, which is followed by cell lysis.[117] Decreasing the extracellular sodium chloride concentration by 50% with an iso-osmotic substitution of mannitol decreased Cl^- influx, cellular swelling, and cellular lysis.[118] Further, hyperosmotic incubation buffer decreased cellular swelling and lysis, but did not prevent.[118] Therefore, Cl^- influx may be the trigger for water influx and additional Na^+ influx which provides the osmotic force necessary to cause cellular swelling and lysis (Figure 4). Increased Cl^- influx has been reported to occur in the late stages of cell injury in renal proximal tubules exposed to a variety of diverse toxicants including t-butylhydroperoxide, bromohydroquinone, mercuric chloride, and tetrafluoroethyl-L-cysteine.[118] The mechanism by which Cl^- influx occurs during toxicant exposure is unknown, however, inhibitors of Ca^{2+}-activated Cl^- channels (IAA-94, niflumic acid, NPPB) block toxicant-induced Cl^- influx and are cytoprotective.[118,119] Interestingly, the cytoprotective actions of glycine and strychnine have been suggested to involve inhibition of Cl^- influx.[117]

Toxicants can directly interact with membrane consitituents to increase membrane permeability to ions. For example, the antifungal amphotericin B binds to cholesterol in the plasma membrane and forms a pore that increases permeability to K^+ and H^+.[120-122] Heavy metals such as gold, silver, copper, and mercury appear to act directly on the plasma membrane to increase permeability to K^+.[123,124] It is unclear how these alterations in K^+ and H^+ permeability lead to cell death. Interestingly, Reeves and Shah[125] have demonstrated that inhibition of K^+ channels decreases hypoxic injury in rat renal proximal tubules.

7.13.4.9 Cytoskeleton Rearrangements

Some of the early morphological changes that occur following toxicant exposure are blebbing of the plasma membrane, loss of the apical brush border, and alterations in membrane polarity. These changes are the result of

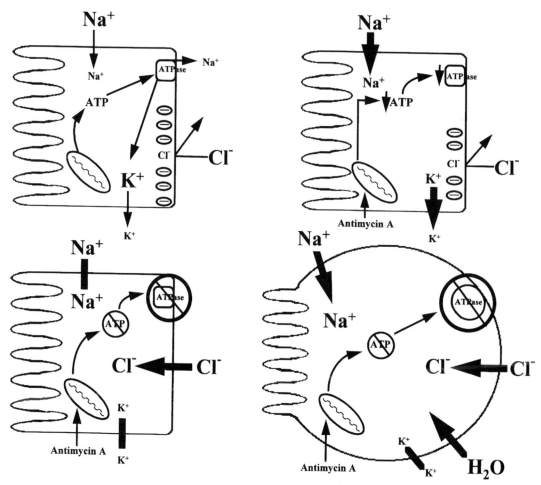

Figure 4 (Upper left panel) Schematic representation of a normal renal cell. The inside of the cell is negative with respect to the outside, which decreases the ability of chloride to enter the cell. (Upper right panel) Inhibition of mitochondrial function, by a compound such as antimycin A that blocks cellular respiration, decreases ATP levels and Na^+,K^+-ATPase activity, increases Na^+ influx and K^+ efflux, and decreases the membrane potential. (Lower left panel) Subsequently, there is an obligatory entry of chloride into the cell (down the concentration gradient) by an unidentified pathway. (Lower right panel) The increase in chloride influx results in water influx, increased Na^+ influx, and cell swelling. These processes provide the osmotic force that ultimately leads to cell lysis (reproduced by permission of Little Brown and Co. from 'Diseases of the Kidney,' 1996).

alterations in cytoskeletal components and cytoskeletal–cell membrane interactions produced by the toxicant. The toxicant can act directly to affect these components or indirectly by interfering with cell processes such as energy metabolism, and calcium and phospholipid homeostasis. For example, treatment of rat proximal tubular cells in primary culture with dichlorovinyl-L-cysteine causes depolymerization of F-actin and an increase in G-actin.[126] These changes were reported to be independent of ATP depletion and due to the formation of reactive intermediates and their interaction with thiol groups on actin and related proteins.

Considerable research has been performed on the effects of ischemia on the proximal tubular cytoskeletal network.[127] While there are distinct differences between ischemic and toxic injury, there are many similarities. Thus, many of the findings in ischemic models of proximal tubular injury may provide insights to the possible effect of toxicants that affect energy production. Under normal physiological conditions, the proximal tubular epithelium is polarized with respect to certain transporters and enzymes. For example, alkaline phosphatase is localized to the brush border membrane, and Na^+,K^+-ATPase to the basolateral membrane. Molitoris[128] reported that *in vivo* ischemic injury and *in vitro* ATP depletion results in the dissociation of the Na^+,K^+-ATPase from the actin cytoskeleton and redistribution to the apical membrane of proximal tubular cells. While the physiological significance of this translocation of the enzyme is unknown, the authors proposed that this may explain the decreased sodium reabsorption observed in ischemic injury. While some morphological and biochemical data have suggested that cytoskeletal alterations play a role in toxicant-induced cell injury, the mechanisms underlying these changes await further studies.

7.13.5 CELLULAR PROTECTIVE MECHANISMS

The renal proximal tubular cell contains several protective mechanisms to counteract damage by reactive intermediates and ROS. The tripeptide glutathione, a major detoxifying agent in the cell, is located in the cytosol, mitochondria, and nucleus.[130] Glutathione can detoxify electrophiles by forming a glutathione conjugate, which is accomplished directly or with the aid of glutathione S-transferases. Renal proximal tubular cells can detoxify bromohydroquinone and related quinone-containing compounds by forming mono- and disubstituted glutathione conjugates.[26] Glutathione can neutralize ROS via glutathione peroxidase and glutathione reductase. For example, organic peroxides are reduced to water and an alcohol by glutathione peroxidase with the formation of glutathione disulfide. Glutathione disulfide is, in turn, reduced to glutathione by glutathione reductase. Catalase and superoxide dismutase can also detoxify ROS by converting superoxide anion to hydrogen peroxide and hydrogen peroxide to water, respectively.

Vitamin E, alpha-tocopherol, is a lipid-soluble antioxidant found within cellular membranes.[130,131] Vitamin E can donate an electron to the peroxyl radical formed during lipid peroxidation, which prevents further lipid peroxidation. From this reaction, vitamin E radical is produced which is recycled back to vitamin E by vitamin C. Vitamin C, ascorbic acid, is also a very effective reducing agent and free radical scavenger.[132,133] Vitamin C can detoxify compounds containing a quinone nucleus, such as bromohydroquinone, but it does so by reducing the toxic bromoquinone and bromoquinone radical back to bromohydroquinone.[26] In the presence of iron, vitamin C can also act as a pro-oxidant.

In experiments aimed at identifying the cytoprotective action of glutathione, it was found that the constitutive amino acid glycine was cytoprotective in numerous models.[134,135] The cytoprotective actions of glycine were found to be shared by a few other amino acids of similar structure, including D- and L-alanine, β-alanine,

and 1-aminocyclopropane-1-carboxylic acid. Glycine has since been shown to be cytoprotective against diverse forms of renal proximal tubular cellular injury, such as anoxia, inhibitors of glycolysis and oxidative phosphorylation, bromohydroquinone, halogenated alkane and alkene cysteine conjugates, and to a lesser degree *t*-butylhydroperoxide and mercuric chloride.[134,136] In addition, work by Schnellmann and co-workers[136–140] demonstrated that the neuronal glycine receptor antagonist strychnine, in a similar manner to glycine, was cytoprotective against a number of diverse toxicants. Thus, strychnine and glycine have been proposed to exert their cytoprotective actions through a ligand–acceptor interaction. Strychnine binds to the plasma membrane of the renal proximal tubule in a saturable and reversible manner at concentrations similar to those necessary for cytoprotection.[138] Further, proteins corresponding to two of the three subunits of the neuronal inhibitory glycine receptor have been identified on the basolateral portion of the renal proximal tubule.[139,140] Thus, it has been proposed that glycine and strychnine may be cytoprotective via binding to a novel glycine receptor in the kidney. The neuronal glycine receptor forms a Cl^- channel and Cl^- influx contributes to the osmotic force for swelling in cellular injury. In conjunction with the observation that glycine and strychnine block Cl^- influx, glycine and strychnine may exert their cytoprotective effects by modulating Cl^- influx. Alternatively, Gores and co-workers have suggested that glycine is cytoprotective in hepatocytes by its ability to inhibit calpains.[141] This hypothesis would require that calpain activity be decreased by glycine during cell injury. Edelstein,[86] however, reported that glycine did not inhibit calpain activity in rat renal proximal tubules exposed to hypoxia. The precise mechanism by which glycine is cytoprotective is still controversial and further studies are needed to definitively identify the cytoprotective mechanism.

Acidosis is not typically considered to be a cellular defense mechanism, but decreasing intracellular pH has been shown to be cytoprotective in several *in vitro* models and has improved our understanding of cell death.[2,142] Reduction of extracellular pH in renal proximal tubules in suspension results in cytoprotection.[119,142–145] Extracellular acidosis can result from the addition of acid to the media or acidification by the cell as observed in renal proximal tubules exposed to anoxia.[144] In this situation, the tubules spontaneously reduce the pH of the surrounding media, possibly due to ATP hydrolysis and proton accumulation. Therefore, during ischemia *in vivo* it is possible that the extracellular acidosis in the area of ischemia may be cytoprotective.

Our understanding of the cytoprotective effect of extracellular acidosis has improved markedly. Rodeheaver and Schnellmann[142] demonstrated that reducing the extracellular pH to 6.4 prevented renal proximal tubular cell death induced by a variety of mitochondrial inhibitors (rotenone, antimycin A, oligomycin, carbonyl cyanide-*p*-trifluoromethoxyphenylhydrazone) and ion exchangers (monensin, nigericin, valinomycin), but exacerbated cell death induced by the oxidants hydrogen peroxide, *t*-butylhydroperoxide, and ochratoxin A. The potentiation of cell death by oxidants was associated with an increase in glutathione disulfide formation, lipid peroxidation, and mitochondrial dysfunction and a decrease in glutathione peroxidase and reductase activities. Therefore, potentiation of oxidant-induced cell death by extracellular acidosis may be due to decreased free radical detoxification.

The cytoprotective effects of extracellular acidosis have previously been demonstrated to be independent of preservation of mitochondrial function or ATP.[143–146] It has been shown that extracellular acidosis could be initiated well after a toxic insult and still be cytoprotective.[119,144,145] When extracellular acidosis was initiated 15 min after the addition of antimycin A or carbonyl cyanide-*p*-trifluoromethoxyphenylhydrazone (a time point after mitochondrial inhibition, ATP depletion, and disruption of sodium and potassium gradients), cell death was completely prevented 45 and 105 min later. Initiation of extracellular acidosis by addition of acid 2 h following the addition of tetrafluor-

oethyl-L-cysteine or *t*-butylhydroperoxide was also cytoprotective up to 2 h later. These data indicate that the cytoprotective effects of acidosis occur late in the process of cell injury. However, the precise mechanism by which acidosis prevents cell death is not known. Extracellular acidosis does not decrease chloride influx or cell swelling, indicating that its effects occur subsequent to these changes.[119] It is possible that acidosis inhibits the activity of the calcium-dependent calpains. Calpain activity decreases as pH decreases and calpains act in the late phase of cell death during the same time that extracellular acidosis exerts cytoprotection. Further studies are necessary to elucidate the precise mechanism of extracellular acidosis cytoprotection.

7.13.6 PARALLEL PATHWAYS

There are many potential sites within a cell for initial toxicant interaction, including DNA, plasma membranes, proteins, lipids, and enzymes. The type and extent of this interaction determines the cascade of events that occur thereafter. There are also various pathways that can lead from initial toxicant exposure to eventual cell injury (reversible) or death (irreversible). Somewhere during this process there is a "point of no return." That is, a point at which the cell is destined to die and no intervention can alter this course (Figure 5). A great deal of research has focused on identifying this point and the overall sequence of events that occur following toxicant exposure. It is doubtful that a single sequence of events can explain the

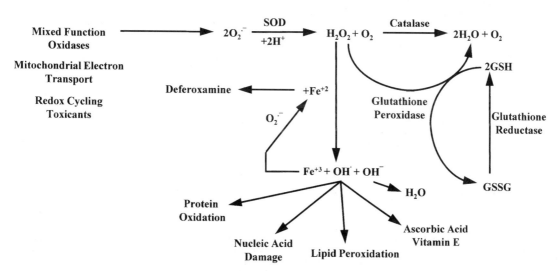

Figure 5 Schematic representation of toxicant-induced renal cell death. A toxicant may activate multiple pathways that lead to cellular injury and cell death. Any one of these representative pathways may be sufficient to cause cell death. The "point of no return" represents that point at which cell injury is not reversible and the cell will die.

damage, considering the number of initial targets available for alkylating species and ROS. A more plausible explanation is that the initial toxic insult leads to the activation of different yet parallel pathways, each of which are capable of leading to cell death. For example, a compound that inhibits a site of the electron transport system in the mitochondria may lead to ROS formation and ATP depletion. Either the ROS or the mitochondrial inhibition could potentially kill the cell and directly affect other important cellular molecules. The depletion of ATP will alter numerous systems and may lead to activation of proteinases and phospholipases. These pathways do not necessarily need to be exclusive as it is possible that some of them may converge. Even a sublethal exposure to a toxicant can have multiple effects on cellular macromolecules and processes. Therefore, it is imperative to consider the complexity of the degradative processes that occur during toxicant exposure when examining the effects of a toxic agent on the renal proximal tubule.

ACKNOWLEDGMENTS

R.G.S was supported by NIH grant ES-04410. G.W.M. was a recipient of a Society of Toxicology Graduate Student Fellowship (Sponsored by Procter and Gamble).

7.13.7 REFERENCES

1. J. M. Weinberg, in 'Diseases of the Kidney,' eds. R. W. Schrier and C. W. Gottschalk, Little Brown, Boston, MA, 1993, pp. 1031–1098.
2. J. M. Weinberg, 'The cell biology of ischemic renal injury.' *Kidney. Int.*, 1991, **39**, 476–500.
3. G. Majno and I. Joris, 'Apoptosis, oncosis, and necrosis. An overview of cell death.' *Am J. Pathol.*, 1995, **146**, 3–15.
4. A. H. Wyllie, 'Apoptosis (the 1992 Frank Rose Memorial Lecture).' *Br. J. Cancer*, 1993, **67**, 205–208.
5. G. B. Corcoran and S. D. Ray, 'The role of the nucleus and other compartments in toxic cell death produced by alkylating hepatotoxicants.' *Toxicol. Appl. Pharmacol.*, 1992, **113**, 167–183.
6. F. von Recklinghausen, 'Untersuchungen uber Rachitis und Osteomalacie,' Gustav Fisher, Jena, Verlag, 1910.
7. M. Erecinska and D. F. Wilson, 'Inhibitors of cytochrome c oxidase.' *Pharmacol. Ther.*, 1982, **8**, 1–20.
8. S. R. Gullans, P. C. Brazy, S. P. Solotoff *et al.*, 'Metabolic inhibitors: effects on metabolism and transport in the proximal tubule.' *Am. J. Physiol.*, 1982, **243**, F133–F140.
9. E. A. Lock, in 'Mechanisms of Injury in Renal Disease and Toxicity,' ed. R. S. Goldstein, CRC Press, Boca Raton, FL, 1994, pp. 173–206.
10. H. Endou, 'Cytochrome P-450 monooxygenase system in the rabbit kidney: its intranephron localization and its induction.' *Jpn. J. Pharmacol.*,
1983, **33**, 423–433.
11. C. J. Henderson, A. R. Scott, C. S. Yang *et al.*, 'Testosterone-mediated regulation of mouse renal cytochrome P-450 isoenzymes.' *Biochem. J.*, 1990, **266**, 675–681.
12. J. H. Smith, 'Role of renal metabolism in chloroform nephrotoxicity.' *Comments Toxicol.*, 1986, **1**, 125–144.
13. S. G. Emeigh Hart, D. R. Koop, D. S. Wyand *et al.*, 'Immunohistochemical localization of acetaminophen (APAP) and cytochrome P450 IIE1 in hepatic and extrahepatic target tissues in the CD-1 mouse.' *Toxicol. Pathol.*, 1990, **18**, 695–703.
14. S. G. Emeigh Hart, W. P. Beierschmitt, D. S. Wyand *et al.*, 'Acetminpohen nephrotoxicity in CD-1 mice. I. Evidence of a role for *in situ* activation in selective covalent binding and toxicity.' *Toxicol. Appl. Pharmacol.*, 1994, **126**, 267–275.
15. A. A. Elfarra, in 'Toxicology of the Kidney,' eds. J. B. Hook and R. S. Goldstein, Raven Press, New York, 1993, pp. 387–414.
16. T. J. Monks, H. H. Lo and S. S. Lau, 'Oxidation and acetylation as determinants of 2-bromocystein-*S*-ylhydroguinone-mediated nephrotoxicity.' *Chem. Res. Toxicol.*, 1994, **7**, 495–502.
17. L. H. Lash, in 'Mechanisms of Injury in Renal Disease and Toxicity,' ed. R. S. Goldstein, CRC Press, Boca Raton, FL, 1994, pp. 207–234.
18. R. K. Zalups, 'Organic anion-transport and action of gamma-glutamyl transpeptidase in kidney linked mechanistically to renal tubular uptake of inorganic mercury.' *Toxicol. Appl. Pharmacol.*, 1995, **132**, 289–298.
19. B. Halliwell and J. M. C. Gutteridge, 'Free Radicals in Biology and Medicine,' 2nd edn., Clarendon Press, Oxford, 1989.
20. H. Sies, 'Oxidative Stress: Oxidants and Antioxidants,' Academic Press, New York, 1991.
21. B. Halliwell, J. M. C. Gutteridge and C. E. Cross, 'Free radicals, antioxidants, and human disease: where are we now?' *J. Lab. Clin. Med.*, 1992, **119**, 598–620.
22. B. O. Lund, D. M. Miller and J. S. Woods, 'Studies on Hg(II)-induced H_2O_2 formation and oxidative stress *in vivo* and *in vitro* in rat kidney mitochondria.' *Biochem. Pharmacol.*, 1993, **45**, 2017–2024.
23. P. D. Walker and S. V. Shah, 'Gentamicin enhanced production of hydrogen peroxide by renal cortical mitochondria.' *Am. J. Physiol.*, 1987, **253**, C495–C499.
24. P. C. Brown, D. M. Dulik and TW Jones, 'The toxicity of menadione (2-methyl-1,4-napthoquinone) and two thioether conjugates studied with isolated renal epithelial cells.' *Arch. Biochem. Biophys.*, 1991, **285**, 187–196.
25. R. G. Schnellmann, '2-Bromohydroquinone-induced toxicity to rabbit renal proximal tubules: evidence against oxidative stress.' *Toxicol. Appl. Pharmacol.*, 1989, **99**, 11–18.
26. R. G. Schnellmann, T. J. Monks, L. J. Mandel *et al.*, '2-Bromohydroquinone-induced toxicity to rabbit renal proximal tubules: the role of biotransformation, glutothione, and covalent binding.' *Toxicol. Appl. Pharmacol.*, 1989, **99**, 19–27.
27. H. Fukino, M. Hirai, Y. M. Hsueh *et al.*, 'Effect of zinc pretreatment on mercuric chloride-induced lipid peroxdation in the rat kidney.' *Toxicol. Appl. Pharmacol.*, 1984, **73**, 395–401.
28. A. Schmid, W. Beuter and L. Mayring, 'Mechanism of action of *S*-(dichlorovinyl)-L-cysteine.' *Zentralbl. Veterinaermed. [A]*, 1983, **30**, 511–520.
29. W. Beuter, C. Cojocel, W. Muller *et al.*, 'Peroxidative

damage and nephrotoxicity of dichlorovinyleysteine in mice.' *J. Appl. Toxicol.*, 1989, **9**, 181–186.

30. Q. Chen, T. W. Jones, P. C. Brown *et al.*, 'The mechanism of cysteine conjugate cytoxicity in renal epithelial cells. Covalent binding leads to thiol depletion and lipid peroxidation.' *J. Biol. Chem.*, 1990, **265**, 21603–21611.

31. C. E. Groves, E. A. Lock and R. G. Schnellmann, 'Role of lipid peroxidation in renal proximal tubule cell death induced by haloalkene cysteine conjugates.' *Toxicol. Appl. Pharmacol.*, 1991, **107**, 54–62.

32. K. Sugihara, S. Nakano, M. Koda *et al.*, 'Stimulatory effect of cisplatin on production of lipid peroxidation in renal tissues.' *Jpn. J. Pharmacol.*, 1987, **43**, 247–252.

33. J. Hannemann and K. Baumann, 'Cisplatin-induced lipid peroxidation and decrease of gluconeogenesis in rat kidney cortex: different effects of antioxidants and radical scavengers.' *Toxicology*, 1988, **51**, 119–132.

34. C. Wang and A. K. Salahudeen, 'Cyclosporine nephrotoxicity: attenuation by an antioxidant-inhibitor of lipid peroxidation *in vitro* and *in vivo*.' *Transplantation*, 1994, **58**, 940–946.

35. A. Lefurgey, P. Ingram and L. J. Mandel, 'Heterogeneity of calcium compartmentation: electron probe analysis of renal tubules.' *J. Membr. Biol.*, 1986, **94**, 191–196.

36. E. Carafoli, 'Intracellular calcium homeostasis.' *Annu. Rev. Biochem.*, 1987, **56**, 395–433.

37. W. R. Jacobs, M. Sgambati, G. Gomez *et al.*, 'Role of cytosolic Ca in renal tubule damage induced by anoxia.' *Am. J. Physiol.*, 1991, **260**, C545–C554.

38. J. M. Weinberg, J. A. Davis, N. F. Roeser *et al.*, 'Role of increased cytosolic free calcium in the pathogenesis of rabbit proximal tubule cell injury and protection by glycine or acidosis.' *J. Clin. Invest.*, 1991, **87**, 581–590.

39. T. Takano, S. P. Soltoff, S. Murdaugh *et al.*, 'Intracellular respiratory dysfunction and cell injury in short-term anoxia of rabbit renal proximal tubules.' *J. Clin. Invest.*, 1985, **76**, 2377–2384.

40. R. G. Schnellmann, 'Calpains and capacitative calcium entry as mediators in renal cell death.' *Fundam. Appl. Toxicol.*, 1997 **36**, 149.

41. R. G. Schnellmann, X. Yang and T. J. Cross, 'Calpains play a critical role in renal proximal tubule (RPT) cell death.' *Can. J. Physiol. Pharmacol.*, 1994, **72**, 44.

42. N. Ueda and S. V. Shah, 'Role of intracellular calcium in hydrogen peroxide-induced renal tubular cell injury.' *Am. J. Physiol.*, 1992, **263**, F214–F221.

43. E. L. Greene and M. S. Paller, 'Calcium on free radicals in hypoxia/reoxygenation injury of renal epithelial cells.' *Am. J. Physiol.*, 1994, **266**, F13–F20.

44. T. W. Jones, A. Wallin, H.Thor *et al.*, 'The mechanism of pentachlorobutadienyl-glutathione nephrotoxicity studied with isolated rat renal epithelial cells.' *Arch. Biochem. Biophys.*, 1986, **251**, 504–513.

45. C. E. Groves, E. A. Lock and R. G. Schnellmann, 'The effects of haloalkene cysteine conjugates on cytosolic free calcium levels in suspensions of rat renal proximal tubules.' *J. Biochem. Toxicol.*, 1990, **5**, 187–192.

46. L. H. Lash and M. W. Anders, 'Cytotoxicity of *S*-(1,2-dichlorovinyl)glutathione and *S*-(1,2-dichlorovinyl)-L-cysteine in isolated rat kidney cells.' *J. Biol. Chem.*, 1986, **261**, 13076–13081.

47. S. Vamvakas, V. K. Sharma, S. S. Sheu *et al.*, 'Perturbations of intracellular calcium distribution in kidney cells by nephrotoxic haloalkenyl cysteine S-conjugates.' *Mol. Pharmacol.*, 1990, **38**, 455–461.

48. B. van de Water, J. P. Zoetewey, H. J. de Bont *et al.*, 'The relationship between intracellular Ca^{2+} and the mitochondrial membrane potential in isolated proximal tubular cells from rat kidney exposed to the nephrotoxin 1,2-dichlorovinyl-cysteine.' *Biochem. Pharmacol.*, 1993, **45**, 2259–2267.

49. Q. Chen, T. W. Jones and J. L. Stevens, 'Early cellular events couple covalent binding of reactive metabolites to cell killing by nephrotoxic cysteine conjugates.' *J. Cell. Physiol.*, 1994, **161**, 293–302.

50. M. W. Smith, I. S. Ambudkar, P. C. Phelps *et al.*, 'HgCl$_2$-induced changes in cytosolic Ca^{2+} of cultured rabbit renal tubular cells.' *Biochim. Biophys. Acta*, 1987, **931**, 130–142.

51. M. W. Smith, P. C. Phelps and B. F. Trump, 'Cytosolic Ca^{2+} deregulation blebbing after HgCl$_2$ injury to cultured rabbit proximal tubule cells as determined by digital imaging microscopy.' *Proc. Natl. Acad. Sci. USA*, 1991, **88**, 4926–4930.

52. J. V. Bonventre, 'Phospholipase A2 and signal transduction.' *J. Am. Soc. Nephrol.*, 1992, **3**, 128–150.

53. D. Portilla, S. V. Shah, P. A. Lehman *et al.*, 'Role of cytosolic calcium-independent plasmalogen-selective phospholipase A2 in hypoxic injury to rabbit proximal tubules.' *J. Clin. Invest.*, 1994, **93**, 1609–1615.

54. S. D. Finkelstein, D. Gilfor and J. L. Farber, 'Alterations in the metabolism of lipids in ischemia of the liver and kidney.' *J. Lipid Res.*, 1985, **26**, 726–734.

55. H. D. Humes, V. D. Nguyen, D. A. Cieslinski *et al.*, 'The role of free fatty acids in hypoxia-induced injury to renal proximal tubule cells.' *Am. J. Physiol.*, 1989, **256**, F688–F696.

56. D. Portilla, L. J. Mandel, D. BarSagi *et al.*, 'Anoxia induces phopholipase A2 activation in rabbit renal proximal tubules.' *Am. J. Physiol.*, 1992, **262**, F354–F360.

57. J. F. Wetzels, X. Wang, P. E. Gengaro *et al.*, 'Glycine protection against hypoxic but not phospholipase A2-induced injury in rat proximal tubules.' *Am. J. Physiol.*, 1993, **264**, F94–F99.

58. R. G. Schnellmann, X. Yang and J. B. Carrick, 'Arachidonic acid release in renal proximal tubule cell injuries and death.' *J. Biochem. Toxicol.*, 1994, **9**, 211–217.

59. D. Bunnachak, A. R. Almeida, J. F. Wetzels *et al.*, 'Ca^{2+} uptake, fatty acid, and LDH release during proximal tubule hypoxia: effects of mepacrine and dibucaine.' *Am. J. Physiol.*, 1994, **266**, F196–F201.

60. M. J. Arends, R. G. Morris and A. H. Wyllie, 'Apoptosis. The role of the endonuclease.' *Am. J. Pathol.*, 1990, **136**, 593–608.

61. J. J. Cohen, R. C. Duke, V. A. Fadok *et al.*, 'Apoptosis and programmed cell death in immunity.' *Annu. Rev. Immunol.*, 1992, **10**, 267–293.

62. M. M. Compton and J. A. Cidlowski, 'Thymocyte apoptosis—a model of programmed cell death.' *Trends Endocrinol. Metab.*, 1992, **3**, 17–23.

63. D. J. McConkey, P. Hartzell, P. Nicotera *et al.*, 'Calcium-activated DNA fragmentation kills immature-thymocytes.' *FASEB J.*, 1989, **3**, 1843–1849.

64. P. Nicotera, G. Bellomo and S. Orrenius, 'Calcium-mediated mechanisms in chemically induced cell death.' *Annu. Rev. Pharmacol. Toxicol.*, 1992, **32**, 449–470.

65. K. Cain, S. H. Inayat-Hussain, L. Kokileva *et al.*, 'DNA cleavage in rat liver nuclei activated by Mg^{2+} or Ca^{2+} is inhibited by a variety of structurally unrelated inhibitors.' *Biochem. Cell Biol.*, 1994, **72**, 631–638.

66. G. R. Bicknell and G. M. Cohen, 'Cleavage of DNA to large kilobase pair fragments occurs in some forms of necrosis as well as apoptosis.' *Biochem. Biophys. Res. Comm.*, 1995, **207**, 40–47.

67. K. Cain, S. H. Inayat-Hussain, L. Kokileva *et al.*,

'Multi-step DNA cleavage in rat liver nuclei is inhibited by thiol reactive agents.' *FEBS Lett.*, 1995, **358**, 255–261.

68. K. Cain, S. H. Inayat-Hussain, J. T. Wolfe *et al.*, 'DNA fragmentation into 200–250 and/or 30–50 kilobase pair fragments in rat liver nuclei is stimulated by Mg^{2+} alone and Ca^{2+}/Mg^{2+} but not by Ca^{2+} alone.' *FEBS Lett.*, 1994, **349**, 385–391.

69. M. Schumer, M. C. Colombel, I. S. Sawczuk *et al.*, 'Morphologic, biochemical, and molecular evidence of apoptosis during the reperfusion phase after brief periods of renal ischemia.' *Am. J. Pathol.*, 1992, **140**, 831–838.

70. N. Ueda and S. V. Shah, 'Endonuclease-induced DNA damage and cell death in oxidant injury to renal tubular epithelial cells.' *J. Clin. Invest.*, 1992, **90**, 2593–2597.

71. R. G. Schnellmann, A. R. Swagler and M. M. Compton, 'Absence of endonuclease activation during acute cell death in renal proximal tubules.' *Am. J. Physiol.*, 1993, **265**, C485–C490.

72. M. Iwata, D. Meyerson, B. Torok-Starb *et al.*, 'An evaluation of renal tubular DNA laddering in response to oxygen deprivation and oxidant injury.' *J. Am. Soc. Nephrol.*, 1994, **5**, 1307–1313.

73. G. R. Bicknell, R. T. Snowden and G. M. Cohen, 'Formation of high molecular mass DNA fragments is a marker of apoptosis in the human leukaemic cell line, U937.' *J. Cell Sci.*, 1994, **107**, 2483–2489.

74. H. Enright, R. P. Hebbel and K. A. Nath, 'Internucleosomal cleavage of DNA as the sole criterion for apoptosis may be artifactual.' *J. Clin. Lab. Med.*, 1994, **124**, 63–68.

75. S. Fukushima, M. A. Davis and B. F. Trump, 'Biochemical evidence of apoptosis in renal tissue resulting from tissue preparation.' *Toxicologist*, 1995, **15**, 298.

76. J. C. Kosek, R. I Mazze and M. J. Cousins, 'Nephrotoxicity of gentamicin.' *Lab. Invest.*, 1974, **30**, 48–57.

77. G. Laurent, B. K. Kishore and P. M. Tulkens, 'Aminoglycoside-induced renal phospholipidosis and nephrotoxicity.' *Biochem. Pharmacol.*, 1990, **40**, 2383–2392.

78. G. J. Kaloyanides, 'Drug–phospholipid interactions: role in aminoglycoside nephrotoxicity.' *Ren. Fail.*, 1992, **14**, 351–357.

79. P. D. Wilson and P. A. Hartz, 'Mechanisms of cyclosporine A toxicity in defined cultures of renal tubules epithelia: a role for cysteine proteases.' *Cell Biol. Int. Rep.*, 1991, **15**, 1243–1258.

80. S. J. Borghoff, B. G. Short and J. A Swenberg, 'Biochemical mechanisms and pathology of alpha 2u-globulin nephropathy.' *Annu. Rev. Pharmacol. Toxicol.*, 1990, **30**, 349–367.

81. L. D. Lehman-McKeeman, in 'Toxicology of the Kidney,' eds. J. B. Hook and R. S. Goldstein, Raven Press, New York, 1993, pp. 477–494.

82. J. A. Swenberg, 'Alpha 2u-globulin nephropathy: review of the cellular and molecular mechanisms involved and their implications for human risk assessment.' *Environ. Health Perspect.*, 1993, **101**, 39–44.

83. R. L. Melnick, 'Mechanistic data in scientific public health decisions.' *Regul. Toxicol. Pharmacol.*, 1992, **16**, 109–110.

84. X. Yang and R. G. Schnellmann, 'Proteinases in renal cell death.' *J. Toxicol. Environ. Health*, 1996 **48**, 319–332.

85. T. C. Saido, H. Sorimachi and K. Suzuki, 'Calpain: new perspectives in molecular diversity and physiology–pathological involvement.' *FASEB J.*, 1994, **8**, 814–822.

86. C. L. Edelstein, E. Wieder, P. Gengaro *et al.*, 'Hypoxia-induced calpain activation in rat proximal tubules.' *J. Am. Soc. Nephrol.*, 1994, **5**, 896.

87. S. S. Sarang and R. G. Schnellmann, 'Measurement of calpain activity *in situ* in renal proximal tubules (RPT) exposed to antimycin A.' *J. Am. Soc. Nephrol.*, 1995, **6**, 1004.

88. L. J. Mandel, 'Metabolic substrates, cellular energy production, and the regulation of proximal tubular transport.' *Annu. Rev. Physiol.*, 1985, **47**, 85–101.

89. S. R. Gullans and S.C Hebert, in 'The Kidney,' eds. B. M. Brenner and F. C. Rector, Jr., Saunders, Philadelphia, PA, 1991, pp. 76–109.

90. R. G. Schnellmann and R. G. Griner, in 'Mechanisms of Injury in Renal Disease and Toxicity,' ed. R. S. Goldstein, CRC Press, Boca Raton, FL, 1994, pp. 247–266.

91. R. G. Schnellmann, in 'Methods in Toxicology,' eds. C. A. Tyson and J. M. Frazier, Academic Press, Florida, 1994, pp. 128–129.

92. C. E. Kobryn and L. J. Mandel, 'Decreased protein phosphorylation induced by anoxia in proximal renal tubules.' *Am. J. Physiol.*, 1994, **267**, C1073–C1079.

93. R. G. Schnellmann, J. W. Griffin and S. S. Sarang, 'Glycine and strychnine do not block antimycin A-induced protein dephosphorylation.' *J.Am. Soc. Nephrol.*, 1995, **6**, 1004.

94. R. G. Schnellmann, 'Protein phosphatase inhibitors prevent antimycin-A induced renal cell death.' *Toxicologist*, 1995, **15**, 296.

95. R. G. Schnellmann, E. A. Lock and L. J. Mandel, 'A mechanism of *S*-(1,2,3,4,4-pentachloro-1,3-butadienyl)-L-cysteine toxicity to rabbit renal proximal tubules.' *Toxicol. Appl. Pharmacol.*, 1987, **90**, 513–521.

96. A. Wallin, T. W. Jones, A. E.Vercesi *et al.*, 'Toxicity of *S*-pentachlorobutadienyl-L-cysteine studied with isolated rat renal cortical mitochondria.' *Arch. Biochem. Biophys.*, 1987, **258**, 365–372.

97. R. G. Schnellmann, T. J. Cross and E. A. Lock, 'Pentachlorobutadienyl-L-cysteine uncouples oxidative phosphorylation by dissipating the proton gradient.' *Toxicol. Appl. Pharmacol.*, 1989, **100**, 498–505.

98. P. J. Hayden and J. L. Stevens, 'Cysteine conjugate toxicity, metabolism, and binding to macromolecules in isolated rat kidney mitochondria.' *Mol. Pharmacol.*, 1990, **37**, 468–476.

99. J. M. Weinberg, P. G. Harding and H. D. Humes, 'Mitochondrial bioenergetics during the initiation of mercuric chloride-induced renal injury. I. Direct effects of *in vitro* mercuric chloride on renal mitochondrial function.' *J. Biol. Chem.*, 1982, **257**, 60–67.

100. J. M. Weinberg, P. D. Harding and H. D. Humes, 'Mitochondrial bioenergetics during the initiation of mercuric chloride-induced renal injury. II. Function of alterations of renal cortical mitochondria isolated after mercuric, chloride treatment.' *J. Biol. Chem.*, 1982, **257**, 68–74.

101. J. A. Gordon and V. H. Gattone, 'Mitochondrial alterations in cisplatin-induced acute renal failure.' *Am. J. Physiol.*, 1986, **250**, F991–F998.

102. R. Safirstein, J. Winston, M. Goldstein *et al.*, 'Cisplatin nephrotoxicity.' *Am J. Kidney. Dis.*, 1986, **8**, 356–367.

103. H. R. Brady, B. C. Kone, M. E. Stromski *et al.*, 'Mitochondrial injury: an early event in cisplatin toxicity to renal proximal tubules.' *Am. J. Physiol.*, 1990, **258**, F1181–F1187.

104. V. G. Lockard, R. D. Phillips, A. W. Hayes *et al.*, 'Citrinin nephrotoxicity in rats: a light and electron microscopic study.' *Exp. Mol. Pathol.*, 1980, **32**, 226–240.

105. M. D. Aleo, R. D. Wyatt and R. G. Schnellmann, 'The role of altered mitochondrial function in citrinin-induced toxicity to rat renal proximal tubule

suspensions.' *Toxicol. Appl. Pharmacol.*, 1991, **109**, 455–463.

106. G. M. Chagas, A. P. Campello and M. L. Kluppel, 'Mechanism of citrinin-induced dysfunction of mitochondria. I. Effects on respiration, enzyme activities and membrane potential of renal cortical mitochondria.' *J. Appl. Toxicol.*, 1992, **12**, 123–129.

107. G. M. Chagas, M. A. Oliveira, A. P. Campello *et al.*, 'Mechanism of citrinin-induced dysfunction of mitochondria. III. Effects on renal cortical and liver mitochondrial swelling.' *J. Appl. Toxicol.*, 1995, **15**, 91–95.

108. G. M. Chagas, M. A. Oliveira, A. P. Campello *et al.*, 'Mechanism of citrinin-induced dysfunction of mitochondria. IV—Effect on Ca^{2+} transport.' *Cell Biochem. Funct.*, 1995, **13**, 53–59.

109. J. H. Moore and B. Truelove, 'Ochratoxin A: inhibition of mitochondrial respiration.' *Science*, 1970, **168**, 1102–1103.

110. S. Suzuki, Y. Kozuka, T. Satoh *et al.*, 'Studies on the nephrotoxicity of ochrotoxin A in rats.' *Toxicol. Appl. Pharmacol.*, 1975, **34**, 479–490.

111. M. D. Aleo, R. D. Wyatt and R. G. Schnellmann, 'Mitochondrial dysfunction is an early event in ochratoxin A but not oosporein toxicity to rat renal proximal tubules.' *Toxicol. Appl. Pharmacol.*, 1991, **107**, 73–80.

112. B. M. Tune, in 'Toxicology of the Kidney,' eds. J. B. Hook and R. S. Goldstein, Raven Press, New York, 1993, pp. 257–282.

113. G. F. Rush and G. D. Ponsler, 'Cephaloridine-induced biochemical changes and cytotoxicity in suspensions of rabbit isolated proximal tubules.' *Toxicol. Appl. Pharmacol.*, 1991, **109**, 314–326.

114. M. D. Aleo, G. O. Rankin, T. J. Cross *et al.*, 'Toxicity of *N*-(3,5-dichlorophenyl) succinimide and metabolites to rat renal proximal tubules and mitochondria.' *Chem. Biol. Interact.*, 1991, **78**, 109–121.

115. A. Leaf, 'On the mechanism of fluid exchange of tissues *in vitro*.' *Biochem. J.*, 1956, **62**, 241–248.

116. A. Leaf, 'Maintenance of concentration gradients and regulation of cell volume.' *Ann. NY Acad. Sci.*, 1959, **72**, 386–404.

117. G. W. Miller and R. G. Schnellmann, 'Cytoprotection by inhibition of chloride channels: the mechanism of action of glycine and strychnine.' *Life Sci.*, 1993, **53**, 1211–1215.

118. G. W. Miller and R. G. Schnellmann, 'Inhibitors of renal chloride transport do not block toxicant-induced chloride influx in the proximal tubule.' *Toxicol. Lett.*, 1995, **76**, 179–184.

119. S. L. Waters and R. G. Schnellmann, 'Extracellular acidosis and chloride channel inhibitors act in the late phase of cellular injury to prevent death.' *J. Pharmacol. Exp. Ther.*, 1996, **278**, 1012–1017.

120. P. R. Steinmetz and R. F. Husted, in 'Nephrotoxic Mechanisms of Drugs and Environmental Toxins,' ed. J. Stein, Plenum, New York, 1982 pp. 95–100.

121. F. Z. Gil and G. Malnic, 'Effect of amphotericin B on renal tubular acidification in the rat.' *Pflugers Arch.*, 1989, **413**, 280–286.

122. M. A. Carlson and R. E. Condon, 'Nephrotoxicity of amphotericin B.' *J. Am. Coll. Surgeons*, 1994, **179**, 361–381.

123. B. C. Kone, M. Kaleta and S. R. Gullans, 'Silver ion (Ag^+)-induced increases in cell membrane K^+ and Na^+ permeability in the renal proximal tubule: reversal by thiol reagents.' *J. Membr. Biol.*, 1988, **102**, 11–19.

124. B. C. Kone, R. M. Brenner and S.R Gullans, 'Sulfhydryl-reactive heavy metals increase cell membrane K^+ and Ca^{2+} transport in renal proximal tubule.' *J. Membr. Biol.*, 1990, **113**, 1–12.

125. W. B. Reeves and S. V. Shah, 'Activation of potassium channels contributes to hypoxic injury in proximal tubules.' *J. Clin. Invest.*, 1994, **94**, 2289–2294.

126. B. Van de Water, J. J. Jaspers, D. H. Maasdam *et al.*, '*In vivo* and *in vitro* detachment of proximal tubular cells and F-actin damage: consequences for renal function.' *Am. J. Physiol.*, 1994, **267**, F888–F899.

127. B. A. Molitoris, in 'Acute Renal Failure,' 3rd edn., eds. J. M. Lazarus and B. M. Brenner, Churchill Livingstone, New York, 1993, pp. 1–32.

128. B. A. Molitoris, 'Na(+)-K(+)-ATPase that redistributes to the apical membrane during ATP depletion remains functional.' *Am. J. Physiol.*, 1993, **265**, F693–F697.

129. J. Vina, 'Glutathione Metabolism and Physiological Functions,' CRC Press, Boca Raton, FL, 1990.

130. D. C. Liebler, 'The role of metabolism in the antioxidant function of vitamin E.' *Crit. Rev. Toxicol.*, 1993, **23**, 147–169.

131. L. Packer and J. Fuchs, 'Vitamin E in Health and Disease,' Marcel Dekker, New York, 1993.

132. R. C. Rose and A. M. Bode, 'Biology of free radical scavengers: an evaluation of ascorbate.' *FASEB J.*, 1993, **7**, 1135–1142.

133. H. E. Sauberlich, 'Pharmacology of vitamin C.' *Annu. Rev. Nutr.*, 1994, **14**, 371–391.

134. J. M. Weinberg, M. A. Venkatachalam, R. Garzo-Quintero *et al.*, 'Structural requirements for protection by small amino acids against hypoxic injury in kidney proximal tubules.' *FASEB J.*, 1990, **4**, 3347–3354.

135. L. J. Mandel, R. G. Schnellmann and W. R. Jacobs, 'Intracellular glutathione in the protection from anoxic injury in renal proximal tubules.' *J. Clin. Invest.*, 1990, **85**, 316–324.

136. G. W. Miller, E. A. Lock and R. G. Schnellmann, 'Strychnine and glycine protect renal proximal tubules from various nephrotoxicants and act in the late phase of necrotic cell injury.' *Toxicol. Appl. Pharmacol.*, 1994, **125**, 192–197.

137. M. D. Aleo and R. G. Schnellmann, 'The neurotoxicants strychnine and bicuculline protect renal proximal tubules from mitochondrial inhibitor-induced cell death.' *Life Sci.*, 1992, **51**, 1783–1787.

138. G. W. Miller and R. G. Schnellmann, 'A novel low-affinity strychnine binding site on renal proximal tubules: role in toxic cell death.' *Life Sci.*, 1993, **53**, 1203–1209.

139. G. W. Miller and R. G. Schnellmann, 'A putative cytoprotective receptor in the kidney: relation to the neuronal strychnine-sensitive glycine receptor.' *Life Sci.*, 1994, **55**, 27–34.

140. G. W. Miller, B. W. Newton and R. G. Schnellmann, 'Immunohistochemical localization of the cytoprotective glycine/strychnine receptor in rabbit kidney cortex.' *Toxicologist*, 1995, **15**, 295.

141. J. C. Nichols, S. F. Bronk, R. L. Mellgren *et al.*, 'Inhibition of nonlysosomal calcium-dependent, proteolysis by glycine during anoxic injury of rat hepatocytes.' *Gastroenterology*, 1994, **106**, 168–176.

142. D. P. Rodeheaver and R. G. Schnellmann, 'Extracellular acidosis ameliorates metabolic-inhibitor-induced and potentiates oxidant-induced cell death in renal proximal tubules.' *J. Pharmacol. Exper. Ther.*, 1993, **265**, 1355–1360.

143. J. V. Bonventre and J. Y. Cheung, 'Effects of metabolic acidosis on viability of cells exposed to anoxia.' *Am. J. Physiol.*, 1985, **249**, C149–C159.

144. J. M. Weinberg, *J. Clin. Invest.*, 'Oxygen deprivation-induced injury to isolated rabbit kidney tubules.' 1985, **76**, 1193–1208.

145. R. G. Schnellmann, R. S. Counts, G. W. Miller *et al.*, 'Temporal aspects of the cytoprotection produced by extracellular acidosis.' *Pharmacologist*, 1993, **35**, 146.

7.14
The Renal Medulla and Distal Nephron Toxicity

PETER H. BACH

University of East London, UK

7.14.1 INTRODUCTION

The medulla and its corticomedullary (deep) and distal nephrons are the least easily accessible part of the kidney and have, therefore, been less well investigated than the rest of the organ. In addition, the inner medulla accounts for a small fraction of the renal mass and therefore requires extensive care to assess its pathology (especially the papilla tip) and it provides little tissue for *in vitro* studies. This, in part, explains why many of the lesions that afflict this region of the kidney are still not adequately investigated or well understood.

7.14.2 THE MEDULLA STRUCTURE AND FUNCTION

An in-depth review of the medulla and distal nephron is beyond the scope of this chapter, but such knowledge is essential to understand how toxicants affect it. The medulla has a number of unique morphological,[1–6] functional,[7–11] and biochemical[12–15] features. The medulla can be divided into the outer medulla, made up of the thin descending and the thick ascending limbs of the loops of Henle, collecting ducts, the vasa recta and a dense capillary network, and the inner medulla, the papilla, containing the thin limbs of the loops of Henle, collecting ducts, the vasa recta, and a diffuse network of capillaries.[1–4] Packed into the spaces between these structures are interstitial cells embedded in a matrix rich in glycosaminoglycans.[2,3] The collecting ducts terminate as the ducts of Bellini around the tip of the papillae. Whereas the mouse, gerbil, rat, guinea pig, rabbit, dog, cat, and primate kidney has only a single papillae, the pig and man have multipapillate kidneys. There are between nine and 20 papilla in each human kidney.[2–4]

Urine concentration is a vital part of medullary function as less than 1% of the glomerular filtrate leaves the kidney as urine (unless there is a state of diuresis), the remainder having been reabsorbed. The concentration of urine is complex, and depends (at least in part) on the countercurrent multiplier system which establishes a steep osmotic gradient across the inner medulla.[2,3,7] The high osmolality is a consequence of the differential permeability of the limbs of the loops of Henle and the collecting ducts to water and ions. The ascending limb is thought to have an active transport mechanism which pumps sodium out of the lumen and into the interstitium, but remains impermeable to water. As a consequence the osmolality decreases in this part of the tubule. The descending limb, however, is freely permeable to water, but not sodium ions. The high ion concentration in the interstitium would draw water out of the descending limb, increasing the osmolality towards the turn of the U-loop. This is probably augmented by urea which leaves the collecting ducts, and enters the descending limb via the interstitium. The collecting ducts regulate the final urine concentration by controlling the amount of water that is reabsorbed. The countercurrent exchange associated with the loops of Henle arising from cortical glomeruli offers an important "barrier" zone which is thought to facilitate solute trapping in, and solvent exclusion from, the inner medulla, and thus help to maintain the hyperosmolality in this "compart-

ment."[2,3,7,8,14–16] There are a number of other factors which control, alter or contribute to the urine concentrating process.

The biological roles played by the medullary interstitial cells remain uncertain, but they contribute to carbohydrate metabolism and lipogenesis in the renal medulla which are specialised processes and linked to the high osmolality in this region of the kidney.[4–8,14–16]

There are, however, aspects of this region of the kidney that require additional commentary to put this chapter into context.

7.14.2.1 The Interstitial and Epithelial Cells

The "Type 1" interstitial cells are an important part of the medulla as they occupy 10–20% of tissue volume in the outer medulla and 40% near the papilla tip.[1–5] They produce the glycosaminoglycan matrix which surrounds them, the functions of which are critical for water reabsorption (see Bach and Bridges[16]). One of the most characteristic features associated with the medullary interstitial cells are the numerous cytoplasmic lipid droplets, which occupy 2–4% of the total cell volume. These droplets are mainly phospholipids and large amounts of unsaturated fatty acids, particularly arachidonic acid in the rat.[17–20] Interstitial cells affect those renal functions (discussed in Bohman and Mandal[5]), which are controlled through prostaglandins. These cells may also help regulate blood pressure with antihypertensive renomedullary lipids.[21]

In addition to the microvasculature and the thin loops of Henle, there are two populations of epithelial cells present.[1–4] The connecting epithelial cells, are known to have some heterogeneity in terms of the "dark" and "light" cells,[4] and, in addition, the papilla tip is covered by a single-cell thick layer of epithelial or urothelial cells, referred to as the covering epithelia.

7.14.2.2 Renal Prostaglandins

The biophysiological roles of renal prostanoids are well defined.[22–27] In the absence of cytochrome P450[28,29] this enzymic system has another important role to play in medullary metabolism, as it is responsible for xenobiotic oxidation through prostaglandin hydroperoxidase[30–34] and the other peroxidases.

7.14.3 ANALGESIC NEPHROPATHY

Analgesic nephropathy is best described as the degenerative renal condition leading to renal failure that follows long-term analgesic abuse. While the clinical syndrome has pyelo-nephritis as one of its hallmarks, the underlying condition is a renal papillary necrosis. Renal papillary necrosis (RPN) in man has been associated with chronic consumption or long-term abuse of mixed analgesics, and clinical doses of nonsteroidal anti-inflammatory drugs (NSAID) and other therapeutic agents[16,35–47] (Table 1). The reader is referred to Chapter 30, this volume, for a more in-depth review of analgesic nephropathy.

There are few clinical symptoms associated with the early development of analgesic asso-ciated RPN.[16,35–47] The progression of renal damage is insidious and renal function may be severely compromised before the condition become obvious. Early symptoms include dependence, emotional instability, anxiety, headaches, introversion, and neurosis; upper gastrointestinal disease, such as peptic ulcera-tion of the stomach and duodenum, dyspepsia, and anaemia as a result of gastrointestinal tract (GIT) bleeding, hemolysis, and iron deficiency. Intermediate symptoms include urinary tract disease, such as bacteriuria, sterile pyuria, nocturia, dysuria, microscopic hematuria, uret-eral colic, lower back pains; urinanalysis shows a defect in ability to concentrate and acidify. The late symptoms, such as hypertension, cardiovascular manifestations, ischemic heart disease, peripheral vascular disease, renal

Table 1 NSAID and other therapeutic agents reported to have caused RPN in humans.

NSAID linked to RPN man	Other medicines linked to RPN in man
Benoxaprofen[39]	Pentazocine and aspirin[46]
Indometacin [39]	Rifampicin[47]
Tolmetin [39]	
Naproxen[39]	
Ketoprofen[39]	
Sulindac[39]	
Diclofenac sodium[40]	
Fenoprofen[41]	
Flurbiprofen[42]	
Indomethacin[43]	
Ibuprofen[44]	
Naproxen[45]	

calculi and bladder stones, and decreased glomerular filtration rate, increased blood urea nitrogen, renal tubular acidosis, and carcinoma of renal tract all can have other causes. Even radiological examination may not identify papillary necrosis (if the necrosed papillae remains *in situ*), and even when loss of the papillae is obvious (in the presence or absence of other degenerative renal changes) the under-lying cause still has to be established.[16,35–38] The epidemiology of the condition is less

Table 2 Chemicals with papillotoxic potential in animals.

Chemical	Species	Ref.
2,2-Bis(bromomethyl)-1,3-propanediol	F344/N rats	64
	B6C3F1 mice	64
Cyclophosphamide (intravesical instillation of acetic acid	Rat	65
D-Ormaplatin (tetraplatin)	Fischer-344	66
Diphenylamine-induced	Syrian hamster, Sprague-Dawley rat and Mongolian Gerbils	67
Ethoxyquin	Fischer-344 rats	68
Formaldehyde	Rats	69
Mesalazine	Cynomolgus monkeys	70
Mesalazine	Dogs and rats	71
Nefiracetam (N-(2,6-dimethyl-phenyl)-2-(2-oxo-1-pyrrolidinyl) acetamide)	Beagle dogs	72
1-Naphthol	Charles River CD1 mice	73
N-Phenylanthranilic acid	Sprague-Dawley rat	74
Phenylbutazone	Male Wistar	75
Phenylbutazone and water deprivation	Horses	76
Triethanolamine	Fischer-344 rats	77
2-Chloroethanamine	Wistar rat	78
3-Bromopropanamine	Wistar rat	78
2-Chloro-N,N-dimethylethanamine	Wistar rat	78
L-Triiodothyronine	SKF Wistar rat	79

certain than it might be because of the diffuse symptoms.[16,48,49]

It is now agreed that the primary lesion of analgesic abuse is in the medulla, where the "fine elements," such as the interstitial cells, endothelia, and loops of Henle are the earliest affected parts, and subsequent degenerative cortical changes lead to renal functional compromise and end-stage renal disease.[16,48,49]

7.14.4 EXPERIMENTAL MODELS OF PAPILLARY NECROSIS

In order to understand the molecular basis of papillary necrosis and the associated nephropathy considerable research has been focused on the development of animal models.[50–100]

7.14.4.1 Analgesic- and NSAID-induced Papillary Necrosis in Animals

The administration of analgesics and NSAIDs to animals does not produce a robust, reproducible model RPN. This may be due, in part, to gastric ulceration that is a frequent and often fatal early consequence of NSAID dosing.[38] More importantly, there appear to be unidentified and uncontrolled factors that make biological variability very large both within and between experimental groups. Nevertheless, suprapharmacological doses of mixed[50] and single analgesics,[51,52] and other drugs and several different chemicals[16,38,50–100] have been used to produce RPN in experimental animals (Table 2).

Table 3   Most popular animal models of renal papillary necrosis.

Papillotoxin	Administration	Advantages	Disadvantages
2-Bromoethanamine hydrobromide	Fresh solution i.v., i.p., s.c., or p.o., but dose needs to be adjusted accordingly	Dose related, rapidly induced lesion (24–48 h) that affects up to the whole medulla in all animals so far investigated. The best described of the model lesions	Metabolism not defined. There are also subtle proximal tubular changes
Diphenylamine	Gavage using a suspension	Dose related lesion that appears to be confined to the papilla tip or the central zone of the medulla. Lesions develop in 7–21 d	Relatively poorly studied. Also causes marked proximal tubule necrosis
Ethyleneimine	Administered by i.v. or i.p. injection	Relatively well studied, but most of the published data is before 1970	Explosive and mutagenic properties have stopped the use of this compound. The unstable nature of the compound also means that there is some uncertainty regarding what dose and compound was administered in published data
N-Phenylanthranilic acid	Gavage as a suspension	Dose related lesion that appears to be confined to the papilla tip or the central zone of the medulla. Lesions develop in 7–21 d	Rather slow to produce the lesion. Relatively poorly studied. Also causes proximal tubule necrosis
Sodium *N*-phenylanthranilate	Given by i.p. injection	Dose related papillary necrosis caused in 24–48 h	Poorly investigated model, that causes some short-lived pharmacological effects of muscular weakness and respiratory depression in rodents

Data from Refs. 16, 53, 56–63, 67, 74, 78, 80, 86–98, 100, 107–111, and from Bach, Gregg, and Hardy using unpublished sources.

7.14.4.2 The Use of Nonanalgesic- and Non-NSAID-induced Models of Papillary Necrosis

The greatest success in the study of RPN has been gained from the use of a limited number (Table 3) of compounds that produce a model lesion in laboratory animals. Administration of these compounds to rats satisfy the requirements of experimental pathology,[16,53] and cause all of the most important physiological and pathological changes described in man.[4,16,38] This includes increased and decreased medullary mucopolysaccharide matrix staining[38,52,57,60,61] and the presence of Oil Red "O" positive lipid material, which is especially marked in epithelial cells.[38,54–56,99] Essentially similar changes are seen in mucopolysaccharide staining[4,48,49] and lipid staining[4,49,62] in the kidneys of human analgesic abusers. These models are highly relevant to facilitate the study of the progression of papillary necrosis from the first cell-specific injury, through a series of secondary changes to marked damage in the cortex.

Ethyleneimine was the first of the prototype compounds studied, but its explosive instability and powerful mutagenic and alkylating activity[63] has stopped its use (see Bach and Bridges[16] for review). 2-Bromoethanamine (BEA), is now the most widely used model, but there are also other compounds[64–79] that have been reported to cause this lesion (Table 2).

7.14.4.3 The Relative Papillotoxicity of Analgesic- and NSAID vs. Model Compounds

The problem of putting much of the experimental data into context is the paucity of data from which the papillotoxic potential of each chemical could be estimated. Carlton and Engelhardt[80] have compared the papillotoxic effects of several chemicals in Syrian hamsters. BEA ($75 \, mg \, kg^{-1}$) caused RPN in all animals, whereas $100–400 \, mg \, kg^{-1}$ mefenamic acid only affected 40% of animals. A few hamsters given $100–400 \, mg \, kg^{-1}$ indomethacin, but none given up to $400 \, mg \, kg^{-1}$ acetaminophen (paracetamol) or up to $600 \, mg \, kg^{-1}$ phenylbutazone developed renal papillary lesions. This suggests a ranking of the papillotoxicity as BEA ≫ mefenamic acid ≫ indomethacin ≫ acetaminophen (paracetamol) or phenylbutazone. The relevance of this ranking to the human risk of developing papillary necrosis is uncertain, as the role of acetaminophen is controversial. Some reports describe papillary lesions in rats dosed with this compound,[51] and the Gunn rat is uniquely sensitive to acetaminophen,[81–84]

and salicylate,[81,82,85] but it is questionable if the use of this strain is relevant to man. The propensity of acetaminophen or salicylate to cause RPN in man does not appear to be high.

7.14.5 2-BROMOETHANAMINE HYDROBROMIDE MODEL OF PAPILLARY NECROSIS

7.14.5.1 Morphological Changes Caused by BEA

BEA has been widely investigated.[16] A single dose causes a dose-related (Figure 1) lesion in 24–48 h that closely parallels the early morphological and functional changes reported in animals dosed with analgesics and NSAIDs and human analgesic abusers.[16,35,48,49,38,53,57] Repeated dosing causes a lesion up to, but not beyond the corticomedullary junction.[57] Morphological changes include altered staining intensity of the proteoglycan–glycosaminoglycan (or mucopolysaccharide) matrix, which increases[56–61] within 2–4 h, becomes diffuse after 8–12 h, and is lost as necrosis develops. The proteoglycan–glycosaminoglycan matrix appears to be totally absent from 24–48 h (Figure 2). A similar series of alterations in the proteoglycan–glycosaminoglycan matrix has been described in human renal tissue from analgesic abusers.[4,48,49]

The first cell type to undergo degenerative changes are the medullary interstitial cells at the tip of the papilla (Figures 1 (bottom) and 3); subsequently, more of the interstitial cells are affected towards the corticomedullary junction (Figure 1 (top)). It is not until about 12 h that collecting duct epithelia and other areas of the distal nephron show degenerative changes and more widespread necrosis is also apparent.

The cortex is also affected, characterized by hydropic changes in proximal tubules [57] and the progressive loss of alkaline phosphatase, γ-glutamyl transpeptidase, and adenosine triphosphatase from the brush border starting at 8 h.[60] Papillary necrosis is associated with the progressive deposition of neutral lipid material in the capillaries, collecting duct, and covering epithelial cells (Figure 4). Lipid staining extends into the outer medulla (Figure 5) which would appear as normal by hematoxylin and eosin (H&E) staining.[38,56,57] The staining of epithelial and microvascular cells for neutral lipid appears to be pathognomonic for RPN, as it also occurs in the pig, baboon, marmoset, and mice following BEA- (Bach and Gregg, unpublished) and aspirin-induced[54] RPN. By contrast chemicals that affect the cortex (e.g., hexachlorobutadiene, aminoglycosides, cisplatin

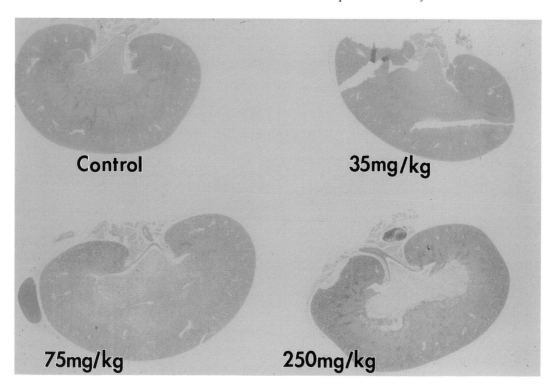

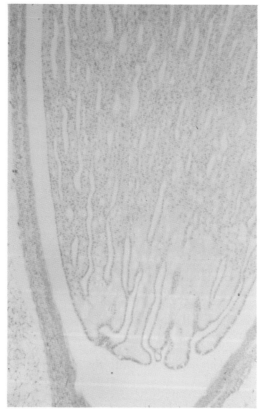

Figure 1 (Top) Rat kidneys 24 h after treatment with BEA. The treatment is control, $35\,mg\,kg^{-1}$ which causes a focal lesion, $75\,mg\,kg^{-1}$ which causes necrosis of the papilla tip and $250\,mg\,kg^{-1}$ which causes necrosis of the entire medulla. H&E ×4 objective. (Bottom) Rat medulla 24 h after treatment with $35\,mg\,kg^{-1}$ BEA which causes a focal lesion of the papilla tip. H&E ×64 objective. (Next page) Rat kidneys 30 d after treatment with $100\,mg\,kg^{-1}$ BEA which causes secondary degenerative changes that affect the cortex. H&E ×64 objective.

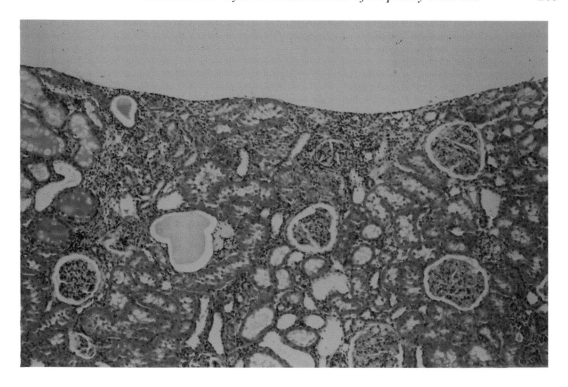

Figure 1 (continued).

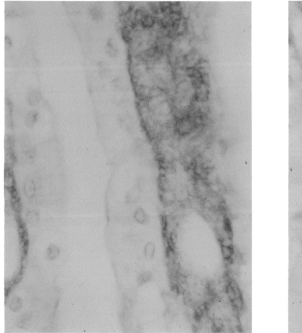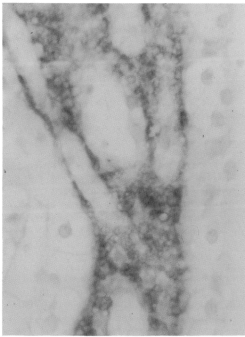

Figure 2 (Left) Control rat kidney showing glycosaminoglycan matrix in between the covering epithelium and collecting ducts, that makes up the mucopolysaccharide, and is stained by Hale's Colloidal Iron ×64 objective. (Right) Rat medulla 6h after treatment with 100 mg kg⁻¹ BEA showing increased staining intensity of mucopolysaccharide and a foamy appearance which precedes loss of staining. Colloidal Iron ×64 objective.

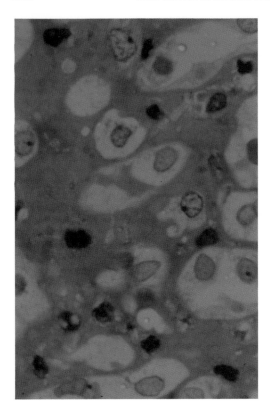

Figure 3 Pyknotic interstitial cell nuclei in the papilla tip, 4 h post BEA (100 mg kg^{-1}), ×1050 magnification.

and polybrene) do not produce specific localized lipid changes in the medullary epithelial cells.[56]

It has generally been assumed that early microvascular changes play a role in medullary anoxia and necrosis, but this does not appear to be the case in BEA-induced RPN. Platelet adhesion is not seen before 8 h after a single dose of BEA. Platelets then increase markedly, but only in those capillaries adjacent to areas where necrosis of the interstitial cells or other fine elements of the medulla had taken place.[61] Colloidal carbon can be used to show microvascular filling, and how this changes as a result of renal injury. After BEA administration there is an early shift of cortical microvascular filling with colloidal carbon to demonstrate shunting of blood to the outer medulla (2–4 h after dosing), whereas at 8–26 h there was a shift of microvascular filling to the inner medulla, which coincided with papillary necrosis. While the necrosed medulla is avascular 48 h after BEA administration, the microvasculature is always patent in the medullary tissue beyond the regions in which necrosis had occurred, confirming an acute medullary necrosis without capillary occlusion.[57] Monastral Blue B staining demonstrates a fully maintained capil-

lary integrity and the absence of any plasma leakage into the interstitium.[38,61]

7.14.5.2 Functional Changes Caused by BEA

The renal functional changes following BEA administration closely parallel those reported for human analgesic abusers[16,53] and include loss of urinary concentrating ability,[57,88–94] electrolyte wasting,[88–94] and severe cortical degeneration which is a late, but consistent, secondary consequence of the medullary lesion.[16,88–93] Specifically, the early changes include increased urea and Na$^+$ excretion and decreased osmolality, glomerular filtration rate, and *p*-aminohippurate clearance,[96] many of which appear to have their genesis within the first few hours after dosing with BEA.

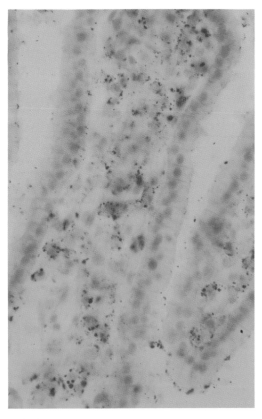

Figure 4 (This page) Control papilla stained with Oil Red "O" showing the lipid droplets present in the renal medullary interstitial cells and absent from the epithelial cells (Oil Red "O" ×25 objective). (Next page, top) The earliest abnormal capillary changes 7 h after BEA (100 mg kg^{-1}) treatment, showing increased lipid in capillaries adjacent to normal capillaries and renal medullary interstitial cells (Oil Red "O" ×25 objective). (Next page, bottom) Mid-papilla 24 h after BEA (100 mg kg^{-1}) treatment showing Oil Red "O" staining capillaries, extensive lipid staining of nephrons and some apparently normal nephron cells (×25 objective).

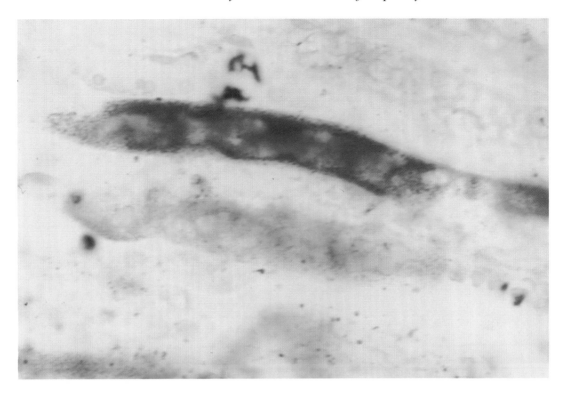

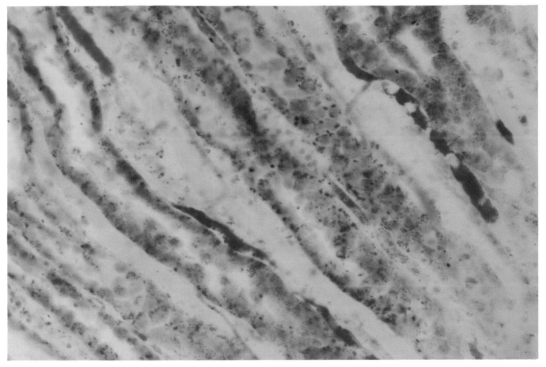

Figure 4 (continued).

7.14.5.3 Strain and Species Response to BEA

BEA targets selectively for the renal medulla, with minimal to slight changes in the cortex within the first few days of treatment in all rodent strains so far investigated: Wistar, Fischer 344, Sprague-Dawley, Holtzman, Donrju, Gunn, Battleboro, and the Munich Wistar Froemter rats.[16] BEA has similar renal effects in the marmoset (Gregg and Bach,

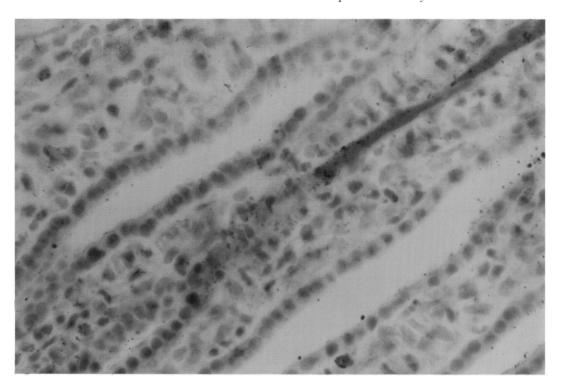

Figure 5 Lipid staining capillaries in cortico-medullary region of the kidney which was morphologically normal by H&E (Oil Red 'O' × 10 objective).

unpublished observation), CD-1, Schnider (Bach and Gregg, unpublished observation), obese, Balb/c and C57Bl/6 mice,[97] Mongolian gerbil,[98] Syrian hamster,[80] domestic pig,[99] and baboon (Bach, unpublished observation).

Interestingly, male MF1-nu/nu/Ola/Hsd nude mice have an atypical response to BEA.[100] Morphological changes in the medulla of these mice closely parallel the early interstitial cell matrix changes, PAS-positive inclusion bodies in the collecting duct cells, followed by collecting duct and interstitial cell necrosis as have been reported in the rat.[56-61] There were, however, marked changes in the cortex including total necrosis of the S_2 and S_3 segments, followed by re-epithelialization of the S_2 segment which precedes the loss of covering epithelium and collecting duct epithelial cells in the papilla. This proximal tubular necrosis was similar to that caused by *p*-aminophenol[101] or hexachloro-1,3-butadiene.[102] Similar to MF1-nu/nu/Ola/Hsd nude mice these changes have also been reported in the Swiss ICR mouse.[59] The significance of the shift of target selectivity to include the cortex in nude and Swiss ICR mice awaits further studies.

7.14.5.4 Metabolism of BEA

There are limited published data on the metabolism of BEA, and then only in the rat.

Following intraperitoneal (i.p.) administration to rats BEA is totally absorbed with an extensive flux into the stomach. The BEA-derived material is distributed to the bladder, stomach, upper GIT, and kidney, and significant amounts are also present in the liver.[103] Murray[104] suggested the cyclization of BEA to ethyleneimine as a mandatory step in the process of RPN, a hypothesis that has been accepted, but not proven. There are some similarities between the kinetics of ethyleneimine[105] in the rat and BEA, but also marked differences. The gastric pH would stabilize BEA and prevent cyclization, but only as long as the molecule is in the stomach. The limited chromatographic data reported[105] exclude metabolic components common to BEA and ethyleneimine. The absence of volatile BEA-derived radiolabeled components in urine and breath of rats (Bach, unpublished data) suggests that no ethyleneimine is excreted *per se*. It cannot, however, exclude the possibility that BEA is cyclized to the proximate or ultimate papillotoxin (ethyleneimine) intracellularly, where it undergoes rapid metabolism once formed.

7.14.5.5 The Mechanism of BEA Target Selective Toxicity

The molecular basis of BEA target selectivity remains controversial, and opinion is divided between vascular injury and direct and selective medullary interstitial cell cytotoxicity.

7.14.5.5.1 Vascular effects

Much of the early literature (see Bach and Bridges;[16] Bach and Gregg[38] for reviews) highlighted microvascular degenerative changes in models of RPN. Many of these were chronically induced lesions where the ability to differentiate microvascular changes as "cause" or "consequence" of RPN is questionable. The acute nature of the BEA-induced lesion does, however, offer the potential to define the relationship between microvascular injury and RPN.

Microangiographic studies in the rat[106] revealed reduced vasa recta perfusion, and tubular injection studies showed unobstructed tubules and collecting ducts supporting vasoconstriction. This study did, however, use high doses of BEA and only assessed the microvasculature at an advanced stage of renal injury. Wolf[98] concluded that RPN in the Mongolian gerbil is due to an ischemic necrosis of the inner medulla that develops secondary to endothelial damage of the vasa recta. The high concentrating capacity of this species may, however, not make this strain atypical of the animals in which RPN has been studied. Ultrastructural studies in the Swiss ICR mouse[59] suggest that endothelial injury is the key factor, but both cortical and papillary injury occurs in this mouse strain. This atypical effect of a BEA-induced proximal tubule necrosis in the Swiss ICR mouse strain raises similar questions regarding the validity of this interpretation. All of these studies have used high doses of BEA, the effects of which produce an injury that affects more than the localized papillary necrosis. Therefore, other concentration-related effects may be involved.

Other investigations using lower doses of BEA and shorter time intervals[38,56,57,60,61] were unable to show any role for microvasculature injury in the pathogenesis of RPN. The only microvascular effects seen were platelet adhesion, well after interstitial necrosis had occurred, and the integrity of the medullary microvascular endothelia remained intact even in areas where necrosis had occurred (see Section 7.14.5.1).

7.14.5.5.2 BEA cytotoxicity and effects on interstitial cells

The mutagenicity of BEA and its analogues 3-bromopropanamine, 2-chloroethanamine, and 2-chloro-*N*,*N*-dimethylethanamine,

correlate with their papillotoxicity. This suggested to Powell and co-workers[78] that a nonenzymically formed direct-acting alkylating species mediates these papillary lesions. They explained the target selectivity of haloalkylamine toxicity by an accumulation in interstitial cells. BEA does not, however, affect all cells similarly. It is much more cytotoxic for 3T3 cells and renal medullary interstitial cells than MDCK and HaK cell lines,[107] differences that have been related to the coincidence of both lipid droplets and peroxidative enzyme activity (see Section 7.14.6). The importance of peroxidative metabolism appears to be supported by the arachidonic-acid-dependent metabolism of BEA to a cytotoxic metabolite in cultured rat medullary interstitial cell cultures.[108] These differences also appear to apply *in vivo*, as BEA has a direct effect on the medullary interstitial cell as the earliest focal lesion.[60,61] It is only subsequently that other fine elements of the medulla, such as the loops of Henle, capillaries, and collecting duct cells are affected. The lesion then spreads to include components up to the cortico–medullary junction.[56–61]

7.14.5.6 The Modulation of BEA Toxicity

A number of factors (Table 4) modulate the lesion caused by BEA and other papillotoxins[86,98,109–112] or affect the subsequent renal changes.[113] These studies have been undertaken to test the hypothesis regarding the role of vascular injury, the formation of reactive intermediates, and the role of the concentrating processes in the genesis of the lesion, but thus far these data have provided little additional insight into the mechanism. The administration of enalpril (an angiotensin-converting enzyme inhibitor) does, however, reduce the chronic degenerative changes that follow the papillary lesion in rats[111] over a 12 month period.

7.14.6 THE MECHANISTIC BASIS OF RENAL PAPILLARY NECROSIS

Data derived from the BEA and other models of RPN, together with what is known about the chronically induced lesion in animals and man have been synthesized into what is now regarded as the "text book explanation" for papillary necrosis. The morphological and histochemical changes[56,57,58,60,61] support distinct pathological changes following a papillotoxic insult. The primary morphological changes occur in the interstitial cells in acute,[16,38,60,61] subchronic,[74] and chronic[16] papillary necrosis. This is followed by damage

Table 4   Modulation of renal papillary necrosis and the subsequent cortical changes.

Modulating chemical	Papillotoxin	Species	Ref.
No effect			
Dimethylsulfoxide	2-Bromoethanamine	Mongolian gerbil	98
Saline induced diuresis	2-Bromoethanamine	Mongolian gerbil	98
Piperonyl butoxide	2-Bromoethanamine	Mongolian gerbil	98
Decreased or reduced severity of papillary necrosis			
Dimethylsulfoxide	Diphenylamine	Syrian hamsters,	107
16,16-Dimethyl prostaglandins E_2	Mefenamate	Rat	108
Reserpine	2-Bromoethanamine	Mongolian gerbil	98
Reserpine	2-Bromoethanamine	Rat	86
Diuresis	2-Bromoethanamine	Rat	109
Diuresis	2-Bromoethanamine	Rat	90–93
Potentiates or increased			
Caffeine	Mefenamate	Rat	110
Ameliorated glomerular sclerosis			
Enalpril	2-Bromoethanamine	Rat	111

to the endothelial cells and loops of Henle, and then collecting duct changes. While the subtle degenerative changes in the proximal tubule of rats are not central to the papillary lesion, the exfoliation of brush border and proximal tubular cells are important components of casts that form in the distal nephron. These appear to contribute to marked tubular dilatation, glomerular sclerosis, and the loss of effective renal parenchyma.

Thus, the renal medullary interstitial cells are the primary target of papillotoxicants[16,38,60,61] which appears to be most likely related to local bioactivation by the many peroxidative enzymes (e.g., hydroperoxidase)[30–32] associated with prostaglandin synthesis[16,114] that are predominant in the medulla. The importance of peroxidative metabolism was first identified by a series of elegant investigations from Zenser and co-workers.[115–117] These and subsequent publications showed that a number of different peroxidases convert phenacetin, *p*-phenetidine, and acetaminophen[115–122] to reactive intermediates that bound covalently to protein and nucleic acid (Figure 6). In addition, Mohandas[125] showed low GSH (glutathione) concentrations, low activities of glutathione reductase, selenium-dependent and selenium-independent glutathione peroxidase, and γ-glutamyl transpeptidase in the inner medulla of rabbits. This high level of peroxidative enzymes in the medulla, compared to low levels of detoxifying enzymes, would be expected to make this region of kidney particularly vulnerable to reactive intermediates that were generated locally.

All of the enzymes involved in prostaglandin synthesis are localized to both the medullary

interstitial and collecting duct cells.[123,124] The ubiquitous distribution of peroxidase activities in the medulla make it most difficult to explain why papillotoxicants affect predominantly the medullary interstitial cells. These cells do, however, contain high levels of polyunsaturated fatty acids[17–20] that would predispose to lipid peroxidation if reactive intermediates were generated locally.[114] The importance of the coincidence of lipid droplets and peroxidative activity[16,114] in cultured interstitial cells as the basis for their injury have been presented above (Section 7.14.5.5.2).

Once the interstitial cells are damaged there are degenerative changes in the other "fine elements" of the medulla, such as the capillaries, loops of Henle, and collecting duct.[126] If the degree of injury is sufficient, the cortex also undergoes degenerative change, but the factors involved in this cascade of events are poorly understood. Interstitial cells have a key role in processes such as the synthesis of prostaglandins,[5,18,22–25] antihypertensive factors,[21] the glycosaminoglycan matrix[16] that surrounds them, and supports the other fine elements of the medulla. The loss of these cells causes quite marked and profound degenerative changes in the cortex, which suggests a significant paracrine function.

The interstitial cells appear be so highly differentiated that there is no repair.[57–61,86–87,97–99,111] This highlights the likely secondary impact of their necrosis, and it is relatively easy to identify a situation in man when exposure to papillotoxins over a number of years will progressively erode the medullary interstitial cell population, and produce a similarly slow development of

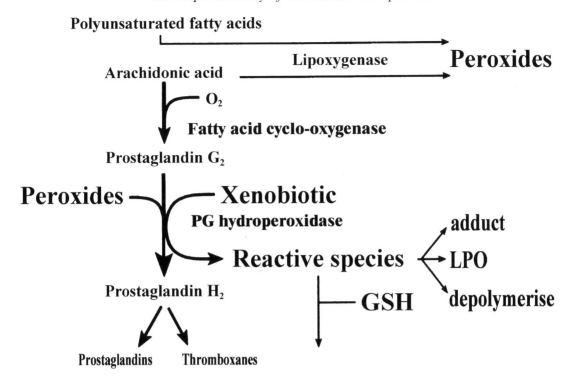

Figure 6 Scheme showing the peroxidase-mediated metabolic activation of chemicals, alternative pathways and possible toxicological consequences.

secondary degenerative changes in the cortex (Figure 7). The presence of multiple papillae in humans, as opposed to only one per kidney in most animals, could serve to reduce the clinical consequences of the toxicity of these compounds in man.

The BEA model provides a basis by which to better understand more subtle criteria that interrelate the different events in the cascade of degenerative changes in the acute, subacute, and then chronic models of RPN. There are a large number of pathophysiological similarities between the acute model of RPN and the chronic analgesic disease in man,[16,38,53] but it may not be valid to extrapolate all of these renal changes to human analgesic abusers. Although this mechanistic explanation[16,114] is attractive, there are a number of anomalies that suggest that the mechanism is more complex. While phenacetin, acetaminophen, and *p*-phenetidine undergo peroxidative activation, none of these compounds have a great propensity to cause RPN in the experimental situation.[115–117] Similarly, while a number of the NSAIDs could undergo peroxidative metabolism there is no data on their metabolism in such systems. There are several chemicals that cause RPN (e.g., L-triiodothyronine, 2,2-bis(bromomethyl)-1,3-propanediol, nefiracetam, D-ormaplatin (tetraplatin), triethanolamine, formaldehyde) that cannot easily be explained by the mechanism described above. This suggests that more than one mechanism underlying RPN may be operative.

7.14.7 THE NEPHROTOXICITY OF GERMANIUM COMPOUNDS

"Anticancer" and "immunostimulatory" health remedies containing germanium are being increasingly widely consumed. There are a number of case reports that highlight the nephrotoxicity of germanium compounds in man,[127,128] these are characterized by increased blood urea nitrogen and serum creatinine, decreased creatinine clearance and proteinuria. In surviving patients renal dysfunction persisted for prolonged periods. Biopsies revealed tubulointerstitial nephropathy with vacuolar distal tubular epithelial degeneration, lipofuscin granules in thick ascending limb of Henle's loop to the distal convoluted tubule accompanying mild tubular atrophy and some desquamation. No glomerular or vascular changes were observed. High germanium concentrations were found in serum and urine and spleen, liver, kidney, adrenal gland, and myocardium.

Germanium dioxide given orally to rats caused renal dysfunctional and histological abnormalities by 4 wk[129] followed by a dose-dependent increase in serum creatinine, BUN,

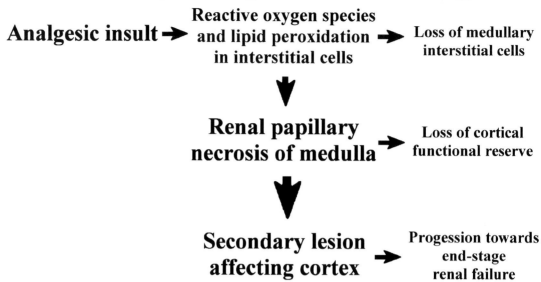

Figure 7 Scheme showing the consequences of renal medullary interstitial cell injury and how this contributes to cortical degeneration.

and serum phosphate and decrease in creatinine clearance, but no proteinuria or other markers of nephrotoxicity[129,130] and after 24 wk increased BUN and serum phosphate, and decreased creatinine clearance.[131] Light microscopy shows vacuolar degeneration and PAS-positive granules in distal tubules in higher-dosage group of germanium dioxide. Electron microscopy reveals electron-dense inclusions in the swollen mitochondrial matrix of the distal tubular epithelium.[129–131] Urinary excretion and renal-tissue content of germanium were elevated following higher dosage of germanium dioxide.[129] Although systemic toxicities were reduced after germanium dioxide was discontinued at week 24, renal tubulointerstitial fibrosis became prominent at week 40 (16 wk after discontinuation). The renal tissue content of germanium was maintained for 40 wk,[131] clearly demonstrated in the mitochondrial matrix of distal tubular epithelial cells using electron probe x-ray microanalysis. Liver dysfunction (increased serum ALT and AST, and hypoproteinemia by the decreased serum total protein and albumin) has also been reported.[129] There are no data on the staging of the lesion or its mechanistic basis.

Although germanium-containing "health" remedies have been withdrawn in developed countries, they are still available in many parts of the world, as health remedies. In addition, these metals are widely used in the semiconductor industry.

7.14.8 NEPHROTOXICITY OF PLATINUM COMPOUNDS

cis-Diamminedichloroplatinum II (cisplatin) causes pathological alterations to the S_3 segment of the rat proximal tubule that has been associated to total renal SH depletion, especially in mitochondrial and cytosolic fractions, where platinum concentrations are the highest.[132]

7.14.8.1 Polyuria and Distal Renal Injury

Platinum nephrotoxicity is, however, more complex as long-term changes include cystic lesions[133] which arise from proximal straight tubules in rats.[134] Distal and collecting tubular necrosis and bursts of cell proliferation and interstitial fibroblast proliferation are also seen.[135] The long-lasting tubular injury and the slowness of repair are consistent with the chronic renal dysfunction. There is a diminished concentration of salt and urea in the papilla as a result of abnormal function of the collecting duct.[136] Single doses of cisplatin

(6 mg kg^{-1}, i.p.) to rats cause focal necrosis in both the proximal and distal tubules and repeated treatment (1 mg kg^{-1}, i.p., twice weekly for 11 wk) cause tubular dilatation, interstitial fibrosis, and thickening of tubular basement membranes.[137] Renal concentration defects arise from impaired sodium chloride transport in the proximal tubule and thick ascending limb of Henle's loop[138] and a vasopressin resistance in the inner medullary collecting duct due to reduced cAMP generation caused by a defect at the level of G proteins.[139]

There are similar changes in humans. High-doses of cisplatin cause a significant impairment of proximal tubular salt and water reabsorption and also affect distal tubular function in patients[140] for 6 months after termination of treatment. Patients treated with cisplatin had impairment of urinary concentrating ability (after intranasal vasopressin) and a urinary acidification defect (after oral ammonium chloride).[141] Histopathology[142] showed degeneration, necrosis, and regenerative changes in the S$_2$ and S$_3$, and distal tubule, enlarged and pleomorphic nuclei in regenerated epithelial cells, and hyperplasia of the collecting duct lining cells. Electron microscopy showed an increased number of cytosomes containing electron-dense granules in all tubular portions and conspicuous nuclear indentation in the thin limb of the loop of Henle and collecting duct. It is not understood why cisplatin causes these distal tubular changes, although it may be due to selective accummulation.[142] If this is the case the reason for selective uptake is unclear.

7.14.9 CONCLUSIONS AND FUTURE RESEARCH DIRECTIONS

There are a range of chemicals that affect the medulla and distal nephron, but the underlying mechanisms are poorly understood. The best studied of these are the broad range of substances that cause papillary necrosis. Despite extensive research on analgesic nephropathy the lesion cannot be diagnosed non-invasively at an early stage in its development. It has therefore not been possible to identify which analgesic or NSAID has the greatest papillotoxic potential, what the risk factors are, and what combination of analgesics and NSAIDs (and/or other therapeutic agents) may be inappropriate. In the absence of sensitive and selective noninvasive tests to identify RPN, little progress can be made in epidemiological studies.

Priorities for future investigations include development of diagnostic techniques to identify the earliest changes, using new technology,

such as magnetic resonance spectroscopy and imaging, and molecular biology to help elucidate the pathogenesis of these lesions and the risk to man of exposure to those compounds that affect the medulla.

ACKNOWLEDGMENTS

The author's research has been supported by The Wellcome Trust, The Kidney Research Fund of Great Britain, Johns Hopkins Center for Alternatives to Animals in Testing, The Smith-Kline Foundation, The European Union, The Cancer Research Campaign and The International Agency for Research on Cancer. David Obatomi and Steve Brant provided critical comments and M. E. van Ek, Nguyen Thi Kim Thanh, and Sarita Mohur helped prepare the manuscript.

7.14.10 REFERENCES

1. E. I. Christensen, A. B. Maunsbach and S. Olsen (eds.), 'Functional Ultrastructure of the Kidney,' Academic Press, London, 1980.
2. D. B. Moffat, 'The Mammalian Kidney,' Cambridge University Press, 1975.
3. D. B. Moffat, 'New ideas on the anatomy of the kidney.' *J. Clin. Pathol.*, 1981, **34**, 1197–1206.
4. W. Kriz and L. Bankir, 'A standard nomenclature for structures of the kidney.' *Am. J. Physiol.*, 1988, **254**, F1–F8.
5. S. O. Bohman and A. K. Mandal (eds.), 'The Renal Papilla and Hypertension,' Plenum, London, 1980.
6. K. V. Lemley and W. Kriz, 'Anatomy of the renal interstitium.' *Kidney Int.*, 1991, **39**, 370–381.
7. B. M. Brenner and F. C. Rector (eds.), 'The Kidney,' 2nd edn., W. B. Saunders, Philadelphia, PA, 1981.
8. A. Garcia-Perez and M. B. Burg, 'Renal medullary organic osmolytes.' *Physiol. Rev.*, 1991, **71**, 1081–1115.
9. F. Morel, 'Methods in kidney physiology: past, present, and future.' *Annu. Rev. Physiol.*, 1992, **54**, 1–9.
10. P. Hansell, 'Evaluation of methods for estimating renal medullary blood flow.' *Ren. Physiol. Biochem.*, 1992, **15**, 217–230.
11. N. G. De Santo, G. Capasso, P. Anastasio *et al.*, 'Renal functional reserve.' *Child Nephrol. Urol.*, 1991, **11**, 140–145.
12. Anonymous, 'Biochemical aspect of renal function.' *Int. J. Biochem.*, 1980, **12**, 1–324.
13. W. G. Guder and B. D. Ross, 'Enzyme distribution along the nephron.' *Kidney Int.*, 1984, **26**, 101–111.
14. A. Garcia-Perez, 'Organic osmolytes in the kidney.' *Semin. Nephrol.*, 1993, **13**, 182–190.
15. M. L. Zeidel, 'Hormonal regulation of inner medullary collecting duct sodium transport.' *Am. J. Physiol.*, 1993, **265**, F159–F173.
16. P. H. Bach and J. W. Bridges, 'Chemically induced renal papillary necrosis and upper urothelial carcinoma. Part 1.' *Crit. Rev. Toxicol.*, 1985, **15**, 217–329.
17. I. Bojesen, 'Quantitative and qualitative analyses of isolated lipid droplets from interstitial cells in renal papillae from various species.' *Lipids*, 1974, **9**,

835–843.

18. I. N. Bojesen, in 'The Renal Papilla and Hypertension,' eds. A. K. Mandal and S. O. Bohman, Plenum, New York, 1980, pp. 121–148.

19. I. N. Bojesen, '*In vitro* and *in vivo* lipogenesis of the rat renal papillae from glucose.' *Biochem. Biophys. Acta*, 1980, **619**, 308–317.

20. I. N. Bojesen, 'The influence of urea on lipogenesis in renal papillae of rats.' *Lipids*, 1980, **15**, 519–523.

21. M. Koltai, D. Hosford, P. Guinotl *et al.*, 'PAF. A review of its effects, antagonists and possible future clinical implications. (Part II).' *Drugs*, 1991, **42**, 174–204.

22. M. J. Dunn and V. L. Hood, 'Prostaglandins and the kidney.' *Am. J. Physiol.*, 1977, **223**, 169–184.

23. M. J. Dunn and E. J. Zambraski, 'Renal effects of drugs that inhibit prostaglandin synthesis.' *Kidney Int.*, 1980, **18**, 609–622.

24. M. J. Dunn, 'Prostaglandins and Bartter's syndrome.' *Kidney Int.*, 1981, **19**, 86–102.

25. W. L. Smith, 'Renal prostaglandin biochemistry.' *Miner. Electrolyte Metab.*, 1981, **6**, 10–26.

26. J. V. Bonventre and R. Nemenoff, 'Renal tubular arachidonic acid metabolism.' *Kidney Int.*, 1991, **39**, 438–449.

27. H. W. Seyberth, A. Leonhardt, B. Tonshoff *et al.*, 'Prostanoids in paediatric kidney diseases.' *Pediatr. Nephrol.*, 1991, **5**, 639–649.

28. M. W. Anders, 'Metabolism of drugs by the kidney.' *Kidney Int.*, 1980, **8**, 636–647.

29. G. F. Rush, J. H. Smith, J. F. Newton *et al.*, 'Chemically induced nephrotoxicity: role of metabolic activation.' *Crit. Rev. Toxicol.*, 1984, **13**, 99–160.

30. B. B. Davis, M. B. Mattammal and T. V. Zenser, 'Renal metabolism of drugs and xenobiotics.' *Nephron*, 1981, **27**, 187–196.

31. B. J. Smith, J. F. Curtis and T. E. Eling, 'Bioactivation of xenobiotics by prostaglandin H synthase.' *Chem. Biol. Interact.*, 1991, **79**, 245–264.

32. D. L. DeWitt, 'Prostaglandin endoperoxide synthase: regulation of enzyme expression.' *Biochim. Biophys. Acta*, 1991, **1083**, 121–134.

33. S. P. Andreoli, 'Reactive oxygen molecules, oxidant injury, and renal disease.' *Pediatr. Nephrol.*, 1991, **5**, 733–742.

34. E. L. Greene and M. S. Paller, 'Oxygen free radicals in acute renal failure.' *Miner. Electrolyte Metabol.*, 1991, **17**, 124–132.

35. R. S. Nanra, 'Clinical and pathological aspects of analgesic nephropathy.' *Br. J. Clin. Pharmacol.*, 1980, **10**, 359S–368S.

36. R. S. Nanra, V. Daniel and M. Howard, 'Analgesic nephropathy induced by common proprietary mixtures.' *Med. J. Aust.*, 1980, **1**, 486–487.

37. L. F. Prescott, 'Analgesic nephropathy: a reassessment of the role of phenacetin and other analgesics.' *Drugs*, 1982, **23**, 75–149.

38. P. H. Bach and N. J. Gregg, 'Experimentally induced renal papillary necrosis and upper urothelial carcinoma.' *Int. Rev. Exp. Pathol.*, 1988, **30**, 1–54.

39. R. C. Allen, R. E. Petty, D. S. Lirenman *et al.*, 'Renal papillary necrosis in children with chronic arthritis.' *Am. J. Dis. Child.*, 1986, **40**, 20–22.

40. S. J. Scott, R. A. Bussey, *et al.*. 'Renal papillary necrosis associated with diclofenac sodium.' *Br. Med. J.*, 1986, **292**, 1050.

41. F. E. Husserl, R. K. Lange and C. M. Kantrow, Jr., 'Renal papillary necrosis and pyelonephritis accompanying fenoprofen therapy.' *JAMA*, 1979, **242**, 1896–1898.

42. E. Colome-Nafria, R. Solans, J. Espinach *et al.*, 'Renal papillary necrosis induced by flurbiprofen.' *DICP*, 1991, **25**, 870–871.

43. H. Mitchell, K. D. Muirden and P. Kincaid-Smith, 'Indomethacin-induced renal papillary necrosis in juvenile chronic arthritis.' *Lancet*, 1982, **2**, 558–559.

44. G. M. Shah, K. K. Muhalwas and R. L. Winer, 'Renal papillary necrosis due to ibuprofen.' *Arthritis Rheum.*, 1981, **24**, 1208–1210.

45. R. J. Caruana and E. L. Semble, 'Renal papillary necrosis due to naproxen.' *J. Rheumatol.*, 1984, **11**, 90–91.

46. K. K. Muhalwas, G. M. Shah and R. L. Winer, 'Renal papillary necrosis caused by long-term ingestion of pentazocine and aspirin.' *JAMA*, 1981, **246**, 867–868.

47. F. M. Lai, K. N. Lai and Y. W. Chong, 'Papillary necrosis associated with rifampicin therapy.' *Aust. N.Z. J. Med.*, 1987, **17**, 68–70.

48. A. Burry, 'Pathology of analgesic nephropathy: Australian experience.' *Kidney Int.*, 1978, **13**, 34–40.

49. F. J. Gloor, 'Changing concepts in pathogenesis and morphology of analgesic nephropathy as seen in Europe.' *Kidney Int.*, 1978, **13**, 27–33.

50. A. W. Macklin and R. J. Szot, 'Eighteen month oral study of aspirin, phenacetin and caffeine in C57BL/6 mice.' *Drug Chem. Toxicol.*, 1980, **3**, 135–163.

51. K. I. Furman, H. Kundig and J. R. Lewin, 'Experimental paracetamol nephropathy and pyelonephritis in rats.' *Clin. Nephrol.*, 1981, **16**, 271–275.

52. E. A. Molland, 'Experimental renal papillary necrosis.' *Kidney Int.*, 1978, **13**, 5–14.

53. P. H. Bach and T. L. Hardy, 'Relevance of animal models to analgesic-associated renal papillary necrosis in humans.' *Kidney Int.*, 1985, **28**, 605–613.

54. E. A. Molland, 'Aspirin damage in the rat kidney in the intact animal and after unilateral nephrectomy.' *J. Pathol.*, 1976, **120**, 43–48.

55. E. A. Molland, in 'Nephrotoxicity, Pathogenesis and Assessment,' eds. P. H. Bach, F. W. Bonner, J. W. Bridges *et al.*, Wiley-Heyden, Chichester, pp. 200–205.

56. P. H. Bach, D. J. Scholey, L. Delacruz *et al.*, 'Renal and urinary lipid changes associated with an acutely induced renal papillary necrosis in rats.' *Food Chem. Toxicol.*, 1991, **29**, 211–219.

57. P. H. Bach, P. Grasso, E. A. Molland *et al.*, 'Changes in the medullary glycosaminoglycan histochemistry and microvascular filling during the development of 2-bromoethanamine hydrobromide-induced renal papillary necrosis.' *Toxicol. Appl. Pharmacol.*, 1983, **69**, 333–344.

58. R. A. Axelsen, 'Experimental renal papillary necrosis in the rat: the selective vulnerability of medullary structures to injury.' *Virchows Arch. A: Pathol. Anat. Histol.*, 1978, **381**, 79–84.

59. D. C. Wolf, J. J. Turek and W. W. Carlton, 'Early sequential ultrastructural renal alterations induced by 2-bromoethylamine hydrobromide in the Swiss ICR mouse.' *Vet. Pathol.*, 1992, **29**, 528–535.

60. N. J. Gregg, E. A. Courtauld and P. H. Bach, 'Enzyme histochemical changes in an acutely-induced renal papillary necrosis.' *Toxicol. Pathol.*, 1990, **18**, 39–46.

61. N. J. Gregg, E. A. Courtauld and P. H. Bach, 'High resolution light microscopic morphological and microvascular changes in an acutely-induced renal papillary necrosis.' *Toxicol. Pathol.*, 1990, **18**, 47–55.

62. A. Munck, F. Lindlar and W. Masshoff, 'Die Pigmentierung der Nierenpapillen und der Schleimhaut der ableitenden Harnwege bei der chronischen sklerosierenden interstitiellen Nephritis ("Phenacetinniere"). (The pigmentation of the renal papillae and the mucosa of the urinary tract in chronic sclerosing interstitial nephritis ("Phenacetin-Niere")).' *Virchows Arch. A: Pathol. Anat.*, 1970, **349**, 323–331.

63. O. C. Dermer and G. E. Ham, 'Ethylenimine and

Other Aziridines,' Academic Press, New York, 1969.

64. M. R. Elwell, J. K. Dunnick, H. R. Brown *et al.*, 'Kidney and urinary bladder lesions in F344/N rats and B6C3F1 mice after 13 weeks of 2,2-bis(bromomethyl)-1,3-propanediol administration.' *Fundam. Appl. Toxicol.*, 1989, **12**, 480–490.

65. T. Okamura, E. M. Garland, R. J. Taylor *et al.*, 'The effect of cyclophosphamide administration on the kidney of the rat.' *Toxicol. Lett.*, 1992, **63**, 261–276.

66. G. J. Kolaja, W. H. Packwood, R. R. Bell *et al.*, 'Renal papillary necrosis and urinary protein alterations induced in Fischer-344 rats by D-ormaplatin.' *Toxicol. Pathol.*, 1994, **22**, 29–38.

67. S. D. Lenz and W. W. Carlton, 'Diphenylamine-induced renal papillary necrosis and necrosis of the pars recta in laboratory rodents.' *Vet. Pathol.*, 1990, **27**, 171–178.

68. G. C. Hard and G. E. Neal, 'Sequential study of the chronic nephrotoxicity induced by dietary administration of ethoxyquin in Fischer-344 rats.' *Fundam. Appl. Toxicol.*, 1992, **18**, 278–287.

69. H. P. Til, R. A. Woutersen, V. J. Feron *et al.*, 'Two-year drinking-water study of formaldehyde in rats.' *Food Chem. Toxicol.*, 1989, **27**, 77–87.

70. S. Nakaura, M. Tsuruta, Y. Tsunenari *et al.*, 'Single dose toxicity study of mesalazine in Cynomolgus monkeys.' *Pharmacometrics*, 1994, **47**, 509–511.

71. K. G. Bilyard, E. C. Joseph and R. Metcalf, 'Mesalazine: an overview of key preclinical studies.' *Scand. J. Gastroenterol. Suppl.*, 1990, **172**, 52–55.

72. T. Sugawara, M. Kato, N. Suzuki *et al.*, 'Thirteen-week oral toxicity study of the new cognition-enhancing agent nefiracetam in dogs.' *Arzneimittelforschung*, 1994, **44**, 217–219.

73. A. Poole and P. Buckley, '1-Naphthol—single and repeated dose (30-day) oral toxicity studies in the mouse.' *Food Chem. Toxicol.*, 1989, **27**, 233–238.

74. T. L. Hardy and P. H. Bach, 'The effects of *N*-phenylanthranilic acid-induced renal papillary necrosis on urinary acidification and renal electrolyte handling.' *Toxicol. Appl. Pharmacol.*, 1984, **75**, 265–277.

75. R. A. Owen and R. Heywood, 'Phenylbutazone-induced nephrotoxicity in the rat.' *Toxicol. Lett.*, 1983, **17**, 117–124.

76. D. E. Gunson and L. R. Soma, 'Renal papillary necrosis in horses afer phenylbutazone and water deprivation.' *Vet. Pathol.*, 1983, **20**, 603–610.

76. D. E. Gunson, 'Renal papillary necrosis in horses.' *J. Am. Vet. Med. Assoc.*, 1983, **182**, 263–266.

77. A. Maekawa, H. Onodera, H. Tanigawa, *et al.*, Lack of carcinogenicity of triethanolamine in F344 rats.' *J. Toxicol. Environ. Health*, 1986, **19**, 345–357.

78. C. J. Powell, P. Grasso, C. Ioannides *et al.*, 'Haloalkylamine-induced renal papillary necrosis: a histopathological study of structure-activity relationships.' *Int. J. Exp. Pathol.*, 1991, **72**, 631–646.

79. S. J. Kennedy and H. B. Jones, in 'Nephrotoxicity: Extrapolation from *In Vitro* to *In Vivo*, and Animals to Man,' eds. P. H. Bach and E. A. Lock, Plenum Press, London, 1989, pp. 611–615.

80. W. W. Carlton and J. A. Engelhardt, 'Experimental renal papillary necrosis in the Syrian hamster.' *Food Chem. Toxicol.*, 1989, **27**, 331–340.

81. R. A. Axelsen, 'Nephrotoxicity of mild analgesics in the Gunn strain of rat.' *Br. J. Clin. Pharmacol.*, 1980, **10**, 309S–312S.

82. M. A. Henry and J. D. Tange, 'Chronic renal lesions in the uninephrectomized Gunn rat after analgesic mixtures.' *Pathology*, 1984, **16**, 278–284.

83. M. A. Henry and J. D. Tange, 'Lesions of the renal papilla induced by paracetamol.' *J. Pathol.*, 1987, **151**, 11–19.

84. G. Thomas and J. D. Tange, 'Experimental pyelonephritis and papillary necrosis in the Gunn rat.' *Pathology*, 1985, **17**, 420–428.

85. N. Mittman, R. Janis and D. Schlondorff, 'Salicylate nephropathy in the Gunn rat: potential role of prostaglandins.' *Prostaglandins*, 1985, **30**, 511–525

86. R. G. Wyllie, G. S. Hill, G. Murray *et al.*, 'Experimental papillary necrosis of the kidney. 3. Effects of reserpine and other pharmacological agents on the lesion.' *Am. J. Pathol.*, 1972, **68**, 235–254.

87. G. S. Hill, R. G. Wyllie, M. Miller *et al.*, 'Experimental papillary necrosis of the kidney. II. Electron microscopic and histochemical studies.' *Am. J. Pathol.*, 1972, **68**, 213–234.

88. S. Sabatini, P. K. Mehta, S. Hayes *et al.*, 'Drug-induced papillary necrosis: electrolyte excretion and nephron heterogeneity.' *Am. J. Physiol.*, 1981, **241**, F14–F22.

89. S. Sabatini, V. Alla, A. Wilson *et al.*, 'The effects of chronic papillary necrosis on acid excretion.' *Pflügers Arch.*, 1982, **393**, 262–268.

90. S. Sabatini, S. Koppera, J. Manaligod *et al.*, 'Role of urinary concentrating ability in the generation of toxic papillary necrosis.' *Kidney Int.*, 1983, **23**, 705–710.

91. S. Sabatini, 'Pathophysiology of drug-induced papillary necrosis.' *Fundam. Appl. Toxicol.*, 1984, **4**, 909–921.

92. S. Sabatini, 'The pathopysiology of experimentally induced renal papillary necrosis.' *Semin. Nephrol.*, 1984, **4**, 27.

93. S. Sabatini, 'Cellular mechanisms of drug-induced papillary necrosis.' *J. Pharmacol. Exp. Ther.*, 1985, **232**, 214–219.

94. R. Vanholder, N. Lameire and W. Eeckhaut *et al.*, 'Renal function studies in an experimental model of papillary necrosis in the rat.' *Arch. Int. Physiol. Biochim.*, 1981, **89**, 63–73.

95. J. A. Arruda, S. Sabatini, P. K. Mehta *et al.*, 'Functional characterization of drug-induced experimental papillary necrosis.' *Kidney Int.*, 1979, **15**, 264–275.

96. M. F. Wilks, B. Schmidt-Nielsen and H. Stolte, 'Early changes in renal function following chemically induced nephropathy.' *Clin. Physiol. Biochem.*, 1986, **4**, 239–251.

97. J. A. Scarlett, N. J. Gregg, S. Nichol *et al.*, in 'Nephrotoxicity: Mechanisms, Early Diagnosis and Therapeutic Management,' eds. P. H. Bach, L. Delacruz, N. J. Gregg *et al.*, Dekker, New York, 1989, pp. 79–84.

98. D. C. Wolf, W. W. Carlton and J. J. Turek, 'Experimental renal papillary necrosis in the Mongolian gerbil (*Meriones unguiculatus*).' *Toxicol. Pathol.*, 1993, **20**, 341–349.

99. N. J. Gregg, M. E. Robbins, J. W. Hopewell *et al.*, 'The effect of acetaminophen on pig kidneys with a 2-bromoethanamine-induced-papillary necrosis.' *Renal Failure*, 1990, **12**, 157–163.

100. N. J. Gregg and P. H. Bach, '2-Bromoethanamine nephrotoxicity in the Nude mouse: an atypical targetting for the renal cortex.' *Int. J. Exp. Pathol.*, 1990, **71**, 659–670.

101. J. M. Davis, K. R. Emslie, R. S. Sweet *et al.*, 'Early functional and morphological changes in renal tubular necrosis due to *p*-aminophenol.' *Kidney Int.*, 1983, **24**, 740–747.

102. E. A. Lock, J. Ishmael and J. B. Hook, 'Nephrotoxicity of hexachloro-1,3-butadiene in the mouse: the effect of age, sex, strain, monoxygenase modifiers and

the role of glutathione.' *Toxicol. Appl. Pharmacol.*, 1984, **72**, 484–494.

103. P. H. Bach, R. Christian, J. R. Baker *et al.*, in 'Mechanisms of Toxicity and Hazard Evaluation,' eds. B. Holmstedt, R. Lauwerys, M. Mercier *et al.*, Elsevier, Amsterdam, 1980, pp. 533–536.

104. G. Murray, R. G. Wyllie, G. S. Hill *et al.*, Experimental papillary necrosis of the kidney. I. Morphology and functional data.' *Am. J. Pathol.*, 1972, **67**, 285–302.

105. G. J. Wright and V. K. Rowe, 'Ethyleneimine: studies of the distribution and metabolism in the rat using carbon-14.' *Toxicol. Appl. Pharmacol.*, 1967, **11**, 575–584.

106. J. T. Cuttino, Jr., F. U. Goss, R. L. Clark *et al.*, 'Experimental renal papillary necrosis in rats: microangiographic and tubular micropuncture injection studies.' *Invest. Radiol.*, 1981, **16**, 107–114.

107. P. H. Bach, C. P Ketley, I. Ahmed *et al.*, 'The mechanisms of target cell injury by nephrotoxins.' *Food Chem. Toxicol.*, 1986, **24**, 775–779.

108. E. M. Grieve, P. H. Whiting and G. M. Hawksworth, 'Arachidonic acid-dependent metabolism of 2-bromoethanamine to a toxic metabolite in rat medullary interstitial cell cultures.' *Toxicol. Lett.*, 1990, **53**, 225–226.

109. S. D. Lenz and W. W. Carlton, 'Decreased incidence of diphenylamine-induced renal papillary necrosis in Syrian hamsters given dimethylsulphoxide.' *Food Chem. Toxicol.*, 1991, **29**, 409–418.

110. G. Elliott, B. A. Whited, A. Purmalis *et al.*, 'Effect of 16,16-dimethyl PGE-2 on renal papillary necrosis and gastrointestinal ulcerations (gastric, duodenal, intestinal) produced in rats by mefenamic acid.' *Life Sci.*, 1986, **39**, 423–432.

111. M. Fuwa and D. Waugh, 'Experimental renal papillary necrosis. Effects of diuresis and antidiuresis.' *Arch. Pathol.*, 1968, **85**, 404–409.

112. P. Champion De Crespigny, T. Hewitson, I. Birchall *et al.*, 'Caffeine potentiates the nephrotoxicity of mefenamic acid on the rat renal papilla.' *Am. J. Nephrol.*, 1990, **10**, 311–315.

113. J. Uemasu, M. Fujiwara, C. Munemura *et al.*, 'Long-term effects of enalapril in rat with experimental chronic tubulo-interstitial nephropathy.' *Am. J. Nephrol.*, 1993, **13**, 35–42.

114. P. H. Bach and J. W. Bridges, 'The role of metabolic activation of analgesics and nonsteroidal anti-inflammatory drugs in the development of renal papillary necrosis and upper urothelial carcinoma.' *Prostaglandins Leukot. Med.*, 1984, **15**, 251–274.

115. T. V. Zenser, M. B. Mattammal and B. B. Davis, 'Demonstration of separate pathways for the metabolism of organic compounds in rabbit kidney.' *J. Pharmacol. Exp. Ther.*, 1979, **208**, 418–421.

116. T. V. Zenser, M. B. Mattammal, W. W. Brown *et al.*, 'Cooxygenation by prostaglandin cyclooxygenase from rabbit inner medulla.' *Kidney Int.*, 1979, **16**, 688–694.

117. B. Andersson, R. Larsson, A. Rahimtula *et al.*, 'Hydroperoxide-dependent activation of *p*-phenetidine catalyzed by prostaglandin synthase and other peroxidases.' *Biochem. Pharmacol.*, 1983, **32**, 1045–1050.

118. D. Ross, R. Larsson, B. Andersson *et al.*, 'The oxidation of *p*-phenetidine by horseradish peroxidase and prostaglandin synthase and the fate of glutathione during such oxidations.' *Biochem. Pharmacol.*, 1985, **34**, 343–351.

119. S. Joshi, T. V. Zenser, M. B. Mattammal *et al.*, 'Kidney metabolism of acetaminophen and phenacetin.' *J. Lab. Clin. Med.*, 1978, **92**, 924–931.

120. J. Mohandas, G. G. Duggin, J. S. Horvath *et al.*, 'Regional differences in peroxidatic activa-

tion of paracetamol (acetaminophen) mediated by cytochrome P450 and prostaglandin endoperoxide synthetase in rabbit kidney.' *Res. Commun. Chem. Pathol. Pharmacol.*, 1981, **34**, 69–80.

121. J. Mohandas, G. G. Duggin, J. S. Horvath *et al.*, 'Metabolic oxidation of acetaminophen (paracetamol) mediated by cytochrome P450 mixed-function oxidase and prostaglandin endoperoxide synthetase in rabbit kidney.' *Toxicol. Appl. Pharmacol.*, 1981, **61**, 252–259.

122. S. D. Nelson, D. C. Dahlin, R. J. Rauckman *et al.*, 'Peroxidase-mediated formation of reactive metabolites of acetaminophen.' *Mol. Pharmacol.*, 1981, **20**, 195–199.

123. W. L. Smith and T. B. Bell, 'Immunohistochemical localization of the prostaglandin-forming cyclooxygenase in renal cortex.' *Am. J. Physiol.*, 1978, **235**, F451–F457.

124. W. L. Smith and G. P. Wilkin, 'Immunochemistry of prostaglandin endoperoxide-forming cyclooxygenase: the detection of the cyclooxygenases in rat, rabbit and guinea pig kidneys by immunofluorescence.' *Prostaglandins*, 1977, **13**, 873–892.

125. J. Mohandas, J. J. Marshall, G. G. Duggin *et al.*, 'Differential distribution of glutathione and glutathione-related enzymes in rabbit kidney. Possible implications in analgesic nephropathy.' *Biochem. Pharmacol.*, 1984, **33**, 1801–1807.

126. P. H. Bach, 'Detection of chemically induced renal injury: the cascade of degenerative morphological and functional changes that follow the primary nephrotoxic insult and evaluation of these changes by *in vitro* methods.' *Toxicol. Lett.*, 1989, **146**, 237–250.

127. B. Hess, J. Raisin, A. Zimmermann *et al.*, 'Tubulointerstitial nephropathy persisting 20 months after discontinuation of chronic intake of germanium lactate citrate.' *Am. J. Kidney. Dis.*, 1993, **21**, 548–552.

128. A. Takeuchi, N. Yoshizawa, S. Oshima *et al.*, 'Nephrotoxicity of germanium compounds: report of a case and review of the literature.' *Nephron*, 1992, **60**, 436–442.

129. T. Sanai, N. Oochi, S. Okuda *et al.*, 'Subacute nephrotoxicity of germanium dioxide in the experimental animal.' *Toxicol. Appl. Pharmacol.*, 1990, **103**, 345–353.

130. T. Sanai, K. Onoyama, S. Osato *et al.*, 'Dose dependency of germanium dioxide-induced nephrotoxicity in rats.' *Nephron*, 1991, **57**, 349–354.

131. T. Sanai, S. Okuda, K. Onoyama *et al.*, 'Chronic tubulointerstitial changes induced by germanium dioxide in comparison with carboxyethylgermanium sesquioxide.' *Kidney Int.*, 1991, **40**, 882–890.

132. M. W. Weiner and C. Jacobs, 'Mechanism of cisplatin nephrotoxicity.' *Fed. Proc.*, 1983, **42**, 2974–2978.

133. D. C. Dobyan, D. Hill, T. Lewis *et al.*, 'Cyst formation in rat kidney induced by *cis*-platinum administration.' *Lab. Invest.*, 1981, **45**, 260–268.

134. J. Zanen, V. Carinci, D. Nonclercq *et al.*, 'Morphometric and tridimensional studies of tubular cystic degeneration in rat kidney following exposure to cisplatin.' *Anal. Cell. Pathol.*, 1993, **5**, 353–366.

135. G. Laurent, V. Yernaux, D. Nonclerq *et al.*, 'Tissue injury and proliferative response induced in rat kidney by *cis*-diamminedichloroplatinum (II).' *Virchows Arch. B: Cell Pathol. Incl. Mol. Pathol.*, 1988, **55**, 129–145.

136. R. Safirstein, P. Miller, S. Dikman, *et al.*, 'Cisplatin nephrotoxicity in rats: defects in papillary hypertoxicity.' *Am. J. Physiol.*, 1981, **10**, F175–F185.

137. D. D. Choie, D. S. Longnecker and A. A. del Campo, 'Acute and chronic cisplatin nephropathy in rats.' *Lab. Invest.*, 1981, **44**, 397–402.

138. A. C. Seguro, M. H. Shimizu, L. H. Kudo *et al.*,

'Renal concentration defect induced by cisplatin. The role of thick ascending limb and papillary collecting duct.' *Am. J. Nephrol.*, 1989, **9**, 59–65.

139. N. L. Wong, V. R. Walker, E. F. Wong *et al.*, 'Mechanism of polyuria after cisplatin therapy.' *Nephron*, 1993, **65**, 623–627.

140. G. Daugaard, U. Abildgaard, N. H. Holstein-Rathlou *et al.*, 'Renal tubular function in patients treated with high-dose cisplatin.' *Clin. Pharmacol. Ther.*, 1988, **44**, 164–172.

141. C. P. Swainson, B. M. Colls and B. M. Fitzharris, '*cis*-Platinum and distal renal tubule toxicity.' *N.Z. Med. J.*, 1985, **98**, 375–378.

142. H. Tanaka, E. Ishikawa, S. Teshima *et al.*, 'Histopathological study of human cisplatin nephropathy.' *Toxicol. Pathol.*, 1986, **14**, 247–257.

7.15
The Tubulointerstitium as a Target

SHARON D. RICARDO and JONATHAN R. DIAMOND
Pennsylvania State College of Medicine, Hershey, PA, USA

7.15.1 INTRODUCTION

Since the mid-1980s many putative mediators of chronic glomerular injury have been identified, including systemic hypertension, glomerular capillary hyperfiltration/hypertension, glomerular hypertrophy, anemia, hyperlipidemia, oxidant stress, up-regulated glomerular peptide growth factor (e.g., transforming growth factor-β (TGF-β), platelet-derived growth factor (PDGF)) expression, glomerular thrombosis and intraglomerular platelet deposition, dysregulated eicosanoid generation, altered glomerular matrix, and proteinase expression (for review see Ref. 1). However, attention has sharply focused on the renal interstitium. Many investigators[2–5] have studied kidney biopsies from patients with a wide variety of chronic glomerular diseases and have clearly shown a strong correlation between the extent of chronic tubulointerstitial damage and the decline in renal function and long-term renal outcome. As aptly indicated by Eddy in an excellent review,[6] "because approximately 80% of renal volume is occupied by tubules, it should not be surprising that chronic tubular damage associated with interstitial fibrosis is so tightly linked to overall renal function."

Although the mechanism(s) to explain the basis for progressive tubulointerstitial damage and the decline in renal function still remains unclear, experimental studies have provided compelling evidence that tubular injury, whether it is related to heavy proteinuria from glomerular disease or from specific tubular insults (e.g., urinary tract obstruction[7]), elicits a cascade of proinflammatory and fibrogenic events. As this review will suggest, the early recruitment of monocytes into the renal interstitium is a pivotal effector mechanism for amplifying the inflammatory response to tubular injury and promoting the later fibrogenic processes that will culminate in interstitial fibrosis and tubular loss.

Progressive interstitial fibrosis is currently regarded as the most important determinant of chronic renal failure (for review see Ref. 8). A histological correlate has been observed between tubular abnormalities and declining filtration rate in renal biopsy specimens from patients with a variety of nephropathies. Investigators have found that impaired renal function including inulin clearance, renal plasma flow rate, and ammonium excretion are more closely related to changes in the renal interstitium than damage to the glomerulus.[9,10] Bohle *et al.*[11,12] performed a series of histological morphometric studies on renal biopsies from patients with glomerulopathies, looking at the consequences of tubulointerstitial changes on renal function. They found that tubular atrophy and the appearance of the tubulointerstitial compartment, such as interstitial volume and extracellular matrix deposition, was an important prognostic marker for loss of renal function.

7.15.2 CELLULAR ACTIONS IN TUBULOINTERSTITIAL FIBROSIS

7.15.2.1 Macrophages

Acute inflammation of the renal tubulointerstitium is associated with an influx of large numbers of immunocompetent cells, particularly macrophages. Macrophages are a rich source of reactive oxygen species (ROS) and other proinflammatory agents, including proteases and eicosanoids. TGF-β, PDGF, interleukin-1 (IL-1), insulin-like growth factor (IGF), and epidermal growth factor (EGF) are examples proinflammatory mediators released by macrophages which may affect extracellular matrix synthesis (for review see Ref. 13).

Macrophage infiltration into the glomerulus and renal interstitium has been found in a number of clinical[14–19] and experimental models such as passive Heymann nephritis,[20] aminonucleoside[21–23] and adriamycin nephropathy,[24] nephrotoxic serum nephritis,[25] renal ablation,[26,27] and streptozotocin-induced diabetic nephropathy.[28] Although macrophages have long been known to produce glomerular injury in acute nephritis, such as antiglomerular basement membrane (GBM) disease,[29] we have fostered the postulate that these infiltrating cells mediate progressive renal disease.[30]

Therapeutic interventions found to reduce macrophage infiltration, such as x-irradiation,[21] administration of antimacrophage serum,[31] liposome encapsulated dichloromethylene diphonate,[32] and an essential fatty acid-deficient diet,[33–35] decreased the severity of immune cell-mediated renal disease in a variety of experimental models. This evidence supports the proposition that macrophages mediate immune-mediated renal injury. This concept will be extended by posing that macrophages also contribute significantly to the "wound-healing" and fibrotic responses observed in tubulointerstitial injury.

Renal interstitial macrophage influx is one of the earliest responses of the kidney to ureteral obstruction.[36] Nagle *et al.*[37,38] performed detailed morphological studies on the changes in the renal interstitium following unilateral ureteral obstruction (UUO). By light microscopy, 24 h after obstruction, there was a subtle widening of the interstitial space. The renal interstitium contained enlarged fibroblasts, mononuclear cells, and margination of polymorphonuclear leukocytes along capillaries with disruption of adjacent tubules. Glomeruli were observed to be congested, but otherwise normal. Cortical proximal tubules exhibited collapsed lumina and fine vacuolation of the cytoplasm. Collecting ducts revealed dilated lumina with flattened epithelium. Ninety-six hours after UUO, the cortical interstitial space was further enlarged, particularly around the proximal tubules and collecting ducts. The interstitium contained a striking increase in the number of mononuclear cells and altered fibroblasts. By 96 h, the proximal tubules exhibited thinning and focal loss of the brush border. Cortical collecting ducts remained widely dilated and contained an occasional cast.

The UUO model is an excellent one because it is nonproteinuric and nonlipidemic, without any direct immune or toxic renal insult (for review see Ref. 39). However, as will be evident, the mechanical disturbance produced by urinary tract obstruction is capable of eliciting a florid macrophage infiltration of the kidney and, if unrelieved, contributes to the cascade of events that ultimately leads to the development of interstitial fibrosis.

Administration of L-arginine to rats decreases macrophage infiltration by inhibiting macrophage chemotaxis, and improves renal function of rats with bilateral ureteral obstruction and puromycin aminonucleoside nephrosis.[40] Similarly, elimination of infiltrating macrophages by whole body x-irradiation of obstructed rats has been found to markedly attenuate thromboxane A_2 production[39] and increase renal plasma flow and glomerular filtration rate,[41] suggesting a role for macrophages in the hemodynamic changes observed in the postobstructed kidney. A significant correlation between renal interstitial macrophage influx and TGF-β1 gene expression in rats with UUO has been observed.[42] Immunohistochemistry revealed localization of this growth factor to mononuclear cells surrounding the obstructed cortical tubules (Figure 1). This phenomenon has also been seen in other models. Eddy and co-workers[43] noted, in acute puromycin aminonucleoside nephrosis, a significant increase in renal TGF-β1 mRNA at the peak of renal interstitial inflammation. Also TGF-β1-positive interstitial monocytes were found only in the interstitium of nephrotic compared to normal rat kidneys.[44] Renal TGF-β1 mRNA levels showed a significant positive correlation with the number of interstitial macrophages. Dietary protein restriction of puromycin aminonucleoside-treated rats appeared to attenuate interstitial fibrosis by diminishing the number of interstitial macrophages present to produce TGF-β1.[45] In experimental diabetic nephropathy produced by streptozotocin, Young *et al.*[28] postulated the early and steadily increasing glomerular macrophage infiltrate is a source of TGF-β1. The production of TGF-β by infiltrating macrophages may represent an important link connecting initial post-obstructive renal inflammation to extracellular matrix accumulation associated with interstitial fibrosis.

7.15.2.2 Renal Tubular Epithelial Cells

Renal tubular epithelial cell injury may initiate a deleterious cascade of events leading to the progression of interstitial fibrosis. Tubular epithelial cells can produce an array of cytokines and growth factors which modulate fibroblast proliferation, extracellular matrix production, and chemoattractants for infiltrating cells (for review see Ref. 46). Proximal tubules have been found to produce the

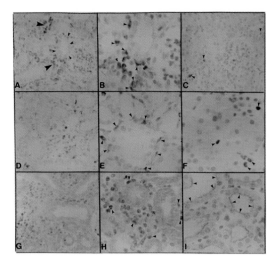

Figure 1 Avidin-biotinylated complex immunohistochemical labeling studies of representative midcoronal sections of obstructed and contralateral unobstructed kidney (CUK) specimens of unilateral obstructed (UUO) rats. (A) labeling with ED-1, a mouse monoclonal IgG anti-rat macrophage antibody that recognizes cytoplasmic antigens in monocytes/macrophages, of an obstructed kidney, 12 h after UUO; large arrowheads, clusters of interstitial macrophages; small arrowheads, cortical tubules (×220). (B) Higher magnification (×440) of cortical tubule in (A) with small arrowheads indicating some of the numerous surrounding interstitial macrophages in a ring-like pattern. (C) Immunostaining with ED-1 antibody of a representative midcoronal section from a CUK specimen at 12 h after UUO; small arrowheads, sparse interstitial macrophage infiltrate in comparison to (A) (×220). (D) Immunostaining with ED-1 antibody at 96 h. Obstructed kidney section revealing a more intense peritubular macrophage infiltrate (×220). (E) Delineation of more intense macrophage infiltrate at higher magnification (×440) in a representative 96 h obstructed kidney specimen. (F) Paucity of interstitial macrophages on ED-1 labeling of a 96 h CUK specimen from a UUO rat (×440). (G) Immunolabeling, utilizing rabbit polyclonal anti-TGF-β antibody on a representative 48 h obstructed kidney section. Note numerous peroxidase-positive peritubular cells labeling positively for this cytokine (×220). (H) Higher magnification (×440) of (G); arrowheads, representative TGF-β1-positive cells that possess nuclear morphology of macrophages; interstitial fibroblasts may represent another cell type that exhibits this immunolabeling pattern. (I) Immunostaining for MCP-1, using a rabbit antimouse MCP-1 antibody on a 96 h obstructed kidney section; this specimen possesses marked apical tubular labeling (small arrowheads) for this chemoattractant peptide at this time point (×440). This apical tubular pattern of MCP-1 immunolocalization was noted as early as 12 h after UUO in obstructed kidneys only and persisted at both the 48 h and 96 h time points.

chemotactic cytokines monocyte chemoattractant peptide (MCP-1) and IL-8,[42,47] as well as the vasoactive peptide endothelin.[48] Growth factors such as, IGF-1, TGF-β, EGF,[49–52] and the cytokines, IL-6 and tumor necrosis factor (TNF), are also generated in response to various stimulants.[52] Renal tubular cells have been found to produce nitric oxide, a vasodilator capable of possessing immunomodulatory and cytotoxic effects, in response to cytokine stimulation.[53,54] McLay *et al.*[54] found that IL-1β, TNF-α, and gamma-interferon (IFN-γ) induced nitric oxide production, secondary to the induction of nitric oxide synthase, from human proximal tubular cells.

Proximal tubular cells may generate potentially damaging ROS or other inflammatory moieties following exposure to oxidant stress.[55] Primary cultures of rabbit proximal tubules were found to generate superoxide and hydrogen peroxide in response to immune activation.[13] Renal dysfunction may result from oxidant injury of the epithelial cells mediated by these toxic oxygen metabolites. Andreoli and McAteer[56] observed that proximal tubules were more susceptible to oxidant injury than distal tubules. Cultured proximal tubular cells were found to respond to injury mediated by ROS with an early decline in ATP levels, while detachment and lytic injury were found to be late reponses to oxidant stress.[56] Oxidant stress has also been found to weaken renal epithelial attachment to the extracellular matrix by altering physiological distribution of integrin receptors.[57]

Alterations of extracellular matrix proteins from proximal tubule epithelium have been found with angiotensin II stimulation. Wolf *et al.*[58] found that exposure of proximal tubules to angiotensin II can induce cellular hypertrophy in proximal tubular cells leading to the secretion of type II collagen. Proximal tubular cells derived from kidneys with autoimmune interstitial nephritis have been found to promote extracellular matrix accumulation by producing collagen types IV, V and also I and III.[59] In summary, renal tubular cells are a rich source of inflammatory compounds and macrophage chemoattractant agents. Tubular dysfunction due to either nonimmune or immune mechanisms may lead to dysregulation of the secretion and synthesis of these inflammatory agents, further contributing to tubulointerstitial injury by enhancing the proinflammatory state of interstitial fibrosis.

7.15.2.3 Interstitial Fibroblasts

Renal interstitial fibrosis frequently occurs in immune and nonimmune kidney disorders and is associated with the overproduction and accumulation of extracellular matrix proteins. The transformation of interstitial fibroblasts into myofibroblasts, resident cells which share both fibroblast and smooth muscle cell features, have been implicated as a source of extracellular matrix deposition. Fibroblasts may enhance collagen and fibronectin synthesis by numerous autocrine and paracrine growth factors. Renal tubular cells and macrophages can release an array of cytokines, such as TGF-β1, fibroblast growth factor (FGF), and PDGF, which are mitogenic for fibroblasts, activating them to proliferate and deposit extracellular matrix proteins in the renal glomerulus and interstitium (for review see Refs. 6, 60–62).

The resident fibroblasts of the kidney have been characterized into three mitotically active progenitor cells.[62,63] These progenitor stem cells can spontaneously differentiate into three postmitotic fibroblasts, with stage VI having the highest biosynthetic activity for collagen and other extracellular matrix proteins, as well as the highest turnover rate for specific intracellular proteins.[62] Muller and co-workers[64–66] have demonstrated that human fibroblasts taken from kidneys with interstitial fibrosis are hyperproliferative and synthesize more collagen than normal renal fibroblasts. The fibrotic kidney fibroblasts were found to be comprised of more hyperproliferative progenitor cells compared to interstitial fibroblasts isolated from normal kidneys.[64] In addition, interstitial fibroblast cultures from the fibrotic kidneys were found to synthesize more total collagen, with significant changes in the relative proportions of the interstitial collagens type I, III, and IV, than normal kidney fibroblasts.[62]

Several chemokines generated by macrophages and tubular epithelial cells may activate proliferative and migratory fibrogenic responses in resident fibroblasts. Tubulointerstitial fibroblasts have been demonstrated to secrete procollagen types I and II, in addition to basement membrane procollagens types IV, V, and laminin, in response to various cytokines.[59,67] TGF-β1 is a potent stimulant for fibroblast proliferation[68] and collagen synthesis.[69] Alvarez *et al.*[69] observed that fibroblasts from the tubulointerstitium of murine kidneys secreted collagen types I and IV in the presence of low concentrations of TGF-β1, IL-1, and EGF. Other paracrine growth factors, such as PDGF[70] and IGF-1,[46] are potent chemoattractants for fibroblasts which can influence fibroblast activation and proliferation.

In addition to activation by chemokines, products of immune activated lymphocytes,[71,72] activation of complement,[73] cleavage of fibronectin,[74] types I, II, III collagens, and collagen-derived peptides,[75] can also enhance fibroblast motility and proliferation. In addition, fibroblast secretion of TGF-β1,[76] and metalloproteinases, collagenase and stromelysin,[77,78] can facilitate the fibroblasts migratory path through extracellular matrix proteins.

Fibroblasts play a critical role in extracellular matrix remodeling and development of renal scarring following ureteral obstruction. Electron microscopic studies of obstructed rabbit kidneys observed that interstitial fibroblasts undergo morphological transformation to form cells containing numerous actin-like cytoplasmic filaments.[79] Accumulation of these α-smooth muscle actin producing cells, presumably myofibroblasts, has been identified at sites of chronic tubulointerstitial injury.[80] Sharma *et al.*[81] demonstrated a diffuse peritubular increase in interstitial collagens I, III, and fibronectin in obstructed rabbit kidneys. An early, transient increase in renal cortical mRNA encoding the α1 monomers of collagen I, III, and IV, indicated increased synthesis of extracellular matrix proteins. *In situ* hybridization studies suggested that cells of the interstitial space, most likely fibroblasts, were involved in the extracellular matrix accumulation associated with obstructive nephropathy.[81]

7.15.3 REACTIVE OXYGEN SPECIES AND PROGRESSIVE RENAL DISEASE

ROS have been implicated as primary mediators in both immune and nonimmune renal diseases (for reviews see Refs. 82–86). In addition to polymorphonuclear leukocytes and monocytes producing ROS, isolated glomeruli, mesangial cells, and tubular epithelial cells are also able to generate these toxic oxygen metabolites.[87–89]

7.15.3.1 ROS Species

Free radicals are defined as any molecular species that contain one or more unpaired electrons.[90] As a consequence, free radicals tend to have paramagnetic properties and are highly reactive and unstable molecules. In addition to being able to peroxidize lipids, ROS can damage proteins and DNA.[91,92] Most of the oxygen consumed by aerobic cells is reduced to water, thereby bypassing ROS generation. However, partial reduction of oxygen can commonly occur during respiration leading to the formation of reactive oxygen intermediates. Oxygen can be univalently reduced to form the free radical superoxide anion. Superoxide anion

is a relatively weak oxidant, not thought to be as toxic as other free radicals but rather a precursor to more reactive oxygen-derived molecules. Superoxide anion is highly unstable and in an aqueous environment will dismutate to produce hydrogen peroxide. Hydrogen peroxide has no unpaired electrons and is not a free radical, however, it is a powerful oxidant which is stable in the absence of trace metals. Hydrogen peroxide, although poorly reactive in aqueous solutions, can cross biological membranes and have direct cytotoxic effects. Oxygen can not be reduced directly by the addition of three electrons, however, superoxide anion and hydrogen peroxide can interact in the presence of a transition metal such as iron to yield the hydroxyl radical via the iron-catalyzed Haber–Weiss reaction. The hydroxyl radical is the most reactive and short-lived free radical produced in biological systems.[93]

7.15.3.2 Antioxidant Systems

Complex antioxidant systems have evolved to metabolize and detoxify ROS, these include: superoxide dismutase (SOD), catalase, and glutathione peroxidase. SOD comprises a group of metalloproteins which catalyze the destruction of superoxide anion by dismutation. They exist as copper–zinc containing enzymes (Cu–ZnSOD), primarily in the cytoplasm, and as manganese-containing proteins, primarily located in the mitochondrial matrix. Catalase reduces hydrogen peroxide to water, whereas selenium-containing glutathione peroxidase protects cells against both hydrogen peroxide and lipid hydroperoxides, toxic end products of lipid peroxidation. Non-enzymatic antioxidants, such as α-tocopherol, ascorbic acid, and carotene, provide important protection of the cell membranes from free radical-induced lipid peroxidation as well as directly scavenging ROS. Metabolic and cellular disruption is observed when ROS are produced in amounts sufficient to overcome normal defense mechanisms or when the defense mechanisms are impaired.

7.15.3.3 ROS-induced Pathogenesis

ROS have been implicated in the pathogenesis of renal injury in a number of experimental models, including passive Heymann nephritis,[94–96] anti-GBM glomerulonephritis,[97–99] puromycin aminonucleoside,[100–102] and adriamycin nephrosis,[103–106] as well as gentamicin- and glycerol-induced acute renal failure.[107,108] Hydrogen peroxide has been implicated as an important mediator of glomerular injury following infusion of phorbol myristate acetate[109] and cobra venom factor.[110,111] Johnson et al.[112–114] postulated a mechanism for hydrogen peroxide-mediated glomerular injury involving the reaction between hydrogen peroxide and halide anions in the presence of neutrophil-derived myeloperoxidase. In the presence of myeloperoxidase, an enzyme found in the primary granules of polymorphonuclear leukocytes, hydrogen peroxide can combine with myeloperoxidase to form an enzyme–substrate complex capable of oxidizing halides. The oxidation of chloride, for example, leads to the production of the powerful oxidant hypochlorous acid. Johnson et al.[113,114] demonstrated that intrarenal infusion of myeloperoxidase followed by nontoxic concentrations of hydrogen peroxide in a chloride-containing solution resulted in proteinuria and glomerular injury. Light and electron microscopy revealed several glomerular morphologic abnormalities including, endothelial cell swelling and glomerular epithelial cell foot process effacement accompanied by halogenation of the GBM.[114] Another study found evidence that the myeloperoxidase–hydrogen peroxide–halide system is activated in an experimental model of immune complex glomerulonephritis.[112] Immune complex glomerulonephritis was induced in rats by in situ perfusion with concanavalin A followed by an anticoncanavalin A antibody. These animals exhibited marked proteinuria and glomerular injury. Rats with glomerulonephritis that received ^{125}I-labeled iodide showed iodination of glomeruli and the glomerular basement membrane. In contrast, iodination of the glomerular basement membrane as well as proteinuria were not observed in neutrophil-depleted rats.[112] These results suggest that in immune complex nephritis the neutrophil-derived myeloperoxidase system is activated in the glomerulus by antigen–antibody complexes. Myeloperoxidase binds to the glomerular capillary wall and subsequently reacts with hydrogen peroxide to produce oxidants such as hypochlorous acid, thereby leading to halogenation of glomerular structural proteins.

7.15.3.4   Reactive Oxygen Species in Obstructive Nephropathy

Macrophage-derived ROS may play an important role in tubulointerstitial inflammation associated with obstructive nephropathy. Modi et al.[115] investigated the effect of pretreatment of probucol, an antioxidant, on normal rats and rats with 24 h unilateral release of bilateral ureteral obstruction. They found

that 3–5 h and three days following release, bilateral ureteral obstructed rats given probucol had significantly reduced renal levels of malondialdehyde, a product of lipid peroxidation, decreased leukocyte infiltration, and decreased renal levels of oxidized glutathione, compared to rats with bilateral ureteral obstruction not receiving probucol. Modi et al.[115] concluded that the protective effects of probucol may be due to either antioxidant properties or to the effect of probucol on leukocyte infiltration.

Ricardo and Diamond[116] have found that decreased mRNA and protein expression of cellular antioxidant enzymes and increased generation of ROS may play an integral role in the development of tubulointerstitial injury and fibrosis associated with experimental hydronephrosis. Ninety-six hours after UUO, levels of total cortical mRNA for catalase and Cu–ZnSOD were decreased 5.5- and 5.0-fold, respectively, compared to the contralateral unobstructed kidney. Levels of both superoxide anion and hydrogen peroxide were significantly increased approximately 10- and 4-fold, respectively, in kidney-slice cultures from UUO rats compared to cultures containing contralateral unobstructed kidney slices.[116] Immunohistochemistry revealed localization of Cu–ZnSOD and catalase to proximal tubules, the staining intensity of which was appreciably diminished in the proximal tubules of obstructed kidneys in comparison to the contralateral unobstructed kidney specimens. The decreased antioxidant mRNA and protein expression in obstructed kidneys may be due either to a direct inactivation by ROS, or regulation by other macrophage-derived inflammatory moieties such as cytokines. Studies have found a correlation between cytokine action and expression of cellular antioxidants. Induction of manganese superoxide dismutase has been observed with IL-1α, IL-1β, and TGF-β in fibroblasts and epithelial cells in vitro.[117,118] We have preliminary evidence suggesting a novel relationship between TGF-β1 activity and catalase. Catalase mRNA levels were found to be down-regulated 14.4- and 2.83-fold following exposure of isolated proximal tubules to TGF-β1 (10 ng), in comparison to either control proximal tubules, or when cultured with 0.1 ng TGF-β1, respectively (unpublished observations). Further evidence supporting a possible paracrine inactivation of tubular antioxidant activity by TGF-β1 and other macrophage-derived inflammatory agents following UUO remains to be elucidated.

Antioxidant enzymes can be inactivated by hydrogen peroxide and other oxygen metabolites.[119,120] A study by Singh et al.[121] found that mRNA levels of Cu–ZnSOD, catalase, and glutathione peroxidase were significantly decreased in rat kidneys following ischemia-reperfusion injury. They concluded that the down-regulation of antioxidant enzymes as a result of oxidative stress may further exacerbate ischemia-reperfusion injury. In a previous study,[122] ischemia-reperfusion injury was found to decrease cytosolic and peroxisomal catalase by both inactivation and proteolysis. During ischemia, the increased cellular levels of hydrogen peroxide and low pH may lead to the conversion of catalase to an inactive iron-bound complex.[119] Cu–ZnSOD can also be irreversibly inactivated by increased levels of hydrogen peroxide.[123] During reperfusion following ischemia, Gulati et al.[122] observed that proteolysis and decreased synthesis also contributed to the decreased catalase activity. They proposed that Ca^{2+} and/or fatty acid-activated proteases may degrade the oxidatively modified catalase. This down-regulation of antioxidant enzymes seen in ischemia-reperfusion injury may be similarly observed in obstructed kidneys, due to the initial vasoconstriction of renal blood flow seen following obstruction with subsequent reperfusion.

ROS may contribute to interstitial fibrosis associated with obstruction by inducing extracellular matrix accumulation. Oxidative stress has previously been found to stimulate collagen $\alpha_1(1)$ gene expression in cultured human fibroblasts.[124] The addition of α-tocopherol to the fibroblast cultures was found to decrease collagen synthesis and gene transcription. The administration of CCl_4, a nephrotoxin which induces lipid peroxidation, to rats stimulates both type I and type IV procollagen synthesis by mesangial cells.[125] A study by Shan et al.[126] investigated how changes in the cellular redox state can influence collagen synthesis from mesangial cells. They found that enhanced levels of glutathione increased the secretion of collagen and mRNA levels of collagen I and IV. Oxidants can also perturb the adhesion of renal tubular epithelial cells to extracellular matrix proteins by altering integrins, cell adhesion molecules.[57] This disruption of the focal contacts between oxidatively modified cells and surrounding matrix may also play a role in the pathogenesis of tubulointerstitial disease.

7.15.4 CHEMOATTRACTANTS

The recruitment of macrophages into the renal interstitium may be due to the release of protein chemoattractants. A number of chemoattractants, such as osteopontin and MCP-1, have been identified. Osteopontin is a sialic acid-rich, phosphoprotein originally isolated in bone.[127] It is expressed in macrophages,[128] many transformed cell lines,[129,130] and in the kidney medulla.[127,131] Increased production of osteopontin has been found in response to various mitogens and growth factors, such as phorbol esters and TGF-β.[132,133] As well as serving as a chemoattractant for monocytes, osteopontin may also promote the early fibrogenic phase of interstitial fibrosis by stimulating fibroblast adhesion.[134,135] Giachelli et al.[136] observed a dramatic increase in osteopontin expression in kidneys from rats with focal tubulointerstitial injury associated with angiotensin II infusion. In situ hybridization and immunohistochemistry revealed osteopontin localization primarily in epithelial cells of the distal tubules, collecting ducts, and Bowman's capsule. Elevated expression of osteopontin in the angiotensin II-infused rats correlated with an early monocyte/macrophage infiltration. Co-localization studies showed that monocytes/macrophages were found almost exclusively in peritubular regions of elevated osteopontin expression, suggesting that osteopontin is an important proinflammatory chemoattractant in tubulointerstitial disease.[136] Pichler et al.[137] observed up-regulated osteopontin expression and increased protein levels in proximal and distal tubules in three experimental models of glomerulonephritis. Osteopontin up-regulation was found to precede histological evidence of tubular injury and correlate with subsequent sites of monocyte/macrophage accumulation.[137] Our group has found up-regulated expression of osteopontin mRNA in the cortex of obstructed rat kidneys as early as 4 h after obstruction.[138] In situ hybridization revealed osteopontin mRNA transcripts localized to both proximal and distal tubules in obstructed kidneys (Figures 2 and 3). A reversal of obstruction and whole body x-irradiation, maneuvers which dramatically reduce macrophage infiltration into the renal cortex, did not effect osteopontin expression, suggesting that persistent up-regulation of osteopontin in obstructed kidneys is important in macrophage accumulation within the peritubular and periglomerular interstitium of the cortex.[138]

The chemoattractant MCP-1, a member of a new family of small cytokines (Scy superfamily/intercrines), is produced by a variety of immune and nonimmune cells, including mononuclear cells,[139] endothelial cells,[140,141] fibroblasts,[142] and smooth muscle cells.[143] Specific pro-inflammatory cytokines, such as IL-1β, TNF-α and INF-γ, have been found to stimulate MCP-1 activity.[47,144] Immunolocalization of MCP-1 to renal tubules has been observed in acute ischemic renal failure in the rat.[145] Rovin

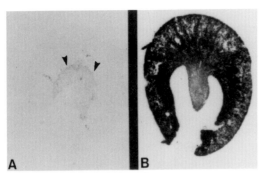

Figure 2 Autoradiographs showing mRNA transcription with ^{35}S-labeled osteopontin probe. (A) Hybridization of a contralateral unobstructed kidney section from a 96 h obstructed rat, showing a normal, low expression of osteopontin in the renal medulla (arrowheads). (B) A 96 h obstructed kidney specimen from the same animal, showing a marked increase in osteopontin mRNA transcription in both the cortex and medulla of the kidney.

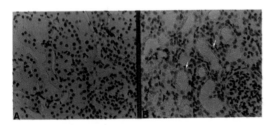

Figure 3 *In situ* hybridization for osteopontin mRNA in obstructed and contralateral unobstructed kidneys from rats, 96 h following UUO. Lightfield photomicrographs of paraffin sections reacted with antisense osteopontin probe from (A) a contralateral unobstructed kidney, showing no osteopontin mRNA transcription, and (B) an obstructed kidney from the same animal. Arrows, fluorescent silver grains represent osteopontin mRNA transcripts localized to the proximal and distal tubules of the renal cortex. Note: glomerular tufts are negative for osteopontin expression ($\times$440).

et al.[55] demonstrated that acute ureteral obstruction results in the release of a specific monocyte chemoattractant that is a lipid. We presented immunohistochemical labeling data demonstrating apical tubular epithelium MCP-1 expression as early as 12 h following ureteral ligation in rats[42] (Figure 1(I)). Northern analysis studies revealed increased expression of MCP-1 mRNA in obstructed kidneys, in comparison to the contralateral control kidneys (Figure 1(I)). These observations suggest that the expression of MCP-1, as a potential chemoattractant mechanism, may be related to the elevated intratubular pressure or some corelated mechanical stress that injures the tubular epithelium as a result of ureteral obstruction.

Other chemoattractants of the small cytokine (Scy superfamily/intercrine) family thought to have proinflammatory or reparative properties and be important in recruitment of inflammatory cells have been described.[146] These include MuRANTES, a macrophage chemoattractant isolated from murine proximal tubules,[146] and HuRANTES, a selective chemoattractant for CD^{4+} lymphocytes and monocytes.[147,148] Wolf *et al.*[149] found increased expression of MuRANTES following stimulation with TNF-α in cultured mesangial cells. Leukotrienes[150] and IL-8[151] may also play a role in the recruitment of circulating macrophages into the renal interstitium leading to induction of tubulointerstitial disease.

7.15.5 CELL ADHESION MOLECULES

Cell adhesion molecule interactions are an important step in the complex cascade of events leading to development of tubulointerstitial fibrosis. Adhesion molecules play a central role in the mediation of inflammatory cell infiltration, cell proliferation, and accumulation of extracellular matrix molecules following renal injury (see Chapter 17, this volume).[152] Monocytes express several membrane-associated adhesion molecules including the immunoglobulin supergene family, intercellular adhesion molecule (ICAM-1, ICAM-2), the selectin family, members of the β1 integrin family (VLA-4, VLA-5), and the β2 or leukocyte integrin family (CD11, CD18).[6] ICAM-1 is a cellular adhesion molecule belonging to the immunoglobulin superfamily which is expressed by many cell types including endothelium and is up-regulated by proinflammatory cytokines.[153,154] Proximal tubular cells are found to express ICAM-1 following stimulation with γ-INF, IL-1, and TNF-α.[155] This ICAM-1 expression has been localized to the apical surface of the proximal tubular epithelial cell.[156,157]

Facilitation of macrophage infiltration into the renal parenchyma and the subsequent development of interstitial injury has been examined in a variety of clinical and experimental models.[62,157–161] Aberrant expression of ICAM-1 has been observed on proximal tubular epithelial cells of patients with focal and segmental glomerulosclerosis, rapidly progressive glomerulonephritis, and mesangioproliferative glomerulonephritis[156–162] (known to be frequently associated with interstitial cellular infiltrates). Muller *et al.*[156] showed increased ICAM-1 expression of proximal tubular epithelial cells, interstitial cells, and peritubular capillaries, in human kidneys with different forms of glomerulonephritis. This increased expression of ICAM-1 was associated with abnormal expression of HLA-DQ and -DP antigens suggesting coordinated induction by lymphocyte-derived cytokines such as IL-1 and IFN-γ.[156]

ICAM-1 plays a major role in T lymphocyte and macrophage adhesion to activated endothelium via binding to lymphocyte function-associated antigen-1 (LFA-1). Hill *et al.*[157] found that ICAM-1 and LFA-1 play a role in the early development of mononuclear leukocytes in experimental anti-GBM glomerulonephritis in the rat. Immunogold labeling studies showed a marked increase in the intensity of ICAM-1 expression in the brush borders of proximal tubules, capillary endothelium, and interstitial fibroblast-like cells, in addition to mononuclear leukocytes showing increased LFA-1 expression.[157] These studies show that ICAM-1 expression is important in the migration and localization of interstitial leukocytes and the development and progression of tubulointerstitial inflammation in the early stages of glomerulonephritis.

7.15.6 GROWTH FACTORS AND CYTOKINES

Growth factors are important in the regulation of cell growth and differentiation. Renal proximal tubular cells and macrophages are a potent source of an array of growth factors, such as TGF-β, IL-1, IL-6, FGF, TNF, and PDGF, which appear integral in the development of tubulointerstitial fibrosis. Growth factors are not only mitogenic for fibroblasts, but can also lead to initiation of macrophage infiltration into the renal interstitium.[163–166]

The polypeptide cytokine TGF-β elicits a variety of responses that promote fibrosis, including stimulation of extracellular matrix genes and up-regulation of TIMP-1 expression.[43,44,60,166,167] TGF-β1 has been reported to induce the production of collagen types I, III, and V by cultured renal fibroblasts,[69] and stimulate proximal tubular epithelial cells to increase production of proteoglycans and type IV collagen.[166,168] TGF-β1 is a strong chemoattractant for monocytes/macrophages which is important in directing an inflammatory response following tubulointerstitial injury. Monocytes exposed to TGF-β1 *in vitro* show augmented gene expression for IL-1, PDGF, FGF, and TNF.[169] The early release of TGF-β1 following renal injury initiates a complex cascade of events that are pivotal to the development of interstitial disease.

The role of up-regulated TGF-β1 expression in tubulointerstitial fibrosis has been observed in a number of experimental models of progressive renal disease, including aminonucleoside[37,163] and adriamycin nephrosis,[24] ureteral obstruction,[42,170,171] streptozotocin-induced diabetic nephropathy,[28,172] and experimental glomerulonephritis.[173–175] An increased TGF-β1 expression in the renal cortex during an early phase of UUO in the rat has been demonstrated[42] (Figure 1). This increase in TGF-β1 mRNA levels during obstructive uropathy appears to be, in part, mediated by angiotensin II.[171]

Macrophage-derived PDGF, IL-1, and TNF may also affect tubular function, thereby playing a role in the progression of tubulointerstitial injury. PDGF is a potent chemoattractant for fibroblasts, promotes extracellular matrix protein synthesis and secretion, and triggers contraction of vascular smooth muscle cells.[163] PDGF is a potent mitogen that can be released from activated platelets, macrophages, endothelial, and mesangial cells and may be important in wound healing and repair processes.[163] Floege *et al.*[176] observed PDGF to be important in the development of sclerosis in a nonimmune model of progressive renal disease. Immunohistochemistry and Northern analysis revealed that proliferation of glomerular cells was accompanied by increased expression of PDGF in rats with severe renal ablation. Increased PDGF receptor sites have been found in chronic transplant rejection, suggesting a role for PDGF in marked tubulointerstitial changes observed in this condition.[177]

TNF and IL-1, also released from macrophages, may be potent mediators of fibroblast proliferation and fibrosis (for reviews see Refs. 178 and 179). In a rabbit hydronephrotic model, release of an IL-1-like monokine was found to stimulate fibroblast proliferation and augment arachidonic acid capacity.[180] Animals with ureteral obstruction fed an essential fatty acid-deficient diet reduced macrophage infiltration and decreased IL-1 production, resulting in diminished fibroblast proliferation. TNFα has a broad spectrum of actions in renal injury. Intravenous infusion of TNFα in the rabbit induced polymorphonuclear leukocyte infiltration into the glomerulus.[181] *In vitro* studies have also found that mesangial cells stimulated with TNFα release the neutrophil chemoattractants IL-8[182] and IP-10, a lipopolysaccharide-and interferon-inducible protein.[183] Both TNF[184] and IL-1[185] can also induce nitric oxide release by rat mesangial cells *in vitro*. As well as inducing vasodilatory actions, nitric oxide has potential immunomodulatory and cytotoxic effects. This induction of nitric oxide synthesis by macrophage-derived cytokines may be involved in the inflammatory response of interstitial fibrosis following renal injury.

7.15.7 METALLOPROTEINASES AND TISSUE INHIBITORS OF METALLOPROTEINASES

The accumulation of extracellular matrix following tubulointerstitial injury represents an imbalance between extracellular matrix deposition and dysregulation of proteases, which degrade matrix components. Proteases are a family of enzymes collectively termed metalloproteinases, capable of degrading both the collagenous and noncollagenous components of the extracellular matrix. Matrix metalloproteinases, such as meprin, stromelysin, collagenase, and gelatinase, are present in renal tissue and play an important role in the regulation of normal tissue remodeling.[186,187] The activity of matrix metalloproteinases is controlled, in part, by inhibitory proteins or tissue inhibitor of metalloproteinases (TIMPs). The TIMPs are composed of a family of three proteins which can bind to metalloproteinases to inhibit extracellular matrix degradation.[188]

The synthesis and secretion of metalloproteinases and their inhibitors, the TIMPs, are regulated by several cytokines and hormones.[189] Macrophage-derived cytokines PDGF, EGF, and αFGF can increase collagenase, transin, and stromelysin mRNA expression and activity.[190,191] IL-1 and phorbol esters which stimulate metalloproteinase activity are also potent activators of TIMP expression.[189] The profibrogenic effect of another macrophage-derived cytokine TGF-β is related to its ability to promote extracellular matrix deposition. TGF-β is a potent repressor of collagenase and stromelysin and can activate increased TIMP-1 expression, particularly in combination with IL-1.[192] The ability of TGF-β to inhibit collagenase synthesis and increase TIMP-1 expression has been demonstrated in human fibroblast cultures.[193] *In situ* hybridization studies have revealed that tissue localization of TIMP-1 transcripts is present at sites of active tissue remodeling, where its expression significantly overlaps with that of TGF-β.[194]

Alterations in the regulation of TIMP expression by TGF-β may contribute to the development of interstitial fibrosis following unrelieved UUO. A marked elevation of TIMP-1 mRNA expression was demonstrated as early as 12 h following UUO.[195] By 96 h after UUO, there was a 30-fold increment in TIMP-1 mRNA in the obstructed kidneys compared to the contralateral unobstructed kidney (Figure 4). Macrophage-derived TGF-β could contribute to the inception of the fibrogenic response following obstruction. We previously observed that macrophage infiltration into the renal interstitium in UUO rats is significantly correlated with a pronounced increment in renal cortical TGF-β1 expression.[42] This markedly increased expression of TGF-β1 following ureteral ligation may induce a profibrogenic state and initiate a cascade of dysregulatory events, including up-regulation of TIMP, leading to postobstructive interstitial fibrosis. On *in situ* hybridization, TIMP-1 mRNA transcripts localized to the peritubular interstitium and peri-adventitial areas of blood vessels (unpublished observations).

Disturbances in metalloproteinase activity and increased TIMP expression have been implicated in variety of other experimental models

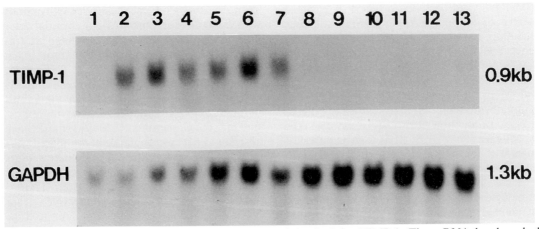

Figure 4 Northern blot analysis of total cortical RNA probed for TIMP-1. The mRNA band probed for TIMP-1 from a representative sham-operated rat kidney (lane 1), obstructed kidneys (lanes 2–7), and contralateral unobstructed kidney (CUK) specimens (lanes 8–13), 48 h after UUO are shown (top) in relation to glyceraldehyde-3-phosphate dehydrogenase (GAPDH) mRNA (bottom). There was a single 0.9 kb transcript for TIMP-1, which is readily apparent in all of the obstructed kidney specimens, but only faintly visible in normal renal cortex or CUK specimens.

of progressive renal disease.[44,174,196–198] Tang et al.[199] assessed the temporal relationship between the accumulation of extracellular matrix, metalloproteinases, and their inhibitors in a rat model of antitubular basement membrane antibody-associated tubulointerstitial nephritis. Increased expression of FGF, PDGF, and TGF-β were observed to correlate with increased extracellular matrix components (including fibronectin, laminin, and collagen types I, III, and IV), a decrease in proteinases transin/stromelysin, and an increase in the proteinase inhibitors TIMP and plasminogen activator inhibitor-1.[199] Interstitial accumulation of collagen I, III and IV, fibronectin and laminin, and increased TIMP-1 expression have also been implicated in the pathogenesis of interstitial fibrosis in chronic puromycin aminonucleoside nephrosis.[44]

The membrane-bound metalloproteinase, meprin, is also implicated in the development of renal injury. Meprin is a metalloendopeptidase located in the brush border membranes of the rat and mouse kidney.[200] Although the function of meprin in normal renal tubules remains unknown, Trachtman et al.[198] proposed that decreased meprin activity in proteinuric conditions may contribute to progressive renal injury as decreased meprin activity was observed in rats with chronic puromycin aminonucleoside nephrosis. In contrast, meprin activity was elevated in streptozocin-induced diabetic nephropathy,[198] an effect which may be secondary to general tubular cell hypertrophy or be triggered by hyperfiltration in diabetic rats. A decreased meprin-β mRNA expression in kidneys has been observed from rats as early as 12 h following UUO.[201] These studies suggest that dysregulation of the renal proximal tubule epithelial cells following ureteral obstruction may lead to diminished meprin activity and hence alterations in extracellular matrix degradation.

7.15.8 ANGIOTENSIN II

Changes in renal hemodynamics and glomerular filtration rate (GFR) occur as a result of complex interactions between vasoconstrictor and vasodilatory agents. Evidence implicating angiotensin II in altered renal hemodynamics leading to tubulointerstitial injury has been observed in obstructive nephropathy,[39,171] puromycin aminonucleoside nephrosis,[202,203] and following angiotensin II infusion.[136,204]

Acute ureteral obstruction causes an initial increase in renal blood flow, due to vasodilator prostaglandins, followed by progressive vasoconstriction of glomerular arterioles, due principally to angiotensin II, leading to a marked decline in GFR.[39] Administration of enalapril, an angiotensin-converting enzyme (ACE) inhibitor, to rats prior to obstruction significantly increases GFR and renal plasma flow.[205] Increased renin secretion, which is inhibited by pretreatment with cyclooxygenase inhibitors,[206] has been observed following ureteral obstruction. This increase in renin secretion leads to increased angiotensin II production.[39] Angiotensin II may also stimulate TGF-β1 mRNA expression following ureteral obstruction. Kaneto et al.[171] reported that treatment of rats with enalapril significantly blunted TGF-β1 mRNA expression following UUO. Angiotensin II has previously been found to modulate proliferation of vascular smooth muscle cells by stimulating TGF-β1 synthesis.[207,208]

A possible role for angiotensin II in initiating tubulointerstitial injury has been observed in angiotensin II infusion studies in rats. Johnson et al.[204] used immunohistochemistry to demonstrate that angiotensin II-mediated hypertension induced vascular, glomerular, tubular, and interstitial cell proliferation, and phenotypic modulation of mesangial and interstitial cells to express α-smooth muscle actin. Angiotensin II-infused rats also developed focal tubulointerstitial injury, associated with an influx of interstitial monocytes, and mild interstitial fibrosis with increased collagen type IV deposition.[204] Angiotensin II infusion has been reported to mediate macrophage infiltration into the renal interstitium.[136] Giachelli et al.[136] observed increased osteopontin mRNA expression and protein localized in the cortical renal tubules following angiotensin II infusion in rats. Angiotensin II may stimulate osteopontin expression of the renal tubules, thereby facilitating monocyte/macrophage accumulation at the sites of tubulointerstitial injury.[136]

7.15.9 CONCLUSION

Utilizing the rat UUO model as a prototype disorder, which culminates in interstitial fibrosis and tubular loss, we offer the following scheme (Figure 5). The mechanical disturbance, resulting from complete ureteral ligation, causes tubular injury/dysfunction. This initial injury state results in a florid proinflammatory and profibrogenic response. First, there are a host of chemoattractant signals elaborated by the tubular epithelium, including MCP-1, osteopontin, ICAM-1 and, perhaps nonpeptide lipid moieties. These chemokines

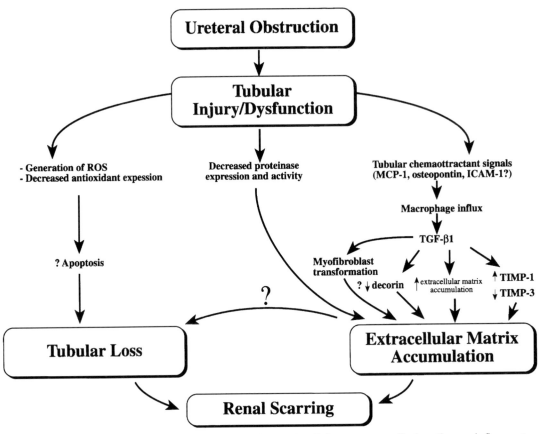

Figure 5 Schematic diagram showing the molecular and cellular events mediating the proinflammatory response to tubular injury/dysfunction, leading to the culmination of extracellular matrix accumulation, tubular loss, and ultimately renal scarring.

mediate macrophage influx into the renal interstitium. One of many macrophage products includes TGF-β1, which causes a myriad of perturbations. Thus far, we have documented, in the UUO model, myofibroblast transformation, increased extracellular matrix expression, increased TIMP-1, and decreased TIMP-3 mRNA expression. We also speculate that the up-regulated TGF-β1 state is associated with a down-regulation of decorin, a member of a class of small dermatan/chondroitin sulfate proteoglycans which can bind to and inhibit the activity of TGF-β1 (as well as the other two mammalian isoforms). Studies are presently underway to determine the mRNA and protein expression of decorin in obstructed vs. unobstructed rat kidneys. The net effect of all of these perturbations is extracellular matrix accumulation and organization and, ultimately interstitial fibrosis. Along the lines of extracellular matrix accumulation, we also postulate that tubular-derived metalloproteinases may be down-regulated with tubular injury/dysfunction post-obstruction. Thus far, we have noted marked

down-regulation of meprin, a proximal tubular brush-border endopeptidase.

Finally, we have observed that coincident with tubular injury/dysfunction postobstruction, there is a dramatic down-regulation of antioxidant enzymes (Cu–ZnSOD and catalase) at both the mRNA and protein levels. In our putative scheme, we offer that down-regulated antioxidant expression may not only result in an amplified generation of ROS but may also contribute to an increased rate of programmed cell death (apoptosis). The process of tubular loss is fundamental to the decline in renal function. It is conceivable that the molecular and cellular events mediating interstitial matrix remodeling during this proinflammatory state may also promote tubular loss. The loss of tubules may develop as a consequence of peritubular capillary occlusion with ensuing ischemia, or due to increased apoptosis from ROS generation with ensuing DNA damage. Clearly, more attention needs to be accorded to linking the processes of tubular cell loss and interstitial extracellular matrix remodeling as major codeterminants of renal scarring.

ACKNOWLEDGMENTS

This study was supported by a Research Grant in Aid from the American Heart Association (Pennsylvania Affiliate) and an Extramural Research Grant from Baxter Healthcare Corporation. J. R. Diamond is an Established Investigator of the American Heart Association. S. D. Ricardo is a Fellow of the National Kidney Foundation and a recipient of the Alyce G. Spector Research Grant from the Kidney Foundation of Central Pennsylvania.

7.15.10 REFERENCES

1. J. R. Diamond, in 'Renal Disease and Toxicology,' ed. R. S. Goldstein, CRC Press, Boca Raton, FL, 1994, chap. 3, pp. 44–61.
2. L. I. Schainuck, G. E. Striker, R. E. Cutler *et al.*, 'Structural-functional correlations in renal disease. II. The correlations.' *Hum. Pathol.*, 1970, **1**, 631–641.
3. R. A. Risdon, J. C. Sloper and H. E. De Wardener, 'Relationship between renal function and histological changes found in renal-biopsy specimens from patients with persistent glomerular nephritis.' *Lancet*, 1968, **2**, 363–366.
4. S. Mackensen-Haen, R. Eissele and A. Bohle, 'Contribution on the correlation between morphometric parameters gained from the renal cortex and renal function in IgA nephritis.' *Lab. Invest.*, 1988, **59**, 239–244.
5. M. Wehrmann, A. Bohle, H. Held *et al.*, 'Long-term prognosis of focal sclerosing glomerulonephritis. An analysis of 250 cases with particular regard to tubulointerstitial changes.' *Clin. Nephrol.*, 1990, **33**, 115–122.
6. A. A. Eddy, 'Experimental insights into the tubulointerstitial disease accompanying primary glomerular lesions.' *J. Am. Soc. Nephrol.*, 1994, **5**, 1273–1287.
7. J. R. Diamond, 'Macrophages and progressive renal disease in experimental hydronephrosis.' *Am. J. Kid. Dis.*, 1995, **26**, 133–140.
8. L. G. Fine, A. C. Ong and J. T. Norman, 'Mechanisms of tubulo-interstitial injury in progressive renal diseases.' *Eur. J. Clin. Invest.*, 1993, **23**, 259–265.
9. L. I. Schainuck, G. E. Striker, R. E. Cutler *et al.*, 'Structural-functional correlations in renal disease. II: The correlations.' *Hum. Pathol.*, 1970, **1**, 631–641.
10. G. E. Striker, L. I. Schainuck, R. E. Cutler *et al.*, 'Structural-functional correlations in renal disease. I. A method for assaying and classifying histopathologic changes in renal disease.' *Hum. Pathol.*, 1970, **1**, 615–630.
11. A. Bohle, S. Mackensen-Haen, H. von Gise *et al.*, 'The consequences of tubulo-interstitial changes for renal function in glomerulopathies. A morphometric and cytological analysis.' *Pathol. Res. Pract.*, 1990, **186**, 135–144.
12. S. Mackensen-Haen, A. Bohle, J. Christensen *et al.*, 'The consequences for renal function of widening of the interstitium and changes in the tubular epithelium of the renal cortex and outer medulla in various renal diseases.' *Clin. Nephrol.*, 1992, **37**, 70–77.
13. M. M. Schwartz, A. K. Bidani and E. J. Lewis, 'Glomerular epithelial cell function and pathology following extreme ablation of renal mass.' *Am. J. Pathol.*, 1987, **126**, 315–324.
14. D. H. Hooke, D. C. Gee and R. C. Atkins, 'Leukocyte analysis using monoclonal antibodies in human glomerulonephritis.' *Kidney Int.*, 1987, **31**, 964–972.
15. T. V. Tuazon, E. E. Schreeberger, A. K. Bhan *et al.*, 'Mononuclear cells in acute allograft glomerulopathy.' *Am. J. Pathol.*, 1987, **129**, 119–132.
16. A. B. Magil and A. H. Cohen, 'Monocytes and focal glomerulosclerosis.' *Lab. Invest.*, 1989, **61**, 404–409.
17. E. Alexopoulos, D. Seron, R. B. Hartley *et al.*, 'Lupus nephritis: correlation of interstitial cells with glomerular function.' *Kidney Int.*, 1990, **37**, 100–109.
18. G. F. Schreiner, 'The role of the macrophage in glomerular injury.' *Semin. Nephrol.*, 1991, **11**, 268–275.
19. D. J. Nikolic-Paterson, H. Y. Lan, P. A. Hill *et al.*, 'Macrophages in renal injury.' *Kidney Int.*, 1994, **45** (Suppl.), S79–S82.
20. A. A. Eddy, G. C. Ho and P. S. Thorner, 'The contribution of antibody-mediated cytotoxicity and immune-complex formation to tubulointerstitial disease in passive Heymann nephritis.' *Clin. Immunol. Immunopathol.*, 1992, **62**, 42–55.
21. J. R. Diamond and I. Pesek-Diamond, 'Sublethal X-irradiation during acute puromycin nephrosis prevents late renal injury: role of macrophages.' *Am. J. Physiol.*, 1991, **260**, F779–F786.
22. C. L. Jones, S. Buch, M. Post *et al.*, 'Renal extracellular matrix accumulation in acute puromycin aminonucleoside nephrosis in rats.' *Am. J. Pathol.*, 1992, **141**, 1381–1396.
23. T. Saito and R. C. Aitkins, 'Contribution of mononuclear leukocytes to the progression of experimental focal glomerular sclerosis.' *Kidney Int.*, 1990, **37**, 1076–1083.
24. K. Tamaki, S. Okuda, T. Ando *et al.*, 'TGF-β1 in glomerulosclerosis and interstitial fibrosis of adriamycin nephropathy.' *Kidney Int.*, 1994, **45**, 525–536.
25. A. A. Eddy, 'Tubulointerstitial nephritis during the heterologous phase of nephrotoxic serum nephritis.' *Nephron*, 1991, **59**, 304–313.
26. J. Floege, C. E. Alpers, M. W. Burns *et al.*, 'Glomerular cells, extracellular matrix accumulation, and the development of glomerulosclerosis in the remnant kidney model.' *Lab. Invest.*, 1992, **66**, 485–496.
27. H. van Goor, M. L. C. van der Horst, V. Fidler *et al.*, 'Glomerular macrophage modulation affects mesangial expansion in the rat after renal ablation.' *Lab. Invest.*, 1992, **66**, 564–571.
28. B. A. Young, R. J. Johnson, C. E. Alpers *et al.*, 'Cellular events in the evolution of experimental diabetic nephropathy.' *Kidney Int.*, 1995, **47**, 935–944.
29. D. J. Nikolic-Paterson, H. Y. Lan, P. A. Hill *et al.*, 'Macrophages in renal injury.' *Kidney Int.*, 1994, **45** (Suppl.), S79–S82.
30. H. Van Goor, G. Ding, D. Kees-Folts *et al.*, 'Biology of disease. Macrophages and renal disease.' *Lab. Invest.*, 1994, **71**, 456–464.
31. K. Matsumoto and M. Hatano, 'Effect of antimacrophage serum on the proliferation of glomerular cells in nephrotoxic serum nephritis in the rat.' *J. Clin. Lab. Immunol.*, 1989, **28**, 39–44.
32. R. Van-Diemen-Steenvoorde, A. Lambers, A. Van der Wal *et al.*, 'Macrophages are responsible for mesangial cell injury and extracellular matrix (ECM) expansion in anti-thy-1 nephritis in rats.' *J. Am. Soc. Nephrol.*, 1991, **2** (Abstract), 585.
33. J. R. Diamond, I. Pesek, S. Ruggieri *et al.*, 'Essential fatty acid deficiency during acute puromycin nephrosis ameliorates late renal injury.' *Am. J. Physiol.*, 1989, **257**, F798–F807.
34. G. F. Schreiner, B. Rovin and J. B. Lefkowith, 'The antiinflammatory effects of essential acid deficiency in experimental glomerulonephritis.' *J. Immunol.*, 1989, **143**, 3192–3199.

35. K. P. Harris, J. B. Lefkowith, S. Klahr *et al.*, 'Essential fatty acid deficiency ameliorates acute renal dysfunction in the rat after administration of the aminonucleoside of puromycin.' *J. Clin. Invest.*, 1990, **86**, 1115–1123.

36. G. F. Schreiner, K. P. Harris, M. L. Purkerson *et al.*, 'Immunological aspects of acute ureteral obstruction: immune cell infiltrate in the kidney.' *Kidney Int.*, 1988, **34**, 487–493.

37. R. B. Nagle, R. E. Bulger, R. E. Cutler *et al.*, 'Unilateral obstructive nephropathy in the rabbit. I. Early morphologic, physiologic, and histochemical changes.' *Lab. Invest.*, 1973, **28**, 456–467.

38. R. B. Nagle, M. E. Johnson and H. R. Jervis, 'Proliferation of renal interstitial cells following injury induced by ureteral obstruction.' *Lab. Invest.*, 1976, **35**, 18–22.

39. S. Klahr, 'New insights into the consequences and mechanisms of renal impairment in obstructive nephropathy. The modulation of macrophage migration and eicosanoid metabolism.' *Am. J. Kid. Dis.*, 1991, **18**, 689–699.

40. A. A. Reyes, B. H. Porras, F. I. Chasalow *et al.*, 'L-arginine decreases the infiltration of the kidney by macrophages in obstructive nephropathy and puromycin-induced nephrosis.' *Kidney Int.*, 1994, **45**, 1346–1354.

41. K. P. G. Harris, G. F. Schreiner and S. Klahr, 'Effect of leukocyte depletion on the function of the postobstructed kidney in the rat.' *Kidney Int.*, 1989, **36**, 210–215.

42. J. R. Diamond, D. Kees-Folts, G. Ding *et al.*, 'Macrophages, monocyte chemoattractant peptide-1, and TGF-β1 in experimental hydronephrosis.' *Am. J. Physiol.*, 1994, **266**, F926–F933.

43. C. L. Jones, S. Buch, M. Post *et al.*, 'Renal extracellular matrix accumulation in acute puromycin aminonucleoside nephrosis in rats.' *Am. J. Pathol.*, 1992, **141**, 1381–1396.

44. C. L. Jones, S. Buch, M. Post *et al.*, 'Pathogenesis of interstitial fibrosis in chronic purine aminonucleoside nephrosis.' *Kidney Int.*, 1991, **40**, 1020–1031.

45. A. A. Eddy, 'Protein restriction reduces transforming growth factor-β and interstitial fibrosis in nephrotic syndrome.' *Am. J. Physiol.*, 1994, **266**, F884–F893.

46. A. C. M. Ong and L. G. Fine, 'Loss of glomerular function and tubulointerstitial fibrosis: cause or effect?' *Kidney Int.*, 1994, **45**, 345–351.

47. R. L. Schmouder, R. M. Strieter and S. L. Kunkel, 'Interferon-γ regulation of human renal cortical epithelial cell-derived monocyte chemotactic peptide-1.' *Kidney Int.*, 1993, **44**, 43–49.

48. D. E. Kohan, 'Endothelin synthesis by rabbit renal tubule cells.' *Am. J. Physiol.*, 1991, **261**, F221–F226.

49. R. Segal and L. G. Fine, 'Polypeptide growth factors and the kidney.' *Kidney Int.*, 1989, **27** (Suppl.), S2–S10.

50. M. V. Rocco, Y. Chen, S. Goldfarb *et al.*, 'Elevated glucose stimulates TGF-β gene expression and bioactivity in proximal tubules.' *Kidney Int.*, 1992, **41**, 107–114.

51. J. Deguchi, T. Kawabata, A. Konodo *et al.*, 'Transforming growth factor-alpha expression of renal proximal tubules in Wistar rats treated with ferric and aluminium nitrilotriacetate.' *Jpn. J. Cancer Res.*, 1993, **84**, 649–655.

52. J. Frank, G. Engel-Blum, H. P. Rodemann *et al.*, 'Human renal tubular cells as a cytokine source: PDGF-B, GM-CSF and IL-6 mRNA expression *in vitro*.' *Exp. Nephrol.*, 1993, **1**, 26–35.

53. B. A. Markewitz, J. R. Michael and D. E. Kohan, 'Cytokine-induced expression of a nitric oxide synthase in rat renal tubule cells.' *J. Clin. Invest.*, 1993, **91**, 2138–2143.

54. J. S. McLay, P. Chatterjee, A. G. Nicolson *et al.*, 'Nitric oxide production by human proximal tubular cells: a novel immunomodulatory mechanism?' *Kidney Int.*, 1994, **46**, 1043–1049.

55. B. H. Rovin, K. P. G. Harris, A. Morrison *et al.*, 'Renal cortical release of a specific macrophage chemoattractant in response to obstruction.' *Lab. Invest.*, 1990, **63**, 213–220.

56. S. P. Andreoli and J. A. McAteer, 'Reactive oxygen molecule-mediated injury in endothelial and renal tubular epithelial cells *in vitro*.' *Kidney Int.*, 1990, **38**, 785–794.

57. J. Gailit, D. Colflesh, I. Rabiner *et al.*, 'Redistribution and dysfunction of integrins in cultured renal epithelial cells exposed to oxidative stress.' *Am. J. Physiol.*, 1993, **264**, F149–F157.

58. G. Wolf, P. D. Killen and E. G. Neilson, 'Intracellular signaling of transcription and secretion of type IV collagen after angiotensin II-induced cellular hypertrophy in culture proximal tubular cells.' *Cell Regul.*, 1991, **2**, 219–227.

59. T. P. Haverty, C. J. Kelly, W. H. Hines *et al.*, 'Characterization of a renal tubular epithelial cell line which secretes the autologous target antigen of autoimmune experimental interstitial nephritis.' *J. Cell. Biol.*, 1988, **107**, 1359–1368.

60. W. A. Border and E. Ruoslahti, 'Transforming growth factor-β in disease: the dark side of tissue repair.' *J. Clin. Invest.*, 1992, **90**, 1–7.

61. G. S. Kuncio, E. G. Neilson and T. Haverty, 'Mechanisms of tubulointerstitial fibrosis.' *Kidney Int.*, 1991, **39**, 550–556.

62. G. A. Muller, J. Markovic-Lipkovski, J. Frank *et al.*, 'The role of interstitial cells in the progression of renal diseases.' *J. Am. Soc. Nephrol.*, 1992, **2**, S198–S205.

63. H. P. Rodemann, K. Bayreuther, P. I. Franz *et al.*, 'Selective enrichment and biochemical characterization of seven human skin fibroblast cell types *in vitro*.' *Exp. Cell. Res.*, 1989, **180**, 84–93.

64. H. P. Rodemann and G. A. Muller, 'Abnormal growth and clonal proliferation of fibroblasts derived from kidneys with interstitial fibrosis.' *Proc. Soc. Exp. Biol. Med.*, 1990, **195**, 57–63.

65. H. P. Rodemann and G. A. Muller, 'Characterization of human renal fibroblasts in health and disease: II. In vitro growth, differentiation and collagen synthesis of fibroblasts from kidneys with interstitial fibrosis.' *Am. J. Kidney Dis.*, 1991, **17**, 684–686.

66. G. A. Muller and H. P. Rodemann, 'Characterization of human renal fibroblasts in health and disease: I. Immunophenotyping of cultured tubular epithelial cells and fibroblasts derived from kidneys with histologically proven intersitial fibrosis.' *Am. J. Kidney Dis.*, 1991, **17**, 680–683.

67. P. Bornstein and H. Sage, 'Regulation of collagen gene expression.' *Prog. Nucleic Acid Res. Mol. Biol.*, 1989, **37**, 67–106.

68. A. E. Postlethwaite, J. Keski-Oja, H. L. Moses *et al.*, 'Stimulation of the chemotactic migration of human fibroblasts by transforming growth factor beta.' *J. Exp. Med.*, 1987, **165**, 251–256.

69. R. J. Alvarez, M. J. Sun, T. P. Haverty *et al.*, 'Biosynthetic and proliferative characteristics of tubulointerstitial fibroblasts probed with paracrine cytokines.' *Kidney Int.*, 1992, **41**, 14–23.

70. A. Knecht, L. G. Fine, K. S. Kleinman *et al.*, 'Fibroblasts of the rabbit kidney culture.: II. Paracrine stimulation of papillary fibroblasts by PDGF.' *Am. J. Physiol.*, 1991, **261**, F292–F299.

71. E. G. Neilson, S. M. Phillips and S. Jimenez, 'Lymphokine modulation of fibroblast proliferation.' *J. Immunol.*, 1982, **128**, 1484–1487.

72. J. A. Elias, B. Freundlich, J. A. Kern *et al.*, 'Cytokine networks in the regulation of inflammation and fibrosis in the lung.' *Chest*, 1990, **97**, 1439–1445.

73. A. E. Postlethwaite, R. Snyderman and A. H. Kang, 'Generation of a fibroblast chemotactic factor in serum by activation of complement.' *J. Clin. Invest.*, 1979, **64**, 1379–1385.

74. H. E. Seppa, K. M. Yamada, S. T. Seppa *et al.*, 'The cell binding fragment of fibronectin is chemotactic for fibroblasts.' *Cell. Biol. Int. Rep.*, 1981, **5**, 813–819.

75. A. E. Postlethwaite, J. M. Seyer and A. H. Kang, 'Chemotactic attraction of human fibroblasts to type I, II, III collagens and collagen-derived peptides.' *Proc. Natl. Acad. Sci. USA*, 1978, **75**, 871–875.

76. A. B. Roberts, K. C. Flanders, P. Kondaiah *et al.*, 'Transforming growth factor β: biochemistry and roles in embryogenesis. Recent tissue repair and remodeling, and carcinogenesis.' *Prog. Horm. Res.*, 1988, **44**, 157–197.

77. K. J. Valle and E. A. Bauer, 'Biosynthesis of the collagenase by human skin fibroblasts in monolayer culture.' *J. Biol. Chem.*, 1979, **254**, 10115–10122.

78. C. M. Alexander and Z. Werb, 'Proteinases and extracellular matrix remodeling.' *Curr. Opin. Cell. Biol.*, 1989, **1**, 974–982.

79. R. B. Nagle, M. R. Kneiser, R. E. Bulger *et al.*, 'Induction of smooth muscle characteristics in renal interstitial fibroblasts during obstructive nephropathy.' *Lab. Invest.*, 1973, **29**, 422–427.

80. C. E. Alpers, K. L. Hudkins, J. Floege *et al.*, 'Human renal cortical interstitial cells with some features of smooth muscle cells participate in tubulointerstitial and crescentic glomerular injury.' *J. Am. Soc. Nephrol.*, 1994, **5**, 201–210.

81. A. K. Sharma, S. M. Mauer, Y. Kim *et al.*, 'Interstitial fibrosis in obstructive nephropathy.' *Kidney Int.*, 1993, **44**, 774–788.

82. J. R. Diamond, 'The role of reactive oxygen species in animal models of glomerular disease.' *Am. J. Kidney Dis.*, 1992, **19**, 292–300.

83. K. A. Nath, A. K. Salahudeen, E. C. Clark *et al.*, 'Role of cellular metabolites in progressive renal injury.' *Kidney Int.*, 1992, **42** (Suppl.), S109–S113.

84. A. C. Alfrey, 'Role of iron and oxygen radicals in the progression of chronic renal failure.' *Am. J. Kidney Dis.*, 1994, **23**, 183–187.

85. R. J. Johnson, D. Lovett, R. I. Lehrer *et al.*, 'Role of oxidants and proteases in glomerular injury.' *Kidney Int.*, 1994, **45**, 352–359.

86. K. A. Nath, M. Fischereder and T. H. Hostetter, 'The role of oxidants in progressive renal injury.' *Kidney Int.*, 1994, **45** (Suppl.), S111–S115.

87. S. V. Shah, 'Light emission by isolated rat glomeruli in response to phorbol myristate acetate.' *J. Lab. Clin. Med.*, 1981, **98**, 46–57.

88. H. H. Radeke, B. Meier, N. Topley *et al.*, 'Interleukin 1-α and tumour necrosis factor-α induce oxygen radical production in mesangial cells.' *Kidney Int.*, 1990, **37**, 767–775.

89. L. Baud and R. Ardaillou, 'Reactive oxygen species: production and role in the kidney.' *Am. J. Physiol.*, 1986, **251**, F765–F776.

90. B. Halliwell and J. M. C. Gutteridge, 'Oxygen toxicity, oxygen radicals, transition metals and disease.' *Biochem. J.*, 1984, **219**, 1–14.

91. B. Chance, H. Sies and A. Boveris, 'Hydroperoxide metabolism in mammalian organs.' *Physiol. Rev.*, 1979, **59**, 527–605.

92. C. E. Vaca, J. Wilhelm and M. Harms-Ringdahl, 'Interactions of lipid peroxidation products with DNA. A review.' *Mutat. Res.*, 1988, **195**, 137–149.

93. W. A. Pryor, 'Oxy-radicals and related species: their formation, lifetimes and reactions.' *Annu. Rev. Physiol.*, 1986, **48**, 657–667.

94. S. V. Shah, 'Evidence suggesting a role for hydroxyl radical in passive Heymann nephritis in rats.' *Am. J. Physiol.*, 1988, **254**, F337–F344.

95. M. A. Rahman, S. S. Emancipator and J. R. Sedor, 'Hydroxyl radical scavengers ameliorate proteinuria in immune complex glomerulonephritis.' *J. Lab. Clin. Invest.*, 1988, **112**, 619–626.

96. T. J. Neale, R. Ullrich, P. Ojha *et al.*, 'Reactive oxygen species and neutrophil respiratory burst cytochrome b_{558} are produced by kidney glomerular cells in passive Heymann nephritis.' *Proc. Natl. Acad. Sci. USA*, 1993, **90**, 3645–3649.

97. A. Rehan, K. J. Johnson, R. C. Wiggins *et al.*, 'Evidence for the role of oxygen radicals in acute nephrotoxic nephritis.' *Lab. Invest.*, 1984, **51**, 396–403.

98. N. W. Boyce and S. R. Holdsworth, 'Hydroxyl radical mediation of immune injury by desferrioxamine.' *Kidney Int.*, 1986, **30**, 813–817.

99. N. W. Boyce, P. G. Tipping and S. R. Holdsworth, 'Glomerular macrophages produce reactive oxygen species in experimental glomerulonephritis.' *Kidney Int.*, 1989, **35**, 778–782.

100. J. R. Diamond, J. V. Bonventre and M. J. Karnovsky, 'A role for oxygen free radicals in aminonucleoside nephrosis.' *Kidney Int.*, 1986, **29**, 478–483.

101. V. Thakur, P. D. Walker and S. V. Shah, 'Evidence suggesting a role for hydroxyl radical in puromycin aminonucleoside-induced proteinuria.' *Kidney Int.*, 1988, **34**, 494–499.

102. S. D. Ricardo, J. F. Bertram and G. B. Ryan, 'Reactive oxygen species in puromycin aminonucleoside nephrosis: *in vitro* studies.' *Kidney Int.*, 1994, **45**, 1057–1069.

103. T. Bertani, A. Poggi, R. Pozzini *et al.*, 'Adriamycin-induced nephrotic syndrome in rats: sequence of pathologic events.' *Lab. Invest.*, 1982, **46**, 16–23.

104. E. G. Mimnaugh, M. A. Trush, M. Bhatnagar *et al.*, 'Enhancement of reactive oxygen-dependent mito-chrondrial membrane lipid peroxidation by the anti-cancer drug adriamycin.' *Biochem. Pharmacol.*, 1985, **34**, 847–856.

105. T. Okasora, T. Takikawa, Y. Utsunomiya *et al.*, 'Suppressive effect of superoxide dismutase on adria-mycin nephropathy.' *Nephron*, 1992, **60**, 199–203.

106. N. Ueda, B. Guidet and S. V. Shah, 'Measurement of intracellular generation of hydrogen peroxide by rat glomeruli *in vitro*.' *Kidney Int.*, 1994, **45**, 788–793.

107. P. D. Walker and S. V. Shah, 'Evidence suggesting a role for hydroxyl radical in gentamicin-induced acute renal failure in rats.' *J. Clin. Invest.*, 1988, **81**, 334–341.

108. K. A. Nath, N. Ueda, P. D. Walker *et al.*, in 'Renal Disease and Toxicology,' ed. R. S. Goldstein, CRC Press, Boca Raton, FL, 1994, chap. 12, pp. 267–278.

109. A. Rehan, K. J. Johnson, R. G. Kunkel *et al.*, 'Role of oxygen radicals in phorbol myristate acetate-induced glomerular injury.' *Kidney Int.*, 1985, **27**, 503–511.

110. A. Rehan, R. C. Wiggins, R. G. Kunkel *et al.*, 'Glomerular injury and proteinuria in rats after intrarenal injection of cobra venom factor. Evidence for the role of neutrophil-derived oxygen free radicals.' *Am. J. Pathol.*, 1986, **123**, 57–66.

111. D. Lotan, B. S. Kaplan, J. S. Fong *et al.*, 'Reduction of protein excretion by dimethyl sulfoxide in rats with passive Heymann nephritis.' *Kidney Int.*, 1984, **25**,

778–788.

112. R. J. Johnson, S. J. Klebanoff, R. F. Ochi *et al.*, 'Participation of the myeolperoxidase-H$_2$O$_2$-halide system in immune complex nephritis.' *Kidney Int.*, 1987, **32**, 342–349.

113. R. J. Johnson, W. G. Couser, E. Y. Chi *et al.*, 'New mechanism for glomerular injury. Myeloperoxidase-hydrogen peroxide-halide system.' *J. Clin. Invest.*, 1987, **79**, 1379–1387.

114. R. J. Johnson, S. J. Guggenheim, S. J. Klebanoff *et al.*, 'Morphologic correlates of glomerular oxidant injury induced by the myeloperoxidase-hydrogen peroxide-halide system of the neutrophil.' *Lab. Invest.*, 1988, **58**, 294–301.

115. K. S. Modi, J. Morrissey, S. V. Shah *et al.*, 'Effects of probucol on renal function in rats with bilateral ureteral obstruction.' *Kidney Int.*, 1990, **38**, 843–850.

116. S. D. Ricardo and J. R. Diamond, 'Reactive oxygen species (ROS) and antioxidant expression in experimental hydronephrosis.' *J. Am. Soc. Nephrol.*, 1995, **6** 401.

117. G. H. W. Wong and D. V. Goeddel, 'Induction of manganese superoxide dismutase by tumor necrosis factor: possible protective mechanism.' *Science*, 1988, **242**, 941–944.

118. G. A. Visner, W. C. Dougall, J. M. Wilson *et al.*, 'Regulation of manganese superoxide dismutase by lipopolysaccharide, interleukin-1, and tumor necrosis factor. Role in the acute inflammatory response.' *J. Biol. Chem.*, 1990, **265**, 2856–2864.

119. Y. Kono and I. Fridovich, 'Superoxide radical inhibits catalase.' *J. Biol. Chem.*, 1982, **257**, 5751–5754.

120. E. Pigeolet, P. Corbisier, A. Houbion *et al.*, 'Glutathione peroxidase, superoxide dismutase and catalase inactivation by peroxides and oxygen derived free radicals.' *Mech. Ageing Dev.*, 1990, **51**, 283–297.

121. I. Singh, S. Gulati, J. K. Orak *et al.*, 'Expression of antioxidant enzymes in rat kidney during ischemia-reperfusion injury.' *Mol. Cell. Biochem.*, 1993, **125**, 97–104.

122. S. Gulati, A. K. Singh, C. Irazu *et al.*, 'Ischemia-reperfusion injury: biochemical alterations in peroxisomes of rat kidney.' *Arch. Biochem. Biophys.*, 1992, **295**, 90–100.

123. S. L. Jewett, S. Cushing, F. Gillespie *et al.*, 'Reaction of bovine-liver copper–zinc superoxide dismutase with hydrogen peroxide. Evidence for reaction with H$_2$O$_2$ and HO$_2^-$ leading to loss of copper.' *Eur. J. Biochem.*, 1989, **180**, 569–575.

124. K. Houglam, D. A. Brenner and M. Chojkier, 'd-α-Tocopherol inhibits collagen α$_1$(1) gene expression in cultured human fibroblasts. Modulation of constitutive collagen gene expression by lipid peroxidation.' *J. Clin. Invest.*, 1991, **87**, 2230–2235.

125. K. Ohyama, J. M. Seyer, R. Raghow *et al.*, 'A factor from damaged rat kidney stimulates collagen biosynthesis by mesangial cells.' *Biochim. Biopys. Acta*, 1990, **1053**, 173–178.

126. Z. Shan, D. Tan, J. Satriano *et al.*, 'Intracellular glutathione influences collagen generation by mesangial cells.' *Kidney Int.*, 1994, **46**, 388–295.

127. C. A. Lopez, J. R. Hoyer, P. D. Wilson *et al.*, 'Heterogeneity of osteopontin expression among nephrons in mouse kidneys and enhanced expression in sclerotic glomeruli.' *Lab. Invest.*, 1993, **69**, 355–363.

128. Y. Miyazaki, M. Setoguchi, S. Yoshida *et al.*, 'The mouse osteopontin gene. Expression in monocyclic lineages and complete nucleotide sequence.' *J. Biol. Chem.*, 1990, **265**, 14432–14438.

129. D. R. Senger, C. A. Peruzzi, C. F. Gracey *et al.*, 'Secreted phosphoproteins associated with neoplastic transformation: close homology with plasma proteins cleaved during blood coagulation.' *Cancer Res.*, 1988,

48, 5770–5774.

130. A. M. Craig, G. T. Bowden, A. F. Chambers *et al.*, 'Secreted phosphoprotein mRNA is induced during multi-stage carcinogenesis in mouse skin and correlates with the metastatic potential of murine fibroblasts.' *Int. J. Cancer*, 1990, **46**, 133–137.

131. K. Yoon, R. Buenaga and G. A. Rodan, 'Tissue specificity and developmental expression of rat osteopontin.' *Biochem. Biophys. Res. Commun.*, 1987, **148**, 1129–1136.

132. A. M. Craig, J. H. Smith and D. T. Denhardt, 'Osteopontin, a transformation-associated cell adhesion phosphoprotein, is induced by 12-tetradecanoyl-phorbol 13-acetate in mouse epidermis.' *J. Biol. Chem.*, 1989, **264**, 9682–9689.

133. S. Kasugai, Q. Zhang, C. M. Overall *et al.*, 'Differential regulation of the 55 and 44 kDa forms of secreted phosphoprotein 1 (SSP-1, osteopontin) in normal and transformed rat bone cells by osteotrophic hormones, growth factors and a tumor promoter.' *Bone Miner.*, 1991, **13**, 235–250.

134. A. Oldberg, A. Franzen and D. Heinegard, 'Cloning and sequence analysis of rat bone sialoprotein (osteopontin) cDNA reveals an Arg-Gly-Asp cell-binding sequence.' *Proc. Natl. Acad. Sci. USA*, 1986, **83**, 8819–8823.

135. M. J. Somerman, L. W. Fisher, R. A. Foster *et al.*, 'Human bone sialoprotein I and II enhance fibroblast attachment *in vitro*.' *Calcif. Tissue Int.*, 1988, **43**, 50–53.

136. C. M. Giachelli, R. Pichler, D. Lombardi *et al.*, 'Osteopontin expression in angiotensin II-induced tubulointerstitial nephritis.' *Kidney Int.*, 1994, **45**, 515–524.

137. R. Pichler, C. M. Giachelli, D. Lombardi *et al.*, 'Tubulointerstitial disease in glomerulonephritis. Potential role of osteopontin (uropontin).' *Am. J. Pathol.*, 1994, **144**, 915–926.

138. J. R. Diamond, D. Kees-Folts, S. D. Ricardo *et al.*, 'Early and persistent up-regulated expression of renal cortical osteopontin in experimental hydronephrosis.' *Am. J. Pathol.*, 1995, **146**, 1455–1466.

139. T. Yoshimura, E. A. Robinson, S. Tanaka *et al.*, 'Purification and amino acid analysis of two human glioma-derived monocyte chemoattractants.' *J. Exp. Med.*, 1989, **169**, 1449–1459.

140. R. M. Strieter, S. L. Kunkel, H. J. Showell *et al.*, 'Endothelial cell gene expression of a neutrophil chemotactic factor by TNF-α, LPS, and IL-1β.' *Science*, 1989, **243**, 1467–1469.

141. A. Sica, J. M. Wang, F. Colotta *et al.*, 'Monocyte chemotactic and activating factor gene expression induced in endothelial cells by IL-1 and tumour necrosis factor.' *J. Immunol.*, 1990, **144**, 3034–3038.

142. T. Yoshimura and E. J. Leonard, 'Secretion by human fibroblasts of monocyte chemoattractant protein-1, the product of gene JE.' *J. Immunol.*, 1990, **144**, 2377–2383.

143. D. T. Graves, Y. L. Jiang, M. J. Williamson *et al.*, 'Identification of monocyte chemotactic activity produced by malignant cells.' *Science*, 1989, **245**, 1490–1493.

144. C. G. Larsen, C. O. Zachariae, J. J. Oppenheim *et al.*, 'Production of monocyte chemotactic and activating factor (MCAF) by human dermal fibroblasts in response to interleukin 1 or tumor necrosis factor.' *Biochem. Biophys. Res. Commun.*, 1989, **160**, 1403–1408.

145. R. Safirstein, J. Megyesi, S. J. Saggi *et al.*, 'Expression of cytokine-like genes JE and KC is increased during renal ischemia.' *Am. J. Physiol.*, 1991, **261**,

F1095–F1101.

146. P. Heeger, G. Wolf, C. Meyers *et al.*, 'Isolation and characterization of cDNA from renal tubular epithelium encoding murine Rantes.' *Kidney Int.*, 1992, **41**, 220–225.

147. T. J. Schall, J. Jongstra, B. J. Dyer *et al.*, 'A human T cell-specific molecule is a member of a new gene family.' *J. Immunol.*, 1988, **141**, 1018–1025.

148. T. J. Schall, K. Bacon, K. J. Toy *et al.*, 'Selective attraction of monocytes and T lymphocytes of the memory phenotype by cytokine RANTES.' *Nature (Lond.)*, 1990, **347**, 669–671.

149. G. Wolf, S. Aberle, F. Thaiss *et al.*, 'TNF induces expression of the chemoattractant cytokine RANTES in cultured mouse mesangial cells.' *Kidney Int.*, 1993, **44**, 795–804.

150. H. R. Brady, A. Papayianni and C. N. Serhan, 'Leukocyte adhesion promotes biosynthesis of lipoxygenase products by transcellular routes.' *Kidney Int.*, 1994, **45** (Suppl.), S90–S97.

151. J. J. Oppenheim, C. O. Zachariae, N. Mukaida *et al.*, 'Properties of the novel proinflammatory supergene "intercrine" cytokine family.' *Annu. Rev. Immunol.*, 1991, **9**, 617–648.

152. J. A. Bruijn and E. de Heer, 'Adhesion molecules in renal disease.' *Lab. Invest.*, 1995, **72**, 387–394.

153. M. L. Dustin, R. Rothlein, A. K. Bhan *et al.*, 'Induction by IL-1 and interferon-γ: tissue distribution, biochemistry, and function of a natural adherence molecule (ICAM-1).' *J. Immunol.*, 1986, **137**, 245–254.

154. R. S. Cotran and J. S. Pober, 'Effects of cytokines on vascular endothelium: their role in vascular and immune injury.' *Kidney Int.*, 1989, **35**, 969–975.

155. A. M. Jevinkar, R. P. Wuthrich, F. Takei *et al.*, 'Differing regulation and function of ICAM-1 and class II antigens on renal tubular cells.' *Kidney Int.*, 1990, **38**, 417–425.

156. G. A. Muller, J. Markovic-Lipkovski and H. P. Rodemann, 'The progression of renal diseases: on the pathogenesis of renal interstitial fibrosis.' *Klin. Wochenschr.*, 1991, **69**, 576–586.

157. P. A. Hill, H. Y. Lan, D. J. Nikolic-Paterson *et al.*, 'ICAM-1 directs migration and localization of interstitial leukocytes in experimental glomerulonephritis.' *Kidney Int.*, 1994, **45**, 32–42.

158. D. C. Brennan, A. M. Jevnikar, F. Takei *et al.*, 'Mesangial cell accessory functions: mediation by intercellular adhesion molecule-1.' *Kidney Int.*, 1990, **38**, 1039–1046.

159. R. P. Wuthrich, A. M. Jevnikar, F. Takei *et al.*, 'Intercellular adhesion molecule-1 (ICAM-1) expression is upregulated in autoimmune murine lupus nephritis.' *Am. J. Pathol.*, 1990, **136**, 441–450.

160. K. Lhotta, H. P. Neumayer, M. Joannidis *et al.*, 'Renal expression of intercellular adhesion molecule-1 in different forms of glomreulonephritis.' *Clin. Sci. (Colch.)*, 1991, **81**, 477–481.

161. A. Dal Canton, G. Fuiano, V. Sepe *et al.*, 'Mesangial expression of intercellular adhesion molecule-1 in primary glomerulosclerosis.' *Kidney Int.*, 1992, **41**, 951–955.

162. D. H. Hooke, D. C. Gee and R. C. Atkins, 'Leukocyte analysis using monoclonal antibodies in human glomerulonephritis.' *Kidney Int.*, 1987, **31**, 964–972.

163. D. A. Kujubu and L. G. Fine, 'Physiology and cell biology update: polypeptide growth factors and their relation to renal disease.' *Am. J. Kidney Dis.*, 1989, **14**, 61–73.

164. A. M. el Nahas, 'Growth factors and glomerular sclerosis.' *Kidney Int.*, 1992, **36** (Suppl.), S15–S20.

165. A. C. M. Ong and L. G. Fine, 'Tubular-derived growth factors and cytokines in the pathogenesis of tubulointerstitial fibrosis: implications for human renal disease progression.' *Am. J. Kidney Dis.*, 1994, **23**, 205–209.

166. K. Sharma and F. N. Ziyadeh, 'The emerging role of transforming growth factor-β in kidney diseases.' *Am. J. Physiol.*, 1994, **266**, F829–F842.

167. M. B. Sporn and A. B. Roberts, 'Transforming growth factor-β: recent progress and new challenges.' *J. Cell Biol.*, 1992, **119**, 1017–1021.

168. H. D. Humes and D. A. Cieslinski, 'Interaction between growth factors and retinoic acid in the induction of kidney tubulogenesis in tissue culture' *Exp. Cell Res.*, 1992, **201**, 8–15.

169. S. M. Wahl, D. A. Hunt, L. M. Wakefield *et al.*, 'Transforming growth factor type β induces monocyte chemotaxis and growth factor production.' *Proc. Natl. Acad. Sci. USA*, 1987, **84**, 5788–5792.

170. G. Ding, H. van Goor, J. Frye *et al.*, 'Transforming growth factor-β expression in macrophages during hypercholesterolemic state.' *Am. J. Physiol.*, 1994, **267**, F937–F943.

171. H. Kaneto, J. Morrissey and S. Klahr, 'Increased expression of TGF-β1 mRNA in the obstructed kidney of rats with unilateral ureteral ligation.' *Kidney Int.*, 1993, **44**, 313–321.

172. J. S. Bollineni and A. S. Reddi, 'Transforming growth factor-β1 enhances glomerular collagen synthesis in diabetic rats.' *Diabetes*, 1993, **42**, 1673–1677.

173. T. Coimbra, R. Wiggins, J. W. Noh *et al.*, 'Transforming growth factor-β production in anti-glomerular basement membrane disease in the rabbit.' *Am. J. Pathol.*, 1991, **138**, 223–234.

174. S. Tomooka, W. A. Border, B. C. Marshall *et al.*, 'Glomerular matrix accumulation is linked to inhibition of the plasmin protease system.' *Kidney Int.*, 1992, **42**, 1462–1469.

175. T. Yamamoto, N. A. Noble, D. E. Miller *et al.*, 'Sustained expression of TGF-β1 underlies development of progressive kidney fibrosis.' *Kidney Int.*, 1994, **45**, 916–927.

176. J. Floege, M. W. Burns, C. E. Alpers *et al.*, 'Glomerular cell proliferation and PDGF expression precede glomerulosclerosis in the remnant kidney model.' *Kidney Int.*, 1992, **41**, 297–309.

177. B. Fellstrom, L. Klareskog, C. H. Heldin *et al.*, 'Platelet-derived growth factor receptors in the kidney: upregulated expression in inflammation.' *Kidney Int.*, 1989, **36**, 1099–1102.

178. L. Baud and R. Ardaillou, 'Tumor necrosis factor alpha in glomerular injury.' *Kidney Int.*, 1994, **45** (Suppl.), S32–S36.

179. M. Ketteler, W. A. Border and N. A. Noble, 'Cytokines and ʟ-arginine in renal injury and repair.' *Am. J. Physiol.*, 1994, **267**, F197–F207.

180. S. M. Spaethe, M. S. Freed, K. De Schryver-Kecksmeti *et al.*, 'Essential fatty acid deficiency reduces the inflammatory cell invasion in rabbit hydronephrosis resulting in suppression of the exagerated eicosanoid production.' *J. Pharmacol. Exp. Ther.*, 1988, **245**, 1088–1094.

181. T. Bertani, M. Abbate, C. Zoja *et al.*, 'Tumor necrosis factor induces glomerular damage in the rabbit.' *Am. J. Pathol.*, 1989, **134**, 419–430.

182. D. J. Kusner, E. L. Leubbers, R. J. Nowinski *et al.*, 'Cytokine- and LPS-induced synthesis of interleukin-8 from human mesangial cells.' *Kidney Int.*, 1991, **39**, 1240–1248.

183. M. Gomez-Chiarri, T. A. Hamilton, J. Egido *et al.*, 'Expression of IP-10, a lipopolysaccharide- and interferon-gamma-inducible protein, in murine mesangial cells in culture.' *Am. J. Pathol.*, 1993, **142**, 433–439.

184. P. A. Marsden and B. J. Ballermann, 'Tumor necrosis factor α activates soluble guanylate cyclase in bovine glomerular mesangial cells via an L-arginine-dependent mechanism.' *J. Exp. Med.*, 1990, **172**, 1843–1852.

185. H. Muhl and J. Pfeilschifter, 'Amplification of nitric oxide synthase expression by nitric oxide in interleukin 1β-stimulated rat mesangial cells.' *J. Clin. Invest.*, 1995, **19**, 1941–1946.

186. J. F. Woessner, Jr., 'Matrix metalloproteinases and their inhibitors in connective tissue remodelling.' *FASEB J.*, 1991, **5**, 2145–2154.

187. M. Davies, J. Martin, G. J. Thomas *et al.*, 'Proteinases amd glomerular matrix turnover.' *Kidney Int.*, 1992, **41**, 671–678.

188. D. E. Kleiner, Jr. and W. G. Stetler-Stevenson, 'Structural biochemistry and activation of matrix metalloproteinases.' *Curr. Biol.*, 1993, **5**, 891–897.

189. A. Mauviel, 'Cytokine regulation of metalloproteinase gene expression.' *J. Cell Biochem.*, 1993, **53**, 288–295.

190. E. A. Bauer, T. W. Cooper, J. S. Huang *et al.*, 'Stimulation of *in vitro* human skin collagenase expression by platelet-derived growth factor.' *Proc. Natl. Acad. Sci. USA*, 1985, **82**, 4132–4136.

191. M. Presta, D. Moscatelli, J. Joseph-Silverstein *et al.*, 'Purification from a human hepatoma cell line of a basic fibroblast-growth factor-like molecule that stimulates capillary endothelial cell plasminogen activator production, DNA synthesis, and migration.' *Mol. Cell Biol.*, 1986, **6**, 4060–4066.

192. S. Chandrasekhar and A. K. Harvey, 'Transforming growth factor-b is a potent inhibitor of IL-1 induced protease activity and cartilage proteoglycan degradation.' *Biochem. Biophys. Res. Commun.*, 1988, **157**, 1352–1359.

193. J. Massague, 'The transforming growth factor-β family.' *Annu. Rev. Cell Biol.*, 1990, **6**, 597–641.

194. S. Nomura, B. L. Hogan, A. J. Wills *et al.*, 'Developmental expression of tissue inhibitor of metalloproteinase (TIMP).' *RNA*, 1989, **105**, 575–583.

195. E. Engelmyer, H. van Goor, D. R. Edwards *et al.*, 'Differential mRNA expression of renal cortical tissue inhibitor of metalloproteinase-1, -2, and -3 in experimental hydronephrosis.' *J. Am. Soc. Nephrol.*, 1994, **5**, 1675–1683.

196. W. H. Barricos and S. V. Shah, 'Proteolytic enzymes as mediators of glomerular injury.' *Kidney Int.*, 1991, **40**, 161–173.

197. J. F. Reckelhoff and C. Baylis, 'Glomerular metalloproteinase activity in the aging rat kidney: inverse correlation with injury.' *J. Am. Soc. Nephrol.*, 1993, **3**, 1835–1838.

198. H. Trachtman, R. Greenwald, S. Moak *et al.*, 'Meprin activity in rats with experimental renal disease.' *Life Sciences*, 1993, **53**, 1339–1344.

199. W. W. Tang, L. Feng, Y. Xia *et al.*, 'Extracellular matrix accumulation in immune-mediated tubulointerstitial injury.' *Kidney Int.*, 1994, **45**, 1077–1084.

200. J. S. Bond, P. E. Butler and R. J. Beynon, 'Metalloendopeptidases of the mouse kidney brush border: meprin and endopeptidase-24.11.' *Biomed. Biochim. Acta*, 1986, **45**, 1515–1521.

201. S. D. Ricardo, J. S. Bond, J. Kaspar *et al.*, 'Down-regulated expression of meprin in experimental hydronephrosis.' *Am. J. Pathol.*, 1996, **36**, F669–F676.

202. J. R. Diamond and S. Anderson, 'Irreversible tubulointerstitial damage associated with chronic aminonucleoside nephrosis. Amelioration by angiotensin I converting enzyme inhibition.' *Am. J. Pathol.*, 1990, **137**, 1323–1332.

203. R. Tanaka, V. Kon, T. Yoshioka *et al.*, 'Angiotensin converting enzyme inhibitor modulates glomerular function and structure by distinct mechanisms.' *Kidney Int.*, 1994, **45**, 537–543.

204. R. J. Johnson, C. E. Alpers, A. Yoshimura *et al.*, 'Renal injury from angiotensin II-mediated hypertension.' *Hypertension*, 1992, **19**, 464–474.

205. M. L. Purkerson and S. Klahr, 'Prior inhibition of vasoconstrictors normalizes GFR in postobstructed kidneys.' *Kidney Int.*, 1989, **35**, 1305–1314.

206. J. L. Blackshear and R. L. Wathen, 'Effects of indomethacin on renal blood flow and renin secretory responses to ureteral occlusion in the dog.' *Miner Electrolyte Metab.*, 1978, **1**, 271–278.

207. G. H. Gibbons, R. E. Pratt and V. J. Dzau, 'Vascular smooth muscle cell hypertrophy vs. hyperplasia. Autocrine transforming growth factor-β1 expression determines growth response to angiotensin II.' *J. Clin. Invest.*, 1992, **90**, 456–461.

208. G. A. Stouffer and G. K. Owens, 'Angiotensin II-induced mitogenesis of spontaneously hypertensive rat-derived cultured smooth muscle cells is dependent on autocrine production of transforming growth factor-β.' *Circ. Res.*, 1992, **70**, 820–828.

7.16
Alterations in Surface Membrane Composition and Fluidity: Role in Cell Injury

BRUCE A. MOLITORIS

Indiana University Medical Center, Indianapolis, IN, USA

7.16.1 INTRODUCTION

Epithelial cells are responsible for the vectorial movement of ions, water, and macromolecules between biological compartments. This is accomplished by having the surface membrane organized into distinct apical and basolateral domains. These domains are distinguishable biochemically, structurally, and physiologically, each containing specific membrane components including ion channels, transport proteins, enzymes, cytoskeletal associations, and lipids. These individual membrane components collectively determine what cellular processes, including absorption, secretion and exchange, the cell will perform. The establishment and maintenance of this specialized organization is a multistage process involving the formation of cell–cell and cell–substratum contacts, and the polarized delivery of plasma membrane components to the appropriate domains. These dynamic processes require the proper organization of cellular cytoskeletal elements. For instance, the increased apical reabsorptive surface area of proximal tubule cells via microvilli is dependent upon the actin cytoskeleton; the stability of cell subratum junctions, such as desmosomes and focal adhesions, is dependent on actin and intermediate filaments (IF); and finally vesicular transport throughout the cell is dependent upon microtubular and actin cytoskeletal elements.

This review will first describe the basic organization of polarized epithelial cells with an emphasis on the differences between apical and basolateral membrane (BML) domains, highlighting the functional and potential pathophysiological significance of these differences. Finally, the role of ischemia-induced changes in surface membrane composition and fluidity and their role in altered cellular function and enhanced susceptibility to injury from nephrotoxicants will be examined.

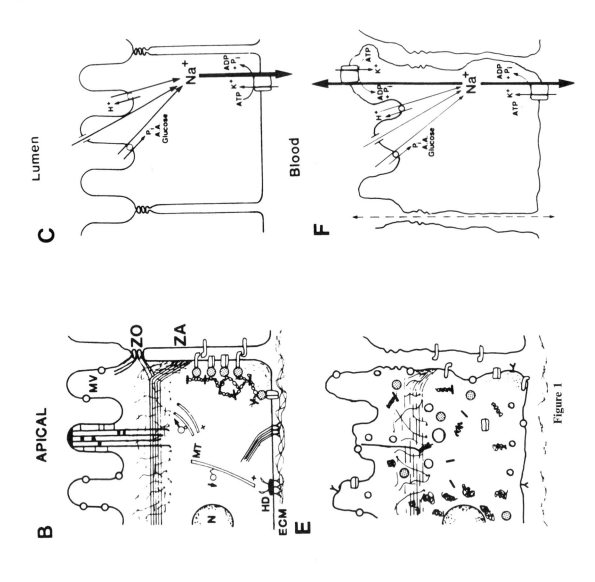

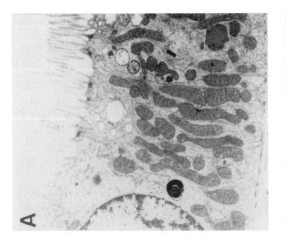

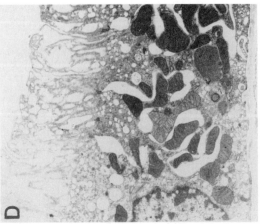

Figure 1

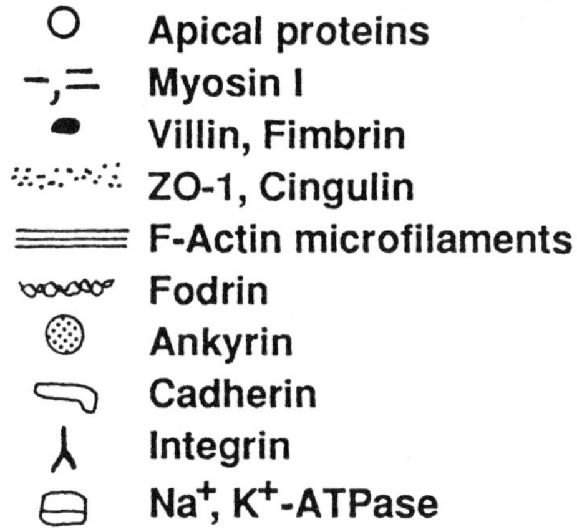

○ **Apical proteins**
⁻,⁼ **Myosin I**
● **Villin, Fimbrin**
⸙⸙⸙⸙ **ZO-1, Cingulin**
≡≡≡ **F-Actin microfilaments**
∿∿∿ **Fodrin**
◉ **Ankyrin**
⟜ **Cadherin**
⅄ **Integrin**
⊜ **Na⁺, K⁺-ATPase**

Figure 1 Effect of ischemia on the actin cytoskeleton and surface membrane polarity in rat renal proximal tubular cells. (A), (B), and (C) show a low power electron micrograph, a drawing of the cytoskeletal architecture, and apical and basolateral sodium transport enzymatic machinery of a rat renal proximal tubular cell under physiologic conditions, respectively. (D), (E), and (F) show the same series of illustrations of a rat renal proximal tubular cell following 20 min of *in vivo* ischemia. Under physiologic conditions (A) the proximal tubule cell is a highly polarized cell with distinctive apical and basolateral membranes, a well defined endocytic system and an apical membrane amplified via microvilli. Following ischemic injury there is disorganization of the cell, a marked increase in intracellular vesicles and loss, fusion and disorganization of the individual microvilli (D). (B) shows the normal surface membrane polarity existing for several transporters and the interactions of the actin cytoskeleton with the surface membrane under physiologic conditions. (E) shows that following ischemic injury there is loss of surface membrane polarity with Na,K-ATPase and integrins moving from the basolateral membrane to the apical membrane, and apical proteins moving to the basolateral domain. This is associated with marked alterations in the actin cytoskeleton and actin cytoskeletal surface membrane interactions. There is cytoplasmic distribution of the actin cytoskeleton from the surface membrane associated with dissociation of actin cytoskeletal membrane proteins such as ankyrin and fodrin throughout the cytoplasm. (C) shows the normal apical and basolateral distribution of sodium transport mediating proteins under physiologic conditions. Following ischemic injury Na⁺K-ATPase redistributes from the basolateral membrane from the apical membrane and then can mediate the pumping of sodium ions from the cell back into the tubular lumen (F).

7.16.2 STRUCTURAL CHARACTERISTICS OF PROXIMAL TUBULE CELLS

The primary function of proximal tubule cells (PTC) is selective reabsorption of mole-cules from the glomerular filtrate. In order to accomplish this task PTC have developed specialized structures for reabsorption and structural stability, and utilize a variety of targeting mechanisms to place receptors and other membrane components at their appropriate site to enable directional transport.[1–4] Figures 1(A)–(C) and Table 1 show the overall characteristics of a proximal tubule cell, some of the key structural elements and the importance of domain specific components for cellular function.

Proper functioning of epithelial cells requires not only structural polarization but also polar distribution of surface membrane components, including enzymes, receptors, ATPases, ion channels, and lipids.[5–7] For example, sodium reabsorption by renal proximal tubular cells is dependent on the polarized delivery of specific carrier proteins such as the Na^+/H^+ antiporter and the Na^+-dependent glucose, amino acid, and phosphate cotransporters to the apical membrane, combined with localization of the Na^+,K^+-ATPase to the basolateral membrane (Figure 1(C)).

Extensive differences in membrane external leaflet lipid composition also exist between the apical and basolateral domains, and result in large physiochemical differences.[8–10] The apical membrane cholesterol:phospholipid (C:PL) and sphingomyelin:phosphatidylcholine (SPH:PC) ratios are high, resulting in high anisotropy and insulating capacity. In contrast, the BLM C:PL and SPH:PC ratios are low, with the phosphatidylcholine (PC) and phosphatidylinositol (PI) contents being relatively high. This results in a membrane that is more fluid (low anisotropy) and across which diffusion can occur more rapidly. Membrane lipid composition and anisotropy also influence the function of intrinsic membrane proteins, for example, Na^+,K^+-ATPase[11,12] and the Na^+-dependent glucose contrasporter.[13]

Apical and BLM domains are separated from each other by the cellular junctional complex. This complex is also responsible for the epithelial cell barrier separating, for proximal tubule cells, the glomerular filtrate and blood compartments. In addition, the structural stability of both the epithelial sheet and individual cells is dependent upon junctional complexes to adhere cells to the extracellular matrix (hemidesmosomes and focal adhesions); to stabilize cell–cell interactions (desmosomes, zonula adherens); to permit communication between cells (gap junctions); and to form both a plasma membrane and extracellular barrier (tight junction).

The tight junction (zonula occludens) encircles the apex of the cell, fusing adjacent cells

Table 1   Surface membrane asymmetry of proximal tubule cells.

Characteristic	Apical membrane		Basolateral membrane	
Proteins				

Enzymes	Leucine aminopeptidase
	Maltase
	Glycosyl-phosphatidylinositol
	  linked proteins, i.e., alkaline
	  phosphatase
			Adenylate cyclase	

Receptors	β_2-Microglobulin receptor GP-330		Insulin Parathyroid hormone EGF Cadherins Integrins Laminin	
ATPases	H^+-ATPase Mg^{2+}-ATPase		Na^+,K^+-ATPase Ca^{2+}-ATPase	
Carriers	Na^+-dependent cotransporters (glucose, Pi, amino acids) Na^+/H^+ antiporter Na^+/sulfate		Cl^-/HCO_3 exchanger Na^+-independent glucose carrier Na^+/HCO_3	
Lipids	*Control (%)*	*Ischemic (%)*	*Control (%)*	
Sphingomyelin	39	30	17	
Phosphatidylcholine (PC)	17	30	37	
Phosphatidylinositol (PI)	2	4.6	6	
Phosphatidylserine (PS)	16	11	9	
Phosphatidylethanolamine	25	22	30	
Sphingomyelin/PC ratio	2.3	1.0	0.5	
Cholesterol/phospholipid	0.9	0.6	0.4	
Physical properties				
Electrical resistance	High		Low	
Membrane fluidity	Low		High	
(Fluorescence polarization)	0.280	0.260	0.230	

Ischemic values represent data from rat renal BBMV following 50 min of *in vivo* ischemia.[61,70]

together and establishing the border between the apical and basolateral domains.[14–16] It forms a barrier to intradomain movement of intrinsic membrane proteins and outer leaflet phospholipids (the "fence" function), and to paracellular movement of solutes between biologic compartments ("gate" function).[17,18]

The zonula adherens consists of a circumferential band of actin microfilaments of mixed polarity encircling the apex of the cell just basal to the zonula occludens.[2,3] Myosin II, α-actin, and tropomyosin have all been identified, and are thought to both cross-link and provide contractile potential for F-actin. This complex is attached to the lateral membrane by an adhesion plaque containing α-actin, zyxin, vinculin, and radixin.[19] These, in turn, link to catenin/cadherin complexes, which mediate cell–cell adhesion. Cadherins are a superfamily of transmembrane glycoproteins that bind with each other in a homophilic manner to mediate Ca^{2+}-dependent cell–cell adhesion.[20] The

family of "classical" cadherins includes at least 12 members, including uvomorulin (E- or epithelial-cadherin). Interactions with catenins appear important both for cell–cell binding and for interactions with the actin cytoskeleton.[21,22] Cadherins are also known to be phosphorylated, and tyrosine kinases may be involved in their regulation.

Desmosomes (macula adherens) lie in a loose band below the zonula adherens and are also distributed along the lateral membrane, forming cell–cell adhesions mediated by cadherins.[23] Hemidesmosomes are distributed along the basal membrane, and form cell–substratum adhesions mediated by integrins.[24]

Integrins are a family of heterodimeric transmembrane glycoproteins that mediate Ca^{2+}-dependent cell–substratum and heterophilic cell–cell binding.[25,26] There are at least 20 members, generated from permutations of pairings between at least 14α- and 8β-subunit isoforms. The pattern of binding to the various components of the extracellular matrix, such as fibronectin, laminin, and various collagens, varies with each integrin family member, but generally involves the Arg-Gly-Asp (R-G-D) tripeptide sequence. Integrins have also been implicated in heterophilic cell–cell adhesion.[27,28] Integrin function appears regulated by both extracellular and intracellular signals. Binding of an extracellular ligand apparently induces conformational changes in the intracellular portion of the β-subunit, which in turn regulate interactions with the actin cytoskeleton.[29] Integrins are also known to be phosphorylated at serine and tyrosine sites. The reader is referred to chapter 16 for more in-depth discussion of integrins.

Gap junctions are present in most tissues and provide a cell–cell channel for passage of molecules less than 5000 daltons.[30,31] These junctions are located along the lateral membrane in proximal tubule cells.

Focal adhesions are distributed along the basal membrane, and anchor cells to the extracellular matrix[19,32] (Figure 1(B)). Actin microfilaments are cross-linked by an α-actinin (in some *in vitro* cases this results in the formation of stress fibers) and linked to the membrane by a protein complex that includes talin, vinculin, and actin microfilament capping proteins. These, in turn, link to integrin heterodimers, which mediate cell–substratum adhesion.

7.16.3   ESTABLISHMENT AND MAINENTANCE OF EPITHELIAL CELL POLARITY

Establishment of a polarized surface membrane begins with cell–cell recognition and contact mediated by cadherins.[13,33–35] This allows for establishment of specific protein domains that serve as foci for assembly of junctional and nonjunctional cytoskeletal structures, contributing to the formation of apical and basolateral domains. Other membrane proteins, such as Na$^+$,K$^+$-ATPase, are subsequently incorporated and organized within this cytoskeletal network.[36,37]

Further genesis and maintenance of a polarized surface membrane depends upon the sorting and targeting of newly-synthesized proteins[4,5,7] and a high degree of fidelity for reutilization of endocytosed membrane components. A variety of pathways appear to be involved with these processes (Figure 2). Apical and basolateral membrane proteins are synthesized in the endoplasmic reticulum, transported through the same Golgi complex and delivered to a trans-Golgi network where sorting occurs. Another method the cell uses to sort is observed

with the Na$^+$,K$^+$-ATPase, which is initially delivered equally to the apical and basolateral membranes of MDCK cells. However, the assembly, with cell–cell contact, of a basolateral actin cortical cytoskeleton allows for selective retention and metabolic stabilization of Na$^+$,K$^+$-ATPase. The end result is a polarized distribution, with higher levels in the basolateral domain.[36,37] A third method used by some cells, notably hepatocytes, to sort apical membrane proteins involves delivery of all or most membrane proteins to the basolateral surface and selective transcytosis of apical proteins to the apical surface.

Establishment of surface membrane lipid polarity is less well understood, in part because of the lack of immunologic techniques and domain-specific markers. Sphingolipids are preferentially transferred from the trans-Golgi network (TGN) to the apical domain, perhaps via vesicles containing coclustered glycosylphosphatidylinositol (GPI)-anchored proteins.[6,38] Possible mechanisms for the sorting of other lipid species are less clear, though lipid transfer proteins may play a role.[39,40] Proteins and lipids delivered to either the apical or basolateral surface are restrained by the zonula occludens "fence" function and by specific interactions with the cortical cytoskeleton. Selective retention in the membrane with exclusion from endocytosis may also play an important role.

The efficient sorting and recycling of endocytosed plasma membrane domain occurs with a high degree of fidelity that is necessary to maintain surface membrane polarity.[41,42] During adsorptive endocytosis, plasma membrane containing receptor-ligand complexes are internalized via clatherin-coated pits (Figure 2 pathway B). Subsequent processing involves sequential loss of clatherin from coated vesicles, fusion of these vesicles with an "early" endosomal compartment and progression to a lysosomal compartment via "late" multivesicular endosomal structures. Endosomal derived membrane components can also recycle back to the membrane of their origin or undergo transcytosis to an alternate surface membrane domain. In polarized epithelial cells this process can occur at both apical and basolateral membrane domains.

It has recently been shown that distinct endosomal systems exist for each plasma membrane (PM) domain.[41–47] An early or "sorting" set of endosomes lies close to either PM domain and receives cargo endocytosed exclusively from that domain. For example, in proximal tubule cells beta-2-microglobulin from apical and basolateral early endosomes was shown to converge in a common, apically oriented, "late" endosomal compartment

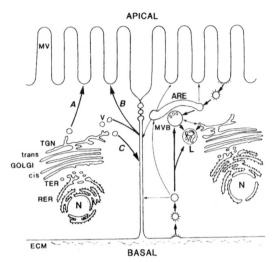

Figure 2 Synthesis, sorting and targeting pathways for membrane components. Proteins are synthesized in the rough endoplasmic reticulum (RER), migrate sequentially through the transitional endoplasmic reticulum (TER) and Golgi apparatus and arrive at the trans-Golgi network (TGN). In some cell types, apically- and basolaterally- targeted components then move via vesicles directly to their corresponding surface membrane domains (pathways A and C, respectively). In other cell types, apically targeted proteins are first delivered to the basolateral domain, then endocytosed and redelivered to the apical domain (pathway B). The nucleus (N), microvilli (MV) and the extracellular matrix (ECM) are also labeled. Right, endocytic pathways. Apical and basolateral receptor mediated endocytosis begins with the budding off of a clatherin-coated pit to form a clatherin coated vesicle. Disassembly of the clatherin coat yields a primary early endosome. From these endosomes membrane components may recycle back to the original domain of origin through the apical recycling endosome (ARE), transfer to lysosomes (L) via the via multivesicular bodies (MVB) or transcytose to the alternate domain.

(multivesicular body) before delivery to lysosomes.[46] However, aminoglycosides internalized via the apical (gentamicin) and basolateral (tobramicin) endocytic pathways converged at the lysosomal level.[47] Translocation of cargo from the early to the common late endosomal compartment appears to take place via a microtubule-dependent mechanism.[48,49] A different pathway has been demonstrated for transcytosis in MDCK cells when a specific marker of basolateral to apical transcytosis (the polymeric IgA immunoglobulin receptor) was used.[43,45] Material endocytosed through the basolateral endosomal system converged with apically endocytosed material at the level of the apical recycling (ARE) or sorting endosome. The material to be transcytosed (degradative)

did not converge with apically endocytosed material in the late apical endosomal compartment. Microtubules were required for this process as well.

Delivery of vesicles to either the apical or basolateral surface is an active process requiring ATP and microtubule (MT)- or actin-based motor molecules.[50] The MT motors kinesin and dynein are responsible for transport to the (+) or (−) end of MTs, respectively. The regulation of this motility and specificity of motor-membrane interactions are poorly understood but could be one mechanism by which targeting occurs. For example, a vesicle destined for the apical surface, toward the (−) end, may preferentially associate and activate dynein or release or inactivate kinesin. Accessory proteins and posttranslational modifications are likely involved in these intricate motility pathways. In addition delivery to the plasma membrane requires movement through or around the cortical actin network. Direct vesicular movement along actin, possibly mediated by a myosin molecule, has been reported and may be necessary for fusion with the plasma membrane to occur.[51] Additional regulators of motility and fusion processes include GTP-binding proteins[52–54] and calcium whose local concentration may be altered by extracellular signals emanating through cadherin or integrin signaling system.

7.16.4   ALTERED SURFACE MEMBRANE POLARITY DURING ISCHEMIA: ROLE OF THE CYTOSKELETON AND FUNCTIONAL SIGNIFICANCE

Tissue ischemia occurs when blood flow is reduced to levels insufficient to maintain cellular energy. Depletion of ATP induces a series of structural, biochemical, and functional deviations. The kidney proximal tubule cell is particularly susceptible to ischemia, and serves as a useful model both *in vivo* and *in vitro*.[55–57] Of central importance appears to be a rapidly occurring, duration-dependent disruption and dissociation of the actin cytoskeleton and associated surface membrane structures.

In vivo, renal ischemia in rats induces disruption of microvillar actin cores and the apical circumferential actin microfilament network, with redistribution of actin from the apical pole to throughout the cytoplasm.[58–62] *In vitro*, using the proximal tubule-derived LLC-PK1 cell line, antimycin A-induced ATP depletion caused disruption of the actin cortical cytoskeleton and redistribution of actin into large cytoplasmic aggregates primarily located in the perinuclear region.[63] Within the first five minutes of ATP depletion, there was a significant conversion of monomeric (G) actin to polymeric (F) actin,[63,64] possibly involving effects on G-actin sequestering proteins and F-actin capping and nucleation-inhibiting proteins. These changes have also been confirmed in mouse proximal tubule cells grown in primary cultures.[56] *In vivo* testing of the role of ATP in maintaining the actin cytoskeleton is also supportive of this hypothesis.[65]

Concurrent with or following this disruption and dissociation of the cortical actin cytoskeleton, the surface membrane undergoes extensive changes (outlined in Figures 1(D)–(F)). These include alterations in microvilli morphology, disruption of cellular junctions, and loss of surface membrane polarity. Apical microvilli are lost by fragmentation with shedding into the lumen and internalization into the cytoplasm.[66,67] Disruption of the zonula occludens leads to increases in intercellular permeability (loss of "gate" function)[8,59,68] and membrane lipids and proteins are free to cross domains (loss of "fence" function). Apical domain sphingomyelin and cholesterol content decreases while PC, PI, and total phospholipid content increases, thus reducing apical membrane fluorescence polarization as a measure of enhanced apical membrane fluidity (Table 1).[9,60,61,69] Na^+,K^+-ATPase, released from the cortical cytoskeleton, redistributes into the apical domain, while the apical marker protein leucine aminopeptidase redistributes into the basolateral domain.[60,64] Whether or not these same events occur during nephrotoxic cellular injury remains to be determined. Alterations in both intermediate filament protein expression[71] and location[72] have also been observed following ischemia. In addition, a recent study also documented the dissociation of ezrin from the actin cytoskeleton during anoxia.[73] Similar investigations using the BS-C-1 cell line and H_2O_2 oxidative stress have revealed disruption of focal adhesions with loss of talin from the basal cell surface.[74] The disruption detaches the associated integrins, allowing them to redistribute into the apical domain. Additional changes occurring to MTs and IFs throughout ischemia ATP depletion have not been analyzed extensively but are suggested by some earlier studies.[75,76] In addition, recent information in a rat model of ischemia shows microtubules in S_3 cells, although not grossly disrupted during ischemia *in vivo*, undergo fragmentation during the first hour of reperfusion. This defect was corrected during 24 h of reperfusion.[77]

The functional ramifications of these changes in cytoskeletal structure and surface membrane polarity are substantial. Loss of microvilli results in a markedly decreased apical membrane surface area. Increases in intercellular permeability allow for increased "backleak" of solutes and water into the interstitial space. Redistribution of membrane lipids changes membrane physiochemical properties, which may affect integral membrane protein function. For example, reduced glucose reabsorption after ischemic injury appears related in part to decreases in the apical SPH:PC ratio. In rat renal brush border membrane vesicles (BBMV) isolated after 15 min of ischemia, the V_{max} for Na^+-dependent glucose transport and the number of phlorizin binding sites decreased dramatically compared to controls.[70] Though this could represent a redistribution of Na^+-dependent glucose "carriers" to the alternate domain, carrier-mediated Na^+-dependent alanine transport was not similarly affected. Moreover, glucose transport in BBMV was highly correlated with the SPH:PC ratio and inversely correlated with membrane fluidity. In addition, recovery from ischemia returned glucose transport to control levels concurrent with the normalization of the apical SPH:PC ratio.[69,70]

Redistribution of membrane proteins to the alternate domain may result in substantial changes in domain-specific functions. Perhaps the best-studied example in this category involves Na^+,K^+-ATPase. Under physiologic circumstances, sodium enters the cell via a variety of apical transport proteins, providing the energy for uptake of a variety of solutes while moving down its electrochemical gradient. Sodium is then transported out of the cell up its electrochemical gradient via basolateral Na^+,K^+-ATPase (Figure 1(C)). Under these circumstances vectorial transport of sodium and other solutes is coupled to ATP utilization. With ischemia-induced redistribution of Na^+,K^+-ATPase to the apical membrane, sodium that has entered the cell may be transported out of the cell across either the apical or the basolateral membrane. Apical transport via Na^+,K^+-ATPase results in a futile cycle, with transport of sodium now uncoupled from ATP utilization. *In vivo* evaluation of sodium reabsorption by micropuncture[78] and by lithium and sodium reabsorption[69] confirmed decreased proximal tubule sodium absorption after ischemia, correlated with apical redistribution of Na^+,K^+-ATPase. Direct evidence came from *in vitro* cell culture studies utilizing LLCPK1 cells.[79] Basolateral redistribution of apical transport proteins could similarly affect cellular transport.

The apical redistribution of integrins may

also have important functional ramifications, as adherence of unattached cells to cell monolayers is increased in cells exhibiting ATP depletion or oxidant stress and this interaction appears dependent on the RGD sequence of integrins.[80] Such cell–cell interactions between cellular material or sloughed cells within a renal tubule lumen may well lead to cellular clumping and lumenal obstruction.[73] Though as yet undocumented, redistribution of cadherins could function similarly. The reader is referred to Chapter 17, this volume, for further details.

The redistribution of surface membrane lipids during ischemia has recently been shown to play an important role in the synergistic interaction between ischemia/hypoxia and aminoglycoside antibiotics to cause acute renal failure in rats.[81] Aminoglycosides bind to both apical and BLM, primarily to acidic phospholipids and especially to phosphatidylinositol (PI).[82] In fact, BLM binding is twice as high as binding to apical membranes. Furthermore, in cell cultural studies (LLCPK$_1$ cells) internalization of aminoglycosides across either domain results in similar cellular toxicity.[47] However, the markedly enhanced rate of endocytosis across the apical domain[83] mediates the predominant uptake of aminoglycosides *in vivo* via this route. Consistent with this observation is the usual pattern of aminoglycoside induced cellular injury in the highly endocytic S$_1$ and S$_2$ cells in rat proximal tubules.

During ischemia PI from the BLM redistributes into the apical domain and gentamicin binding increases proportionately (Table 1).[84] This results in markedly increased cellular uptake of gentamicin and abnormal intracellular distribution of gentamicin, especially in S$_3$ cells.[84] The combination of mild ischemia or hypoperfusion and aminoglycoside administration leads to a change in the pattern of cellular injury seen in the rat.[85] Under physiologic conditions aminoglycosides induce injury primarily to S$_1$ and S$_2$ cells, those with the highest rates of endocytic internalization. Following mild ischemia this pattern of cellular injury changes so that S$_3$ cells are primarily involved. This is consistent with S$_3$ cells being most sensitive to ischemia, due to reduced reperfusion post ischemic injury, and this in turn resulting in enhanced aminoglycoside binding and internalization.[84]

Reestablishment of apical and basolateral membrane polarity occurs during the recovery phase of ischemic acute renal failure and has been demonstrated for leucine aminopeptidase, Na$^+$,K$^+$-ATPase, and apical and BLM lipids.[60,69,77,78] The mechanisms responsible for restitution of surface membrane polarity have not been determined, but involve actual remodeling of surface membrane domains in previously damaged cells and cellular proliferation and differentiation of newly proliferating tubular epithelial cells.[66,76] The rate at which repolarization occurs is dependent on the severity of the injury. Fifteen minutes of ischemia (mild injury) required only 24–48 h of reperfusion, while 50 min of ischemia (moderate to severe injury) required several days for re-establishment of apical and basolateral membrane polarity.[69] In both cases, return of structural polarity (cytoskeleton) is a prerequisite for re-establishment of surface membrane polarity.

The study of factors involved in and responsible for cellular recovery from ischemia is being pursued with vigor, but remains in its infancy.[86,87] Epidermal growth factor,[88] insulin-like growth factor,[89] hepatocyte growth factor,[90] and transforming growth factor-α[91] have all been shown to accelerate recovery. Heat shock proteins have been shown to accumulate during ischemia.[92–95] Recovery from ischemia may be aided by an acidic environment, which was shown to stabilize the actin cytoskeleton during *in vitro* ATP depletion.[96] Finally, the addition of glycine, which acts as a cytoprotectant during ischemia, may facilitate a quicker recovery following ischemia.[97,98] However, the underlying cellular mechanisms by which these agents provide for enhanced recovery remain to be determined. For instance, such basic questions as whether the disrupted and dissociated actin cytoskeletal components are reutilized or if synthesis of new proteins is required remain unanswered.

7.16.5   CONCLUSIONS

In summary, ischemia in kidney proximal tubule cells induces a rapidly-occurring, duration-dependent disruption of the actin cytoskeleton. This, in turn, leads to disruption of associated surface membrane structures and untethering of integral membrane proteins. Lipids and proteins move to alternate domains, and cell–cell and cell-substrate contacts are lost. The cell is no longer able to perform vectorial transport, and cellular and organ-level dysfunction ensues.

ACKNOWLEDGMENTS

This work was supported by National Institute of Diabetes and Digestive and Kidney Diseases Grant DK-41126 (to B.A. Molitoris) and grants from the Veterans Affairs Research Service and the American

Heart Association. B.A. Molitoris is an AHA
Established Investigator.

7.16.6 REFERENCES

1. J. Leiser and B. A. Molitoris, 'Disease Processes in epithelia, the role of the actin cytoskeleton and altered surface membrane polarity.' *Biochem. Biophys. Acta*, 1993, **1225**, 1–13.
2. N. S. Mooseker, 'Organization, chemistry and assembly of the cytoskeletal apparatus of the intestinal brush border.' *Ann. Rev. Cell Biol.*, 1985, **1**, 209–241.
3. A. Bretscher, 'Microfilament structure and function in the cortical cytoskeleton.' *Annu. Rev. Cell Biol.*, 1991, **7**, 337–374.
4. R. W. Mays, K. A. Beck and W. J. Nelson, 'Organization and function of the cytoskeleton in polarized epithelial cells, a component of the protein sorting machinery.' *Curr. Opin. Cell Biol.*, 1994, **6**, 16–24.
5. H. R. Pelham and S. Munro, 'Sorting of membrane proteins in the secretory pathway.' *Cell*, 1993, **75**, 603–605.
6. G. van Meer and K. N. J. Burger, 'Sphingolipid trafficking-sorted out?' *Trends Cell Biol.*, 1992, **2**, 332–337.
7. K. Matter and I. Mellman, 'Mechanisms of cell polarity, sorting and transport in epithelial cells.' *Curr. Opin. Cell Biol.*, 1994, **6**, 545–554.
8. G. F. Carmel, F. Rodriquez, S. Carriere *et al.*, 'Composition and physical properties of lipids from plasma membrane of dog kidney.' *Biochim Biophys Acta*, 1985, **818**, 149–157.
9. B. A. Molitoris and C. Hoilien, 'Static and dynamic components of renal cortical brush border and basolateral fluidity: role of cholesterol.' *J. Membr. Biol.*, 1987, **99**, 165–172.
10. B. A. Molitoris and F. R. Simon, 'Renal cortical brush-border and basolateral membranes, cholesterol and phospholipid composition and relative turnover.' *J. Membr. Biol.*, 1985, **83**, 207–215.
11. P. L. G. Chong, P. A. G. Fortes and D. M. Jameson, 'Mechanisms of inhibition of (Na,K)-ATPase by hydrostatic pressure studied with fluorescent probes.' *J. Biol. Chem.*, 1985, **260**, 14484–14490.
12. E. Sutherland, B. S. Dixon, H. L. Leffert *et al.*, 'Biochemical localization of hepatic surface-membrane Na⁺K⁺ATPase activity depends on membrane lipid fluidity.' *Proc. Natl. Acad. Sci. USA*, 1988, **85**, 8673–8677.
13. B. A. Molitoris and W. J. Nelson, 'Alterations in the establishment and maintenance of epithelial cell polarity as a basis for disease processes.' *J. Clin. Invest.*, 1990, **85**, 3–9.
14. M. Cereijido, L. Gonzalez-Mariscal, R. G. Contreras *et al.*, 'The making of a tight juntion.' *J. Cell Science, Suppl.*, 1993, **17**, 127–132.
15. B. Gumbiner, 'Structure, biochemistry and assembly of epithelial tight junctions.' *J. Am. J. Physiol.*, 1987, **253**, C749–C758.
16. S. Citi, 'The molecular organization of tight junctions.' *J. Cell Biol.*, 1993, **121**, 485–489.
17. J. L. Madara, 'Tight junction dynamics, is paracellular transport regulated?' *Cell*, 1988, **53**, 497–498.
18. L. J. Mandel, R. Bacallao and G. Zampighi, 'Uncoupling of the molecular "fence" and paracellular "gate" functions in epithelial tight junctions.' *Nature*, 1993, **361**, 552–555.
19. E. J. Luna and A. L. Hitt, 'Cytoskeleton–plasma membrane interactions.' *Science*, 1992, **258**, 955–964.
20. M. Takeichi, 'Cadherin cell adhesion receptors as a morphogenetic regulator.' *Science*, 1991, **251**, 1451–1455.
21. M. Ozawa and R. Kemler, 'Molecular organization of the uvomorulin–catenin complex.' *J. Cell Biol.*, 1992, **116**, 989–996.
22. B. M. Gumbiner and P. D. McCrea, 'Catenins as mediators of the cytoplasmic functions of cadherins.' *J. Cell Science, Suppl.*, 1993, **17**, 155–158.
23. P. K. Legan, J. E. Collins and D. R. Garrod, 'The molecular biology of desmosomes and hemidesmosomes, What's in a name?' *Bioessays*, 1992, **14**, 385–393.
24. M. J. William, P. E. Hughes, T. E. O'Toole *et al.*, 'The inner world of cell adhesion: integrin cytoplasmic domains.' *Trends Cell Biol.*, 1994, **4**, 109–112.
25. R. O. Hynes, 'Integrins, versatility, modulation, and signaling in cell adhesion.' *Cell*, 1992, **69**, 11–25.
26. F. M. Pavalko and C. A. Otey, 'Role of adhesion molecule cytoplasmic domains in mediating interactions with the cytoskeleton.' *Proc. Soc. Exp. Biol. Med.*, 1994, **205**, 282–293.
27. M. G. Lampugnani, M. Resnati, E. Dejana *et al.*, 'The role of integrins in the maintenance of endothelial monolayer integrity.' *J. Cell Biol.*, 1991, **112**, 479–490.
28. M. S. Goligorsky, W. Lieberthal, L. Racusen *et al.*, 'Integrin receptors in renal tubular epithelium, new insights into pathophysiology of acute renal failure.' *Am. J. Physiol.*, 1993, **264**, F1–F8.
29. A. A. Reszka, Y., Hayashi and A. F. Horwitz, 'Identification of amino acid sequences in the integrin β1 cytoplasmic domain implicated in cytoskeletal association.' *J. Cell Biol.*, 1992, **117**, 1321–1330.
30. E. C. Beyer, D. L. Paul and D. A. Goodenough, 'Connexin family of gap junction proteins.' *J. Membr. Biol.*, 1990, **116**, 187–194.
31. M. E. Finbow and J. D. Pitts, 'Is the gap junction channel—the connexon—made of connexin or ductin?' *J. Cell Sci.*, 1993, **106**, 463–471.
32. K. Burridge, K. Fath, T. Kelly *et al.*, 'Focal adhesions and transmembrane junctions between extracellular matrix and the cytoskeleton.' *Annu. Rev. Cell Biol.*, 1988, **4**, 487–525.
33. W. J. Nelson, R. W. Hammerton, A. Z. Wang *et al.*, 'Involvement of the membrane-cytoskeleton in development of epithelial cell polarity.' *Semin. Cell Biol.*, 1990, **1**, 359–371.
34. L. M. Wiley, G. M. Kidder and A. J. Watson, 'Cell polarity and development of the first epithelium.' *Bioessays*, 1990, **12**, 67–73.
35. H. McNeil, M. Ozawa, R. Kemler *et al.*, 'Novel function of the cell adhesion uvomorulin as an inducer of cell surface polarity.' *Cell*, 1990, **62**, 309–316.
36. R. W. Hammerton, K. A. Krezeminski, R. W. Mays *et al.*, 'Mechanism for regulating cell surface distribution of Na⁺,K(+)-ATPase in polarized epithelial cells.' *Science*, 1991, **254**, 847–850.
37. W. J. Nelson and R. W. Hammerton, 'A membrane-cytoskeletal complex containing Na⁺,K⁺-ATPase, ankyrin, and fodrin in Madin–Darby canine kidney (MDCK) cells: implications for the biogenesis epithelial cell polarity.' *J. Cell Biol.*, 1989, **108**, 893–902.
38. G. Van Meer and K. Simons, 'Lipid polarity and sorting in epithelial cells.' *J. Cell Biochem.*, 1988, **36**, 51–58.
39. M. Koval and R. E. Pagano, 'Lipid recycling between the plasma membrane and intracellular compartments, transport and metabolism of fluorescent sphingomyelin analogues in cultured fibroblasts.' *J. Cell Biol.*, 1989, **108**, 2169–2181.
40. J. W. Nichols, 'Binding of fluorescent-labeled phosphatidylcholine to rat liver nonspecific lipid transfer protein.' *J. Biol. Chem.*, 1987, **262**, 14172–14177.
41. R. G. Parton, K. Prydz, M. Bomsel *et al.*, 'Meeting of the apical and basolateral endocytic pathways of the

Madin–Darby canine kidney cell in late endosomes.' *J. Cell Biol.*, 1989, **109**, 3259–3272.

42. M. Bomsel, K. Prydz, R. G. Parton *et al.*, 'Endocytosis in filter-grown Madin–Darby canine kidney cells.' *J. Cell Biol.*, 1989, **109**, 3243–3258.

43. G. Apodoca, L. A. Katz and K. E. Mostov, 'Receptor-mediated transcytosis of IgA in MDCK cells via apical recycling endosomes.' *J. Cell Biol.*, 1994, **125**, 67–86.

44. R. Parton, S. A. Kuznetsov *et al.*, 'Microtubule- and motor-dependent fusion *in vitro* between apical and barolateral endocytic vesicles from MDCK cells.' *Cell*, 1990, **62**, 719–731.

45. M. Barroso and E. J. Sztul, 'Basolateral to apical transcytosis in polarized cells is indirect and involves BFA and trimeric G protein sensitive passage through the apical endosome.' *J. Cell Biol.*, 1994, **124**, 83–100.

46. M. Cohen, D. P. Sundin, R. Dahl *et al.*, 'Convergence of apical and basolateral endocytic pathways for β-2 microglobulin in LLC-PK1 cells.' *Am. J. Physiol.*, 1995, **268**, F829–F838.

47. D. M. Ford, R. H. Dahl, C. A. Lamp *et al.*, 'Apically and basolaterally internalized aminoglycosides colocalize in LLC-PK1 lysosomes and alter cell function.' *Am. J. Physiol.*, 1994, **266**, C52–C57.

48. P. P. Breitfeld, W. C. McKinnon, C. E. Mostov *et al.*, 'Effect of orocodazole on vesicular traffic to the apical and basolateral surfaces of polarized MDCK cells.' *J. Cell Biol.*, 1990, **111**, 2365–2373.

49. W. Hunziker, P. Male and I. Mellman, 'Differential microtubule requirements for transcytosis in MDCK cells.' *EMBO J.*, 1990, **9**, 3515–3525.

50. C. Achler, D. Filmer, C. Merte *et al.*, 'Role of microtubules in polarized delivery of apical membrane proteins to the brush border of the intestinal epithelium.' *J. Cell Biol.*, 1989, **109**, 179–189.

51. K. R. Fath and D. R. Burgess, 'Golgi-derived vesicles from developing epithelial cells bind actin filaments and possess myosin-I as cytoplasmically oriented peripheral membrane protein.' *J. Cell Biol.*, 1993, **120**, 117–127.

52. P. Novick and P. Brennwald, 'Friends and family, the role of the Rab GTPases in vesicular traffic.' *Cell*, 1993, **75**, 597–601.

53. M. Barroso and E. S. Sztul, 'Basolateral to apical transcytosis in polarized cells is indirect and involves BFA and trimeric G protein sensitive passage through the apical endosome.' *J. Cell Biol.*, 1994, **124**, 83–100.

54. S. R. Pfeffer, 'Rab GTPases, master regulators of membrane trafficking.' *Curr. Opin. Cell Biol.*, **6**, 522–526.

55. E. M. Fish and B. A. Molitoris, 'Alterations in epithelial polarity and the pathogenesis of disease states.' *N. Engl. J. Med.*, 1994, **330**, 1580–1588.

56. V. M. Kroshian, A. M. Sheridan and W. Lieberthal, 'Functional and cytoskeletal changes induced by sublethal injury in proximal tubular epithelial cells.' *Am. J. Physiol.*, 1994, **266**, F21-F30.

57. J. M. Weinberg, 'The cell biology of ischemic renal injury.' *Kidney Int.*, 1991, **39**, 476–500.

58. P. S. Kellerman, R. A. F. Clark, C. A. Hoilien *et al.*, 'Role of microfilaments in the maintenance of proximal tubule structural and functional integrity.' *Am. J. Physiol.*, 1990, **259**, F279–F285.

59. B. A. Molitoris, S. A. Falk and R. H. Dahl, 'Ischemia-induced loss of epithelial polarity. Role of the tight junction.' *J. Clin. Invest.*, 1989, **84**, 1334–1339.

60. B. A. Molitoris, C. A. Hoilien, R. H. Dahl *et al.*, 'Characterization of ischemia-induced loss of epithelial polarity.' *J. Membr. Biol.*, 1988, **106**, 233–242.

61. B. A. Molitoris, P. D. Wilson, R. W. Schrier *et al.*, 'Ischemia induces partial loss of surface membrane polarity and accumulation of putative calcium ionophores.' *J. Clin. Invest.*, 1985, **76**, 2097–2105.

62. P. S. Kellerman and R. T. Bogusky, 'Microfilament disruption occurs very early in ischemic proximal tubule injury.' *Kidney Int.*, 1992, **42**, 896–902.

63. B. A. Molitoris, A. C. Geerdes and J. R. McIntosh, 'Dissociation and redistribution of Na(+), K(+)-ATPase from its surface membrane actin cytoskelton complex during cellular ATP depletion.' *J. Clin. Invest.*, 1991, **88**, 462–469.

64. B. A. Molitoris, R. Dahl and A. Geerdes, 'Cytoskeleton disruption and apical redistribution of proximal tubule Na(+)-K(+)-ATPase during ischemia.' *Am. J. Physiol.*, 1992, **263**, F488–F495.

65. P. S. Kellerman, 'Exogenesis in adenosine triphosphate (ATP) preserves proximal tubule microfilament structure and function *in vivo* in a maleic acid model of ATP depletion.' *J. Clin. Invest.*, 1993, **92**, 1940–1949.

66. B. Glaumann, H. Glaumann, I. K. Berezesky *et al.*, 'Studies on the cellular recovery from injury II. Ultrastructural studies on the recovery of the pars convoluta of the proximal tubule of the rat kidney from temporary ischemia.' *Virchows Arch. B Cell Pathol.*, 1977, **24**, 1–18.

67. M. A. Venkatachalam, D. B. Jones, H. G. Rennke *et al.*, 'Mechanism of proximal tubule brush border loss and regeneration following mild renal ischemia.' *Lab. Invest.*, 1981, **45**, 355–365.

68. P. E. Canfield, A. E. Geerdes and B. A. Molitoris, 'Effect of reversible ATP depletion on tight-junction integrity in LLC-PK1 cells.' *Am. J. Physiol.*, 1991, **261**, F1038–F1045.

69. D. M. Spiegel, P. D. Wilson and B. A. Molitoris, 'Epithelial polarity following ischemia, a requirement for normal cell function.' *Am. J. Physiol.*, 1989, **256**, F430–F436.

70. B. A. Molitoris and R. Kinne, 'Ischemia induces surface membrane dysfunction. Mechanism of altered Na$^+$-dependent glucose transport.' *J. Clin. Invest.*, 1987, **80**, 647–654.

71. R. Moll, C. Hage and W. Thoenes, 'Expression of intermediate filament proteins in fetal and adult human kidney: modulations of intermediate filament patterns during development and in damaged tissue.' *Lab. Invest.*, 1991, **65**, 74–86.

72. R. Witzgall, D. Brown, C. Schwarz *et al.*, 'Localization of proliferating cell nuclear antigen, vimentin, c-Fos, and clusterin in the postischemic kidney. Evidence for a heterogeneous genetic response among nephron segments, and a large pool of mitotically active and dedifferentiated cells.' *J. Clin. Invest.*, 1994, **93**, 2175–2188.

73. J. Chen, R. B. Doctor and L. J. Mandel, 'Cytoskeletal dissociation of ezrin during renal anoxia: role in microvillar injury.' *Am. J. Physiol.*, 1994, **267**, C784-C795.

74. J. Gailit, D. Colflesh, I. Rabiner *et al.*, 'Redistribution and dysfunction of integrins in cultured renal epithelial cells exposed to oxidative stress.' *Am. J. Physiol.*, 1993, **264**, F149–F157.

75. P. J. Hollenbeck, A. D. Bershadsky, O. Y. Pletjushkina *et al.*, 'Intermediate filament collapse is an ATP-dependent and actin-dependent process.' *J. Cell Sci.*, 1989, **92**, 621–631.

76. A. D. Bershadsky and V. I. Gelfand, 'ATP-dependent regulation of cytoplasmic microtubule disassembly.' *Proc. Natl. Acad. Sci. USA*, 1981, **78**, 3610–3613.

77. M. Abbate, J. V. Bonventre and D. Brown, 'The microtubule network of renal epithelial cells is disrupted by ischemia and reperfusion.' *Am J. Physiol.*, 1994, **267**, F971–F978.

78. B. A. Molitoris, L. K. Chan, J. I. Shapiro *et al.*, 'Loss of epithelial polarity, a novel hypothesis for reduced proximal tubule Na$^+$ transport following ischemic injury.' *J. Membr. Biol.*, 1989, **107**, 119–127.

79. B. A. Molitoris, 'Na(+)-K(+)-ATPase that redistributes to apical membrane during ATP depletion remains functional.' *Am. J. Physiol.*, 1993, **265**, F693–697.

80. M. S. Goligorsky and G. F. DiBona, 'Pathogenetic role of Arg-Gly-Asp-recognizing integrins in acute renal failure.' *Proc. Natl. Acad. Sci.*, 1993, **90**, 5700–5704.

81. R. A. Zager, 'Gentamicin nephrotoxicity in the setting of acute renal hypoperfusion.' *Am. J. Physiol.*, 1988, **254**, F574–F581.

82. H. D. Humes, 'Aminoglycoside nephrotoxicity.' *Kidney Int.*, 1988, **22**, 900–911.

83. E. I. Christensen and S. Nielsen, 'Structural and functional features of protein handling in the kidney proximal tubule. *Semin. Nephrol.*, 1991, **11**, 414–439.

84. B. A. Molitoris, C. Meyer, R. Dahl *et al.*, 'Mechanism of ischemia-enhanced aminoglycoside binding and uptake by proximal tubule cells.' *Am. J. Physiol.*, 1993, **363**, F907–F916.

85. D. M. Spiegel, P. F. Shanley and B. A. Molitoris, 'Mild ischemia predisposes the S_3 segment to gentamicin toxicity.' *Kidney Int.*, 1990, **38**, 459–464.

86. R. Bacallao and L. G. Fine, 'Molecular events in the organization of renal tubular epithelium, from nephrogenesis to regeneration.' *Am. J. Physiol.*, 1989, **257**, F913–F924.

87. F. G. Toback, 'Regeneration after acute tubular necrosis.' *Kidney Int.*, 1992, **41**, 226–246.

88. H. D. Humes, D. A. Cieslinski, T. M. Coimbra *et al.*, 'Epidermal growth factor enhances renal tubule cell regeneration and repair and accelerates the recovery of renal function in postischemic acute renal failure.' *J. Clin. Invest.*, 1989, **84**, 1757–1761.

89. H. Ding, J. D. Kopple, A. Cohen *et al.*, 'Recombinant human insulin-like growth factor-I accelerates recovery and reduces catabolism in rats with ischemic acute renal failure.' *J. Clin. Invest.*, 1993, **91**, 2281–2287.

90. S. B. Miller, D. R. Martin, J. Kissane *et al.*, 'Hepatocyte growth factor accelerates recovery from acute ischemic renal injury in rats.' *Am. J. Physiol.*, 1994, **266**, F129–F134.

91. R. Reiss, A. J. Funke, D. A. Cielinski *et al.*, 'Transforming growth factor-alpha (TGFα) accelerates renal repair and recovery following ischemic injury to the kidney.' *Kidney Int.*, 1990, **37**, 492 (Abstract).

92. S. K. van Why, F. Hildebrandt, T. Ardito *et al.*, 'Induction of intracellular localization of HSP-72 after renal ischemia.' *Am. J. Physiol.*, 1992, **263**, F769–F775.

93. R. Mestril, S. H. Chi, M. R. Sayen *et al.*, 'Isolation of a novel inducible rat heat-shock protein (HSP70) gene and its expression during ischemia/hypoxia and heat shock.' *Biochem. J.*, 1994, **298**, 561–569.

94. R. Mestril, S. H. Chi, M. R. Sayen *et al.*, 'Expression of inducible stress protein 70 in rat heart myogenic cells confers protection against simulated ischemia-induced injury.' *J. Clin. Invest.*, 1994, **93**, 759–767.

95. A. Enami, J. H. Schwartz and S. C. Borkan, 'Transient ischemia or heat stress induced a cytoprotectant protein in rat kidney.' *Am. J. Physiol.*, 1991, **260**, F479–F485.

96. E. M. Fish and B. A. Molitoris, 'Extracellular acidosis minimizes actin cytoskeletal alterations during ATP depletion.' *Am. J. Physiol.*, 1994, **267**, F566–F572.

97. J. M. Weinberg, M. A. Venkatachalam, R. Garzo-Quintero *et al.*, 'Structural requirements for protection by small amino acids against hypoxic injury in kidney proximal tubules.' *FASEB J.*, 1990, **4**, 3347–3354.

98. A. R. Almeida, J. F. Wetzels, D. Bunnachak *et al.*, 'Acute phosphate depletion and *in vitro* rat proximal tubule injury: protection by glycine and acidosis.' *Kidney Int.*, 1992, **41**, 1494–1500.

7.17
Cell Adhesion Molecules in Renal Injury

MICHAEL S. GOLIGORSKY, EISEI NOIRI, and JAMES GAILIT
State University of New York at Stony Brook, NY, USA

VICTOR ROMANOV
NCI-Frederick Cancer Research and Development Center, Frederick, MD, USA

and

HUGH R. BRADY
University College Dublin, Ireland

7.17.1　INTRODUCTION

Adhesion molecules mediating cell–matrix and cell–cell attachment belong to five families: integrins, cadherins (uvomorulin, or E-CAM, L-, K-, and N-CAMs), Ig-like molecules (ICAMs, VCAM-1, and PECAM-1), selectins (L-, P-, and E-selectin), and carbohydrate ligands for selectins (sialyl lewis and GlyCAM-1). This chapter summarizes general information on various adhesion molecules, their normal and pathological expression and distribution in the kidney, as well as some potential clinical applications of the acquired knowledge on cell adhesion molecules in preventing or alleviating renal injury.

7.17.2　INTEGRINS

The integrins are noncovalently bound heterodimeric glycoproteins composed of α and β subunits. At present, 15 distinct α subunits and eight β subunits have been identified on the

Table 1　Combinations of integrin subunits.

Subunits		Ligands (recognition sequence)	Distribution
β_1	α_1	Laminin, collagen I, IV (DGEA)	Broad
	α_2	Laminin, collagen, $\alpha_3\beta_1$ (RGD ?)	Broad
	α_3	Laminin, collagen, fibronectin, epiligrin, entactin, $\alpha_2\beta_1$ (EILDV, RGD ?)	Broad
	α_4	Fibronectin (CS-1), VCAM (RGD)	B/T lymph, M
	α_5	Fibronectin (RGD), Invasin	Neural crest cells
	α_6	Laminin, merosin, kalinin, invasin	Broad
	α_7	Laminin	Broad
	α_8	?	?
	α_9	?	?
	α_v	Fibronectin, vitronectin (RGD)	Epithelial cells
β_2	α_L	ICAM-1, 2, 3	L
	α_M	iC3b, fibrinogen, factor X, ICAM-1	G, M, lymph
	α_X	iC3b, fibrinogen (GPRP)	M, G, lymph
α_v	β_3	α_{IIb} Fibrinogen, fibronectin, vWF, vitronectin (RGD, KQAGDV)	Platelets
		Vitronectin, fibrinogen, vWF, fibronectin, adenovirus penton base, denatured collagen, laminin, tenascin, osteopontin, thrombospondin (RGD)	Endothelial cells, tumor cells
	β_1	Fibronectin, vitronectin (RGD)	Epithelial cells
	β_5	Vitronectin, adenovirus penton base (RGD)	Epithelial cells, carcinomas
	β_6	Fibronectin (RGD)	?
	β_8	?	?
β_4	α_6	Laminin, kalinin	Epithelial cells
β_7	α_4	Fibronectin (CS-1), VCAM, MadCAM (EILDV)	B/T lymph, M
	α_{IEL}	?	

L, leukocytes; G, granulocytes; M, macrophages; lymph, lymphocytes.

protein level. Established combinations of the known integrin subunits into heterodimers and their known ligands are presented in Table 1. Members of this large family of receptors share several common features. Both subunits have a single hydrophobic transmembrane domain, relatively short cytoplasmic tails, and massive extracellular domains. The extracellular domains are compactly folded by virtue of disulfide bonding, associated together, and both chains contribute to the formation of the binding domain. All α subunits contain a seven-fold repeat of a homologous segment, the last three or four repeats of which are likely to contribute to divalent cation binding. The β subunits contain a fourfold cysteine-rich repeat responsible for the folding via internal disulfide groups. Some α subunits, such as the α_1 and α_2, contain a 180 amino acid segment, termed the I domain which, probably, imparts some specificity to the receptors. Cytoplasmic domains of the β subunits are indispensable in connecting the receptors to the cytoskeleton via talin and α-actinin,[1] as detailed below.

7.17.2.1 Extracellular Domain: Ligand Recognition and Interaction

The discovery that an 11 kDa fragment of fibronectin effectively supported cell adhesion[2] and identification of this activity with the tripeptide sequence Arg-Gly-Asp (RGD)[3] provided not only a powerful tool for affinity chromatographic purification of several integrins, but also elucidated relevant recognition sites of the extracellular matrix. While RGD turned out to be one of several recognition sequences (Table 1), the importance of an essential aspartic acid residue has been revealed. Chemical cross-linking and, more importantly, photoaffinity labeling studies have identified N-terminal domains involved in ligand recognition on each subunit and convincingly demonstrated the role of both subunits in ligand binding.[4,5] Using anticomplementary peptide strategy, a peptide sequence APL (complementary to RGD) was identified as a common recognition motif on all known integrin α-subunits.[6] Three contact sites between the receptor and the ligand are predicted—one on the β-subunit and two on the α-subunit.[7]

As indicated above, all α subunits contain sequences homologous to the EF-hand of calcium-binding proteins, with the common distinction—integrins lack the essential aspartic acid residue which is present in position 12 of EF-hand calcium-binding site. It has been hypothesized, therefore, that the aspartic acid

on integrin ligands, by providing the missing cation-coordinating residue, forms a ternary complex with the receptor-bound divalent ion.[8] The cation displacement model emphasizes the role of initial receptor–ligand binding via coordinating aspartic residue, followed by destabilization of a divalent ion bond to EF-hand, extrusion of a divalent ion, and completion of ligand–receptor interaction.

The well-established ability of some integrins, best exemplified by the $\alpha IIb/\beta_3$ platelet integrin, to exist in the inactive and active states has recently been conceptualized on the basis of the above role of aspartic residue in coordination of divalent ion-binding EF-hand. It has been proposed that the residues 68–77 (GSGDS sequence which is analogous to the GRGDS adhesive motif) represent a "nested ligand"of the β_3 subunit. Its interaction with a divalent ion bound to an EF-hand of the αIIb subunit results in the acquisition of a "self-occupied" inactive state. When presented with the RGD sequence, the GSGDS "nested ligand" is displaced, resulting in an active conformational state of the receptor (Figure 1). Recent demonstration of cation binding to the β_3 (118–131) fragment, assessed by terbium luminescence and mass spectrometry, and its displacement by RGD-containing ligands supports this hypothesis.[9] This mechanism, in addition to the cytoplasmic domain-triggered conformational changes of the extracellular integrin domain (see below) may be responsible for the affinity modulation and the conversion from dormancy to active state of integrin receptors.

The fact that all integrins require cations for binding activity provides a completely new perspective for viewing the toxicity of metals. The apparent specificity and affinity of an integrin for its ligand(s) can be modulated by cations. For example, $\alpha_5\beta_1$ integrin binds to

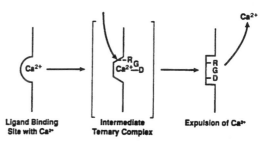

Figure 1 The cation displacement model of ligand binding. The cation binding sequences of integrins are homologous to the EF-hand calcium-binding loops of calmodulin, but lack the coordinating residue. It has been suggested that this coordination position can be filled by the aspartic residue in integrin ligands. See details in the text (reproduced by permission of Academic Press from "Integrins," 1994).[10]

fibronectin in the presence of either Ca^{2+} or Mg^{2+}, but the binding affinity is greater in the presence of Mg^{2+}, and is further increased in the presence of other cations, such as Co^{2+} or Mn^{2+}.[11] The effects of Mn^{2+} on different integrins have been investigated, and the enhanced binding it produces has been exploited in experiments on integrin purification and on molecular mechanisms of ligand recognition. Very little effort, however, has been made to determine whether effects of cations on integrins contribute to the pathophysiology of metal toxicity *in vivo*. For example, it is known that chronic exposure to Mn^{2+} can cause severe neurological disorders. It appears that the neurotoxicity of Mn^{2+} may be partially explained by the recent observation that Mn^{2+} promotes neurite outgrowth, possibly through a direct effect on integrin–extracellular matrix interactions.[12] Thus, some metals may produce toxic effects by promoting inappropriate interactions between cell adhesion molecules and matrix and potentially contributing to toxic sequelae. Metals-induced nephrotoxicity should be evaluated under this angle in the future.

7.17.2.2 Intracellular Domain: Cytoskeletal Interactions

In contrast to the extended extracellular domains, the cytoplasmic tails of integrins are short (except for the β_4 subunit). Numerous studies with truncated cytoplasmic domains of integrins have demonstrated the essential role of the β subunit in establishing integrin–cytoskeletal interactions.[13–15] Using the β_1 subunit as a prototype, three potentially important cytoplasmic regions have been identified. The sequence HDRREFAKFEKE, denoted as cyto-1 region, appears to be essential for the targeting of β_1 subunit to focal adhesions.[16] Four amino acid sequences NPIY and NPKY (cyto-2 and cyto-3, respectively) also participate in localizing the integrin to focal adhesions, but also, quite unexpectedly, represent a concensus signal for clathrin coated pit-mediated internalization of membrane proteins;[17] the significance of this finding awaits elucidation. Above all, the NPxY motif has been found to represent an alternative binding site for Shc (Margoulis, personal communication). Interestingly, these three regions are conserved in many integrin's β subunits. LaFlamme *et al.*[18] constructed chimeric receptors consisting of the extracellular and transmembrane domains of the human IL-2 receptor connected to the intracellular domain of either β_1, β_3, $\beta_3 B$, or β_5 subunit. While the $\beta_3 B$-

containing chimera was expressed diffusely on the cell surface, other chimeric receptors were localized at focal adhesions of transfected human fibroblasts. Expressed at higher levels, β_1 and β_3 chimeras functioned as dominant negative mutants and inhibited endogenous integrin functions in cell spreading and migration. These elegant studies convincingly demonstrate the role of cytoplasmic domains of β subunits in regulating integrin clustering at focal adhesions and mediating cell spreading and locomotion. The cytoplasmic tail of the β_4 chain is unusually large and contains four type III fibronectin-like modules. To examine its role, Spinardi *et al.*[19] expressed full-length and truncated human β_4 cDNAs in epithelial cells which form hemidesmosomes *in vitro*. These investigators have demonstrated that chimeras lacking almost the entire extracellular domain were unable to associate with the α_6 subunit, but were not required for the targeting of this subunit to hemidesmosomes. In contrast, the intracellular domain was required for the assembly of the $\alpha_6\beta_4$ integrin into hemidesmosomes, and at least two type III repeats appeared to be necessary for the interaction with cytoskeletal components of hemidesmosomes.

The functions of α subunits cytoplasmic domains have been extensively studied using genetic engineering approaches. Using the $\alpha_2\beta_1$ collagen/laminin receptor, Chan *et al.*[20] constructed chimeras with cytoplasmic domains of the α_4 and α_5 subunits. These studies revealed different roles of individual cytoplasmic domains in cell migration and contraction of collagen gel. The role of the α_{IIb} subunit was examined by truncating its cytoplasmic tail.[21] This deletion resulted in a constitutively activated $\alpha_{IIb}\beta_3$ receptor. The substitution of α_{IIb} subunit cytoplasmic tail with that of the α_5 subunit similarly resulted in a constitutively active integrin receptor. The similar genetic approach was utilized by Hamler and co-workers[22] to create chimeric VLA-2 and VLA-4 receptors. While deletion of the α chain cytoplasmic domains leads to the loss of constitutive and phorbol ester-stimulated activity, adding 5–7 amino acids after the conserved GFFKR motif restores maximal adhesive activity of VLA-2 and VLA-4 to collagen and VCAM-1, respectively. In addition, these truncated receptors showed perturbations in the sensitivity to divalent ions, but did not affect the intrinsic ability to interact with specific ligands.

7.17.2.3 Focal Adhesions and Involvement of Integrins in Signal Transduction

Focal adhesions are highly specialized domains of the plasma membrane which, on

Proposed Model Of Fibroblast Focal Adhesion In Vitro

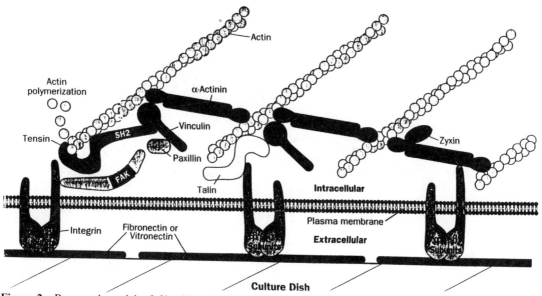

Figure 2 Proposed model of fibroblast focal adhesion *in vitro*. The schema depicts some of the protein interactions on the cytoplasmic side of focal adhesions. Cytoplasmic domains of β integrins bind to either talin or α-actinin. Vinculin associates with talin, α-actinin, tensin, and paxillin. Tensin can also bind to actin filaments (reproduced by permission of Academic Press from "Integrins," 1994).[10]

the extracellular side, form the closest contact with matrix proteins, and, on the cytoplasmic face, represent the sites where converging actin filaments terminate and interact with integrins (Figure 2). In these complex structures, several cytoskeletal proteins are participating in anchoring actin to integrins: vinculin, talin, α-actinin, fimbrin, tensin, paxillin, and zyxin,[23] whereas several other focal adhesion-associated components express enzymatic or yet unidentified activities. Several proteins are substrates for PKC phosphorylation (vinculin, talin, tensin, filamin), and tyrosine kinase phosphorylation (vinculin, talin, tensin, paxillin). Using shearing and quick-freezing procedures, Samuelsson *et al.*[23] in evaluating the cytoplasmic surface and the three-dimensional organization of focal contacts and associated actin bundles, observed that type I aggregates contained β_1 integrins, vinculin, talin, and anchoring actin filaments, whereas type II aggregates did not contain vinculin and talin and were not associated with the actin cytoskeleton. These findings suggest that type I aggregates are relatively stable and represent the classical focal adhesions, whereas type II aggregates containing the β_1 integrin subunits are unanchored, significantly more mobile, and represent the pool of integrins that can be recruited to form new focal adhesions. Ligation of these integrins with the particular epitopes on the extracellular matrix triggers distinct cell signaling events and remodeling of cell shape.

The establishment of focal adhesions via integrin–extracellular ligand binding initiates a cascade of signaling events (Figure 3). The formation of cell–matrix and cell–cell contacts triggers the following cellular responses: rapid activation of the Na^+/H^+ exchanger and cell alkalinization,[24,25] elevation of cytosolic calcium concentration (possibly, due to the activation of a 50 kDa β_3 integrin-associated protein, presumed to represent an integrin-regulated calcium channel in endothelial cells and neutrophils),[26,27] delayed stimulation of a K^+ current,[28] and a series of tyrosine phosphorylation reactions mediated via activation of focal adhesion kinase (FAK).[29,30] Conversely, the reduction of cell–substrate adhesion and eventual detachment of cells from their matrix is associated with activation of a phosphotyrosine phosphatase and decreased tyrosine phosphorylation of focal adhesions.[31] The diversity of signaling mechanisms triggered by the establishment or dissolution of focal adhesions can provide the means for the conformational changes of proteins comprising the focal adhesions (e.g., the β_1 integrin, paxillin, and tensin, all containing phosphotyrosine) and explain the phenomena of affinity modulation of integrins toward the extracellular matrix proteins, outside the cell, and toward the cytoskeletal elements, inside the cell.[14]

The list of the substrates for tyrosine kinase phosphorylation is continuously growing, and presently includes vinculin, talin, tensin,

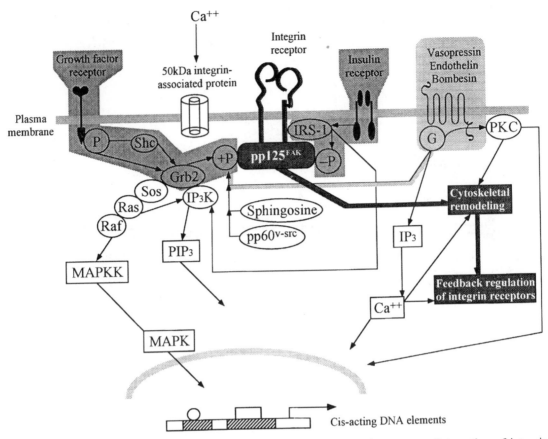

Figure 3 Integrin signaling cascades and interactions with other signaling systems. Interaction of integrin receptors with focal adhesion kinase leading to cell remodeling and, possibly, to feedback regulation of integrins is depicted in red. Potential interactions between integrin receptors and various growth factors (e.g., epidermal growth factor, EGF, or insulin and insulin receptor substrate-1, IRS-1) are outlined in green and blue, respectively. Actions of serpentine G-protein-coupled receptors on the elements of integrin signaling are colored yellow. See text for details (reproduced by permission of Blackwell Scientific from *Kidney Int.*, 1995, **48**, 1375–1385).

paxillin, integrins, pp125[FAK], cadherins, and catenins. pp125[FAK] represents so far the best studied tyrosine kinase associated with focal adhesions. This protein is autophosphorylated during cell adhesion (the cytoplasmic domain of β subunits is necessary for this reaction), but it also undergoes autophosphorylation in response to pp60[v-src], bradykinin, endothelin, and sphingosine, resulting in the increase in its tyrosine kinase activity. Dephosphorylation of pp125[FAK] may be accomplished in part via activation of the SH2 domain-containing protein tyrosine phosphatase Syp (SHPTP2) which is activated by the insulin receptor substrate-1 (IRS-1). This IRS-1 *per se* can be associated with integrin receptor $\alpha_V\beta_3$,[32] thus providing the long-sought link between growth factors and integrins. In Rat-1 fibroblasts expressing insulin receptor, insulin treatment resulted in dephosphorylation of pp125[FAK], whereas in fibroblasts transfected with insulin receptor cDNAs lacking the C-terminal twin tyrosine

phosphorylation sites, the degree of pp125[FAK] phosphorylation was unaffected by insulin.[33] In turn, Grb2 and Pi(3)-kinase are two known substrates of pp125[FAK].[34,35] Collectively, these studies interconnect integrins with receptors for hormones and growth factors, and implicate the state of pp125[FAK] phosphorylation in modulation of the Ras/MAPK and PIP$_3$ signal transduction pathways.

7.17.3 CADHERINS

Cell–cell adhesion molecules (CAMs) belong to two families: calcium-dependent molecules are members of a cadherin family and calcium-independent adhesion molecules belong to an immunoglobulin (Ig) superfamily (see below). The 30 known cadherins are subdivided into N-(neuronal), L- (liver isoform found in chicken and identical to E-cadherin = uvomorulin), and P- (placental) CAMs. General structure of different cadherins retains more than 50% homology: they are comprised of 3–5 internal repeats approximately 100 amino acids long, an N-terminal signal sequence, a single membrane-spanning domain, and a highly conserved cytoplasmic tail.[36] The amino-terminal 113 region is a determinant of binding specificity, with the sequence containing the HAV motif being essential for the homophilic interactions (see below). A cluster of cysteine residues is located extracellularly close to the membrane-spanning region.

The function of cadherins has been best demonstrated experimentally by expressing E-cadherin in L-cells, which usually do not express this adhesion molecule. Normally non-adherent, these cells acquired the polarized adherent appearance of epithelial cells.[37] Conversely, antibodies to cadherins dissociate adherent cells and result in a loss of epithelial cell polarity. In a mixed suspension of cells expressing different types of cadherins, co-aggregation of uniform cells occurs by homophilic interactions via cadherins, and antibodies against cadherins prevent such an aggregation.[38] The recent nuclear magnetic resonance study of the CAD1 domain of E-cadherin (amino acids 1–104) and presentation of solution structure of this domain shed light on the remarkable homophilic interactions of this class of adhesion molecules.[39] CAD domains, which share similar folding topology with CD2 and CD4 extracellular domains of Ig CAMs, are arranged as a rigid string of β-barrels separated by Ca^{2+}-binding pockets formed by a DAD sequence. Such binding of Ca^{2+} at articulation points between CAD tandems should provide rigidity to the structure and may explain the sensitivity of cadherins to proteolysis when calcium is chelated. Homophilic interaction is accomplished through binding of HAV motif to one of the CAD cadherin domains of the opposing cell, providing a cadherin zipper-like contact.

C-terminal domains of cadherins are highly conserved among different members of the class and they are required for the association with α- (homologue of vinculin) or β-catenin, and plakoglobin (γ-catenin). Deletions in the C-terminal portion of cadherins resulted in a loss of adhesion function.[36] Proper functioning of cadherins is a prerequisite for the polarized distribution of various membrane proteins, including the Na,K-ATPase, in epithelia.[40] It appears that cadherin's binding to the cortical actin bundles (but not to stress fibers) is required for their proper positioning at *zonula adherens* and functioning in cell–cell interaction. A curious homology between catenins and a product of a Drosophila segment polarity gene armadillo, as well as a product of a tumor suppressor gene (APC) linked to colon cancer, may reflect an important role of these intracellular linker-proteins in signal transduction processes related to positional information and tumor suppression.[41–49] Altogether, morphoregulatory functions of cadherins provide the basis for the complex arrangement of cellular populations in a defined architecture characteristic of an organ. Any disruption of such an architecture in pathological conditions may engage cadherins. To date, no evidence on the cadherins function in the course of nephrotoxic injury is available, and future studies of these adhesion molecules in renal injury should bring important missing information.

7.17.4 DISTRIBUTION OF INTEGRINS IN THE KIDNEY: PHYSIOLOGY AND PATHOPHYSIOLOGY

Integrins are abundantly expressed in the kidney (Figure 4). Immunohistochemical localization of different subunits along the mammalian nephron has revealed that proximal tubules express subunits α_6, α_1, and, marginally, α_3; that distal tubules express subunits α_1, α_2, α_3, and α_6, and that the β_1 subunits are present along the entire tubule.[50–52] The following relevant combinations of these and several

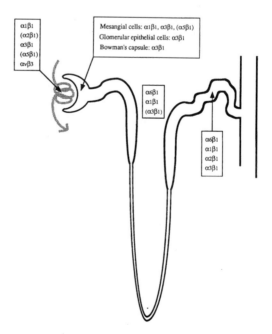

Figure 4 Distribution of integrin receptors along the nephron. Schematic summary of immunohistochemical localization of different integrin receptors in the mammalian kidney.

other subunits may exist in the kidney: $\alpha_1\beta_1$ and $\alpha_2\beta_1$, $\alpha_3\beta_1$, $\alpha_6\beta_1$, $\alpha_6\beta_4$, $\alpha_v\beta_3$, and $\alpha_v\beta_6$.

7.17.4.1 Redistribution of Integrins in Nephrotoxic and Ischemic Renal Injury

Microscopic observations of intravitally stained or fixed BSC-1 cells subjected to oxidant stress yielded an intriguing observation on the loss of the normal highly polarized distribution of several integrin receptor subunits.[53] Using confocal microscopy, it was observed that cells growing as a monolayer exhibited focal adhesions and integrins predominantly at the basolateral membranes. Following oxidative stress or anoxia (Lieberthal, personal communication) this normal pattern underwent a rapid transformation from a highly polarized to an ostensibly random distribution between the apical and basolateral membranes.

Epithelial cell polarity is one of the major attributes of this cellular barrier between the external and internal milieu. The best studied precedent in epithelial cells is related to the anoxia-induced loss of the polar distribution of Na,K-ATPase.[54] Other membrane proteins can undergo repolarization or depolarization in response to stress. The phenomenon of redistribution of the normally polarized integrins is known to occur under many different circumstances, including keratinocytes and corneal epithelial cells undergoing maturation and wound healing, respectively,[55,56] or malignant transformation of epithelial cells.[57–59] The mechanisms governing this stress response are, probably, diverse and include cytoskeletal disruption, phosphorylation events, activation of proteases, modification of matrix proteins, and production of endogenous antiadhesins and repulsins (see below).

The functional consequences of the loss of polarized distribution of integrins in renal tubular epithelial cells are twofold. First, cell detachment from the matrix is a well-established consequence of the dysfunction of several integrins. There is emerging evidence that disruption of epithelial cell–matrix interactions leading to the detachment of these anchorage-dependent cells, or *anoikis* (Greek term for homelessness), induces apoptotic cell death.[60–62] Hence, renal tubular cell detachment from the matrix is a likely consequence of integrin redistribution from the basal cell surface. Second, the redistribution of integrins from the basal to the apical cell membrane predisposes *in situ* cells to homo- and, possibly,

heterophilic interactions with binding sites on the dislodged cells or with matrix fragments.

Our recent immunohistochemical studies of ischemic rat kidneys were aimed to investigate the topography of RGD binding sites and integrin receptors in the ischemic rat kidneys (Romanov *et al.*, unpublished observations). Two RGD peptides were synthesized: a cyclic biotinylated (Bt)RGD peptide and a linear RGD peptide (GRGDSP) labeled with rhodamine green (RhoG). In control, Bt-RGD staining was undetectable, whereas RhoG-GRGDSP staining faintly decorated the basolateral aspect of the proximal tubular cells in a punctated fashion. In contrast, ischemic kidneys showed binding to the basolateral and apical aspects of proximal tubules, peritubular capillaries, and desquamated cells within tubular lumen. The most conspicuous staining of ischemic kidneys was obtained with antibodies to the β_1 (labeling of the apical aspect of proximal and distal tubules, as well as desquamated cells obstructing tubular lumen) and the α_v (glomeruli, tubular epithelia, intima of blood vessels stained faintly, while the obstructing cellular conglomerates showed intense staining) subunits. Application of antibodies against the α_3 subunit showed no staining in control and a faint staining, which had the same topography as the β_1 subunit, of ischemic kidneys. Clearly, expression of RGD binding sites and $\beta1$ integrin subunits along the apical aspect of tubular epithelia and on the surface of desquamated cells is in concert with the hypothesis on the pathogenetic role of RGD-recognizing integrins in tubular obstruction. Unexpectedly, expression of RGD binding sites along the intimal surface of blood vessels in ischemic kidneys was demonstrated, suggesting an additional abnormality of integrin receptors in vascular endothelial cells. These observations are consistent with the proposed role of integrins in triggering the cascade of events leading to cellular detachment and aggregation of detached cells, as schematically summarized in Figure 5.

The role of integrins in ARF has also been confirmed in *in vivo* experiments in rats with ischemic ARF by monitoring the proximal tubular pressure.[63] We have consistently observed the elevation of proximal tubular pressure, as an index of tubular obstruction, in rats after the release of renal artery occlusion. When a linear RGD peptide was injected into the renal artery upon reperfusion, however, the elevation of proximal tubular pressure was curtailed, an effect which was not reproduced by an inactive RGE peptide or by a vehicle alone.[63] These data are consistent with our proposed model of tubular obstruction in ARF.

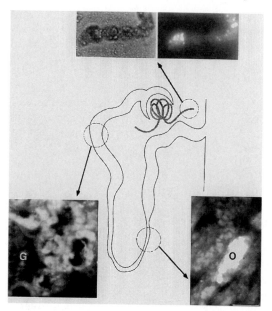

Figure 5 Schematic summary of RGD binding sites in the ischemic rat kidney. Specialized structures along the nephron are magnified to denote RGD binding sites on the intimal surface of resistance arteries (a bright-field and fluorescence view of the microdissected afferent arteriole stained with rhodamine-labeled RGD peptide), immunodetectable β₁ subunits in the proximal nephron on the surface of desquamating or desquamated cells, and RGD binding sites on the surface of epithelial cell and cellular debris impacting the loop of Henle and the distal nephron, as revealed with the rhodamine-labeled RGD peptide (reproduced by permission of Blackwell Scientific from *Kidney Int.*, in press).

It is important to bear in mind that all these interactions leading to cellular agglomeration were inhibited by RGD peptides.

7.17.5 POTENTIAL THERAPEUTIC USE OF RGD PEPTIDES

Arg-Gly-Asp (RGD) sequence is the most common domain contained in a variety of matrix proteins and serves as the recognition site for diverse integrin receptors.[3,64,65] Matrix proteins containing this particular or a related sequence include fibronectin, vitronectin, laminin, thrombospondin, tenascin, von Willebrand factor, osteopontin, and bone sialoprotein 1.[66–70] It has also been discovered that Zn-α_2-glycoprotein contains the RGD-valine sequence and is abundantly expressed in the proximal and distal tubular epithelium,[71] thus expanding the family of the adhesive RGD-containing proteins. Integrin receptors that recognize this sequence include the entire α_V family, $\alpha_3\beta_1$, $\alpha_5\beta_1$, and $\alpha_{IIb}\beta_3$

integrins,[72–74] as well as several other receptors under specific conditions, including $\alpha_2\beta_1$[74] and $\alpha_4\beta_1$.[75] Synthetic RGD peptides have been extensively exploited in *in vitro* studies of cell–matrix and cell–cell interactions.[63,76–78]

There is burgeoning evidence that synthetic RGD peptides have therapeutic potential in diverse pathological situations. Several laboratories have explored their potential in preventing the metastatic spread of B16 melanoma cell line[79,80] and demonstrated that RGD peptides significantly reduce the formation of lung colonies in mice injected with B16 melanoma cells. These peptides, as well as their natural analogues, disintegrins (see below), have been extensively studied as potential antithrombotic agents acting by inhibition of platelet aggregation.[81–84] Recent attempts to engineer the molecule of hirudin containing the RGD sequence and the effect of this chimeric protein on platelet aggregation have been reported.[85]

7.17.5.1 Effects of Cyclic RGD Peptides in Ischemic Acute Renal Failure in Rats

The above mentioned studies from our laboratory have suggested that RGD peptides may prevent tubular obstruction in the ischemic model of ARF.[63,86,87] Therefore, we examined *in vivo* effects of RGD peptides in ischemic ARF in rats and post-transplant ARF in pigs and assessed their potential as diagnostic tools

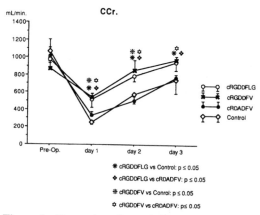

Figure 6 Dynamics of creatinine clearance in rats with ischemic ARF. Rats were subjected to 45 min renal ischemia. After the release of renal artery clamp, either of the indicated cyclic peptides was injected systemically. Creatinine clearance was measured prior to the ischemia (Pre-Op) and daily afterwards for 3 days. Note the accelerated recovery of creatinine clearance in the animals which received cyclic RGD peptides (reproduced by permission of Blackwell Scientific from *Kidney Int.*, 1994, **46**, 1050–1058).[88]

for the detection of tubular obstruction.[88] Experiments were performed in Sprague-Dawley rats subjected to 45 min renal ischemia followed by contralateral nephrectomy. The dynamics of creatinine clearance (C_{Cr}) in the postischemic period are summarized (Figure 6). Cyclic RGDDFLG and RGDDFV-treated groups showed a significantly lesser retention of Cr, as well as an accelerated recovery of C_{Cr} on postoperative day 1, compared to both control groups (animals treated with either the inactive cyclic peptide or with the vehicle). In addition, a significant difference in C_{Cr} between cyclic RGDDFLG or RGDDFV-treated and cyclic RDADFV-treated groups was observed on postoperative day 3. These data indicate that the recovery of renal function after ischemic injury occurred faster in cyclic RGDDFV-treated (2 days) or RGDDFLG-treated (3 days) groups, compared to animals treated with either the inactive peptide or with the vehicle ($p < 0.05$). When a single injection of cyclic RGD peptides was performed 2 h after ischemia, it resulted in functional protection comparable to that achieved with the injection immediately after the release of renal artery. The effectiveness of cyclic RGD peptides was partly diminished when injected 8 h after ischemia. The rank order for the cyclic peptides in ameliorating ischemic ARF and accelerating recovery of C_{Cr} was as follows: cyclic RGDDFV ≥ RGDDFLG ≫ RDADFV.

Histologic examination of kidneys obtained from rats treated with the active peptide or with the vehicle revealed striking differences in the degree of tubular obstruction. Forty-eight hours after acute ischemia, peptide-treated kidneys showed no tubular dilatation and a very mild tubular obstruction, compared with the severely dilatated and obstructed tubules in the vehicle group. Occasional necrotic and dividing tubular epithelial cells *in situ* were observed with equal frequency in both groups. These data suggest that cyclic RGD peptides do not affect significantly the fate of tubular cells, for example, lethal injury or mitogenesis, but rather act by preventing tubular obstruction, a conclusion consistent with the other findings.

The effects of a single infusion of the cyclic RGD peptide, RGDDFV, immediately before transplantation of 24 h cryopreserved pig kidneys have also been examined. This procedure improved functional parameters of the grafts.[89]

The observed *in vivo* effects of cyclic RGD peptides in renal ischemia and transplantation are possibly mediated by their tubular effects. Given the molecular weights of these cyclic peptides being less than 900 Da, and their electrical charge being close to neutral, it is reasonable to assume that the peptides should readily undergo glomerular filtration.[90,91] Thus, a single systemic administration of cyclic peptides *in vivo* ameliorated acute ischemic renal failure, probably through their inhibitory action on cell–cell conglomeration in the tubular lumen.

Recently, several strategies to manipulate adhesion molecules during the course of ischemic injury have been proposed. One approach explores the therapeutic effects of anti-ICAM-1 and anti-LFA-1 antibodies. It has been demonstrated that these antibodies are effective in protection against acute ischemic myocardial reperfusion [92,93] and in renal ischemia.[94] Most probably, these effects are confined to the inhibition of leukocyte migration and/or blood coagulation in the renal microvascular bed, thus resulting in improved renal hemodynamics. The approach described in the present work utilizing small, filterable and less immunogenic peptides is directed toward inhibition of tubular obstruction by the viable desquamated epithelial cells,[95,96] thus aiming to improve urodynamics.

7.17.5.2 Tissue Distribution of RGD Peptides and Their Metabolic Clearance

To evaluate the metabolism of RGD peptides, Tc-99m-GRGDSPC was injected in a tail artery of normal Sprague-Dawley rats and rats subjected to the 45 min renal ischemia followed by the contralateral nephrectomy. The clearance of radiolabeled RGD from the blood in both groups of animals was rapid and indistinguishable between the control and ARF rats. In control, renal accumulation of radioactivity by 10 min postinjection was $12.47 \pm 0.47\%$ (both kidneys) of the administered dose, with the gut and liver retaining $6.58 \pm 0.99\%$ and $5.19 \pm 0.49\%$ of the injected dose, respectively. In ARF group, renal accumulation accounted for $3.53 \pm 1.24\%$ (one kidney), while the gut and liver accumulation accounted for $6.72 \pm 0.79\%$ and $3.31 \pm 0.44\%$, respectively. The other organs (brain, lungs, heart, spleen, and stomach) contained less than 1% of the injected dose in both groups of animals. By 180 min postinjection, the gut content of 99m-Tc amounted to $8.62 \pm 0.92\%$ injected dose in controls and $19.5 \pm 02.16\%$ in ARF animals, suggesting that in ARF the gut may be an important route for the elimination of 99m-Tc-GRGDSPC. When the data were expressed per gram wet tissue weight, the kidneys retained the largest portion of the injected radioactivity at 10 min postinjection: $8.42 \pm 0.18\%/g$ wet weight in control and $4.42 \pm 1.88\%/g$ wet weight in ARF rats).

When the results are expressed as the ratio of the activities of 99m-Tc-GRGDSPC/111-In-DTPA, thus normalizing the amount of the accumulated RGD peptide to the glomerular filtration rate,[97] it is clear that postischemic kidneys retain about three times more of the filtered 99m-Tc-labeled RGD peptide compared to the normal kidney. Taking into account previous findings of surface expression of integrins on desquamation of tubular epithelial cells, the above data suggest specific binding of the injected 99m-Tc-labeled RGD peptide to these exposed integrins. Based on these results, it is quite plausible that the ratio 99m-Tc-RGD:111-In-DTPA may serve as a sensitive diagnostic parameter for detection of tubular obstruction in this and other pathological conditions.

7.17.6 OVERVIEW OF LEUKOCYTE TRAFFICKING IN HOST DEFENCE AND INFLAMMATION

Leukocyte adhesion to endothelial cells is a central event in their recruitment during host defense, inflammation, ischemia reperfusion and other vascular events.[98-107] Leukocyte–endothelial cell adhesion is supported by interaction of cell surface adhesion molecules that are members of larger superfamilies of cell surface receptors involved in immunosurveillance, inflammation, hemostasis, wound healing, morphogenesis, maintenance of tissue architecture, atherogenesis, and tumor metastasis. Figure 7 summarizes the prominent structural features of the major leukocyte adhesion molecules and presents a consensus model for their involvement in physiologic leukocyte migration in host defense. Most leukocyte–endothelial cell adhesion is mediated though coordinated interactions of four major families of leukocyte adhesion molecules: selectins and their ligands, some of which are sialomucins, and the integrins and their ligands, many of which are members of the immunoglobulin (Ig) superfamily of cell surface molecules. Initial adhesion of leukocytes to endothelium occurs in venules and is supported by binding of selectins on leukocytes or endothelial cells to their glycosylated molecular partners. Selectin-mediated adhesion allows leukocytes to roll on endothelium where they are subject to soluble and cell-associated activation signals from endothelium and extravascular tissue. Activated leukocytes are immobilized on endothelium through interaction of leukocyte integrins with endothelial cell ligands. Following immobilization, leukocytes undergo shape change and diapedese between endothelial cells and through an extracellular matrix to the site of inflammation. Phagocytic leukocytes ingest and destroy invading microorganisms predominantly through release of reactive oxygen species, lysosomal enzymes, and other cytotoxic products into phagolysosomes with relative sparing of host tissue. Recruited leukocytes are themselves a rich source of chemoattractants, cytokines, and chemokines, and amplify the inflammatory response.

7.17.6.1 Classification, Biochemistry, and Distribution of Leukocyte Adhesion Molecules

7.17.6.1.1 Selectins

Three selectins have been characterized: P-selectin (platelet selectin; also expressed by endothelial cells), E-selectin (endothelial cell selectin), and L-selectin (leukocyte selectin) (Figure 7).[98-107] These molecules are 40–60% homologous at the nucleotide and amino acid level and share several features: a 120 amino acid NH_2 terminal C-type (calcium dependent) lectin-binding domain, a conserved epidermal growth factor-like domain of $\sim$30 amino acids, several 60 amino acid repeat sequences that share homology with complement regulatory proteins, a single membrane spanning domain and a short intracellular COOH-terminal domain.[103,106] The genes encoding P-, E-, and L-selectin are localized to a cluster on the long arm of chromosome 1.[103,106]

Selectins support leukocyte–endothelial cell and leukocyte–platelet adhesion. L-selectin (CD62L, LAM-1, Mel-14) is constitutively expressed by most leukocytes, mediates adhesion of lymphocytes to endothelial cells of high endothelial venules (HEV) of lymphoid tissue and regulates physiologic homing of lymphocytes from blood to lymphatics. Adhesion of leukocytes to cytokine-activated endothelium derived from other vascular beds, including glomeruli,[108] can be attenuated by anti-L-selectin mAb, suggesting that L-selectin also regulates leukocyte trafficking in non-lymphoid tissue during inflammation. P-selectin (CD62P; PADGEM; GMP-140) is stored in alpha granules of platelets and Weibel–Palade bodies of endothelial cells and mobilized to the cell surface upon activation. P-selectin supports adhesion of granulocytes, monocytes, some lymphocyte subsets, and carcinoma cells. E-selectin (CD62E; ELAM-1) is only expressed by endothelial cells following induction by cytokines and supports adhesion of granulocytes, monocytes, some memory T-lymphocytes, natural killer cells and some carcinoma cells.

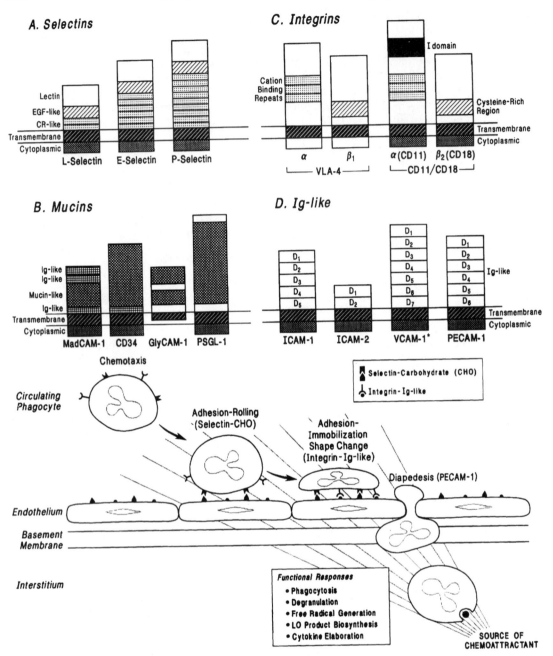

Figure 7 Structural features of major leukocyte adhesion molecules and overview of their roles in leukocyte trafficking during inflammation. *Upper panel:* Some structural features of the four major families of leukocyte adhesion molecules. See text for explanation. Abbreviations: VLA-4, very late activation antigen-4; MadCAM, mucosal addressin cell adhesion molecule; GlyCAM-1, glycosylation-dependent cell adhesion molecule; PSGL-1, P-selectin glycoprotein ligand; ICAM, intercellular adhesion molecule; VCAM-1, vascular cell adhesion molecule-1; PECAM-1; platelet endothelial cell adhesion molecule-1; EGF-like, epidermal growth factor-like domain; CR-like, complement regulatory protein-like domain; Ig-like, immunoglobulin-like; D, domain. *Lower panel:* Leukocytes migrate towards the site of inflammation up concentration gradients of soluble and/or cell-associated chemoattractants (chemotaxis), adhere to endothelial cells (margination/adhesion), and disrupt tight junctions and migrate between endothelial cells (diapedesis). Initial adhesion of leukocytes and endothelial cells is supported by interaction of leukocyte or endothelial cell selectins with carbohydrate-containing ligands, many of which are mucins. Selectin-mediated adhesion supports leukocyte rolling on endothelium where they are subject to activation signals from endothelium and extravascular tissue. Upon activation, leukocytes are immobilized by interaction of leukocyte integrins with endothelial cell ligands such as ICAM-1 and VCAM-1, members of the Ig-like superfamily. The molecular basis for diapedesis has not been established fully, but appears to involve PECAM-1, an Ig-like molecule localized at intercellular junctions (reproduced by permission of Blackwell Scientific from *Kidney Int.*, 1994, **45**, 1285–1300).[99]

7.17.6.1.2 Mucins and other selectin ligands

The ligands for the selectins have been the subject of heated debate and readers are referred to several extensive reviews of this controversial topic.[100,103,105,106] Three ligands for L-selectin have been characterized: MadCAM-1, CD34, and GlyCAM-1 (Figure 7). These structurally related sialomucins contain a high density of O-linked sugars on their extracellular domains that probably extend beyond the glycoprotein and polysaccharide glycocalyx of the cell. MadCAM-1 (mucosal addressin cell adhesion molecule-1) is the most extensively characterized ligand for L-selectin and is preferentially localized to mucosal lymphoid tissue. Cloning of the gene for MadCAM-1 also demonstrated Ig-like domains similar to those of Ig-like adhesion receptors for leukocyte integrins, and MadCAM-1 is also a ligand for the $\alpha_4\beta_7$ integrin involved in lymphocyte homing in Peyer's patches. This economy of adhesive function (i.e., binding sites for L-selectin and $\alpha_4\beta_7$ integrin on the same molecule) may have evolved to facilitate the physiologic trafficking of lymphocytes within mucosal lymphoid tissue.

The structural characteristics of the leukocyte ligands for P- and E-selectin are less well-defined and appear to be a group of diverse and complex macromolecules that share certain anionic carbohydrate structures.[100,103,105,106] Sialyl Lewis^x (SLe^x) is a sialylated, fucosylated tetrasaccharide that decorates many proteins. Congenital deficiency of SLe^x results in the immunodeficiency syndrome leukocyte adhesion deficiency type II. This moiety is an important ligand for both P- and E-selectin and is probably presented as a residue on sialomucins or other fucosylated molecules. Candidate molecules for the latter role include P-selectin glycoprotein ligand (PSGL-1) and E-selectin ligand-1 (ESL-1). PSGL-1 bears a sialylated extracellular domain that contains three potential sites for N-linked glycosylation and a region of 15 decameric repeats with potential sites for O-linked glycosylation. Enzymatic disruption of O-linked, but not N-linked moieties results in loss of P-selectin binding. ESL-1 is a 150 kDa fucosylated glycoprotein expressed by myeloid cells whose predicted amino acid sequence is 94% identical to a cystein-rich fibroblast growth factor receptor.

7.17.6.1.3 Integrins

The integrins that appear most important in leukocyte–endothelial adhesion are the very late activation antigen-4 β_1 integrin ($\alpha_4\beta_1$; CD49d/CD29; VLA-4), the CD11/CD18 family of β_2 integrins and $\alpha_4\beta_7$ integrin.[98,99,107]

VLA-4 is constitutively expressed by lymphocytes, monocytes, basophils, and eosinophils, but not neutrophils and is a ligand for the inducible Ig-like vascular cell adhesion molecule-1 (VCAM-1), fibronectin and possibly other matrix components. Three heterodimeric β_2 integrins have been characterized and classified as CD11a/CD18, CD11b/CD18, and CD11c/CD18; CD11 and CD18 refer to α and β subunits, respectively (Figure 7). CD18 is a 678 amino acid protein encoded by a single gene on chromosome 21 and is characterized by a highly conserved cytoplasmic tail of 46 amino acids with several potential phosphorylation sites, a highly conserved membrane-spanning domain, and a longer extracellular region which contains a conserved cysteine-rich region. Point mutations in the latter region cause an immunodeficiency syndrome, termed leukocyte adhesion deficiency type 1, characterized by impaired neutrophil migration to the extravascular space and recurrent life-threatening bacterial infections. The CD11a, CD11b, and CD11c α subunits are comprised of 1063, 1136, and 1144 amino acids, respectively, and are encoded by distinct genes in a cluster on chromosome 16. Notable features of the CD11 subunits include short nonhomologous cytoplasmic regions containing potential phosphorylation sites, highly conserved membrane spanning regions, a longer extracellular region which contains several cation-binding repeats, and an "I" (interactive) domain which is relatively highly conserved across integrin subfamilies and may be an important adhesion domain. CD11a is constitutively expressed by granulocytes, monocytes, and lymphocytes, whereas CD11b and CD11c are expressed by granulocytes and monocytes, but not lymphocytes (Table 1). The major ligands for CD11a/CD18 are the Ig-like intercellular adhesion molecules (ICAM) 1 and 2. ICAM-1 is also a ligand for CD11b/CD18; however, CD11b/CD18 also supports leukocyte adhesion to cellular and acellular substrates by mechanisms that are independent of ICAM-1. The ligand(s) for CD11c/CD18 and role of this molecule in leukocyte migration have not been defined.

7.17.6.1.4 Immunoglobulin-like molecules and other integrin ligands

The immunoglobulin superfamily of cell surface proteins is divided into C1-type surface proteins which function in antigen recognition and the C2-type which mediate complement binding and cell adhesion. Five C2-type Ig-like molecules support leukocyte–endothelial cell interactions: ICAM-1, ICAM-2, VCAM-1, PECAM-1, and MadCAM-1 (Figure 7).[98,102,107] These molecules contain one or more regions homologous to Ig, each consisting of a disulfide-bridged loop containing antiparallel β-pleated strands arranged into two sheets, a transmembrane domain and a short cytoplasmic tail. ICAM-1 is encoded for by chromosome 19 and contains five tandem immunoglobulin domains (D1–D5). ICAM-1 is constitutively expressed at low levels by endothelial cells and its expression is enhanced upon exposure to cytokines such as tumor necrosis factor (TNF) and interleukin-1 (vide infra). Cytokines also induce *de novo* expression of ICAM-1 by other cell-types, including mesangial cells and renal epithelial cells. ICAM-1 is also expressed by lymphocytes and some other leukocytes, and homotypic and heterotypic lymphocyte adhesion mediated by interaction of ICAM-1 and CD11a/CD18 facilitates other important lymphocyte functions including antigen recognition, lymphocyte costimulation, and cytotoxicity. ICAM-2 is a noninducible Ig-like adhesion molecule, encoded by chromosome 17, that is constitutively expressed at high levels by endothelial cells. ICAM-2 (CD102) is a glycoprotein of molecular weight 60 kDa, contains two Ig-like domains and is also a ligand for CD11a/CD18, but not other β_2 integrins. Low levels of ICAM-2 are detected on mononuclear cells, including lymphocytes and NK cells. Because resting T lymphocytes express ICAM-2, but minimal levels of ICAM-1, it has been suggested that ICAM-2 may play a central role in the initial interaction of lymphocytes with antigen-presenting cells.

VCAM-1 (CD106; INCAM-110) is a 110 kDa glycoprotein that contains seven Ig-like domains, a 22 amino acid transmembrane domain and a short 19 amino acid cytoplasmic tail (Figure 7). A second form of VCAM-1, consisting of six Ig-like domains, is generated in some tissues by alternate splicing. VCAM-1 is absent or expressed at low levels by resting endothelial cells and is induced by exposure to cytokines. The latter also induce VCAM-1 on other cell-types, including mesangial cells, renal epithelial cells and vascular smooth muscle cells. VCAM-1 is constitutively expressed by parietal epithelial cells of Bowman's capsule and its function in this location is unclear. VCAM-1 supports adhesion of eosinophils, basophils, monocytes, and lymphocytes, but not neutrophils, through interaction with VLA-4. PECAM-1 (CD31) is a 130 kDa integral membrane glycoprotein expressed by platelets, some leukocytes and endothelial cells. PECAM-1 is constitutively expressed preferentially at tight junctions of endothelial cells

and appears to regulate leukocyte diapedesis through a homotypic adhesion mechanism.

7.17.7 REGULATION OF LEUKOCYTE ADHESION BY CHEMOATTRACTANTS, CYTOKINES, AND CHEMOKINES

To prevent indiscriminate and chaotic leukocyte adhesion in health, leukocytes and endothelial cells are maintained in an "anti-adhesive phenotype." Inflammatory mediators provoke a "pro-adhesive" phenotypic switch by increasing the avidity and/or expression of preformed adhesion molecules and triggering *de novo* synthesis of adhesion molecules through activation of gene transcription.[98,107] Table 2 summarizes some mechanisms and mediators involved in this phenotypic transformation. A diverse array of peptide, lipid, and carbohydrate mediators provoke rapid leukocyte adhesion to endothelium through modulation of preformed molecules. These include bacterial cell wall peptides, complement components such as C5a, lipid-derived stimuli such as LTB_4 and PAF, and cytokines and chemokines such as MCP-1, IL-8, TNF, and GM-CSF. Most activate leukocytes through engagement of specific cell surface receptors and recruitment of an orchestrated cascade of signal transduction events that ultimately result in mobilization of preformed molecules (e.g., CD11b/CD18, CD11c/CD18) from "specific" tertiary granules to the cell surface and/or enhanced avidity of constitutively expressed molecules (e.g., CD11/CD18 integrins, L-selectin) for cognate ligands. Endothelial cells are also subject to rapid modulation of their adhesiveness for leukocytes, predominently through changes in the expression of preformed P-selectin. Thrombin, histamine, LTC_4, LTD_4, hydrogen peroxide, and the membrane attack complex of complement (MAC) can each provoke rapid mobilization of P-selectin from Weibel–Palade bodies to the cell surface. The signal transduction events that subserve these responses are still being appreciated.

Prolonged exposure (hours to days) of endothelial cells and many other cell-types augments their adhesiveness for leukocytes by activating adhesion molecule gene transcription (Table 2). Cytokines also promote leukocyte recruitment by triggering endothelial biosynthesis of soluble and cell-associated stimuli for leukocyte adhesion such as IL-8, MCP-1, and PAF. The mechanism(s) by which cytokines regulate gene transcription is currently under investigation. ICAM-1 biosynthesis is regulated, at least in part, by protein kinase C and

Table 2 Regulation of leukocyte adhesion by chemoattractants, cytokines, chemokines, and other inflammatory mediators.

Response	Sites of action	Mechanisms	Molecules	Some stimuli
Sec–min	Leukocytes	Increase in avidity and/or expression of preformed adhesion molecules	CD11/CD18, VLA-4, L-selectin	C5a, LTB$_4$, PAF IL-8, MCP-1, TNFα bacterial wall peptides E-selectin, ANCA
	Platelets	Increased expression of preformed adhesion molecules	P-selectin	ADP, thrombin histamine
	Endothelial	Increased expression of preformed adhesion molecules	P-selectin	Histamine, thrombin H$_2$O$_2$, MAC, LTC$_4$, LTD$_4$
	Mesangial	Increase in cell adhesiveness	?ligand(s)	LTD$_4$
Hours–days	Leukocytes	*De novo* synthesis of adhesion molecules	CD11/CD18, L-selectin	GM-CSF
	Endothelial	*De novo* synthesis of adhesion molecules	ICAM-1, VCAM-1, E-selectin, ligands for L-selectin, ?P-selectin	TNFα, IL-1β, IFc IL-4, endotoxin
	Mesangial, smooth muscle epithelial	*De novo* synthesis of adhesion molecules	ICAM-1, VCAM-1	TNFα, IL-1β, IFc IL-4, endotoxin

Source: Brady.[99]
LT, leukotriene; PAF, platelet activating factor; IL, interleukin; MCP, monocyte chemotactic peptide; TNF, tumor necrosis factor; HETE, hydroxyeicosatetraenoic acid; ADP, adenosine diphosphate; H$_2$O$_2$, hydrogen peroxide; MAC, membrane attack complex of complement; GM-CSF, granulocyte-macrophage colony-stimulating factor; IF, interferon; ANCA, antineutrophil cytoplasmic.

triggered by binding of the transcription regulatory protein AP-1 to a binding site on the ICAM-1 promoter. In contrast, VCAM-1 and E-selectin appear to be induced predominently through the actions of the pleiotropic transcription regulatory factor NFkB. The mechanism(s) by which cytokines activate NFkB is incompletely understood, but may involve receptor-triggered changes in cell free radical generation, fatty acids levels and/or redox state.

7.17.8 LEUKOCYTE ADHESION MOLECULES AND KIDNEY DISEASES

A host of complementary *in vivo* and *in vitro* studies have defined the patterns of leukocyte adhesion molecules by renal cells.[108–151] In normal kidney, ICAM-1 is expressed at low levels on the luminal surface of endothelial cells of large vessels, glomeruli, and peritubular capillaries, by some cells in the mesangium,

on the luminal surface of some parietal epithelial cells of Bowman's capsule, on the brush border of proximal tubule cells, and by fibroblast-like interstitial cells.[109–127] VCAM-1 is constitutively expressed by parietal epithelial cells of Bowman's capsule and occasional endothelial cells of large vessels and peritubular capillaries.[125,128–132] P-selectin and E-selectin are not usually detected in normal kidney. Albeit based on a limited number of reports, ICAM-2, PECAM-1, and the mucin CD34 appear to be expressed by renal microvascular endothelial cells under basal conditions.[99]

7.17.8.1 Glomerular and Tubulointerstitial Injury

Glomerular expression of ICAM-1 is typically upregulated in humans with active crescentic glomerulonephritis, mesangioproliferative glomerulonephritis, IgA nephropathy, Henoch–Schonlein purpura, proliferative grades of lupus nephritis, and in rats with experimental nephrotoxic serum nephritis.[109–118] In these diseases, striking ICAM-1 expression is usually also detected on the luminal surface of proximal tubules, distal convoluted tubule and collecting duct cells, and on fibroblast-like interstitial cells.[109–118] In general, the level of expression correlates with disease activity and the intensity of the local leukocyte infiltrate. Variable levels of ICAM-1 have been reported in patients with minimal change disease, focal segmental glomerulosclerosis and membranous nephropathy. Soluble forms of ICAM-1 circulate in normal blood and elevated circulating levels of ICAM-1 have been reported in acute glomerulonephritis. It is, as yet, unclear whether the latter phenomenon represents another level of regulation of leukocyte adhesion or merely reflects accumulation of soluble ICAM-1 in blood because of impairment of glomerular filtration.[134]

Prominent VCAM-1 expression has been reported on proximal tubule cells of patients with vasculitis and crescentic nephritis, lupus nephritis, IgA nephropathy, and allergic interstitial nephritis.[128,129] As with ICAM-1, VCAM-1 levels tend to correlate with the degree of active inflammation and the intensity of the surrounding leukocytic infiltrate. Interestingly, VCAM-1 expression is not usually striking on vascular endothelial cells in either glomerulonephritis or tubulointerstitial nephritis.[128,129] It should be noted that upregulation of VCAM-1 has also been reported in proximal tubules of patients with diabetic nephropathy, amyloid, gouty nephropathy, minimal change disease, and membranous nephropathy–diseases that are not usually associated with leukocytic infiltration of the kidney. These findings probably highlight the fact that both increased adhesion molecule expression and local generation of chemoattractants are required for leukocyte recruitment. The stimulus for induction of ICAM-1 and VCAM-1 on tubular epithelium in patients with glomerulonephritis is unclear. Potential mechanisms include primary involvement of the tubulointerstitium in the disease process and/or the "down-stream" actions of glomerular cytokines that reach the tubulointerstitium via blood, urine, or tissue diffusion.

The pattern of expression of P- and E-selectin in glomerulonephritis and tubulointerstitial inflammation is largely undefined. Prominent glomerular expression of P-selectin has been noted in a murine model of complement-independent immune complex glomerulonephritis,[147] but not in a complement-dependent rat model.[148] A preliminary report suggests that E-selectin is expressed by glomerular endothelial cells in some patients with acute glomerulone-

phritis, lupus nephritis, and IgA nephropathy, but not in patients with focal segmental glomerulosclerosis or membranous nephropathy.[134] E-selectin expression on interstitial venules is another common finding in patients with lupus nephritis and marked interstitial inflammation.[134]

The anti-inflammatory efficacy of mAb against different adhesion molecules has been assessed in experimental models of glomerulonephritis, tubulointerstitial disease, allograft rejection, and renal ischemia-reperfusion injury (Table 3).[93,135,142–151] It should be noted that mAb, in addition to direct inhibition of leukocyte–endothelial cell interactions, may potentially confer antiinflammatory activity via several other mechanisms.[2,136–142] These actions include inhibition of adhesion-dependent antigen presentation, transcellular eicosanoid generation, and cytotoxicity, and should be considered when interpreting the results of interventional studies on antiadhesion therapy (Figure 8).[136–142] Whereas many studies have demonstrated a protective effect, it is evident that the profile of adhesion molecules that support leukocyte recruitment can vary markedly depending on the type of experimental model, species, and time of study. In nephrotoxic serum nephritis in Long–Evans rats, a classic model of immune complex-mediated glomerulonephritis, mAb against CD18, CD11b, and ICAM-1 inhibited neutrophil infiltration and proteinuria, whereas mAb against CD11a and E-selectin were not beneficial.[135] Interestingly, mAb against VLA-4 also blocked neutrophil recruitment, even though this one integrin does not play a direct role in neutrophil trafficking.[135] MAb against CD11a and ICAM-1 also attenuate proteinuria and crescent formation in crescentic glomerulonephritis in Wistar–Kyoto rats, a model of rapidly progressive glomerulonephritis.[145,146] Importantly, inflammation was significantly reduced even when mAb were administered after disease was established, an important consideration if antiadhesion therapy is to be useful therapeutically in humans. In contrast to these studies of glomerulonephritis in rats, anti-CD18 mAb does not afford protection in nephrotoxic serum nephritis in rabbits, despite evidence of engagement of leukocyte CD18 integrins and inhibition of leukocyte trafficking to other sites (Table 3). Experiences with anti-P-selectin mAb have varied. Anti-P-selectin mAb confered significant protection in a complement-independent model of murine nephrotoxic serum nephritis,[147] but did not influence neutrophil recruitment in the complement-dependent concanavalin-A/ferritin model of immune complex nephritis in rats.[148]

Table 3 Efficacy of some adhesion-blocking monoclonal antibodies in kidney diseases.

Renal disease	Species (strain)	Molecule	Protection	Ref.
Nephrotoxic serum nephritis	Rat (Long–Evans)	CD18	yes	135
		CD11a	no	
		CD11b	yes	
		VLA-4	yes	
		ICAM-1	yes	
		E-selectin	no	
Nephrotoxic serum nephritis	Rat (Lewis)	CD11b	yes	143
Nephrotoxic serum nephritis	Rat (Wistar)	ICAM-1	yes	144
Nephrotoxic serum nephritis	Mouse (C57/Bl10)	P-selectin	yes	147
Nephrotoxic serum nephritis	Rabbit (New Zealand White)	CD18	no	b
		ICAM-1	no	
Nephrotoxic serum nephritis	Rabbit (New Zealand White)	CD18	no	b
Con A/ferritin nephritis	Rat (Sprague-Dawley)	P-selectin	no	148
Crescentic glomerulonephritis	Rat (Wistar–Kyoto)	CD11a	yes	146
		ICAM-1	yes	
		CD11a + ICAM-1	yes	
Crescentic glomerulonephritis	Rat (Wistar–Kyoto)	CD11a + ICAM-1	yes	145
Tubulointerstitial nephritis	Mouse (CBA/Ca, kdkd)	ICAM-1	yes	149
Acute allograft rejection	Monkey (Cynomolgus)	ICAM-1	yes	150
Acute allograft rejection	Human	ICAM-1	yes	151
Ischemia-reperfusion	Rabbit (New Zealand White)	CD18	no	154
Ischemia-reperfusion	Rabbit (New Zealand White)	CD18	no	a
		ICAM-1	no	
Ischemia-reperfusion	Rat (Sprague-Dawley)	ICAM-1	yes	95

Source: Brady.[99]
[a]J. Neuringer and H. R. Brady, unpublished observations. [b]Y. M. O'Meara, D. J. Salant, H. R. Brady, P. G. Tipping, L. J. Cornthwaite, and S. R. Holdsworth, XII International Congress of Nephrology, Jerusalem, Israel, June 1993.

Because mAb may potentially affect immune function by multiple mechanisms, many of which are unrelated to leukocyte–endothelial cell interactions, these observations will have to be confirmed using other interventions. Along these lines, details regarding the course of experimental glomerulonephritis in adhesion molecule deficient mice ("knockout mice") are eagerly awaited.

Harning *et al.* monitored the efficacy of anti-ICAM-1 mAb on hereditary autoimmune tubulointerstitial nephritis in the kdkd variant of the CBA/Ca mice (Table 3).[149] These animals develop progressive and ultimately lethal tubulointerstitial nephritis beginning approximately 4 weeks after birth. Nephritis is associated with upregulation of ICAM-1 expression in the renal interstitium, on infiltrating leukocytes, and on the basolateral surface of renal tubule epithe-lium. Anti-ICAM-1 mAb caused a marked reduction in leukocyte infiltration, tubular injury, and proteinuria in this model; however, survival was unchanged.

7.17.8.2 Allograft Rejection

The distribution of ICAM-1 in acute allograft rejection in humans differs markedly from that observed in glomerulonephritis.[119–127] Typically, there is striking induction of ICAM-1 on the luminal surface of proximal tubule cells, on some distal tubule and collecting duct cells, and on infiltrating leukocytes. The glomeruli are spared. Marked changes in ICAM-1 levels on vascular endothelium are not usually detected, perhaps being

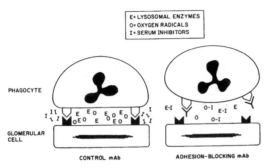

Figure 8 Leukocyte adhesion promotes leukocyte-mediated injury to renal glomerular mesangial cells in culture. Some potential mechanisms by which leukocyte adhesion could promote cytotoxicity during glomerular inflammation. Adherent leukocytes release increased quantities of toxic oxygen radicals (O) and lysosomal enzymes (E) upon activation than leukocytes in suspension. In addition, adhesion facilitates release of these toxic mediators in close proximity to resident glomerular cells with relative exclusion of endogenous inhibitors (I) present in extracellular fluid. By inference, mAb or other agents that inhibit adhesion may not only confer protection by blocking leukocyte recruitment to sites of inflammation, but also inhibit leukocyte-mediated cytotoxicity within a local inflammatory milieu (reproduced by permission of Blackwell Scientific from *Kidney Int.*, 1994, **45**, 1285–1300).[99]

veiled by constitutive expression of ICAM-1 at this site. Impressive *de novo* focal expression of VCAM-1 has been reported towards the basolateral surface of proximal tubule cells and on occasional distal tubules.[125,130,131] In addition, striking induction of VCAM-1 is characteristically seen on the endothelium of peritubular capillaries, venules and arterioles, but not on glomerular cells. Again, ICAM-1, VCAM-1, and E-selectin expression tends to be most intense in areas of leukocyte infiltration, implicating these molecules in leukocyte recruitment. The authors are not aware of reports of P-selectin expression in rejection. The endothelial distribution of E-selectin tends to parallel that of ICAM-1 and VCAM-1, but is a less consistent finding.[125] It should be noted that the absence of marked P- and E-selectin expression in some patients with otherwise active glomerulonephritis, tubulointerstitial nephritis or allograft rejection may reflect the transience of P- and E-selectin expression or extensive endothelial injury, and does not preclude a role for this molecule in leukocyte recruitment. Indeed, focal absence of PECAM-1, which is constitutively expressed by normal renal endothelium, has been reported in acute allograft rejection,[125,126] suggesting that the absence of these molecules may indeed be a marker for extensive endothelial injury.

Two studies have reported efficacy of anti-ICAM-1 mAb in acute renal allograft rejection (Table 3).[150,151] Anti-ICAM-1 mAb markedly delayed the onset of rejection and improved recipient survival in Cynomolgus monkeys when employed as the sole immunosuppressive agent. This protective action was associated with partial inhibition of T lymphocyte and monocyte infiltration of the tubulointerstitium. Anti-ICAM-1 mAb also reversed established rejection in allograft recipients maintained on subtherapeutic doses of cyclosporine, albeit without a dramatic reduction in the intensity of the leukocyte infiltrate. These results in nonhuman primates have been confirmed in a phase 1 clinical trial of anti-ICAM-1 in high risk human allograft recipients.[151]

7.17.8.3 Ischemia-reperfusion

Neutrophils play a pivotal role in the pathophysiology of ischemia-reperfusion injury in many vascular beds and mAb that inhibit selectin- or β_2-integrin-mediated adhesion are protective in several experimental models.[107] The role of neutrophils in renal ischemic-reperfusion injury is less clear (Table 3).[95,152–154] Whereas there is some indirect evidence for involvement of neutrophils, it is difficult to detect neutrophils by immunohistochemical methods in kidneys following ischemia-reperfusion injury and the interpretation of tissue myeloperoxidase content, another well-established index of neutrophil infiltration, can be difficult in renal tissue due to the abundant hydrogen peroxide produced by ischemic tubular epithelium. Pharmacologic strategies to deplete animals of neutrophils prior to induction of ischemia or to inhibit leukocyte–endothelial cell adhesion have yielded conflicting results. In the latter regard, anti-ICAM-1 mAb confered marked functional protection in rats, whereas anti-CD18 and anti-ICAM-1 mAb did not confer functional or morphologic protection in two studies of ischemic ARF in rabbits (Table 3). Given the high incidence of ischemic ARF in clinical practice, it is likely that this will be an area of intensive investigation in the near future.

7.17.8.4 Dialysis Membrane Incompatibility

Exposure of circulating blood to cellulosic hemodialysis membranes results in rapid upregulation of CD11/CD18 on granulocytes and monocytes, and shedding of granulocyte L-selectin.[155–159] These events occur in parallel with complement activation, rapid modulation of leukocyte adhesiveness for endothelium, sequestration of leukocytes in the pulmonary vasculature, and hypoxemia: the classic features of the dialysis "first-use syndrome." Leukocyte activation is less marked when newer synthetic biocompatible membranes are employed. Interestingly, the use of cellulosic membranes for treatment of oliguric ARF is associated with a poorer prognosis when compared with patients treated with biocompatible membranes, possibly because activated leukocytes accumulate in the kidney and exacerbate renal injury.

7.17.9 FUTURE STRATEGIES FOR INHIBITION OF LEUKOCYTE ADHESION IN INFLAMMATORY DISEASES

7.17.9.1 Soluble Adhesion Receptors and Antisense Oligonucleotides

Whereas mAb are clearly capable of inhibiting leukocyte–endothelial cell interactions and leukocyte recruitment during inflammation *in vitro* and *in vivo*, they are nevertheless imperfect tools for routine treatment of humans because of their antigenicity and need for parenteral administration. These drawbacks have prompted a furious search for safe, convenient, and efficacious antiadhesive agents. Promising strategies include the use of soluble forms or analogues of adhesion molecules, and antisense oligonucleotides that inhibit *de novo* adhesion molecule biosynthesis. Synthetic oligosaccharides that scavenge binding sites on selectins and prevent their interactions with mucins and other glycosylated, fucosylated ligands have already proven efficacious in lung models of acute inflammation [160] and are currently being tested in other systems (including human diseases). Phosphorothioate oligodeoxynucleotide antisense compounds also appear promising, although as yet not extensively tested in inflammatory models.[161] These molecules are relatively resistant to degradation by virtue of chemical modification of one or more substituents on the internucleotide phosphate and are bioavailable when given subcutaneously and by other parenteral routes.

7.17.9.2 Endogenous "Stop Signals" for Leukocyte Recruitment

Inflammation is frequently self-limiting, raising the possibility that endogenous "stop signals" are generated during the evolution of inflammatory responses that ultimately inhibit further leukocyte recruitment and limit tissue injury. Dissection of the molecular components of the endogenous "braking systems" may ultimately yield novel and relatively nontoxic strategies for therapeutic intervention in human disease. Potential braking signals include nitric oxide, prostacyclin, transforming growth factor beta, and products of the 12/15-lipoxygenase pathways.[98,99,107] The latter are particularly interesting as there is an expanding body of evidence suggesting that products of the 5-lipoxygenase (leukotrienes) and 15-lipoxygenase systems (15-HETE, lipoxins) may evoke counter-regulatory actions on neutrophil trafficking in a manner akin to the actions of prostacyclin and thromboxane on platelet aggregability and vascular tone.[162–167] 15-Hydroxyeicosatetraenoic acid (15-HETE) is a 15-lipoxygenase-derived eicosanoid whose levels are frequently elevated at sites of inflammation and which has impressive anti-inflammatory activity in experimental models of inflammation, including glomerulonephritis, and human psoriasis.[163] 15(S)-HETE is rapidly esterified into neutrophil cell membrane inositol-containing lipids to levels that approximate those of arachidonic acid where it can influence neutrophil receptor function and signal transduction events. Intriguingly, 15(S)-HETE-remodeled neutrophils display markedly impaired neutrophil adhesion and migration across endothelium in response to chemoattractants and cytokines,[164,165] suggesting that this compound may be an important inhibitor of neutrophil trafficking *in vivo*.

Lipoxins are a more recent addition to the families of bioactive eicosanoids and can also function as endogenous leukotriene antagonists and inhibitors of neutrophil recruitment in inflammation.[162] Lipoxins are generated in nanogram quantities through transcellular biosynthetic routes during platelet–neutrophil interactions. Unlike leukotrienes, lipoxins provoke little or no chemotaxis, adhesion, degranulation, or free radical generation at submicromolar concentrations. In contrast, both LXA$_4$ and LXB$_4$ inhibit leukotriene-induced neutrophil chemotaxis, adhesion to endothelial cells, and migration across confluent endothelial cell monolayers in most *in vitro* assay systems and in acute glomerulonephritis *in vivo*.[167] Definitive proof that lipoxins are important regulators of neutrophil–endothelial cell interactions will require the development of selective pharmacologic agents to further study lipoxin biosynthesis or bioactivity. Manipulation of these exciting mediator systems may offer new therapeutic approaches for modulating neutrophil trafficking and vascular tone during multicellular vascular events in human disease.

7.17.10 SUMMARY

Recent years have produced solid evidence that diversity of adhesion molecules is matched by the diversity of pivotal roles they play in various physiological and pathological processes. Abberant distribution and/or function of adhesion molecules is becoming a well recognized phenomenon in renal injury. There are initial attempts to use the knowledge on adhesion molecules to correct certain abnormalities of their distribution and function in disease, as was briefly summarized above. With further clarification of disfunctions in adhesion molecules, therapies targeted toward them will become more diversified.

ACKNOWLEDMENTS

Studies reported in this chapter were supported in part by NIH grants DK41573, DK45695 (MSG), DK44380 and a RAG award from the VA (HB). EN was a Fellow of the NKF.

7.17.11 REFERENCES

1. C. A. Otey, G. B. Vasquez, K. Burridge *et al.*, 'Mapping of the α-actinin binding site within the β_1 integrin cytoplasmic domain.' *J. Biol. Chem.*, 1993, **268**, 21193–21197.
2. M. D. Pierschbacher, E. G. Hagman and E. Ruoslahti, 'Location of the cell-attachment site in fibronectin with monoclonal antibodies and proteolytic fragments of the molecule.' *Cell*, 1981, **26**, 259–267.
3. M. D. Pierschbacher and E. Ruoslahti, 'The cell attachment activity of fibronectin can be duplicated by small synthetic fragments of the molecule.' *Nature*, 1984, **309**, 30–33.
4. J. W. Smith and D. A. Cheresh, 'Integrin (αvβ3)–ligand interaction. Identification of a heterodimeric RGD binding site on the vitronectin receptor.' *J. Biol. Chem.*, 1990, **265**, 2168–2172.
5. S. E. D'Souza, M. H. Ginsberg, T. A. Burke *et al.*, 'Localization of an Arg-Gly-Asp recognition site within an integrin adhesion receptor.' *Science*, 1988, **242**, 91–93.
6. D. B. Taylor and T. K. Gartner, 'A peptide corresponding to GPIIbα 300–312, a presumptive fibrinogen gamma-chain binding site on the platelet integrin GPIIb/IIIa, inhibits the adhesion of platelets to at least four adhesive ligands.' *J. Biol. Chem.*, 1992, **267**, 11729–11733.
7. I. Segal, 'Enzyme Kinetics,' Wiley, New York, 1975.
8. J. W. Smith and D. A. Cheresh, 'Labeling of integrin αvβ3 with [58]Co(III) Evidence of metal ion coordination sphere involvement in ligand binding.' *J. Biol. Chem.*, 1991, **266**, 11429–11432.
9. S. E. D'Souza, T. A. Haas, R. S. Piotrowicz *et al.*, 'Ligand and cation binding are dual functions of a discrete segment of the integrin β3 subunit: cation displacement is involved in ligand binding.' *Cell*, 1994, **79**, 659–667.
10. D. A. Cheresh and R. P. Mecham (eds.), 'Integrins,' Academic Press, San Diego, CA, 1994.
11. J. Gailit and E. Ruoslahti, 'Regulation of the fibronectin receptor affinity by divalent cations.' *J. Biol. Chem.*, 1988, **263**, 12927–12933.
12. W. H. Lin, D. Higgins, M. Pacheco *et al.*, 'Manganese induces spreading and process outgrowth in rat pheochromocytoma (PC12) cells.' *J. Neurosci. Res.*, 1993, **34**, 546–561.
13. J. Solowska, J. L. Guan, E. E. Marcantonio *et al.*, 'Expression of normal and mutant avian integrin subunits in rodent cells.' *J. Cell. Biol.*, 1989, **109**, 853–861.
14. S. E. LaFlamme, S. K. Akiyama and K. M. Yamada, 'Regulation of fibronectin receptor distribution.' *J. Cell. Biol.*, 1992, **117**, 437–447.
15. B. Geiger, D. Salomon, M. Takeichi *et al.*, 'A chimeric N-cadherin/β1-integrin receptor which localizes to both cell–cell and cell–matrix adhesions.' *J. Cell. Sci.*, 1992, **103**, 943–951.
16. A. A. Reszka, Y. Hayashi and A. F. Horwitz, 'Identification of amino acid sequences in the integrin β1 cytoplasmic domain implicated in cytoskeletal association.' *J. Cell. Biol.*, 1992, **117**, 1321–1330.
17. W. J. Chen, J. L. Goldstein and M. S. Brown, 'NPXY, a sequence often found in cytoplasmic tails, is required for coated pit-mediated internalization of the low density lipoprotein receptor.' *J. Biol. Chem.*, 1990, **265**, 3116–3123.
18. S. E. LaFlamme, L. A. Thomas, S. S. Yamada *et al.*, 'Single subunit chimeric integrins as mimics and inhibitors of endogenous integrin functions in receptor localization, cell spreading and migration, and matrix assembly.' *J. Cell. Biol.*, 1994, **126**, 1287–1298.
19. L. Spinardi, Y. L. Ren, R. Sanders *et al.*, 'The β4 subunit cytoplasmic domain mediates the interaction of α6β4 integrin with the cytoskeleton of hemidesmosomes.' *Mol. Biol. Cell.*, 1993, **4**, 871–884.
20. B. M. Chan, P. D. Kassner, J. A. Schiro *et al.*, 'Distinct cellular functions mediated by different VLA integrin α subunit cytoplasmic domains.' *Cell*, 1992, **68**, 1051–1060.
21. T. E. O'Toole, D. Mandelman, J. Forsyth *et al.*, 'Modulation of the affinity of integrin αIIbβ3 (GPIIb-IIIa) by the cytoplasmic domain of α_{IIb}.' *Science*, 1991, **254**, 845–847.
22. P. Kassner, S. Kawaguchi and M. E. Hemler, 'Minimum α chain cytoplasmic tail sequence needed to support integrin-mediated adhesion.' *J. Biol. Chem.*, 1994, **269**, 19859–19867.
23. S. J. Samuelsson, P. W. Luther, D. W. Pumplin *et al.*, 'Structures linking microfilament bundles to the membrane at focal contacts.' *J. Cell. Biol.*, 1993, **122**, 485–496.
24. M. A. Schwartz, E. J. Cragoe, Jr. and C. P. Lechene, 'pH regulation in spread cells and rounded cells.' *J. Biol. Chem.*, 1990, **265**, 1327–1332.
25. S. I. Galkina, G. F. Sud'Ina and L. B. Margolis, 'Cell–cell contacts alter intracellular pH.' *Exp. Cell. Res.*, 1992, **200**, 211–214.
26. M. E. Jaconi, J. M. Theler, W. Schlegel *et al.*, 'Multiple elevations of cytosolic-free Ca^{2+} in human neutrophils: initiation by adherence receptors of the integrin family.' *J. Cell. Biol.*, 1991, **112**, 1249–1257.
27. M. A. Schwartz, E. J. Brown and B. Fazeli, 'A 50-kDa integrin-associated protein is required for integrin-regulated calcium entry in endothelial cells.' *J. Biol. Chem.*, 1993, **268**, 19931–19934.

28. A. Becchetti, A. Arcangeli, M. R. Del Bene *et al.*, 'Response to fibronectin–integrin interaction in leukaemia cells: delayed enhancing of a K^+ current.' *Proc. R. Soc. London Ser. B*, 1992, **248**, 235–240.

29. M. D. Schaller, C. A. Borgman, B. S. Cobb *et al.*, 'Pp125^{FAK}, a structurally unique protein tyrosine kinase associated with focal adhesions.' *Proc. Natl. Acad. Sci. USA*, 1992, **89**, 5192–5196.

30. L. Kornberg, H. S. Earp, J. T. Parsons *et al.*, 'Cell adhesion or integrin clustering increases phosphorylation of a focal adhesion-associated tyrosine kinase.' *J. Biol. Chem.*, 1992, **267**, 23439–23442.

31. P. A. Maher 'Activation of phosphotyrosine phosphatase activity by reduction of cell–substrate adhesion.' *Proc. Natl. Acad. Sci. USA*, 1993, **90**, 11177–11181.

32. K. Vuori and E. Ruoslahti, 'Association of insulin receptor substrate-1 with integrins.' *Science*, 1994, **266**, 1576–1578.

33. T. S. Pillay, T. Sasaoka and J. M. Olefsky, 'Insulin stimulates the tyrosine dephosphorylation of pp125 focal adhesion kinase' *J. Biol. Chem.*, 1995, **270**, 991–994.

34. D. D. Schlaepfer, S. K. Hanks, T. Hunter *et al.*, 'Integrin-mediated signal transduction linked to Ras pathway by Grb2 binding to focal adhesion kinase.' *Nature*, 1994, **372**, 786–791.

35. H. C. Chen and J. L. Guan, 'Association of focal adhesion kinase with its potential substrate phosphatidylinositol 3-kinase.' *Proc. Natl. Acad. Sci. USA*, 1994, **91**, 10148–10152.

36. M. Takeichi, 'Cadherins: a molecular family important in selective cell–cell adhesion.' *Annu. Rev. Biochem.*, 1990, **59**, 237–252.

37. A. Nose, A. Nagafuchi and M.Takeichi, 'Expressed recombinant cadherins mediate cell sorting in model systems.' *Cell*, 1988, **54**, 993–1001.

38. M. Takeichi, T. Atsumi, C. Yoshida *et al.*, 'Selective adhesion of embryonal carcinoma cells and differentiated cells by Ca^{2+}-dependent sites.' *Dev. Biol.*, 1981, **87**, 340–350.

39. M. Overduin, T. S. Harvey, S. Bagby *et al.*, 'Solution structure of the epithelial cadherin domain responsible for selective cell adhesion.' *Science*, 1995, **267**, 386–389.

40. B. A. Molitoris and W. J. Nelson, 'Alterations in the establishment and maintenance of epithelial cell polarity as a basis for disease processes.' *J. Clin. Invest.*, 1990, **85**, 3–9.

41. P. D. McCrea, C. W. Turck and B. Gmbiner, 'A homolog of the armadillo protein in Drosophila (plakoglobin) associated with E-cadherin.' *Science*, 1991, **254**, 1359–1361.

42. M. Peifer and E. Weischaus, 'The segment polarity gene armadillo encodes a functionally modular protein that is the Drosophila homolog of human plakoglobin.' *Cell*, 1990, **63**, 1167–1176.

43. M. Peifer, P. D. McCrea, K. J. Green *et al.*, 'The vertebrate adhesive junction proteins β-catenin and plakoglobin and the Drosophila segment polarity gene armadillo form a multigene family with similar properties.' *J. Cell. Biol.*, 1992, **118**, 681–691.

44. M. Peifer, 'The product of the Drosophila segment polarity gene armadillo is part of a multi-protein complex resembling the vertebrate adherens junction.' *J. Cell. Sci.*, 1993, **105**, 993–1000.

45. R. S. Bradley, P. Cowin and A. M. C. Brown, 'Expression of Wnt-1 in PC12 cells results in modulation of plakoglobin and E-cadherin and increased cellular adhesion.' *J. Cell. Biol.*, 1993, **123**, 1857–1865.

46. L. Hinck, W. J. Nelson and J. Papkoff, 'Wnt-1 modulates cell–cell adhesion in mammalian cells by stabilizing β-catenin binding to the cell adhesion protein cadherin.' *J. Cell. Biol.*, 1994, **124**, 729–741.

47. P. D. McCrea, W. M. Brieher and B. M. Gumbiner, 'Induction of a secondary body axis in Xenopus by antibodies to β-catenin.' *J. Cell. Biol.*, 1993, **123**, 477–484.

48. L. K. Su, B. Vogelstein and K. W. Kinzler, 'Association of the APC tumor suppressor protein with catenins.' *Science*, 1993, **262**, 1734–1737.

49. B. Rubinfeld, B. Souza, I. Albert *et al.*, 'Association of the APC gene product with β-catenin.' *Science*, 1993, **262**, 1731–1734.

50. M. Korhonen, J. Ylanne, L. Laitinen *et al.*, 'The α1–α6 subunits of integrins are characteristically expressed in distinct segments of developing and adult human nephron.' *J. Cell. Biol.*, 1990, **111**, 1245–1254.

51. M. Korhonen, J. Ylanne, L. Laitinen *et al.*, 'Distribution of β1 and β3 integrins in human fetal and adult kidney.' *Lab. Invest.*, 1990, **62**, 616–625.

52. E. E. Simon and J. A. McDonald, 'Extracellular matrix receptors in the kidney cortex.' *Am. J. Physiol.*, 1990, **259**, F783–F792.

53. J. Gailit, D. Colflesh, I. Rabiner *et al.*, 'Redistribution and disfunction of integrins in cultured renal epithelial cells exposed to oxidative stress.' *Am. J. Physiol.*, 1993, **264**, F149–F157.

54. E. M. Fish and B. A. Molitoris, 'Alterations in epithelial polarity and the pathogenesis of disease states.' *New. Engl. J. Med.*, 1994, **330**, 1580–1588.

55. M. Guo, L. T. Kim, S. K. Akiyama *et al.*, 'Altered processing of integrin receptors during keratinocyte activation.' *Exp. Cell. Res.*, 1991, **195**, 315–322.

56. M. A. Kurpakus, V. Quaranta and J. C. R. Jones, 'Surface relocation of alpha 6 beta 4 integrins and assembly of hemidesmosomes in an *in vitro* model of wound healing.' *J. Cell. Biol.*, 1991, **115**, 1737–1750.

57. S. Dedhar and R. Saulnier, 'Alterations in integrin receptor expression on chemically transformed human cells: specific enhancement of laminin and collagen receptor complexes.' *J. Cell. Biol.*, 1990, **110**, 481–489.

58. M. Korhonen, L. Laitinen, J. Ylanne *et al.*, 'Integrin distribution in renal cell carcinomas of various grades of malignancy.' *Am. J. Pathol.*, 1992, **141**, 1161–1171.

59. J. D. Knox, A. E. Cress, V. Clark *et al.*, ' Differential expression of extracellular matrix molecules and the α6-integrins in the normal and neoplastic prostate.' *Am. J. Pathol.*, 1994, **145**, 167–174.

60. S. M. Frisch and H. Francis, 'Disruption of epithelial cell–matrix interactions induces apoptosis.' *J. Cell. Biol.*, 1994, **124**, 619–626.

61. J. E. Meredith, Jr., B. Fazeli and M. A. Schwartz, 'The extracellular matrix as a cell survival factor.' *Mol. Biol. Cell.*, 1993, **4**, 953–961.

62. E. Ruoslahti and J. C. Reed, 'Anchorage dependence, integrins, and apoptosis.' *Cell*, 1994, **77**, 477–478.

63. M. S. Goligorsky and G. F. DiBona, 'Pathogenetic role of Arg-Gly-Asp-recognizing integrins in acute renal failure.' *Proc. Natl. Acad. Sci. USA*, 1993, **90**, 5700–5704.

64. M. D. Pierschbacher and E. Ruoslahti, 'Variants of the cell recognition site of fibronectin that retain attachment-promoting activity.' *Proc. Natl. Acad. Sci. USA*, 1984, **81**, 5985–5988.

65. E. Ruoslahti and M. D. Pierschbacher, 'New perspectives in cell adhesion: RGD and integrins.' *Science*, 1987, **238**, 491–497.

66. S. K. Akiyama, K. Nagata and K. M. Yamada, 'Cell surface receptors for extracellular matrix components.' *Biochim. Biophys. Acta.*, 1990, **1031**, 91–110.

67. K. M. Yamada, 'Adhesive recognition sequences.' *J. Biol. Chem.*, 1991, **266**, 12809–12812.

68. E. Ruoslahti, 'Integrins.' *J. Clin. Invest.*, 1991, **87**, 1–5.

69. R. O. Hynes, 'Integrins: versatility, modulation, and signaling in cell adhesion.' *Cell*, 1992, **69**, 11–25.

70. P. Joshi, C. Y. Chung, I. Aukhil *et al.*, 'Endothelial cells adhere to the RGD domain and the fibrinogen-like terminal knob of tenascin.' *J. Cell. Sci.*, 1993, **106**, 389–400.

71. M. Takagaki, K. Honke, T. Tsukamoto *et al.*, 'Zn-α2-glycoprotein is a novel adhesive protein.' *Biochem. Biophys. Res. Commun.*, 1994, **201**, 1339–1347.

72. E. Ruoslahti, N. A. Noble, S. Kagami *et al.*, 'Integrins.' *Kidney Int.*, 1994, **44** (Suppl.), s17–s22.

73. S. M. Albelda and C. A. Buck, 'Integrins and other cell adhesion molecule.' *FASEB. J.*, 1990, **4**, 2868–2880.

74. P. M. Cardarelli, S. Yamagata, I. Taguchi *et al.*, 'The collagen receptor a2β1, from MG-63 and HT1080 cells, interacts with a cyclic RGD peptide.' *J. Biol. Chem.*, 1992, **267**, 23159–23164.

75. P. Sanchez-Aparicio, C. Dominquez-Jimenez and A. Garcia-Pardo, 'Activation of the α4β1 integrin through the β1 subunit induces recognition of the RGDS sequence in fibronectin.' *J. Cell. Biol.*, 1994, **126**, 271–279.

76. M. J. Elices, L. A. Urry and M. E. Hemler, 'Receptor functions for the integrin VLA-3: fibronectin, collagen, and laminin binding are differentially influenced by Arg-Gly-Asp peptide and by divalent cations.' *J. Cell. Biol.*, 1991, **112**, 169–181.

77. M. D. Pierschbacher and E. Rouslahti, 'Influence of stereochemistry of the sequence Arg-Gly-Asp-Xaa on binding specificity in cell adhesion.' *J. Biol. Chem.*, 1987, **262**, 17294–17298.

78. J. P. Kim, K. Zhang, J. D. Chen *et al.*, 'Mechanism of human keratinocyte migration on fibronectin: unique roles of RGD site and integrins.' *J. Cell. Physiol.*, 1992, **151**, 443–450.

79. M. J. Humphries, K. Olden and K. M. Yamada, 'A synthetic peptide from fibronectin inhibits experimental metastasis of murine melanoma cells.' *Science*, 1986, **233**, 467–470.

80. H. Kumagai, M. Tajima, Y. Ueno *et al.*, 'Effect of cyclic RGD peptide on cell adhesion and tumor metastasis.' *Biochem. Biophys. Res. Commun.*, 1991, **177**, 74–82.

81. E. F. Plow, J. C. Loftus, E. G. Levin *et al.*, 'Immunologic relationship between platelet membrane glycoprotein GPIIb/IIIa and cell surface molecules expressed by a variety of cells.' *Proc. Natl. Acad. Sci. USA*, 1986, **83**, 6002–6006.

82. I. F. Charo, L. A. Fitzgerald, B. Steiner *et al.*, 'Platelet glycoproteins IIb and IIIa: evidence for a family of immunologically and structurally related glycoproteins in mammalian cells.' *Proc. Natl. Acad. Sci. USA*, 1986, **83**, 8351–8355.

83. A. M. Krezel, G.Wagner, J. Seymour-Ulmer *et al.*, 'Structure of the RGD protein decorsin: conserved motif and distinct function in leech proteins that affect blood clotting.' *Science*, 1994, **264**, 1944–1947.

84. Y. Tomiyama, E. Brojer, Z. M. Ruggeri *et al.*, 'A molecular model of RGD ligands. Antibody D gene segments that direct specificity for the integrin αIIbβ3.' *J. Biol. Chem.*, 1992, **267**, 18085–18092.

85. A. Knapp, T. Degenhardt and J. Dodt, 'Hirudisins. Hirudin-derived thrombin inhibitors with disintegrin activity.' *J. Biol. Chem.*, 1992, **267**, 24230–24234.

86. J. Oliver, 'Correlation of structure and function and mechanisms of recovery in acute tubular necrosis.' *Am. J. Med.*, 1953, **15**, 535–557.

87. J. Oliver, M. MacDowell and A. Tracy, 'The pathogenesis of acute renal failure associated with traumatic and toxic injury. Renal Ischemia, nephrotoxic damage, and the ischemic episode.' *J. Clin. Invest.*, 1951, **30**, 1307–1351.

88. E. Noiri, J. Gailit, D. Sheth *et al.*, 'Cyclic RGD peptide ameliorate ischemic acute renal failure in rats.' *Kidney Int.*, 1994, **46**, 1050–1058.

89. T. Forest, E. Noiri, D. A. Reim *et al.*, 'Moderation of ARF by cyclic R6D peptides in autologously transplanted pig kidneys.' *J. Am. Soc. Nephrol.*, 1994, **5**, 897.

90. H. G. Rennke, Y. Patel and M. A. Venkatachalam, 'Glomerular filtration of proteins: clearance of anionic, neutral and cationic horseradish peroxidase in the rat.' *Kidney Int.*, 1978, **13**, 278–288.

91. J. P. Caulfield and M. G. Farquhar, 'The permeability of glomerular capillaries to graded dextrans. Identification of the basement membrane as the primary filtration barrier.' *J. Cell. Biol.*, 1974, **63**, 883–903.

92. X. L. Ma, D. J. Lefer, A. M. Lefer *et al.*, 'Coronary endothelium and cardiac protective effects of a monoclonal antibody to intercellular adhesion molecule-1 in myocardial ischemia and reperfusion.' *Circulation*, 1992, **86**, 937–946.

93. K. Youker, C. W. Smith, D. C. Anderson *et al.*, 'Neutrophil adherence to isolated adult cardiac myocytes. Induction by cardiac lymph collected by ischemia and reperfusion.' *J. Clin. Invest.*, 1992, **89**, 602–609.

94. K. J. Kelly, W. W. Williams, Jr., R. B. Colvin *et al.*, 'Antibody to intercellular adhesion molecule 1 protects the kidney against ischemic injury.' *Proc. Natl. Acad. Sci. USA*, 1994, **91**, 812–816.

95. L. C. Racusen, B. A. Fivush, Y. L. Li *et al.*, 'Dissociation of tubular cell detachment and tubular cell death in clinical and experimental "acute tubular necrosis".' *Lab. Invest.*, 1991, **64**, 546–556.

96. M. Graber, B. Lane, R. Lamia *et al.*, 'Bubble cells: renal tubular cells in the urinary sediment with characteristic viability.' *J. Am. Soc. Nephrol.*, 1991, **1**, 999–1004.

97. K. Kubota, H. L. Atkins, D. Anaise *et al.*, 'Quantitative evaluation of renal excretion on the dynamic DTPA renal scan.' *Clin. Nucl. Med.*, 1989, **14**, 8–12.

98. T. M. Carlos and J. M. Harlan, 'Leukocyte-endothelial adhesion molecules.' *Blood*, 1994, **84**, 2068–2101.

99. H. R. Brady, 'Leukocyte adhesion molecules and kidney diseases.' *Kidney Int.*, 1994, **45**, 1285–1300.

100. A. Varki, 'Selectin ligands.' *Proc. Natl. Acad. Sci. USA*, 1994, **91**, 7390–7397.

101. R. R. Lobb and M. E. Hemler, 'The pathophysiologic role of alpha 4 integrins *in vivo*.' *J. Clin. Invest.*, 1994, **94**, 1722–1728.

102. T. A. Springer, 'Traffic signals for lymphocyte recirculation and leukocyte emigration: the multistep paradigm.' *Cell*, 1994, **76**, 301–314.

103. M. P. Bevilacqua and R. M. Nelson, 'Selectins.' *J. Clin. Invest.*, 1993, **91**, 379–387.

104. A. J. H. Gearing and W. Newman, 'Circulating adhesion molecules in disease.' *Immunol. Today*, 1993, **14**, 506–512.

105. Y. Shimizu and S. Shaw, 'Cell adhesion. Mucins in the mainstream.' *Nature*, 1993, **366**, 630–631.

106. L. A. Lasky, 'Selectins: interpretors of cell-specific carbohydrate inflammation during inflammation.' *Science*, 1992, **258**, 964–969.

107. J. M. Harlan and D. Y. Liu (eds.) 'Adhesion: Its Role in Inflammatory Disease,' Freeman, New York, 1992.

108. H. R. Brady, O. Spertini, W. Jimenez *et al.*, 'Neutrophils, monocytes and lymphocytes bind to cytokine-activated kidney glomerular endothelial cells

through L-selectin (LAM-1) *in vitro*.' *J. Immunol.*, 1992, **149**, 2437–2444.

109. G. A. Bishop and B. M. Hall, 'Expression of leukocyte and lymphocyte adhesion molecules in the human kidney.' *Kidney Int.*, 1989, **36**, 1078–1085.

110. G. A. Muller, J. Markovic-Lipkovski and C. A. Muller, 'Intercellular adhesion molecule-1 expression in human kidneys with glomerulonephritis.' *Clin. Nephrol.*, 1991, **36**, 203–208.

111. K. Lhotta, H. P. Neumayer, M. Joannidis *et al.*, 'Renal expression of intercellular adhesion molecule-1 in different forms of glomerulonephritis.' *Clin. Sci.*, 1991, **81**, 477–481.

112. J. Markovic-Lipkovski, C. A. Muller, T. Risler *et al.*, 'Mononuclear leukocytes, expression of HLA class II antigens and intercellular adhesion molecule 1 in focal segmental glomerulosclerosis.' *Nephron*, 1991, **59**, 286–293.

113. A. Dal Canton, G. Fuiano, V. Sepe *et al.*, 'Mesangial expression of intercellular adhesion molecule-1 in primary glomerulosclerosis.' *Kidney Int.*, 1992, **41**, 951–955.

114. R. Waldherr, M. Eberlein-Gonska, I. L. Noronha *et al.*, 'TNF alpha and ICAM-1 expression in renal disease.' *J. Am. Soc. Nephrol.*, 1990, **1**, 544 (abstract).

115. R. P. Wuthrich, A. M. Jevnikar, F. Takei *et al.*, 'Intercellular adhesion molecule-1 (ICAM-1) expression is upregulated in autoimmune murine lupus nephritis.' *Am. J.Pathol.*, 1990, **136**, 441–450.

116. C. Diaz Gallo, A. M. Jevnikar, D. C. Brennan *et al.*, 'Autoreactive kidney-infiltrating T-cell clones in murine lupus nephritis.' *Kidney Int.*, 1992, **42**, 851–859.

117. P. A. Hill, H. Y. Lan, D. J. Nikolic-Paterson *et al.*, 'The ICAM-1–LFA-1 interaction in glomerular leukocytic accumulation in anti-GBM glomerulonephritis.' *Kidney Int.*, 1994, **45**, 700–708.

118. P. A. Hill, H. Y. Lan, D. J. Nikolic-Paterson *et al.*, 'ICAM-1 directs migration and localization of interstitial leukocytes in experimental glomerulonephritis.' *Kidney Int.*, 1994, **45**, 32–42.

119. R. J. Faull and G. R. Russ, 'Tubular expression of intercellular adhesion molecule-1 during renal allograft rejection.' *Transplantation*, 1989, **48**, 226–230.

120. K. Kanagawa, H. Ishikura, C. Takahashi *et al.*, 'Identification of ICAM-1 positive cells in the nongrafted and transplanted rat kidney—an immunohistochemical and ultrastructural study.' *Transplantation*, 1991, **52**, 1057–1062.

121. T. Matsuno, K. Sakagami, S. Saito *et al.*, 'Expression of intercellular adhesion molecule-1 and perforin in kidney allograft rejection.' *Transplant. Proc.*, 1992, **24**, 1306–1307.

122. N. Yoshizawa, T. Oda and H. Nakamura, 'Expression of ICAM-1 and HLA-DR by tubular cells and immune cell infiltration in human renal allografts.' *Transplant. Proc.*, 1992, **24**, 1308–1309.

123. W. Moolenaar, J. A. Bruijn, E. Schrama *et al.*, 'T-cell receptors and ICAM-1 expression in renal allografts during rejection.' *Transplant. Int.*, 1991, **4**, 140–145.

124. C. B. Andersen, H. Blaehr, S. Ladenfoged *et al.*, 'Expression of intercellular adhesion molecule-1 in human renal allografts and cultured human tubular cells.' *Nephrol. Dial. Transplant.*, 1992, **7**, 147–154.

125. C. Brockmeyer, M. Ulbrecht, D. J. Schendel *et al.*, 'Distribution of cell adhesion molecules (ICAM-1, VCAM-1, ELAM-1) in renal tissue during allograft rejection.' *Transplantation*, 1993, **55**, 610–615.

126. S. V. Fuggle, J. B. Sanderson, D. W. R. Gray *et al.*, 'Variation in expression of endothelial adhesion molecules in pretransplanted and transplanted kidneys—correlation with intragraft events.' *Transplantation*, 1993, **55**, 117–123.

127. E. von Willebrand, R. Loginov, K. Salmela *et al.*, 'Relationship between intercellular adhesion molecule-1 and HLA class II expression in acute cellular rejection of human kidney allografts.' *Transplant. Proc.*, 1993, **25**, 870–871.

128. D. Seron, J. S. Cameron and D. O. Haskard, 'Expression of VCAM-1 in normal and diseased kidney.' *Nephrol. Dial. Transplant.*, 1991, **6**, 917–922.

129. R. P. Wuthrich, 'Vascular cell adhesion molecule-1 (VCAM-1) expression in murine lupus nephritis.' *Kidney Int.*, 1992, **42**, 903–914.

130. D. M. Briscoe, J. S. Pober, W. E. Harmon *et al.*, 'Expression of vascular cell adhesion molecule-1 in human renal allografts.' *J. Am. Soc. Nephrol.*, 1992, **3**, 1180–185.

131. C. E. Alpers, K. L. Hudkins, C. L. Davis *et al.*, 'Expression of vascular cell adhesion molecule-1 in human allograft rejection.' *J. Am. Soc. Nephrol.*, 1992, **3**, 851 (abstract).

132. Y. Lin, J. A. Kirby, K. Clark *et al.*, 'Renal allograft rejection: induction and function of adhesion molecules on cultured epithelial cells.' *Clin. Exp. Immunol.*, 1992, **90**, 111–116.

133. H. Redl, H. P. Dinges, W. A. Buurman *et al.*, 'Expression of endothelial leukocyte adhesion molecule-1 in septic but not traumatic/hypovolemic shock in the baboon.' *Am. J. Pathol.*, 1991, **139**, 461–466.

134. H. Yokoyama, N. Tomosugi, M. Takaeda *et al.*, 'Glomerular expression of cellular adhesion molecules and serum TNF alpha and soluble ICAM-1 levels in human glomerulonephritis.' *J. Am. Soc. Nephrol.*, 1992, **3**, 669 (abstract).

135. M. S. Mulligan, K. J. Johnson, R. F. Todd, III *et al.*, 'Requirements for leukocyte adhesion molecules in nephrotoxic nephritis.' *J. Clin. Invest.*, 1993, **91**, 577–587.

136. D. C. Brennan, A. M. Jevnikar, F. Takei *et al.*, 'Mesangial cell accessory functions: mediation by intercellular adhesion molecule-1.' *Kidney Int.*, 1990, **38**, 1039–1046.

137. M. G. Suranyi, G. A. Bishop, C. Clayberger *et al.*, 'Lymphocyte adhesion molecules in T cell-mediated lysis of human kidney cells.' *Kidney Int.*, 1991, **39**, 312–319.

138. M. D. Denton, P. A. Marsden, F. W. Luscinskas *et al.*, 'Cytokine-induced phagocyte adhesion to human mesangial cells: role of CD11/CD18 integrins and ICAM-1.' *Am. J. Physiol.*, 1991, **261**, F1071–F1079.

139. H. R. Brady, M. D. Denton, W. Jimenez *et al.*, 'Chemoattractants provoke monocyte adhesion to human mesangial cells and mesangial cell injury mechanism.' *Kidney Int.*, 1992, **42**, 480–487.

140. H. R. Brady and C. N. Serhan, 'Adhesion promotes transcellular leukotriene biosynthesis during neutrophil–glomerular endothelial cell interactions: inhibition by antibodies against CD18 and L-selectin.' *Biochem. Biophys. Res. Commun.*, 1992, **186**, 1307–1314.

141. H. R. Brady, A. Papayianni and C. N. Serhan, 'Leukocyte adhesion promotes lipoxygenase products by transcellular routes.' *Kidney Int.*, 1994, **45** (Suppl.), s90–s97.

142. H. R. Brady, S. Lamas, A. Papayianni *et al.*, 'Lipoxygenase product formation and cell adhesion during neutrophil–glomerular endothelial cells.' *Am. J. Physiol.*, 1995, **268**, F1–F12.

143. X. Wu, J. Pippin and J. B. Lefkowith, 'Attenuation of immune-mediated glomerulonephritis with an anti-CD11b monoclonal antibody.' *Am. J. Physiol.*, 1993, **264**, F715–F721.

144. J. Wada, H. Makino, K. Shikata *et al.*, 'Role of intercellular adhesion molecule-1 in nephrotoxic

serum nephritis.' *J. Am. Soc. Nephrol.*, 1992, **3**, 647 (abstract).

145. K. Kawasaki, E. Yaoita, T. Yamamoto *et al.*, 'Antibodies against intercellular adhesion molecule-1 and lymphocyte function-associated antigen-1 prevent glomerular injury in rat experimental crescentic glomerulonephritis.' *J. Immunol.*, 1993, **150**, 1074–1083.

146. K. Nishikawa, Y. J. Guo, M. Miyasaki *et al.*, 'Antibodies to intercellular adhesion molecule 1/ lymphocyte function-associated antigen-1 prevent crescent formation in rat autoimmune glomerulonephritis.' *J. Exp. Med.*, 1993, **177**, 667–677.

147. P. G. Tipping, X. R. Huang, M. C. Berndt *et al.*, 'A role for P selectin in complement-independent neutrophil-mediated glomerular injury.' *Kidney Int.*, 1994, **46**, 79–88.

148. A. Papayianni, C. N. Serhan, M. L. Phillips *et al.*, 'Transcellular biosynthesis of lipoxin A4 during adhesion of platelets and neutrophils in experimental immune complex glomerulonephritis.' *Kidney Int.*, 1995, **47**, 1295–1302.

149. R. Harning, J. Pelletier, G. Van *et al.*, 'Monoclonal antibody to MALA-2 (ICAM-1) reduces acute autoimmune nephritis in kdkd mice.' *Clin. Immunol. Immunopathol.*, 1992, **64**, 129–134.

150. A. B. Cosimi, D. Conti, F. L. Delmonico *et al.*, '*In vivo* effects of monoclonal antibody to ICAM-1 (CD54) in nonhuman primates with renal allografts. *J. Immunol.*, 1990, **144**, 4604–4612.

151. C. E. Haug, R. B. Colvin, F. L. Delmonico *et al.*, 'A phase I trial of immunosuppression with anti-ICAM-1 (CD54) mAb in renal allograft recipients.' *Transplantation*, 1993, **55**, 766–772.

152. S. L. Linas, P. F. Shanley, D. Whittenburg *et al.*, 'Neutrophils accentuate ischemia-reperfusion injury in isolated perfused rat kidneys.' *Am. J. Physiol.*, 1988, **255**, F728–F735.

153. P. O. Hellberg and T. O. Kallskog, 'Neutrophil-mediated post-ischemic tubular leakage in the rat kidney.' *Kidney Int.*, 1989, **36**, 555–561.

154. M. A. Thornton, R. Winn, C. E. Alpers *et al.*, 'An evaluation of the neutrophil as a mediator of *in vivo* renal ischemic-reperfusion injury.' *Am. J. Pathol.*, 1989, **135**, 509–515.

155. M. A. Arnaout, R. M. Hakim, R. F. Todd, III *et al.*, 'Increased expression of an adhesion-promoting surface glycoprotein in the granulocytopenia of hemodialysis.' *N. Engl. J. Med.*, 1985, **312**, 457–462.

156. A. K. Cheung, M. Hohnholt and J. Gilson, 'Adherence of neutrophils to hemodialysis membranes: role of complement receptors.' *Kidney Int.*, 1991, **40**, 1123–1133.

157. J. Himmelfarb, P. Zaoui and R. Hakim, 'Modulation of granulocyte LAM-1 and MAC-1 during dialysis— a prospective randomized controlled trial.' *Kidney Int.*, 1992, **41**, 388–395.

158. A. K. Cheung, C. J. Parker and M. Hohnholt, 'Beta 2 integrins are required for neutrophil degranulation induced by hemodialysis membranes.' *Kidney Int.*, 1993, **43**, 649–660.

159. S. Stuard, M-P. Carreno, J-L. Poignet *et al.*, 'A major role for CD62P/CD15s interaction in leukocyte margination during hemodialysis.' *Kidney Int.*, 1995, **48**, 93–102.

160. M. S. Mulligan, J. C. Paulson, S. De Frees, *et al.*, 'Protective effects of oligosaccharides in P-selectin-dependent lung injury.' *Nature*, 1993, **364**, 149–151.

161. C. F. Bennett, T. P. Condon, S. Grimm *et al.*, 'Inhibition of endothelial cell adhesion molecule expression with antisense oligonucleotides.' *J. Immunol.*, 1994, **152**, 3530–3540.

162. C. N. Serhan, 'Lipoxin biosynthesis and its impact in inflammatory and vascular events.' *Biochim. Biophys. Acta*, 1994, **1212**, 1–25.

163. A. A. Spector, J. A. Gordon and S. A. Moore, 'Hydroxyeicosatetraenoic acids (HETEs).' *Prog. Lipid. Res.*, 1988, **27**, 271–323.

164. S. Takata, M. Matsubara, P. G. Allen *et al.*, 'Remodeling of neutrophil phospholipids with 15(S)-hydroxyeicosatetraenoic acid inhibits leukotriene B4-induced neutrophil migration across endothelium.' *J. Clin. Invest.*, 1994, **93**, 499–508.

165. S. Takata, A. Papayianni, M. Matsubara *et al.*, '15-hydroxyeicosatetraenoic acid inhibits neutrophil migration across cytokine-activated endothelium.' *Am. J. Pathol.*, 1994, **145**, 541–549.

166. D. B. Fischer, J. W. Christman, and K. F. Badr, 'Fifteen-*S*-hydroxyeicosatetraenoic acid (15-*S*-HETE) specifically antagonizes the chemotactic action and glomerular synthesis of leukotriene B4 in the rat.' *Kidney Int.*, 1992, **41**, 1155–1160.

167. H. R. Brady, A. Papayianni and C. N. Serhan, 'Potential vascular roles for lipoxins in the "stop programs" of host defense and inflammation.' *Trends Cardiovasc. Med.*, 1995, **5**, 186–192.

7.18
Immunological Response of the Kidney

GERALD C. GROGGEL

University of Nebraska Medical Center, Omaha, NE, USA

7.18.1   INTRODUCTION

Immunologic responses in the kidney are a major cause of chronic renal failure. The immunologic response in the kidney can involve either humoral or cellular immune responses to a variety of antigens, both endogenous and exogenous.[1] The humoral response involves antibodies which are deposited in the kidney by two mechanisms: (i) the deposition of circulating immune complexes or (ii) the *in situ* binding of antibodies to antigens within the kidney.[1] The antigens present within the kidney can either be a structural element of the kidney, such as the glomerular basement membrane, or an antigen that is trapped or planted in the kidney by a variety of mechanisms. Antibodies can also interact with soluble antigens to form immune complexes and these immune complexes which escape clearance by the mononuclear phagocytic system can circulate and accumulate within the kidney, particularly the glomerulus. These immune complexes can then undergo continuous rearrangement. Once antibody is deposited in the kidney by either mechanism, a variety of other mediators can be activated. Recently, a new mechanism for antibodies in inducing immunologic responses in the kidney has been identified. Antibodies to constituents of neutrophil granules have been identified as antineutrophil cytoplasmic antibodies (ANCA)[2] and may induce immune injury by activating neutrophils. Certain diseases such as Wegener's granulomatosis have active inflammatory changes within the glomerulus without glomerular immune deposits. Many of these diseases are associated with the presence of this antibody in the circulation. The second type of immune reaction within the kidney is the cell mediated immune response, which involves activated T cells, monocytes and macrophages.[1] Antigen-specific T cells can respond to their specific antigens in a delayed type hypersensitivity reaction with release of lymphokines to recruit monocytes and macrophages. A second reaction by T cells, can involve cytotoxic T cells, which can directly damage cells.

After initiation of the immune response, either cellular or humoral, a variety of mediators of inflammation are recruited.[3] These mediators of immune injury will vary depending on the initial immune response as well as the location within the kidney in which the immune response is occurring. The two major sites where immune responses can occur within the kidney are the glomerulus and the tubulointerstitium. The response of the kidney, particularly the glomerulus, to these mediators is rather uniform, and includes (i) cellular proliferation and hypertrophy, (ii) the synthesis and deposition of extracellular matrix within the glomerulus, which can lead to sclerosis and scarring, and (iii) expression of new antigens on the glomerular cells.[4] These responses often involve the synthesis and secretion of multiple growth factors. The major emphasis of this review will be on the mediators of the glomerular immune response.

Both the humoral and cellular immune responses as well as the identification of mediators of inflammation have been studied most extensively in experimental animal models of glomerular and tubulointerstitial injury. Most of this chapter will refer to these experimental studies, but there is accumulating evidence for these immune responses in the pathogenesis of human renal disease as well. The best example of this is antiglomerular basement membrane nephritis where deposits of antibody to the glomerular basement membrane (GBM) produce linear IgG deposition in glomeruli by immunofluorescence microscopy.

7.18.2 MECHANISMS OF IMMUNE DEPOSIT FORMATION

A variety of classifications of immunologic renal disease have been devised. Dixon originally proposed a classification of glomerulonephritis based on pathogenetic mechanisms.[5] He proposed two basic forms of glomerulonephritis: (i) induced by circulating antibodies reacting with fixed structural antigens of the GBM and producing linear staining by immunofluorescent microscopy and (ii) the other induced by the deposition in the glomerulus of circulating immune complexes and producing granular immune deposits by immunofluorescent microscopy (Figure 1). This approach was used for many years, but this classification has had to be expanded. Recent work has identified an *in situ* mechanism for the development of immune deposits within the glomeruli.[6,7] Thus granular immune deposits can result from trapping of circulating immune complexes, but also can result from antigen or antibody binding to a glomerular structure and initiating immune complex formation *in situ*.[6] Although antibodies reacting with an intrinsic glomerular antigen can result in either linear or granular types of deposits (Figure 1), immune injury may occur in the absence of visible deposits. In this case, injury is thought to be mediated by cells rather than by immunoglobulins.

7.18.3 SITES OF GLOMERULAR IMMUNE DEPOSITS

There are alternative classifications to those based on the mechanisms of immune deposit

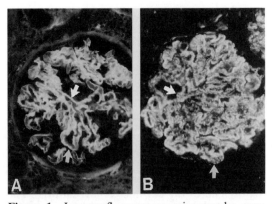

Figure 1 Immunofluorescence micrographs comparing linear and granular staining. A. Smooth linear deposits of anti-GBM antibody (arrows) are present along the glomerular basement membrane in a patient with anti-GBM glomerulonephritis. B. Irregular granular deposits of IgG (arrows) are present along the glomerular capillary walls (fluorescein isothiocyanate conjugated antihuman IgG; original magnification ×832).

formation. One of these uses the histologic appearance to divide glomerular disorders into inflammatory and noninflammatory types.[3] Alternatively, glomerular immune injury can be classified according to the site of the immune deposits and injury.[8] Deposits may lie within the subepithelial space, the subendothelial space, the GBM itself or the mesangium (Figure 2.) The site of the immune deposit in the glomerulus is an important determinant of the functional and histological lesion which results as well as the mediators involved in the immune injury. Such a classification, based on the site of deposits, will be used for this chapter.

7.18.3.1 Subepithelial Immune Deposits

Deposits in the subepithelial space are characteristic of human membranous nephropathy.

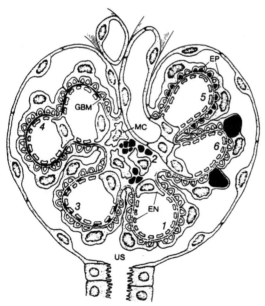

Figure 2 Schematic representation of a single glomerulus illustrating the various sites of immune complex deposits. Loop 1 represents a normal glomerular capillary wall with endothelial cells (EN), glomerular basement membrane (GBM) and epithelial cells (EP). MC represents mesangial cells and US is the urinary space. Shown in 2 are immune complex deposits within the mesangial matrix as found in IgA nephropathy. Loop 3 demonstrates subendothelial immune deposits such as are found in membranoproliferative glomerulonephritis. Shown in loop 4 are the linear deposits along the GBM found in anti-GBM glomerulonephritis. In loop 5 are the diffuse subepithelial deposits of membranous nephropathy. Shown in loop 6 are the subepithelial humps found in poststreptococcal glomerulonephritis.

At this site, deposits are formed primarily if not entirely by *in situ* mechanisms.[7] The antigen can be either an endogenous, insoluble glomerular antigen, or an exogenous, soluble antigen which is deposited in the glomerulus on the basis of some physicochemical characteristic. *In situ* formation of immune deposits was initially identified in a rat model of human membranous nephropathy, Heymann nephritis.[9] Rats immunized with a crude renal cortical extract, Fx1A, form anti-Fx1A antibodies which deposit along the capillary wall in a granular pattern. In this model, antibody interacts with antigens on the surface of the glomerular epithelial cell (GEC). The best characterized of these antigens is a glycoprotein, GP330, which has a molecular weight of 330 kDa and is also found on the brush border of the renal proximal tubule cell.[10] If rats are injected with a heterologous antibody to this antigen, passive Heymann nephritis, they deposit complement fixing IgG and develop proteinuria over a period of 4–5 d.[11] The heterologous IgG and rat C3 stain in a granular pattern. Subepithelial electron dense deposits are seen by electron microscopy. It appears that once the immune complexes form on the GEC, patching and shedding of the immune complexes from the cell surface occurs.[12,13] This phenomenon most likely produces the finely granular appearance seen. The deposits may reform on a local or *in situ* basis.

Exogenous antigen may also bind to glomerular structures and initiate subepithelial deposit formation. Positively charged antigens, such as a chemically modified cationic bovine serum albumin (BSA), can bind to anionic sites of the capillary wall.[14] This has been demonstrated in a chronic serum sickness model in the rabbit in which the animal receives daily injections of cationic BSA.[14] In this model, granular deposits of cationic BSA, rabbit IgG and C3 are seen and subepithelial deposits are demonstrated by electron microscopy.[14] Certain monoclonal antibodies to DNA have been shown to bind directly to glomerular structures in the mouse and produce deposits.[15]

Histologically, Heymann nephritis is very similar to human membranous nephropathy. This membranous nephropathy may be produced by an autoantibody to a native GEC antigen or to foreign antigens which are structurally similar to antigens in the glomerular epithelial cell. Unfortunately, to date these antigens have not been identified in human membranous nephropathy.

Subendothelial deposits may dissolve and the individual reactants move across the capillary wall to a subepithelial location where they reform immune complexes.[16] Cationic ferritin has been shown to localize to the subendothelial space where antibody then binds to the ferritin. This complex dissolves and migrates across the GBM, to reform in the subepithelial space.[16] It appears unlikely that circulating immune complexes are able to localize directly to the subepithelial space.[7] Studies which have reported subepithelial localization of circulating immune complexes have involved low-avidity antibodies or non-covalently linked immune complexes.[17,18] Both of these could undergo intravascular dissociation and subsequent reformation in the subepithelial space by an *in situ* mechanism.

7.18.3.2　Glomerular Basement Membrane

A second site of immunologic injury in the glomerulus is the GBM. This is the first model to demonstrate direct antibody binding to GBM antigens and was originally described by Masugi in the 1930s.[19] In this model, rats which have been administered serum from rabbits immunized with rat kidney develop glomerulonephritis with linear deposits of IgG and complement along the glomerular capillary wall.[20] The target antigen has been demonstrated to be a component of the GBM. This *in situ* formation of immune deposits occurs in both a heterologous phase, in which injected heterologous antibody immediately binds to the GBM, and an autologous phase, in which the heterologous IgG acts as a planted antigen and the animal responds with antibody to this foreign protein in 7–10 d.[20]

Work has demonstrated that the antigen in anti-GBM nephritis is localized to the non-collagenous globular domain of the α-3 chain of the type IV collagen molecule in the GBM.[21] This antigen is missing from the GBM of many patients with some forms of glomerulonephritis of the Alport type.[22] This auto-antibody may also react with epitopes on the alveolar basement membrane in the lungs and produce Goodpasture's syndrome.

Anti-GBM nephritis is the best example of human glomerulonephritis in which the pathogenesis has been well defined. Circulating anti-GBM antibodies have been demonstrated in patients with this disorder and IgG antibody eluted from the kidneys of these patients has also been isolated. Antibodies from either of these sources are able to transfer the disease to animals. Reoccurrence of this disease has been demonstrated in transplanted kidneys. Interestingly, patients with Alport's syndrome, who lack this antigen in their native kidneys, may develop anti-GBM nephritis in their transplanted kidney.

7.18.3.3　Subendothelial Immune Deposits

A third site of injury is the subendothelial space where deposits are present between the endothelial cell and the GBM. Deposits in this location are usually associated with a severe proliferative and inflammatory response within the glomerulus and are usually seen in association with deposits within the mesangium. Subendothelial deposits may form *in situ* such as when cationic antigen binds to anionic sites on the endothelial cell.[16] Lectins, such as concanavalin A, may act as antigen and bind to sugar residues in the GBM.[23] Endothelial cell angiotensin converting enzyme can act as an antigen to produce subendothelial deposits.[24] Antibodies to the endothelial cell have been found in systemic lupus erythematosus, a disease where subendothelial deposits are frequently found.[25]

Deposits in this location may also result from the trapping of circulating immune complexes.[26] A number of factors influence this process including the charge,[27] size,[26] avidity, and complement-fixing ability[28] of the immune complexes; glomerular properties such as charge and permeability of the capillary wall and the clearance function of the mesangium;[29] and systemic factors such as the function of the mononuclear phagocyte system,[30] renal blood flow,[31] and the number of complement receptors on the erythrocyte.[28]

Hebert and colleagues have shown in the primate that the erythrocyte possesses a complement receptor, CR1, which is important in binding immune complexes.[28,32] This receptor binds and transports C3b-containing immune complexes to the mononuclear phagocyte system, primarily in the liver.[32] If this erythrocytic system is defective in such a way that the immune complexes do not fix complement, immune complexes will persist in the circulation and deposit in the glomerulus, usually in a subendothelial location.[33] Deposits in this location are in direct contact with the circulation and may be enlarged by binding other circulating complexes, antigens, or antibodies or may also be reduced or solubilized by circulating antigen excess.[34] Deposits in this location are readily able to recruit circulating cells which can mediate tissue damage.

The classic experimental model of this type of glomerular injury is serum sickness. In acute serum sickness, rabbits are immunized with one dose of a foreign protein, bovine serum albumin (BSA), form anti-BSA antibodies and develop glomerulonephritis with subendothelial and mesangial deposits of BSA and anti-BSA after 1–2 weeks.[35] Immune complex glomerulonephritis can also be induced in animal models by passive infusion of preformed immune complexes. Such complexes predominantly deposit in a subendothelial or mesangial location, not in the subepithelial space. These observations suggest that for subepithelial immune complexes to form, the antigen and antibodies must independently cross the GBM and then complex in the subepithelial space.

A variety of both exogenous and endogenous antigens have been identified in glomerular immune complex deposits in humans. These include tumor antigens, thyroid antigens, ribonucleoproteins, DNA, autologous Ig, and microbial antigens such as Streptococcal antigens, hepatitis B and C, *Treponema pallidum*, and *Schistosma mansoni*.

7.18.3.4 Mesangial Immune Deposits

Another site of immune deposits within the glomerulus is the mesangium. Deposits are found here alone or in association with subendothelial deposits. Since the mesangium is separated from the systemic circulation only by the glomerular endothelial cell with its wide fenestrae,[36] it is readily accessible to circulating macromolecules including immune complexes.[36] The complexes may be degraded locally or pass out of the glomerulus through the hilum into the renal lymphatics. Thus, circulating immune complexes may readily deposit in the mesangium. However, deposits may also form here by an *in situ* mechanism. In classic studies by Mauer *et al.*, heat-aggregated human IgG was injected into rabbits and deposited in the mesangium of their kidneys.[37] The kidneys were removed and transplanted into normal rabbits, which received rabbit anti-human IgG and formed immune deposits *in situ*. The rabbits developed a proliferative glomerulonephritis with deposits of human IgG, rabbit IgG and rabbit complement.

Thy 1 is a glycoprotein located on the cell surface of certain cells, particularly the thymus. Thy 1.1-like antigen is found on the cell membrane of mesangial cells. When rats are injected with antibodies to this antigen, they develop mesangial deposits of heterologous IgG and rat C3, mesangial cell lysis, and a proliferative mesangial glomerulonephritis which is complement but not leukocyte dependent and is sometimes associated with proteinuria.[38,39] This represents another model in which the immune deposits form *in situ* in the mesangium.

7.18.3.5 Glomerular Injury without Immune Deposits

Lastly, renal immune injury may occur without any glomerular immune deposits. Some examples of human immunologic renal disease without deposits include minimal change disease and type II membranoproliferative glomerulonephritis. In these types of diseases, the immunologic injury may be produced by sensitized cells such as T cells rather than by the classic antigen–antibody complex. In minimal change disease in humans, the proteinuria is thought to be due to lymphokines and cytokines produced by the T cells.

Bhan and co-workers originally showed that rats with subnephritogenic amounts of rabbit anti-GBM IgG only developed disease after receiving T cells from donors who had been sensitized to rabbit IgG.[40] B-cell deficient chickens have been used to identify an antibody-independent role for T cells in glomerulonephritis.[41] B-cell deficient chickens immunized with GBM developed a proliferative glomerulonephritis without immune deposits.[41] This lesion was transferred to other chickens using only sensitized lymphocytes.[42] But T cells were not identified in the glomerular lesions.

7.18.4 TUBULOINTERSTITIAL IMMUNE INJURY

Tubulointerstitial nephritis can also result from immunologic injury.[43] Antibodies can bind to intrinsic tubular and interstitial antigens and produce immune deposits and injury.[44] T-cell mediated injury may also occur in the interstitium.[43]

Rats immunized with a heterologous tubular antigen preparation can develop antitubular basement membrane (TBM) antibodies similar to those seen in anti-GBM nephritis with linear deposits along the TBM.[43] The antibodies are associated with an interstitial nephritis and mononuclear cell infiltrate in disease-susceptible strains. Antigen-specific T cells appear to be induced and mediate the injury either by recruitment of macrophages or by direct cell cytotoxicity. Tamm–Horsfall protein is found in certain cells along the nephron and can act as an antigen to form immune deposits and interstitial nephritis. Certain strains, such as Lewis rats, lack the TBM antigen. When immunized with TBM antigen, they develop a cellular immune interstitial nephritis without immunoglobulin, which can be easily transferred with sensitized T cells.[45] Thus the tubulointerstitial areas of kidney can be the site at which both humoral and cellular components of the immune system produce immunologic injury.

Acute and chronic interstitial nephritis in humans appear to be mediated by cellular immunity and can be induced by drugs, infections or systemic autoimmune disorders. Interstitial nephritis frequently accompanies glomerular disorders and correlates better with the progression of the renal disease than does the glomerular disease.

7.18.5 MEDIATORS OF IMMUNE INJURY

Much recent research has been aimed at identifying those factors which mediate immunologic injury in the kidney. Functional manifestations of this injury include changes in glomerular filtration rate and proteinuria reflecting defects in glomerular permeability. Structural manifestations include GBM thickening, cellular proliferation, necrosis, thrombosis, crescent formation, and sclerosis.

A variety of mediators of immune glomerular injury have been identified.[3] Shown in Figure 3 are the pathways by which the glomerular immune response leads to injury to the capillary wall. These include (i) cellular elements consisting of polymorphonuclear cells, mononuclear cells including monocytes and macrophages, lymphocytes, platelets and also mesangial cells[1] and (ii) soluble elements including complement, particularly the membrane attack complex, fibrinogen, prostanoids, leukotrienes, platelet activating factor, interleukins, tumor necrosis factor, angiotensin II, and antineutrophil cytoplasmic antibodies. Some of these soluble mediators are covered in other chapters and will not be discussed here.

The contribution of each of these mediators will vary depending on the particular disease, and the stage and severity of the disease. Roles for these mediators have been most extensively studied in antibody-mediated experimental models of glomerular immune injury. Understanding the mediators involved may allow therapeutic interventions in various types of immune renal injuries without complete understanding of the underlying immune response.

7.18.5.1 Antibody

Antibody alone may produce injury without involving any other mediators. In nephrotoxic nephritis in the guinea-pig, anti-GBM antibody produces proteinuria without the involvement of complement or cells.[46] In the isolated perfused rat kidney, anti-GBM antibody can produce changes in glomerular permeability

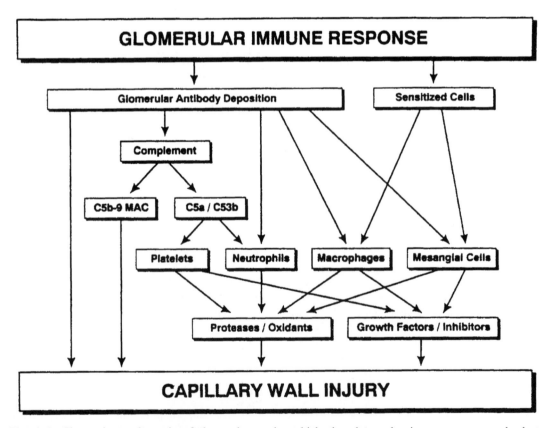

Figure 3 Shown is a schematic of the pathways by which the glomerular immune response leads to capillary wall injury through a variety of mediators (after Couser[3]).

without other mediators.[47] In passive Heymann nephritis, administration of the $F(ab')_2$ or $F(ab')$ fragments of heterologous anti-Fx1A antibody produces proteinuria in the rat without fixing complement.[48] Two separate monoclonal antibodies to the GEC membrane or to the GEC slit diaphragm have been shown to cause proteinuria in the rat independent of other mediators.[49,50]

The mechanism whereby antibody alone produces injury in these models is unknown but may be similar to that by which puromycin aminonucleoside damages the GEC and produces proteinuria. Injury to the GEC could include detachment of the GEC, which could then lead to alterations in permeability. Alternatively, the GEC could be stimulated to produce substances which could damage the glomerular capillary wall. GEC in culture have been shown to produce a neutral proteinase with activity against type IV collagen and the GBM.[51]

7.18.5.2 Complement

The usual mechanism whereby immune deposits and antibody are thought to initiate injury is through the activation of the complement system.[52,53] The complement system can be activated by either the classical pathway, which is initiated by IgG or IgM, or by the alternative pathway, which can be initiated by IgA or nonimmunologically by polysaccharides, bacteria, and other substances.[54] Both pathways lead to the formation of a C3 convertase to form C3b.[54] This then leads to the formation of a C5 convertase to form C5a and C5b. Generation of C5b from either pathway initiates self-assembly of C5b-9, which is the membrane attack complex.

The classic role for complement in immune injury is the generation of chemotactic factors, particularly C5a, which attract neutrophils to the site of immune deposits.[55] Neutrophils may also be attracted by immune adherence mechanisms involving C3b. The neutrophils then produce injury as discussed later in this chapter. The best examples of complement–neutrophil mediated glomerulonephritis are the heterologous and autologous phases of anti-GBM nephritis.[55] Depletion of complement in this model prevents neutrophil recruitment and injury.[55] Neutrophil depletion also prevents injury and repletion of neutrophils produces proteinuria.[56] There may be neutrophil-mediated glomerulonephritis which is not dependent on complement. Work has also shown that complement may be important for the recruitment of platelets to sites of immune deposits.

A new role for complement in noninflammatory immunologic injury has been identified[57] which does not involve the recruitment of neutrophils but rather the formation of the membrane attack complex (C5b-9)(MAC). A role for the MAC was first shown in passive Heymann nephritis in the rat, a model of human membranous nephritis.[57] After administration of heterologous antibody to Fx1A, immune deposits form in the subepithelial space as a result of the antibody binding to an antigen on the GEC. The antibody activates complement and the MAC is formed. Depletion of complement in this model with cobra venom factor prevents any injury without affecting antibody deposition.[57] In contrast, neutrophil-depletion has no effect on this injury.[57] This was the first evidence for a cell-independent effect of complement in mediating renal immune injury. In rabbits deficient in the sixth component of complement and thus unable to form the MAC, injury is prevented in a model of membraneous nephropathy.[58] Similarly in the isolated perfused kidney, injury can be prevented following antibody to Fx1A deposition when perfused with C6 or C8 deficient serum but proteinuria develops immediately with perfusion of normal serum containing all the complement components.[59] The participation of the MAC in glomerular disease is not confined to membranous nephropathy. Proteinuria is reduced and renal function preserved in the heterologous phase of anti-GBM nephritis in rabbits deficient of C6.[60] Many studies using an antibody to a neoantigen of the MAC have shown deposits of the MAC at sites of immune injury in models which are complement dependent.[61] There is a substantial body of evidence identifying a role for the MAC in mediating glomerular immune injury.

The MAC results from proteolytic cleavage of C5 to form C5b.[62] C5b combines with C6 and C7 to form the C5b67 complex which can bind to cell membranes or bind to S-protein (vitronectin) in the plasma. C5b67 on cell membranes binds C8 and multiple C9 molecules and inserts into the lipid bilayer and this leads to lysis of the cell, particularly if it is nonnucleated such as erythrocytes. If S-protein binds to C5b67, this complex can bind C8 and C9 but cannot insert into membranes.

The mechanism whereby the MAC produces injury in the kidney is unknown. It may be that C5b-9 is activating cells such as the GEC in Heymann nephritis and this leads to the release of inflammatory mediators. Antibody to GEC combined with sublytic concentrations of complement leads to noncytolytic injury *in vitro*.[63] This sublytic C5b-9 attack also leads to increased intracellular calcium, activation of phospholipase C, increased IP2, IP3, diacylglycerol and phosphatidic acid and release of PGF_2 and thromboxane.[64] It has also been shown that treatment of glomerular mesangial cells in culture with sublytic amounts of C5b-9 leads to release of H_2O_2 and superoxide anion.[65] A second possible mechanism for MAC-induced injury would be stimulation of production of extracellular matrix components (such as type IV collagen when exposed to C5b-9[66]) by GEC. Lastly, activation and formation of MAC on GEC may cause detachment of GEC which would lead to proteinuria.

Interestingly, in the autologous phase of anti-GBM nephritis where rabbits are immunized with sheep IgG and then receive a subnephrotoxic dose of sheep anti-rabbit GBM, rabbits deficient in C6 appeared to have more severe disease.[67] At 5 d after injection of the anti-GBM antibody, C6-deficient rabbits had the same degree of proteinuria, much greater loss of renal function and much more severe histologic appearance than normocomplementemic controls. The C6-deficient animals also had persistence of their immune deposits as measured by quantitating sheep IgG in isolated glomeruli.[67] Thus in this model the complement deficient animals had more severe injury because of failure to clear their immune deposits. This suggests a role for the MAC in the solubilization and clearance of glomerular immune deposits. Thus complement may play more than one role in immune-mediated renal injury.

Another important function of complement in immune injury is its role in the removal of immune complexes from the circulation via the complement CR1 receptor, which is specific for C3b.[32,68] In primates, immune complexes are transported to the fixed macrophage system by erythrocytes which have the CR1 receptors.[32,68] Immune complexes which activate complement have C3b bound and this interacts with its receptor, CR1, on the red blood cells.[68] The erythrocyte-bound immune complexes are carried to the liver where they are removed and the erythrocyte returned to the circulation. The binding of the immune complexes to the erythrocytes prevents their deposition in organs such as the kidneys. When complement is depleted, a large proportion of injected complexes are deposited in the kidney rather than being cleared by the liver.[33] Transport of immune complexes by erythrocytes thus has a protective function which ensures their safe delivery to the mononuclear phagocyte system where they are eliminated. These findings may explain why patients with complement deficiencies have a higher incidence of immune complex-mediated diseases.[68]

7.18.5.3 Neutrophils

The site of immune injury plays an important role in determining the presence of neutrophils.[8] In the subendothelial and mesangial areas, neutrophils are accessible to the immune deposits and can localize to these areas in response to a variety of mediators including C5a, platelet activating factor and products of the lipoxygenase pathway. Neutrophils may adhere to tissue via receptors for the Fc portion of the immunoglobulin molecule and C3b. A variety of leukocyte adhesion molecules have been described including integrins, intracellular adhesions molecules, vascular cell adhesion molecules and selectin[69] (see Chapter 16, this volume for an in-depth review). Neutrophils then release proteinases and oxidants to produce tissue injury. Neutrophils are not accessable to the subepithelial space.

The importance of the polymorphonuclear leukocyte (PMN) has been established in a number of different models of immune glomerular injury. Perhaps best documented is the participation of PMNs in nephrotoxic nephritis. Cochrane et al. showed that there was a rapid influx of PMNs into rat glomeruli which peaked at about 2 h after injection of nephrotoxic serum (NTS).[55] Rats depleted of PMNs had markedly lower levels of proteinuria following NTS administration, and the inhibition persisted throughout the first week of nephritis. Naisch et al. extended these observations, documenting the participation of PMNs in the autologous phase of nephrotoxic nephritis.[70] Rabbits depleted of PMNs by either nitrogen mustard or anti-polymorphonuclear cell serum had a reduction in autologous phase induced-proteinuria, fibrin deposition, and azotemia. A correlation between fibrin deposition and PMN accumulation was also noted by Hooke et al. who used monoclonal antibodies to characterize the cellular infiltration in a number of forms of glomerular and interstitial disease.[71] PMNs, normally found in small numbers along with monocytes in glomeruli, were increased in postinfectious glomerulonephritis and, to a lesser degree, in crescentic glomerulonephritis. In a model less commonly thought of as involving PMNs, *in situ* immune complex glomerulonephritis, Cook et al. documented an infiltration of PMNs within 30 min of complex formation.[72] The infiltration was superseded in the subsequent 24 h by an accumulation of macrophages.

It seems then that PMNs can be important either because of their predominance in an infiltrating cell population or their early appearance in an otherwise evolving infiltrate. They may be of functional importance in producing proteinuria, azotemia, or fibrin deposition. A surprising finding in a study by Cochrane *et al.* was that early on PMNs produced negligible glomerular destruction.[55] Rather, ultrastructural studies showed that PMNs displaced endothelial cells away from the basement membrane. Even though endothelial cell apposition was restored after 12–24 h, accompanying proteinuria persisted. The importance of PMNs in decreasing glomerular filtration was emphasized by Tucker *et al.* in a rat model of anti-GBM nephritis.[73] They showed that glomerular ultrafiltration coefficient and single nephron glomerular filtration rate did not decrease in rats depleted of PMNs prior to injection of anti-GBM antibody.[73]

Despite the importance of PMN infiltration in causing renal dysfunction, it is their central role in producing other mediators of tissue injury that has drawn most attention by renal pathophysiologists. Specifically, PMNs are well known for their ability to produce reactive oxygen free species (ROS) and tissue proteinases, substances that have gained increasing attention in contributing to the pathogenesis of glomerular injury.

The model in which there is strongest evidence for direct participation of reactive oxygen species is the heterologous phase of nephrotoxic nephritis. Adachi *et al.* administered nephrotoxic serum to rats followed by the daily administration of superoxide dismutase (SOD). The treated rats had less proteinuria than control rats.[74] Less mesangial thickening and less capillary obliteration were observed in the SOD-treated animals. Rehan *et al.* injected nephrotoxic serum directly into the renal artery of the rat, producing significant proteinuria which was prevented by treatment with catalase.[75] There was an inverse relationship between catalase dose and proteinuria. The catalase-treated animals did not manifest the same degree of endothelial cell injury or epithelial cell foot process fusion present in controls. However, catalase infusion did not improve the glomerular filtration rate of treated animals. In a model of immune complex disease in which neutrophils were exposed to aggregated IgG, neutrophils were shown to elaborate superoxide anion.[76]

7.18.5.4 Proteases

Proteases are produced by inflammatory cells and mediate tissue damage. There are four major classes of proteases, which are named for the chemical species at their enzymatic core: serine, cysteine and aspartic proteases, and metalloenzymes (Table 1).[77] These species differ in their cells of origin, their intralysosomal or extracellular secretion, and the pH at which optimal activity takes place (Table 1).[77]

The evidence that proteases contribute to kidney disease is derived from three types of studies: (i) *in vitro* studies demonstrating the capacity of various classes of proteases to digest different substances found within the kidney, (ii) *in vivo* studies in which proteases are either isolated in various models of glomerular disease or administered to experimental animals with subsequent demonstration of glomerular disease or dysfunction, and (iii) studies examining the ability of protease inhibitors to prevent histologic or functional parameters of induced disease.

Since they are the predominant species within the azurophil granules of PMNs and are most active at neutral pH, the serine proteases including elastase, Cathepsin G and plasmin are considered to be of major importance in inflammatory reactions. Davies *et al.* demonstrated substantial release of hydroxyproline when the GBM was incubated with either elastase or Cathepsin G.[78] Serine protease inhibitors have been shown to be effective in preventing neutrophil mediated detachment of endothelial cell monolayers from culture wells.[79] Shah *et al.* reported that a metalloproteinase, activated by a product of the myeloperoxidase–hydrogen peroxide–halide system, mediated degradation of GBM by neutrophils stimulated with PMA.[80] Moreover, Lovett *et al.* demonstrated that a neutral metalloproteinase capable of digesting GBM was also produced by glomerular mesangial cells in culture.[81] This proteinase was further found to specifically degrade both soluble and insoluble type IV collagen, a major component of the GBM. Baricos *et al.* purified lysosomal

Table 1 Classes of proteases.

Protease class	Examples	pH range	Location
Serine	Elastase Cathepsin G Plasmin	7–9	Intra and extracellular
Cysteine	Cathepsin B and L	3–7	Intralysosomal
Aspartic	Cathepsin D	2–6	Intralysosomal
Metallo- enzymes	Gelatinase Type IV collagenase	5–9	Intra and extracellular

Source: Baricos and Shah.[77]

cysteine proteases, Cathepsin B and L, from human kidney and found that Cathepsin L degraded GBM at acid pH.[82] They further demonstrated the presence of Cathepsin B in isolated glomeruli in the absence of PMNs and found these cysteine proteinases similarly capable of GBM degradation at acid pH.

Johnson *et al.* demonstrated that injection of the serine proteases, elastase and Cathepsin G into the renal arteries caused dose-dependent proteinuria.[83] Neither invasion by white blood cells nor proliferation of resident glomerular cells was seen. Despite the significant proteinuria, no glomerular histologic changes were noted. Davin *et al.* studied the role of neutral proteinases and metalloproteinases in two models of glomerular disease, accelerated nephrotoxic serum nephritis (NSN) and puromycin aminonucleoside nephrosis (PAN).[84] In both models, urinary excretion of neutral proteinases was greatly increased and parallelled the excretion of type IV collagen and laminin. They demonstrated that whereas serine proteinases predominated in normal urine, both serine and metalloproteinases were found in infiltrative and noninfiltrative disease models. Schrijver *et al.* studied PMN-dependent, complement-independent NSN in Beije mice in which the PMNs are deficient in both elastase and Cathepsin G.[85] The nephrotoxic serum caused only a small increment in proteinuria at the highest doses despite an influx of glomerular PMNs comparable to that of "normal" mice with NSN, suggesting that proteinuria could be attributed to serine neutral proteinases in this model. Despite the difference in proteinuria, damage to endothelial cells was comparable to controls. A specific cysteine proteinase inhibitor, trans-epoxysuccinyl-2-leucylamido-(4-guanidine) butane (E-64) was studied in a complement and neutrophil independent model of anti-GBM antibody disease.[86] E-64, administered prior to and following anti-GBM antibody administration, significantly reduced the degree of proteinuria. The specific activity of Cathepsin B and L in the glomeruli was found to be decreased. This study is particularly compelling in that a protease inhibitor was effective in ameliorating glomerulonephritis which is not dependent on cellular infiltration.

Plasmin is a serine protease which is capable of degrading extracellular matrix as well as fibrin.[87] Plasmin is generated from plasminogen by two plasminogen activators, tissue-type plasminogen activator (t-PA) and urokinase-type plasminogen activator (u-PA).[87] There are inhibitors of the system, the most important being plasminogen activator inhibitor 1 (PAI-1).[87] PAI-1 is able to inhibit both

t-PA and u-PA. PAI-1 is often found in extracellular matrix in association with vitronectin. Plasmin is able to degrade extracellular matrix directly but it is also able to activate metalloproteases.[88]

All of the components of this system are produced within the glomerulus. Mesangial cells synthesize t-PA and PAI-1 and express the u-PA receptor on their cell surface.[89,90] PAI-1 is found in the mesangial cell matrix.[91] The GEC synthesizes primarily u-PA and PAI-1 and expresses both the u-PA receptor and a plasminogen/plasmin receptor.[92,93] Components of this system are upregulated by a number of cytokines, including tumor necrosis factor α, transforming growth factor-β and interleukin 1. Cultured mesangial cells were demonstrated to degrade extracellular matrix only in the presence of plasminogen.[94]

In models of glomerular immune injury, roles for the plasminogen/plasmin system have been identified. In the model of antithymocyte serum glomerulonephritis in the rat, the isolated glomeruli had decreased plasminogen activator activity, increased synthesis of PAI-1 and increased PAI-1 deposition in the matrix.[95] When an antibody to transforming growth factor-β was administered to these animals, there was a decrease in glomerular PAI-1 deposition.[95] In an accelerated model of anti-GBM glomerulonephritis in the rat, there was increased expression of glomerular PAI-1 mRNA[96] and increased PAI-1 bioactivity with no change in expression of t-PA and u-PA mRNA. In a rat model of anti-TBM antibody-associated tubulointerstitial nephritis, there was increased expression of PAI-1 mRNA which correlated with increased PAI-1 biologic activity.[97] The net effect of these increases of PAI-1 in these models would be a decrease in plasmin formation, which could lead to an accumulation of extracellular matrix.

7.18.5.5 Mononuclear Cells

Mononuclear cells, particularly monocytes and macrophages, are also important mediators of immune injury.[98] Monocytes and macrophages have been identified in the glomeruli in certain models of glomerulonephritis and particularly in proliferative types.[98] Schreiner *et al.* first showed that depletion of monocytes was protective in a model of accelerated anti-GBM nephritis.[99] In models associated with crescent formation, macrophages are an important component of the crescents. In a model of crescentic nephritis, macrophage depletion prevented glomerular fibrin deposition and markedly decreased procoagulant activity.[100] In this same model, glomerular macrophage accumulation preceded glomerular fibrin deposition and coincided with increased glomerular procoagulant activity.[101] The macrophage is an important effector cell in immune injury because it is able to synthesize and release a number of inflammatory mediators.[102] These include a procoagulant tissue-factor-like activity, reactive oxygen species, tumor necrosis factor, proteolytic enzymes, prostaglandins, polypeptide growth factors, and complement components. Macrophages are able to respond to cytokines produced by sensitized lymphocytes. Monocyte accumulations in glomeruli particularly correlate with fibrin deposition. Thus these mononuclear cells may play many roles in immune injury.

A role for lymphocytes, and especially T lymphocytes, is suspected in glomerular immune injury although no clear role in producing injury has yet been identified. The infiltration of lymphocytes very early in the development of anti-GBM nephritis has been described.[103] Anti-GBM nephritis in the chicken can be mediated by T lymphocytes[41] suggesting that cell mediated immunity alone, in the absence of antibody, can induce glomerulonephritis. T lymphocytes are able to produce a number of products such as lymphokines, which control the immune response and could act to stimulate macrophages and recruit other T cells and B cells.

7.18.5.6 Platelets

Platelets have long been known to be important in hemostasis but recently they have been shown to be effector cells in glomerular immune injury.[104] When activated, platelets release many inflammatory mediators including growth factors, proteases, vasoactive amines, thromboxane, cationic proteins, and platelet activating factor.[104] Platelets have been found within the glomeruli in many types of glomerulonephritis.

In subendothelial proliferative glomerulonephritis involving concanavalin A, very early platelet accumulation was identified before neutrophils appeared. Complement depletion with cobra venom factor completely prevented platelet infiltration and platelet-depletion significantly reduced albuminuria.[105,106] Similarly, in mesangial proliferative glomerulonephritis induced by anti-Thy 1 antibody, platelet depletion reduced the degree of mesangial cell proliferation and improved renal function,[107] and complement depletion with cobra venom factor markedly reduced glomerular platelet accumulation.[108] In this mesangial

proliferative model, both platelet and complement depletion reduced glomerular platelet-derived growth factor (PDGF) and PDGF mRNA expression.[109] The expression of PDGF receptor and mRNA for this protein was also reduced.[109] This suggests that platelets may stimulate mesangial cells to release growth factors which may act in an autocrine fashion to stimulate mesangial cell proliferation.

Platelets may have several roles in glomerular immune injury.[104] They may lead to glomerular thrombosis by augmenting tissue factor release from monocytes and macrophages or to glomerulosclerosis by releasing growth factors such as PDGF and transforming growth factor-β. They also may affect glomerular hemodynamics through the release of vasoactive substances which may mediate mesangial cell contraction.

7.18.5.7 Mesangial Cells

The last cellular mediator of renal immunologic injury is the glomerular mesangial cell.[110] The mesangium consists of extracellular matrix and intrinsic smooth muscle-like contractile mesangial cells.[111,112] There is a second cell type in the mesangium which is bone-marrow derived and makes up less than 5% of the cells.[111] The mesangium is covered only by a fenestrated endothelium and no basement membrane so that circulating macromolecules such as immune complexes can easily enter into the mesangium.[112] The mesangium also regulates glomerular filtration rate by controlling the surface area available for filtration through contraction and relaxation of the mesangial cells.[113] Mesangial cells can respond to and produce a number of biologically active substances.[111]

Mesangial cell proliferation is a feature of many types of glomerular immune injury in humans. These diseases include IgA nephropathy, Henoch Schonlein purpura, lupus nephritis, mesangial proliferative glomerulonephritis such as found in steroid-resistant nephrotic syndrome, and membranoproliferative glomerulonephritis. It appears that the mesangial cell may play a pivotal role in many types of immune injury.

Mesangial cells are able to respond to a variety of stimuli including growth factors such as PDGF, epidermal growth factor, transforming growth factor, insulin-like growth factor 1, insulin, thrombin, interleukin 1, interleukin 6, tumor necrosis factor, fibronectin, endothelin, prostaglandins, angiotensin II, vasopressin, lipids, immune complexes, MAC of complement, heparan sulfate proteoglycan, and aggregated immunoglobulin.[110,111] Mesangial cells respond to these stimuli by proliferating, undergoing hypertrophy, contracting, migrating, changing matrix metabolism and synthesizing a variety of inflammatory mediators. These mediators include neutral proteinases, plasminogen activators, plasminogen activator inhibitors, interleukin 1, platelet activating factor, PDGF, insulin-like growth factor 1, basic fibroblast growth factor, eicosanoids and oxygen radicals.[111,114] Mesangial cells are also able to synthesize many components of extracellular matrix.[115]

Deregulated growth of mesangial cells may be important in glomerulosclerosis and progressive renal disease.[116] Heparan sulfate, a component of the glomerular basement membrane and mesangial matrix, has been demonstrated to be a regulator of mesangial cell growth.[117] When the glycosaminoglycan form of heparan sulfate is added to mesangial cells in culture, their growth is significantly inhibited.[117] The effect is reversible and unique to heparan sulfate since chondroitin sulfate, another glycosaminoglycan found in the glomerulus, had no effect. This inhibitory action of heparan sulfate requires certain structural characteristics, particularly a low degree of sulfation.[117] In both immune and nonimmune glomerular injury, the structure of glomerular heparan sulfate is altered.[118,119] Thus, changes in glomerular heparan sulfate structure produced in immunologic injury of the glomerulus may be important in the development of progressive renal disease. Heparan sulfate has also been shown to stimulate the synthesis of certain extracellular matrix components including laminin, fibronectin and type V collagen.[115]

In mesangial proliferative glomerulonephritis induced by anti-Thy 1, there is increased production of PDGF and its receptor by mesangial cells.[109] Administration of a neutralizing antibody to PDGF caused a significant decrease in mesangial cell proliferation. Increased gene expression of PDGF has been demonstrated in several proliferative glomerular immune lesions in humans.[120] In an immune complex glomerulonephritis model in the rat, the antithymocyte serum glomerulonephritis, mesangial cells synthesize and release a GBM degrading neutral proteinase.[121]

7.18.5.8 Antineutrophil Cytoplasmic Antibodies

A new, soluble mediator of immune injury has been identified in humans. This is a group of autoantibodies referred to as antineutrophil cytoplasmic antibodies (ANCA).[122] These autoantibodies react with constituents of neutrophil azurophilic granules and monocyte lysosomes.[2] There are two types of ANCAs as defined by immunofluorescent staining patterns of ethanol-fixed neutrophils exposed to ANCA.[2] One is characterized by a cytoplasmic staining pattern and is reactive with the granule constituent, serine proteinase 3. The other type manifests a perinuclear pattern of staining and the antigen is usually myeloperoxidase. ANCAs have been found in the serum of patients with vasculitis either involving the kidney alone or other organs as well.[2]

There is experimental evidence that these autoantibodies are able to induce neutrophils to undergo an oxidative burst causing the release of reactive oxygen species.[123] Priming neutrophils with cytokines such as tumor necrosis factor enhances this effect by inducing antigens on the cell surface which react with ANCA. Antimyeloperoxidase autoantibodies were demonstrated to stimulate neutrophils to damage endothelials cells in culture.[124]

7.18.6 CONCLUSION

In summary, this review has attempted to provide an overview of the mechanisms of immune injury in the kidney, emphasizing mediators of the immunologic injury. The next few years should lead to greater understanding of immune renal injury at the cellular and molecular levels and possibly to newer therapeutic interventions.

7.18.7 REFERENCES

1. C. B. Wilson, in 'The Kidney,' 4th edn., eds. B. M. Brenner and F. C. Rector, Saunders, Philadelphia, PA, 1991, pp. 1062–1181.
2. R. J. Falk and J. C. Jennette, 'Wegener's granulomatosis, systemic vasculitis and antineutrophil cytoplasmic autoantibodies.' *Annu. Rev. Med.*, 1991, **42**, 459–469.
3. W. G. Couser, 'Mediation of immune glomerular injury.' *J. Am. Soc. Nephrol.*, 1990, **1**, 13–29.
4. R. J. Johnson, 'The glomerular response to injury: progression or resolution?' *Kidney Int.*, 1994, **45**, 1769–1782.
5. F. J. Dixon, 'The pathogenesis of glomerulonephritis.' *Am. J. Med.*, 1968, **44**, 493–498.
6. W. G. Couser and D. J. Salant, '*In situ* immune complex formation and glomerular injury.' *Kidney Int.*, 1980, **17**, 1–13.
7. W. G. Couser, 'Mechanisms of glomerular injury in immune-complex disease.' *Kidney Int.*, 1985, **28**, 569–583.
8. D. J. Salant, S. Adler, C. Darby *et al.*, 'Influence of antigen distribution on the mediation of immunologic glomerular injury.' *Kidney Int.*, 1985, **27**, 938–950.
9. W. G. Couser, D. R. Steinmuller, M. M. Stilmant *et al.*, 'Experimental glomerulonephritis in the iso-

lated perfused rat kidney.' *J. Clin. Invest.*, 1978, **62**, 1275–1287.

10. D. Kerjaschki and M. G. Farquhar, 'The pathogenetic antigen of Heymann nephritis is a membrane glycoprotein of the renal proximal tubule brush border.' *Proc. Natl. Acad. Sci. USA*, 1982, **79**, 5557–5581.

11. D. J. Salant, C. Darby and W. G. Couser, 'Experimental membranous glomerulonephritis in rats. Quantitative studies of glomerular immune deposit formation in isolated glomeruli and whole animals.' *J. Clin. Invest.*, 1980, **66**, 71–81.

12. G. Camussi, J. R. Brentjens, B. Noble *et al.*, 'Antibody-induced redistribution of Heymann's antigen on the surface of cultured glomerular visceral epithelial cells: possible role in the pathogenesis of Heymann glomerulonephritis.' *J. Immunol.*, 1985, **135**, 2409–2416.

13. J. R. Brentjens and G. Andres, 'Interaction of antibodies with renal cell surface antigens.' *Kidney Int.*, 1989, **35**, 954–968.

14. W. A. Border, H. J. Ward, E. S. Kamil *et al.*, 'Induction of membranous nephropathy in rabbits by administration of an exogenous cationic antigen.' *J. Clin. Invest.*, 1982, **69**, 451–461.

15. M. P. Madaio, J. Carlson, J. Cataldo *et al.*, 'Murine monoclonal anti-DNA antibodies bind directly to glomerular antigens and form immune deposits.' *J. Immunol.*, 1987, **138**, 2883–2889.

16. A. Vogt, R. Rohrbach, F. Shimizu *et al.*, 'Interaction of cationized antigen with rat glomerular basement membrane: *in situ* immune complex formation.' *Kidney Int.*, 1982, **22**, 27–35.

17. F. G. Germuth, Jr., E. Rodriguez, C. A. Lorelle *et al.*, 'Passive immune complex glomerulonephritis in mice: models for various lesions found in human disease. II. Low avidity complexes and diffuse proliferative glomerulonephritis with subepithelial deposits.' *Lab. Invest.*, 1979, **41**, 366–371.

18. G. R. Gallo, T. Caulin-Glaser and M. E. Lamm, 'Charge of circulating immune complexes as a factor in glomerular basement membrane localization in mice.' *J. Clin. Invest.*, 1981, **67**, 1305–1313.

19. M. Masugi, 'Uber die experimentelle Glomerulonephritis durch das spezifische Antinierenserum. Ein Beitrag zur Pathogense der diffusen Glomerulonephritis.' *Beitr. Pathol.*, 1934, **92**, 429–466.

20. C. B. Wilson and R. C. Blantz, 'Nephroimmunopathology and pathophysiology, Editorial Review.' *Am. J. Physiol.*, 1985, **248**, F319–F331.

21. J. Wieslander, J. F. Barr, R. J. Butkowski *et al.*, 'Goodpasture antigen of the glomerular basement membrane: localization to noncollagenous regions of type IV collagen.' *Proc. Natl. Acad. Sci. USA*, 1984, **81**, 3838–3842.

22. R. C. McCoy, H. K. Johnson, W. J. Stone *et al.*, 'Absence of nephritogenic GBM antigen(s) in some patients with hereditary nephritis.' *Kidney Int.*, 1982, **21**, 642–652.

23. S. M. Globus and C. B. Wilson, 'Experimental glomerulonephritis induced by in situ formation of immune complexes in the glomerular capillary wall.' *Kidney Int.*, 1979, **16**, 148–157.

24. S. Matsuo, A. Fukatsu, M. L. Taub *et al.*, 'Glomerulonephritis induced in the rabbit by antiendothelial antibodies.' *J. Clin. Invest.*, 1987, **79**, 1798–1811.

25. D. B. Clines, A. P. Lyss, M. Reeber *et al.*, 'Presence of complement-fixing anti-endothelial antibodies in systemic lupus erythematosus.' *J. Clin. Invest.*, 1984, **73**, 611–625.

26. M. Mannik, 'Pathophysiology of circulating immune complexes.' *Arthritis Rheum.*, 1982, **25**, 783–787.

27. T. Caulin-Glasser, G. R. Gallo and M. E. Lamm, 'Nondissociating cationic immune complexes can deposit in glomerular basement membrane.' *J. Exp. Med.*, 1983, **158**, 1561–1572.

28. J. B. Cornacoff, L. A. Hebert, W. L. Smead *et al.*, 'Primate erythrocyte-immune complex-clearing mechanism.' *J. Clin. Invest.*, 1983, **71**, 236–247.

29. S. M. Mauer, A. J. Fish, E. B. Blau *et al.*, 'I. Kinetic studies of macromolecular uptake in normal and nephrotic rats.' *J. Clin. Invest.*, 1972, **51**, 1092–1101.

30. U. Barcelli, R. Rabbmacher, Y. M. Ooi *et al.*, 'Modification of glomerular immune complex deposition in mice by activation of the reticuloendothelial system.' *J. Clin. Invest.*, 1981, **67**, 20–27.

31. L. A. Hebert, C. L. Allhiser and S. M. Koethe, 'Some hemodynamic determinants of immune complex trapping by the kidney.' *Kidney Int.*, 1978, **14**, 452–465.

32. L. A. Hebert and G. Cosio, 'The erythrocyte-immune complex–glomerulonephritis connection in man.' *Kidney Int.*, 1987, **31**, 877–885.

33. F. J. Waxman, L. A. Hebert, J. B. Cornacoff *et al.*, 'Complement depletion accelerates the clearance of immune complexes from the circulation of primates.' *J. Clin. Invest.*, 1984, **74**, 1329–1340.

34. A. O. Haakenstad, G. E. Striker and M. Mannik, 'Removal of glomerular immune complex deposits by excess antigen in chronic mouse model of immune complex disease.' *Lab. Invest.*, 1983, **48**, 323–331.

35. A. J. Fish, A. F. Michael, R. L. Vernier *et al.*, 'Acute serum sickness nephritis in the rabbit. An immune deposit disease.' *Am. J. Pathol.*, 1966, **49**, 997–1022.

36. A. F. Michael, W. F. Keane, L. Raji *et al.*, 'The glomerular mesangium.' *Kidney Int.*, 1980, **17**, 141–154.

37. M. S. Mauer, D. E. R. Sutherland, R. J. Howard *et al.*, 'The glomerular mesangium. 3. Acute immune mesangial injury: a new model of glomerulonephritis.' *J. Exp. Med.*, 1973, **137**, 553–570.

38. T. Yamamoto and C. B. Wilson, 'Quantitative and qualitative studies of antibody-induced mesangial cell damage in the rat.' *Kidney Int.*, 1987, **32**, 514–525.

39. T. Yamamoto and C. B. Wilson, 'Complement dependence of antibody-induced mesangial cell injury in the rat.' *J. Immunol.*, 1987, **138**, 3758–3765.

40. A. K. Bhan, E. E. Schneeberger, A. B. Collins *et al.*, 'Evidence for a pathogenic role of a cell-mediated immune mechanism in experimental glomerulonephritis.' *J. Exp. Med.*, 1978, **148**, 246–260.

41. W. K. Bolton, F. L. Tucker and B. C. Sturgill, 'New avian model of experimental glomerulonephritis consistent with mediation by cellular immunity. Nonhumorally mediated glomerulonephritis in chickens.' *J. Clin. Invest.*, 1984, **73**, 1263–1276.

42. W. K. Bolton, M. Chandra, T. M. Tyson *et al.*, 'Transfer of experimental glomerulonephritis in chickens by mononuclear cells.' *Kidney Int.*, 1988, **34**, 598–610.

43. C. B. Wilson, 'Study of the immunopathogenesis of tubulointerstitial nephritis using model systems.' *Kidney Int.*, 1989, **35**, 938–953.

44. C. B. Wilson, 'Nephritogenic tubulointerstitial antigens.' *Kidney Int.*, 1991, **39**, 501–517.

45. K. M. Bannister, T. R. Ulich and C. B. Wilson, 'Induction, characterization, and cell transfer of autoimmune tubulointerstitial nephritis.' *Kidney Int.*, 1987, **32**, 642–651.

46. W. G. Couser, M. M. Stilmant, and N. B. Jermanovich, 'Complement-independent nephrotoxic nephritis in the guinea pig.' *Kidney Int.*, 1977, **11**, 170–180.

47. W. G. Couser, C. Darby, D. J. Salant *et al.*, 'Anti-GBM antibody-induced proteinuria in isolated perfused rat kidney.' *Am. J. Physiol.*, 1985, **249**, F241–F250.

48. D. J. Salant, M. P. Madaio, S. Adler *et al.*, 'Altered glomerular permeability induced by F(ab)₂ and Fab

antibodies to rat renal tubular epithelial antigen.'
Kidney Int., 1981, **21**, 36–43.

49. D. L. Mendrick and H. G. Rennke, 'Induction of proteinuria in the rat by a monoclonal antibody against SGP-115/107.' *Kidney Int.*, 1988, **33**, 818–830.

50. M. Orikasa, K. Matsui, T. Oite *et al.*, 'Massive proteinuria induced in rats by a single intravenous injection of a monoclonal antibody.' *J. Immunol.*, 1988, **141**, 807–814.

51. R. Johnson, H. Yamabe, Y. P. Chen *et al.*, 'Glomerular epithelial cells secrete a glomerular basement membrane-degrading metalloproteinase.' *J. Am. Soc. Nephrol.*, 1992, **2**, 1388–1397.

52. W. G. Couser, P. J. Baker and S. Adler, 'Complement and the direct mediation of immune glomerular injury: a new perspective.' *Kidney Int.*, 1985, **28**, 879–890.

53. A. V. Cybulsky, R. J. Quigg, and D. J. Salant, in 'Immunopathology of Renal Disease,' eds. B. M. Brenner and J. Stein, Churchill Livingstone, New York, 1988, pp. 57–86.

54. R. D. Schreiber and H. J. Muller-Eberhard, in 'Immunologic Mechanisms Of Renal Disease,' eds. B. Brenner and J. Stein, Churchill Livingstone, New York, 1979, pp. 67–105.

55. C. G. Cochrane, E. R. Unanue and F. J. Dixon, 'A role of polymorphonuclear leukocytes and complement in nephrotoxic nephritis.' *J. Exp. Med.*, 1965, **122**, 99–116.

56. P. M. Henson, 'Pathologic mechanisms in neutrophil-mediated injury.' *Amer. J. Path.*, 1972, **68**, 593–612.

57. D. J. Salant, S. Belok, M. P. Madaio *et al.*, 'A new role for complement in experimental membranous nephropathy in rats.' *J. Clin. Invest.*, 1980, **66**, 1339–1350.

58. G. C. Groggel, S. Adler, H. G. Rennke *et al.*, 'Role of the terminal complement pathway in experimental membranous nephropathy in the rabbit.' *J. Clin. Invest.*, 1983, **72**, 1948–1957.

59. A. V. Cybulsky, H. G. Rennke, I. D. Feintzeig *et al.*, 'Complement-induced glomerular epithelial cell injury. Role of the membrane attack complex in rat membranous nephropathy.' *J. Clin. Invest.*, 1986, **77**, 1096–1107.

60. G. C. Groggel, D. J. Salant, C. Darby *et al.*, 'Role of terminal complement pathway in the heterologous phase of antiglomerular basement membrane nephritis.' *Kidney Int.*, 1985, **27**, 643–651.

61. R. J. Falk, A. P. Dalmasso, Y. Kim *et al.*, 'Neoantigen of the polymerized ninth component of complement. Characterization of a monoclonal antibody and immunohistochemical localization in renal disease.' *J. Clin. Invest.*, 1983, **72**, 560–573.

62. H. J. Muller-Eberhard, 'Molecular organization and function of the complement system.' *Annu. Rev. Biochem.*, 1988, **57**, 321–347.

63. R. J. Quigg, A. V. Cybulsky, J. B. Jacobs *et al.*, 'Anti-Fx1A produces complement-dependent cytotoxicity of glomerular epithelial cells.' *Kidney Int.*, 1988, **34**, 43–52.

64. A. V. Cybulsky, D. J. Salant, R. J. Quigg *et al.*, 'Complement C5b-9 complex activates phospholipases in glomerular epithelial cells.' *Am. J. Physiol.*, 1989, **257**, F826–F836.

65. S. Adler, P. J. Baker, R. J. Johnson *et al.*, 'Complement membrane attack complex stimulates production of reactive oxygen metabolites by cultured rat mesangial cells.' *J. Clin. Invest.*, 1986, **77**, 762–767.

66. I. Torbohm, M. Schonermark, A. M. Wingen *et al.*, 'C5b-8 and C5b-9 modulate the collagen release of human glomerular epithelial cells.' *Kidney Int.*, 1990, **37**, 1098–1104.

67. G. C. Groggel and D. A. Terreros, 'Role of the terminal complement pathway in accelerated autologous

anti-glomerular basement membrane nephritis.' *Am. J. Pathol.*, 1990, **136**, 533–540.

68. J. A. Schifferli, Y. C. Ng and D. K. Peters, 'The role of complement and its receptor in the elimination of immune complexes.' *New Eng. J. Med.*, 1986, **315**, 488–495.

69. H. R. Brady, 'Leukocyte adhesion molecules and kidney diseases.' *Kidney Int.*, 1994, 45, 1285–1300.

70. P. F. Naisch, N. M. Thomson, I. J. Simpson *et al.*, 'The role of polymorphonuclear leukocytes in the autologous phase of nephrotoxic nephritis.' *Clin. Exp. Immunol.*, 1975, **22**, 102–111.

71. D. H. Hooke, D. C. Gee and R. C. Atkins, 'Leucocyte analysis using monoclonal antibodies in human glomerulonephritis.' *Kidney Int.*, 1987, **31**, 964–972.

72. H. T. Cook, J. Smith and V. Cattell, 'Isolation and characterization of inflammatory leukocytes from glomeruli in an in situ model of glomerulonephritis in the rat.' *Am. J. Pathol.*, 1987, **126**, 126–136.

73. B. J. Tucker, L. C. Gushwa and C. B. Wilson *et al.*, 'Effect of leucocyte depletion on glomerular dynamics during active glomerular immune injury.' *Kidney Int.*, **28**, 1985, 28–35.

74. T. Adachi, M. Fukuta, Y. Ito *et al.*, 'Effect of superoxide dismutase on glomerular nephritis.' *Biochem. Pharmacol.*, 1986, **35**, 341–345.

75. A. Rehan, K. J. Johnson, R. C. Wiggins *et al.*, 'Evidence for the role of oxygen radicals in acute nephrotoxic nephritis.' *Lab. Invest.*, 1984, **51**, 396–403.

76. R. B. Johnson, Jr. and J. E. Lehmeyer, 'Elaboration of toxic oxygen by-products by neutrophils in a model of immune complex disease.' *J. Clin. Invest.*, 1976, **57**, 836–841.

77. W. H. Baricos and S. V. Shah, 'Proteolytic enzymes as mediators of glomerular injury.' *Kidney Int.*, 1991, **40**, 161–173.

78. M. Davies, A. J. Barrett, J. Travis *et al.*, 'The degradation of human glomerular basement membrane with purified lysosomal proteinases: evidence for the pathogenic role of the polymorphonuclear leukocyte in glomerulonephritis.' *Clin. Sci. Mol. Med.*, 1978, **54**, 233–240.

79. J. M. Harlan, P. D. Killen, L. A. Harker *et al.*, 'Neutrophil-mediated endothelial injury *in vitro* mechanisms of cell detachment.' *J. Clin. Invest.*, 1981, **68**, 1394–1403.

80. S. V. Shah, W. H. Baricos and A. Basci, 'Degradation of human glomerular basement membrane by stimulated neutrophils. Activation of a metalloproteinase(s) by reactive oxygen metabolites.' *J. Clin. Invest.*, 1987, **79**, 25–31.

81. D. H. Lovett, B. R. Sterzel, M. Kashgarian *et al.*, 'Neutral proteinase activity produced *in vitro* by cells of the glomerular mesangium.' *Kidney Int.*, 1983, **23**, 342–349.

82. W. H. Baricos, S. L. Cortez, Q. C. Le *et al.*, 'Glomerular basement membrane degradation by endogenous cysteine proteinases in isolated rat glomeruli.' *Kidney Int.*, 1990, **38**, 395–401.

83. R. J. Johnson, W. G. Couser, C. E. Alpers *et al.*, 'The human neutrophil serine proteinases, elastase, and cathepsin G, can mediate glomerular injury *in vivo*.' *J. Exp. Med.*, 1988, **168**, 1169–1174.

84. J. C. Davin, M. Davies, J. M. Foidart *et al.*, 'Urinary excretion of neutral proteinases in nephrotic rats with a glomerular disease.' *Kidney Int.*, 1987, **31**, 32–40.

85. G. Schrijver, J. Schwalkwijk, J. C. M. Robben *et al.*, 'Antiglomerular basement membrane nephritis in Beije mice. Deficiency of leukocytic neutral proteinases prevents the induction of albuminuria in the heterologous phase.' *J. Exp. Med.*, 1989, **169**, 1435–1448.

86. W. H. Baricos, S. E. O'Connor, S. L. Cortez *et al.*,

'The cysteine proteinase inhibitor, E-64, reduces proteinuria in an experimental model of glomerulonephritis.' *Biochem. Biophys. Res. Commun.*, 1988, **155**, 1318–1323.

87. J. D. Vassalli, A. P. Sappino and D. Belin, 'The plasminogen activator/plasmin system.' *J. Clin. Invest.*, 1991, **88**, 1067–1072.

88. C. S. He, S. M. Wilhelm, A. P. Pentland *et al.*, 'Tissue cooperation in a proteolytic cascade activating human interstitial collagenase.' *Proc. Natl. Acad. Sci. USA*, 1989, **86**, 2632–2636.

89. R. Lacave, E. Rondeau, S. Ochi *et al.*, 'Characterization of a plasminogen activator and its inhibitor in human mesangial cells.' *Kidney Int.*, 1989, **35**, 806–811.

90. G. Nguyen, X. M. N. Li, M. N. Peraldi *et al.*, 'Receptor binding and degradation of urokinase-type plasminogen activator by human mesangial cells.' *Kidney Int.*, 1994, **46**, 208–215.

91. J. Hagege, M. N. Peraldi, E. Rondeau *et al.*, 'Plasminogen activator inhibitor-1 deposition in the extracellular matrix of cultured human mesangial cells.' *Am. J. Pathol.*, 1992, **141**, 117–128.

92. E. Rondeau, S. Ochi, R. Lacave *et al.*, 'Urokinase synthesis and binding by glomerular epithelial cells in culture.' *Kidney Int.*, 1989, **36**, 593–600.

93. L. Becquemont, G. Nguyen, M. N. Peraldi *et al.*, 'Expression of plasminogen/plasmin receptors on human glomerular epithelial cells.' *Am. J. Physiol.*, 1994, **267**, F303–F310.

94. A. P. Wong, S. L. Cortez and W. H. Baricos, 'Role of plasmin and gelatinase in extracellular matrix degradation by cultured rat mesangial cells.' *Am. J. Physiol.*, 1992, **263**, F1112–F1118.

95. S. Tomooka, W. A. Border, B. C. Marshall *et al.*, 'Glomerular matrix accumulation is linked to inhibition of the plasmin protease system.' *Kidney Int.*, 1992, **42**, 1462–1469.

96. L. Feng, W. W. Tang, D. J. Loskutoff *et al.*, 'Dysfunction of glomerular fibrinolysis in experimental antiglomerular basement membrane antibody glomerulonephritis.' *J. Am. Soc. Nephrol.*, 1993, **3**, 1753–1764.

97. W. W. Tang, L. Feng, Y. Xia *et al.*, 'Extracellular matrix accumulation in immune-mediated tubulointerstitial injury.' *Kidney Int.*, 1994, **45**, 1077–1084.

98. G. F. Schreiner, 'The role of the macrophage in glomerular injury.' *Semin. Nephrol.*, 1991, **11**, 268–275.

99. G. F. Schreiner, R. S. Cotran, V. Pardo *et al.*, 'A mononuclear cell component in experimental immunological glomerulonephritis.' *J. Exp. Med.*, 1978, **147**, 369–384.

100. S. R. Holdsworth and P. G. Tipping, 'Macrophage-induced glomerular fibrin deposition in experimental glomerulonephritis in the rabbit.' *J. Clin. Invest.*, 1987, **76**, 1367–1374.

101. P. G. Tipping and S. R. Holdsworth, 'The participation of macrophages, glomerular procoagulant activity, and factor VIII in glomerular fibrin deposition. Studies on anti-GBM antibody induced glomerulonephritis in rabbits.' *Am. J. Pathol.*, 1986, **124**, 10–17.

102. C. F. Nathan, 'Secretory products of macrophages.' *J. Clin. Invest.*, 1987, **79**, 319–326.

103. J. I. Kreisberg, D. B. Wayne and M. D. Karnovsky, 'Rapid and focal loss of negative charge associated with mononuclear cell infiltration early in nephrotoxic serum nephritis.' *Kidney Int.*, 1979, **16**, 290–300.

104. R. J. Johnson, 'Platelets in inflammatory glomerular injury.' *Semin. in Nephrol.*, 1991, **11**, 276–284.

105. R. J. Johnson, C. E. Alpers, P. Pritzl *et al.*, 'Platelets mediate neutrophil-dependent immune complex nephritis in the rat.' *J. Clin. Invest.*, 1988, **82**, 1225–1235.

106. R. J. Johnson, C. E. Alpers, C. Pruchno *et al.*, 'Mechanisms and kinetics for platelet and neutrophil localization in immune complex nephritis.' *Kidney Int.*, 1989, **36**, 780–789.

107. R. J. Johnson, R. L. Garcia, P. Pritzl *et al.*, 'Platelets mediate glomerular cell proliferation in immune complex nephritis induced by anti-mesangial cell antibodies in the rat.' *Am. J. Pathol.*, 1990, **136**, 369–374.

108. R. J. Johnson, P. Pritzl, H. Iida *et al.*, 'Platelet–complement interactions in mesangial proliferative nephritis in the rat.' *Am. J. Pathol.*, 1991, **138**, 313–321.

109. H. Iida, R. Seifert, C. E. Alpers *et al.*, 'Platelet-derived growth factor (PDGF) and PDGF receptor are induced in mesangial proliferative nephritis in the rat.' *Proc. Nat. Acad. Sci. USA*, 1991, **88**, 6560–6564.

110. H. E. Abboud, 'Resident glomerular cells in glomerular injury: mesangial cells.' *Semin. Nephrol.*, 1991, **11**, 304–311.

111. P. Mene, M. S. Simonson and M. J. Dunn, 'Physiology of the mesangial cell.' *Physiol. Rev.*, 1989, **69**, 1347–1424.

112. D. Schlondorff, 'The glomerular mesangial cell: an expanding role for a specialized pericyte.' *FASEB J.*, 1987, **1**, 272–281.

113. J. I. Kreisberg, M. Venkatachalam and D. Troyer, 'Contractile properties of cultured glomerular mesangial cells.' *Am. J. Physiol.*, 1985, **249**, F457–F463.

114. R. B. Sterzel and D. H. Lovett, in 'Immunopathology Of Renal Disease,' eds. C. B. Wilson, B. M. Brenner and J. H. Stein, Churchill Livingstone, New York, 1988, pp. 137–173.

115. G. C. Groggel and M. L. Hughes, 'Heparan sulfate stimulates extracellular matrix component synthesis by mesangial cells.' *Nephron.*, 1995, **71**, 197–202.

116. L. J. Striker, E. P. Peten, S. J. Elliot *et al.*, 'Mesangial cell turnover: effect of heparin and peptide growth factors.' *Lab. Invest.*, 1991, **64**, 446–456.

117. G. C. Groggel, G. M. Marinides, P. Hovingh *et al.*, 'Inhibition of rat mesangial cell growth by heparan sulfate.' *Am. J. Physiol.*, 1990, **258**, F259–F265.

118. G. C. Groggel, P. Hovingh, W. A. Border *et al.*, 'Changes in glomerular heparan sulfate in puromycin aminonucleoside nephrosis.' *Am. J. Pathol.*, 1987, **128**, 521–527.

119. G. C. Groggel, J. Stevenson, P. Hovingh *et al.*, 'Changes in heparan sulfate correlate with increased glomerular permeability.' *Kidney Int.*, 1988, **33**, 517–523.

120. L. Gesualdo, M. Pinzani, J. J. Floriano *et al.*, 'Platelet-derived growth factor expression in mesangial proliferative glomerulonephritis.' *Lab. Invest.*, 1991, **65**, 160–167.

121. D. H. Lovett, R. J. Johnson, H. P. Marti *et al.*, 'Structural characterization of the mesangial cell type IV collagenase and enhanced expression in a model of immune complex-mediated glomerulonephritis.' *Am. J. Pathol.*, 1992, **141**, 85–98.

122. R. J. Falk and J. C. Jennette, 'Anti-neutrophil cytoplasmic autoantibodies with specificity for myeloperoxidase in patients with systemic vasculitis and idiopathic necrotizing and crescentic glomerulonephritis.' *New Eng. J. Med.*, 1988, **318**, 1651–1657.

123. R. J. Falk, R. S. Terrell, L. A. Charles *et al.*, 'Antineutrophil cytoplasmic autoantibodies induce neutrophils to degranulate and produce oxygen radicals *in vitro*.' *Proc. Natl. Acad. Sci. USA*, 1990, **87**, 4115–4119.

124. B. Ewert, J. C. Jennette and R. Falk, 'Anti-myeloperoxidase antibodies stimulate neutrophils to damage human endothelial cells.' *Kidney Int.*, 1992, **41**, 375–383.

7.19
Carcinogenic Response of the Kidney

SCOT L. EUSTIS

US SmithKline Beecham Pharmaceuticals, King of Prussia, PA, USA

7.19.1 INTRODUCTION

Renal carcinogenesis in humans traditionally has been investigated with epidemiological studies, pathological evaluation of surgical or autopsy specimens, and surrogate studies with animal models. Due to the inherent insensitivity of epidemiological methods and the limitations of standard pathological techniques, studies with animal models have contributed greatly to our understanding of the pathogenesis and histogenesis of renal neoplasia. Almost all types of kidney neoplasms occurring in humans also have been observed in many species of animals, and most have been experimentally induced, particularly in rodents, by exposure to chemicals, hormones, radiation, or viruses. Since there are many comprehensive reviews of renal carcinogenesis describing the spontaneous occurrence, morphology and experimental induction of renal neoplasms in animals, these topics will not be covered in detail here.[1-8]

The most widely practiced method for determining the potential carcinogenicity (renal or other) of a chemical or prospective drug and for defining the relative human risk is the rodent carcinogenicity bioassay. Rodent bioassays, however, are of value only if the mechanisms of cancer induction in the surrogate species are similar to the mechanisms in man. It has become increasingly clear that the induction of cancer is polygenic; that is, the initiation, promotion and progression of cancer is modulated by several classes of genes. These include, for example, the genes which determine the response (i.e., metabolism, distribution, and elimination) of the organism to xenobiotics, such as those genes coding for P450, glutathione transferase, and transacetylase, the genes which encode for DNA repair enzymes, and genes which encode for important growth or cell cycle regulatory proteins (oncogenes, tumor suppressor genes, cyclin/cyclin dependent kinases, cell death proteins, etc.). It is also clear that individual and species susceptibility to spontaneous, chemical or radiation carcinogenesis is related to genetic polymorphisms in various gene superfamilies.[9-12] In contrast to random outbred populations, such as humans, where allelic diversity is maintained through recombination, inbred populations of rodents have highly characteristic gene patterns where heterozygosity (allelic diversity) is greatly diminished or lost. This raises the possibility of carcinogenic effects in rodents (in response to drug or chemical administration) that are highly genotype-specific.[13]

Technical advances in molecular biology and methods of DNA analysis combined with mapping of the human genome have made it possible to study renal carcinogenesis in humans in a more direct and detailed manner. These rapid advances in understanding the genetic basis of human renal neoplasia pose a major challenge to toxicologists who must relate findings in experimental animals to humans during the processes of hazard identification and risk assessment. For significant advances to be made in the field of risk assessment, toxicologists must attain a similar level of understanding of the the molecular genetics of renal neoplasia in experimental animals. The purpose of this review is to give the reader a brief summary of findings regarding the genetic basis of renal cancer in humans and animals, particularly the rat.

7.19.2 CLASSIFICATION AND HISTOGENESIS OF RENAL NEOPLASMS

It is important to remember that cellular phenotype (state of differentiation, form, location and function), as determined by the pattern of gene expression (genotype), is of singular importance in determining the morphologic types of neoplasms, frequency of spontaneous occurrence, and susceptibility to carcinogenic agents and/or promotional influences (risk factors). The pattern of gene expression in a population of cells determines the rate of cell turnover, cellular receptors and responsiveness to promotional influences (steroids, estrogens, nonsteroidal hormones, and autocrine/paracrine growth factors), the capacity of DNA repair enzymes to excise and repair altered DNA, the capacity to metabolize xenobiotics, the capacity of glutathione transferases to conjugate electrophilic intermediates, and other functional characteristics. Therefore, an understanding of the various tumor types and their histogenesis is a necessary foundation for understanding the etiology, risk factors, mechanism of induction, and molecular genetics of renal cancers.

A great deal of effort has been made to describe the morphology of renal neoplasms and develop an appropriate classification system based primarily on purported histogenesis or cellular origin.[6,14-16] Neoplasms of the kidney may originate from epithelial cells of the nephron or renal tubule (metanephrogenic blastema in origin), collecting ducts (mesonephric ureteral bud), or pelvic urothelium (also mesonephric ureteral bud), from mesenchymal cells of the interstitium and supporting tissues (fibroblasts, vascular smooth muscle, endothelium, etc.), and from cells of the embryonic metanephric blastema (Table 1).

Unlike other internal organs, the epithelium of the nephron is derived embryologically from the mesoderm rather than the endoderm. Therefore, cells of the metanephric blastema are considered to have potential for differentiation into the nephron epithelium (tubules) as well as the secondary mesenchyme (connective tissues) of the kidney. Conceptually, this has led to some confusion in descriptive terminology and controversy regarding the classification of certain embryonal-like neoplasms in the rat (e.g., confusion of renal mesenchymal tumor with nephroblastoma).

7.19.2.1 Renal Cell Neoplasms

Tumors occurring in the renal cortex and presumed to arise from the renal tubule are

Table 1 Classification and histogenesis of renal neoplasms.

Histogenesis	Benign neoplasm	Malignant neoplasm
Epithelium		
Nephron (renal tubule)	Renal cell adenoma	Renal cell carcinoma
Collecting duct	Oncocytoma	Malignant oncocytoma ??
	Collecting duct adenoma	Collecting duct carcinoma
Pelvic urothelium	Transitional cell papilloma	Transitional cell carcinoma
Mesenchyme		
Endothelium	Hemangioma	Hemangiosarcoma
Smooth muscle	Leiomyoma	Leiomyosarcoma
Fibrocyte/fibroblast	Fibroma	Fibrosarcoma
Stem cell ??	Lipoma	Liposarcoma, mesenchymal tumor, rat
Metanephric blastema		Nephroblastoma

generally called renal cell adenoma or carcinoma. Two variants believed to arise from the collecting ducts, oncocytoma and collecting duct adenoma or carcinoma, have distinguishing morphological, biological, and/or cytogenetic characteristics. The precise segment of the nephron from which renal cell tumors arise is debated and has been shown to vary with species, agent (chemical class, radiation, or viral), and mechanism of induction. Sequential studies of experimental animals given potent genotoxic carcinogens demonstrate a morphological progression from small intratubular foci of proliferating cells (hyperplasia/dysplasia), through intermediate-size tumor nodules effacing the tubule of origin, to large masses replacing much of the kidney.[17–20] Although histological criteria for the diagnosis of preneoplastic lesions (i.e., hyperplasia or dysplasia), adenoma and carcinoma have been established, an absolute distinction on a morphological basis is not clear in either humans or animals. In general, the potential for metastasis is correlated with size of the tumor mass, although exceptions occur.

Renal cell adenoma and carcinoma are commonly subclassifed according to cytological characteristics and growth pattern as described below. The reader is cautioned that these subcategories are not absolute, as cytology and growth pattern may vary within any given tumor. Further, corroborating ultrastructural studies have been based on the examination of limited numbers of tumors. In experimental animals, these cytological characteristics have been described for tumors induced by potent genotoxic carcinogens, and it is not clear to what extent these associations exist for other spontaneous or induced renal tumors. Nevertheless, when strict diagnostic criteria are adhered too, certain cytological types have distinguishing biological characteristics, molecular genetic changes and histogenesis, particularly in humans.

Pathologists have described clear, basophilic, acidophilic (eosinophilic), oncocytic, and chromophobic cell types based on their affinity for histologic dyes (hematoxylin and eosin). Ultrastructural studies have shown that these staining characteristics are related to the relative abundance of intracellular nonparticulate glycogen (clear cells), free and bound ribosomes (basophilic cells), abnormally large mitochondria (oncocytic cells), or smooth endoplasmic reticulum and/or normal mitochondria (acidophilic cells). Chromophobic cells lack the distinguishing abundance of glycogen, mitochondria, ribosomes, or endoplasmic reticulum observed in the other cell types. Chromophobe cells of at least some human renal tumors contain an abundance of small cytoplasmic vesicles.

In humans, the majority of renal cell carcinomas (approximately two-thirds to three-fourths) are comprised predominantly of clear cells.[16] While there has been a long held belief that human clear cell tumors originate from the proximal tubule, ultrastructural and immunohistochemical studies provide conflicting evidence suggesting origin from either proximal or distal tubule.[21–27]

In rodents, renal cell tumors induced by dimethylnitrosamine, diethylnitrosamine, N-(4'-fluoro-4-biphenylyl)acetamide, N-nitrosomorpholine, and lead acetate were shown to consist primarily of basophilic or, rarely, chromophobe cells arising from the proximal tubule.[17–20] In contrast to the rat, chromophobe cell tumors of humans are believed to originate from the intercalated cells of the collecting ducts, based on their characteristic cytoplasmic vesicles and immunoreactivity to epithelial membrane antigen and carbonic anhydrase C.[28] Further, chromophobe (collecting duct) tumors of humans exhibit chromosomal changes that differ from those of clear cell or papillary renal cell tumors.

The collecting duct is also believed to be the site of origin of chemically induced oncocytomas in the rat as well as oncocytomas in humans, based on ultrastructural characteristics, lectin histochemistry and positive reactivity for carbonic anhydrase C and band 3 anion exchanger.[20,29–33] Carcinogen induced renal cell tumors in rats consisting principally of clear and acidophilic cells were also demonstrated to arise from the outer medullary collecting ducts.[34]

Renal cell neoplasms, like those of other organs, are also subclassified according to growth pattern. These categories typically include solid (or compact), acinar, tubular, papillary, and cystic.[16] There is evidence that human papillary tumors, like oncocytomas, have a distinctive set of genetic and chromosomal abnormalities and differing molecular pathogenesis.

7.19.2.2 Transitional Cell Neoplasms

The pathological characteristics of transitional cell neoplasms are similar in humans and other animals. Renal transitional cell neoplasms originate from the pelvic urothelium and are similar in histological characteristics to the transitional cell neoplasms arising from the urinary bladder mucosa. The tumors may be exophytic, that is, protrude into the pelvic lumen, or endophytic, displacing or invading the papillae and medulla. Experimentally induced transitional cell tumors of rats show a continuum from small foci of urothelial hyperplasia to benign papillary or nodular tumors (papilloma or adenoma) and finally to large carcinomas with cellular atypia and invasion of the renal medulla or adjacent structures.[35] With tumors induced by potent carcinogens, the benign stages may not be clearly evident and very small lesions may exhibit malignant characteristics such as cellular atypia and invasion. While attempts are usually made to distinguish the relatively benign from the malignant tumors in animal studies, some medical pathologists consider all transitional cell tumors in humans to be malignant, albeit of varying grades of malignancy.

7.19.2.3 Nephroblastoma (Wilms' Tumor)

Nephroblastoma is generally, but not always, observed in young animals including humans and many laboratory (rodents, rabbit), domesticated (dog, cat, cattle, chickens, etc.) and wild animals.[36,37] Since cytologic characteristics, predominance in young animals, and association with congenital malformations in humans, the nephroblastoma is presumed to arise from the metanephric blastema. Nephroblastoma in lower animals such as the rat generally consists of undifferentiated embryonal cells similar to the metanephric blastema, with partial organoid differentiation forming tubule- and glomerulus-like structures. In contrast, nephroblastoma in humans (Wilms' tumor) has a wider range of histologic characteristics including blastemal cells, epithelial structures, and neoplastic mesenchymal tissue.[38] Wilm's tumor in man is associated with a spectrum of hamartomatous and/or preneoplastic lesions (nodular blastema, metanephric hamartoma, etc.), sometimes termed the nephroblastomatosis complex, as well as congenital malformations of other organs.[39,40]

7.19.2.4 Mesenchymal Neoplasms

Primary benign or malignant neoplasms of mesenchymal tissue are generally uncommon in humans and other animals, including rodents. These tumors are believed to arise from secondary mesenchyme (adult connective and supporting tissues), except perhaps for those sarcomas associated with some cases of Wilm's tumor in humans which may arise from embryonal blastema (see above).

Lipomatous tumors of the rat occur infrequently and are usually seen in aged animals approximately 2 yr or older.[41] Since adipocytes are not normally present in the kidney, their origin and histogenesis is uncertain. It is speculated that they may arise from a medullary interstitial cell capable of synthesizing triacylglycerol. Benign tumors consisting of fully differentiated adipocytes have been termed lipoma, lipomatous hamartoma, and adipocyte metaplasia, while the malignant tumors are termed liposarcoma. Malignant potential seems to correlate somewhat with size, but metastases are generally uncommon. Whether lipomatous tumors of the rat have any similarities in histogenesis with medullary fibroma and/or angiomyolipoma of humans is uncertain.

7.19.2.5 Renal Mesenchymal Tumor of the Rat

As noted above, there is some confusion and controversy regarding the classification, particularly in the rat, of a neoplasm that has been termed "renal mesenchymal tumor."[41] While chemical-induced renal mesenchymal tumors of the rat have characteristics that distinguish them from nephroblastoma, it is less clear from the literature how spontaneously occurring tumors of similar morphology in the rat or other animal species differ from nephroblastoma. Sequential studies of dimethylnitrosamine induced renal mesenchymal tumors (rat) indicate an origin from interstitial mesenchymal cells, often near the corticomedullary junction.[3] As defined, these tumors consist of undifferentiated mesenchymal cells and varying proportions of differentiated or partially differentiated mesenchymal cell types (fibroblasts, osteoblasts, endothelial cells, smooth muscle, and skeletal muscle). Thus, the rat mesenchymal tumor may contain components resembling other malignant mesenchymal tumors including fibrosarcoma, hemangiosarcoma, osteosarcoma, leiomyosarcoma or rhabdomyosarcoma. Although the undifferentiated mesenchymal cells of the rat tumor may resemble the "blastemal cells" of nephroblastoma, there is no organoid differentiation or formation of primitive nephrons consisting of tubule- or glomerulus-like structures. Glomeruli, tubules, and urothelium seen within rat mesenchymal tumors are interpreted as pre-existing structures. Since the rat mesenchymal tumor is characterized principally by mesenchymal differentiation, it has been compared with the mesoblastic nephroma of childhood (human) which also exhibits a heterogenous population of neoplastic mesenchymal cells (fibroblasts and smooth muscle).[3,42]

7.19.3 MOLECULAR GENETICS

7.19.3.1 Renal Cell Carcinoma of Humans

Renal cell carcinoma in humans occurs in both sporadic and inherited forms, including von Hippel–Lindau (VHL) disease, "pure" familial renal cell carcinoma associated with constitutional (germline) translocations between chromosomes 3 and 6 [t(3;6)(p13;q25.1)] or between chromosomes 3 and 8 [t(3;8) (p14;q24)], hereditary papillary renal cell carcinoma, and possibly other familial types.[43–46] Renal cancer also occurs in patients with tuberous sclerosis, patients with autosomal dominant polycystic kidney disease, and patients with acquired cystic disease of chronic dialysis.[47–49]

The majority of all renal cell carcinomas occurring sporadically, in patients with VHL disease, or in patients with constitutional chromosome 3 translocations are clear cell tumors with a solid or "acinar" growth pattern. In kidneys of patients with VHL, there is a morphologic continuum from benign cysts with atypical cells, which occur in over half of VHL patients, to renal cell carcinoma.[50] Renal cell tumors associated with acquired cystic disease and autosomal dominant polycystic kidney disease also show a spectrum of hyperplastic and neoplastic lesions apparently originating from the proximal tubules, but the tumors tend to have a papillary architecture.[51]

7.19.3.1.1 Tumor suppressor genes

Advances in DNA analysis combined with high-resolution cytogenetic techniques are rapidly expanding our knowledge of the molecular and genetic basis of renal cell neoplasia. While only a limited number of renal cell neoplasms have been analyzed in depth, chromosomal and DNA alterations clearly distinguish subtypes of renal neoplasms.[52,53] A high proportion (>90% in some studies) of sporadic and inherited nonpapillary renal cell carcinomas (primarily clear cell type) are characterized by abnormalities of the short arm of chromosome 3 (3p).[54] In contrast, many papillary renal cell carcinomas exhibit trisomy of chromosome 17, without rearrangements in 3p.[55] Chromophobe renal cell carcinomas and oncocytomas similarly reveal distinct chromosomal alterations without 3p rearrangements. The molecular genetics of the latter three tumor types are discussed separately below.

Cytogenetic studies (karyotyping and chromosomal banding) of sporadic and VHL disease-associated nonpapillary clear cell tumors have revealed consistent loss of the 3p13-pter region, often as a result of unbalanced translocations between 3p13 and 5q22, interstitial or terminal deletions of 3p, or monosomy of chromosome 3.[56,57] In kindreds with constitutional (germline) translocations involving the short arm of chromosome 3 to chromosome 8 [t(3;8)(p14.24.13)] or 3 to 6 [t(3;6)(p13;q25.1)], the renal tumors were shown to retain the normal chromosome 3 homologue while the derivative chromosome 8 or 6 carrying the translocated 3p-pter segment was lost.[58] In another kindred without a constitutional translocation, renal tumors exhibited translocations of chromosome 3 to chromosome 11.[59] The breakpoints of translocations in these three kindreds cluster in the 3p13–p14 region. These chromosomal abnormalities suggested the existence of one or more tumor suppressor genes located on chromosome 3 distal to 3p13 in the development of renal cell carcinoma. According to current hypothesis, both alleles of a tumor suppressor gene must be inactivated before a tumor can develop. In the sporadic form, tumor initiation is believed to be the result of acquired (somatic) mutations involving the same suppressor gene locus on each of the two homologous chromosomes within the cell (e.g., one allele contributed by each parent). In the inherited forms, one mutation is inherited (germline) while the second or somatic mutation is acquired.

Restriction fragment length polymorphism (RFLP) analysis, conducted by hybridization of labeled oligonucleotide probes to DNA extracted from tumor and normal tissue, has been used to identify allelic loss (loss of heterozygosity) on chromosome 3p in renal cell carcinoma, and more precisely define the location of the putative tumor suppressor gene. Loss of alleles (terminal or interstitial deletions of nucleotide sequences) in the distal portion of 3p was demonstrated in most (up to 90%) sporadic nonpapillary clear cell carcinomas, as well as other histologic variants (e.g., granular, mixed cell, or sarcomatoid cell types). The minimal regions defined by commonly deleted loci extended between 3p13 and 3p14.3 (3p13–14.3), 3p21–24, 3p24–25, and 3p21.[60–64] Loss of 3p25–26 was often coincident with loss of one or more of the other 3p regions. These findings indicated that one or more critical genes were located in these regions. Loss of heterozygosity at 3p loci was not correlated with tumor stage, indicating the singular importance of allelic loss at the tumor suppressor gene(s) in early tumor development rather than tumor progression.

Through genetic linkage analysis, physical mapping, and positional DNA cloning strategies, the critical gene responsible for VHL disease has been localized to 3p25–26.[65–69] Analysis of blood or lymphoblastoid cell line DNA from patients with VHL demonstrated germline mutations in the *VHL* gene distributed among all three exons containing the reported open reading frame.[68,70,71] The mutations included point mutations (nucleotide transitions and transversions), deletions or insertions resulting in frameshifts of the coding sequence, splice-site mutations, and missense and nonsense mutations. As noted above, renal cell carcinomas from VHL patients have consistently demonstrated loss of the region 3p13-pter, including the region of the VHL gene, from a single copy of chromosome 3, similar to sporadic renal cell carcinomas.[57] Further, it has been shown that the 3p alleles lost were the wild-type (normal) alleles inherited from the non-*VHL* gene-carrying parent.[72] Consistent with these findings, the

spectrum of mutations identified in DNA extracted from VHL renal cell carcinomas were similar to the germline mutations. Together, these findings provide evidence that (i) the inherited predisposition to VHL is associated with germline mutations in a tumor suppressor gene at 3p25–26, and (ii) initiation of renal cell carcinoma is the result of inactivating somatic mutations and/or loss of the wild-type normal *VHL* gene allele.

Using PCR amplification of the three exons comprising the VHL cDNA, reverse trascriptase PCR with VHL gene-specific primers, single strand conformational polymorphism analysis, and nucleotide sequencing, somatic mutations in the VHL gene have also been identified in 57% (56 of 98) sporadic renal cell carcinomas in one study, 33% (10 of 30 tumors) in another, and in cell lines derived from sporadic renal cell carcinoms.[70,71] Due to the inherent sensitivity limits of these techniques, the actual proportion of sporadic renal cell carcinomas with mutations in the VHL gene may be much higher. The mutations included point mutations (nucleotide transitions and transversions) and deletions or insertions resulting in frameshifts of the coding sequence. Mutations in the *VHL* gene were not identified in over 180 carcinomas originating in other tissues including breast, pancreas, lung, colon, ovary, prostate, uterus, and bladder in one study, nor in 120 carcinomas in another study.[70,71] These findings indicate that somatic mutation with loss of heterozygosity at the VHL gene is important in the initiation of sporadic nonpapillary renal cell carinomas.

As noted above, cytogenetic studies have also implicated a number of other chromosomes and possibly tumor suppressor genes in the progression of renal cell carcinomas.[53,54] Loss of chromosome 6q, 8p, 9, 14q, and 17p sequences and trisomy of 5q sequences are frequently seen in nonpapillary renal cell carcinomas, while loss of the Y chromosome and trisomy of chromosomes 3q, 7, 8, 12, 16, 17, and 20 are observed in papillary renal cell tumors. Chromophobe cell carcinomas (collecting duct tumors) exhibit loss of chromosomes 1, 2, 6, 10, 13, 17, and 21. While it is not clear which of these chromosomal abnormalities may contribute to tumor progression, several of the frequently affected chromosomes are known to contain tumor suppressor genes (Table 2).

The *p53* tumor suppressor gene, which is localized to the short arm of chromosome 17 (17p13.1), is one of most commonly mutated genes in human cancer, and its potential involvement in renal cell cancer has been investigated more thoroughly than others.[73] Since mutation of one allele and loss of the

Table 2 Chromosomal location of tumor suppressor genes in humans.

Suppressor gene	Chromosome	Neoplasms
VHL	3p25	Renal cell carcinoma, pheochromocytoma
RCC*	3p14	Renal cell carcinoma
WT1	11p13	Wilms' tumor
TSC1	9q34.3	Tuberous sclerosis
TSC2	16p13.3	Tuberous sclerosis
TP53 (*p53*)	17p13.1	Many
APC/MCC	5q21–22	Colon, stomach, pancreatic cancer
DCC	18q21-qter	Colon cancer
BRCA1	17q21	Breast cancer
WT2	11p15.5	Beckwith–Wiedemann syndrome
NF1	17q11.2	Sarcoma, glioma
NF2	22q12	Schwannoma
RB1	13q14	Retinoblastoma, sarcoma
NB1	1p36	Neuroblastoma
MLM	9p21	Melanoma
BCNS	9q31	Medulloblastoma, skin carcinoma
MEN1	11q13	Pituitary, parathyroid, and islet cell tumor
nm23	17q21	Many
MTS1 (p16)	9q21	Many

wild-type (normal) allele is the most commonly observed alteration, renal cell tumors and tumor cell lines have been examined for loss of heterozygosity or allelic deletion at 17p13.1 or other closely associated loci. In studies by various investigators, loss of heterozygosity at 17p13.1, the *p53* locus, or 17p13.3 has been detected in 10–60% of informative nonpapillary renal cell carcinomas or cell lines evaluated.[74–76] In one of the more comprehensive evaluations of cell lines derived from advanced (high grade) carcinomas, loss of heterozygosity was observed in 14/29 (48%), *p53* mutations were seen in 11/33 (33%), and 6/9 with loss of heterozygosity also had mutations in the remaining allele.[75] Together, these findings suggest that *p53* is important in the progression (development of malignant characteristics) of nonpapillary renal cell carcinoma.

Loss of heterozygosity or allelic deletion has also been observed at or near loci of the *BRCA1*, *Rb*, *NF2*, *DCC*, *nm23-H1*, and *MTS1(p16)* genes in small numbers of nonpapillary and collecting duct (chromophobe) carcinomas, but the significance of these findings is unknown.[77–82]

7.19.3.1.2 Oncogenes and growth factors

Although the importance of tumor suppressor gene(s) on chromosome 3 has been clearly established for renal cell carcinoma, the relative importance of dominantly acting onco-genes, through gain of function mutations, is unknown. The protein products of oncogenes directly participate in cell growth, proliferation and differentiation as growth factors (TGF-α), growth factor receptors (*erb*B1), components of signal transduction (*ras*), transcription regulators (*fos*, *jun*), and cell cycle regulators.

Transforming growth factor-α (TGF-α), which binds to the epidermal growth factor (EGF) receptor (the *erb*B1 oncogene product), may support the growth of renal cell carcinomas through an autocrine mechanism involving activation of tyrosine kinase activity. Over-expression of TGF-α and the EGF receptor has been demonstrated in early and late stage carcinomas.[83–86] Moreover, high expression of the EGF receptor gene and low expression of HER2/*neu* (*erb*B2) gene were observed in grade 2 and 3 renal cell carcinomas, while the inverse was seen in normal kidney tissue (low EGF receptor gene and high HER2/*neu* gene).[87] In papillary renal cell tumors, there was no correlation of *erb*B genes (*erb*B1, *erb*B2, and *erb*B3) expression and trisomy of chromosomes 7, 12, and 17 (*erb*B genes are mapped to these chromosomes).[54]

The ras genes code for membrane-associated guanine nucleotide binding proteins important in signal transduction, and mutations or over-expression of these genes are associated with many tumors in humans and animals. However, in a survey of 41 primary renal cell tumors and 10 tumor cell lines, no mutations in K-*ras* or N-*ras* genes were seen, while mutations in

H-*ras* were detected in only 2%.[88] Over-expression of K-*ras* mRNA has been observed in some, but not all, renal cell carcinomas examined.[89] These findings suggest that *ras* oncogenes are not critical factors in the genesis of renal cell carcinoma.

The expression of oncogenes c-*myc* and c-*fos*, which code for nuclear transcription factors, were shown to correlate with malignancy grade of renal cell carcinomas as determined from histological characteristics.[87,90,91] The expression of interleukin-6 and basic fibroblastic growth factor in renal carcinomas was reported to show some correlation with poor survival.[92,93]

7.19.3.1.3 DNA methylation and renal cancer

There is a steadily growing body of evidence indicating that DNA methylation plays a fundamental, albeit not primary, role in regulating gene transcription, and further, that alterations in DNA methylation may be important events in the induction and/or progression of neoplasia, including renal cancer. Vertebrate DNA is normally highly methylated, but stably nonmethylated sequences occur. A fraction of nonmethylated DNA comprising approximately 1% of the genome is relatively high in guanine and cytosine content, is nonmethylated at CpG dinucleotide sequences in germ-line and adult tissues, and occurs as discrete islands (referred to as CpG islands) usually 1–2 kilobase pairs (kbp) long.[94,95] A large proportion of CpG islands are associated with genes (including "house-keeping" and tissue-specific genes) and encompass the 5′ ends where transcription begins, as well as their upstream promoter regions. Association with the 5′ domain and promoter region of genes suggests that CpG islands may be preferred sites of interaction between DNA and DNA-binding proteins (e.g., transcription factors).

Methylation of cytosine in CpG islands may render genes transcriptionally inactive, as it does on the inactive homologue of the X chromosome and by controlling differential expression of paternal and maternal alleles of "imprinted genes." Aberrant methylation may also increase the mutability of the affected genes.[96] Cytosine is prone to deamination, and when methylated gives rise to thymine. The latter can escape DNA repair mechanisms in *E. coli*, and possibly in the vertebrate genome as well. It is also speculated that methylation of CpG islands may result in replication delays and contribute to the formation of "fragile sites" which are breakpoints for chromosomal rearrangement and recombination.[96]

Aberrant methylation (hypermethylation) of CpG islands may contribute to the genesis and/or progression of renal cell neoplasia by transcriptional inactivation of tumor suppressor genes or by increasing the mutability of suppressor genes or oncogenes. Methylation of a normally nonmethylated CpG island in the 5′ region of the *VHL* gene was found in 5 of 26 (19%) spontaneous (nonfamilial) renal cell carcinomas examined.[97] Four of the carcinomas had lost one copy of *VHL* (LOH), while one had two methylated alleles. Moreover, four of the carcinomas also had no detectable *VHL* mutations, whereas one had a point (missense) mutation of the retained allele. None of the five carcinomas expressed the *VHL* gene. These findings suggest that a fraction of renal cell carcinomas are caused by methylation of the CpG island of the *VHL*-tumor suppressor gene resulting in transcriptional inactivation.

Other studies suggest that methylation of CpG islands on chromosome 17p, which contains the *p53* suppressor gene, precedes and may be the cause of allelic loss (LOH) and *p53* mutations in some renal carcinomas.[98] As noted above, there is evidence that 17p allele loss and *p53* mutations occur infrequently in early stage renal cell carcinomas, but may appear during progression in late-stage carcinomas. Hypermethylation at locus D17S5 (a marker locus on 17p) was observed in all late-stage renal cell carcinomas (11 total) with 17p allele loss, and 6 of the 11 carcinomas also had *p53* mutations.[98] Hypermethylation of D17S5 was also seen at high incidence in early-stage tumors (50%) which exhibited low incidences of 17p allelic loss (13%) and no *p53* mutations (0%). These findings suggest that hypermethylation of 17p loci preceeds and may contribute to both 17p allelic loss and *p53* gene mutation in renal cell carcinoma.

7.19.3.1.4 Defective DNA mismatch repair

Studies suggest that mutations in genes encoding DNA mismatch repair enzymes may play an important role in the progression of human renal cell carcinoma, as they do in hereditary nonpolyposis colon cancer and perhaps other cancers.[99–102] The DNA mismatch repair system ensures the precision and fidelity of chromosome replication and genetic recombination which are intrinsically imperfect.[103–105] Replication and recombination errors produce base-pairing anomalies within the DNA helix such as single-nucleotide insertions, transversions, and A → G transitions. These are recognized and excised by the mismatch repair system to prevent spontaneous mutations in mammalian cells.

Defects in DNA mismatch repair and the resultant genomic instability are detected by the examination of cellular DNA for instability of microsatelite repeats. Microsatellites are normal regions of the genome, generally considered noncoding, composed of one to six bases of repeated DNA sequence. While polymorphic between individuals, the pattern within an individual is set at birth and is the same in any issue from the individual. Alterations in the number of mono-, di-, and trinucleotide repeats comprising the "microsatellites" within tumor tissues of an individual are thought to be a manifestation of replication errors and increased mutation rate.[106]

In mammalian cells, microsatellite instability and defective DNA mismatch repair has been shown to be associated with four different inherited mutant genes, each homologous to bacterial mismatch DNA repair genes.[103] In hereditary nonpolyposis colon cancer, defects in DNA mismatch repair result in mutation rates that are more than 100 times that of normal human cells.[100] Genomic instability of microsatellite repeats were identified in 9/36 (25%) renal cell carcinomas, a rate similar to that (11.6–28%) reported in colorectal cancer.[99] Moreover, genomic instability was seen more frequently in high stage (pT₃) and/or high grade (grade 3) carcinomas than in low stage/grade tumors. Although microsatellite repeats are indicative of genomic instability and defective DNA repair, it is the subsequent mutations that occur in other genes (oncogenes, growth factors/receptor genes, suppressor genes, etc.) that are of singular importance in the process of carcinogenesis.

7.19.3.2   Papillary Renal Cell Carcinoma

Sporadic tumors with a papillary or tubulopapillary growth pattern comprise about 10% of renal cell neoplasms with a 6:1 to 8:1 ratio of men to women.[52] Renal tumors occurring in patients with autosomal dominant polycystic kidney disease or with acquired cystic disease of chronic dialysis are often papillary rather than clear cell types. Typically, the kidneys of affected patients often have multifocal or bilateral papillary tumors and frequent papillary tubular changes believed to be preneoplastic lesions. A hereditary form of papillary renal cell carcinoma with autosomal dominant inheritance has been recognized and described.[106]

The genetic changes seen in sporadic papillary tumors are distinct from those seen in nonpapillary renal cell carcinomas. Papillary tumors commonly exhibit loss of the Y chromosome and trisomies of chromosomes 3q, 7, 8, 12, 16, 17, and 20.[107] Small (2–5.5 mm diameter) benign tumors showed trisomy of chromosomes 7 and 17 and loss of the Y chromosome, while larger, usually malignant, tumors exhibited these changes and trisomy of 3q, 8, 12, 17, and 20. Rearrangement of 3p and trisomy of 5q22-qter, common chromosomal alterations found in nonpapillary renal cell carcinomas, were not seen. Another subset of sporadic papillary renal cell carcinomas was reported to have a translocation of chromosomes X and 1 [t(X;1)(p11.2;q21)] as a consistent feature.[108] The breakpoint Xp11.2 has been reported in synovial sarcomas, and a member of a helix-loop-helix family of transcription factors (TFE3) has been localized to Xp11.2.[109] This finding suggests a role for this transcription factor in some papillary renal cell carcinomas.

It has been reported that 12 sporadic papillary renal cell carcinomas did not exhibit loss of heterozygosity on 3p or mutations in the *VHL* gene.[69] In addition, the hereditary form with autosomal dominant inheritance was not linked to polymorphic markers on 3p and there was no loss of heterozygosity at loci on 3p in papillary renal tumors from affected family members.[105] Together, these cytogenetic and molecular studies indicate that tumor suppressor genes on 3p such as the *VHL* gene are not involved in the development of papillary renal cell tumors.

The evidence for involvement of the *p53* tumor suppressor gene in papillary renal cell carcinoma is not clear. In one series of papillary renal cell tumors, one allele of the *p53* gene was frequently duplicated (trisomy of chromosome 17) and the gene was overexpressed in nearly all the tumors evaluated.[54] However, analysis of *p53* exons 2–11 failed to detect mutations of *p53* gene. Since the authors did not report on the malignancy or tumor stage, these observations do not preclude a role for *p53* in the progression of papillary renal cell tumors.

7.19.3.3   Collecting Duct (Chromophobe Cell) Carcinoma of Humans

As noted previously, chromophobe renal cell carcinoma is a recognized variant comprising about 5% of renal cell tumors, and is believed to arise from cortical collecting tubules or ducts.[110] They apparently are characterized by a specific combination of chromosome loss including 1, 2, 6, 10, 13, 17, and 21, and gross rearrangement of mitochondrial DNA.[81] Small numbers of chromophobe renal cell carcinomas have been examined for allelic loss on chromosome 3p, but the results were inconsistent with only half of the informative tumors exhibiting loss of heterozygosity.[111] Further, no mutations in the VHL gene were seen in six tumors evaluated.[112] In a series of 19 chromophobe cell (collecting duct) carcinomas, loss of chromosome 17 was seen in 13/17 (76%) tumors and mutations of *p53* were seen in 20%.[80,112] These findings suggest that the molecular pathogenesis of chromophobe cell carcinomas differs from that of other renal cell carcinomas, perhaps reflecting its origin from collecting ducts instead of proximal tubules.

7.19.3.4   Oncocytoma of Humans

Approximately 5% of all human renal neoplasms are oncocytomas.[113,114] While most occur sporadically, there are rare occurrences of bilateral oncocytomas or families with multiple members affected.[115] The tumors are generally benign, but some have been reported to have malignant potential on the basis of DNA ploidy and karyotype and a few have metastasized.[116] In contrast to renal cell carcinomas, oncocytomas lack consistent cytogenetic or karyotypic changes and do not exhibit rearrangement of chromosome 3.[117] Molecular analysis using probes to detect polymorphic loci over the length of 3p also failed to show loss of alleles from the short arm.[118] Subsets of these tumors exhibit loss of the Y chromosome as well as chromosome 1, or a balanced translocation between 11q13 and other chromosomes.[52] Restriction analysis of mitochondrial DNA has shown an additional 40 bp fragment not seen in normal DNA or other renal tumors, which may contribute to the abnormal morphology of mitochondria in oncocytomas.[119]

7.19.3.5   Renal Cell Carcinoma of the Rat

7.19.3.5.1   *Tumor suppressor genes*

Aside from renal cell carcinoma of humans, little is known about the molecular genetics of this neoplasm in species other than the rat. Interestingly, there are intriguing similarities in the biology and molecular genetics of renal cell carcinoma between the rat and human. Renal cell carcinoma of the rat also occurs in both spontaneous and inherited forms. The hereditary form occurs in a line of Long-Evans rats (Eker rat) that carries a single gene mutation predisposing to multiple/bilateral renal cell carcinomas.[120,121] The mutation has an autosomal dominant pattern of inheritance with 100% penetrance. Heterozygous carriers

develop a spectrum of renal tubule lesions from atypical tubules to hyperplasia, adenoma and carcinoma, which appear in rats as early as 4 months of age. Tumors of the uterus and spleen also develop, but at lower penetrance. The homozygous state, however, results in embryonic death at day 9–10 of gestation, suggesting that the gene has a pleiotropic effect in development and carcinogenesis.

Through linkage analysis, the Eker gene has been localized to rat chromosome 10q12, thus eliminating several known or putative human suppressor genes which map to different chromosomes in the rat, for example, *VHL* (rat chromosome 4), *RB1* (rat chromosome 15), *WT1* (rat chromosome 3), and adenomatous polyposis coli gene (*APC*) (rat chromosome 18).[122] The Eker gene has failed to show linkage with homologous genes on human chromsome 3p, and the rat homolog of the *VHL* gene has been mapped to 4q42.[123] Further, analysis of eight cell lines derived from independent renal tumors from Eker rats failed to detect mutations in the *VHL* gene at 4q42.[124]

Rat chromosome band 10q12, where the gene predisposing the Eker rat to renal tumors is localized, was shown to be syntenic with human chromosome band 16p13.3, the site of the tuberous sclerosis 2 (*TSC2*) gene.[125] Patients with tuberous sclerosis are predisposed to multiple harmartomas and neoplasia including renal cell carcinoma. A specific rearrangement of the rat TSC2 gene (r*Tsc2*) was shown to cosegregate with heterozygous carriers of the Eker mutation, and give rise to an aberrant transcript that deletes the 3′ end of the mRNA containing the region of *rap1*GAP homology. The protein of *rap1* (p21^{rap1}) is an effector of the GTPase family of proteins involved in signal transduction pathways regulating cell growth and differentiation. Loss of heterozygosity (e.g., loss of the wild-type or normal allele) at the r*Tsc2* locus was detected in 72% of primary tumors or cell lines while the mutation (aberrant r*Tsc2* fragment) was retained in each, further evidence that r*Tsc2* functions as a classical tumor suppressor gene.[125] It was presumed that other subtle intragenic mutations were present in tumors that did not show loss of heterozygosity.

Although the *VHL* tumor suppressor gene seems to play a primary role in the development of both hereditary and spontaneous renal cell carcinomas in humans, it is not known if the r*Tsc2* gene of rats has a similar role in the development of spontaneous tumors.

Studies of radiation and chemical carcinogenesis in rats carrying the Eker mutation provide interesting insight into the process of renal carcinogenesis. Exposure of hetero-zygous carriers to sublethal ionizing radiation produces an 11- to 12-fold enhancement of tumor multiplicity with a linear dose–response relationship, while normal rats exhibited a quadratic dose–response relationship to radia-tion.[126] This suggests that in heterozygotes two events (one inherited, one somatic) are necessary to produce renal tumors, also consistent with the two-hit hypothesis and involvement of one or more tumor suppressor genes.

In another study, 6/10 (60%) of spontaneous renal tumors in hybrid F1 rats with the *Eker* mutation showed loss of the wild type (normal) allele, with individual tumors exhibiting differ-ent patterns of allelic loss even within the same kidney. The latter finding indicates independent clonal origin of the renal tumors. In contrast, none of the nine *N*-ethyl-*N*-nitrosourea induced tumors examined had allelic loss, suggesting that *N*-ethyl-*N*-nitrosourea produced intra-genic mutations in the normal allele.[127] Ethyl-nitrosourea is a well-characterized mutagen that acts by direct ethylation of DNA, and substitutions at A:T base pairs appear to con-stitute a high proportion of ethylation-induced mutations *in vivo*.

N-Ethyl-*N*-nitrosourea, given at day 15 of gestation to pregnant female rats lacking the Eker mutation, typically induce Wilms' tumors in the pups around 5 months of age rather than renal cell tumors. In this model of transplacen-tal carcinogenesis, the pups of Eker rats given *N*-ethyl-*N*-nitrosourea developed high inci-dences of renal cell tumors beginning as early as 1 week of age, and no increase in the incidences of Wilms' tumors.[128] These findings suggest that the r*Tsc2* gene may control terminal differentiation of renal tubule cells, while the Wilms' tumor gene acts at an earlier stage controlling the differentiation of meta-nephric stem cells. This is also supported by studies with dimethylnitrosamine, which typi-cally induces both renal cell and mesenchymal cell tumors in the rat. Dimethylnitrosamine treated rats carrying the Eker mutation have a 70-fold increase in susceptibility to the devel-opment of renal cell tumors as a result of the mutation.[129] In contrast, the Eker mutation imparted no increase in susceptibility to the development of renal mesenchymal tumors, suggesting that the r*Tsc2* gene does not play a significant role in the differentiation of renal stromal cells.

The *p53* tumor suppressor gene seems to have a limited role in the development of renal cell tumors in rats. No mutations in *p53* cDNA were identified in seven tumor-derived cell lines or in six spontaneous renal tumors from Eker rats, and mutations were seen in only 2 of 15

dimethylnitrosamine-induced tumors, one a $C \rightarrow T$ transition at codon 140 in an adenoma, and the other a $C \rightarrow A$ transversion at codon 281 in a transitional cell carcinoma.[130,131]

7.19.3.5.2 *Oncogenes and growth factors*

Like renal cell carcinomas in humans, dominantly acting oncogenes seem to have limited importance in the development of this neoplasm in the rat, although studies are extremely limited. Rat renal cell tumors do not exhibit a high frequency of ras oncogene activation, but overexpression of TGF-α and EGF receptor mRNA transcripts has been observed in cell lines derived from tumors in the Eker rat.[132,133] It is not known when dur-ing tumor development altered expression of TGF-α first occurs, or whether it is important in the initiation of neoplasia or progression.

7.19.3.6 Nephroblastoma (Wilms' Tumor) of Humans

Wilms' tumor is one of the most frequent childhood cancers, accounting for about 10% of pediatric malignancies.[134] It is believed to arise from the metanephric blastema *in utero*, with a peak incidence occurring in children 2–4 yr of age. Wilms' tumor occurs in both familial and sporadic forms with the latter comprising >90% of the cases.[135] The familial form is inherited as an autosomal dominant trait with varying penetrance. About 8% of Wilms' tumors are also associated with other develop-mental abnormalities in the WAGR complex (bilateral aniridia, defects in the genitourinary tract and mental retardation), the Denys–Drash syndrome (DDS; nephropathy and gona-dal dysgenesis) and Beckwith–Wiedemann syndrome (BWS; fetal overgrowth, hemihyper-trophy and embryonal tumors).[136,137]

Genetic studies have linked the development of Wilms' tumor to two distinct loci on chromo-some 11 (11p13 and 11p15), and 30–40% of Wilms' tumors exhibit loss of heterozygosity at either 11p13 or 11p15.[137–140] Pure familial Wilms' tumor (not associated with WAGR, DDS or HWS) apparently does not involve genes on chromosome 11, and there are con-flicting reports regarding an association with chromosome 16q where loss of heterozygosity has been detected. Other evidence suggests chromosome 7 may contain a suppressor gene predisposing to Wilms' tumor. A Wilms' tumor patient was shown to have a constitutional balanced translocation between chromosomes 1 and 7 [t(1;7)(q42;p15)], and cytogenetic evaluation of the tumor revealed an acquired (somatic) abnormality of the other chromosome 7.[141] In addition, a number of chromosome 7 abnormalities (trisomy, allelic loss, etc.) have been seen in Wilms' tumors from other patients. These observations reflect the genetic complexity of Wilms' tumor and suggest that more than one tumor suppressor gene and deregulation of other genes contribute to tumor pathogenesis.

Deletion and linkage analysis and positional DNA cloning led to the isolation of a gene (*WT1*) at 11p13 which has been confirmed as one of the principal genes predisposing to Wilms' tumor. Relatively large deletions at 11p13, visible by high resolution microscopy and cytogenetic methods, are found in only a small subset (1%) of Wilms' tumors, but smaller deletions encompassing the *WT1* gene are detectable in about 10% by Southern blotting. However, through DNA isolation and cloning of the *WT1* gene, point mutations and small intragenic deletions have also been described. Many of the mutations are either missense mutations within the zinc fingers of *WT1*, predicted to disrupt *WT1* DNA binding, or translational frameshifts resulting in production of truncated polypeptide. Since many of the alterations occur within the zinc finger domain, this region may be a "hotspot" for mutations. *WT1* mutations have also been identified in persistent nephrogenic rests.[142]

The *WT1* gene encompasses 10 exons spanning approximately 50 kb of DNA in 11p13 close to the aniridia gene (*AN2*, *PAX6*), and generates four transcripts through alternative splice mechanims.[143,144] The four splice variants are expressed in a ratio that is relatively constant in normal kidneys, Wilms' tumors and in other *WT1* expressing tissues during development. The *WT1* protein product appears to play a major role in the switch between proliferation and differentiation pathways during embryonal development of the urogenital system.

The expression pattern and *WT1* motifs are consistent with a tissue-specific transcription regulator and share homology with the early growth response genes *EGR-1* and *EGR-2*. The *WT1* product was shown to function as a transcriptional repressor on promoters containing the *EGR-1* site, and based on transcriptional repression by *WT1* of reporter constructs in transfection assays, it was proposed that the *WT1* gene product also acts as a transcriptional repressor for insulin-like growth factor 2 gene (*IFG-2*), insulin-like growth factor 1 receptor gene (*IGF-1R*), *Pax-2* and *-8*, and platelet

derived growth factor chain A (PDGF-A). Based on sequence analysis, insulin receptor, epidermal growth factor (EGF) receptor, *raf*, K-*ras*, and *myc* oncogenes may also be downstream target genes for *WT1*.

IGF-2 and PDGF-A are growth factors which stimulate proliferation of a variety of cell types. *IGF-2* is expressed in blastemal cells of fetal kidney, but expression is repressed when cells differentiate. *IGF-2* mRNA levels in Wilms' tumor are 10- to 60-fold greater than normal kidney tissue, suggesting that mutations in the *WT1* gene result in failure to repress transcription of *IGF-2*. Interestingly, the IGF-2 gene is located at 11p15, a region sometimes exhibiting loss of heterozygosity. *In vitro* data also suggest that the *WT1* gene represses expression of the *IGF-1R* promoter, and elevated expression of *IGF-1R* mRNA in Wilms' tumor correlated inversely with *WT1* expression. Like *IGF-2*, PDGF-A mRNA levels are elevated in Wilms' tumor, corresponding to levels of expression in fetal rather than adult tissue. *IGF-2*, *IGF-1R* and PDGF-A are capable of contributing to carcinogenesis through autocrine growth stimulation, and the increased expression of these genes in Wilms' tumor may reflect loss of functional *WT1* as a result of mutations. *WT1* may act antagonistically with *EGR1*, which activates expression of the growth promoting genes. Wilms' tumor is generally not associated with *p53* mutations, but a number of studies have demonstrated *p53* mutations in a high percentage of anaplastic variants, a subtype associated with poor prognosis.[145,146]

As early as 1972, a two-step mutation model was proposed to account for the incidence of sporadic and heritable forms.[147] It was proposed that sporadic unilateral Wilms' tumor was the result of two postzygotic (somatic) mutations, while the hereditary form was the result of a germ-line (constitutional) mutation and a somatic mutation resulting in functional loss of a tumor suppressor gene. Following the first mutation (constitutional or postzygotic), homozygosity at the gene can be achieved during mitotic recombination or loss and reduplication, or the second allele can be altered by point mutation or intragenic deletions. However, this model does not seem to fit all Wilms' tumor cases, in that *WT1* is mutated in only some cases with many of the tumors retaining at least one normal functional allele. This is supported by the observation that *WT1* mRNA is overexpressed in many Wilms' tumors. Even if *WT1* acts as a classical tumor suppressor in some instances, other tumor suppressor genes located at 11p15, 16q, or perhaps on chromosome 7 may also predispose to tumor development.

7.19.3.7 Nephroblastoma and Renal Mesenchymal Tumor of the Rat

7.19.3.7.1 *Tumor suppressor genes*

Investigation of the molecular pathogenesis of nephroblastoma in animals has been limited by the rarity of spontaneous occurrence and lack of a suitable model in experimental animals. Mutant $Sey^{Dey}/+$ heterozygous mice lack the region of chromosome 2 encompassing the mouse homologue of the aniridia (*AN2*) and *WT1* genes, but do not develop nephroblastomas.[148] Targeted inactivation of the mouse *WT1* locus causes death of mutant homozygous embryos with failure to develop kidneys and gonads, reflecting the importance of these genes in urogenital development. The absence of nephroblastoma in heterozygous $Sey^{Dey}/+$ mice is an indication of the genetic complexity of this tumor and possible requirement for inactivation of more than one suppressor gene. As mentioned above, a second tumor locus in human Wilms' tumor is located at 11p15 where a single chromosome nondisjunction or recombination could effect both genes. In mice, the chromosome region syntenic with human 11p15 is located on chromosome 7.[149]

In newborn rats, the administration of the alkylating agent, *N*-nitroso-*N'*-methylurea, produced embryonal kidney tumors after 4–8 months of which 55% were purely mesenchymal (rat mesenchymal tumor) and 30% were nephroblastomas.[150] Seven of 18 tumors (four nephroblastomas and three mesenchymal tumors) were shown to contain point mutations in *WT1* cDNA. Each of the four nephroblastomas had a T → A transversion in codon 111 predicted to be a Phe → Tyr substitution in the mutated *WT1* protein. CG → TA transitions at codons 128, 364, or 372 were observed in three mesenchymal tumors. Codons 111 and 128 lie in the regulatory domain of *WT1* that is required for transcriptional suppression of the PDGFA chain promoter. Of the 18 *N*-nitroso-*N'*-methylurea induced tumors, seven (including four nephroblastomas) exhibited two- to sevenfold elevations of *IGF-2* mRNA, and each of the nephroblastomas had elevated *WT1* mRNA.

The principal DNA alkylation sites affected by the *N*-nitroso-alkylating agents such as *N*-nitroso-*N'*-methylurea are the O^6 position of deoxyguanine and the O^4 and O^2 positions of deoxythymidine. The C → T transitions observed are consistent with the formation of O^6-methylguanine adducts by *N*-nitroso-*N'*-methylurea, while the T → A transversion may be related to O^2-deoxythymidine DNA adducts which are inefficiently repaired and persistent.[151]

7.19.3.7.2 *Oncogenes and growth factors*

Mutations in members of the *ras* gene family are frequently observed in spontaneous and chemically induced tumors in animals, including the kidney. Activating mutations in the ras genes commonly occur in either of two "hotspot" regions, codons 12 and 13 in exon 1 or codons 59–61 in exon 2. Point mutations resulting in GGT → GAT transition at codon 12 of the K-*ras* gene was observed in 90% of spontaneous and 100% of *N*-nitrosoethylurea induced nephroblastomas in rats.[152] Several authors have also reported K-*ras* mutations (G → A transition in codon 12) in 35–90% of rat mesenchymal tumors and 75% of renal cell tumors examined.[152–156] Transforming activity of K-*ras* sequences from these tumors have also been identified by the NIH 3T3 transfection assay.

Kidney tumors induced in rats by the alkylating *N*-nitroso compounds also exhibited a high incidence of G → A transition mutations in the *p53* gene, which were almost exclusively in codons 264 and 213 of exon 6.[155] Similar mutations in codons 264 and 213 were also seen in spontaneous nephroblastomas in WAB/Not rats, but at a lower frequency. However, none of 10 nickel subsulfide/iron induced rat renal mesenchymal tumors and only one of 10 methyl(methoxymethyl)nitrosamine induced tumors contained *p53* point mutations in the region of exons 4–10.[157] This finding suggests that mutated *p53* may not be essential to the development and/or progression of these neoplasms.

Rat renal mesenchymal tumors induced by dimethylnitrosamine were shown to have elevated levels of *fos* mRNA, but the overexpression was not found to arise by gene amplification or gene arrangement.[158] The *fos* gene product is a transcription factor that may be involved in traversal of $G_0 → G_1$ in the cell cycle. Deregulation of the *fos* gene in tumor cells may be important for maintaining the rate of cell division.

7.19.4 SUMMARY

Technical advances in molecular biology and methods of DNA analysis combined with mapping of the human and animal genomes are rapidly changing the nature of investigative carcinogenesis. Elucidating the molecular genetics of neoplasia in experimental animals must keep pace with that of humans if significant advancements are to be made in hazard identification and risk assessment. Tumor suppressor genes (*VHL* at 3p25–26 in humans; *rTSC2* at 10q12 in rats) rather than dominant acting oncogenes seem to play a principal role in the genesis of hereditary and/ or sporadic renal cell carcinoma. The *p53* suppressor gene may be involved in the progression, but not initiation, of this neoplasm. Mutations in genes encoded DNA mismatch repair enzymes and hypermethylation of normally nonmethylated CpG dinucleotide sequences also seem to be involved in the progression of human renal cell carcinoma. The *WT1* gene is one of the principal genes involved in the development of nephroblastoma (Wilms' tumor) in humans and rats. The protein product of *WT1* seems to act as a transcriptions repressor of a number of downstream genes, including insulin-like growth factor (IFG-1), insulin-like growth receptor (IGF-1R), PAx-2 and -8, platelet derived growth factor chain A (PDGF-A), *ras*, K-*ras*, and *myc*. The molecular genetics of other kidney neoplasms (collecting duct carcinoma, oncocytoma, transitional cell carcinoma, etc.) in humans or animals is not well understood.

7.19.5 REFERENCES

1. M. Guerin, I. Chouroulinkov and M. R. Riviere, in 'The Kidney: Morphology, Biochemistry and Physiology,' eds. C. Rouiller and A. F. Mueller, Academic Press, New York, 1969, pp. 199–268.
2. J. M. Hamilton, 'Renal carcinogenesis.' *Adv. Cancer Res.*, 1975, **22**, 1–56.
3. G. C. Hard, 'Experimental models for the sequential analysis of chemically-induced renal carcinogenesis.' *Toxicol. Pathol.*, 1986, **14**, 112–122.
4. Y. Hiasa and N. Ito, 'Experimental induction of renal tumors.' *Crit. Rev. Toxicol.*, 1987, **17**, 279–343.
5. M. M. Lipsky, Jr. and B. F. Trump, 'Chemically induced renal epithelial neoplasia in experimental animals.' *Int. Rev. Exp. Pathol.*, 1988, **30**, 357–383.
6. P. Bannasch and H. Zerban, in 'Contemporary Issues of Surgical Pathology,' ed. J. N. Eble, Churchill Livingstone, New York, 1990, pp. 1–34.
7. D. Dietrich and J. A. Swenberg, in 'Toxicology of the Kidney,' eds. J. B. Hook and R. S. Goldstein, Raven Press, New York, 1993, pp. 495–537.
8. N. Konishi and Y. Hiasa, in 'Carcinogenesis,' eds. M. P. Waalkes and J. M. Ward, Raven Press, New York, 1994, pp. 123–159.
9. J. R. Idle, 'Is environmental carcinogenesis modulated by host polymorphism?' *Mutat. Res.*, 1991, **247**, 259–266.
10. D. W. Nebert, D. R. Nelson, M. J. Coon *et al.*, 'The P450 superfamily: update on new sequences, gene mapping and recommended nomenclature.' *DNA Cell Biol.*, 1991, **10**, 1–14.
11. R. M. Evans, 'The steroid and thyroid hormone receptor superfamily.' *Science*, 1988, **240**, 889–895.
12. J. B. Storer, T. J. Mitchell and R. J. M. Fry, 'Extrapolation of the relative risk of radiogenic neoplasms across mouse strains and to man.' *Radiat. Res.*, 1988, **114**, 331–353.
13. R. W. Tennant, 'Stratification of rodent carcinogenicity bioassay results to reflect relative human hazard.' *Mutat. Res.*, 1993, **286**, 111–118.
14. G. C. Hard, C. L. Alden, E. F. Stula *et al.*, 'Proliferative Lesions of the Kidney of Rats,' Society of Toxicologic Pathologists. Standard System of Nomenclature and Diagnostic Pathology, 1993.
15. F. K. Mostofi, I. A. Sesterhenn and L. H. Lobin (eds.), 'Histological Typing of Kidney Tumours. International Histological Classification of Tumours,' World Health Organization, Geneva, 1981, vol. 25.
16. W. Thoenes, S. Storkel and H. J. Rumpelt, 'Histopathology and classification of renal cell tumors (adenomas, oncocytomas and carcinomas). The basic cytological and histopathological elements and their use in diagnostics.' *Pathol. Res. Pract.*, 1986, **181**, 125–143.
17. G. C. Hard and W. H. Butler, 'Morphogenesis of epithelial neoplasms induced in the rat kidney by dimethylnitrosamine.' *Cancer Res.*, 1971, **31**, 1496–1505.
18. J. H. Dees, B. M. Heatfield, M. D. Reuber *et al.*, 'Adenocarcinomas of the kidney. III. Histogenesis of renal adenocarcinomas induced in rats by N-(4'-fluoro-4-biphenylyl)acetamide.' *J. Natl. Cancer Inst.*, 1980, **64**, 1537–1545.
19. P. Bannasch, R. Krech and H. Zerban, 'Morphogenese und Mikromorphologie epithelialer Nierentumoren bei Nitrosomorpholin-vergifteten Ratten: IV. Tubularen Lasionen und basophile Tumoren.' [Morphogenesis and micromorphology of epithelial tumors induced in the rat kidney by nitrosomorpholine. IV. Tubular lesions and basophilic tumors.] *J. Cancer Res. Clin. Oncol.*, 1980, **98**, 243–265.
20. E. Nogueira, 'Rat renal carcinogenesis after simultaneous exposure to lead acetate and N-nitrosodiethylamine.' *Virchows Arch. B Cell Pathol. Incl. Mol. Pathol.*, 1987, **53**, 365–374.
21. C. M. Oberling, M. Riviere and F. Haguenau, 'Ultrastructure of clear cells in renal carcinoma and its importance for the demonstration of their renal origin.' *Nature*, 1960, **186**, 402.
22. E. R. Fisher and B. Horvat, 'Comparative ultrastructural study of so-called renal adenoma and carcinoma.' *J. Urol.*, 1972, **108**, 382–386.
23. R. Seljelid and J. L. E. Ericsson, 'Electron microscopic observations on specializations of the cell surface in renal clear cell carcinoma.' *Lab. Invest.*, 1965, **14**, 435.
24. H. Braunstein and J. U. Adelman, 'Histochemical study of the enzymatic activity of human neoplasms. II. Histogenesis of renal cell carcinoma.' *Cancer*, 1966, **19**, 935–938.
25. A. C. Wallace and R. C. Nairn, 'Renal tubular antigens in kidney tumors.' *Cancer*, 1972, **29**, 977–981.
26. F. F. L. Gu, S. L. Cai, B. J. Cai *et al.*, 'Cellular origin of renal cell carcinoma: an immunohistochemical study on monoclonal antibodies.' *Scand. J. Urol. Nephrol. Suppl.*, 1991, **138**, 203–206.
27. C. Cohen, P. A. McCue and P. B. Derose, 'Histogenesis of renal cell carcinoma and renal oncocytoma. An immunohistochemical study' *Cancer*, 1988, **62**, 1946–1951.
28. S. Storkel, P. V. Steart, D. Drenckhahn *et al.*, 'The human chromophobe cell renal cell carcinoma: its probable relation to intercalated cells of the collecting duct.' *Virchows Arch. B Cell Pathol. Incl. Mol. Pathol.*, 1988, **56**, 237–245.
29. E. Nogueira and P. Bannasch, 'Oncocytic transformation of the rat renal collecting duct epithelium by

various carcinogens.' *J. Cancer Res. Clin. Oncol.*, 1987, **113**, 19.

30. E. Nogueira and P. Bannasch, 'Cellular origin of rat renal cell oncocytoma.' *Lab. Invest.*, 1988, **59**, 337–343.

31. H. Zerban, E. Nogueira, G. Riedasch *et al.*, 'Renal oncocytoma: origin from the collecting duct.' *Virchows Arch. B Cell Pathol. Incl. Mol. Pathol.*, 1987, **52**, 375–387.

32. M. Ortmann, M. Vierbuchen, G. Koller *et al.*, 'Renal oncocytoma. I. Cytochrome *c* oxidase in normal and neoplastic renal tissue as detected by immunohistochemistry—a valuable aid to distinguish oncocytomas from renal cell carcinomas.' *Virchows Arch. B Cell Pathol. Incl. Mol. Pathol.*, 1988, **56**, 165–173.

33. S. Storkel, B. Pannen, W. Thoenes *et al.*, 'Intercalated cells as a probable source for the development of renal oncocytoma.' *Virchows Arch. B Cell Pathol. Incl. Mol. Pathol.*, 1988, **56**, 185–189.

34. E. Nogueira, F. Klimek, E. Weber *et al.*, 'Collecting duct origin of rat renal clear cell tumors.' *Virchows Arch. B Cell Pathol. Incl. Mol. Pathol.*, 1989, **57**, 275–283.

35. E. Nogueira, A. Cardesa and U. Mohr, 'Experimental models of kidney tumors.' *J. Cancer Res. Clin. Oncol.*, 1993, **119**, 190–198.

36. G. C. Hard, in 'Wilms' Tumor, Clinical and Biological Manifestations,' eds. C. Pochedly and E. S. Baum, Elsevier, New York, 1984, pp. 147–167.

37. G. C. Hard, in 'Wilms' Tumor, Clinical and Biological Manifestations,' eds. C. Pochedly and E. S. Baum, Elsevier, New York, 1984, pp. 169–189.

38. J. L. Bennington and J. R. Beckwith, 'Atlas of Tumor Pathology: Tumors of the Kidney, Renal Pelvis and Ureter,' Armed Forces Institute of Pathology, Bethesda, MD, 1975.

39. K. E. Bove and A. J. McAdams, 'The nephroblastomatosis complex and its relationship to Wilm's tumor: a clinicopathological treatise.' *Perspect. Pediatr. Pathol.*, 1976, **3**, 185–223.

40. R. W. Miller, J. F. Fraumeni and M. D. Manning, 'Association of Wilm's tumor with aniridia, hemihypertrophy and other congenital malformations.' *N. Engl. J. Med.*, 1964, **270**, 922.

41. G. C. Hard, in 'Pathology of Tumors of Laboratory Animals,' eds. V. S. Turusov and U. Mohr, International Agency for Research on Cancer, Lyon, 1990, pp. 301–324.

42. R. P. Bolande, A. J. Brough and R. J. Izant, Jr., 'Congenital mesoblastic nephroma of infancy. A report of eight cases and their relationship to Wilms' tumor.' *Pediatrics*, 1967, **40**, 272–278.

43. G. M. Glenn, P. L. Choyke, B. Zbar *et al.*, 'von Hippel–Lindau disease: clinical review and molecular genetics.' *Problems Urol.*, 1990, **4**, 312.

44. A. J. Cohen, F. P. Li, S. Berg *et al.*, 'Hereditary renal-cell carcinoma associated with a chromosomal translocation.' *N. Engl. J. Med.*, 1979, **301**, 592–595.

45. F. P. Li, H. J. Decker, B. Zbar *et al.*, 'Clinical and genetic studies of renal cell carcinomas in a family with a constitutional chromosome 3;8 translocation. Genetics of familial renal carcinoma.' *Ann. Intern. Med.*, 1993, **118**, 106–111.

46. G. Kovacs, P. Brusa and W. De Riese, ' Tissue-specific expression of a constitutional 3;6 translocation: development of multiple bilateral renal-cell carcinomas.' *Intern. J. Cancer*, 1989, **43**, 422–427.

47. J. Bernstein and T. O. Robbins, 'Renal involvement in tuberous sclerosis.' *Ann. NY Acad. Sci.*, 1991, **615**, 36–49.

48. J. R. Gregoire, V. E. Torres, K. E. Holley *et al.*, 'Renal epithelial hyperplastic and neoplastic proliferation in autosomal dominant polycystic kidney disease.' *Am. J. Kidney Dis.*, 1987, **9**, 27–38.

49. P. N. Bretan, Jr., M. P. Busch, H. Hricak *et al.*, 'Chronic renal failure: a significant risk factor in the development of acquired renal cysts and renal cell carcinoma. Case reports and review of the literature.' *Cancer*, 1986, **57**, 1871–1879.

50. D. Solomon and A. Schwartz, 'Renal pathology in von Hippel–Lindau disease.' *Hum. Pathol.*, 1988, **19**, 1072–1079.

51. J. J. Grantham and E. Levine, 'Acquired cystic kidney: replacing one kidney disease with another.' *Kidney Int.*, 1985, **28**, 99–105.

52. G. Kovacs, 'Molecular differential pathology of renal cell tumours.' *Histopathology.*, 1993, **22**, 1–8.

53. G. Kovacs, 'Molecular cytogenetics of renal cell tumors.' *Adv. Cancer Res.*, 1993, **62**, 89–124.

54. G. Kovacs, 'The value of molecular genetic analysis in the diagnosis and prognosis of renal cell tumors.' *World J. Urol.*, 1994, **12**, 64–68.

55. G. Kovacs, 'Papillary renal cell carcinoma. A morphological and cytogenetic study of 11 cases.' *Am. J. Pathol.*, 1989, **134**, 27–34.

56. G. Kovacs and S. Frisch, 'Clonal chromosomal abnormalities in tumor cells from patients with sporadic renal cell carcinomas.' *Cancer Res.*, 1989, **49**, 651–659.

57. G. Kovacs and H. F. Kung, 'Nonhomologous chromatid exchange in hereditary and sporadic renal cell carcinomas.' *Proc. Natl. Acad. Sci. USA*, 1991, **88**, 194–198.

58. S. Pathak, L. C. Strong, R. E. Ferrell *et al.*, 'Familial renal cell carcinoma with a 3;11 chromosome translocation limited to tumor cells.' *Science*, 1982, **217**, 939–941.

59. P. Anglard, K. Tory, H. Brauch *et al.*, 'Molecular analysis of genetic changes in the origin and development of renal cell carcinoma.' *Cancer Res.*, 1991, **51**, 1071–1077.

60. P. Anglard, E. Trahan, S. Liu *et al.*, 'Molecular and cellular characterization of human renal cell carcinoma cell lines.' *Cancer Res.*, 1992, **52**, 348–356.

61. K. Yamakawa, R. Morita, E. Takahashi *et al.*, 'A detailed deletion mapping of the short arm of chromosome 3 in sporadic renal cell carcinoma.' *Cancer Res.*, 1991, **51**, 4707–4711.

62. A. H. van der Hout, P. van der Vlies, C. Wijmenga *et al.*, 'The region of common allelic losses in sporadic renal cell carcinoma is bordered by the loci D3S2 and THRB.' *Genomics*, 1991, **11**, 537–542.

63. H. Brauch, S. Pomer, T. Hieronymus *et al.*, 'Genetic alterations in sporadic renal-cell carcinoma: molecular analyses of tumor suppressor gene harboring chromosomal regions 3p, 5q and 17p.' *World J. Urol.*, 1994, **12**, 162–168.

64. B. R. Seizinger, G. A. Rouleau, L. J. Ozelius *et al.*, 'von Hippel–Lindau disease maps to the region of chromosome 3 associated with renal cell carcinoma.' *Nature*, 1988, **332**, 268–269.

65. B. R. Seizinger, D. I. Smith, M. R. Filling-Katz *et al.*, , 'Genetic flanking markers refine diagnostic criteria and provide insights into the genetics of von Hippel–Lindau disease.' *Proc. Natl. Acad. Sci. USA*, 1991, **88**, 2864–2868.

66. S. Hosoe, H. Brauch, F. Latif *et al.*, 'Localization of the von Hippel–Lindau disease gene to a small region of chromosome 3.' *Genomics*, 1990, **8**, 634–640.

67. F. Latif, K. Tory, J. Gnarra *et al.*, 'Identification of the von Hippel–Lindau disease tumor suppressor gene.' *Science*, 1993, **260**, 1317–1320.

68. F. M. Richards, M. E. Phipps, F. Latif *et al.*, 'Mapping the von Hippel–Lindau disease tumour suppressor gene: identification of germline deletions

by pulsed field gel electrophoresis.' *Hum. Mol. Genet.*, 1993, **2**, 879–882.

69. J. R. Gnarra, K. Tory, Y. Weng *et al.*, 'Mutations of the *VHL* tumour suppressor gene in renal carcinoma.' *Nat. Genet.*, 1994, **7**, 85–90.

70. J. M. Whaley, J. Naglich, L. Gelbert *et al.*, 'Germ-line mutations in the von Hippel–Lindau tumor-suppressor gene are similar to somatic von Hippel–Lindau aberrations in sporadic renal cell carcinoma.' *Am. J. Hum. Genet.*, 1994, **55**, 1092–1102.

71. K. Tory, H. Brauch, M. Linehan *et al.*, 'Specific genetic change in tumors associated with von Hippel–Lindau disease.' *J. Natl. Cancer Inst.*, 1989, **81**, 1097–1101.

72. A. J. Levine, J. Momand and C. A. Finlay, 'The *p53* tumour suppressor gene.' *Nature*, 1991, **351**, 453–456.

73. J. C. Presti, Jr., P. H. Rao, Q. Chen *et al.*, 'Histopathological, cytogenetic and molecular characterization of renal cortical tumors.' *Cancer Res.*, 1991, **51**, 1544–1552.

74. R. E. Reiter, P. Anglard, S. Liu *et al.*, 'Chromosome 17p deletions and *p53* mutations in renal cell carcinoma.' *Cancer Res.*, 1993, **53**, 3092–3097.

75. K. Foster, P. A. Crossey, P. Cairns *et al.*, 'Molecular genetic investigation of sporadic renal cell carcinoma: analysis of allele loss on chromosomes 3p, 5q, 11p, 17 and 22.' *Br. J. Cancer*, 1994, **69**, 230–234.

76. J. Ishikawa, H. J. Xu, S. X. Hu *et al.*, 'Inactivation of the retinoblastoma gene in human bladder and renal cell carcinomas.' *Cancer Res.*, 1991, **51**, 5736–5743.

77. A. Leone, O. W. McBride, A. Weston *et al.*, 'Somatic allelic deletion of *nm23* in human cancer.' *Cancer Res.*, 1991, **51**, 2490–2493.

78. J. D. Brooks, G. S. Bova, F. F. Marshall *et al.*, 'Tumor suppressor gene allelic loss in human renal cancers.' *J. Urol.*, 1993, **150**, 1278–1283.

79. M. R. Speicher, B. Schoell, S. du Manoir *et al.*, 'Specific loss of chromosomes 1, 2, 6, 10, 13, 17, and 21 in chromophobe renal cell carcinomas revealed by comparative genomic hybridization.' *Am. J. Pathol.*, 1994, **145**, 356–364.

80. P. Cairns, L. Mao, A. Merlo *et al.*, 'Rates of *p16* (*MTS1*) mutations in primary tumors with 9p loss.' *Science*, 1994, **265**, 415–417.

81. U. Bergerheim, M. Nordenskjold, and V. P. Collins, 'Deletion mapping in human renal cell carcinoma.' *Cancer Res.*, 1989, **49**, 1390–1396.

82. P. E. Petrides, S. Bock, J. Bovens *et al.*, 'Modulation of pro-epidermal growth factor, pro-transforming growth factor-α and epidermal growth factor receptor gene expression in human renal carcinomas.' *Cancer Res.*, 1990, **50**, 3934–3939.

83. L. G. Gomella, E. R. Sargent, T. P. Wade *et al.*, 'Expression of transforming growth factor-α in normal human adult kidney and enhanced expression of transforming growth factors α and β1 in renal cell carcinoma.' *Cancer Res.*, 1989, **49**, 6972–6975.

84. L. G. Gomella, P. Anglard, E. R. Sargent *et al.*, 'Epidermal growth factor receptor gene analysis in renal cell carcinoma.' *J. Urol.*, 1990, **143**, 191–193.

85. J. H. Mydlo, J. Michaeli, C. Cordon-Cardo *et al.*, 'Expression of transforming growth factor α and epidermal growth factor receptor messenger RNA in neoplastic and nonneoplastic human kidney tissue.' *Cancer Res.*, 1989, **49**, 3407–3411.

86. U. Weidner, S. Peter, T. Strohmeyer *et al.*, 'Inverse relationship of epidermal growth factor receptor and *HER2/neu* gene expression in human renal cell carcinoma.' *Cancer Res.*, 1990, **50**, 4504–4509.

87. D. M. Nanus, I. R. Mentle, R. J. Motzer *et al.*, 'Infrequent *ras* oncogene point mutations in renal cell carcinoma.' *J. Urol.*, 1990, **143**, 175–178.

88. D. J. Slamon, J. B. deKernion, I. M. Verma *et al.*,

'Expression of cellular oncogenes in human malignancies.' *Science*, 1984, **224**, 256–262.

89. M. Yao, T. Shuin, H. Misaki *et al.*, 'Enhanced expression of c-*myc* and epidermal growth factor receptor (c-*erb*B-1) genes in primary human renal cancer.' *Cancer Res.*, 1988, **48**, 6753–6757.

90. T. Kinouchi, S. Saiki, T. Naoe *et al.*, 'Correlation of c-*myc* expression with nuclear pleomorphism in human renal cell carcinoma.' *Cancer Res.*, 1989, **49**, 3627–3630.

91. D. M. Nanus, B. J. Schmitz-Drager, R. J. Motzer *et al.*, 'Expression of basic fibroblast growth factor in primary human renal tumor: correlation with poor survival.' *J. Natl. Cancer Inst.*, 1993, **85**, 1597–1599.

92. J. Takenawa, Y. Kaneko, M. Fukumota *et al.*, 'Enhanced expression of interleukin-6 in primary human renal cell carcinomas.' *J. Natl. Cancer Inst.*, 1991, **83**, 1668–1672.

93. A. Bird, 'The essentials of DNA methylation.' *Cell*, 1992, **70**, 5–8.

94. A. P. Bird, 'CpG islands as gene markers in the vertebrate nucleus.' *Trends Genet.*, 1987, **3**, 342.

95. C. Laird, E. Jaffe, G. Karpen *et al.*, 'Fragile sites in human chromosomes as regions of late-replicating DNA.' *Trends Genet.*, 1987, **3**, 274.

96. J. G. Herman, F. Latif, Y. Weng *et al.*, 'Silencing of the *VHL* tumor-suppressor gene by DNA methylation in renal carcinoma.' *Proc. Natl. Acad. Sci. USA*, 1994, **91**, 9700–9704.

97. M. Makos, B. D. Nelkin, R. E. Reiter *et al.*, 'Regional DNA hypermethylation at D17S5 precedes 17p structural changes in the progression of renal tumors.' *Cancer Res.*, 1993, **53**, 2719–2722.

98. T. Uchida, C. Wada, C. Wang *et al.*, 'Genomic instability of microsatellite repeats and mutations of H-, K-, and N-*ras*, and *p53* genes in renal cell carcinoma.' *Cancer Res.*, 1994, **54**, 3682–3685.

99. L. A. Aaltonen, P. Peltomaki, F. S. Leach *et al.*, 'Clues to the pathogenesis of familial colorectal cancer.' *Science*, 1993, **260**, 812–816.

100. M. Koi, A. Umar, D. P. Chauhan *et al.*, 'Human chromosome 3 corrects mismatch repair deficiency and microsatellite instability and reduces *N*-methyl-*N'*-nitro-*N*-nitrosoguanidine tolerance in colon tumor cells with homozygous *hMLH1* mutation.' *Cancer Res.*, 1994, **54**, 4308–4312.

101. J. R. Eshleman and S. D. Markowitz, 'Microsatellite instability in inherited and sporadic neoplasms.' *Curr. Opin. Oncol.*, 1995, **7**, 83–89.

102. P. Modrich, 'Mechanisms and biological effects of mismatch repair.' *Annu. Rev. Genet.*, 1991, **25**, 229–253.

103. P. Modrich, 'Mismatch repair, genetic stability and cancer.' *Science*, 1994, **266**, 1959–1960.

104. P. Modrich, 'Mismatch repair, genetic stability and tumor avoidance.' *Philos. Trans. R. Soc. Lond. B Biol. Sci.*, 1995, **347**, 89–95.

105. R. Parsons, G. M. Li, M. J. Longley *et al.*, 'Hypermutability and mismatch repair deficiency in RER+ tumor cells.' *Cell*, 1993, **75**, 1227–1236.

106. B. Zbar, K. Tory, M. Merino *et al.*, 'Hereditary papillary renal cell carcinoma.' *J. Urol.*, 1994, **151**, 561–566.

107. G. Kovacs, L. Fuzesi, A. Emanuel *et al.*, 'Cytogenetics of papillary renal cell tumors.' *Genes Chromosomes Cancer*, 1991, **3**, 249–255.

108. A. M. Meloni, R. M. Dobbs, J. E. Pontes *et al.*, 'Translocation (X;1) in papillary renal cell carcinoma. A new cytogenetic subtype.' *Cancer Genet. Cytogenet.*, 1993, **65**, 1–6.

109. P. S. Henthorn, C. C. Stewart, T. Kadesch *et al.*, 'The gene encoding human TFE3, a transcription factor that binds the immunoglobulin heavy-chain enhancer, maps to Xp11.22.' *Genomics*, 1986, **11**, 374–378.

110. W. Thoenes, S. Storkel and H. J. Rumpelt, 'Human chromophobe cell carcinoma.' *Virchows Arch. B Cell Pathol. Incl. Mol. Pathol.*, 1985, **48**, 207–217.

111. A. Kovacs, S. Storkel, W. Thoenes *et al.*, 'Mitochondrial and chromosomal DNA alterations in human chromophobe renal cell carcinomas.' *J. Pathol.*, 1992, **167**, 273–277.

112. T. Shuin, K. Kondo, S. Torigoe *et al.*, 'Frequent somatic mutations and loss of heterozygosity of the von Hippel–Lindau tumor suppressor gene in primary human renal cell carcinomas.' *Cancer Res.*, 1994, **54**, 2852–2855.

113. M. J. Klein and Q. J. Valensi, 'Proximal tubular adenomas of kidney with so-called oncocytic features: a clinicopathologic study of 13 cases of a rarely resported neoplasm.' *Cancer*, 1976, **38**, 906–914.

114. G. M. Farrow, in 'Urological Pathology.' ed. W. M. Murphy, W. B. Saunders, Philadelphia, PA, pp. 445–451.

115. T. N. Fairchild, D. H. Dail and G. E. Brannen, 'Renal oncocytoma-bilateral, multifocal.' *Urology*, 1983, **22**, 355–359.

116. M. M. Lieber, K. M. Tomera and G. M. Farrow, 'Renal oncocytoma.' *J. Urol.*, 1981, **125**, 481–485.

117. T. B. Crotty, K. M. Lawrence, C. A. Moertel *et al.*, 'Cytogenetic analysis of six renal oncocytomas and a chromophobe cell renal carcinoma. Evidence that -Y, -1 may be a characteristic anomaly in renal oncocytomas.' *Cancer Genet. Cytogenet.*, 1992, **61**, 61–66.

118. H. Brauch, K. Tory, W. M. Linehan *et al.*, 'Molecular analysis of the short arm of chromosome 3 in five renal oncocytomas.' *J. Urol.*, 1990, **143**, 622–624.

119. G. Kovacs, C. Welter, L. Wilkens *et al.*, 'Renal oncocytoma. A phenotypic and genotypic entity of renal parenchymal tumors.' *Am. J. Pathol.*, 1989, **134**, 967–971.

120. R. Eker and J. Mossige, 'A dominant gene for renal adenomas in the rat.' *Nature*, 1961, **189**, 858.

121. J. I. Everitt, T. L. Goldsworthy, D. C. Wolf *et al.*, 'Hereditary renal cell carcinoma in the Eker rat: a rodent familial cancer syndrome.' *J. Urol.*, 1992, **148**, 1932–1936.

122. R. S. Yeung, K. H. Buetow, J. R. Testa *et al.*, 'Susceptibility to renal carcinoma in the Eker rat involves a tumor suppressor gene on chromosome 10.' *Proc. Natl. Acad. Sci. USA*, 1993, **90**, 8038–8042.

123. C. M. Aldaz, R. S. Yeung, F. Latif *et al.*, 'Colocalization of the rat homolog of the von Hippel–Lindau (*Vhl*) gene and the plasma membrane Ca^{2+} transporting ATPase isoform 2 (*Atp2b2*) gene to rat chromosome bands $4q41.3 \rightarrow 42.1$.' *Cytogenet. Cell Genet.*, 1995, **71**, 253–256.

124. C. Walker, Y.-T. Ahn, J. Everitt *et al.*, 'Renal cell carcinoma development in the rat independent of alterations at the *VHL* gene locus.' *Mol. Carcinog.*, 1996, **15**, 154–161.

125. R. S. Yeung, G.-H. Xiao, F. Jin *et al.*, 'Predisposition to renal carcinoma in the Eker rat is determined by germ-line mutation of the tuberous sclerosis 2 (*TSC2*) gene.' *Proc. Natl. Acad. Sci. USA*, 1994, **91**, 11413–11416.

126. O. Hino, A. J. P. Klein-Szanto, J. J. Freed *et al.*, 'Spontaneous and radiation-induced renal tumors in the Eker rat model of dominantly inherited cancer.' *Proc. Natl. Acad. Sci. USA*, 1993, **90**, 327–331.

127. Y. Kubo, H. Mitani and O. Hino, 'Allelic loss at the predisposing gene locus in spontaneous and chemically induced renal cell carinoma in the Eker rat.' *Cancer Res.*, 1994, **54**, 2633–2635.

128. O. Hino, H. Mitani and A. G. Knudson, 'Genetic predisposition to transplacentally induced renal cell carcinomas in the Eker rat.' *Cancer Res.*, 1993, **53**, 5856–5858.

129. C. Walker, T. L. Goldsworthy, D. C. Wolf *et al.*, 'Predisposition to renal cell carcinoma due to alteration of a cancer susceptibility gene.' *Science*, 1992, **255**, 1693–1695.

130. L. Recio, J. J. Freed, A. G. Knudson *et al.*, 'Analysis of tumor suppressor genes in hereditary rat renal cell carcinoma.' *Proc. Am. Assoc. Cancer Res.*, 1992, **34**, A2311.

131. G. Horesovsky, L. Recio, J. Everitt *et al.*, 'Low frequency of *p53* mutations in spontaneous and dimethylnitrosamine induced renal tumors in Eker rats.' *Proc. Am. Assoc. Cancer Res.*, 1993, **34**, A653.

132. L. Recio, S. C. Lane, J. Ginsler *et al.*, 'Analysis of ras DNA sequences in rat renal cell carcinoma.' *Mol. Carcinog.*, 1991, **4**, 350–353.

133. C. Walker, J. Everitt, J. J. Freed *et al.*, 'Altered expression of transforming growth factor-α in hereditary rat renal cell carcinoma.' *Cancer Res.*, 1991, **51**, 2973–2978.

134. J. L. Young, Jr. and R. W. Miller, 'Incidence of malignant tumors in US children.' *J. Pediatr.*, 1975, **86**, 254–258.

135. J. F. Fraumeni, Jr. and A. G. Glass, 'Wilms' tumor and congenital aniridia.' *JAMA*, 1968, **206**, 825–828.

136. R. W. Miller, J. F. Fraumeni, Jr. and M. D. Manning, 'Association of Wilms' tumor with aniridia, hemihypertrophy and other congenital malformations.' *N. Engl. J. Med.*, 1964, **270**, 922.

137. K. A. Williamson and V. Von Heyningen, 'Towards an understanding of Wilms' tumour.' *Int. J. Exp. Pathol.*, 1994, **75**, 147–155.

138. J. Pelletier, 'Molecular genetics of Wilms' tumor: insights into normal and abnormal renal development.' *Can. J. Oncol.*, 1994, **4**, 262–272.

139. G. G. Re, D. J. Hazen-Martin, D. A. Sens *et al.*, 'Nephroblastoma (Wilms' tumor): a model system of aberrant renal development.' *Semin. Diagn. Pathol.*, 1994, **11**, 126–135.

140. M. J. Coppes and B. R. G. Williams, 'The molecular genetics of Wilms' tumor.' *Cancer Invest.*, 1994, **12**, 57–65.

141. H. P. Wilmore, G. F. J. White, R. T. Howell *et al.*, 'Germline and somatic abnormalities of chromosome 7 in Wilms' tumor.' *Cancer Genet. Cytogenet.*, 1994, **77**, 93–98.

142. S. Park, A. Bernard and K. E. Bove, 'Inactivation of WT1 in nephrogenic rests, genetic precursors to Wilms' tumour.' *Nat. Genet.*, 1993, **5**, 363–367.

143. K. M. Call, T. Glaser, C. Y. Ito *et al.*, 'Isolation and characterization of a zinc finger polypeptide gene at the human chromosome 11 Wilms' tumor locus.' *Cell*, 1990, **60**, 509–520.

144. M. Gessler, A. Poutska, W. Cavenee *et al.*, 'Homozygous deletion in Wilms' tumors of a zinc-finger gene identified by chromosome jumping.' *Nature*, 1990, **343**, 774–778.

145. N. Bardeesy, D. Falkoff, M. J. Petruzzi *et al.*, 'Anaplastic Wilms' tumour, a subtype displaying poor prognosis, harbours *p53* gene mutations.' *Nat. Genet.*, 1994, **7**, 91–97.

146. N. Bardeesy, J. B. Beckwith and J. Pelletier, 'Clonal expansion and attenuated apoptosis in Wilms' tumors are associated with *p53* gene mutations.' *Cancer Res.*, 1995, **55**, 215–219.

147. A. G. Knudson, Jr. and L. C. Strong, 'Mutation and cancer: a model for Wilms' tumor of the kidney.' *J. Natl. Cancer Inst.*, 1972, **48**, 313–324.

148. T. Glaser, J. Lane and D. Housman, 'A mouse model of the aniridia—Wilms' tumor deletion syndrome.' *Science*, 1990, **250**, 823.

149. D. A. Meyers, T. H. Beaty, N. E. Maestri *et al.*, 'Multipoint mapping studies of six loci on chromosome 11.' *Hum. Hered.*, 1987, **37**, 94.

150. P. M. Sharma, M. Bowman, B.-F. Yu *et al.*, 'A rodent model for Wilms' tumor: embryonal kidney neoplasms induced by *N*-nitroso-*N'*-methylurea.' *Proc. Natl. Acad. Sci. USA*, 1994, **91**, 9931.

151. A. K. Basu and J. M. Essigmann, 'Site specifically modified oligodeoxynucleotides as probes for the structural and biological effects of DNA—damaging agents.' *Chem. Res. Toxicol.*, 1988, **1**, 1.

152. H. Ohgaki, P. Kleihues and G. C. Hard, 'Ki-*ras* mutations in spontaneous and chemically induced renal tumors of the rat.' *Mol. Carcinogen.*, 1991, **4**, 455.

153. S. Sukumar, A. Perantoni, C. Reed *et al.*, 'Activated K-*ras* and N-*ras* oncogenes in primary renal mesenchymal tumors induced in F344 rats by methyl-(methoxymethyl)nitrosamine.' *Mol. Cell Biol.*, 1986, **6**, 2716.

154. S. Sukumar, '*Ras* oncogenes in chemical carcinogenesis.' *Curr. Top. Microbiol. Immunol.*, 1990, **148**, 93.

155. A. O. Perantoni, C. D. Reed, M. Watatani *et al.*, 'Multiple 12th codon mutations in the K-*ras* oncogenes in renal mesenchymal tumors induced in newborn F344 rats by methyl(methoxymethyl)nitrosamine (DMN-OMe).' *Proc. Am. Assoc. Cancer Res.*, 1990, **31**, 137.

156. H. Ohgaki, G. C. Hard, N. Hirota *et al.*, 'Selective muation of condons 204 and 213 of the *p53* gene in rat tumors induced by alkylating *N*-nitroso compounds.' *Cancer Res.*, 1992, **52**, 2995.

157. C. M. Weghorst, K. H. Dragnev, G. S. Buzard *et al.*, 'Low incidence of point mutations detected in the *p53* tumor suppressor gene from chemically induced rat renal mesenchymal tumors.' *Cancer Res.*, 1994, **54**, 215.

158. S. K. Shore, G. C. Hard and P. N. Magee, 'Deregulation and overexpression of c-*fos* proto-oncogene in rat renal cell lines and primary tumors induced by dimethylnitrosamine.' *Oncogene*, 1988, **3**, 567.

7.20
Carcinogenic Response of the Bladder

TERRY V. ZENSER, VIJAYA M. LAKSHMI, and BERNARD B. DAVIS
St. Louis University and VA Medical Center, MO, USA

7.20.1 INTRODUCTION

Bladder cancer represents approximately 7% of human malignancies and is the third most prevalent cancer type in men 60 years and older.[1,2] In 1995, there will be an estimated 50 500 new cases and 11 200 deaths due to bladder cancer in the United States.[3] It is of historical significance that bladder cancer was one of the first human cancers to be identified as being caused by a specific occupational chemical agent, aromatic amines. The first symptom in 90% of bladder cancer patients is hematuria. There are several unique features of bladder cancer which provide opportunities for its study and the development of chemoprevention strategies: the tissue is accessible for observation and biopsy, patients with successful resection of the primary tumor are at high risk for recurrence and/or progression, and the relatively short time period for the latter to occur. The following describes the carcinogenic response of the bladder.

7.20.2 BLADDER CELL PHYSIOLOGY AND MORPHOLOGY OF PRENEOPLASTIC AND NEOPLASTIC LESIONS

7.20.2.1 Bladder Cell Physiology

Epithelial cancer in the urinary bladder arises in the transitional epithelium. This epithelium has a unique structure and function and resides in a unique environment. Chapter 3, this volume contains an in-depth review of the functional anatomy of the bladder. A prominent, and perhaps the most important function of this epithelium, is to provide a barrier to prevent back-diffusion of substances from the urine into the blood stream. Impermeability to solute and solvent then is an important characteristic of these cells.[4]

The urothelium is exposed to urine on its mucosal surface and is therefore somewhat unique when compared to the internal environment of most other tissues. Prominent are the variations in pH from 4.5 to 7.5 and solute concentration from 50 to 1200 mOsm L^{-1}. However, it is also apparent that the function of the kidney to excrete certain toxicants into the urine at the same time as subserving volume and solute homeostasis can result in high urinary concentrations of certain toxicants and xenobiotics.

A solute that is filtered at the glomerulus will have a urine:plasma concentration ratio ranging from 0.2 to 4.0, if there is no movement of freely filterable solute out of the renal tubule. If there is relative impermeability of the tubular

epithelium to solute or tubular secretion, the solute concentration ratio can increase more than 10-fold. Therefore, urothelial cells can be exposed to very high concentrations of certain chemicals.

The cellular structure of the mucosal surface of the urinary bladder is eight cells thick. The innermost layer, the one lining the bladder lumen, is composed of cells which are multinucleated in character. Tight intercellular junctions characterize the subluminal layers.

These specific features of the urothelium, that is, exposure to a highly unique environment, the structural characteristics of multinucleated cells lining the lumen, and the physiological impermeability of the cells, may be important when considering the pathogenesis of malignant degeneration of these cells.

7.20.2.2 Morphology of Preneoplastic Lesions

Urothelial cancer is characterized by the frequency of premalignant lesions and by the high incidence of multifocal disease. It is also associated with cofactors which would tend to disrupt the highly specialized architecture of the urothelium and thus impair the important characteristic of impermeability of the membrane to the contents of the urine. Chemical and mechanical factors such as exposure to the products of tobacco and urinary tract obstruction may increase the susceptibility to chemically-induced bladder cancer. An important potential mechanism for this effect would be a decreased structural integrity of the barrier maintaining urinary solute and solvent.[5,6]

There are premalignant histological changes which are associated with damage to the structural integrity of the urothelium (Figures 1 and 2). A common change is a hyperplastic

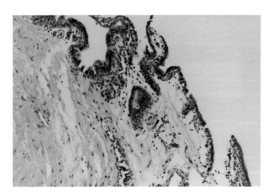

Figure 1 Normal urinary bladder mucosa and wall. The mucosa is lined by a single layer of normal transitional cells abutting submucosal loose connective tissue (H&E stain, original magnification 200×).

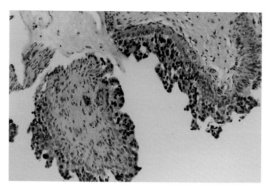

Figure 2 Precancerous or dysplastic change of urinary bladder mucosa. Mucosal lining cells are larger than the normal ones with hyperchromatic nuclei and many mitoses (H&E stain, original magnification 200×).

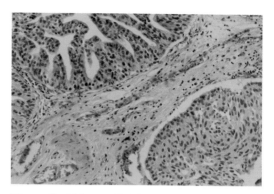

Figure 3 Urinary bladder transitional cell carcinoma. Tumor cells are polygonal and multilayered with areas of papillary pattern (upper portion) and syncytial pattern (lower portion). Nuclei are hyperchromatic and there are mitotic cells (H&E stain, original magnification 200×).

reaction which by definition involves an increase in normal appearing cells. Cellular polarity is maintained, but it appears as if there may be some compromise of the permeability barrier which is dependent upon not only cellular but intercellular organization. There may also be a dysplastic reaction or a loss of complete differentiation of the luminal cells. Dysplasia involves a proliferation of cells that are not fully differentiated to transitional epithelium. This may range from a loss of polarity to changes in the intracellular organization and may include nuclear crowding and changes in nucleolar structure. Metaplasia to squamous or glandular epithelium also occurs. Leukoplakia (white, thickened patches of mucous membrane) is also considered premalignant.

7.20.2.3 Morphology of Neoplastic Lesions

Carcinoma *in situ* of the urinary bladder can be defined as an intraepithelial neoplasm in which the malignant cells are confined to the bladder mucosa (Figure 3). Melicow described it as areas of flat carcinoma of the epithelium surrounding visible tumor.[7] Descriptions of the frankly anaplastic cells and the diffuse multifocal nature of carcinoma *in situ* have also appeared. These changes are complex and represent significant changes in the biological function of transitional epithelium. When changes increase susceptibility to malignant degeneration, they should be considered premalignant. Leukoplakia should be considered as premalignant and advanced dysplasia as carcinoma *in situ*.

Urinary bladder cancer is primarily a disease of the urothelium, with greater than 90% of the

lesions identified as transitional cell carcinomas. The lesions range from carcinoma *in situ* as discussed above to invasive tumors. Invasive tumors are then classified according to the extent of invasion of submucosa, muscle, perivesicular tissue or involving other pelvic structures. Two cell types make up the remainder of the carcinomas, adenocarcinoma (less than 5%), and squamous cell (5–10%). In areas where Schistosomal infections are prominent in the etiology of bladder cancer, the percentage of squamous cell lesions will increase to greater than 50%.[5,6]

7.20.3 ETIOLOGIC AGENTS

7.20.3.1 Aromatic Amines: Occupational Exposure

Bladder cancer was one of the first neoplasms to be associated with occupational exposure to chemicals (Table 1). The observation by Rehn, in 1895, that several of his patients with bladder cancer worked in the same synthetic dyestuffs

Table 1 Aromatic amines and related chemicals suspected of being human bladder carcinogens.

Benzidine
2-Naphthylamine
4-Aminobiphenyl
o-Toluidine
4,4′-Methylene bis(2-chloroaniline)
4-Chloro-*o*-toluidine
4,4′-Methylene dianiline
Benzidine-derived azo dyes
Phenacetin-containing compounds

factory, led to the suspicion that aromatic amines cause bladder cancer.[8] Unfortunately, it proved difficult to reproduce this lesion in experimental animals. However, in 1938, Hueper and Wiley were successful in inducing very similar bladder tumors by administering 2-naphthylamine to dogs.[9] For this reason, many studies have used dogs to evaluate the carcinogenic potential of aromatic amines. Benzidine, 2-naphthylamine, and 4-aminobiphenyl are all specific occupational arylamines listed as carcinogenic in humans and associated with exceptionally high risks of inducing bladder cancer.[10] One study reported that all 15 workers distilling 2-naphthylamine developed bladder cancer,[11] indicating that in exceptional situations in which the exposure to these carcinogens is high, individual susceptibility is irrelevant. Dyestuff users also have an increased risk of bladder cancer (Table 2). For example, Kimono painters, many of whom ingest benzidine-based dyes by licking their brush, had seven times the expected incidence of bladder cancer.[12] These dyes may initiate cancer because of contamination by the parent arylamine in the final product and/or due to azo reduction of the dye to the parent arylamine in the gastrointestinal tract.[13]

Workers in several other occupations, including chemical,[14] rubber,[15] and leather[16] industries, are also exposed to these carcinogenic amines. Workers exposed to o-toluidine,[17,18] 4,4'-methylene bis(2-methylaniline),[17] or 4-chloro-o-toluidine[19] were found to have an increased risk of developing bladder cancer. 1-Naphthylamine and aniline are not considered carcinogenic.

Exposure to combustion gases and soot from coal,[20] or to motor vehicle exhaust[21] also increase the relative risk of bladder cancer. Major chemical components of these gases and soot are polycyclic aromatic hydrocarbons, including nitropolycyclics, which can be biologically reduced to their corresponding N—OH

Table 2   Occupations associated with human bladder cancer.

Textile workers
Dye workers
Tire and rubber workers
Chemical workers
Leather workers
Bootblacks
Painters
Truck drivers
Drill press operators
Aluminum workers

arylamine derivatives[22,23] and could then initiate the carcinogenic process. An increased incidence of bladder cancer is also observed in the aluminum industry,[24] and the chemicals involved may be arylamines and/or polycyclic aromatic hydrocarbons. Arylamine contaminants, that is, phenyl-a-naphthylamine, may also be responsible for the excess risk of bladder cancer observed by exposure to cutting oils and cutting fluids.[25] Identifying the specific etiologic agent causing bladder cancer in an occupational setting is difficult, because workers are exposed to numerous compounds.

7.20.3.2   Coffee Drinking

Coffee has been associated with an increased incidence of bladder cancer. A population-based, case-control study conducted in Massachusetts demonstrated an unexpected increased bladder cancer risk for coffee drinking.[26] However, subsequent studies have indicated little or no association, suggesting that coffee is not a bladder carcinogen. Interestingly, cigarette smoking is a strong risk factor for bladder cancer and a strong correlate of coffee drinking.[27]

7.20.3.3   Artificial Sweeteners

Studies in rodents suggested a potential for artificial sweeteners to cause bladder cancer in humans. Sodium cyclamate was first reported in 1969 to cause bladder cancer in mice. Saccharin was also shown to be a bladder carcinogen in rats; however, bladder cancer was observed only in rats fed a diet containing at least 2.5% sodium saccharin. If saccharin is administered (up to 7.5% of the diet) as the acid rather than the sodium salt, it is not carcinogenic.[28] Saccharin does not bind DNA. However, it does increase the proliferative rate of rat bladder epithelial cells,[29] which appears to be secondary to the formation of silicate precipitates and/or microcrystals in urine. Precipitate and crystal formation is enhanced by urinary pH >6.5, increasing urinary concentrations of sodium, and large amounts of protein.[30] The urinary protein requirement was rather specific, with the rat $\alpha_{2\mu}$-globulin being the most effective. Male rats are more susceptible than female rats, and the mouse, hamster, and monkey appear resistant to effects of sodium saccharin on bladder.[31] These extensive studies with saccharin indicate the species-specific nature of this response and that it would not be a human carcinogen. This has been confirmed by numerous case–control studies.[32,33]

7.20.3.4 Drugs

Cyclophosphamide treatment and analgesic abuse result in an increased risk of bladder cancer. When used as an alkylating agent to treat both malignant and nonmalignant diseases in the early 1950s, cyclophosphamide-associated increases in bladder cancer were observed.[34,35] Cyclophosphamide produces bladder tumors in both rats and mice. Heavy consumption of phenacetin-containing analgesics are linked to renal pelvis, ureter, and bladder cancer in human.[36] However, acetaminophen, a popular analgesic, shows no increased risk for these cancers.[37]

7.20.3.5 Infectious Agents

Infections may be etiologic agents in bladder cancer. In Egypt, bladder cancer may develop in patients with bilharziasis, a parasitic infection.[38] This parasitic infection is thought to contribute to the different types of bladder tumors observed in Egypt compared to the United States.[39] In the United States, most bladder cancers are transitional cell carcinomas (90%) with a few squamous cell carcinomas (6%). In Egypt, squamous cell carcinomas comprise a much larger proportion of bladder cancer (67%) than transitional cell carcinomas (23%). Bladders from infected individuals exhibit an increased inflammatory and regenerative response along with the generation of nitrosamines, which should contribute to increased risk.

Urinary tract infection is a significant risk factor for the development of bladder cancer. Although risk is greater with chronic infection, it also increases with the number of episodes of acute urinary tract infection and multiplies with cigarette smoking.[40] Using the heterotopically transplanted rat urinary bladder model, chronic inflammation (induced by repeated intravesical instillation of killed *Escherichia coli*) was shown to enhance N-methyl-N-nitrosourea-initiated bladder carcinogenesis.[41] Chronic inflammation beginning as late as 18 weeks after N-methyl-N-nitrosourea treatment resulted in a significant increase in tumors by 31 weeks. Besides causing increased cell proliferation, urinary tract infections may also generate nitrosamines. Thus, the combined effect of both of these events may contribute to increased risk.

While these effects of infection described above appear to be related to promotional aspects of the carcinogenic process, certain viruses may cause mutations. There is increasing evidence of a role for certain strains of human papilloma virus in the etiology of urogenital cancers, including urinary bladder.[42,43] A model for bladder carcinogenesis has been proposed in which human papilloma virus is thought to immortalize bladder epithelium in a manner analogous to SV40 binding of pRB and p53.[44]

7.20.3.6 Tobacco

Cigarette smoking is consistently demonstrated as a cause of bladder cancer. Compared to nonsmokers, smokers are at a two- to threefold higher risk of developing bladder cancer.[45-48] Smoking unfiltered cigarettes may increase the risk of bladder cancer (35–50%) compared to filtered cigarettes.[48] Deep inhaling may also put smokers at greater risk.[46] In general, risk increases with a longer duration of smoking and increasing intensity of smoking.[45,47,48] Cessation of cigarette smoking results in approximately a 30–60% reduction in risk.[49] Tobacco smoke contains a myriad of chemicals including aromatic amines.[50] While aromatic amines are thought to be responsible for the bladder cancer associated with smoking, other chemicals may also contribute. Black tobacco is more carcinogenic than blond tobacco.[47] Pipe and cigar smokers and users of snuff and chewing tobacco do not appear to have a greater risk of developing bladder cancer.[45,46]

7.20.3.7 Urologic Agents

Urine is an important etiologic factor in bladder cancer. Urine appears to do more than just deliver the carcinogen to the bladder. Using the heterotopically transplanted rat urinary bladder model, bladders initiated with N-butyl-N-(hydroxybutyl)nitrosamine produced more tumors if they were instilled with urine than saline.[51] One of the factors in urine contributing to this response was epidermal growth factor, which appears to act by maintaining a proliferative (promotive) response.[52] Urinary retention is also a factor insofar as medical conditions that cause retention, such as benign prostatic hypertrophy, are associated with increased risk.[53] The upper hemisphere of the bladder (dome) has less contact with urine than other parts of the bladder and is less likely to have tumors.[54]

7.20.3.8 Radiation

Radiation can cause bladder cancer. Treatment of patients with high-dose ^{131}I for thyroid cancer results in increased incidence of bladder cancer.[55] Atomic bomb survivors in Japan demonstrated increased bladder cancer-related deaths.[56] Women receiving therapeutic pelvic ionizing radiation for dysfunctional uterine bleeding appear to have an increased risk of bladder cancer.[57]

7.20.3.9   Hair Dyes

Hair dyes are mutagenic and cause bladder tumors in experimental animals. However, persons exposed to these dyes, that is, hairdressers and barbers, do not demonstrate an increased risk of bladder cancer.[58]

7.20.3.10   Familial Predisposition

Familial predisposition to bladder cancer has been demonstrated. However, the familial risk appears to be associated with environmental factors such as cigarette smoking, suggesting genetic and environmental interactions.[59]

7.20.3.11   Dietary Factors

The role of dietary factors in bladder cancer is not clear. Decreasing levels of serum selenium have been associated with an increasing trend in risk.[60] Fatty foods have also been associated with bladder cancer risk.[61] Alcohol drinking has not been consistently associated with increased risk, possibly due to positive findings confounded by smoking.[62] Dietary supplements such as retinoids protect against bladder cancer in animals, but not consistently in humans.[63] Chlorination by-products in water are not associated with bladder cancer, although initial ecological studies suggested a link.[64] High levels of arsenic in water are associated with an increased incidence of bladder cancer.[65]

7.20.3.12   Agents in Experimental Animals

Several chemicals induce bladder cancer in rodents and were used to develop models for studying bladder carcinogenesis. N-Butyl-N-(hydroxybutyl)nitrosamine   (BBN),[66,67] N-[4-(5-nitro-2-furyl)-2-thiazolyl]formamide (FANFT),[68]   and N-methyl-N-nitrosourea (MNU)[69] are the chemicals most commonly used. All of these chemicals induce a high incidence of bladder cancer in rodents and have been useful for studying factors modify-

ing bladder carcinogenesis, that is, two-stage carcinogenesis. These chemicals induce neoplasia in susceptible species with histologic characteristics similar to those reported for humans, most tumors being transitional cell carcinomas. A major difference between these chemicals is their route and method of administration. MNU is a direct acting carcinogen and therefore must be instilled directly into the bladder. Usually, only a single dose of MNU is given. BBN is administered in the drinking water. For a low level of initiation, BBN is given at a dose of 0.01% in drinking water for 4 weeks. BBN induces papillary and nodular hyperplasia, which is considered a preneoplastic lesion.[67] Susceptible species include rat, mouse, hamster, and guinea pig. FANFT induces bladder cancer in rat, mouse, hamster, and dog with similar results observed in males and females, but is not carcinogenic in guinea pig. Rats fed 0.2% FANFT for 10 or more weeks develop a high incidence of lesions (90–100%), which are irreversible and eventually result in bladder carcinomas within one year.[68] However, hyperplastic lesions developing through 6 weeks of FANFT feeding appear to be reversible if FANFT feeding is discontinued. FANFT is not immunosuppressive. For both the FANFT and BBN models, tumor yield was dependent upon carcinogen dose and duration of treatment.

4,4′-Methylene dianiline and 4,4′-methylene bis(2-chloroaniline), both structural analogues of benzidine, produce bladder tumors in experimental animals. These compounds are used as curing agents for certain resins and plastics, and may possibly cause bladder cancer in humans.[70]

7.20.4   EPIDEMIOLOGY IN HUMANS

7.20.4.1   Occupational Exposure

With the development of modern techniques of epidemiology, such as cohort and follow-up study, Case *et al.*[71] clearly demonstrated the increased incidence of bladder cancer in workers involved in the manufacture of aromatic amines. As a result, the manufacture of carcinogenic aromatic amines has been prohibited in Western countries, but unfortunately not in developing countries. As mentioned above, occupations with strong evidence of exposure to aromatic amines and risk of bladder cancer are dye workers, aromatic amine manufacturing workers, rubber workers, leather workers, painters, aluminum workers, and truck drivers.

7.20.4.2 Age, Race, and Sex

The incidence and mortality due to bladder cancer increases sharply with age. About 70% of the cases occur among individuals 65 yr or older.[72,73] At all ages, whites have a higher (approximately twofold) incidence than blacks. This racial difference in incidence is not related to social class. Hispanics and Asians have a slightly lower incidence than blacks; American Indians have a very low rate. The incidence in males is about three to four times greater than females. This sex difference is not entirely explained by differences in cigarette smoking and occupational exposure.[74]

Age, race, and sex affect the stage of bladder cancer.[75] Although bladder cancer before the age of 40 is uncommon, in younger patients most tumors are low-grade, papillary, noninvasive, and transitional-cell carcinomas. In patients 40–44 yr, greater than 80% have a local tumor, compared to approximately 60% of patients over 84 years. The proportion of localized tumors for white men, white women, black men, and black women are 73, 69, 59, and 49%, respectively. Whites and blacks have a similar risk for more advanced tumors.

7.20.4.3 Geographic Variation

The incidence of bladder cancer worldwide varies about 10-fold.[76] Relatively low rates are found in eastern Europe and several areas of Asia, whereas in western Europe and North America, higher rates occur. Geographic variation may reflect local practices in registration of tumor type, that is, papillomas vs. benign. Urban areas have a higher mortality rate than rural areas in the United States and other countries.[76,77] In the United States, bladder cancer incidence and mortality are higher in parts of the northeast and upper midwest and lower in the South.[78]

7.20.4.4 Time of Exposure

Similar results are observed for cigarette smoking and occupational exposure. The earlier an individual is exposed, the greater the risk. Risk decreases with increasing time since the last exposure. The similarity between these events suggests a similar mechanism of carcinogenesis for smoking and occupational exposure.[79] A similar etiologic agent, that is, aromatic amines, for cigarette smoking and occupational exposure may also be suggested by the these results. As the average age of bladder cancer diagnosis is approximately 68, there appears to be a significant latent period between initial exposure and diagnosis.

7.20.5 BIOMARKERS

7.20.5.1 *N*-Acetylator Phenotype/Genotype (Marker of Risk)

N-Acetylation is thought to play an important role in aromatic amine metabolism. *N*-acetyltransferase (NAT) enzymes, NAT1 and NAT2,[80] catalyze the *N*-acetylation of bicyclic aromatic amines, while heterocyclic amines are poor substrates.[81,82] Because primary aromatic amines are easier to oxidize than their acetylated amide products, *N*-acetylation is thought to compete with *N*-oxidation.[83] Thus, *N*-acetylation is considered as a detoxification step for aromatic amines.

Rapid and slow acetylator phenotypes are present in humans, with the slow acetylator phenotype occurring at frequencies of 10–90% in various ethnic populations.[84] Slow acetylators are thought to generate higher levels of *N*-hydroxy products than rapid acetylators.[83] This is consistent with the higher incidence of bladder cancer observed in slow acetylators.[85,86] A number of studies have correlated the slow acetylator phenotype with an increased formation of macromolecular adducts by 4-aminobiphenyl, an arylmonoamine.[87–89] For the aryldiamine, benzidine, an increased incidence of bladder cancer was not observed in slow acetylator Chinese workers.[90] In studies in which acetylator phenotype was correlated with the incidence of bladder cancer, the workers were exposed to a mixture of arylmonoamines and aryldiamines.[91] As explained in Section 7.19.6.1.1, the fate of *N*-acetylated aryldiamines, such as benzidine, and arylmonoamines, like 4-aminobiphenyl, may be different.

NATs can also activate aromatic and heterocyclic amines.[83] *N*-Hydroxy (*N*-OH) metabolites can undergo transferase catalyzed *O*-acetylation resulting in the formation of *N*-acetoxy derivatives. *N,O*-Transacetylation of hydroxamic acids also yields *N*-acetoxy esters. This reaction is catalyzed by microsomal enzymes which do not require acetyl CoA and are sensitive to paraoxon inhibition. *N*-Acetoxy esters are unstable and can be transformed through heterolytic fission into highly reactive nitrenium ions which attack DNA, especially the nucleophilic C-8 site on 2′-deoxyguanosine.[92] Bladder cells from a variety of species, including man, contain NAT and *N,O*-acetyltransferase activities.[93–98] Therefore, these transferase enzymes may facilitate end-organ activation of carcinogens to form DNA adducts.

7.20.5.2 Cytochrome P450 1A2 (Marker of Risk)

Aromatic amines are thought to be involved in the initiation of bladder cancer in certain groups of occupationally exposed individuals. Pioneering work by James and Elizabeth Miller showed that these carcinogens are oxidized to *N*-OH intermediates that react with DNA to form adducts.[92] The cytochrome P450 most responsible for this oxidation in humans is cytochrome P450 1A2.[99] Individuals with the slow acetylator phenotype and rapid cytochrome 1A2 phenotype exhibit the highest level of 4-aminobiphenyl hemoglobin adduct.[100] 4-Aminobiphenyl is a recognized human bladder carcinogen found in cigarette smoke.[50]

7.20.5.3 Carcinogen Macromolecule Adducts (Marker of Exposure)

Depending upon their structure and location, DNA adducts may cause replication errors, spontaneous deamination and depurination, and DNA damage. Formation of the guanine C-8 aromatic amine adducts in bacteria, mammalian cells, and rodents correlates with induction of mutations and tumorigenesis.[101,102] The guanine C-8 adduct of 4-aminobiphenyl was identified in biopsy samples of human bladder from smokers.[103] Levels of this adduct in exfoliated urothelial cells correlated significantly with the level of 4-aminobiphenyl hemoglobin adduct and the number of cigarettes smoked.[104] In addition, the exfoliated cell guanine C-8 adduct of 4-aminobiphenyl correlated with the mutagenicity of an individual's urine. This suggests that aromatic amine DNA adducts may be involved in the initiation of bladder cancer. The level of 4-aminobiphenyl hemoglobin adduct is higher in slow acetylators.[87]

7.20.5.4 *p53* Gene (Marker of Tumor Development)

Mutations of the *p53* gene are the most common genetic defect in human tumors.[105] The *p53* gene functions as a tumor-suppressor gene and more specifically as a cell-cycle regulator.[106] When DNA damage occurs, levels of p53 protein increase, arresting the cell cycle and allowing time for repair.

Mutations in the *p53* gene appear to be an early event in the formation of carcinoma *in situ* and much less frequent in noninvasive papillary tumors.[107] Mutation of the *p53* gene and nuclear accumulation of p53 protein are associated with the grade and stage of bladder cancer.[108] In bladder cancer patients treated by radical cystectomy, nuclear accumulation of p53 protein identifies transitional-cell carcinomas with a propensity for progression that is independent of tumor grade and stage.[109] The high risk of progression despite radical cystectomy would suggest benefit from adjuvant therapy. Thus, nuclear accumulation of p53 protein may be an important biomarker in the multistep progression of bladder cancer.

7.20.6 UNDERLYING MECHANISMS

7.20.6.1 Genotoxic

7.20.6.1.1 Genetic alterations mediated by aromatic amines

A model for aromatic amine-induced bladder cancer is described in which four competing enzymatic pathways are considered to play a role in carcinogenesis (Figure 4).

The oxidation of aromatic amine carcinogens to *N*-hydroxy derivatives is thought to be a necessary event in their initiation of bladder cancer.[92] The liver is the proposed site for oxidation. Once formed, *N*-hydroxy metabolites must reach their target organ, the bladder. It is proposed that *N*-glucuronide conjugates are formed which are excreted from the liver, filtered by the kidney, and accumulate in urine within the lumen of the bladder. The acidic pH of urine hydrolyzes these glucuronide conjugates to their *N*-hydroxy amines.[110] As indicated above, cytochrome P450 1A2 (*N*-oxidation) is a potential biomarker to assess the risk of aromatic amine-induced bladder cancer.

Peroxidatic metabolism of benzidine and other carcinogenic amines by prostaglandin H synthase to form DNA adducts has been demonstrated.[111-113] This enzyme is present in the bladder.[111,114,115] Arylhydroxamic acids are oxidized by a single electron process to give the nitroxyl free radical intermediate which is thought to undergo a bimolecular disproportionation to the corresponding nitroso derivative and *O*-acetyl ester of the hydroxamic acid.[116] The latter forms DNA adducts. Thus, peroxidation provides a potential mechanism for activation of carcinogenic amines within the bladder.

N-Acetylation is thought to detoxify aromatic amines.[83] This is because primary amines are easier to oxidize than amides. Acetylator phenotype/genotype (*N*-acetylation) is a potential biomarker for aromatic amine-induced bladder cancer.

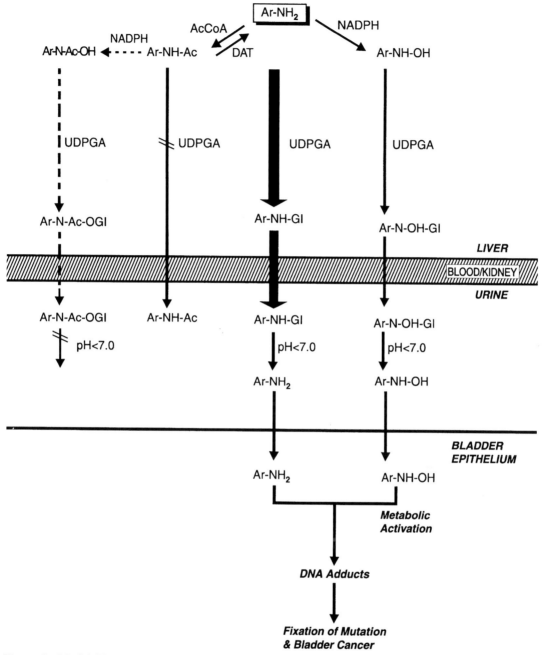

Figure 4 Model illustrating the role of oxidation, glucuronidation, acetylation, and deacetylation pathways in aromatic amine-induced bladder cancer. See text for details. Abbreviations: arylamine, Ar-NH$_2$; *N*-hydroxyarylamine, Ar-NH-OH; *N*-acetylarylamine, Ar-NH-Ac; *N*-hydroxy-*N*-acetylarylamine, Ar-*N*-Ac-OH; arylamine *N*-glucuronide, Ar-NH-Gl; *N*-hydroxyarylamine *N*-glucuronide, Ar-*N*-OH-Gl; *O*-glucuronide, Ar-*N*-Ac-OGl; acetyl CoA, AcCoA; deacetylation, DAT; UDP-glucuronic acid, UDPGA.

Glucuronidation represents a major pathway for metabolizing carcinogenic aromatic amines.[117,118] UDP-glucuronosyltransferases consist of isozymes in two distinct gene families which metabolize many endogenous and xenobiotic compounds. Polymorphism in humans may depend upon age, genetic incidence, and exposure to inducers (i.e., phenobarbital and tobacco). Since glucuronidation results in inactivation and excretion, *N*-glucuronidation competes with *N*-oxidation. While *N*-glucuronides of primary arylamines and *N*-OH-arylamines are formed, arylamides are not *N*-glucuronidated.[119] Therefore, *N*-glucuronidation and *N*-acetylation are also competing pathways. The fate of *N*-acetylated aryldiamines, such as benzidine, and arylmonoamines, like 4-aminobiphenyl, may be different.

In contrast to 4-aminobiphenyl, acetylation of benzidine yields a product (N-acetylbenzidine) that is still susceptible to N'-oxidation and N'-glucuronidation.[119–121]

Deacetylase (carboxylesterases) can dramatically influence the extent of acetylated product formation. A human liver arylacetamide deacetylase has been purified and characterized.[122,123] Microsomal N,O-acetyltransferases also exhibit deacetylase activity. Dogs, which are susceptible to aromatic amine carcinogenesis,[124,125] lack N-acetylation activity,[126,127] but contain microsomal N,O-acetyltransferases (deacetylases).[96,128]

The role of oxidation, glucuronidation, acetylation, and deacetylation pathways in aromatic amine-induced bladder cancer is illustrated in Figure 4. Aromatic amine (Ar-NH$_2$) is metabolized by oxidation (NADPH), N-glucuronidation using UDP-glucuronic acid (UDPGA) as cosubstrate, N-acetylation using acetyl CoA (AcCoA) as cosubstrate, or deacetylation (DAT). N-Acetyl-aromatic amine (Ar-NH-Ac) is not glucuronidated, but can be oxidized, to a limited extend, to N-hydroxy-N-acetyl-aromatic amine (Ar-N-Ac-OH). The latter hydroxamic acid can be metabolized to an O-glucuronide (Ar-N-Ac-OGl) which is not acid labile. Deacetylation converts the N-acetylated product back to the aromatic amine. Thus, glucuronidation, acetylation, oxidation, and deacetylation are competing pathways in aromatic amine metabolism. Hepatic glucuronidation results in detoxification and excretion. N-Glucuronides of aromatic amines (Ar-NH-Gl) and N-hydroxy-aromatic amines (Ar-N-OH-Gl) can be hydrolyzed by acidic urine to their corresponding arylamines which can then enter the bladder and undergo further metabolism to initiate carcinogenesis. Bladder epithelium contains O-acetyltransferase and N,O-acetyltransferase activities[93,96,129] capable of activating N-OH arylamines to N-acetoxy esters which can react with DNA. In addition, prostaglandin H synthase is present in bladder epithelium[111] and is capable of activating arylamines[113,130] and their hydroxamic acids.[116] During cell replication, DNA adducts may cause mutations which become fixed and participate in the carcinogenic process.

7.20.6.1.2 Multistep accumulation of genetic alterations

Cancer is thought to be the result of multiple mutational events. These genetic changes may occur by endogenous or exgenous chemicals, or by physical phenomena. A mechanism for exogenous chemicals (aromatic amines) is described above. Endogenous mechanisms include spontaneous deamination and DNA adducts derived from lipid oxidation. Physical damage to DNA can occur by ionizing radiation.

There is a growing body of evidence demonstrating the importance of carcinogenic DNA adducts in the pathogenesis of cancer by genotoxic agents in humans.[131] An increased risk of developing liver cancer is associated with aflatoxin DNA adducts.[132] Furthermore, an increased risk of developing lung cancer is associated with a tendency to form polycyclic aromatic hydrocarbon DNA adducts.[133] Carcinogenic DNA adducts have been correlated with chromosomal aberrations and somatic cell mutations[134–137] which themselves are mechanisms for activation of certain oncogenes involved in carcinogenesis in animals and man.[138]

A number of specific genes and chromosomal loci have been implicated in bladder cancer. The H-ras oncogene was first detected in a bladder tumor cell line[139] and subsequently found in about 10–15% of bladder tumors.[140] The loss of heterozygosity of chromosomes 9, 11, and 17 suggests that tumor suppressor genes may be involved.[44] A partial loss of chromosome 9 occurs in transitional cell carcinomas at all stages, while deletions on chromosome 17 are predominately observed in invasive tumors. Nuclear accumulation of p53 protein may be a prognostic indicator of bladder tumor progression.[109] There is increasing evidence of a role for certain strains of human papilloma virus in the etiology of urogenital cancers, including urinary bladder.[42,43] The accumulation of these genetic changes within individual cells results in genetic instability which enhances the probability of additional genetic damage. To become tumorigenic, a selective growth advantage must occur along with clonal expansion.

7.20.6.2 Nongenotoxic

There are a number of agents which are not metabolically activated to reactive intermediates, do not interact with DNA directly, are not genotoxic in short-term screens, for example, Ames test, and yet increase the incidence of cancer in experimental animals.[141] These agents are termed "nongenotoxic" and include a class of compounds referred to as tumor promoters. The balance between cell proliferation and apoptosis, programmed cell death,[142] directly affects the survival and growth of initiated cells, and preneoplastic and tumor cell populations. Nongenotoxic agents can effect carcinogenesis by increasing proliferation, decreasing apoptosis, or both in specific target cell populations. Although the mechanism by which chemicals induce cell growth alterations are not completely known, nongenotoxic changes in growth factor signaling and cell communication may be critical in influencing the balance of cell growth processes.

Foreign bodies may be nongenotoxic agents. Implantation of pellets of various materials such as glass, cholesterol, or wax into the bladder lumen of rodents produces cancer.[143] Urinary calculi can be generated by certain chemicals such as uracil, melamine, and sodium saccharin.[30,144] Formation of calculi depends upon physical properties (e.g., solubility). Nongenotoxic substances may act by a receptor-mediated mechanism. Epidermal growth factor is one example for the bladder.[52] 2,3,7,8-Tetrachlorodibenzo-p-dioxin (TCDD)-mediated tumor promotion in rat liver is not caused by increasing cell proliferation, but rather by substantially inhibiting apoptosis.[145] Modifiers of apoptosis regulation in rodent liver are intrinsic and extrinsic factors which include peroxisome proliferators, growth factors, cytotoxic agents (furan), food deprivation, TCDD, phenobarbital, and DNA damage.[142]

Cytokines may be involved in the enhancement of bladder cancer observed with chronic inflammation.[40] Interleukin 6 enhanced N-methyl-N-nitrosourea-initiated transformation of MYP3 cells, an anchorage-dependent nontumorigenic rat bladder epithelial cell line.[146] This effect of interleukin 6 was not due to differences in its receptor mRNA expression between parental and transformed cells, but may be due to transformants having better binding affinity to interleukin 6 than the parental cells, or that there are differences in the subsequent signal-transducing mechanisms.

Both necrosis and apoptosis induce nongenotoxic effects. Chemicals can induce necrosis and regenerative hyperplasia. Necrosis and apoptosis are two different categories of cell death separated by morphology, time of appearance, and tissue.[147] In contrast to necrosis, apoptosis is a genetically programmed process that generally requires gene transcription and translation. Apoptosis presents opportunities for the development of drugs, methods, and procedures for the prevention of bladder cancer.

7.20.7 REFERENCES

1. A. R. Feldman, L. Kessler, M. H. Myers *et al.*, 'The prevalence of cancer. Estimates based on the Connecticut Tumor Registry.' *N. Engl. J. Med.*, 1986, **315**, 1394–1397.

2. E. Silverberg and J. Lubera, 'Cancer statistics, 1987.' *CA Cancer J. Clin.*, 1987, **37**, 2–19.

3. P. A. Wingo, T. Tong and S. Bolden, 'Cancer statistics, 1995.' *CA Cancer J. Clin.*, 1995, **45**, 8–30.

4. T. Berl and R. W. Schrier, in 'Renal and Electrolyte Disorders,' ed. R. W. Schrier, Little, Brown and Company, Boston/Toronto, 1986, pp. 1–78.

5. W. R. Fair, Z. Y. Fuks and H. I. Scher, in 'Cancer: Principles & Practice of Oncology,' eds. T. DeVita, Jr., S. Hellman and S. A. Rosenberg, J. B. Lippincott, Philadelphia, PA, 1993, pp. 1052–1059.

6. M. T. Macfarlane, R. A. Figlin and J. B. deKernion, in 'Cancer Medicine,' eds. J. F. Holland, E. I. Frei, R. C. Bast, Jr. *et al.*, Lea & Febiger, Philadelphia/London, 1993, pp. 1546–1559.

7. M. N. Melicow, 'Histological study of vesical urothelium intervening between gross neoplasm in total cystectomy.' *J. Urol.*, 1952, **68**, 261–279.

8. L. Rehn, 'Blasengeschwulste bei fuchsinarbeitern.' *Arch. Klin. Chir.*, 1895, **50**, 588–600.

9. W. C. Hueper and F. H. Wiley, 'Experimental production of bladder tumors in dogs by administration of beta-naphthylamine.' *J. Ind. Hyg. Toxicol.*, 1938, **20**, 46–84.

10. 'IARC Monographs on the Evaluation of Carcinogenic Risks to Humans,' International Agency for Research on Cancer, Lyon, 1987.

11. R. A. M. Case, 'Tumours of the urinary tract as an occupational disease in several industries.' *Ann. R. Coll. Surg. Engl.*, 1966, **39**, 213–235.

12. O. Yoshida, T. Harada, T. Kato and M. Miyakawa, 'Bladder cancer in workers in the dyeing industry.' *Igaku No Ayumi (Jpn.)*, 1971, **79**, 421–422.

13. D. L. Morgan, J. K. Dunnick, T. Goehl *et al.*, Summary of the National Toxicology Program benzidine dye initiative.' *Environ. Health Perspect.*, 1994, **102**, 63–78.

14. P. A. Schulte, K. Ringen, G. P. Hemstreet *et al.*, 'Risk factors for bladder cancer in a cohort exposed to aromatic amines.' *Cancer*, 1986, **58**, 2156–2162.

15. R. A. M. Case and M. E. Hosker, 'Tumour of the urinary bladder as an occupational disease in the rubber industry in England and Wales.' *Br. J. Prev. Soc. Med.*, 1954, **8**, 39–50.

16. P. Cole, R. Hoover and G. H. Friedell, 'Occupation and cancer of the lower urinary tract.' *Cancer*, 1972, **29**, 1250–1260.

17. G. F. Rubino, G. Scansetti, G. Piolatto *et al.*, 'The carcinogenic effect of aromatic amines: an epidemiologic study on the role of *o*-toluidine and 4,4′-methylene bis(2-methylaniline) in inducing bladder cancer in man.' *Environ. Res.*, 1982, **27**, 241–254.

18. E. Ward, A. Carpenter, S. Markowitz *et al.*, 'Excess number of bladder cancers in workers exposed to *ortho*-toluidine and aniline.' *J. Natl. Cancer Inst.*, 1991, **83**, 501–506.

19. M. J. Stasik, 'Carcinomas of the urinary bladder in a 4-chloro-*o*-toluidine cohort.' *Int. Arch. Occup. Environ. Health*, 1988, **60**, 21–24.

20. G. Steineck, N. Plato, S. E. Norell *et al.*, 'Urothelial cancer and some industry-related chemicals: an evaluation of the epidemiologic literature.' *Am. J. Ind. Med.*, 1990, **17**, 371–391.

21. D. T. Silverman, R. N. Hoover, T. J. Mason *et al.*, 'Motor exhaust-related occupations and bladder cancer.' *Cancer Res.*, 1986, **46**, 2113–2116.

22. P. A. Schulte, E. Ward, M. Boeniger *et al.*, in 'Carcinogenic and Mutagenic Responses to Aromatic Amines and Nitroarenes,' eds. C. M. King, L. J. Romano and D. Schuetzle, Elsevier Science, New York, 1988, pp. 23–35.

23. M. R. Guerin and M. V. Buchanan, in 'Carcinogenic and Mutagenic Responses to Aromatic Amines and Nitroarenes,' eds. C. M. King, L. J. Romano and D. Schuetzle, Elsevier Science, New York, 1988, pp. 37–45.

24. G. Theriault, C. Tremblay, S. Cordier *et al.*, 'Bladder cancer in the aluminum industry.' *Lancet*, 1984, **1**, 947–950.

25. P. Vineis and S. Di Prima, 'Cutting oils and bladder cancer.' *Scand. J. Work Environ. Health*, 1983, **9**, 449–450.

26. P. Cole, 'Coffee-drinking and cancer of the lower urinary tract.' *Lancet*, 1971, **1**, 1335–1337.

27. P. Hartge, R. Hoover, D. W. West *et al.*, 'Coffee drinking and risk of bladder cancer.' *J. Natl. Cancer Inst.*, 1983, **70**, 1021–1026.

28. S. M. Cohen, L. B. Ellwein, T. Okamura *et al.*, 'Comparative bladder tumor promoting activity of sodium saccharin, sodium ascorbate, related acids, and calcium salts in rats.' *Cancer Res.*, 1991, **51**, 1766–1777.

29. R. Hasegawa and S. M. Cohen, 'The effect of different salts of saccharin on the rat urinary bladder.' *Cancer Lett.*, 1986, **30**, 261–268.

30. S. M. Cohen, M. Cano, R. A. Earl *et al.*, 'A proposed role for silicates and protein in the proliferative effects of saccharin on the male rat urothelium.' *Carcinogenesis*, 1991, **12**, 1551–1555.

31. Anonymous, in 'IARC Monographs on the Evaluation of the Carcinogenic Risk of Chemicals to Humans,' eds. Anonymous, IARC, Lyon, 1980, pp. 111–170.

32. O. Moller-Jensen, J. B. Knudsen, B. L. Sorensen *et al.*, 'Artificial sweeteners and absence of bladder cancer risk in Copenhagen.' *Int. J. Cancer*, 1983, **32**, 577–582.

33. I. I. Kessler and J. P. Clark, 'Saccharin, cyclamate, and human bladder cancer. No evidence of an association' *JAMA*, 1978, **240**, 349–355.

34. Anonymous (ed.), 'International Agency for Research on Cancer: Evaluation of the Carcinogenic Risk of Chemicals to Humans. Some Antineoplastic and Immunosuppressive Agents,' IARC, Lyon, 1981, vol. 26.

35. L. A. Levine and J. P. Richie, 'Urological complications of cyclophosphamide.' *J. Urol.*, 1989, **141**, 1063–1069.

36. Anonymous (ed.), 'International Agency for Research on Cancer: Evaluation of the Carcinogenic Risk of Chemicals to Humans. Some Pharmaceutical Drugs,' IARC, Lyon, 1980, vol. 24.

37. M. McCredie and J. H. Stewart, 'Does paracetamol cause urothelial cancer or renal papillary necrosis?' *Nephron*, 1988, **49**, 296–300.

38. A. R. Ferguson, 'Associated bilharziosis and primary malignant disease of the urinary bladder with observations in a series of forty cases.' *J. Pathol. Bacteriol.*, 1911, **16**, 76–94.

39. H. N. Tawfik, in 'Unusual Occurrences as Clues to Cancer Etiology,' eds. R. W. Miller, S. Watanabe, J. F. Fraumeni, Jr. *et al.*, Scientific Societies Press, Tokyo, 1989, pp. 197–209.

40. A. F. Kantor, P. Hartge, R. N. Hoover *et al.*, 'Urinary tract infection and risk of bladder cancer.' *Am. J. Epidemiol.*, 1984, **119**, 510–515.

41. K. Kawai, H. Kawamata, S. Kemeyama *et al.*, 'Persistence of carcinogen-altered cell population in rat urothelium which can be promoted to tumors by chronic inflammatory stimulus.' *Cancer Res.*, 1994, **54**, 2630–2632.

42. K. Anwar, M. Phil, H. Naiki *et al.*, 'High frequency of human papillomavirus infection in carcinoma of the urinary bladder.' *Cancer*, 1992, **70**, 1967–1973.

43. Y. F. Shibutani, M. P. Schoenberg, V. L. Carpiniello *et al.*, 'Human papillomavirus associated with bladder cancer.' *Urology*, 1992, **40**, 15–17.

44. C. A. Reznikoff, C. Kao, E. M. Messing *et al.*, 'A molecular genetic model of human bladder carcinogenesis.' *Semin. Cancer Biol.*, 1993, **4**, 143–152.

45. J. D. Burch, T. E. Rohan, G. R. Howe *et al.*, 'Risk of bladder cancer by source and type of tobacco exposure: a case-control study.' *Int. J. Cancer*, 1989, **44**, 622–628.

46. P. Cole, R. R. Monson, H. Haning *et al.*, 'Smoking and cancer of the lower urinary tract.' *N. Engl. J. Med.*, 1971, **284**, 129–134.

47. J. Clavel, S. Cordier, L. Boccon-Gibod, *et al.*, 'Tobacco and bladder cancer in males: increased risk for inhalers and smokers of black tobacco.' *Int. J. Cancer*, 1989, **44**, 605–610.

48. P. Hartge, D. Silverman, R. Hoover *et al.*, 'Changing cigarette habits and bladder cancer risk: a case-control study.' *J. Natl. Cancer Inst.*, 1987, **78**, 1119–1125.

49. E. L. Wynder and S. D. Stellman, 'Comparative epidemiology of tobacco-related cancers.' *Cancer Res.*, 1977, **37**, 4608–4622.

50. C. Patriankos and D. Hoffmann, 'Chemical studies on tobacco smoke.' *J. Anal. Toxicol.*, 1979, **3**, 150–154.

51. R. Oyasu, Y. Hirao and K. Izumi, 'Enhancement by urine of urinary bladder carcinogenesis.' *Cancer Res.*, 1981, **41**, 478–481.

52. H. Momose, H. Kakinuma, S. Y. Shariff *et al.*, 'Tumor-promoting effect of urinary epidermal growth factor in rat urinary bladder carcinogenesis.' *Cancer Res.*, 1991, **51**, 5487–5490.

53. S. Mommsen and A. Sell, 'Prostatic hypertrophy and venereal disease as possible risk factors in the development of bladder cancer.' *Urol. Res.*, 1983, **11**, 49–52.

54. O. Parkash and H. Kiesswetter, 'The role of urine in the etiology of cancer of the urinary bladder.' *Urol. Int.*, 1976, **31**, 343–348.

55. C. J. Edmonds and T. Smith, 'The long-term hazards of the treatment of thyroid cancer with radioiodine.' *Br. J. Radiol.*, 1986, **59**, 45–51.

56. National Reserach Council (US), 'Health Effects of Exposure to Low Levels of Ionizing Radiation Effects BEIR V,' National Academy Press, Washington, DC, 1990.

57. P. D. Inskip, R. R. Monson, J. K. Wagoner *et al.*, 'Cancer mortality following radiotherapy for uterine bleeding.' *Radiat. Res.*, 1990, **123**, 331–344.

58. P. Hartge, R. Hoover, R. Altman *et al.*, 'Use of hair dyes and risk of bladder cancer.' *Cancer Res.*, 1982, **42**, 4784–4787.

59. A. F. Kantor, P. Hartge, R. N. Hoover *et al.*, 'Familial and environmental interactions in bladder cancer risk.' *Int. J. Cancer*, 1985, **35**, 703–706.

60. K. J. Helzlsouer, G. W. Comstock and J. S. Morris, 'Selenium, lycopene, α-tocopherol, β-carotene, retinol, and subsequent bladder cancer.' *Cancer Res.*, 1989, **49**, 6144–6148.

61. J. Claude, E. Kunze, R. Frentzel-Beyme *et al.*, 'Lifestyle and occupational risk factors in cancer of the lower urinary tract.' *Am. J. Epidemiol.*, 1986, **124**, 578–589.

62. D. B. Thomas, C. N. Uhl and P. Hartge, 'Bladder cancer and alcoholic beverage consumption.' *Am. J. Epidemiol.*, 1983, **118**, 720–727.

63. H. A. Tyler, R. G. Notley, F. A. W. Schweitzer *et al.*, 'Vitamin A status and bladder cancer.' *Eur. J. Surg. Oncol.*, 1986, **12**, 35–41.

64. K. S. Crump and H. A. Guess, 'Drinking water and cancer: review of recent epidemiological findings and assessment of risks.' *Annu. Rev. Public Health*, 1982, **3**, 339–357.

65. M. M. Wu, T. L. Kuo, Y. H. Hwang *et al.*, 'Dose–response relation between arsenic concentration in well water and mortality from cancers and vascular diseases.' *Am. J. Epidemiol.*, 1989, **130**, 1123–1132.

66. N. Ito, Y. Hiasa, A. Tamai *et al.*, 'Histogenesis of urinary bladder tumors indicated by *N*-butyl-*N*-(4-hydroxybutyl)nitrosamine in rats.' *Gann*, 1969, **60**, 401–410.

67. N. Ito, S. Fukushima, T. Shirai *et al.*, 'Modifying factors in urinary bladder carcinogenesis.' *Environ. Health Perspect.*, 1983, **49**, 217–222.

68. S. M. Cohen, in 'Carcinogenesis: A Comprehensive Survey,' ed. G. T. Bryan, Raven Press, New York, 1978, pp. 171–231.

69. R. M. Hicks and J. S. J. Wakefield, 'Rapid induction of bladder cancer in rats with *N*-methyl-*N*-nitrosourea I. Histology.' *Chem. Biol. Interact.*, 1972, **5**, 139–152.

70. P. A. Schulte, K. Ringen, G. P. Hemstreet *et al.*, 'Occupational cancer of the urinary tract.' *Occup. Med.*, 1987, **2**, 85–107.

71. R. A. M. Case, M. W. Hosker, D. B. McDonald *et al.*, 'Tumors of the urinary bladder in workmen engaged in the manufacture and use of certain dyestuff intermediates in the British chemical industry.' *Br. J. Ind. Med.*, 1954, **11**, 75–104.

72. W. P. D. Logan (ed.), 'Cancer mortality by occupation and social class, 1851–1971,' IARC, Lyon, 1982.

73. H. Seidman, M. H. Mushinski, S. K. Gelb *et al.*, 'Probabilities of eventually developing or dying of cancer—United States, 1985.' *CA Cancer J. Clin.*, 1985, **35**, 36–56.

74. P. Hartge, E. B. Harvey, W. M. Linehan *et al.*, 'Unexplained excess risk of bladder cancer in men.' *J. Natl. Cancer Inst.*, 1990, **82**, 1636–1640.

75. C. Schairer, P. Hartge, R. N. Hoover *et al.*, 'Racial differences in bladder cancer risk: a case-control study.' *Am. J. Epidemiol.*, 1988, **128**, 1027–1037.

76. C. Muir, J. Waterhouse, T. Mack *et al.* (eds.), 'Cancer Incidence in Five Continents,' IARC, Lyon, IARC Publ. No. 88, 1987, vol. V.

77. W. J. Blot and J. F. Fraumeni, Jr., 'Geographic patterns of bladder cancer in the United States.' *J. Natl. Cancer Inst.*, 1978, **61**, 1017–1023.

78. T. J. Mason, 'Atlas of Cancer Mortality for US Counties: 1950–1969.' National Institutes of Health, 1975.

79. P. Vineis, 'Epidemiological models of carcinogenesis: the example of bladder cancer.' *Cancer Epidemiol., Biomarkers Prev.*, 1992, **1**, 149–153.

80. D. M. Grant, M. Blum, M. Beer *et al.*, 'Monomorphic and polymorphic human arylamine *N*-acetyltransferases: a comparison of liver isozymes and expressed products of two cloned genes.' *Mol. Pharmacol.*, 1990, **39**, 184–191.

81. R. F. Minchin, P. T. Reeves, C. H. Teitel *et al.*, '*N*- and *O*-acetylation of aromatic and heterocyclic amine carcinogens by human monomorphic and polymorphic acetyltransferases expressed in *COS-1* cells.' *Biochem. Biophys. Res. Commun.*, 1992, **185**, 839–844.

82. D. W. Hein, M. A. Doll, T. D. Rustan *et al.*, 'Metabolic activation and deactivation of arylamine carcinogens by recombinant human NAT1 and polymorphic NAT2 acetyltransferases.' *Carcinogenesis*, 1993, **14**, 1633–1638.

83. D. W. Hein, 'Acetylator genotype and arylamine-induced carcinogenesis.' *Biochim. Biophys. Acta*, 1988, **948**, 37–66.

84. D. P. Evans, '*N*-acetyltransferase.' *Pharmacol. Ther.*, 1989, **42**, 157–234.

85. S. Mommsen, N. M. Barfod and J. Aagaard, '*N*-acetyltransferase phenotypes in the urinary bladder carcinogenesis of a low risk population.' *Carcinogenesis*, 1985, **6**, 199–201.

86. G. M. Lower, Jr., T. Nilsson, C. E. Nelson *et al.*, '*N*-acetyltransferase phenotype and risk in urinary bladder cancer: approaches in molecular epidemiology. Preliminary results in Sweden and Denmark.' *Environ. Health Perspect.*, 1979, **29**, 71–79.

87. P. Vineis, N. Caporaso, S. R. Tannenbaum *et al.*, 'Acetylation phenotype, carcinogen–hemoglobin adducts, and cigarette smoking.' *Cancer Res.*, 1990, **50**, 3002–3004.

88. P. Vineis, H. Bartsch, N. Caporaso *et al.*, 'Genetically based *N*-acetyltransferase metabolic polymorphism and low-level environmental exposure to carcinogens.' *Nature*, 1994, **369**, 154–156.

89. M. C. Yu, P. L. Skipper, K. Taghizadeh *et al.*, 'Acetylator phenotype, aminobiphenyl–hemoglobin adduct levels, and bladder cancer risk in white, black, and Asian men in Los Angeles, California.' *J. Natl. Cancer Inst.*, 1994, **86**, 712–716.

90. R. B. Hayes, W. Bi, N. Rothman *et al.*, '*N*-acetylation phenotype and genotype and risk of bladder cancer, in benzidine-exposed workers.' *Carcinogenesis*, 1993, **14**, 675–678.

91. R. A. Cartwright, R. W. Glasham, H. J. Rogers *et al.*, 'Role of *N*-acetyltransferases phenotypes in bladder carcinogenesis: a pharmacogenetic epidemiological approach to bladder cancer.' *Lancet*, 1982, **2**, 842–845.

92. J. A. Miller and E. C. Miller, in 'Biological Reactive Intermediates,' eds. D. J. Jollow, J. J. Kocsis, R. Snyder *et al.*, Plenum, New York, 1977, pp. 6–24.

93. W. G. Kirlin, A. Trinidad, T. Yerokun *et al.*, 'Polymorphic expression of acetyl coenzyme A-dependent arylamine *N*-acetyltransferase and acetyl coenzyme A-dependent *O*-acetyltransferase-mediated activation of *N*-hydroxyarylamines by human bladder cytosol.' *Cancer Res.*, 1989, **49**, 2448–2454.

94. S. Swaminathan and C. A. Reznikoff, 'Metabolism and nucleic acid binding of *N*-hydroxy-4-acetylaminobiphenyl and *N*-acetoxy-4-acetylaminobiphenyl by cultured human uroepithelial cells.' *Cancer Res.*, 1992, **52**, 3286–3294.

95. S. M. Frederickson, J. F. Hatcher, C. A. Reznikoff *et al.*, 'Acetyl transferase-mediated metabolic activation of *N*-hydroxy-4-aminobiphenyl by human uroepithelial cells.' *Carcinogenesis*, 1992, **13**, 955–961.

96. J. F. Hatcher and S. Swaminathan, 'Microsome-mediated transacetylation and binding of *N*-hydroxy-4-aminobiphenyl to nucleic acids by hepatic and bladder tissues from dog.' *Carcinogenesis*, 1992, **13**, 1705–1711.

97. D. W. Hein, 'Genetic polymorphism and cancer susceptibility: evidence concerning acetyltransfrases and cancer of the urinary bladder.' *Bioessays*, 1988, **9**, 200–204.

98. S. J. Land, K. Zukowski, M. S. Lee *et al.*, 'Metabolism of aromatic amines: relationships of *N*-acetylation, *O*-acetylation, *N,O*-acetyltransfer and deacetylation in human liver and urinary bladder.' *Carcinogenesis*, 1989, **10**, 727–731.

99. M. A. Butler, M. Iwasaki, F. P. Guengerich *et al.*, 'Human cytochrome P-450$_{PA}$ (P-450IA2), the phenacetin *O*-deethylase, is primarily responsible for the hepatic 3-demethylation of caffeine and *N*-oxidation of carcinogenic arylamines.' *Proc. Natl. Acad. Sci. USA*, 1989, **86**, 7696–7700.

100. H. Bartsch, C. Malaveille, M. Friesen *et al.*, 'Black (air-cured) and blond (flue-cured) tobacco cancer risk. IV. Molecular dosimetry studies implicate aromatic amines as bladder carcinogens.' *Eur. J. Cancer*, 1993, **29A**, 1199–1207.

101. F. A. Beland and F. F. Kadlubar, 'Formation and persistence of arylamine DNA adducts *in vivo*.' *Environ. Health Perspect.*, 1985, **62**, 19–30.

102. M. C. Poirier and F. A. Beland, 'DNA adduct measurements and tumor incidence during chronic carcinogen exposure in animal models: implications for DNA adduct-based human cancer risk assessment.' *Chem. Res. Toxicol.*, 1992, **5**, 749–755.

103. G. Talaska, A. Z. S. S. al-Juburi and F. F. Kadlubar, 'Smoking-related carcinogen-DNA adducts in biopsy samples of human urinary bladder: identification of *N*-(deoxyguanosin-8-yl)-4-aminobiphenyl as a major adduct.' *Proc. Natl. Acad. Sci. USA*, 1991, **88**, 5350–5354.

104. G. Talaska, M. Schamer, P. Skipper *et al.*, 'Detection of carcinogen–DNA adducts in exfoliated urothelial cells of cigarette smokers: association with smoking, hemoglobin adducts, and urinary mutagenicity.' *Cancer Epidemiol., Biomarkers Prev.*, 1991, **1**, 61–66.

105. M. Hollstein, D. Sidransky, B. Vogelstein *et al.*, 'p53 mutations in human cancers.' *Science*, 1991, **253**, 49–53.

106. D. P. Lane, 'Cancer. p53, guardian of the genome.' *Nature*, 1992, **358**, 15–16.

107. C. H. I. Spruck, III, P. F. Ohneseit, M. Gonzalez-Zulueta *et al.*, 'Two molecular pathways to transitional cell carcinoma of the bladder.' *Cancer Res.*, 1994, **54**, 784–788.

108. D. Esrig, C. H. Spruck, III, P. W. Nichols *et al.*, 'p53 nuclear protein accumulation correlates with mutations in the p53 gene, tumor grade, and stage in bladder cancer.' *Am. J. Pathol.*, 1993, **143**, 1389–1397.

109. D. Esrig, D. Elmajian, S. Groshen *et al.*, 'Accumulation of nuclear p53 and tumor progression in bladder cancer.' *N. Engl. J. Med.*, 1994, **331**, 1259–1264.

110. F. F. Kadlubar, J. A. Miller and E. C. Miller, 'Hepatic microsomal *N*-glucuronidation and nucleic acid binding of *N*-hydroxy arylamines in relation to urinary bladder carcinogenesis.' *Cancer Res.*, 1977, **37**, 805–814.

111. R. W. Wise, T. V. Zenser, F. F. Kadlubar *et al.*, 'Metabolic activation of carcinogenic aromatic amines by dog bladder and kidney prostaglandin H synthase.' *Cancer Res.*, 1984, **44**, 1893–1897.

112. F. F. Kadlubar, C. B. Frederick, C. C. Weis *et al.*, 'Prostaglandin endoperoxide synthetase-mediated metabolism of carcinogenic aromatic amines and their binding to DNA and protein.' *Biochem. Biophys. Res. Commun.*, 1982, **108**, 253–258.

113. Y. Yamazoe, T. V. Zenser, D. W. Miller *et al.*, 'Mechanism of formation and structural characterization of DNA adducts derived from peroxidative activation of benzidine.' *Carcinogenesis*, 1988, **9**, 1635–1641.

114. A. Danon, T. V. Zenser, D. L. Thomasson *et al.*, 'Eicosanoid synthesis by cultured human urothelial cells: potential role in bladder cancer.' *Cancer Res.*, 1986, **46**, 5676–5681.

115. T. J. Flammang, Y. Yamazoe, R. W. Benson *et al.*, 'Arachidonic acid-dependent peroxidative activation of carcinogenic arylamines by extrahepatic human tissue microsomes.' *Cancer Res.*, 1989, **49**, 1977–1982.

116. H. Bartsch, J. A. Miller and E. C. Miller, '*N*-acetoxy-*N*-acetylaminoarenes. One-electron nonenzymatic and enzymatic oxidation products of various carcinogenic aromatic acethydroxamic acids.' *Biochim. Biophys. Acta*, 1972, **273**, 40–51.

117. G. J. Dutton (ed.), 'Glucuronidation of Drugs and Other Compounds,' CRC Press, Boca Raton, FL, 1980.

118. G. J. Mulder, M. W. H. Coughtrie and B. Burchell, in 'Conjugation Reactions In Drug Metabolism: An Integrated Approach,' ed. G. J. Mulder, Taylor & Francis, New York, 1990, pp. 51–105.

119. S. R. Babu, V. M. Lakshmi, F. F. Hsu *et al.*, '*N*-acetyl-benzidine-*N'*-glucuronidation by human, dog, and rat liver.' *Carcinogenesis*, 1993, **14**, 2605–2611.

120. C. B. Frederick, C. C. Weis, T. J. Flammang *et al.*, 'Hepatic *N*-oxidation, acetyl-transfer and DNA-binding of the acetylated metabolites of the carcinogen, benzidine.' *Carcinogenesis*, 1985, **6**, 959–965.

121. S. R. Babu, V. M. Lakshmi, T. V. Zenser *et al.*, 'Glucuronidation of *N*-acetylbenzidine by human liver.' *Drug Metab. Dispos.*, 1994, **22**, 922–927.

122. M. R. Probst, P. Jeno and U. A. Meyer, 'Purification and characterization of a human liver arylacetamide deacetylase.' *Biochem. Biophys. Res. Commun.*, 1991, **177**, 453–459.

123. M. R. Probst, M. Beer, D. Beer *et al.*, 'Human liver arylacetamide deacetylase. Molecular cloning of a novel esterase involved in the metabolic activation of arylamine carcinogens with high sequence similarity to hormone-sensitive lipase.' *J. Biol. Chem.*, 1994, **269**, 21650–21656.

124. T. J. Haley, 'Benzidine revisited: a review of the literature and problems associated with the use of benzidine and its congeners.' *Clin. Toxicol.*, 1975, **8**, 13–42.

125. J. L. Radomski, 'The primary aromatic amines: their biological properties and structure-activity relationships.' *Annu. Rev. Pharmacol. Toxicol.*, 1979, **19**, 129–157.

126. G. M. Lower, Jr. and G. T. Bryan, 'Enzymatic *N*-acetylation of carcinogenic aromatic amines by liver cytosol of species displaying different organ susceptibilities.' *Biochem. Pharmacol.*, 1973, **22**, 1581–1588.

127. L. A. Poirier, J. A. Miller and E. C. Miller, 'The *N*- and ring-hydroxylation of 2-acetylamino-fluorene and the failure to detect *N*-acetylation of 2-aminofluorene in the dog.' *Cancer Res.*, 1963, **23**, 790–800.

128. T. Sone, K. Zukowski, S. J. Land *et al.*, 'Characteristics of a purified dog hepatic microsomal *N,O*-acyltransferase.' *Carcinogenesis*, 1994, **15**, 595–599.

129. S. M. Frederickson, E. M. Messing, C. A. Reznikoff *et al.*, 'Relationship between *in vivo* acetylator phenotypes and cytosolic *N*-acetyltransferase and *O*-acetyltransferase activities in human uroepithelial cells.' *Cancer Epidemiol., Biomarkers Prev.*, 1994, **3**, 25–32.

130. T. V. Zenser, M. B. Mattammal, H. J. Armbrecht *et al.*, 'Benzidine binding to nucleic acids mediated by the peroxidative activity of prostaglandin endoperoxide synthetase.' *Cancer Res.*, 1980, **40**, 2839–2845.

131. K. Hemminki, 'DNA adducts, mutations and cancer.' *Carcinogenesis*, 1993, **14**, 2007–2012.

132. G. Qian, R. K. Ross, M. C. Yu *et al.*, 'A follow-up study of urinary markers of aflatoxin exposure and liver cancer risk in Shanghai, People's Republic of China.' *Cancer Epidemiol., Biomarkers Prev.*, 1994, **3**, 3–10.

133. D. Tang, R. M. Santella, A. M. Blackwood *et al.*, 'A molecular epidemiological case-control study of lung cancer.' *Cancer Epidemiol., Biomarkers Prev.*, 1995, **4**, 341–346.

134. G. Talaska, W. W. Au, J. J. B. Ward, Jr. *et al.*, 'The correlation between DNA adducts and chromosomal aberrations in the target organ of benzidine exposed, partially-hepatectomized mice.' *Carcinogenesis*, 1987, **8**, 1899–1905.

135. T. R. Fox, A. M. Schumann, P. G. Watanabe *et al.*, 'Mutational analysis of the H-*ras* oncogene in spontaneous C57BL/6 × C3H/He mouse liver tumors and tumors induced with genotoxic and nongenotoxic hepatocarcinogens.' *Cancer Res.*, 1990, **50**, 4014–4019.

136. D. M. Grant, P. D. Josephy, H. L. Lord *et al.*, '*Salmonella typhimurium* strains expressing human arylamine *N*-acetyltransferases: metabolism and mutagenic activation of aromatic amines.' *Cancer Res.*, 1992, **52**, 3961–3964.

137. W. B. Melchior, Jr., M. M. Marques and F. A. Beland, 'Mutations induced by aromatic amine DNA adducts in pBR322.' *Carcinogenesis*, 1994, **15**, 889–899.

138. J. L. Bos, 'The ras gene family and human carcinogenesis.' *Mutat. Res.*, 1988, **195**, 255–271.

139. C. Shih and R. A. Weinberg, 'Isolation of a transforming sequence from a human bladder carcinoma cell line.' *Cell*, 1982, **29**, 161–169.

140. J. Fujita, S. K. Srivastava, M. H. Kraus *et al.*, 'Frequency of molecular alterations affecting ras protooncogenes in human urinary tract tumors.' *Proc. Natl. Acad. Sci. USA*, 1985, **82**, 3849–3853.

141. S. M. Cohen and L. B. Ellwein, 'Genetic errors, cell proliferation, and carcinogenesis.' *Cancer Res.*, 1991, **51**, 6493–6505.

142. C. B. Thompson, 'Apoptosis in the pathogenesis and treatment of disease.' *Science*, 1995, **267**, 1456–1462.

143. G. T. Bryan, 'Pellet implantation studies of carcinogenic compounds.' *J. Natl. Cancer Inst.*, 1969, **43**, 255–261.

144. T. Shirai, E. Ikawa, S. Fukushima, T. Masui *et al.*, 'Uracil-induced urolithiasis and the development of reversible papillomatosis in the urinary bladder of F344 rats.' *Cancer Res.*, 1986, **46**, 2062–2067.

145. S. Stinchcombe, A. Buchmann, K. W. Boch *et al.*, 'Inhibition of apoptosis during 2,3,7,8-tetrachlorodibenzo-*p*-dioxin-mediated tumour promotion in rat liver.' *Carcinogenesis*, 1995, **16**, 1271–1275.

146. M. Okamoto, H. Kawamata, K. Kawai *et al.*, 'Enhancement of transformation *in vitro* of a nontumorigenic rat urothelial cell line by interleukin 6.' *Cancer Res.*, 1995, **55**, 4581–4585.

147. G. Majno and I. Joris, 'Apoptosis, oncosis, and necrosis. An overview of cell death.' *Am. J. Pathol.*, 1995, **146**, 3–15.

7.21

Glutathione Status and Other Antioxidant Defense Mechanisms

LAWRENCE H. LASH

Wayne State University School of Medicine, Detroit, MI, USA

7.21.1 INTRODUCTION

7.21.1.1 Redox Balance and Oxidative Stress in Cells

7.21.1.1.1 Definition of redox balance

Redox pairs whose midpoint potentials fall between these two extreme values are the most effective, if present at sufficient concentrations in cells, in regulating intracellular redox homeostasis.

Just as processes in nature tend towards higher entropy and disorder without the input of energy, redox status tends toward oxidation without the input of reducing agents. These reducing agents serve as donors of electrons and hydrogen atoms, with electrons being on carbon, oxygen, nitrogen, sulfur, and phosphorus atoms. The intracellular milieu is highly reduced under physiological conditions. Maintenance of this state, however, requires a constant supply of reductants, which can act through both enzymatic and nonenzymatic pathways to reduce oxidants.

Disturbances in the ability of cells to maintain redox homeostasis can result primarily from three causes: (i) a deficit in the supply of reductants, (ii) an excess in the supply of oxidants, or (iii) inhibition of any of the enzymes that are involved in regulating redox status. Enzymatic processes that are critical to redox homeostasis can be inhibited by cofactor depletion or by the presence of a specific inhibitor. In either case, cellular redox status will gradually become more oxidized with time, until the inhibitor is removed or new enzyme is synthesized. In many cases, no apparent signs of cellular dysfunction or injury are observed when a specific enzyme that is involved in the regulation of redox status is inhibited or otherwise not functioning at optimal activity levels. However, subsequent exposure of cells to an oxidant or a chemical that generates an oxidant under these conditions, will often result in greater dysfunction or toxicity than would occur in normal cells.

The situation that results from an imbalance of redox status in favor of oxidants is termed an "oxidative stress." Reactive oxygen species (ROS) are a major factor in production of oxidative stress, and are generated by the partial reduction of molecular oxygen (Figure 1). During mitochondrial respiration, molecular oxygen normally undergoes a four-electron reduction to water. Partial reduction can occur, however, at various steps, such as the ubiquinone-cytochrome b step, yielding the one-electron reduced form, the superoxide anion radical ($O_2^{-\cdot}$). Since $O_2^{-\cdot}$ is an anion, it cannot readily cross biological membranes. In

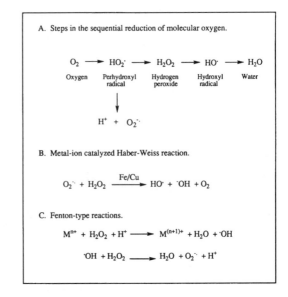

A. Steps in the sequential reduction of molecular oxygen.

$$O_2 \longrightarrow HO_2^{\cdot} \longrightarrow H_2O_2 \longrightarrow HO^{\cdot} \longrightarrow H_2O$$

Oxygen · Perhydroxyl radical · Hydrogen peroxide · Hydroxyl radical · Water

$$H^+ + O_2^{-\cdot}$$

B. Metal-ion catalyzed Haber–Weiss reaction.

$$O_2^{-\cdot} + H_2O_2 \xrightarrow{Fe/Cu} HO^{\cdot} + {}^-OH + O_2$$

C. Fenton-type reactions.

$$M^{n+} + H_2O_2 + H^+ \longrightarrow M^{(n+1)+} + H_2O + {}^{\cdot}OH$$

$${}^{\cdot}OH + H_2O_2 \longrightarrow H_2O + O_2^{-\cdot} + H^+$$

Figure 1 Redox reactions of molecular oxygen producing free radicals and reactive species.

contrast, two-electron reduction of molecular oxygen to generate hydrogen peroxide, which can dismutate via metal-catalyzed Haber–Weiss or Fenton-type reactions to form the hydroxyl radical, generates reactive species that can readily cross biological membranes and produce widespread cellular injury. In addition to these ROS, molecular oxygen can be electronically activated by, for example, ultraviolet light or x rays, to an excited state, producing singlet oxygen ($^1\Delta O_2$). Concentrations of these reactive species are normally maintained at very low intracellular levels, but can increase to supraphysiological levels if redox homeostasis is not maintained.

7.21.1.1.2 Cellular markers of oxidative stress

"Oxidative stress" has been the subject of extensive studies in both the clinical and basic sciences since the mid-1980s. Numerous toxicological and pathological conditions have been associated with oxidative stress, including the decrease in tissue function and cellular degeneration that occurs in aging.[1] The mitochondrion is an excellent example of a subcellular organelle whose function is closely linked to maintenance of redox balance. A large number of mitochondrial enzymes, including dehydrogenases and transport ATPases, contain critical sulfhydryl groups that must be maintained in the reduced form for proper function.[2] Furthermore, redox-sensitive components in the electron transport chain, such as iron ions on heme prosthetic groups and iron–sulfur centers, may be oxidized during a redox imbalance, thereby producing mitochondrial dysfunction.

Kidney cells have a high basal metabolic rate, reflected by its high rate of oxygen consumption, thus making them dependent on mitochondrial function. As mitochondria are the primary intracellular sites of oxygen consumption, they may also be primary sites of generation of ROS. Although normal electron transport in mitochondria involves four-electron reduction of molecular oxygen to water, partial reduction reactions occur even under physiological conditions, causing release of $O_2^{-\cdot}$ and hydrogen peroxide. Toxic or pathological conditions, such as oxidative stress, that lead to an impairment of mitochondrial function, can increase release of ROS. Renal cells possess several membrane transporters for regulation of ion homeostasis and metabolite distribution that are directly or indirectly dependent on mitochondrial ATP and hence, on redox status (Figure 2).

The primary consumer of cellular ATP is the $(Na^+ + K^+)$-stimulated ATPase (EC 3.6.1.3) on the basolateral plasma membrane, which catalyzes efflux of two Na^+ ions and uptake of three K^+ ions, thus establishing and maintaining the electrochemical gradient for Na^+ and K^+ ions across renal cellular plasma membranes. Other primary active transport systems that are critical to renal cellular function and are redox-dependent include the Ca^{2+}-ATPases on the plasma membrane and the endoplasmic reticulum. Oxidation or alkylation of the sulfhydryl groups on the various cellular ATPases, which can readily occur in oxidative stress, inhibits their activity and compromises the ability of the cell to regulate ion and metabolite homeostasis. Measurement of intracellular concentrations of Na^+, K^+, or Ca^{2+} ions can, therefore, serve as markers of cellular dysfunction.

Other transport systems whose mechanism involves coupling of substrate transport to the gradient of Na^+ ions include the organic anion secretory system on the basolateral plasma membrane (which is characteristically inhibited by probenecid and *p*-aminohippurate (PAH)) and Na^+-coupled transport systems for uptake of glucose and amino acids on the brush-border plasma membrane.

Ames and co-workers[1] argue that oxidative damage plays a major role in the mitochondrial dysfunction that occurs progressively in aging. Several mitochondrial functions decline with age due to the intrinsic rate of proton leakage across the mitochondrial inner membrane, decreased membrane fluidity, and decreased levels of cardiolipin. Hence, the baseline redox status would be viewed as tending over time to a more oxidized state as the efficiency of regulatory processes declines. This tendency would, therefore, lead to enhanced generation of ROS and other pro-oxidant species. Alterations in mitochondrial efficiency and function have obvious consequences for cellular energetics, causing a decrease in the ability of the cell to generate ATP for biosynthetic reactions and active transport processes (cf. Figure 2).

Oxidative stress can cause peroxidation of membrane lipids, which can be assessed by measurement of lipid peroxidation by-products. Polyunsaturated fatty acids of membrane lipids are susceptible to attack by free radicals, including those derived by the partial reduction of molecular oxygen. As shown in Figure 3, lipid peroxidation can proceed through four steps: (i) an initiation reaction in which a free radical abstracts a hydrogen atom from a fatty acid residue of a lipid molecule, which forms a conjugated diene after an intramolecular rearrangement, (ii) an oxidation reaction in which molecular oxygen reacts with the conjugated diene to form a peroxy radical derivative, (iii) a propagation reaction in which the lipid radical reacts with another cellular constituent, and (iv) a degradation reaction, which results in breakdown of the fatty acid residue to several characteristic products. Many of these products, such as alkanes and malondialdehyde, can be measured in renal tissue to indicate occurrence of an oxidative stress. Depending on the specific fatty acid residue that is peroxidized, a large variety of complex products can be produced, including alkanals, alkenals, ketones, hydroxy acids, and hydroxyaldehydes. Several of these products can be measured in urine or exhaled breath with noninvasive methods, thereby being of utility in clinical studies of oxidative injury.

Other cellular consequences of oxidative stress include glutathione (GSH) and protein sulfhydryl oxidation, DNA single and double strand breaks, and modified DNA bases such as 8-hydroxy 2-deoxyguanosine. Changes in these parameters depend on a balance between reductants and oxidants and can thus serve as markers of redox imbalance. Cells contain several classes of antioxidants that are either normal constituents of the cell or are obtained in the diet. Additionally, enzymatic mechanisms exist for reduction of oxidants, thereby maintaining the reduced intracellular environment.

Several studies in the 1990s have suggested that oxidative stress may be the underlying factor in or cause of apoptosis or programmed cell death.[3,4] In this manner, alterations in

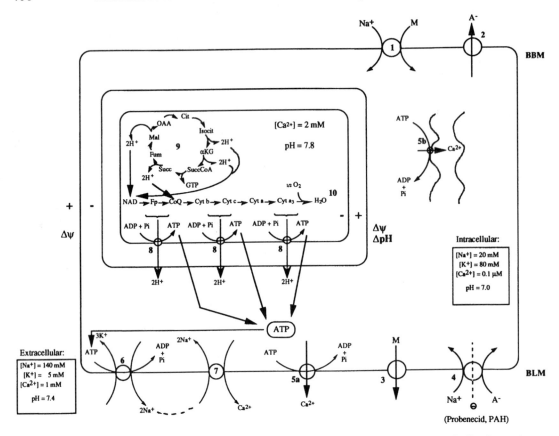

Figure 2 Energy-dependent functions in a renal epithelial cell. 1, Na⁺-dependent metabolite (e.g., glucose, amino acids) uptake across the BBM; 2, anion (A⁻) efflux system on the BBM; 3, facilitated transport system on BLM for efflux of metabolites; 4, Na⁺-coupled anion uptake system on BLM (probenecid- and PAH-sensitive); 5a, Ca²⁺-dependent ATPase on BLM; 5b, Ca²⁺-dependent ATPase on endoplasmic reticulum; 6, (Na⁺K⁺)-stimulated ATPase on BLM; 7, Na⁺–Ca²⁺ exchanger on BLM; 8, F₀F₁-ATPase on mitochondrial inner membrane; 9, citric acid cycle; 10, mitochondrial electron transport chain. Abbreviations: BBM, brush-border membrane; BLM, basolateral membrane; PAH, *p*-aminohippurate; Fp, flavoprotein; CoQ, coenzyme Q or ubiquinone; Cyt, cytochrome; Cit, citrate; Isocit, isocitrate; αKG, α-Ketoglutarate; SuccCoA, succinyl-CoA, Succ, succinate; Fum, fumarate; Mal, malate; OAA, oxaloacetate.

redox status may function as an initial signaling mechanism to produce cell death that occurs during organogenesis or exposure to low doses of various toxicants. The proto-oncogene product bcl-2, which is a mitochondrial protein, is thought to inhibit apoptosis by acting as an antioxidant.[5] Mutations in the bcl-2 gene, which either produce nonfunctional protein or no protein, can cause an oxidative stress that promotes apoptosis.

7.21.1.2 Nonenzymatic Antioxidant Defense Mechanisms

7.21.1.2.1 *Glutathione*

Glutathione (GSH) is a tripeptide containing L-glutamate, L-cysteine, and glycine, is the predominant nonprotein thiol in all mammalian cells, and is found at concentrations of 1–5 mM. The structure of GSH shown in Figure 4 high-lights the two critical features of the molecule that enable it to have such a wide variety of critical cellular functions. Since GSH is not synthesized by the ribosomal protein synthetic machinery, but is synthesized by two enzymes that are primarily of cytosolic origin (see Section 7.21.2.1.1), the peptide bond between the L-glutamyl and L-cysteinyl residues exists between the α-amino group of L-cysteine and the γ-carboxyl group of L-glutamate. The isopeptide bond makes the GSH molecule resistant to protease-catalyzed degradation. The second critical feature of the GSH molecule is the sulfhydryl or thiol group, which gives GSH the properties of a nucleophile and a reductant. Both these chemical properties allow GSH to function, both nonenzymatically and as a cosubstrate or cofactor in specific enzyme-catalyzed reactions (see Section 7.21.1.3.1), to prevent the accumulation of oxidants or reactive electrophiles that can produce cellular injury.

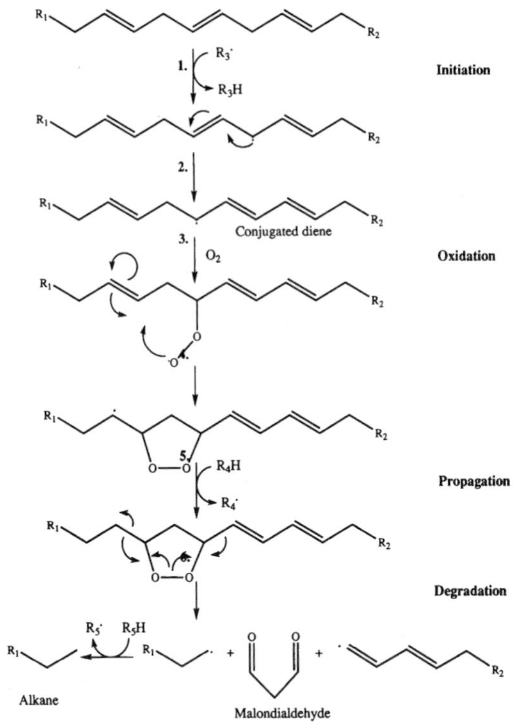

Figure 3 Lipid peroxidation reactions. R_1–R_5 are alkyl groups.

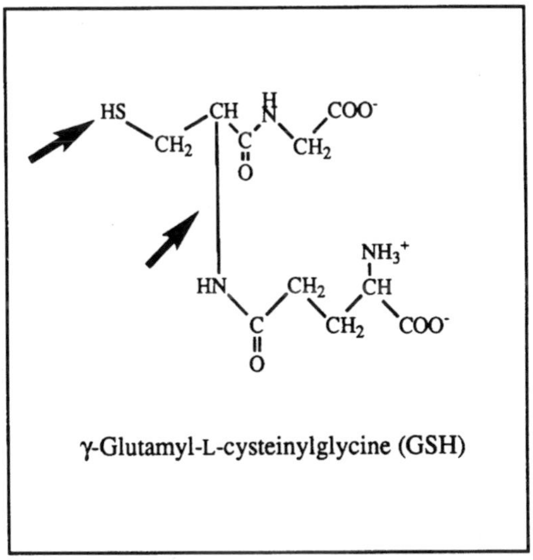

Figure 4 Structure of glutathione.

The ability of the GSH molecule to undergo reversible oxidation to glutathione disulfide (GSSG) allows it to participate in redox reactions with most physiological redox couples. The reaction has a midpoint electrochemical or redox potential (E'_0 of the GSH/GSSG pair) of -0.23 V. In addition to the disulfide form, which is the predominant form of glutathione besides GSH, several other oxidized forms of glutathione are present in cells, albeit at relatively low concentrations (Table 1). Of particular importance in a consideration of oxidative stress and GSH status is the presence of mixed disulfides of GSH and protein, which may be involved in regulation of intermediary metabolism and enzyme activity.[6]

7.21.1.2.2 α-Tocopherol

α-Tocopherol (αT; vitamin E) is a lipid-soluble vitamin that is believed to play a role in protection of membranes against oxidative damage. αT reacts with a free radical, which may be lipid-derived, to convert it to a nonradical, and, therefore, nonreactive form. αT is converted to the chromanoxyl radical, which is relatively stable, but cannot function in scavenging other free radicals. To prevent oxidative stress from occurring, αT must be regenerated from the free radical form. Although several reductants, including ascorbate and ubiquinol, have been suggested to function in the regeneration of functional αT, GSH appears to be the most likely physiological reductant. Burk[7] showed that GSH prevents lipid peroxidation of microsomal membranes *in vitro*, but only in the presence of adequate concentrations of αT.[8] In rats fed a vitamin E-deficient diet, enhanced lipid peroxidation is observed in liver, kidney, lung, heart, and skeletal muscle,[9] consistent with the general importance of αT in maintenance of cellular redox status.

The above-mentioned data were taken to suggest that an αT chromanoxyl radical reductase activity, that is at least partly dependent on GSH, is present in mammalian tissues. Additional studies by Packer and co-workers[10] suggested the presence of multiple mechanisms for regeneration of functional αT from the chromanoxyl radical. These mechanisms include the quenching of radical formation by both ascorbate and thiols (dihydrolipoate and GSH) in liposomes, low-density lipoproteins, and biological membranes from liver microsomes, mitochondria, and submitochondrial particles. Furthermore, the operation of multiple electron transport mechanisms, involving NADPH- and NADH-dependent enzymes in microsomes and NADH- or succinate-dependent enzymes in mitochondria, were demonstrated, indicating that the initial hypothesis of a single vitamin E radical reductase as a single regenerating system was an oversimplification.

7.21.1.2.3 Ascorbate

Ascorbic acid is a water-soluble antioxidant that can scavenge both hydroxyl radicals and

Table 1 Forms of glutathione found in biological systems.

			Sulfur valence number			
-2	*-1*	*0*	*+1*	*+2*	*+3*	*+4*
GSH (thiol)	GSSR (disulfide)	GSOH (sulfenic acid)	G-SO-S-R (thiosulfinate)	GSO_2^- (sulfinic acid)	$G-SO_2S^-$ (thiosulfonate)	GSO_3^- (sulfonic acid)
G-S-R (thioether; sulfide)	GSSH (persulfide)		G-SO-R (sulfoxide)	$G-SO_2$-R (sulfone)	G-SO2S-R (thiosulfonate)	
G-S-CO-R (thiol ester)	$GSSO_3^-$ (thiosulfate ester)					

R represents various groups to which the sulfur can be bound, and can be another molecule of GSH or another organic molecule. Sulfur is linked to these groups through a carbon bond. Modified from L. H. Lash, 'Functions of the Basal-Lateral Membrane in Thiol Metabolism,' Ph.D. dissertation, Emory University, 1984.

superoxide anions and can also react with thiyl radicals and hypochlorous acid. Although most animals can synthesize ascorbic acid from glucose, humans, other primates, guinea pigs, and fruit bats have lost one of the enzymes necessary in the biosynthetic pathway and must obtain ascorbic acid from the diet. Ascorbic acid is normally present in high concentrations (millimolar range) in several tissues, including the eye, kidney, spinal cord, lung, and adrenal cortex. Human plasma contains 50–200 µM ascorbic acid, which is 10- to 50-fold higher than reported concentrations of GSH, suggesting that ascorbic acid may play an important role in certain extracellular redox reactions.

Ascorbic acid undergoes reversible oxidation, first to the semidehydroascorbic acid radical and then, by dismutation, to dehydroascorbic acid. One chemical that may promote the oxidation of ascorbic acid is the αT chromanoxyl radical. There is much controversy about the regeneration of ascorbic acid from its oxidized forms. GSH is generally accepted to be the primary intracellular reductant for dehydroascorbic acid. The controversy, however, lies in the issue of whether or not the regeneration occurs enzymatically or nonenzymatically.[11] Although evidence of enzymatic reduction of dehydroascorbate has been reported for a number of years, some investigators have concluded that these findings are either artifact or that any enzymatic activity that may be present is quantitatively insignificant. In the mid-1990s, Maellaro *et al.*[12] reported the purification of a GSH-dependent dehydroascorbate reductase activity from rat liver, although the physiological importance of this enzymatic activity is unclear.

7.21.1.2.4 Other nonenzymatic antioxidants

Numerous other antioxidants are found in cells and extracellular fluids. Some are naturally-occurring and are normal constituents of cells, whereas others are ingested in the diet. β-Carotene (vitamin A) and uric acid are potent singlet oxygen quenchers. Plasma proteins such as ceruloplasmin are also redox active and may play a role in maintenance of plasma redox status. In the diet, flavonoids, which are plant antioxidants (e.g., rutin, quercetin), can provide antioxidant defense in the gastrointestinal tract. Synthetic chemicals that are commonly added to foods, such as the preservatives butylated hydroxytoluene (BHT) and butylated hydroxyanisole (BHA) may similarly provide antioxidant defense.

7.21.1.3 Enzymatic Antioxidant Defense Mechanisms

7.21.1.3.1 GSH-dependent enzymes

GSH is a critical element of cellular defense against oxidants and reactive electrophiles chiefly due to the presence of the thiol group on the cysteinyl residue. Reactivity of the thiol group, however, is dependent on formation of the thiolate anion (GS^-). The pK_a for the ionization of the thiol group is influenced by the chemical properties of the adjacent atoms in the molecule. For simple aliphatic thiols, the pK_a varies from 9 to 11; the pK_a for GSH is 9.2. In contrast, aliphatic aminothiols, such as cysteine and cysteamine, have pK_a values near eight due to the inductive effect of the nearby positively charged ammonium group; aromatic or heteroaromatic thiols have lower pK_a values, some as low as 6 to 7. Hence, although the proportion of the thiol present as the reactive thiolate anion varies considerably between different thiols, the mechanism underlying their reactivity is the same, namely, generation of the thiolate anion and subsequent nucleophilic attack on electrophiles. Although the alkaline pK_a of the thiol group of GSH would indicate that only 1.6% of total GSH molecules would be in the deprotonated form, the proportion of thiolate anion is increased when GSH binds to the active site of an enzyme that requires GSH to be in the thiolate form, thereby increasing the efficiency of the reaction.

The reactions of GSH that are important for cellular defense against oxidative stress can thereby be divided into functions of GSH as a nucleophile and as a reductant (Figure 5). The GSH S-transferases (GSTs), which in the kidney are a family of isozymes that are believed to be localized exclusively in the cytosol, catalyze the formation of thioether conjugates of GSH with a vast number of electrophilic compounds. Although the tissue with the highest activity of GST is the liver, where the isozymes comprise up to 5% of total cytosolic protein, the kidneys also contain significant amounts of the enzyme. In general, GSH S-conjugates are converted by additional renal enzymes to mercapturic acids (i.e., N-acetylcysteine conjugates), which are highly polar and are excreted readily into the urine, thereby completing a detoxification function. Many of the xenobiotic substrates for the GSTs are generated by the action of cytochrome P450s, whereas others, such as halogenated hydrocarbons, are direct substrates. Besides xenobiotics, GSH conjugation of endogenous chemicals also occurs (e.g., leukotriene formation), indicating the widespread general importance of this metabolic pathway. Whereas most conjugation reactions of GSH with electrophiles and other compounds results in detoxication, in some cases, however, formation of a GSH S-conjugate is the initial step in a bioactivation mechanism that produces direct-acting epi-sulfonium compounds or acts as a precursor for cysteine S-conjugates that are substrates for the cysteine conjugate β-lyase (β-lyase) (EC 4.4.1.13), a pyridoxal phosphate-containing enzyme that metabolizes certain types of cysteine S-conjugates to reactive and toxic acylating species (see Section 7.21.2.6).

As a reductant, GSH functions in a number of critical reactions. Thioltransferases catalyze thiol–disulfide exchange reactions and are widely distributed in mammalian tissues and are found predominantly in cytosolic fractions.[13] Several activities are present and each has a broad substrate specificity. GSH functions as a reductant in the lysis of disulfide bonds in both low-molecular weight compounds and in proteins. Substrates include disulfides of coenzyme A, cystine, cystamine, homocystine, albumin, lysozyme, and ribonuclease. The microsomal thioltransferase is most active towards protein disulfide bonds, and may be important in post-translational processing. In these thioltransferase reactions, GSH acts catalytically to alter the thiol–disulfide status of proteins. In this way, thioltransferases may have important regulatory functions.[6,14,15]

Since peroxides and other ROS are requisite components of life in an aerobic environment, nonenzymatic and enzymatic mechanisms have evolved to detoxify these species. Besides catalase (see Section 7.21.1.3.3), GSH peroxidase (GPX) is the other major enzyme that can detoxify hydrogen peroxide. In addition to acting on hydrogen peroxide, GPX also detoxifies organic peroxides, indicating that peroxides that result from lipid peroxidation may be reduced in this manner. The major form of GPX contains selenium and is found in both cytosolic and mitochondrial matrix compartments of most mammalian cells. Although the enzyme has a broad specificity for peroxide substrate, it is specific for GSH as the hydrogen donor. A selenium-independent GPX, which is a catalytic activity of a GST isozyme, is also present and acts solely on organic peroxides.[16,17] Although this activity can increase substantially in the livers of selenium-deficient rats, its role under normal physiological conditions is unclear.[18]

GSH S-Transferases:

$$R\text{-}X + GSH \longrightarrow R\text{-}SG + H\text{-}X$$

Thioltransferases:

$$RSSR' + GSH \rightleftharpoons RSSG + R'SH$$

$$RSSG + GSH \rightleftharpoons GSSG + R'SH$$

GSH Peroxidase:

$$2\,GSH + ROOH \longrightarrow GSSG + ROH + H_2O$$

GSSG Reductase:

$$GSSG + NADPH + H^+ \rightleftharpoons 2\,GSH + NADP^+$$

Figure 5 Enzymatic reactions of GSH as a nucleophile and a reductant. R-X is an electrophile with X^- as the leaving group. RSSR$'$ is either a low-molecular weight, inter- or intramolecular protein, or a mixed disulfide of protein with a low-molecular weight compound.

A critical element in the ability of the cell to utilize GSH for detoxication of peroxides and various electrophiles is the ability to regenerate GSH from the GSSG that is generated during the GPX reaction or thiol–disulfide exchange reactions. This is accomplished by the enzyme GSSG reductase (GRD), which is a flavoprotein that in liver exists as a dimer of molecular weight 125 kDa and contains one molecule of FAD per subunit.[19] Although the enzyme is highly specific for GSSG, some activity is exhibited with the mixed disulfides of GSH and γ-glutamylcysteine and of GSH and coenzyme A as substrates. NADPH is obligatorily utilized as a reductant.

The interplay between these various reactions of peroxide metabolism and GSH oxidation and GSSG reduction is illustrated in the concept of the GSH redox cycle (Figure 6). Three sets of reactions, catalyzed by GPX, GRD, and an NADPH-generating enzyme, are linked to form the redox cycle. This system can be experimentally manipulated by inhibiting GRD with 1,3-(2-bis-chloro)-1-nitrosourea (BCNU), by depleting GSH, or by limiting substrate supply, thereby limiting NADPH generation and GSH regeneration. In each case, interference with any step in the cycle makes a cell more susceptible to injury from oxidative stress.

7.21.1.3.2 Superoxide dismutase

Superoxide dismutase (SOD) is found in both the cytosolic and mitochondrial matrix fractions of most cells as a copper- and manganese-containing enzyme, respectively. It catalyzes the dismutation of $O_2^{-\cdot}$ to hydrogen peroxide according to the following reaction:

$$2O_2^{-\cdot} + 2H^+ \rightarrow H_2O_2 + O_2$$

The role of $O_2^{-\cdot}$ in toxicity can be assessed by inhibition of the enzyme with diethyldithiocarbamate. If $O_2^{-\cdot}$ is the reactive species involved in a toxic process, then inhibition of the enzyme should enhance toxicity.

7.21.1.3.3 Catalase

Catalase is a heme-containing protein found in peroxisomes, which are most abundant in liver cells and renal proximal tubules. Catalase detoxifies hydrogen peroxide by converting it to water and oxygen, according to the following reaction:

$$2H_2O_2 \rightarrow 2H_2O + O_2$$

As with $O_2^{-\cdot}$, the action of hydrogen peroxide in a toxic mechanism can be demonstrated by inhibition of catalase, using either azide or 1,3-aminotriazole. Enhancement of toxicity in the presence of one of these inhibitors is consistent with a role for hydrogen peroxide in the process of toxicity.

7.21.1.3.4 NADPH:quinone oxidoreductase (DT-diaphorase)

NADPH:quinone oxidoreductase (EC 1.6.99.2), commonly known as DT-diaphorase, is found in cytosolic and mitochondrial matrix fractions, and is inhibited by dicumarol. The enzyme catalyzes the two-electron reduction of quinones to hydroquinones, utilizing NADPH as the source of reducing equivalents, according to the following reaction:

$$NAD(P) + H^+ + Quinone$$
$$\rightarrow NAD(P)^+ + Hydroquinone$$

DT-diaphorase is important toxicologically because of the large number of quinone-containing or quinone-like chemicals in nature that are either used therapeutically or are present in the environment. One of the most studied and best characterized examples is the redox-cycling quinone menadione (Figure 7). Menadione produces cytotoxicity in a variety of tissues and cell types, including renal proximal tubular (PT) and distal tubular (DT) cells,[20] by a mechanism that involves generation of $O_2^{-\cdot}$ and is called redox cycling. The fully oxidized form (quinone) can undergo a one-electron, NADPH-dependent reduction via a flavoprotein to produce the semiquinone. The semiquinone can react with molecular oxygen, regenerating the quinone and producing $O_2^{-\cdot}$. DT-diaphorase plays a detoxication role by bypassing the semiquinone and reducing the quinone by two electrons to the hydroquinone, which can redox cycle, but does so more slowly than the semiquinone. Inhibition of DT-diaphorase by dicumarol enhances menadione-induced generation of $O_2^{-\cdot}$ and cytotoxicity.

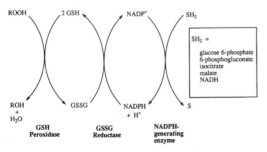

Figure 6   GSH redox cycle.

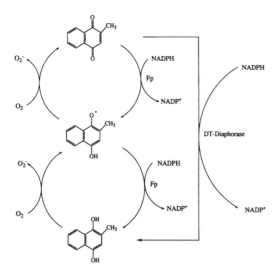

Figure 7 Redox cycling reactions of menadione.
Fp = flavoprotein such as NADPH:cytochrome
P450 reductase.

7.21.2 RENAL GSH STATUS

7.21.2.1 GSH Turnover and the γ-Glutamyl Cycle

7.21.2.1.1 GSH synthesis

As with most endogenous chemicals whose concentrations within cells are closely regulated, GSH is synthesized and degraded by separate pathways, thus allowing for more efficient control of the two processes. GSH is synthesized from its precursor amino acids, L-glutamate, L-cysteine, and glycine, in two ATP-dependent reactions that occur predominantly in the cytosolic fraction of most cells. In the first reaction, catalyzed by γ-glutamylcysteine synthetase (GCS) (EC 6.3.2.2), L-glutamate and L-cysteine are linked by a peptide bond between the γ-carboxy group of L-glutamate and the α-amino group of L-cysteine to form γ-glutamyl cysteine:

L-Glutamate + L-cysteine + ATP

$$\xrightarrow[\text{Mg}^{2+}]{} \gamma\text{-Glutamyl-L-cysteine} + \text{ADP} + P_i$$

The enzyme exhibits K_m values for ATP, L-glutamate, and L-cysteine of 0.2, 1.8, and 0.3 mM, respectively.[21] GCS is relatively specific for L-glutamate, but is also active when L-cysteine is replaced by α-aminoisobutyrate ($K_m = 3$ mM). The reaction proceeds through an enzyme-bound γ-glutamyl-phosphate intermediate, which accounts for the requirement for ATP. Both GCS and glutamine synthetase are inhibited by methionine sulfoximine, which is phosphorylated by ATP and then strongly, but noncovalently, binds to the enzyme active site.[22] Subsequent synthesis of higher homologues of methionine sulfoximine (i.e., prothionine and buthionine sulfoximine) produced more potent and selective inhibitors of γ-glutamylcysteine synthetase.[23,24]

The second step in GSH biosynthesis is catalyzed by GSH synthetase (EC 6.3.2.3):

γ-Glutamyl-L-cysteine + glycine + ATP

$$\xrightarrow[\text{Mg}^{2+}]{} \text{GSH} + \text{ADP} + P_i$$

The enzyme is specific for glycine and exhibits K_m values for ATP, γ-glutamylcysteine, and glycine of 0.36, 0.16, and 2.3 mM, respectively.[21] In addition to γ-glutamylcysteine, γ-glutamyl-α-aminoisobutyrate and γ-glutamylalanine are substrates, forming ophthalmic acid and norophthalmic acid, respectively, both of which normally occur *in vivo*. The catalytic mechanism proceeds through formation of an enzyme-bound acyl phosphate intermediate, similar to the mechanism of GCS.

Two factors that are critical for the regulation of GSH biosynthesis are the concentrations of cysteine and GSH. Normally, glutamate and glycine are present at concentrations that are significantly higher than the respective K_m values of the two enzymes for these substrates, so that these two amino acids are not rate-limiting. However, intracellular cysteine is present at concentrations that are much lower than those of the other two precursors, and are probably no higher than the K_m of GCS for cysteine (i.e., 0.3 mM). Additionally, free cysteine is cytotoxic to mammalian cells because of its facile (pK_a of thiol group = 8.3) auto-oxidation to cystine, which is insoluble in aqueous medium.[25,26] Hence, cysteine is typically the rate-limiting substrate for GSH synthesis. GSH also regulates its own synthesis by feedback inhibition of GCS ($K_i = 2.3$ mM), thereby preventing intracellular GSH concentrations from exceeding a preset level. However, investigators have developed several methods to overcome this limitation, and this is discussed in Section 7.21.4.1.

7.21.2.1.2 GSH degradation

GSH degradation is initiated by γ-glutamyltransferase (GGT) (EC 2.3.2.2), which is a glycoprotein in the brush-border membrane of certain epithelial cells, including the proximal tubules of the kidney. In most mammals the kidneys have the highest GGT activity of all tissues, although the ratio of GGT activity in the kidney to that in the liver varies considerably among species, with the rat having the highest kidney:liver ratio.[27] GGT can catalyze either hydrolysis of the γ-glutamyl group (reaction a) or transfer of the γ-glutamyl group to an amino acid or dipeptide acceptor (reaction b):

$$\text{GSH} + H_2O \xrightarrow{a} \text{L-Glutamate} + \text{L-cysteinylglycine}$$

or

$$\text{GSH} + \text{amino acid}$$

$$\xrightarrow{b} \gamma\text{-Glutamyl-amino acid} + \text{L-cysteinylglycine}$$

Due to the latter reaction, the enzyme is often called "γ-glutamyl transpeptidase." In either case, the cysteinylglycine that is formed is then hydrolyzed by either cysteinylglycine dipeptidase (EC 3.4.13.6) or aminopeptidase M (EC 3.4.11.2), both of which are enzymatic activities on the renal proximal tubular brush-border membrane:

$$\text{L-Cysteinylglycine} + H_2O \rightarrow \text{L-Cysteine} + \text{glycine}$$

The cysteine and glycine thus generated are reabsorbed by Na^+-dependent transport systems on the brush-border membrane (Figure 8). The γ-glutamyl group attached to an amino acid or dipeptide acceptor is transported across the brush-border membrane into the renal cell for conversion to glutamate (see below). GSH can then be resynthesized intracellularly from its precursors.

Some dipeptides and most neutral amino acids are good acceptors for the γ-glutamyl group in the transpeptidation, whereas branched-chain, acidic, and basic amino acids are poor acceptors and D-amino acids and L-proline are inactive.[21] Besides GSH, several *S*-substituted derivatives of GSH are substrates for hydrolysis or transpeptidation, including GSH *S*-conjugates of various endogenous and exogenous chemicals. Hence, GGT plays a critical role in the metabolism of GSH *S*-conjugates for their conversion to mercapturates (detoxication function) or for their conversion to the corresponding cysteine *S*-conjugates that are then substrates for the β-lyase (bioactivation function; see Section 7.21.2.6).

There has been some controversy on the relative physiological importance of hydrolysis versus transpeptidation in the GGT reaction. One study showed that at pH 7.4, and in the presence of a mixture of amino acids at concentrations found in plasma, the two reactions each accounted for 50% of the total activity of GGT.[28] Other studies, however,

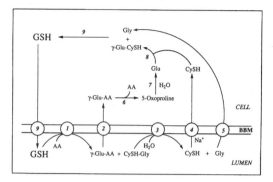

Figure 8 Reactions of the γ-glutamyl cycle in renal proximal tubular cells. Reactions: 1, γ-glutamyltransferase; 2, uptake of γ-glutamyl-amino acid into renal proximal tubular cell; 3, cysteinylglycine dipeptidase activity; 4 and 5, Na^+-dependent uptake of cysteine and glycine; 6, γ-glutamylcyclotransferase; 7, 5-oxoprolinase; 8, γ-glutamylcysteine synthetase; 9, GSH synthetase. Abbreviations: BBM, brush-border membrane; AA, amino acid; CySH-Gly, cysteinylglycine; CySH, cysteine; gly, glycine; γ-Glu-AA, γ-glutamyl-amino acid; γ-Glu-CySH, γ-glutamylcysteine; Glu, glutamate.

have indicated a smaller role for transpeptidation and concluded that hydrolysis of GSH is the primary physiological function of the enzyme.[29] This controversy is also central to that involving the physiological function of the so-called γ-glutamyl cycle, which will be discussed in Section 7.21.2.1.3.

Several inhibitors of GGT have been developed, and these have been useful tools in the study of GSH and GSH *S*-conjugate metabolism. Among the most effective inhibitors are γ-glutamyl derivatives of hydrazines, such as γ-glutamyl-(*o*-carboxy)-phenylhydrazide, and diazo compounds, such as 6-diazo-5-oxo-L-norleucine.[21] The combination of serine and borate is a potent transition state analogue of γ-glutamyltransferase and, hence, is an effective inhibitor as well.[30] Acivicin [L-(αS,5S)-α-amino-3-chloro-4,5-dihydro-5-isoxazoleacetic acid; AT-125] has been used as a specific inhibitor of GGT as it forms an irreversible adduct with the enzyme active site, although acivicin becomes a nonspecific alkylating agent at high concentrations.[31,32]

7.21.2.1.3 Other enzymes of the γ-glutamyl cycle

The turnover of GSH consists of three principal phases: (i) efflux of GSH from the cell into the lumen by transport across the brush-border membrane, (ii) degradation of GSH into its constituent amino acids in the lumen and reuptake of the component amino acids, and (iii) intracellular resynthesis of GSH

(Figure 8). The controversial aspect of this cycle concerns the function of GGT as a transpeptidase, as discussed above, and the role of γ-glutamyl-amino acid or γ-glutamyl-peptide formation in the uptake of these γ-glutamyl acceptors into renal cells and other epithelial cells that possess GGT activity.

Meister[21] proposed in the mid-1970s that the "γ-glutamyl cycle" could play an important role in the transport of certain amino acids and dipeptides. Once transported into the renal proximal tubular cell, the γ-glutamyl-amino acid is a substrate for the enzyme γ-glutamyl-cyclotransferase (EC 2.3.2.4), which releases the amino acid and forms 5-oxoproline. 5-Oxoproline is then hydrolyzed to glutamate by the enzyme 5-oxoprolinase (EC 3.5.2.9). Patients with a genetic deficiency in 5-oxoprolinase exist and have elevated tissue levels of 5-oxoproline, while those with a deficiency in GSH synthetase exist and excrete large amounts of 5-oxoproline in the urine due to conversion of the accumulated γ-glutamylcysteine to cysteine and 5-oxoproline.[21] Although there is evidence consistent with the function of the "γ-glutamyl cycle" in amino acid transport, there is no clear evidence to demonstrate that it is anything more than a physiologically minor route of transport.

7.21.2.2 Interorgan GSH Metabolism

7.21.2.2.1 Hepatic GSH turnover

The liver continuously releases both GSH and GSSG into the plasma and the bile.[33,34] Since the enzymes involved in GSH degradation are found primarily in brush-border membranes of epithelial cells, such as those of the jejunum, renal proximal tubule, and biliary tract, the hepatocyte does not degrade GSH but translocates it via the bile or plasma to other tissues for its turnover. Since degradation of GSH occurs primarily in extrahepatic tissues, the liver is viewed as being a source of either GSH or cysteine for other tissues (see Section 7.21.2.2.2). As shown in Figure 9, GSH efflux occurs across both the sinusoidal and canalicular membranes, releasing GSH into the plasma and the bile, respectively. Quantitatively, transport of GSH across the sinusoidal membrane into the plasma is the primary route. Under conditions of oxidative stress, however, efflux of GSSG from the hepatocyte into the bile can greatly increase.[33]

The two hepatic membrane transport systems for GSH differ. The sinusoidal membrane system is membrane potential-dependent, is

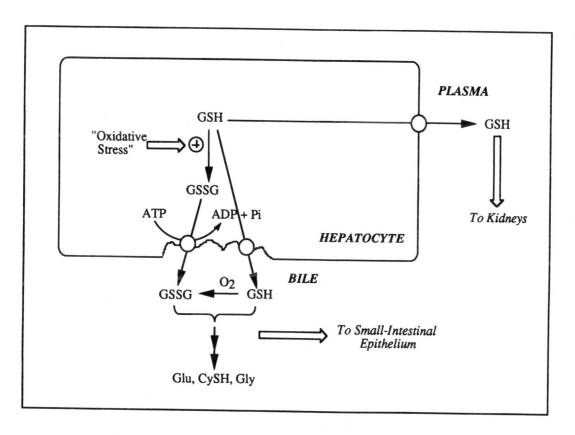

Figure 9 Transport processes for GSH and GSSG in the hepatocyte.

saturable, and is inhibited by methionine, indicating that intracellular methionine plays a role in regulating the efflux of GSH from the hepatocyte.[33-35] Transport is also inhibited by GSSG and GSH *S*-conjugates, suggesting that they are transported across the sinusoidal membrane by the same system as GSH. In contrast, the transport of GSH across the canalicular membrane is distinct from that of GSSG and GSH *S*-conjugates. GSH efflux into the bile occurs by passive diffusion whereas that of GSSG and GSH *S*-conjugates is catalyzed by a transport ATPase. Once in the bile, GSH, GSSG, or GSH *S*-conjugates are degraded by biliary GGT and dipeptidases or are delivered to the small-intestinal epithelium, where GGT and dipeptidases degrade them.

7.21.2.2.2 Enterohepatic and renal–hepatic circulation of GSH and GSH-derived metabolites

The concept that GSH and GSH *S*-conjugates undergo interorgan metabolism was developed from the observations that GGT and the dipeptidases that degrade GSH, GSSG, and GSH *S*-conjugates exhibit tissue-specific distributions and that GSH is found in plasma.[33,36] The liver is the primary source of plasma GSH, which is delivered via enterohepatic and renal-hepatic circulations, to sites of uptake or degradation (Figure 10). The GSH that is secreted from the liver into the bile is delivered to the small-intestinal epithelium either as intact GSH or as the constituent

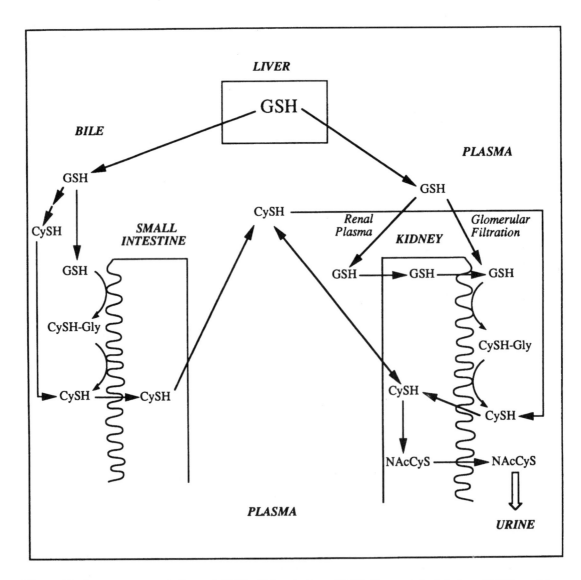

Figure 10 Interorgan metabolism of GSH. Abbreviations: CySH-Gly, cysteinylglycine; CySH, cysteine; Gly, glycine; NAcCys, *N*-acetylcysteine.

amino acids after degradation in the bile. Ultimately, cysteine is either utilized for intestinal synthesis of GSH or proteins or is secreted into plasma. The GSH that is secreted from the liver into the plasma is primarily extracted by the kidneys. GSH S-conjugates are processed in an identical manner, except that they are predominantly secreted from the liver by the biliary route due to their polarity and molecular weight.

As shown in the scheme, there are essentially three mechanisms by which the cysteinyl moiety of hepatic GSH that is secreted into the bile or the plasma is extracted by the kidneys: (i) the cysteine from the small-intestinal epithelium that is derived from biliary GSH is taken up across the basolateral plasma membrane, (ii) plasma GSH undergoes glomerular filtration, degradation by brush-border membrane enzymes, and uptake into proximal tubular cells of the constituent amino acids, and (iii) plasma GSH enters the renal peritubular plasma and is actively transported in the proximal tubular cells.

There are two potential functions of this interorgan system of GSH metabolism. The first function was proposed in the mid-1970s; GSH is envisioned as serving as a transport-stabilized form of cysteine. This seems to be a logical function, since cysteine readily undergoes autoxidation, is cytotoxic, and it is generally the rate-limiting component in biochemical processes that require it as a substrate. The liver, then, would secrete GSH to supply other tissues, which have the capability to degrade GSH, with cysteine primarily for protein synthesis. The second function has been hypothesized since the mid-1980s, and involves the utilization of extracellular GSH for detoxication or protection of certain epithelial cells from oxidants and other reactive electrophiles. The protective mechanism involves uptake of the intact tripeptide into the cells by transport across the basolateral plasma membrane, leading to an increase in intracellular GSH concentration and an enhanced detoxication ability. Besides the renal proximal tubular cell,[37] which has been the most extensively studied cell type that can transport GSH from the extracellular space into the cell, this protective function has been characterized in small-intestinal epithelial cells,[38] pulmonary alveolar type II cells,[39] and retinal pigment epithelium.[40]

7.21.2.3 Renal Handling of Plasma GSH

7.21.2.3.1 *Plasma GSH and glomerular filtration*

As discussed above, the liver continuously secretes GSH into the plasma and this efflux process quantitatively accounts for hepatic GSH turnover. Besides acting as a transport form of cysteine, which is one of the suggested functions of the process of interorgan GSH metabolism, plasma GSH may be important in the regulation of the thiol-disulfide status of plasma proteins and of cellular proteins, such as enzymes, hormone receptors, and transporters, on certain plasma membranes.[41] Exogenously added GSH reacts rapidly with plasma proteins, forming reversible mixed disulfides, indicating that such interactions can occur *in vivo* and may be an important component of the interorgan processing of GSH and may be critical for the maintenance of plasma redox status during exposure to an oxidative stress. These reactions are extremely rapid, with exogenously added GSH having a half-life of approximately 4 min.[41,42]

The kidney is the primary organ that extracts plasma GSH. During a single pass through the kidneys, approximately 80% of plasma GSH is removed.[43,44] As shown in Figure 11, only 30% of the total plasma GSH is extracted by glomerular filtration, leaving the majority (i.e., 50% of total) to be extracted by a mechanism(s) on the basolateral plasma membrane.

7.21.2.3.2 *Turnover of intracellular or filtered GSH*

The filtered GSH enters the tubular lumen and is metabolized by GGT (Figure 11, reaction 3) and dipeptidase (Figure 11, reaction 4) on the brush-border plasma membrane to the constituent amino acids, which are then reabsorbed by Na^+-dependent transport processes. GSH S-conjugates will be similarly metabolized to glutamate, glycine, and the corresponding cysteine conjugate, followed by reabsorption by the proximal tubular cell.

Intracellular GSH must be transported out of the proximal tubular cell into the lumen to be accessible to the active site of GGT (Figure 11, reaction 2).[44] This transport has been characterized in isolated brush-border membrane vesicles, and was found to be membrane potential-dependent but Na^+ independent.[45] Since the transport of GSH across the brush-border membrane is not coupled to that of another charged species and GSH has a net charge of -1 at physiological pH, the transport process would involve the net translocation of one negative charge across the membrane. Since the lumen is positive relative to the intracellular milieu, the normal direction of transport would be efflux from the cell into the lumen. Although the transporter is theoretically

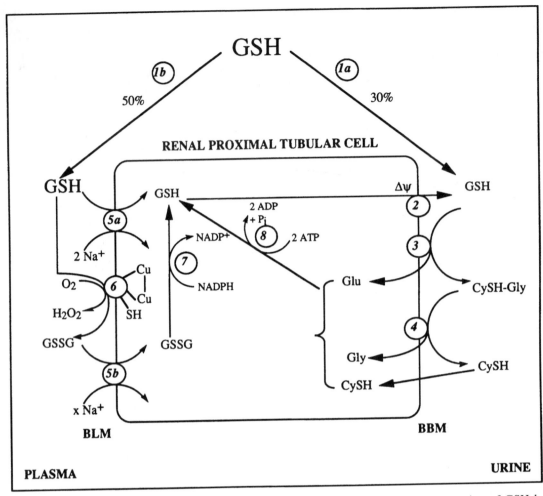

Figure 11   Handling of GSH by renal proximal tubular cell. Reactions: 1a and 1b, extraction of GSH by glomerular filtration and basolateral mechanisms, respectively; 2, efflux of GSH into lumen; 3, γ-glutamyl-transferase; 4, cysteinylglycine dipeptidase activity; 5a, Na^+-coupled uptake of GSH across the basolateral membrane; 5b, Na^+-coupled uptake of GSSG across the basolateral membrane; 6, thiol oxidase; 7, GSSG reductase; 8, resynthesis of GSH by γ-glutamylcysteine synthetase and GSH synthetase. Abbreviations: BBM, brush-border membrane; BLM, basolateral membrane; Glu, glutamate; CySH, cysteine; Gly, glycine; CySH-Gly, cysteinylglycine.

bidirectional, under most conditions that are physiologically attainable, the brush-border membrane transport of GSH will be unidirectional.

7.21.2.3.3   Basolateral extraction

Most of the renal extraction of plasma GSH occurs by processes localized to the basolateral plasma membrane. Studies with the isolated perfused rat kidney demonstrated that the primary mechanism of extraction involved transport of intact GSH across the basolateral plasma membrane into the cell.[46–50] This was a novel finding because the dogma had been that uptake of intact GSH did not occur in any tissues but rather, that apparent uptake was

actually due to GGT.[51–53] GGT activity has been detected in the renal microvascular compartment,[54] suggesting that degradation could indeed play a role in basolateral extraction of GSH.

Studies in isolated basolateral plasma membrane vesicles from rat renal cortex unambiguously identified a GSH transporter that is distinct from GGT-mediated degradation of GSH (Figure 11, reaction 5a).[55,56] GSH uptake is coupled to the Na^+ ion gradient. Although the transport is membrane potential-dependent, the presence of a K^+ ion gradient is not sufficient to stimulate active uptake and intravesicular accumulation of GSH. Rather, only the presence of a Na^+ ion gradient can accomplish this. Further analysis showed that uptake of two Na^+ ions are coupled to the uptake of each GSH molecule.[56] Since GSH has a net charge of -1 at physiological pH, the transport involves the translocation of one positive charge across the membrane. Since the intracellular milieu is electrically negative relative to the plasma, the physiological direction of transport is uptake into the cell. That transport of GSH is not mediated by GGT was demonstrated by the occurrence of GSH uptake despite irreversible inhibition of GGT by acivicin[31] and by analysis of the distribution of transport and degradation activities in membrane fractions obtained by Percoll density-gradient centrifugation.[55,56] Further, the GSH transporter has been partially purified and reconstituted.[57]

In addition to mediating the uptake of GSH, the transporter can also catalyze uptake of GSSG (Figure 11, reaction 6), several γ-glutamyl amino acids, and GSH S-conjugates.[56,58] The stoichiometry of the coupling between GSSG and Na^+ transport has not been determined. Since GSSG has a net charge of -2 at physiological pH, coupling of two Na^+ ions to each GSSG molecule would not provide an electrostatic driving force for uptake. However, rates of GSSG uptake by renal basolateral plasma membrane vesicles were severalfold lower than those of GSH, suggesting that the disulfide form is poorly transported due to a lower driving force, since GSH S-conjugates and γ-glutamyl compounds are taken up at rates that are comparable to that for GSH.[56,58]

GSH in the peritubular plasma can have one additional fate: GSH can be oxidized to GSSG by a copper-containing thiol oxidase on the basolateral plasma membrane. The reaction catalyzed is oxidation of two molecules of GSH with molecular oxygen as electron acceptor, forming one molecule of GSSG and hydrogen peroxide. The background on the discovery of the thiol (glutathione) oxidase activity and its distinction from GGT was reviewed in the early 1980s,[59] and the enzyme has been purified to apparent homogeneity from porcine kidney.[60] Although the physiological role of this enzymatic activity is not clear, it may be important for maintenance of the thiol-disulfide status of plasma and plasma membrane proteins.

7.21.2.4 Renal Cellular Heterogeneity of GSH and GSH-dependent Detoxication

GSH metabolism exhibits marked differences along the nephron.[61–64] Although the intracellular content of GSH is only approximately 30% lower in the most distal portions of the rat or rabbit nephron than in the proximal tubules, striking differences are found in the levels of the various GSH-dependent enzymes. For example, GRD, GPX, and GST activities are present predominantly in the proximal tubules and are much lower or nearly absent from the thick ascending limb or distal tubules.[63,64] These differences may contribute to the intrinsic susceptibility of different nephron segments to oxidative stress and injury. Indeed, isolated distal tubular cells from rat renal cortex are much more susceptible *in vitro* to injury from oxidants, such as H_2O_2 and *t*-butyl hydroperoxide (*t*BH), than are isolated proximal tubular cells from rat renal cortex.[62] Hence, the inherent capability of a cell population to metabolize reactive species or oxidants will play a role in determining their susceptibility *in vitro* to certain forms of injury.

7.21.2.5 Intracellular Compartmentation of GSH: Derivation of the Renal Mitochondrial Pool

In addition to cellular heterogeneity, subcellular heterogeneity and compartmentation of GSH within cells play a critical role in the cellular response to oxidative stress. As described above (see Section 7.21.1.1.1), regulation of mitochondrial redox status is critical to maintenance of cellular energetics and, hence, cellular function. The mitochondrial GSH pool in renal proximal tubular cells comprises approximately 30% of total cellular GSH and has a concentration of approximately 5 mM, which is the same as that in the cytosol.[65]

Most or all of the GCS and GSH synthetase activities in the cell are localized in the cytosol. The absent to minimal *de novo* GSH biosynthesis within mitochondria suggests that GSH might be transported into the organelle from the cytosolic compartment. If a transport system exists, it must be somehow energy-dependent or charge- or membrane potential-compensated, since GSH has a net charge of -1 to -2 at the pH of the cytosol or mitochondrial matrix (pH ≈ 7.0 or 7.8, respectively) and since a pH gradient (ΔpH as much as 0.8) and membrane potential ($\Delta\psi$ as high as $-200\,mV$; matrix negative relative to cytosol) are present across the mitochondrial inner membrane.

Indeed, a carrier-mediated mechanism for uptake of GSH by renal cortical mitochondria has been identified.[66,67] The mechanism of uptake appears to be an electroneutral exchange of divalent GSH (net charge $= -2$) with dicarboxylic acids, such as succinate or malate.[68] Although it is likely that renal mitochondrial GSH transport is a catalytic property of one of the known anion transporters in the mitochondrial inner membrane, it may be a novel transporter that has a broad substrate specificity. The renal mitochondrial GSH transport theoretically should be reversible, although this is energetically unfavorable. Indeed, GSH efflux into the cytosol occurs, but is much slower than uptake.[68]

The toxicological importance of this transport system is illustrated by the observation that *t*BH-induced mitochondrial injury is prevented by the increased uptake of intact GSH by renal cortical mitochondria.[66] Hence, by supplying the mitochondria with GSH, these organelles can be protected from injury due to oxidative stress.

7.21.2.6　Renal Metabolism of GSH S-Conjugates

The metabolism of most GSH *S*-conjugates is similar to that of GSH and GSSG (cf. Figures 8 and 10). Thus, the GSH *S*-conjugate is metabolized by γ-glutamyltransferase and dipeptidases to the corresponding cysteine *S*-conjugate. Cysteine *S*-conjugates of most xenobiotics are then metabolized to mercapturates (*N*-acetylcysteine conjugate), which are highly polar and are generally excreted in the urine. By this "classical" route of metabolism, GSH conjugation serves an important detoxication function in removing reactive electrophiles from cells.

In contrast to this route, several classes of xenobiotics, such as many halogenated hydrocarbons, are conjugated with GSH and are converted to cysteine *S*-conjugates that are substrates for enzymes, such as the β-lyase, that generate reactive metabolites that are nephrotoxic and, in some cases, nephrocarcinogenic.

One such halogenated hydrocarbon that has received and continues to receive much attention is trichloroethylene (Tri) (see Refs. 69 and 70 for reviews) (Figure 12). Tri is conjugated with GSH by the GSTs. Although the liver contains the highest activity of GSH conjugation activity of any tissue, GSH *S*-conjugates of Tri and similar compounds exhibit primarily nephrotoxicity and no hepatotoxicity. This is explained in part by the fact that the GSH *S*-conjugates, once formed in the liver, are exported to the kidneys for further metabolism and generation of the reactive metabolite. In addition, Lash and co-workers have shown in the mid-1990s that Tri conjugation with GSH (Figure 12, reaction 1) also may occur in isolated renal proximal tubular cells,[71] indicat-

ing that the bioactivation process may, to some extent, occur entirely within the target organ. The GSH *S*-conjugate of Tri, *S*-(1,2-dichlorovinyl)glutathione (DCVG), is then metabolized to the cysteine *S*-conjugate *S*-(1,2-dichlorovinyl)-L-cysteine (DCVC). Formation of the cysteine *S*-conjugate serves as a branch point, as either detoxication by *N*-acetylation (Figure 12, reaction 3a) or bioactivation by one of four routes involving three enzymes, can occur.

The most studied and best characterized bioactivation route involves the β-lyase, which converts DCVC to an unstable, reactive thiol 1,2-dichlorovinylthiol (DCVSH), either directly via a β-elimination reaction (Figure 12, reaction 4a) or indirectly via a transamination reaction (Figure 12, reaction 4b). Alternative bioactivation reactions for DCVC include those that are catalyzed by the L-amino acid oxidase (EC 1.4.3.2; Figure 12, reaction 5) and by the cysteine conjugate *S*-oxidase (Figure 12, reaction 6). Renal L-amino acid oxidase shows marked species-dependent differences in its abundance, and is probably only quantitatively significant in rats. In rats, DCVC bioactivation by this route has been estimated to account for at most 35% of the total.[69,72] DCVSH may also be detoxified by methylation, forming a thiomethyl derivative that is readily excreted, typically in the feces (Figure 12, reaction 9).

The quantitative significance of the sulfoxidation pathway is unknown at present, although the product DCVC sulfoxide has been shown to be a more potent renal toxicant than DCVC, both *in vivo* and in isolated renal proximal tubular cells.[73] DCVC sulfoxide formation is catalyzed by a microsomal enzyme that is a flavin monooxygenase (FMO)-like protein.[74,75]

7.21.3　OXIDANT INJURY IN RENAL CELL POPULATIONS

In studying mechanisms of renal cellular injury and factors that contribute to susceptibility to injury from oxidative stress, the heterogeneity of the nephron must be considered. For example, energy metabolism differs along the rat nephron (Table 2). While the proximal tubule, medullary thick ascending limb, and distal convoluted tubule all exhibit relatively high rates of mitochondrial oxygen consumption, the proximal tubules exhibit low rates of glycolysis and high rates of gluconeogenesis, whereas the other two cell types exhibit the opposite pattern with regard

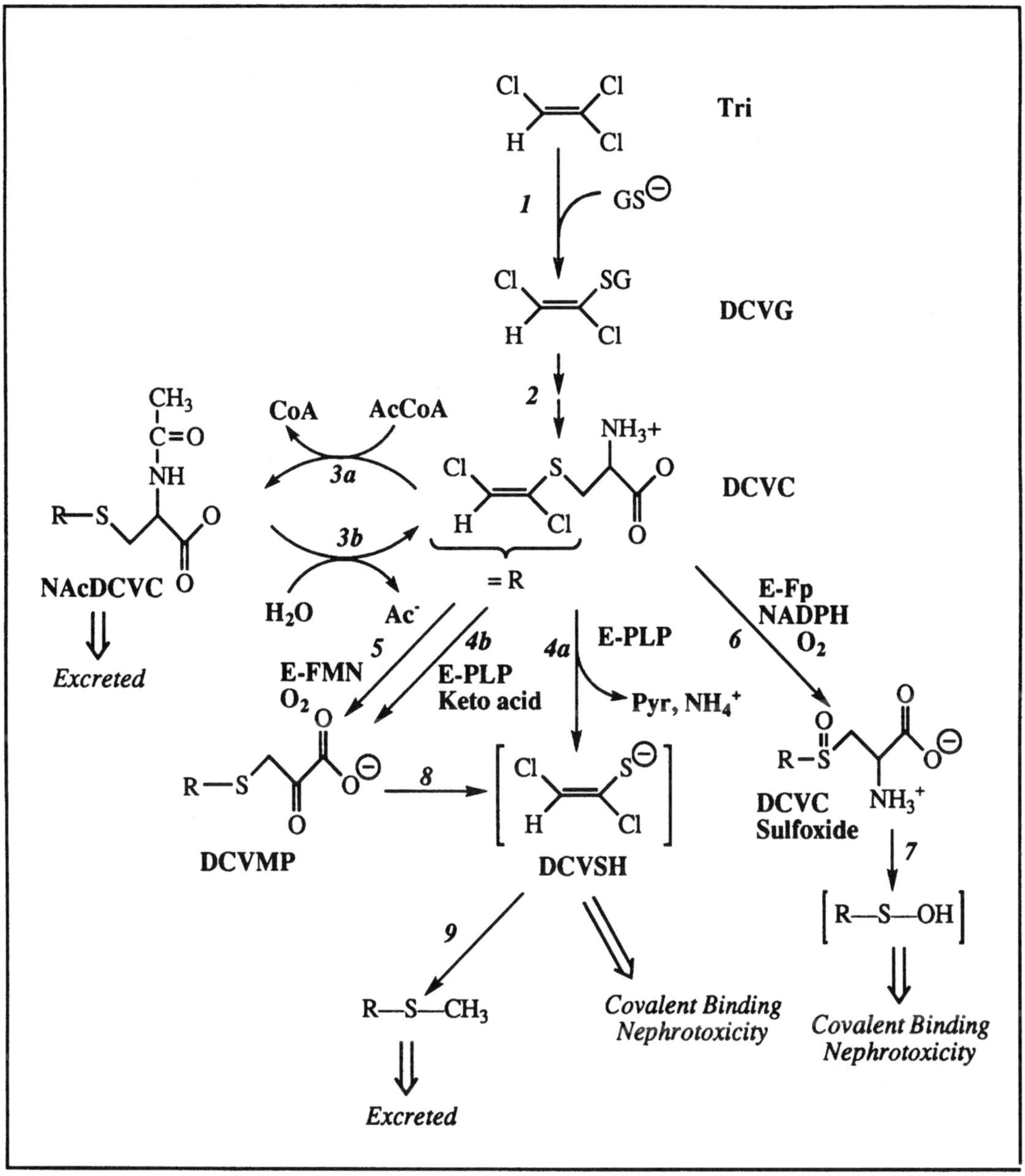

Figure 12 Metabolism of trichloroethylene by the GSH conjugation pathway. Reactions: 1, GSH S-transferase; 2, γ-glutamyltransferase and cysteinylglycine dipeptidase; 3a, cysteine conjugate N-acetyltransferase; 3b, deacetylase; 4a, cysteine conjugate β-lyase: β-elimination; 4b, cysteine conjugate β-lyase: transamination; 5, L-amino acid (L-hydroxy acid) oxidase; 6, cysteine conjugate S-oxidase; 7 and 8, *retro*-Michael reaction; 9, thiomethyltransferase. Abbreviations: Tri, trichloroethylene; DCVG, S-(1,2-dichlorovinyl)glutathione; DCVC, S-(1,2-dichlorovinyl)-L-cysteine; NAcDCVC, S-(1,2-dichlorovinyl)-N-acetyl-L-cysteine; AcCoA, acetyl-coenzyme A; CoA, coenzyme A; Ac-, acetate; E-FMN, enzyme-bound flavin mononucleotide; E-PLP, enzyme-bound pyridoxal phosphate; Pyr, pyruvate; E-Fp, flavoprotein; DCVSH, 1,2-dichlorovinylthiol; DCVMP, S-(1,2-dichlorovinyl)-3-mercaptopropionic acid.

to carbohydrate metabolism. Of toxicological importance, the distribution of active transport systems and certain drug metabolism enzymes differs along the nephron, indicating that accumulation and metabolism of potential toxicants will differ in each cell population.

Relative activities of detoxication enzymes will also contribute to differences in cellular susceptibility to oxidants and other toxic chemicals. Numerous *in vivo* and *in vitro* studies have demonstrated that biochemical differences such as those summarized above contribute in a

Table 2 Biochemical heterogeneity of the rat nephron.

	Nephron region		
Parameter	*Proximal tubule*	*Medullary thick ascending limb*	*Distal convoluted tubule*
Brush border microvilli	Numerous, long	Few, short	Few or none
Oxygenation *in vivo*	High	Low	Low
Energy metabolism			
Glycolysis	Low	High	High
Gluconeogenesis	Low	Low	Low
Mitochondrial QO_2	High	High	High
Active transport systems	Organic ion Na^+-coupled amino acid, hexose	Ion transport	Ion transport
GSH metabolism and transport	High activities	Low activities	Low activities
Cytochrome P450	High	Absent	Low
Cysteine conjugate β-lyase	High	Absent	Low

mechanistically relevant way to differences in susceptibility of cells to chemical and pathological injury.[61,62,76]

7.21.3.1 Proximal Tubular Cells

Renal proximal tubular cells are the primary target cells for a vast number of xenobiotics due to their anatomical position as the first renal epithelial cell type to be exposed to filtered chemicals, to the presence of transport systems that readily facilitate uptake and accumulation of chemicals within the cell, and to the large number of bioactivation enzymes that can generate reactive species.

One example of a selective proximal tubular toxicant is cephaloridine (CPH).[77] The reader is referred to Chapter 24, this volume, for a more in-depth review of CPH nephrotoxicity. The thiophene ring may be metabolized by renal cytochrome P450 to an epoxide that can react with GSH, thereby producing GSH depletion and oxidative stress, and the pyridinium ring, in analogy to paraquat, may redox cycle, producing O_2^-. In support of oxidative stress as a mechanism of CPH-induced proximal tubular toxicity, CPH causes lipid peroxidation and GSH oxidation *in vivo* and *in vitro*.[78] Additionally, CPH-induced cellular injury is partially prevented by preincubation of isolated proximal tubular cells with either GSH or α-tocopherol.[78]

Oxidative stress has been implicated in the mechanism of action of numerous other prox-imal tubular toxicants. DCVC inhibits several sulfhydryl-dependent processes in isolated rat renal cortical cells[73,79] and causes GSH oxidation, lipid peroxidation, and inhibition of sulfhydryl-dependent enzymes in suspensions of isolated rat renal cortical mitochondria.[80] Inorganic mercury also produces lipid peroxidation and GSH oxidation in *in vitro* renal systems.[81–83] Hence, several classes of proximal tubular toxicants act by generation of an oxidative stress.

ROS may also play a role in cellular injury due to ischemia/reperfusion. Paller and co-workers[84] have shown that primary cultures of rat renal proximal tubular cells exhibit a marked release of O_2^-, hydrogen peroxide, and hydroxyl radical after reoxygenation of hypoxic cultures, an effect that was accompanied by lipid peroxidation and cell death. Injury was largely prevented by addition of SOD, catalase, dimethylthiourea, deferoxamine, or inhibitors of cytochrome P450.[84,85] They hypothesized that labile iron is released from the P450 heme moiety and that this iron serves as a catalyst for Fenton-type reactions, producing hydroxyl radical.

In contrast, Borkan and Schwartz[86] concluded that ROS are not responsible for hypoxia/reoxygenation-induced injury in freshly isolated suspensions of rat proximal tubules. They found that while *t*BH caused lipid peroxidation and cell injury, hypoxia/reoxygenation was not associated with lipid peroxidation and SOD, allopurinol, and catalase did not attenuate the injury. This disparity highlights the difficulties in making conclusions about the importance of oxidative stress based on *in vitro* experimentation alone. Differences in the *in vitro* models used, such as freshly isolated cells or tubules versus primary cell cultures, may allow different biochemical mechanisms to predominate under appropriate conditions.

7.21.3.2 Distal Tubular Cells

Unlike proximal tubular cells, there are not many chemicals that are selectively cytotoxic to distal tubular cells. This is because presumably most chemicals that undergo glomerular filtration or that are present in the renal circulation, come into contact first with the proximal tubules. However, pathological conditions that may involve oxidative stress, such as ischemia and reperfusion, are not cell type selective in the same way as many xenobiotics. Rather, each nephron segment will exhibit a characteristic susceptibility to injury from these conditions based in part on their biochemical and physiological properties.

To investigate inherent susceptibilities of cells from different nephron segments, we employed Percoll density-gradient centrifugation to prepare suspensions of isolated cells from the proximal tubular and distal tubular regions and have evaluated the susceptibility of these two cell populations to various types of chemical toxicants and pathological conditions.[76] Exposure of proximal tubular and distal tubular cells from rat kidney to the oxidant *t*BH caused a comparable degree of oxidation of GSH to GSSG. Preincubation of cells with 5 mM GSH increased GSH content in proximal tubular cells but not that in distal tubular cells. Although the percent of cellular GSH oxidized was equivalent, distal tubular cells were much more susceptible than proximal tubular cells to cytotoxicity from *t*BH. Preincubation of either cell population with GSH significantly decreased cytotoxicity due to oxidant exposure. Similar patterns are observed with hydrogen peroxide or the redox-cycling quinone menadione.[62] This same pattern of greater susceptibility of distal tubular cells to thiol oxidants is also observed when these freshly isolated cells are placed into primary culture.[87]

In addition to agents that produce oxidation of thiols, freshly isolated distal tubular cells are also more sensitive to agents that alkylate thiols, such as methyl vinyl ketone and allyl alcohol.[20] Hence, this cell population exhibits an extreme sensitivity to both thiol oxidants and thiol alkylating agents.

7.21.3.3 Glomerular Epithelial Cells

In the mid-1990s, Shah[88] reviewed the role of ROS in glomerular injury. One mechanism involves the infiltration of polymorphonuclear leukocytes and monocytes, which release proteases and oxidants, producing a local inflammatory response and glomerulonephritis.[88,89] Neutrophil-derived products that are involved in glomerular injury include superoxide anion, hydrogen peroxide, products of the myeloperoxidase–hydrogen peroxide–chloride reaction system, hydroxyl radical, possibly singlet oxygen, serine and metalloproteinases, several cationic proteins (e.g., defensins, lysozyme), and phospholipase-derived products. Glomerular epithelial cells serve as sources of reactive oxygen metabolites in noninflammatory forms of glomerular disease, such as occurs in many types of chemically induced glomerular injury. Mesangial cells produce oxidants in response to many types of stimuli, such as immune complexes, complement, and cytokines.

Physiological manifestations of glomerular disease include decreased glomerular filtration rate, proteinuria, and morphological changes in glomerular epithelial cells and mesangial cells.[88] Glomerular basement membrane structure is altered by reactions mediated by oxygen metabolites. Exposure of basement membrane or extracellular matrix proteins to ROS may make them more susceptible to degradation by proteases. Increased production of prostaglandins and thromboxanes also occurs in certain stages of glomerulonephritis. Although ROS are normally regarded as cytotoxic and direct mediators of cellular injury, at low concentrations, these oxygen metabolites may alter cellular regulatory mechanisms, producing alterations in cellular regulatory functions rather than cellular necrosis and cell death.

7.21.4 MODULATION OF OXIDANT INJURY IN RENAL CELLS

7.21.4.1 GSH Status

Numerous studies of oxidant injury in renal cells have shown that alterations in intracellular GSH status are important determinants of cellular injury.[37,62,66,87,90–98] Increases in cellular GSH content, by administration of exogenous GSH or *N*-acetylcysteine, or prior decreases in renal cellular GSH content with diethylmaleate, prevents or exacerbates cellular injury, respectively.[91–93]

Some controversy exists, however, on the role of intact GSH as compared with glycine in the apparent GSH-dependent protection of renal cells from hypoxia or ischemia. Weinberg and co-workers[94,96,99] and Mandel *et al.*[97] have concluded that the GSH-dependent protection from cellular injury due to oxygen deprivation is due to degradation of GSH and the consequent formation of glycine, and not from an increase in the intracellular content of GSH. Although glycine (either added directly or derived from GSH degradation) and other small, neutral amino acids such as alanine afford significant protection from several forms of chemical and pathological injury, other data from Hagen *et al.*[37] and Paller and Patten[98] have shown clearly that an increase in intracellular GSH content, such as by uptake of GSH from the extracellular medium via the Na^+-coupled transport system on the basolateral plasma membrane, is a mechanism of GSH-dependent protection and is distinct from protection due to degradation of GSH and release of glycine. As with many other controversial situations, it appears that under appropriate conditions, multiple protective and toxicity mechanisms can be operative. Clearly, therefore, modulation of intracellular GSH status can have profound effects on the response of renal cells to oxidative stress.

Segmental differences in the handling of GSH may influence the ability of GSH to protect the different cell types from toxicity and their inherent susceptibility to oxidative injury. Uptake of exogenous GSH is much more active in rat renal proximal tubular cells than in distal tubular cells. GSH uptake by proximal tubular cells occurs at rates that are severalfold more rapid and elevated intracellular levels of GSH are retained for longer periods of time than in distal tubular cells. This suggests that the proximal tubular cells can more effectively utilize exogenous GSH for protection against oxidative stress. An additional factor that may be important is that activities of GRD and GPX are much higher in proximal tubular cells,[61–63] indicating that proximal tubular cells have a greater capacity to reduce peroxides and to regenerate GSH from GSSG. This conclusion is supported by the observation of a similar degree of GSH oxidation induced by *t*BH in proximal tubular and distal tubular cells but significantly greater cytotoxicity in distal tubular cells.[62]

Chronic administration of a subtoxic dose of methyl mercury[100] or acute administration of a subtoxic dose of inorganic mercury[101] (and L. H. Lash and R. K. Zalups, unpublished data) increase renal cellular GSH content through upregulation of GCS. This alteration in GSH homeostasis may then modify susceptibility to subsequent exposures to methyl mercury or inorganic mercury or to other oxidants.

7.21.4.2 Captopril and Other Thiol Compounds

Besides GSH, other thiol-containing compounds have been useful in protecting renal cells from oxidative injury. Captopril [1-(3-mercapto-2-methyl-1-oxopropyl)-1-proline] is a thiol-containing compound that is an angiotensin converting enzyme inhibitor. Chavez *et al.*[102] found that *in vivo* or *in vitro* administration of captopril to rats or rat renal cortical mitochondria, respectively, prevented the mitochondrial inhibitory effects of inorganic mercury. Andreoli[103] showed that captopril could scavenge hydrogen peroxide generated by hypoxanthine/xanthine oxidase in cultured normal human renal cortical (NHK-C) cells, thereby diminishing cellular injury.

Methimazole is commonly used to treat hyperthyroidism, is a substrate for the flavin-containing monooxygenase, and protects kidneys, both *in vivo* and *in vitro*, from injury due to DCVC.[73,104] It was presumed that the mechanism of protection involved competitive inhibition of DCVC bioactivation by the cysteine conjugate *S*-oxidase. Subsequent studies by Sausen *et al.*[104] showed that the protective effects of methimazole are more widespread as *in vivo* treatment of rats protected them from nephrotoxicity due to CPH, DCVC, 2-bromohydroquinone, and cisplatin. Methimazole protected against nephrotoxicity by acting as an antioxidant.

Other thiol-containing scavengers of ROS, such as dimethylthiourea and dimethyl sulfoxide, have been used to protect isolated perfused kidneys of rats from oxidant injury due to ischemia/reperfusion[105] and glucose/glucose oxidase or hydrogen peroxide.[106]

7.21.4.3 Allopurinol and Nucleotide Metabolism

Alterations in adenine nucleotide status, which occur rapidly during periods of oxygen depletion, are thought to play a role in oxidative injury by providing substrate for xanthine oxidase (Figure 13). Hypoxia, anoxia, or ischemia cause degradation of ATP to hypoxanthine. During oxidative stress, such as occurs in reperfusion after ischemia, xanthine dehydrogenase is converted to xanthine oxidase. The dehydrogenase activity catalyzes oxidation of hypoxanthine to xanthine or xanthine to uric acid, using NAD^+ as the electron acceptor. During oxidative stress, however, the dehydro-

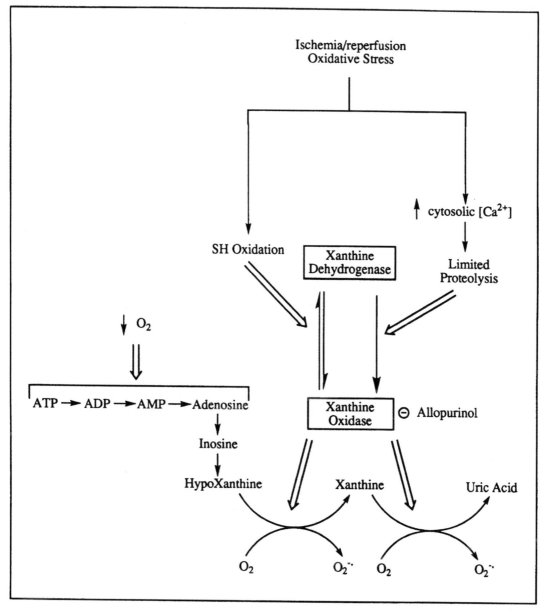

Figure 13 Role of adenine nucleotide degradation and xanthine oxidase in cellular injury from oxidative stress.

genase can be converted to the oxidase by two mechanisms, one involving sulfhydryl group oxidation and one involving activation of Ca^{2+}-dependent proteases.[107] The first mechanism, sulfhydryl oxidation, is reversible by treatment with thiol reductants such as dithiothreitol, and was observed to occur prior to the second mechanism,[107] which is irreversible. Treatment with the xanthine dehydrogenase (oxidase) inhibitor allopurinol has often been observed to protect against reoxygenation or reperfusion injury, implicating the hypoxanthine/xanthine oxidase mechanism in cellular injury.

In contrast to this conclusion, Doctor and Mandel,[108] using suspensions of freshly isolated proximal tubules from rats, found no protective effect of allopurinol in anoxia or hypoxia and reoxygenation. They concluded that although xanthine oxidase is activated during re-oxygenation, there is no critical role for ROS in reoxygenation injury as the proximal tubules were capable of detoxifying them. The injury that was observed during both the oxygen deprivation and the reoxygenation phases was concluded to be due to non-oxidative processes.

7.21.4.4 Calcium Antagonists

As partially illustrated in Figure 2, renal cells have multiple mechanisms for regulating intracellular Ca^{2+} ion status. Increases in cytosolic Ca^{2+} ion concentrations have been associated with numerous toxic and pathologic changes in cells. Many of the regulatory processes for Ca^{2+} ion distribution are affected by intracellular thiol/disulfide status. Hence, oxidative stress that results in thiol oxidation should influence cellular ability to transport and regulate Ca^{2+} ions. Cellular energetics (i.e., ATP supply) is also critical to the ability of renal cells to regulate Ca^{2+} ion homeostasis. Hence, inhibition of mitochondrial function, which will alter cellular ATP generation, will also diminish the cell's ability to regulate Ca^{2+} ions.

Green and Paller[109] studied the role of Ca^{2+} ions and calmodulin in mediating hypoxia/reoxygenation injury and free radical production in isolated renal proximal tubular cells. Their studies showed that decreases in extracellular Ca^{2+} ion concentrations attenuated lethal cell injury. This was associated with a decrease in $O_2^{-\cdot}$ production. Conversely, use of ionomycin, a Ca^{2+} ionophore which increases cytosolic Ca^{2+} ion concentrations, potentiated cellular injury. Calmodulin inhibitors, such as trifluoperazine and calmidazolium, decreased cellular injury. Ueda and Shah[110] investigated the role of changes in intracellular Ca^{2+} ions on hydrogen peroxide-induced injury in LLC-PK$_1$ cells, which are derived from the proximal tubular region. Studies with Ca^{2+} ionophores, Ca^{2+} chelators, voltage-gated Ca^{2+} channel blockers, and incubations of cells in media containing different concentrations of free Ca^{2+} ions showed a close relationship between hydrogen peroxide-induced cytotoxicity, as assessed by trypan blue exclusion, and free Ca^{2+} ion concentration.

In addition to cytotoxic effects of alterations in intracellular free Ca^{2+} concentrations, changes in Ca^{2+} homeostasis can also alter gene expression. Maki et al.[111] studied the effects of hydrogen peroxide and $O_2^{-\cdot}$ generation by the xanthine/xanthine oxidase system on cellular Ca^{2+} ion concentrations and induction of oncogenes in primary cultures of rat renal proximal tubular cells. Time-dependent changes in induction of c-*fos*, c-*jun*, and c-*myc* occurred as a consequence of oxidative stress. These responses were inhibited by SOD but not by catalase, suggesting that generation of $O_2^{-\cdot}$ directly altered expression of the mRNA for these oncogenes that have a role in regulation of cellular proliferation and differentiation. Furthermore, chelation of extracellular Ca^{2+} ions by EGTA or of intracellular Ca^{2+} ions by Quin 2/AM markedly decreased expression of c-*fos* and two inhibitors of protein kinase C also decreased expression of c-*fos*, indicating that Ca^{2+} ions and protein kinase C activation, changes that are initially provoked by oxidative stress, regulate oncogene expression. These gene regulatory changes are further supported by studies showing that oxidant-induced DNA damage in LLC-PK$_1$ cells correlated with a rise in intracellular Ca^{2+} ion concentrations.[112]

7.21.4.5 Iron-dependent Oxidative Injury

Besides Ca^{2+} ions, intracellular concentrations of iron play a role in the expression of oxidant-induced cellular injury. One mechanism of action of iron involves participation in metal-catalyzed Haber–Weiss and Fenton reactions, as described in Section 7.21.1.1.1. Several studies have shown a relationship between the disposition of iron and oxidant-induced cellular injury. For example, Wu and Paller[113] showed that mild, subacute renal iron loading of rats increased susceptibility to ischemic injury. These findings may be relevant to certain renal diseases in humans, where there is evidence of cellular iron accumulation in various proteinuric and chronic renal diseases. Conversely, extracellular iron chelators protect primary cultures of rat renal proximal tubular cells from hypoxia/reoxygenation injury.[114] Golconda et al.[112] demonstrated a relationship between iron status and cellular Ca^{2+} ion concentrations. They suggested that oxidant-induced DNA damage may be diminished by iron chelators due to their secondary ability to limit the increases in Ca^{2+} ion concentrations that occur in oxidative stress.

There is much interest in the ability of prior administration of subtoxic doses of toxicants to confer resistance to tissue damage. Vogt et al.[115] found that prior administration of endotoxin to rats conferred resistance to the glycerol model of acute renal failure. Their data and previous studies suggest that the administration of endotoxin causes an upregulation of antioxidant defenses. In particular, elevated levels of heme oxygenase and the iron transport protein ferritin occur, indicating that cellular levels of free iron are an important factor that determines susceptibility to oxidative stress.

7.21.4.6 Endonuclease Inhibitors

Endonucleases are critical cellular enzymes that appear to be important in regulation of gene expression and programmed cell death or apoptosis. Several studies have linked alterations in endonuclease activity, which is also influenced by changes in intracellular Ca^{2+} ion concentrations, with either regulatory changes in gene expression or DNA damage. Ueda and Shah[116] characterized hydrogen peroxide-induced DNA damage in LLC-PK$_1$ cells. Within minutes of treatment of cells with hydrogen peroxide, DNA damage (single- and double-strand breaks) and cell death occurred. Several endonuclease inhibitors, such as aurin-tricarboxylic acid, Evans blue, and zinc ions, prevented the occurrence of DNA strand breaks, the resultant DNA fragmentation, and cell death. This protection was not associated with prevention of oxidant-induced increases in cytosolic free Ca^{2+} ion concentrations, indicating that endonuclease activation is a more proximal event to initiation of DNA damage and cell death.

ACKNOWLEDGMENTS

Research from the author's laboratory is supported by a grant from the National Institute of Diabetes and Digestive and Kidney Diseases (R01-DK40725) and a Cooperative Agreement from the US Environmental Protection Agency (CR-822240-01–1). L. H. Lash is a recipient of a Research Career Development Award from the National Institute of Diabetes and Digestive and Kidney Diseases (K04-DK02090).

7.21.5 REFERENCES

1. M. K. Shigenaga, T. M. Hagen and B. N. Ames, 'Oxidative damage and mitochondrial decay in aging.' *Proc. Natl. Acad. Sci. USA*, 1994, **91**, 10771–10778.
2. D. P. Jones and L. H. Lash, in 'Mitochondrial Dysfunction, Methods in Toxicology,' ed. L. H. Lash and D. P. Jones, Academic Press, San Diego, CA, 1993, vol. 2, pp. 1–7.
3. P. A. Sandstrom, M. D. Mannie and T. M. Buttke, 'Inhibition of activation-induced death in T cell hybridomas by thiol antioxidants: oxidative stress as a mediator of apoptosis.' *J. Leukocyte Biol.*, 1994, **55**, 221–226.
4. T. A. Sarafian and D. E. Bredesen, 'Is apoptosis mediated by reactive oxygen species?' *Free Radic. Res.*, 1994, **21**, 1–8.
5. D. J. Kane, T. A. Sarafian, R. Anton *et al.*, 'Bcl-2 inhibition of neural death: decreased generation of reactive oxygen species.' *Science*, 1993, **262**, 1274.
6. D. M. Ziegler, 'Role of reversible oxidation–reduction of enzyme thiols-disulfides in metabolic regulation.' *Annu. Rev. Biochem.*, 1985, **54**, 305–329.
7. R. F. Burk, 'Glutathione-dependent protection by rat liver microsomal protein against lipid peroxidation.' *Biochim. Biophys. Acta*, 1983, **757**, 21–28.
8. K. E. Hill and R. F. Burk, 'Influence of vitamin E and selenium on glutathione-dependent protection against microsomal lipid peroxidation.' *Biochem. Pharmacol.*, 1984, **33**, 1065–1068.
9. J. A. Awad, J. D. Morrow, K. E. Hill *et al.*, 'Detection and localization of lipid peroxidation in selenium-

and vitamin E-deficient rats using F2-isoprostanes.' *J. Nutr.*, 1994, **124**, 810–816.
10. V. E. Kagan and L. Packer, 'Light-induced generation of vitamin E radicals: assessing vitamin E regeneration.' *Methods Enzymol.*, 1994, **234**, 316–320.
11. B. S. Winkler, S. M. Orselli and T. S. Rex, 'The redox couple between glutathione and ascorbic acid: a chemical and physiological perspective.' *Free Radic. Biol. Med.*, 1994, **17**, 333–349.
12. E. Maellaro, B. Del Bello, L. Sugherini *et al.*, 'Purification and characterization of glutathione-dependent dehydroascorbate reductase from rat liver.' *Biochem. J.*, 1994, **301**, 471–476.
13. D. B. Wetlaufer, V. P. Saxena, A. K. Ahmed *et al.*, 'Protein thiol–disulfide interchange and interfacing with biological systems.' *Adv. Exp. Med. Biol.*, 1977, **86A**, 43–50.
14. K. Axelsson and B. Mannervik, 'General specificity of cytoplasmic thioltransferase (thiol:disulfide oxidoreductase) from rat liver for thiol and disulfide substrates.' *Biochim. Biophys. Acta*, 1980, **613**, 324–336.
15. B. Mannervik and K. Axelsson, 'Role of cytoplasmic thioltransferase in cellular regulation by thiol–disulfide exchange.' *Biochem. J.*, 1980, **190**, 125–130.
16. R. A. Lawrence and R. F. Burk, 'Glutathione peroxidase activity in selenium-deficient rat liver.' *Biochem. Biophys. Res. Commun.*, 1976, **71**, 952–958.
17. R. A. Lawrence and R. F. Burk, 'Species, tissue, and subcellular distribution of non-selenium-dependent glutathione peroxidase.' *J. Nutr.*, 1978, **108**, 211–215.
18. K. E. Hill and R. F. Burk, 'Effect of selenium deficiency and vitamin E deficiency on glutathione metabolism in isolated rat hepatocytes.' *J. Biol. Chem.*, 1982, **257**, 10668–10672.
19. I. Carlberg and B. Mannervik, 'Purification and characterization of the flavoenzyme glutathione reductase from rat liver.' *J. Biol. Chem.*, 1975, **250**, 5475–5480.
20. L. H. Lash and E. B. Woods, 'Cytotoxicity of alkylating agents in isolated rat kidney proximal tubular and distal tubular cells.' *Arch. Biochem. Biophys.*, 1991, **286**, 46–56.
21. A. Meister and S. S. Tate, 'Glutathione and related γ-glutamyl compounds: biosynthesis and utilization.' *Annu. Rev. Biochem.*, 1976, **45**, 559–604.
22. O. W. Griffith and A. Meister, 'Differential inhibition of glutamine and γ-glutamylcysteine synthetases by α-alkyl analogs of methionine sulfoximine that induce convulsions.' *J. Biol. Chem.*, 1978, **253**, 2333–2338.
23. O. W. Griffith, M. E. Andersen and A. Meister, 'Inhibition of glutathione biosynthesis by prothionine sulfoximine (*S-n*-propyl homocysteine sulfoximine), a selective inhibitor of γ-glutamylcysteine synthetase.' *J. Biol. Chem.*, 1979, **254**, 1205–1210.
24. O. W. Griffith and A. Meister, 'Potent and specific inhibition of glutathione synthesis by buthionine sulfoximine (*S-n*-butyl homocysteine sulfoximine).' *J. Biol. Chem.*, 1979, **254**, 7558–7560.
25. G. Saez, P. J. Thornalley, H. A. O. Hill *et al.*, 'The production of free radicals during the autoxidation of cysteine and their effect on isolated rat hepatocytes.' *Biochim. Biophys. Acta*, 1982, **719**, 24–31.
26. J. Vina, G. T. Saez, D. Wiggins *et al.*, 'The effect of cysteine oxidation on isolated hepatocytes.' *Biochem. J.*, 1983, **212**, 39–44.
27. C. A. Hinchman and N. Ballatori, 'Glutathione-degrading capacities of liver and kidney in different species.' *Biochem. Pharmacol.*, 1990, **40**, 1131–1135.
28. R. D. Allison and A. Meister, 'Evidence that trans-peptidation is a significant function of γ-glutamyl transpeptidase.' *J. Biol. Chem.*, 1981, **256**, 2988–2992.
29. T. M. McIntyre and N. P. Curthoys, 'Comparison of the hydrolytic and transfer activities of rat renal

γ-glutamyltranspeptidase.' *J. Biol. Chem.*, 1979, **254**, 6449–6504.

30. S. S. Tate and A. Meister, 'Serine–borate complex as a transition-state inhibitor of γ-glutamyl transpeptidase.' *Proc. Natl. Acad. Sci. USA*, 1978, **75**, 4806–4809.

31. D. J. Reed, W. W. Ellis and R. A. Meck, 'The inhibition of γ-glutamyl transpeptidase and glutathione metabolism of isolated rat kidney cells by L-(αS,5S)-α-amino-3-chloro-4,5-dihydro-5-isoxazoleacetic acid (AT-125; NSC-163501).' *Biochem. Biophys. Res. Commun.*, 1980, **94**, 1273–1277.

32. C. S. Schasteen, N. P. Curthoys and D. J. Reed, 'The binding mechanism of glutathione and the anti-tumor drug L-(αS,5S)-α-amino-3-chloro-4,5-dihydro-5-isoxazoleacetic acid (AT-125; NSC-163501) to γ-glutamyltransferase.' *Biochem. Biophys. Res. Commun.*, 1983, **112**, 564–570.

33. L. H. Lash, D. P. Jones and M. W. Anders, 'Glutathione homeostasis and glutathione S-conjugate toxicity in the kidney.' *Rev. Biochem. Toxicol.*, 1988, **9**, 29.

34. T. Y. Aw, M. Ookhtens, C. Ren *et al.*, 'Kinetics of glutathione efflux from isolated rat hepatocytes.' *Am. J. Physiol.*, 1986, **250**, G236–G243.

35. T. Y. Aw, M. Ookhtens and N. Kaplowitz, 'Inhibition of glutathione efflux from isolated rat hepatocytes by methionine.' *J. Biol. Chem.*, 1984, **259**, 9355–9358.

36. T. M. McIntyre and N. P. Curthoys, 'The interorgan metabolism of glutathione.' *Int. J. Biochem.*, 1980, **12**, 545–551.

37. T. M. Hagen, T. Y. Aw and D. P. Jones, 'Glutathione uptake and protection against oxidative injury in isolated kidney cells.' *Kidney Int.*, 1988, **34**, 74–81.

38. L. H. Lash, T. M. Hagen and D. P. Jones, 'Exogenous glutathione protects intestinal epithelial cells from oxidative injury.' *Proc. Natl. Acad. Sci. USA*, 1986, **83**, 4641–4645.

39. T. M. Hagen, L. A. Brown and D. P. Jones, 'Protection against paraquat-induced injury by exogenous GSH in pulmonary alveolar type II cells.' *Biochem. Pharmacol.*, 1986, **35**, 4537–4542.

40. P. Sternberg, Jr., P. C. Davidson, D. P. Jones *et al.*, 'Protection of retinal pigment epithelium from oxidative injury by glutathione and precursors.' *Invest. Ophthalmol. Vis. Sci.*, 1993, **34**, 3661–3668.

41. L. H. Lash and D. P. Jones, 'Distribution of oxidized and reduced forms of glutathione and cysteine in rat plasma.' *Arch. Biochem. Biophys.*, 1985, **240**, 583–592.

42. M. E. Anderson and A. Meister, 'Dynamic state of glutathione in blood plasma.' *J. Biol. Chem.*, 1980, **255**, 9530–9533.

43. D. Häberle, A. Wahllander and H. Sies, 'Assessment of the kidney function in maintenance of plasma glutathione concentration and redox state in anaesthetized rats.' *FEBS Lett.*, 1979, **108**, 335–40.

44. O. W. Griffith and A. Meister, Translocation of intracellular glutathione to membrane-bound γ-glutamyl transpeptidase as a discrete step in the γ-glutamyl cycle: glutathionuria after inhibition of transpeptidase.' *Proc. Natl. Acad. Sci. USA*, 1979, **76**, 268–272.

45. M. Inoue and Y. Morino, 'Direct evidence for the role of the membrane potential in glutathione transport by renal brush-border membranes.' *J. Biol. Chem.*, 1985, **260**, 326–331.

46. K. Ormstad, T. Lästbom and S. Orrenius, 'Translocation of amino acids and glutathione studied with the perfused kidney and isolated renal cells.' *FEBS Lett.*, 1980, **112**, 55–59.

47. K. Ormstad, T. Lästbom and S. Orrenius, 'Evidence for different localization of glutathione oxidase and γ-glutamyl transferase activities during extracellular glutathione metabolism in isolated perfused kidney.' *Biochim. Biophys. Acta*, 1982, **700**, 148–153.

48. B. B. Rankin and N. P. Curthoys, 'Evidence for renal paratubular transport of glutathione.' *FEBS Lett.*, 1982, **147**, 193–196.

49. B. B. Rankin, W. Wells and N. P. Curthoys, 'Rat renal peritubular transport and metabolism of plasma [³⁵S]glutathione.' *Am. J. Physiol.*, 1985, **249**, F198–F204.

50. R. D. Scott and N. P. Curthoys, 'Renal clearance of glutathione measured in rats pretreated with inhibitors of glutathione metabolism.' *Am. J. Physiol.*, 1987, **252**, F877–F882.

51. M. E. Anderson, R. J. Bridges and A. Meister, 'Direct evidence for interorgan transport of glutathione and that the nonfiltration mechanism for glutathione utilization involves γ-glutamyl transpeptidase.' *Biochem. Biophys. Res. Commun.*, 1980, **96**, 848–853.

52. W. A. Abbott, R. J. Bridges and A. Meister, 'Extracellular metabolism of glutathione accounts for its disappearance from the basolateral circulation of the kidney.' *J. Biol. Chem.*, 1984, **259**, 15393–15400.

53. M. Inoue, S. Shinozuka and Y. Morino, 'Mechanism of renal peritubular extraction of plasma glutathione: the catalytic activity of contraluminal γ-glutamyltransferase is prerequisite to the apparent peritubular extraction of plasma glutathione.' *Eur. J. Biochem.*, 1986, **157**, 605–609.

54. P. D. Dass, R. P. Misra and T. C. Welbourne, 'Presence of γ-glutamyltransferase in the renal microvascular compartment.' *Can. J. Biochem.*, 1981, **59**, 383–386.

55. L. H. Lash and D. P. Jones, 'Transport of glutathione by renal basal-lateral membrane vesicles.' *Biochem. Biophys. Res. Commun.*, 1983, **112**, 55–60.

56. L. H. Lash and D. P. Jones, 'Renal glutathione transport. Characteristics of the sodium-dependent system in the basal-lateral membrane.' *J. Biol. Chem.*, 1984, **259**, 14508–14514.

57. T. M. Hagen and D. P. Jones, 'Reconstitution and partial purification of the Na⁺-GSH symporter of rat kidney cortex.' *FASEB J.*, 1988, **2**, A752 (abstract).

58. L. H. Lash and D. P. Jones, 'Uptake of the glutathione conjugate S-(1,2-dichlorovinyl)glutathione by renal basal-lateral membrane vesicles and isolated kidney cells.' *Mol. Pharmacol.*, 1985, **28**, 278–282.

59. L. H. Lash, D. P. Jones and S. Orrenius, 'The renal thiol (glutathione) oxidase. Subcellular localization and properties.' *Biochim. Biophys. Acta*, 1984, **779**, 191–200.

60. L. H. Lash and D. P. Jones, Purification and properties of the membranal thiol oxidase from porcine kidney.' *Arch. Biochem. Biophys.*, 1986, **247**, 120–130.

61. L. H. Lash, 'Susceptibility to toxic injury in different nephron cell populations.' *Toxicol. Lett.*, 1990, **53**, 97–104.

62. L. H. Lash and J. J. Tokarz, 'Oxidative stress in isolated proximal and distal tubular cells.' *Am. J. Physiol.*, 1990, **259**, F338–F347.

63. W. G. Guder and B. D. Ross, 'Enzyme distribution along the nephron.' *Kidney Int.*, 1984, **26**, 101–111.

64. J. Mohandas, J. J. Marshall, G. G. Duggin *et al.*, 'Differential distribution of glutathione and glutathione-related enzymes in rabbit kidney. Possible implications in analgesic nephropathy.' *Biochem. Pharmacol.*, 1984, **33**, 1801–1807.

65. R. G. Schnellmann, S. M. Gilchrist and L. J. Mandel, 'Intracellular distribution and depletion of glutathione in rabbit renal proximal tubules.' *Kidney Int.*, 1988, **34**, 229–233.

66. T. B. McKernan, E. B. Woods and L. H. Lash, 'Uptake of glutathione by renal cortical mitochondria.' *Arch. Biochem. Biophys.*, 1991, **288**, 653.

67. R. G. Schnellmann, 'Renal mitochondrial glutathione transport.' *Life Sci.*, 1991, **49**, 393–398.

68. Z. Chen and L. H. Lash, 'Properties of glutathione uptake and efflux in renal cortical mitochondria.' *Toxicologist*, 1995, **15**, 301 (abstract).

69. M. W. Anders, L. H. Lash, W. Dekant *et al.*, 'Biosynthesis and metabolism of glutathione *S*-conjugates to toxic forms.' *Crit. Rev. Toxicol.*, 1988, **18**, 311.

70. I. W. Davidson and R. P. Beliles, 'Consideration of the target organ toxicity of trichloroethylene in terms of metabolite toxicity and pharmacokinetics.' *Drug Metab. Rev.*, 1991, **23**, 493–599.

71. L. H. Lash, Y. Xu, A. A. Elfarra *et al.*, 'Glutathione-dependent metabolism of trichloroethylene in isolated liver and kidney cells of rats and its role in mitochondrial and cellular toxicity.' *Drug Metab. Dispos.*, 1995, **23**, 846–853.

72. J. L. Stevens, P. B. Hatzinger and P. J. Hayden, 'Quantitation of multiple pathways for the metabolism of nephrotoxic cysteine conjugates using selective inhibitors of L-α-hydroxy acid oxidase (L-amino acid oxidase) and cysteine conjugate β-lyase.' *Drug Metab. Dispos.*, 1989, **17**, 297–303.

73. L. H. Lash, P. J. Sausen, R. J. Duescher *et al.*, 'Roles of cysteine conjugate β-lyase and *S*-oxidase in nephrotoxicity: studies with *S*-(1,2- dichlorovinyl)-L-cysteine and *S*-(1,2-dichlorovinyl)-L-cysteine sulfoxide.' *J. Pharmacol. Exp. Ther.*, 1994, **269**, 374–383.

74. P. J. Sausen and A. A. Elfarra, 'Cysteine conjugate *S*-oxidase. Characterization of a novel enzymatic activity in rat hepatic and renal microsomes.' *J. Biol. Chem.*, 1990, **265**, 6139–6145.

75. P. J. Sausen, R. J. Duescher and A. A. Elfarra, 'Further characterization and purification of the flavin-dependent *S*-benzyl-L-cysteine *S*-oxidase activities of rat liver and kidney microsomes.' *Mol. Pharmacol.*, 1993, **43**, 388–396.

76. L. H. Lash and J. J. Tokarz, 'Isolation of two distinct populations of cells from rat kidney cortex and their use in the study of chemical-induced toxicity.' *Anal. Biochem.*, 1989, **182**, 271–279.

77. B. M. Tune, 'The nephrotoxicity of cephalosporin antibiotics—Structure-activity relationships.' *Comments Toxicol.*, 1986, **1**, 145.

78. L. H. Lash, J. J. Tokarz and E. B. Woods, 'Renal cell type specificity of cephalosporin-induced cytotoxicity in suspensions of isolated proximal tubular and distal tubular cells.' *Toxicology*, 1994, **94**, 97–118.

79. L. H. Lash and M. W. Anders, 'Cytotoxicity of *S*-(1,2-dichlorovinyl)glutathione and *S*-(1,2-dichlorovinyl)-L-cysteine in isolated rat kidney cells.' *J. Biol. Chem.*, 1986, **261**, 13076–13081.

80. L. H. Lash and M. W. Anders, 'Mechanism of *S*-(1,2-dichlorovinyl)-L-cysteine- and *S*-(1,2-dichlorovinyl)-L-homocysteine-induced renal mitochondrial toxicity.' *Mol. Pharmacol.*, 1987, **32**, 549–556.

81. B. O. Lund, D. M. Miller and J. S. Woods, 'Mercury-induced H_2O_2 production and lipid peroxidation *in vitro* in rat kidney mitochondria.' *Biochem. Pharmacol.*, 1991, **42**, Suppl., S181–S187.

82. D. M. Miller and J. S. Woods, 'Redox activities of mercury–thiol complexes: implications for mercury-induced porphyria and toxicity.' *Chem. Biol. Interact.*, 1993, **88**, 23–35.

83. L. H. Lash and R. K. Zalups, 'Mercuric chloride-induced cytotoxicity and compensatory hypertrophy in rat kidney proximal tubular cells.', *J. Pharmacol. Exp. Ther.*, 1992, **261**, 819–829.

84. M. S. Paller and T. V. Neumann, 'Reactive oxygen species and rat renal epithelial cells during hypoxia and reoxygenation.' *Kidney Int.*, 1991, **40**, 1041–1049.

85. M. S. Paller and H. S. Jacob, 'Cytochrome P450 mediates tissue-damaging hydroxyl radical formation during reoxygenation of the kidney.' *Proc. Natl. Acad. Sci. USA*, 1994, **91**, 7002–7006.

86. S. C. Borkan and J. H. Schwartz, 'Role of oxygen free radical species in *in vitro* models of proximal tubular ischemia.' *Am. J. Physiol.*, 1989, **257**, F114–F125.

87. L. H. Lash, J. J. Tokarz and D. M. Pegouske, 'Susceptibility of primary cultures of proximal tubular and distal tubular cells from rat kidney to chemically induced toxicity.' *Toxicology*, 1995, **103**, 85–103.

88. S. V. Shah, 'The role of reactive oxygen metabolites in glomerular disease.' *Annu. Rev. Physiol.*, 1995, **57**, 245–262.

89. R. J. Johnson, D. Lovett, R. I. Lehrer *et al.*, 'Role of oxidants and proteases in glomerular injury.' *Kidney Int.*, 1994, **45**, 352–359.

90. R. Baliga, M. Baliga and S. V. Shah, 'Effect of selenium-deficient diet in experimental glomerular disease.' *Am. J. Physiol.*, 1992, **263**, F56–F61.

91. M. S. Paller, 'Renal work, glutathione and susceptibility to free radical-mediated postischemic injury.' *Kidney Int.*, 1988, **33**, 843–849.

92. R. C. Scaduto, Jr., V. H. Gattone II, L. W. Grotyohann *et al.*, 'Effect of an altered glutathione content on renal ischemic injury,' *Am. J. Physiol.*, 1988, **255**, F911–F921.

93. S. O. Slusser, L. W. Grotyohann, L. F. Martin *et al.*, 'Glutathione catabolism by the ischemic rat kidney.' *Am. J. Physiol.*, 1990, **258**, F1546–F1553.

94. M. S. Paller, 'Hydrogen peroxide and ischemic renal injury: effect of catalase inhibition.' *Free Radic. Biol. Med.*, 1991, **10**, 29–34.

95. J. M. Weinberg, J. A. Davis, M. Abarzua *et al.*, 'Cytoprotective effects of glycine and glutathione against hypoxic injury to renal tubules.' *J. Clin. Invest.*, 1987, **80**, 1446–1454.

96. J. M. Weinberg, J. A. Davis, M. Abarzua *et al.*, 'Relationship between cell adenosine triphosphate and glutathione content and protection by glycine against hypoxic proximal tubule cell injury.' *J. Lab. Clin. Med.*, 1989, **113**, 612–632.

97. L. J. Mandel, R. G. Schnellmann and W. R. Jacobs, 'Intracellular glutathione in the protection from anoxic injury in renal proximal tubules.' *J. Clin. Invest.*, 1990, **85**, 316–324.

98. M. S. Paller and M. Patten, 'Protective effects of glutathione, glycine, or alanine in an *in vitro* model of renal anoxia.' *J. Am. Soc. Nephrol.*, 1992, **2**, 1338–1344.

99. J. M. Weinberg, D. N. Buchanan, J. A. Davis *et al.*, 'Metabolic aspects of protection by glycine against hypoxic injury to isolated proximal tubules.' *J. Am. Soc. Nephrol.*, 1991, **1**, 949–958.

100. J. S. Woods, H. A. Davis and R. P. Baer, 'Enhancement of γ-glutamylcysteine synthetase mRNA in rat kidney by methyl mercury.' *Arch. Biochem. Biophys.*, 1992, **296**, 350–353.

101. R. K. Zalups and L. H. Lash, 'Effects of uninephrectomy and mercuric chloride on renal glutathione homeostasis.' *J. Pharmacol. Exp. Ther.*, 1990, **254**, 962–970.

102. E. Chavez, C. Zazueta, A. Osnornio *et al.*, 'Protective behavior of captopril on Hg^{2+}-induced toxicity on kidney mitochondria. *In vivo* and *in vitro* experiments.' *J. Pharmacol. Exp. Ther.*, 1991, **256**, 385–390.

103. S. P. Andreoli, 'Captopril scavenges hydrogen peroxide and reduces, but does not eliminate, oxidant-induced cell injury.' *Am. J. Physiol.*, 1993, **264**, F120–F127.

104. P. J. Sausen, A. A. Elfarra and A. J. Cooley, 'Methimazole protection of rats against chemically induced kidney damage *in vivo*.' *J. Pharmacol. Exp. Ther.*, 1992, **260**, 393–401.

105. S. L. Linas, D. Whittenburg and J. E. Repine, 'O₂

metabolites cause reperfusion injury after short but not prolonged renal ischemia.' *Am. J. Physiol.*, 1987, **253**, F685–F691.

106. S. L. Linas, P. F. Shanley, C. W. White *et al.*, 'O$_2$ metabolite-mediated injury in perfused kidneys is reflected by consumption of DMTU and glutathione.' *Am. J. Physiol.*, 1987, **253**, F692–F701.

107. T. G. McKelvey, M. E. Höllwarth, D. N. Granger *et al.*, 'Mechanisms of conversion of xanthine dehydrogenase to xanthine oxidase in ischemic rat liver and kidney.' *Am. J. Physiol.*, 1988, **254**, G753–G760.

108. R. B. Doctor and L. J. Mandel, 'Minimal role of xanthine oxidase and oxygen free radicals in rat renal tubular reoxygenation injury.' *J. Am. Soc. Nephrol.*, 1991, **1**, 959–969.

109. E. L. Greene and M. S. Paller, 'Calcium and free radicals in hypoxia/reoxygenation injury of renal epithelial cells.' *Am. J. Physiol.*, 1994, **266**, F13–F20.

110. N. Ueda and S. V. Shah, 'Role of intracellular calcium in hydrogen peroxide-induced renal tubular injury.' *Am. J. Physiol.*, 1992, **263**, F214–F221.

111. A. Maki, I. K. Berezesky, J Fargnoli *et al.*, 'Role of [Ca^{2+}]$_i$ in induction of *c-fos*, *c-jun*, and *c-myc* mRNA in rat PTE after oxidative stress.' *FASEB J.*, 1992, **6**, 919–924.

112. M. S. Golconda, N. Ueda and S. V Shah, 'Evidence suggesting that iron and calcium are interrelated in oxidant-induced DNA damage.' *Kidney Int.*, 1993, **44**, 1228–1234.

113. Z. L. Wu and M. S. Paller, 'Iron loading enhances susceptibility to renal ischemia in rats.' *Ren. Fail.*, 1994, **16**, 471–480.

114. M. S. Paller and B. E. Hedlund, 'Extracellular iron chelators protect kidney cells from hypoxia/reoxygenation.' *Free Radic. Biol. Med.*, 1994, **17**, 597–603.

115. B. A. Vogt, J. Alam, A. J. Croatt *et al.*, 'Acquired resistance to acute oxidative stress: Possible role of heme oxygenase and ferritin.' *Lab. Invest.*, 1995, **72**, 474–483.

116. N. Ueda and S. V. Shah, 'Endonuclease-induced DNA damage and cell death in oxidant injury to renal tubular epithelial cells.' *J. Clin. Invest.*, 1992, **90**, 2593–2597.

7.22

Molecular Responses to Oxygen Deprivation and Ischemia

RALPH WITZGALL
Universität Heidelberg, Germany

and

JOSEPH V. BONVENTRE
Harvard Medical School, Boston, MA, USA

7.22.1 INTRODUCTION

Oxidative phosphorylation is the most common pathway for ATP generation in the animal kingdom. Animal cells, however, are often exposed to periods of relative or absolute oxygen deprivation. Some species can generate energy in the absence of oxygen (anaerobiosis) to support energy demand of the entire organism or provide partial energy support for the entire organism throughout the animal's entire life cycle or in some stages of it.[1] Furthermore, within a given organism at a given time some cells may generate energy anaerobically while others require oxygen. Organisms that are dependent upon oxygen for ATP generation have developed responses, both adaptive and maladaptive, to the environmental influence that oxygen deprivation represents. Lower organisms, such as prokaryotes, are very well adapted to this kind of "stress" and have developed special "defense" mechanisms to deal with it. Some eukaryotes, such as yeast, have also developed mechanisms permitting

429

survival when exposed to low oxygen pressure. The adaptive mechanisms become more complex when one looks at phylogenetically highly developed animals like vertebrates. Turtles and whales can dive for 30 minutes or longer without any difficulty,[2] whereas, 30 minutes of anoxia is incompatible with survival of the great majority of mammals. In turtles this ability to dive for long periods is the result of their ability to generate energy by anaerobiosis in all tissues including the brain. By contrast, diving mammals, such as the Waddell seal, reduce their metabolic demands when diving underwater. There is significant variability in the length of time of ischemia/hypoxia that can be tolerated by a particular organ such as kidney, muscle, and brain. Though this varying sensitivity of cells and tissues to oxygen deprivation seems to be a very fundamental feature of biological organisms, the reasons for it are incompletely understood.[1,3]

The term "ischemia" derives from the Greek "ischo," to keep back, and "haima," blood, and is defined as: localized tissue anemia due to obstruction of the inflow of arterial blood.[4] When one considers ischemic injury, it is important to recognize that the cellular and tissue response mechanisms and the character of the injury during the ischemic phase are very different from those of the postischemic or recovery phase. In the ischemic period there is decreased blood flow and impaired delivery of oxygen and nutrients as well as increased accumulation of metabolic waste products in the tissue.

In this review we will focus on the response of the kidney to ischemia. Some of the mechanisms and responses can be generalized to other organs like the heart, brain, and liver, while other processes are specific for an organ and cannot be generalized.[5–8] A brain-specific process, for example, is the involvement of excitotoxic amino acids in ischemic injury. Beneficial effects of NMDA- and non-NMDA-receptor antagonists[9–11] are specific for that organ.[12] By contrast, reactive oxygen radicals and phospholipase A_2 are likely to play a damaging role in brain, kidney, and other organs.[13–15]

While studies *in vitro* have added greatly to our insight into cellular mechanisms of injury, extrapolating findings to an organ *in vivo* is not always without concerns because of the inability to accurately mimic *in vivo* conditions *in vitro*. What makes a cell die (and when a cell is dead) is not always well defined. Release of lactate dehydrogenase, uptake of vital dyes, or morphological changes are used to measure cell death. But does an increase in cell membrane permeability define the "point of no return?" It

is generally acknowledged that there are two fundamentally different ways that a cell can die, necrosis and apoptosis (also called "programmed cell death" or "cell suicide").[16] A cell becomes necrotic when it is so profoundly damaged that too many of its control and regulatory circuits are not functioning any more. Thus, the cell cannot maintain its integrity and has no alternative but to die. This process of cell death is usually associated with cell and mitochondrial swelling and results in inflammation. A cell undergoing apoptosis destroys itself by turning on a special genetic program which leads to cell death.[17–20] By contrast with necrosis, the apoptotic cell is characterized by nuclear chromatin condensation, less cell swelling, and phagocytosis of the apoptotic cell without inflammation. We will return to the molecular aspects of apoptosis later in this chapter.

7.22.2 GENETIC RESPONSE OF BACTERIA TO THEIR (AN)AEROBIC ENVIRONMENT

7.22.2.1 Fnr and OxyR in *Escherichia coli*

In our discussion of gene control mechanisms we will first discuss bacteria before describing processes apparent in yeast and metazoans. We start with bacteria because of their lower phylogenetic rank. This does this mean, however, that bacteria have developed less complex genetic programs in their strategies to cope with hypoxia, as we will see as we discuss their processes. In order to survive under both low and high oxygen tension, *Escherichia coli* has developed different strategies to deal with either situation. Examples of regulatory genes necessary for life under aerobic conditions are the *arcA/arcB* and *oxyR* genes; genes necessary for life under anaerobic conditions are *narX/narL* and *fnr*.

Escherichia coli, as a facultative anaerobic bacterium, uses O_2, and other substrates like fumarate or nitrate, as electron acceptor.[21] Fnr, the product of the *Fnr* gene, is essential for the induction of genes involved in anaerobic energy generation, primarily by acting as a transcriptional activator.[22,23] Fnr is believed to contain a redox/O_2-sensitive region which, upon sensing an anaerobic environment, results in conversion of the protein into an active conformation for DNA binding. Some genes regulated by Fnr include nitrate reductase (*narGHJI*), fumarate reductase (*frdABCD*), trimethylamine-*N*-oxide/dimethyl sulfoxide reductase (*dmsABC*), and nitrite reductase (fdnGHI).[23] Fnr can also repress the expression of genes involved in oxidative metabolism.

The *fnr* gene has been cloned and sequenced and was shown to exhibit homologies with other *E. coli* transcriptional regulator proteins.[24] The protein contains four cysteine residues, three of which are clustered in an eight amino acid long region at the NH$_2$-terminus with the fourth near the center of the molecule. Replacement of any of these cysteines with serine completely inactivates the Fnr protein. When the significance of these cysteines was investigated by reactivity with the alkylating agent iodoacetate, the protein from the bacteria grown under anaerobic conditions was not alkylated to a greater extent, as might have been expected, but to the same extent as the protein from aerobically grown bacteria (2.4 residues in either case).[25] What was different, however, was the time required for half-maximal alkylation. The "anaerobic" protein was much more slowly alkylated ($t_{1/2,max} = 50$ min) than the "aerobic" protein ($t_{1/2,max} = 6$ min). Phenanthroline, a metal-chelating agent, enhanced the reactivity of the cysteines, when the bacteria were grown under anaerobic conditions.[25] Fe^{2+} and Cu^{2+} ions were able to reverse this effect. Co^{2+} could only partially reverse it, while Mg^{2+}, Ca^{2+}, and Na$^+$ had no effect at all. The presence of phenanthroline also lowered both the activity and the expression level of fumarate reductase. Based on these results and mutagenesis studies a model can be put forward in which the deoxygenated Fnr protein would bind a Fe^{2+} ion, which would be oxidized under aerobic conditions and would render Fnr inactive. Fnr is believed to bind to Fnr-regulated promoters in response to anaerobiosis.[23]

Another mechanism by which an organism can protect itself against hypoxia and reoxygenation is by increasing its defense mechanisms against reactive oxygen species. Ames and co-workers examined a *Salmonella typhimurium* mutant which was resistant to H$_2$O$_2$.[26] Pretreatment of wild-type *S. typhimurium* with 60 mM H$_2$O$_2$ for 60 min made the bacteria resistant to subsequent treatment with 10 mM H$_2$O$_2$. The development of resistance could be blocked by chloramphenicol, indicating that protein synthesis was necessary. Pretreatment with H$_2$O$_2$ also conferred increased resistance to other chemical oxidants and exposure to 50 °C. Analysis by two-dimensional protein gels showed that the synthesis of 30 proteins was induced. The mutants (called *oxyR1*), which are selected at a frequency of about one in a million, constitutively expressed nine of twelve early H$_2$O$_2$-inducible proteins. The proteins which were expressed at a higher level, were shown to be catalase (45–50 times greater activity than in wild-type bacteria), peroxidase,

superoxide dismutase, glutathione reductase, NAD(P)H-dependent hydroperoxide reductase, and heat shock proteins.

The *E. coli oxyR* was subsequently cloned.[27] In its NH$_2$-terminal region the protein shows a region with high homology to other regulatory proteins in *E. coli*, probably representing a helix-turn-helix motif of DNA-binding proteins. A closer look at the mechanism, by which OxyR turns on the oxidative stress-defense genes, revealed that the level of OxyR was not altered after treatment with H$_2$O$_2$.[28] Also, strains that overexpress OxyR do not show increased expression of defense-genes. These findings suggest a posttranslational modification of OxyR. Oxidized, but not reduced, OxyR can activate transcription. Indeed, when the OxyR protein was prepared from the bacteria either without dithiothreitol (DTT), or with 1 mM DTT only, OxyR was able to activate transcription *in vitro*, whereas it was no longer able to do so in the presence of 100 mM DTT. When the DTT was removed by gel filtration, the OxyR-protein was again active.[28] The increased transcriptional activity of the oxidized form of OxyR is likely not due to a higher DNA-binding affinity, but to a conformational change of the already bound protein. An OxyR-mutant (Ser199 changed to Cys) in 1 mM DTT shows the same footprint as the wild-type protein in 100 mM.[28] The recognition sequence for the OxyR protein was also determined. The oxidized protein recognizes four adjacent ATAG nucleotide elements, whereas the reduced protein recognizes two sets of two ATAG nucleotide elements separated by one helical turn. Depending on the redox state of the cell, OxyR positions itself differently on the promoter and functions either as a transcriptional activator or repressor.[29]

7.22.2.2 FixL and FixJ in *Rhizobium meliloti*

Nitrogen fixation, as developed by some bacteria (e.g., *Klebsiella pneumoniae*, *Rhizobium meliloti*), is a highly complex biological process.[30] There are at least 23 genes necessary for nitrogen fixation in *R. meliloti* (*fix* and *nif* genes) and low oxygen tension is a critical signal for induction of the expression of most of these genes.[31] One of the key proteins in this process is NifA, the product of the *nifA* gene, which controls a wide variety of genes necessary for nitrogen fixation. In this section we will focus on *R. meliloti*, probably one of the best understood organisms in this respect. In the 1980s it was found that the *nifA* gene is O$_2$-regulated. Bacteria, grown under microaerobic conditions, showed a strong induction of NifA activity even in the presence of fixed nitrogen sources. The peak activity occurred at 1% O_2. The *nifA* gene induction was not due to low energy levels in the cells, because treatment of the bacterial cultures with sodium azide or 2,4-dinitrophenol did not imitate the effects of low oxygen. This suggested a "direct" influence of low oxygen tension to upregulate *nifA* activity.[32] Soon after these initial observations, additional experiments identified two other genes, *fixL* and *fixJ*, as playing a role further upstream of *nifA* in the cascade of gene activation events.[31] *fixL* and *fixJ* were cloned, sequenced and found to encode proteins with homology to already known proteins. FixJ displayed strong similarity to bacterial regulatory proteins, whereas FixL had characteristics of a membrane protein.[31] When *fixL* and *fixJ* were induced artificially by IPTG, activation of a *nifA* promoter/*lacZ* gene-hybrid was demonstrated under microaerobic conditions. When FixJ was expressed alone (i.e., not with FixL) the *nifA* promoter was activated independently of the oxygen pressure. Co-expression of FixL with FixJ completely changed the response. Under microaerobic conditions, expression of both proteins led to a twofold higher increase of *nifA* promoter-activity than in the presence of FixJ alone. When the cultures were grown in the presence of greater amounts of oxygen, FixL led to a sevenfold decrease in *nifA* promoter activity compared to expression of FixJ only.[33] These results strongly suggested that FixL served as an oxygen sensor, that FixJ could activate genes independently of FixL, and that FixL would transmit information regarding the O_2 status of the cell through FixJ.

Further progress towards a better understanding of the regulatory mechanisms were made by overexpressing a truncated FixL protein (called FixL*) and a full-length FixJ protein in *E. coli*.[34] Both proteins could be separated by gel filtration, indicating that they were not tightly associated. However, FixL* was able to autophosphorylate and to phosphorylate FixJ. When subjected to spectrophotometry, FixL* exhibited the typical spectrum of heme-proteins and it also changed its absorption spectrum according to the oxygen pressure in a reversible way.[34] Studies *in vitro* demonstrated a basal transcriptional activator activity of purified, recombinant FixJ. Phosphorylation of FixJ by FixL* greatly enhanced the activity of FixJ as a transcriptional activator. Anaerobiosis markedly increased transcription by RNA polymerases from *R. meliloti* or *E. coli*, when both FixJ and FixL* were present. It can therefore be proposed that FixL, through its hydrophobic domain(s), is anchored in the cell membrane, where it can sense the oxygen tension by virtue of its heme moiety or can sense other signals related to oxygen availability such as redox potential of the cells.[35] It then transmits this information through its kinase domain to FixJ, which in turn will bind to its promoters and activates transcription.

7.22.3 GENETIC RESPONSE OF YEAST TO THE (AN)AEROBIC ENVIRONMENT

Saccharomyces cerevisiae can grow both under aerobic and anaerobic conditions. Among the genes which are differentially regulated by the availability of oxygen we will focus on those coding for cytochrome c, in particular the iso-1-cytochrome c gene. Cytochrome c exists in two major forms in *S. cerevisiae*, iso-1-cytochrome c (coded for by the *CYC1* gene), accounting for 95% of the cytochrome c activity, and iso-2-cytochrome c (encoded by the *CYC7* gene), accounting for the remaining 5%. The cellular levels of both proteins change with the supply of catabolites and oxygen. Glucose leads to a decreased level of both isoforms, whereas aerobic conditions cause just the opposite phenomenon.[36–38] After determination of amino acid sequences, the genes for both isoforms were cloned.[39] The clones revealed further information on the 5'- and 3'-noncoding regions. Guarente and Mason[40] demonstrated that heme could activate *CYC1* and defined the *cis*-acting sequences necessary for heme regulation.[41] They defined an upstream activation site, UAS (-312/-258 from the most upstream transcription initiation site), which conferred inducibility on *CYC1* by heme. That this region is also involved in the regulation by oxygen was suggested by Northern blot analysis and mutagenesis experiments.[42] In yeast grown anaerobically no CYC1-transcript could be detected, whereas a 0.7 kb mRNA could easily be seen under aerobic conditions, suggesting at least a partial regulation at a transcriptional level. When a construct was used in which the region between −180 and −580 from the most upstream transcription initiation site was deleted, two effects were noticed.[42] First, at a low oxygen level a CYC1-transcript was present, albeit at a lower level then in aerobically grown wild-type cells. This revealed that the shut-off mechanism, which was employed in wild-type cells, was not functional in the mutants. In addition, no matter whether the yeast were grown under aerobic or anaerobic conditions, the level of CYC1 mRNA was the same.

These findings were confirmed and subsequently extended. Glucose repressed both the level of CYC1 and the related CYC7 mRNA, indicating how the levels of iso-1- and iso-2-cytochrome c in the cell might be regulated by sugars.[41,43] Furthermore, both the levels of CYC1 and CYC7 mRNAs were low when the cells were cultivated in a nitrogen atmosphere.[43] These observations led to an examination of the nucleotide sequences which mediated these responses in both genes. It was also of interest to evaluate whether there were different *cis*-acting sequences for the induction by O_2 and heme, and the repression by glucose. Indeed, the UAS of *CYC1* was further dissected into two subsites, UAS1 and UAS2. UAS1 is centered at about -275, whereas UAS2 could be localized between -229 and -211.[44] Both UAS1 and UAS2 mediate repression by glucose; however, only UAS1 was necessary to confer inducibility by heme[44] (and therefore oxygen[40,41]). Heme activator protein 1 (HAP1) was proposed to activate transcription via UAS1, whereas HAP2 only effected UAS2.[44] When the HAP2 locus was cloned, it was found that the steady-state level of HAP2 transcript was very low, that its level increased under the influence of nonfermentable carbon sources, and that heme did not exert any influence on the level of HAP2 mRNA.[45]

After closer investigation of UAS1 it became apparent that two regions in UAS1 were required for the expression of *CYC1*. Single base changes in region A between -288 and -285 or region B at -266 and -265 resulted in marked reductions in activity.[46] Region B is specifically bound by a factor as shown in gel-shift assays. This factor is not present in yeast grown anaerobically or in the absence of heme.[47] Region A seems to harbor the *cis*-acting sequence responsible for catabolite repression but not that for heme-induction.[46] By footprint analysis the binding-site of HAP1 in the *CYC1* gene could be exactly defined to lie between -269 and -247.[48] In the same report the authors also demonstrated a HAP1 binding site in the *CYC7* gene, which surprisingly had no obvious similarity with that in the *CYC1* gene.

HAP1 has been cloned and sequenced.[48–50] The protein displays some very interesting features. It consists of several domains which show homologies to other gene families. At the NH_2-terminus lies a region which closely resembles a zinc finger of the C4 class.[51] Adjacent to this domain but located further to the COOH-terminus a repeat of 12 glutamine and one glutamic acid residues can be found, resembling the opa-repeat of *Drosophila* transcription factors.[52] The function of the opa-sequence is not known, whereas zinc fingers are implicated in DNA-binding. Adjacent to the

opa-sequence are seven repeats of a (somewhat variable) KCPVDH motif. Finally, the COOH-terminus of HAP1 is highly acidic and therefore might represent a transcriptional activation domain.[53]

Guarente and co-workers verified the functional significance of these proposed domains. Residues 1 through 247 (which include the zinc finger and opa-motif) were sufficient to act through UAS1, as shown by complementing a HAP1-deficient yeast strain. No heme response was seen when residues 247–444 (KCPVDH motif) were deleted (nondeleted construct showed a fourfold induction by heme), although the baseline-level of activation through UAS1 was increased by 54. By using electrophoretic mobility gel-shift experiments the DNA-binding region of HAP1 was defined as encompassing residues 1 through 99 (in case of UAS1) and 1 through 148 (in case of *CYC7*), respectively, which means that the opa-repeat is not necessary to bind specifically to DNA and that the zinc finger motif is sufficient.[54] Mutation of Cys-64 or Cys-81 abolished the DNA-binding activity.

To elucidate further the nature of the results with the KCPVDH mutants, gel shifts were performed with the deletion mutants. A mutant consisting of residues 1–444 bound to UAS1 or *CYC7* in a heme-dependent manner, whereas mutant 1–244 bound no matter whether heme was present or not. Based on these results it was suggested that the seven KCPVDH repeats mask the DNA-binding activity of HAP1 in the absence of heme, whereas in the presence of heme the zinc finger is unmasked, HAP1 can bind in the promoter region of *CYC1* or *CYC7* and activate the genes.

7.22.4 GENETIC RESPONSE OF METAZOANS TO OXYGEN DEPRIVATION AND ISCHEMIA

7.22.4.1 Genetic Response *In Vivo*

After providing a brief summary of the genetic response of brain, heart, and liver to ischemia we will devote most attention to the response of the kidney to oxygen deprivation and ischemia.

7.22.4.1.1 *Ischemia in brain, heart, liver*

There is a growing literature reporting induction of various genes with ischemia in brain,[55–58] heart,[59–61] and liver.[62,63] In this brief chapter it is not possible to do justice to this expanding field. Only some of the reported findings are, therefore, presented. In the mid-1980s data were reported on the molecular nature of proteins expressed after ischemia in the brain. Focal cerebral ischemia in rats, produced by unilateral occlusion of the common carotid artery and middle cerebral artery for 2 h with subsequent recovery for 50 min, induced various proteins depending on the cerebral blood flow (CBF).[64] At a CBF of 50–60 mL $100\,g^{-1}\,min^{-1}$, the increased synthesis of actin and proteins of M_r 27 kDa, 34 kDa, 73 kDa, 79 kDa (identified according to their electrophoretic migration pattern after pulse labeling) was noticed. Blood flows of 40–50 mL $100\,g^{-1}\,min^{-1}$ caused the induction of additional 55 kDa and 70 kDa proteins. Blood flows of 15–25 mL $100\,g^{-1}\,min^{-1}$ induced yet another protein of 40 kDa molecular weight.

It was not until the mid-1980s that our knowledge about the identity of proteins and mRNAs induced after ischemia increased markedly. The mRNA encoding Krox-20, also known as Egr-2, is expressed rapidly and transiently in the ischemic cortex perfused by the right middle cerebral artery. The expression of Egr-2/Krox-20 is controlled, at least in part, at the transcriptional level as shown by nuclear run-on assays. *Egr-2/Krox-20* is an immediate-early response gene, other examples of which include *c-fos*, *c-jun*, *jun B*, and *jun D*. As in the case of Egr-2/Krox-20, mRNAs coding for the proteins c-Fos, c-Jun, and Jun B were expressed rapidly and transiently after ischemia, although mRNA encoding c-Jun was increased to a much lower extent when compared to the others. Elevated mRNA levels, in each case, were due, at least in part, to transcriptional upregulation. The level of Jun D mRNA did not change at all. For the description of the expression of other proteins and mRNAs (e.g., NGFI-B, TIS1, TIS11, TIS21, Hsp70, GFAP) in different models of cerebral ischemia the reader is referred to other publications.[65–67]

The laboratory of Witzgall and Bonventre has reported that the heat-shock response protected cultured neurons against amino acid excitotoxicity, which is believed to be the effector mechanism of ischemic injury to the brain *in vivo*.[68] Glutamate-induced excitotoxicity was inhibited by preheating the cells to 42.2 °C for 20 min. This protection was afforded to cells subsequently exposed to glutamate 3 h or 24 h after they were heated. Protection required new protein synthesis.

In hearts of mice exposed to hypobaric decompression of 256 mm Hg at 7% O_2 the synthesis of 85 kDa and 95 kDa proteins steadily increased over a period of 16 h, while other proteins of 71 kDa and 79 kDa molecular weight reached peak levels at 3 h and returned to baseline at 16 h.[69] Total heart transcriptional activity was increased, also. Knowlton *et al.*

were able to demonstrate increased levels of heat shock protein 70 (Hsp70) mRNA and protein in a rabbit model of either a single 5 min minute coronary occlusion and subsequent reperfusion or repetitive occlusion/reperfusion cycles.[70] Those results were confirmed in the pig and extended to confirm increases in c-Fos, c-Jun, Jun B, Egr-1, HBGF-1, and TGF-1 mRNAs, whereas the mRNA levels of c-Myc remained almost unchanged. Similar to the findings in ischemic brain tissue, c-Jun mRNA expression was increased only slightly. Marber *et al.* have reported that overexpression of the rat inducible 70 kDa heat stress protein in a transgenic mouse increases the resistance of the heart to ischemic injury.[71]

The heat shock response is triggered with ischemia-reperfusion in the liver[72] and heart.[73] Fujio *et al.* compared the proteins generated by *in vitro* translation assays of liver extracts of rats undergoing whole body hyperthermia (42 °C, 15 min), two-thirds partial hepatectomy or liver ischemia for 1 h.[74] *In vitro* translation assays showed increased synthesis of proteins of molecular mass of 70 (peak at 2.5 h),[74] and 100 kDa (peak at 6 h) after hyperthermia. The 70, 71, and 85 but not the 100 kDa proteins were increased in level, though with a different time course after ischemia (maximum level of p70 and p71 at 2.5 h, of p85 at 6 h) and partial hepatectomy (p70, p71, and p85 peaked at 6 h and returned to normal levels at 24 h). By differential screening of a shock liver cDNA library from swine (4 h of cardiogenic shock and 4 h of recovery) an induction of metal-lothionein was found.[75] c-Fos mRNA is rapidly (30 min) and transiently expressed after 1 h of ischemia to the liver due to increased transcriptional activity.[76] Furthermore, Hsp70 mRNA could be detected as early as 2 h and as late as 7 h of reperfusion, whereas c-myc mRNA remained at the same level as in sham-operated rats (comparable to postischemic porcine myocardium). The mRNA encoding Grp[78], another member of the Hsp70 family, appeared to be increased very late in the recovery phase, 24 h after re-establishment of circulation. Tacchini and co-workers confirmed the expression of c-Fos and Hsp70 mRNA in postischemic rat liver and furthermore demonstrated increased levels of c-Jun and heme-oxygenase mRNA. The increased expression of Hsp70 mRNA was probably due to the binding of the heat shock factor to the heat shock element in the promoter region of the *Hsp70* gene.[76]

7.22.4.1.2 *Ischemia in the kidney*

When one considers the genetic events that occur in the postischemic kidney it is important to consider them in the context of the cellular events that characterize the period of injury and repair that occurs as a consequence of ischemia. As indicated previously, cells can die by two processes, necrosis or apoptosis.

(i) Apoptosis

One of the ways in which a cell can die is via a genetic program of cell death that has been called apoptosis. Apoptosis is a complex biological process by which cells destroy themselves. The apoptotic cell undergoes condensation of cytoplasmic organelles, condensation and fragmentation of the chromatin, budding of the cell membrane, and activation of an endonuclease which leads to the characteristic oligonucleosomal pattern of chromosomal DNA degradation.[17,18] Apoptosis plays a role in embryonic development, metamorphosis, involution and atrophy of glands, and in lymphoid germinal centers.[17]

Pathological evidence of apoptosis has been found in the postischemic animal kidney[77–79] and apoptosis has been described in clinical ARF in man.[6] Apoptosis seems to be particularly prevalent in post-transplant ARF, where it coexists with necrosis.[80] Gobé et al. described the morphological changes in rat kidneys that occurred with unilateral renal artery partial occlusion.[79] No apoptotic cells could be detected in the contralateral, nonclamped kidneys, whereas in the partially ischemic kidneys the number of apoptotic cells steadily increased up to eight days after the partial occlusion was imposed. Between eight and 28 days the number of apoptotic cells remained fairly constant. During that time the injured kidney also became atrophic. In the acute phase (2–8 days of clamping), both necrotic and apoptotic phenomena could be detected (in particular in the proximal tubule), whereas in the chronic phase (10–28 days) necrosis was negligible.

Depending on the length of the ischemic period, apoptosis could also be demonstrated in an acute model of ischemic acute renal failure in the rat. A 5 min period of ischemia produced only few apoptotic bodies at 24 h and 48 h after ischemia. After ischemic periods of 30 and 45 min the picture was more dramatic. Not only could apoptotic bodies be detected by 12 h after ischemia, but their number also had increased compared to shorter periods of ischemia.[78] Those morphological findings were corroborated by the appearance of oligonucleosomal DNA fragments. A study *in vitro* showed that H_2O_2 was able to induce a phenomenon similar to apoptosis in LLC-PK$_1$ cells.[81] It can therefore be speculated that H_2O_2, generated in the postischemic period, contributes to the induction of apoptosis *in vivo*.

The genetic program of apoptosis is complex and only incompletely understood. The expression of several genes has been associated with induction of apoptosis in a number of cell types.[19,82] The B-cell lymphoma gene-2 (*bcl*-2) encodes an antiapoptotic protein that is able to inhibit apoptosis induced by many stimuli. Many genes that play roles in control of proliferation, such as *p53*, *c-myc*, *Rb-1*, *E1A*, *cyclin D1*, *c-fos*, and *p34^{cdc2}*, also have been implicated in the control of apoptosis, suggesting potential commonality of signaling pathways between proliferation and apoptosis. Overexpression of either the gene encoding interleukin-1β-converting enzyme (ICE)[83] or the related protease, Nedd-2/Ich-1[84] is proapoptotic. While the expression patterns of genes involved in apoptosis has not been described in the postischemic kidney it is clear that *bcl-2*[85,86] and *nedd-2*[84] are important for the apoptosis that normally occurs in the developing kidney.

(ii) Genes whose expression is altered with ischemia/reperfusion

A large number of genes have been found to be altered in expression in response to ischemia/ reperfusion (Table 1). A number of growth factors, including platelet-derived growth factor (PDGF), transforming growth factor-β (TGF-β), and heparin-binding epidermal growth factor (EGF)-like growth factor, are increased in the postischemic tissue and have been implicated in the repair process. c-Fos mRNA is expressed to elevated levels in the postischemic kidney.[87,88] The immediate-early gene induction was apparent either by clamping one renal artery for 10 or 30 min (higher levels were achieved by longer ischemia),[87] or by bilateral clamping for 50 min.[88] In either case, c-Fos appeared early (1 h after the onset of reperfusion[87,88]) and transiently (undetectable at 4 h postischemia[88]). In addition to *c-fos* another immediate-early gene, *Egr-1*, was found to be induced in the postischemic kidney,[87,88] although the results were slightly different from those obtained with *c-fos*. Whereas with *c-fos* the duration of the ischemic period had an impact on the level of c-Fos mRNA, Egr-1 mRNA reached very high levels one hour after the onset of reperfusion even after relatively short periods of ischemia.[87] Egr-1 mRNA levels declined toward baseline at a rate much slower than that of *c-fos* mRNA.[87,88] mRNAs encoded by two other immediate-early genes, *JE* and *KC*, were also expressed markedly higher in postischemic

Table 1   Some genes whose kidney expression is modulated with ischemia/reperfusion.

Growth factors

Heparin binding EGF-like growth factor	↑
HGF	↑
PDGF	↑
TGFβ	↑
Growth hormone	↓
IGF-1	↓
prepro EGF	↓

Transcription factors

Egr-1	↑
c-Fos	↑
Kid-1	↓

Cytokines:

JC	↑
KE	↑

Cell surface proteins

Intercellular adhesion molecule-1 (ICAM-1)	↑
c-met	↑
E-selectin	↑
IGF-1 binding protein	↓

Cytoskeletal proteins

actin	↓
vimentin	↑

Defense proteins

HSP-70	↑
superoxide dismutase	↓

Others

Clusterin	↑
Histone H2b	↑
Lipocortin (annexin 1)	↑
renin	↓

---

kidneys than in controls.[89] These latter two immediate-early genes encode cytokines which are secreted and may act in an autocrine or paracrine fashion in the postischemic tissue.

Another important class of genes that are upregulated by ischemia/reperfusion are those that encode proteins that regulate cell–cell adhesion. Intercellular adhesion molecule-1 (ICAM-1) mRNA levels are increased with ischemia reperfusion of the kidney.[90] Protein expression is upregulated primarily on the endothelium of the outer medulla.[91] The importance of this expression was demonstrated by finding that treatment with antibodies against ICAM-1 protected the rat kidney against ischemic injury.[91] In addition, animals that had been genetically engineered to produce no ICAM-1 protein were protected against ischemic injury.[91]

It has been reported[92] that the mRNA levels of a member of the HSP 70 gene family are increased soon after an ischemic insult to the kidney. mRNA levels are maximal at 3 h after clamping the renal artery for 40 min. Similar findings were subsequently reported by Van Why *et al.*[93] who localized HSP-72 to the apical membrane of proximal tubule cells at 15 min of reoxygenation after 45 min of ischemia.[93] The protein is distributed throughout the cytoplasm in a vesicular pattern from 2–6 h of reperfusion when microvilli begin to recover, and is located away from the apical membrane at 24 h of reperfusion. The early localization of HSP-72 to the apical part of the proximal cell positions this protein at the site of the cell undergoing rapid changes during the early period of ischemia. There is rapid loss of apical microvilli with ischemia and loss of cell polarity.[94]

It has been reported that the heat shock response protects the pig kidney against warm ischemia.[95] In this study, donor kidneys were heated to 42.5 °C *in vivo* for 15 min, allowed to recover for 4–6 h, and then exposed to 90 min of ischemia at 37–38 °C. The kidneys were then removed, flushed with Euro-Collins solution at 4 °C and stored at this temperature for 20 h prior to transplantation into a littermate. Compared with controls the heat shock group had significantly greater survival rates and improved renal function on day 6 after the allograft was placed into the recipient. The cellular mechanisms responsible for protection postheat shock are not known although the multiple cellular events that heat shock proteins can influence suggest multiple possible mechanisms for this protection.[96]

Up to this point we have discussed a number of genes whose expression is increased with ischemia and reperfusion. There are, however, other genes whose expression is decreased. These include prepro epidermal growth factor (preproEGF), Tamm Horsfall protein, renin and superoxide dismutase.[88,97,98] Interestingly, the number of binding sites for EGF in the proximal tubule, but not the glomerulus, of the postischemic kidney was fourfold higher than in controls. Similarly, the number of binding sites for hepatocyte growth factor (HGF) receptor, encoded by the *c-Met* gene was also increased.[99]

Entry of cells into the cell cycle 24–48 h postreperfusion has been demonstrated by incorporation of ^{3}H-thymidine,[100] increased synthesis of histone H2b mRNA,[101] or immunocytochemical detection of ribonucleotide reductase[99] and PCNA.[102] mRNA levels of a kidney-specific DNA-binding protein of the zinc-finger family, Kid-1, is decreased following ischemic and folic acid-induced injury.[103] This gene is downregulated under other conditions where mitogenesis is induced, for example, early in postnatal renal development. It is interesting that the Kruppel-associated box-A (KRAB-A) in the non-zinc-finger region of the Kid-1 protein confers transcriptional repressor activity.[104] Kid-1 may act, therefore, by inhibiting the expression of genes that encode proteins important for growth arrest or differentiation.

The response of the kidney to ischemic injury is not homogeneous. Most of the damage occurs in the S3 segment of the proximal tubule, whereas other structures in the kidney are remarkably resistant to ischemia. In order to understand the role of genetic programs in the injury and repair process it is therefore crucial to determine where in the kidney molecular changes take place. As has already been mentioned, the number of binding sites for EGF increases in the postischemic kidney. Those binding sites have been localized (tentatively) to the regenerating tubules. Although the levels of EGF produced by the postischemic kidney fall, perhaps other growth factors, whose levels increase, act upon this receptor or other receptors present in the surviving epithelial cells of the regenerating tubules. Most of the surviving cells in the S3 segment of the proximal tubule enter the cell cycle as judged by immunocytochemical detection of PCNA and appearance of mitotic figures in many cells of the regenerating tubule. In the postischemic kidney, the expression of vimentin,[102] a protein found in cells of mesodermal origin and found in metanephrogenic mesenchyme early in renal development but not in normal epithelia, suggests that the dividing cells also dedifferentiate.

Many of the postischemic molecular changes, however, occur in structures other than the S3 segment of the proximal tubule and one therefore has to be careful when interpreting Northern or Western blots from whole organs. Egr-1 protein is located at greatly elevated levels in the nuclei of cells in the thick ascending limb and cortical and medullary collecting ducts.[105] c-Fos protein shows a similar pattern of expression in the postischemic kidney as Egr-1; in addition it also is expressed in the S3 segment. JE antigen can be detected in the apical regions of the thick ascending limb. Clusterin, a protein which may play a role in apoptosis or cell–cell adhesion, is detected both in the S3 segment and the thick ascending limb with strikingly different staining patterns.[102]

7.22.4.2   Genetic Response to Anoxia *In Vitro*

The number of articles describing gene induction events in mammalian cells after hypoxia (Table 2) has steadily increased. The

Table 2   Some mammalian genes whose expression is modified by hypoxia in cell culture.

c-Fos ↑
c-Jun ↑
endothelial leukocyte adhesion molecule-1 (ELAM-1) ↑
endothelin ↑
erythropoetin ↑
GADD 153 ↑
glucose regulated protein 78 ↑
growth arrest and DNA damage (GADD) 45 ↑
heme oxygenase ↑
HSP-70 ↑
intercellular adhesion molecule-1 (ICAM-1) ↑
interleukin-1α ↑
interleukin-8 ↑
lactate dehydrogenase ↑
p53 ↑
phosphoglycerate kinase 1 ↑
platelet derived growth factor A and B ↑
ref-1/HAP1 ↑
TGF-β1 ↑
tyrosine hydroxylase ↑
vascular endothelial growth factor ↑
phosphoenolpyruvate carboxykinase ↓
placental growth factor ↓

first publications provided only indirect information regarding the nature of these genes. Sciandra *et al.*[106] compared the protein pattern (via one- and two-dimensional protein gel electrophoresis) in CHO (Chinese hamster ovary) cells after glucose deprivation and anoxia (<0.4% O_2). They found new synthesis of 68 kDa and 89 kDa proteins 4 h after the onset of anoxia, and of 76 kDa and 97 kDa proteins at a later time-point (>12 h of anaerobiosis). By their analyses, the 68 kDa and 89 kDa proteins could also be detected in heat-shocked CHO cells and the 76 kDa and 97 kDa proteins were detected in glucose-deprived cells. These four proteins were synthesized at an even higher level after restoration of oxygen, and returned to normal cellular levels approximately 24 h after reoxygenation (though each different protein returned to baseline levels with different time kinetics). Similar observations were made by other groups.

This work provided only indirect evidence for genetic control of expression of the genes encoding these proteins and did not provide the identity of the proteins. By more direct means the identity of some of them could be established and new genes were added to a growing list of hypoxia-inducible genes. Of several protooncogenes tested, only *c-fos* was shown to be inducible in primary human skin fibroblasts with exposure to 95% N_2/5% CO_2. No induction could be demonstrated for *c-Abl*, c-ErbB, c-Fes, c-Fgr, c-Mos, c-Myc, c-Ha-Ras, c-Sis, and c-Src.[107] 30 min, 1 h, and 2 h of oxygen deprivation yielded approximately the same level of c-fos induction with maximum levels achieved 30–60 min after reoxygenation. By contrast, 6 h or longer of oxygen deficiency markedly suppressed c-Fos expression. Ausserer et al. showed that the higher mRNA levels of c-Jun during hypoxia resulted both from transcriptional activation of its gene and from message stabilization in SiHa cells, a human squamous cell carcinoma cell line.[108]

An interesting set of experiments was performed to examine the response to hypoxia of mRNA levels of glycolytic enzymes in rat skeletal muscle. Rat skeletal muscle myoblasts grown under conditions of 93% N_2/2% O_2/5% CO_2 show a higher overall transcription rate when compared to cells grown at higher oxygen tensions. The transcription rates for triose phosphate isomerase, aldolase, pyruvate kinase, and lactate dehydrogenase increased by day 1 and reached a steady-state by day 3, while that for glyceraldehyde-3-phosphate dehydrogenase and cytochrome c dehydrogenase decreased.[109] In endothelial cells, however, higher levels of glyceraldehyde-3-phosphate dehydrogenase mRNA were found with hypoxia and this increase was shown to be due to an increased transcriptional rate. Kadowaki and Kitagawa[110] provided evidence that mRNA levels for several mitochondrial proteins, including NADH dehydrogenase subunits 3 and 4/4L, cytochrome c oxidase subunit III, and H+ ATPase subunit 6/6L, decreased slowly during hypoxia.

Other mRNAs shown to be induced during hypoxia *in vitro* include those coding for oxygen-regulated protein ORP 80/glucose-regulated protein 78,[111] ORP 33 (which is identical to heme oxygenase),[112] Gadd (growth arrest and DNA damage) 45 and Gadd 153.[113] Increased protein levels during or after hypoxia were demonstrated for the cell adhesion molecules ELAM-1 and ICAM-1,[114] the tumor suppressor p53,[115] and ferritin.[116] In the latter case no changes in mRNA levels were found.

The interpretation of gene induction in *in vitro* studies becomes more complicated due to possible secondary effects by cytokines. If primary cultures of human umbilical vein endothelial cells are exposed to 1% or 0% oxygen, the mRNA level of PDGF(platelet-derived growth factor)-B increases by at least eightfold and tenfold, respectively, after 3 d of culture under low oxygen tension.[117] The main reason for this increase in mRNA levels appears to be a higher transcription rate though the half-life of the PDGF-B mRNA was slightly

prolonged under 1% oxygen (90 min vs. 60 min in 21% oxygen). The level of PDGF-A mRNA was not affected, so that it is likely that increased amounts of PDGF-BB homodimers, and not AB heterdimers, are produced. Other investigators reported the increased expression, under hypoxic conditions, of VEGF (vascular endothelial growth factor),[118,119] endothelin,[120] interleukin-1α,[114] and interleukin-8.[121] As was the case for PDGF-B, the increased expression of endothelin and interleukin-8 was due at least to some extent to a higher transcriptional rate.

A number of other genes have been found to be induced with hypoxia in tissue culture. Genes encoding the glycolytic enzymes, lactic dehydrogenase and phosphoglycerate kinase-1, are upregulated.[122–124] Although the control of glycolysis is primarily at the level of posttranslational modifications of glycolytic enzymes, upregulation of these genes might enhance energy production from glycolysis in the hypoxic cell.

Hypoxia also results in upregulation of genes which have been implicated in regulation of vascular tone, wound healing, angiogenesis, proliferation, antiproliferation, and embryonic development. These include genes encoding growth factors such as PDGF-A and B, TGF-β1, vascular endothelial growth factor (VEGF), and endothelin. Interleukin-1α,[125] tyrosine hydroxylase,[126,127] and HSP70[128] are also upregulated. By contrast, hypoxia results in a decrease in placental growth factor mRNA levels.[129] It has been reported that hypoxia-responsive transcription-activating *cis* elements are present in the 5′ flanking region of the VEGF, LDH, and phosphoglycerate kinase-1 genes. Hypoxia increases the mRNA levels of VEGF, as it does tyrosine hydroxylase, by increased transcription as well as increased RNA stability.

7.22.4.3 Potential Gene Regulatory Mechanisms

Induction of gene expression in response to ischemia likely results as a direct consequence of one or more of the metabolic changes in the cell that occur as a result of the ischemic insult. In addition to low oxygen tension, ischemia/ reperfusion is associated with the generation of reactive oxygen species (ROS), severe acidification of the intracellular environment, and ATP depletion. ROS are believed to play an important role in the pathophysiology of ischemic injury to many organs including the kidney.[8,130–135] As has already been discussed, heme oxygenase mRNA and protein is induced by hypoxia. In addition, this gene is induced by hydrogen peroxide.[136] Reactive oxygen species also induce the hypoxia-induced mRNAs encoding c-Fos,[137] c-Jun, Egr-1, and the ischemia-induced JE mRNA.[138,139] In each case the induction results at least to some extent from transcriptional activation. Mesangial cells, glomerular endothelial and glomerular epithelial cells respond to a challenge with hydrogen peroxide by expressing increased amounts of manganese superoxide dismutase (Mn-SOD) but not Cu/Zn-SOD.[140] Again, this upregulation involves transcriptional activation. A different mode of action must be assumed for the induction of surface GMP-140 (granule membrane protein-140) on human endothelial cells, where H_2O_2 treatment results in the translocation of GMP-140 from intracellular stores to the cell membrane.[141]

This section describes several potential mechanisms by which transcriptional activation might be achieved in the postischemic tissue. We will use as examples the transcription factor, heat shock factor (HSF), the regulatory protein, Ref-1 (also known as APE or HAP1, not to be confused with yeast HAP1), the antioxidant response element (ARE), K^+-channels, the hypoxia-inducible factor 1 (HIF-1), NF-κB, and second messengers.

New synthesis of transcription factors is not necessarily required to upregulate the transcription of other genes. For example, the *hsp70* gene is activated by binding of a normally inactive form of heat shock factor (HSF) to a heat shock element (HSE) in its promoter region.[142,143] The classical activation event is heat, but hypoxia,[144] low pH (5.8 to 6.4), nonionic detergents like NP-40, Triton X-100, urea (0.5–1.0 M), H_2O_2, and Ca^{2+} can result in enhanced HSF binding to HSE, as demonstrated by electrophoretic mobility shift assays.[145,146] Low pH, and Ca^{2+}, in addition to H_2O_2 (described previously), play a major role in ischemic tissue injury.[6–8]

Mestril *et al.* cloned the cDNA and gene from the rat which coded for an inducible form of hsp70. This gene contains two heat shock elements in its promoter region, both of which are necessary for induction by hypoxia, but only one suffices for induction by heat.[147] The same laboratory was able to demonstrate a partial protection against hypoxia by stably overexpressing the human form of inducible hsp70 in rat embryonic cardiocytes.[148]

A good deal of attention has been directed towards understanding the control of transcription factors by the cellular redox potential. The first evidence for this regulatory mechanism derived from the finding that the binding of the transcription factor AP-1 to its recognition site was dependent on its redox status. Reduction of

a conserved cysteine residue in the DNA binding domain of both c-Fos and c-Jun enhances their DNA-binding activity *in vitro*. The reduction results from the action of a 35.5 kDa protein named Ref-1 (also known as APE/HAP1).[149] The cDNA coding for Ref-1 was subsequently cloned and shown to be identical to an already known cDNA coding for an endonuclease repair enzyme, HAP1.[150] Ref-1 is a multifunctional nuclear protein and not only stimulates the DNA binding activity of Fos–Jun heterodimers and Jun–Jun homodimers (but not Fos–Fos homodimers), but also of a number of other transcription factors such as ATF-1, ATF-2, CREB, NF-κB, and myb. Ref-1 is a redox factor which reduces a critical cysteine residue in the DNA binding domain of Fos and Jun. When the level of Ref-1 was decreased by expressing a Ref-1 antisense RNA, the cells were more susceptible to damage by hypoxia and hyperoxia.[151] Consistent with that notion is the induction of Ref-1 by hypoxia, which is due to transcriptional activation of its gene.[152]

The rat genes encoding the glutathione S-transferase Ya subunit and NAD(P)H:quinone reductase can be activated by H_2O_2. At least part of this activation can be traced back to the sequence 5′-GTGACnnnGC-3′ in the promoter regions of these genes, termed the antioxidant response element (ARE). Promoter constructs containing the ARE and a reporter gene are responsive to incubation of cells with H_2O_2.[153–155] Similar results have been obtained in the case of the murine gene coding for metallothionein-I.[156]

A very different mechanism for how cells respond to low oxygen is demonstrated by the genetic modulation of K^+ channels. Type I cells in the carotid body from rabbits contain K^+ channels, whose open probability is decreased by reduced O_2 tension, whereas the single-channel conductance is unaltered by O_2 tension. When O_2 tension was between 70 and 150 mmHg, the open probability was proportional to the applied O_2 tension.[157]

Ischemia results in loss of cell potassium, accumulation of sodium, and depolarization of membrane potential.[158] As it is known that depolarization of cells can induce transcription of some genes (e.g., *Egr-1*[159]), it is possible that membrane depolarization would be a plausible mechanism for induction of genes during/after hypoxia. The role of membrane depolarization in the upregulation of gene transcription with oxygen deprivation, however, has yet to be demonstrated.

Hypoxia-inducible factor 1 (HIF-1) is felt to play a role in the activation of the erythropoietin gene during low oxygen tension. Subsequently it was found that HIF-1 activity is also induced in a variety of other cell lines which do not produce erythropoietin in response to hypoxia.[160] The genes coding for several glycolytic enzymes activated by hypoxia were also shown to contain HIF-1 binding sites. Those binding sites were recognized by a hypoxia-inducible protein. Reporter constructs containing those elements responded to low oxygen tension.[124]

The transcription factor NF-κB usually resides in the cytosol where it is bound to an inhibitor, IκB.[161] Upon activation of the cell NF-κB is released from IκB and moves to the nucleus where it binds to DNA and activates its target genes. It appears that the involvement of oxidative stress is a common theme in the activation of NF-κB. Many agents activating NF-κB such as interleukin-1 or ultraviolet light increase intracellular oxidative stress. Several chemically distinct antioxidants are able to prevent the activation of NF-κB. In addition, NF-κB is posttranslationally activated by low concentrations of hydrogen peroxide.[161] Interestingly, low oxygen tension can activate NF-κB. In Jurkat T-cells 0.02% oxygen results in the tyrosine phosphorylation and subsequent degradation of IκB. Inhibition of tyrosine phosphorylation of IkB prevented its degradation and activation of NF-κB.[162]

The final topic in this section discusses other important intermediates in signal transduction. H_2O_2, acidosis, and hypoxia each lead to increases in mRNAs for c-Fos, c-Jun, Egr-1, JE, c-Myc, Gadd45, and Gadd153. The expression of c-Fos, c-Jun, Jun B, and Egr-1 mRNAs could be induced by low pH even when protein kinase C was downregulated by prior incubation with phorbol ester or when a rise of intracellular calcium was prevented by exposing the cells to BAPTA.[163] Herbimycin A, however, a tyrosine kinase inhibitor, was able to block induction of those mRNAs by acidosis.[163]

Exposure of rat proximal tubular epithelial cells to EGTA (an extracellular chelator of Ca^{2+}), Quin 2/AM (an intracellular chelator of Ca^{2+}), or protein kinase inhibitors relatively specific for protein kinase C (H-7 and staurosporine) resulted in a modest inhibition of induction of c-Fos mRNA by H_2O_2, whereas 2-aminopurine (a protein kinase inhibitor with a broad spectrum) almost completely abolished the response to H_2O_2.[138] Similar to the induction by low pH, c-Fos and c-Jun mRNA induction by H_2O_2 in a mouse osteoblast cell line could not be blocked by downregulation of protein kinase C, although incubation with the relatively specific protein kinase C inhibitor H-7 partially prevented expression.[139] In rat aortic smooth muscle cells phospholipase A_2 (PLA_2) may be important for the induction of

c-Fos mRNA by H_2O_2. The PLA_2 inhibitors mepacrine and *p*-bromophenacylbromide were able to block expression of c-Fos mRNA, whereas exposure of those cells to arachidonic acid increased c-Fos mRNA. NDGA (nordihydroguaiaretic acid), a phospholipase A_2 and lipoxygenase/cytochrome P450 inhibitor, blocked c-Fos expression, but indomethacin, a cyclooxygenase inhibitor, did not.[137] Gadd45 and Gadd153 expression during hypoxia can be blocked efficiently by the protein kinase inhibitors H-7 and 2-aminopurine, but not by the tyrosine kinase inhibitor, genestein.[113]

Most of these studies provided only indirect insight into the nature of the upstream proteins involved in the induction of genes such as *c-fos* or *c-jun*. The activation of the p54 stress-activated protein kinase (SAPK) after renal ischemia/reperfusion or ATP depletion in MDCK cells was demonstrated in the mid-1990s in our laboratory.[164] As *c-jun* is one of the immediate-early response genes activated early in the post-ischemic period,[159] Pombo *et al.* evaluated whether p54 SAPK, which phosphorylates c-Jun within the N-terminal transactivation domain, is activated in response to ischemia. SAPKs are a subfamily of the extracellular signal-regulated kinases (ERKs). SAPK kinase activity was assayed using GST-c-Jun(1–135), containing the N-terminal transactivation domain of c-Jun, as substrate. SAPKs were not activated by ischemia alone, but reperfusion for as little as 5 min was associated with a 4.6-fold increase in kinase activity. Kinase activity was increased 7.6-fold at 20 min following reperfusion and remained elevated at 90 min of reperfusion (4.9-fold). In contrast, activity of the related ERK-1 and -2 was increased only 1.4-fold and only at the 5 min reperfusion timepoint. SAPKs accounted for the majority of the N-terminal c-Jun kinase activity of kidney at 5 min following reperfusion. ATP repletion following ATP depletion in MDCK renal epithelial cells, was associated with a 15- to 29-fold activation of SAPKs with a similar time course of activation to that seen in the kidney after ischemia and reperfusion. Thus SAPKs, possibly by *trans*-activating c-Jun may function in the early stages of the complex genetic response to ischemia, ultimately leading to mitogenesis and/or cytoprotection and repair.

7.22.4.4 The Erythropoietin Gene

Hematopoiesis is one of the most complex biological control mechanisms in man. The generation of the different populations of blood cells is very tightly regulated, a necessary feature given the diversity of action of each of the different cell types. One can discern three lineages in the bone marrow, the erythropoietic lineage, the thrombopoietic and the granulo-cyte-macrophage lineage. A wide variety of growth factors are involved in the regulation of the different lineages. Some of them exert their effects very early in differentiation and therefore influence more than one cell type, whereas others act very late in development and have very specific effects.[166] Erythropoietin (Epo) is a cytokine which promotes growth and differentiation in the red cell lineage. Though it was known for a long time that erythropoietin is produced in the kidney, the identity of the erythropoietin-producing cell in the kidney was clarified only in the late 1980s.[167,168] Koury *et al.* found that the more severe the anemia in an animal the higher is the number of erythro-poietin-producing cells and not, as one might expect, the level of mRNA per cell.[169] Transgenic mice containing a transgene under control of regulatory sequences from the human or murine erythropoietin gene expressed the transgene either in peritubular or proximal convoluted cells.

The investigation of the regulation of erythropoietin production was hampered by the lack of a good *in vitro* model until Goldberg *et al.* recognized that two human hepatoma cell lines, Hep3B and HepG2, constitutively express erythropoietin.[170] Under the influence of hypoxia and cobalt these cells can be stimulated to synthesize more of this cytokine (Hep3B: 18-fold and sixfold increase by hypoxia and cobalt, respectively; HepG2: threefold increase by either agent). Erythropoietin production was increased both at the mRNA and protein level.[170] With decreases in oxygen tension to approximately 5% O_2 a markedly higher steady-state-level of mRNA was observed. In three subsequent reports the underlying reasons for this upregulation were established. Schuster *et al.* exposed rats to subcutaneous injections of cobalt and to anemia/hypoxia, by lowering their hematocrits to 18% to 24% and by placing them into a hypobaric chamber (0.4 ATM).[171] Within 4 h of the injection with cobalt chloride, a significant level of transcription from the erythropoietin gene was noticed (baseline transcription was undetectable). The same was true for the hypoxic/anemic animals. Similar experiments were performed with the Hep3B cell line.[172] Nuclear extracts prepared from Hep3B cells grown at 1% O_2 provided a consistently richer source for driving erythropoietin transcription *in vitro* than those prepared from Hep3B cells grown at 21% O_2 or from HeLa cells. In HeLa cells it did not matter whether they were grown under hypoxic conditions or not. This

suggested that, although there may be a basal level of erythropoietin gene transcription in quite a variety of cell types, only in some cells is transcription inducible by hypoxia.

The nature of the oxygen sensor responsible for upregulation of the erythropoietin gene remains unclear. Besides hypoxia and cobalt, nickel and manganese stimulate transcription from the Epo gene. Protein synthesis is necessary for both the induction by hypoxia and cobalt ions.[173] It has been suggested that a rapidly turning over heme protein is a key component of transcriptional regulation because nickel, cobalt, and manganese can substitute for iron in the porphyrin ring.[173] These ions, in contrast to iron, cannot bind O_2, so that the heme protein would be locked in its deoxy-state and subsequently turn on transcription. New protein synthesis is required because these ions cannot be incorporated into an already existing heme protein. The view of Goldberg *et al.* of a heme protein as a common pathway through which all these agents act is supported by these observations: (i) there is no additive effect for hypoxia plus cobalt, hypoxia plus nickel, or cobalt plus nickel; (ii) carbon monoxide (CO) to a great extent abolishes Epo induction by hypoxia (CO binds tightly to reduced heme protein and locks it in the oxy-state); (iii) CO does not abrogate the effects of nickel or cobalt (CO binds only to reduced iron in the porphyrin ring); and (iv) inhibitors of heme synthesis also decrease the response of Hep3B cells to either hypoxia, nickel or cobalt.

The understanding of the effects of oxygen on Epo regulation is further complicated by the fact that posttranscriptional events play a role. When Hep3B cells are grown at 1% O_2 and are transferred subsequently to 21% O_2, the steady-state mRNA levels of Epo decline by 50% 90–120 min after reoxygenation.[174] This represents a maximal estimate of the actual half-life of the mRNA because new mRNA synthesis was not blocked. When mRNA synthesis was blocked by actinomycin D, however, instead of shortening the steady-state half-life of Epo mRNA, it was prolonged to 7–8 h (similar findings were made with cycloheximide). This suggests that a protein with a high turnover rate plays a role in the degradation of Epo mRNA.

The *cis*-elements involved in the regulation of the *Epo* gene appear to be located both at the 5′- and 3′-end of the gene.[173] When one compares the murine and human *Epo* genes, strong homology in the coding region is found, which is not unusual. More surprisingly, however, there is the same degree of homology in parts of the noncoding sequences: (i) 140 bp upstream of the transcription initiation site; (ii) two regions in the first of four introns; and

(iii) about 100–220 bp $3'$ to the stop codon. Constructs, which utilized the conserved sequences located in the $5'$-end of the gene, conferred hypoxia- as well as cobalt-responsiveness to a vector with a growth hormone reporter gene. The addition of the conserved $3'$ noncoding sequence increased the growth hormone level even further.[174] The specificity of the results were supported by the finding that CO partly abrogated the hypoxia inducibility of the constructs and that no inducibility was found in COS7 cells in contrast to Hep3B cells (COS7 cells neither produces Epo constitutively nor under hypoxia). It was argued that the COS7 cells did not express the necessary *trans*-acting factors, but care has to be taken with this interpretation, because COS7 cells are derived from African green monkey and there may exist a species barrier (vector constructs were made from the human *Epo* gene). HIF-1, the hypoxia-inducible factor-1, which has already been mentioned above, is newly synthesized in hypoxic Hep3B cells and binds to an enhancer region in the $3'$-end of the human erythropoietin gene. Studies in the mid-1990s in transgenic mice reveal that *cis* DNA sequences required for induction in the kidney are located more than 9.5 kb $5'$ to the gene.[175]

Additional studies provided more evidence about the nature of the *trans*-acting factors involved in the regulation of erythropoietin protein expression. A cytosolic Epo RNA binding protein was identified in several organs including liver, kidney, brain, and spleen. Whereas this binding activity did not change after hypoxia in liver or kidney, an increased binding activity was observed in brain and spleen with hypoxia.

Though there are still many questions left unanswered, for example, why the inducibility of the growth hormone reporter gene is less than that of the endogenous Epo gene and what are the negative regulatory elements, the above experiments offer an exciting insight into the regulation of a hypoxia-response gene in mammals.

tion is not "unexpected" in these lower organisms, and they have developed strategies to deal with it. It is assumed that some sort of coordinated events are also involved in the response of higher organisms, including man, to ischemic injury. A steadily increasing number of ischemia/hypoxia-induced genes have been identified, but almost nothing is known about their regulation, or about the network in which these genes interact. It will require a great deal more basic research to come to a better understanding of the underlying molecular events in ischemia. Many questions remain. Why is one region of an organ more susceptible than the other one? What is the tissue distribution of gene expression? Can the different tolerance level to ischemia of different cell types be explained on the basis of differential gene expression? There are only very few clues as to the answers to these questions[105] and much more has to be done.

Understanding the molecular basis of cell injury and the mechanisms brought to bear by the cell and organ to enhance survival will lead to new strategies for protection of the cells and tissue. Perhaps the genetic response measured in experimental animals does not occur optimally in humans, in which case potentiation of the response might be therapeutically useful. Growth factors might serve to be useful therapeutically to enhance recovery.[100,176–181] Interference with cytokines important in the ischemic injury might be effective therapeutically, as experiments with delivery of the IL-1 receptor antagonist by gene therapy to the brain have suggested.[182] Alternatively other cytokines may be effective therapeutically, if it is found that it is possible to trigger a genetic response that will help to minimize the cell injury or potentiate cell division or repair. It is hoped further research will bring a better insight into the basic mechanisms of ischemic injury and new strategies to prevent and treat.

7.22.5 CONCLUSIONS AND OUTLOOK

In this chapter we have tried to provide a brief overview of the molecular events that occur in response to anoxic, hypoxic, and ischemic injury. Study of bacteria and yeast indicates how these organisms adapt to hypoxia and what regulatory mechanisms are brought to bear in this adaptation. Of course those findings cannot easily be extrapolated to vertebrates and mammals, although there are many examples of parallel mechanisms existing in yeast and mammalian cells. Oxygen depriva-

7.22.6 REFERENCES

1. P. W. Hochachka, P. L. Lutz, T. Sick *et al.*, 'Surviving Hypoxia, Mechanisms of Control and Adaptation,' CRC Press, Boca Raton, FL, 1993.
2. P. L. Lutz and P. W. Hochachka, in 'Surviving Hypoxia, Mechanisms of Control and Adaptation,' eds. P. W. Hochachka, P. L. Lutz, T. Sick *et al.*, CRC Press, Boca Raton, FL, 1993, pp. 459–469.
3. P. W. Hochachka, 'Defense strategies against hypoxia and hypothermia.' *Science*, 1986, **231**, 234–241.
4. 'Webster's Seventh New Collegiate Dictionary,' Merriam, Springfield, MA, 1963.
5. J. M. Weinberg, 'The cell biology of ischemic renal injury.' *Kidney Int.*, 1991, **39**, 476–500.

6. J. V. Bonventre, 'Mechanisms of ischemic acute renal failure.' *Kidney Int.*, 1993, **43**, 1160–1178.

7. J. V. Bonventre, in 'Textbook of Molecular Medicine,' ed. J. L. Jamison, Blackwell Science, Cambridge, MA, 1996, in press

8. R. Thadhani, M. Pascual and J. V. Bonventre, 'Acute renal failure.' *New Engl. J. Med.*, 1996, **334**, 1448–1460.

9. J. C. Grotta, C. M. Picone, P. T. Ostrow *et al.*, 'CGS-19725, a competitive NMDA receptor antagonist, reduces calcium–calmodulin binding and improves outcome after global cerebral ischemia.' *Ann. Neurol.*, 1990, **27**, 612–619.

10. K. M. Raley-Susman and P. Lipton, '*In vitro* and protein synthesis in the rat hippocampal slice: the role of calcium and NMDA receptor activation.' *Brain Res.*, 1990, **515**, 27–38.

11. M. J. Sheardown, E. Nielsen, A. J. Hansen *et al.*, '2,3-Dihydroxy-6-nitro-7-sulfamoyl-benzo(F)quinoxaline: a neuroprotectant for cerebral ischemia.' *Science*, 1990, **247**, 571–574.

12. J. V. Bonventre and W. J. Koroshetz, 'Phospholipase A_2 (PLA_2) activity in gerbil brain: characterization of cytosolic and membrane-associated forms and effects of ischemia and reperfusion on enzymatic activity.' *J. Lipid Mediat.*, 1993, **6**, 457–471.

13. H. S. Basaga, 'Biochemical aspects of free radicals.' *Biochem. Cell Biol.*, 1990, **68**, 989–998.

14. B. Halliwell and J. M. C. Gutteridge, 'Role of free radicals and catalytic metal ions in human disease, an overview.' *Methods Enzymol.*, 1990, **186**, 1–85.

15. S. W. Werns and B. R. Lucchesi, 'Free radicals and ischemic tissue injury.' *Trends Pharmacol. Sci.*, 1990, **11**, 161–166.

16. J. F. R. Kerr, A. H. Wyllie and A. R. Currie, 'Apoptosis: a basic biological phenomenon with wide-ranging implications in tissue kinetics.' *Br. J. Cancer*, 1972, **26**, 239–257.

17. J. Searle, J. F. R. Kerr and C. J. Bishop, 'Necrosis and apoptosis: distinct modes of cell death with fundamentally different significance.' *Pathol. Annu.*, 1982, **17**, 229–259.

18. A. H. Wyllie, J. F. R. Kerr and A. R. Currie, 'Cell death, the significance of apoptosis.' *Int. Rev. Cytol.*, 1980, **68**, 251–306.

19. H. Steller, 'Mechanisms and genes of cellular suicide.' *Science*, 1995, **267**, 1445–1449.

20. W. C. Earnshaw, 'Apoptosis: lessons from *in vitro* systems.' *Trends Cell Biol.*, 1995, **5**, 217–220.

21. B. A. Haddock and C. W. Jones, 'Bacterial respiration.' *Bacteriol. Rev.*, 1977, **41**, 47–99.

22. D. J. Shaw and J. R. Guest, 'Nucleotide sequence of the *fnr* gene and primary structure of the Fnr protein of *Escherichia coli*.' *Nucleic Acids Res.*, 1982, **10**, 6119–6130.

23. S. B. Melville and R. P. Gunsalus, 'Isolation of an oxygen-sensitive FNR protein of *Escherichia coli*: interaction at activator and repressor sites of FNR-controlled genes.' *Proc. Natl. Acad. Sci. USA*, 1996, **93**, 1226–1231.

24. D. J. Shaw and J. R. Guest, 'Amplification and product identification of the *fnr* gene of *Escherichia coli*.' *J. Gen. Microbiol.*, 1982, **128**, 2221–2228.

25. M. Trageser and G. Unden, 'Role of cysteine residues and of metal ions in the regulatory functioning of FNR, the transcriptional regulator of anaerobic respiration in *Escherichia coli*.' *Mol. Microbiol.*, 1989, **3**, 593–599.

26. M. F. Christman, R. W. Morgan, F. S. Jacobson *et al.*, 'Positive control of a regulon for defenses against oxidative stress and some heat-shock proteins in *Salmonella typhimurium*.' *Cell*, 1985, **41**,

753–762.

27. K. Tao, K. Makino, S. Yonei *et al.*, 'Molecular cloning and nucleotide sequencing of *oxyR*, the positive regulatory gene of a regulon for an adaptive response to oxidative stress in *Escherichia coli*: homologies between OxyR protein and a family of bacterial activator proteins.' *Mol. Gen. Genet.*, 1989, **218**, 371–376.

28. G. Storz, L. A. Tartaglia and B. N. Ames, 'Transcriptional regulator of oxidative stress-inducible genes; direct activation by oxidation.' *Science*, 1990, **248**, 189–194.

29. M. B. Toledano, I. Kullik, F. Trinh *et al.*, 'Redox-dependent shift of OxyR-DNA contacts along an extended DNA-binding site, a mechanism for differential promoter selection.' *Cell*, 1994, **78**, 897–909.

30. R. A. Dixon, 'The genetic complexity of nitrogen fixation. The ninth Fleming lectures.' *J. Gen. Microbiol.*, 1984, **130**, 2745–2755.

31. M. David, M. L. Daveran, J. Batut *et al.*, 'Cascade regulations of *nif* gene expression in *R. meliloti*.' *Cell*, 1988, **54**, 671–683.

32. G. Ditta, E. Virts, A. Palomares *et al.*, 'The *nifA* gene of *R. melliloti* is oxygen regulated.' *J. Bacteriol.*, 1987, **169**, 3217–3223.

33. P. de Philip, J. Batut and P. Boistard, '*Rhizobium meliloti* Fix L is an oxygen sensor and regulates *R. meliloti nifA nixK* genes differently in *Escherichia coli*.' *J. Bacteriol.*, 1990, **172**, 4255–4262.

34. M. A. Gilles-Gonzalez, G. S. Ditta and D. R. Helinski, 'A haemoprotein with kinase activity encoded by the oxygen sensor of *Rhizobium meliloti*.' *Nature*, 1991, **350**, 170–172.

35. E. K. Monson, G. S. Ditta and D. R. Helinski, 'The oxygen sensor protein, *FixL*, of *R. meliloti*. Role of histidine residues in heme binding, phosphorylation, and signal transduction.' *J. Biol. Chem.*, 1995, **270**, 5243–5250.

36. C. H. Chin, 'Effect of aeration on the cytochrome systems of the resting cells of brewer's yeast.' *Nature*, 1950, **165**, 926–927.

37. R. S. Criddle and G. Schatz, 'Promitochondria of anaerobically grown yeast. I. Isolation and biochemical properties.' *Biochemistry*, 1969, **8**, 322–334.

38. F. Sherman and J. W. Stewart, 'Genetics and biosynthesis of cytochrome c.' *Annu. Rev. Genet.*, 1971, **5**, 257–296.

39. D. L. Montgomery, D. W. Leung, M. Smith *et al.*, 'Isolation and sequence of the gene for iso-2-cytochrome c in *Saccharomyces cerevisiae*.' *Proc. Natl. Acad. Sci. USA*, 1980, **77**, 541–545.

40. L. Guarente and T. Mason, 'Heme regulates transcription of the *CYC1* gene of *S. cerevisiae* via an upstream activation site.' *Cell*, 1983, **32**, 1279–1286.

41. H. Hörtner, G. Ammerer, E. Hartter *et al.*, 'Regulation of synthesis of catalases and iso-l-cytochrome c in *Saccharomces cerevisiae* by glucose, oxygen and heme.' *Eur. J. Biochem.*, 1982, **128**, 179–184.

42. C. V. Lowry, J. L. Weiss, D. A. Walthall *et al.*, 'Modulator sequences mediate oxygen regulation of *CYC1* and a neighboring gene in yeast.' *Proc. Natl. Acad. Sci. USA*, 1983, **80**, 151–155.

43. T. M. Laz, D. F. Pietras and F. Sherman, 'Differential regulation of the duplicated isocytochrome c genes in yeast.' *Proc. Natl. Acad. Sci. USA*, 1984, **81**, 4475–4479.

44. L. Guarente, B. Lalonde, P. Gifford *et al.*, 'Distinctly regulated tandem upstream activation sites mediate catabolite repression of the *CYC1* gene of *S. cerevisiae*.' *Cell*, 1984, **36**, 503–511.

45. J. L. Pinkham and L. Guarente, 'Cloning and molecular analysis of the HAP2 locus: a global regulator of respiratory genes in *Saccharonyces cerevisiae*.' *Mol. Cell Biol.*, 1985, **5**, 3410–3416.

46. B. Lalonde, B. Arcangioli and L. Guarente, 'A single *Saccharomyces cerevisiae* upstream activation site (UAS1) has two distinct regions essential for its activity.' *Mol. Cell Biol.*, 1986, **6**, 4690–4696.

47. B. Arcangioli and B. Lescure, 'Identification of proteins involved in the regulation of yeast iso-l-cytochrome c expression by oxygen.' *EMBO J.*, 1985, **4**, 2627–2633.

48. K. Pfeifer, T. Prezant and L. Guarente, 'Yeast HAP1 activator binds to two upstream activation sites of different sequence.' *Cell*, 1987, **49**, 19–27.

49. F. Creusot, J. Verdiere, M. Gaisne *et al.*, 'CYP1 (HAP1) regulator of oxygen-dependent gene expression in yeast. I. Overall organization of the protein sequence displays several novel structural domains.' *J. Mol. Biol.*, 1988, **204**, 263–276.

50. J. Verdiere, M. Gaisne, B. Guiard *et al.*, 'CYP1 (HAP1) regulator of oxygen-dependent gene expression in yeast. II. Missense mutation suggests alternative Zn fingers as discriminating agent of gene control.' *J. Mol. Biol.*, 1988, **204**, 277–282.

51. A. Klug and D. Rhodes, ''Zinc fingers': a novel protein motif for nucleic acid recognition.' *Trends Biochem. Sci.*, 1987, **12**, 464–469.

52. K. A. Wharton, B. Yedvobnick, V. G. Finnerty *et al.*, 'opa: a novel family of transcribed repeats shared by the Notch locus and other developmentally regulated loci in *D. melanogaster*.' *Cell*, 1985, **40**, 55–62.

53. P. J. Mitchell and R. Tjian, 'Transcriptional regulation in mammalian cells by sequence-specific DNA binding proteins.' *Science*, 1989, **245**, 371–378.

54. K. Pfeifer, K. S. Kim, S. Kogan *et al.*, 'Functional dissection and sequence of yeast HAP1 activator.' *Cell*, 1989, **56**, 291–301.

55. X. Wang and G. Z. Feurerstein, 'Induced expression of adhesion molecules following focal brain ischemia.' *J. Neurotrauma*, 1995, **12**, 825–832.

56. A. C. Yu, Y. L. Lee, W. Y. Fu *et al.*, 'Gene expression in astrocytes during and after ischemia.' *Prog. Brain Res.*, 1995, **105**, 245–253.

57. Y. Collaco-Moraes, B. S. Aspey, J. S. de Belleroche *et al.*, 'Focal ischemia causes an extensive induction of immediate early genes that are sensitive to MK-801.' *Stroke*, 1994, **25**, 1855–1860.

58. K. Kogure and H. Kato, 'Arterial gene expression in cerebral ischemia.' *Stroke*, 1993, **24**, 2121–2127.

59. A. Herskowitz, S. Choi, A. A. Ansari *et al.*, 'Cytokine mRNA expression in postischemia/reperfused myocardium.' *Am. J. Pathol.*, 1995, **146**, 419–428.

60. A. Yao, T. Takahashi, T. Aoyagi *et al.*, 'Immediate-early gene induction and MAP kinase activation during recovery from metabolic inhibition in cultured cardiac myocytes.' *J. Clin. Invest.*, 1995, **96**, 69–77.

61. A. S. Wechsler, J. C. Entwistle, III, T. Yeh, Jr. *et al.*, 'Early gene changes in myocardial ischemia.' *Ann. Thorac. Surg.*, 1994, **58**, 1282–1284.

62. A. Bernelli-Zazzera, G. Cairo, L. Schiaffonati *et al.*, 'Stress proteins and reperfusion stress in the liver.' *Ann. NY Acad. Sci.*, 1992, **663**, 120–124.

63. S. Goto, I. Matsumoto, N. Kamada *et al.*, 'The induction of immediate-early genes in postischemic and transplanted livers in rats. Its relation to organ survival.' *Transplantation*, 1994, **58**, 840–845.

64. M. Jacewicz, M. Kiessling and W. A. Pulsinelli, 'Selective gene expression in focal cerebral ischemia.' *J. Cereb. Blood Flow Metab.*, 1986, **6**, 263–272.

65. G. An, T. N. Lin, J. S. Liu *et al.*, 'Expression of *c-fos* and *c-jun* family genes after focal cerebral ischemia.' *Ann. Neurol.*, 1993, **33**, 457–464.

66. R. M. Gubits, R. E. Burke, G. Casey-McIntosh *et al.*, 'Immediate-early gene induction after neonatal hypoxia-ischemia.' *Mol. Brain Res.*, 1993, **18**, 228–238.

67. T. Neumann-Haefelin, C. Wießner, P. Vogel *et al.*,

68. 'Differential expression of the immediate-early genes *c-fos*, *c-jun*, *junB*, and *NCFI-B* in the rat brain following transient forebrain ischemia.' *J. Cereb. Blood Flow Metab.*, 1994, **14**, 206–216.

68. G. Rordorf, W. J. Koroshetz and J. V. Bonventre, 'Heat shock protects cultured neurons from glutamate toxicity.' *Neuron*, 1991, **7**, 1043–1051.

69. G. Howard and T. E. Geoghegan, 'Altered cardiac tissue gene expression during acute hypoxic exposure.' *Mol. Cell. Biochem.*, 1986, **69**, 155–160.

70. A. A. Knowlton, P. Brecher and C. S. Apstein, 'Rapid expression of heat shock protein in the rabbit after brief cardiac ischema.' *J. Clin. Invest.*, 1991, **87**, 139–147.

71. M. S. Marber, R. Mestril, S. H. Chi *et al.*, 'Over-expression of the rat inducible 70 kDa heat stress protein in a transgenic mouse increases the resistance of the heart to ischemic injury.' *J. Clin. Invest.*, 1995, **95**, 1446–1456.

72. G. Cairo, L. Bardella, L. Schiaffonati *et al.*, 'Synthesis of heat shock proteins in rat liver after ischemia and hyperthermia.' *Hepatology*, 1985, **5**, 357–361.

73. R. W. Currie, 'Effects of ischemia and perfusion temperature on the synthesis of stress-induced (heat shock) proteins in isolated and perfused rat hearts.' *J. Mol. Cell Cardiol.*, 1987, **19**, 795–808.

74. N. Fujio, T. Hatayama, H. Kinoshita *et al.*, 'Induction of mRNAs for heat shock proteins in livers of rats after ischemia and partial hepatectomy.' *Mol. Cell Biochem.*, 1987, **77**, 173–177.

75. T. G. Buchman, D. E. Cabin, J. M. Porter *et al.*, 'Change in hepatic gene expression after shock/rescusitation.' *Surgery*, 1989, **106**, 283–291.

76. L. Schiaffonati, E. Rappocciolo, L. Tacchini *et al.*, 'Reprogramming of gene expression in postischemic rat liver: induction of proto-oncogenes and *hsp70* gene family.' *J. Cell Physiol.*, 1990, **143**, 79–87.

77. R. Beeri, Z. Symon, M. Brezis *et al.*, 'Rapid DNA fragmentation from hypoxia along the thick ascending limb of rat kidneys.' *Kidney Int.*, 1995, **47**, 1806–1810.

78. M. Schumer, M. C. Colombel, I. S. Sawczuk *et al.*, 'Morphologic, biochemical, and molecular evidence of apoptosis during the reperfusion phase after brief periods of renal ischemia.' *Am. J. Pathol.*, 1992, **140**, 831–838.

79. G. C. Gobé, R. A. Axelsen and J. W. Searle, 'Cellular events in experimental unilateral ischemic renal atrophy and in regeneration after contralateral nephrectomy.' *Lab. Invest.*, 1990, **63**, 770–779.

80. S. Olsen, J. F. Burdick, P. A. Keown *et al.*, 'Primary acute renal failure ("acute tubular necrosis") in the transplanted kidney: morphology and pathogenesis.' *Medicine (Baltimore)*, 1989, **68**, 173–187.

81. N. Ueda and S. V. Shah, 'Endonuclease-induced DNA damage and cell death in oxidant injury to renal tubular epithelial cells.' *J. Clin. Invest.*, 1992, **90**, 2593–2597.

82. B. A. Osborne and L. M. Schwartz, 'Essential genes that regulate apoptosis.' *Trends Cell Biol.*, 1994, **4**, 394–399.

83. M. Miura, H. Zhu, R. Rotello *et al.*, 'Induction of apoptosis in fibroblasts by IL-1β-converting enzyme, a mammalian homolog of the *C. elegans* cell death gene *ced-3*.' *Cell*, 1993, **75**, 653–660.

84. S. Kumar, M. Kinoshita, M. Noda *et al.*, 'Induction of apoptosis by the mouse *Nedd-2* gene, which encodes a protein similar to the product of the *Caenorhabditis elegans* cell death gene *ced-3* and the mammalian IL-1β-converting enzyme.' *Genes Devel.*, 1994, **8**, 1613–1626.

85. D. J. Veis, C. M. Sorenson, J. R. Shutter *et al.*, 'Bcl-2-deficient mice demonstrate fulminant lymphoid apoptosis, polycystic kidneys, and hypopigmented hair.' *Cell*, 1993, **75**, 229–240.

86. C. M. Sorenson, S. A. Rogers, S. J. Korsmeyer *et al.*, 'Fulminant metanephric apoptosis and abnormal kidney development in bc1-2-deficient mice.' *Am. J. Physiol.*, 1995, **268**, F73–F81.

87. A. J. Ouellette, R. A. Malt, V. P. Sukhatme *et al.*, 'Expression of two "immediate-early" genes, *Egr-1* and *c-fos*, in response to renal ischemia and during compensatory renal hypertrophy in mice.' *J. Clin. Invest.*, 1990, **85**, 766–771.

88. R. Safirstein, P. M. Price, S. J. Saggi *et al.*, 'Changes in gene expression after temporary renal ischemia.' *Kidney Int.*, 1990, **37**, 1515–1521.

89. R. Safirstein, J. Megyesi, S. J. Saggi *et al.*, 'Expression of cytokine-like genes *JE* and *KC* is increased during renal ischemia.' *Am. J. Physiol.*, 1991, **261**, F1095–F1101.

90. K. J. Kelly, W. W. Williams, Jr., R. B. Colvin *et al.*, 'Intercellular adhesion molecule-1-deficient mice are protected against ischemic renal injury.' *J. Clin. Invest.*, 1996, **97**, 1056–1063.

91. K. J. Kelly, W. W. Williams, Jr., R. B. Colvin *et al.*, 'Intercellular adhesion molecule-1 deficient mice are protected against renal ischemia.' *J. Clin. Invest.*, 1996, **97**, 1056–1063.

92. B. S. Polla, N. Mili, Y. R. Donati *et al.*, 'Les proteines du choc thermique: quelles implcations en nephrologie? [Heat-shock proteins: what implications in nephrology?].' *Nephrologie*, 1991, **12**, 119–123.

93. S. K. Van Why, F. Hildebrandt, T. Ardito *et al.*, 'Induction and intracellular localization of HSP-72 after renal ischemia.' *Am. J. Physiol.*, 1992, **263**, F769–F775.

94. E. M. Fish and B. A. Molitoris, 'Alterations in epithelial polarity and the pathogenesis of disease states.' *N. Engl. J. Med.*, 1994, **330**, 1580–1588.

95. G. A. Perdrizet, H. Kaneko, T. M. Buckley *et al.*, 'Heat-shock and recovery protects renal allografts from warm ischemic injury and enhances HSP72 production.' *Transplant. Proc.*, 1993, **25**, 1670–1673.

96. S. Lindquist, 'The heat-shock response.' *Annu. Rev. Biochem.*, 1986, **55**, 1151–1191.

97. R. L. Safirstein, A. Zelent and P. Price, 'Reduced renal prepro-epidermal growth factor mRNA and decreased EGF excretion in ARF.' *Kidney Int.*, 1989, **36**, 810–815.

98. R. L. Safirstein and J. V. Bonventre, in 'Molecular Nephrology,' eds. D. Schlöndorff and J. V. Bonventre, Marcel Dekker, New York, 1995, pp. 839–854.

99. G. L. Matejka and E. Jennische, 'IGF-I binding and IGF-I mRNA expression in the postischemic regenerating rat kidney.' *Kidney Int.*, 1992, **42**, 1113–1123.

100. H. D. Humes, D. A. Cieslinski, T. M. Coimbra *et al.*, 'Epidermal growth factor enhances renal tubule cell regeneration and repair and accelerates the recovery of renal function in postischemic acute renal failure.' *J. Clin. Invest.*, 1989, **84**, 1757–1761.

101. M. E. Rosenberg and M. S. Paller, 'Differential gene expression in the recovery from ischemic renal injury.' *Kidney Int.*, 1991, **39**, 1156–1161.

102. R. Witzgall, D. Brown, C. Schwarz *et al.*, 'Localization of proliferating cell nuclear antigen, vimentin, c-Fos, and clusterin in the postischemic kidney. Evidence for a heterogenous genetic response among nephron segments, and a large pool of mitotically active and dedifferentiated cells.' *J. Clin. Invest.*, 1994, **93**, 2175–2188.

103. R. Witzgall, E. O'Leary, R. Gessner *et al.*, 'Kid-1, a putative renal transcription factor: regulation during ontogeny and in response to ischemia and toxic injury.' *Mol. Cell Biol.*, 1993, **13**, 1933–1942.

104. R. Witzgall, E. O'Leary, A. Leaf *et al.*, 'The Kruppel-associated box-A (KRAB-A) domain of zinc finger proteins mediates transcriptional repression.' *Proc. Natl. Acad. Sci. USA*, 1994, **91**, 4514–4518.

105. J. V. Bonventre, V. P. Sukhatme, M. Bamberger *et al.*, 'Localization of the protein product of the immediate early growth response gene, *egr-1*, in the kidney after ischemia and reperfusion.' *Cell Regul.*, 1991, **2**, 251–260.

106. J. J. Sciandra, J. R. Subjeck and C. S. Hughes, 'Induction of glucose-regulated proteins during anaerobic exposure and of heat-shock proteins after reoxygenation.' *Proc. Natl. Acad. Sci. USA*, 1984, **81**, 4843–4847.

107. Y. Deguchi, S. Negoro and S. Kishimoto, '*c-fos* expression in human skin fibroblasts by reperfusion after oxygen deficiency: a recovery change of human skin fibroblasts after oxygen deficiency stress.' *Biochem. Biophys. Res. Commun.*, 1987, **149**, 1093–1098.

108. W. A. Ausserer, B. Bourrat-Floeck, C. J. Green *et al.*, 'Regulation of *c-jun* expression during hypoxic and low-glucose stress.' *Mol. Cell Biol.*, 1994, **14**, 5032–5042.

109. K. A. Webster, 'Regulation of glycolytic enzyme RNA transcriptional rates by oxygen availability in skeletal muscle cells.' *Mol. Cell Biochem.*, 1987, **77**, 19–28.

110. T. Kadowaki and Y. Kitagawa, 'Hypoxic depression of mitochondrial mRNA levels in HeLa cell.' *Exp. Cell Res.*, 1991, **192**, 243–247.

111. D. E. Roll, B. J. Murphy, K. R. Laderoute *et al.*, 'Oxygen regulated 80 kDa protein and glucose regulated 78 kDa protein are identical.' *Mol. Cell Biochem.*, 1991, **103**, 141–148.

112. B. J. Murphy, K. R. Laderoute, S. M. Short *et al.*, 'The identification of heme oxygenase as a major hypoxic stress protein in Chinese hamster ovary cells.' *Br. J. Cancer*, 1991, **64**, 69–73.

113. B. D. Price and S. K. Calderwood, 'Gadd45 and Gadd153 messenger RNA levels are increased during hypoxia and after exposure of cells to agents which elevate the levels of the glucose-regulated proteins.' *Cancer Res.*, 1992, **52**, 3814–3817.

114. R. Shreeniwas, S. Koga, M. Karakurum *et al.*, 'Hypoxia-mediated induction of endothelial cell interleukin-l-alpha. An autocrine mechanism promoting expression of leukocyte adhesion molecules on the vessel surface.' *J. Clin. Invest.*, 1992, **90**, 2333–2339.

115. T. G. Graeber, J. F. Peterson, M. Tsai *et al.*, 'Hypoxia induces accumulation of p53 protein, but activation of a Gl-phase checkpoint by low-oxygen conditions is independent of p53 status.' *Mol. Cell Biol.*, 1994, **14**, 6264–6277.

116. Y. Qi and G. Dawson, 'Hypoxia specifically and reversibly induces the synthesis of ferritin in oligodendrocytes and human oligodendrogliomas.' *J. Neurochem.*, 1994, **63**, 1485–1490.

117. S. Kourembanas, R. L. Hannan and D. V. Faller, 'Oxygen tension regulates the expression of the platelet-derived growth factor-B chain gene in human endothelial cells.' *J. Clin. Invest.*, 1990, **86**, 670–674.

118. A. Minchenko, T. Bauer, S. Salceda *et al.*, 'Hypoxic stimulation of vascular endothelial growth factor expression *in vitro* and *in vivo*.' *Lab. Invest.*, 1994, **71**, 374–379.

119. D. Shweiki, A. Itin, D. Soffer *et al.*, 'Vascular endothelial growth factor induced by hypoxia may mediate hypoxia-initiated angiogenesis.' *Nature*, 1992, **359**, 843–845.

120. S. Kourembanas, P. A. Marsden, L. P. McQuillan *et al.*, 'Hypoxia induces endothelin gene expression and secretion in cultured human endothelium.' *J. Clin. Invest.*, 1991, **88**, 1054–1057.

121. M. Karakurum, R. Shreeniwas, J. Chen *et al.*, 'Hypoxic induction of interleukin-8 gene expression in human endothelial cells.' *J. Clin. Invest.*, 1994, **93**, 1564–1570.

122. J. D. Firth, B. L. Ebert, C. W. Pugh *et al.*, 'Oxygen-regulated control elements in the phosphoglycerate

kinase 1 and lactate dehydrogenase A genes: similarities with the erythropoietin 3' enhancer.' *Proc. Natl. Acad. Sci. USA*, 1994, **91**, 6496–6500.

123. J. D. Firth, B. L. Ebert and P. J. Ratcliffe, 'Hypoxic regulation of lactate dehydrogenase A. Interaction between hypoxia-inducible factor 1 and cAMP response elements.' *J. Biol. Chem.*, 1995, **270**, 21021–21027.

124. G. L. Semenza, P. H. Roth, H. M. Fang *et al.*, 'Transcriptional regulation of genes encoding glycolytic enzymes by hypoxia-inducible factor 1.' *J. Biol. Chem.*, 1994, **269**, 23757–23763.

125. E. T. Clark, T. R. Desai, K. L. Hynes *et al.*, 'Endothelial cell response to hypoxia-reoxygenation is mediated by IL-1.' *J. Surg. Res.*, 1995, **58**, 675–681.

126. M. F. Czyzyk-Krzeska and J. E. Beresh, 'Characterization of the hypoxia-inducible protein binding site within the pyrimidine-rich tract in the 3'-untranslated region of the tyrosine hydroxylase mRNA.' *J. Biol. Chem.*, 1996, **271**, 3293–3299.

127. M. L. Norris and D. E. Millhorn, 'Hypoxia-induced protein binding to O_2-responsive sequences on the tyrosine hydroxylase gene.' *J. Biol. Chem.*, 1995, **270**, 23774–23779.

128. J. C. Copin, E. Pinteaux, M. Ledig *et al.*, '70 kDa heat shock protein expression in cultured rat astrocytes after hypoxia: regulatory effect of almitrine.' *Neurochem. Res.*, 1995, **20**, 11–15.

129. J. M. Gleadle, B. L. Ebert, J. D. Firth *et al.*, 'Regulation of angiogenic growth factor expression by hypoxia, transition metals, and chelating agents.' *Am. J. Physiol.*, 1995, **268**, C1362–C1368,

130. L. Baud and R. Ardaillou, 'Involvement of reactive oxygen species in kidney damage.' *Br. Med. Bull.*, 1993, **49**, 621–629.

131. E. L. Greene and M. S. Paller, 'Oxygen free radicals in acute renal failure.' *Miner. Electrolyte Metab.*, 1991, **17**, 124–132.

132. S. L. Linas, D. Whittenburg, J. E. Repine *et al.*, 'Ischemia increases neutrophil retention and worsens acute renal failure: role of oxygen metabolites and ICAM 1. *Kidney Int.*, 1995, **48**, 1584–1591.

133. J. M. McCord, 'Oxygen-derived free radicals in postischemic tissue injury.' *N. Engl. J. Med.*, 1985, **312**, 159–163.

134. M. S. Paller, J. R. Hoidal and T. F. Ferris, 'Oxygen free radicals in ischemic acute renal failure.' *J. Clin. Invest.*, 1984, **74**, 1156–1164.

135. G. M. Rubanyi and P. M. Vanhoutte, 'Oxygen-derived free radicals, endothelium, and responsiveness of vascular smooth muscle.' *Am. J. Physiol.*, 1986, **250**, H815–H821.

136. S. M. Keyse and R. M. Tyrrell, 'Heme oxygenase is the major 32 kDa stress protein induced in human skin fibroblasts by UVA radiation, hydrogen peroxide, and sodium arsenite.' *Proc. Natl. Acad. Sci. USA*, 1989, **86**, 99–103.

137. G. N. Rao, B. Lasségue, K. K. Griendling *et al.*, 'Hydrogen peroxide-induced *c-fos* expression is mediated by arachidonic acid releases: role of protein kinase C.' *Nucleic Acids Res.*, 1993, **21**, 1259–1263.

138. A. Maki, I. K. Berezesky, J. Fargnoli *et al.*, 'Role of $[Ca^{2+}]_i$ in induction of *c-fos*, *c-jun*, and *c-myc* mRNA in rat PTE after oxidative stress.' *FASEB J.*, 1992, **6**, 919–924.

139. K. Nose, M. Shibanuma, K. Kikuchi *et al.*, 'Transcriptional activation of early-response genes by hydrogen peroxide in a mouse osteoblastic cell line.' *Eur. J. Biochem.*, 1991, **201**, 99–106.

140. T. Yoshioka, T. Homma, B. Meyrick *et al.*, 'Oxidants induce transcriptional activation of manganese superoxide dismutase in glomerular cells.' *Kidney Int.*, 1994, **46**, 405–413.

141. K. D. Patel, G. A. Zimmerman, S. M. Prescott *et al.*, 'Oxygen radicals induce human endothelial cells to express GMP-140 and bind neutrophils.' *J. Cell Biol.*, 1991, **112**, 749–759.

142. M. Bienz and H. R. B. Pelham, 'Mechanisms of heat-shock gene activation in higher eukaryotes.' *Adv. Genet.*, 1987, **24**, 31–72.

143. P. K. Sorger, 'Heat-shock factor and the heat-shock response.' *Cell*, 1991, **65**, 363–366.

144. A. J. Giaccia, E. A. Auger, A. Koong *et al.*, 'Activation of the heat-shock transcription factor by hypoxia in normal and tumor cell lines *in vivo* and *in vitro*.' *Int. J. Radiat. Oncol. Biol. Phys.*, 1992, **23**, 891–897.

145. J. Becker, V. Mezger, A. M. Courgeon *et al.*, 'Hydrogen peroxide activates immediate binding of a *Drosophila* factor to DNA heat-shock regulatory element *in vivo* and *in vitro*.' *Eur. J. Biochem.*, 1990, **189**, 553–558.

146. D. D. Mosser, P. T. Kotzbauer, K. D. Sarge *et al.*, '*In vitro* activation of heat shock transcription factor DNA-binding by calcium and biochemical conditions that affect protein conformation.' *Proc. Natl. Acad. Sci. USA*, 1990, **87**, 3748–3752.

147. R. Mestril, S. H. Chi, M. R. Sayen *et al.*, 'Isolation of a novel inducible rat heat-shock protein (*HSP70*) gene and its expression during ischaemia/hypoxia and heat shock.' *Biochem. J.*, 1994, **298**, 561–569.

148. R. Mestril, S. H. Chi, M. R. Sayen *et al.*, 'Expression of inducible stress protein 70 in rat heart myogenic cells confers protection against simulated ischemia-induced injury.' *J. Clin. Invest.*, 1994, **93**, 759–767.

149. S. Xanthoudakis and T. Curran, 'Identification and characterization of Ref-1, a nuclear protein that facilitates AP-1 DNA-binding activity.' *EMBO J.*, 1992, **11**, 653–665.

150. S. Xanthoudakis, G. Miao, F. Wang *et al.*, 'Redox activation of Fos–Jun DNA binding activity is mediated by a DNA repair enzyme.' *EMBO J.*, 1992, **11**, 3323–3335.

151. L. J. Walker, R. B. Craig, A. L. Harris *et al.*, 'A role for the human DNA repair enzyme HAP1 in cellular protection against DNA damaging agents and hypoxic stress.' *Nucleic Acids Res.*, 1994, **22**, 4884–4889.

152. K. S. Yao, S. Xanthoudakis, T. Curran *et al.*, 'Activation of AP-1 and of a nuclear redox factor, Ref-1, in the response of HT29 colon cancer cells to hypoxia.' *Mol. Cell Biol.*, 1994, **14**, 5997–6003.

153. L. V. Favreau and C. B. Pickett, 'Transcriptional regulation of the rat NAD(P)H:quinone reductase gene. Characterization of a DNA-protein interaction at the antioxidant responsive element and induction by 12-*O*-tetradecanoylphorbol-13-acetate.' *J. Biol. Chem.*, 1993, **268**, 19875–19881.

154. T. H. Rushmore and C. B. Pickett, 'Transcriptional regulation of the rat glutathione *S*-transferase Ya subunit gene. Characterization of a xenobiotic-responsive element controlling inducible expression by phenolic antioxidants.' *J. Biol. Chem.*, 1990, **265**, 14648–14653.

155. T. H. Rushmore, M. R. Morton and C. B. Pickett, 'The antioxidant responsive element. Activation by oxidative stress and identification of the DNA consensus sequence required for functional activity.' *J. Biol. Chem.*, 1991, **266**, 11632–11639.

156. T. Dalton, R. D. Palmiter and G. K. Andrews, 'Transcriptional induction of the mouse metallothionein-I gene in hydrogen peroxide-treated Hepa cells involves a composite major late transcription factor/antioxidant response element and metal response promoter elements.' *Nucleic Acids Res.*, 1994, **22**, 5016–5023.

157. E. K. Weir and S. L. Archer, 'The mechanism of acute hypoxic pulmonary vasoconstriction: the tale of two channels.' *FASEB J.*, 1995, **9**, 183–189.

158. J. Mason, F. Beck, A. Dorge *et al.*, 'Intracellular electrolyte composition following renal ischemia.' *Kidney Int.*, 1981, **20**, 61–70.

159. V. P. Sukhatme, X. M. Cao, L. C. Chang *et al.*, 'A zinc finger-encoding gene coregulated with *c-fos* during growth and differentiation, and after cellular depolarization.' *Cell*, 1988, **53**, 37–43.

160. G. L. Wang and G. L. Semenza, 'General involvement of hypoxia-inducible factor 1 in transcriptional response to hypoxia.' *Proc. Natl. Acad. Sci. USA*, 1993, **90**, 4304–4308.

161. S. Grimm and P. A. Baeuerle, 'The inducible transcription factor NF-κB: structure–function relationship of its protein subunits.' *Biochem. J.*, 1993, **290**, 297–308.

162. A. C. Koong, E. Y. Chen and A. J. Giaccia, 'Hypoxia causes the activation of NFκB through the phosphorylation of IκBα on tyrosine residues.' *Cancer Res.*, 1994, **54**, 1425–1430.

163. Y. Yamaji, O. W. Moe, R. T. Miller *et al.*, 'Acid activation of immediate-early genes in renal epithilial cells.' *J. Clin. Invest.*, 1994, **94**, 1297–1303.

164. C. M. Pombo, J. V. Bonventre, J. Avruch *et al.*, 'The stress-activated protein kinases are major *c-jun* amino-terminal kinases activated by ischemia and reperfusion.' *J. Biol. Chem.*, 1994, **269**, 26546–26551.

165. R. L. Safirstein, 'Gene expression in nephrotoxic and ischemic acute renal failure.' *J. Am. Soc. Nephrol.*, 1994, **4**, 1387–1395.

166. T. M. Dexter and E. Spooncer, 'Growth and differentiation in the hemopoietic systems.' *Annu. Rev. Cell Biol.*, 1987, **3**, 423–441.

167. S. T. Koury, M. C. Bondurant and M. J. Koury, 'Localization of erythropoietin synthesizing cells in murine kidneys by *in situ* hybridization.' *Blood*, 1988, **71**, 524–527.

168. C. Lacombe, J. L. Da Silva, P. Bruneval *et al.*, 'Peritubular cells are the site of erythropoietin synthesis in the murine hypoxic kidney.' *J. Clin. Invest.*, 1988, **81**, 620–623.

169. S. T. Koury, M. J. Koury, M. C. Bondurant *et al.*, 'Quantitation of erythropoietin-producing cells in kidneys of mice by *in situ* hybridization: correlation with hematocrit, renal erythropoietin mRNA, and serum erythropoietin concentration.' *Blood*, 1989, **74**, 645–651.

170. M. A. Goldberg, G. A. Glass, J. M. Cunningham *et al.*, 'The regulated expression of erythropoietin by two human hepatoma cell line.' *Natl. Acad. Sci. USA*, 1987, **84**, 7972–7976.

171. S. J. Schuster, E. V. Badiavas, P. Costa-Giomi *et al.*, 'Stimulation of erythropoietin gene transcription during hypoxia and cobalt exposure.' *Blood*, 1989, **73**, 13–16.

172. P. Costa-Giomi, J. Caro and R. Weinmann, 'Enhancement by hypoxia of human erythropoietin gene transcription *in vitro*.' *J. Biol. Chem.*, 1990, **265**, 10185–10188.

173. M. A. Goldberg, S. P. Dunning and H. F. Bunn, 'Regulation of the erythropoietin gene: evidence that the oxygen sensor is a heme protein.' *Science*, 1988, **243**, 1412–1415.

174. M. A. Goldberg, C. C. Gaut and H. F. Bunn, 'Erythropoietin mRNA levels are governed by both the rate of gene transcription and posttranscriptional events.' *Blood*, 1991, **77**, 271–277.

175. D. L. Porter and M. A. Goldberg, in 'Molecular Nephrology,' eds. D. Schlöndorff and J. V. Bonventre, Marcel Dekker, New York, 1995, pp. 551–560.

176. S. B. Miller, D. R. Martin, J. Kissane *et al.*, 'Insulin-like growth factor I accelerates recovery from ischemic acute tubular necrosis in the rat.' *Proc. Natl. Acad. Sci. USA*, 1992, **89**, 11876–11880.

177. S. B. Miller, D. R. Martin, J. Kissane *et al.*, 'Rat models for clinical use of insulin-like growth factor I in acute renal failures.' *Am. J. Physiol.*, 1994, **266**, F949–F956.

178. S. B. Miller, D. R. Martin, J. Kissane *et al.*, 'Hepatocyte growth factor accelerates recovery from acute ischemic renal injury in rats.' *Am. J. Physiol.*, 1994, **266**, F129–F134.

179. A. M. Lefer, P. Tsao, N. Aoki *et al.*, 'Mediation of cardioprotection by transforming growth factor-beta.' *Science*, 1990, **249**, 61–64.

180. W. Lieberthal, A. M. Sheridan and C. R. Valeri, 'Protective effect of atrial natriuretic factor and mannitol following renal ischemia.' *Am. J. Physiol.*, 1990, **258**, F1266–F1272.

181. J. Norman, Y. K. Tsau, A. Bacay *et al.*, 'Epidermal growth factor accelerates functional recovery from ischaemic acute tubular necrosis in the rat: role of the epidermal growth factor receptor.' *Clin. Sci. (Colch.)*, 1990, **78**, 445–450.

182. A. L. Betz, G. L. Yang and B. L. Davidson, 'Attenuation of stroke size in rats using an adenoviral vector to induce overexpression of interleukin-l receptor antagonist in brain.' *J. Cereb. Blood Flow Metab.*, 1995, **15**, 547–551.

7.23
Gene Expression Following Nephrotoxic Exposure

ROBERT L. SAFIRSTEIN
University of Texas Medical Branch at Galveston, TX, USA

7.23.1 INTRODUCTION

A diverse group of compounds damage the kidney, yet the kidney responds to them at the molecular level in a limited way. This is especially surprising in view of the heterogeneous nature of renal cellular structure. Nonetheless, the molecular response mounted by the kidney is similar whether the toxin damages DNA, inhibits mitochondrial respiration, or binds to cytosolic macromolecules. This suggests that the renal cells respond to cellular stress through a common pathway. It is not so clear, however, what the ultimate outcome of this stress pathway is. Whether it is protective and reparative or whether it contributes to cell damage and organ dysfunction may depend on the specific cells expressing the changes, the cellular burden of the toxin, or the duration of exposure.

Renal cells exposed to nephrotoxins undergo at least three fates in response to a nephrotoxin. A cell can remain indifferent to it and appear, histologically at least, normal. Other cells may undergo cell death, either by frank necrosis, or by apoptosis. Finally, new cells may replace those lost and enter a regenerative phase. Whether the kidney survives the insult intact or whether it will be damaged irreversibly will depend on the relative contribution of each these processes. Insight into the molecular responses that serve each of these aspects of cell fate are emerging from studies of cells in culture, but drawing parallels between these observations *in vitro* to those *in vivo* is fraught with hazard. Given the kidney's cellular diversity and the growing appreciation that the reaction of a cell to toxin is very cell specific, it will be necessary to characterize the fate of the renal cells expressing the response in detail. It is also apparent that while many nephron segments superficially appear indifferent to a nephrotoxin, they actively respond to the toxin at least at the molecular level.

This chapter will summarize the molecular responses noted after nephrotoxin exposure *in vivo* as well as *in vitro*. Special attention will be placed on those observations *in vitro* that bear a reasonable relationship to the situation in the animal. An attempt will be made to place these responses within the context of the known fate of these cells, wherever the site of the molecular response has been unequivocally established. Possible mechanisms and transduction pathways for gene activation will be discussed. Chronic toxicity and the molecular antecedents to these events will be summarized, but the data on this issue is sparse and confined to only a few models. While the study of the molecular response to ischemia/reperfusion injury of the kidney has matured greatly over the last few years, much less is known about these events following nephrotoxic damage. For that reason, the attempts to place the molecular responses of the kidney to nephrotoxic damage within some biologic context will be, by necessity, highly speculative and for the most part incomplete.

7.23.2 MOLECULAR RESPONSES AFTER ACUTE EXPOSURE: THE STRESS RESPONSE IN THE KIDNEY

7.23.2.1 Genes Involved in the Acute Response

Prominent among the products expressed in the kidney after nephrotoxic damage are those representative of the mammalian stress response,[1,2] which consist of transcription factors, proteases, chemokines, metallothienein, and structural proteins. The expression of the transcription factors *c-fos* and *c-jun* has been observed in a wide variety of acute exposures including cisplatin,[3] mercury chloride,[4] cadmium chloride,[5] folic acid,[6] and nephrotoxic cysteine conjugates.[4,7] The characteristics of their expression conforms to a typical immediate early (IE) gene response as the expression of the genes is short-lived and superinduced by cycloheximide. Other stress-induced genes such as *HSP-70*,[8,9] *clusterin*,[10–12] *egr-1*,[5] *IL-6*,[13] and *gadd 153*[14] have also been observed (Table 1).

Coincident with this response to stress is the downregulation of several kidney specific genes, including epidermal growth factor (EGF)[3] and Tamm–Horsfall protein.[17] These genes are usually expressed to a high degree by the mature kidney only. This suggests a loss of the differentiated phenotype of the kidney during acute injury. Consistent with this notion of dedifferentiation during toxin-induced damage is the upregulation of vimentin,[11,17,26,27] which is not ordinarily expressed by tubular elements of the mature kidney.

As DNA synthesis increases after a wide variety of nephrotoxic insults,[33] many investigators have explored the possibility that the kidney produces a growth factor following nephrotoxic challenge. Two growth factors produced in the kidney, EGF and insulin-like growth factor-1 (IGF-1), are downregulated after all insults examined (Table 1). Fibroblast growth factor-1 (FGF-1) expression is upregulated during *S*-(1,2-dichlorovinyl)-L-cysteine (DCVC) exposure,[19] but has not been examined in other forms of renal failure. Increased transforming growth factor β (TGF-β) expression has been observed in cisplatin[3] and cyclosporine-induced[15] renal injury. Finally, increased hepatocyte growth factor (HGF) expression has been observed in mercury chloride-induced renal failure.[20] These growth factors might promote survival or regeneration of renal tubular epithelial cells, as the expression of these factors precedes the maximum changes in renal DNA synthesis.

There is a paucity of information on the intranephronal sites in which these changes in gene expression take place. As the IE gene response can serve proliferative, apoptotic, or protective roles, depending, among other things, upon the cell in which this response takes place, it will be necessary to determine the outcome of the cell in which the IE response occurs. That the site of expression cannot be deduced from the histological appearance of the cell is exemplified by the somewhat unexpected changes in prepro epidermal growth factor (ppEGF) expression. ppEGF is expressed in the thick ascending limb and distal tubule in the normal kidney,[34] a site that is histologically intact during nephrotoxin-induced damage. Although cisplatin, gentamicin, and folic acid damage the proximal tubule, ppEGF is downregulated during each of these insults (Table 1). Also, in ischemia/reperfusion, which induces a typical IE response, many of these genes are also expressed predominantly in the distal nephron and spatially separated from the cells undergoing necrosis and regeneration.[35] Thus, the predominant site of the molecular response to ischemia occurs in cells which appear indifferent to the stress. An important area of future research, therefore, will be to identify the cells expressing the genes and then determining whether the cell survives intact or whether it undergoes cell death or regeneration.

7.23.2.2 Mechanism of Gene Expression

Most of the observations on the mechanism of gene expression during nephrotoxic damage have been made in cell culture. Almost no

Table 1   Gene expression following nephrotoxic injury.

Category	Model	Direction	Ref.
Cytokines			
TGF-β	Cyclosporine	↑	15
	Cisplatin	↑	3
Interleukin-	Cadmium	↑	13
Growth factors (GF)			
Insulin-like GF	Folate	↑	16
Epidermal GF	Cisplatin	↓	3
	Cyclosporine	↓	17
	Aminoglycoside	↓	17
	Folate	↓	18
Fibroblast GF	Nephrotoxic cysteine conjugates	↑	19
Hepatocyte GF	Mercury chloride	↑	20
Matrix components			
Collagen	Cyclosporine	↑	21
			22
Fibronectin	Purine amino nucleoside nephrosis	↑	23
Procoagulants	Mercury chloride	↑	24
Transcripton factors			
c-fos, c-myc	Folic acid	↑	6,25
c-fos	Cisplatin	↑	3
c-fos, c-myc	Nephrotoxic cystein conjugates	↑	7
c-myc, c-fos	Mercury chloride	↑	4
c-fos, c-jun, c-myc	Cadmium chloride	↑	5
gadd153	Nephrotoxic cysteine conjugates	↑	14
Miscellaneous			
Vimenten	Cyclosporine	↑	11
	Mercuric chloride	↑	26
	Daunomycin	↑	26
	Nephrotoxic cysteine conjugates	↑	27
		↑	17
Osteopontin	Cyclosporine	↑	15
HSPs	Mercury chloride	↑	8
	Folic acid	↑	9
	Iron	↑	28
Clusterin	Gentamicin	↑	10
	Cyclosporine	↑	11
	Glycerol	↑	12
Tamm−Horsfall	Gentamicin	↓	17
Prostaglandin synthetase	Folic acid	↓	29
Glut 1	Gentamicin	↑	30
Vasoactive compounds			
Kininogen	Dichromate	↑	31
Endothelin-1	Cyclosporine	↑	32

information is available *in vivo*. As the kinetics of *c-fos* and *c-jun* expression *in vivo* mimic the IE response *in vitro*, induction of these genes may be transcriptional and independent of new protein synthesis. Consistent with this notion, at least *in vitro*, is the observation that the expression of *c-fos* and *c-jun* induction provoked by cadmium is inhibited by actinomycin and superinduced by cycloheximide.[5] *HSP-70* induction after exposure to DCVC is also inhibited by actinomycin,[14] and DCVC induction of *c-fos* is independent of protein synthesis.[7] These findings suggest that gene activation occurs by modification of some preformed transcription factors following nephrotoxic exposure. Some of the genes induced in the response to cadmium, such as metallothionein and heme oxygenase[36] have metal responsive elements within them, but others, such as *c-jun*, c-*myc*, and *HSP-70* do not, so that metal-sensing *cis*-elements cannot be responsible for all of the molecular changes induced, even by metals. Other *cis*-acting elements that appear to be activated by heavy metals include both the antioxidant response element (ARE) and the phorbol ester response element (TRE).[37]

Searching for the transduction pathways responsible for the transcriptional activation of these genes, several laboratories have explored possible roles for intracellular calcium and protein kinase activity in the response. The cadmium-induced IE gene response is partially inhibited by chelating intracellular calcium and inhibiting kinase activity.[5] Induction of *c-fos* and *HSP-70* by DCVC, however, was not affected by inhibition of the observed increases in intracellular calcium, but was sensitive to the redox state of intracellular thiols.[7,14] Not all of the responses were suppressed by such treatment, as the expression of *gadd 153* was actually potentiated under reducing conditions. FK506 induction of plasminogen activator inhibiting (PAI) protein is dependent on prior *c-jun* activation.[38] This pattern of gene expression, prominent activation of *c-jun*, and independent roles for intracellular calcium and protein kinase activity is similar to the pattern of gene expression following activation of the stress response induced by UV light and oxidative stress, and suggest the participation of the stress kinase pathway (Figure 1).

Exposure of cells to environmental stress, including toxin exposure, seems to initiate a genetically programmed series of events resembling those after growth factor or phorbol ester treatment of cells in culture.[1,2] Thus, the expression of genes subsequent to oxygen stress, or heavy metal exposure, bears a close resemblance to those induced by growth factors. While the stress response is similar to that initiated by growth factors, the consequence of the genetically programmed events initiated by environmental stress is antiproliferative, rather than proliferative. In the stress response, gene activation depends on the activation of pre-existent transcription factors by phosphorylation via a transduction pathway served by members of the mitogen activated protein kinases (MAPKs) family, known as stress-activated kinases, or SAPKs, and p38 MAPK.[39,40] These kinases are substrates for and are activated by a distinct group of kinases that are activated by cellular stress. These upstream regulators are not activated by growth factors, so that the pathways are functionally distinct. While many of the downstream targets of the stress and growth factor pathways are the same, many of the substrates for each are distinct. This substrate specificity may help to explain why there may be different outcomes even when there is overlap in gene expression. Due to the similarities in gene induction between the stress-activated pathway *in vitro* and that observed *in vivo* in the kidney, it is highly likely that this group of kinases are also activated after nephrotoxic stress.

7.23.3 MOLECULAR RESPONSES TO CHRONIC EXPOSURE

The common feature of almost all forms of renal injury that end in chronic renal failure is the presence of fibrosis. In glomerulonephritis, the role of antecedent inflammation and resident and infiltrating renal cells has been amply demonstrated.[41] The molecular determinants of this inflammatory and profibrotic response include the expression of chemokines and cytokines, such as MCP-1, Rantes, TGF-β, and PDGF among others, as well as adhesion and extracellular matrix proteins. Acute injury to the kidney by nephrotoxins also provokes similar changes in gene expression, as demonstrated above (Table 1), but in this case the expression is regulated and short-lived when the proximate cause of the injury is removed. Only a few studies have addressed the issue of chronic stress and gene expression in the kidney. Cyclosporine nephrotoxicity, especially in combination with a low salt diet, is the best studied example of chronic exposure at this level.[42]

Cyclosporine induces increased expression of several genes typical of the mammalian stress response both *in vitro* and *in vivo*. Increased expression of *c-fos*, and *TGF-β*,[43] *HSP-70*,[44] and plasminogen activator inhibitor-1 (*PAI-1*)[38] has been observed. In fibroblasts studied *in vitro*, *PAI-1* expression was dependent on activating protein-1 (AP-1) driven transcriptional activation.[38] Many of these same responses, including *c-fos*, *TGF-β*, and *PAI-1* expression occur after long-term exposure to rats in which the kidney

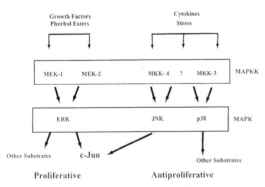

Figure 1 Model for the induction of *c-jun* by both the proliferative or stress-induced pathways. Arrows denote activation of downstream targets by modification of phosphorylation sites.

develops fibrosis and renal failure.[42] Increased collagen gene induction is also noted after long-term cyclosporine exposure.[43] To date, no detailed analysis of the sites of synthesis have been carried out so that the contribution of vascular, tubular, or interstitial cells in the process cannot be determined. An open question is what ultimately is the fate of cells which express the stress-related genes over a prolonged period of time, and whether more chronic expression leads to a different cellular outcome than short-term expression.

7.23.4 POSSIBLE FUNCTIONS OF THE RENAL MOLECULAR RESPONSE

A framework in which to view the mechanism of gene activation after nephrotoxic stress and the possible outcomes of the response is shown in Figure 2. A nephrotoxin, by damaging DNA, binding to intracellular proteins, generating oxidant stress, or disrupting organelle function, would presumably activate this pathway and begin the cascade of gene changes summarized above. The mechanisms of activation are unknown. In ischemia, the stress-kinase pathway has already been shown to be active and responsible for the majority of *c-jun* activation,[45] so that it is highly likely that this pathway will also be activated during nephrotoxic damage. Whether this pathway serves a cytoprotective function, or whether the cells lose viability and die by apoptosis or necrosis, is not known and will perhaps depend on the dose of the toxin and the cell in which the response is provoked. The outcome of the response will probably be different if the toxin is applied briefly or whether the stress is continuously applied. Since the pathway is genetically programmed, it will also be affected by the specific structure of the host genome. This might help to explain the different sensitivities

of the host to a nephrotoxin. It will be necessary to identify the cells that participate in the response as well as correlate the response to functional, cytoprotective, or cytoreductive events in order to place the response in some biologic context. Also a cell expressing the stress response may recruit other cells into the response by the production of substances that have autocrine, juxtacrine, or even endocrine activities. Characterization of the transduction pathway stimulated by nephrotoxic exposure will help in this regard, since one pathway is regenerative while the other is not. Due to the distinct substrate specificity of the different genes, this will also help to identify the important transcription factors involved in the response. Ultimately, definitive information about the role of any specific pathway or gene product will depend on experiments that either eliminate or upregulate the pathway in as targeted a way as possible. Thus, tools of molecular biology and genetics will be increasingly necessary to understand the response of the kidney to nephrotoxic injury.

7.23.5 REFERENCES

1. P. Herrlich, H. Ponta and H. J. Rahmsdorf, 'DNA damage-induced gene expression: signal transduction and relation to growth factor signaling.' *Rev. Physiol. Biochem. Pharmacol.*, 1992, **119**, 187–223.
2. N. J. Holbrook and A. J. Fornace, Jr., 'Response to adversity: molecular control of gene activation following genotoxic stress.' *New Biol.*, 1991, **3**, 825–833.
3. R. Safirstein, A. Z. Zelent and P. M. Price, 'Reduced renal prepro-epidermal growth factor mRNA and decreased EGF excretion in ARF.' *Kidney Int.*, 1989, **36**, 810–815.
4. S. Vamvakas, D. Bittner and U. Koster, 'Enhanced expression of the protooncogenes *c-myc* and *c-fos* in normal and malignant renal growth.' *Toxicol. Lett.*, 1993, **67**, 161–172.
5. M. Matsuoka and K. M. Call, 'Cadmium-induced expression of immediate early genes in LLC-PK1 cells.' *Kidney Int.*, 1995, **48**, 383–389.
6. B. Cowley, Jr., F. L. Smardo, Jr., J. J. Grantham *et al.*, 'Elevated *c-myc* protooncogene expression in autosomal recessive polycystic kidney disease.' *Proc. Natl. Acad. Sci. USA*, 1987, **84**, 8394–8398.
7. K. Yu, Q. Chen, H. Liu *et al.*, 'Signalling the molecular stress response to nephrotoxic and mutagenic cysteine conjugates: differential roles for protein synthesis and calcium in the induction of *c-fos* and *c-myc* mRNA in LLC-PK1 cells.' *J. Cell. Physiol.*, 1994, **161**, 303–311.
8. P. L. Goering, B. R. Fisher, P. P. Chaudhary *et al.*, 'Relationship between stress protein induction in rat kidney by mercuric chloride and nephrotoxicity.' *Toxicol. Appl. Pharmacol.*, 1992, **113**, 184–191.
9. L. Bardella and R. Comolli, 'Differential expression of *c-jun* and *c-fos* and *hsp70* mRNAs after folic acid and ischemia-reperfusion injury; effect of antioxidant treatment.' *Exp. Nephrol.*, 1994, **2**, 158–165.
10. W. K. Aulitzky, P. N. Schlegel, D. F. Wu *et al.*, 'Measurement of urinary clusterin as an index of nephrotoxicity.' *Proc. Soc. Exp. Biol. Med.*, 1992, **199**, 93–96.

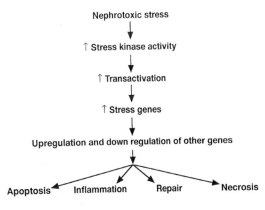

Figure 2 Pathophysiology of nephrotoxic injury and possible cellular fates.

11. I. A. Darby, T. Hewitson, C. Jones *et al.*, 'Vascular expression of clusterin in experimental cyclosporine nephrotoxicity.' *Nephron*, 1995, **3**, 234–239.

12. K. A. Nath, J. Dvergsten, R. Correa-Rotter *et al.*, 'Induction of clusterin in acute and chronic oxidative renal disease in the rat and its dissociation from cell injury.' *Lab. Invest.*, 1994, **71**, 209–218.

13. F. Kayama, T. Yoshida, M. R. Elwell *et al.*, 'Cadmium-induced renal damage and proinflammatory cytokines: possible role of IL-6 in tubular epithelial cell regeneration.' *Toxicol. Appl. Pharmacol.*, 1995, **134**, 26–34.

14. Q. Chen, K. Yu and J. L. Stevens, 'Regulation of the cellular stress response by reactive electrophiles. The role of covalent binding and cellular thiols in transcriptional activation of the 70-kilodalton heat shock protein gene by nephrotoxic cysteine conjugates.' *J. Biol. Chem.*, 1992, **267**, 24322–24327.

15. R. H. Pichler, N. Franceschini, B. A. Young *et al.*, 'Pathogenesis of cyclosporine nephropathy: roles of angiotensin II and osteopontin.' *J. Am. Soc. Nephrol.*, 1995, **6**, 1186–1196.

16. M. K. Hise, L. Li, N. Mantzouris *et al.*, 'Differential mRNA expression of insulin-like growth factor system during renal injury and hypertrophy.' *Am. J. Physiol.*, 1995, **269**, F817–F824.

17. E. J. Nouwen, W. A. Verstrepen, N. Buyssens *et al.*, 'Hyperplasia, hypertrophy, and phenotypic alterations in the distal nephron after acute proximal tubular injury in the rat.' *Lab. Invest.*, 1994, **70**, 479–493.

18. B. D. Cowley and S. Gudapaty, 'Temporal altera-tions in regional gene expression after nephrotoxic renal injury.' *J. Lab. Clin. Med.*, 1995, **125**, 187–199.

19. G. Zhang, T. Ichimura, J. A. Maier *et al.*, 'A role for fibroblast growth factor type-1 in nephrogenic repair. Autocrine expression in rat kidney proximal tubule epithelial cells *in vitro* and in the regenerating epithelium following nephrotoxic damage by *S*-(1,1,2,2-tetrafluoroethyl)-L-cysteine *in vivo*.' *J. Biol. Chem.*, 1993, **268**, 11542–11547.

20. T. Igawa, K. Matsumoto, S. Kanda *et al.*, 'Hepatocyte growth factor may function as a renotropic factor for regeneration in rats with acute renal injury.' *Am. J. Physiol.*, 1993, **265**, F61–F69.

21. G. Wolf, P. D. Killen and E. G. Neilson, 'Cyclosporin A stimulates transcription and procollagen secretion in tubulointerstitial fibroblasts and proximal tubular cells.' *J. Am. Soc. Nephrol.*, 1990, **1**, 918–922.

22. C. C. Nast, S. G. Adler, A. Artishevsky *et al.*, 'Cyclosporine induces elevated procollagen α1 (I) mRNA levels in the rat renal cortex.' *Kidney Int.*, 1991, **39**, 631–638.

23. C. L. Jones, S. Buch, M. Post *et al.*, 'Pathogenesis of interstitial fibrosis in chronic purine aminonucleoside nephrosis.' *Kidney Int.*, 1991, **40**, 1020–1031.

24. A. Kanfer, D. de Prost, C. Guettier *et al.*, 'Enhanced glomerular procoagulant activity and fibrin deposition in rats with mercuric chloride-induced autoimmune nephritis.' *Lab. Invest.*, 1987, **57**, 138–143.

25. B. D. Cowley, Jr., L. J. Chadwick, J. J. Grantham *et al.*, 'Sequential protooncogene expression in regenerating kidney following acute renal injury.' *J. Biol. Chem.*, 1989, **264**, 8389–8393.

26. H. J. Grone, K. Weber, E. Grone *et al.*, 'Coexpression of keratin and vimentin in damaged and regenerating tubular epithelia of the kidney.' *Am. J. Pathol.*, 1987, **129**, 1–8.

27. A. Wallin, G. Zhang, T.W. Jones *et al.*, 'Mechanism of the nephrogenic repair response. Studies on proliferation and vimentin expression after 35S-1,2-dichloro-vinyl-L-cysteine nephrotoxicity *in vivo* and in cultured proximal tubule epithelial cells.' *Lab. Invest.*, 1992, **66**, 474–484.

28. A. Fukuda, T. Osawa, H. Oda *et al.*, 'Oxidative stress response in iron-induced acute nephrotoxicity: enhanced expression of heat shock protein 90.' *Biochem. Biophys. Res. Commun.*, 1996, **219**, 76–81.

29. R. M. Ingerowski, F. Haux and F. Bruchhausen, 'Decreased prostaglandin synthetase activity during kidney regeneration after folic acid or 2,4,5-triamino-6-styrylpyrimidine application.' *Nauyn Schmiedebergs Arch. Pharmacol.*, 1977, **298**, 157–162.

30. J. H. Dominguez, C. C. Hale and M. Qulali, 'Studies of renal injury. I. Gentamicin toxicity and expression of basolateral transporters.' *Am. J. Physiol.*, 1996, **39**, F245–F253.

31. G. Bompart, A. Colle, M. L. Dos Reiss *et al.*, 'Increase in renal and urinary low- and high-molecular weight kininogens during chromate-induced acute renal failure in the rat: evidence for renal kininogen production.' *Nephron*, 1993, **65**, 612–618.

32. Z. A. Abassi, F. Pieruzzi, F. Nakhoul *et al.*, 'Effects of cyclosporine A on the synthesis, excretion and metabolism of endothelin in the rat.' *Hypertension*, 1996, **27**, 1140–1148.

33. F. Toback, 'Control of Renal Regeneration After Acute Tubule Necrosis,' Springer, New York, 1984, pp. 748–762.

34. E. C. Salido, L. Barajas, J. Lechago *et al.*, 'Immunocytochemical localization of epidermal growth factor in mouse kidney.' *J. Histochem. Cytochem.*, 1986, **34**, 1155–1160.

35. J. Megyesi, J. Di Mari, N. Udvarhelyi *et al.*, 'DNA synthesis is dissociated from the immediate-early gene response in the post-ischemic kidney.' *Kidney Int.*, 1995, **48**, 1451–1458.

36. J. Alam, J. Cai and A. Smith.'Isolation and characterization of the mouse heme oxygenase-1 gene. Distal 5′ sequences are required for induction by heme or heavy metals.' *J. Biol. Chem.*, 1994, **269**, 1001–1009.

37. T. Prestera, P. Talalay, J. Alam *et al.*, 'Parallel induction of heme oxygenase-1 and chemoprotective phase 2 enzymes by electrophiles and antioxidants: regulation by upstream antioxidant-responsive elements (ARE).' *Mol. Med.*, 1995, **1**, 827–837.

38. H. Goldberg, 'FK506 increases plasminogen activator inhibitor-1 (PAI-1) gene expression.' *J. Am. Soc. Nephrol.*, 1995, **6**, 998.

39. J. M. Kyriakis, P. Banerjee, E. Nikolakaki *et al.*, 'The stress-activated protein kinase subfamily of c-jun kinases.' *Nature*, 1994, **369**, 156–160.

40. R. J. Davis, 'The mitogen-activated protein kinase signal transduction pathway.' *J. Biol. Chem.*, 1993, **268**, 14553–14556.

41. H. Okada, F. Strutz, T. Danoff *et al.*, 'Possible pathogenesis of renal fibrosis.' *Kidney Int.*, 1995, **49**, S37–S38.

42. F. S. Shihab, T. Andoh, A. Tanner *et al.*, 'Fibrosis in chronic renal cyclosporine toxicity: preferential expression in medulla.' *J. Am. Soc. Nephrol.*, 1995, **6**, 1004

43. S. Saggi, T. Andoh, S. Rajaraman *et al.*, 'Salt depletion enhances renal c-fos and TGF-beta in chronic cyclosporin nephrotoxicity.' *J. Am. Soc. Nephrol.*, 1995, **6**, 1003.

44. C. Yuan, E. Bohen and M. Carome, 'Cyclosporine induces *hsp70* gene expression and protects against subsequent CYA toxicity in LLC-PK1 cells.' *J. Am. Soc. Nephrol.*, 1995, **6**, 1067.

45. C. M. Pombo, J. V. Bonventre, J. Avruch *et al.*, 'The stress-activated protein kinases are major c-jun amino-terminal kinases activated by ischemia and reperfusion.' *J. Biol. Chem.*, 1994, **269**, 26546–26551.

7.24
Nephrotoxicity of β-Lactam Antibiotics

BRUCE M. TUNE

Stanford University School of Medicine, CA, USA

7.24.1 INTRODUCTION

The β-lactam antibiotics (β-lactams) are modified microbial exotoxins whose natural function is to prevent the growth of competing bacteria.[1] The major β-lactam groups (penicillins, cephalosporins, penems) are cyclic dipeptides (Figure 1) with two common properties:[1,2] (i) structural homology with bacterial cell wall peptidoglycan precursors; and (ii) closure of their peptide regions into reactive lactam rings. The dipeptide configuration of the β-lactams fits the active sites of a series of cell wall-synthesizing carboxypeptidases and related bacterial proteins. However, instead of causing hydrolysis of the peptide bond, the interaction promotes ring opening, covalent binding of the antibiotic to the enzyme, and arrest of cell wall synthesis.[1]

The naturally occurring β-lactams have high target specificity for a limited group of bacterial proteins, and are therefore nontoxic to the microorganisms that produce them. For the same reason, and because of very limited exposure, mammalian cells are also unaffected by environmental β-lactams. The β-lactams developed for clinical use have been pharmaceutically modified to increase their antimicrobial spectra. Bactericidal potency has been enhanced by increasing one or more of the following properties: resistance to β-lactamases, ability to pass through cell wall porins, affinity for molecular targets, and reactivity or acylating potential.[2]

Advances in technology have allowed the development of β-lactams that were too reactive to be isolated in the early decades of the antibiotic era.[3] The result of this progress has been the introduction of more broadly effective antibiotics, but, unfortunately, also an increase of toxic side effects. Although the penicillins are almost completely free of cytotoxicity, several cephalosporins and the majority of penems developed so far are nephrotoxic.[3–7] This toxicity has been severe enough with certain cephalosporins to preclude their clinical use and with two broad-spectrum penems to require their combined administration with nephroprotective transport inhibitors.[3,7]

7.24.1.1 Terminology

The present discussion of the dose-related nephrotoxicity of the β-lactams will not include idiosyncratic or allergic reactions (interstitial nephritis, etc.).[8] In this analysis, nephrotoxicity is defined as the production of acute proximal tubular cell necrosis in the *in situ* kidney,[3,9–11] with decreased glomerular filtration at fully toxic dosage.[12–14] Isolated case reports of acute renal failure following the use of a particular β-lactam, regardless of histopathologic findings, are not accepted as proof of nephrotoxicity. Cytotoxicity will refer to the production of altered cell metabolism, cell injury, or cell death *in vitro*, whether or not the conditions of exposure reflect those causing nephrotoxicity *in vivo*.

7.24.1.2 Models Studied

There has been limited clinical investigation of β-lactam nephrotoxicity.[15–17] Studies in humans[18–20] and nonhuman primates[3,9] have shown important similarities to results in other laboratory animals: (i) restriction of cellular necrosis to the proximal renal tubule,[9,10,21] (ii) prevention of toxicity by inhibitors of organic anion secretion,[3,7,11,22–24] and (iii) augmentation of nephrotoxicity by minimally toxic aminoglycoside regimens.[12,14]

Most laboratory studies have used rabbits or rats. The rabbit offers the advantage of greater sensitivity to the β-lactams.[3,9,25,26] For example, the ND-50 (50% nephrotoxic dose, the

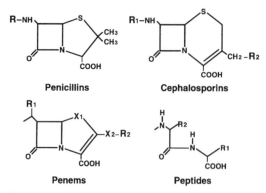

Figure 1 Basic structures of the penicillins, cephalosporins, and penems. In the penems X_1 is sulfur; in the carbapenems X_1 is carbon. The thienamycins are carbapenems with a sulfur at X_2. The equivalent region of a dipeptide illustrates the 90° strained configuration resulting from closure of the amide into a lactam ring.

amount producing tubular necrosis in one-half of the animals) of cephaloridine (Cld) is 90–140 mg kg^{-1} in the rabbit; the ND-50 of Cld is 1000–1400 mg kg^{-1} in the rat,[9,25] an amount that delivers solute loads large enough to reduce nephrotoxic injury.[27,28] The rabbit has the disadvantage of developing diarrhea after multiple doses of broad-spectrum β-lactams.[5,12] Other species have not been extensively studied. Guinea pigs (ND-50 of Cld = 200–300 mg kg^{-1}),[9,25,29] may develop lethal diarrhea after a series of doses, even with narrow-spectrum β-lactams;[29,30] monkeys (ND-50 of Cld = 200–300 mg kg^{-1}, vs. ≥6 g day^{-1} in humans)[9,16,25,31] are very expensive; and mice (ND-50 of Cld = 600–3100 mg kg^{-1}, depending on strain)[9,25,32] have kidneys too small for many of the methods used in rabbits and rats. The large size and relative resistance of dogs and cats (ND-50 of Cld > 1000–1500 mg kg^{-1}),[9,25] make them unsuitable for studies of β-lactam nephrotoxicity.

A number of models of toxicity have been reviewed for the present discussion: the *in situ* kidney,[9,10,22,23,25,33–37] biochemical alterations in tissues or subcellular fractions from *in vivo*-exposed kidneys,[4,11,13,38–45] and *in vitro* models such as renal cortical slices,[23,39,46,47] proximal tubule suspensions,[48–50] tubular cells in culture,[51,52] and isolated subcellular fractions.[41,53,54]

In interpreting the results of various methods of study, gradations of cytotoxicity of different β-lactams derived only from *in vitro* exposure[47,51,52] are not assumed to reflect their nephrotoxic potentials *in vivo*. Moreover, *in vitro* tubular cell injury is accepted as relevant only when it is produced at β-lactam concentrations and times of exposure similar to those that cause nephrotoxicity *in vivo* and develops before generalized cytolytic damage.

7.24.1.3 Nephrotoxic β-Lactams

With the possible exception of BLP-1654,[55] a ureidopenicillin that was not released for clinical use, the penicillins have shown essentially no direct nephrotoxicity. However, a number of cephalosporins and penems are toxic to the renal tubule.[4] When the most toxic of these are given to rabbits, single doses only slightly above the therapeutic range cause a highly selective necrosis of the proximal tubule.[3,9–11] Studies of interactions between mildly nephrotoxic cephalosporin regimens and minimally damaging aminoglycoside,[12] ischemic,[56] or endotoxemic[14,57] injuries have shown significant toxic synergies, most severely with cephalosporin–aminoglycoside–endotoxin combinations.[4]

The cephalosporins have been available for study for more than 25 years, and therefore provide the basis for much of the present discussion. Cld and cephaloglycin (Cgl) (single ND-50s in the female New Zealand white rabbit = 90 and 60 mg kg^{-1}, respectively) have been the most thoroughly studied.[4] Limited investigations have been done with cefaclor (Ccl) (rabbit ND-50 ≥ 800 mg kg^{-1}) and cefazolin (rabbit ND-50 = 800–1200 mg kg^{-1}),[12,24,58,59] and toxicity has also been reported with cefamandole[26] and cefpimizole.[60]

Imipenem (Imip), the first penem released for clinical use, is marketed in combination with cilastatin, an inhibitor of its tubular cell metabolism, transport, and toxicity.[3] When administered without cilastatin, Imip is about as toxic as Cld in rabbits and monkeys.[3,43] Panipenem is comparably nephrotoxic and is therefore used in combination with betamipron, another nephroprotective transport inhibitor.[7] Some nephrotoxicity has been found with at least two other penems under development.[5,6]

The results of toxicologic tests of medications under development are made available to regulatory agencies before the drugs are tested in humans. However, detailed studies of toxicity may never be published in scientific journals. It is therefore almost impossible to provide a complete and accurately ranked list of all of the nephrotoxic β-lactam antibiotics.

7.24.2 MECHANISMS OF TOXICITY

The nephrotoxic β-lactams produce proximal tubular cell necrosis in the intact animal within 16–24 h after parenteral administration of single toxic doses.[3,9–11] Ultrastructural damage develops over 1–5 h after administration of Cld to rabbits, with loss of brush border through endocytosis of microvilli, formation of subapical membranous whorls and uncoated vesicles, decrease of basolateral cell interdigitations, and rounding of mitochondria.[10] It is not certain whether the early mitochondrial alterations represent a change in structure or a loss of their vertical orientation, but there is evidence of mitochondrial calcification by 24 h after administration of Cgl to guinea pigs.[21] Although tubular necrosis is most widespread in the S2 segment,[10,11,23,24] glucosuria is an early finding,[5,31] indicating injury to the S1 segment as well.[61] N-Acetyl-β-D-glucosaminidase, γ-glutamyl transpeptidase, lactate dehydrogenase, and other tubular cell enzymes are released into the urine,[5,62] and glomerular filtration is decreased,[5,31,57] by 1–2 days. The value of enzymuria as a predictor of acute renal failure is unclear.[5]

Cld causes little or no cumulative nephrotoxicity when given in a series of subtoxic doses separated by enough time to allow its clearance from the tubular cell.[11,31] Similar serial dosage of Cgl[11] or Imip[3] has a limited tendency to produce cumulative injury. This pattern is in contrast to aminoglycoside nephrotoxicity, which is minimal after a single large dose but accumulates with continuing administration and progressive intracellular sequestration.[63] As in the case of gentamicin, however, continued administration of Cld in moderately toxic dosage allows regeneration of proximal tubular cells with relative resistance to further injury by the antibiotic.[64,65]

Several mechanisms of toxicity have been proposed from studies with Cld, the most widely investigated nephrotoxic cephalosporin: (i) at the cellular level, concentrative uptake by an organic anion secretory transporter;[23] and (ii) at the molecular level, (a) production of a highly reactive intermediate, through epoxidation of the R_1 substituent thiophene by a cytochrome P450 mixed function oxidase (MFO),[66] (b) lipid peroxidation, through redox cycling of electrons in the R_2 pyridinium,[13] (c) acylation and inactivation of tubular cell proteins, through the spontaneous reactivity of the β-lactam ring,[67] and (d) respiratory toxicity,[48] through acylation of mitochondrial anionic substrate transporters,[44] and possibly also through reversible inhibition of mitochondrial fatty acylcarnitine uptake.[68]

7.24.2.1 Tubular Transport

Like the penicillins, many of the cephalosporins are secreted across the proximal tubule,[69] beginning with transport into the tubular cell by a basolateral oxoglutarate/ organic anion (hippurate) antiporter.[70–72] The resulting intracellular concentrations can be orders of magnitude greater than in any other cell type;[11,34] this is particularly true in rabbits,[24] which are 10-fold more sensitive to β-lactam nephrotoxicity than rats and mice, and are closest to humans and nonhuman primates in this sensitivity.[4] Most studies of transport have therefore been done in rabbits;[11,23,29,34,35,53,59,73,74] studies in guinea pigs, rats, and mice, however, have confirmed the relationship between transport and toxicity.[22,24,29,32,75]

7.24.2.1.1 Cephalothin and cephaloridine

The range of concentrative uptake is illustrated by these two early (R_1 = 2-thienylacetyl)

cephalosporins (Figure 1): cephalothin (Cpt) (R_2 = acetoxy) and Cld (R_2 = pyridinyl). Early testing in laboratory animals[31,76] and subsequent clinical experience[16] with Cpt revealed little or no renal toxicity. In contrast, Cld, released shortly after Cpt, was found to be nephrotoxic in several laboratory species[9,25] and in clinical use.[16]

Cpt, a simple anionic cephalosporin, is rapidly secreted across the tubular cell. Its low and transient intracellular concentrations may explain Cpt's minimal toxicity to the kidney.[35,77] In contrast, Cld has a quaternary nitrogen in the pyridinium side ring at the 3-methyl (R_2) position. Although Cld is rapidly transported into the tubular cell at the basolateral side, its movement from cell to tubular fluid is greatly restricted compared to normally secreted organic anions, probably because it is a zwitterion.[35,39,53,70,78] Due to this restricted luminal membrane transport, peak tubular cell concentrations of Cld in the rabbit are twice, and areas-under-the-curve of concentration and time (AUC) five times, those of any other cephalosporin after equivalent dosage.[77]

The importance of the transport–toxicity relationship of Cld has been demonstrated in several ways: (i) correlations of more than 10-fold interspecies, intersex, and interstrain differences of cortical uptake with variations of nephrotoxicity;[24,32,53] (ii) increasing toxicity in the newborn rabbit as tubular cell transport matures;[24,73] (iii) prevention of toxicity by organic anion transport inhibitors, like probenecid or p-aminohippurate, but only when enough of the inhibitor is given to produce a sustained reduction of cortical uptake;[29,74] and (iv) increased tissue concentrations and nephrotoxicity after treatment with an inhibitor of organic cation transport that reduces Cld efflux from the tubular cell.[39,78]

7.24.2.1.2 Ceftazidime

The extreme differences between the intracellular sequestration of Cld, Cpt, and other cephalosporins, best illustrated by their cortical AUCs (Table 1), are important contributors to the large differences in their nephrotoxicity. The net effects of differing basolateral transport into the cell, luminal-side efflux into the tubular fluid, and possibly reabsorptive transport in the case of the aminocephalosporins[79] are important determinants of their widely different AUCs, as illustrated schematically in Figure 2.

It should be noted that the rate of tubular secretion is not a reliable indicator of tubular cell concentrations or AUCs. Cld undergoes

Table 1   Renal cortical uptake and nephrotoxicity of cephalosporins in the rabbit.

	Cortical conc. (μg g^{-1})[a,b]	Cortex/serum ratios[b,c]	Cortical half-lives (h)[b]	Cortical AUC[d]	ND-50 (mg kg^{-1})[e]	Refs.
Cld	2837 $\pm$ 53	16.1 $\pm$ 0.6	1.32 $\pm$ 0.25	6069 $\pm$ 957	90	14,36,57,74
Cgl	1088 $\pm$ 56	6.6 $\pm$ 1.0	0.65 $\pm$ 0.06	1287 $\pm$ 99	60	11,12,14,56,59,67
Clx	1323 $\pm$ 258	7.7 $\pm$ 1.6	0.42	1129	>1500	77
Ccl	499 $\pm$ 48	6.5 $\pm$ 1.0	0.46	459	$\geq$800	37
Ctz	406 $\pm$ 42	1.4 $\pm$ 0.1	0.68	497	>1500	77

[a]Peak (or steady-state) concentrations measured in cortex (μg g^{-1} wet tissue) after infusion (0.5 h) of 100 mg kg^{-1} BW of the cephalosporins into female NZW rabbits. Values (means $\pm$ SEM) for cephaloridine (Cld) and cephaloglycin (Cgl) are derived from several previously published means; values for cephalexin (Clx), cefaclor (Ccl), and ceftazidime (Ctz) are from single studies (five to six animals in each group). [b]The following comparisons are statistically significant (>95% by ANOVA): cortical concentrations (conc.), Cld > Cgl and Clx > Ccl, and Ctz; cortex/serum ratios (C/S), Cld > all others, Cgl, Clx, and Ccl > Ctz; cortical half-lives, Cld > Cgl. [c]C/S inulin = 1.4 $\pm$ 0.1 (from means of seven studies), not significantly different from C/S Ctz; intracellular Ctz is therefore probably minimal. [d]Areas under the curves of cortical concentration and time = (0.25 × peak cortical concentrations) + (peak cortical concentrations × cortical half-lives/ ln 0.5). [e]50% nephrotoxic dose, the amount producing tubular necrosis in 50% of the animals.

little or no secretion into the tubular fluid, but has uniquely high and prolonged intracellular concentrations.[35,77,80] Although the cellular mechanisms of ceftazidime (Ctz) transport have not been fully determined, it undergoes both secretory and reabsorptive movement, with minimal net secretion[81] and little or no concentrative uptake in the tubular cell.[77] Ctz has an R_2 substituent pyridinium identical to that of Cld, but its transport by the proximal tubule is very different, most likely as a result of its unusual anionic R_1 substituent group.

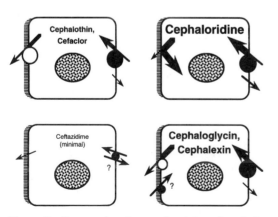

Figure 2   Proposed patterns of uptake of cephalosporin antibiotics by proximal renal tubular cells. Differences in active transport from blood-to-tubular cell (shown on the right-hand side in each example) and movement from cell to luminal fluid (left-hand side), possibly including luminal-side reabsorptive transport of cephaloglycin and cephalexin, result in wide variations of intracellular AUCs, schematically represented by print size.

7.24.2.1.3   Cephaloglycin, cefaclor, and cephalexin

These aminocephalosporins (R_1 = phenylglycyl) were developed for their acid-stability and absorption across the gastrointestinal tract. The R_1 amino group is cationic at physiologic pH.[82] However, in the present discussion the phenylglycyl cephalosporins will be designated anionic for the purpose of contrasting them with Cld, which has a fixed cationic R_2 pyridinium.[2] They may undergo simultaneous secretory and reabsorptive transport in the proximal tubule.[79]

Cgl (R_2 = acetoxy), the most nephrotoxic of the commercially released cephalosporins, is no longer in clinical use. Cgl is secreted across the tubular cell, with a cortical AUC one-fifth of that of Cld but three times that of Ccl (Table 1). Cgl nephrotoxicity is prevented by decreasing intracellular concentrations with probenecid,[11] and augmented when cellular concentrations are increased by transient ureteral occlusion.[59]

Ccl (entire 3-CH$_2$—R_2 group = Cl), which is mildly nephrotoxic, is efficiently secreted across the tubular cell. The resulting limited intracellular sequestration of Ccl (Table 1) could account for its low toxicity compared to that of Cgl. In support of this interpretation is the very large increase of nephrotoxicity when tubular cell Ccl concentrations are increased by a brief period of ureteral obstruction.[59]

Cephalexin (Clx) (R_2 = H), presents a striking departure from a simple relationship between uptake and toxicity. This cephalosporin is secreted across the tubular cell, with cortical concentrations and AUCs comparable to those of Cgl (Table 1). Clx, however, is virtually nontoxic to the kidney, leading to the conclusion that, while concentrative uptake is necessary, it not sufficient for the production of tubular necrosis.

7.24.2.2 Molecular Mechanisms

7.24.2.2.1 *MFO-metabolite hypothesis*

Cobaltous chloride and piperonyl butoxide, which block and compete for cytochrome P450-dependent MFO activity, respectively, reduce the nephrotoxicity of Cld in the mouse and the rat.[66] It was therefore proposed that MFO activity produces a toxic metabolite, through epoxidation of the thiophene ring in the R_1-thienylacetyl side group common to Cld and Cpt. Cpt undergoes so little intracellular sequestration compared to Cld[35,67] that it need not produce significant nephrotoxicity by this mechanism.

The validity of this hypothesis came into early question for two reasons: (i) there was no demonstration of NADPH-dependent, MFO-mediated binding of Cld to rat renal microsomes, as shown with other MFO-activated cytotoxins;[66] and (ii) other nephrotoxic cephalosporins do not have a thiophene or similar side ring.[11]

Evidence against this mechanism of toxicity in the rabbit kidney followed: (i) cobaltous chloride does not protect the rabbit against Cld nephrotoxicity;[83] (ii) although piperonyl butoxide prevents Cld nephrotoxicity in the rabbit, it reduces the cortical uptake of the cephalosporin by enough to explain this effect;[83,84] (iii) Cgl binding to rabbit renal cortical microsomes, in contrast to that resulting from MFO activation, is decreased by NADPH and is not reduced by piperonyl butoxide;[67] and (iv) piperonyl butoxide does not prevent Cgl from producing tubular necrosis in the rabbit.[83]

7.24.2.2.2 *Lipid peroxidation*

Hook and co-workers,[13] studying the rat and the rabbit, observed that Cld depleted reduced glutathione (GSH) in the renal cortex, while increasing its oxidized form (GSSG). Cld also caused the formation of conjugated dienes, as evidence of cell membrane lipid peroxidation, and selenium and vitamin E deficiency, which potentiate the development of oxidative injury, increased Cld nephrotoxicity.[13] The authors suggested that redox cycling in Cld's pyridinium ring was the vehicle of electron transfer causing peroxidative injury.

Table 2 Side group substituents of several cephalosporins.[a]

Cephalosporin[b]	R_1 side group	R_2 side group
Cephaloglycin	phenylglycyl	acetoxy
Cephaloridine	thienylacetyl	pyridinyl
Cefaclor	phenylglycyl	chlorine[c]
Cephalothin	thienylacetyl	acetoxy
Cephalexin	phenylglycyl	hydrogen

[a] See Figure 1 for core structure. [b] Decreasing order of nephrotoxicity in the rabbit, ranging from severe to absent.[77] [c] Entire $-CH_2-R_2$ replaced by $-Cl$.

These observations have been extended: (i) Cld and NADPH undergo reduction and oxidation, respectively, in the presence of active renal cortical microsomes *in vitro*;[41] (ii) Cld induces malondialdehyde (MDA), superoxide anion, and hydrogen peroxide production in cortical microsomes or mitochondria,[41,43] and MDA production in renal cortical slices;[85,86] (iii) Cld-induced MDA production is prevented by various antioxidants and free radical scavengers;[49,85,86] (iv) cefotaxime and Clx, which are not nephrotoxic, cause little or no oxidative stress;[41,44] and (v) Cld-induced MDA production is decreased by cobaltous chloride in the rat, suggesting a role of a P450-dependent reductase in the generation of lipid peroxides.[85]

Several observations are difficult to reconcile with lipid peroxidation being a primary mechanism of injury by all nephrotoxic β-lactams: (i) *in vitro* exposure to Cpt, which lacks a pyridinium ring, produces as much MDA in cortical slices as does Cld;[47] (ii) GSH depletion by Cld is not prevented by antioxidants;[86] and (iii) other nephrotoxic β-lactams do not have pyridinyl (or thienylacetyl) side groups (Table 2) and do not cause significant oxidative injury.[43,44]

7.24.2.2.3 *β-Lactam reactivity*

The very similar renal cortical AUCs and very different toxicities of Cgl and Clx indicate that they have different toxic potentials at a subcellular level, which could only result from the single difference between them—their R_2 substituents. However, examination of the side groups of several toxic and nontoxic cephalosporins (Table 2) does not suggest common or even similar R_1 or R_2 groups as a direct cause of toxicity. It therefore seems likely that: (i) the core structure of the toxic cephalosporins participates in the molecular events leading to tubular necrosis, (ii) the R_2 side groups influence the toxic potential of the core, and (iii) the side groups can exert this effect independent of their effects on cellular AUCs. These conditions also describe the antibiotic properties of the cephalosporins,[2] which suggests shared properties in their antimicrobial and nephrotoxic chemistry.

(i) Antimicrobial action

The β-lactam antibiotics are modified cyclic dipeptides (Figure 1) with three-dimensional configurations resembling the D-alanyl-D-alanine carboxy terminal sequence of bacterial cell wall precursor peptidoglycans.[87] After being accepted as substrates of a number of carboxypeptidases and other membrane-bound proteins essential for bacterial replication, the β-lactams acylate these proteins, blocking cell wall formation and proliferation.

Indelicato and Yamana and co-workers[88-92] have compared the reactivities of a variety of penicillins and cephalosporins in a model designed to quantify their ability to acylate bacterial proteins. The order of their reactivity (Ctz > Ccl > Cld ≥ Cgl ≥ Cpt ≥ cefazolin ≫ several penicillins > Clx and other 3-methyl-cephalosporins) is determined by the side groups, generally more by the R_2 than by the R_1 substituents.[90] The more reactive cephalosporins have R_2 groups with comparatively unstable bonds to the core. The leaving of these groups causes intramolecular electron shifts that destabilize the β-lactam ring, favoring acylation of the molecular targets.[90] Reactivity contributes to the antimicrobial potencies of the antibiotics, although resistance to β-lactamases, bacterial cell wall penetration, and target affinity are also important.[1]

(ii) Nephrotoxic potential

The rank order of nephrotoxicity of the β-lactams [Cgl > Cld ≫ Ccl > cefazolin > Cpt ≫ the penicillins, Clx, and Ctz (no toxicity)][11,12,24,31,51,59,75,93,94] correlates partially with the order of their reactivity. The relative stability of the penicillins and the 3-methyl-cephalosporins (like Clx), which is largely a result of the absence of leaving groups in the R_2 position, fits well with their uniform lack of toxicity. However, several of the more reactive cephalosporins are not nephrotoxic. A more convincing model of toxicity emerges when intracellular sequestration (Table 1) is also considered. For example, Ctz, although very highly reactive, undergoes essentially no concentrative uptake by the tubular cell. Ccl, which

is also highly reactive, has a low renal cortical AUC (7% and 37% of the AUCs of Cld and Cgl, respectively). Cpt, although fairly reactive, also has very low tubular cell concentrations.[67]

The single exception to this correlation of nephrotoxicity with reactivity and the AUC is Cld. The combined high reactivity and high tubular cell AUC of Cld would not predict a lower toxicity than that of Cgl. An explanation for this discrepancy may be found in Cld's pyridinium ring. The presence of a fixed cation so near to the carboxyl could limit Cld's access to its anion-receptive molecular targets (see Section 7.24.3.2), thereby requiring the high and prolonged cellular concentrations needed to produce nephrotoxicity. This hypothesis is supported by the following observations: (i) Cld inhibition of mitochondrial anionic substrate transport is less than that of Cgl after *in vitro* and *in vivo* exposure;[68,95] and (ii) brief *in vivo* exposure of the tubular cell, terminated by an intravenous bolus of probenecid, produces less injury with Cld than with similar brief exposure after an equally toxic dose of Cgl.[11]

Browning *et al.*[56] compared the covalent binding of three β-lactams to tubular cell proteins of the rabbit kidney: Cgl (highly nephrotoxic), Cpt (minimally nephrotoxic), and benzylpenicillin (no nephrotoxicity). Several hours after *in vivo* administration, covalently bound Cgl in nuclear, mitochondrial, microsomal, and cytosolic proteins was 3.5 times greater than Cpt and 7 times greater than penicillin. Binding after 1 h of *in vitro* incubation was similar. The order of binding of the three β-lactams is consistent with the hypothesis that cephalosporin nephrotoxicity is initiated by a process of tubular cell uptake, followed by spontaneous acylation of one or more target proteins. The binding of Cgl was so diffuse among subcellular fractions that a specific target could not be identified by this method. As in the antimicrobial process, acylation by the cephalosporins may impair the functions of multiple proteins. It is not assumed, however, that the function of every acylated protein is affected.

7.24.2.3 Molecular Targets

7.24.2.3.1 Glucose metabolism: gluconeogenesis

Goldstein *et al.*[46] showed a reduction of gluconeogenesis by Cld in rat renal cortical slices within 0.25 h of *in vitro* exposure. Tenfold higher Cld concentrations have a similar effect on isolated hepatocytes.[96] This cytotoxic effect involves inhibition of glucose-6-phosphatase;[46,96] reduction of tubular cell and/or mitochondrial carnitine transport may also have a role.[68] Although reduced gluconeogenesis should not produce cellular necrosis by itself, it is one of the most rapidly developing biochemical effects of Cld in the tubular cell.

7.24.2.3.2 Cytochrome P450 activity

Cojocel et al.[97] have measured a decrease of cytochrome P450 activity in rat renal cortex within 3 h of injection of Cld. Following this decrease was a reduction of 50–53 kDa proteins and the appearance of a 44 kDa protein in renal cortical microsomes; this effect, however, was observed only after 3 days of exposure.[98] The mechanisms and functional significance of these observations remain to be determined.

7.24.2.3.3 Mitochondrial metabolism

Mitochondrial injury was evaluated as a possible mechanism of Cld nephrotoxicity because: (i) the basolateral transport of organic anions[70] and Cld[34] may produce higher concentrations in the cytoplasm surrounding mitochondria than in other regions of the cell,[99] and (ii) ultrastructural changes develop in mitochondria within a few hours after Cld administration.[10] Studies in rabbit renal cortical mitochondria, using glutamate, malate, and succinate as substrates, showed decreased respiration after nephrotoxic doses of Cld or Cgl (in vivo exposure) and during in vitro exposure of untreated mitochondria to concentrations found in the renal cortex.[11,38,48] Respiration in renal tubular suspensions was inhibited within 2 h, but not after 5–10 min of in vitro exposure to Cld.[48]

There are several lines of evidence that mitochondrial injury may contribute to cephalosporin nephrotoxicity: (i) Cld and Cgl (nephrotoxic) produce respiratory toxicity in vivo, while Clx (nontoxic) does not;[11] (ii) respiratory toxicity evolves over 0.5–1 h after Cgl administration, as intracellular antibiotic concentrations decline substantially;[11,75] (iii) Cld decreases the ATP content of cortical slices and renal tubules by 1–1.5 h;[46,49,50] (iv) ultrastructural damage to mitochondria and other membranous structures evolves after the development of respiratory toxicity, in a pattern[10] that resembles the lesions produced by ischemia[100] and cyanide;[101] (v) the respiratory toxicity of Cgl, measured 1 h after administration, is augmented by a nephrotoxic-potentiating regimen of aminoglycoside that has no respiratory toxicity by itself;[12] and (vi) renal mitochondrial calcification is seen 24 h after Cgl intoxication.[21]

7.24.3 MECHANISMS OF MITOCHONDRIAL INJURY

7.24.3.1 Respiration

The nephrotoxic cephalosporins produce similar patterns of respiratory toxicity in mitochondria exposed either in vitro or in vivo:[38] (i) with the naturally occurring substrates, respiration with glutamate and malate is inhibited less than respiration with succinate; (ii) respiration with the artificial electron donors ascorbate and tetramethylphenylenediamine is relatively unaffected; and (iii) state 3 (ATP-generating) respiration is reduced more than the slower, ADP-independent, respiration.

However, the in vivo and in vitro effects of the cephalosporins on respiration differ in certain important ways.[102] Toxicity to respiration in mitochondria exposed in vitro can be produced by a variety of penicillins and cephalosporins and is immediate in onset. In contrast, toxicity from in vivo exposure, measured in subsequently isolated mitochondria, occurs only with the nephrotoxic cephalosporins and develops more slowly. Moreover, early in vitro toxicity is reversed either by removal of the cephalosporin or by increasing concentrations of metabolic substrate, while in vivo toxicity is not.

7.24.3.2 Transport

Several observations suggest an effect of the β-lactams on mitochondrial substrate transporters: (i) the natural ability of the antibiotics to acylate and inactivate membrane-bound proteins[87] and to acylate tubular cell proteins;[67] (ii) the greater toxicity of the β-lactams to respiration with succinate, which donates electrons to the mid-region of the cytochrome chain, than to respiration with NAD-dependent substrates, which donate electrons proximal to that locus—consistent with an action outside the electron transport chain;[11,38] and (iii) most importantly, the comparatively indiscriminate in vitro respiratory toxicity, which in its early stages can be reversed by substrate excess. This competitive inhibition provided evidence that many β-lactams may act as inhibitors of mitochondrial anionic substrate transport, but in vivo nephrotoxicity develops only with the comparatively sequestered and reactive (i.e., nephrotoxic) ones, which acylate these carriers and cause irreversible injury to substrate uptake.

Further studies demonstrated correlations between decreased respiration and decreased uptake of succinate in mitochondria exposed *in vivo* to nephrotoxic doses of Cld, Cgl, or Imip.[42–44] Succinate efflux was unaffected in each case, indicating that reduced net uptake was the result of decreased substrate entry. *In vitro* exposure of mitochondria to Cgl produced similar results.[42] Several observations support the conclusion that these effects are specific to β-lactam-induced nephrotoxic injury *in vivo*:[45] (i) Clx, which is not nephrotoxic, does not decrease either the uptake of or respiration with succinate; (ii) respiratory chain poisons reduce respiration but not uptake; and (iii) finally, the uptake of phosphate and ADP, which are carried into the mitochondrion by their own transporters, and which have critical roles in mitochondrial respiration, are relatively unaffected by the nephrotoxic β-lactams.

The evolution of *in vitro* toxicity was studied in rabbit kidney mitochondria subjected to continuing exposure to Clx, Cgl, and Imip over the range of concentrations found in the rabbit renal cortex after *in vivo* exposure.[54] The two highly reactive (nephrotoxic) β-lactams produced a time- and concentration-dependent transition to irreversible injury to both respiration with and the uptake of succinate by the mitochondrion, while producing little or no injury to ADP transport.

7.24.3.2.1 *Oxoglutarate carrier*

Preliminary studies[4] have been done with the purified oxoglutarate (ketoglutarate) transport protein. *In vitro* Cgl reduces oxoglutarate transport in normal mitochondria and in proteoliposomes prepared with the isolated carrier from normal mitochondria. On two-dimensional gel electrophoresis of the carrier protein, *in vivo* Cgl pretreatment produces a new 34 kDa protein alongside the 34 kDa native protein, shifted from a p*I* of approximately 6.7 to one of around 6.5. Acid phosphatase does not eliminate the new protein spot, ruling against phosphorylation as a cause of the altered p*I*. Although there are other possible explanations for this new protein, the small change of isoelectric point without a significant difference in molecular weight is consistent with β-lactam acylation of an amino group in the transporter.

7.24.3.2.2 *Cytotoxic potential*

The question remained whether the scope and severity of mitochondrial injury by the nephrotoxic β-lactams was sufficient to kill the tubular cell. In support of this mechanism, reductions of uptake and state 3 respiration with different dicarboxylic and tricarboxylic substrates of the Krebs cycle (succinate, glutamate, malate, oxoglutarate, and citrate) range from 35% to 75% in mitochondria isolated from renal cortex 1 h after toxic β-lactam administration,[4,11,38] and there are probably greater reductions in the 25–50% of tubular cells killed by the doses used in these studies.[29,75] Moreover, ATP-generating respiration (state 3 minus state 4) is more reduced than state 3, and ATP concentrations are decreased by Cld within 0.5–1 h of the onset of respiratory toxicity.[46,49]

7.24.3.3 Cephaloridine vs. the Anionic β-Lactams

Although the attack by all three nephrotoxic β-lactams on mitochondrial substrate transport, and the virtual absence of oxidative injury with Cgl and Imip,[42–44] seemed to establish respiratory injury as a unifying mechanism of the nephrotoxicity of all β-lactams, several properties of Cld challenged this generalization: (i) the augmentation of Cld nephrotoxicity by potentiators of oxidative injury[13] and its reduction by antioxidants;[13,49,86,103] (ii) the slower emergence of Cld's respiratory toxicity compared to the onset of oxidative injury in isolated renal tubules;[49,50] (iii) Cld's limited *in vivo* toxicity to mitochondrial respiration with pyruvate,[45] an important intermediate of amino acid and carbohydrate catabolism and a precursor of Krebs cycle intermediates;[104] and (iv) Cld's higher ND-50 compared to Cgl, even though it has uniquely high cortical concentrations and is comparably reactive.[77] Studies have provided evidence that Cld's R_2 pyridinium, which ironically is responsible for its unusual cellular uptake and probably also its oxidative potential, may limit its toxicity to mitochondrial anionic substrate carriers.

7.24.3.3.1 *Monocarboxylates: pyruvate and short-chain fatty anions*

Studies of the mitochondrial metabolism of monocarboxylic substrates revealed greater differences between Cld and the anionic β-lactams than were apparent from earlier work with the dicarboxylates. Evidence had been provided for a toxic effect of the β-lactams on the catabolism of fatty acids: (i) Cgl toxicity in the rabbit kidney was significantly increased by 24 h of fasting,[4] which increases dependence upon fatty acid metabolism;[105] and (ii) this augmentation of injury was prevented by intravenous glucose, but not by an identical volume and osmolar infusion of saline.[4] Protocols were therefore designed to compare the effects of Cld and the anionic β-lactams on respiration with, and the uptake of, pyruvate (three-carbon) and the short-chain fatty anions butyrate (four-carbon) and valerate (five-carbon).[95,106] These three anions are precursors of acetyl-coenzyme A[104,107] and are important substrates of proximal tubular cell respiration.[108–110]

Whereas *in vivo*, Cld, Cgl, and Imip were comparably toxic to mitochondrial dicarboxylate metabolism, their effects on the mitochondrial metabolism of the monocarboxylic substrates were not the same. Cgl and Imip were as toxic to respiration with the three monocarboxylates after 1 h of *in vivo* exposure as they were to respiration with succinate. However, Cld was only slightly toxic to respiration with valerate and had no effect on respiration with pyruvate or butyrate. The effects of the three nephrotoxic β-lactams on monocarboxylic substrate uptake paralleled their effects on respiration with the monocarboxylates. Clx, studied as a control anionic cephalosporin, affected butyrate metabolism in a manner that was also not seen in earlier studies. *In vivo* Clx stimulated both the uptake and oxidation of butyrate,[95] an action that might explain Clx's protective effect against the toxic cephalosporins at doses insufficient to reduce their cortical concentrations.[36]

The mitochondrial effects of *in vivo* Cld, Cgl, Imip, and Clx on transport and respiration with different anionic substrates are summarized in Table 3. The influence of each β-lactam on substrate uptake and metabolism, whether inhibitory, nontoxic, or stimulatory, is qualitatively the same. This cumulative experience provides strong evidence that the toxicity of the β-lactams to mitochondrial respiration with anionic substrates is caused by an attack on the substrate carriers of the inner membrane.

Unlike the consistently parallel *in vivo* effects of the β-lactams on uptake and respiration (Table 3), *in vitro* exposure does not always have the same effects on monocarboxylate metabolism as *in vivo* exposure. For example, Clx has inhibitory *in vitro* and stimulatory *in vivo* effects on butyrate uptake and butyrate-mediated respiration, and even on respiration with the carnitine ester of the long-chain fatty anion palmitate,[95] while Imip stimulates respiration with butyrate or valerate *in vitro* but inhibits it after *in vivo* exposure of more than 15 min.[106] These different effects may be a consequence of altered substrate affinities as a result of β-lactam–carrier interactions, analogous to the effects of certain inhibitors on mitochondrial pyruvate and ADP/ATP transporters.[111–113] The results with the nephrotoxic and nontoxic β-lactams suggest that reversible interactions may either inhibit or stimulate transport, while acylation of the carrier causes the irreversible injury that blocks substrate uptake and thereby contributes to nephrotoxicity.[68,95,106]

7.24.3.3.2   *Zwitterionic substrates: the long-chain fatty acylcarnitines*

Before they can undergo beta-oxidation in the mitochondrial inner matrix, long-chain fatty acids like palmitate must undergo esterification to carnitine and be transported by the acylcarnitine–carnitine antiporter (carnitine carrier) across the inner membrane.[114,115] Further protocols were designed to compare the effects of Cld and Cgl on the transport of carnitine and respiration with palmitoylcarnitine.[68,95]

Cld has significant structural homology with carnitine, with a carboxyl separated from a quaternary nitrogen by three carbons in relatively tetrahedral configurations (Figure 3). As a result, Cld is a potent inhibitor of both the tubular reabsorption and the mitochondrial uptake of carnitine and its fatty-acyl esters,

Table 3   *In vivo* effects of β-lactam antibiotics on mitochondrial anionic substrate uptake and respiration.

	Cld	Cgl	Imip	Clx	Refs.
Succinate[a]	↓↓↓/↓↓	↓↓↓/↓↓↓	↓↓/↓↓	–/–	11,38,42–45,75
Pyruvate[a]	–/–	↓/↓↓	↓↓↓/↓↓		45,106
Butyrate[a]	–/–	↓↓↓/↓↓↓	↓↓/↓↓↓	↑↑/↑↑↑	95,106

[a]*In vivo* exposure; irreversible reduction (↓) or stimulation (↑) of transport/respiration: –, no change; ↓ or ↑, 20–24% change; ↓↓ or ↑↑, 25–49% change; ↓↓↓ or ↑↑↑, ≥50% change.

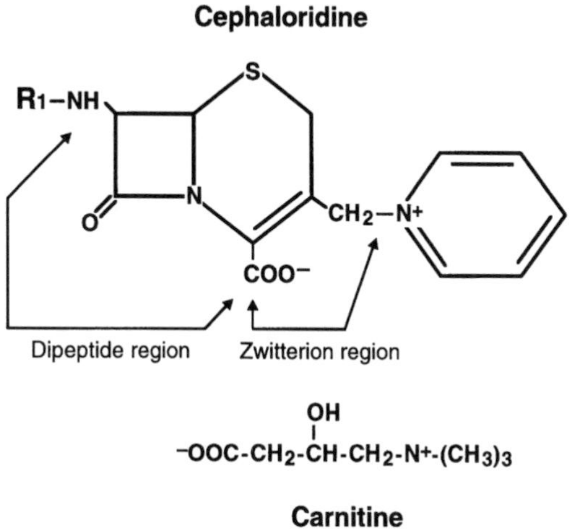

Cephaloridine

Dipeptide region   Zwitterion region

$$OOC-CH_2-CH-CH_2-N^+-(CH_3)_3$$
$$|$$
$$OH$$

Carnitine

Figure 3 Molecular structure of cephaloridine, showing the protein-acylating dipeptide region common to all β-lactams and implicated in their irreversible attack on mitochondrial anionic substrate transporters, and the zwitterionic region, which is responsible for cephaloridine's competitive inhibition of carnitine transport. Carnitine is shown below for comparison. The other nephrotoxic β-lactams studied have similar dipeptide homology, but do not have the zwitterionic region.

causing massive acylcarnitinuria and a doubling of intramitochondrial free carnitine in the intact kidney.[68] Presumably because the zwitterionic region of Cld is separate from its β-lactam ring, preventing acylation of the carnitine carriers, toxicity to carnitine and acylcarnitine transport is reversed during isolation of *in vivo*-exposed mitochondria.[68] It is not known whether the reversible effect on the carnitine carriers is of sufficient magnitude or duration to contribute to Cld nephrotoxicity.

Cgl lacks Cld's structural homology with carnitine and therefore has little or no effect on either the mitochondrial or the tubular transport of carnitine or acylcarnitines.[95] Although Cld and Cgl have very different effects on the carnitine carriers, both are comparably toxic to respiration with palmitoylcarnitine after *in vivo* exposure.[68] This single case of impaired respiration in the absence of irreversible injury to the corresponding carrier by either cephalosporin suggests separate mechanisms of altered mitochondrial palmitoylcarnitine metabolism, such as lipid peroxidation by Cld and depletion of mitochondrial Krebs-cycle intermediates by Cgl.

7.24.3.3.3 *Conclusions*

In vivo Cgl and Imip have essentially the same toxicity to the mitochondrial metabolism of all

of the metabolic substrates tested. Cld causes similar injury to respiration with dicarboxylic substrates, but at considerably greater AUCs than those of Cgl. Moreover, even with its higher intracellular concentrations, *in vivo* Cld has little or no early effect on the mitochondrial metabolism of pyruvate and the short-chain fatty anions. Cld's zwitterionic charge may hinder its ability to acylate monocarboxylic and other anionic carriers, causing it to be less nephrotoxic than might be predicted from its uniquely high intracellular concentrations, its high reactivity, and its singular ability to produce oxidative injury. This important difference from the anionic β-lactams may also explain the failure of Cld, in contrast to Cgl and the less nephrotoxic anionic cephalosporins, to augment the tubular necrosis caused by aminoglycosides[12,116] or by renal ischemia.[56]

7.24.4 NEW LACTAM ANTIBIOTICS

7.24.4.1 Cephalosporins

Following recognition of the influence of Cld's pyridinium on its transport and toxicity,[35,78] there was a hiatus of several years before the release of another cephalosporin with a side group containing a quaternary nitrogen,[93] and a further delay before a similar zwitterionic cephalosporin was introduced.[60] With the recognition of the facilitation by cationic substituents of penetration through bacterial cell wall porins, interest in the R_2 pyridinyl substituents has been renewed.[117]

7.24.4.1.1 *Cefpimizole*

This pyridinyl cephalosporin has significant nephrotoxicity in laboratory animals.[60] There are no published data regarding either the mechanisms of its toxicity or its potential risk to humans.

7.24.4.1.2 *DQ-2556: a new mechanism of injury?*

This cephalosporin combines R_1 and R_2 substituents separately associated with breadth of antibacterial spectrum in other antibiotics of its class, including an R_1 group quaternary nitrogen in the same position as that of Cld and Ctz. DQ-2556 is excreted by the kidney mainly through glomerular filtration, with a small component of probenecid-inhibitable tubular secretion.[118] Intravenous doses of DQ-2556 in the range of 1200 mg kg^{-1} cause acute constriction of the small branches of the renal arteries in rats (but not in rabbits), followed by the development of necrosis of proximal convoluted and straight tubules.[119] Calcium channel blockade, which does not reduce Cld nephrotoxicity,[120] protects against these reactions.[119] DQ-2556 inhibits *N*-methylnicotinamide transport in brush border membrane vesicles of rat, but not rabbit, kidneys.[121]

The mechanisms of the vasoconstrictive nephrotoxicity of DQ-2556 are not known. Although its cationic structure and transport properties suggest a possible neuromimetic effect, a partially protective effect of probenecid seems to contradict this interpretation.[118]

7.24.4.2 Penems

Although the patterns of toxicity of Imip to mitochondrial substrate uptake and metabolism indicate important similarities to the toxic anionic cephalosporins, experience with the penems is too limited to permit the conclusion that they and the cephalosporins are nephrotoxic by identical mechanisms.

Imip differs from the cephalosporins in several respects: (i) although it is secreted by the proximal tubule, Imip is rapidly hydrolyzed by cytoplasmic and brush border dehydropeptidase-I (DHP-I) to inactive metabolites;[3,122] (ii) Imip is very highly reactive, which may allow acylation of mitochondrial anion transporters (or other target proteins) with comparatively low AUCs;[2,123,124] (iii) Imip is marketed in combination with cilastatin, a DHP-I inhibitor that also blocks its tubular secretion and nephrotoxicity;[3] and (iv) in an *in vitro* system that bypasses the effect on secretory transport, cilastatin, which has structural homology with the β-lactams, partially blocks the toxicity of Imip, but not that of Cgl, to mitochondrial succinate metabolism.[54] DHP-I activity does not increase the mitochondrial toxicity of Imip *in vitro*, ruling against the formation of a toxic by-product of hydrolysis.[54]

Toxic screening has revealed varying degrees of nephrotoxicity of several new penems in laboratory animals.[5-7] Panipenem toxicity is severe enough that it has been released in combination with an inhibitor of its renal transport and toxicity.[7] Rates of hydrolysis of these new penems by renal DHP-I vary greatly between molecular and animal species.[5,125-127] Currently published data are insufficient to establish correlations between susceptibility to DHP and nephrotoxicity.[5,125-127]

7.24.4.3 γ-Lactams

γ-Lactam antibiotics have the basic structure of secreted organic anions[128,129] and a mechanism of antibacterial action much like that of the β-lactams.[91,128] The most potent γ-lactams tested are more reactive than the most reactive cephalosporins.[91] We may, therefore, have in this new class of antibiotics the opportunity to test further the proposed roles of uptake and reactivity in producing nephrotoxicity.

7.24.5 FUTURE DIRECTIONS

Early molecular changes in β-lactam-induced tubular cell injury have been measured within 0.25–1 h after exposure to nephrotoxic doses *in vivo* or appropriate concentrations *in vitro*.[11,13,46,86,106] Although suspensions of proximal renal tubules have been useful in studies of the cellular mechanisms of Cld nephrotoxicity,[48-50] the lack of physiologic movement of the organic anions from cell to luminal fluid limits the utility of these collapsed tubules for studying the more typically secreted β-lactams.[70] Several newer approaches hold further potential promise for investigation of the early molecular events in β-lactam nephrotoxicity.

7.24.5.1 Tubular Cells in Culture

Proximal tubular cells have been studied in both primary cultures and established cell lines. Primary cultures are more likely to preserve physiologic functions, like the secretion of *p*-aminohippurate (PAH).[130] However, primary cultures multiply for a limited number of generations and are therefore difficult to grow in the monolayers needed to measure secretory movement into and across the cell. Continuous cell lines are more reliably grown in confluent monolayers, but are more likely to have lost certain basic functions.[131,132]

Two continuous proximal tubular cell lines have been used to study β-lactam toxicity. Gstraunthaler *et al.*[52] compared three cephalosporins in a porcine kidney cell line (LLC-PK1) and concluded that their *in vitro* cytotoxicity and *in vivo* nephrotoxicity were similar. However, the data are preliminary, and LLC-PK1 cells lack organic anion transport capacity,[131,132] which has been shown to be a necessary first step in β-lactam nephrotoxicity.[11,22-24,49,73,74] Williams *et al.*[51] tested a more convincing *in vitro* model of β-lactam injury, using a continuous rabbit proximal tubular cell line (LLC-RK1). However, the cells were not grown on porous membranes to facilitate basolateral uptake. Moreover, antibiotic concentrations and times of exposure (AUCs) were very high and the rank order of cytotoxicity only approximated that of nephrotoxicity.

A continuous proximal tubular cell line from the opossum kidney (OK) grown on porous membranes has recently been shown to secrete PAH across the basolateral and luminal membranes.[133,134] Although the secretory uptake and subsequent efflux of PAH are carrier mediated (inhibited by probenecid), they do not generate the high cellular concentrations seen in perfused tubules or the *in situ* kidney.[34,35,70] It is possible that luminal-side efflux is enhanced by the large volume of apical bathing medium, and that concentrations can be manipulated to establish cellular concentrations closer to those in the intact kidney, making the OK cell a suitable *in vitro* model for the study of β-lactam toxicity. The OK cell also has organic cation transport capacity[92] and might also be useful for the study of zwitterionic cephalosporins. Although it does not proliferate and differentiate as rapidly, the human proximal tubular cell line HK-2 may prove useful as a tool for toxicologic study of the β-lactams.[135]

7.24.5.2 Subcellular Probes

Advances in nucleomagnetic resonance[136–139] and microspectrofluorimetry[140,141] hold great promise for application as subcellular probes of the biochemical mechanisms of β-lactam toxicity. Used in combination with appropriate models with cultured tubular cells, these techniques may permit finer analysis of currently suspected mechanisms of injury and recognition of others not yet detected.

7.24.6 CLINICAL TOXICITY

7.24.6.1 Prevention

7.24.6.1.1 Monotherapy and β-lactam combinations

Due to their widely varying renal cortical AUCs and ring reactivities, the β-lactam antibiotics have an extremely broad range of nephrotoxic potential. Laboratory animal screening by the pharmaceutical industry has fortunately prevented the medical use of several highly toxic β-lactams—with the exception of Imip and panipenem, which are marketed in combination with nephroprotective transport inhibitors.[3,7]

The present risk of nephrotoxic injury from monotherapy with β-lactams is, therefore, very low. None of the penicillins in use is known to cause direct toxicity to the kidney, whether given alone or in combination with other drugs. Renal toxicity is also of little or no concern with the currently available cephalosporins or the Imip–cilastatin combination alone, or with double β-lactam regimens.[142,143]

7.24.6.1.2 Nephrotoxic synergies

The risks are not as low, however, with multiple drug combinations. The treatment of serious infections may require more than one nephrotoxic antibiotic, often in the presence of septicemia and/or nephrotoxic anticancer regimens. Acute renal failure is relatively common in such cases, and it is difficult to predict the likelihood, or identify with certainty the causes, of renal injury. Renal failure may be prevented by early treatment of infections, attention to the choice and dosage of potentially nephrotoxic antibiotics, and maintenance of hemodynamic stability, renal perfusion, and urinary solute excretion.[57,74,144]

Additive aminoglycoside–Cpt nephrotoxicity has been demonstrated in humans.[18–20] Toxic synergies between aminoglycosides and several mildly toxic cephalosporin regimens have been shown in rabbits,[12,37] and this combined injury is further augmented by mild renal ischemia[56] or endotoxemia.[14,57] Studies in rats have either shown no additive aminoglycoside–cephalosporin toxicity[145,146] or a protective effect of large doses of a nontoxic cephalosporin or penicillin against an aminoglycoside.[27,147] It is possible that the resistance of rats to cephalosporin nephrotoxicity[9,25,30] is responsible for the negative results in this species, and that the sodium load resulting from the large β-lactam doses was protective.[27,28] Fortunately, controlled studies have documented the clinical safety of combinations of nontoxic cephalosporins with aminoglycosides.[94,142]

The nephrotoxic antibiotics (aminoglycosides, vancomycin, mildly toxic β-lactams) are commonly used together with or following other medications that are toxic to the kidney. Drugs that may produce acute tubular necrosis in simultaneous or sequential combinations include: aminoglycosides, vancomycin, cisplatin, ifosfamide, cyclosporine, amphotericin B, and the toxic β-lactams.[148] Although the interactions of combinations of these nephrotoxicants have not been established in every case, the known toxic synergies between several pairs of drugs from this list indicates a potential for a variety of toxic interactions.

7.24.6.1.3 Diuretics

The potent diuretics furosemide and ethacrynic acid have variably been described as additive or protective in the development of nephrotoxic injury.[144] β-Lactam nephrotoxicity has figured prominently in this controversy. Early clinical reports of Cld toxicity identified diuretic use as a correlate of the development of tubular necrosis, together with old age and serious cardiovascular disease.[16] However, there is no direct clinical evidence of a nephrotoxic synergy between the diuretics and the β-lactams. Studies in laboratory animals alternatively suggested potentiation of[149,150] or protection against[151] Cld nephrotoxicity by the loop diuretics; these studies were complicated by the additional renal insults of glycerol injection or severe dehydration.

In rabbits given furosemide, or its vehicle, followed immediately by a Ringer's infusion sufficient to maintain hydration and sustain a diuresis for 4 h, both furosemide-Ringer's- and vehicle-Ringer's-treated animals showed slight protection against a toxic dose of Cld.[74] It has been concluded from these and other studies that: (i) proximal tubular solute (sodium or mannitol) diuresis can be protective, and hypovolemia can be potentiating, in nephrotoxic injury, and (ii) the loop diuretics may contribute to this injury indirectly, through the production of dehydration and/or a post-diuretic oliguria.[144,148]

7.24.6.1.4 Development of nontoxic lactam antibiotics

A number of bacteria have developed multiple-antibiotic resistance, in some cases to β-lactams that have been effective for up to half a century.[152,153] While the β-lactams, and possibly the γ-lactams,[91,128] are promising groups for new and effective therapies for these infections, the urgency for development may compromise the screening-out of potentially nephrotoxic antibiotics.

The testing required by pharmaceutical regulatory agencies does not generally include toxicologic studies with potentiating factors. Careful attention will, therefore, be required as new cephems, penems, and other lactam ring antibiotics are used under conditions of risk. Although transport inhibitors can be used to protect against injury from new lactams with nephrotoxic potential, they may elevate blood levels[69] and thereby increase the likelihood of nephrotoxicity if eliminated before the antibiotic.[154] Moreover, evidence has been presented that the associated inhibition of β-lactam transport across the choroid plexus[155] can increase CNS concentrations and predispose to neurotoxicity.[156]

7.24.6.2 Treatment

Antibiotic-induced acute renal failure may be oliguric or nonoliguric. With proper management the renal failure is typically reversible.[157] While aggressive fluid resuscitation can limit the severity of renal injury from hypotensive, endotoxemic, and nephrotoxic insults,[57,144] there will be a risk of overhydration if renal failure develops. The renal failure resulting from β-lactams and aminoglycosides follows a time course comparable to acute tubular necrosis from other causes.[16,157] Recovery from combined endotoxemic or ischemic and nephrotoxic insults is less predictable. There is no evidence that dialytic removal of the antibiotics will hasten renal recovery, but clinical stability and good nutrition are likely to be salutary, as in other causes of acute renal failure.

7.24.7 SUMMARY AND CONCLUSIONS

Several of the cephem and penem antibiotics produce acute tubular necrosis when given in large single doses. Concentrative uptake by the tubular cell, resulting from the balance of basolateral secretory transport and subsequent net movement into the tubular lumen, makes the proximal renal tubule the principal target of toxicity.[158] Differences among the separate components of transport result in a wide range of intracellular concentrations and AUCs, and thereby contribute to the highly variable nephrotoxicity of the different β-lactams in humans and laboratory animals.[24,32,53,75,159]

Cld, the most thoroughly studied cephalosporin, has several effects on the rat and/or rabbit proximal tubular cell: (i) lipid peroxidation, resulting either from electron shifts in its R_2 substituent pyridinium,[13,41] or possibly from its R_1 thiophene;[47,68] (ii) irreversible injury to the mitochondrial uptake of and respiration with Krebs-cycle substrates and their amino acid precursors, through cephem-ring acylation of anionic substrate carriers;[38,44] (iii) reversible inhibition of tubular cell and mitochondrial carnitine transport, and long-chain fatty acylcarnitine-mediated respiration, caused by Cld's zwitterionic region;[68] (iv) inhibition of gluconeogenesis;[46] and (v) reduction of cytochrome P450 activity.[97] Lipid peroxidation and inhibition of carnitine transport are effects of Cld alone; mitochondrial toxicity is a property of Cld, Cgl, and Imip; the effects of Cgl and Imip on gluconeogenesis and P450 activity have not been studied.

The most nephrotoxic β-lactams do not share common or similar side groups to account for their nephrotoxicity.[77] They do, however, share reactive β-lactam (cephem or penem) rings[2] capable of acylating proteins within the tubular cell.[67] The side groups can affect the cellular transport of the β-lactams and/or the reactivity involved in their attack on intracellular targets.[77] An important example of the effects of the substituents is the case of cephalexin (nontoxic) and Cgl (the most nephrotoxic cephalosporin), which differ only in the presence of an R_2 hydrogen and an R_2 acetoxy, respectively. The acetoxy leaving group conveys to Cgl a small increase in the cortical AUC, but a very large increase in cephem ring reactivity and toxicity.[158]

The pathogenic role of acylation is indicated by the combined influence on nephrotoxicity of the cortical AUCs and reactivities of several cephalosporins[4] and the demonstration of nephrotoxicity in the majority of penems,[3,5–7] which are more reactive than the cephalosporins.[2] The importance of mitochondrial substrate carriers as targets of acylation is supported by (i) the toxicity of Cld, Cgl, and Imip to mitochondrial substrate uptake and respiration[42–44] in the absence of significant oxidative injury from Cgl or Imip,[43,44] and (ii) the direct demonstration of Cgl toxicity to the isolated mitochondrial oxoglutarate carrier.[148]

The several toxic effects of Cld, the evidence that oxidative injury is pathogenic in its nephrotoxicity,[13,106] and the toxicity of Cgl and Imip to the mitochondrial transport and metabolism of every anionic substrate studied[42,43,106] suggest that the β-lactams are nephrotoxic through more than one molecular mechanism. Studies of the consequences of acylation of other tubular cell proteins will probably reveal additional molecular targets. Finally, considering their substantial concentrative uptake by the tubular cell, and their highly variable and complex side group substituents, it will not be surprising to learn of other important cytotoxic effects of β-lactam or γ-lactam antibiotics in the kidney.

ACKNOWLEDGMENT

This work was sponsored by a grant from the National Institutes of Health (USPHS #DK 33814).

7.24.8 REFERENCES

1. D. J. Waxman and J. L. Strominger, 'Penicillin-binding proteins and the mechanism of action of beta-lactam antibiotics.' *Annu. Rev. Biochem.*, 1983, **52**, 825–869.
2. J. R. E. Hoover, in 'Handbook of Experimental Pharmacology. II. Beta-lactam Antibiotics,' eds. A. L. Demain and N. A. Solomon, Springer, Berlin, 1983, vol. 67, pp. 119–245.
3. J. Birnbaum, F. M. Kahan, H. Kropp *et al.*, 'Carbapenems, a new class of beta-lactam antibiotics. Discovery and development of imipenem/cilastatin.' *Am. J. Med.*, 1985, **78**, Suppl. 6A, 3–21.
4. B. M. Tune, in 'Toxicology of the Kidney,' eds. J. B. Hook and R. S. Goldstein, Raven Press, New York, 1993, pp. 257–281.
5. J. C. Topham, L. B. Murgatroyd, D. V. Jones *et al.*, 'Safety evaluation of meropenem in animals: studies on the kidney.' *J. Antimicrob. Chemother.*, 1989, **24**, Suppl. A, 287–306.
6. M. Brughera, G. Scampini, M. L. Ferrari *et al.*, 'Toxicologic profile of FCE 22101 and its orally available ester FCE 22891.' *J. Antimicrob. Chemother.*, 1989, **23**, Suppl C, 129–135.
7. Y. Hirouchi, H. Naganuma, Y. Kawahara *et al.*, 'Preventive effect of betamipron on nephrotoxicity and uptake of carbapenems in rabbit renal cortex.' *Jpn. J. Pharmacol.*, 1994, **66**, 1–6.
8. M. Barza, 'The nephrotoxicity of cephalosporins: an overview.' *J. Infect. Dis.*, 1978, **137**, Suppl. S60–S73.
9. R. M. Atkinson, J. P. Currie, B. Davis *et al.*, 'Acute toxicity of cephaloridine, an antibiotic derived from cephalosporin C.' *Toxicol. Appl. Pharmacol.*, 1966, **8**, 398–406.
10. F. Silverblatt, M. Turck and R. Bulger, 'Nephrotoxicity due to cephaloridine: a light- and electron-microscopic study in rabbits.' *J. Infect. Dis.*, 1970, **122**, 33–44.
11. B. M. Tune and D. Fravert, 'Cephalosporin nephrotoxicity. Transport, cytotoxicity and mitochondrial toxicity of cephaloglycin.' *J. Pharmacol. Exp. Ther.*, 1980, **215**, 186–190.
12. J. P. Bendirdjian, D. J. Prime, M. C. Browning *et al.*, 'Additive nephrotoxicity of cephalosporins and aminoglycosides in the rabbit.' *J. Pharmacol. Exp. Ther.*, 1981, **218**, 631–685.
13. C. H. Kuo, K. Maita, S. D. Slieght *et al.*, 'Lipid peroxidation: a possible mechanism of cephaloridine-induced nephrotoxicity.' *Toxicol. Appl. Pharmacol.*, 1983, **67**, 78–88.
14. B. M. Tune and C. Y. Hsu, 'Augmentation of antibiotic nephrotoxicity by endotoxemia in the rabbit.' *J. Pharmacol. Exp. Ther.*, 1985, **234**, 425–430.
15. R. D. Foord, 'Cephaloridine and the kidney.' *Progr. Antimicrob. Anticancer Ther.*, 1969, **1**, 597–604.
16. R. D. Foord, 'Cephaloridine, cephalothin and the kidney.' *J. Antimicrob. Chemother.*, 1975, **1**, 119–133.
17. R. D. Foord, 'Aspects of clinical trials with ceftazidime worldwide.' *Am. J. Med*, 1985, **79**, Suppl. 2A, 110–113.
18. J. Klastersky, C. Hensgens and L. Debusscher, 'Empiric therapy for cancer patients: comparative study of ticarcillin-tobramycin, ticarcillin-cephalothin, and cephalothin-tobramycin.' *Antimicrob. Agents Chemother.*, 1975, **7**, 640–645.

19. J. C. Wade, C. R. Smith, B. G. Petty *et al.*, 'Cephalothin plus an aminoglycoside is more nephrotoxic than methicillin plus an aminoglycoside.' *Lancet*, 1978, **2**, 604–606.

20. S. C. Schimpff, H. Gaya, J. Klastersky *et al.*, 'Three antibiotic regimens in the treatment of infection in febrile granulocytopenic patients with cancer.' EORTC International Antimicrobial Therapy Project Group, *J. Infect. Dis.*, 1978, **137**, 14–29.

21. F. Silverblatt, 'Pathogenesis of nephrotoxicity of cephalosporins and aminoglycosides: a review of current concepts.' *Rev. Infect. Dis.*, 1982, **4**, S360–S365.

22. K. J. Child and M. G. Dodds, 'Nephron transport and renal tubular effects of cephaloridine in animals.' *Br. J. Pharmacol.*, 1967, **30**, 354–370.

23. B. M. Tune, 'Effect of organic acid transport inhibitors on renal cortical uptake and proximal tubular toxicity of cephaloridine.' *J. Pharmacol. Exp. Ther.*, 1972, **181**, 250–256.

24. B. M. Tune, 'Relationship between the transport and toxicity of cephalosporins in the kidney.' *J. Infect. Dis.*, 1975, **132**, 189–194.

25. J. S. Welles, W. R. Gibson, P. N. Harris *et al.*, 'Toxicity, distribution, and excretion of cephaloridine in laboratory animals.' *Antimicrob. Agents Chemother.*, 1965, **5**, 863–869.

26. J. S. Wold, J. S. Welles, N. V. Owen *et al.*, 'Toxicologic evaluation of cefamandole naftate in laboratory animals.' *J. Infect. Dis.*, 1978, **137**, S51–S59.

27. P. Dellinger, T. Murphy, V. Pinn *et al.*, 'Protective effect of cephalothin against gentamicin-induced nephrotoxicity in rats.' *Antimicrob. Agents Chemother.*, 1976, **9**, 172–178.

28. R. D. Adelman, in 'Nephrotoxicity, Ototoxicity of Drugs,' ed. J.-P. Fillastre, Editions INSERM, Rouen, 1982, pp. 213–224.

29. B. M. Tune, K. Y. Wu and R. L. Kempson, 'Inhibition of transport and prevention of toxicity of cephaloridine in the kidney. Dose-responsiveness of the rabbit and the guinea pig to probenecid.' *J. Pharmacol. Exp. Ther.*, 1977, **202**, 466–471.

30. J. S. Welles, in 'Cephalosporins and Penicillins. Chemistry and Biology,' ed. E. H. Flynn, Academic Press, New York, 1972, pp. 581–608.

31. R. L. Perkins, M. A. Apicella, I. S. Lee *et al.*, 'Cephaloridine and cephalothin: comparative studies of potential nephrotoxicity.' *J. Lab. Clin. Med.*, 1968, **71**, 75–84.

32. D. A. Pasino, K. Miura, R. S. Goldstein *et al.*, 'Cephaloridine nephrotoxicity: strain and sex differences in mice.' *Fundam. Appl. Toxicol.*, 1985, **5**, 1153–1160.

33. K. J. Child and M. G. Dodds, 'Mechanism of urinary excretion of cephaloridine and its effects on renal function in animals.' *Br. J. Pharmacol.*, 1966, **26**, 108–119.

34. B. M. Tune and M. Fernholt, 'Relationship between cephaloridine and *p*-aminohippurate transport in the kidney.' *Am. J. Physiol.*, 1973, **225**, 1114–1117.

35. B. M. Tune, M. Fernholt and A. Schwartz, 'Mechanism of cephaloridine transport in the kidney.' *J. Pharmacol. Exp. Ther.*, 1974, **191**, 311–317.

36. B. M. Tune, M. C. Browning, C. Y. Hsu *et al.*, 'Prevention of cephalosporin nephrotoxicity by other cephalosporins and by penicillins without significant inhibition of renal cortical uptake.' *J. Infect. Dis.*, 1982, **145**, 174–180.

37. B. M. Tune, in 'Nephrotoxic Mechanisms of Drugs and Environmental Toxins,' ed. G. Porter, Plenum, New York, 1982, pp. 151–164.

38. B. M. Tune, K. Y. Wu, D. Fravert *et al.*, 'Effect of cephaloridine on respiration by renal cortical mitochondria.' *J. Pharmacol. Exp. Ther.*, 1979, **210**, 98–100.

39. J. S. Wold and S. A. Turnipseed, 'The effect of renal cation transport inhibitors on the *in vivo* and *in vitro* accumulation and efflux of cephaloridine.' *Life Sci.*, 1980, **27**, 2559–2564.

40. C. H. Kyo, K. Maita, S. D. Sleight *et al.*, 'Lipid peroxidation: a possible mechanism of cephalosporin-induced nephrotoxicity.' *Toxicol. Appl. Pharmacol.*, **67**, 78–88.

41. C. Cojocel, J. Hannemann and K. Baumann, 'Cephaloridine-induced lipid peroxidation by reactive oxygen species as a possible mechanism of cephaloridine nephrotoxicity.' *Biochim. Biophys. Acta*, 1985, **834**, 402–410.

42. B. M. Tune, R. K. Sibley and C. Y. Hsu, 'The mitochondrial respiratory toxicity of cephalosporin antibiotics. An inhibitory effect on substrate uptake.' *J. Pharmacol. Exp. Ther.*, 1988, **245**, 1054–1059.

43. B. M. Tune, D. Fravert and C.-Y. Hsu, 'Thienamycin nephrotoxicity. Mitochondrial injury and oxidative effects of imipenem in the rabbit kidney.' *Biochem. Pharmacol.*, 1989, **38**, 3779–3783.

44. B. M. Tune, D. Fravert and C. Y. Hsu, 'Oxidative and mitochondrial toxic effects of cephalosporin antibiotics in the kidney. A comparative study of cephaloridine and cephaloglycin.' *Biochem. Pharmacol.*, 1989, **38**, 795–802.

45. B. M. Tune and C. Y. Hsu, 'The renal mitochondrial toxicity of cephalosporins: specificity of the effect on anionic substrate uptake.' *J. Pharmacol. Exp. Ther.*, 1990, **252**, 65–69.

46. R. S. Goldstein, L. R. Contardi, D. A. Pasino *et al.*, 'Mechanisms mediating cephaloridine inhibition of gluconeogenesis.' *Toxicol. Appl. Pharmacol.*, 1987, **87**, 297–305.

47. C. Cojocel, U. Göttsche, K. L. Tölle *et al.*, 'Nephrotoxic potential of first-, second-, and third-generation cephalosporins.' *Arch. Toxicol.*, 1988, **62**, 458–464.

48. B. M. Tune, K. Y. Wu, D. Longerbeam *et al.*, in 'Proceedings of the Seventh International Congress of Nephrology,' ed. G. Lemieux, Karger, Basel, 1978, pp. 279–287.

49. G. F. Rush and G. D. Ponsler, 'Cephaloridine-induced biochemical changes and cytotoxicity in suspensions of rabbit isolated proximal tubules.' *Toxicol. Appl. Pharmacol.*, 1991, **109**, 314–326.

50. G. F. Rush, R. A. Heim, G. D. Ponsler *et al.*, 'Cephaloridine-induced renal pathological and biochemical changes in female rabbits and isolated proximal tubules in suspension.' *Toxicol. Pathol.*, 1992, **20**, 155–168.

51. P. D. Williams, D. A. Laska, L. K. Tay *et al.*, 'Comparative toxicities of cephalosporin antibiotics in a rabbit kidney cell line (LLC-RK1).' *Antimicrob. Agents Chemother.*, 1988, **32**, 314–318.

52. G. Gstraunthaler, D. Steinmassl and W. Pfaller, 'Renal cell cultures: a tool for studying tubular function and nephrotoxicity.' *Toxicol. Lett.*, 1990, **53**, 1–7.

53. P. D. Williams, M. J. Hitchcock and G. H. Hottendorf, 'Effect of cephalosporins on organic ion transport in renal membrane vesicles from rat and rabbit kidney cortex.' *Res. Commun. Chem. Pathol. Pharmacol.*, 1985, **47**, 357–369.

54. B. M. Tune and C. Y. Hsu, 'The renal mitochondrial toxicity of beta-lactam antibiotics: *in vitro* effects of cephaloglycin and imipenem.' *J. Am. Soc. Nephrol.*, 1990, **1**, 815–821.

55. B. B. Williams, R. D. Cushing and A. M. Lerner, 'Letter. Severe combined nephrotoxicity of BL-P1654 and gentamicin.' *J. Infect. Dis.*, 1974, **130**, 694–695.

56. M. C. Browning, C. Y. Hsu, P. L. Wang *et al.*, 'Interaction of ischemic and antibiotic-induced injury in the rabbit kidney.' *J. Infect. Dis.*, 1983, **147**, 341–351.

57. B. M. Tune, C. Y. Hsu and D. Fravert, 'Mechanisms of the bacterial endotoxin-cephalosporin toxic synergy and the protective action of saline in the rabbit kidney.' *J. Pharmacol. Exp. Ther.*, 1988, **244**, 520–525.

58. J. S. Wold, in 'Toxicology of the Kidney,' ed. J. B. Hook, Raven Press, New York, 1981, pp. 251–266.

59. P. L. Wang, D. J. Prime, C. Y. Hsu *et al.*, 'Effects of ureteral obstruction on the toxicity of cephalosporins in the rabbit kidney.' *J. Infect. Dis.*, 1982, **145**, 574–581.

60. S. Hashimoto, H. Ishii, T. Fujimoto *et al.*, 'Acute and subacute toxicity studies of AC-1370 sodium in mice and rats.' *Toxicol. Lett.*, 1984, **23**, 135–140.

61. B. M. Tune and M. B. Burg, 'Glucose transport by proximal renal tubules.' *Am. J. Physiol.*, 1971, **221**, 580–585.

62. D. T. Plummer and E. O. Ngaha, in 'Nephrotoxicity. Interaction of Drugs with Membrane Systems Mitochondria-lysosomes,' ed. J.-P. Fillastre, Masson, New York, 1978, pp. 175–191.

63. G. J. Kaloyanides, in 'Diseases of the Kidney,' 5th edn., eds. R. Schrier and C. Gottschalk, Little, Brown, Boston, MA, 1992, pp. 1131–1163.

64. D. N. Gilbert, D. C. Houghton, W. M. Bennett *et al.*, 'Reversibility of gentamicin nephrotoxicity in rats: recovery during continuous drug administration.' *Proc. Soc. Exp. Biol. Med.*, 1979, **160**, 99–103.

65. E. D. Wachsmuth, 'Adaptation to nephrotoxic effects of cephaloridine in subacute rat toxicity studies.' *Toxicol. Appl. Pharmacol.*, 1982, **63**, 446–460.

66. R. J. McMurtry and J. R. Mitchell, 'Renal and hepatic necrosis after metabolic activation of 2-substituted furans and thiophenes, including furosemide and cephaloridine.' *Toxicol. Appl. Pharmacol.*, 1977, **42**, 285–300.

67. M. C. Browning and B. M. Tune, 'Reactivity and binding of beta-lactam antibiotics in rabbit renal cortex.' *J. Pharmacol. Exp. Ther.*, 1983, **226**, 640–644.

68. B. M. Tune and C. Y. Hsu, 'Toxicity of cephaloridine to carnitine transport and fatty acid metabolism in rabbit renal cortical mitochondria: structure-activity relationships.' *J. Pharmacol. Exp. Ther.*, 1994, **270**, 873–880.

69. J. M. Brogard, F. Comte and M. Pinget, 'Pharmacokinetics of cephalosporin antibiotics.' *Antibiot. Chemother.*, 1978, **25**, 123–162.

70. B. M. Tune, M. B. Burg and C. S. Patlak, 'Characteristics of *p*-aminohippurate transport in proximal renal tubules.' *Am. J. Physiol.*, 1969, **217**, 1057–1063.

71. K. J. Ullrich, G. Rumrich and S. Klöss, 'Contraluminal organic anion transport in the proximal renal tubule: V. Interaction with sulfamoyl- and phenoxy diuretics, and with β-lactam antibiotics.' *Kidney Int.*, 1989, **36**, 78–88.

72. K. J. Ullrich, G. Rumrich, C. David *et al.*, 'Bisubstrates: substances that interact with both, renal contraluminal organic anion and organic cation transport systems.' *Pflügers Arch.*, 1993, **425**, 300–312.

73. J. S. Wold, R. R. Joost and N. V. Owen, 'Nephrotoxicity of cephaloridine in newborn rabbits: role of the renal anionic transport system.' *J. Pharmacol. Exp. Ther.*, 1977, **201**, 778–785.

74. B. M. Tune, K. Y. Wu, D. F. Longerbeam *et al.*, 'Transport and toxicity of cephaloridine in the kidney. Effect of furosemide, *p*-aminohippurate and saline diuresis.' *J. Pharmacol. Exp. Ther.*, 1977, **202**, 472–478.

75. B. M. Tune and D. Fravert, 'Mechanisms of cephalosporin nephrotoxicity: a comparison of cephaloridine and cephaloglycin.' *Kidney Int.*, 1980, **18**, 591–600.

76. C. C. Lee, E. B. Herr and R. C. Anderson, 'Pharmacological and toxicological studies on cephalothin.' *Clin. Med.*, 1963, **70**, 1123–1138.

77. B. M. Tune, 'The nephrotoxicity of cephalosporin antibiotics. Structure–activity relationships.' *Commun. Toxicol.*, 1986, **1**, 145–170.

78. J. S. Wold, S. A. Turnipseed and B. L. Miller, 'The effect of renal cation transport on cephaloridine nephrotoxicity.' *Toxicol. Appl. Pharmacol.*, 1979, **47**, 115–122.

79. K. I. Inui, T. Okano, M. Takano *et al.*, 'Carrier-mediated transport of amino-cephalosporins by brush border membrane vesicles isolated from rat kidney cortex.' *Biochem. Pharmacol.*, 1983, **32**, 621–626.

80. B. Tune, 'Effects of L-carnitine on the renal tubular transport of cephaloridine.' *Biochem. Pharmacol.*, 1995, **50**, 562–564.

81. C. Carbon, F. Dromer, N. Brion *et al.*, 'Renal disposition of ceftazidime illustrated by interferences with probenecid, furosemide, and indomethacin in rabbits.' *Antimicrob. Agents Chemother.*, 1984, **26**, 373–377.

82. T. Yamana and A. Tsuji, 'Comparative stability of cephalosporins in aqueous solution: kinetics and mechanisms of degradation.' *J. Pharm. Sci.*, 1976, **65**, 1563–1574.

83. B. M. Tune, C. H. Kuo, J. B. Hook *et al.*, 'Effects of piperonyl butoxide on cephalosporin nephrotoxicity in the rabbit. An effect on cephaloridine transport.' *J. Pharmacol. Exp. Ther.*, 1983, **224**, 520–524.

84. C. H. Kuo, B. M. Tune and J. B. Hook, 'Effect of piperonyl butoxide on organic anion and cation transport in rabbit kidneys.' *Proc. Soc. Exp. Biol. Med.*, 1983, **174**, 165–171.

85. C. Cojocel, K. H. Laeschke, G. Inselmann *et al.*, 'Inhibition of cephaloridine-induced lipid peroxidation.' *Toxicology*, 1985, **35**, 295–305.

86. R. S. Goldstein, D. A. Pasino, W. R. Hewitt *et al.*, 'Biochemical mechanisms of cephaloridine nephrotoxicity: time and concentration dependence of peroxidative injury.' *Toxicol. Appl. Pharmacol.*, 1986, **83**, 261–270.

87. D. J. Tipper and J. L. Strominger, 'Mechanism of action of penicillins: a proposal based on their structural similarity to acyl-D-alanyl-D-alanine.' *Proc. Natl. Acad. Sci. USA*, 1965, **541**, 1133–1141.

88. J. M. Indelicato, T. T. Norvilas, R. R. Pfeiffer *et al.*, 'Substituent effects upon the base hydrolysis of penicillins and cephalosporins. Competitive intramolecular nucleophilic amino attack in cephalosporins.' *J. Med. Chem.*, 1974, **17**, 523–527.

89. T. Yamana, A. Tsuji, K. Kanayama *et al.*, 'Comparative stabilities of cephalosporins in aqueous solution.' *J. Antibiot. (Tokyo)*, 1974, **27**, 1000–1002.

90. J. M. Indelicato, A. Dinner, L. R. Peters *et al.*, 'Hydrolysis of 3-chloro-3-cephems. Intramolecular nucleophilic attack in cefaclor.' *J. Med. Chem.*, 1977, **20**, 961–963.

91. J. M. Indelicato and C. E. Pasini, 'The acylating potential of γ-lactam antibacterials: base hydrolysis of bicyclic pyrazolidinones.' *J. Med. Chem.*, 1988, **31**, 1227–1230.

92. C. E. Pasini and J. M. Indelicato, 'Pharmaceutical properties of loracarbef: the remarkable solution stability of an oral 1-carba-1-dethiacephalosporin antibiotic.' *Pharm. Res.*, 1992, **9**, 250–254.

93. K. Capel-Edwards and D. A. H. Pratt, 'Renal tolerance of ceftazidime in animals.' *J. Antimicrob. Chemother.*, 1981, **8**, Suppl. B, 241–245.

94. Anonymous, 'Ceftazidime combined with a short or long course of amikacin for empirical therapy of gram-negative bacteremia in cancer patients with granulocytopenia.' The EORTC International Antimicrobial Therapy Cooperative Group, *N. Engl. J. Med.*, 1987, **317**, 1692–1698.

95. B. M. Tune and C. Y. Hsu, 'Toxicity of cephalosporins to fatty acid metabolism in rabbit renal cortical mitochondria.' *Biochem. Pharmacol.*, 1995, **49**, 727–734.

96. R. S. Goldstein, P. F. Smith, J. B. Tarloff *et al.*, 'Biochemical mechanisms of cephaloridine nephrotoxicity.' *Life Sci.*, 1988, **42**, 1809–1816.

97. C. Cojocel, W. Kramer and D. Mayer, 'Depletion of cytochrome P-450 and alterations of activities of drug metabolizing enzymes induced by cephaloridine in the rat kidney cortex.' *Biochem. Pharmacol.*, 1988, **37**, 3781–3785.

98. W. Kramer, C. Cojocel and D. Mayer, 'Specific alterations of rat renal microsomal proteins induced by cephaloridine.' *Biochem. Pharmacol.*, 1988, **37**, 4135–4140.

99. G. A. Currie, P. J. Little and S. J. McDonald, 'The localisation of cephaloridine and nitrofurantoin in the kidney.' *Nephron*, 1966, **3**, 282–288.

100. M. A. Venkatachalam, D. B. Bernard, J. F. Donohoe *et al.*, 'Ischemic damage and repair in the rat proximal tubule: differences among the S1, S2, and S3 segments.' *Kidney Int.*, 1978, **14**, 31–49.

101. B. F. Trump and R. E. Bulger, 'Studies of cellular injury in isolated flounder tubules. IV. Electron microscopic observations of changes during the phase of altered homeostasis in tubules treated with cyanide.' *Lab. Invest.*, 1968, **18**, 731–739.

102. J. P. Bendirdjian, D. J. Prime, M. C. Browning *et al.*, in 'Nephrotoxicity, Ototoxicity of Drugs,' ed. J. P. Fillastre, Editions INSERM, Rouen, 1982, pp. 303–319.

103. P. J. Sausen, A. A. Elfarra and A. J. Cooley, 'Methimizole protection of rats against chemically induced kidney damage *in vivo*.' *J. Pharmacol. Exp. Ther.*, 1992, **260**, 393–401.

104. L. Stryer, in 'Biochemistry,' ed. L. Stryer, Freeman, New York, 1988, pp. 373–397.

105. M. J. Weidemann and H. A. Krebs, 'The fuel of respiration in rat kidney cortex.' *Biochem. J.*, 1969, **112**, 149–166.

106. B. Tune, 'Effects of nephrotoxic beta-lactam antibiotics on the mitochondrial metabolism of monocarboxylic substrates.' *J. Pharmacol. Exp. Ther.*, 1995, **274**, 194–199.

107. L. Stryer, in 'Biochemistry,' ed. L. Stryer, Freeman, New York, 1988, pp. 469–494.

108. K. L. Klein, M. S. Wang, S. Torikai *et al.*, 'Substrate oxidation by isolated single nephron segments of the rat.' *Kidney Int.*, 1981, **20**, 29–35.

109. K. Y. Jung, S. Uchida and H. Endou, 'Nephrotoxicity assessment by measuring cellular ATP content. I. Substrate specificities in the maintenance of ATP content in isolated nephron segments.' *Toxicol. Appl. Pharmacol.*, 1989, **100**, 369–382.

110. R. S. Balaban and L. J. Mandel, 'Metabolic substrate utilization by rabbit proximal tubule. An NADH fluorescence study.' *Am. J. Physiol.*, 1988, **254**, F407–F416.

111. G. Paradies and S. Papa, 'Substrate regulation of the pyruvate-transporting system in rat liver mitochondria.' *FEBS Lett.*, 1976, **62**, 318–321.

112. G. Paradies and S. Papa, 'On the kinetics and substrate specificity of the pyruvate translocator in rat liver mitochondria.' *Biochim. Biophys. Acta*, 1977, **462**, 333–346.

113. P. V. Vignais, G. Brandolin, F. Boulay *et al.*, in 'Anion Carriers of Mitochondrial Membranes,' eds. A. Azzi, K. A. Nalecz, M. J. Nalecz *et al.*, Springer, Berlin, 1989, pp. 133–146.

114. A. Tzagoloff, in 'Mitochondria,' ed. A. Tzagoloff, Plenum, New York, 1982, pp. 39–60.

115. B. Alberts, D. Bray, J. Lewis *et al.*, in 'Molecular Biology of the Cell,' Garland, New York, 1989, pp. 342–404.

116. D. Dolislager, D. Fravert and B. M. Tune, 'Interaction of aminoglycosides and cephaloridine in the rabbit kidney.' *Res. Commun. Chem. Pathol. Pharmacol.*, 1979, **26**, 13–23.

117. H. Nikaido, W. Liu and E. Y. Rosenberg, 'Outer membrane permeability and beta-lactamase stability of dipolar ionic cephalosporins containing methoxyamino substituents.' *Anitmicrob. Agents Chemother.*, 1990, **34**, 337–342.

118. K. Matsubayashi, S. Shintani, K. Yoshida *et al.*, 'Nonlinear pharmacokinetics of DQ-2556, a new 3-quaternary ammonium cephalosporin antibiotic, in rats caused by non-Michaelis–Menten type, dose-dependent renal clearance.' *J. Pharm. Sci.*, 1994, **83**, 186–192.

119. M. Kato, M. Yoshida, H. Shimada *et al.*, 'Nephrotoxicity of a new cephalosporin, DQ-2556, in rats.' *Fundam. Appl. Toxicol.*, 1992, **18**, 532–539.

120. M. C. Browning, 'Effect of verapamil on cephaloridine nephrotoxicity in the rabbit.' *Toxicol. Appl. Pharmacol.*, 1990, **103**, 383–388.

121. P. P. Sokol, 'Effect of DQ-2556, a new cephalosporin, on organic ion transport in renal plasma membrane vesicles from the dog, rabbit and rat.' *J. Pharmacol. Exp. Ther.*, 1992, **255**, 436–441.

122. H. Kropp, J. G. Sundelof, R. Hajdu *et al.*, 'Metabolism of thienamycin and related carbapenem antibiotics by the renal dipeptidase, dehydropeptidase.' *Antimicrob. Agents Chemother.*, 1982, **22**, 62–70.

123. J. S. Kahan, F. M. Kahan, R. Goegelman *et al.*, 'Thienamycin, a new beta-lactam antibiotic. I. Discovery, taxonomy, isolation and physical properties.' *J. Antibiot. (Tokyo)*, 1978, **32**, 1–11.

124. H. Kropp, J. G. Sundelof, J. S. Kahan *et al.*, 'MK0787 (*N*-formimidoyl thienamycin): evaluation of *in vitro* and *in vivo* activities.' *Antimicrob. Agents Chemother.*, 1980, **17**, 993–1000.

125. R. Battaglia, M. S. Benedetti, E. Frigerio *et al.*, 'The disposition and urinary metabolism of ^{14}C-labelled FCE 22891, a pro-drug of FCE 22101, in animals.' *J. Antimicrob. Chemother.*, 1990, **25**, 133–139.

126. S. R. Norrby, L. A. Burman, D. Sassella *et al.*, 'Pharmacokinetics in healthy subjects of FCE 22101 and its acetoxymethyl ester, FCE 22891: effect of co-administration of imipenem/cilastatin on the renal metabolism of FCE 22101.' *J. Antimicrob. Chemother.*, 1990, **25**, 371–383.

127. M. Hikida, K. Kawashima, M. Yoshida *et al.*, 'Inactivation of new carbapenem antibiotics by dehydropeptidase-I from porcine and human renal cortex.' *J. Antimicrob. Chemother.*, 1992, **30**, 129–134.

128. N. E. Allen, J. N. Hobbs Jr., D. A. Preston *et al.*, 'Antibacterial properties of the bicyclic pyrazolidinones.' *J. Antibiot. (Tokyo)*, 1990, **43**, 92–99.

129. J. V. Møller and M. I. Sheikh, 'Renal organic anion transport system: pharmacological, physiological, and biochemical aspects.' *Pharmacol. Rev.*, 1982, **34**, 315–358.

130. I. S. Yang, J. M. Goldinger, S. K. Hong *et al.*, 'Preparation of basolateral membranes that transport *p*-aminohippurate from primary cultures of rabbit kidney proximal tubule cells.' *J. Cell. Physiol.*, 1988, **135**, 481–487.

131. J. J. Mertens, J. G. Weijnen, W. J. van Doorn *et al.*, 'Differential toxicity as a result of apical and basolateral treatment of LLC-PK1 monolayers with *S*-(1,2,3,4,4-pentachlorobutadienyl)glutathione and *N*-acetyl-(1,2,3,4,4-pentachlorobutadienyl)-L-cysteine.' *Chem. Biol. Interact.*, 1988, **65**, 283–93.

132. T. K. Ip, P. Aebischer and P. M. Galletti, 'Cellular control of membrane permeability. Implications for a bioartificial renal tubule.' *ASAIO Trans.*, 1988, **34**, 351–355.

133. R. Hori, M. Okamura, A. Takayama *et al.*, 'Transport of organic anion in OK kidney epithelial cell line.' *Am. J. Physiol.*, 1993, **264**, F975–F980.

134. M. Takano, K. Hirozane, M. Okamura *et al.*, '*p*-Aminohippurate transport in apical and basolateral membranes of the OK kidney epithelial cells.' *J. Pharmacol. Exp. Ther.*, 1994, **269**, 970–975.

135. M. J. Ryan, G. Johnson, J. Kirk *et al.*, 'HK-2: an immortalized proximal tubule epithelial cell line from normal adult human kidney.' *Kidney Int.*, 1994, **45**, 48–57.

136. M. P. Harrison, D. V. Jones, R. J. Pickford *et al.*, 'β-Hydroxybutyrate: a urinary marker of imipenem-induced nephrotoxicity in the cynomolgus monkey detected by high field ^{1}H NMR spectroscopy.' *Biochem. Pharmacol.*, 1991, **41**, 2045–2049.

137. T. Schutz, R. González-Méndez, D. C. Nabseth *et al.*, 'A study of nephrotoxin-induced acute tubular necrosis with ^{31}P magnetic resonance spectroscopy.' *Magn. Reson. Med.*, 1991, **18**, 159–168.

138. L. B. Murgatroyd, R. J. Pickford, I. K. Smith *et al.*, '^{1}H-NMR spectroscopy as a means of monitoring nephrotoxicity as exemplified by studies of cephaloridine.' *Hum. Exp. Toxicol.*, 1992, **11**, 35–41.

139. M. L. Anthony, K. P. R. Gartland, C. R. Beddell *et al.*, 'Cephaloridine-induced nephrotoxicity in the Fischer 344 rat: proton NMR spectroscopic studies of urine and plasma in relation to conventional clinical chemical and histopathological assessments of nephronal damage.' *Arch. Toxicol.*, 1992, **66**, 525–537.

140. G. T. A. McEwan, C. D. A. Brown, B. H. Hirst *et al.*, 'Characterisation of volume-activated ion transport across epithelial monolayers of human intestinal T84 cells.' *Pflügers Arch., Eur. J. Physiol.* 1993, **423**, 213–220.

141. D. T. Thwaites, C. D. Brown, B. H. Hirst *et al.*, 'Transepithelial glycylsarcosine transport in intestinal Caco-2 cells mediated by expression of H$^+$-coupled carriers at both apical and basal membranes.' *J. Biol. Chem.*, 1993, **268**, 7640–7642.

142. J. H. Joshi, K. A. Newman, B. W. Brown *et al.*, 'Double beta-lactam regimen compared to an aminoglycoside/beta-lactam regimen as empiric antibiotic therapy for febrile granulocytopenic cancer patients.' *Support Care Cancer*, 1993, **1**, 186–194.

143. D. J. Winston, W. G. Ho, D. A. Bruckner *et al.*, 'Beta-lactam antibiotic therapy in febrile granulocytopenic patients. A randomized trial comparing cefoperazone plus piperacillin, ceftazidime plus piperacillin, and imipenem alone.' *Ann. Intern. Med.*, 1991, **115**, 849–859.

144. N. G. Levinsky, D. B. Bernard and P. A. Johnston, in 'Acute Renal Failure,' eds. B. M. Brenner and J. M. Lazarus, Saunders, Philadelphia, PA, 1983, pp. 712–722.

145. W. O. Harrison, F. J. Silverblatt and M. Turck, 'Gentamicin nephrotoxicity: failure of three cephalosporins to potentiate injury in rats.' *Antimicrob. Agents Chemother.*, 1975, **8**, 209–215.

146. F. C. Luft, V. Patel, M. N. Yum *et al.*, 'Nephrotoxicity of cephalosporin-gentamicin combinations in rats.' *Antimicrob. Agents Chemother.*, 1976, **9**, 831–839.

147. R. Bloch, F. C. Luft, L. I. Rankin *et al.*, 'Protection from gentamicin nephrotoxicity by cephalothin and carbenicillin.' *Antimocrob. Agents Chemother.*, 1979, **15**, 46–49.

148. B. M. Tune, V. M. Reznik and S. A. Mendoza, in 'Pediatric Nephrology,' 3rd edn., eds. M. A. Holliday, T. M. Barratt and E. A. Avner, Williams and Wilkins, Baltimore, MD, 1994, pp. 1212–1226.

149. M. G. Dodds and R. D. Foord, 'Enhancement by potent diuretics of renal tubular necrosis induced by cephaloridine.' *Br. J. Pharmacol.*, 1970, **40**, 227–236.

150. A. L. Linton, R. R. Bailey and D. I. Turnbull, 'Relative nephrotoxicity of cephalosporin antibiotics in an animal model.' *Can. Med. Assoc. J.*, 1972, **107**, 414–416.

151. R. R. Bailey, R. Natale, D. I. Turnbull *et al.*, 'Protective effect of furosemide in acute tubular necrosis and acute renal failure.' *Clin. Sci.*, 1973, **45**, 1–17.

152. J. T. Smith and C. S. Lewis, 'Mechanisms of antimicrobial resistance and implications for epidemiology.' *Vet. Microbiol.*, 1993, **35**, 233–242.

153. B. G. Spratt, 'Resistance to antibiotics mediated by target alterations.' *Science*, 1994, **264**, 388–393.

154. B. M. Tune and C. Y. Hsu, in 'Prevention in Nephrology,' eds. M. E. De Broe and G. A. Verpooten, Kluwer, Dordrecht, 1991, pp. 39–49.

155. E. H. Barany, 'The liver-like anion transport system in rabbit kidney, uvea and chroroid plexus. I. Selectivity of some inhibitors, direction of transport, possible physiological substrates.' *Acta Physiol. Scand.*, 1973, **88**, 412–429.

156. S. E. Schliamser, O. Cars and S. R. Norrby, 'Neurotoxicity of beta-lactam antibiotics: predisposing factors and pathogenesis.' *J. Antimicrob. Chemother.*, 1991, **27**, 405–425.

157. D. J. Prime and B. M. Tune, in 'Pediatrics Update,' ed. A. J. Moss, Elsevier/North-Holland, New York, 1981, pp. 265–285.

158. B. M. Tune, in 'Renal Disposition and Nephrotoxicity of Xenobiotics,' eds. M. W. Anders, W. Dekant, D. Henschler *et al.*, Academic Press, Orlando, FL 1993, pp. 249–267.

159. C. R. Ross and P. D. Holohan, 'Transport of organic anions and cations in isolated renal plasma membranes.' *Annu. Rev. Pharmacol. Toxicol.*, 1983, **23**, 65–85.

7.25
Amphotericin B

JOSE F. BERNARDO and ROBERT A. BRANCH
University of Pittsburgh, PA, USA

7.25.1 INTRODUCTION

During the past two decades systemic mycoses have been increasingly prominent as causes of disease, particularly in severely ill and immunocompromised patients. This is the result of an increase in the frequency of community acquired fungal infections and the increased incidence of opportunistic and nosocomial fungal infections, particularly in patients with decreased immune response due to AIDS, organ transplants, cancer, or prosthetic devices.[1] *Candida* species now constitute the fourth most common organism isolated from hospitalized patients.[2] The incidence of disseminated fungal infections in patients with hematologic malignancies is 20–40%, with a fatality rate as high as 70%.[3] In a report from The National Cancer Institute, it was estimated that 43% of patients dying from acute leukemia had systemic fungal infection at autopsy.[4] *Candida albicans* is the most common causative agent of oral candidiasis in patients with AIDS, while cryptococcosis, which is diagnosed in 5–9% of cases, is associated with the highest mortality rate.[5] These trends are likely to increase rather than decrease in the foreseeable future. Thus, management of a systemic fungal infection will remain an important consideration in the treatment of severely ill and immunosuppressed patients.[6]

Since its introduction in 1956, and despite the availability of newer agents, amphotericin B (AmB) continues as the most effective systemic therapy for serious fungal infections.[7] This agent is not only useful for fungal infections but is also a recognized second-choice therapy for cutaneous, mucocutaneous, and visceral leishmaniasis.[8] Mucocutaneous leishmaniasis is a polymorphic disease of the skin and mucous membranes that can cause late sequelae due to chronic necrotizing lesions involving nasopharyngeal tissues, and is prevalent in Asia, Africa, and South America. Visceral leishmaniasis, a chronic systemic disease, is usually fatal if left untreated. Formerly considered a disease of the developing world, it is becoming more frequently diagnosed in patients with AIDS.[9] In both conditions, patients not responding to antimonial therapy can be successfully treated with AmB. Therefore, management of these parasitic protozoa diseases constitutes additional indications for AmB administration.

Despite its clinical efficacy, the usefulness of AmB, is compromised by the frequent occurrence of several infusion-related and dose-dependent adverse effects that often necessitate changes in, or premature discontinuation of, therapy. Infusion-related changes include fever, chills, nausea, vomiting, anorexia, headache, bronchospasm, hypotension, and anaphylaxis, whereas dose-related events include anaemia, cardiotoxicity, and renal failure with potassium and magnesium wasting. The most limiting adverse effect in clinical practice, however, is nephrotoxicity. Thus, despite being the most efficacious antifungal drug available, AmB's narrow therapeutic index continues to be its limiting factor.[10]

AmB is a member of the polyene macrolide class of antibiotics. The molecule consists of a large macrolide lactone ring and a sugar moiety: one side of the ring is composed of a rigid lipophilic chain of seven conjugated double bonds, and the opposite side of the ring contains a similar number of hydroxylated carbon atoms (Figure 1). Thus, the molecule is amphipatic, and this feature of its structure is believed to be important in its mechanism of action.[11] The major action of AmB is believed to be on the cell membrane of fungal and mammalian cells. The drug binds to sterols in the cell membrane and induces formation of aqueous pores. This results in impairment of barrier function, leading to the leakage of potassium, change in intracellular concentration of free calcium and subsequently metabolic disruption and cell death.[11,12]

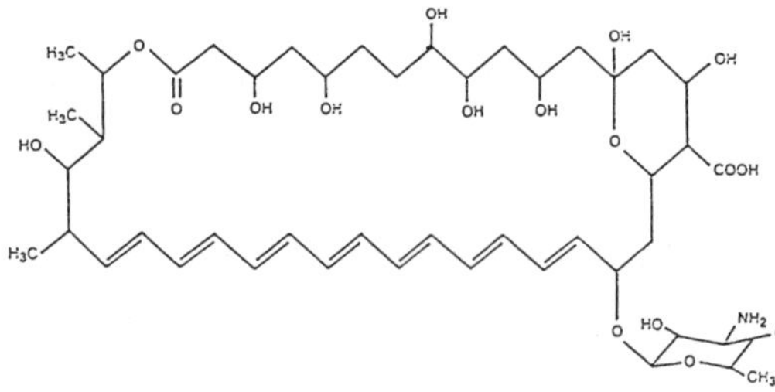

Figure 1 Chemical structure of amphotericin B.

The cellular events that follow this membrane effect are complex and depend on a variety of factors, such as the growth phase of the cells, the dose, and the mode of AmB administration.[13] Some studies suggest that cell lethality is not simply a consequence of changes in permeability of membranes, and that formation of active oxygen species may play a role in the lytic or lethal actions of AmB.[14,15]

7.25.2 CLINICAL MANIFESTATIONS OF NEPHROTOXICITY

The major limitation associated with AmB therapy is its potential to induce disturbances in both glomerular and tubular renal function.[10] The glomerular changes are mostly functional and dependent on hemodynamic changes at the level of renal microcirculation, whereas the tubular changes are thought to be related to direct binding of AmB to cell membranes, that explain abnormalities in water, potassium, and magnesium metabolism. The most alarming clinical manifestation is azotemia; however, renal tubular acidosis, decreased concentrating ability of the kidney, and electrolyte disturbances such as urinary potassium/magnesium wasting leading to hypokalemia and hypomagnesemia are also important manifestations and may dominate the clinical picture (Table 1).

Table 1 Presenting features of renal dysfunction associated with amphotericin B-induced nephrotoxicity.

Glomerular	Increased serum creatinine
	Decreased creatinine clearance
	Increased BUN
Tubulointerstitial	Hypostenuria
	Renal tubular acidosis
	Hypokalemia
	Hypomagnesemia

7.25.2.1 Azotemia

The incidence of AmB-induced renal dysfunction reported in the literature ranges between 50% and 90%. The high variability can be explained by several factors, among which are the case-definition of nephrotoxicity used, the dose of AmB, the coadministration of other nephrotoxic agents, age and other proposed risk factors. Shortly after its introduction, a survey of 56 patients treated between 1956 and 1963, demonstrated that 83% of patients showed increased levels of serum creatinine.[16] Studies since 1980 indicate that in almost every patient treated with AmB, the glomerular filtration rate (GFR) falls approximately 40% within the first 2–3 weeks of therapy and then stabilizes at around 20–60% of the baseline rate throughout the subsequent course of treatment.[17,18] The clinical features are summarized in Table 2. In general, azotemia is a transient event occurring during therapy and renal function usually returns to the baseline values after discontinuation of the drug. Whenever necessary, the temporal discontinuation of therapy for a few days usually allows renal function to recover enough to permit administration of the full course of therapy. In rare cases, however, permanent renal damage persists after cessation of therapy, and it is not clear whether this is related to total dose or individual predisposition to toxicity. The relationship of the cumulative dose of AmB to the development of nephrotoxicity is still controversial. Earlier studies suggested that greater cumulative doses of AmB (e.g., 3–4 g) were associated with a higher risk of nephrotoxicity.[19] This implied that the patient likelihood of a rise in the serum creatinine concentration increases in proportion to the length of therapy and hence to the cumulative dose. However, our experience indicates that there is no time dependence in the onset of nephrotoxicity (Figure 2).[20] Azotemia can be observed after a cumulative dose as low as

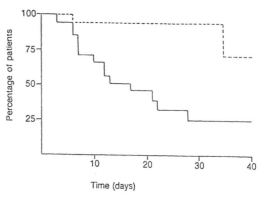

Figure 2 Estimated proportion of patients retaining normal renal function during therapy with amphotericin B. Patients received amphotericin B with (– – –, $n = 17$) or without (——, $n = 21$) parenteral salt supplementation due to coadministration of ticarcillin (reproduced by permission of Springer from *Klin. Wochenschr.*, 1987, **65**, 500–506).

100 mg, particularly when associated with sodium and/or volume deficits. In one study, larger cumulative doses have been associated with irreversible renal impairment. In the study reported by Winn, 88% of patients who had received cumulative doses exceeding 5 g were found to have persistent renal dysfunction.[21]

The development of azotemia is modified by various factors. The frequency of dosing may be one of them, as administration of drug on alternate days reduces the incidence of nephrotoxicity.[22,23] The presence of sodium depletion is a strong potentiating risk factor for development of azotemia; more importantly, sodium supplementation ameliorates this side effect.[10,20] Age also seems to be important, as a greater incidence in the older age group is apparent.[19,24] A case control study in the late 1980s revealed that higher daily doses, concomitant diuretic use, and abnormal baseline renal function are all significant independent risk factors for AmB nephrotoxicity.[25]

7.25.2.2 Urinary Concentrating Defect

Many studies have demonstrated that AmB almost invariably induces loss of the concentrating ability of the kidney.[16,26,27] This abnormality occurs early in the course of therapy and is independent of dose. In a detailed mechanistic study, Barbour *et al.* evaluated a small number of patients who developed hyposthenuria and showed a defect in free water reabsorption under maximal stimuli of a high water load. These results were interpreted as indicating a tubular functional abnormality with failure of the vasopressin response in the medullary collecting

Table 2 Typical features of amphotericin B-induced renal failure.

May occur at any time after starting therapy
Reversible
Tubular abnormalities appear early
Asymptomatic increase in serum creatinine or BUN
Hypokalemia
Pre-existing sodium depletion aggravates the condition
Can be managed by discontinuation or dose reduction of amphotericin B
Can be prevented with sodium supplementation

tubule.[28] The mechanisms for this defect are not fully understood. The impairment in concentrating ability probably reflects direct tubular toxicity since it occurs in the absence of a decrease in GFR, and is temporally unrelated to azotemia. It appears that the underlying mechanisms include the partial inhibition of water permeability (an arginine-vasopressin dependent mechanism) and the dissipation of the medullary tonicity due to decreased urea transport.[29]

7.25.2.3 Electrolyte Disturbances

Excessive renal wasting of potassium and magnesium leading to electrolyte disorders are both expected consequences and common adverse effects in patients receiving AmB.[10,25] Although hypokalemia has been emphasized in prior studies, its impact on patient management and on the course of other manifestations of AmB nephrotoxicity have not been well examined. Potassium depletion is expressed not only through systemic effects (muscle weakness, fatigue, cramps, rhabdomyolysis, and myoglobinuria), but also through renal abnormalities (impairment of concentrating ability, defect of urinary acidification, renal insufficiency, and abnormal sodium reabsorption).[30] It is conceivable that these effects may influence or contribute to the nephrotoxicity of AmB.

In clinical practice potassium depletion is suspected by the presence of hypokalemia. In the absence of potassium supplements, approximately 75% of patients will develop hypokalemia during a course of treatment with AmB.[31] If the requirements of potassium supplementation to maintain a plasma potassium level at the lower limit of the normal range are considered markers of a potassium losing diathesis, the incidence of potassium depletion has been reported to be 90% or more.[18,32] Since patients who require AmB are often severely ill and unable to tolerate oral potassium supplementation, it may be necessary to administer large intravenous doses of potassium chloride over relatively long intervals ($6–7\,h\,d^{-1}$) to avoid severe potassium depletion. Therefore, the maintenance of such continuous intravenous infusion may interfere with the administration of other multiple parenteral therapies.[18]

The mechanisms by which urinary potassium wasting takes place are not clear. Selective distal tubular epithelial toxicity seems to be, at least in part, responsible for the profound potassium wasting.[33] Alternatively, a recent study has shown that AmB also affects sodium flux in both the distal and transverse human colon, suggesting a change in sodium/potassium exchange to result in potassium loss.[34]

Magnesium wasting is also a consequence of AmB administration.[35,36] Magnesium balance is probably decreased in all patients, but clinically relevant magnesium depletion is often overlooked and only occurs when the urinary loss is high and not replaced. Similar to the case of potassium, if the presence of magnesium depletion is evaluated not only by the serum level but also by using more specific tests, the incidence of magnesium depletion may increase. Barton et al.[35] showed renal magnesium wasting leading to mild to moderate hypomagnesemia after relatively small cumulative doses. The lowest serum level and the largest fractional excretion of magnesium was observed after a cumulative dose of approximately 500 mg; no further changes were observed despite continued AmB treatment.[35] Since the follow-up did not include magnesium balance studies, the possible role of either reduced intake or altered gastrointestinal absorption could not be ascertained. This abnormality was fully reversible by stopping therapy, evidenced by the normal serum and urinary magnesium levels measured 1 yr after discontinuation of therapy.

The mechanisms for the observed AmB-induced renal magnesium wasting are not clear; however, increased urinary excretion of magnesium despite its reduced filtered load suggests a tubular defect in magnesium reabsorption.[35] Refractoriness to correction of hypokalemia may also occur and potassium replacement may not be successful unless magnesium deficiency is corrected first.

7.25.2.4 Renal Tubular Acidosis

Renal tubular acidosis is also an overlooked abnormality because the acidifying defect usually occurs without systemic acidosis.[33] It has been reported to frequently occur in patients receiving total doses of 0.5–1 g or more, and is generally reversible if therapy is discontinued.[7] In our experience, this is an underestimate, as it is one of the earlier manifestations of tubular toxicity, since all patients developed an inability to acidify the urine in response to an acid load after 2 weeks of therapy and a cumulative dose of 300 mg of AmB.[32] This defect appears to be a specific tubular effect of AmB because defects in acid secretion, attributed to increased passive permeability of the luminal membrane to hydrogen ions, have been demonstrated in the isolated turtle bladder.[37,38] Additionally, the impaired excretion of titratable acids is greater than can be accounted for by a depression of GFR.[33,39] It is also probable that distal renal tubular acidosis is a contributing pathogenic mechanism for urinary losses of potassium and magnesium.[31,33,39]

7.25.3 PATHOLOGICAL FINDINGS

Despite the almost complete compromise of whole kidney function, histological changes associated with AmB therapy are minimal, occurring mostly in the tubulointerstitial structures. Tubular damage primarily involves the distal convoluted tubule and the ascending limb of the loop of Henle.[35] Morphologic changes include fragmentation and thickening of basement membranes, necrosis and vacuolization of distal tubular epithelium, and after prolonged therapy being maintained in the presence of azotemia, nephrocalcinosis.[16,26,33,40] The scarce glomerular changes include calcific foci, along with hypercellularity and vacuolization of smooth-muscle cells in small arteries and arterioles.[27,33] In a detailed study that included functional evaluation and morphometric and electron microscopy, the effect of sodium depletion on the kidney histological changes after multiple-dose AmB administration was evaluated in rats.[41] Salt-depleted rats had smaller kidneys than control rats. In control AmB-treated rats, cortical changes were largely confined to the medullary ray. In contrast, in salt depleted rats that received AmB, similar changes extended to the medulla. The histologic changes were more extensive in the presence of salt depletion and included focal rupture and calcification of the thick ascending limbs and occasional involvement of contiguous glomerulus structures, such as macula densa segments, with atrophic changes in the thick ascending limb in the inner stripe. The S2/S3 proximal tubules did not seem to be involved in the calcification process. No AmB-induced vascular alterations were noted in any of the groups. Since the changes occurred in areas known to be most vulnerable to hypoxia and in areas rich in oxygen, the results supported the dual pathogenetic mechanisms for AmB nephrotoxicity, that is, direct cytotoxicity and vasoconstriction.[41]

7.25.4 MECHANISMS OF NEPHROTOXICITY

As is common for the majority of nephrotoxicants, the possible sites of nephron involvement should be identified before mechanisms can be proposed to account for renal cell injury.[42] Based upon structural and functional changes, AmB is known to cause acute renal vasoconstriction and to damage preferentially the distal tubular epithelium, but the exact mechanisms mediating its nephrotoxicity have not been clearly defined. The initial event is thought to involve binding of AmB to membrane sterols in the renal vasculature and epithelial cells to alter membrane permeability. This interaction may then trigger other cellular events that result in activation of second messenger systems, release of mediators, or activation of renal homeostatic mechanisms. It is, therefore, possible that the membrane effect *per se* is not the sole factor that determines the extent of change in renal function. Furthermore, factors which interact with these secondary responses and mechanisms may modify the net effect of AmB on renal functions.

7.25.4.1 Effects on Cell Membranes

AmB is believed to bind to cell membranes, induce micropore formation, influence ionic flux, and thereby change intracellular concentrations of ions.[43,44] It is generally accepted that the presence of membrane sterols is a necessary requirement for the formation of the micropores, leading to the leakage of potassium, essential metabolites and, eventually, cell necrosis.[12] Studies have demonstrated that at lower doses the first alterations that occur are functional in nature and relate to sodium-dependent processes leading to alteration of the Na^+/K^+ gradients. Joly *et al.* evaluated the noxious effects of free AmB on primary cultures of renal proximal tubules by measuring the changes in sodium-dependent uptake as a characteristic function of proximal tubular cells, changes in cell permeability and cell death.[45] After defining that cell necrosis, as evaluated by LDH release, increased significantly after incubation with $40 \mu M$ AmB, the studies were conducted using lower AmB concentrations to evaluate functional but not lethal effects of free AmB. It was important that the concentrations evaluated were close to systemic concentrations of patients treated with conventional doses of AmB (range $0.5-2.2 \mu M$).[46] The phosphate and α-methylglucopyranoside uptakes (markers of sodium-dependent processes) were significantly reduced (50%) by $2.5 \mu M$ and $5 \mu M$ AmB, respectively, and the effect was dose-dependent up to $20 \mu M$. Similarly potassium release, a marker of cell permeability, was increased at $2.5 \mu M$ AmB and further increased with higher AmB concentrations. The results suggest parallel alterations of sodium-dependent uptake and potassium release by AmB which occur prior to cytotoxicity.

In a study from our laboratory, isolated cultured mesangial cells were used to measure the effect of AmB on intracellular calcium concentrations and evaluate mechanisms that contribute to the intracellular response.[47] Rat glomeruli were isolated and mesangial cells were cultured and grown to confluence in the absence of epithelial cells. Using fura-2 as a probe to measure intracellular calcium concentrations, an AmB concentration-dependent increase in free calcium concentration was demonstrated. The increase in free cytosolic calcium concentration by AmB was inhibited by removal of extracellular calcium, sodium, and diltiazem (20 µM) (Figure 3). The role of intracellular calcium changes following AmB has been corroborated by a study that demonstrated AmB-induced smooth muscle constriction in perfused afferent arterioles and isometrically contracting rings of rabbit aorta and renal artery, effects which were also prevented in calcium-free medium and by calcium channel blockers.[48] The dependence of the response on extracellular sodium and calcium is consistent with the hypothesis that an interaction occurs between AmB and cell membrane sterols leading to pore formation which increases their permeability. This results in sodium diffusing down its concentration gradient leading to depolarization of the membrane and opening of calcium channels. Thus at low concentrations, AmB-induced membrane perturbation initiates cell responses via a calcium channel-mediated mechanism.

The dynamics of the subsequent events that follow the formation of micropores and the cell damaging effect of AmB had not been established by the mid-1990s. Rapid increased ion conductance and efflux of potassium may generate sudden changes in cell volume with consequent cell necrosis. However, inhibition of enzymes or oxidative damage also seems to be important in the cell lethal action of AmB. Interestingly, Ramos *et al.* using the parasitic protozoa *Leishmania mexicana* have demonstrated a protective effect of magnesium against the lethal effects of AmB.[12] The study evaluated the effect of AmB on the viability of promastigote forms and promastigotes that were transformed to amastigote-like forms by elevating the temperature of the culture medium. The fluorescence of complexes of ethidium bromide (EB) with nucleic acid as an indicator of cell permeability was continuously monitored. Changes in fluorescence were related to complete permeabilization induced by digitonin. An increase in the magnesium content of the incubation medium prevented the AmB-induced incorporation of EB into *Leishmania* promastigotes to a greater extent than it was into heat-transformed cells, suggesting that other membrane factors subsequent to AmB binding were also affected. This effect was apparent even when magnesium was added following the preexposure of *Leishmania* promastigotes to AmB. AmB at 0.5 µM induced a time-dependent increase in the fluorescence that was modified by the addition of Mg^{2+} (10 mM) (Figure 4). These observations were used to infer that AmB alters endogenous anionic components at the membrane surface level. It is of interest that this effect has also been found

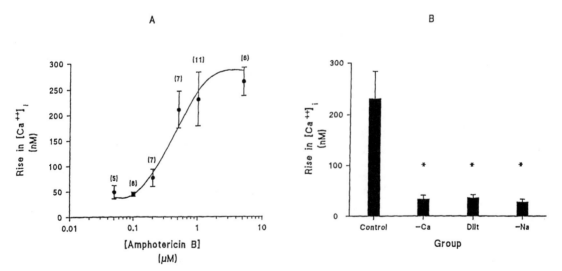

Figure 3 Concentration-dependent increase in intracellular calcium levels in cultured glomerular mesangial cells (A) and its inhibition by removal of Ca^{2+} and Na^+ ions from the medium (−Ca and −Na, respectively), and by addition of 20 µM diltiazem (Dilt) (B) (reproduced by permission of Elsevier Science B.V. from *Eur. J. Pharmacol.*, 1992, **226**, 79–85).

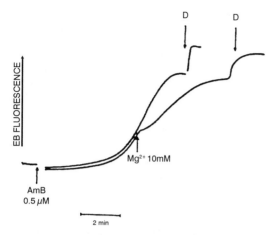

Figure 4 Effect of adding Mg^{2+} after the addition of AmB on the incorporation of ethidium bromide into *Leishmania* promastigotes. The Mg^{2+} concentration was 10 mM; the AmB concentration was 0.5 μM. Maximal EN incorporation was determined by adding digitonin (arrow D) (reproduced by permission of the American Society of Microbiology from *Antimicrob. Agents Chemother.*, 1990, **34**, 1584–1589).

in *S. cerevisiae* and *C. albicans*, suggesting a common mechanism underlying the action of AmB on both parasitic protozoa and fungi.

7.25.4.2 Whole Animal Studies

7.25.4.2.1 Acute studies

The infusion of AmB, intravenously or into the renal artery, induces a short term reduction in renal blood flow (RBF) and an increase in renal vascular resistance in both rats and dogs.[49–51] This response is transient, with a rapid return toward the baseline values after stopping the infusion. A similar pattern is observed with GFR; however, despite the return of RBF to baseline in the postinfusion period, there is usually a progressive fall in GFR after stopping the infusion, indicating a further reduction in filtration fraction. Renal micropuncture studies in our laboratory have confirmed that AmB in doses that did not change systemic blood pressure, but did cause a 40% reduction in whole kidney RPF, a 32% reduction in GFR, and a 35% reduction in SNGFR, decreased the glomerular capillary ultrafiltration coefficient (K_f), as well as increased afferent and efferent arteriolar resistances, consistent with a direct effect on mesangial and/or vascular smooth muscle contraction (Figure 5).[52] Pressures in both the glomerulus and tubule decreased to a similar extent, so that the pressure gradient across the

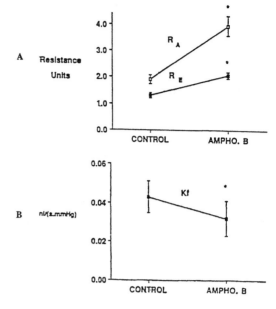

Figure 5 Effect of infusion of amphotericin B (AMPHO) into the renal artery (0.05 mg kg^{-1} min^{-1}) on RA and RE (A) and on the K_f (B) in four rats. Values are means ± SEM. *$P < 0.05$ (reproduced by permission of Williams and Wilkins from *J. Pharmacol. Exp. Ther.*, 1990, **253**, 34–37).

glomerulus remained unchanged. Previous micropuncture studies demonstrated a similar vasoconstriction of the afferent arteriole but also an increased permeability of the tubular epithelium to inulin.[53] Thus, the reduction in GFR after acute AmB infusions can be attributed to contraction of afferent and efferent arteriolar smooth muscle cells and of glomerular mesangial cells, as well as increased tubular permeability with back-leak into the interstitial space. These micropuncture studies are consistent with measurements of RBF and GFR that have been obtained after acute AmB infusions in rats and dogs.[49,51,54–56]

Animal experiments have provided evidence for a protective effect of sodium supplementation on the acute renal effects of AmB. Using *in situ* perfused rat kidneys we have shown that prior salt loading delayed and partially blocked the fall in RBF (Figure 6), and prevented the reduction in GFR (Table 3).[57] The results are consistent with prior observations of salt dependence in the dog, that showed enhancement of the acute renovascular response in the presence of salt depletion.[54] Initially, inhibition of the tubuloglomerular feedback (TGF) mechanism was proposed as an explanation, since physiological and pharmacological interventions that blocked TGF, namely salt loading and administration of furosemide, theophylline

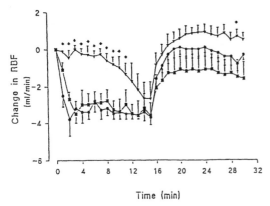

Figure 6 Change in RBF in the normal salt (control, ●, $n = 7$), salt depleted ($-$NaCl, ■, $n = 7$) and salt loaded ($+$NaCl, ▼, $n = 6$) groups, during and after a 15 min infusion of amphotericin B at $0.025\,\mathrm{mg\,kg^{-1}\,min^{-1}}$ intrarenal artery. Each point represents the mean $\pm$ SEM. Repeated measures ANOVA: $P < 0.0001$ comparing $+$NaCl to control and $-$NaCl; control and $-$NaCl no significant difference (reproduced by permission of the American Society of Microbiology from *Antimicrob. Agents Chemother.*, 1991, **35**, 2509–2514).

or calcium channel blockers, inhibited the acute renal effects of AmB.[50–52,54,58–64] However, the potential role for TGF contributing to the acute renal response to AmB, was definitively excluded in direct micropuncture studies showing that AmB-induced reduction in single nephron GFR was the same irrespective of whether the TGF reflex was intact or interrupted.[48] Furthermore, distal tubular chloride concentrations were not increased by AmB. Thus, the signal for TGF was unchanged by AmB. Finally, AmB did not increase the sensitivity of the TGF mechanism to other stimuli.

The precise mechanisms of the contractile responses to AmB have not been identified.

Theoretically, the drug can act either directly on the glomerular vasculature or via an indirect mechanism involving secondary mediators. The acute glomerular response to AmB is also blocked by diltiazem (Table 3), consistent with the role of calcium gated channels suggested by experiments in mesangial cell cultures.[47] Neither renal denervation nor angiotensin II receptor blockade prevent the renal vasoconstriction or the reduction in GFR.[56] Endothelin does not appear to be involved in the acute responses to AmB.[65] Potassium depletion does not modify the acute renovascular response.[66] A role for thromboxane A_2 is suggested based upon partial inhibition of the AmB-induced vasoconstriction and reduction in GFR by pretreatment with ibuprofen or a thromboxane receptor antagonist.[67]

During the early 1990s, an alternative hypothesis of nitric oxide (NO) as a potential modulator was developed from the observation that salt loading enhances endogenous NO.[68] If this hypothesis is valid, then increased availability of substrate for NO, that is, L-arginine, might be expected to diminish the renovascular response. We have shown that an acute excess of L-arginine confers protection against the acute renovascular changes after a single i.v. administration of AmB in the rat, whereas the inhibitor of NO, N^G-nitro-L-arginine methyl ester (L-NAME) potentiates the response. These observations support this hypothesis and suggest that AmB could be acting through mechanisms related to NO metabolism.[69]

7.25.4.2.2 *Multiple-dose studies*

Animal models of multiple-dose nephrotoxicity have also shown that similar interventions

Table 3 Effect of salt supplement or calcium channel blockade on the creatinine clearance (mL min^{-1}) in rats treated with AmB.[a]

Group	Number	Baseline	AmB infusion	Post-AmB infusion
Control	6	0.83 ± 0.08	0.40 ± 0.09^{b}	0.35 ± 0.07^{b}
Sodium depleted	5	0.74 ± 0.15	0.56 ± 0.18^{b}	0.40 ± 0.05^{b}
Sodium loaded	7	1.13 ± 0.18	0.96 ± 0.17^{c}	1.34 ± 0.25^{c}
Diltiazem	6	0.65 ± 0.07	0.60 ± 0.02^{c}	0.60 ± 0.08^{c}

[a]Each value represents the mean $\pm$ standard error of the mean at baseline, during and after a 15 min infusion of amphotericin B at $0.025\,\mathrm{mg\,kg^{-1}\,min^{-1}}$ intra-arterially (i.a.). Control group of rats were fed normal-salt diet. Sodium-depleted group of rats received low-salt diet (Na composition, $<0.05\%$; Ralston Purina Co., St. Louis, MO) and tap water for at least 1 week. Sodium loaded was achieved by subcutaneous injection of deoxycorticosterone acetate 1 week prior to the day of the experiment, followed by maintenance on a normal-salt diet and 1.0% saline drinking water. Diltiazem ($20\,\mu\mathrm{g\,kg^{-1}\,min^{-1}}$, i.a.) was started 30 min prior to amphotericin B and continued throughout the experiment in rats maintained on normal salt intake. [b]$P < 0.05$ compared to baseline. [c]$P < 0.05$ compared to same period in control group.

that modulate the acute renal response can modify the nephrotoxicity of AmB. Rats co-treated with sodium bicarbonate sustain smaller reductions in GFR compared with control rats treated with AmB alone for 3 weeks.[39] Oral NaCl supplementation also attenuates the decrease in GFR and the elevation in renovascular resistance induced by daily administration of AmB over 3 weeks.[61,63] In addition, renal impairment following a 7 d course of AmB in rats was less severe when theophylline was co-administered.[55] These interventions also attenuate the acute renal responses to AmB, suggesting that similar mechanisms contribute to its multiple-dose nephrotoxicity. Similar logic would suggest that salt supplementation and theophylline are protecting the kidney by a mechanism unrelated to TGF. It is, however, possible that the latter does contribute to AmB nephrotoxicity but only at later stages of therapy, when severe damage to the tubules may have taken place. Interestingly, the protection by salt loading is associated with lower concentrations of drug in the kidney despite similar concentrations in plasma and liver tissue.[61] This raises an alternative possible mechanism of protection by salt loading involving a pharmacokinetic interaction with AmB which limits its uptake into the kidney.

Since the early 1990s, several interesting alternative hypotheses have been raised. Cell death induced by AmB in the medullary thick ascending limb is prevented by ouabain.[70] A reasonable explanation for this observation is that ouabain, by inhibiting transport, decreases the oxygen demand of an area of the nephron, which has a limited oxygen supply. It is conceivable that AmB-induced renal vasoconstriction and ischemia to this section of the nephron would enhance the cell death produced by a direct toxic action of AmB on those cells. Thus, any maneuver that improves renal perfusion, or decreases oxygen demand, would be expected to be protective. Additional support for this hypothesis has also been obtained. Methimazole, a free radical scavenger, has been shown to ameliorate cephaloridine and cisplatin-induced nephrotoxicity.[71] In a preliminary study, we have evaluated the effect of concomitant administration of methimazole on AmB-induced reductions in GFR using an established model of AmB nephrotoxicity in rats.[72] Methimazole provided a renoprotective effect to the glomerular response after multiple-dose AmB administration. Since reactive oxygen metabolites have been implicated in several models of toxic renal failure,[71,73,74] it is conceivable that these mechanisms may also be operating responses after exposure to AmB and blunted by methimazole.

Studies on the effects of chronic administration of calcium channel blockers on multiple-dose AmB induced nephrotoxicity have been discordant. In the rat, nifedipine does not offer a significant protective effect,[75] but diltiazem blunts the increase in serum creatinine and the decreases in GFR and RPF.[62] It is possible that these differences relate to the heterogeneity of calcium channels and the differential activity of calcium channel blockers on them.

A study in 1992 demonstrated that coadministration of 5-flucytosine with AmB to rats, which is commonly used clinically to obtain a synergistic antifungal effect, protects against acute and multiple-dose nephrotoxicity.[76] The mechanisms by which flucytosine influences the renal response to AmB are not clear but may relate to (i) its administration in 0.9% NaCl, which itself is protective, (ii) a renal vasodilator effect of flucytosine that antagonizes AmB-induced vasoconstriction, and (iii) reduction in renal uptake of AmB.

Potassium depletion *per se* induces changes in tubular and glomerular renal function; that is, impaired ability to concentrate and acidify urine, altered sodium chloride reabsorption and an increase in renovascular resistance to reduce glomerular function.[30,77] Each of these effects resembles the effect of AmB itself. Thus, hypokalemia and AmB could have additive or synergistic effects in influencing the renal function. The influence of increased urinary potassium losses leading to potassium depletion and hypokalemia during multiple-dose AmB administration on renal function has been evaluated.[66] The response to a single i.v. administration of AmB was first evaluated in hypokalemic and normokalemic rats. The study showed that the acute glomerular response of RBF and RVR to AmB was not modified by potassium depletion, and there were no differences in changes in GFR. In contrast to the glomerular response, AmB treatment to potassium-depleted rats caused a greater increase in fractional sodium excretion than in rats receiving potassium supplementation, suggesting a greater tubulotoxic effect (Figure 7). The implication that potassium depletion enhances AmB-induced sodium loss was further evaluated during multiple-dose AmB administration. The renal effect of repeated administration of AmB in potassium depletion was evaluated in rats. Under potassium-depleted conditions, there was an enhancement of AmB-induced decrease in GFR.[78] After 3 d of treatment with AmB, potassium-depleted rats had GFR values that were 40–50% lower than rats without depletion. These results suggest that potassium depletion *per se* enhances the renal responsiveness to AmB.

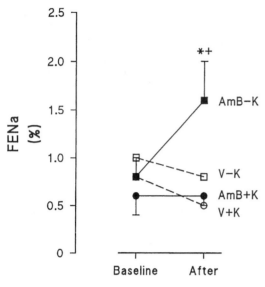

Figure 7 Fractional Na excretion (FENa) before and after a 30 min infusion of either amphotericin B at $0.03\,mg\,kg^{-1}\,min^{-1}$ i.v. or its vehicle in the normal potassium group (●, AmB; $n = 10$; and ○, V, $n = 6$) and potassium-depleted group of rats (■, AmB-K, $n = 10$; and □, V-K, $n = 6$). Means ± SEM. Repeated measures ANOVA revealed $P < 0.05$. *$P < 0.05$ compared with the baseline period; $+P < 0.05$ compared with AmB (reproduced by permission of S. Karger from *Nephron*, 1995, **70**, 235–241).

7.25.5 APPROACHES TO REDUCE NEPHROTOXICITY

If nephrotoxicity is a limiting factor in the therapeutic use of AmB, then an increased understanding of the mechanisms involved in determining the change and extent of change in renal function might be used to lead to the development of strategies to broaden its therapeutic index.

7.25.5.1 Other Analogues of AmB

Several derivatives of AmB were developed including methyl ester compounds (AmB methyl ester and *d*-ornithyl AmB methyl ester) and DAPEG AmB (reaction of AmB with the *N*-(3-dimethylaminopropyl)-*N'*-ethyl carbodiimide).[79] These relatively water soluble compounds, particularly the AmB methyl ester, appeared to be less nephrotoxic. However, the appearance of neurotoxicity and ototoxicity after a limited clinical trial resulted in the termination of studies with these related salts.

7.25.5.2 Sodium Supplementation as a Modulator of Nephrotoxicity

The relevance of salt loading to minimize AmB is becoming recognized in clinical practice. Studies in animals which demonstrate a renal protective effect of salt loading on AmB-induced nephrotoxicity have provided the rational basis to evaluate this intervention in humans. Initial reports of a protective effect with salt loading, including those from our clinical service, were anecdotal.[16,80] These observations were confirmed in a retrospective study. This study revealed that only 2 of 17 patients (12%) receiving coadministration of ticarcillin (with its obligatory sodium load of $150\,mEq\,d^{-1}$) had a nephrotoxic response to AmB, compared with 14 of 21 patients (67%) not receiving ticarcillin.[20] In a companion phase II study, the benefit of routine intravenous saline (1 L of 0.9% saline) was assessed prospectively in leukemic patients receiving a 28 d course of AmB for persistent fever of unknown origin.[20] Only 2 of 20 patients (10%) developed mild renal dysfunction which did not necessitate interruption of therapy. The full course of high-dose AmB was successfully completed in all patients, including four with mild renal impairment prior to therapy.

It is difficult to study the influence of salt loading on AmB nephrotoxicity in patients receiving the drug in the USA due to the diversity of the patient population being treated. Many patients are severely ill with systemic diseases that diminish baseline renal function, are immunocompromised, or are receiving concomitant potentially nephrotoxic drug therapy. The ability of monotherapy with AmB to eradicate the parasite in mucocutaneous leishmaniasis offers a unique opportunity to design a study to evaluate the influence of salt supplementation on the renal function during AmB therapy in a prospective, randomized, placebo controlled trial.[32] In patients with mucocutaneous leishmaniasis in Per, pretreatment of AmB with dextrose resulted in increased mean serum creatinine over time, whereas pretreatment with saline prevented this response (Figure 8). Similarly, creatinine clearance decreased in the dextrose group, but remained unchanged in the saline group, resulting in a significantly different response over time. All patients responded to therapy with AmB with remission of the disease as assessed clinically and by histologic examination, with no difference in the therapeutic response between the two groups. This study confirmed the protective effect of salt loading against AmB nephrotoxicity. Based upon these

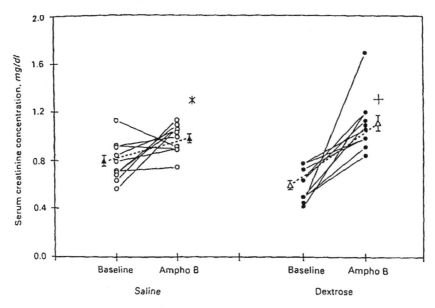

Figure 8 Baseline and maximum serum creatinine concentrations achieved in patients receiving AmB (150 mg kg^{-1} for 10 weeks) with either 1 L of saline or 1 L of 5% dextrose in water. * $P < 0.05$ compared to baseline; + $P < 0.0001$ compared to baseline. Comparison of the maximal change in serum creatinine in the two groups gave a $P = 0.01$ (reproduced by permission of Blackwell Science Publishing Inc. from *Kidney Int.*, 1991, **40**, 302–308).

studies, we have recommended routine salt supplementation when feasible with administration of AmB.

However, the protective effect of salt supplementation resulted in enhanced renal wasting of potassium in response to AmB. Greater amounts of potassium supplements were apparent in patients treated with AmB and saline. Even though there was no major change in GFR, there was a substantial difference in the amount of potassium supplements needed to maintain a serum potassium level in the low normal range between the groups. Serum potassium was significantly decreased within the first 2 weeks of AmB treatment. The saline loaded group required significantly higher amounts of supplements to maintain a normal serum potassium level (Figure 9). Conventionally, as in this study, potassium supplements are given to maintain serum potassium at 3.0 mEq L^{-1}. Although at the lower range of normal values, this level already indicates a total body potassium deficit and the supplements are designed to avoid severe potassium depletion. Whether it is more beneficial to prevent potassium depletion by using potassium salts from the start of a therapeutic course remains to be evaluated.

7.25.5.3 Liposomes

The incorporation of AmB into liposomes was designed to decrease the untoward effects

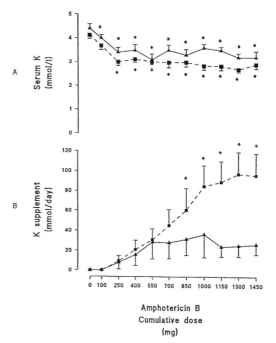

Figure 9 Serum K levels (A) and K supplements given to maintain serum K levels at or above 3 mmol L^{-1} (B) in patients receiving AmB with either 1 L of 0.9% NaCl (■) or 1 L of 5% dextrose in water (▲). Notice the difficulty in maintaining serum K levels despite significantly higher amounts of supplements in the former group. * $P < 0.05$ compared with baseline (reproduced by permission of Blackwell Science Publishing Inc. from *Kidney Int.*, 1991, **40**, 302–308).

of this drug, to enhance its therapeutic activity, and to provide site specific delivery of high doses of the drug.[81,82] AmB is a highly lipophilic drug. At a neutral pH, AmB is practically insoluble in water. It is conventionally formulated as a mixed micellar dispersion with deoxycholate as detergent (Fungizone, Bristol–Myers Squibb). It is not always effective and it is highly toxic. In clinical situations where AmB was ineffective, poor bioavailability was suspected. However, the possibility of serious toxic effects does not allow the use of higher, perhaps more efficacious, doses than the maximal tolerable dosage of $0.7–1.0 \, \text{mg kg}^{-1} \text{d}^{-1}$.[83]

The narrow therapeutic index of conventional AmB led to the development of new formulations that utilize either liposomes or complex phospholipids as drug carriers. Since the appearance of early reports demonstrating that disseminated candidiasis in rats could be successfully treated with AmB incorporated into liposomes and that similarly encouraging observations were apparent in patients with systemic fungal disease, a striking number of publications have appeared on this topic in the medical literature.

Liposomes are microscopic vesicles consisting of one or more phospholipid membranes surrounding a discrete water compartment. The lipid layer is composed of amphipatic phospholipids whose hydrophobic tails associate on addition to water, while the polar hydrophilic heads align toward the bulk of the water phase. A variety of liposomes with unique physical and chemical structure can be manufactured by altering nonpolar and polar groups. Excellent reviews were published in the early 1990s.[82–84]

Different mixtures using several lipoproteins and combined in different ratios have appeared during the 1980s. As has been reviewed, there are at least five formulations that have been tested in man. They include L-AmpB liposomes, AmB lipid complex (ABLC), intralipid AmB, AmB colloidal dispersion (ABCD) and AmBisome (Table 4). Since the 1980s, these preparations have been extensively examined in *in vitro*, whole animal, and clinical studies.[3,45,46,81,85–107]

Whether the ultimate goal of the incorporation of AmB into liposomes, that is, to reduce the nephrotoxic effects, has being reached is not clearly established. A comparison of conventional AmB with a new formulation of the drug should answer the following questions: (i) do the two formulations have the same or different actions? (ii) if they have the same action, what is the dose ratio for their antifungal and toxic effects on mammalian systems, especially nephrotoxicity? and (iii) is there a selective advantage in the relative dose ratios of efficacy and toxicity to indicate a wider therapeutic margin, that is, is the dose ratio liposomal formulation of AmB/conventional AmB lower for the antifungal effect compared to the dose ratio liposomal formulation of AmB/conventional AmB for the nephrotoxic effect?

A review of the literature suggests that liposomal formulations of AmB and conventional AmB have a similar action on fungal and mammalian cells. Very few studies, however, have established a dose ratio for antifungal and nephrotoxic effects. In most, only one aspect of the activity or only one formulation was studied. The multitude of formulations prepared by either a particular investigator or by pharmaceutical companies have made it very difficult to compare relative advantages and disadvantages. Thus, comparisons between studies are difficult, and inferences should be made with caution. There is evidence that the fungicidal activity of liposomal formulations of AmB is influenced by several properties of the

Table 4   Physical and chemical features of several liposomal formulations.

Formulation	Chemical composition	Particle size
Fungizone	Amphotericin B 50 mg and deoxycholate 41 mg	
L-AmpB	DMPC/DMPG[a] in 7:3 molar ratio and AmB at a concentration of 5 mol%	>1 μm in diameter
ABLC	DMPC/DMPG in 7:3 molar ratio and AmB at a concentration of 33 mol%	1.6–11.1 μM 90% between 1.6 and 6 μM
AmBisome	Phosphatidylcholine, cholesterol disteroylphosphatidylglycerol and AmB in a molar ratio of 2:1:0.8:0.4	<100 nm unilamellar vesicles
ABCD	Cholesteryl sulfate and AmB in a 1:1 molar ratio	Discoidal complexes of 150 nm in diameter

[a]DMPC:DMPG: dimyristoyl phosphatidylcholine/dimyristoyl phosphatidylglycerol.

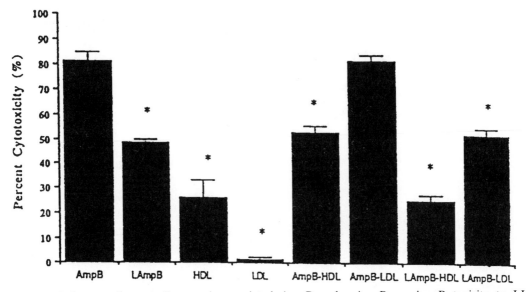

Figure 10 Influence of serum lipoprotein-associated AmpB and L-AmpB on AmpB toxicity to LLC PK1 renal cells. Percent cytotoxicity of LLC PK1 cells in serum-free medium at 37 °C containing various treatments: AmpB (20 μg mL^{-1}), L-AmpB (20 μg mL^{-1} of AmpB), high-density lipoproteins (0.5 mg protein mL^{-1}), low-density lipoproteins (0.5 mg mL^{-1}), mixtures of AmpB (20 μg mL^{-1}) with high-density lipoproteins (AmpB-HDL) or low-density lipoproteins (AmpB-LDL) and mixtures of L-AmpB (20 μg mL^{-1}) with high-density lipoproteins (L-AmpB-HDL) or low-density lipoproteins (L-AmpB-LDL). (Mean ± standard deviation; $n = 3$; $*P < 0.05$ vs. AmpB) (reproduced with permission of the Pharmaceutical Society of Great Britain from *J. Pharm. Pharmacol.*, 1991, **43**, 831–835).

liposomes including lipid composition, physical size, the molar ratio of lipids, and the presence or absence of sterols.[85–87] Thus, each formulation has to be considered as a separate entity. Furthermore, the tests used to assess *in vivo* toxicity have rarely examined renal function adequately. The testing of any new formulation requires one to obtain the acute LD$_{50}$ of the drug. It should be noted that this is an acute response to the drug and does not relate to the nephrotoxic potential of chronic therapy.[86,88–90] Finally, Phase II–III studies in humans are complicated by difficulties in the diagnosis of fungal infection, the underlying clinical condition of the patients, and frequent concomitant use of other nephrotoxic agents.

7.25.5.3.1 Cell studies

The binding of AmB to various compounds or formulations may result in reduced bioavailability of free AmB with a consequent reduction in toxicity to mammalian and/or fungal cells. Thus, the different formulations may act as a reservoir for free AmB. Since it is recognized that AmB has a higher affinity for ergosterol (the main sterol in fungal cell membranes) than for cholesterol (that is found in human cells) it is possible that the reduced amounts of free AmB are sufficient to be toxic to fungal cells and not to mammalian

cells. Ralph *et al.* demonstrated that L-AmpB is generally less active than AmB on yeast cells, and has a slower onset of action, suggesting that this formulation acts as a reservoir for free AmB which is the active moiety.[91] Others found either an equivalent or lower efficacy for ABCD.[86,89,92] These differences may be attributed to the different preparations used and/or to the different strains of fungi examined.

The selective rate of transfer may be related not only to changes in the sterol components of the cell membranes but also to different levels of expression of lipoprotein receptors in the target cells. Much has been learned from the *in vitro* studies focused on addressing the relationship between either free AmB or liposomal formulations of AmB with both high- and low-density lipoproteins (HDLs, LDLs). In a series of elegant studies by Lopez-Berenstein and coworkers, it was shown that AmB predominantly associates with HDL and that this effect is enhanced when it is incorporated into positively or negatively charged liposomes.[108,109] Wasan *et al.* have evaluated the influence of HDLs and LDLs on the toxicity of AmB to fungal and renal cells. The minimum inhibitory concentration of AmB and L-AmpB on *Candida albicans* fungal cells was not unchanged whether or not HDLs or LDLs were added to the incubation plates. However, HDL-associated AmB was less toxic than free AmB to LLC PK1 cells (mammalian renal tubule cells) while LDL-

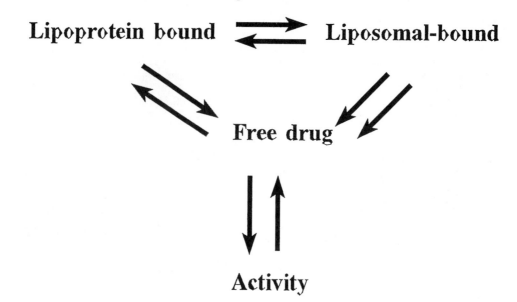

Figure 11 Hypothesis for serum distribution of AmB entrapped into liposomes. After intravenous administration, AmB may exist in several serum compartments. The biologic activity and, therefore, the toxicity is determined by free AmB. The dynamics of the events have not yet been clearly established.

associated AmB was as toxic as free AmB. In addition, L-AmpB and both HDL- and LDL-associated L-AmpB were less toxic to LLC PK1 cells than AmB (Figure 10).[110] The examination for the presence of HDL and LDL receptors in the LLC PK1 cells revealed high- and low-affinity LDL receptors but only a low-affinity HDL receptor. After trypsinization of the LLC PK1 renal cells, to reduce the LDL receptor, LDL-associated AmB was also less toxic than free AmB. Thus, the reduced level of toxicity of HDL-associated AmB and of L-AmpB may be explained by the low level of expression of HDL receptor in LLC PK1.[110] Taking this information into account, it appears that AmB in liposomal formulations may exist in a complex system that includes free drug, lipoprotein-bound drug and liposomal bound. Since the dynamic of the equilibrium is not known yet, it is very difficult to conclude whether higher doses of AmB in liposomal formulations result in comparable free drug concentration as the conventional preparation in the lower dose (Figure 11). These results also support the notion that the relative distribution of AmB among the serum lipoproteins seems to be a major factor influencing the therapeutic index of AmB incorporated into liposomes.

An *in vitro* evaluation of a therapeutic ratio between mammalian and fungal cells is required to evaluate whether or not there is a clinical advantage conferred by the incorporation of AmB into liposomes. In one of the few studies which calculated a concentration ratio for the actions of L-AmpB and conventional AmB on mammalian and fungal cells, Juliano *et al.* compared the *in vitro* toxicity of these preparations in both *Candida albicans* and mammalian erythrocytes.[87] While the two formulations were equipotent in their effects on ion fluxes in yeast cells (indicating formation of membrane pores in fungi), only AmB induced such an effect in erythrocytes, despite achieving concentrations of L-AmpB that were between 10- and 20-fold higher than those of AmB. The time required to achieve this effect in fungi was the same for the two formulations, which suggested that L-AmpB did not constitute a slow-release form of the drug. The reason for the reduced toxicity of L-AmpB is proposed to be a preferential transfer of AmB from liposomes to fungal cells compared with its transfer to mammalian cells.

Few studies have directly compared the antifungal and toxic effects of the two formulations to confirm a wider therapeutic index. A greater toxicity to kidney epithelial cell structures of AmB was apparent when compared with L-AmpB. LLCPK1 renal cells were exposed to short exposure times (2 h), with different formulations of L-AmpB. They exhibited different EC_{50} values, the most potent having an EC_{50} 13 times that of AmB.[93] Taken in concert with their previous study,[101] this indicates that the toxic concentration ratio of L-AmpB /AmB (13–20:1) is higher than its antifungal concentration ratio (1:1). These acute studies require cautious interpretation as longer exposure of LLCPK1 cells to L-AmpB (1–2 d), which is more representative of what occurs clinically, resulted in profound toxic effects at concentrations similar to those of AmB (L-AmpB/AmB = 1:1), manifested by changes in cellular transport processes and in morphology.[93] This finding raises questions as to the applicability and relevance of results derived from short-term *in vitro* experiments to whole animal or clinical situations.

7.25.5.3.2 Whole animal studies

Several formulations of AmB entrapped into liposomes have been extensively tested in whole animal models of fungal disease. Animal studies suggest that liposomal formulations of AmB are effective in the treatment of fungal infections, but usually require higher doses than AmB.[85,88,89,96,97,105,106] The lack of concomitant assessment of renal function in many of these studies makes it impossible to determine a dose ratio, although most reports confirm that the drug was well tolerated.

Detailed studies concerning pharmacokinetic parameters and tissue distribution have been published.[83,90,111] However, the interpretation of AmB serum levels obtained after systemic administration of these formulations is complicated by the fact that most of the assays do not discriminate between free drug, that is, biologically active, protein-bound and liposome-bound drug. In addition, as a result of the particle size, a significant proportion of AmB in serum settles as a sediment during the centrifugation process.[83] Therefore, the free AmB fraction in serum may differ extensively between the formulations not only due to factual differences, but also due to artifactual processes. In general, the tissue distribution and pharmacokinetics differed from the conventional drug. A greater volume of distribution and greater systemic clearance is apparent when AmB is incorporated into liposomes, suggesting substantial penetration into many organs. They accumulate in the tissues of the reticuloendothelial system as expected for a particulate formulation.[112] After parenteral administration, the total concentrations in liver and spleen were higher for the liposomal formulations than with conventional AmB. Concentrations of AmB in the kidney are dose-dependent for both formulations, but with lower values for liposomal formulations at equivalent doses. For both AmBisome and ABLC, the difference in biodistribution of these drugs may partly explain the reduced kidney toxicity of these formulations.[90,111] Comparable or slightly higher kidney concentrations of AmB are found in animals treated with higher doses of AmB using either AmBisome ($5\,mg\,kg^{-1}$) or ABLC ($10\,mg\,kg^{-1}$) when compared to that obtained after $1\,mg\,kg^{-1}$ of conventional AmB.

It is undisputed that liposomal formulations can usually be given at higher doses than AmB deoxycholate and are less toxic. Initial reports in a mouse model using a multiple-dose dosing regimen of liposomal AmB described only transient changes in liver enzymes and no abnormalities in renal function as evaluated by changes in serum urea or creatinine. In a report where the acute nephrotoxicity of the two formulations was examined in rabbits, AmB induced a fall in GFR and a rise in urinary sodium and potassium excretion rates, while L-AmpB, at 2.5 times the dose, did not affect these parameters.[94] Here again, antifungal activity was not assessed. These results should, however, be interpreted cautiously since unexpectedly L-AmpB increased the excretion of *N*-acetylglucosaminidase (NAG) in these animals, a parameter that was not altered by AmB deoxycholate. The authors suggested that increased delivery of AmB to the renal tubular cells by the liposomes led to an interaction with lysosomes and provided an additional mechanism of injury. In support of this conclusion is the finding that the tubular toxicity (abnormalities in urinary concentrating ability, natriuresis, kaliuresis and index of NAG) induced by repeated administration of AmB and L-AmpB over 5 d were similar when a high enough dose of L-AmpB (2.4 times that of AmB) was used.[95] Since kidney tissue levels of AmB in animals receiving a single dose of liposome AmB were equal to or greater than those obtained in animals treated with the same dose of free AmB, the conclusion was that the protective effect of the liposomes was not related to changes in pharmacokinetic of the drug in the target organ. The differential toxicity could have been related to a limited interaction between the liposomal formulation of AmB and the cell membrane.[95]

7.25.5.3.3 Human studies

The results from initial clinical studies are encouraging, but none of these reports can be considered as definitive evidence for a wider therapeutic index. Most of the published studies were part of the early assessment of safety of the particular formulation involved. The administration of the liposomal formulations of AmB is still based mainly on the basis of failed prior therapy and they are usually used on a compassionate basis in very complex clinical conditions.

In terms of clinical effectiveness, the literature provides numerous anecdotal examples of patients who failed to respond to conventional AmB, or those who developed nephrotoxicity on it, that were either complete or partial responders to AmB in liposomes, and did not sustain or even had recovery of renal function while on therapy.[3,81,98–104,106] The administration of L-AmpB was also devoid of the other adverse effects associated with AmB, and was generally well-tolerated compared with previous studies using AmB. In several hematologic conditions associated with neutropenia, the eradication of the disease while being treated with a liposomal formulation of AmB was also temporally associated with the remission of the underlying disease and with the recovery of neutrophil counts. Therefore, the relative contribution of the drug, neutrophil recovery and disease status on the efficacy to eradicate the disease was uncertain.[113]

Despite not being strictly a liposomal formulation, the preparation of Fungizone in Intralipid 20% instead of in glucose as is conventionally recommended, was associated with notable reduction in clinical and renal toxicity.[114] HIV positive patients with oral candidiasis were considered for treatment. The 22 subjects had comparable characteristics regarding HIV status infection and candidiasis, renal function and concomitant use of other nephrotoxic drugs. They were randomized to receive Fungizone prepared either in Intralipid 20% or in glucose. Four out of 11 AmB–glucose treatments were discontinued due to renal abnormalities or intolerable acute effects, whereas all 11 AmB–Intralipid 20% were treated without serious problems. Four out of the seven patients that completed the AmB–glucose regimen had at least one serum creatinine exceeding $1.5 \, \text{mg} \, \text{dL}^{-1}$ vs. one out of 11 patients that completed the AmB–Intralipid 20% regimen. Clinical side effects were noted in 36/38 infusions with AmB–glucose but only in 10/44 with AmB–Intralipid 20%. All patients responded to therapy with AmB as assessed by an oral candidiasis score. Thus, marked reduction in clinical and renal effects without lower efficacy was accomplished by using this formulation.

The limitations conferred by the usual complex clinical settings surrounding the use of AmB in liposomal formulations make it difficult to establish randomized clinical trials. However, the clinical conditions of visceral leishmaniasis and the prophylactic use of AmB during severe neutropenia may offer the opportunity to compare renal toxic effects due to these formulations.

7.25.6 CONCLUSION

AmB remains the most effective antifungal and antileishmanial agent. Nephrotoxicity is a well recognized dose-limiting complication, leading to interruption or discontinuation of the therapy. It is commonly expressed as azotemia and decreased GFR, however, tubular abnormalities are also important. The underlying mechanisms include direct vasoconstrictor effects and direct cytotoxicity, as a reflection of its action on cell membranes leading to alteration of cell permeability. These effects are amenable to being modulated. In the clinical setting, the use of salt supplementation lowers incidence and severity of nephrotoxicity; however, careful attention to potassium disorders should be encouraged. The information accumulated supports the notion that liposomal formulations of AmB have a wider therapeutic index. The mechanism of protection remains to be elucidated.

7.25.7 REFERENCES

1. D. W. Denning, 'Epidemiology and pathogenesis of systemic fungal infections in the immunocompromised host.' *J. Antimicrob. Chemother.*, 1991, **28**, 1–16.
2. F. Meunier and B. Wong, 'Overview of management of fungal infections: Part 1.' *Clin. Infect. Dis.*, 1993, **17**, S492–S493.
3. V. J. Wiebe and M. W. De Gregorio, 'Liposome-encapsulated amphotericin B: promising new treatment for disseminated fungal infections.' *Rev. Infect. Dis.*, 1988, **10**, 1097–1101.
4. G. P. Bodey, 'Fungal infections complicating acute leukemia.' *J. Chronic Dis.*, 1966, **19**, 667–687.
5. D. Shürmann, B. de Matos Marques, T. Grünewald et al., 'Safety and efficacy of liposomal amphotericin B in treating AIDS-associated disseminated cryptococcosis.' *J. Infect. Dis.*, 1991, **164**, 620–622.
6. R. J. Hay, 'Overview of the treatment of disseminated fungal infections.' *J. Antimicrob. Chemother.*, 1991, **28**, 17–25.
7. H. A. Gallis, R. H. Drew and W. W. Pickard, 'Amphotericin B: 30 years of clinical experience.' *Rev. Infect. Dis.*, 1990, **12**, 308–329.
8. Anonymous, in 'Control of Communicable Disease in Man,' ed. A. S. Benenson, APHA, Washington, 1990, pp. 238–243.
9. R. Dietze, E. P. Milan, J. D. Berman et al., 'Treatment of Brazilian kala-azar with a short course of amphocil (amphotericin B cholesterol dispersion).' *Clin. Infect. Dis.*, 1993, **17**, 981–986.
10. R. Sabra and R. A. Branch, 'Amphotericin B nephrotoxicity.' *Drug. Saf.*, 1990, **5**, 94–108.
11. D. W. Warnock, 'Amphotericin B: an introduction.' *J. Antimicrob. Chemother.*, 1991, **28**, 27–38.
12. H. Ramos, J. Milhaud, E. Cohen et al., 'Enhanced action of amphotericin B on *Leishmania mexicana* resulting from heat transformation.' *Antimicrob. Agents Chemother.*, 1990, **34**, 1584–1589.
13. J. Brajtburg, W. G. Powderly, G. S. Kobayashi et al., 'Amphotericin B: current understanding of mechanisms of action.' *Antimicrob. Agents Chemother.*, 1990, **34**, 183–188.
14. J. Brajtburg, S. Elberg, D. R. Schwartz et al., 'Involvement of oxidative damage in erythrocyte lysis induced by amphotericin B.' *Antimicrob. Agents Chemother.*, 1985, **27**, 172–176.

15. M. L. Sokol-Anderson, J. Brajtburg and G. Medoff, 'Amphotericin B-induced oxidative damage and killing of *Candida albicans.' J. Infect. Dis.*, 1986, **154**, 76–83.

16. W. T. Butler, J. E. Bennett, D. W. Alling *et al.*, 'Nephrotoxicity of amphotericin B, early and late events in 81 patients.' *Ann. Intern. Med.*, 1964, **61**, 175–187.

17. G. Medoff and G. S. Kobayashi, 'Strategies in the treatment of systemic fungal infections.' *N. Engl. J. Med.*, 1980, **302**, 145–155.

18. J. S. Clements Jr. and J. E. Peacock Jr., Amphotericin B revisited: reassessment of toxicity.' *Am. J. Med.*, 1990, **88**, 22N–27N.

19. R. P. Miller and J. H. Bates, 'Amphotericin B toxicity. A follow-up report of 53 patients.' *Ann. Intern. Med.*, 1969, **71**, 1089–1095.

20. R. A. Branch, E. K. Jackson, E. Jacqz *et al.*, 'Amphotericin-B nephrotoxicity in humans decreased by sodium supplements with coadministration of ticarcillin or intravenous saline.' *Klin. Wochenschr.*, 1987, **65**, 500–506.

21. W. A. Winn, 'Coccidioidomycosis and amphotericin B.' *Med. Clin. North. Am.*, 1963, **47**, 1131–1148.

22. M. L. Littman, P. L. Horowitz and J. G. Swadey, 'Coccidioidomycosis and its treatment with amphotericin B.' *Am. J. Med.*, 1958, **24**, 568–592.

23. S. I. Rubin, D. R. Krawiec, H. Gelberg *et al.*, 'Nephrotoxicity of amphotericin B in dogs: a comparison of two methods of administration.' *Can. J. Vet. Res.*, 1989, **53**, 23–28.

24. R. S. Stein, K. Albridge, R. K. Lenox *et al.*, 'Nephrotoxicity in leukemic patients receiving empirical amphotericin B and aminoglycosides.' *South. Med. J.*, 1988, **81**, 1095–1099.

25. M. A. Fisher, G. H. Talbot, G. Maislin *et al.*, 'Risk factors for Amphotericin B-associated nephrotoxicity.' *Am. J. Med.*, 1989, **87**, 547–552.

26. C. W. Holeman and H. Einstein, 'The toxic effects of amphotericin B in man.' *Calif. Med.*, 1963, **99**, 90–93.

27. W. E. Bullock, R. G. Luke, C. E. Nuttall *et al.*, 'Can mannitol reduce amphotericin B nephrotoxicity? Double-blind study and description of a new vascular lesion in kidneys.' *Antimicrob. Agents Chemother.*, 1976, **10**, 555–563.

28. G. L. Barbour, K. D. Straub, B. L. O'Neal *et al.*, 'Vasopressin-resistant nephrogenic diabetes insipidus. A result of amphotericin B therapy.' *Arch. Intern. Med.*, 1979, **139**, 86–88.

29. Y. Yano, J. L. Monteiro and A. C. Seguro, 'Effect of amphotericin B on water and urea transport in the inner medullary collecting duct.' *J. Am. Soc. Nephrol.*, 1994, **5**, 68–74.

30. B. D. Rose, in 'Clinical Physiology of Acid–Base and Electrolyte Disorders,' ed. B. D. Rose, McGraw-Hill, New York, 1989, pp. 715–756,

31. J. L. Burgess and R. Birchall, 'Nephrotoxicity of amphotericin B with emphasis on changes in tubular function.' *Am. J. Med.*, 1972, **53**, 77–84.

32. A. Llanos, J. Cieza, J. Bernardo *et al.*, 'Effect of salt supplementation on amphotericin B nephrotoxicity.' *Kidney Int.*, 1991, **40**, 302–308.

33. D. K. McCurdy, M. Frederic and J. R. Elkinton, 'Renal tubular acidosis due to amphotericin B.' *N. Engl. J. Med.*, 1968, **278**, 124–130.

34. J. H. Sellin and R. De Soigne, 'Ion transport in human colon *in vitro.' Gastroenterology*, 1987, **93**, 441–448.

35. C. H. Barton, M. Pahl, N. D. Vaziri *et al.*, 'Renal magnesium wasting associated with amphotericin B therapy.' *Am. J. Med.*, 1984, **77**, 471–474.

36. J. B. Douglas and J. K. Healy, 'Nephrotoxic effects of amphotericin B, including renal tubular acidosis.' *Am. J. Med.*, 1969, **46**, 154–162.

37. P. R. Steinmetz and L. R. Lawson, 'Defect in urinary acidification induced *in vitro* by amphotericin B.' *J. Clin. Invest.*, 1970, **49**, 596–601.

38. F. Z. Gil and G. Malnic, 'Effect of amphotericin B on renal tubular acidification in the rat.' *Pflugers Arch.*, 1989, **413**, 280–286.

39. T. H. Gouge and V. T. Andreoli, 'An experimental model of amphotericin B nephrotoxicity with renal tubular acidosis.' *J. Lab. Clin. Med.*, 1971, **78**, 713–724.

40. F. J. Takacs, Z. M. Tomkiewicz and J. P. Merrill, 'Amphotericin B nephrotoxicity with irreversible renal failure.' *Ann. Intern. Med.*, 1963, **59**, 716–724.

41. S. N. Heyman, I. E. Stillman, M. Brezis *et al.*, 'Chronic amphotericin nephropathy: morphometric, electron microscopic and functional studies.' *J. Am. Soc. Nephrol.*, 1993, **4**, 69–80.

42. G. A. Porter, in 'Nephrotoxicity in the Experimental and Clinical Situation,' eds. P. H. Bach and E. A. Lock, Martinus Nijhoff Publishers, Dordrecht, 1987, pp. 613–641.

43. T. E. Andreoli, 'On the anatomy of amphotericin B-cholesterol pores in lipid bilayer membranes.' *Kidney Int.*, 1973, **4**, 337–345.

44. A. Vertut-Croquin, J. Bolard, M. Chabbert *et al.*, 'Differences in the interaction of the polyene antibiotic amphotericin B with cholesterol- or ergosterol-containing phospholipid vesicles. A circular dichroism and permeability study.' *Biochemistry*, 1983, **22**, 2939–2944.

45. V. Joly, L. Saint-Julien, C. Carbon *et al.*, 'Interactions of free and liposomal amphotericin B with renal proximal tubular cells in primary culture.' *J. Pharmacol. Exp. Ther.*, 1990, **255**, 17–22.

46. N. Collette, P. van der Auwera, A. P. Lopez *et al.*, 'Tissue concentrations and bioactivity of amphotericin B in cancer patients treated with amphotericin B-deoxycholate.' *Antimicrob. Agents Chemother.*, 1989, **33**, 362–368.

47. R. Sabra and R. A. Branch, 'Effect of amphotericin B on intracellular calcium levels in cultured glomerular mesangial cells.' *Eur. J. Pharmacol.*, 1992, **226**, 79–85.

48. B. P. Sawaya, H. Weihprecht, W. R. Campbell *et al.*, 'Direct vasoconstriction as a possible cause for amphotericin B-induced nephrotoxicity in rats.' *J. Clin. Invest.*, 1991, **87**, 2097–2107.

49. W. T. Butler, G. J. Hill, C. F. Szwed *et al.*, 'Amphotericin B renal toxicity in the dog.' *J. Pharmacol. Exp. Ther.*, 1964, **143**, 47–56.

50. J. F. Gerkens, H. T. Heidemann, E. K. Jackson *et al.*, 'Aminophylline inhibits renal vasoconstriction produced by intrarenal hypertonic saline.' *J. Pharmacol. Exp. Ther.*, 1983, **225**, 611–615.

51. H. T. Heidemann, G. F. Gerkens, E. K. Jackson *et al.*, 'Effect of aminophylline on renal vasoconstriction produced by amphotericin B in the rat.' *Naunyn. Schmiedebergs Arch. Pharmacol.*, 1983, **324**, 148–152.

52. R. Sabra, K. Takahashi, R. A. Branch *et al.*, 'Mechanisms of amphotericin B-induced reduction of the glomerular filtration rate: a micropuncture study.' *J. Pharmacol. Exp. Ther.*, 1990, **253**, 34–37.

53. J. T. Cheng, R. T. Witty, R. R. Robinson *et al.*, 'Amphotericin B nephrotoxicity: increased renal resistance and tubule permeability.' *Kidney Int.*, 1982, **22**, 626–633.

54. J. F. Gerkens and R. A. Branch, 'The influence of sodium status and furosemide on canine acute amphotericin B nephrotoxicity.' *J. Pharmacol. Exp. Ther.*, 1980, **214**, 306–311.

55. H. T. Heidemann, M. Bolten and G. Inselmann, 'Effect of chronic theophylline administration on amphotericin B nephrotoxicity in rats.' *Nephron*, 1991, **59**, 294–298.

56. J. P. Tolins and L. Raij, 'Adverse effect of amphotericin B administration on renal hemodynamics in the rat. Neurohumoral mechanisms and influence of calcium channel blockade.' *J. Pharmacol. Exp. Ther.*, 1988, **245**, 594–599.

57. R. Sabra and R. A. Branch, 'Mechanisms of amphotericin B-induced decrease in glomerular filtration rate in rats.' *Antimicrob. Agents Chemother.*, 1991, **35**, 2509–2514.

58. J. Schnermann, 'Regulation of single nephron filtration rate by feedback—facts and theories.' *Clin. Nephrol.*, 1975, **3**, 75–81.

59. K. Thurau, 'Modification of angiotensin-mediated tubulo-glomerular feedback by extracellular volume.' *Kidney Int., Suppl.*, 1975, S202–S207.

60. F. S. Wright and J. Schnermann, 'Interference with feedback control of glomerular filtration rate by furosemide, triflocin and cyanide.' *J. Clin. Invest.*, 1974, **53**, 1695–1708.

61. A. Ohnishi, T. Ohnishi, W. Stevenhead *et al.*, 'Sodium status influences chronic amphotericin B nephrotoxicity in rats.' *Antimicrob. Agents Chemother.*, 1989, **33**, 1222–1227.

62. J. P. Tolins and L. Raij, 'Chronic amphotericin B nephrotoxicity in the rat: protective effect of calcium channel blockade.' *J. Am. Soc. Nephrol.*, 1991, **2**, 98–102.

63. J. P. Tolins and L. Raij, 'Chronic amphotericin B nephrotoxicity in the rat, protective effect of prophylactic salt loading.' *Am. J. Kidney Dis.*, 1988, **11**, 313–317.

64. C. J. Kuan, R. A. Branch and E. K. Jackson, 'Effect of an adenosine receptor antagonist on acute amphotericin B nephrotoxicity.' *Eur. J. Pharmacol.*, 1990, **178**, 285–291.

65. S. N. Heyman, B. A. Clark, N. Kaiser *et al.*, '*In-vivo* and *in-vitro* studies on the effect of amphotericin B on endothelin release.' *J. Antimicrob. Chemother.*, 1992, **29**, 69–77.

66. J. F. Bernardo, S. Murakami, R. A. Branch *et al.*, 'Potassium depletion potentiates amphotericin-B-induced toxicity to renal tubules.' *Nephron*, 1995, **70**, 235–241.

67. W. D. Hardie, J. Ebert, M. Frazer *et al.*, 'The effect of thromboxane A2 receptor antagonism on amphotericin B-induced renal vasoconstriction in the rat.' *Prostaglandins*, 1993, **45**, 47–56.

68. P. J. Schultz and J. P. Tolins, 'Adaptation to increased dietary salt intake in the rat. Role of endogenous nitric oxide.' *J. Clin. Invest.*, 1993, **91**, 642–650.

69. J. Bernardo, S. Murakami, K. Osaka *et al.*, 'The effect of L-arginine (L-Arg) on amphotericin B (AmB)-induced nephrotoxicity.' *J. Am. Soc. Nephrol.*, 1994, **5**, 917 (abstract).

70. M. Brezis, S. Rosen, P. Silva *et al.*, 'Polyene toxicity in renal medulla: injury mediated by transport activity.' *Science*, 1984, **224**, 66–68.

71. P. J. Sansen, A. A. Elfarra and A. J. Cooley, 'Methimazole protection of rats against chemically induced kidney damage *in vivo*.' *J. Pharmacol. Exp. Ther.*, 1992, **260**, 393–401.

72. K. Osaka, J. Bernardo, S. Murakami *et al.*, 'Protective role of methimazole (M) on amphotericin B (AmB)-induced nephrotoxicity.' *J. Am. Soc. Nephrol.*, 1994, **5**, 929 (abstract).

73. M. S. Paller, J. R. Hoidal and T. F. Ferris, 'Oxygen free radicals in ischemic acute renal failure in the rat.' *J. Clin. Invest.*, 1984, **74**, 1156–1164.

74. T. Nakajima, A. Hishida and A. Kato, 'Mechanisms for protective effects of free radical scavengers on gentamicin-mediated nephropathy in rats.' *Am. J. Physiol.*, 1994, **266**, F425–F431.

75. A. Soupart and G. Decaux, 'Nifedipine and amphotericin B nephrotoxicity in the rat.' *Nephron*, 1989, **52**, 278–280.

76. H. T. Heidemann, K. H. Brune, R. Sabra *et al.*, 'Acute and chronic effects of flucytosine on amphotericin B nephrotoxicity in rats.' *Antimicrob. Agents Chemother.*, 1992, **36**, 2670–2675.

77. S. L. Linas and D. Dickmann, 'Mechanism of decreased renal blood flow in the potassium-depleted conscious rat.' *Kidney Int.*, 1982, **21**, 757–764.

78. J. F. Bernardo, S. Murakami, R. Sabra *et al.*, in 'Proceedings of the 6th Congress on Nephrotoxicity and Nephrocarcinogenicity, The Netherlands, 1994,' ed. Leiden University, The Netherlands, 1994, p. 37.

79. P. D. Hoeprich, 'Clinical use of amphotericin B and derivatives: lore, mystique and fact.' *Clin. Infect. Dis.*, 1992, **14**, S114–S119.

80. H. T. Heidemann., G. F. Gerkens, W. A. Spickard *et al.*, 'Amphotericin B nephrotoxicity in humans decreased by salt repletion.' *Am. J. Med.*, 1983, **75**, 476–481.

81. G. Lopez-Berestein, V. Fainstein, R. Hopfer *et al.*, 'Liposomal amphotericin B for the treatment of systemic fungal infections in patients with cancer: a preliminary study.' *J. Infect. Dis.*, 1985, **151**, 704–710.

82. C. Gates and R. J. Pinney, 'Amphotericin B and its delivery by liposomal and lipid formulations.' *J. Clin. Pharm. Ther.*, 1993, **18**, 147–153.

83. R. Janknegt, S. de Marie, I. A. J. M. Bakker-Wondenberg *et al.*, 'Liposomal and lipid formulations of amphotericin B. Clinical pharmacokinetics.' *Clin. Pharmacokinet.*, 1992, **23**, 279–291.

84. G. Gregoriadis, 'Overview of liposomes.' *J. Antimicrob. Chemother.*, 1991, **28**, 39–48.

85. N. I. Payne, R. F. Cosgrove, A. P. Green *et al.*, '*In-vivo* studies of amphotericin B liposomes derived from proliposomes: effect of formulation on toxicity and tissue disposition of the drug in mice. *J. Pharm. Pharmacol.*, 1987, **39**, 24–28.

86. F. C. Szoka, Jr., D. Milholland and M. Barza, 'Effect of lipid composition and liposome size on toxicity and *in vitro* fungicidal activity of liposome-intercalated amphotericin B.' *Antimicrob. Agents Chemother.*, 1987, **31**, 421–429.

87. R. L. Juliano, C. W. M. Grant, K. R. Barber *et al.*, 'Mechanism of the selective toxicity of amphotericin B incorporated into liposomes.' *Mol. Pharmacol.*, 1987, **31**, 1–11.

88. K. V. Clemons and D. A. Stevens, 'Comparative efficacy of amphotericin B colloidal dispersion and amphotericin B deoxycholate suspension in treatment of murine coccidiodomycosis.' *Antimicrob. Agents Chemother.*, 1991, **35**, 1829–1833.

89. J. S. Hostetler, K. V. Clemons, L. H. Hanson *et al.*, 'Efficacy and safety of amphotericin B colloidal dispersion compared with those of amphotericin B deoxycholate suspension for treatment of disseminated murine cryptococcosis.' *Antimicrob. Agents Chemother.*, 1992, **36**, 2656–2660.

90. R. T. Proffitt, A. Satorius, S. M. Chiang *et al.*, 'Pharmacology and toxicology of a liposomal formulation of amphotericin B (AmBisome) in rodents.' *J. Antimicrob. Chemother.*, 1991, **28**, 49–61.

91. E. D. Ralph, A. M. Khazindar, K. R. Barber *et al.*, 'Comparative *in vitro* effects of liposomal amphotericin B, amphotericin B-deoxycholate and free amphotericin B against fungal strains determined by using MIC and minimal lethal concentration susceptibility studies and time-kill curves.' *Antimicrob. Agents Chemother.*, 1991, **35**, 188–191.

92. E. Anaissie, V. Paetznick, R. Proffitt *et al.*, 'Comparison of the *in vitro* antifungal activity of free and liposome-encapsulated amphotericin B.' *Eur. J. Clin. Microbiol. Infect. Dis.*, 1991, **10**, 665–668.

93. H. J. Krause and R. L. Juliano, 'Interactions of liposome-incorporated amphotericin B with kidney epithelial cell cultures.' *Mol. Pharmacol.*, 1987, **34**, 286–297.

94. V. Joly, F. Dromer, J. Barge *et al.*, 'Incorporation of amphotericin B (AMB) into liposomes alter AMB-induced acute nephrotoxicity in rabbits.' *J. Pharmacol. Exp. Ther.*, 1989, **251**, 311–316.

95. P. Longuet, V. Joly, P. Amirault *et al.*, 'Limited protection by small unilamellar liposomes against the renal tubular toxicity induced by repeated amphotericin B infusion in rats.' *Antimicrob. Agents Chemother.*, 1991, **35**, 1303–1308.

96. J. A. Gondal, R. P. Swartz and A. Rahman, 'Therapeutic evaluation of free and liposome-encapsulated amphotericin B in the treatment of systemic candidiasis in mice.' *Antimicrob. Agents Chemother.*, 1989, **33**, 1544–1548.

97. G. Lopez-Berestein, R. Mehta, R. L. Hopfer *et al.*, 'Treatment and prophylaxis of disseminated infection due to *Candida albicans* in mice with liposome-encapsulated amphotericin B.' *J. Infect. Dis.*, 1983, **147**, 939–945.

98. G. Lopez-Berestein, G. P. Bodey, L. S. Frankel *et al.*, 'Treatment of hepatosplenic candidiasis with liposomal-amphotericin B.' *J. Clin. Onc.*, 1987, **5**, 310–317.

99. G. Lopez-Berestein, 'Liposomes as carriers of antifungal drugs.' *Ann. NY Acad. Sci.*, 1988, **544**, 590–597.

100. G. Lopez-Berestein, G. P. Bodey, V. Fainstein *et al.*, 'Treatment of systemic fungal infections with liposomal amphotericin B.' *Arch. Intern. Med.*, 1989, **149**, 2533–2536.

101. F. Meunier, J. P. Sculier, A. Coune *et al.*, 'Amphotericin B encapsulated in liposomes administered to cancer patients.' *Ann. NY Acad. Sci.*, 1988, **544**, 598–610.

102. A. Llanos-Cuentas, J. Chang, J. Cieza *et al.*, in 'Program and Abstracts of the 30th Interscience Conference on Antimicrobial Agents and Chemotherapy, Atlanta, GA, 1990,' ed. American Society for Microbiology, 1990, Abstract 568.

103. F. Meunier, H. G. Prentice and O. Ringdén, 'Liposomal amphotericin B (AmBisome): safety data from a Phase II/III clinical trial.' *J. Antimicrob. Chemother.*, 1991, **28**, 83–91.

104. O. Ringdén, F. Meunier, J. Tollemar *et al.*, 'Efficacy of amphotericin B encapsulated in liposomes (AmBisome) in the treatment of invasive fungal infections in immunocompromised patients.' *J. Antimicrob. Chemother.*, 1991, **28**, 73–82.

105. R. L. Taylor, D. M. Williams, P. C. Craven *et al.*, 'Amphotericin B in liposomes: a novel therapy for histoplasmosis.' *Am. Rev. Respir. Dis.*, 1982, **125**, 610–611.

106. F. Meunier, 'New methods for delivery of antifungal agents.' *Rev. Infect. Dis.*, 1989, **11**, S1605–S1612.

107. J. R. Graybill, P. C. Craven, R. L. Taylor *et al.*, 'Treatment of murine crytococcosis with liposome-associated amphotericin B.' *J. Infect. Dis.*, 1982, **145**, 748–752.

108. K. M. Wasan, G. A. Brazeau, A. Keyhani *et al.*, 'Roles of liposome composition and temperature in distribution of amphotericin B in serum lipoproteins.' *Antimicrob. Agents Chemother.*, 1993, **37**, 246–250.

109. K. M. Wasan, M. G. Rosenblum, L. Cheung *et al.*, 'Influence of lipoproteins on renal cytotoxicity and antifungal activity of amphotericin B.' *Antimicrob. Agents Chemother.*, 1994, **38**, 223–227.

110. K. M. Wasan and G. Lopez-Berestein, 'Modification of amphotericin B's therapeutic index by increasing its association with serum high-density lipoproteins.' *Ann. NY Acad. Sci.*, 1994, **730**, 93–106.

111. S. J. Olsen, M. R. Swerdel, B. Blue *et al.*, 'Tissue distribution of amphotericin B lipid complex in laboratory animals.' *J. Pharm. Pharmacol.*, 1991, **43**, 831–835.

112. G. Poste, 'Liposome targetting *in vivo*: problems and opportunities.' *Biol. Cell.*, 1983, **47**, 19–37.

113. R. Chopra, S. Blair, J. Strang *et al.*, 'Liposomal amphotericin B (AmBisome) in the treatment of fungal infections in neutropenic patients.' *J. Antimicrob. Chemother.*, 1991, **28**, 93–104.

114. P. Y. Chavanet, I. Garry, N. Charlier *et al.*, 'Trial of glucose versus fat emulsion in preparation of amphotericin for use in HIV infected patients with candidiasis.' *BMJ*, 1992, **305**, 921–925.

7.26
Aminoglycoside Nephrotoxicity

CONSTANTIN COJOCEL

Hoechst Marion Roussel, Frankfurt am Main, Germany

7.26.1 INTRODUCTION

The use of aminoglycoside (AG) is limited by concerns about toxicity, primarily nephrotoxicity, and ototoxicity. The kidney is especially vulnerable to toxic injury because of its rich blood supply and capacity to concentrate toxicants within tubule epithelial cells and the interstitium via various transport processes and renal counter-current exchange. It is also an important site of xenobiotic metabolism and is able to transform parent compounds into reactive toxic metabolites.[1]

Although AG are not metabolized in the renal tubular cells, intracellular bioactivation of AG appears to play a role in their toxic injury.[2] As with ischemic acute tubular necrosis, nephrotoxicants impair glomerular filtration rate (GFR) by causing intrarenal vasoconstriction, direct injury to tubule epithelium, and tubule obstruction. The contribution of each pathogical and biochemical mechanism to the development of acute renal failure varies among different agents. Whereas intrarenal vasoconstriction triggers the cyclosporin-induced acute renal failure,[3] direct epithelial cell toxicity appears to be the primary cause of acute tubular necrosis induced by AG.[4]

Early symptoms of AG nephrotoxicity observed in the period from 24 h up to the fifth day of AG treatment are excretion of myeloid bodies in the urinary sediment,[5] an increase in the excretion of brush border and lysosomal enzymes, and proteinuria.[6–10] Characteristic symptoms of nephrotoxicity after 7–10 days of AG treatment, such as an increase in serum creatinine, a decrease in urinary osmolality, and tubule cell necrosis, can be regarded as late symptoms of AG nephrotoxicity.[9,11–13]

Although newer antibiotics such as extended-spectrum β-lactams and fluorinated quinolones have been developed, AG continue to be used widely in clinical practice, despite their toxicity and narrow therapeutic ratio, due to their rapid bactericidal action, post-antibiotic effect, and other pharmacodynamic parameters which allow infrequent dosing schedules, and to their common synergistic pharmacological action when used in combination with β-lactam antibiotics.

In this chapter, the differential nephrotoxic effects and the mechanisms of aminoglycoside nephrotoxicity are reviewed.

7.26.2 AMINOGLYCOSIDE UTILIZATION

Following the discovery of streptomycin,[14] AG antibiotics have found wide application in the treatment of bacterial infections. Despite numerous new antibiotics used in clinical management of infectious diseases, aminoglycoside (AG) antibiotics are still widely prescribed because of proven efficacy against life-threatening aerobic Gram-negative infections, especially when resistance to β-lactam antibiotics is suspected. AG are also active against many staphylococci, and certain microbacteria.[15]

AG belongs to the group of antibacterial active substances which inhibit ribosomal protein synthesis by binding to the 30S ribosomal subunit and interfere with translation of mRNA after active uptake into bacteria.[16] In their studies on the *in vitro* inhibition of protein biosynthesis, Benveniste and Davies[17] showed that tobramycin, kanamycin B, gentamicin C 1a, and sisomicin were all equally active on an equimolar concentration basis. Additional AG such as amikacin and netilmicin are available for the treatment of life-threatening bacterial infections. The intracellular mechanism of action of AG differs from that of β-lactam antibiotics which inhibit cell wall synthesis. The alterations in cell wall integrity caused by β-lactams can, in many cases, facilitate AG transport across bacterial cell wall. This accounts for the synergy that is frequently observed when the two types of antibacterial agents are combined.

AG consist, for the most part, of a central 2-deoxystreptamine molecule with glycosidic linkages to two or more amino sugars. AG react basically ($pK_a \geq 8$), are highly water-soluble, have low lipid solubility, and low capacity to penetrate membranes. The number of amino groups per AG molecule, and their cationic structure, appears to be correlated to the nephrotoxic potential.[18,19] As highly polar cations, AG are absorbed poorly from the intestinal tract but absorbed rapidly from intramuscular and subcutaneous site of injection. AG solutions are most stable at a pH of between six and eight and are traditionally administered by intermittent infusion 2–3 times a day.

Two main families of AG antibiotics can, in principle, be distinguished: the kanamycins and the gentamicins. Kanamycins A, B, and C, amikacin, and tobramycin belong to the kanamycin group, whereas gentamicins C1, C1a, and C2, sisomicin, and netilmicin are assigned to the gentamicin group. Davies and Courvalin[20] divided AG into different generations according to their molecular structure and antibacterial mode of action. Thus the first generation of AG is represented by kanamycin and kanamycin derivatives and the second generation by gentamicin, sisomicin, and tobramycin, which have a broader antibacterial effect than kanamycin. Amikacin and netilmicin are third-generation AG antibiotics which exhibit increased activity against microorganisms resistant to second-generation AG. Amikacin is a derivative of kanamycin A, whereas netilmicin is a direct derivative of sisomicin or gentamicin C 1a.[21] However both netilmicin and amikacin show a distinctly broader antibacterial spectrum than the parent substance.

Extensive use of AG in hospital has led to a substantial increase of the bacterial resistance to AG, which is quantitatively related to their use. Bacterial resistance to AG is caused predominantly by AG-modifying enzymes which are capable of adenylating, phosphorylating, or acetylating the target AG molecule resulting in loss of antimicrobial activity. AG resistance of 378 strains of Gram-negative bacteria isolated from urine was investigated.[22] The observed resistance to gentamicin was 16%, to tobramycin 11%, to netilmicin 3%, and to amikacin 4%. A ten-year study[23] reported a reduction in overall AG resistance after introduction of high-level amikacin use: amikacin 3.8% to 3.2%, gentamicin 12% to 6.4%, and tobramycin 9.5% to 4.8%. From a clinical point of view, more careful use of AG should restrict the development of resistance.

7.26.3 CLINICAL INCIDENCE AND SEVERITY

A number of studies have defined patient populations at high risk for development of acute renal failure (ARF) during a course of AG treatment.[24–26] ARF is characterized by a rapid decline in GFR, perturbation of extracellular fluid volume, electrolyte and acid-base homoeo-stasis, and retention of nitrogenous waste from protein catabolism.[27] AG-induced ARF is typically nonoliguric, without any decline in the urinary volume, and frequently occurs 7–10 days during the course of AG treatment.

Patients who are septic and hypotensive are more likely to suffer oliguric renal failure ($<400 \, \text{ml} \, \text{d}^{-1}$) early in the course of AG treatment. ARF is typically asymtomatic and often is detected by an acute rise in serum creatinine and blood urea nitrogen. ARF is usually reversible since the kidney is able to recover from almost complete loss of function. Acute tubular necrosis (ATN), one of the ARF forms, is caused by AG.[5] The overall mortality rate from ATN (from all causes) is about 50% and has changed little between 1960 and the 1990s.[27] Mortality rates differ, however, depending on the cause of ARF, being about 30% in nephrotoxicant-induced ARF. Most patients who survive begin renal recovery within 10–21 days and regain sufficient renal function to live normal lives. ARF is irreversible in about 5% of patients, probably due to complete cortical necrosis.[28]

The high incidence of associated nephrotoxicity represents an important concern in the clinical use of AG. In various types of populations of patients treated with AG, the incidence of nephrotoxicity has ranged from less than 2% to almost 50%.[29,30] Relatively healthy patients receiving a short course of an AG may be expected to have a low incidence of nephrotoxicity, whereas hypotensive, septic patients and those receiving a prolonged, high-dose course of AG therapy for endocarditis or osteomyelites are likely to have a much higher incidence of renal injuries.

However, the overall incidence of AG-induced nephrotoxicity in large unselected populations has been estimated to occur in 5–10% of patients.[24,31–33] Another factor affecting the incidence of nephrotoxicity is the criteria used to assess renal damage. In early clinical trials, nephrotoxicity was defined by the presence of cylinduria and other urinary sediment changes, tubular and/or glomerular proteinuria, or small changes of the serum creatinine. However, such changes do not directly represent significant alterations of the GFR. Well-designed clinical trials in the mid-1990s have used significant changes in the serum creatinine as the marker for the renal damage and consequently report lower incidence of renal injuries than did prior studies that used more sensitive but less specific tests. In many studies, nephrotoxicity was defined as an increase in serum creatinine of more than 50% or $35 \, \mu\text{mol} \, \text{L}^{-1}$ above the base line.[34–36]

Finally, the specific AG used may influence the incidence of nephrotoxicity. The nephrotoxic potentials of the AG available in the mid-1990s, have been widely debated. Clearly, in any given study, multiple factors can influence the incidence of nephrotoxicity. Patients who are aged, debilitated, or have preexisting renal insufficiency have a higher incidence of nephrotoxicity.[21,28,37] Even when plasma AG concentrations are mentioned within the recommended ranges, 10–15% of treated patients showed clinically detectable decreases in renal function.[37,38] In a large study of hospital acquired acute renal failure, 11% of over 2000 cases were attributable to AG.[39]

7.26.4 RISK FACTORS

A higher incidence of nehrotoxicity occuring under treatment with AG is dependent on various factors: the AG administered dose, frequent dosing intervals, prolonged duration of treatment, presence of hypotension, hypovolemia, shock, liver and/or renal disease, patient age, gender, obesity, the quality of the monitoring of the treatment, and previous treatment with AG.[24–26,37,38,40]

A number of other therapeutic agents may potentiate the nephrotoxicity when given concurrently with an AG.[41–46]

The results of many studies raise questions about the impact of some identified risk factors. It has been proposed, for example, that previous AG treatment might increase risk only if the therapy was recent, allowing rapid saturation of the renal cortex tissue as a result of the prolonged tissue half-life of AG. Similarly, although both high peak and trough serum levels of AG have been claimed to correlate with nephrotoxicity, in many AG treated patients serum levels are poor predictors of renal outcome.[47] However, in many studies, large total doses of AG given for prolonged time intervals appeared to correlate with nephrotoxicity. Of those risk factors associated with the patient's status at the time of AG administration, volume depletion, hypotension, shock, and liver disease appear to be most consistently involved in the development of AG nephrotoxicity. Any preexisting renal disease and the advanced age of the patient may increase risk of nephrotoxicity only if the dose of the AG is not adjusted appropriately according to the diminished renal function of such patients.

Gender of both human and animals also has to be considered as a risk factor in AG-induced nephrotoxicity. Although the female gender proved to be a risk factor in some clinical studies,[24,33,48] this was not confirmed by other studies either in human or animal models, and males showed nephrotoxic effects when treated with AG.[35,45,46,49–52]

Many Gram-negative infections capable of inducing a prologed endotoxemia could lead to synergistic toxic effects with AG.[53] Interactions between AG and endotoxin, a lipopolysacchride (LPS) constituent of the cell wall of Gram-negative bacteria, released locally or systemically are very complex.

In patients septic shock, secondary to endotoxin release into the circulation, or liver failure (no detoxification of endotoxin) is associated with an increased incidence of AG nephrotoxicity.[24] Endotoxin modifies the accumulation kinetics of gentamicin and tobramycin in renal cortex and isolated renal tubules resulting in an increased accumulation of [^{3}H]tobramycin and toxicity in proximal tubules[53–55] from endotoxemic rats. Taken together, the results of animal studies demonstrate that renal infection, Gram-negative sepsis, and endotoxin potentiate AG nephrotoxicity.

The mechanism by which endotoxin and AG act synergistically to alter renal function is complex. LPS can disturb the hemodynamic status of the host, activate immunological and inflammatory processes, and cause injuries to cell organelles. Endotoxin is able to stimulate the release of vasoactive hormones and mediators of inflammation such as tumor necrosis factor-alpha (TNF-α) and interleukin-1. TNF-α is an important mediator during shock and renal damage caused by LPS.[56] TNF-α release results in a reduction of the vasodilator PGI$_2$ production by vascular endothelium. In addition, TNF-α promotes adhesion receptor expression on granulocytes and endothelial cells, which, along with the elaboration of inflammatory eicosanoids and TXA2, promotes vascular stasis and thrombosis. It appears that TNF-α plays a role in amphotericin nephrotoxicity[57,58] and gentamicin nephrotoxicity.[53]

Pentoxifylline (PTX), a hemorrheologically active methylxantine derivative, stimulates the synthesis and release of endothelial prostacyclin PGI$_2$ while reducing vascular hyperviscosity.[59,60] PTX has been shown to inhibit TNF-α production via the inhibition of TNF-mRNA.[61,62] In a rat study and in bone marrow patients receiving amphotericin B, PTX protected against amphotericin-induced nephrotoxicity and promptly reversed renal dysfunction and allowed continued administration of the immunosuppressive agent cyclosporine A and amphotericin B.[62,63] Similarly, PTX could be protective against nephrotoxicty caused by AG after treatment of endotoxemic patients following Gram-negative infections.

7.26.4.1 Combination Therapy

Concurrent use of nephrotoxic agents such as vancomycin has been claimed to increase the risk of AG nephrotoxicity. Cephaloridine and other cephalosporins are able to cause dose-related damage to the proximal tubule, and a number of studies have shown synergistic nephrotoxicity between AG and cephalosporins.[28] However, in animal studies it has been difficult to demonstrate synergistic nephrotoxicity between these antibiotics.[64] In fact, a protective effect against AG nephrotoxicity from concurrent cephalosporin use has been reported in some animal studies.[42] Likewise, AG nephrotoxicity has not been shown to be increased in a significant manner by concurrent cephalosporin use in humans.[28]

Vancomycin, another potentially nephrotoxic drug, causes renal injury in less than 5% of treated patients.[65,66] A number of studies, however, suggest that as many as 35% of patients treated with the combination of vancomycin and an AG will experience nephrotoxicity, confirming animal data of synergistic renal damage between the two drugs.[41,43,65,67,68] The risk of nephrotoxicity, in patients treated with combined vancomycin and an AG is enhanced by various factors such as increased age, male sex, liver disease, concurrent amphotericin therapy, neutropenia, and peritonitis.[66] Length of therapy and serum concentrations of AG and vancomycin are also associated with increased nephrotoxicity. The effect of vancomycin on tobramycin binding to rat renal brush border membrane was investigated.[69] This study showed that binding of tobramycin to the brush-border membrane was enhanced by preincubation of membrane vesicles with vancomycin because of the increase in the number of the negatively charged binding sites on the membrane surface.

Loop diuretics and other potent diuretics also may enhance AG nephrotoxicity. Based on the results from animal models, it appears that AG nephrotoxicity is greater with diuretics that induce volume depletion.[37]

Other therapeutic agents associated with an increased risk of AG nephrotoxicity include amphotericine B, cisplatin, cyclosporine, parathyroid hormone nonsteroidal anti-inflammatory drugs, and radiographic contrast agents.[52,70,71]

7.26.4.2 Differential Nephrotoxicity of Aminoglycosides

One of the most debated issues concerns the differential nephrotoxicity of the AG.[41,43–46] While streptomycin has only minimal nephrotoxic potential, neomycin is assessed as one of the most nephrotoxic of these agents. Of the widely clinically used AG, gentamicin has been shown to be more nephrotoxic than tobramycin or netilmicin in numerous *in vitro* and *in vivo* animal studies.[28,35,37,72] Amikacin appears to be less nephrotoxic than gentamicin. Kanamycin and sisomicin with a nephrotoxic potential at least comparable to that of gentamicin are not often clinically used.[73]

Direct measurement of the AG concentration in the renal cortex in patients with normal renal function showed a somewhat lower renal accumulation of tobramycin compared to gentamicin. In patients with diseased kidneys,[74] a higher cortical accumulation of tobramycin was shown when compared to gentamicin. Despite similar cortical concentrations[75] and an equal number of amino groups per AG molecule, clinical and experimental findings showed netilmicin to be less nephrotoxic than gentamicin.[76]

A number of studies show that AG exert differential neprotoxic injuries at different sites in the nephron.[21,23,31,77–81] While the results of some studies suggest that gentamicin has higher nephrotoxicity than tobramycin,[82] the results of other studies show that gentamicin and tobramycin are of essentially equal nephrotoxicity, but are more nephrotoxic than kanamycin and amikacin.[76,83] The latter two cause only slight functional and structural alterations, though kanamycin appears to be more nephrotoxic than amikacin.

In contrast, at the same dosage, netilmicin has practically no effect on protein filtration, reabsorption, and degradation, or on the glomerular ultrastructure.[76,77] Higher doses of amikacin, kanamycin, and netilmicin are nonetheless linked with functional and structural changes.[84,85]

In low-risk patients, there is probably little significant difference in the nephrotoxic potential between gentamicin, tobramycin, or netilmicin. However, in patients at high risk for nephrotoxicity, even small differences in nephrotoxic potential between individual drugs may potentiate other risk factors and add to the burden of renal damage.

7.26.4.3 Fractionation of the Daily Dose

Both animal and human studies have shown that the AG concentration in the renal cortex is significantly lower when administered as a single daily dose than when the same dose is divided and administered more frequently.[73]

7.26.5 DETECTION AND TREATMENT

The clinical pattern of AG nephrotoxicity has been well studied, both in humans and in animal models.[32,37] In the early stage of AG nephrotoxicity, initial appearance of brush-border and lysosomal enzymes in the urine and granular casts in the urinary sediment is followed by the development of polyuria and nephrogenic diabetes insipidus. A subsequent decline in the GFR is associated with a rise in both blood urea nitrogen and plasma creatinine concentration. In common with other forms of ATN, the urinary sodium concentration (greater than $40\,mEq\,L^{-1}$) and fractional excretion of sodium (greater than 1% of the filtered load) are typically high.[32,52,76]

The most frequently used parameter for detection of nephrotoxicity is the rise in serum creatinine. This, however, has the disadvantage of being unable to register early and less extensive damage to the kidney.[87]

Kidney damage is already clinically manifest after 24 hours and can be detected in the form of phospholipiduria accompanied by increased excretion in the urine of brush-border and/or lysosomal marker enzymes such as alanine aminopeptidase and *N*-acetyl-β-glucosaminidase, respectively.[8–10] Increased excretion of low-molecular-weight proteins (LMWP) such as β_2-microglobulin and lysozyme (muramidase) is observed after 3–5 days.[74,76,88] Hence, a decrease in tubular reabsorption of LMWP, measured as increased protein excretion, is also a sensitive and early parameter of gentamicin-induced tubular damage since under normal conditions 99% of glomerularly filtered protein such as lysozyme undergoes almost complete tubular reabsorption.[89] Thus, daily determination of LMWP and/or lysosomal enzymes in urine allows early recognition of tubular damage.[17,74,78,88,90] The lysosomal enzyme *N*-acetyl-β-D-glucosaminidase (NAG) cannot be filtered by the glomerulus because of its molecular size, hence NAG activity measured in urine should be indicative of the increased lability of the lysosomal membrane and enzyme release from kidney tubules.[90–93]

After 7 days of gentamicin treatment, the decrease in the concentrating capability of the kidney, measured as a decrease in urinary osmolality and increasing proximal tubule cell necrosis,[21] is paralleled by a significant increase in sodium and potassium excretion.[94] After 10 days of gentamicin treatment there is a significant increase in the serum creatinine concentration which occurs at the point at which proximal tubule cell necrosis is at a maximum.[21] When gentamicin therapy is prolonged to 14–28 days, the serum creatinine concentration increases, whereas proximal tubule cell necrosis shows a clear decrease, as does the gentamicin concentration in the kidney tissue.[21]

It should be noted that regeneration of the renal cortex and reversibility of kidney function can progress even in the presence of continued AG admnistration and toxic drug levels in the renal cortex.[94,95]

During gentamicin treatment lasting more than two weeks, rats showed clear toxic renal effects between the seventh and 14th day but with clear regression of the pathological findings and the impairment of functional parameter in the second and third week, in spite of continuing AG treatment.[94] A reversal of the pathophysiological renal injuries to the control conditions occurs after both low and higher doses of AG.

As long as the threshold concentration in the rat kidney parenchyma is not reached, recovery of renal structure and function takes place without necrosis and regeneration.[96] At higher

dosages and periods of treatment exceeding 10 days, necrosis and regeneration processes occur concomitantly.

After gentamicin treatment, an increase in the incorporation of [^{3}H]-thymidine into proximal tubule cells was observed.[97] During these processes, necrotic epithelial cells are excreted in the urine. This may provide an explanation for the decrease in the gentamicin concentration detected in kidney parenchyma during or after treatment. It means that at high gentamicin doses, no correlation between renal AG concentration and nephrotoxicity can be found after the start of the necrosis-regeneration process. Moreover, the new epithelial cells possess a lower accumulation ability and a greater tolerance towards AG. Using transmission electron microscopy (TEM) detection of urinary sediments from patients and animals,[5] it has been shown that tobramycin induces earlier but less severe AG nephrotoxicity than gentamicin and that renal ultrastructural changes may occur in the absence of major renal dysfunction.[5]

Gentamicin-induced displacement of LMWP from the binding sites of the proximal tubule leads to an almost total inhibition of tubular protein reabsorption.[22] This can be detected by measuring the quotient protein clearance GFR, which in the case of lysozyme rose from 0.23 (control conditions) to one, after 20 min of gentamicin infusion.[22] These results are in good agreement with other research, suggesting the rapid development of nephrotoxic side effects after i.v. administration of gentamicin.[96]

7.26.5.1 Once-daily Dosing vs. Multiple Dosing

In the acute clinical care of patients with life-threatening bacterial infections, two types of administration are common for the daily AG dose: continuous infusion as single dose and fractionation of the dose. Traditional AG dosage regimens, usually every 8–12 hours, were designed to maintain the serum concentration above the minimum inhibitory concentration for the majority of the dosage interval, with the intention of preventing regrowth of organisms. It was assumed that the bactericidal action would be optimal once the minimum bactericidal concentration was reached, and that high peak AG concentrations would correlate with toxicity and not necessarily with improved antibacterial activity. Although routine monitoring of serum AG concentrations is an accepted practice, there is only weak evidence to support it.[47] Previous clinical studies[98,99] had already demonstrated for amikacin, gentamicin, and tobramycin as well as for netilmicin the lack of any obvious correlation between peak and trough serum concentrations and the onset of toxicity.

With traditional daily dosing, inaccurate timing of AG dose administration and inaccurate drawing of serum drug concentrations can lead to misinterpretation or a delay in interpreting serum concentrations, resulting potentially in suboptimal dosing and increased nephrotoxicity.[100] Experimental data concerning the pharmacodynamics of AG support the concept of using a once-daily dose, rather than fractionating the daily dose to two or three portions given every 8 or 12 hours.[100–102]

AG eradicate bacteria at a rate proportional to the peak serum drug concentration attained. This is commonly referred to as concentration-dependent killing. For optimal antibacterial action, AG must be present in sufficiently high concentration at the infection site.

High peak levels of AG result in a more rapid and efficient bactericidal effect against susceptible organisms. AG also inhibit the growth of susceptible organisms even after the serum concentration has fallen below the MIC. This phenomenon is referred to as the postantibiotic effect (PAE). The duration of the PAE for AG with Gram-negative bacteria has been reported to be from 3–24 hours *in vivo*, depending on the peak concentration and the organism. The mechanism of PAE is believed to be irreversible drug binding to bacterial ribosomal subunits and cause nonlethal damage.[103,104]

In vitro and *in vivo* animal studies have shown that microorganisms exhibit a temporary down-regulation of AG uptake which persists for several hours after each administration. If the first exposure is sublethal, the bacterial cell down-regulates its uptake of AG in a process known as adaptive resistance. Surviving bacteria stop further internalizing of AG and become refractory to subsequent drug for a period of time. Exposure of organisms to another AG dose during this temporary period of adapative resistance results in diminished killing.[105,106] Longer dosing intervals take advantage of the PAE of AG and avoid this adaptive resistance phenomenon.

Occurrence and the degree of toxic tubular injury appear to be correlated to the intracellular AG accumulation in the renal cortex.[107]. In the absence of tubular necrosis, the degree of AG nephrotoxicity at lower AG dosages is dependent on the intracellular AG concentration in the renal parenchyma.[96,108]

Since the uptake of AG into renal tubular cells is a saturable process,[96,108] the more important determinant of nephrotoxicity is the duration of AG exposure to the kidney. Administering the total daily dose as one infusion reduces the duration of exposure of renal cells to AG and thus results in less accumulation and nephrotoxicity.

An increasing number of experimental and clinical studies show that single daily dosing regimens are as effective as multiple daily doses, have a lower incidence of nephrotoxicity and delay the development of nephrotoxicity.[109–111]

The data from more than 20 clinical trials on the safety and efficacy of AG given either once daily or as divided doses were assessed in three independent meta-analyses.[35,36,112] Taken together, the results of these meta-analyses show that once-daily administration of AG in patients without pre-existing renal impairment is as effective as multiple daily dosing or there is a small but significant difference in clinical efficacy in favor of once-daily administration.[36] Furthermore, the results of these meta-analyses either showed no difference in toxicity between once-daily and multiple daily dosing or that the single daily dose had a lower risk of nephrotoxicity and there was no significant difference in mortality.[35]

Thus, once-daily dosing of AG enables maximum clinical efficacy by achieving optimum concentration-dependent bactericidal activity and the postantibiotic effect, and avoids first-exposure adaptive resistance. Considering the additional convenience and reduced cost,[113] once daily dosing should be the preferred mode of AG administration to certain patient populations.

Patients receiving an AG should be well hydrated to avoid volume depletion, and should be potassium and magnesium repleted, especially when concurrent treatment with potent diuretics is required. If GFR declines, one should consider replacement of the AG with a different, but equally efficacious antibiotic. AG-induced renal injury can continue and develop to severe ARF, even if treatment was discontinued at the time the serum creatinine begins to rise.

Since even a small decrease in GFR can be a signal of progressive renal injury, recognizing incipient renal damage is essential. Already established AG-induced ARF must be clinically managed as any other type of ARF, maintaining appropriate fluid and electrolyte levels and, when necessary, dialysis.[27]

7.26.6 PATHOPHYSIOLOGY OF AMINOGLYCOSIDE NEPHROTOXICITY

Numerous animal studies, as well as several studies of AG-induced injuries to human kidneys, have examined the pathophysiological changes involved in AG nephrotoxicity.[34,114–117] AG-induced glomerular and tubular damage is dependent on a number of factors, including dosage and duration of treatment, the age and gender of the patient, preexisting renal disease, endotoxemia, hypomagnesemia, and hypokalemia.[118–122]

After administration of AG over several days, the rapidity of AG accumulation in the renal cortex and the severity of subsequent pathophysiological changes of the glomerulus and tubules depend to a great extent on dose, dosage regimen (undivided or divided into 2–3 portions), dosage interval, and period of treatment. After a 4 day gentamicin treatment (undivided dose), the threshold concentration is reached both in humans[123] and in rats[124,125] regardless of the dose or the mode of administration. Subdivision of the daily dose of gentamicin led to a more rapid accumulation of gentamicin in the rat renal cortex and to saturation of renal gentamicin accumulation after 4 days. The plateau for renal gentamicin accumulation did not change after 7 days of treatment and remained relatively unchanged on continuous treatment of rats over 14, 21, or 28 days.[108] However, increased AG accumulation in the renal cortex does not necessarily imply increased toxicity.[126,127] Therefore, the induction of pathophysiological changes cannot be inferred automatically from cumulative AG potency.

Ultrastructural injuries to the tubular epithelium and the endothelium of the glomerular capillaries are characteristic features of AG nephrotoxicity. Although proximal tubular necrosis is the major renal consequence of aminoglycoside nephrotoxicity, decreased GFR is the major cause of morbidity.[31,37,86]

7.26.6.1 Changes in Glomerular Ultrastructure and Permeability

It is recognized generally that water and solute pass the glomerular capillary wall through an extracellular pathway which consists of endothelial fenestrae (EF), basement membrane, the pores of slit diaphragms, and the filtration slits between foot-like processes of the podocytes (Figure 1). Filtration of macromolecules moving through this pathway is restricted not only by the rate of glomerular plasma flow and osmotic presssure across the capillary wall, but also by their size, shape, and surface charge, which will interact with the strong negatively charged filtration pathway.[128,129] Positively charged substances, such

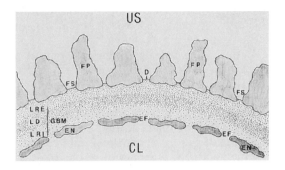

Figure 1 Schematic description of the filtration barrier. EN, endothelium; EF, endothelial fenestrae. Glomerular basement membrane (GBM) with LRI, lamina rara interna; LD, lamina densa; and LRE, lamina rara externa. FP, foot processes of podocytes; FS, filtration slits; D, slit diagram; CL, capillary lumen; US, urinary space.

as the low-molecular weight protein lysozyme, can cross the filtration barrier (Figure 2) more easily than those of neutral or negative charge with the same molecular size.[128–131]

7.26.6.1.1 Changes after acute aminoglycoside treatment

In animal studies, after a single dose of gentamicin (30 or 60 mg kg^{-1}), no ultrastructural changes of the glomerular capillary wall layers could be established.[76,132] However, acute AG treatment does alter glomerular permeability. The fractional clearance of a low molecular weight protein (LMWP) is expressed by the ratio of its urinary clearance and GFR. Under the conditions of complete inhibition of tubular lysozyme reabsorption by sodium iodoacetate, the fractional clearance of lysozyme (C_{LY}/GFR) is a measure of the glomerular sieving coefficient for lysozyme (GSC_{LY}). Perfusion of the isolated rat kidney with gentamicin increased GSC_{LY} from 0.8 to 1.0.[12] The results of these and other experimental studies[133] indicate that the acute administration of polycationic substances increases glomerular permeability to proteins. Interestingly, a mean value of 0.87 has been found for the glomerular ultrafiltration ratio of AG,[134] a value close to the control value of 0.8 found for GSC_{LY}.

As the GSC_{LY} was measured as the C_{LY}/GFR, the gentamicin-induced reduction in GFR, accompanied by an increase in lysozyme clearance in the intact animal, may in part explain the increase in GSC_{LY} from 0.8 to 1.0 following intravenous infusion of a single dose of gentamicin to rats or gentamicin perfusion of the isolated rat kidney.[22] Various polycations are

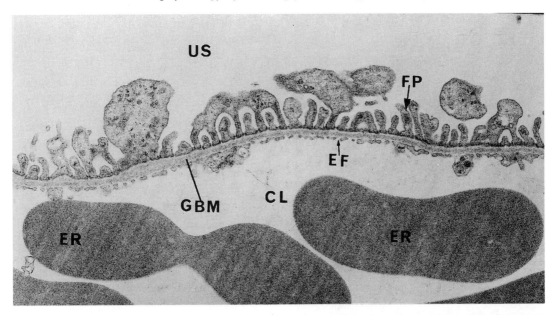

Figure 2 Transmission electron micrograph of a glomerular capillary from a control rat displaying the filtration barrier. CL, capillary lumen; EF, endothelial fenestrae; GBM, glomerular basement membrane; FP, foot processes of podocyte; US, urinary space; ER, erythrocytes. (Original magnification ×16 300).

able to interact with glomerular basement membrane anionic sites and have been shown to induce alterations in glomerular permselectivity.[128–131] Therefore, it is likely that the administration i.v. to rats of the polycationic gentamicin prior i.v. infusion of the cationic lysozyme[12] enabled occupation of glomerular anionic binding sites by gentamicin and consequent loss in the ability of the glomerulus to restrict filtration of lysozyme.

7.26.6.1.2 *Changes after chronic aminoglycoside treatment*

There is accumulating evidence that not only the renal tubules but also the glomerulus may be the target of chronic AG nephrotoxicity.

(i) Ultrasructural alterations

Contradictory results with respect to AG-induced changes to the endothelial fenestrae of the glomerular capillaries were obtained by various research groups following AG treatment of rats for several days.[82,132,135] The results of a number of studies indicated little or no morphological change to the glomerulus as a result of treatment with AG.[135–137] In other studies, however, AG treatment of rats (30 mg kg^{-1} d^{-1} for 7, 14, or 21 days) resulted in injury of the glomerular capillary endothelium and the myoepitheloid cells of the juxtamedullary

apparatus, and in alterations of the glomerular permeability.[76,132]

Morphometric analysis of the ultrastructural changes revealed by TEM and scanning electron microscopy (SEM)[76,132,138] showed a reduction in size and a significant decrease in the density of EF to about 80% of control following treatment of rats with gentamicin or tobramycin for 7 days (Figures 3 and 4). No changes to the glomerular ultrastructure were detected after an identical treatment with netilmicin, amikacin or kanamycin. However, the glomerular epithelial cells of the visceral layer, including the slit diaphragm between podocytic foot processes, were free of any remarkable injuries. Similarly, no changes of the mesangial cells or of the basement membrane were observed.[132]

It is unlikely that AG are taken up into the endothelial cells to compromise cell function directly, since no radioactivity has been detected in the endothelium after administration of labeled gentamicin.[139] However, it is possible that polycationic AG could neutralize negatively charged sites around the EF and thus lead to a reduction in diameter or eventual collapse of the fenestrae.

The results of TEM showed a partial or total loss of cytoplasmic granules from the myoepitheloid cells in the afferent arterial wall of the juxtaglomerular apparatus after treatment of rats with gentamicin and tobramycin; only slight changes were observed in netilmicin-treated rats.[132] Since these results did not

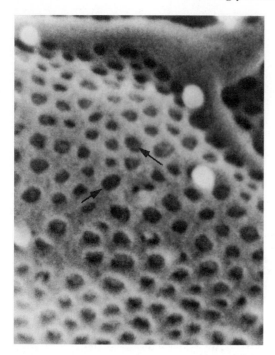

Figure 3 Scanning electron micrograph of the luminal surface of a glomerular capillary from a control rat. Notice the large endothelial fenestrae (arrows). (Original magnification ×35 000).

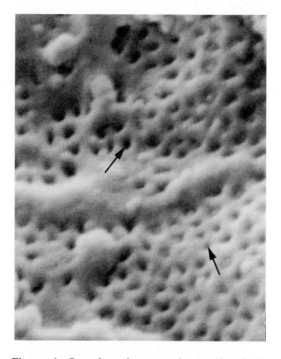

Figure 4 Scanning electron micrograph of the luminal surface of a glomerular capillary from a rat treated with 30 mg kg^{-1} d^{-1} tobramycin for 7 days. Notice the marked decrease in number and size of endothelial fenestrae (arrows). (Original magnification ×28 000).

show disintegration of the granules, elevated exocytotic excretion processes, or activation of rough endoplasmic reticulum, it is uncertain whether the depletion of cytoplasmic granules from the myoepitheloid cells after AG treatment was related to activation of the renin-angiotensin system.

(ii) Changes of the glomerular filtration rate and ultrafiltration coefficient

The glomerular filtration of a molecule is a very complex process, and is dependent on the filtration properties of the glomerular layers, the physicochemical properties of the molecule, and haemodynamic factors. The results of a number of studies[140,141] showed a dose-dependent decrease in GFR after several days of gentamicin treatment.

After AG treatment of rats (30 mg kg^{-1} d^{-1}, divided in two portions) for 7 days, there was a clear decrease in GFR (Table 1).[76,94]

A significant decrease in the ultrafiltration coefficient (K_f) after treatment with gentamicin of rats (40 mg kg^{-1} d^{-1}) was established by Baylis *et al.*,[140] K_f being computed as the product of hydraulic permeability and surface area available for filtration.[142] According to Baylis *et al.*,[140] the gentamicin-induced decrease in K_f could explain at least in part the reduction in GFR. In turn, a decrease in the number and diameter of the endothelial fenestrae signifies a decrease in the glomerular filtration area, and may be contributory to the decrease in K_f. Mesangial cell contraction and relaxation potentially modify glomerular hemodynamics by regulating the surface area available for filtration and the glomerular ultrafiltration coefficient.[142,143] Because angiotensin II and vasopressin can cause mesangial cell contraction, their increased release can contribute to the impairment of renal function and finally to ARF. The decrease of K_f may represent a primary alteration of the glomerular filtration barrier or a secondary response to tubular injury. However, in animal studies K_f may be decreased by as much as 50% without altering single nephron GFR.[144]

The results of studies on the possible role of the renin-angiotensin system in AG-induced glomerular injury are contradictory.[132,138,145,146] Glomerular dysfunctions observed in animal models of ARF, where alterations of the renin-angiotensin system have been well documented, are produced after a short interval following treatment.[147–150] However, AG-induced glomerular dysfunction is a

Table 1 Glomerular filtration rate (GFR) in aminoglycoside treated rats.[a]

Duration of treatment	7 days	14 days	21 days
GFR	mL min^{-1} g^{-1} *kidney*	mL min^{-1} g^{-1} *kidney*	mL min^{-1} g^{-1} *kidney*
Control	0.95 ± 0.04 (*n* = 5)	1.02 ± 0.07 (*n* = 5)	0.96 ± 0.11 (*n* = 6)
Netilmicin	0.51 ± 0.14[b] (*n* = 7)	0.78 ± 0.13[b] (*n* = 6)	0.80 ± 0.13[b] (*n* = 7)
Kanamycin	0.43 ± 0.08[b] (*n* = 5)	0.56 ± 0.03[b] (*n* = 6)	0.64 ± 0.06[b] (*n* = 5)
Gentamicin	0.45 ± 0.18[b] (*n* = 7)	0.42 ± 0.11[b] (*n* = 6)	0.48 ± 0.09[b] (*n* = 7)

Source: Cojecel *et al.*[94]
[a]All experimental data are given as mean value ± SD, with the number (*n*) of rats treated given in brackets. [b]Statistically significantly different to the control ($p < 0.05$).

rather late manifestation during the course of the toxicity in man and experimental animals.

For other types of renal insufficiency, the AG-induced alterations in glomerular structure and function may be a feedback response to tubular damage.[147–150]

Accumulation of adenosine in ischemia because of ATP breakdown induces a vasoconstriction which can cause a decrease in renal blood flow and in GFR. In experimental ischemic ARF adenosine antagonists such as theophylline increase renal blood flow and GFR.

(iii) Changes in glomerular sieving of proteins

A correlation appears to exist between the lowering of the glomerular filtration area (and consequent decrease in K_f) and the reduction in both the GFR and glomerular protein filtration after treatment of rats with AG for 7 days.[76] AG treatment decreased the GSC$_{LY}$ from a control value of 0.8 to values of 0.6 and 0.5 in kidneys from gentamicin- and tobramycin-treated rats, respectively (Table 2).[76] Kanamycin slightly decreased the GSC$_{LY}$ whereas amikacin and netilmicin had no effect.

7.26.6.2 Ultrastructural and Functional Tubule Injuries

7.26.6.2.1 *Ultrastructural alterations of the kidney tubule*

In animal models, gentamicin-induced ATN primarily affects the S1 and S2 segments of the proximal tubule. However, gentamicin administration in the presence of concurrent ischemia[151] or mild renal ischemia preceding gentamicin dosing[152] caused significant injury to the S3 segment of the proximal tubule.[153] These data suggest that clinically unrecognized transient renal ischemia may greatly predispose patients to rapidly occuring AG-induced nephrotoxicity.

The earliest structural changes of the proximal tubule demonstrated by TEM include an increase in the size, number, and structure of the lysosomes. Cytosegresomes, which are altered secondary lysosomes containing electron-dense lamellated whorl-like configurations known as myeloid bodies, are present in significant numbers in the cytoplasm of the proximal tubule cells.

With further progression of ARF, alterations of the brush-border membrane microvilli appear (Figures 5 and 6) along with vacuolation of the cytoplasm and alterations of the

Table 2 Glomerular sieving coefficient of lysozyme (GSC$_{LY}$) in kidneys from rats treated with aminoglycosides for 7 days.

GSC$_{LY}$	GSC$_{LY}$[a]
Control	0.08 ± 0.03
Amikacin	0.77 ± 0.05
Netilmicin	0.78 ± 0.08
Kanamycin	0.70 ± 0.02[b]
Gentamicin	0.60 ± 0.09[b]
Tobramycin	0.47 ± 0.15[b]

[a]The GSC$_{LY}$ was determined as lysozyme clearance/GFR in the isolated perfused rat kidney. All experimental data are given as mean value ± SD. [b]Statistically significantly different from the control ($p < 0.05$).

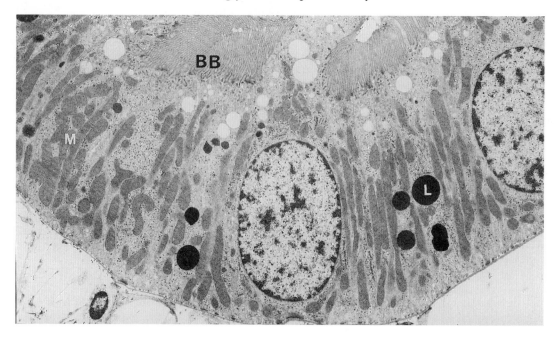

Figure 5 Transmission electron micrograph of renal tubular cells from a control rat. BB, brush-border; L, lysosomes; M, mitochondria. (Original magnification ×6300.)

cisternae of the endoplsmic reticulum. TEM studies show localized brush-border microvilli losses and an increase in the number of lysosomal cytosegresomes.[154] At this stage, the lumen of the renal tubules contains fragments of degenerating cells, myeloid bodies, and cytoplasmic debris.

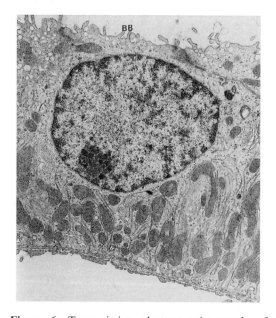

Figure 6 Transmission electron micrograph of renal proximal tubular cells from a rat treated with 30 mg kg^{-1} d^{-1} gentamicin for 7 days. Notice the loss of brush-border (BB). M, mitochondria. (Original magnification ×15 150).

In the following stages, mitochondrial swelling becomes evident and epithelial cell necrosis occurs. Finally, the tubular lumen is filled with sloughed degenerating cells and cellular debris. Apical bleb formation was observed by TEM 24 hours after injection of 100 mg kg^{-1} gentamicin to rats[155] and had a focal distribution at 24 and 48 hours. Shedding of blebs containing subcellular organelles including lysosomes into the lumen could explain the concomitant detection of phospholipids, lysosomal enzymes, and brush-border enzymes in the urine. Renal tubular injuries result ultimately in intratubular obstruction, backleak of glomerular filtrate through damaged tubular epithelium, and reduction in GFR.[86]

Intratubular obstruction appears to be one of the major mechanisms leading to a decline in GFR in both ischemic and nephrotoxic ARF. In the cases when intraluminal blockage to urine flow is diffuse enough to involve most functioning nephrons, a relevant decline in renal function occurs.[157–159] Micropuncture studies of gentamicin-induced ARF revealed evidence of reestablishment of normal single nephron GFR when microperfusion of the tubule dislodged obstructing casts or debris.[158]

Shedding of the necrotic cells into the tubular lumen not only reduces renal excretory function by obstructing urine flow but also by leaving gaps along the tubular epithelia through which glomerular ultrafiltrate can re-enter the circulation. Results of studies using microinjection techniques suggest that there is a correlation between the magnitude of backleak and the extent of tubular injury present in ATN.[157] However, there is little evidence for backleak of filtrate in gentamicin-induced ARF.[140]

After a single dose of gentamicin ($80 \, mg \, kg^{-1}$ body weight) in rats, ultrastructural changes to the proximal tubule epithelial cells such as brush-border losses, increase in number and multimembranous restructuring of lysosomes, formation of myeloid bodies, expansion of the endoplasmic reticulum, and mitochondrial swelling are already induced after 80 min.[160,161] Similar ultrastructural changes such as an increase in the lysosome count and in the lysosome diameter, formation of myeloid bodies, and the lysosomal accumulation of undigested cell membranes occur following application of much smaller gentamicin doses ($1-1.5 \, mg \, kg^{-1} \, d^{-1}$) over 2 or 4 days in rats[160] and in humans, respecitvely.[162]

Regeneration of the tubule epithelial cells could be detected by electron microscopy already three days after the end of a 10 days treatment of rats with $40 \, mg \, kg^{-1} \, d^{-1}$ gentamicin.[163] An almost complete regeneration of the tubular epithelial cells occurred 7 days after the end of treatment; after four weeks no differences from the control kidneys could be detected.[163]

7.26.6.2.2 *Alterations of tubular transport processes*

AG nephrotoxicity is often associated with depression of tubular transport of electrolyte, calcium, magnesium and glucose reabsorption, and alterations of tubular protein and organic ion transport. The mechanisms by which AG affect amino acid and fatty acid transport into proximal tubule have not been investigated sufficiently. After 4 days of gentamicin administration to rats, the uptake of α-aminoisobutiric acid across the basolateral membrane was not affected in the early course of AG nephrotoxicity.[164]

(i) *Glucose transport*

The results of earlier studies described development of glucosuria in the course of AG nephrotoxicity.[164] Futher studies suggested that impairment of the glucose uptake at the brush-border membrane contributes to the development of glucosuria.[165] Both gentamicin and netilmicin are able to inhibit Na^+-dependent D-glucose transport in the rabbit renal brush-border membrane vesicles.[166] However, in this study netilmicin had a weaker effect on glucose transport and, in contrast to gentamicin, was not able to affect membrane fluidity.

(ii) *Alterations of the organic ion transport*

While the organic base transport was decreased in both brush-border[167] and basolateral membrane[164,168] of the proximal tubule, the organic acid transport at the basolateral membrane was increased within 24 hours after administration of a single dose of gentamicin.[169] Stimulation of the organic acid transport system by AG appears to be unspecific since other nephrotoxins have similar effects.[9] The mechanism of the interaction between AG and the organic base transport system is not clear since the uptake of polycationic AG into tubule cells is known to occur by endocytosis.

(iii) *Decrease of sodium and potassium reabsorption*

It is noteworthy that potassium and sodium excretion are considered as sensitive and early parameters of gentamicin nephrotoxicity.[170] According to Giebisch et al.,[171] the presence of strongly cationic substances in the distal tubule lumen may promote increased potassium excretion. High plasma or perfusate concentrations of cations such as AG or lysozyme cause increased sodium and potassium excretion,[76,174] as does increased potassium excretion after i.v. infusion of the cationic amino acids L-lysine or L-arginine.[172,173]

While the potassium content in humans, and most probably in rats also, is $50 \, mmol \, kg^{-1}$ body weight, only 2% of this total potassium content is found in the extracellular space. Maintenance of the overall potassium content and its distribution between the intracellular and extracellular domains is of great importance to normal cellular function.[175] Small changes in the ratio of intracellular potassium/extracellular potassium may cause serious disturbances in muscle and heart function. The percentage potassium excretion amounted to 5.9% in the control rats, 16.9% after netilmicin treatment, 28.5% after kanamycin treatment, and 37.3% after gentamicin treatment.[94] Administration of $2 \, mg \, kg^{-1} \, d^{-1}$ amiloride did prevent a decrease in plasma potassium concentration, but failed to offer any protection against gentamicin nephrotoxicity.[176]

Perfusion of the isolated rat kidney, with $1 \, mg \, mL^{-1}$ gentamicin, caused an increase in sodium and potassium excretion to 17% and 78% of the filtered amount, respectively.[22,76] This is a considerable electrolyte loss despite the fact that at the same GFR and the same plasma or perfusate concentration of potassium or sodium, the isolated perfused kidney clearly shed more potassium and sodium than the kidney *in situ*, after acute administration of lysozyme or gentamicin.[76,174] It is also conceivable that the accumulation of the cationic substances such as lysozyme and/or gentamicin could displace potassium from the renal cells and hence cause increased potassium excretion. Usually, chronic AG treatment causes a clear increase in potassium excretion but only a minor increase in sodium excretion.[177] After a 7 day treatment with gentamicin, kanamycin, or netilmicin ($30\,mg\,kg^{-1}\,d^{-1}$), the percentage sodium excretion remained in the normal range, not exceeding 1% of the amount of glomerularly filtered sodium.[76,177]

(iv) Impairment of tubular transport of calcium and magnesium

Acute AG infusion increased urinary excretion of calcium and magnesium in rats.[178] Chronic administration of gentamicin in man resulted in renal wasting of magnesium and hypomagnesemia.[179] Decrease of magnesium content in the renal cortex following treatment of dogs with gentamicin[180] suggests that impairment of magnesium transport into tubule cells contributed to the renal wasting of magnesium.

Since calcium has been shown to be involved in a variety of cell membrane functions including semipermeability, hormone action, and enzyme activity, displacement of calcium from the cell membranes may be crucial in AG-induced functional alterations of renal tubules. Whereas a decrease in the calcium transport at the basolateral membrane was observed after a single dose administration of gentamicin, no such changes were seen in the brush-border.[181]

(v) Alterations of renal protein reabsorption

After the passage across the glomerular basement membrane into the glomerular ultrafiltrate, renal proximal tubule cells are responsible for endocytic reabsorption and lysosomal degradation of filtered endogenous and exogenous proteins. Phospholipids, particularly in the crypt region of the brush-border membrane appear to have common anionic binding sites on the surface of the membrane for various cationic substances such as AG, lysine, arginine, and lysozyme.[80,165,182–185] Lysozyme and gentamicin bind to the acidic phospholipids, particularly phosphatidylinositol, and subsequently are internalized into tubule cells by endocytosis and stored in lysosomes.[190] Experimental evidence gathered from a number of studies showed for the first time

that acute and chronic administration of AG to rats decreases renal reabsorption, as well as accumulation and degradation of lysozyme.[12,76–78,94] Further *in vitro* studies using the MDCK kidney cell line[186] or the isolated perfused kidney,[76,187] and *in vivo* studies were carried out to evaluate the effects of gentamicin on renal handling of proteins using lysozyme and other proteins such as rat β2-microglobulin, human β2-microglobulin, retinol binding protein, cytochrome C, or horseradish peroxidase (HRP) as probes.[12,76,78,94,187–189] The results of these studies indicated that gentamicin affects the tubular endocytosis of proteins, the subcellular internalization pathway,[76,89,93,94,177,186] and subsequent lysosomal degradation.[12,76,77,189]

In an elegant ultrastructural study[190] it has been shown that lysosomes altered by the treatment with gentamicin do not fuse or fuse less effectively with incoming pinocytotic vesicles. In the same study, more than half the lysosomes from control rats contained the globular protein HRP while only one-third of lysosomes from gentamicin treated rats stored HRP; lysosomes overloaded with phospholipid stored little HRP.[190] These findings provide a cytological explanation for the AG-induced inhibition of lysosomal degradation of proteins reabsorbed by proximal tubular cells.[12,76,77,131]

The i.v. administration of a single dose of gentamicin to intact animals led, within 20 min of the start of gentamicin infusion, to an almost complete inhibition of tubular lysozyme reabsorption.[12] Similarly, a concentration-dependent inhibition of lysozyme reabsorption by addition of gentamicin to the perfusate was observed in the isolated perfused rat kidney.[12,76,187] An interaction between cationic lysozyme and polycationic AG for the anionic binding sites at the proximal tubule brush-border could explain the inhibition of tubular protein reabsorption.

After endocytotic absorption, gentamicin reaches the lysosomes within 60 min and accumulates there in high concentrations.[18,19,139] Lysosomal accumulation and long-term storage of gentamicin alter the permeability of the lysosomal membrane[91] and possibly disrupts lysosomal digestive capacity by reducing the enzymatic activity of lysosomal phospholipases and proteases.[139,191–193]

AG differ in their potential to inhibit protein reabsorption in the rat kidney.[78] While netilmicin and amikacin had practically no effect on renal protein reabsorption, kanamycin caused a distinct decrease in protein reabsorption by the kidney. Treatment of rats with gentamicin or tobramycin over several days caused a reduction in the amount of lysozyme filtered,[76] which resulted in a decrease in lysozyme reabsorption to about 50% of the amount filtered, both in the intact animal and in the isolated perfused rat kidney (Figure 7).[76,94,177] This may be due to the reduction in GFR and/or GSC_{LY} which led to a decrease in the amount of the filtered lysozyme.[94,132,154] This decrease in AG-induced renal protein reabsorption explains the reduction in renal protein accumulation (Figure 8),[76,94,154,177] and to a certain degree, the decrease in renal protein degradation.[76]

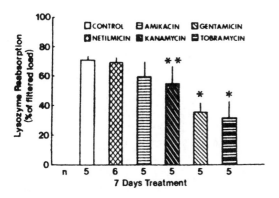

Figure 7 Effect of aminoglycosides on renal reabsorption of lysozyme after 7 days of treatment of rats with 30 mg kg day. n, Number of isolated kidneys. *Significantly different from control. **Significantly different from gentamicin and tobramycin (after Cojocel *et al.*[76]).

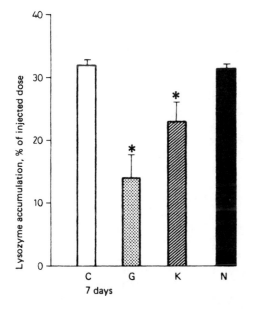

Figure 8 Effect of aminoglycosides on renal accumulation of lysozyme. Measurements were carried out by using ^{125}I-lysozyme as tracer. C, control; G, gentamicin; K, Kanamycin; N, netilmicin. *Significantly different from control (after Cojocel *et al.*[94]).

7.26.6.2.3 *Alterations of phospholipid and protein catabolism*

(i) Phospholipid degradation

Lysosomes are involved in the intracellular traffic of membrane related to endocytosis and lysosomal phospholipases are believed to play an important role in this process.[194] AG treatment and subsequent lysosomal accumulation of AG decrease the activity of sphingomyelinase and lysosomal phospholipases A1, A2 and C, and phopatidylinositol specific phospholipase C.[94,193,195–197] This alteration is characterized structurally by an enlargement of lysosomes and accumulation of lamellar, osmiophillic material (myloid bodies) due to the fast turnover of phospholipids and inhibition of their intralysosomal degradation. The increase in number and size of lysosomes and intralysosomal accumulation of myeloid bodies (phospholipidosis) is considered an early event in the development of AG-induced nephrotoxicity.[197]

Exposure of rat kidneys to AG caused a significant decrease in renal protein degradation as it was demonstrated in the isolated kidney and renal cortical slices models by using ^{125}I-lysozyme or non-labeled lysozyme and measuring either of TCA-soluble radioactivity or the amino acid tyrosine as products of renal protein degradation.[76,77,93]

Autoradiographic studies showed that gentamicin enters lyosomes, the site of protein degradation, one hour after i.v. injection.[139] Catabolism of ^{125}I-lysozyme by isolated rat kidneys perfused without gentamicin becomes evident after the onset of perfusion.[12] A concentration- and time-dependent decrease in the renal protein degradation was detected.[12] The almost complete abolishment of renal protein degradation at perfusate concentration of gentamicin of $0.5\,mg\,mL^{-1}$ and higher could be due primarily to inhibition of tubular reabsorption of lysozyme by gentamicin and additionally to the inhibition of the enzymatic activity of lysosomal proteases responsible for the lysosomal catabolism of lysozyme such as cathepsin B.[191,193,198]

After AG treatment of rats for 7 days, renal protein degradation was investigated using the isolated perfused kidney.[76] The results of these studies showed that AG may be subdivided into two groups based on their effects on renal protein degradation. The highly nephrotoxic AG, gentamicin, and tobramycin induced the strongest decreases in renal protein degradation, whereas netilmicin, amikacin, and kanamycin appear to have a weak or no effect (Figure 9).[76]

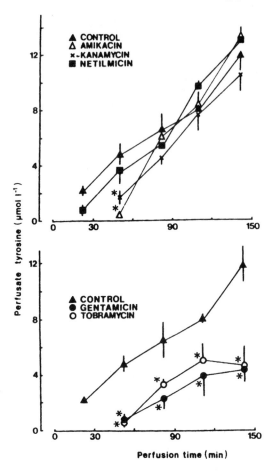

Figure 9 Renal degradation of lysozyme after aminoglycoside treatment of rats $(30\,mg\,kg^{-1}\,day^{-1})$ for 7 days. Renal protein degradation was measured as the net increase in tyrosine concentration due to the addition of lysozyme to the perfusate and subsequent lysosomal degradation. Values represent five or more isolated kidneys. *Significantly different from control (after Cojocel *et al.*[76]).

7.26.6.2.4 Alterations of mitochondrial function

In vitro exposure of isolated renal mitochondria to AG induced inhibition of state 3 and stimulation of state 4 respiration but also inhibition of dinitrophenol (DNP)-uncoupled respiration.[199–201] Impaired mitochondrial function has also been shown following *in vivo* AG administration. Administration of gentamicin $(40\,mg\,kg^{-1}\,d^{-1})$ for 7 days resulted in inhibition of state 3 and DNP-uncoupled respiration while state 4 respiration was not affected significantly.[202] Interestingly, injections of low dose gentamicin $(3\,mg\,kg^{-1}\,d^{-1}$ for 28 days) stimulated oxygen consumption of renal mitochondria while injection of higher doses of gentamicin $(40\,mg\,kg^{-1}\,d^{-1}$ for 28

days) decreased oxygen consumption.[203] *In vitro* studies have shown gentamicin to competitively inhibit mitochondrial uptake of calcium along with competitive displacement of magnesium from the inner mitochondrial membrane.[204]

7.26.7 MECHANISMS OF NEPHROTOXICITY

Numerous research groups have focused their efforts on elucidation of the mechanism of AG nephrotoxicity. The results of a number of studies indicate that polycationic AG exert toxic effects both on the glomeruli and kidney tubules[82,84,132,138,140,176] and that they have considerable potential for the alteration of cellular structure and function.[13,37] The primary molecular mechanism which triggers the cascade of morphophysiological changes responsible for the development of cellular necrosis is still unknown. The findings of various research groups suggest that AG-induced injuries to the plasma cell membranes could be a highly likely candidate for one of the factors which triggers AG-induced tubule cell necrosis.[1,9,37] In spite of important progress having been made, the relationship between molecular structure and the nephrotoxic effects of AG has yet to be settled.

7.26.7.1 Interactions Between AG and Various Cationic Substances During Renal Endocytosis

Cationic polypeptides, proteins, cationic AG are internalized by the proximal tubule cells via endocytosis. Interactions between various substances during renal endocytotic absorption has been investigated in various studies.[18,19,205–209] For example, the i.v. injection of arginine resulted in a temporary increase in the excretion of protein in human urine.[210] Further studies have shown that both tubular lysozyme reabsorption and binding of cationic lysozyme at the brush-border membrane are inhibited by cationic amino acids and cationic proteins, whereas neutral amino acids and neutral proteins have no effect.[205,211] Using proteins with modifications to their overall charge it was shown that the more cationic the protein, the more strongly it is absorbed.[131,212]

Binding of the cationic peptide aprotinin at the brush-border membrane is competitively inhibited by AG, whereby the inhibitory potency of a specific AG appears to be related to the number of free amino groups attached to the AG molecule.[18,19] These phenomena are consistent with the presence, in the crypts of the brush-border membranes, of negatively charged binding sites[213–215] for which positively charged substances compete.[18,19,182,185,208]

The overall charge of the AG molecule at physiological pH plays an important part in the cellular uptake, accumulation, and toxicity of these antibiotics. On the basis of several studies, a correlation can be established between AG nephrotoxicity and the number of amino groups attached to the AG molecule.[5,92,201] The classification of AG according to their nephrotoxic potency and the number of amino groups shows the following pattern: neomycin > gentamicin = tobramycin > streptomycin.[19]

Neomycin, with six, possesses the highest number of amino groups; streptomycin, with three, the lowest. The AG used clinically, gentamicin, netilmicin, and tobramycin, possess five amino groups, whereas amikacin and kanamycin have only four. Netilmicin is less nephrotoxic than gentamicin or tobramycin, which have the same number of amino groups and the same accumulation in the renal cortex; the origin of this discrepancy is still unknown. It should also be noted that differences in nephrotoxicity were established even among the three gentamicin derivatives C1, C1A, and C2.[31] The results of these studies suggest that the differences in nephrotoxicity of these three gentamicin derivatives may be explained by the different positioning of a methyl group in the molecule. Thus, other factors apart from the overall charge such as the molecular weight and conformation also appear to play a part for the binding affinity and nephrotoxicity of AG.

7.26.7.2 Binding to the Brush-border Membrane

AG binding to phosphatidylinositol is considered to be the first step in the cellular absorption of AG and in the inhibition of phospholipase activity.[165,213] As is to be expected from its polycationic structure, the AG molecule can bind to the negative binding site(s) of the phosphatidylinositol molecule.[213] Although this binding is primarily electrostatic in nature, molecular conformation has an important role in the penetration of the AG molecule into the lipid layer of the membrane.[213] Deep penetration of the gentamicin molecule into the lipid layer of the membrane is accompanied by high phospholipase inhibition, whilst poor penetrative ability of streptomycin is associated with low phospholipase inhibition. An exception to this behavior is exhibited by amikacin, for which a lower energy of penetration relative to kanamycin A was calculated.[213] Both these AG show equal inhibitory activity towards phospholipases.

Data indicate that AG binding is qualitatively similar for brush-border membrane vesicles (BBMV) obtained from human and rat kidney.[48] However, a lack of correlation between AG binding affinity to BBMV and their nephrotoxic potetial has been reported.[165] Furthermore, *in vitro* binding affinity of AG does not correlate with *in vivo* rat strain differences in nephrotoxicity.[48] Similarly, no differences were observed between AG binding to BBMV from diabetic and nondiabetic rats.[214] It appears that AG binding to BBMV is a necesssary but not a determining factor of nephrotoxicity.[165]

An important common biochemical alteration to the composition of the luminal and basolateral plasma membranes is the occurrence of an appreciable increase in the phospholipid content of these membranes after treatment with AG. Of particular interest is the increase in phosphatidylinositol (the probable membrane receptor for AG). There may exist a causal link between these changes to the phospholipid composition of the plasma membrane and consequent alterations to membrane function and the AG-induced inhibition of phospholipid metabolism.

7.26.7.3 Injuries of the Plasma Cell Membranes

Early functional, biochemical, and structural damage occurs during AG treatment, highlighting the luminal plasma cell membrane of the proximal tubule cells as one of the most important sites of cellular AG interactions. It seems possible that the cascade of damage of AG nephrotoxicity is triggered at the luminal plasma membrane.[9,86] Although AG are mainly absorbed on the luminal side of the tubule cell, the findings of various research groups demonstrate AG transport into the tubule cells through the basolateral membrane.[118] Various AG-induced functional changes to the basolateral membrane have already been described,[9] including, among others, a decrease in the enzymatic activities of Na^+,K^+-ATPase and adenylate cyclase.[216,217] The inhibition of Na^+,K^+-ATPase with consequent impairment of membrane function may be regarded as a contributory biochemical mechanism to AG toxicity.

The ability of various tissues to maintain a potassium concentration gradient between the extracellular space and the intracellular space is also dependent on various other factors such as the hydrogen ion balance, and the plasma concentrations of insulin, epinephrine, and aldosterone.[175,218,219] The decrease in the enzymatic activity of cortical and medullary Na^+,

K^+-ATPase and Mg^{2+}-ATPase after a 7 day gentamicin treatment[141] may offer a partial explanation of the increases in sodium and potassium excretion.

From the histological findings and the decrease in activity of Mg^{++}-ATPase and Na^+, K^+-ATPase after seven days of gentamicin treatment, it is fair to assume that the reduction in ATPase activity is a consequence of the gentamicin-induced parenchymal lesions.[141]

Taken together, the results of this study and those of of Purnell *et al.*,[176] indicate that the electrolyte loss is a consequence, rather than the cause, of AG-induced nephrotoxicity. The mechanism by which gentamicin nephrotoxicity is potentiated by potassium or magnesium depletion remains unclear.[176,220,221]

7.26.7.4 Injuries on Lysosome Function

The intralysosomal accumulation of AG leads to a decrease in the enzymatic activity of phospholipases and proteases such as cathepsins[191,222] and to a reduction in the lysosomal digestion of proteins[77] and phospholipids.[195] The likely mechanism responsible for reduction of cathepsin B and L activities is an inhibition in generation of active intralysosomal cathepsins by gentamicin.[193] The process of generation of active lysosomal enzymes, including phospholipases and sphingomyelinase, involves biosynthesis of enzymes precursors, transport to and uptake into lysosomes where proteolytic processing leads to generation of mature active enzymes.[223] AG may inhibit one or more of these steps.

Gentamicin has been shown to inhibit microsomal protein synthesis in the renal cortex[224] and to decrease protein catabolism by the kidney.[76] Cathepsins B and L and cysteine-proteinases are involved in lysosomal catabolism of endogenous and exogenous proteins[198] and proteolytic activation of lysosomal enzyme precursors.[223] Consequently, a decrease in catepsin activity may lead to decreased activities of lysosomal enzymes involved in phospholipid degradation. Therefore, the AG-induced accumulation of phospholipids and formation of myeloid bodies (phospholipidosis) in the renal cortex[160,225] may be at least in part due to AG inhibition of cathepsin activity.

AG-induced phospholipidosis has been proposed as a possible mechanism for AG nephrotoxicity.[13] According to this hypothesis, the disruption of normal lysosomal digestive function leads to cellular depletion of important substrates necessary for cell membrane regeneration and the maintenance of various metabolic processes and the function of the various cell organelles.

Alternatively, the intralysosomal accumulation of AG may lead to changes in the permeability of the lysosomal membrane, causing damage to other cell organelles (e.g., mitochondria) through release of lysosomal hydrolases. However, the results of clinical experiments show no direct relationship between the changes in lysosomal membrane permeability induced *in vitro* and the potency of AG with respect to clinical nephrotoxicity.[226,227]

7.26.7.5 Alterations of the Mitochondrial Function

AG cause changes not only to the plasma membrane and the lysosomes, but also to the mitochondria. Swelling of the mitochondria after treatment with AG in rats was detected using TEM.[160,161] In *in vitro* studies, gentamicin induced inhibition of mitochondrial oxidative phosphorylation.[199,200] As pinocytosis and intracellular protein degradation are energy-consuming processes,[228,229] impairment of mitochondrial respiration might contribute to the reduction in pinocytotic protein reabsorption and intralysosomal protein degradation.

Results of studies showing a reduction in renal gluconeogenesis induced by treatment with gentamicin[230] are clearly indicative of impairment of mitochondrial energy-releasing processes. However, no correlation could be made between the potency of an AG with respect to *in vitro* inhibition of mitochondrial respiration and its clinical nephrotoxicity.[19]

7.26.7.6 Reactive Oxygen Species and Lipid Peroxidation

The reactive oxygen species formed as a consequence of intracellular bioactivation of AG may result in lipid peroxidation and denaturation of proteins and enzymes and leads to nephrotoxic injury.

The results of a number of studies suggest that AG-induced lipid peroxidation in the kidney tissue[231,232] is a consequence and not a pivotal mechanism in the pathogenesis of AG nephrotoxicity.[233–235] Whether AG exert their nephrotoxicity via a free radical mechanism is still controversial.[231,234,236] AG are considered to be redox-inactive compounds.[234,237]

Data suggest that conversion of AG to a redox-active form may require the participation of transition metals such as iron.[2,238,239] The results of an *in vitro* study indicate that gentamicin promotes the oxidation of arachidonic acid in the presence of iron salts, in a similar manner to that of bleomycin.[12] It was suggested that gentamicin, iron, and reactive oxygen species form a ternary complex facilitate the electron transfer.[12] The ligated iron is the site of redox activation of the drug, while the organic moiety has no redox-active residue. If Fe^{3+} is the starting oxidation state, in the presence of thiols, NAD(P)H or ascorbate, it is reduced to Fe^{2+}. Addition of one more electron to the ternary complex will form the the "active gentamicin" which, in the presence of oxygen and availability of iron, is able to facilitate formation of reactive oxygen species such as superoxide and hydoxyl radical which subsequently may cause lipid peroxidation and nephrotoxicity.

Under *in vivo* conditions, the availability of physiological iron is critical for the rate of electron transfer and the rate of reactive oxygen radical production. *In vivo*, most of the iron is bound to heme and nonheme proteins and does not directly catalyze the generation of reactive oxygen species. Ferritin, which is present at the subcellular level in the cytosol and endoplasmatic reticulum, appears to be a source for iron in its ferric form *in vivo*.[240] Superoxide generated by xantine oxidase is capable of reductive release of F^{3+} from ferritin, a process which is dependent on the activity of microsomal NADPH-cytochrome P450 reductase.[240] Another source needed for the formation of reactive oxygen species are iron-rich mitochondria. The results of experiments with scavengers of reactive oxygen species indicate that gentamicin-induced iron release from mitochondria is mediated by hydrogen peroxide.[241] Thus, mitochondria should be considered as a potential source of iron for the generation of more reacting oxygen species responsible for initiation of lipid peroxidation and nephrotoxicity.

7.26.7.7 Effects of Nitric Oxide

Increase of gentamicin nephrotoxicity after concomitant treatment of rats with gentamicin and a specific inhibitor (N^G-nitro-L-arginine methyl ester) of nitric oxide (NO) synthesis[242] is in good agreement with the increases of nitrite production and of basal cGMP in glomeruli isolated from gentamicin treated rats.[243] These results indicate that gentamicin treatment increases glomerular NO synthesis in the kidney.[242] Interestingly, the results of earlier studies showed that plasma renin activity was greater in gentamicin treated rats than in control rats.[138] Results of experiments in the isolated perfused rat kidney[244] or in mouse

juxtaglomerular cells[245] exposed to elevated NO levels suggest a stimulatory effect of NO on renin secretion.

Other studies showed that the renin release was not significantly affected by either stimulation or inhibition of NO syntase.[246] The same studies also indicate that the interaction of NO and the granular cells of juxtamedular apparatus results in cGMP-mediated inhibition of renin secretion.[247] These contradictory results may be due to the fact that in the juxtaglomerular apparatus of the intact kidney NO is probably generated by several cell types.[248,249] It is suggested that NO originating from endothelial cells may directly inhibit renin secretion while NO generated by macula densa may stimulate renin secretion by some indirect mechanism.[247] Increased gentamicin nephrotoxicity following inhibition of NO synthesis[242,243] suggests that endogenously released NO plays a protective role in gentamicin nephrotoxicity. This nephroprotective role of NO may be due, on one hand, to its vasodilation effect and participation in the regulation of the renin release,[246] and, on the other hand, to its involvement in the defense against cellular damage induced by reactive oxygen species.[250]

Based on the available data on AG nephrotoxicity, it is highly unlikely that any of the individual mechanisms of cellular toxicity discussed above is solely responsible for nephrotoxicity, but rather AG nephrotoxicity is more likely to arise through the combined effect of several biochemical lesions. AG are in a position simultaneously to stimulate phospholipid synthesis,[9] inhibit both the degradation of phospholipids and proteins,[77,195,225] and synthesis of microsomal proteins.[251] The subsequent elimination of myeloid bodies and necrotic cells leads to the formation of tubular casts, obstruction of the tubular lumen, and failure of renal function.[15,252]

On the basis of these findings, AG nephrotoxicity may be viewed as the net result of alterations of a variety of metabolic pathways.

7.26.8 SPECIES DIFFERENCES IN RENAL RESPONSE TO AG NEPHROTOXICITY

AG can lead to oto- and nephrotoxic reactions, and in rare instances, can also trigger reversible neuromuscular, allergic, hepatic, and haematological reactions. In humans it is frequently impossible to distinguish between symptoms due to illness and AG-induced side effects. To elucidate various aspects of AG toxicity, comparative toxicity studies in experimental animal models has proven worthwhile. Ototoxicity trials relating to vestibular function are carried out mainly in cats, those relating to cochlear function are carried out predominantly in guinea pigs, and nephrotoxicity studies are performed frequently in mice, rats, rabbits, and dogs.

Treatment of rats or mice with 10–80 mg kg^{-1} of gentamicin[256] resulted in alterations to all nephrotoxicity and biochemical parameters in rats but not in mice, with the exception that a significant increase in urinary activity of *N*-acetyl-beta-D-glucosaminidase at 80 mg kg^{-1} was noted in mice. In the same study, the concentration of gentamicin in the cytosol from rat whole kidney was 3.4-fold higher when compared with that in the mouse kidney after 80 mg kg^{-1} gentamicin. While necrosis occured in the rat renal proximal tubules, no pathological changes were observed in the mouse kidney.[256]

The rat is a suitable animal model for studies of AG nephrotoxicity in humans since pharmacokinetics and toxicology of AG are remarkably similar in rats and man.[257]

Electron microscope investigations on the glomerular capillaries in humans and in rats indicate the presence of the same basic structural and functional features in both species. The endothelium with its fenestrae, the basal membrane, and the visceral membrane with filtration clefts between the podocyte foot processes are all present in both species and represent a similar ultrastructure.[253,255,258]

Along the proximal and distal tubules, the height of the brush-border and the distribution, density, and shape of the mitochondria and lysosomes are similar in rats and humans.[254,255] The normal distribution of enzymes along the nephron is similar in rats and humans and includes marker enzymes of the brush-border membrane such as alkaline phosphatase, alanine aminopeptidase, and α-glucosidase, enzymes of the basolateral membrane such as Na$^+$,K$^+$-ATPase, enzymes which are mainly localized in the lysosomes such as acid phosphatase, β-glucuronidase, and NAG[19,77,259–262] and enzymes invoved in drug metabolism.[263–265] Common characteristics of AG binding to cell membranes[48] or of the renal transport system for glucose,[266–268] amino acids,[269] and low-molecular-weight proteins[76,93,174,270–273] in rats and humans have also been described.

The similarities between rats and humans in their pharmacokinetics and susceptibility towards AG nephrotoxicity enable extrapolation of knowledge gained from this animal model to elucidation of the mechanism(s) of AG nephrotoxicity in humans. In combination studies, cephem antibiotics significantly lowered gentamicin nephrotoxicity and renal gentamicin concentrations in rats.[274] In contrast, results of human studies with cephem antibiotics and gentamicin suggested either no interaction or a synergistic effect.[64,275]

Results of *in vivo* studies suggest rat strain differences in AG nephrotoxicity; specifically Fischer rats appear to be more sensitive than Sprague-Dawley rats[276] although other studies showed that Sprague-Dawley rats were as responsive as Fischer rats to AG nephrotoxicity.[277] However, these results do not correlate well with the results of *in vitro* studies designed to investigate the strain differences in binding of tobramycin[266] or other AG[165] to renal brush-border membrane. Taken together, these results suggest that the binding of AG to renal brush-border membrane may be a necessary but not determining factor in the pathogenesis of AG nephrotoxicity.[165,266]

7.26.9 PREVENTION OF AG-INDUCED RENAL DAMAGE

Correction of the risk factors associated with AG nephrotoxicity is essential to the prevention of nephrotoxicity. Of these factors, correction of volume depletion and/or congestive heart failure and diminished renal perfusion is of primary importance. Furthermore, adjustment of dosage to compensate for the pre-existing renal dysfunction and awareness of the decreased GFR that may be present in elderly patients despite apparently normal levels of plasma creatinine may be important in preventing AG nephrotoxicity. To minimize the risk of renal injury, using once-daily dosing and the shortest appropriate course of AG as well as avoiding concurrent treatment with other nephrotoxic agents are also advisable.

Occurrence of ARF from AG treatment may be prevented by early treatment of serious infections, together with maintenance of hemodynamic stability, renal perfusion, and urinary solute excretion. While fluid resuscitation can limit the severity of renal injury from hypotensive, endotoxemic, and nephrotoxic insults, overhydration should be avoided in those patients developing renal failure. Once established, AG-induced ARF follows a time course comparable to ATN of different origins.[27] Recovery from combined endotoxemic or ischemic and nephrotoxic injuries is less predictible. While there is no evidence that dialysis will speed up renal recovery, clinical stability and good nutrition are likely to improve recovery, as is also the case with other types of ARF.[27]

7.26.9.1 Radical Scavangers

The results of earlier studies suggested that treatment of rats with hydroxy radical scavangers protected against gentamicin-induced renal failure.[232] However, in contrast to these observations, superoxide dismutase or hydroxyl radical scavangers had no effect on gentamicin-induced iron mobilization[241] and conferred no protection against the development of ARF and proximal tubular necrosis.[278] Catalase, an effective detoxifying enzyme of hydrogen peroxide, prevented gentamicin-induced iron mobilization from mitochondria[241] thus impairing formation of the gentamicin-iron-complex, the toxic form of "active gentamicin".[12] Interestingly, glutathione[279] and iron chelators[280] appear to have protective effects against gentamicin-induced ototoxicity.

7.26.9.2 Poly-L-aspartic Acid

The results of numerous studies have shown that poly-L-aspartic acid (PAA) prevents gentamicin nephrotoxicity[281–283] in spite of renal cortical gentamicin accumulation which is greater in PAA-treated rats than in control rats.[284] Thus, it is possible that PAA prevents gentamicin-induced renal injuries, perhaps by complexing with gentamicin. Alternatively, PAA is thought to sequester gentamicin within the lysosomal compartment and thereby reduce cellular toxicity.[285] However, while PAA pretreatment significantly decreased the magnitude of gentamicin calciuria, it did not prevent furosemide calciuresis.[286] Combining PAA and concomitant administration of large gentamicin doses with less frequent dosing schedules (once daily) would increase bacterial efficacy while minimizing nephrotoxicity.[281] There is evidence that the protection confered by PAA is related to its ability to serve as an anionic substrate for the cationic AG and thereby prevent AG from interacting electrostatically with membrane anionic phopholipids.[278]

7.26.9.3 Pyridoxal-5′-phosphate

Combined administration of pyridoxal-5′-phosphate ($250\,\mathrm{mg\,kg^{-1}\,d^{-1}}$) with gentamicin ($60\,\mathrm{mg\,kg^{-1}\,d^{-1}}$) for 14 days[287] prevented enzymuria, depression of renal cortical enzymes and the development of phospholipidosis. When a higher dose of gentamicin was used ($100\,\mathrm{mg\,kg^{-1}\,d^{-1}}$ for 6 days), $250\,\mathrm{mg\,kg^{-1}\,d^{-1}}$ pyridoxal-5′-phosphate did not prevent enzymuria, the increase in renal cortical phospholipid and malondialdehyde, or the depression of

catalase but blunted the decrease in creatinine clearance and the incidence in tubular cell necrosis.[278] Pyridoxal-5′-phosphate is also an anionic compound that has been shown to form a complex with AG[288] and thus provide some protection against AG nephrotoxicity.

7.26.9.4 HWA-448

HWA-448 is the structural analogue of the hemorrheologic agent pentoxifylline. It has been shown to reduce renal injuries caused by gentamicin or amphotercin-B *in vivo*.[289,290] Similarly, it showed protective effects against gentamicin-induced decrease in the specific activities of cellular leukin aminopeptidase, Na$^+$, K$^+$-ATPase and *N*-acetyl glucosaminidase, and prevented gentamicin-induced phospholipidosis in LLC-PK$_1$ renal cell monolayers.[291] Interestingly, it has been shown that HWA-448 decreases cellular toxicity by a mechanism independent of plasma membrane binding or intacellular gentamicin accumulation.[291] The specific intracellular mechanism of renal protection provided by HWA-448 against AG nephrotoxicity is not known.

7.26.9.5 Calcium Antagonists

Clinical observations suggest that a diet rich in calcium reduces AG nephrotoxicity.[292,293] Gentamicin or tobramycin administration results in a dose-related, reversible calciuresis. Magnesuria of much less magnitude is simultaneously observed.[294,295] Gentamicin calciuria is also characterized by a notable absence of concomitant sodium or potassium wasting.[296] Polycationic gentamicin competes at physiologic pH with calcium for electrostatic binding to renal brush-border membrane.[293] Ca-ATPase is another possible site of gentamicin interference with intracellular events involved in the transtubular transport of calcium.[286] Nephroprotective action of dietary calcium loading has been thought to stabilize cellular membranes by competitive inhibition of gentamicin binding to renal cell membranes.[293] There is a growing body of evidence supporting the notion that calcium antagonists confer a renal protective effect.[297] The mechanism by which verapamil exerts its renal protective effect[298] still remains to be elucidated.[286] Further agents such as lipopetide daptomycin[299] and alpha-lipoic acid[300] exert renal protective effects against gentamicin nephrotoxicity.

7.26.10 REFERENCES

1. G. J. Kaloyanides, 'Metabolic interactions between drugs and renal tubulointerstitial cells: role in nephrotoxicity.' *Kidney Int.*, 1991, **39**, 531–540.
2. E. M. Priuska and J. Schacht, Formation of free radicals by gentamicin and iron and evidence for an iron/gentamicin complex. Biochem. Pharmacol., 1995, **50**(11), 1749–1752.
3. P. H. Whiting, 'Mechanisms underlying cyclosporin A nephrotoxicity.' *Toxicol. Lett.*, 1990, **53**, 69–73.
4. H. H. Rasmussen and L. S. Ibels, 'Acute renal failure. Multivariate analysis of causes and risk factors.' *Am. J. Med.*, 1982, **73**, 211–218.
5. A. K. Mandal and W. M. Bennett, 'Transmission electron microscopy of urinary sediment in the assessment of aminoglycoside nephrotoxicity in the rat.' *Nephron*, 1988, **49**, 67–73.
6. U. Burchardt, G. Schinköthe, G. Müller *et al.*, 'Ausscheidungskinetik von Enzymen und Protein mit dem Harn bei Applikation therapeutischer Gentamicindosen.' [Urinary excretion kinetics of enzymes and proteins in the application of therapeutic doses of gentamycin.] *Wochenschr. Schweiz. med.*, 1978, **108**, 1541–1545.
7. U. Burchardt, H. Krosch, G. Müller *et al.*, 'Changes of urinary enzymes excretion after drug application.' *Curr. Probl. Clin. Biochem.*, 1979, **9**, 183–191.
8. C. Josepovitz, E. Pastoriza-Munoz, D. Timmerman *et al.*, 'Inhibition of gentamicin uptake in rat renal cortex *in vivo* by aminoglycosides and organic polycation.' *J. Pharmacol. Exper. Ther.*, 1982, **223**, 314–321.
9. G. J. Kaloyanides, 'Aminoglycoside-induced functional and biochemical defects in the renal cortex.' *Fundam. Appl. Toxicol.*, 1984, **4**, 930–943.
10. A. W. Mondorf, J. Hendus, J. Breier *et al.*,'Tubular-toxic effects of aminoglycosides and their combinations with cephalosporins.' *Curr. Probl. Clin. Biochem.*, 1979, **9**, 192–200.
11. W. M. Bennett, 'Aminoglycoside nephrotoxicity. Experimental and clinical considerations.' *Mineral Electrolyte Metab.*, 1981, **6**, 277–286.
12. C. Cojocel and J. B. Hook, 'Effects of acute exposures to gentamicin on renal handling of proteins.' *Toxicology*, 1983, **28**, 347–356.
13. G. J. Kaloyanides and E. Pastoriza-Munoz, 'Aminoglycoside nephrotoxicity.' *Kidney Int.*, 1980, **18**, 571–582.
14. A. Schatz, E. Bugie and S. A. Waksman, 'Streptomycin, a substance exhibiting activity against gram-positive and gram-negative bacteria.' *Proc. Soc. Exp. Biol. Med.*, 1944, **55**, 449–450.
15. R. S. Edson and C. L. Terrell, 'The aminoglycosides.' *Mayo Clin. Proc.*, 1991, **66**, 1158–1164.
16. J. Davies, R. Benveniste, K. Kvitek *et al.*, 'Biologic effects of molecular manipulation.' *J. Infect. Dis.*, 1969, **119**, 351–354.
17. R. Beneviste and J. Davies, 'Structure-activity relationships among the aminoglycoside antibiotics: role of hydroxyl and amino groups.' *Antimicrob. Agents Chemother.*, 1973, **4**, 402–409.
18. M. Just, G. Endmann and E. Habermann, 'The renal handling of polybasic drugs. 1. Gentamicin and aprotinin in intact animals.' *Naunyn Schmiedebergs Arch. Pharmacol.*, 1977, **300**, 57–66.
19. M. Just and E. Habermann, 'The renal handling of polybasic drugs. 2. *In vitro* studies with brush border and lysosomal preparations.' *Naunyn Schmiedebergs Arch. Pharmacol.*, 1977, **300**, 67–76.
20. J. Davies and P. Courvalin, 'Mechanisms of resistance to aminoglycosides.' *Am. J. Med.*, 1977, **62**, 868–872.

21. G. G. Jackson, 'The key role of aminoglycosides in antibacterial therapy and prophylaxis.' *J. Antimicrob. Chemother.*, 1984, **13**, Suppl. A, 1–7.
22. J. Kallova, L. Langsadl, A. Bencakova *et al.*, 'Resistance to aminoglycosides among Gram-negative bacteria isolated from urine in Slovakia.' *Antiinfective Drug Chemother.*, 1995, **3**(13), 189–192.
23. D. N. Gerding, T. A. Larson, R. A. Huges *et al.*, 'Aminoglycoside resistence and aminoglycoside usage: ten years of experience in one hospital.' *Antimicrob. Agents Chemother.*, 1991, **35**, 1284–1290.
24. R. D. Moore, C. R. Smith, J. J. Lipsky *et al.*, 'Risk factors for nephrotoxicity in patients treated with aminoglycosides.' *Ann. Inter. Med.*, 1984, **100**, 352–357.
25. C. L. Sawyers, R. D. Moore, S. A. Lemer *et al.*, 'A model for predicting nephrotoxicity in patients treated with amnoglycosides.' *J. Infect. Dis.*, 1986, **153**, 1062–1068.
26. Y. W. Lam, C. J. Arana, L. R. Shikuma *et al.*, 'The clinical utility of a published nomogram to predict aminoglycoside nephrotoxicity.' *JAMA* 1986, **255**, 639–642.
27. H. R. Brady and B. M. Brenner, in 'Harrison's Principles of Internal Medicine,' eds. K. J. Isselbacher, E. Braunwald, J. D. Wilson *et al.*, 1994, McGraw-Hill, New York, pp. 1265–1274.
28. G. B. Appel and H. C. Neu, 'The nephrotoxicity of antimicrobial agents.' *N. Engl. J. Med.*, 1977, **296**, 722–728.
29. O. C. Tablan, M. P. Reyes, W. F. Rintelmann *et al.*, 'Renal and auditory toxicity of high-dose, prolonged therapy with gentamicin and tobramycin in pseudomonas endocarditis.' *J. Infect. Dis.*, 1984, **149**, 257–263.
30. T. F. Keys, S. B. Kurtz, J. D. Jones *et al.*, 'Renal toxicity during therapy with gentamicin or tobramycin.' *Mayo Clin. Proc.*, 1981, **56**, 556–559.
31. W. M. Bennett, 'Aminoglycoside nephrotoxicity.' *Nephron*, 1983, **35**, 73–77.
32. G. B. Appel and H. C. Neu, 'Gentamicin in 1978.' *Ann. Intern. Med.*, 1978, **89**, 528–538.
33. A. Whelton, 'Therapeutic initiatives for the avoidance of aminoglycoside toxicity.' *J. Clin. Pharmacol.*, 1985, **25**, 67–81.
34. J. Blaser, H. P. Simmen, U. Thurnheer *et al.*, 'Nephrotoxicity, high frequency ototoxicity, efficacy and serum kinetics of once versus thrice daily dosing of netilmicin in patients with serious infections.' *J. Antimicrob. Chemother.*, 1995, **36**, 803–814.
35. M. Barza, J. P. Ioannidis, J. C. Capelleri *et al.*, 'Single or multiple daily doses of aminoglycosides: a meta-analysis.' *BMJ*, 1996, **312**, 338–345.
36. W. J. Munckhof, M. L. Grayson and J. D. Turnidge, 'A meta-analysis of studies on the safety and efficacy of aminoglycosides given either once daily or as divided doses.' *J. Antimicrob. Chemother.*, 1996, **37**, 645–663.
37. H. D. Humes, J. M. Weinberg and T. C. Knaus, 'Clinical and psychophysiologic aspects of aminoglycoside nephrotoxicity: an overview.' *Am. J. Kidney Dis.*, 1982, **2**, 5–29.
38. G. R. Matzke, R. L. Lucarotti and H. S. Shapiro, 'Controlled comparison of gentamicin and tobramycin nephrotoxicity.' *Am. J. Nephrol.*, 1983, **3**, 11–17.
39. S. H. Hou, D. A. Bushinsky, J. B. Wish *et al.*, 'Hospital-acquired renal insufficiency: a prospective study.' *Am. J. Med.*, 1983, **74**, 243–248.
40. G. B. Corcoran, D. E. Salazar and J. J. Schentag, 'Excessive aminoglycoside nephrotoxicity in obese patients.' *Am. J. Med.*, 1988, **85**, 279.

41. G. B. Appel, D. B. Given, L. R. Levine *et al.*, 'Vancomycin and the kidney.' *Am. J. Kidney Dis.*, 1986, **8**, 75–80.

42. R. Kojima, M. Ito and Y. Suzuki, 'Studies on the nephrotoxicity of aminoglycoside antibiotics and protection from these effects (7): Effect of latamoxef on binding of tobramycin to brush border membranes isolated from rat kidney cortex.' *Jpn. J. Pharmacol.*, 1989, **51**, 465–473.

43. J. S. Wold and S. A. Turnipseed, 'Toxicology of vancomycin in laboratory animals.' *Rev. Infect. Dis.*, 1981, **3**, 224–235.

44. R. D. Meyer, 'Risk factors and comparisons of clinical nephrotoxicity of aminoglycosides.' *Am. J. Med.*, 1986, **80**, (6B), 119–125.

45. D. A. Evans, J. Buring, S. Mayrent *et al.*, 'Qualitative overview of randomized trials of aminoglycosides.' *Am. J. Med.*, 1986, **80**, (6B), 39–43.

46. L. A. Cone, 'A survey of prospective controlled clinical trials of gentamicin, tobramycin, amikacin and netilmicin.' *Clin. Ther.*, 1982, **5**, 155–162.

47. J. P. McCormack and P. J. Jewesson, 'A critical reevaluation of the therapeutic range of amino-. glycosides.' *Clin. Infect. Dis.* 1992, **14**, 320–339.

48. J. H. Todd and G. H. Hottendorf, 'Renal brush border membrane vesicle aminoglycoside binding and nephrotoxicity.' *J. Pharmacol. Exp. Ther.*, 1995, **274**(1), 258–263.

49. S. A. Lerner, B. A. Schmitt, R. Seligsohn *et al.*, 'Comparative study of ototoxicity and nephrotoxicity in patients randomly assigned treatment with amikacin and gentamicin.' *Am. J. Med.*, 1986, **80**, (6B), 98–104.

50. M. E. Plaut, J. J. Schentag and W. J. Jusko, 'Aminoglycoside nephrotoxicity: comparative assessment in critical ill patients.' *J. Med.*, 1979, **10**, 257–266.

51. G. D. Kumin, 'Clinical nephrotoxicity of tobramycin and gentamicin. A prospective study.' *JAMA*, 1980, **244**, 1808–1810.

52. G. Kahlmeter and J. I. Dahlager, 'Aminoglycoside toxicity–a review of clinical studies published between 1975 and 1982.' *J. Antimicrob. Chemother.*, 1984, **13**, Suppl A, 9–22.

53. P. Auclaire, D. Tarhif, D. Beauchamp *et al.*, 'Prolonged endotoxemia enances the renal injuries induced by gentamicin in rats.' *Antimicrob. Agents Chemother.*, 1990, **34**(5), 889–895.

54. M. G. Bergeron, Y. Bergeron and Y. Marais, 'Autoradiography of tobramycin uptake by the proximal and distal tubules of normal and endotoxin-treated rats.' *Antimicrob. Agents Chemother.*, 1986, **29**, 1005–1009.

55. M. J. Bergeron, C. Lessard and A. Turcotte, 'In-vitro uptake of gentamicin and tobramycin by renal tubules in the presence or absence of *Escherichia coli* endo-. toxin.' *Antimicrob. Chemother.*, 1986, **18**, 375–380.

56. K. J. Tracey, S. F. Lowry and A. Cerami, 'Cachectin: a hormone that triggers acute shock and chronic cachexia.' *J. Infect. Dis.*, 1988, **157**, 413–420.

57. P. P. Nawroth and D. M. Stern, 'Modulation of endothelial cell hemostatic properties by tumor necrosis factor.' *J. Exp. Med.*, 1986, **163**, 740–745.

58. E. Holler, H. J. Kolb, A. Moller *et al.*, 'Increased serum levels of tumor necrosis factor alpha precede major complications of bone marrow transplantation.' *Blood*, 1990, **75**, 1011–1016.

59. R. Matzky, H. Darius and K. Schror, 'The release of prostaglandine (PGI$_2$) by petoxifylline from human vascular tissue.' *Arzneimittelforschung*, 1982, **32**, 1315–1318.

60. W. J. Novick, Jr., G. Sulivan and G. Mandell, 'New pharmacological studies with pentoxifylline.' *Biorheology*, 1990, **27**, 449–454.

61. R. M. Strieter, D. G. Remick, P. A. Ward *et al.* 'Cellular and molecular regulation of tumor necrosis factor alpha production by pentoxifylline.' *Biochem. Biophys. Res. Commun.*, 1988, **155**, 1230–1236.

62. J. A. Bianco, F. R. Appelbaum, J. Nemunaitis *et al.*, 'Phase I-II trial of pentoxifylline for the prevention of transplant-related toxicities following bone marrow transplantation.' *Blood*, 1991, **78**(5), 1205–1211.

63. K. M. Wasan, K. Vadiei, G. Lopez-Berestein *et al.*, 'Pentoxifylline in amphotericn B toxicity rat model.' *Antimicrob. Agents. Chemother.*, 1990, **34**, 241–244.

64. F. C. Luft, in 'The Aminoglycosides,' eds. A. Whelton and H. C. Neu, Dekker, New York, 1982, pp. 387–399.

65. M. J. Rybak, J. J. Frankowski, D. J. Edwards *et al.*, 'Alanine aminopeptidase and beta$_2$-microglobulin excretion in patients receiving vancomycin and gentamicin.' *Antimicrob. Agents Chemother.*, 1987, **31**(10), 1461–1464.

66. D. J. Pauly, D. M. Musa, M. R. Lestico *et al.*, 'Risk of nephrotoxicity with combination vancomycin-aminoglycoside antibiotic therapy.' *Pharmacotherapy*, 1990, **10**(6), 378–382.

67. C. A. Wood, S. J. Kohlhepp, P. W. Kohnen *et al.*, 'Vancomycin enhancement of experimental tobramycin nephrotoxicity.' *Antimicrob. Agents Chemother.*, 1986, **30**(1), 20–24.

68. N. J. Saunders, 'Vancomycin administration and monitoring reappraisal.' *J. Antimicrob. Chemother.*, 1995, **36**, 279–282.

69. Y. Yano, A. Hiraoka and T. Oguma, 'Enhancement of tobramycin binding to rat renal brush border membrane by vancomycin.' *J. Pharmacol. Exper. Ther.*, 1995, **274**(2), 695–699.

70. M. Iwamori, N. Tayama, Y. Nomura *et al.*, 'Hormone-dependent enhancment in binding of oto- and nephrotoxic aminoglycoside antibiotics.' *Acta Otolaryngol. Suppl (Stockh.)*, 1994, **514**, 117–121.

71. M. B. Goetz and J. Sayers, 'Nephrotoxicity of vancomycin and aminoglycoside therapy separately and in combination.' *J. Antimicrob. Chemother.*, 1993, **32**, 325–334.

72. G. Kahlmeter, T. Hallberg and C. Kamme, 'Gentamicin and tobramycin in patients with various infection-nephrotoxicity.' *J. Antimicrob. Chemother.*, 1978, **4** Suppl A, 47–52.

73. S. L. Preston and L. L. Briceland, 'Single daily dosing of aminoglycosides.' *Pharmacotherapy*, 1995, **15**(3), 297–316.

74. J. J. Schentag, T. A. Sutfin, M. E. Plaut *et al.*, 'Early detection of aminoglycoside nephrotoxicity with urinary β-microglobulin.' *J. Med.*, 1978, **9**, 201–210.

75. F. C. Luft, M. N. Yum and S. A. Kleit, 'Comparative nephrotoxicities of netilmicin and gentamicin in rats.' *Antimicrob. Agents Chemother.*, 1976, **10**, 845–849.

76. C. Cojocel, N. Dociu, K. Maita *et al.*, 'Effects of aminoglycosides on glomerular permeability, tubular reabsorption and intracellular catabolism of the cationic low-molecular-weight protein lysozyme.' *Toxicol. Appl. Pharmacol.*, 1983, **68**, 96–109.

77. C. Cojocel, J. H. Smith, K. Maita *et al.*, 'Renal protein degradation: a biochemical target of specific nephrotoxicants.' *Fundam. Appl. Toxicol.*, 1983, **3**, 278–284.

78. C. Cojocel and J. B. Hook, 'Differential effect of aminoglycoside treatment on glomerular filtration and renal reabsorption of lysozyme in rats.' *Toxicology*, 1981, **22**, 261–267.

79. K. Solez, L. C. Racusen and S. Olsen, 'The pathology of drug toxicity.' *J. Clin. Pharmacol.*, 1983, **23**, 484–490.

80. A. Whelton and K. Solez, 'Aminoglycoside nephrotoxicity–a tale of two transports. *J. Lab. Clin. Med.*, 1982, **99**, 148–155.

81. A. Whelton and K. Solez, 'Pathophysiologic mechanisms in aminoglycoside nephrotoxicity.' *J. Clin. Pharmacol.*, 1983, **23**, 453–460.

82. F. C. Luft and A. P. Evan, 'Comparative effects of tobramycin and gentamicin on glomerular ultrastructure.' *J. Infect. Dis.*, 1980, **142**, 910–913.

83. G. C. McCormick, E. Weinberg, G. Briziarelli et al., 'Comparative toxicity of gentamicin and tobramycin in rats at low multiples of the human therapeutic dose.' *Toxicol. Appl. Pharmacol.*, 1982, **63**, 194–200.

84. F. C. Luft, G. R. Aronoff, A. P. Evan et al., 'The effect of aminoglycoside on glomerular endothelium: a comparative study.' *Res. Comm. Chem. Pathol. Pharmacol.*, 1981, **34**, 89–95.

85. E. H. Weinberg, W. E. Field, W. D. Gray et al., 'Preclinical toxicologic studies of netilmicin.' *Arzneimittelforschung*, 1981, **31**, 816–822.

86. H. D. Humes, 'Aminoglycoside nephrotoxicity.' *Kidney Int.*, 1988, **33**, 900–911.

87. R. M. Reichley, D. J. Ritchie and T. C. Bailey, 'Analysis of various creatinine clearance formulas in predicting gentamicin elimination in patients with low serum creatinine.' *Pharmacotherapy*, 1995, **15**(5), 625–630.

88. G. Nicot, J. P. Charmes, J. P. Vallette et al., 'Signes urinaire precoces de nephrotoxicite des aminosides.' [Early urinary markers of aminoglycoside nephrotoxicity]. *Therapie*, 1980, **35**, 369–374.

89. C. Cojocel and K. Baumann, 'Renal handling of endogenous lysozyme in the rat.' *Ren. Physiol.*, 1983, **6**, 258–265.

90. R. Gibey, J. L. Dupond, D. Alber et al., 'Predictive value of urinary *N*-acetyl-beta-D-glucosaminidase (NAG), alanine-aminopeptidase (AAP) and beta-2-microglobulin B2M in evaluating nephrotoxicity of gentamicin.' *Clin. Chim. Acta*, 1981, **116**, 25–34.

91. G. Viotte, J. P. Morin, M. Godin et al., 'Changes in the renal function of rats treated with cephoxitin and a comparison with other cephalosporins and gentamicin.' *J. Antimicrob. Chemother.*, 1981, **7**, 537–550.

92. V. Patel, F. C. Luft, M. N. Yum et al., 'Enzymuria in gentamicin-induced kidney damage.' *Antimicrob. Agents Chemother.*, 1975, **7**, 364–369.

93. C. Cojocel, N. Docius, K. Maita et al., 'Renal ultrastructural and biochemical injuries induced by aminoglycosides.' *Environ. Health Persp.*, 1984, **57**, 293–299.

94. C. Cojocel, N. Docius, E. Ceacmacudis et al., 'Nephrotoxic effects of aminoglycoside treatment on renal protein reabsorption and accumulation.' *Nephron*, 1984, **37**, 113–119.

95. D. N. Gilbert, D. C. Houghton, W. M. Bennett et al., 'Reversibility of gentamicin nephrotoxicity in rats: recovery during continuous drug administration.' *Proc. Soc. Exp. Biol. Med.*, 1979, **160**, 99–103.

96. R. A. Giuliano, G. J. Paulus, G. A. Verpooten et al., 'Recovery of cortical phospholipidosis and necrosis after acute gentamicin loading in rats.' *Kidney Int.*, 1984, **26**, 838–847.

97. G. Laurent, P. Maldague, M. B. Carlier et al., 'Increased renal DNA synthesis *in vivo* after administration of low doses of gentamicin to rats.' *Antimicrob. Agents Chemother.*, 1983, **24**, 586–593.

98. J. Klastersky, R. F. Meunier-Carpentier, L. Coppens-Kahan et al., 'Clinical and bacteriological evaluation of netilmicin in gram-negative infections.' *Antimicrob. Agents Chemother.*, 1977, **12**, 503–509.

99. C. R. Smith, K. L. Baughman, C. O. Edwards et al., 'Controlled comparison of amikacin and gentamicin.' *New. Engl. J. Med.*, 1977, **296**, 349–353.

100. S. C. Li, L. L. Ioannides-Demos, W. J. Spicer, et al. 'Prospective audit of aminoglycoside usage in a general hospital with assessments of clinical processes and adverse clinical outcomes.' *Med. J. Aust.*, 1989, **151**, 224–232.

101. D. N. Gilbert, 'Once-daily aminoglycoside therapy.' *Antimicrob. Agents Chemother.*, 1991, **35**, 399–405.

102. M. E. Levison, 'New dosing regimens for aminoglycoside antibiotics.' *Ann. Intern. Med.*, 1992, **117**, 693–694.

103. W. A. Craig and B. Vogelman, 'The postantibiotic effect.' *Ann. Intern. Med.*, 1987, **106**, 900–902.

104. B. S. Vogelman and W. A. Craig, 'Postantibiotic effect.' *J. Antimicrob. Chemother.*, 1985, **15**, 37–46.

105. G. G. Jackson, G. L. Daikos and V. T. Lolans, 'First-exposure effect of netilmicin n bacterial susceptibility as a basis for modifing the dosage regimen of aminoglyoside antibiotics.' *J. Drug Dev.*, 1988, **1**, Suppl. 3, 49–54.

106. G. L. Daikos, G. G. Jackson, V. T. Lolans et al., 'Adaptive resistance to aminoglycoside antibiotics fom first-exposure down-regulation.' *J. Infect. Dis.*, 1990, **162**, 414–420.

107. H. Mattie, W. A. Craig and J. C. Pechere, 'Determinants of efficacy and toxicity of aminoglycosides.' *J. Antimicrob. Chemother.*, 1989, **24**, 281–293.

108. G. R. Aronoff, S. T. Pottratz, M. E. Brier et al., 'Aminoglycoside accumulation kinetics in rat rental parenchyma.' *Antimicrob. Agents Chemother.*, 1983, **23**, 74–78.

109. W. M. Bennett, C. E. Plamp, III, D. N. Gilbert et al., 'The influence of dosage regimen on experimental gentamicin nephrotoxicity: dissociation of peak serum levels from renal failure.' *J. Infect. Dis.*, 1979, **140**, 576–580.

110. S. H. Powell, W. L. Thomson, A. A. Luthe et al., 'Once-daily vs. continuous aminoglycoside dosing: efficacy and toxicity in animal and clinical studies of gentamicin, netilmicin and tobramycin.' *J. Infect. Dis.*, 1983, **147**, 918–932.

111. J. M. Prins, H. R. Büller, E. J. Kuijper et al., 'Once versus thrice daily gentamicin in patients with serious infections.' *Lancet*, 1993, **341**, 355–339.

112. A. M. Galloe, N. Graudal, H. R. Christensen et al., 'Aminoglycosides: single or multiple daily dosing? A meta-analysis on efficacy and safety.' *Eur. J. Cli. Pharmacol.*, 1995, **48**(1), 39–43.

113. P. Periti, 'Pharmacoeconomic evaluation of once-daily aminoglycoside treatment.' *J. Chemother.*, 1995, **7**(4), 380–394.

114. A. Whelton, G. G. Carter, T. J. Craig et al., 'Comparison of the intrarenal disposition of tobramycin and gentamicin: therapeutic and toxicologic answers.' *J. Antimicrob. Chemother.*, 1978, **4**, Suppl. A, 13–22.

115. W. M. Bennett, F. Luft and G. A. Porter, 'Pathogenesis of renal failure due to aminoglycosides and contrast media used in roentgenography.' *Am. J. Med.*, 1980, **69**, 767–774.

116. F. Silverblatt, 'Pathogenesis of nephrotoxicity of cephalosporins and aminoglycosides: a review of current concepts.' *Rev. Infect. Dis.*, 1982, **4**, S360–365.

117. S. Kacew and M. G. Bergeron, 'Pathogenic factors in aminoglycoside-inducd nephrotoxicity.' *Toxicol. Lett.*, 1990, **51**, 241–259.

118. W. M. Bennett, C. E. Plamp, III, W. C. Elliott et al., 'Effect of basic amino acids and aminoglycosides on ^{3}H-gentamicin uptake in cortical slices of rat and human kidney.' *J. Lab. Clin. Med.*, 1982, **99**, 156–162.

119. R. E. Cronin, 'Aminoglycoside nephrotoxicity: pathogenesis and prevention.' *Clin. Nephrol.*, 1979, **11**, 251–256.

120. M. M. Hansen and K. Kaaber, 'Nephrotoxicity in

combined cephalothin and gentamicin therapy.' *Acta Med. Scand.*, 1977, **201**, 463–467.

121. D. N. McMartin and S. G. Engel,'Effect of aging on gentamicin nephrotoxicity and pharmacokinetics in rats.' *Res. Commun. Chem. Pathol. Pharmacol.*, 1982, **38**, 193–207.

122. A. P. Panwalker, J. B. Malow, V. M. Zimelis *et al.*, 'Netilmicin: clinical efficacy, tolerance and toxicity.' *Antimicrob. Agents Chemother.*, 1978, **13**, 170–176.

123. M. E. De Broe, G. J. Paulus, G. A. Verpooten *et al.*, 'Early effects of gentamicin, tobramycin and amikacin on the human kidney.' *Kidney Int.*, 1984, **25**, 643–652.

124. R. A. Giuliano, D. E. Pollet, G. A. Verpooten *et al.*, 'Influence of dose regimen on renal accumulation of aminoglycosides.' *Arch. Int. Pharmacodyn. Ther.*, 1982, **260**, 277–279.

125. R. Marre, M. Abraham, H. Freiesleben *et al.*, 'Netilmicin und tobramycin: vergleichende tierexperimentelle untersuchungen zur pharmakokinetik, nierenverträglichkeit und therapeutischen effektivität.' *Arzneimittelforschung*, 1979, **29**, 940–945.

126. A. Whelton, G. G. Carter, T. J. Craig *et al.*, 'Comparison of the intrarenal disposition of tobramycin and gentamicin: therapeutic and toxicologic answers.' *J. Antimicrob. Chemother.*, 1978, **4** Suppl., 13–22.

127. P. D. Williams, G. H. Hottendorf and D. B. Bennett, 'Inhibition of renal membrane binding and nephrotoxicity of aminoglycosides.' *J. Pharmacol. Exp. Ther.*, 1986, **237**(3), 919–925.

128. B. M. Brenner, T. H. Hostetter and H. D. Humes, 'Glomerular permselectivity: barrier function based on discrimination of molecular size and charge.' *Am. J. Physiol.*, 1978, **234**, F455–F460.

129. S. M. Shea and A. B. Morrison, 'A stereological study of the glomerular filter in the rat. Morphometry of the slit diaphragm and basement membrane.' *J. Cell. Biol.*, 1975, **67**, 436.

130. M. A. Venkatachalam and H. G. Rennke, 'The structural and molecular basis of glomerular filtration.' *Circ. Res.*, 1978, **43**, 337–347.

131. K. Baumann, C. Cojocel and W. Pape, in 'The Pathogenicity of Cationic Proteins,' eds, P. P. Lambert, P. Bergmann and R. Beauwens, Raven Press, New York, 1983, pp. 249–260.

132. K. Maita, C. Cojocel, N. Docius *et al.*, 'Effects of aminoglycosides on glomerular ultrastructure.' *Pharmacology*, 1984, **29**, 292–300.

133. V. M. Vehaskari, E. R. Root, F. G. Germuth Jr. *et al.*, 'Glomerular charge and urinary protein excretion: effects of systemic and intrarenal polycation infusion in the rat.' *Kidney Int.*, 1982, **22**, 127–135.

134. E. Pastoriza-Munoz, D. Timmermann, S. Feldman *et al.*, 'Ultrafiltration of gentamicin and netilmicin *in vivo*.' *Pharmacol. Exp. Ther.*, 1982, **220**, 604–608.

135. P. S. Avasthi, A. P. Evan, J. W. Huser *et al.*, 'Effect of gentamicin on glomerular ultrastructure.' *J. Lab. Clin. Med.*, 1981, **98**, 444–454.

136. D. C. Houghton, C. E. Plamp, III, J. M. DeFehr *et al.*, 'Gentamicin and tobramycin nephrotoxicity. A morphologic and functional comparison in the rat.' *Am. J. Pathol.*, 1978, **93**, 137–152.

137. F. C. Luft, V. Patel, M. N. Yum *et al.*, 'Experimental aminoglycoside nephrotoxicity.' *J. Lab. Clin. Med.*, 1975, **86**, 213–220.

138. F. C. Luft, G. R. Aronoff, A. P. Evan *et al.*, 'The renin-angiotensin system in aminoglycoside-induced acute renal failure.' *J. Pharmacol. Exp. Ther.*, 1982, **220**, 433–439.

139. F. G. Silverblatt and C. Kuehn, 'Autoradiography of gentamicin uptake by the rat proximal tubule cell.' *Kidney Int.*, 1979, **15**, 335–345.

140. C. Baylis, H. R. Rennke and B. M. Brenner, 'Mechanisms of the defect in glomerular ultrafiltration associated with gentamicin administration.' *Kidney Int.*, 1977, **12**, 344–353.

141. A. Sugarman, R. S. Brown, P. Silva *et al.*, 'Features of gentamicin nephrotoxicity and effect of concurrent cephalothin in the rat.' *Nephron*, 1983, **34**, 239–247.

142. P. Mene, M. S. Simonson and M. J. Dun, 'Physiology of mesangial cells.' *Physiol. Rev.*, 1989, **69**(4), 1347–1424.

143. H. Latta and S. Fligiel, 'Mesangial fenestrations, sieving, filtration and flow.' *Lab. Invest.*, 1986, **52**, 591–598.

144. V. Savin, L. Karniski, F. Cuppage *et al.*, 'Effect of gentamicin on isolated glomeruli and proximal tubules of the rabit.' *Lab. Invest.*, 1985, **52**(1), 93–102.

145. R. C. Blantz, 'The glomerulus, passive filter or regulatory organ?' *Klin. Wochenschr.*, 1980, **58**, 957–964.

146. N. Schor, L. Ichikawa, H. G. Rennke *et al.*, 'Pathophysiology of altered glomerular function in aminoglycoside-treated rats.' *Kidney Int.*, 1981, **19**, 288–296.

147. G. F. DiBona and L. L. Sawin, 'The renin-angiotensin system in acute renal failure in the rat.' *Lab. Invest.*, 1971, **25**, 528–532.

148. D. E. Oken, S. C. Cotes, W. Flamenbaum *et al.*, 'Active and passive immunization to angiotensin in experimental acute renal failure.' *Kidney Int.*, 1975, **7**, 12–18.

149. G. B. Ryan, D. Alcorn, J. P. Coghlan *et al.*, 'Ultrastructural morphology of granule release from juxtaglomerular myoepithelioid and peripolar cells.' *Kidney Int.*, 1982, Suppl.12, S3–S8.

150. P. Weber, E. Held, E. Uhlich *et al.*, 'Reaction constants of renin in juxtaglomerular apparatus and plasma renin activity after renal ischemia and hemorrhage.' *Kidney Int.*, 1975, **7**, 331–341.

151. R. A. Zager, 'Gentamicin nephrotoxicity in the seting of acute renal hypoperfusion.' *Am. J. Physiol.*, 1988, **254**, F574–F581.

152. D. M. Spiegel, P. F. Shanley and B. A. Molitoris, 'Mild ischemia predisposes the S3 segment to gentamicin toxicity.' *Kidney Int.*, 1990, **38**, 459–464.

153. B. A. Molitoris, C. Meyer, R. Dahl *et al.*, 'Mechanism of ischemia-enhanced aminoglycoside binding and uptake by proximal tubule cells.' *Am. J. Physiol.*, 1993, **264**, F907–F916.

154. C. Cojocel and J. B. Hook, 'Aminoglycoside nephrotoxicity.' *Trends Pharmacol.*, 1983, **4**, 174–179.

155. G. J. Kaloyanides, 'Aminoglycoside-induced functional and biochemical defects in the renal cortex.' *Fundam. Appl. Toxicol.*, 1984, **4**, 930–943.

156. H. D. Humes and J. M. Weinberg, 'Alterations in renal tubular cell metabolism in acute renal failure.' *Miner. Electrolyte Metab.*, 1983, **9**, 290–305.

157. J. F. Donohoe, M. A. Venkatachalam, D. B. Bernard *et al.*, 'Tubular leakage and obstruction after renal ischemia: structural-functional correlations.' *Kidney Int.*, 1978, **13**, 208–222.

158. J. Neugarten, H. S. Aynedjian and N. Bank, 'Role of tubular obstruction in acute renal failure due to gentamicin.' *Kidney Int.*, 1983, **24**, 230–335.

159. D. C. Houghton, C. E. Plamp, III and J. M. DeFehr, 'Gentamicin and tobramycin nephropathy. A morphologic and functional comparison in the rat.' *Am. J. Path.*, 1978, **93**, 137–152.

160. J. C. Kosek, R. I. Mazze and M. J. Cousins, 'Nephrotoxicity of gentamicin.' *Lab. Invest.*, 1974, **30**, 48–57.

161. J. Vera-Roman, T. P. Krishnakantha and F. E. Cuppage, 'Gentamicin nephrotoxicity in rats. I. Acute biochemical and ultrastructural effects.' *Lab. Invest.*, 1975, **33**, 412–417.

162. M. E. De Broe, G. J. Paulus, G. A. Verpooten *et al.*, 'Early effects of gentamicin, tobramycin and amikacin on the human kidney.' *Kidney Int.*, 1984, **25**, 643–652.

163. D. C. Houghton, M. Hartnett, M. Campbell-Boswell

et al., 'A light and electron microscopic analysis of gentamicin nephrotoxicity in rats.' *Am. J. Pathol.*, 1976, **82**, 589–612.

164. W. M. Kluwe and J. B. Hook, 'Analysis of gentamicin uptake by rat renal cortical slices.' *Toxicol. Appl. Pharmacol.*, 1978, **45**, 531–539.

165. M. Sastrasinh, T. C. Knauss, J. M. Weinberg *et al.*, 'Identification of the aminoglycoside binding site of renal brush border membranes.' *J. Pharmacol. Exp. Ther.*, 1982, **222**, 350–358.

166. H. Nakahama, T. Moriyama, M. Horio *et al.*, 'Netilmicin affects Na$^+$-dependent D-glucosetransport and the membrane fluidity of rabbit renal brush-border membrane vesicles: a comparison with gentamicin.' *Toxicol. Lett.*, 1990, **53**, 201–202.

167. W. M. Bennett, C. E. Plamp, III, R. A. Parker *et al.*, 'Alterations in organic ion transport induced by gentamicin nephrotoxicity in the rat.' *J. Lab. Clin. Med.*, 1980, **95**, 32–39.

168. B. J. Smyth, J. H. Todd, J. E. Bylander *et al.*, 'Selective exposure of human proximal tubule cells to gentamicin provides evidence for basolateral component of toxicity.' *Toxicol. Lett.*, 1994, **74**, 1–13.

169. L. Cohen, R. Lapkin and G. J. Kaloyanides, 'Effect of gentamicin on the renal function in the rat.' *J. Pharmacol. Exp. Ther.*, 1975, **193**, 264–273.

170. J. E. Riviere, E. J. Hinsman, G. L. Coppoc *et al.*, 'Single dose gentamicin nephrotoxicity in the dog: early functional and ultrastructural changes.' *Res. Commun. Chem. Pathol. Pharmacol.*, 1981, **33**, 403–418.

171. G. Giebisch, R. M. Klose and G. Malnic, 'Renal tubular potassium transport.' *Bull. Schweiz. Akad. Med. Wiss.*, 1967, **23**, 287–312.

172. H. W. Dickerman and W. G. Walker, 'Effect of cationic amino acid infusion on potassium metabolism *in vivo*.' *Am. J. Physiol.*, 1964, **206**, 403–408.

173. D. Y. Mason, D. T. Howes, C. R. Taylor *et al.*, 'Effect of human lysozyme (muramidase) on potassium handling by the perfused rat kidney. A mechanism for renal damage in human monocytic leukaemia.' *J. Clin. Path.*, 1975, **28**, 722–727.

174. C. Cojocel N. Docius and K. Baumann, 'Early nephrotoxicity at high plasma concentrations of lysozyme in the rat.' *Lab. Invest.*, 1982, **46**, 149–157.

175. R. A. DeFronzo and M. Bia, in 'The Kidney: Physiology and Pathophysiology,' eds. D. W. Seldin and G. Giebisch, Raven Press, New York, 1985, pp. 1179–1206.

176. J. Purnell, D. C. Haughton, G. A. Porter *et al.*, 'Effect of amiloride on experimental gentamicin nephrotoxicity.' *Nephron*, 1985, **40**, 166–170.

177. C. Cojocel, N. Docius, E. Ceacmacudis *et al.*, 'Effects of aminoglycosides on renal protein reabsorption and accumulation.' *Contrib. Nephrol.*, 1984, **42**, 196–201.

178. E. Pastoriza-Munoz, D. Timmerman and G. J. Kaloyanides, 'Tubular wasting of cations associated with netilmicin infusion.' *Clin. Res.*, 1983, **31**, 483A.

179. R. Patel and A. Savage, 'Symptomatic hypomagnesemia associated with gentamicin therapy.' *Nephron*, 1979, **23**, 50–52.

180. R. E. Cronin, R. E. Bulger, P. Southern *et al.*, 'Natural history of aminoglycoside nephrotoxicity in the dog.' *J. Lab. Clin. Med.*, 1980, **95**, 463–474.

181. P. D. Williams, M. E. Trimble, L. Crespo *et al.*, 'Inhibition of renal Na$^+$,K$^+$-adenosine triphosphatase by gentamicin.' *J. Pharmacol. Exp. Ther.*, 1984, **231**, 248–253.

182. G. Beyer, F. Bode and K. Baumann, 'Binding of lysozyme to brush border membranes of rat kidney.' *Biochim. Biophys. Acta*, 1983, **732**, 372–376.

183. F. Bode, H. Pockrand-Hemstedt, K. Baumann *et al.*, 'Analysis of the pinocytic process in rat kidney. 1. Isolation of pinocytic vesicles from rat kidney cortex.' *J. Cell Biol.*, 1974, **63**, 998–1008.

184. F. Bode, K. Baumann and R. Kinne, 'Analysis of the pinocytic process in rat kidney. 11. Biochemical composition of pinocytic vesicles compared to brush border microvilli, lysosomes and basolateral plasma membranes.' *Biochim. Biophys. Acta*, 1976, **433**, 294–310.

185. H. Schöttke, R. Schwartz and K. Baumann, 'Effect of low-molecular-weight proteins on protein lysozyme binding to isolated brush-border membranes of rat kidney.' *Biochim. Biophys. Acta*, 1984, **770**, 210–215.

186. G. S. Vince and R. T. Dean, 'Endocytosis of proteins by kidney tubule cells: inhibition by the aminoglycoside gentamicin.' *Biochem. Pharmacol.*, 1986, **35**(18), 3182–3184.

187. B. E. Sumpio, I. H. Chaudry and A. E. Baue, 'Reduction of the drug-induced nephrotoxicity by ATP-MgCl$_2$. II. Effects on gentamicin-treated isolated perfused kidneys.' *J. Surgical Res.*, 1985, **38**, 438–445.

188. A. Bernard, C. Viau, A. Ouled *et al.*, 'Effects of gentamicin on the renal uptake of endogenous and exogenous proteins in conscious rats.' *Toxicol. Appl. Pharmacol.*, 1986, **84**, 431–438.

189. H. Smaoui, M. Schaeverbeke, J. P. Mallie *et al.*, 'Transplacental effects of gentamicin on endocytosis in rat renal proximal tubule cells.' *Pediatr. Nephrol.*, 1994, **8**(4), 447–450.

190. L. Giurgea-Marion, G. Toubeau, G. Laurent *et al.*, 'Impairment of lysosome-pinocytotic vesicle fusion in rat kidney proximal tubules after treatment with gentamicin at low doses.' *Toxicol. Appl. Pharmacol.*, 1986, **86**, 271–285.

191. J. P. Morin, J. P. Viotte, A., Vandewalle *et al.*, 'Gentamicin-induced nephrotoxicity: a cell biology approach.' *Kidney Int.*, 1980, **18**, 583–590.

192. G. Aubert-Tulkens, F. Van Hoof and P. Tulkens, 'Gentamicin-induced lysosomal phospholipidosis in cultured rat fibroblasts. Quantitative ultrastructural and biochemical study.' *Lab. Invest.*, 1979, **40**, 481–491.

193. C. J. Olbricht, M. Fink and E. Gutjahr, 'Alterations in lysosomal enzymes of the proximal tubule in gentamycin nephrotoxicity.' *Kidney Int.*, 1991, **39**, 636–646.

194. P. Tulkens and A. Trouet, 'The uptake and intracellular accumulation of aminoglycoside antibiotics in lysosomes of cultured fibroblasts.' *Biochem. Pharmacol.*, 1978, **27**, 415–424.

195. G. Laurent, M. B. Carlier, B. Rollman *et al.*, 'Mechanisms of aminoglycoside-induced lysosomal phospholipidosis: *in vitro* and *in vivo* studies with gentamicin and amikacin.' *Biochem. Pharmacol.*, 1982, **31**, 3861–3870.

196. K. Y. Hostetler and L. B. Hall, 'Inhibition of kidney lysosomal phospholipases A and C by aminoglycoside antibiotics: possible mechanisms of aminoglycoside toxicity. *Proc. Nat. Acad. Sci. USA*, 1982, **79**, 1663–1667.

197. J. P. Fillastre and M. Godin, in 'Mechanism of Injury in Renal Disease and Toxicity,' ed. R. S. Goldstein, CRC Press, Boca Raton, FL, 1994, pp. 123–147.

198. T. Yuzuriha, K. Katayama and T. Fujita, 'Studies on biotransformation of lysozyme. II. Tissue distribution of 1311-labelled lysozyme and degradation in kidney after intravenous injections in rats.' *Chem. Pharm. Bull. (Tokyo)*, 1975, **23**, 1315–1322.

199. J. M. Weinberg, P. G. Harding and H. D. Humes, 'Mechanisms of gentamicin-induced dysfunction of renal cortical mitochondria. II. Effects on mitochondrial monovalent cation transport.' *Arch. Biochem. Biophys.*, 1980, **205**, 232–239.

200. J. M. Wienberg, and H. D. Humes, 'Mechanisms of

gentamicin-induced dysfunction of renal cortical mitochondria. I. Effects on mitochondrial respiration.' *Arch. Biochem. Biophys.*, 1980, **205**, 221–231.

201. J. M. Weinberg, C. F. Simmons, Jr. and H. D. Humes, 'Alterations of mitochondrial respiration induced by aminoglycoside antibiotics.' *Res. Comm. Chem. Pathol. Pharmacol.*, 1980, **27**, 521–531.

202. C. F. Simmons, Jr., R. T. Boguski and H. D. Humes, 'Inhibitory effects of gentamicin on renal mitochonrial oxidative phosphorilation.' *J. Pharmacol. Exp. Ther.*, 1980, **214**, 709–715.

203. F. E. Cuppage, K. Setter, P. Sullivan *et al.*, 'Gentamicin nephrotoxicity. II. Physiological, biochemical and morphological effects of prolonged administration to rats.' *Virchows Arch. B Cell Pathol.*, 1977, **24**, 121–138.

204. M. Sastrasinh, J. M. Weinberg and H. D. Humes *et al.*, 'The effects of gentamicin on calcium uptake by renal mitochondria.' *Life Sci.*, 1982, **30**, 2309–2315.

205. C. Cojocel, M. Franzen-Sieveking, G. Beckmann *et al.*, 'Inhibition of renal accumulation of lysozyme (basic low molecular weight protein) by basic proteins and other basic substances.' *Pflügers Arch.*, 1981, **390**, 211–215.

206. M. Just, A. Röckel, A. Stanjek *et al.*, 'Is there any transtubular reabsorption of filtered proteins in rat kidney?' *Naunyn Schmiedebergs Arch. Pharmacol.*, 1975, **289**, 229–236.

207. M. Just and E. Habermann, 'Interactions of a protease inhibitor and other peptides with isolated brush border membranes from rat renal cortex.' *Naunyn Schmiedebergs Arch. Pharmacol.*, 1973, **280**, 161–176.

208. H. J. Ryser, 'Uptake of protein by mammalian cells: an underdeveloped area. The penetration of foreign proteins into mammalian cells can be measured and their functions explored.' *Science*, 1968, **159**, 390–396.

209. B. E. Sumpio and T. Maack, 'Kinetics, competition and selectivity of tubular absorption of proteins. *Am. J. Physiol.*, 1982, **243**, F379–F392.

210. C. E. Mogensen, E. Vittinghus and K. Solling, 'Increased urinary excretion of albumin, light chains, and B2-microglobulin after intravenous arginine administration in normal man.' *Lancet*, 1975, **27**, 581–583.

211. P. D. Ottosen, K. M. Madsen, F. Bode *et al.*, 'Inhibition of protein reabsorption in the renal proximal tubule by basic amino acids.' *Ren. Physiol.*, 1985, **8**, 90–99.

212. E. I. Christensen, H. G. Renke and F. A. Carone, 'Renal tubular uptake of protein: effect of molecular charge.' *Am. J. Physiol.*, 1983, **244**, F436–F441.

213. R. Brasseur, G. Laurent, J. M. Ruysschaert, 'Interactions of aminoglycoside antibiotics with negatively charged lipid layers. Biochemical and conformational studies.' *Biochem. Pharmacol.*, 1984, **33**, 629–637.

214. C. Josepovitz, R. Levine, T. Farruggella *et al.*, '[³H]-netilmicin binding constants and phospholipid composition of renal plasma membranes of normal and diabetic rats.' *J. Pharmacol. Exp. Ther.*, 1985, **233**, 298–303.

215. Y. S. Kanwar and M. G. Farquhar, 'Anionic sites in the glomerular basement membrane. *In vivo* and *in vitro* localization to the laminae rarae by cationic probes.' *J. Cell Biol.*, 1979, **81**, 137–153.

216. P. D. Williams, P. D. Holohan and C. R. Ross, 'Gentamicin nephrotoxicity. I. Acute biochemical correlates in rats.' *Toxicol. Appl. Pharmacol.*, 1981, **61**, 234–242.

217. P. D. Williams, P. D. Holohan and C. R. Ross, 'Gentamicin nephrotoxicity. II. Plasma membrane changes.' *Toxicol. Appl. Pharmacol.*, 1981, **61**, 243–251.

218. R. A. DeFronzo, R. Lee, A. Jones *et al.*, 'Effect of insulinopenia and adrenal hormone deficiency on acute potassium tolerance.' *Kidney Int.*, 1980, **17**, 586–594.

219. M. J. Field and G. J. Giebisch, 'Hormonal control of renal potassium excretion.' *Kidney Int.*, 1985, **27**, 379–387.

220. K. R. Brinker, R. E. Bulgar, D. C. Bobyan *et al.*, 'Effect of potassium depletion on gentamicin nephrotoxicity.' *J. Lab. Clin. Med.*, 1981, **98**, 292–301.

221. L. I. Rankin, H. Krous, A. W. Fryer *et al.*, 'Enhancement of gentamicin nephrotoxicity by magnesium depletion in the rat.' *Miner. Electrolyte Metab.*, 1984, **10**, 199–203.

222. Y. Matsuzawa and K. Y. Hostettler, 'Inhibition of lysosomal phospholipase A and phospolipase C by chloroquine and 4,4'-bis(diethylaminoethoxy), A, β-diethyldiphenylethane.' *J. Biol. Chem.*, 1980, **255**, 5190–5194.

223. E. I. Christensen and A. B. Maunsbach, 'Effects of dextran on lysosomal ultrastructure and protein digestion in renal proximal tubule.' *Kidney Int.*, 1979, **16**, 301–311.

224. W. M. Bennett, L. M. Mela-Riker, D. C. Houghton *et al.*, 'Microsomal protein synthesis inhibition: an early manifestation of gentamicin nephrotoxicity.' *Am J. Physiol.*, 1988, **255**, F265–F269.

225. S. Feldman, M. Y. Wang and G. J. Kaloyanides, 'Aminoglycosides induce a phospholipidosis in the renal cortex of the rat: an early manifestation of nephrotoxicity.' *J. Pharmacol. Exp. Ther.*, 1982, **220**, 514–520.

226. J. P. Morin, J. Fresel, J.-P. Fillastre *et al.*, in 'Nephrotoxicity: Interaction of Drugs with Membrane Systems: Mitochondria-Lysosomes,' ed. J. P. Fillastre, Masson, New York, 1978, pp. 253–263.

227. R. Lüllman-Rauch, in 'Lysosomes in Biology and Pathology,' eds. J. T. Dingle, P. J. Jacques and I. H. Shaw, North Holland, Amsterdam, 1979, vol. 6, pp. 49–130.

228. J. D. Etlinger and A. L. Goldberg, 'A soluble ATP-dependent proteolytic system responsible for the degradation of abnormal proteins in reticulocytes.' *Proc. Natl. Acad. Sci. USA*, 1977, **74**, 54–58.

229. W. J. Reville, D. E. Goll, M. H. Stromer *et al.*, 'A Ca²⁺-activated protease involved in myofibrilar protein turnover. Subcellular localization of the protease in porcine skeletal muscle.' *J. Cell Biol.*, 1976, **70**, 1–8.

230. W. M. Kluwe and J. B. Hook, 'Functional nephrotoxicity of gentamicin in the rat.' *Toxicol. Appl. Pharmacol.*, 1978, **45**, 163–175.

231. L. S. Ramsammy, K. Y. Ling, C. Josepovitz *et al.*, 'Effect of gentamicin and lipid peroxidation in rat renal cortex.' *Biochem. Pharmacol.*, 1985, **34**, 3895–3900.

232. P. D. Walker and S. V. Shah, 'Evidence suggesting a role for hydroxyl radical in gentamicin-induced acute renal failure in rats.' *J. Clin. Invest.*, 1980, **81**, 334–341.

233. L. S. Ramsammy, C. Josepovitz, K. Y. Ling *et al.*, 'Effects of diphenyl-phenylenediamine on gentamicin-induced lipid peroxidation and toxicity in rat renal cortex.' *J. Pharmacol. Exp. Ther.*, 1986, **238**, 83–88.

234. J. D. Swann and D. Acosta, 'Failure of gentamicin to elevate cellular malodialdehyde content or increase generation of intracellular reactive oxygen species in primary cultures of renal cortical epithelial cells.' *Biochem. Pharmacol.*, 1990, **40**(7), 1523–1526.

235. T. H. Ben Ismail, B. H. Ali and A. A. Bashir, 'Influence of iron, deferoxamine and ascorbic acid on gentamicin-induced nephrotoxicity in rats.' *Gen. Pharmacol.*, 1994, **25**, 1249–1252.

236. L. S. Ramsammy, C. Josepivitz, K. Y. Ling *et al.*, 'Failure and inhibition of lipid peroxidation by vitamin E to protect against gentamicin nephrotoxicity in the rat.' *Biochem. Pharmacol.*, 1987, **36**, 2125–2132.

237. C. Cojocel, J., Hannemann and K. Baumann, 'Cephaloridine-induced lipid peroxidation initiated by reactive oxygen species as a possible mechanism of cephaloridine nephrotoxicity.' *Biochim. Biophys, Acta*, 1985, **834**, 402–410.

238. H. C. Sutton, 'Efficiency of chelated iron compounds as catalysts for the Haber–Weiss reaction.' *Free Radic. Biol. Med.*, 1985, **1**, 195–202.

239. S. E. Kays, W. A. Crowell and M. A. Johnson, 'Iron supplementation increases gentamicin nephrotoxicity in rats.' *J. Nutr.*, 1991, **121**, 1869–1875.

240. C. E. Thomas and S. D. Aust, 'Reductive release of iron from ferritin by cation free radicals of paraquat and other bipyridyls.' *J. Biol. Chem.*, 1986, **261**(28), 13064–13070.

241. N. Ueda, B. Guidet and S. V. Shah, 'Gentamicin-induced mobilization of iron from cortical mitochondria.' *Am J. Physiol.*, 1993, **265**, F435–F439.

242. L. Rivas-Cabanero, A. Rodriguez-Barbero, M. Arevalo *et al.*, 'Effect of N^G-nitro-L-arginine methyl ester on nephrotoxicity induced by gentamicin in rats.' *Nephron*, 1995, **71**, 203–207.

243. L. Rivas-Cabanero, A. Montero and J. M. Lopez-Novoa, 'Increased glomerular nitric oxide synthesis in gentamicin-induced renal failure.' *Eur. J. Pharmacol.*, 1994, **270**, 119–121.

244. H. Scholz and A. Kurtz, 'Involvement of endothelium-derived relaxing factor in the pressure control of renin secretion from isolated perfused kidney.' *J. Clin. Invest.*, 1993, **91**, 1088–1094.

245. A. Kurtz, B. Kaissling, R. Bussse *et al.*, 'Endothelial cells modulate renin secretion from isolated mouse juxtaglomerular cells.' *J. Clin. Invest.*, 1991, **88**, 1147–1154.

246. S. G. Greenberg, X. R. He, J. B. Schnermann *et al.*, 'Effect of nitric oxide on renin secretion. I. Studies in isolated juxtaglomerular granular cells.' *Am. J. Physiol.*, 1995, **268**, F948–F952.

247. X. R. He, S. G. Greenberg, J. P Briggs *et al.*, 'Effect of nitric oxide on renin secretion. II. Studies in the juxtaglomerular apparatus.' *Am. J. Physiol.*, 1995, **268**, F953–F959.

248. K. Ujiie, L. Yuen, L. Hogarth *et al.*, 'Localization and regulation of endothelial NO synthase mRNA expression in rat kidney.' *Am. J. Physiol.*, 1994, **267**, F296–F302.

249. C. S. Wilcox, W. J. Welch, F. Murad *et al.*, 'Nitric oxide synthase in macula densa regulates glomerular capillary pressure.' *Proc. Natl. Acad. Sci. USA*, 1992, **89**, 11993–11997.

250. D. A. Wink, I. Hanbauer, M. C. Krishn *et al.*, 'Nitric oxide protects against cellular damage and cytoxicity from reactive oxygen species.' *Proc. Natl. Acad. Sci. USA*, 1993, **90**, 9813–9817.

251. W. C. Buss, M. K. Piatt and R. Kauten, 'Inhibition of mammalian microsomal protein synthesis by aminoglycoside antibiotics.' *J. Antimicrob. Chemother.*, 1984, **14**, 231–241.

252. J. Neugarten, H. S. Aynedjian and N. Bank, 'Role of tubular obstruction in acute renal failure due to gentamicin.' *Kidney Int.*, 1983, **24**, 330–335.

253. W. Bargmann, 'Histologie und Mikroskopische Anatomie des Menschen,' Georg Thieme Verlag, Stuttgart, 1967.

254. A. B. Maunsbach, 'Observations on the segmentation of the proximal tubule in the rat kidney. Comparison of results from phase contrast, fluorescence and electron microscopy.' *J. Ultrastruct. Res.*, 1966, **16**, 239–258.

255. W. Thoenes, 'Die Mikromorphologie des Nephrons in ihrer Beziehung zur Funktion. Teil 1: Funktionseinheit: Glomerulum-Proximales und distales Konvolut.' *Klin. Wochenschr.*, 1961, **39**, 504–518.

256. S. Suzuki, S. Takamura, J. Yoshida *et al.*, 'Comparison of gentamicin nephrotoxicity between rats and mice.' *Comp. Biochem. Physiol. C Pharmacol. Toxicol. Endocrinol.*, 1995, **112**(1), 15–28.

257. G. H. Hottendorf, in 'Nephrotoxicity Ototoxicity of Drugs,' ed. J. P. Fillastre, University of Rouen, 1996, pp. 257–268.

258. P. M. Andrews, in ' Biochemical Research Applications of Scanning Electron Microscopy,' eds. G. M. Hodges and R. C. Hallowes, Academic Press, New York, 1979, vol. I, pp. 273–306.

259. F. C. Luft and V. Patel, in 'Nephrotoxicity,' eds. J. P. Fillastre, Masson, New York, 1978, pp. 168–174.

260. A. W. Mondorf, J. Breier, J. Hendus *et al.*, 'Effect of aminoglycosides on proximal tubular membranes of the human kidney.' *Eur. J. Clin. Pharmacol.*, 1978, **13**, 133–142.

261. W. Pape, R. Kochmann, G. Kochmann *et al.*, 'Excretion of neutral a-glucosidase, determined with a continuous assay, and of acid–glucosidase in the urine of human reference subjects.' *J. Clin. Chem. Clin. Biochem.*, 1983, **21**, 511–517.

262. M. C. Sanchez-Bernal, J. Martin-Barrientos and J. A. Cabezas, 'Effect of tobramycin and gentamicin on the activity of some glycosidases in rat serum and urine.' *Comp. Biochem. Physiol.*, 1984, **79**, 401–405.

263. J. P. Miguet, H. Allemand and D. Vuitton, 'Les tests *in vivo* d'induction enzymatique medicamenteuse chez l'homme.' [*In vivo* tests for the evaluation of hepatic enzyme inductions in humans.] *Gastroenterol. Clin. Biol.*, 1981, **5**, 798–811.

264. B. Decouvelaere, B. Terlain and A. Bieder, 'Biotransformation de la métapramine chez trois espèces animaux (rat, lapin, chien) et chez l'homme.' [Biotransformation of metapramine in three animal species (dog, rabbit, rat) and in man.] *Therapie*, 1982, **37** 249–257.

265. J. A. Boutin, B. Antione, A. M. Batt *et al.*, 'Heterogeneity of hepatic microsomal UDP-glucuoronosyltransferase activities: comparison between human and mammalian species activities.' *Chem. Biol. Interact.*, 1984, **52**, 173–184.

266. M. Sastrasinh, J. M. Weinberg and H. D. Humes, 'The effects of gentamicin on glucose transport by isolated renal brush border membranes.' *Clin. Res.*, 1982, **30**, 462A.

267. P. P. Frohnert, B. Höhmann, R. Zwiebel *et al.*, 'Free flow micropuncture studies of glucose transport in the rat nephron.' *Pflügers Arch.*, 1970, **315**, 66–85.

268. H. E. Renschler, 'Die anwendung enzymatischer methoden zur bestimmung von insulin.' *Klin. Wschr.*, 1963, **41**, 615–618.

269. L. M. Pepe, P. D. McNamara, J. W. Foreman *et al.*, 'Preservation of brush border transport systems for proline and alpha-methyl-D-glucoside from rat, dog and human kidney.' *Lab. Invest.*, 1982, **47**, 611–617.

270. C. Gauthier, H. Nguyen-Simonnet, C. Vincent *et al.*, 'Renal tubular absorption of β2-microglobulin.' *Kidney Int.*, 1984, **26**, 170–175.

271. T. Maack, V. Johnson, S. T. Kau *et al.*, 'Renal filtration, transport and metabolism of low-molecular-weight-proteins: a review.' *Kidney Int.*, 1979, **16**, 251–270.

272. W. Pruzanski and M. E. Platts, 'Serum and urinary proteins, lysozyme (muramidase), and renal dysfunc-

tion in mono- and myelomonocytic leukemia.' *J. Clin. Invest.*, 1970, **49**, 1694–1708.

273. J. J. Schentag and M. E. Plaut, 'Patterns of urinary β2-microglobulin excretion by patients treated with aminoglycosides.' *Kidney Int.*, 1980, **17**, 654–661.

274. K. Furuhama and T. Onodera, 'The influence of cephem antibiotics on gentamicin nephrotoxicity in norml, acidotic, dehydrated, and unilaterally nephrectomized rats.' *Toxicol. Appl. Pharmacol.*, 1986, **86**, 430–436.

275. A. W. Mondorf, in 'The Aminoglycosides,' eds. A. Whelton and H. C. Neu, Dekker, New York, 1982, pp. 283–301.

276. M. K. Reinhard, G. H. Hottendorf and E. D. Powell, 'Differences in the sensitivity of Fisher and Sprague-Dawley rats to aminoglycoside nephrotoxicity.' *Toxicol. Pathol.*, 1991, **19**(1), 66–71.

277. H. O. Garland, T. J. Birdsey, C. G. Davidge *et al.*, 'Effects of gentamicin, neomycin and tobramycin on renal calcium and magnesium handling in two rat strains.' *Clin. Exp. Pharmacol. Physiol.*, 1994, **21**(2), 109–115.

278. G. J. Kaloyanides, L. Ramsammy and C. Josepovitz, in 'Nephrotoxicity,' eds. P. H. Bach, N. J. Gregg, M. F. Wilks *et al.*, Dekker, 1991, pp. 99–104.

279. S. L. Garetz, R. A. Altschuler and J. Schacht, 'Attenuation of gentamicin ototoxicity by glutathione in the guinea pig *in vivo*.' *Hearing Res.*, 1994, **77**, 81–87.

280. B. Song and J. Schacht, 'Protective effects of iron chelators on gentamicin ototoxicity.' *Inner Ear Biol. Abst.*, 1995, **32**, O–8.

281. S. K. Swan, D. N. Gilbert, S. J. Kohlhepp *et al.*, 'Pharmacologic limits and the protective effect of polyaspartic acid on experimental gentamicin nephrotoxicity.' *Antimicrob. Agents. Chemother.*, 1993, **37**, 347–348.

282. T. Whittem, R. G. Schnellmann and D. C. Ferguson, 'Poly-L-aspartic acid does but triiodothyrosine does not protect against gentamicin-induced cytotoxicity in the porcine kidney cell line LLC-PK1.' *J. Pharmacol. Exp. Ther.*, 1992, **262**, 834–840.

283. S. J. Kohlhepp, D. N. McGregor, S. J. Cohen *et al.*, 'Determinants of the *in vitro* interaction of polyaspartic acid and aminoglycoside antibiotics.' *J. Pharmacol. Exp. Ther.*, 1992, **263**, 464–1470.

284. S. K. Swan, D. N. Gilbert, S. J. Kohlhepp *et al.*, 'Duration of the protective effect of polyaspartic acid on experimental gentamicin nephrotoxicity.' *Antimicrob. Agents Chemother.*, 1992, **36**, 2556–2558.

285. B. K. Kishore, Z. Kallay, P. Lambricht *et al.*, 'Mechanism of protection afforded by polyaspartic acid against gentamicin-induced phospholiüidosis. I. Polyaspartic acid binds gentamicin and displaces it from negatively charged phospholipid layers *in vitro*.' *J. Pharmacol. Exp. Ther.*, 1990, **255**, 867–874.

286. W. C. Elliot and D. S. Patchin, 'Effects and interactions of gentamicn,polyaspartic acid and diuretics on urine calcium concentration. *J. Pharmacol. Exp. Ther.*, 1995, **273**, 280–284.

287. S. Kacew, 'Inhibition of gentamicin-induced nephro-

toxicity by pyridoxal-5'-phosphate in the rat.' *J. Pharmacol. Exp. Ther.*, 1989, **248**, 360–366.

288. R. C. Keniston, S. Cabellon Jr. and K. S. Yarbrough, 'Pyridoxal-5'-phosphate as an antidote for cyanide, spermine, gentamicin and dopamine toxicity: an *in vivo* study.' *Toxicol. Appl. Pharmacol.*, 1987, **88**, 443–441.

289. R. E. Cronin, M. E. Burton and S. E. Demian, 'Protective effects of pentoxifylline analogue (HWA-448) but not pentoxifylline against gentamicin nephrotoxicity.' *J. Am. Soc. Nephrol.*, 1990, **1**, 595a (Abstract).

290. D. R. Luke, K. M. Wasan, R. R. Verani *et al.*, 'Attenuation of amphotericin-B nephrotoxicity in the candidiasis rat model.' *Nephron*, 1991, **59**, 139–144.

291. D. M. Ford, R. E. Thieme, C. A. Lamp *et al.*, 'HWA-448 reduces gentamicin toxicity in LLC-PK$_1$ cells.' *J. Pharmacol. Exp. Ther.*, 1995, **274**, 29–33.

292. M. L. Quarum, D. C. Houghton, D. N. Gilbert *et al.*, 'Increasing dietary calcium moderates experimental gentamicin nephrotoxicity.' *J. Lab. Clin. Med.*, 1984, **103**, 104–114.

293. H. D. Humes, M. Sastrsinh and J. M. Weinberg, 'Calcium is a competitive inhibitor of gentamicin-renal membrane binding interactions and dietary calcium supplementation protects against gentamicin nephrotoxicity.' *J. Clin. Invest.*, 1984, **73**, 134–147.

294. S. B. Chahwala and E. S. Harpur, 'Gentamicin-induced hypercalciuria in the rat.' *Acta Pharmacol. Toxicol. (Copenh.)*, 1983, **53**, 358–362.

295. J. E. Foster, E. S. Harpur and H. O. Garland, 'An investigation of the acute effect of gentamicin on the renal handling of electrolytes in the rat.' *J. Pharmacol. Exp. Ther.*, 1992, **261**, 38–43.

296. U. Finsterer and R. Rotzer, 'Renal effekte von Dopamin bei gesunden Erwachsenen unter besonderen Berucksichtigung der Exkretion von Phosphat und kalzium.' [Renal effects of dopamine in healthy adults with special references to the excretion of phosphte and calcium]. *Infusionther. Klin. Ernahr.*, 1986, **13**, 222–230.

297. H. H. Neumayer, J. Gellert and F. C. Luft, 'Calcium antagonists and renal protection.' *Renal Fail.*, 1993, **15**(3), 353–358.

298. D. J. Kazierad, G. J. Wojcik, D. E. Nix *et al.*, 'The effect of verapamil on the nephrotoxic potential of gentamicin as measured by urinary enzyme excretion in healthy voluteers.' *J. Clin. Pharmacol.*, 1995, **35**(2), 196–201.

299. K. Gurnani, H. Khouri, M. Couture *et al.*, 'Molecular basis of the inhibition of gentamicin nephrotoxicity by daptomycin; an infrared spectroscopic investigation.' *Biochim. Biophys. Acta*, 1995, **1237**(1), 86–94.

300. P. Sandhya, S. Mohandass and P. Varalakshmi, 'Role of D L alpha-lipoic acid in gentamicin induced nephrotoxicity.' *Mol. Cell. Biochem.*, 1995, **145**(1), 11–17.

7.27
Antineoplastic Agents

GRUSHENKA H. I. WOLFGANG and MARTHA E. I. LEIBBRANDT
Chiron Corporation, Emeryville, CA, USA

7.27.1 INTRODUCTION

The use of chemotherapy in the management of neoplastic disease is well established. The number of drugs approved for the treatment of cancer is now in excess of 50.[1] Many of these drugs carry a risk of renal complications. In addition, antineoplastic regimens frequently include drug combinations which may increase the potential for nephrotoxic effects.[2] Numerous reviews and book chapters have been published which address the nephrological consequences of cancer therapy[3-12] and strategies for managing these therapies.[13-15] The intent of this chapter is to discuss the antineoplastic agents which cause renal injury most frequently, those with unique or known mechanisms of action, and to evaluate whether preclinical animal experiments can predict renal complications in humans.

Renal failure in cancer patients is a common problem. In addition to renal failure from antineoplastic therapy, renal complications may result from the tumor mass itself, tumor products, immunologic responses to the tumor, or treatment with nephrotoxic antibiotics, radiation, or radiocontrast media. Any underlying renal disease may predispose patients to increased nephrotoxic injury from antineoplastic agents. The spectrum of renal disease includes acute renal failure (prerenal, intrinsic, and post-renal), chronic renal failure, and specific tubular dysfunction (reviewed by Rieselbach and Garnick[3]). With advances in cancer treatment many patients are surviving longer; this extended survival has allowed individuals to manifest with cumulative and delayed toxic side effects.

Renal function in patients undergoing antineoplastic therapy is most frequently estimated by blood urea nitrogen, serum creatinine, and creatinine clearance. Patients undergoing treatment may have problems with emesis/starvation, muscle wasting, and repeated infections; these conditions will substantially affect creatinine clearance. Thus, during evaluation for potential nephrotoxicity it is very important to measure renal hemodynamics as well as the standard tubular functional parameters (reviewed in Daugaard and Abildgaard[16]).

Nephrotoxicity following chemotherapy has been observed with alkylating agents, antitumor antibiotics, antimetabolites, and biologics. Table 1 lists specific agents with demonstrated nephrotoxicity along with an indication of the risk of such toxicity. The dose and duration of treatments, the type of renal effects, and the suggested treatment or prevention modalities are also presented in Table 1. The structures of selected antineoplastic agents are shown in Figure 1.

7.27.2 ALKYLATING DRUGS

7.27.2.1 Cisplatin

Cisplatin (*cis*-diamminedichloroplatinum (II)) has proved highly effective for the treatment of solid tumors including those of the testis, ovary, bladder, head and neck, lung, breast, and bladder.[17] Clinically, nephrotoxicity is dose-limiting and manifests as polyuria, acute or chronic renal failure, or chronic hypomagnesemia. Since cisplatin is one of the most effective drugs in its class, modulation of its nephrotoxicity has become an important issue for clinicians and scientists. A complete understanding of its mechanism of nephrotoxicity would, in addition, aid in the development of new platinum antineoplastic agents. Of the hundreds of platinum compounds developed, few have proved as effective as cisplatin, and a lower potential for renal damage has generally led to enhanced neuro- and ototoxicities. Carboplatin, for example, is less nephrotoxic than cisplatin, but has an increased potential for myelosuppressive side effects. The search

for an orally active platinum compound has led to the study of JM216, a platinum complex which is not nephrotoxic in rodent models.[18] The effectiveness of JM216 in human patients is under investigation, and it is hoped that such a drug will obviate the need for hospitalization during treatment and the hydration which must accompany cisplatin therapy in order to reduce nephrotoxicity.

7.27.2.1.1 Clinical features

(i) Incidence, severity, and time frame

Acute changes in renal function following cisplatin therapy in the absence of hydration occurred in as many as 36% of patients.[19] With successive courses of cisplatin treatment of $50 \, mg \, m^{-2}$ or greater, renal functional impairment became more severe, suggesting cumulative damage to the renal tubules. Nephrotoxicity, detected as increased blood urea nitrogen (BUN), occurred during the second week of therapy.[20] Pretreatment hydration has reduced toxicity to as little as 5%, although most patients experience some degree of renal functional impairment.[21]

(ii) Pathophysiology

Hypomagnesemia has been reported in 40–100% of patients receiving cisplatin chemotherapy,[22,23] and deficiencies in magnesium have been linked to morphological changes in the kidney structure itself.[24] Mavichak *et al.*[25] suggest that decreases in serum magnesium may arise from changes in the thick ascending loop of Henle. Regardless of its etiology, hypomagnesemia may be used as an indicator of cisplatin toxicity and targeted for treatment. Alterations in other urinary electrolytes such as calcium and sodium are usually reversed once underlying hypomagnesemia is corrected.[26,27]

Since creatinine is freely filtered by the glomerulus, serum levels are often used to assess renal function. When serum creatinine increases, it suggests that glomerular filtration has fallen as a result of physiological or structural changes to cells of the nephron. Decreased glomerular filtration rate (GFR) was seen in patients 12 months[28,29] and as many as 52 months following treatment with cisplatin.[30] Some clinicians used increases in serum creatinine greater than 30% to indicate nephrotoxicity in patients treated with $80 \, mg \, m^{-2}$ cisplatin.[31] However, serum creatinine may not be the most sensitive measure of GFR since certain patients with low basal

Table 1 Antineoplastic agents with nephrotoxic potential.

Class	Compound	Risk	Type	Clinical dose	Treatment/prevention
Alkylating agents	Cisplatin	High	Acute renal failure	$50–100\,mg\,m^{-2}$ every 4 weeks	Hydration, diuresis, sodium thiosulfate, DDTC, WR-2721
	Carboplatin	Low	Acute renal failure	$360\,mg\,m^{-2}$ every 4 weeks	
	Cyclophosphamide	Low	Cystitis	$40–50\,mg\,kg^{-1} \times 2–5$ days	Mesna, hydration
	Ifosfamide	Intermediate	Cystitis, Fanconi's syndrome	$1.2–2.5\,g\,m^{-2} \times 5$ days	Mesna, hydration
	Streptozotocin	High	Acute renal failure	$500\,mg\,m^{-2} \times 5$ days every 3 weeks	Limit single dose $<1.5\,g\,m^{-2}$
	Carmustine	Low	Chronic renal failure	$150–200\,mg\,m^{-2}$ every 6 weeks	Stop at cumulative dose of $1200\,mg\,m^{-2}$
	Lomustine	High	Chronic renal failure	$130\,mg\,m^{-2}$ every 6 weeks	Stop at cumulative dose of $1200\,mg\,m^{-2}$
	Semustine	High	Chronic renal failure	$150–200\,mg\,m^{-2}$ every 6 weeks	Stop at cumulative dose of $1200\,mg\,m^{-2}$
Antitumor antibiotics	Mitomycin C	High	HUS, Chronic renal failure	$10\,mg\,m^{-2}$ every 6–8 weeks	Stop at cumulative dose of 60 mg
	Mithramycin	High	Acute renal failure	$25–50\,\mu g\,kg^{-1} \times 5$ days	Every other day dosing; monitor daily
	Doxorubicin	Low	Acute renal failure	$60–75\,mg\,m^{-2}$ every 3 weeks	
	Daunorubicin	Low		$45\,mg\,m^{-2} \times 3$ days	
Antimetabolite	Methotrexate, high dose	High	Acute renal failure	$>1\,g\,m^{-2}$	Urinary alkalinization, hydration, Leucovorin rescue
	5-Azacytidine	Low	Acute renal failure	$200\,mg\,m^{-2} \times 5$ days	Monitor renal function
Other	Gallium nitrate	Intermediate	Hypocalcemia, Hypophosphatemia	$200\,mg\,m^{-2} \times 5$ days	Maintain urine flow, avoid doses $>300\,mg\,m^{-2}\,d^{-1} \times 7$ days

HUS, hemolytic uremic syndrome.

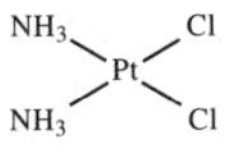

Cisplatin

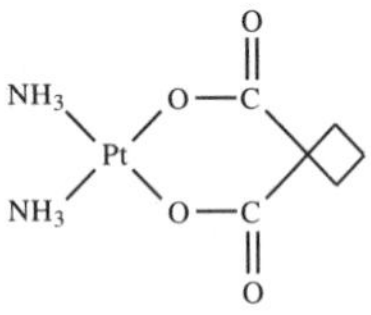

Carboplatin

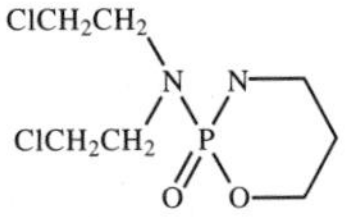

Cyclophosphamide

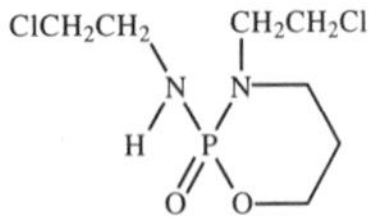

Ifosfamide

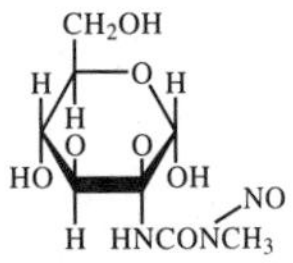

Streptozotocin

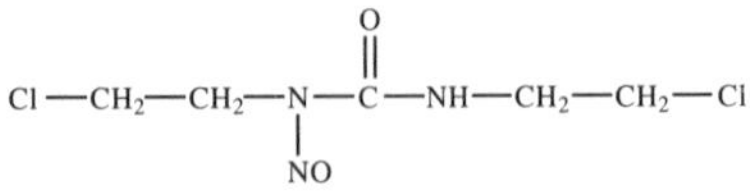

Carmustine

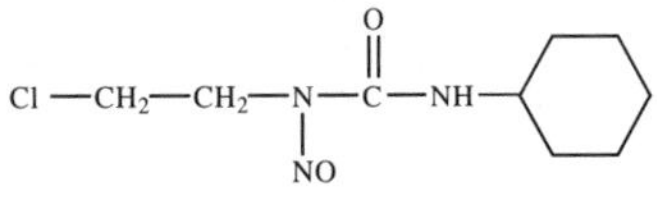

Lomustine

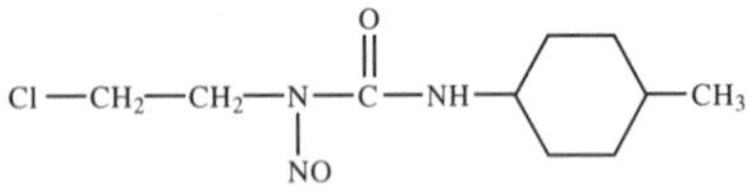

Semustine

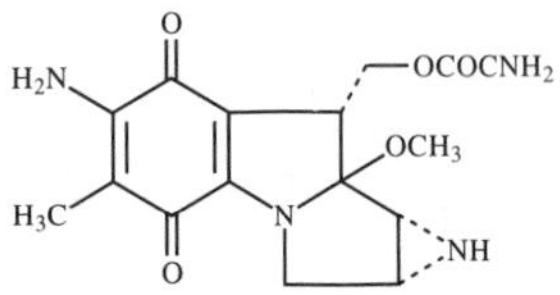

Mitomycin C

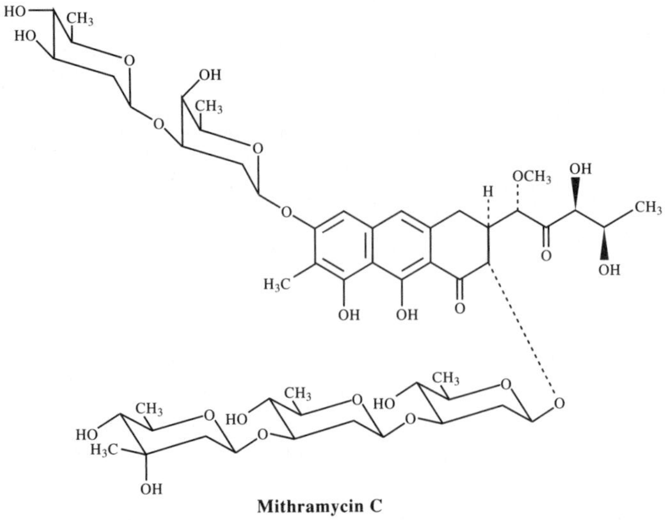

Mithramycin C

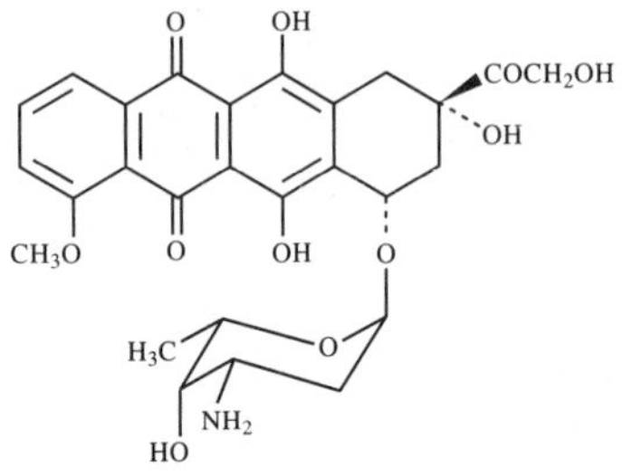

Doxorubicin

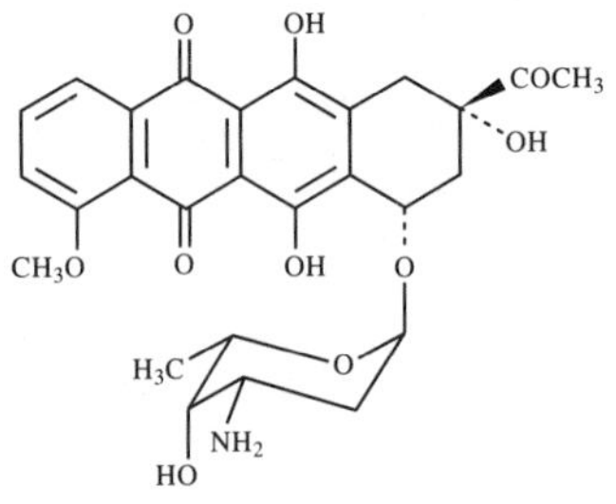

Daunorubicin

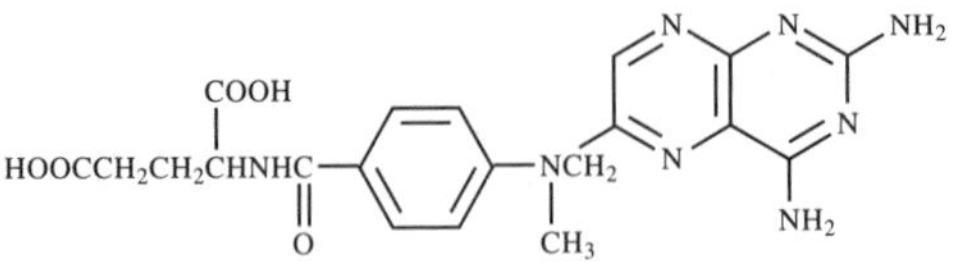

Methotrexate

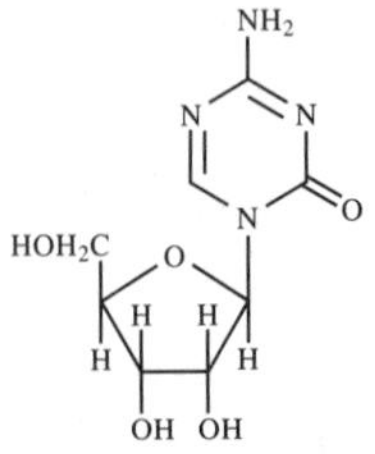

5-Azacytidine

Figure 1 Structures of selected antineoplastic agents.

(pretreatment) creatinine levels may exhibit no noticeable increase when treated with cisplatin.[32]

Although serum creatinine and creatinine clearance remain the most commonly used measures of nephrotoxicity (in more than 90% of clinical studies), more specific measures of renal function have also been tested. For example, increases in urinary excretion of β_2 microglobulin indicated damage to the proximal tubule and appeared approximately 5 d following treatment.[33,34] Urinary excretion of alanine aminopeptidase (a brush border enzyme of proximal tubules) and *N*-acetyl-β-D glucosamidase (a lysosomal enzyme found in proximal tubule cells) were elevated following cisplatin treatment, indicating the sensitivity of the proximal tubule to this compound.[33]

Although cisplatin is filtered by glomerular filtration,[3] nephrotoxicity involves primarily the distal parts of the proximal tubule or the distal nephron itself. Early deaths in patients receiving cisplatin without adequate hydration showed that tubular dilatation, coagulative necrosis, and interstitial edema were the most prominent findings, and damage occurred primarily in the distal tubules and collecting ducts.[35] In another study, disruption of tubule cell brush border and significant degeneration of mitochondria were correlated with renal dysfunction.[20] Since studies on the morphological changes accompanying human nephrotoxicity are few and contradictory, implicating the distal tubule[35] and collecting ducts,[36] proximal and distal tubules,[28] proximal convoluted tubules[37] and all of the proximal and distal tubules and collecting ducts,[38] it is difficult to select a single relevant species for studying cisplatin nephrotoxicity. Preclinical animal studies with cisplatin predicted that renal damage would be a predominant side effect in humans, but as discussed below, the site of damage in dogs and rats is proximal to the lesion found in humans.

7.27.2.1.2 Animal studies

Despite species differences in the anatomy of the kidney (uni- vs. multipapillary) and renal concentrating ability, animal models of cisplatin nephrotoxicity have proved useful for studying the factors influencing human response to platinum-based compounds. The differences which exist between human and animal cisplatin-induced nephrotoxicity appear to involve the specific cell type affected by cisplatin; the time frame of development and severity of the lesions are relatively similar across species. As the rat is the most commonly used species for studies on cisplatin nephro-

toxicity, we will focus our review on this animal, referring to others where relevant studies exist.

Rats treated with a single i.v. dose of cisplatin ($4-6 \, \text{mg} \, \text{kg}^{-1}$) displayed renal toxicity in $3-6 \, \text{d}$.[27,39] Doses of $10 \, \text{mg} \, \text{kg}^{-1}$ or higher caused significant morbidity or death within $4 \, \text{d}$.[33] Renal toxicity in rat was detected as in humans by measurement of increased BUN and decreased GFR,[39,40] as well as vasopressin-resistant polyuria.[41] Decreased inulin clearance was observed $3 \, \text{d}$ following treatment,[41] and may be a more reliable measure of nephrotoxicity than BUN which responds more slowly to alterations in GFR. In rats, the effects on kidney function arose from acute proximal tubule impairment, leading to a decrease in fractional and absolute proximal tubule reabsorption, specifically in the pars convoluta and pars recta.[42] Similar effects were observed in dogs treated with 2.5, 5,[43] or $10 \, \text{mg} \, \text{kg}^{-1}$ cisplatin.[44] Renal dysfunction in monkeys treated with a total of $6.25 \, \text{mg} \, \text{kg}^{-1}$ cisplatin over a five-day period was more severe than that observed in dogs, although renal lesions in both species regressed entirely within $124 \, \text{d}$.[43] In pigs, on the other hand, bolus injections of $1.5-2.5 \, \text{mg} \, \text{kg}^{-1}$ cisplatin had no effect on GFR for up to $24 \, \text{wk}$.[45]

Histopathologic evaluation of kidneys from cisplatin-treated animals has suggested that the proximal tubule is the region most sensitive to the effects of this complex. Selective damage to the proximal tubules is manifest as widespread tubular necrosis and degeneration, predominantly in the third segment (S_3) or pars recta of the proximal tubule, located in the cortico-medullary junction (Figure 2).[40,46–50] The lumens of damaged tubules are often filled with acidic protein casts. Despite the specificity of the lesion for S_3, some have suggested that the minimal damage to the S_1 and S_2 segments of the proximal tubule may be physiologically important in the early stages of renal failure.[49] No changes were reported in the glomeruli of rats treated with cisplatin. The extensive damage to the proximal tubule of rat correlated in most studies with functional changes (i.e., peak elevation of BUN). In contrast, a study in rabbits given a single i.v. dose of cisplatin ($5 \, \text{mg} \, \text{kg}^{-1}$) showed marked degeneration of tubules in the outer cortex while those in the corticomedullary junction displayed only slight necrosis involving tubular dilatation and protein casts.[51] However, the reported increase in BUN and sparing of glomeruli in rabbits was similar to that seen in cisplatin-treated rats and dogs. In pigs, the Bowman's capsule of glomeruli were thickened by cisplatin treatment, but there was little degeneration of

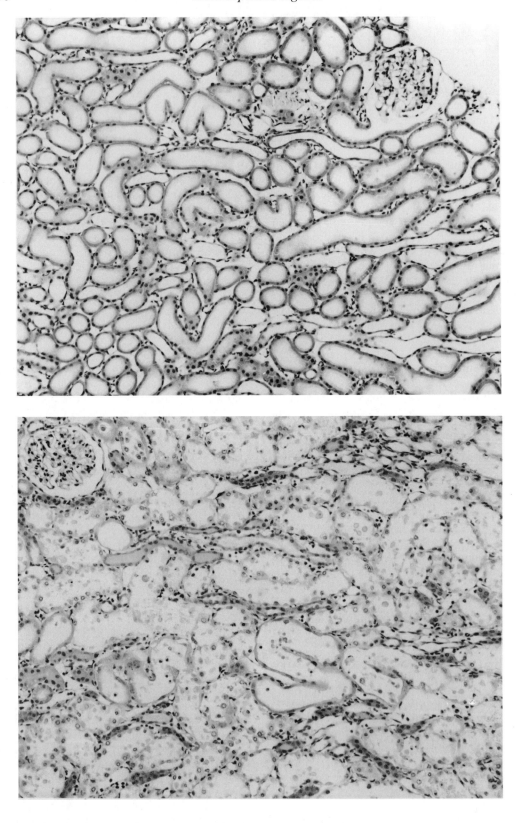

Figure 2 Histopathology of kidneys from a (top) control and (bottom) cisplatin-treated rat. Animals were sacrificed in day 4 following an single intravenous dose of saline or 6.5 mg kg^{-1} cisplatin. Kidneys were perfusion fixed, process, and stained with hematoxylin and eosin (400 s). Marked diffuse proximal tubular necrosis is clearly evident in the cisplatin-treated kidney.

tubular epithelia.[45] It appears that species specificity is an important component of the nephrotoxicity induced by cisplatin. In some respects, the anatomical differences, of unipapillary and multipapillary kidneys may contribute to differential toxicity. Despite such differences, however, cisplatin induces a lesion whose severity and time-frame of development are very similar across species, and this explains why preclinical ani-mal studies correctly predicted that nephrotoxicity would be a key feature of human exposure.[52]

The most thorough ultrastructural studies of cisplatin-induced nephrotoxicity have been performed in the rat. In animals with physiological evidence of acute renal failure (decreased creatinine clearance, increased BUN), loss and thinning of proximal tubule brush border, cellular swelling, chromatin condensation, cytoplasmic vacuoles, rounded mitochondria, and increased number of lysosomes have been reported.[49,53,54] Dispersed ribosomes and nucleolar segregation in the absence of changes in mitochondrial ultrastructure have also been reported 24–96 h following a single i.p. (10 mg kg^{-1}) injection of cisplatin.[49] Approximately 7 d following challenge with a moderate dose of cisplatin, the kidneys of most animals begin to regenerate.[40,50] However, recovery does not appear to be complete in all animals, and persistent cysts (nodular foci) can be seen in rats as many as 6–15 months following treatment.[53]

Although DNA adduct formation by intercalation of cisplatin with nucleotide residues[55] is the classical mechanism for the drug's antineoplastic effect, studies in rat suggest that renal DNA may not be the primary site of drug binding and, hence, toxicity. The relative mitotic quiescence of renal tubule cells compared to tumor cells may partially explain the different location of intracellular binding. In rats given 6 mg kg^{-1} cisplatin, 70–75% of the platinum measured in the kidney was in the cytosol.[56,57] Of the 25–30% present in the nucleus, 5–10-fold more platinum was associated with chromosomal proteins than with DNA.[57] Since the highest platinum concentrations were measured in the microsomal fraction of both liver and kidney, the interaction of cisplatin with cytosolic proteins and the translational apparatus may play a more critical role in the etiology of nephrotoxicity than the formation of DNA adducts.

7.27.2.1.3 *Mechanism of injury*

To define the mechanism of cisplatin's nephrotoxicity, extensive experiments *in vivo*, and in renal cortical slices, primary cultures, or suspensions of renal proximal tubule cells and cell lines displaying proximal tubule cell characteristics (e.g., OK, MDCK, LLC-PK$_1$) have been performed. The mechanism of cisplatin's preferential damage in the kidney remains elusive. The contribution of effects arising from overwhelming cellular stress as compared to those arising from alteration of a specific subcellular target are not easily determined. We discuss some of the most likely targets/mechanisms of cisplatin's renal toxicity: lysosomes, transport of organic ions, lipid peroxidation, mitochondria, macromolecular syntheses, and the role of thiol groups.

(i) *Lysosomes*

The acidic, intracellular compartment of lysosomes contains acid phosphatase which degrades many proteins engulfed by the lysosome. Several studies suggest that cisplatin alters lysosomal function, but a consensus on the degree and direction of that change has not been reached. In one *in vivo* study, the number of lysosomes and the level of acid phosphatase increased,[58] while in another the specific degradation of lysozyme decreased in the renal cortical slices of rats treated with 5 mg kg^{-1} cisplatin.[59] *In vitro*, the accumulation of neutral red dye by the lysosomes of cultured rat renal proximal tubule cells decreased following 10 μM cisplatin treatment.[60] The type of specific lysosomal changes initiated by cisplatin remains poorly defined and more studies are needed to establish the primacy of cisplatin's effects on this organelle.

(ii) *Organic ion transport*

The uptake of the organic ions *p*-aminohippuric acid (PAH) and tetraethylammonium (TEA) may be used to evaluate the effect of cisplatin on organic ion transport systems in the renal tubule. In rat renal cortical slices, cisplatin reduced the accumulation of PAH and TEA at concentrations ranging from 100 μM to 3 mM.[39,61–63] There is little understanding of the specifics of cisplatin's uptake into cells, although it is unlikely that it competes for both cationic and anionic transport pathways. Cisplatin transport has been comprehensively discussed by Gately and Howell[64] who suggest that cisplatin may enter cells through both passive and facilitated diffusion pathways, such that its kinetics are determined by external drug concentration. This model has been proposed for tumor cell uptake of cisplatin and its relevance to renal handling of this drug will not be known until appropriate studies have been conducted.

(iii) Lipid peroxidation and mitochondria

Cisplatin treatment causes oxidative stress in exposed tissues which can lead to lipid peroxidation. Lipid peroxidation caused by free radicals may also occur when cellular defense mechanisms are compromised, permitting the activity of cytotoxins. Mitochondrial function is often discussed in concert with lipid peroxidation because of their mutual requirement for oxygen. The hypothesis that cisplatin injures renal tubule cells by interfering with mitochondrial function is one of the most thoroughly researched areas of cisplatin's nephrotoxicity.

Lipid peroxidation (as measured by increased levels of malondialdehyde) has been reported following *in vivo*[65] and *in vitro*[62,66–68] administration of cisplatin. Some of these reports have correlated lipid peroxidation with the inhibition of mitochondrial function by cisplatin.[67–70] One hypothesis proposes that the injury of mitochondria by platinum compounds leads to lipid peroxidation and cell death. However, in some cases these toxic effects are also correlated with the loss of membrane integrity,[67] suggesting an advanced state of cellular dysfunction. Isolated mitochondria were also vulnerable to lipid peroxidation when treated with cisplatin.[66] Although mitochondrial function is inhibited by cisplatin, this effect appears in some studies only at a threshold of $100\,\mu M$ cisplatin (or about 3–10-fold blood levels in patients),[67,71] or with concentrations of $2\,mM$ in rat renal cortical slices.[66,68] In primary cultures of renal proximal tubule cells, other cellular functions were inhibited before mitochondrial succinic dehydrogenase[51,72] or the mitochondrial conversion of 3-(4,5-dimethylthiazol-2-yl)-2,5-diphenyltetrazolium bromide (MTT) to formazan[60] were decreased. Lipid peroxidation may be the consequence rather than the cause of cisplatin's nephrotoxicity, thus explaining why it is seen at different concentrations and different time points in different types of renal cell preparations. Studies on the mitochondria of intact cells which use lower concentrations of drug and measure cellular functions in addition to ATP production will define the temporal sensitivity of renal mitochondria to cisplatin.

(iv) Macromolecular synthesis

Transcription and translation are integral processes in the life of all cells. Due to their centrality to cell homeostasis, they are important potential targets for cisplatin. The antitumor effect of this planar drug is generally believed to involve binding to guanine bases in the DNA, causing cross-linking and interference with cell division.[55,73,74] Cisplatin inhibited DNA synthesis in renal proximal tubule cells as well as in tumor cells,[51,72,75] although Yasumasu *et al.*[76] reported only a transient depression in DNA synthesis at 8–24 h which was reversed by 72 h.

Cisplatin's inhibition of RNA synthesis[51,77] was hardly surprising given this drug's ability to corrupt the DNA template by adduct formation. Some authors[78,79] have suggested that cisplatin inhibits transcription by interfering with the polymerase enzyme, rather than the DNA template itself. Ultimately, the cisplatin–polymerase complex could reduce DNA replication as effectively as a cisplatin–DNA complex. However, in quiescent cells, such as those of the renal proximal tubule, interference with mRNA synthesis would prove more immediately toxic than decreases in *de novo* DNA synthesis. The suggestion[80] that cisplatin interferes with the opening of repressive chromatin structures and blocks transcription factor binding directly may explain its organ- and cell-specific toxicity. In the renal proximal tubule, the association of cisplatin with transcriptional complexes in active regions of the genome would be different than in other organs because of tissue-specific gene expression. Toxicity could arise if gene products vital to cytoprotection (drug transport or deactivation) were affected.

Rosenberg and Sato[74,81] reported that cisplatin interferes with the formation of the 80S ribosomal subunit leading to a reduction in ribosome-catalyzed protein synthesis.[51,60,72,77] Ultrastructural studies indicated degranulation of rough endoplasm reticulum (loss of ribosomes) and dispersion of ribosomes in cultured proximal tubule cells[60] or rat kidney[49] following cisplatin treatment. The mitochondria in both these studies appeared normal. Progressive changes in nucleolar structure involving condensation, separation of granular and fibrillar components, and finally disintegration were observed in rat proximal tubule cells,[60] similar to changes reported in the neurons of rats treated with cisplatin[82] and in liver and heart cells subsequent to adriamycin treatment.[83] Such findings agree with a mechanism of toxicity involving inhibition of protein synthesis through interference with nucleolar ribosome biogenesis.

(v) Binding to thiol groups

Interactions between toxic agents and the thiol groups of glutathione (GSH) or metallothionein have long been proposed as cytoprotective mechanisms. Since cisplatin is based around a platinum metal moiety, it was thought that its nephrotoxicity could be mediated by the metal-binding, sulfur-rich protein metallothionein, in much the same way as cadmium. Metallothionein mRNA expression levels were increased in mice treated with cisplatin,[84] but not in the LLC-PK$_1$ cell line.[85] Cisplatin preincubated with GSH, which represents the largest intracellular nonprotein thiol pool, was less toxic to cultured renal cells[85] and caused less severe side effects (elevation of BUN, weight loss) when administered *in vivo*.[86,87] However, the relationship between GSH and cisplatin may not be as simple and direct as these studies would imply, since significant decreases in protein-bound thiol groups were observed without changes in GSH[88] and GSH-dependent factors were not inactivated[89] after cisplatin treatment. Finally, some antioxidants (including penicillamine and *N*-acetylcysteine) which protect thiol groups from oxidation do not prevent nephrotoxicity, while others, such as sodium selenite and β-hydroxylethylrutoside, may be protective through direct inactivation of cisplatin itself rather than their antioxidant properties.

(vi) Tissue platinum and toxicity

Patients with higher renal platinum levels displayed greater toxicity than those with lower platinum.[90] Higher platinum levels were measured in liver compared to kidney.[90,91] This is in agreement with the preferential distribution of platinum to the liver of rats following treatment with a single i.v. dose of cisplatin.[57] Rats treated with cisplatin had higher levels of platinum in the kidney after 72 h compared to liver,[56,92] although at timepoints up to and including 48 h post-treatment, platinum levels in liver were comparable to kidney.[56] Mice treated for 24 h had similar platinum distribution.[84] These results suggest that the kidney clears platinum less efficiently than the liver, and longer exposure may be responsible for the enhanced sensitivity of renal epithelia. *In vitro* experiments have confirmed a relationship between platinum concentrations and toxicity,[63,72] although platinum concentration alone does not necessarily determine toxicity. The subcellular distribution of platinum offers clues as to which organelles may have the

lowest (mitochondria, plasma membrane) or highest (endoplasmic reticulum, nucleus) platinum levels, possibly predisposing the latter to site-specific toxicity.

(vii) Summary

In summary, cisplatin's nephrotoxicity may involve mitochondrial and lysosomal dysfunction, transcriptional and translational inhibition, binding to vital homeostatic proteins such as polymerases, thiols, and histones, lipid peroxidation or simply platinum concentration (Table 2). The evidence suggests that lipid peroxidation occurs in cells whose function has been severely compromised. A working theory which encompasses many of the *in vitro* and *in vivo* observations might involve cisplatin binding to proteins in the cytosol, endoplasmic reticulum, and nucleus, thereby inhibiting ribosome formation and activity, and blocking transcription of mRNA. This, in turn, would decrease protein synthesis which would interfere with cellular machinery such as mitochondria and lysosomes, leading to lipid peroxidation and loss of viability. Whether levels of specific proteins or areas of the genome make some cells fractionally more sensitive to cisplatin should be investigated. Basal gene expression, response to toxic stimuli, and the physiology of anatomical location and drug concentration undoubtedly work in concert to facilitate cisplatin's toxicity to proximal and distal tubules. Once the intracellular target for this drug is defined, appropriate measures can be taken to avoid dose-limiting toxicity.

7.27.2.1.4 Modulation of toxicity

Since cisplatin remains one of the most effective antineoplastic agents available for the treatment of multiple tumor types, strategies for modulating its dose-limiting nephrotoxicity have been actively investigated. These include aggressive hydration and splitting of doses, as well as strategies for mediating the cytotoxic effects of cisplatin on the renal tubule. Modulation of nephrotoxictiy has been discussed in depth in other reviews.[4,93,94]

(i) Reducing systemic toxicity

In 1973, a study by Higby *et al.*[95] reported that 100% of patients given a bolus i.v. dose of 50 mg m^{-2} cisplatin had renal toxicity. Fractionation of the dose of cisplatin into five administrations reduced toxicity to 30%.[96,97]

Table 2 Suggested mechanisms of cisplatin nephrotoxicity.

Target	Ref.
Lysosomes	58–60
Organic ion transport	39,61–64
Lipid peroxidation	62,65,66,68,70
Mitochondria	67–70
DNA, RNA synthesis	51,72,75–79
Protein synthesis	51,60,63,72,74,77,81
Binding to thiol groups	85–87
Platinum levels/ distribution	47,57,63,72

Administration of cisplatin in smaller doses over longer infusion periods reduced nephrotoxicity to as little as 5% of patients.[98,99] Systemic toxicity can also be diminished by delivering the drug more directly to the tumor, thereby minimizing the exposure of nontarget tissues. For example, ovarian cancer responds well to i.p. administration of cisplatin, although this mode of administration does not reduce toxicity completely since the drug must still be eliminated by the kidney.

(ii) Modulating elimination of cisplatin

Since cisplatin's toxicity is due, at least in part, to the exposure of the kidney during elimination, another strategy for overcoming proximal tubule cell damage is to reduce the length of time that the sensitive cell type is exposed to cisplatin, or the concentration of cisplatin in the tubule. Hydration therapy with isotonic saline is one of the most common adjuncts to cisplatin therapy. It reduced toxicity in 40% of patients to as little as 5%,[100] probably by preventing a drop in the glomerular filtration rate. Ozols *et al.*[101] observed that cisplatin dissolved in 3% saline reduced nephrotoxicity even further. Use of hypertonic saline may minimize formation of cisplatin's toxic hydrolysis products through elevation of extracellular chloride ion concentration. Mannitol diuresis can also reduce toxicity, but only in the initial cycle of treatment.[100] Like hydration with saline, mannitol must be given prior to or during cisplatin treatment to be effective. Some authors have suggested that mannitol's protective effects have less to do with its diuretic properties than with a possible role in preventing platinum binding in the renal tubule,[102] particularly since plasma concentrations of platinum actually increased with adjuvant mannitol treatment.[103]

(iii) Antidotes

While modulators of elimination such as mannitol may prevent acute renal failure, they are surprisingly ineffective at modulating the cumulative toxicity that occurs with repeated cycles of treatment. Antidotes, however, effectively reduce cumulative toxicity by counteracting or antagonizing the action of cisplatin at a cellular level. Many of the antidotes exploit cisplatin's attraction for thiol-containing cellular factors. Sodium thiosulfate, for example, may inactivate toxic platinum species through binding.[104,105] It successfully reduced the clinical nephrotoxicity of cisplatin[106–108] and inactivated the drug in blood.[109,110] However, in another study, sodium thiosulfate increased toxicity and inhibited the antineoplastic activity of cisplatin.[111]

Sodium diethyldithiocarbamate (DDTC) is a chelating agent which may preferentially remove platinum from cysteine and methionine without altering the stability of DNA adducts. DDTC removed platinum bound to all factors other than two guanine residues thought to be involved in cisplatin's antitumor effects.[112] Cisplatin's antineoplastic activity was not affected by DDTC,[111–113] but DDTC treatment did reverse its inhibition of protein synthesis.[114] Despite the fact that DDTC has been reported to reduce nephrotoxicity without affecting the pharmacokinetics of cisplatin when given at 2.5 g m^{-2} or 5 g m^{-2},[115,116] DDTC itself has been reported to cause hypertension, burning sensation, agitation, and diaphoresis which may limit its usefulness in certain patients.

WR-2721 was developed as a radiation protector. It is hydrolyzed *in vivo* into WR-1065, a diaminothiol which purportedly accumulates in normal but not tumor cells.[117–119] As an active chelator, WR-1065 may prevent cisplatin from binding to key cellular components in the kidney. Paradoxically, WR-2721 itself may reduce glutathione synthesis by inhibiting γ-glutamylcysteine synthetase. This inhibition combined with dose-limiting emesis and hypotension require that its dose be carefully controlled to prevent enhancement of cisplatin's systemic toxicities. Finally, WR-2721 treatment alone inhibited metastatic melanoma and decreased cisplatin's myelosuppressive, neurotoxic and ototoxic side effects.[120]

GSH and the small metal-binding protein metallothionein together comprise a significant proportion of the cellular thiol pool. Although GSH reacts pharmacologically with cisplatin at a slow rate, it rapidly inactivates cisplatin's toxic hydrolysis products.[121] When infused 15 min before a clinical dose of 90 mg m^{-2} cisplatin, it reduced nephrotoxicity, and was not itself associated with unwanted side effects.[122] However, experiments in mice[123] and rats[124] suggest that the modulation of cisplatin's toxicity by GSH may involve more than the mechanism of thiol binding; pharmacological depletion of GSH did not always increase cisplatin-induced renal damage. Metals which induce metallothionein protein expression reduced toxicity,[125,126] but the role of metallothionein as a chelator of platinum is difficult to distinguish from other properties of metallothionein, or the role of other gene products whose expression may be increased by metal treatment. By obviating the need for additional chemical treatments, transgenic animals which over- or underexpressed GSH or metallothionein would be ideal models for investigating the role of the cellular thiol pool in the modulation of cisplatin nephrotoxicity.

One of the most interesting classes of drugs being used to reduce cisplatin renal damage are steroids and a related compound, urinastatin. Steroids may prevent the release of degradative enzymes by stabilizing renal tubular lysosomal membranes.[127] ORG-2766, a peptide analogue of alpha-MSH_{4-10}, displayed none of the hormonal effects associated with steroid use and protected animals from nephrotoxicity.[128] The proteinase inhibitor, urinastatin, may also prevent renal lysosomal enzyme release.[129] Clinical trials are proposed to determine the usefulness of urinastatin as a chemoprotective agent when given twice a day for three days of each treatment cycle.[129]

As the mechanism of cisplatin nephrotoxicity is more completely defined, the development of protective strategies should expand to encompass specific cellular targets, thereby minimizing interference with its antineoplastic activity. Already, greater understanding of the role of the proximal tubule and its response to cisplatin exposure is giving rise to novel compounds. Further investigation of the gene expression which defines the proximal tubule phenotype will also facilitate development of specific strategies for protection.

7.27.2.1.5 *Platinum analogues*

(i) Carboplatin

Carboplatin (diammine (1,1-cyclobutanedicarboxylato) platinum(II)) is a second-generation platinum compound which has greater chemical stability and lower reactivity with proteins than cisplatin. These characteristics may explain its lower nephrotoxicity[130,131] despite the fact that its elimination is primarily renal.[132] Since it is less reactive against nucleophilic sites on DNA, higher doses of carboplatin are required to achieve DNA adduct levels comparable to cisplatin. Like cisplatin, carboplatin probably exerts its antitumor activity through irreversible DNA binding. Unlike cisplatin, its dose-limiting side effect is myelosuppression, particularly thrombocytopenia.[133,134] However, renal dysfunction and neurotoxicity are evident when doses reach 2.5-fold the maximum tolerated dose of $400\,mg\,m^{-2}$.[135,136]

At pharmacologic doses, plasma platinum levels 24 h following carboplatin treatment are similar to those after cisplatin.[137,138] Tumor tissue levels of platinum in human patients[139] and in nude mice[140] were also comparable for carboplatin and cisplatin. This suggests that drug distribution and platinum tissue concentration are not simple predictors for determining the relative toxicity or activity of platinum complexes. For example, greater concentrations of platinum were measured in the renal cortex of dogs treated with carboplatin when compared with cisplatin,[141] and intrarenal comparison showed no significant difference in distribution between the cortex and medulla of rats treated with these compounds.[76] The lower nephrotoxicity of carboplatin compared to cisplatin is, therefore, not the result of proportionately less accumulation of platinum, agreeing with studies on cisplatin's toxicity in different tissues where absolute platinum levels were poor indicators of cytotoxicity.

A number of studies comparing the nephrotoxicity of these two compounds in rats *in vivo*[142,143] or *in vitro* in rat renal cortical slices[62,63,144] and primary cultures of renal proximal tubule cells[60,72] have confirmed carboplatin's lower nephrotoxicity. Relative to cisplatin, carboplatin is less potent in inhibiting DNA and protein synthesis, and decreasing organic ion transport and gluconeogenesis. It has been suggested that carboplatin is excreted primarily through glomerular filtration because it associates less readily with plasma proteins. If this were true, it would bypass the renal proximal tubule which is so sensitive to cisplatin.[76,145]

As with cisplatin, the mechanism of carboplatin's renal toxicity is not clear. However, reports comparing their differential toxicity in the kidney indicate that carboplatin induces the same effects as cisplatin although to a lesser degree. Inhibition of protein synthesis, organic ion transport, neutral red uptake, and mitochondrial conversion of MTT progress in a similar, but less rapid, manner as cisplatin-induced toxicity.[60,63] Carboplatin treatment also led to condensation of the nucleolus and degranulation of rough endoplasmic reticulum in human renal proximal tubule cells (unpublished observation), which agrees with the effects of cisplatin on the ultrastructure of rat renal proximal tubule cells.[60]

(ii) Other platinum analogues

A detailed review of some of the cisplatin analogues under development has been published elsewhere.[146] In order to qualify for marketing approval in the United States, a new compound must show activity in cancers unresponsive to cisplatin (gastrointestinal disease, for example), reduce nephrotoxicity, neurotoxicity, or ototoxicity, be formulated for oral administration and absorption, be active in tumors resistant to cisplatin and not display new chronic systemic toxicities. No third-generation platinum analogues have yet met these criteria. Many of them trade increased potential neurotoxictiy (oxaliplatin and ormaplatin) or myelosuppression (lobaplatin, 254-S, and DWA 2114R) for reduced nephrotoxicity. However, the possible ability of ormaplatin to overcome cisplatin resistance may encourage its use in conjunction with appropriate neuroprotective agents.

An orally-active platinum compound of the ammine/amine platinum(IV) dicarboxylate family, JM-216, is being awaited with interest because of its reported effectiveness against cisplatin-resistant cell lines.[147,148] It was not nephrotoxic in rodent models despite severe leucopenia[149] and has shown no nephrotoxicity in clinical trials to date. Probably as a result of its oral route of administration, JM-216 accumulates primarily in the liver and causes hepatotoxicity, particularly with repeated dosing. Using another approach to circumvent what may be a structure–toxicity relationship between *cis* isomers and renal function, Kell and *et al.*[150] evaluated the efficacy of a *trans* platinum complex and showed it to have antitumor activity *in vitro* and *in vivo*. The toxicity of this *trans* platinum complex is still being investigated.

The goal of third- and fourth-generation platinum compounds has been to overcome cisplatin resistance and provide oral bioavailability. Drugs such as 254-S and oxaliplatin show efficacy, but no general advantages over cisplatin (first-generation) and carboplatin (second-generation). One of the difficulties in developing drugs displaying good antitumor activity in the absence of nephro- or neurotoxicities is that the parent compound, cisplatin, is still so poorly understood. In order to delineate those structural characteristics which confer specific toxicities, one must have a thorough understanding of the mechanism through which platinum complexes enter the cell, are metabolized, are distributed, and finally degraded. Thus, studies which further expand our knowledge of cisplatin's nephrotoxicity *in vitro* and *in vivo* will facilitate the development of new compounds with fewer side effects and more desirable pharmacokinetic properties.

7.27.2.2 Ifosfamide/Cyclophosphamide

Ifosfamide and cyclophosphamide are oxazaphosphorines, which are highly effective alkylating agents used in the management of a variety of malignant diseases. Both compounds are metabolized by hepatic microsomal oxidation to 4-hydroxy (4-OH) metabolites and subsequently decompose to alkylating mustard derivatives and acrolein.[151,152] Unlike cyclophosphamide, ifosfamide undergoes considerable oxidation of chloroethyl side chains with liberation of chloroacetaldehyde.

One major side effect of both agents is hemorrhagic cystitis. The incidence is dose-dependent and ranges from 4% to 36% for cyclophosphamide and 40% to 50% for ifosfamide.[10] However, with adequate hydration and/or concomitant administration of mesna, a synthetic thiol compound which detoxifies reactive metabolites in the urinary tract,[153,154] the incidence of this complication has been much reduced. With ifosfamide and mesna the incidence of cystitis has been reduced to less than 5%.[155–158] While both compounds can cause hemorrhagic cystitis, only ifosfamide is nephrotoxic. Nephrotoxicity may manifest as glomerular, proximal tubular, or distal tubular dysfunction (reviewed by Skinner *et al.*[159]). Proximal tubular impairment is most common and usually presents as Fanconi's syndrome. As ifosfamide is used in pediatric malignancies, nephrotoxicity has emerged as a serious adverse event.[159–163] The nephrotoxicity appears to be dose-related, may occur after one or several courses of therapy,[158,162,164,165] and is more likely to occur in patients having prior or concomitant cisplatin therapy.[166–168] The overall incidence of renal tubular dysfunction appears to be between 10% and 20%, with only 1–3% of cases showing severe clinical symptoms.[169] Although mesna does not prevent the renal toxicity, the nephrotoxicity in many cases is reversible.[170] Kamen *et al.*[171] have questioned whether ifosfamide should be used in preference to cyclophosphamide when considering toxicity, efficacy, and cost of therapy.

The findings from preclinical studies of ifosfamide in mice, rats, dogs, and monkeys have been summarized by Barnett.[172] Irritation of the bladder was noted after single high-dose administration in rats (200 mg kg^{-1}; p.o.) and dogs (66 mg kg^{-1}; i.v.). Hematuria and/or hemorrhagic urinary bladders were observed in multiple dose studies in mice, rats, dogs, and monkeys. In a long-term study in rats (dosing once every 3 wk for 6 months), i.p. doses of 25, 50, or 100 mg kg^{-1} resulted in incidences of cystitis of 31, 100, and 97%, respectively. Renal pathology was only noted in dogs given high-dose ifosfamide, and in a single monkey that died during a 14 d study. Reversible elevations in BUN were also observed in the 14 d monkey study.

Two *in vitro* systems have been used to investigate the mechanism of ifosfamide-induced toxic effects. Mohrmann *et al.*[173–175] have induced an experimental Fanconi syndrome in LLC-PK$_1$ cells, showing that metabolites of ifosfamide impair transport rates of D-glucose, phosphate, and L-alanine. Although ifosfamide can impair thymidine uptake in these cells, the 4-OH metabolite was more toxic than the parent, and acrolein was the most toxic metabolite. Acrolein also inhibited RNA synthesis and decreased total protein synthesis. Interestingly, chloroacetaldehyde also reduced transport, total protein, and thymidine uptake in these cells. In an isolated perfused rat kidney model,[176] chloroacetaldehyde (210 µM) decreased fractional reabsorption of sodium, glucose, inorganic phosphate, and sulfate, as well as PAH clearance. In this model, neither ifosfamide nor acrolein (at 470 µM) had effects on functional parameters. These *in vitro* studies suggest that the chloroacetaldehyde metabolite may play a role in the development of renal tubular damage after ifosfamide therapy.

The preclinical studies predicted the hemorrhagic cystitis which was subsequently seen in clinical trials. There was a suggestion from the large animal studies that renal toxicity would occur, but more sensitive measures of renal toxicity such as urinary enzyme and fractional excretion were not measured in the preclinical studies. *In vitro* studies of metabolism and toxicity have suggested that the nephrotoxicity is due to the metabolites acrolein and chloroacetaldehyde but the molecular mechanism of tubular cell damage is yet to be elucidated.

7.27.2.3 Nitrosoureas

The nitrosoureas are a group of lipid-soluble alkylating agents including streptozotocin, carmustine (BCNU), lomustine (CCNU), semustine (methyl-CCNU), and chlorozotocin (DCNU).

Streptozotocin is diabetogenic in rodents, dogs, and monkeys and has been used clinically to treat pancreatic islet cell tumors. Nephro-

toxicity has proven to be the dose-limiting toxicity of this agent, with some abnormality of renal function occurring in as many as 65% of patients.[177–179] Toxicity manifests as proteinuria, reduction in creatinine clearance, Fanconi syndrome (phosphaturia, glycosuria, and amino aciduria), and azotemia.[180] Tubular and glomerular injury, as well as interstitial fibrosis, present pathologically.[181]

Lesions in the proximal convoluted renal tubules (necrosis, vacuolization, tubular degeneration) were evident following single doses of streptozotocin in mice (250 mg kg^{-1}), dogs (6.3 mg kg^{-1}), and monkeys (80 mg kg^{-1}). In multidose studies, elevations in BUN, glucosuria, and renal lesions were seen after five daily doses of ≥ 3.1 mg kg^{-1} in dogs and ≥ 20 mg kg^{-1} in monkeys.[182] When doses are expressed on the basis of body surface area, toxicity was similar in both dogs and monkeys. Cumulative doses of 2400–2500 mg m^{-2} were lethal in these species; in comparison, 2500 mg m^{-2} is given to patients clinically (as 500 mg m^{-2} × 5 d).

Unlike streptozotocin, renal failure is cumulative and delayed following treatment with semustine, carmustine and lomustine. Semustine has proved to be the most toxic of the three compounds.[183,184] However, no nephrotoxicity has been observed with semustine at cumulative doses below 1400 mg m^{-2}.[184] At cumulative doses above 1400 mg m^{-2}, the incidence of nephrotoxicity is 14–38% with a median time to onset of 27 months. The most common manifestation of nephrotoxicity is a slowly rising serum creatinine. Renal tissue from biopsy or autopsies of patients treated with carmustine and/or semustine had findings of tubular atrophy, interstitial fibrosis, and glomerular sclerosis.[184,185] Carmustine alone rarely causes renal toxicity, with only mild elevations in serum creatinine being observed in patients.[185] Only a few cases of nephrotoxicity have been reported following long-term lomustine administration.[183]

Semustine, carmustine, and lomustine all caused renal toxicity in dogs and monkeys, with monkeys being the more sensitive species.[186,187] Following single i.v. infusions of lomustine at doses of 20 mg kg^{-1} or higher, azotemia and progressive renal lesions including fatal chronic interstitial nephritis were observed in monkeys.[187] Animal deaths were seen beginning 16 d after treatment. Renal lesions were still evident in survivors 100 d after dosing. High doses of semustine in monkeys resulted in severe irreversible damage to the proximal convoluted tubules.[184] Dogs receiving high doses demonstrated azotemia.[186] Single doses of semustine (21–63 mg kg^{-1}) produced dose-related elevations in BUN in mice.[188] In mice, acute tubular necrosis of the proximal tubules was detected on day 3; the lesions were most severe on day 6, by day 21 cortical fibrosis and atrophy were present. Rats given single doses of 20–100 mg kg^{-1} of semustine showed dose-related decreases in PAH as early as 1 h after treatment. Decreases in PAH uptake were still evident 28 d after dosing with 20–80 mg kg^{-1}, indicating chronic effects. Animals receiving 100 mg kg^{-1} and 140 mg kg^{-1} semustine did not survive the duration of the study. Changes in other measures of toxicity (enzymuria, proteinuria, polyuria, and alkalinuria) were delayed from 1 to 6 d after dosing, some measures increased in severity throughout the 28 d period.[189] In a separate study, congestion of the papillary vasculature and papillary necrosis were observed within 7 d of a single high dose of semustine (250 mg kg^{-1}).[190] Chlorozotocin was more toxic than semustine in rats, and produced acute tubular necrosis followed by papillary necrosis.[190] Long-term testing in animals has not been conducted with these compounds.

In general, man is not as sensitive to nephrotoxic effects of nitrosoureas as animals, although similar nephrotoxic effects have been noted. Although the compounds are rapidly metabolized and the route of elimination is primarily renal, the mechanism of nitrosourea-induced renal toxicity is not well defined and there is currently no explanation for the delayed onset of toxicity.

7.27.3   ANTIBIOTICS

7.27.3.1   Mitomycin-C

Mitomycin-C is an antitumor antibiotic isolated from *Streptomyces caespitosus*. The toxicity most often manifested in humans is a hemolytic uremic syndrome (HUS), although renal failure or microangiopathic hemolytic anemia occasionally occur. A number of reviews of the renal effects of this agent in patients have been published.[191–196] In approximately 10% of patients, mitomycin-C toxicities are lethal. Patients present with proteinuria, microscopic hematuria, anemia, and thrombocytopenia. The renal side effects have a latency period of up to 15 months, most developing HUS within 5 months of the last dose of mitomycin. The incidence of HUS appears to be dose-related, with cumulative doses of 60 mg being associated with development of the syndrome.[194] Pathology of HUS includes nuclear atypias, mesangiolysis, arteriolar fibrin thrombi and intimal hyperplasia, thickening

of the glomerular capillary wall, and glomerular sclerosis.[197,198] The mechanism of vascular damage is unknown but may involve either direct toxicity (alkylation) or immune complex mediated toxicity.[198] Although numerous treatments have been employed, including corticosteroids, plasma exchange, dialysis, furosemide, aspirin, dipyridamole, and immunopheresis, most have been ineffective.[193] Careful monitoring of the patient after treatment is recommended.

The LD$_{50}$ of mitomycin-C varies from 1 to 2.5 mg kg^{-1} in mice, rats, cats, dogs, and monkeys.[199] Preclinical toxicological investigations demonstrated necrotizing nephrosis in rhesus monkeys following doses of 0.5–2 mg kg^{-1}; however, similar lesions were not noted in rats or dogs.[200] Repeated doses of mitomycin-C induced hydronephrosis and papillomatous hyperplasia of the epithelia of the uretopelvic junction in mice.[201] In rats, urinary enzyme leakage was reported following a single high dose[202] and morphologic changes of capillary collapse and degenerative changes in the glomerular endothelial cell, as well as tubular necrosis, were demonstrated after direct injection into the left kidney of rats.[203] Although early animal studies indicated that the kidney may be a target, most lesions in animals are described as tubular lesions and are not similar to HUS seen in humans.

7.27.3.2   Mithramycin

Mithramycin is an antitumor antibiotic derived from *Streptomyces tanashiensis* which inhibits DNA-dependent RNA synthesis by binding to DNA. Mithramycin (plicamycin) is used to treat hypercalcemia of malignancy at low doses. Nephrotoxicity was observed in the past, when used at doses of 25–50 µg kg^{-1} for 5 d to treat malignancies.[204] Increases in BUN or decreased creatinine clearance were seen in 40% of patients and proteinuria in 78% of patients. Eleven percent of patients died of renal complications; histopathologic findings included renal tubular swelling with hydropic degeneration, necrosis of both proximal and distal renal tubules, and tubular atrophy. Minimal renal functional changes have been reported using single low-dose (25 µg kg^{-1}) treatment for hypercalcemia.[205]

Radiolabeled drug studies in mice have shown high concentrations of mithramycin in the kidney.[206] Gilsdorf *et al.*[207] demonstrated decreases in urine volume, PAH clearance, and sodium excretion following intravenous administration of mithramycin (0.5 mg kg^{-1}) in dogs.

No change was evident in creatinine clearance, suggesting tubular rather than glomerular effects.

7.27.3.3 Doxorubicin/Daunorubicin

Doxorubicin (adriamycin) is an anthracyline antibiotic that is used for the treatment of solid as well as disseminated tumors. Its antitumor mechanism is inhibition of RNA and DNA synthesis through intercalation into DNA. Nephrotoxicity following doxorubicin treatment appears to be species specific, being reported only in rats and rabbits.[208] Nephrotoxicity most commonly follows long-term administration and is characterized by glomerular damage, although vacuolization of the tubular epithelium, protein casts, and interstitial fibrosis also occur.[208–210] In rats, a single dose of 7.5 mg kg^{-1} of doxorubicin induces a nephrotic syndrome characterized by acute glomerular injury and sustained proteinuria.[210] Epithelial cell detachment, cytoplasmic vacuolization, glomerular epithelial cell alterations, glomerular anionic charge loss, and foot process fusion have been observed. It has been suggested that platelet activating factor (PAF) or thromboxane B_2 may be involved in adriamycin-induced nephrotoxicity.[211] PAF is synthesized by mesangial cells and may cause loss of glomerular anionic charge. Bertelli *et al.*[212] demonstrated that doxorubicin is cytotoxic without entering cells, suggesting that the cell membrane is the subcellular target. In the same study, free radicals were partially implicated in doxorubicin-induced damage of glomerular epithelial cells.

The major side effect of doxorubicin is cardiomyopathy which is dose-dependent and leads to congestive heart failure. The cardiomyopathy occurs primarily at cumulative doses above 400 mg m^{-2}. Prerenal azotemia accompanies the diminished cardiac output in these patients.[5] It is likely that nephrotoxicity is not seen in patients because the risk of cardiomyopathy limits doses to nonnephrotoxic levels.

Daunorubicin is an anthracycline which is structurally related to doxorubicin. Several animal studies have indicated that daunorubicin is a nephrotoxin.[7,213,214] A single i.v. dose of 20 mg kg^{-1} daunorubucin in rats caused loss of glomerular foot processes by 4 d, glomerular basement membrane thickening by 4 wk, and resulted in chronic glomerulonephritis with persistent nephrotic syndrome.[213] Renal tumors were observed one year following drug treatment.[214] However, like doxorubicin, little nephrotoxic potential has been observed in man.

7.27.4 ANTIMETABOLITES

7.27.4.1 Methotrexate

Methotrexate, an inhibitor of folate synthesis, is one of the earliest and most commonly used agents in cancer chemotherapy. Methotrexate is most often administered with leucovorin, a reduced folate, to decrease side effects of this cytotoxic drug. However, high-dose regimens (>1 g m^{-2}) have been associated with a high incidence of nephrotoxicity, even with the use of leucovorin rescue.[5,215] Prior to the recognition that maintenance of a high urinary volume and high pH was essential in high-dose methotrexate treatment, renal toxicity was observed in up to 60% of patients and deaths due to renal failure were observed.[215–217]

Methotrexate is excreted primarily by the kidneys and much of the dose is excreted unchanged. At high doses (>50 mg kg^{-1}), a 7-OH metabolite is also excreted in significant amounts. While methotrexate has low solubility in aqueous medium, the 7-OH metabolite has even more limited solubility than methotrexate and may contribute to nephrotoxicity at high doses.[218] Decreases in renal function can dramatically increase plasma concentrations of methotrexate and can exacerbate toxicity. Little systemic toxicity has been associated with plasma levels below 1×10^{-7} M at 48 h after drug administration, whereas a high percentage of patients with plasma levels of 1×10^{-7} M or above exhibited toxicities (usually myelosuppression).[215,219,220] By monitoring plasma methotrexate levels, toxicity can be minimized, doses of leucovorin can be minimized, and therapeutic efficacy can be maximized.

The pathogenesis of nephrotoxicity is thought to be due to precipitation of methotrexate or a metabolite within the distal tubule, which in turn causes intrarenal obstructive nephropathy. At physiological pH, methotrexate is fully ionized, but in the acidic environment of the urine the solubility is reduced.[221,222] Precipitated material (drug) has been observed in tubules of a patient who died from renal failure; the concentration of methotrexate was higher in the medulla than in the cortex.[215] Patients who are dehydrated and/or excrete acidic urine are thus at a high risk of nephrotoxicity. Direct tubular toxicity has also been proposed as a mechanism for methotrexate-induced renal damage since renal failure has been reported without evidence of precipitation. Condit *et al.*[217] noted marked necrosis of the convoluted tubules in three patients who died of nephrotoxicity. A direct effect on glomerular hemodynamics has been suggested due to effects on GFR (decreased inulin and creatinine clearances).[223,224]

The frequency of renal failure can be decreased by maintaining hydration and urinary alkalinization (with i.v. or oral bicarbonate) during methotrexate treatment. Plasma levels of methotrexate should be monitored following treatment and leucovorin should be continued until plasma levels are low.

7.27.4.2 5-Azacytidine

Preclinical studies indicated that this pyrimidine analogue may have nephrotoxic potential. Elevations in BUN and renal morphological changes were observed in dogs and monkeys following i.v. administration of 5-azacytidine.[225] The mouse was found to be less sensitive. Renal effects in clinical trials have been observed primarily in studies where 5-azacytidine has been used in combination therapy,[226] where nephrotoxicity manifests as tubular transport defects and systemic acidosis. No nephrotoxicity was observed in patients treated with 5-azacytidine alone.[225]

7.27.5 OTHER

7.27.5.1 Gallium Nitrate

Gallium nitrate is a hydrated nitrate salt of the heavy metal gallium. It was initially investigated as an antineoplastic agent, but is approved for treatment of cancer-related hypercalcemia. The primary dose-limiting toxicity of gallium nitrate in cancer chemotherapy trials was nephrotoxicity.[227–229] Elevations in BUN and creatinine have been reported in 12.5% of cases.[230,231] Unlike cisplatin, the nephrotoxicity is not cumulative, can be prevented with hydration, and is usually reversible.

In rats, the mechanism of toxicity appears to be the result of deposition of a gallium–calcium–phosphorus complex in the tubules.[232] High doses (100 mg kg^{-1}) in rats caused tubular necrosis and tubular obstruction; the glomeruli did not appear to be affected.

7.27.5.2 Biologics

Biologics are often cleared by renal excretion, but reports of renal abnormalities with protein therapeutics are rare. Oliguria, reduction of fractional sodium excretion, and azotemia has been reported following high-dose interleukin-2 therapy.[233] The renal complications observed with this cytokine are believed to be of prerenal origin.[5] Nephrotic syndrome and proteinuria have been reported in rare cases after treatment with interferon-α.[234]

7.27.6 CONCLUSIONS

Renal complications following cancer chemotherapy have been demonstrated with an array of agents (Table 1). A decrease in the incidence of renal side effects can be achieved by careful monitoring of patients, novel dosing regimens, and prophylactic measures. Preclinical animal studies in many cases predicted the potential for renal toxicity. Despite extensive research, the mechanism of toxicity for most nephrotoxic antineoplastic agents is not well understood. Further studies on mechanisms of toxicity should allow the development of new chemotherapeutics with fewer renal side effects.

7.27.7 REFERENCES

1. Anonymous, 'Drugs of Choice for Cancer Chemotherapy.' *Med. Lett. Drugs Ther.*, 1995, **37**, 25–32.
2. R. S. Finley, 'Drug interactions in the oncology patient.' *Semin. Oncol. Nurs.*, 1992, **8**, 95–101.
3. R. E. Rieselbach and M. B. Garnick, in 'Diseases of the Kidney,' eds. R. W. Schrier and C. W. Gottschalk, Little, Brown, Boston, MA, 1993, pp. 1165–1186.
4. R. F. Borch, in 'Toxicology of the Kidney,' eds. J. B. Hook and R. S. Goldstein, Raven Press, New York, 1993, pp. 283–301.
5. R. G. Narins, M. Carley, E. J. Bloom *et al.*, 'The nephrotoxicity of chemotherapeutic agents.' *Semin. Nephrol.*, 1990, **10**, 556–564.
6. E. Cobos and R. R. Hall, 'Effects of chemotherapy on the kidney.' *Semin. Nephrol.*, 1993, **13**, 297–305.
7. R. B. Weiss and D. S. Poster, 'The renal toxicity of cancer chemotherapeutic agents.' *Cancer Treat. Rev.*, 1982, **9**, 37–56.
8. N. J. Vogelzang, 'Nephrotoxicity from chemotherapy: prevention and management.' *Oncology (Huntingt.)*, 1991, **5**, 97–105.
9. W. P. Patterson and G. P. Reams, 'Renal toxicities of chemotherapy.' *Semin. Oncol.*, 1992, **19**, 521–528.
10. W. Kreusser, R. Herrmann, W. Tschöpe *et al.*, 'Nephrological complications of cancer therapy.' *Contrib. Nephrol.*, 1982, **33**, 223–238.
11. F. Ries and J. Klastersky, 'Nephrotoxicity induced by cancer chemotherapy with special emphasis on cisplatin toxicity.' *Am. J. Kidney Dis.*, 1986, **8**, 368–379.
12. H. G. Healy and A. R. Clarkson, 'Renal complications of cytotoxic therapy.' *Aust. NZ J. Med.*, 1983, **13**, 531–539.
13. R. L. Schilsky, 'Renal and metabolic toxicities of cancer chemotherapy.' *Semin. Oncol.*, 1982, **9**, 75–83.
14. D. Wujcik, 'Current research in side effects of high-dose chemotherapy.' *Semin. Oncol. Nurs.*, 1992, **8**, 102–112.
15. D. K. Jorkasky and I. Singer, 'Drug-induced tubulointerstitial nephritis: special cases.' *Semin. Nephrol.*, 1988, **8**, 62–71.
16. G. Daugaard and U. Abildgaard, 'Evaluation of nephrotoxicity secondary to cytostatic agents.' *Crit. Rev. Oncol. Hematol.*, 1992, **13**, 215–240.
17. M. Rozencweig, D. D. von Hoff, M. Slavik *et al.*, 'cis-Diamminedichloroplatinum(II). A new anticancer drug.' *Ann. Intern. Med.*, 1977, **86**, 803–812.
18. M. J. McKeage, S. E. Morgan, F. E. Boxall *et al.*, 'Lack of nephrotoxicity of oral ammine/amine platinum(IV) dicarboxylate complexes in rodents.' *Br. J. Cancer*, 1993, **67**, 996–1000.
19. A. W. Prestayko, J. C. D'Aoust, B. F. Issell and S. T. Crooke, 'Cisplatin (cis-diamminedichloroplatinum(II)).' *Cancer Treat. Rev.*, 1979, **6**, 17–39.

20. A. H. Rossof, R. E. Slayton and C. P. Perlia, 'Preliminary clinical experience with *cis*-diamminedichloroplatinum (II) (NSC-119875 CACP).' *Cancer*, 1972, **30**, 1451–1456.

21. G. Daugaard, U. Abildgaard, N. H. Holstein-Rathlou *et al.*, 'Renal tubular function in patients treated with high-dose cisplatin.' *Clin. Pharmacol. Ther.*, 1988, **44**, 164–172.

22. R. L. Schilsky and T. Anderson, 'Hypomagnesemia and renal magnesium wasting in patients receiving cisplatin.' *Ann. Intern. Med.*, 1979, **90**, 929–931.

23. R. L. Schilsky, A. Barlock and R. F. Ozols, 'Persistent hypomagnesemia following cisplatin chemotherapy for testicular cancer.' *Cancer Treat. Rep.*, 1982, **66**, 1767–1769.

24. M. Lam and D. J. Adelstein, 'Hypomagnesemia and renal magnesium wasting in patients receiving cisplatin.' *Am. J. Kidney Dis.*, 1986, **8**, 164–169.

25. V. Mavichak, C. M. L. Cappin, N. L. M. Wong *et al.* 'Renal magnesium wasting and hypocalciuria in chronic *cis*-platinum nephropathy in man.' *Clin. Sci. (Colch.)*, 1988, **75**, 203–207.

26. J. B. Hill and A. Russo, 'Cisplatin-induced hypomagnesemic hypocalcemia (Letter).' *Arch. Intern. Med.*, 1981, **141**, 1100.

27. J. P. Fillastre and G. Raguenez-Viotte, 'Cisplatin nephrotoxicity.' *Toxicol. Lett.*, 1989, **46**, 163–175.

28. M. Dentino, F. C. Luft, M. N. Yum *et al.*, 'Long-term effect of *cis*-diamminedichloride platinum (CDDP) on renal function and structure in man.' *Cancer*, 1978, **41**, 1274–1281.

29. A. C. Friedman and E. M. Lautin, '*cis*-Platinum(II) diamine dichloride: another cause of bilateral small kidneys.' *Urology*, 1980, **16**, 584–586.

30. P. Fjeldborg, J. Sorensen and P. E. Helkjaer, 'The long-term effect of cisplatin on renal function.' *Cancer*, 1986, **58**, 2214–2217.

31. A. B. Campbell, S. M. Kalman and C. Jacobs, 'Plasma platinum levels: relationship to cisplatin dose and nephrotoxicity.' *Cancer Treat. Rep.*, 1983, **67**, 169–172.

32. P. J. Hilton, Z. Roth, S. Lavender *et al.*, 'Creatinine Clearance in patients with proteinuria.' *Lancet*, 1969, **2**, 1215–1216.

33. B. R. Jones, R. B. Bhalla, J. Mladek *et al.*, 'Comparison of methods of evaluating nephrotoxicity of *cis*-platinum.' *Clin. Pharmacol. Ther.*, 1980, **27**, 557–562.

34. P. K. Buamah, A. Howell, H. Whitby *et al.*, 'Assessment of renal function during high-dose *cis*-platinum therapy in patients with ovarian carcinoma.' *Cancer Chemother. Pharmacol.*, 1982, **8**, 281–284.

35. J. C. Gonzalez-Vitale, D. M. Hayes, E. Cvitkovic *et al.*, 'The renal pathology in clinical trials of *cis*-platinum(II) diamminedichloride.' *Cancer*, 1977, **39**, 1362–1371.

36. C. Kanaka, O. H. Oetliker and M. G. Bianchetti, 'Chronic cisplatin tubulopathy in humans and animals: clear-cut discrepant findings.' *Nephron*, 1991, **59**, 693.

37. W. T. Hardaker Jr., R. A. Stone and R. McCoy, 'Platinum nephrotoxicity.' *Cancer*, 1974, **34**, 1030–1032.

38. H. Tanaka, E. Ishikawa, S. Teshima *et al.*, 'Histopathological study of human cisplatin nephropathy.' *Toxicol. Pathol.*, 1986, **14**, 247–257.

39. R. S. Goldstein, B. Noordewier, J. T. Bond *et al.*, '*cis*-Dichlorodiammineplatinum nephrotoxicity: time course and dose response of renal functional impairment.' *Toxicol. Appl. Pharmacol.*, 1981, **60**, 163–175.

40. J. M. Ward and K. A. Fauvie, 'The nephrotoxic effects of *cis*-diamminedichloroplatinum(II) (NSC-119875) in male F344 rats.' *Toxicol. Appl. Pharmacol.*, 1976, **38**, 535–547.

41. R. Safirstein, P. Miller and J. B. Guttenplan, 'Uptake and metabolism of cisplatin by rat kidney.' *Kidney Int.*, 1984, **25**, 753–758.

42. G. Daugaard, N. H. Holstein-Rathlou and P. P. Leyssac, 'Effect of cisplatin on proximal convoluted and straight segments of the rat kidney.' *J. Pharmacol. Exper. Ther.*, 1988, **244**, 1081–1085.

43. U. Schaeppi, I. A. Heyman, R. W. Fleischman *et al.*, '*cis*-Dichlorodiammineplatinum(II) (NSC-119 875) preclinical toxicologic evaluation of intravenous injection in dogs, monkeys, and mice.' *Toxicol. Appl. Pharmacol.*, 1973, **25**, 230–241.

44. G. Daugaard, U. Abildgaard, N. H. Holstein-Rathlou *et al.*, 'Acute effect of cisplatin on renal hemodynamics and tubular function in dog kidneys.' *Ren. Physiol.*, 1986, **9**, 308–316.

45. M. E. C. Robbins, T. B. Bywaters, R. S. Jaenke *et al.*, 'Long-term studies of cisplatin-induced reductions in porcine renal functional reserve.' *Cancer Chemother. Pharmacol.*, 1992, **29**, 309–315.

46. D. C. Dobyan, J. Levi, C. Jacobs *et al.*, 'Mechanism of *cis*-platinum nephrotoxicity. II. Morphologic observations.' *J. Pharmacol. Exp. Ther.*, 1980, **213**, 551–556.

47. D. D. Choie, D. S. Longnecker and A. A. del Campo, 'Acute and chronic cisplatin nephropathy in rats.' *Lab. Invest.*, 1981, **44**, 397–402.

48. S. Chopra, J. S. Kaufman, T. W. Jones *et al.*, '*cis*-Diamminedichloroplatinum-induced acute renal failure in the rat.' *Kidney Int.*, 1982, **21**, 54–64.

49. T. W. Jones, S. Chopra, J. S. Kaufman *et al.*, '*cis*-Diamminedichloroplatinum(II)-induced acute renal failure in the rat. Correlation of structural and functional alterations.' *Lab. Invest.*, 1985, **52**, 363–373.

50. K. Nosaka, J. Nakada and H. Endou, Cisplatin-induced alterations in renal structure, ammoniagenesis, and gluconeogenesis of rats.' *Kidney Int.*, 1992, **41**, 73–79.

51. L. K. Tay, C. L. Bregman, B. A. Masters *et al.*, 'Effects of *cis*-diamminedichloroplatinum(II) on rabbit kidney *in vivo* and on rabbit renal proximal tubule cells in culture.' *Cancer Res.*, 1988, **48**, 2538–2543.

52. B. J. Foster, B. J. Harding, M. K. Wolpert-DeFilippes *et al.*, 'A strategy for the development of two clinically active cisplatin analogs: CBDCA and CHIP.' *Cancer Chemother. Pharmacol.*, 1990, **25**, 395–404.

53. D. C. Dobyan, D. Hill, T. Lewis *et al.*, 'Cyst formation in rat kidney induced by *cis*-platinum administration.' *Lab. Invest.*, 1981, **45**, 260–268.

54. D. Nonclercq, G. Toubeau, P. Tulkens *et al.*, 'Renal tissue injury and proliferative response after successive treatments with anticancer platinum derivatives and tobramycin.' *Virchows. Arch. B. Cell. Pathol. Incl. Mol. Pathol.*, 1990, **59**, 143–158.

55. J. M. Pascoe and J. J. Roberts, 'Interactions between mammalian cell DNA and inorganic platinum compounds. I. DNA interstrand cross-linking and cytotoxic properties of platinum(II) compounds.' *Biochem. Pharmacol.*, 1974, **23**, 1359–1365.

56. D. D. Choie, A. A. Del Campo and A. M. Guarino, 'Subcellular localization of *cis*-dichlrodiammineplatinum(II) in rat kidney and liver.' *Toxicol. Appl. Pharmacol.*, 1980, **55**, 245–252.

57. R. Parti and W. Wolf, 'Quantitative subcellular distribution of platinum in rat tissues following i.v. bolus and i.v. infusion of cisplatin.' *Cancer Chemother. Pharmacol.*, 1990, **26**, 188–192.

58. S. K. Aggarwal, 'A histochemical approach to the mechanism of action of cisplatin and its analogues.' *J. Histochem. Cytochem.*, 1993, **41**, 1053–1073.

59. C. Cojocel, J. H. Smith, K. Maita *et al.*, 'Renal protein degradation: a biochemical target of specific nephrotoxicants.' *Fundam. Appl. Toxicol.*, 1983, **3**, 278–284.

60. M. E. I. Leibbrandt, G. H. I. Wolfgang, A. L. Metz et al., 'Critical subcellular targets of cisplatin and related platinum analogs in rat renal proximal tubule cells.' *Kidney Int.*, 1995, **48**, 761–770.

61. R. S. Goldstein and G. H. Mayor, 'Minireview: the nephrotoxicity of cisplatin.' *Life Sci.*, 1983, **32**, 685–690.

62. J. Hannemann and K. Baumann, ' Nephrotoxicity of cisplatin, carboplatin, and transplatin. A comparative *in vitro* study.' *Arch. Toxicol.*, 1990, **64**, 393–400.

63. M. E. I. Leibbrandt and G. H. I. Wolfgang, 'Differential toxicity of cisplatin, carboplatin, and CI-973 correlates with cellular platinum levels in rat renal cortical slices.' *Toxicol. Appl. Pharmacol.*, 1995, **132**, 245–252.

64. D. P. Gately and S. B. Howell, 'Cellular accumulation of the anticancer agent cisplatin: a review.' *Br. J. Cancer*, 1993, **67**, 1171–1176.

65. K. Sugihara, S. Nakano and M. Gemba, 'Effect of cisplatin on *in vitro* production of lipid peroxides in rat kidney cortex.' *Jpn. J. Pharmacol.*, 1987, **44**, 71–76.

66. J. G. Zhang and W. E. Lindup, 'Role of mitochondria in cisplatin-induced oxidative damage exhibited by rat renal cortical slices.' *Biochem. Pharmacol.*, 1993, **45**, 2215–2222.

67. H. R. Brady, M. L. Zeidel, B. C. Kone et al., 'Differential actions of cisplatin on renal proximal tubule and inner medullary collecting duct cells.' *J. Pharmacol. Exp. Ther.*, 1993, **265**, 1421–1428.

68. S. J. McGuiness and M. P. Ryan, 'Mechanism of cisplatin nephrotoxicity in rat renal proximal tubule suspensions.' *Toxicol. In Vitro*, 1994, **8**, 1203–1212.

69. J. A. Gordon and V. H. Gattone II, 'Mitochondrial alterations in cisplatin-induced acute renal failure.' *Am. J. Physiol.*, 1986, **250**, F991–F998.

70. H. R. Brady, B. C. Kone, M. E. Stromski et al., 'Mitochondrial injury: an early event in cisplatin toxicity to renal proximal tubules.' *Am. J. Physiol.*, 1990, **258**, F1181–F1187.

71. R. G. Fish, M. D. Shelley, J. Badman et al., 'Platinum accumulation in adult cancer patients receiving cisplatin, plasma and urinary platinum concentrations on first drug exposure.' *Drug Invest.*, 1994, **7**, 175–182.

72. F. Courjault, D. Leroy, I. Coquery et al., 'Platinum complex-induced dysfunction of cultured renal proximal tubule cells. A comparative study of carboplatin and transplatin with cisplatin.' *Arch. Toxicol.*, 1993, **67**, 338–346.

73. A. Eastman, 'The formation, isolation, and characterization of DNA adducts produced by anticancer platinum complexes.' *Pharmacol. Ther.*, 1987, **34**, 155–166.

74. J. M. Rosenberg and P. H. Sato, 'Cisplatin inhibits *in vitro* translation by preventing the formation of complete initiation complex.' *Mol. Pharmacol.*, 1993, **43**, 491–497.

75. K. Uchida, Y. Tanaka, T. Nishimura et al., 'Effect of serum on inhibition of DNA synthesis in leukemia cells by *cis*- and *trans*-[Pt(NH$_3$)$_2$Cl$_2$].' *Biochem. Biophys. Res. Commun.*, 1986, **138**, 631–637.

76. T. Yasumasu, T. Ueda, J. Uozumi et al., 'Comparative study of cisplatin and carboplatin on pharmacokinetics, nephrotoxicity, and effect on renal nuclear DNA synthesis in rats.' *Pharmacol. Toxicol.*, 1992, **70**, 143–147.

77. H. C. Harder and B. Rosenberg, 'Inhibitory effects of antitumor platinum compounds on DNA, RNA, and protein syntheses in mammalian cells *in vitro*.' *Int. J. Cancer*, 1970, **6**, 207–216.

78. W. Bernard, in 'Advances in Cytopharmacology, vol. I, First International Symposium on Cell Biology and Cytopharmacology,' eds. F. Clementi and B. Ceccarelli, Raven Press, New York, 1971, pp. 50–59.

79. D. Miller, D. M. A. Minahan, M. E. Friedman et al., 'Effects of *cis*- and *trans*-platinum complexes on RNA transcription.' *Chem. Biol. Interact.*, 1983, **44**, 311–316.

80. J. S. Mymryk, E. Zaniewski and T. K. Archer, 'Cisplatin inhibits chromatin remodeling, transcription factor binding, and transcription from the mouse mammary tumor virus promoter *in vivo*,' *Proc. Natl. Acad. Sci. USA*, 1995, **92**, 2076–2080.

81. J. Rosenberg and P. Sato, 'Messenger RNA loses the ability to direct *in vitro* peptide synthesis following incubation with cisplatin.' *Mol. Pharmacol.*, 1988, **33**, 611–616.

82. K. Tomiwa, C. Nolan and J. B. Cavanagh, 'The effects of cisplatin on rat spinal ganglia: a study by light and electron microscopy and by morphometry.' *Acta Neuropathol. (Berlin)*, 1986, **69**, 295–308.

83. J. A. Merski, I. Daskal and H. Busch, 'Effects of adriamycin on ultrastructure of nucleoli in the heart and liver cells of the rat.' *Cancer Res.*, 1976, **36**, 1580–1584.

84. G. Singh and J. Koropatnick, 'Differential toxicity of *cis* and *trans* isomers of dichlorodiammineplatinum.' *J. Biochem. Toxicol.*, 1988, **3**, 223–233.

85. T. J. Montine and R. F. Borch, 'Role of endogenous sulfur-containing nuelophiles in an *in vitro* model of *cis*-diamminedichloroplatinum(II)-induced nephrotoxicity.' *Biochem. Pharmacol.*, 1990, **39**, 1751–1757.

86. C. L. Litterst, F. Bertolero and J Uozumi, in 'Biochemical Mechanisms of Platinum Antitumor Drugs,' eds. D. C. H. McBrien and R. F. Slater, IRL Press, Oxford, 1986, pp. 171–198.

87. F. Zunino, G. Pratesi, A. Micheloni et al., 'Protective effect of reduced glutathione against cisplatin-induced renal and systemic toxicity and its influence on the therapeutic activity of the antitumor drug.' *Chem. Biol. Interact.*, 1989, **70**, 89–101.

88. J. Levi, C. Jacobs, M. McTigue et al., 'Effects of *cis*-diamminedichloroplatinum on renal sulfhydryl groups.' *Kidney Int.*, 1978, **14**, 728.

89. N. P. E. Vermeulen and G. S. Baldew, 'The role of lipid peroxidation in the nephrotoxicity of cisplatin.' *Biochem. Pharmacol.*, 1992, **44**, 1193–1199.

90. D. H. Stewart, N. Z. Mikhael, A. A. Nanji et al., 'Renal and hepatic concentrations of platinum: relationship to cisplatin time, dose, and nephrotoxicity.' *J. Clin. Oncol.*, 1985, **3**, 1251–1256.

91. P. Tothill, H. S. Klys, L. M. Matheson et al., 'The long-term retention of platinum in human tissues following the administration of cisplatin or carboplatin for cancer chemotherapy,' *Eur. J. Cancer*, 1992, **28A**, 1358–1361.

92. K. S. Blisard, D. A. Harrington, D. A. Long et al., 'Relative lack of toxicity of transplatin compared with cisplatin in rodents.' *J. Comp. Pathol.*, 1991, **105**, 367–375.

93. D. R. Gandara, E. A. Perez, V. Wiebe et al., 'Cisplatin chemoprotection and rescue: pharmacologic modulation of toxicity.' *Semin. Oncol.*, 1991, **18**, 49–55.

94. V. Pinzani, F. Bressolle, I. J. Haug et al., 'Cisplatin-induced renal toxicity and toxicty-modulating strategies: a review.' *Cancer Chemother. Pharmacol.*, 1994, **35**, 1–9.

95. D. J. Higby, H. J. Wallace Jr., J. F. Holland et al., '*cis*-Diamminedichloroplatinum (NSC-119875): a phase I study.' *Cancer Chemother. Rep.*, 1973, **57**, 459–463.

96. L. H. Einhorn and J. Donohue, '*cis*-Diamminedichloroplatinum, vinblastine, and bleomycin combination chemotherapy in disseminated testicular cancer.' *Ann. Intern. Med.*, 1977, **87**, 293–298.

97. D. M. Hayes, E. Cvitkovic, R. B. Golbey *et al.*, 'High-dose *cis*-platinum diamminedichloride: amelioration of renal toxicity by mannitol diuresis.' *Cancer*, 1977, **39**, 1372–1381.

98. C. Jacobs, J. R. Bertino, D. R. Goffinet *et al.*, '24-hour infusion of *cis*-platinum in head and neck cancers.' *Cancer*, 1978, **42**, 2135–2140.

99. P. Salem, M. Khalyl, K. Jabboury *et al.*, '*cis*-Diamminedichloroplatinum(II) by 5-days continuous infusion. A new dose schedule with minimal toxicity.' *Cancer*, 1984, **53**, 837–840.

100. M. Al-Sarraf, W. Fletcher, N. Oishi *et al.*, 'Cisplatin hydration with and without mannitol diuresis in refractory disseminated malignant melanoma: a southwest oncology group study.' *Cancer Treat. Rep.*, 1982, **66**, 31–35.

101. R. F. Ozols, B. J. Corden, J. Jacob *et al.*, 'High-dose cisplatin in hypertonic saline.' *Ann. Intern. Med.*, 1984, **100**, 19–24.

102. J. D. Bitran, R. K. Desser, A. A. Billings *et al.* 'Acute nephrotoxicity following *cis*-dichlorodiammine-platinum.' *Cancer*, 1982, **49**, 1784–1788.

103. R. J. Belt, K. J. Himmelstein, T. F. Patton *et al.*, 'Pharmacokinetics of non-protein-bound platinum species following administration of *cis*-diamminedichloroplatinum(II).' *Cancer Treat. Rep.*, 1979, **63**, 1515–1521.

104. M. Shea, J. A. Koziol and S. B. Howell, 'Kinetics of sodium thiosulfate, a cisplatin neutralizer.' *Clin. Pharmacol. Ther.*, 1989, **35**, 419–425.

105. O. R. Leeuwenkamp, W. J. F. Van der Vijgh, J. P. Neijt *et al.*, 'Reaction kinetics of cisplatin and its monoaquated species with the (potential) renal protecting agents (di)mensa and thiosulphate. Estimation of the effect of protecting agents on the plasma and peritoneal AUCs of CDDP.' *Cancer Chemother.*, 1990, **27**, 111–114.

106. C. E. Pfeifle, S. B. Howell, R. D. Felthouse *et al.*, 'High-dose cisplatin with sodium thiosulfate protection.' *J. Clin. Oncol.*, 1985, **3**, 237–244.

107. A. Hirosawa, H. Niitani, K. Hayashibara *et al.*, Effects of sodium thiosulfate in combination therapy of *cis*-dichlorodiammineplatinum and vindesine.' *Cancer Chemother. Pharmacol.*, 1989, **23**, 255–258.

108. M. Markman, R. D'Aquistino, N. Iannotti *et al.*, 'Phase I trial of high-dose intravenous cisplatin with simultaneous intravenous sodium thiosulfate.' *J. Cancer Res. Clin. Oncol.*, 1991, **117**, 151–155.

109. Y. Iwamoto, T. Kawano, M. Ishizawa *et al.*, 'Inactivation of *cis*-dichlorodiammineplatinum(II) in blood and protection of its toxicity by sodium thiosulfate in rabbits.' *Cancer Chemother. Pharmacol.*, 1985, **15**, 228–232.

110. R. Abe, T. Akiyoshi and T. Baba, 'Two-route chemotherapy using cisplatin and its neutralizing agent, sodium thiosulfate, for intraperitoneal cancer.' *Oncology*, 1990, **47**, 422–426.

111. T. B. Felder, K. Wasserman, R. Shah, *et al.*, 'Effect of diethyldithiocarbamate (DDTC) and sodium thiosulfate on the cytotoxicity and pharmacology of cisplatin.' *Proc. Am. Assoc. Cancer Res.*, 1987, **28**, 31.

112. D. L. Bodenner, P. C. Dedon, P. C. Keng *et al.*, 'Selective protection against *cis*-diamminedichloroplatinum(II)-induced toxicity in kidney, gut, and bone marrow by diethyldithiocarbamate.' *Cancer Res.*, 1986, **46**, 2751–2755.

113. R. F. Borch, J. C. Katz, P. H. Lieder *et al.*, 'Effect of diethyldithiocarbamate on tumor response to *cis*-platinum in a rat model.' *Proc. Natl. Acad. Sci. USA*, 1980, **77**, 5541–5444.

114. T. J. Montine and R. F. Borch, 'Quiescent LLC-PK1 cells as a model for *cis*-diamminedichlroplatinum(II) nephrotoxicity and modulation by thiol rescue agents.' *Cancer Res.*, 1988, **48**, 6017–6024.

115. R. Qazi, A. Y. C. Change, R. F. Borch *et al.*, 'Phase I clinical and pharmacokinetic study of diethyl-dithiocarbamate as a chemoprotector from toxic effects of cisplatin.' *J. Natl. Cancer Inst.*, 1988, **80**, 1486–1488.

116. J. M. Berry, C. Jacobs, B. Sikic *et al.*, 'Modification of cisplatin toxicity with diethyldithiocarbamate.' *J. Clin. Oncol.*, 1990, **8**, 1585–1590.

117. P. M. Calabro-Jones, J. A. Aguilera, J. F. Ward, *et al.*, 'Uptake of WR-2721 derivatives by cells in culture: identification of the transported form of the drug.' *Cancer Res.*, 1988, **48**, 3634–3640.

118. G. D. Smoluk, R. C. Fahey, P. M. Calabro-Jones *et al.*, 'Radioprotection of cells in culture by WR-2721 and derivatives: form of the drug responsible for protection.' *Cancer Res.*, 1988, **48**, 3641–3647.

119. L. M. Shaw, D. Glover, A. Turrisi *et al.*, 'Pharmacokinetics of WR-2721.' *Pharmacol. Ther.*, 1988, **39**, 195–201.

120. D. Glover, S. Grabelsky, K. Fox *et al.*, 'Clinical trials of WR-2721 and *cis*-platinum.' *Int. J. Radiat. Oncol. Biol. Phys.*, 1989, **16**, 1201–1204.

121. P. C. Dedon and R. F. Borch, 'Characterization of the reactions of platinum antitumor agents with biologic and nonbiologic sulfur-containing nucleophiles.' *Biochem. Pharmacol.*, 1987, **36**, 1955–1964.

122. S. Oriana, S. Böhm, G. B. Spatti *et al.* 'A preliminary clinical experience with reduced glutathione as Protector against cisplatin toxicity.' *Tumori*, 1987, **73**, 337–340.

123. M. E. Anderson, A. Naganuma and A. Meister, 'Protection against cisplatin toxicity by administration of glutathione ester.' *FASEB J.*, 1990, **4**, 3251–3255.

124. R. D. Mayer, K. E. Lee and A. T. Cockett, 'Improved use of buthionine sulfoximine to prevent cisplatin nephrotoxicity in rats.' *J. Cancer Res. Clin. Oncol.*, 1989, **115**, 418–422.

125. A. Naganuma, M. Satoh and N. Imura, 'Prevention of lethal and renal toxicity of *cis*-diamminedichloroplatinum(II) by induction of metallothionein synthesis without compromising its antitumor activity in mice.' *Cancer Res.*, 1987, **47**, 983–987.

126. P. J. Boogaard, A. Slikkerveer, J. F. Nagelkerke *et al.*, The role of metallothionein in the reduction of cisplatin-induced nephrotoxicity by Bi^{3+} pretreatment in the rat *in vivo* and *in vitro*. Are antioxidant properties of metallothionein more relevant than platinum binding?' *Biochem. Pharmacol.*, 1991, **41**, 369–375.

127. S. Umeki, M. Watanabe, S. Yagi *et al.*, 'Supplemental fosfomycin and/or steroids that reduce cisplatin-induced nephrotoxicity.' *Am. J. Med. Sci.*, 1988, **295**, 6–10.

128. S. Tognella, 'Pharmacological interventions to reduce platinum-induced toxicity.' *Cancer Treat. Rev.*, 1990, **17**, 139–142.

129. S. Umeki, K. Tsukiyama, N. Ukimoto *et al.*, 'Urinastatin (Kunitz-type proteinase inhibitor) reducing cisplatin nephrotoxicity.' *Am. J. Med. Sci.*, 1989, **298**, 221–226.

130. W. C. Rose and J. E. Schurig, 'Preclinical antitumor and toxicologic profile of carboplatin.' *Cancer Treat. Rev.*, 1985, **12** Suppl. A, 1–19.

131. W. J. F. van der Vijgh, 'Clinical pharmacokinetics of carboplatin.' *Clin. Pharmacokinet.*, 1991, **21**, 242–261.

132. Z. H. Siddik, M. Jones, F. E. Boxall *et al.*, 'Comparative distribution and excretion of carboplatin and cisplatin in mice.' *Cancer Chemother. Pharmacol.*, 1988, **21**, 19–24.

133. A. H. Calvert, S. J. Harland, D. R. Newell *et al.*, 'Early clinical Studies with *cis*-diammine-1,1-cyclobutane dicarboxylate platinum II.' *Cancer Chemother. Pharmacol.*, 1982, **9**, 140–147.

134. G. A. Curt, J. J. Grygiel, B. J. Corden *et al.*, 'A phase I and pharmacokinetic study of diamminecyclobutanedicarboxylatoplatinum (NSC 241240).' *Cancer Res.*, 1983, **43**, 4470–4473.

135. R. Canetta, K. Bragman, L. Smaldone *et al.*, 'Carboplatin: current status and future prospects.' *Cancer Treat. Rev.*, 1988, **15**, Suppl. B, 17–32.

136. T. C. Shea, M. Flaherty, A. Elias *et al.*, 'A phase I clinical and pharmacokinetic study of carboplatin and autologous bone marrow support.' *J. Clin. Oncol.*, 1989, **7**, 651–661.

137. J. B. Vermorken, W. J. F. van der Vijgh, I. Klein *et al.*, 'Pharmacokinetics of free and total platinum species after short-term infusion of cispaltin.' *Cancer Treat. Rep.*, 1984, **68**, 505–513.

138. F. Elferink, W. J. F. van der Vijgh, I. Klein *et al.*, 'Pharmacokinetics of carboplatin after i.v. administration.' *Cancer Treat. Rep.*, 1987, **71**, 1231–1237.

139. B. Hecquet, A. Caty, C. Fournier *et al.*, 'Comparison of platinum concentrations in human head and neck tumours following administration of carboplatin, iproplatin or cisplatin.' *Bull. Cancer (Paris)*, 1987, **74**, 433–436.

140. E. Boven, W. J. F. van der Vijgh, M. M. Nauta *et al.*, 'Comparative activity and distribution of five platinum analogues in nude mice bearing human ovarian carcinoma xenografts.' *Cancer Res.*, 1985, **45**, 86–90.

141. W. J. F. van der Vijgh, P. Lelieveld, I. Klein *et al.*, in 'Proceedings of the 13th International Congress of Chemotherapy,' eds. K. H. Spitzy and K. Karrer, H. Egermann, Vienna, 1983, pp. 57–59.

142. D. Nonclercq, G. Toubeau, G. Lauren *et al.*, 'Tissue injury and repair in the rat kidney after exposure to cisplatin and carboplatin.' *Exp. Mol. Pathol.*, 1989, **51**, 123–140.

143. G. H. I. Wolfgang, M. A. Dominick, K. M. Walsh *et al.*, 'Comparative nephrotoxicity of a novel platinum compound, cisplatin, and carboplatin in male Wistar rats.' *Fundam. Appl. Toxicol.*, 1994, **22**, 73–79.

144. J. S. Phelps, A. J. Gandolfi, K. Brendel *et al.*, 'Cisplatin nephrotoxicty: *in vitro* studies with precision-cut rabbit renal cortical slices.' *Toxicol. Appl. Pharmacol.*, 1987, **90**, 501–512.

145. Z. H. Siddik, S. E. Dible, F. E. Boxall *et al.*, in 'Biochemical Mechanisms of Platinum Antitumor Drugs,' eds. D. C. H. McBrien and R. F. Slater, IRL Press, Oxford, 1986, pp. 171–198.

146. R. B. Weiss and M. C. Christian, 'New cisplatin analogues in development. A review.' *Drugs*, 1993, **46**, 360–377.

147. A. M. Casazza, W. C. Rose, C. Comereski *et al.*, 'JM216: a novel orally active platinum complex.' *Proc. Am. Assoc. Cancer Res.*, 1992, **33**, 536.

148. L. R. Kelland, B. A. Murrer, G. Abel *et al.*, 'Ammine/amine platinum(IV) dicarboxylates: a novel class of platinum complex exhibiting selective cytotoxicity to intrinsically cisplatin-resistant human ovarian carcinoma cell lines.' *Cancer Res.*, 1992, **52**, 822–828.

149. M. J. McKeage, S. E. Morgan, R. E. Boxall *et al.*, 'Lack of nephrotoxicity of orally administered mixed ammine/amine platinum(IV) dicarboxylate complexes in rodents.' *Ann. Oncol.* (Suppl.), 1992, **3**, 111.

150. L. R. Kelland, C. F. J. Barnard, K. J. Mellish *et al.*, 'A novel *trans*-platinum coordination complex possessing *in vitro* and *in vivo* antitumor activity.' *Cancer Res.*, 1994, **54**, 5618–5622.

151. N. E. Sladek, 'Metabolism of oxazaphosphrines,' *Pharmacol. Ther.*, 1988, **37**, 301–355.

152. K. L. Dechant, R. N. Brogden, T. Pilkington *et al.*, 'Ifosfamide/mesna. A review of its antineoplastic activity, pharmacokinetic properties, and therapeutic efficacy in cancer.' *Drugs*, 1991, **42**, 428–467.

153. N. Brock, J. Pohl, J. Stekar *et al.*, 'Studies on the urotoxicity of oxazaphosphorine cytostatics and its prevention. III. Profile of action of sodium 2-mercaptoethane sulfonate (mesna).' *Eur. J. Cancer Clin. Oncol.*, 1982, **18**, 1377–1387.

154. N. Brock, P. Hilgard, J. Pohl *et al.*, 'Pharmacokinetics and mechanism of action of detoxifying low-molecular weight thiols.' *J. Cancer Res. Clin. Oncol.*, 1984, **108**, 87–97.

155. W. P. Brade, K. Hedrich, and M. Varini, 'Ifosfamide-pharmacology, safety and therapeutic potential.' *Cancer Treat. Rev.*, 1985, **12**, 1–47.

156. M. E. Scheulen, N. Niederle, K. Bremer *et al.*, 'Efficacy of ifosfamide in refractory malignant diseases and uroprotection by mesna: results of a clinical phase II study with 151 patients.' *Cancer Treat. Rev.*, 1983, **10** Suppl. A, 93–101.

157. P. Y. Holoye, B. S. Glisson, J. S. Lee *et al.*, 'Ifosfamide with mesna uroprotection in the management of lung cancer.' *Am. J. Clin. Oncol.*, 1990, **13**, 148–155.

158. M. Zalupski and L. H. Baker, 'Ifosfamide.' *J. Natl. Cancer Inst.*, 1988, **80**, 556–566.

159. R. Skinner, I. M. Sharkey, A. D. J. Pearson *et al.*, 'Ifosfamide, mesna, and nephrotoxicity in children.' *J. Clin. Oncol.*, 1993, **11**, 173–190.

160. R. Shore, M. Greenberg, D. Geary *et al.*, 'Iphosphamide-induced nephrotoxicity in children.' *Pediatr. Nephrol.*, 1992, **6**, 162–165.

161. A. Suarez, H. McDowell, P. Niaudet *et al.*, 'Long-term follow-up of ifosfamide renal toxicity in children treated for malignant mesenchymal tumors: an International Society of Pediatric Oncology Report.' *J. Clin. Oncol.*, 1991, **9**, 2177–2182.

162. C. D. Burk, I. Restaino, B. S. Kaplan *et al.*, 'Ifosfamide-induced renal tubular dysfunction and rickets in children with Wilms tumor.' *Pediatrics*, 1990, **117**, 331–335.

163. D. Heney, J. Wheeldon, P. Rushworth *et al.*, 'Progressive renal toxicity due to ifosfamide.' *Arch. Dis. Child.*, 1991, **66**, 966–970.

164. W. P. Patterson and A. Khojasteh, 'Ifosfamide-induced renal tubular defects.' *Cancer*, 1989, **63**, 649–651.

165. R. Skinner, A. D. J. Pearson, L. Price *et al.*, 'Nephrotoxicity after ifosfamide.' *Arch. Dis. Child.*, 1990, **65**, 732–738.

166. M. P. Goren, R. K. Wright, C. B. Pratt *et al.*, 'Potentiation of ifosfamide neurotoxicity, hematotoxicity, and tubular nephrotoxicity by prior *cis*-diamminedichloroplatinum(II) therapy.' *Cancer Res.*, 1987, **47**, 1457–1460.

167. R. Rossi, S. Danzebrink, D. Hillebrand *et al.*, 'Ifosfamide-induced subclinical nephrotoxicity and its potentiation by *cis*-platinum.' *Med. Pediatr. Oncol.*, 1994, **22**, 27–32.

168. R. M. Rossi, C. Kist, U. Wurster *et al.*, 'Estimation of ifosfamide/*cis*-platinum-induced renal toxicity by urinary protein analysis.' *Pediatr. Nephrol.*, 1994, **8**, 151–156.

169. M. Brandis, K. von der Hardt, R. B. Zimmerhackl *et al.*, 'Cytostatics-induced tubular toxicity.' *Clin. Invest.*, 1993, **71**, 855–857.

170. G. Sutton, 'Ifosfamide and mesna in epithelial ovarian carcinoma.' *Gynecol. Oncol.*, 1993, **51**, 104–108.

171. B. A. Kamen, E. Frenkel and O. M. Colvin, 'Ifosfamide: should the honeymoon be over?' *J. Clin. Oncol.*, 1995, **13**, 307–309.

172. D. Barnett, 'Preclinical toxicology of ifosfamide.' *Semin. Oncol.*, 1982, **9**, 8–13.

173. M. Mohrmann, S. Ansorge, U. Schmich *et al.*, 'Toxicity of ifosfamide, cyclophosphamide and their metabolites in renal tubular cells in culture.' *Pediatr. Nephrol.*, 1994, **8**, 157–163.

174. M. Mohrmann, A. Pauli, M. Ritzer *et al.*, 'Inhibition of sodium-dependent transport systems in LLC-PK$_1$ cells by metabolites of ifosfamide.' *Renal Physiol. Biochem.*, 1992, **15**, 289–301.

175. M. Mohrmann, A. Pauli, H. Walkenhorst *et al.*, 'Effect of ifosfamide metabolites on sodium-dependent phosphate transport in a model of proximal tubular cells (LLC-PK$_1$) in culture.' *Ren. Physiol. Biochem.*, 1993, **16**, 285–298.

176. M. J. Zamlauski-Tucker, M. E. Morris and J. E. Springate, 'Ifosfamide metabolite chloroacetaldehyde causes fanconi syndrome in the perfused rat kidney.' *Toxicol. Appl. Pharmacol.*, 1994, **129**, 170–175.

177. L. Sadoff, 'Nephrotoxicity of streptozotocin.' *Cancer Chemother. Rep.*, 1970, **54**, 457–459.

178. L. E. Broder and S. K. Carter, 'Pancreatic islet cell carcinoma. II. Results of therapy with streptozotocin in 52 patients.' *Ann. Intern Med.*, 1973, **79**, 108–118.

179. R. B. Weiss, 'Streptozocin: a review of its pharmacology, efficacy, and toxicity.' *Cancer Treat. Rep.*, 1982, **66**, 427–438.

180. P. S. Schein, M. J. O'Connell, J. Blom *et al.* 'Clinical antitumor activity and toxicity of streptozotocin (NSC-85998).' *Cancer*, 1974, **34**, 993–1000.

181. R. L. Myerowitz, G. P. Sartiano, and T. Cavallo, 'Nephrotoxic and cytoproliferative effects of streptozotocin: report of a patient with multiple hormone-secreting islet cell carcinoma.' *Cancer*, 1976, **38**, 1550–1555.

182. B. S. Levine, M. C. Henry, C. D. Port *et al.*, 'Toxicological evaluation of streptozotocin (NSC 85998) in mice, dogs, and monkeys.' *Drug Chem. Toxicol.*, 1980, **3**, 201–212.

183. R. B. Weiss and B. F. Issell, 'The nitrosoureas: carmustine (BCNU) and lomustine (CCNU).' *Cancer Treat. Rev.*, 1982, **9**, 313–330.

184. R. B. Weiss, J. G. Posada, Jr., R. A. Kramer *et al.*, 'Nephrotoxicity of semustine.' *Cancer Treat. Rep.*, 1983, **67**, 1105–1112.

185. R. G. Schacht, H. D. Feiner, G. R. Gallo *et al.*, 'Nephrotoxicity of nitrosoureas.' *Cancer*, 1981, **48**, 1328–1334.

186. V. T. Oliverio, 'Toxicology and pharmacology of the nitrosoureas.' *Cancer Chemother. Rep.*, 1973, **4**, 13–20.

187. U. Schaeppi, R. W. Fleischman, R. S. Phelan *et al.*, 'CCNU (NSC-79037): preclinical toxicologic evaluation of a single intravenous infusion in dogs and monkeys.' *Cancer Chemother. Rep.*, 1974, **5**, 53–64.

188. E. P. Denine, S. D. Harrison, Jr. and J. C. Peckham, 'Qualitative and quantitative toxicity of sublethal doses of methyl-CCNU in BDF$_1$ mice.' *Cancer Treat. Rep.*, 1977, **61**, 409–417.

189. R. A. Kramer and M. R. Boyd, 'Nephrotoxicity of 1-(2-chloroethyl)-3-(*trans*-4-methylcyclohexyl)-1-nitrosourea (MeCCNU) in the Fischer-344 rat.' *J. Pharmacol. Exp. Ther.*, 1983, **227**(2), 409–414.

190. R. A. Kramer, M. R. Boyd and J. H. Dees, 'Comparative nephrotoxicity of 1-(2-chloroethyl)-3-(*trans*-4-methylcyclohexyl)-1-nitrosourea (MeCCNU) and chlorozotocin: functional–structural correlations in the Fischer-344 rat.' *Toxicol. Appl. Pharmacol.*, 1986, **82**, 540–550.

191. J. Verweij, M. E. L. van der Burg and H. M. Pinedo, 'Mitomycin C-induced hemolytic uremic syndrome. Six case reports and a review of the literature on renal, pulmonary, and cardiac side effects of the drug.' *Radiother. Oncol.*, 1987, **8**, 33–41.

192. J. Verwey, J. de Vries and H. M. Pinedo, 'Mitomycin C-induced renal toxicity, a dose-dependent side effect?' *Eur. J. Cancer Clin. Oncol.*, 1987, **23**, 195–199.

193. J. B. Lesesne, N. Rothschild, B. Erickson *et al.*, 'Cancer-associated hemolytic-uremic syndrome: analysis of 85 cases from a national registry.' *J. Clin. Oncol.*, 1989, **7**, 781–789.

194. T. S. Ravikumar, R. Sibley, K. Reed *et al.*, 'Renal toxicity of mitomycin-C.' *Am. J. Clin. Oncol.*, 1984, **7**, 279–285.

195. R. W. Hamner, R. Verani and E. J. Weinman, 'Mitomycin-associated renal failure. Case report and review.' *Arch. Intern. Med.*, 1983, **143**, 803–807.

196. S. J. Rabadi, J. D. Khandekar and H. J. Miller, 'Mitomycin-induced hemolytic uremic syndrome: case presentation and review of literature.' *Cancer Treat. Rep.*, 1982, **66**, 1244–1247.

197. F. Ries, 'Nephrotoxicity of chemotherapy.' *Eur. J. Cancer Clin. Oncol.*, 1987, **24**, 951–953.

198. S. Jain and A. E. Seymour, 'Mitomycin C associated hemolytic uremic syndrome.' *Pathology*, 1987, **19**, 58–61.

199. S. T. Crooke and W. T. Bradner, 'Mitomycin C: a review,' *Cancer Treat. Rev.*, 1976, **3**, 121–139.

200. F. S. Philips, H. S. Schwartz and S. S. Sternberg, 'Pharmacology of mitomycin-C. I. Toxicity and pathologic effects.' *Cancer Res.*, 1960, **20**, 1354–1361.

201. M. Matsuyama, K. Suzumori and T. Nakamura, 'Biological studies of anticancer agents. I. Effects of prolonged intraperitoneal injections of mitomycin C on urinary organs.' *J. Urol.*, 1964, **92**, 618–620.

202. J. Verweij, S. Kerpel-Fronius, M. Stuurman *et al.*, 'Mitomycin C-induced organ toxicity in Wistar rats: a study with special focus on the kidney.' *J. Cancer Res. Clin. Oncol.*, 1988, **114**, 137–141.

203. C. Blanco, M. L. Sainz-Maza, F. Garijo *et al.*, 'Kidney cortical necrosis induced by mitomycin-C: a morphologic experimental study.' *Ren. Fail.*, 1992, **14**, 31–39.

204. B. J. Kennedy, 'Metabolic and toxic effects of mithramycin during tumor therapy.' *Am. J. Med.*, 1970, **49**, 494–503.

205. J. P. Fillastre, J. Maillot, M. A. Canonne *et al.*, 'Renal function and alterations in plasma electrolyte levels in normocalcaemic and hypercalcaemic patients with malignant diseases, given an intravenous infusion of mithramycin.' *Chemotherapy*, 1974, **20**, 280–295.

206. B. J. Kennedy, M. Sandberg-Wolheim, M. Loken *et al.*, 'Studies with tritiated mithramycin in C3H mice.' *Cancer Res.*, 1967, **27**, 1534–1538.

207. R. B. Gilsdorf, J. W. Yarbro and A. S. Leonard, 'Influence of mithramycin on RNA synthesis and renal function in the canine kidney—a trial study.' *J. Surg. Oncol.*, 1970, **2**, 89–96.

208. F. S. Philips, A. Gillodega, H. Marquardt *et al.*, Some observations on the toxicity of adriamycin (NSC-123127).' *Cancer Chemother. Rep.*, 1975, **6**, 177–181.

209. L. F. Fajardo, J. R. Eltringham, J. R. Stewart *et al.*, 'Adriamycin nephrotoxicity.' *Lab Invest.*, 1980, **43**, 242–253.

210. T. Bertani, A. Poggi, R. Pozzoni *et al.*, 'Adriamycin-induced nephrotic syndrome in rats: sequence of pathologic events.' *Lab Invest.*, 1982, **46**, 16–23.

211. J. Egido, A. Robles, A. Ortiz *et al.*, 'Role of platelet-activating factor in adriamycin-induced nephropathy in rats.' *Eur. J. Pharmacol.*, 1987, **138**, 119–123.

212. R. Bertelli, F. Ginevri, R. Gusmano *et al.*, 'Cytotoxic effect of adriamycin and agarose-coupled adriamycin on glomerular epithelial cells: role of free radicals.' *In Vitro Cell Dev. Biol.*, 1991, **27A**, 799–804.

213. S. S. Sternberg, F. S. Philips and A. P. Cronin, 'Renal tumors and other lesions in rats following a single intravenous injection of daunomycin.' *Cancer Res.*, 1972, **32**, 1029–1036.

214. S. S. Sternberg, 'Cross-striated fibrils and other ultrastructural alterations in glomeruli of rats with daunomycin nephrosis.' *Lab. Invest.*, 1970, **23**, 39–51.

215. D. D. Von Hoff, J. S. Penta, L. J. Helman *et al.*, 'Incidence of drug-related deaths secondary to high-dose methotrexate and citrovorum factor administration.' *Cancer Treat. Rep.*, 1977, **61**, 745–748.

216. S. W. Pitman, L. M. Parker, M. H. N. Tattersall *et al.*, 'Clinical trial of high-dose methotrexate (NSC-740) with citrovorum factor (NSC-3590)—toxicologic and therapeutic observations'. *Cancer Chemother. Rep.*, 1975, **6**, 43–49.

217. P. T. Condit, R. E. Chanes and W. Joel, 'Renal toxicity of methotrexate.' *Cancer*, 1969, **23**, 126–131.

218. S. A. Jacobs, R. G. Stoller, B. A. Chabner *et al.*, '7-Hydroxymethotrexate as a urinary metabolite in human subjects and rhesus monkeys receiving high-dose methotrexate.' *J. Clin. Invest.*, 1976, **57**, 534–538.

219. R. G. Stoller, K. R. Hande, S. A. Jacobs *et al.*, 'Use of plasma pharmacokinetics to predict and prevent methotrexate toxicity.' *N. Engl. J. Med.*, 1977, **297**, 630–634.

220. R. G. Stoller, H. G. Kaplan, F. J. Cummings *et al.*, 'A clinical and pharmacological study of high-dose methotrexate with minimal leucovorin rescue.' *Cancer Res.*, 1979, **39**, 908–912.

221. K. R. Hande, R. C. Donehower and B. A. Chabner, in 'Clinical Pharmacology of Antineoplastic Drugs,' ed. H. M. Pinedo, Elsevier, New York, 1978, pp. 97–114.

222. R. M. Fox, 'Methotrexate nephrotoxicity.' *Clin. Exp. Pharm. Physiol. Suppl.*, 1979, **5**, 43–45.

223. S. B. Howell and J. Carmody, 'Changes in glomerular filtration rate associated with high-dose methotrexate therapy in adults.' *Cancer Treat. Rep.*, 1977, **61**, 1389–1391.

224. H. T. Abelson, M. T. Fosburg, G. P. Beardsley *et al.*, 'Methotrexate-induced renal impairment: clinical studies and rescue from systemic toxicity with high-dose leucovorin and thymidine.' *J. Clin. Oncol.*, 1983, **1**, 208–216.

225. D. D. Von Hoff and M. Slavik, '5-Azacytidine—a new anticancer drug with significant activity in acute myelobalstic leukemia,' *Adv. Pharmacol. Chemother.*, 1977, **14**, 285–326.

226. B. A. Peterson, A. J. Collins, N. J. Vogelzang *et al.*, '5-Azacytidine and renal tubular dysfunction.' *Blood*, 1981, **57**, 182–185.

227. R. P. Warrell Jr., C. J. Coonley, D. J. Straus *et al.*, 'Treatment of patients with advanced malignant lymphoma using gallium nitrate administered as a seven-day continuous infusion.' *Cancer*, 1983, **51**, 1982–1987.

228. R. P. Warrell and R. S. Bockman, 'Gallium in the treatment of hypercalcemia and bone metastasis.' *Important Adv. Oncol.*, 1989, 205–220.

229. I. H. Krakoff, R. A. Newman and R. S. Goldberg, 'Clinical toxicologic and pharmacologic studies of gallium nitrate.' *Cancer*, 1979, **44**, 1722–1727.

230. T. E. Hughes and L. A. Hansen, 'Gallium nitrate.' *Pharmacother.*, 1992, **26**, 354–362.

231. T. G. Hall and R. A. Schaiff, 'Update on the medical treatment of hypercalcemia of malignancy.' *Clin. Pharm.*, 1993, **12**, 117–125.

232. R. A. Newman, A. R. Brody and I. H. Krakoff, 'Gallium nitrate (NSC-15200) induced toxicity in the rat: a pharmacologic, histopathologic, and micro-analytical investigation.' *Cancer*, 1979, **44**, 1728–1740.

233. A. Belldegnum, D. E. Webb, H. A. Austin III *et al.*, 'Effects of interleukin-2 on renal function in patients receiving immunotherapy for advanced cancer.' *Am. Intern. Med.*, 1987, **106**, 817–822.

234. B. H. Ault, F. B. Stapleton, L. Gaber *et al.*, 'Acute renal failure during therapy with recombinant human gamma interferon.' *N. Engl. J. Med.*, 1988, **319**, 1397–1400.

7.28
The Pathogenesis and Prevention of Radiocontrast-induced Renal Dysfunction

GEORGE L. BAKRIS

Rush University, Chicago, IL, USA

7.28.1 INTRODUCTION

At least 10% of all acute renal failure (ARF) cases observed in the hospital setting are caused by radiographic studies involving the use of various radiocontrast media (RCM).[1,2] Moreover, the third most common cause of hospital-acquired ARF is a consequence of RCM.[2]

Given this clinical problem, insight into the mechanisms of renal dysfunction as well as defining the population at risk is of paramount importance. This information can then guide physicians toward the development of a prophylactic strategy to prevent renal failure in high-risk patients requiring RCM studies.

Since the 1960s investigators have used various animal models to define the mechanism(s) that contribute to the development of RCM-induced renal dysfunction. In addition, clinical studies have attempted to outline possible prophylactic measures that ameliorate or prevent this problem. This chapter presents an overview of the history, risk factors, and mechanisms involved in renal dysfunction following RCM administration. It also focuses on both clinical and laboratory studies that compare the renal effect of hyper- and low-osmolar RCM. Lastly, a review of studies that examine strategies to prevent RCM-induced renal dysfunction are discussed.

7.28.2 HISTORY

Radiocontrast agents were first used for *in vivo* angiographic studies in the early 1920s.[3] Agents used during this period included strontium bromide, thorium dioxide, and sodium bromide. Although excellent RCM, both the strontium and thorium derivatives, unfortunately, were associated with an increased incidence of malignant diseases. Furthermore, these agents had long decay times and thus were associated with prolonged radioactivity many years after their use.[3,4] Organic diiodinated preparations were initially used during the early 1920s when Swick demonstrated that these compounds were reliable agents for intravenous (i.v.) urographic studies.[5] These hyperosmolar iodinated compounds produced the first case reports of renal dysfunction following their use in 1931.[6] Consequently, they were replaced in the mid-1950s with triiodinated compounds such as sodium diatrizoate; these triiodinated agents were found to be less toxic but more viscous than the diiodinated compounds.

The major chemical difference between the triiodinated RCM and those previously used was the presence of three atoms of iodine per molecule as opposed to one or two. As a result of increasing the number of iodine atoms per compound, these triiodinated RCM did not provide an ideal imaging substance. This was due to a lack of water solubility secondary to the high osmotic composition. It is this hyperosmolarity that is a primary cause of renal failure.[7–13] Moreover, the acute change in serum osmolality induced by hyperosmolar RCM results in a number of vascular and peripheral nerve alterations, including vasodilation, the sensation of heat and pain, as well as involuntary movements. These effects may be mediated by adenosine or other local mediators (see below).

Investigators in the early 1970s developed a lower (nonionic) osmotic and, hence, less toxic RCM. These agents are termed nonionic because an organic side chain has replaced the carboxyl group and, hence, these agents do not ionize in solution. They also differ from the ionic RCM in the number of osmotic particles per iodine atom (roughly 50% less than ionic agents), which account for the low osmolarity.[5,14]

Since the mid-1980s, attempts to develop nontoxic RCMs have led to the development of gadolinium-DTPA (Gd-DTPA) and carbon dioxide (DTPA, diethyltriaminepentaacetic acid).[15,16] Gd-DTPA is used primarily with magnetic resonance imaging technology to evaluate central nervous system problems.[15] Conversely, carbon dioxide is substituted for conventional RCM for evaluation of arterial beds.[16] The historic evolution of RCM is summarized in Table 1.

7.28.3 RENAL PHARMACOLOGY AND PHYSIOLOGY

The ionic and nonionic RCMs used in medical practice are the 2,4,6-tri-iodinated benzoic acid derivatives (Figure 1). The major

Table 1   The historic evolution of clinically used radiocontrast media.

Agent	Year first used
Sodium iodide	1918,[a] 1923
Stontium bromide	1923
Thorium dioxide	1923
Monoidodinated compounds Lopax and others	1929
Diiodinated compounds Diodrast, Skiodan, Diodon, and others	Early 1930s
Triiodinated compounds Sodium diatrizoate, diatrizoate meglumine, etc.	Mid-1950s
Low-osmolar (nonionic) compounds Metrizamide, iohexol, iopamidol and others	Late 1970s, Early 1980s
Carbon dioxide (CO_2)	1982
Gadolinium-DTPA/dimeglumine	1984

[a]Initially suggested for use as radiocontrast agent.

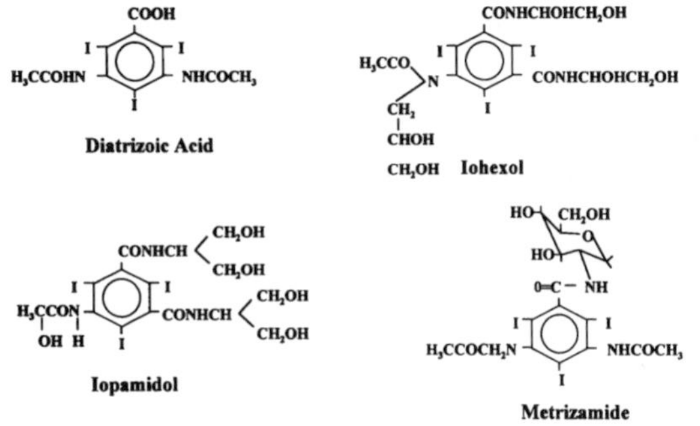

Figure 1 The chemical structures of both hyperosmolar and low-osmolar RCM commonly used in clinical practice.

difference between ionic and nonionic RCM lies in their divergent osmolarities and not in their iodine content or viscosity. A comparison of osmolarities among the RCMs is summarized in Table 2.

The most widely used hyperosmolar RCM are the diatrizoate derivatives, which are water soluble and have a low pK_a due to their carboxyl group.[17,18] They exist as anions in biological systems and, hence, are only in the extracellular space.[18] In addition, they are not significantly protein bound and almost entirely (99%) renally excreted; the remaining 1% of their excretion is accounted for by gastrointestinal and biliary losses, which become more pronounced in renal failure.[19,20]

Injection of RCM results in a rapid equilibration across capillary membranes with the exception of an intact blood–brain barrier; however, RCM can enter the cerebrospinal fluid through fenestrae in the choroid plexus.[21] During the first phase of distribution following RCM administration, there is a marked increase in intravascular osmolality which causes a rapid fluid shift across capillary membranes toward the hypertonic (intravascular) compartment.[19,21,22] Furthermore, the kinetics of RCM distribution depend primarily on the following factors: (i) organ blood flow, (ii) capillary density and permeability, and (iii) interstitial diffusion distances.[21–24]

Triiodinated compounds as well as DTPA are freely filtered by the glomerulus and neither secreted nor reabsorbed by the tubules.[24,25] Consequently, these agents are similar to inulin, a marker of glomerular filtration rate (GFR). Indeed, iohexol, a low osmolar RCM has been shown to provide an accurate measurement of GFR.[26] Thus, after bolus i.v. injection of various RCM, an equilibration of these agents occurs in the extracellular space and simultaneous excretion by glomerular filtration ensues in subjects with normal renal function.

The amount of RCM entering the tubular fluid is determined predominantly by two factors: plasma concentration and GFR.[22,24] In persons with normal renal function, the plasma half-life of RCM is between 30 min and 60 min.[22,24] Also, the peak time for excretion of RCM is about 3 min following i.v. injection. Peak urine iodine concentrations occur approximately 1 h after RCM administration.[27–29]

Urine flow rates increase following administration of meglumine salts but not with sodium salts. This results in higher urinary concentrations of the sodium-associated RCM as compared to the meglumine derivatives.[30] Although the initial concentration of RCM in the tubule is the same as the plasma the urinary concentration of RCM increases five- to 10-fold as a result of proximal tubular sodium and water reabsorption. This increased concentration may help explain, in part, enhanced toxicity to this portion of the renal tubule.

Table 2 Physical properties of commonly used radiocontrast media.

	Osmolarity (mOsm)	Iodine (%)	Viscosity 37° ($M_pA.S.$)
Ionic			
Iothalamate meglumine	1217	28	4.0
Sodium diatrizoatea	1470	30	2.5
Meglumine-sodium diatrizoate	1690	37	9.1
Iothalamate-sodium	1965	40	4.5
Nonionic			
Iotrol[a]	300	30	9.1
Metrizamide	450	28	5.0
Iopamidol[a]	570	28	3.8
Iohexol	620	28	4.8

Range of molecular weights for radiocontrast media mentioned above is 636 (diatrizoate) to 1626 (Iotrol). [a]Most commonly used radiocontrast agents.

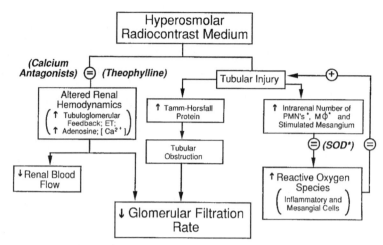

Figure 2 A proposed model that amalgamates the proposed mechanisms involved in RCM-induced nephrotoxicity. RCM both reduces renal blood flow and causes tubular injury. In the high-risk patient, for example, renal insufficiency with or without diabetes, the persistent vasoconstriction coupled with direct tubular injury leading to transient intratubular obstruction in a low-flow state combine to yield insurmountable insults to the kidney and, hence, development of renal failure. SOD, superoxide dismutase; PMN, polymorphonuclear leukocytes; Mf, macrophages; = inhibits; +potentiates; *SOD does not affect renal hemodynamic responses to RCM but preserves GFR.

Proximal tubular sodium reabsorption following RCM treatment is independent of hydration and is influenced by diuretics and osmotic load.[20,31] Furthermore, in states of dehydration, the concentration of RCM can be further increased in the collecting duct system secondary to the increased levels of antidiuretic hormone, which is increased in volume-depleted states. It is conceivable, therefore, that the concentration of RCM in the ultrafiltrate leaving the proximal tubule may be up to 50 or 100 times that entering the tubule.[20–24,27–31]

Individuals with normal renal function excrete all but 1% of the injected RCM through the kidneys. In patients with renal dysfunction between 25% and 36% of the RCM administered is eliminated by extrarenal routes, such as the gastrointestinal and biliary systems, prior to dialysis.[34,35] A number of studies demonstrate that the mean half-life of RCM is roughly doubled in dialysis patients when compared to normals.[32–36] In patients with end-stage renal disease RCM is removed by dialysis, since these agents are not protein bound and possess relatively low molecular weights. Last, the clearance of these agents by hemodialysis, at blood flow rates between 172 and 250 mL min^{-1}, varies from 65% to 80% following a 4 h treatment.

7.28.4 MECHANISMS OF NEPHROTOXICITY

Nephrotoxicity secondary to radiocontrast agents can be broadly grouped under two categories: renal hemodynamic effects (vasoconstriction) and cellular injury. A summary of factors implicated in the hemodynamic and tubular changes that follow RCM injection are

Table 3 Potential mediators of the hemodynamic and tubular effects of RCM-induced renal injury and their prevention.

Potential mediators	*Preventive*
Hemodynamic (vasoconstriction)	
Adenosine	Theophylline
Calcium	Calcium channel blockers
Endothelin	BQ123
Tubular injury	
Increased reactive oxygen species	SOD, H$_2$O$_2$, allopurinol
Increased cell metabolism	??

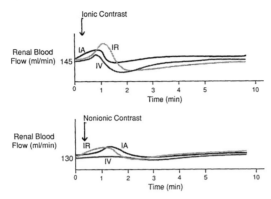

Figure 3 Representative effects of a hyperosmolar, sodium diatrizoate, and low-osmolar, iopamidol, RCM on renal blood flow in the normal dog. IA, intra-arterial (femoral artery); IV, intravenous; IR, intrarenal.

summarized in Table 3. An overall schematic representation integrating these factors and the resulting consequences of RCM administration is summarized in Figure 2. The principal factor that leads to both the hemodynamic and tubular changes following RCM injection relates to its osmotic state.[37,38,41,42]

The variability of renal vascular effects that follow injection of RCM depends upon the dosage and route of administration.[18,20,23] Figure 3 demonstrates the time course and variability of renal vascular responsiveness between an intrarenal, intraarterial (thoracic aorta), and i.v. bolus injection of an ionic and nonionic RCM in the normal dog. These observations on renal vascular effects of RCM are also supported by other investigators.[37–40] As can be noted, administration of RCM characteristically gives a biphasic renal blood flow (RBF) response.

In addition to the renal vascular changes altered peripheral arterial flows are also depen-

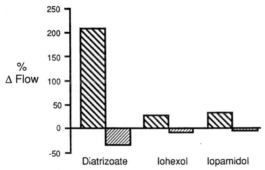

Figure 4 The percent change in blood flow in both the renal and femoral arteries in a euvolemic dog following either high- or low-osmolar RCM injection. Note: Results obtained from two separate dogs. Light bars, femoral artery; dark bars, renal artery.

dent on the osmotic concentration of the RCM utilized. The contrast between femoral and RBF responses following administration of either a high- or low-osmolar RCM are depicted in Figure 4. Clearly, less dramatic renal hemodynamic and peripheral arterial effects are seen with nonionic (low osmolar) RCM. The mechanism for these differential effects on blood flow relates to adenosine release (see below).

A number of studies have searched for possible mechanisms that account for the renal vasoconstrictor response that follows RCM administration.[18,43–52] These studies demonstrate that intraarterial injection of a hyperosmolar RCM or hypertonic saline result in a biphasic RBF response, which is characterized by a transient increase followed by a prolonged decrease in RBF. This decrease in RBF is approximately 30–40% from baseline and correlates positively with the osmolality of the solution injected.[53–55] These osmosis-dependent changes in RBF are shown in Figure 3.

There is uniform recovery from vasoconstriction that follows hyperosmolar RCM injection in normal dogs. Conversely, in three volume-depleted dogs with a 1&5/6 nephrectomy there is no recovery from this reduction in RBF even after 6 h of observation (Figure 5). This effect on RBF supports the clinical observation that volume-depleted subjects with renal insufficiency are at high risk of development of acute renal failure following RCM.

Hyperosmolar RCM also causes cellular injury to the renal tubule. *In vitro* studies have evaluated the effects of hyperosmotic RCM on proximal tubular cell integrity and metabolism.[56,57] These studies clearly demonstrate that hyperosmolar but not low-osmolar RCM cultured with proximal tubular cells kills them within a period of a few minutes.[56,57] Moreover, *in vitro* studies indicate that low-osmolar RCMs cause only minor changes in the cellular integrity of human mesangial cells.[58] Clinical evidence to confirm the findings of tubular injury come from both animal and human reports, which document the presence of granular and hyaline casts in the urine within 2 h following injection of a hyperosmolar RCM.[53,56] Moreover, renal biopsies from dogs given intrarenal injection of hyperosmolar RCM demonstrate the presence of polymorphonuclear leukocytes and macrophages in the interstitium of the kidney within 3 h of injection (see below).

The remaining discussion in this section focuses on specific factors implicated in both altered renal hemodynamics and tubular injury associated with RCM administration.

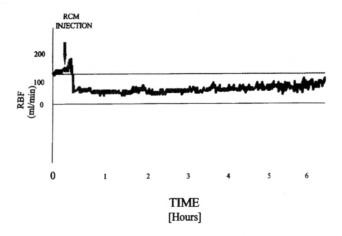

Figure 5   The renal blood flow response of a volume-depleted dog with a 1&5/6 nephrectomy following intrarenal injection of a hyperosmolar RCM, Renograffin.

7.28.4.1   Hemodynamic

7.28.4.1.1   Calcium

A number of studies have investigated the vasoconstrictor response following RCM administration by attempting to ablate this response utilizing inhibition of the renin–angiotensin system, α-adrenergic antagonists, vasodilators, as well as vasodilating prostaglandins, such as PGE_2.[43–52] None of these agents attenuated the vasoconstriction associated with RCM. The first evidence to demonstrate obliteration of this vasocontrictor response was provided in 1984 with the use of caclium channel blockers (CCBs).[46] Infusion of non-hypotensive doses of either two different CCBs or a calcium chelator into the renal artery of three different groups of euvolemic dogs markedly blunted the vasoconstrictor response following intrarenal injection of an ionic RCM.[46] These observations were later corroborated by a group of French investigators.[47] The clinical importance of this observation is demonstrated in two separate studies where dihydropyridine CCBs prevented the decline in GFR following RCM injection (see below).

In addition to blocking vasoconstriction, CCBs also inhibit renal autoregulatory mechanisms, that is, tubuloglomerular feedback (TGF) and afferent arteriolar "myogenic" responsiveness.[59–61] The delivery of a high osmotic load associated with RCM to the macula densa in the area of the distal tubule results in stimulation of the TGF processes, and hence, reduces GFR. Thus, CCBs would theoretically attenuate any decreases in GFR seen after RCM injection. No published studies are available that specifically examine this autoregulatory process following RCM administration. However, hypertonic saline and other hypertonic solutions have been examined with regard to this autoregulatory process; these studies provide support for the hypothesis that hyperosmolar RCM stimulate the TGF mechanism and, in turn, reduce GFR.[54] CCBs or calcium chelators inhibit this response and act to attenuate these renal hemodynamic effects. That all studies referenced in this section were performed in normal animal models should be noted. Hence, they cannot necessarily be extrapolated to different pathophysiological conditions, such as severe renal insufficiency or diabetes, where autoregulation is impaired.

7.28.4.1.2   Adenosine

Adenosine is a well-known vasodilator in the peripheral circulation; in the renal cortex it is a vasoconstrictor.[62–65] One of the mediators of its release is a hypertonic ultrafiltrate delivered to the macula densa in the area of the distal tubule as seen with hyperosmolar RCM. To test the hypothesis that the renal hemodynamic response to adenosine infusion is similar to the vasoconstrictor and recovery phase following RCM injection. Studies were done to evaluate whether adenosine is a mediator of the renal vasoconstrictor response following intrarenal hyperosmolar RCM injection. Normal dogs were treated with RCM followed by vehicle or the adenosine antagonist, theophylline, or the adenosine agonist, dipyridamole.[45] These agents did not alter baseline RBF or arterial pressure. The results clearly demonstrate that adenosine has a primary role in mediating the vasoconstrictor response associated with RCM administration (Figure 6). These results were

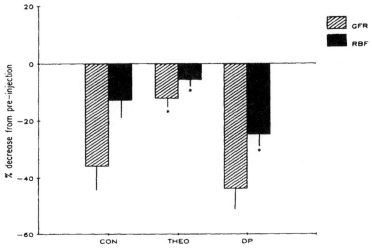

Figure 6 The renal hemodynamic changes following intrarenal injection of a hyperosmolar RCM, Renograffin, in the presence and absence of an adenosine agonist, dipyridamole, and an adenosine antagonist, theophylline (after Arend *et al.*[45]).

subsequently corroborated and have been reviewed.[47,66]

Adenosine also plays a critical role in the maintenance of renal autoregulatory responses, specifically TGF.[62,65] Specifically, inhibition of adenosine activity via an adenosine (A_1) receptor antagonist or its production via theophylline attenuates the TGF mechanism.[62,63] Thus, TGF-associated vasoconstriction that follows RCM injection may be mediated, in part, by an increase in renal adenosine activity.

7.28.4.1.3 Endothelin

Margulis *et al.* first demonstrated an increase in both plasma and urinary endothelin concentration following of RCM administration in the dog.[67] Unlike other autocoids, such as adenosine, whose release is dependent on RCM tonicity, endothelin release is not.[42,68] *In vivo* experiments in rats as well as endothelial cell culture, using similar iodine concentrations with different osmolalities, indicated that neither iodine nor osmotic concentration of the RCM alters endothelin release;[68] thus, the mechanism for increased endothelin concentrations following RCM is unclear. One exception, however, is the lack of endothelin release with ioversol.[42] This is postulated to result from a lack of endothelial cell stimulation; however, there are no published data to support this hypothesis. Additional studies in rats have shown that a continuous infusion of the ETA (endothelin A) receptor antagonist, BQ123, totally obliterates the renal vasoconstrictor effects of both low- and hyperosmolar RCM.[69]

These observations taken together with previous studies suggest that the vasoconstriction observed immediately after RCM administration is mediated by both adenosine and endothelin. It is unclear, however, which is the primary initiator of this response. It appears that both autocoids have important and perhaps additive effects in mediating the reduction in RBF and GFR. Yet from a practical standpoint, since CCBs block the effects of endothelin and adenosine inhibitors, such as theophylline, block the vasoconstrictor response to RCM, there is no clinical need to develop new compounds that affect only endothelin. Moreover, the available clinical data (discussed later in the chapter) supports the concept that theopylline, in both the presence and absence of a calcium antagonist, protects against RCM nephrotoxicity. There is no data supporting an additive protective role of theophylline and CCBs.

7.28.4.2 Tubular Injury

7.28.4.2.1 Reactive oxygen species

Reactive oxygen species (ROS) include the superoxide anion, hydrogen peroxide, hydroxyl radical, and singlet oxygen. These molecules are released from renal cells in response to various stimuli. A number of reviews describe the effects of these species on renal cells.[70,71] Even with very short half-lives, ROS are implicated in contributing to renal demise under various experimental conditions, for example, ischemic renal failure, acute glomerulonephritis, and toxic renal diseases.[70–72]

ROS are released by a variety of different cells including polymorphonuclear leukocytes, macrophages, and glomerular mesangial cells.[72,73] They help primarily by destroying bacteria and other invasive organisms in biological systems. Their production can be inhibited by glucocorticoids and specific scavengers such as superoxide dismutase (SOD), glutathione, and dimethyl sulfoxide.[72,74,75]

In vitro investigations utilizing electron spin resonance techniques demonstrate that exposure of human mesangial cells to an ionic RCM (diatrizoate sodium) produces an increase in ROS (Figure 7), including both superoxide and hydroxyl radical species. These substances can be suppressed with SOD and dimethyl sulfoxide, respectively.[71,72]

In normal, mildly volume-depleted dogs the oxygen free radical scavenger, SOD, partially blocked the fall in GFR following hyperosmolar RCM treatment, but had no effect on RBF.[48] In addition, renal biopsies, performed within 3 h of intrarenal RCM administration confirmed a large influx of polymorphonuclear leukocytes and macrophages in both the glomerular and tubular areas.[48] It is hypothesized that these cells, in addition to mesangial cells, release ROS to instigate the tubular injury initially induced by the hyperosmotic properties of RCM (Figure 2).

Additional studies support the concept that generation of ROS is dependent on the volume status of an animal. Yoshioka *et al.* found that the proximal tubular content of SOD was much lower in volume-depleted rats when compared with euvolemic animals.[76] Furthermore, the volume-depleted group showed the greatest declines in GFR following ionic radiocontrast administration. Collectively, these data support the notion that ROS may play an important role in the genesis of tubular injury following hyperosmolar RCM treatment. The potential

clinical benefit of free radical scavengers, however, has not been shown in any clinical trial.

7.28.4.2.2 *Markers of tubular injury (Tamm–Horsfall Protein)*

Tamm–Horsfall protein is a large glycoprotein normally found only in the ascending limb of the loop of Henle extending into the very early portion of the distal tubule and, thus, is a marker of tubular injury.[77] This protein is released following tubular destruction to this area of the nephron and forms the matrix of urinary casts.[78,79] The solubility of the Tamm–Horsfall protein in the urine depends on the pH, salt concentration, and concentration of proteins in the urine, as well as other factors.[77]

A number of studies demonstrate increases in urinary levels of Tamm–Horsfall protein following hyperosmolar RCM administration.[80–82] Given that RCM administration generates ROS by cells in the glomerulus, as well as those in the renal circulation, multiple experiments were undertaken to evaluate the role of ROS in contributing to increases in urinary Tamm–Horsfall protein following hyperosmolar RCM administration.[81] Studies conducted in volume-depleted dogs demonstrate that SOD clearly attenuates the increase in urinary Tamm–Horsfall protein and blunts declines in GFR following hyperosmolar RCM administration (Figure 8). In addition, these studies demonstrated fewer urinary casts following RCM injection in the kidney that received SOD.[81]

Preliminary studies in euvolemic dogs that received intrarenal injections of the nonionic, low-osmolar RCM, iopamidol, demonstrate a much smaller increase in urinary Tamm–

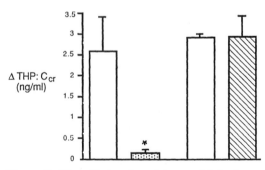

Figure 8 The effects of an intrarenal injection of the hyperosmolar RCM sodium diatrizoate, on urinary Tamm–Horsfall protein (THP). Data expressed as mean ± SEM and represent change in ration of Tamm–Horsfall protein to creatinine clearance (C_{cr}). *$P < 0.05$ compared with baseline. Open bars, baseline; dotted bars, SOD; light cross-hatch bars, heat-inactivated SOD.

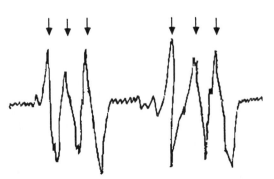

Figure 7 The electron spin resonance spectra following a hyperosmolar RCM, sodium diatrizoate, administered to human mesangial cells. Arrows indicate generation of hydroxyl radicals (after Baud *et al.*[73]).

Horsfall protein excretion as well as ROS generation when compared to hyperosmolar RCM (unpublished observations). Furthermore, there were relatively smaller declines in RBF and GFR with the low-osmolar agents. Taken together these studies support the concept that ROS contribute to the genesis of tubular injury. This is evidenced by increased levels of urinary Tamm–Horsfall protein after RCM injection. The clinical implications of these findings have not been explored.

7.28.4.2.3 *Renal metabolism*

Numerous studies have investigated the effects of high- and low-osmolar RCM on cellular enzyme activity, prostaglandin production, sodium-potassium-adenosine triphosphatase (Na-K-ATPase) activity and renal oxygen metabolism.[82–91] In general, these studies indicate that there is no reliable urinary enzymatic indicator of RCM injury with the possible exception of proximal tubular enzymes alanine aminopeptidase and γ-glutamyl transpeptidase.[83] These two enzymes were found in significantly higher concentrations in the urine of patients as early as 24 h following nonionic RCM studies. Interestingly, these enzymes were elevated in the absence of any significant increase in serum creatinine. Furthermore, this trend was present with both ionic (diatrizoate) and nonionic (iohexol) contrast agents. Unfortunately, these studies do not provide follow-up data (>72 h post-RCM administration) to assess the clinical predictability of these enzyme markers on the development of renal dysfunction.

Although Na-K-ATPase activity is clearly inhibited by hyperosmolar RCM, the mechanism for this inhibition is unclear.[84] Data (from our laboratory) suggests that ROS generation that follows RCM treatment may play a role in Na-K-ATPase inhibition. Two separate investigators documented depression of Na-K-ATPase activity in both the heart and ischemic kidney associated with increased oxygen free radical generation.[91,92] To eliminate the effects of renal vasoconstriction on Na-K-ATPase activity, *in vitro* studies were performed on canine renal slices and both Na-K-ATPase activity as well as mitochondrial ATPase activity were measured after exposure to either ionic (diatrizoate sodium) or nonionic (iohexol) RCM in the presence and absence of SOD. RCM treatment significantly reduced ouabain-sensitive Na-K-ATPase, but not mitochondrial ATPase in dog renal slices (Figure 9). Moreover, this reduction in Na-K-ATPase

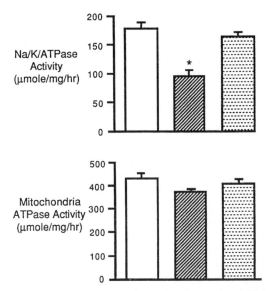

Figure 9 The effects of the hyperosmolar RCM, diatrizoate sodium, on Na-K-ATPase and mitochondrial ATPase activity in normal dog kidney. *$P < 0.05$ compared with baseline. Open bars, baseline; dark cross-hatched (middle) bars, contrast medium alone; dotted bars, contrast medium with superoxide dismutase.

activity by RCM was attenuated in the presence of SOD.

The observation that RCM decreases Na-K-ATPase activity was also noted in *in vivo* dog studies. Natriuresis following RCM administration (diatrizoate sodium) was lower in dogs pretreated with SOD when compared to controls.[81] Thus, the mechanism of Na-K-ATPase inhibition associated with RCM may be related to the ability of RCM to generate ROS. ROS would then cause either direct damage or inhibition of the Na-K-ATPase pump or inhibition of enzymatic activity secondary to tubular damage or necrosis.

The effects of RCM as well as other hypoxic factors on the medullary thick ascending limb of Henle's loop have also been evaluated.[51,52,88,89] In a unilaterally nephrectomized rat model that was treated with indomethacin and given a hyperosmolar RCM (iothalamate), a significant fall was noted in creatinine clearance after 24 h.[51] In addition, histologic studies in this model demonstrate that the loop of Henle had the greatest amount of histologic injury. Consequently, these and other investigators have speculated that the loop of Henle, because of its very poor oxygen supply but high oxygen demand, is the most susceptible area of the nephron to hypoxic injury following ionic RCM injection.[51,52,88]

Last, numerous studies have evaluated the modulatory role of prostaglandins of the E series (PGEs) on renal hemodymanics following RCM administration.[50,86,90] Prostaglandin E_2 has been shown to prevent hypoxic injury to the medullary area of the kidney by inhibiting oxygen consumption.[89] Further studies in normal dogs, however, demonstrate that prostaglandin inhibition does not result in RCM-induced renal dysfunction regardless of the osmotic content of the agent used.[86,90] Conversely, prostaglandin inhibition in rabbits with a decreased renal mass clearly potentiates the vasoconstrictor response that follows RCM injection.[50] These disparate effects are presumably due to the modulatory role of PGEs themselves. PGEs are important primarily in pathophysiologic states, such as heart failure, diabetes or renal insufficiency, when vasoconstrictors, such as endothelin, are increased. Moreover, there are species differences in the PGE response to vasoconstrictors. Thus, studies evaluating the effects of various RCM on renal physiology need to be performed in pathophysiologically relevant models, such as the volume-depleted dog with a 1&5/6 nephrectomy, to attain clinically meaningful data.

7.28.5   MODELS OF CONTRAST NEPHROTOXICITY

The renal hemodynamic effects of RCM must emulate the clinical or pathophysiological setting of RCM-induced renal failure to be meaningful. Evaluation of the renal hemodynamic effects of RCM have been largely performed in normal dogs, rabbits, or rats. Therefore, any assertions from these studies to renal hemodynamic alterations in pathophysiological states (i.e., congestive heart failure, diabetic nephropathy, cirrhosis) are, at best, speculative.

To date, only one clinically relevant animal model has been developed to investigate the renal hemodynamic effects of radiocontrast agents. This dog model was also used to study possible prophylactic measures that would prevent detrimental renal hemodynamic changes following RCM administration. Margulies and co-workers evaluated the renal hemodynamic effects of the RCM Vascoray in dogs with congestive heart failure (CHF).[49] CHF was induced after 8 d of pacing the dogs at a ventricular rate of 250 beats per minute. Radiocontrast agents were given as an i.v. infusion in the presence and absence of an infusion of atrial natriuretic factor. This study demonstrates that an i.v. infusion of RCM in the setting of heart failure results in a significant and more persistent decline in GFR than under normal circumstances. These investigators noted that the decline in renal function occurred only in the group of dogs with CHF and that this decline was attenuated by atrial natriuretic factor infusion. This is the only published pathophysiological model in the dog that allows for study of altered renal hemodynamics following RCM administration. Additional studies in man by the same investigators also noted that intraarterial infusion of atrial natriureti peptide (ANP) attenuated declines in renal hemodynamics that follow arteriography.[93]

Another model of contrast nephrotoxicity is described in the rabbit by Vari and co-workers utilizing a 1&5/6 nephrectomized rabbit model that has received large doses of nonsteroidal antiinflammatory agents followed by large bolus injections of hyperosmolar RCM.[50] This model has potential usefulness in the investigation of renal hemodynamics; however, no therapeutic interventions were performed to assess "protection" from RCM-induced renal dysfunction. Such a model, however, does focus directly on the pathophysiology of end-stage renal disease and the possible protective physiological mechanisms involved in maintaining renal hemodynamics following RCM injection. As such, this study provides an insight into the mechanisms of RCM-induced nephrotoxicity in the presence of renal disease. Unfortunately, patients with significant preexisting renal dysfunction are at much higher risk for exacerbation of this problem following RCM administration. Thus, this model would be very appropriate to study therapeutic interventions.

Unfortunately, the rat is particularly refractory to renal hemodynamic changes following large doses of ionic RCM. Furthermore, some studies report that data derived from this model are difficult, if not impossible, to reproduce.[50] Hence, with the exception of the 1&5/6 nephrectomy or the CHF model in the dog, and possibly the renal failure model in the rabbit, there are no meaningful pathophysiological animal models by which to study possible therapeutic interventions and their effects on renal function.

7.28.6   RISK FACTORS

It is difficult to arrive at an actual incidence of RCM-associated renal dysfunction or failure, largely due to varying definitions among the various studies. A number of studies report the incidence of renal failure following either venous or angiographic procedures to range from 0%, in a random population, to 92% in a high-risk population, that is, patients with diabetes or heart failure with renal insufficiency.[1,2,6,7,11,13,94–139] A review of these studies illustrates the importance of uniform definitions and clear identification of patient populations in order to meaningfully compare one study with another.

In a classic study by Byrd and Sherman, a number of clinical risk factors are outlined to define patients who are at risk of developing RCM-induced nephrotoxicity.[10] From this study, subjects with advanced age, prior renal insufficiency (serum creatinine $>1.6\,mg\,dL^{-1}$), dehydration, and hyperuricemia all had a greater than 50% risk for developing contrast nephrotoxicity. Those who received multiple-contrast exposure within 24 h or had diabetes mellitus had a 33% or 37% greater risk of developing nephrotoxicity, respectively. Since the mid-1980s, a number of well-done prospective and retrospective studies have evaluated some of these risk factors in patients undergoing arteriographic studies.[13,95–139] In addition, some studies have compared the possible differences in renal effects following hyperosmolar vs. low-osmolar RCM administration in patients with these and other risk factors.[13,106–119,122]

A study by Parfrey *et al.* documents that the population at highest risk for developing RCM-induced nephrotoxicity following angiographic study were diabetic patients with renal insufficiency.[106] These investigators found that given equal hydration status, the presence of preexisting renal insufficiency alone resulted in a greater than fourfold higher risk for the development of acute renal failure when compared to the normal population. Interestingly, this was not true for diabetes alone. A large multicenter trial, The Iohexol Cooperative Study confirmed these observations.[140] This study of 1196 patients who underwent coronary angiography demonstrated that preexisting evidence of renal insufficiency (serum creatinine $>1.6\,mg\,dL^{-1}$) was by far the most powerful risk factor for predicting renal failure following RCM injection. However, the presence of diabetes in subjects with renal insufficiency added to the risk profile. Moreover, this study clearly demonstrated that use of a low-osmolar RCM instead of conventional RCM reduces the incidence of renal impairment in this highest risk group by more than 50%, that is, the diabetic with renal impairment. Thus, these and other smaller studies clearly support the concept that the preexisting level of renal function is the strongest risk factor for predicting renal dysfunction following RCM treatment.

A review of several prospective angiographic studies demonstrates that the probability of developing renal impairment following an RCM study is a function of the preexisting level of renal function (Figure 10). Moreover, if concomitant conditions, such as heart failure,

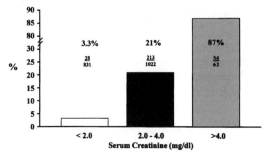

Figure 10 The percent of people with preexisting renal insufficiency that developed renal failure following an angiographic study with hyperosmolar RCM. Percentages are stratified by level of renal dysfunction.[13,94,140,142]

diabetes, cirrhosis, or volume depletion are present, the risk of renal dysfunction is even greater.[140] For these groups of patients low osmolar RCM are recommended.[13]

While management of patients that develop renal failure from RCM is clearly supportive and may require a temporary need for dialysis, no study has addressed the issue of the need for dialysis following a radiographic procedure in a "dialysis patient." It is generally accepted clinical practice to dialyze a patient with end-stage renal disease (ESRD) immediately following a RCM-requiring procedure to eliminate the hyperosmolar RCM. Note that about one-third of any given RCM is metabolized and eliminated prior to initiation of dialysis. Therefore, based on case reports and clinical experience, there is no need for dialysis following a procedure unless the patient has one of the conditions listed in Table 4. Since the low-osmolar RCM would be predicted to cause less

Table 4 Pre-existing conditions that require dialysis in ESRD patients following imaging procedures with radiocontrast medium.

Volume overload secondary to:
  (i)   Congestive heart failure
  (ii) Severe hypoalbuminemia (serum albumin $<2.0\,g\,d^{-1}$)
  (iii) Aggressive pretest hydration

Stages III or IV hypertension (not reducible with medication)[a]

Greater than 300 mL of hyperosmolar radiocontrast medium used[b]

[a]Most people are hypertensive secondary to volume overload. In the absence of hypoalbuminemia or severe renal dysfunction, they will respond to antihypertensive medication. [b]Note this value was extrapolated from case reports where clinical problems occurred with this amount of RCM used. This value applies to people with body weights 55–85 kg. Obviously if subjects weight more they can tolerate more RCM. The converse is also true.

fluid shifts and side effects, they are preferred in patients with ESRD. However, until a prospective, randomized study is performed that assesses the validity and cost effectiveness of hemodialysis following a RCM requiring procedure, this issue will remain unresolved.

7.28.7   ATHEROEMBOLIC RENAL DISEASE

One of the complications of a RCM treatment is the development of atheroembolic-induced renal failure. Patients with the highest probability of developing this problem are those with at least one of the following pre-existing conditions: (i) smoking, (ii) peripheral vascular disease, (iii) age >50, and (iv) poorly controlled hypertension with subsequent progressive renal disease.[142-147] Roughly two-thirds of the reported cases occur following a stimulus that triggers showers of microemboli from a damaged aorta.[148,149]

The pathogenesis of atheroembolic renal disease is attributed to occlusion of small arteries by fatty-type material from eroded plaques in the diseased aorta. These plaques then travel into the renal artery to the arcuate and interlobular arterioles. At this anatomic point these plaques occlude the vessel and result in ischemia to the glomerular tuft. Over time, this ischemia results in scarring and nephron loss.

Clinically, skin mottling ("livedo reticularis") of the lower extremities or "blue toes" is a common feature of this process.[142-147] However, variable presentations including abdominal pain, gastrointestinal bleeding, or pancreatitis may also occur.[142,147] Laboratory clues to the diagnosis include a transient eosinophilia, hypo-complementemia, and increased sedimentation rate.[143,147,150]

The clinical distinction between RCM-induced nephrotoxicity and atheroembolic renal disease lies in the time course. RCM-induced nephrotoxicity is noted usually within hours, while atheroembolic renal disease has a time course of days to weeks after the procedure.[147] In order to minimize the risk of atheroemboli in high-risk subjects some authors suggest a brachial rather than ileofemoral approach to cardiac catheterizations.[151] In this way the aorta, the repository for atheromatous plaques, is avoided.

7.28.8   PROPHYLAXIS AGAINST NEPHROTOXICITY

To achieve true prophylaxis against RCM induced nephrotoxicity, the treatment needs to precede the injection of RCM by a reasonable time period. Numerous clinical studies have evaluated various therapeutic interventions shown to prevent declines in renal function in animal models following RCM. Pharmacologic interventions include: (i) loop diuretics at various time periods throughout the contrast study; (ii osmotic diuretics, such as mannitol; (iii) CCBs prior to RCM administration; and finally (iv) theophylline prior to a contrast procedure to block adenosine. Other interventions evaluated include ANP and dopamine infusions. These will not be discussed as the number of total subjects studied is too few and results are conflicting.[152] A summary of all prophylactic interventions shown in clinical studies to reduce the incidence of RCM-induced renal impairment are noted in Table 5.

7.28.8.1   Hydration

Dehydration is known to magnify the risk of RCM-induced renal dysfunction in patients who already have other risk factors for this problem.[153] Thus, pretreatment with normal saline should always be considered, since it reduces the risk of nephrotoxicity by RCM. Three prospective studies in people with either diabetes or mild renal insufficiency failed to support the concept that hydration reduces risk of RCM nephrotoxicity.[95,96,100] Conversely, in another prospective study saline infusion (0.9%) did prevent renal failure following angiography when administered at a rate of $550 \, \text{mL} \, \text{h}^{-1}$ during the procedure.[154] Strangely, these same investigators had reported one year earlier that dehydration was not a risk factor for RCM-induced renal dysfunction. A closer scrutiny of their previous study demonstrates numerous methodologic flaws,

Table 5   Summary of recommended interventions to reduce the incidence of RCM-induced renal dysfunction.[a]

Intervention	Refs.
Hydration	10,153–155,156
Low-osmolar RCM	13,107–111,141,157
Calcium channel blockers	158,159
Theophylline	160,161
Dose reduction	94,110,156,157,162

[a]Note also that use of minimal contrast volume, avoidance of repeated contrast injections, and discontinuance of prostaglandin antagonists (NSAIDs), all have a role in minimizing exacerbation of pre-existing renal dysfunction following RCM injections. Furosemide is not of benefit. Dopamine and mannitol have a questionable prophylactic role and are not recommended for routine use.

including a lack of aggressive rehydration. A subsequent prospective randomized clinical study by Soloman and colleagues further supports the concept that adequate hydration prior to a contrast procedure minimizes renal dysfunction.[154] Moreover, neither furosemide nor mannitol augmented protection against RCM-induced nephrotoxicity over saline alone.[155] Thus, dehydration will exacerbate renal dysfunction especially if other risk factors are present.

7.28.8.2 Mannitol

Some studies have evaluated the effects of mannitol at various time periods before, during, and after angiographic studies.[126,130,163,164] Mannitol prevents ischemic renal failure in the dog and maintains GFR during controlled renal hypoperfusion studies in rats.[163,165,166] In a study by Old *et al.*, an infusion of 500 mL of 5% mannitol prevented RCM-induced declines in renal function in six patients with renal insufficiency (mean baseline serum creatinine = $2.5 \, mg \, dL^{-1}$). A comparable group of five patients who did not receive mannitol experienced a significant increase in serum creatinine from $2.3 \, mg \, dL^{-1}$ to $3.7 \, mg \, dL^{-1}$ following exposure to hyperosmolar RCM.[164] In another study, Anto *et al.* studied 37 patients, all with chronic renal insufficiency, who were given 1500 mL of one-half normal saline plus 250 mL of 20% mannitol within one hour of the procedure.[126] Of these patients 21% exhibited increases in serum creatinine. The authors conclude that mannitol offers some protection against RCM-induced renal dysfunction. This conclusion is based on the fact that a similar group of 40 patients, who did not receive this therapy, experienced a 70% incidence of renal failure. Other studies by Shafi *et al.* demonstrate similar findings.[130] These studies must be regarded as preliminary as they have small sample sizes and are published only as abstracts.

Problems may develop with the use of mannitol in this clinical setting if not closely monitored. Mannitol is generally given prior to a radiographic procedure; however, if the procedure is delayed and mannitol infusion not stopped, a diuresis would ensue prior to angiographic study and the patient could conceivably be relatively dehydrated at the start of the procedure. Hence, the patient's risk for nephrotoxicity may potentially be increased secondary to dehydration. This information, taken together with the study by Soloman *et al.*, suggests that mannitol should not be a first-line agent for prophylaxis against RCM-induced renal dysfunction.

The questionable renal preserving effect of mannitol may be related to systemic increases in ANP. A study in patients with coronary artery disease documents significant increases in plasma ANP levels during mannitol infusion.[167] This increase in ANP correlated with an increase in RBF. Unfortunately, an inhibitor of ANP was not given to fully assess this possible mechanism on renal function.

7.28.8.3 Diuretics

A number of studies have used i.v. furosemide alone and in combination with mannitol for prophylaxis against RCM-induced nephrotoxicity. Porush *et al.* studied 17 patients with an average serum creatinine of $3.8 \, mg \, dL^{-1}$ who underwent i.v. RCM studies and received i.v. furosemide (dose equal to 4000 divided by the creatinine clearance) one hour prior to the procedure.[163] There were significantly fewer subjects that developed a decrease in renal function in the furosemide group than would have been predicted following RCM treatment. However, other studies suggest that furosemide may actually increase the risk of RCM nephrotoxicity. Soloman *et al.* demonstrated in a group of high-risk subjects, that furosemide increased the probability of RCM induced renal dysfunction when compared to hydration alone.[155] Moreover, Weinstein and colleagues showed similar increases in risk of RCM-induced renal failure when patients were pretreated with furosemide.[169] Consequently, furosemide should not be given to patients undergoing a procedure that requires use of hyperosmolar RCM since it may increase their risk of renal dysfunction.

7.28.8.4 Calcium Channel Blockers

Two separate clinical studies evaluated the prophylatic effects of CCBs on RCM-induced decreases in renal function.[158,159] These studies are based on the initial observation made in 1984 that CCBs obliterate decreases in RBF and GFR following intrarenal RCM administration.[46] Russo *et al.* investigated the possible prophylactic effects of CCBs following high- vs. low-osmotic RCM in 30 male patients undergoing i.v. pyelography for hematuria and proteinuria.[158] RCM was infused and the calcium antagonist nifedipine was given only to the group that received a hyperosmolar RCM. The results demonstrate that 10 mg of sublingual nifedipine 5 min before injection significantly increased GFR and RBF above baseline as early as 1 h after the procedure.

These increases in renal hemodynamics were statistically significant compared to subjects who did not receive nifedipine. Unfortunately, this study only measured renal hemodynamics for 2 h following the procedure.

Neumayer *et al.*, in a prospective, randomized, double-blind clinical trial also demonstrated a prophylactic effect of the calcium antagonist, nitrendipine, against development of RCM induced renal dysfunction.[159] These investigators studied the renal hemodynamic and tubular enzyme effects of the low-osmolar RCM, iopamidol, in 35 patients undergoing renal arteriography, in the presence and absence of the nitrendipine. A clear blunting in GFR decline was noted in the nitrendipine group, with a return to baseline GFR after 48 h, whereas a persistent 20% decline in GFR was observed in the 48 h following RCM administration alone. Interestingly, urinary protein excretion also increased in the group receiving RCM alone, but not in the nitrendipine group. However, the baseline levels of urinary protein excretion in the nitrendipine group were roughly 2.5 times that of the control group. This is in keeping with previous studies documenting that the dihydropyridine compounds increase proteinuria in various pathophysiological states, such as diabetes.[169] Thus, CCBs may have an important role in attenuating declines in renal hemodynamics following either high- or low-osmolar RCM.

7.28.8.5   Adenosine (A$_1$) Receptor Antagonists

Two clinical studies have evaluated the effects of the adenosine antagonist, theophylline, for prophylactic use against RCM-induced-nephrotoxicity.[160,161] In one randomized, prospective double-blind study, subjects with GFRs below 75 mL min^{-1} were given either theophylline or placebo 45 min before injection with a low-osmolar RCM.[160] While the number of participants was small ($N = 35$), the theophylline group had no significant reduction in renal hemodynamics compared to a 22% decrease in GFR with low-osmolar RCM alone. A second prospective randomized clinical study of 93 patients with coronary artery disease and serum creatinine of <2.0 mg dL^{-1} produced similar results with theophylline.[161] Specifically, theophylline, given at a dose of 2.9 mg kg^{-1} every 12 h for four doses prior to the RCM study, prevented renal failure in a high-risk group that received low-osmolar RCM and were adequately hydrated. These studies coupled with animal studies support the concept that adenosine inhibition prevents RCM induced nephrotoxicity.

7.28.8.6   Dose Adjustment

The dose of RCM is important to monitor in order to prevent renal toxicity. Patients with severe renal dysfunction should have an adjustment in the dose of RCM administered. Cigarroa and co-workers provide guidelines for dosing with RCM to reduce the risk of RCM-induced nephrotoxicity in patients with renal disease.[162] These investigators demonstrate that appropriate dose reductions of RCM can prevent significant increases in RCM-induced renal dysfunction. The formula suggested is as follows:

$$\text{Radiocontrast medium limit} = \frac{5\,\text{mL contrast kg}^{-1}\,(\text{max} = 300\,\text{mL})}{\text{serum creatinine (mg dL}^{-1})}$$

The type of RCM (hyper- or low-osmolar) used is also of importance in preventing nephrotoxicity. A recent meta-analysis of 31 clinical trials clearly demonstrates that low-osmolar agents are the drugs of choice for any high-risk patient.[13] Moreover, the majority of prospective studies that examine the cardiac effects of low-osmolar RCM suggest that they are clearly beneficial. This benefit was defined by fewer arrhythmias, less hypotension, and less interaction with CCBs when compared to hyperosmolar agents.[141,170–174]

As a result of these metaanalyses and clinical trials, the question as to whether everyone should receive low-osmolar RCM has been posed. An early evaluation of the cost-benefit ratio between low- and hyperosmolar agents did not support the use of low-osmolar RCM in the population at high-risk for renal dysfunction.[175] However, a safety and cost effectiveness study determined that low-osmolar agents have far fewer adverse reactions and would potentially reduce the costs associated with morbidity in a high-risk population.[157] Moreover, these authors concluded that hyperosmolar RCM should be used in normal individuals since they did not have a significantly higher morbidity associated with this cheaper RCM.

7.28.9   SUMMARY

Our understanding of the mechanisms and pathophysiology of RCM-induced renal dysfunction has only minimally expanded since the mid-1980s.[156] However, large clinical trials

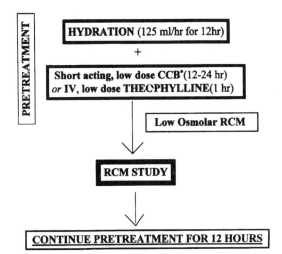

PRETREATMENT

HYDRATION (125 ml/hr for 12hr)

+

Short acting, low dose CCB*(12-24 hr)
or IV, low dose THEOPHYLLINE(1 hr)

Low Osmolar RCM

RCM STUDY

CONTINUE PRETREATMENT FOR 12 HOURS

Figure 11 A general strategy to prevent RCM-induced nephrotoxicity in high-risk patients. Short-acting CCBs are preferred since it is possible to reach a blood level quickly. Antihypertensive doses are not required. 120 mg of either diltiazem or verapamil in divided doses is suggested. Long-acting agents, such as amlodipine, may be used in heart failure, but the drug should be started at least 3 d in advance of the procedure. The theophylline dose needed for prophylaxis is at subtherapeutic blood levels for asthma.[161] Hydration with normal saline should be given to all those without heart failure at the rate recommended.

have now clearly demonstrated that patients with preexisting renal insufficiency and diabetes, heart failure, or other conditions affecting renal hemodynamics are at very high risk for RCM-induced renal dysfunction. These studies also demonstrate that pretreatment with either theophylline, CCBs, and use of low-osmolar RCM reduce the risk of RCM-induced nephrotoxicity in this high-risk group. Moreover, other clinical studies have demonstrated no benefit from furosemide, mannitol, or dopamine. A number of specific issues remain inadequately studied in high-risk subjects; and these include: (i) the use of CCBs alone; (ii) the expanded use of carbon dioxide arteriography; and (iii) the potentially additive effects of theophylline with CCBs and other related topics. At present, the best prophylaxis to prevent RCM-induced nephrotoxicity includes: use of a low-osmolar RCM in a patient who has received adequate hydration along with either theophylline or CCBs given sufficiently prior to RCM injection. A proposed scheme is illustrated in Figure 11.

7.28.10 REFERENCES

1. S. H. Hou, D. A. Bushinsky, J. B. Wish *et al.* 'Hospital-acquired renal insufficiency: a prospective study.' *Am. J. Med.*, 1983, **74**, 243–258.

2. N. Shusterman, B. L. Strom, T. G. Murray *et al.* 'Risk factors and outcome of hospital-acquired acute renal failure. Clinical epidemiologic study.' *Am. J. Med.*, 1987, **83**, 65–71
3. E. D. Osborne, C. G. Sutherland, A. J. Scholl *et al.* 'Roentgenography or urinary tract during excretion of sodium iodide.' *JAMA*, 1923, **80**, 368–372.
4. P. Silpananta, F. F. Illescas and H. Sheldon, 'Multiple malignant neoplasms 40 years after angiography with Thorotrast.' *Can. Med. Assoc. J.*, 1983, **128**, 289–292.
5. D. Sutton in 'Textbook of Radiology and Imaging,' 4th edn., ed. D. Sutton, Churchill-Livingstone, Edinburgh, 1987, pp. 692–699.
6. E. P. Pendergrass, G. W. Chamberlin, E. W. Godfrey *et al.*, 'A survey of deaths and unfavorable sequelae following the administration of contrast media.' *Am. J. Radiol.*, 1942, **48**, 741–762.
7. R. D. Alexander, S. L. Berkes and G. Abuelo, 'Contrast media-induced oliguric renal failure.' *Arch. Intern. Med.*, 1978, **138**, 381–384.
8. Z. Ansari and D. S. Baldwin, 'Acute renal failure due to radiocontrast agents.' *Nephron*, 1976, **17**, 28–40.
9. O. Bartley, U. Bengtsson and G. Cedarbon, 'Renal function before and after urography and angiography with large doses of contrast media.' *Acta Radiol.*, 1969, **8**, 9–12.
10. L. Byrd and R. L. Sherman, 'Radiocontrast-induced acute renal failure: a clinical and pathophysiologic review.' *Medicine (Baltimore)*, 1979, **58**, 270–279.
11. J. A. Diaz-Buxo, R. D. Wagoner, R. R. Hattery *et al.* 'Acute renal failure after excretory urography in diabetic patients.' *Ann. Intern. Med.*, 1975, **83**, 155–158.
12. F. A. Krumlovsky, N. Simon, S. Santhanam *et al.*, 'Acute renal failure. Association with administration of radiographic contrast material.' *JAMA*, 1978, **239**, 125–127.
13. B. J. Barrett and E. J. Carlisle. 'Metaanalysis of the relative nephrotoxicity of high- and low-osmolality iodinated contrast media.' *Radiology*, 1993, **188**, 171–178.
14. R. F. Spataro, 'Newer contrast agents for urology.' *Radiol. Clin. North Am.*, 1984, **22**, 365–380.
15. H. J. Weinmann, R. C. Brasch, W. R. Press *et al.*, 'Characteristics of gadolinium-DTPA complex: a potential NMR contrast agent.' *Am. J. Roentgenol.*, 1984, **142**, 619–624
16. J. M. Seeger, S. Self, T. R. S. Harward *et al.*, 'Carbon dioxide gas as an arterial contrast agent.' *Ann. Surgery*, 1993, **217**, 688–698.
17. M. A. Bettmann, 'Angiographic contrast agents: conventional and new media compared.' *Am. J. Roentgenol.*, 1982, **139**, 787–794.
18. G. H. Mudge, 'Nephrotoxicity of urographic radiocontrast drugs.' *Kidney Int.*, 1980, **18**, 540–552.
19. P. Schiantarelli, F. Peroni, P. Tirone *et al.* 'Effects of iodinated contrast media on erythrocytes. I. Effects of canine erythrocytes on morphology.' *Invest. Radiol.*, 1973, **8**, 199–204.
20. L. B. Talner, 'Urographic contrast media in uremia. Physiology and pharmacology.' *Radiol. Clin. North Am.*, 1972, **10**, 421–432.
21. M. R. Sage, 'Kinetics of water-soluble contrast media in the central nervous system.' *Am. J. Roentgenol.*, 1983, **141**, 815–824.
22. T. W. Morris and H. W. Fischer, 'The pharmacology of intravascular radiocontrast media.' *Annu. Rev. Pharmacol. Toxicol.*, 1986, **26**, 143–160.
23. P. B. Dean, L. Kivisaari and M. Kormano, 'The diagnostic potential of contrast enhancement pharmacokinetics.' *Invest. Radiol.*, 1978, **13**, 533–540.
24. F. A. Burgener and D. J. Hamlin, 'Contrast enhance-

ment in abdominal CT: bolus vs. infusion.' *Am. J. Roentgenol.*, 1981, **137**, 351–358.

25. J. Haustein, H. P. Niendorf, G. Krestin *et al.*, 'Renal tolerance of gadolinium-DTPA/dimeglumine in patients with chronic renal failure.' *Invest Radiol.*, 1992, **27**, 153–156.

26. M. V. Rocco, V. M. Buckalew, L. Cole *et al.*, 'Measurement of GFR with a single intravenous injection of nonradioactive iohexol and a single blood sample.' *J. Am. Soc. Nephrol.*, 1993, **4**(3), 323.

27. G. H. Mudge, 'The maximal urinary concentration of diatrizoate.' *Invest. Radiol.*, 1980, **15**, S67–S78.

28. R. F. Spataro, H. W. Fischer and L. Boglan, 'Urography with low-osmolality contrast media: comparative urinary excretion of Iopamidol, Hexabrix, and Diatrizoate.' *Invest. Radiol.*, 1982, **17**, 494–500.

29. R. W. Katzberg, R. C. Pabico, T. W. Morris *et al.*, 'Effects of contrast media on renal function and subcellular morphology in the dog.' *Invest. Radiol.*, 1986, **21**, 64–70.

30. G. T. Benness, 'Urographic contrast agents. A comparison of sodium and methylglucamine salts.' *Clin. Radiol.*, 1970, **21**, 150–156.

31. W. R. Cattel, I. K. Fry, A. G. Spencer *et al.* 'Excretion urography. I. Factors determining the excretion of Hypaque.' *Br. J. Radiol.*, 1967, **40**, 561–571.

32. J. Bahlmann and H. L. Kruskemper, 'Elimination of iodine-containing contrast media by hemodialysis.' *Nephron.*, 1973, **10**, 250–255.

33. H. G. Von Baars, J. Schabel and M. Weiss, 'Kontrastmittei-Elimination durch Hamodialyze, (contrast medium elimination by hemidialysis).' *Z. Urol. Nephrol.*, 1984, **77**, 473–481.

34. P. Ackrill, C. S. McIntosh, C. Nimmon *et al.*, 'A comparison of the clearance of urographic contrast medium (sodium diatrizoate) by peritoneal and hemodialysis.' *Clin. Sci. Mol. Med.*, 1976, **50**, 69–74.

35. N. Milman and E. Christensen, 'Elimination of diatrizoate by peritoneal dialysis in renal failure.' *Acta Radiol. Diagn. (Stockh.)*, 1974, **15**, 265–272.

36. A. Waaler, M. Svaland, P. Faurchald *et al.*, 'Elimination of iohexol, a low-osmolar nonionic contrast medium, by hemodialysis in patients with chronic renal failure.' *Nephron*, 1990, **56**, 81–85.

37. T. W. Morris, R. W. Katzberg and H. W. Fischer, 'A comparison of the hemodynamic responses to metrizamide and meglumine/sodium diatrizoate in canine renal angiography.' *Invest. Radiol.*, 1978, **13**, 74–78.

38. L. H. Norby and G. F. DiBona, 'The renal vascular effects of meglumine diatrizoate.' *J. Pharmacol. Exp. Ther.*, 1975, **193**, 932–940.

39. L. B. Talner and A. J. Davidson, 'Renal hemodynamic effects of contrast media.' *Invest. Radiol.*, 1968, **3**, 310–317.

40. W. J. H. Caldicott, N. K. Hollenberg and H. L. Abrams, 'Characteristics of response of renal vascular bed to contrast media. Evidence for vasoconstriction induced by renin–angiotensin system.' *Invest. Radiol.*, 1970, **5**, 539–547.

41. R. B. Harvey, 'Vascular resistance changes produced by hyperosmotic solutions.' *Am. J. Physiol.*, 1960, **199**, 31–34.

42. S. N. Heyman, B. A. Clark, L. Cantley, *et al.*, 'Effects of ioversol vs. iothalamate on endothelin release and radiocontrast nephropathy.' *Invest. Radiol.*, 1993, **28**, 313–318.

43. R. W. Katzberg, G. Schulman, L. G. Meggs *et al.*, 'Mechanism of the renal response to contrast medium in dogs. Decrease in renal function

due to hypertonicity.' *Invest. Radiol.*, 1983, **18**, 74–80.

44. T. Larson, K. Hudson, J. I. Mertz *et al.*, 'Renal vasoconstrictive response to contrast medium. The role of sodium balance and the renin–angiotensin system.' *J. Lab. Clin. Med.*, 1983, **101**, 385–391.

45. L. J. Arend, G. L. Bakris, J. C. Burnett, Jr. *et al.*, 'Role of intrarenal adenosine in the renal hemodynamic response to contrast media.' *J. Lab. Clin. Med.*, 1987, **110**, 406–411.

46. G. L. Bakris and J. C. Burnett, Jr., 'A role for calcium in radiocontrast-induced reduction in renal hemodynamics.' *Kidney Int.*, 1985, **27**, 465–468.

47. G. Deray, F. Martinez, P. Cacoub *et al.*, 'A role for adenosine, calcium, and ischemia in radiocontrast-induced intrarenal vasoconstriction.' *Am. J. Nephrol.*, 1990, **10**, 316–322.

48. G. L. Bakris, N. Lass, A. O. Gaber *et al.*, 'Radiocontrast medium-induced declines in renal function: a role for oxygen free radicals.' *Am. J. Physiol.*, 1990, **258**, F115–F120.

49. K. B. Margulies, L. J. McKinley, P. G. Cavero *et al.*, 'Induction and prevention of radiocontrast-induced nephropathy in dogs with heart failure.' *Kidney Int.*, 1990, **38**, 1101–1108.

50. R. C. Vari, L. A. Natarajan, S. A. Whitescarver *et al.*, 'Induction, prevention and mechanisms of contrast medium-induced acute renal failure.' *Kidney Int.*, 1988, **33**, 699–707.

51. S. N. Heyman, M. Brezis, Z. Greenfeld *et al.*, 'Protection role of feurosemide and saline in radiocontrast-induced acute renal failure in the rat.' *Am. J. Kidney Dis.*, 1989, **14**, 377–385.

52. S. N. Heyman, M. Brezis, C. A. Reubinoff *et al.*, 'Acute renal failure with selective medullary injury in the rat.' *J. Clin. Invest.*, 1988, **82**, 401–412.

53. G. H. Mudge, F. A. Meier and K. K. Ward, in 'Acute Renal Failure,' eds. K. Solez and A. Whelton, Marcel Dekker, New York, 1984, pp. 361–388.

54. J. G. Gerber, R. A. Branch, A. S. Nies *et al.*, 'Influence of hypertonic saline on canine renal blood flow and renin release.' *Am. J. Physiol.*, 1979, **237**, F441–F446.

55. R. W. Katzberg, T. W. Morris and F. A. Burgerer *et al.*, 'Renal renin and hemodynamic responses to selective renal artery catherization and angiography.' *Invest. Radiol.*, 1977, **12**, 381–388.

56. H. D. Humes, D. A. Hunt and M. D. White, 'Direct toxic effect of the radiocontrast agent diatrizoate on renal proximal tubule cells.' *Am. J. Physiol.*, 1987, **252**, F246–F255.

57. J. M. Messana, D. A. Cieslinski, V. D. Nguyen *et al.*, 'Comparison of the toxicity of the radiocontrast agents, iopanidol and diatrizoate to rabbit renal proximal tubule cells *in vitro*.' *J. Pharmacol. Exp. Ther.*, 1988, **244**, 1139–1144

58. S. Bhandaru and G. L. Bakris, 'Effects of high- and low-osmolar radiocontrast medium on mesangial cell growth and cell integrity (abstract).' *J. Invest. Med.*, 1995, **43**, 103.

59. L. G. Navar, W. J. Champion and C. E. Thomas, 'Effects of calcium channel blockade on renal vascular resistance responses to changes in perfusion pressure and angiotensin-converting enzyme inhibition in dogs.' *Circ. Res.*, 1986, **58**, 874–881.

60. P. G. Baer and L. G. Navar, 'Renal vasodilatation and uncoupling of blood flow and filtration rate autoregulation.' *Kidney Int.*, 1973, **4**, 12–21.

61. R. Muller-Suur, H. U. Gutsche, H. J. Schurek *et al.*, 'Acute reversible inhibition of tubuloglomerular feedback mediated afferent vasoconstriction by the calcium antagonist verapamil.' *Curr. Probl. Clin. Biochem.*, 1976, **6**, 291–298.

62. H. Osswald, W. S. Spielman and F. G. Knox, 'Mechanism of adenosine-mediated decreases in glomerular filtration rate in dogs.' *Circ. Res.*, 1978, **43**, 465–469.

63. H. Tagawa and A. J. Vander, 'Effects of adenosine compounds on renal function and renin secretion in dogs.' *Circ. Res.*, 1970, **26**, 327–338.

64. J. E. Hall, J. P. Granger and R. L. Hester, 'Interactions between adenosine and angiotensin II in controling glomerular filtration.' *Am. J. Physiol.*, 1985, **248**, F340–F346.

65. W. S. Spielman and C. I. Thompson, 'A proposed role for adenosine in the regulation of renal hemodymanics and renin release.' *Am. J. Physiol.*, 1982, **242**, F423–F435.

66. H. Osswald, C. Gleiter and B. Muhlbauer, 'Therapeutic use of theophylline to antagonize renal effects of adenosine.' *Clin. Nephrol.*, 1995, **43** (Suppl. 1), S33–S37.

67. K. Margulis, F. L. Hildebrand, D. M. Heublein *et al.*, 'Intraarterial atrial natriuretic factor attenuates radiocontrast-induced nephropathy in humans.' *J. Am. Soc. Nephrol.*, 1991, **2**, 1041.

68. S. N. Heyman, B. A. Clark, N. Kaiser *et al.*, 'Radiocontrast agents induce endothelin release *in vivo* and *in vitro*.' *J. Am. Soc. Nephrol.*, 1992, **3**, 58–65.

69. S. Oldroyd, S. J. Lee, J. Haylor *et al.*, 'Role for endothelin in the renal responses to radiocontrast media in the rat.' *Clin. Sci. (Colch.)*, 1994, **87**, 427–434.

70. L. Baud and R. Ardaillou, 'Reactive oxygen species: production and role in the kidney.' *Am. J. Physiol.*, 1986, **251**, F765–F776.

71. C. E. Cross, B. Halliwell, E. T. Borish *et al.*, 'Oxygen radicals and human disease.' *Ann. Intern. Med.*, 1987, **107**, 526–545.

72. S. V. Shah, 'Role of reactive oxygen metabolites in experimental glomerular disease.' *Kidney Int.*, 1989, **35**, 1093–1106.

73. L. Baud, J. Hagege, J. Sraer *et al.*, 'Reactive oxygen production by cultured rat glomerular mesangial cells during phagocytosis is associated with stimulation of lipoxygenase activity.' *J. Exp. Med.*, 1983, **158**, 1836–1852.

74. J. M. Messana, D. A. Cieslinski, R. P. O'Connor *et al.*, 'Glutathione protects against exogenous oxidant injury to rabbit renal proximal tubules.' *Am. J. Physiol.*, 1988, **255**, F874–F884.

75. R. C. Scaduto, Jr., V. H. Gattone II, L. W. Grotyohann *et al.*, 'Effect of an altered glutathione content on renal ischemic injury.' *Am. J. Physiol.*, 1988, **255**, F911–F921.

76. T. Yoshioka, A. Fogo and J. K. Beckman, 'Reduced activity of antioxidant enzymes underlies contrast media-induced renal injury in volume depletion.' *Kidney Int.*, 1992, **41**, 1008–1015.

77. J. R. Hoyer and M. W. Seiler, 'Pathophysiology of Tamm–Horsfall protein.' *Kidney Int.*, 1987, **16**, 279–289.

78. R. Patel, J. K. McKenzie and E. G. McQueen, 'Tamm–Horsfall urinary microprotein and tubular obstruction by casts in acute renal failure.' *Lancet*, 1964, **1**, 41–46.

79. R. H. Schwartz, W. E. Berdon, J. Wagner *et al.*, 'Tamm–Horsfall urinary microprotein precipitation by urographic contrast agents: *in vitro* studies.' *Am. J. Roentgenol. Radium Ther. Nucl. Med.*, 1970, **108**, 698–701.

80. W. E. Berdon, R. H. Schwartz, J. Becker *et al.*, 'Tamm–Horsfall proteinuria. Its relationship to prolonged nephrograms in infants and children and to renal failure following intravenous urography in adults with multiple myeloma.' *Radiology*, 1969, **92**, 714–722.

81. G. L. Bakris, A. O. Gaber and J. D. Jones, 'Oxygen free radical involvement in urinary Tamm–Horsfall protein excretion after intrarenal injection of contrast medium.' *Radiology*, 1990, **175**, 51–60.

82. G. S. Nicot, L. J. Merle, J. P. Charmes *et al.*, 'Transient glomerula proteinuria, enzymuria and nephrotoxic reaction induced by radiocontrast media.' *JAMA*, 1984, **252**, 2432–2434.

83. Z. Parvez, S. Ramamursthy, N. B. Patel *et al.*, 'Enzyme markers of contrast media-induced renal failure.' *Invest. Radiol.*, 1990, **25**, Suppl. 1, S133–S134.

84. J. Lang and E. C. Lasser, 'Inhibition of adenosine triphosphate and carbonic anhydrase by contrast media.' *Invest. Radiol.*, 1975, **10**, 314–316.

85. L. B. Talner, H. N. Rushmer and M. N. Coel, 'The effect of renal artery injection of contrast material on urinary enzyme excretion.' *Invest. Radiol.*, 1972, **7**, 311–322.

86. R. J. Workman, M. I. Schaff, R. V. Jackson *et al.*, 'Relationship of renal hemodynamics and functional changes following intravascular contrast to the renin–angiotensin system and renal prostacycline in the dog.' *Invest. Radiol.*, 1983, **18**, 160–166.

87. F. R. DeRubertis and P. A. Craven, 'Effects of osmolality and oxygen availability on soluble cyclic AMP-dependent protein kinase activity of rat renal inner medulla.' *J. Clin. Invest.*, 1978, **62**, 1210–1221.

88. M. Brezis, S. N. Rosen and F. H. Epstein, 'The pathophysiologic implications of medullary hypoxia.' *Am. J. Kidney Dis.*, 1989, **13**, 253–258.

89. S. Lear, P. Silva, V. E. Kelley *et al.*, 'Prostaglandin E_2 inhibits oxygen consumption in rabbit medullary thick ascending limb.' *Am. J. Physiol.*, 1990, **258**, F1372–F1378.

90. G. Lund, S. Einzig, J. Rysavy *et al.*, 'Effect of prostaglandin inhibition on the renal vascular response to ionic and nonionic contrast media in the dog.' *Acta Radiol. Diagn. (Stockh.)*, 1984, **25**, 407–410.

91. K. Kato, M. Kato, T. Matsuoka *et al.*, 'Depression of membrane-bound Na^+-K^+-ATPase activity induced by free radicals and by ischemia of kidney.' *Am. J. Physiol.*, 1988, **254**, C330–C337.

92. M. S. Kim and T. Akera, 'O_2 free radicals: cause of ischemia-reperfusion injury to cardiac Na^+-K^+-ATPase.' *Am. J. Physiol.*, 1987, **252**, H252–H257.

93. K. Margulies, L. J. McKinley, R. Allgren *et al.*, 'Intraarterial atrial natriuretic factor attenuates radiocontrast-induced nephropathy in human (abstract).' *J. Am. Soc. Nephrol.*, 1991, **2**, 666.

94. A. S. Berns, 'Nephrotoxicity of contrast media.' *Kidney Int.*, 1989, **36**, 730–740.

95. R. L. Eisenberg, W. O. Bank and M. W. Hedgcock, 'Renal failure after major angiography.' *Am. J. Med.*, 1980, **68**, 43–46.

96. S. Kumar, J. D. Hull, S. Lathi *et al.*, 'Incidence of renal failure after angiography.' *Arch. Intern. Med.*, 1981, **141**, 1268–1270.

97. B. A. Barnes, A. S. Shaw, A. Leaf *et al.*, 'Oliguria following diagnostic translumbar aortography.' *N. Engl. J. Med.*, 1955, **252**, 113–117.

98. F. R. Stark and J. W. Coburn, 'Renal failure following methylglucamine diatrizoate (Renografin) aortography: report of a case with unilateral renal artery stenosis.' *J. Urol.*, 1966, **96**, 848–851.

99. F. K. Port, R. D. Wagoner and R. E. Fulton, 'Acute renal failure after angiography.' *Am. J. Roentgenol. Radium Ther. Nucl. Med.*, 1974, **121**, 544–550.

100. J. A. D'Elia, R. E. Gleason, M. Alday *et al.*, 'Nephrotoxicity from angiographic contrast material. A prospective study.' *Am. J. Med.*, 1982, **72**, 719–725.

101. R. A. Mason, L. A. Arbeit and F. Giron, 'Renal dysfunction after arteriography.' *JAMA*, 1985, **253**, 1001–1004.

102. S. T. Cochran, W. S. Wong and D. J. Roe, 'Predicting angiography-induced acute renal function impairment: clinical risk model.' *Am. J. Roentgenol.*, 1983, **14**, 1027–1033.

103. L. A. Weinrauch, R. W. Healy, O. S. Leland, Jr. *et al.*, 'Coronary angiography and acute renal failure in diabetic azotemic nephropathy.' *Ann. Intern. Med.*, 1977, **86**, 56–59.

104. C. P. Taliercio, R. E. Vlietstra, L. D. Fisher *et al.*, 'Risks for renal dysfunction with cardiac angiography.' *Ann. Intern. Med.*, 1986, **104**, 501–504.

105. R. D. Swartz, J. E. Rubin, B. W. Leeming *et al.*, 'Renal failure following major angiography.' *Am. J. Med.*, 1978, **65**, 31–37.

106. P. S. Parfrey, S. M. Griffiths, B. J. Barrett *et al.*, Contrast material-induced renal failure in patients with diabetes mellitus, renal insufficiency or both. A prospective controlled study.' *N. Engl. J. Med.*, 1989, **320**, 143–149.

107. S. J. Schwab, M. A. Hlatky, K. S. Pieper *et al.*, 'Contrast nephrotoxicity: a randomized controled trial of nonionic and ionic radiographic contrast agent.' *N. Engl. J. Med.*, 1989, **320**, 149–153.

108. C. J. Davidson, M. Hlatky, K. G. Morris *et al.*, 'Cardiovascular and renal toxicity of a nonionic radiographic contrast agent after cardiac catheterization. A prospective trial.' *Ann. Intern. Med.*, 1989, **110**, 119–124.

109. M. L. Kinnison, N. R. Powe and E. P. Steinberg, 'Results of randomized controlled trials of low- vs. high-osmolality contrast media.' *Radiology*, 1989, **170**, 381–389.

110. A. S. Gomes, J. F. Lois, J. D. Baker *et al.*, 'Acute renal dysfunction in high-risk patients after angiography: comparison of ionic and nonionic contrast media.' *Radiology*, 1989, **170**, 65–68.

111. U. Albrechtsson, B. Hultberg, H. Larusdottir *et al.*, 'Nephrotoxicity of ionic and nonionic contrast media in aortofemoral angiography.' *Acta Radiol. Diagn. (Stockh.)*, 1985, **26**, 615–618.

112. C. Tornquist and S. Holtas, 'Renal angiography with iohexol and metrizoate.' *Radiology*, 1984, **150**, 331–334.

113. C. R. Bird, B. P. Drayer, R. Velja *et al.*, 'Safety of contrast media in cerebral angiography; iopamidol vs. methylglucamine iothalamate.' *Am. J. Roentgenol.*, 1984, **5**, 801–803.

114. B. A. Sacks, H. P. Ellison, S. Bartek *et al.*, 'A comparison of hexabrix and Renografin 60 in peripheral arteriography.' *Am. J. Roentgenol.*, 1983, **140**, 975–977.

115. K. H. Barth and M. A. Mertens, 'A double-blind comparative study of Hexabrix and Renografin-76 in aortography and visceral arteriography.' *Invest. Radiol.*, 1984, **19**, S323–S325.

116. P. Roy, P. Robillard, C. L'Homme *et al.*, 'Iohexol: a new nonionic agent in adult peripheral arteriography.' *J. Can. Assoc. Radiol.*, 1985, **36**, 113–117.

117. M. Pathria, S. Somers and G. Gill, 'Pain during angiography: a randomized double-blind trial comparing ioxaglate and ditrizoate.' *J. Can. Assoc. Radiol.*, 1987, **38**, 32–34.

118. M. E. Gale, A. H. Robbins, R. J. Hamburger *et al.*, 'Renal toxicity of contrast agents: iopamidol, iothalamate, and diatrizoate.' *Am. J. Roentgenol.*, 1984, **142**, 333–335.

119. G. A. Khoury, J. C. Hopper, Z. Varghese *et al.*, 'Nephrotoxicity of ionic and nonionic contrast material in digital vascular imaging and selective renal arteriography.' *Br. J. Radiol.*, 1983, **56**, 631–635.

120. S. Harkonen and C. M. Kjellstrand, 'Exacerbation of diabetic renal failure following intravenous pyelography.' *Am. J. Med.*, 1977, **63**, 939–946.

121. B. E. VanZee, W. E. Hoy, T. E. Talley *et al.*, 'Renal injury associated with intravenous pyelography in nondiabetic and diabetic patients.' *Ann. Intern. Med.*, 1978, **89**, 51–54.

122. D. Peltz, A. J. Fox and F. Vinuela, 'Clinical trial of Iohexol vs. Conray 60 for cerebral angiography.' *Am. J. Roentgenol.*, 1984, **5**, 565–568.

123. E. P. Pendergrass, P. J. Hodges, R. J. Trondreau *et al.*, 'Further consideration of deaths and unfavorable sequelae following administration of contrast media in urography in the United States.' *Am. J. Roentgenol.*, 1955, **74**, 262–287.

124. A. Kamdar, P. Weidman, D. L. Makoff *et al.*, 'Acute renal failure following intravenous use of radiographic contrast dyes in patients with diabetes mellitus.' *Diabetes*, 1977, **26**, 643–649.

125. V. K. Pillay, P. C. Robbins, F. D. Schwartz *et al.*, 'Acute renal failure following intravenous urography in patients with long-standing diabetes mellitus and azotemia.' *Radiology*, 1970, **95**, 633–636.

126. H. R. Anto, S. Y. Chou, J. G. Porush *et al.*, 'Infusion intravenous pyelography and renal function. Effects of hypertonic mannitol in patients with chronic renal insufficiency.' *Arch. Intern. Med.*, 1981, **141**, 1652–1656.

127. S. Harkonen and C. M. Kjellstrand, 'Intravenous pyelography in nonuremic diabetic patients.' *Nephron*, 1979, **24**, 268–270.

128. A. Carvallo, T. A. Rakowski, W. P. Argy Jr. *et al.*, 'Acute renal failure following drip infusion pyelography.' *Am. J. Med.*, 1975, **65**, 38–45.

129. J. L. Terwel, R. Moxen, J. M. Onaindia *et al.*, 'Renal function impairment caused by intravenous urography. A prospective study.' *Arch. Intern. Med.*, 1981, **141**, 1271–1274.

130. T. Shafi, C. Y. Chou, J. G. Porush *et al.*, 'Infusion intravenous pyelography and renal function. Effects in patients with chronic renal insufficiency.' *Arch. Intern. Med.*, 1978, **138**, 1218–1221.

131. S. D. Shieh, S. R. Hirsch, B. R. Boshell *et al.*, 'Low risk of contrast media-induced acute renal failure in nonazotemic type-2 diabetes mellitus.' *Kidney Int.*, 1982, **21**, 739–743.

132. L. A. Bergman, M. R. Ellison and G. Dunea, 'Acute renal failure after drip-infusion pyelography.' *N. Engl. J. Med.*, 1978, **279**, 1277.

133. S. Harkonen and C. Kjellstrand, 'Contrast nephropathy.' *Am. J. Nephrol.*, 1981, **7**, 69–77.

134. J. A. Diaz-Buxo, R. D. Wagoner, R. R. Hattery *et al.*, 'Acute renal failure after excretory urography in diabetic patients.' *Ann. Intern. Med.*, 1975, **83**, 155–158.

135. W. B. Schwartz, A. Hurwit and A. Ettinger, 'Intravenous urography in the patient with renal insufficiency.' *N. Engl. J. Med.*, 1963, **269**, 277–283.

136. B. C. Cramer, P. S. Parfrey, T. A. Hutchison *et al.*, 'Renal function following infusion of radiologic contrast material. A prospective controled study.' *Arch. Intern. Med.*, 1985, **145**, 87–89.

137. O. Sunnegardh, S. O. Hietala, S. Wirell *et al.*, 'Systemic, pulmonary and renal hemodynamic effects of intravenously infused iopentol.' *Acta Radiol.*, 1990, **31**, 395–399.

138. K. J. Berg, F. Kolmannskog, P. E. Lillevold *et al.*, 'Iopentol in patients with chronic renal failure: its effects on renal function and its use as glomerular filtration rate parameter.' *Scand. J. Clin. Lab. Invest.*, 1992, **52**, 27–33.

139. B. J. Barrett, P. S. Parfrey, H. M. Vavasour *et al.*, 'Contrast nephropathy in patients with impaired renal function: high vs. low-osmolar media.' *Kidney Int.*, 1992, **41**(5), 1274–1279.

140. M. R. Rudnick, S. Goldfarb, L. Wexler *et al.*, 'Nephrotoxicity of ionic and nonionic contrast

media in 1196 patients: a randomized trial. The Iohexol Cooperative Study.' *Kidney Int.*, 1995, **47**, 254–261.

141. R. Katholi, G. J. Taylor, W. T. Woods *et al.*, 'Nephrotoxicity of nonionic low-osmolality vs. ionic high-osmolality contrast media: a prospective double-blind randomized comparison in human beings.' *Radiology*, 1993, **186**, 183–187.

142. J. P. Kassirer, 'Atheroembolic renal disease.' *N. Engl. J. Med.*, 1969, **280**, 812–818.

143. M. C. Smith, M. K. Ghose and A. R. Henry, 'The clinical spectrum of renal cholesterol embolization.' *Am. J. Med.*, 1981, **71**, 174–180.

144. A. Meyrier, P. Buchet, P. Simon *et al.*, 'Atheromatous renal disease.' *Am. J. Med.*, 1988, **85**, 139–146.

145. H. G. Colt, R. J. Begg, J. Saporito *et al.*, 'Cholesterol emboli after cardiac catheterization. Eight cases and a review of the literature.' *Medicine (Baltimore)*, 1988, **67**, 389–400.

146. H. S. Rosman, T. P. Davis, D. Reddy *et al.*, 'Cholesterol embolization: clinical findings and implications.' *J. Am. Coll. Cardiol.*, 1990, **15**, 1296–1299.

147. J. W. Coburn and K. L. Agre, in 'Diseases of the Kidney,' 5th edn., eds. R. W. Schrier and C. W. Gottschalk, Little Brown, Boston, MA, 1993, pp. 2119–2135.

148. A. Om, S. Ellahham and G. DiSciasco, 'Cholesterol embolism: an underdiagnosed clinical entity.' *Am. Heart J.*, 1992, **124**, 1321–1326.

149. W. Thurlbeck and B. Castleman, 'Atheromatous emboli to kidneys after aortic surgery.' *N. Engl. J. Med.*, 1957, **257**, 444–447.

150. B. S. Kasinath, H. L. Corwin, A. K. Bidani *et al.*, 'Eosinophilia in the diagnosis of atheroembolic renal disease.' *Am. J. Nephrol.*, 1987, **7**, 173–177.

151. Anonymous, 'Case records of the Massachusetts General Hospital. Weekly clinopathological exercises. Case 38-1993. Renal failure and a painful toe in a 70-year-old man after an acute myocardial infarct.' *N. Engl. J. Med.*, 1993, **329**, 948–955.

152. B. J. Barrett, 'Contrast nephrotoxicity.' *J. Am. Soc. Nephrol.*, 1994, **5**(2), 125–137.

153. P. J. Dudzinski, A. F. Petrone, M. Persoff *et al.*, 'Acute renal failure following high dose excretory urography in dehydrated patients.' *J. Urol.*, 1971, **106**, 619–628.

154. R. L. Eisenberg, W. O. Bank and M. W. Hedgcock, 'Renal failure after major angiography can be avoided with hydration.' *Am. J. Roentgenol.*, 1981, **136**, 859–861.

155. R. Soloman, C. Werner, D. Mann *et al.*, 'Effects of saline, mannitol, and furosemide to prevent acute decreases in renal function induced by radiocontrast agents.' *N. Eng. J. Med.*, 1994, **331**, 1416–1420.

156. R. O. Berkseth and C. M. Kjellstrand, 'Radiologic contrast-induced nephropathy.' *Med. Clin. North Am.*, 1984, **68**, 351–370.

157. E. P. Steinberg, R. D. Moore, N. R. Powe *et al.*, 'Safety and cost effectiveness of high-osmolality as compared with low-osmolality contrast material in patients undergoing cardiac angiography.' *N. Engl. J. Med.*, 1992, **326**, 425–430.

158. D. Russo, A. Testa, L. Della Volpe *et al.*, 'Randomized prospective study on renal effects of two different contrast media in humans: protective role of a calcium channel blocker.' *Nephron*, 1990, **55**, 254–257.

159. H. H. Neumayer, W. Junge, A. Kufner *et al.*,

'Prevention of radiocontrast media-induced nephrotoxicity by the calcium channel blocker nitrendipine: a prospective randomized clinical trial.' *Nephrol. Dial. Transplant.*, 1989, **4**, 1030–1036.

160. C. M. Erley, S. H. Duda, S. Schlepekow *et al.*, 'Adenosine antagonist theophylline prevents reduction of glomerular filtration rate after contrast media application.' *Kidney Int.*, 1994, **45**, 1425–1431.

161. R. E. Katholi, G. J. Taylor, W. P. McCann *et al.*, 'Nephrotoxicity from contrast media: attenuation with theophylline.' *Radiology*, 1995, **195**, 17–22.

162. R. G. Cigarroa, R. A. Lange, R. H. Williams *et al.*, 'Dosing of contrast material to prevent contrast nephropathy in patients with renal disease.' *Am. J. Med.*, 1989, **86**, 649–652.

163. J. G. Porush, S. Y. Chou, H. R. Anto *et al.*, in 'Acute Renal Failure,' ed. H. E. Eliahou, John Libbey, London, 1982, pp. 161–167.

164. C. W. Old, C. M. Duarte, L. M. Lehrmer *et al.*, 'A prospective evaluation of mannitol in the prevention of radiocontrast acute renal failure.' *Clin. Res.*, 1981, **29**, 472 (abstract).

165. R. E. Cronin, A. M. Erickson, A. De Torrente *et al.*, 'Norepinephrine-induced acute renal failure: a reversible ischemic model of acute renal failure.' *Kidney Int.*, 1978, **14**, 187–190.

166. C. R. Morris, E. A. Alexander, F. J. Bruns *et al.*, 'Restoration and maintenance of glomerular filtration by mannitol during hypoperfusion of the kidney.' *J. Clin. Invest.*, 1972, **51**, 1555–1564.

167. B. R. C. Kurnik, L. S. Weisberg, I. M. Cuttler *et al.*, 'Effects of atrial natriuretic peptide versus mannitol on renal blood flow during radiocontrast infusion in chronic renal failure.' *J. Lab. Clin. Med.*, 1990, **116**, 27–36.

168. J. M. Weinstein, S. Heyman and M. Brezis, 'Potential deleterious effect of furosemide in radiocontrast nephropathy.' *Nephron*, 1992, **62**, 413–415.

169. B. K. Demarie and G. L. Bakris, 'Effects of different calcium antagonists of proteinuria associated with diabetes mellitus.' *Ann. Int. Med.*, 1990, **113**, 987–988.

170. N. B. Aron, D. A. Feinfeld, A. T. Peters *et al.*, 'Acute renal failure associated with ioxaglate, a low-osmolality radiocontrast agent.' *Am. J. Kidney Dis.*, 1989, **13**, 189–193.

171. C. B. Higgins, M. Kuber and R. A. Slutsky, 'Interaction between verapamil and contrast media in coronary arteriography: comparison of standard ionic and new nonionic media.' *Circulation*, 1983, **68**, 628–635.

172. K. H. Gerber, C. B. Higgins, Y. S. Yuh *et al.*, 'Regional myocardial hemodynamic and metabolic effects of ionic and nonionic contrast media in normal and ischemic states.' *Circulation*, 1982, **65**, 1307–1314.

173. G. Lund Einzigs, J. Rysavy, B. Borgwardt *et al.*, 'Role of ischemia in contrast-induced renal damage: an experimental study.' *Circulation*, 1984, **69**, 783–789.

174. D. L. Morris, J. A. Wisenski, E. W. Gertz *et al.*, 'Potentiation by nifedipine and diltiazem of the hypotensive response after contrast angiography.' *J. Am. Coll. Cardiol.*, 1985, **6**, 785–791.

175. H. W. Fischer, R. F. Spataro and R. M. Rosenberg, 'Medical and economic considerations in using a new contrast medium.' *Arch. Intern. Med.*, 1988, **146**, 1717–1721.

7.29
Cyclosporine

JUNE MASON

Sandoz Pharmaceuticals, East Hanover, NJ, USA

and

LEON C. MOORE

State University of New York at Stony Brook, NY, USA

7.29.1 INTRODUCTION

Since its introduction in the early 1980s, the clinical assessment of cyclosporine has undergone a radical transformation. Initially, as the wonder drug that dramatically changed the outcome of organ transplantation, it was viewed with excitement. Soon after, when the side affects associated with high doses began to become apparent, its long-term safety was questioned. Now, after prolonged use at lower doses both in transplantation and in autoimmune-type diseases, such as rheumatoid arthritis, psoriasis, and nephrotic syndrome, it is viewed as one of many drugs that may be dangerous if not used cautiously, but is safe if watchfully dosed.

The features of cyclosporine's side effects that make it so interesting are the presence of a clearly defined set of pathophysiological and pathomorphological findings, some reversible and some not, which affect predominately the kidney but also other organs. Within the kidney the lesions that are observed affect all three tissue types, the vessels, the tubules, and the interstitium, without it being evident whether these effects are interrelated, whether some effects are primary and others secondary or whether all effects are independent.

The goal of this chapter is to give a summary of the renal side effects that cyclosporine and other immunosuppressants have. The major focus is to present new findings or new explanations for old findings, in preference to discussing each of the pathophysiological or toxicological findings in detail. As such, it is not intended to review all the clinical or experimental findings in depth but to summarize the principle observations and refer to previous review articles for details and original references. Thus, although not exhaustive, this review, by complementing and extending those before it, should provide a comprehensive coverage of the field from a contemporary perspective.

For this chapter the format that has been chosen is to survey the pathophysiological effects before discussing the morphological effects. For each theme, a short summary of the major clinical findings will be given that includes a discussion of any mechanistic studies. Next, a detailed account of the relevant experimental findings in animals will be given, placing emphasis on the newer findings or interpretations. Finally, in Section 7.29.8, the major features of each topic are highlighted and integrated into a broad overview.

7.29.2 HEMODYNAMIC EFFECTS

7.29.2.1 Clinical Observations

One of the first observations made about cyclosporine was its ability to impair renal function. This was sometimes manifest as frank oliguric renal failure during its early, high-dose usage, and later as an almost inevitable rise in serum creatinine or urea. The rise in serum creatinine is now known to largely reflect a reversible vasoconstriction that causes a decrease in glomerular filtration. The evidence that this renal dysfunction reflects vasoconstriction rather than tissue damage comes from the fact that the rise in serum creatinine is reversible[1,2] and is not progressive, even after prolonged treatment in organ recipients.[3,4]

In contrast to chronic renal failure, where the reduction in glomerular filtration is much more severe, cyclosporine-induced decreases in glomerular filtration rate can be well predicted from the rise in serum creatinine and urea.[5] This is because glomerular filtration rate is only modestly affected and does not fall to less than half of normal. In autoimmune diseases, where rejection is not a threat and doses can be decreased as needed, serum creatinine is monitored closely. By preventing it from rising more than 30–50% above baseline values, any danger of subsequent structural damage to the kidney can be minimized.[6]

The clinical picture that accompanies the reduction of renal function is one of prerenal azotemia. Serum urea is always more elevated than serum creatinine[7] and the excretion of sodium, relative to that filtered (fractional excretion) is usually very low.[7] This setting, together with the salt-sensitive nature of the renal dysfunction, implies that circulating fluid volume may be reduced. In support, some measurements of plasma volume using labeled erythrocytes show it to be reduced, although others using labeled albumin show it to be increased.[7] However, the latter findings must be interpreted cautiously, as albumin will give erroneously high estimates of plasma volume if it leaks from the vessels, as in animal studies.[8]

Mechanistic studies in humans have shown that the cause of the increase in serum creatinine and urea is a reduction in glomerular filtration rate (GFR) accompanied by a reduction in renal plasma flow (RPF) or renal blood flow (RBF).[7] This combination of parallel decreases in renal perfusion and glomerular filtration in the absence of decreased systemic blood pressure, means that vasoconstriction, predominately of the afferent vessels, is the cause of reduced renal function. Sophisticated studies of glomerular dynamics show that decreased permeation through the glomerular membrane is not generally involved.[7]

Earlier theories about single causes of vasoconstriction, such as decreased prostaglandin or increased thromboxane synthesis, have been largely surpassed. Newer theories propose that endothelial damage leads to increased endothelin release and decreased production of nitric oxide (NO) from L-arginine. Plasma levels of endothelin may be raised in transplant recipients.[9] In addition, the renal vasodilation that follows L-arginine infusion is blunted during cyclosporine therapy.[10] Hence, endothelial damage is thought to cause renal vasoconstriction by decreasing the production of nitric oxide, thereby reducing the response to many vasodilators, as well as by increasing endothelin release.

The effects of vasoconstriction are not confined to the kidney and are seen in the systemic circulation as an increase in systolic and diastolic blood pressure. Sympathetic activity can be increased in the clinical setting,[11] but it is unclear whether this effect is primary or is secondary to other effects, such as volume depletion. The rise in blood pressure, like the renal dysfunction, is increasingly believed to result primarily from endothelial damage. It is responsive to dose reduction and will reverse completely upon drug discontinuation in those patients with autoimmune diseases who acquired an elevated blood pressure whilst taking cyclosporine.[1]

However, the rise in arterial pressure, particularly in patients prone to it because of their underlying condition, is a cause for concern. Despite wide variations in the prevalence and severity of hypertension that reflect pre-existing cardiovascular or renal insufficiency or concomitant therapy, calcium channel blockers or angiotensin-converting enzyme (ACE) inhibitors are felt to be the most appropriate antihypertensive therapy. Nevertheless, the search for the most specific antagonist of renal and systemic vasoconstriction persists, and agents that block the production or action of endothelin or enhance the activity of the nitric oxide system may hold promise in this regard.

7.29.2.2 Animal Findings

In animal models, cyclosporine lowers renal function and raises blood pressure just as in humans. Whereas early studies failed to show that cyclosporine consistently raised blood pressure, careful telemetry studies in conscious rats clearly showed the development of a dose-dependent moderate hypertension after acute or chronic treatment.[12] Invasive methods or tail-cuff plethysmography may be unable to detect moderate hypertension (15–30 mmHg) reliably in cyclosporine-treated rats, as both methods themselves induce changes in blood pressure during measurement.[13]

Renal vasoconstriction reduces glomerular filtration rate and renal blood flow in rats. Initial studies, using acute administration of high doses, indicated mainly preglomerular vasoconstriction together with a reduction in the ultrafiltration coefficient. Later studies using chronic treatment and lower doses,

confirmed the preglomerular vasoconstriction but not the decrease in filtration coefficient.[7] Subsequent studies have shown a direct vasoconstrictive action of cyclosporine on isolated afferent arterioles[14,15] and in renal microvessels of kidneys perfused *ex vivo*.[16] Interestingly, arteriolar responses were seen at bath concentrations much lower than blood concentrations needed for immunosuppression *in vivo*, an effect possibly related to the use of protein-free perfusate solutions that minimize cyclosporine binding.[14]

Early research efforts did not allow the systemic and renal vasoconstriction to be associated with a single vasoactive system or substance. Disturbances in the renin–angiotensin system,[7] in thromboxane synthesis,[7] in prostaglandin production,[7] and in the sympathetic nervous system[7] all seemed to be involved to some extent in the responses to either acute or chronic cyclosporine exposure.[17] This apparent multifactorial character of the hemodynamic dysfunction was later reconfirmed in a single study that compared the effects of several agents and excluded differences in technique or dosage as confounding factors.[18]

Much research has focused on two important effects of chronic cyclosporine treatment that promise to supply a more unified explanation of pathogenesis of the renal hemodynamic dysfunction: endothelial injury, caused either by acute or chronic treatment with cyclosporine[7,12,19] or its intravenous vehicle,[7,20] and a direct sensitization of the vascular smooth muscle cells in the renal and systemic circulation.

7.29.2.2.1 Endothelial effects

Endothelial injury results in three events that have important effects on renal and systemic hemodynamics. The first is a reduction in intravascular volume that is attributable to an alteration in capillary permeability to protein. As shown in Figure 1, cyclosporine-treated rats exhibit reduced plasma volume which results from increased leakage of albumin from the vasculature,[8] and not from overt renal protein loss.[7] Circulatory hypovolemia is completely consistent with the sensitivity of renal function to low dietary salt intake, and the improvement produced by high salt intake or acute repletion of plasma or blood volume.[7]

Reduced blood volume is a powerful stimulus for activation of the sympathetic nervous and the renin–angiotensin systems, both of which can elicit systemic and renal vasoconstriction. There is conflicting evidence concerning a direct action of cyclosporine to activate the sympathetic nervous system.[12,17,21]

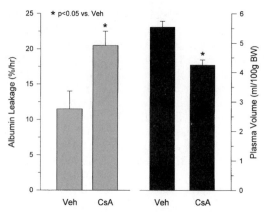

Figure 1 Left: effect of cyclosporine (CsA, $25\,\text{mg}\,\text{kg}^{-1}\text{d}^{-1}$ i.p., 21 d) on the leakage of ^{125}I-albumin from the plasma compartment in the rat. Right: corresponding effect of cyclosporine on intravascular plasma volume, estimated after correction for albumin leakage (mean ± SE, $n = 6$/group) (after Moore et al.[8]).

Although treatment with sympathetic antagonists improved cyclosporine-induced renal vasoconstriction in some studies, renal dyfunction is observed in both rats and humans with denervated kidneys,[17] suggesting that sympathetic activation cannot be the sole mediator of the renal vasoconstriction.

The second important consequence of endothelial injury by cyclosporine is the release of endothelin-1, a potent vasoconstrictor. Exposure to cyclosporine has been shown to elevate plasma endothelin levels,[22–24] induce endothelin synthesis in cultured human endothelial cells[25] and in systemic vessels,[26] and raise endothelin receptor density in renal vessels.[22,27] Renal function is improved after endothelin inactivation by receptor blockade or with anti-endothelin-1 antibodies.[22–24] Further, cyclosporine-induced vasoconstriction has been shown to be endothelin-mediated in renal arterioles perfused *in vitro*,[14] although platelet-activating factor or thromboxane A_2 are required in addition to produce afferent or efferent arteriolar constriction, respectively.[15]

The third major consequence of cyclosporine-induced endothelial injury is impaired synthesis of NO (nitric oxide or endothelium-derived relaxing factor). A suppression of endothelium-dependent vasodilation, which depends upon NO release, was seen in renal[28,29] and systemic vessels,[30–32] while the response to vasodilators that do not depend upon this release, was not affected.[33] In rats, the renal vasoconstriction *in vivo* and in perfused vessels *ex vivo* can be reversed by excess L-arginine, the substrate for NO synthase.[34–36] Direct measurements in afferent arterioles from cyclosporine-treated rats show both diminished NO release during acetylcholine-induced vasodilation, and reduced NO synthase immunoreactivity in the renal microvascular endothelium.[37]

7.29.2.2.2 *Vascular smooth muscle effects*

There is considerable evidence that cyclosporine also directly sensitizes vascular smooth muscle cells and contractile glomerular mesangial cells to a variety of constrictor agents. The normal response to constrictors in these contractile cells is a rise in intracellular calcium, caused by enhanced calcium influx and increased calcium mobilization from intracellular stores. After exposure to cyclosporine, the resting cytosolic calcium concentration is raised, calcium influx is elevated, and intracellular calcium stores are increased.[38,39] Thus, the threshold to vasoconstrictor stimulation is lowered and the resulting response is enhanced, particularly to angiotensin II and vasopressin, two vasoactive peptides that are elevated in cyclosporine-treated rats.[17]

The sensitization of vascular smooth muscle and mesangial cells can be blocked by atrial natriuretic peptide,[40,41] an effect that might explain its beneficial effects on renal hemodynamic dysfunction and systemic hypertension in cyclosporine-treated animals and humans.[42–44] Interestingly, endothelin may also mediate some of these abnormalities in vascular smooth muscle cells, as endothelin receptor blockade can blunt the hyperresponsiveness to vasopressin in glomerular mesangial cells.[45] As both vascular smooth muscle cells and mesangial cells can produce endothelin,[46,45] sensitization by endothelin may be independent of endothlial injury.

In the kidney, as elsewhere in the circulation, NO is an important modulator of the vascular reactivity to vasoactive agents. In cyclosporine-treated animals, there is an inhibition of this vasodilatory system, coupled with enhanced endothelin secretion and sensitization of the vascular smooth muscle cells. Together, this causes a marked shift in the balance between dilatory and constrictive forces, such that even normal levels of vasoactive agents can suffice to cause substantial vasoconstriction. Hence, these new insights begin to provide a unified mechanism to explain the consistent findings that inhibition of a wide variety of endogenous vasoconstrictors can partially ameliorate cyclosporine-induced renal hemodynamic dysfunction.

7.29.3 TUBULAR EPITHELIAL EFFECTS

7.29.3.1 Clinical Observations

The one feature of cyclosporine's effects on the kidney that distinguishes this drug from almost all other nephrotoxic agents is the fact that it has remarkably little effect on the ability of the tubular epithelium to perform transport. Generally, with nephrotoxins that target the tubules, the reabsorption of sodium, which makes up the bulk of the substances to be reabsorbed, is diminished. Yet with cyclosporine, the fractional excretion of sodium is very low, suggesting that reabsorptive function is unusually well preserved. Once again, this indicates that the cause of renal dysfunction is primarily vascular and not tubular in nature.[7]

There are some minor disturbances in tubular transport, the most obvious of which is an impairment of magnesium transport. Lowered serum magnesium levels are seen in virtually all patients treated with cyclosporine, and the increase in fractional excretion of the ion makes it quite clear that this is caused by a reduction in tubular reabsorption, rather than a failure of intestinal absorption.[7] The exact site of the defect and the exact cause are unknown but the resulting hypomagnesemia with low-dose therapy is generally benign and only rarely warrants magnesium supplementation. However, high doses of cyclosporine and concomitant use of diuretics may cause magnesium levels to fall more severely, and the value of magnesium supplementation in averting any danger of seizures and hypertensive episodes has been debated.

Potassium transport may be impaired and lead to mild hyperkalemia, sometimes in association with hyporeninemic hypoaldosteronism. This condition may resemble a type IV tubular acidosis, which suggests that the defect is in the distal tubule.[7] Neither the hyperkalemia nor the associated acidosis is of a magnitude that warrants clinical intervention, except under unusual circumstances, and is generally so discrete that it often remains undetected. The somewhat curious decrease in the secretion of kallikrein and Tamm–Horsfall protein, that are synthesized in the distal tubule, are of no clinical relevance.[7]

Further impairments in transport are seen with some weak organic acids. The extraction of *p*-aminohippuric acid (PAH) has been reported to be lowered.[7] PAH is used to measure renal plasma flow and reduced tubular secretion could result in underestimation of renal perfusion. Of more significance is the decreased clearance of uric acid, which can lead to a rise in serum levels.[7] The resultant hyperuricemia is generally modest and of no concern clinically, as it does not generally lead to gout. However, clinically relevant hyperuricemia can result, particularly in kidney recipients in whom diuretics are used, and can on occasion result in the development of gout.[7]

This collection of minor disturbances in tubular reabsorptive and secretory functions raises the suspicion that there must be a minor disturbance in tubular sodium and water reabsorption, perhaps masked in whole kidney studies because the amount of salt and water delivered to the tubules is lowered by the decrease in filtration rate. Subtle disturbances in distal salt and water reabsorption are indeed suggested by the decrease in tubular diluting or concentrating ability seen in some studies,[7] which could provide a mechanism for some of the other effects seen.

7.29.3.2 Animal Findings

The findings in animal studies are in good agreement with those in humans and show only minor changes in renal tubular function. Because in animals tubular reabsorption can be studied independent of the filtration and quantity of filtered water and solutes so delivered to the tubular epithelium, more specific statements can be made about the degree of tubular reabsorption or secretion than in humans. Whereas tubular function is adequately preserved for reabsorption of the reduced filtered load of water and solutes, closer inspection confirms a reduction in sodium reabsorption, not detectable in studies of the whole kidney.

Although in intact animals the fractional excretion of sodium in the urine is very low,[7] cyclosporine can impair the urinary diluting and concentrating mechanism.[7] It has also been shown to reduce the diluting ability of single nephrons and, hence, the absolute amount of sodium transport in the thick ascending limb and distal tubular segment.[7] However, this effect was only demonstrated *in vivo* when the tubule was perfused at rates above those present normally. In contrast, in whole animals when a lowered filtration rate reduces tubular fluid and sodium delivery, reduced urine concentrating ability, indicative of reduced tubular sodium reabsorption, is not evident.[7,20]

Subsequent evidence has confirmed these findings by studying the effects of acute cyclosporine exposure on Na,K-ATPase activity in isolated tubular segments. Sodium pump activity was found to be inhibited by cyclosporine in the more distal but not proximal segments, that is in the medullary thick ascending limb and cortical and outer medullary collecting ducts but not in the proximal convoluted S2 segment.[47] This finding can then explain why tubular reabsorption is well preserved in the whole kidney, as the bulk of it occurs in the proximal segments, yet is also reduced in the more distal segments, as shown by the reduction in concentrating and diluting ability.[7]

This decrease in Na,K-ATPase activity in more distal segments could further explain the decreased tubular potassium secretion and magnesium reabsorption that result in hyperkalemia and hypomagnesemia.[7] Potassium secretion in the distal segments is driven by the activity of Na,K-ATPase, so that when it is depressed, potassium secretion falls, and all the more so if tubular flow rate is simultaneously decreased.[48] Similarly, magnesium reabsorption in the thick ascending limb is driven by the transtubular electrical potential established by the activities of the sodium pump and Na,K,2Cl cotransporter,[48] such that when Na,K-ATPase activity is reduced, distal magnesium reabsorption is depressed.

7.29.4 RENIN–ANGIOTENSIN SYSTEM

7.29.4.1 Clinical Observations

The renin–angiotensin system is notoriously difficult to evaluate under clinical conditions because of the effect of cardiac or renal insufficiency and associated therapy on blood pressure and vascular volume. Not surprisingly, the findings have been contradictory and equivocal. In general, plasma renin has been found to be low and the response to several types of stimulation has been seen to be blunted.[7] Similarly, aldosterone concentrations can be lower than normal, particularly during stimulation by low salt diet.[7] Hence, it has generally been assumed that the renin–angiotensin system is depressed by cyclosporine.

Other observations suggest the contrary; whilst renin levels were low or normal, prorenin levels were elevated, and the juxtaglomerular apparatus (JGA) that produces renin was enlarged.[49] Also, in renal biopsies the number of renin-containing cells was greater during cyclosporine therapy than after its withdrawal.[50] Similarly, although the JGA was not visibly enlarged in renal biopsies taken during cyclosporine therapy, the number of renin-containing cells outside it was increased, new cells having been recruited for this task.[51] Hence, there is good reason to suspect that cyclosporine stimulates renin production and also influences its processing in some way.

However, this pathophysiology of renin is unlikely to be the main cause of either the renal or the peripheral vasoconstriction. There is little evidence that systemic angiotensin II levels are raised, as neither the use of receptor blockers nor of ACE inhibitors to reduce angiotensin II synthesis has been particularly successful in reversing the renal dysfunction. Also, receptor blockers are not able to lower the raised blood pressure caused by cyclosporine. Hence, a major role for angiotensin II as the agent responsible for the vasoconstriction, is not under discussion, even though ACE inhibitors are often used to control cyclosporine-induced hypertension.

7.29.4.2 Animal Findings

Early studies in both experimental animals and humans, which showed that inhibition of the renin–angiotensin system could not fully reverse or prevent the renal functional disturbance seen with cyclosporine, seemed to indicate that renin and angiotensin were not involved in its pathogenesis. However, there is mounting evidence in animals that renin and angiotensin may both play central roles in the development of the morphological changes associated with cyclosporine treatment and that they may also contribute to the renal dysfunction that invariably develops.

Cyclosporine treatment in rats has a number of effects on the renin–angiotensin system. It has long been known to elevate plasma renin activity and raise the renin content of the kidney.[7,52] It is known that these changes are associated with the transformation of smooth muscle cells along the afferent arteriole into renin-containing cells, as illustrated in Figure 2. Messenger RNA levels for the angiotensin II type I receptor are decreased in the kidney,[52] and increased in other vascular beds,[53] the latter possibly explaining the beneficial effects of angiotensin-converting enzyme inhibition on cyclosporine-induced hypertension. Finally, cyclosporine blunts angiotensin II-stimulated aldosterone secretion.[7]

The effects of cyclosporine on renin release are complex and likely involve both primary and secondary phenomena. Primary effects have been shown in kidney slices and isolated juxtaglomerular cells, in which renin secretion is stimulated by high cyclosporine concentrations.[7,54] The secondary effects arise from any direct cyclosporine-induced activation of the sympathetic nervous system, which stimulates renin release, or by a decrease of circulating plasma volume, which also activates the sympathetic nervous system and stimulates renin release.

Some studies indicate that cyclosporine may also interfere with intracellular renin processing

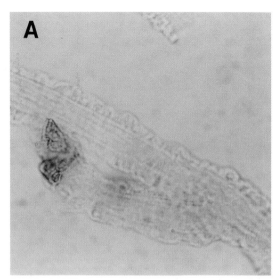
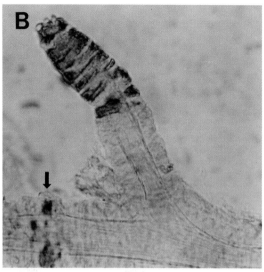

Figure 2 Cyclosporine A-induced recruitment of new renin-containing cells in the afferent arteriole. Shown are arterioles microdissected from HCl macerated kidneys and subsequently immunostained for rat renin.[87] The glomeruli become detached during dissection, but the arterioles remain largely intact. (A) typical control arteriole with renin deposits (brown peroxidase stain) confined to the distal juxtaglomerular segment. (B) arteriole from a cyclosporine-treated rat ($25\,mg\,kg^{-1}d^{-1}$ i.m. 21 d) showing extensive recruitment of renin-containing cells along the arteriole, and in the interlobular artery (arrow).

and affect its production, storage, and secretion. Plasma concentrations of prorenin, the immediate precursor of renin, are elevated by cyclosporine.[7] In isolated juxtaglomerular cells, cyclosporine increases renin production, suggesting increased gene expression, and prorenin accumulation, additionally suggesting an abnormality in prorenin conversion.[54] In rat kidneys, cyclosporine causes the slow accumulation of an acidic renin isoform.[55] This isoform is normally present in only very small amounts and has a low rate of secretion following

stimulation compared to normal renin isoforms.[56] Thus, renin synthesis, processing, and release are disturbed.

An intriguing aspect of the action of cyclosporine to stimulate the renin–angiotensin system is that it occurs at a low-dose, and one that is close to that at which immunosuppression is achieved in the rat.[57] As illustrated in Figure 3, impaired renal function, as reflected by the rise in serum creatinine, develops at a higher dose than that needed to raise plasma renin. Thus, activation of the renin–angiotensin system may be a primary effect and may share the same mechanism of action as immunosuppression. This direct stimulation of the renin–angiotensin system could then be reinforced at higher doses by other, indirect effects, such as sympathetic activation and renal vasoconstriction.

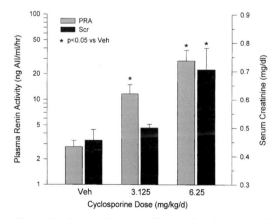

Figure 3 Comparison on effects of cyclosporine-A treatment (21 d, p.o.) on plasma renin activity and plasma creatinine concentration in the rat. $n = 6$ rats/dose and group. Plasma renin activity increases at a lower cyclosporine dose than serum creatinine (after Kaskel *et al.*[57]).

7.29.5 STRUCTURAL CHANGES AND TISSUE DAMAGE

7.29.5.1 Clinical Observations

The final definition of the morphological changes to the kidney caused by cyclosporine resulted from painstaking studies of biopsies taken after kidney transplantation to determine whether renal dysfunction was caused by rejection or toxicity. The prevailing belief that toxicity is characterized by tubular necrosis, as with classical nephrotoxins, hampered this

574

work, as did the presence of other changes resulting from ischemia and rejection. However, careful comparison of kidney biopsies from transplanted kidneys with those from the native kidneys of heart recipients or patients with autoimmune diseases finally enabled the cyclosporine effects to be separated from the others.

The morphological changes seen in humans were identified when the doses of cyclosporine routinely in use were very high. As doses have diminished, so have the incidence and severity of cyclosporine-induced nephropathy. Initial findings in heart recipients[49,58] suggested a high rate of chronic renal damage leading to a high incidence of end-stage renal failure. Subsequently, the incidence of cyclosporine-induced renal damage and the number of graft losses it causes was shown to be reduced in kidney recipients receiving low rather than high doses of cyclosporine.[4,59] Finally, in a large series of patients with autoimmune diseases, the risk of developing cyclosporine-associated renal damage was confirmed to be dose-dependent and very small with low doses.[6]

The morphological changes that are seen affect the tubules, the interstitium and the vessels. Whether these changes are interconnected or independent has not been fully established. The changes to the tubules were the ones most frequently seen when high doses were used. These consist of isometric vacuolization, giant mitochondria, lysosomal inclusion bodies, sporadic single-cell necroses, and microcalcifications, all predominantly in the proximal tubule.[7,60] These changes appear to be reversible upon drug withdrawal and are of no functional significance, other than to alert the clinician to the possibility of over-dosing. Tubular atrophy is another lesion but this is thought to be secondary to vascular damage.[7,60]

The interstitial alterations are more rare than the tubular ones and involve the development of a fibrotic lesion. Interstitial fibrosis is common in many types of renal injury and cyclosporine stimulates matrix protein deposition in several tissues.[17,19] Nonetheless, because of its distribution, the fibrosis is viewed as mostly being secondary to vascular damage. Instead of being diffuse, the fibrosis is typically striped, the pattern both matching the tissue region normally supplied by one afferent arteriole and matching the area that contains the atrophied tubules.[60] Thus, this striped fibrosis, illustrated in Figure 4, is believed to represent the local ischemic damage caused by the obliterative arteriolopathy.

The changes to the vessels are more discrete and involve the afferent arterioles and sometimes the smallest arteries. Typical injuries are

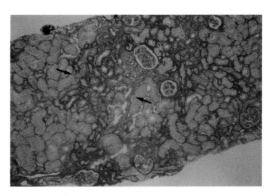

Figure 4 Human biopsy section (van Gieson's stain) showing cyclosporine-induced striped interstitial fibrosis (arrows) consistent with localized ischemia subsequent to loss of individual nephrons (reprinted by permission of *The New England Journal of Medicine*, G. Fentren and M. J. Mihatsch, 'Risk factors for cyclosporine-induced nephropathy in patients with autoimmune diseases. International kidney biopsy registry of cydosporine in autoimmune diseases.' *N. Engl. J. Med.*, 1992, **326**, 1654–1660, Copyright 1992, Massachusetts Medical Society).

illustrated in Figure 5. Damage is seen in the endothelium first, as vacuoles, lysosomal inclusion bodies and ultimately as occasional single-cell necroses and an interruption of the endothelial lining. Single-cell necroses of the myocytes and their replacement by lumpy protein deposits (hyaline) lead to a characteristic circular or nodular pattern of eosinophilic material in the arteriolar wall. The resultant narrowing of the lumen and development of protein or fibrin thrombi can lead to occlusion of the vessel and the collapse and atrophy of the attached glomerulus.[7,60] The thrombotic and proliferative changes reminiscent of the hemolytic uremic syndrome that can lead to segmental, focal glomerular sclerosis are considered to be a separate lesion.[59]

This obliterative arteriolopathy, once thought to be irreversible, has been shown to regress upon dose reduction or drug withdrawal. Patients with biopsy-proven arteriolar damage, which had been graded according to its severity, were discontinued from cyclosporine therapy and biopsied again. In the majority of the patients the severity of the arteriolar lesions decreased after dose reduction or discontinuation of cyclosporine. In addition, the vessels showed clear evidence of a remodeling process, as the lumena were patent and the hyaline deposits in the arteriolar wall were absent, having been replaced by irregularly-arranged smooth muscle cells.[61]

However, the great mystery is why vasculopathy is confined to the afferent arteriole and smallest arteries. As the only physiological

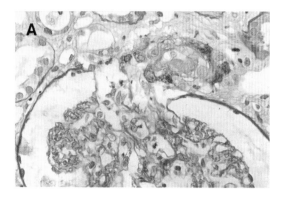

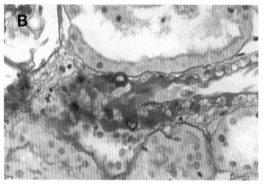

Figure 5 (A) human biopsy section (PAS stain) showing cyclosporine arteriolopathy with hyaline deposits intruding into the lumen. The brown immunoperoxidase stain shows the location of renin-containing granular cells. The micrograph was kindly supplied by M. Mihatsch. (B) Axial section of human afferent arteriole with cyclosporine arteriolopathy. Again, the injury is co-localized with renin-containing cells. The presence of renin-positive cells upstream from the juxtaglomerular apparatus is consistent with recruitment of new renin-containing cells ((B) reproduced by permission of Dustri-Verlag Dr. Karl Feistle from *Clin. Nephrol.*, 1995, **43**, 226–231).

function unique to these vessels is the ability to synthesize and store renin, the distribution of renin containing-cells and of arteriolopathy was examined in biopsy specimens. As shown in Figure 5, renin-positive cells were found outside the juxtaglomerular region, extending up the afferent arteriole with exactly the same distribution as the necrotic myocytes. With increasing severity of arteriolopathy, the number of renin-positive cells was noted to decrease. Thus, it was concluded that the renin-producing, smooth muscle cells along the arteriole were probably the ones to become necrotic and be replaced by protein deposits.[51]

7.29.5.2 Animal Findings

Tubular morphological changes similar to those seen in human biopsies, vacuolization,

single-cell necroses, and microcalcifications are relatively easy to replicate in rats.[7] However, it has been much harder to induce the vascular and interstitial injury in experimental animals. Early attempts to reproduce interstitial or vascular injury used excessive doses or hypertensive rats but neither the pattern of interstitial damage nor the location of the vascular injury mirrored those seen in humans.[17] Thus the adequacy of the rat as a model for the most serious of the side effects induced by cyclosporine was long called into question.

However, subsequent studies have clearly shown that chronic treatment with moderate doses of cyclosporine, when coupled with dietary salt restriction, produces tubulo-interstitial and vascular injuries in rats similar in many respects to those seen in humans. Within weeks of treatment, an injury to the distal arterioles develops,[62] as illustrated in Figure 6. It consists of smooth muscle cell hyperplasia and the accumulation of lumpy, eosinophilic deposits in the vascular wall.[63] This injury was greatly reduced by blockade of the angiotensin II type I receptor[63] or converting enzyme inhibition,[64] but not by other antihypertensive agents with other sites of action,[63] nor by endothelin receptor blockade.[23]

The concurrent development of striped tubulointerstitial fibrosis was also demonstrated in the salt-restricted rat model,[65,66] as shown in Figure 7. Besides tubular injury, consisting of tubular dilatation, basement membrane thickening, cell swelling and sloughing, the surrounding interstitium was fibrotic. The interstitial fibrosis but not the tubular injury could be reduced by angiotensin II type I receptor blockade. Both tubular cells and interstitium showed an increase in type IV collagen mRNA.[63] Increased expression of osteopontin, a macrophage chemoattractant, was seen most commonly in tubular cells and interstitial fibroblasts, whereas tubular epithelial cells showed an increase expression of transforming growth factor-β,[63] a potent stimulant of extracellular matrix production associated with renal fibrosis in a variety of diseases.[67]

However, two key issues still require clarification. The first is whether the interstitial injuries in rats and humans share the same etiology. The interstitial fibrosis in humans is localized and appears to be a consequence of afferent arteriolar occlusion. In the rat, the interstitial fibrosis is much more diffuse, and arteriolar occlusion is not common. The second, related issue is the extent to which the interstitial fibrosis is dependent on the ischemia that results from the action of salt-restriction to amplify the depression in renal perfusion caused by cyclosporine. Renal perfusion in

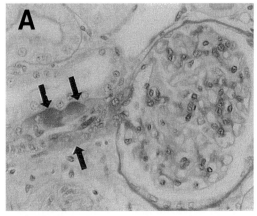

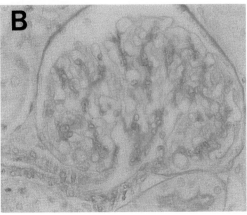

Figure 6 (A) Cyclosporine arteriolopathy in a salt-depleted (0.05% sodium diet), cyclosporine-treated rat (15 mg kg^{-1}d^{-1}, p.o., 35 d) showing lumpy, eosi-nophillic hyaline deposits. (B) Section from a cyclosporine-treated, salt-depleted rat treated with an angiotensin II receptor blocker (Losartan, 10 mg kg^{-1}d^{-1}), which significantly reduced the incidence of cyclosporine arteriolopathy (repro-duced by permission of Williams and Wilkins from R. H. Pichler, N. Franceschini, B. A. Young *et al.* 'Pathogenesis of cyclosporine nephropathy: roles of angiotensin II and osteopontin.' *J. Am. Soc. Nephrol.*, 1995, **6**, 1186–1196).

this model may be profoundly depressed and marked hypotension can develop, particularly in studies using severe salt restriction.[68] Ische-mia alone, induced either by renal arterial clamping[69] or by prolonged angiotensin II infusion[70] is known to cause interstitial changes virtually identical to those seen with cyclospor-ine in salt-restricted rats.

In this regard, a subsequent study has shown that dietary L-arginine supplementation can markedly reduce the renal functional impair-ment and the structural changes seen in both the interstitium and the vessels.[34] Although suggestive of a role for endothelial injury and NO in the structural injuries caused by cyclos-porine, the beneficial effects could also reflect a

lessening of renal ischemia. Renal vasodilation with dietary L-arginine could result both from enhanced NO production and stimulated release of glugagon, a renal vasodilator. More studies will indicate if the salt-depletion model lives up to its promise to supply mechanistic insight into cyclosporine-induced tissue injury.

7.29.6 CELLULAR OR BIOCHEMICAL CHANGES

7.29.6.1 General Observations

The mechanisms of immunosuppression afforded by cyclosporine requires it to bind to form a pentameric complex with calcineurin, calmodulin, calcium, and cyclophilin, its major binding protein. As cyclophilins are ubiquitous and abundant proteins found throughout nature, cyclosporine may exert effects in differ-ent tissues, reflecting the action that cyclophilin has in these cells. The most abundant cyclophilin is the A form but there are other forms, C and D, with a more restricted tissue distribution.[71]

Cyclophilins are all peptidyl-proline *cis–trans* isomerases (PPIases or rotamases) that catalyze *cis–trans* isomerization of the peptide bond. This rotamase activity is believed to be important in protein folding, an essential component of protein trafficking within the cell and across the cell membrane.[72] Perhaps, this is the reason that during cyclosporine therapy epithelial cells involved in protein transport develop empty cytosolic vacuoles, derived from the endoplasmic reticulum. Such vacuoles can be seen in the salivary gland, exocrine and endocrine pancreas,[73] and proxi-mal tubule.[7] It is possible that these vacuoles once contained protein that, because it is improperly folded, cannot be fully chaperoned across the cell or cell membranes. Inhibition of rotamase activity may also underlie the cyclo-sporine-induced accumulation of abnormal renin isoforms in juxtaglomerular cells.[55]

Cyclosporine acts on T lymphocytes by inhibiting a calcium-dependent signal transduc-tion pathway that requires calcineurin activa-tion. Calcineurin, the phosphatase that is inhibited when bound to the cyclosporine–cyclophilin complex, is a calmodulin-binding enzyme that regulates gene transcription in many cells. The basis of the selective action on T cells appears to be that their calcineurin levels are sufficiently low to be strongly inhibited at low cyclosporine concentrations.[74] Cyclospor-ine and other immunosuppressants which inhibit calcineurin, are increasingly being used as probes to identify cellular processes that utilize this pathway. Alterations found in the

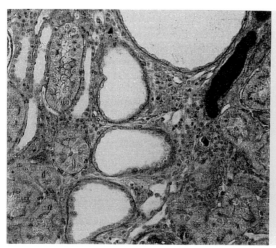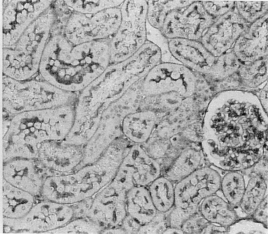

Figure 7 Left: tubulo-interstitial fibrosis in a cortical section (400×) from a salt-depleted, cyclosporine-treated rat (15 mg kg^{-1}d^{-1}, p.o., 35 d). Marked interstitial thickening and tubular dilation is evident. Right: section from a normal control rat showing typical narrow interstitial regions (reprinted by permission of Blackwell Science, Inc., from B. A. Young, E. A. Burdmann, R. J. Johnson *et al.*, 'Cellular proliferation and macrophage influx precede interstitial fibrosis in cyclosporine nephrotoxicity. *Kidney Int.*, 1995, **48**, 439–448).

same dose range as needed for immunosuppression are usually considered to be calcineurin-mediated.

One such cyclosporine side effect that is known to be caused by calcineurin inhibition is inhibited distal tubular sodium transport.[75,76] Another good candidate is the alteration in renin synthesis and secretion by the juxtaglomerular cells. In these cells, an increase in cellular calcium inhibits the secretion as well as synthesis via one or more calmodulin-mediated pathways.[77] If this pathway subsequently involves calcineurin, its inhibition by cyclosporine would eliminate an inhibitory control on renin production and secretion, and provide a mechanism to explain the accumulation of renin within the juxtaglomerular cells.[52,54] Further, intracellular calcium regulates many intracellular events in many different cell types, including cell contraction, hormone synthesis and release, and electrolyte transport. More studies will be needed to determine the extent to which cyclosporine's side effects can be attributed to its disruption of the calcium–calmodulin–calcineurin signaling pathway.

A number of functional changes to mitochondria have been reported. These include a number of alterations in metabolic function, such as a reduction in the quantity of the F1-ATPase subunit, a decrease in oxidative phosphorylation, a fall in malate dehydrogenase activity, and reduced gluconeogenesis.[60] However, to what extent these effects, seen mostly *ex vivo* during exposure to high concentrations, represent a decrease in function *in vivo* is unclear. There are, in addition, changes in membrane properties, such as an alteration of the inner mitochondrial membrane permeability to calcium and to protein[60] but their significance is also unclear.

7.29.7 SIMILAR IMMUNOSUPPRESSANTS

7.29.7.1 Molecular Mechanism of Action

There are two other immunosuppressants which confer selective immunosuppression, rather than inhibiting lymphocyte proliferation by interfering with cellular metabolism. These are tacrolimus (FK 506) and sirolimus (rapamycin), which are polyketides produced by bacterial organisms, rather than fungal ones. Tacrolimus and sirolimus are structural analogues that share the same intracellular binding protein, an immunophilin that is analogous in function to cyclophilin, but they act in different places in the cascade of events that leads first to T-cell activation then to lymphocyte growth and proliferation.

Tacrolimus and cyclosporine, despite their different binding proteins (immunophilins), share the mechanism of inhibiting the calcium-dependent signal transduction following T-cell receptor activation by antigen-presenting cells. The interruption of signal transduction inhibits the expression of "early" gene products regulating the synthesis of cytokines such as interleukin-2 (IL-2), a major growth factor. Sirolimus acts further down the activation cascade by interrupting the calcium-independent signal transduction that follows IL-2 receptor activation. This, in turn, inhibits the expression of "late" gene products after growth factor stimulation by IL-2, which control cell division.[78]

All these lipophilic immunosuppressants enter the cell by diffusion and bind in the cytosol to their own immunophilin. This inhibits the rotamase activity of the immunophilins but does not itself cause immunosuppression.[78] Further actions of sirolimus are unclear, but cyclosporine and tacrolimus both act to inhibit calcineurin, a calcium–calmodulin-activated serine–threonine phosphatase. They do so by presenting their bound immunophilin for further binding to the calcineurin–calmodulin complex.[79] Calcineurin inhibition prevents part of nuclear factor of activated T cells (NF-AT) from migrating to and binding within the nucleus to initiate transcription of the IL-2 gene.[78–80]

7.29.7.2 Renal Side Effects

Common features and differences in the actions of the three immunosuppressants provide some insight into the origin of the renal side effects. Initial observations indicated that tacrolimus was relatively free of side effects. However, later observations showed renal functional impairment, generalized vasoconstriction, some increase in systemic blood pressure,[81,82] as well as renal morphological changes. Most significant is the finding that the renal morphological changes seen in man, once thought to be unique to cyclosporine, are replicated in every detail in patients treated with tacrolimus.[83] This indicates that the similar side effects are probably the result of the similar mode of action.

This view is consistent with the observations made with sirolimus, which, although sharing the same immunophilin as tacrolimus, has a completely different mode of action. Animal studies have indicated that sirolimus neither causes the typical morphological changes in rodents seen with cyclosporine or tacrolimus nor does it depress renal function.[84–86] Clinical findings have not yet been fully established, as sirolimus is still undergoing clinical evaluation. However, the expectation is that nephrotoxicity will not be pronounced in humans because sirolimus does not inhibit calcineurin, thereby preserving calcium-dependent pathways in the tubular and vascular cells.

The suggestion that calcineurin could be involved in the development of side effects was developed from observations made in other tissues. Both tacrolimus and cyclosporine block the degranulation of mast cells, neutrophils, basophils and cytotoxic T lymphocytes.

Transcription events are not involved in this blockade but as a rise in cellular calcium is an essential part of the signaling pathway, this suggested that calcineurin inhibition by tacrolimus or cyclosporine binding could be responsible.[80] If so, a similar inhibition of other calcineurin-regulated events may then underlie toxicity.

Consistent with this view is the finding that the calcineurin-dependent Na/K-pump in the renal tubules can be inhibited by tacrolimus and cyclosporine.[47,76] As shown in Figure 8, a low concentration of cyclosporine (600 ng mL^{-1}) inhibits Na,K-ATPase activity in the medullary thick ascending limb and cortical collecting tubule (distal segments) but not the S2 segment (proximal segment).[47] Tacrolimus (6 ng mL^{-1}) has virtually identical effects in the proximal S2 segment and cortical collecting ducts.[76] The basis for the differential sensitivity of Na,K-ATPase to cyclosporine or tacrolimus is that 10 times less calcineurin activity is found in the affected distal segments than in the nonaffected proximal segment.[75] Correspondingly, inhibition of Na,K-ATPase activity in the S2 segment could only be achieved with much higher doses of cyclosporine (25 μg mL^{-1}) or tacrolimus (250 ng mL^{-1}).[75] Hence, cells with low calcineurin activity, but with key functions regulated by calcineurin, such as distaltubular cells or T lymphocytes, are particularly sensitive to low concentrations of cyclosporine or tacrolimus.

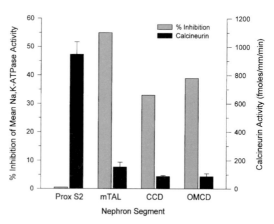

Figure 8 Relationship between inhibition of Na,K-ATPase activity by cyclosporine-A (600 ng mL^{-1}) and calcineurin activity. Data is shown for the proximal tubular S2 segment, medullary thick ascending limb (mTAL), cortical collecting duct (CCD), cortical connecting tubules (CNT), and outer medullary collecting ducts (OMCD). Tacrolimus (6 ng mL^{-1} has similar effects on Na,K-ATPase activity in S2 and CCD segments).[76] Figure drawn by authors from data published by Tumlin and co-workers.[47,75]

7.29.8 SUMMARY AND CONCLUSIONS

Significant progress has been made since the early 1990s with regard to identifying the common mechanisms that underlie several pathological events in the kidney, and confirming that some of the side effects are related to the mechanism of immunosuppression. Figure 9 summarizes the major pathophysiological pathways involved in cyclosporine nephrotoxicity.

In both the systemic and renal circulation, endothelial injury seems to be the most significant effect of cyclosporine on the vasculature. It results in plasma volume depletion, increased activity of the endothelin system, decreased NO synthesis, and alterations in prostaglandin and thromboxane release. These changes, together with a direct sensitization of vascular smooth muscle cells, result in renal and systemic vasoconstriction which depress renal function and raise arterial blood pressure.

The effects of cyclosporine on tubular function are modest and, in general, are not sufficient to compromise tubular processing of the glomerular filtrate. Animal studies do reveal subtle changes in tubular reabsorption. Decreases in distal nephron sodium transport, in particular, provide mechanisms to explain the reductions in potassium excretion and in magnesium reabsorption that lead to the characteristic mild hyperkalemia and hypomagnesemia observed.

Clinical and animal findings indicate that the renin–angiotensin system is activated and disrupted by cyclosporine and tacrolimus, although the evidence is much more consistent in animal studies than in clinical investigations. The potential significance of these changes in renin processing lies in the fact that they occur in the juxtaglomerular region, the site of the vascular lesion in man, which is only found in the distal preglomerular vasculature, not in any other vessels, and not in any other organ. Hence, if an alteration in renin processing were to lead to cell necrosis, this could explain the unusually specific distribution of the vascular lesion seen with cyclosporine.

Whilst no information exists about the calcineurin-dependence of renin synthesis or release, the fact that tacrolimus and cyclosporine both stimulate the renin–angiotensin system at low doses just sufficient for immunosuppression suggests that calcineurin may be involved in renin processing. If renin stimulation is, indeed, related to the development of arteriolopathy, this means that it cannot be prevented with immunosuppressants of this type, other than by careful monitoring and appropriate dose reduction. However, it opens up the possibility that other immunosuppressants with other modes of action need not cause arteriolopathy.

The establishment of a salt-restriction rat model that reproduces many of the cyclosporine-induced vascular and interstitial changes seen in man represents a major advance. The findings in this model suggest that activation of the renin–angiotensin system is a key factor in the etiology of vascular and interstitial damage. However, questions remain about the mechanism of interstitial injury in the rat, and its relevance to the injury in humans. More studies will indicate if this model lives up to its initial promise and is able to supply the much-needed pathological mechanism that underlies structural damage in humans.

Although very little is known for certain about the degree to which specific biochemical changes underlie some of the cellular changes observed, the suspicion remains that the biochemical processes involved in the immunosuppressive action of cyclosporine and certain tissue-specific side effects are interrelated. Side effects, such as altered mitochondrial function, remain of uncertain significance. The supposition that the side effects are related to the distribution of different types of cyclophilin has not been substantiated. Nonetheless, inhibition of cyclophilin, which is a rotamase, may underlie some deficiencies in protein transport.

The fact that tacrolimus has a renal side effect profile almost identical to cyclosporine, particularly if sirolimus remains free of these major side effects, strongly suggests that the interference with the calcium–calmodulin–calcineurin-dependent pathways is responsible for some of the specific side effects seen, as has been demonstrated for cyclosporine and tacrolimus inhibition of renal tubular Na,K-ATPase

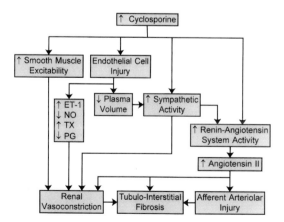

Figure 9 Diagram summarizing the major pathophysiologic pathways involved in cyclosporine nephrotoxicity. Abbreviations: ET-1, endothelin-1; NO, nitric oxide; TX, thromboxane A_2; PG, vasodilatory prostaglandins.

activity. Final evaluation of this possibility depends upon much more information becoming available about the regulatory role of calcineurin in those cell types adversely affected by cyclosporine and tacrolimus.

7.29.9 REFERENCES

1. G. Feutren, K. Abeywickrama, D. Friend *et al.*, 'Renal function and blood pressure in psoriatic patients treated with cyclosporine A.' *Br. J. Dermatol.*, 1990, **122**, Suppl. 36, 57–69.
2. A. A. M. J. Hollander, J. L. C. M. van Saase, A. M. M. Kootte *et al.*, 'Beneficial effects of conversion from cyclosporine to azathioprine after kidney transplantation.' *Lancet*, 1995, **345**, 610–614.
3. D. H. Van Buren, J. F. Burke and R. M. Lewis, 'Renal function in patients receiving long-term cyclosporine therapy.' *J. Am. Soc. Nephrol.*, 1994, **4**, S17–S22.
4. G. Thiel, T. Fellmann, J. Rosman *et al.*, 'Long-term safety profile of sandimmune in renal transplantation.' *Transplant Proc.*, 1992, **24**, 71–77
5. J. Mason and L. C. Moore, 'Indirect assessment of renal dysfunction in patients taking cyclosporin A for autoimmune diseases.' *Br. J. Dermatol.*, 1990, **122**, Suppl. 36, 79–84.
6. G. Feutren and M. J. Mihatsch, 'Risk factors for cyclosporine-induced nephropathy in patients with autoimmune diseases. International Kidney Biopsy Registry of Cyclosporine in Autoimmune Diseases [see comments].' *N. Engl. J. Med.*, 1992, **326**, 1654–1660.
7. J. Mason, 'The pathophysiology of Sandimmune (cyclosporine) in man and animals.' *Pediatr. Nephrol.*, 1990, **4**, 554–574.
8. L. C. Moore, J. Mason, L. Feld *et al.*, 'Effect of cyclosporine on endothelial albumin leakage in rats', *J. Am. Soc. Nephrol.*, 1992, **3**, 51–57.
9. M. Grieff, R. Loertscher, S. A. Shoharb *et al.*, 'Cyclosporine-induced elevation in circulating endothelin-1 in patients with solid organ transplants.' *Transplantation*, 1993, **56**, 880–884.
10. R. S. Gaston, S. D. Schlessinger, P. W. Sanders *et al.*, 'Cyclosporine inhibits the renal response to L-arginine in human kidney transplant recipients.' *J. Am. Soc. Nephrol.*, 1995, **5**, 1426–1433.
11. U. Scherrer, S. F. Vissing, B. J. Morgan *et al.*, 'Cyclosporine-induced sympathetic activation and hypertension after heart transplantation.' *N. Engl. J. Med.*, 1990, **323**, 693–699.
12. A. Rego, R. Vargas, S. Cathapermal *et al.*, 'Systemic vascular effects of cyclosporin A treatment in normotensive rats.' *J. Pharmacol. Exp. Ther.*, 1991, **259**, 905–915.
13. A. K. Bidani, K. A. Griffin, M. Picken *et al.*, 'Continuous telemetric blood pressure monitoring and glomerular injury in the rat remnant kidney model.' *Am J. Physiol.*, 1993, **265**, F391–F398.
14. D. M. Lanese and J. D. Conger, 'Effects of endothelin receptor antagonist on cyclosporine-induced vasoconstriction in isolated rat renal arterioles.' *J. Clin. Invest.*, 1993, **91**, 2144–2149.
15. D. M. Lanese, S. A. Falk and J. D. Conger, 'Sequential agonist activation and site-specific mediation of acute cyclosporine constriction in rat renal arterioles.' *Transplantation*, 1994, **58**, 1371–1378.
16. I. T. Bloom, F. R. Bentley and R. N. Garrison, 'Acute cyclosporine-induced renal vasoconstriction is mediated by endothelin-1.' *Surgery*, 1993, **114**, 480–488.
17. L. C. Racusen and K. Solez,. in 'Toxicology of the Kidney,' 2nd, edn., eds. J. B. Hook and R. S. Goldstein, Raven Press, New York, 1993, chap. 12.
18. J. D. Conger, G. E. Kim and J. B. Robinette, 'Effects of ANG II, ETA, and TxA2 receptor antagonists on cyclosporin A renal vasoconstriction.' *Am. J. Physiol.*, 1994, **267**, F443–F449.
19. J. B. Kopp and P. E. Klotman, 'Cellular and molecular mechanisms of cyclosporine nephrotoxicity.' *J. Am. Soc. Nephrol.*, 1990, **1**, 162–179.
20. A. Tibell, M. Larsson and A. Alvestrand, 'Dissolving intravenous cyclosporin A in a fat emulsion carrier prevents acute renal side effects in the rat.' *Transplant. Int.*, 1993, **6**, 69–72.
21. B. J. Morgan, T. Lyson, U. Scherrer *et al.*, 'Cyclosporine causes sympathetically mediated elevations in arterial pressure in rats.' *Hypertension*, 1991, **18**, 458–466.
22. V. Kon and M. Awazu, 'Endothelin and cyclosporine nephrotoxicity.' *Ren. Fail.*, 1992, **14**, 345–350.
23. T. E. Hunley, A. Fogo, S. Iwasaki *et al.*, 'Endothelin A receptor mediates functional but not structural damage in chronic cyclosporine nephrotoxicity.' *J. Am. Soc. Nephrol.*, 1995, **5**, 1718–1723.
24. A. Fogo, S. E. Hellings, T. Inagami *et al.*, 'Endothelin receptor antagonism is protective in *in vivo* acute cyclosporine toxicity.' *Kidney Int.*, 1992, **42**, 770–774.
25. T. E. Bunchman and C. A. Brookshire, 'Cyclosporine-induced synthesis of endothelin by cultured human endothelial cells.' *J. Clin. Invest.*, 1991, **88**, 310–314.
26. P. A. Phillips, K. A. Rolls, L. M. Burrell *et al.*, 'Vascular endothelin responsiveness and receptor characteristics *in vitro* and effects of endothelin receptor blockade *in vivo* in cyclosporin hypertension.' *Clin. Exp. Pharmacol. Physiol.*, 1994, **21**, 223–226.
27. S. Iwasaki, T. Homma and V. Kon, 'Site specific regulation in the kidney of endothelin and its receptor subtypes by cyclosporine.' *Kidney Int.*, 1994, **45**, 592–597.
28. D. Diederich, Z. Yang and T. F. Luscher, 'Chronic cyclosporine therapy impairs endothelium-dependent relaxation in the renal artery of the rat', *J. Am. Soc. Nephrol.*, 1992, **2**, 1291–1297.
29. M. Verbeke, J. Van de Voorde, L. de Ridder *et al.*, 'Functional analysis of vascular dysfunction in cyclosporin treated rats.' *Cardiovasc. Res.*, 1994, **28**, 1152–1156.
30. M. J. Gallego, A. L. Garcia Villalon, A. J. Lopez Farre *et al.*, 'Mechanisms of the endothelial toxicity of cyclosporin A. Role of nitric oxide, cGMP, and Ca^{2+}.' *Circ. Res.*, 1994, **74**, 477–484.
31. D. Diederich, J. Skopec, A. Diederich *et al.*, 'Cyclosporine produces endothelial dysfunction by increased production of superoxide.' *Hypertension*, 1994, **23**, 957–961.
32. K. Sudhir, J. S. Macgregor, T. DeMarco *et al.*, 'Cyclosporine impairs release of endothelium-derived relaxing factors in epicardial and resistance coronary arteries.' *Circulation*, 1994, **90**, 3018–3023.
33. M. J. Gallego, A. Lopez Farre, A. Riesco *et al.*, 'Blockade of endothelium-dependent responses in conscious rats by cyclosporin A: effect of L-arginine.' *Am. J. Physiol.*, 1993, **264**, H708–H714.
34. M. P. Gardner, T. F. Andoh and W. M. Bennett, 'Dietary L-arginine supplementation reduces the functional and structural manifestations of chronic cyclosporine (CSA) nephrotoxicity (abstract).' *J. Am. Soc. Nephrol.*, 1995, **6**, 997.
35. I. T. Bloom, F. R. Bentley, D. A. Spain *et al.*, 'An experimental study of altered nitric oxide metabolism as a mechanism of cyclosporin-induced renal vasoconstriction.' *Br. J. Surg.*, 1995, **82**, 195–198.

36. L. De Nicola, S. C. Thomson, L. M. Wead *et al.*, 'Arginine feeding modifies cyclosporine nephrotoxicity in rats.' *J. Clin. Invest.*, 1993, **92**, 1859–1865.

37. J. D. Conger, S. A. Falk, M. Friedemann *et al.*, 'Reduced renovascular nitric oxide (NO) generation with chronic cyclosporine a (CSA) treatment (abstract).' *J. Am. Soc. Nephrol.*, 1995, **6**, 995.

38. R. Locher, R. Huss, and W. Vetter, 'Potentiation of vascular smooth muscle cell activity by cyclosporin A.' *Eur. J. Clin. Pharmacol.*, 1991, **41**, 297–301.

39. D. Bokemeyer, U. Friedrichs, A. Backer *et al.*, 'Cyclosporine A enhances total cell calcium independent of Na, K-ATPase in vascular smooth muscle cells.' *Clin. Invest.*, 1994, **72**, 992–995.

40. D. Bokemeyer, H. J. Kramer and H. Meyer-Lehnert, 'Atrial natriuretic peptide blunts the cellular effects of cyclosporine in smooth muscle.' *Hypertension*, 1993, **21**, 166–172.

41. H. Meyer-Lehnert, D. Bokemeyer, U. Friedrichs *et al.*, 'Cellular signaling by cyclosporine A in contractile cells: interactions with atrial natriuretic peptide.' *Clin. Invest.*, 1993, **71**, 153–160.

42. C. C. Lang, I. S. Henderson, R. Mactier *et al.*, 'Atrial natriuretic factor improves renal function and lowers systolic blood pressure in renal allograft recipients treated with cyclosporin A.' *J. Hypertens.*, 1992, **10**, 483–488.

43. C. C. Lang, A. M. Choy, T. H. Pringle *et al.*, 'Renal, hemodynamic and neurohormonal effects of atrial natriuretic factor in cardiac allograft recipients treated with cyclosporin A.' *Am. J. Cardiol.*, 1993, **72**, 1083–1084.

44. C. Bagnis, G. Deray, M. Dubois *et al.*, 'Prevention of acute cyclosporin nephrotoxicity by verapamil and atrial natriuretic factor in the rat.' *Nephrol. Dial. Transplant.*, 1994, **9**, 1143–1148.

45. H. Meyer-Lehnert, U. Friedrich, D. Bokemeyer *et al.*, 'Cyclosporine A induced calcium accumulation in glomerular mesangial cells: role of endothelin (abstract).' *J. Am. Soc. Nephrol.*, 1994, **5**, 926.

46. D. Bokemeyer, U. Friedrichs, A. Backer *et al.*, 'Atrial natriuretic peptide inhibits cyclosporin A-induced endothelin production and calcium accumulation in rat vascular smooth muscle cells.' *Clin. Sci.*, 1994, **87**, 383–387.

47. J. A. Tumlin and J. M. Sands, 'Nephron segment-specific inhibition of Na^+,K^+-ATPase activity by cyclosporine A.' *Kidney Int.*, 1993, **43**, 246–251.

48. B. D. Rose, 'Clinical Physiology of Acid–Base and Electrolyte Disorders,' 4th edn., McGraw-Hill, New York, 1994.

49. B. D. Myers, R. Sibley, L. Newton *et al.*, 'The long-term course of cyclosporine-associated chronic nephropathy.' *Kidney Int.*, 1988, **33**, 590–600.

50. D. S. Gardiner, M. A. Watson, B. J. Junor *et al.*, 'The effect of conversion from cyclosporin to azathioprine on renin-containing cells in renal allograft biopsies.' *Nephrol. Dial. Transplant.*, 1991, **6**, 363–367.

51. E. H. Strom, R. Epper and M. J. Mihatsch, 'Ciclosporin-associated arteriolopathy: the renin producing vascular smooth muscle cells are more sensitive to ciclosporin toxicity.' *Clin. Nephrol.*, 1995, **43**, 226–231.

52. A. Tufro-McReddie, R. A. Gomez, L. L. Norling *et al.*, 'Effect of CsA on the expression of renin and angiotensin type I receptor genes in the rat kidney.' *Kidney Int.*, 1993, **43**, 615–622.

53. J. Iwai, Y. Kanayama, N. N. Inoue *et al.*, 'Increased endothelial gene expression of angiotensin AT1A receptor in cyclosporine induced hypertensive rats.' *Eur. J. Pharmacol.*, 1993, **248**, 341–344.

54. A. Kurtz, R. Della Bruna, and K. Kuhn, 'Cyclosporine A enhances renin secretion and production in isolated juxtaglomerular cells.' *Kidney Int.*, 1988, **33**, 947–953.

55. L. L. Norling, A. Tufro-McReddie, R. A. Gomez *et al.*, 'Accumulation of acidic renin isoforms in kidneys of cyclosporine A-treated rats.' *J. Am. Soc. Nephrol.*, 1996, **7**, 331.

56. J. A. Opsahl, P. A. Abraham, J. G. Shake *et al.*, 'Role of renin isoelectric heterogeneity in renal storage and secretion of renin.' *J. Am. Soc. Nephrol.*, 1993, **4**, 1054–1063.

57. F. J. Kaskel, D. Casellas and L. C. Moore, 'Dissociation between renin–angiotensin system activation and hemodynamic dysfunction induced by cyclosporines a and g in the rat (abstract).' *J. Am. Soc. Nephrol.*, 1995, **6**, 999.

58. B. D. Myers and L. Newton, 'Cyclosporine-induced chronic nephropathy: an obliterative microvascular renal injury', *J. Am. Soc. Nephrol.*, 1991, **2**, S45–S52.

59. M. J. Mihatsch, K. Morozumi, E. H. Storm *et al.*, 'Renal transplant morphology after long-term therapy with cyclosporine.' *Transplant Proc.*, 1995, **27**, 39–42.

60. M. J. Mihatsch, F. Gudat, B. Ryffel *et al.*, in 'Renal Pathology: With Clinical and Functional Correlations,' eds. C. C. Tischer and B. M. Brenner, J.B. Lippincott Co., Philadelphia, PA, 1994, chap. 52.

61. K. Morozumi, G. Thiel, F. W. Albert *et al.*, 'Studies on morphological outcome of cyclosporine-associated arteriolopathy after discontinuation of cyclosporine in renal allografts.' *Clin. Nephrol.*, 1992, **38**, 1–8.

62. B. A. Young, E. A. Burdmann, R. J. Johnson *et al.*, 'Cyclosporine A induced arteriolopathy in a rat model of chronic cyclosporine nephropathy.' *Kidney Int.*, 1995, **48**, 431–438.

63. R. H. Pichler, N. Franceschini, B. A. Young *et al.*, 'Pathogenesis of cyclosporine nephropathy: roles of angiotensin II and osteopontin.' *J. Am. Soc. Nephrol.*, 1995, **6**, 1186–1196.

64. E. A. Burdmann, T. F. Andoh, C. C. Nast *et al.*, 'Prevention of experimental cyclosporin-induced interstitial fibrosis by losartan and enalapril.' *Am. J. Physiol.*, 1995, **38**, F491–F499.

65. W. M. Bennett, 'The nephrotoxicity of immunosuppressive drugs.' *Clin. Nephrol.*, 1995, **43**, S3–S7.

66. E. A. Burdmann, B. Young, T. F. Andoh *et al.*, 'Mechanisms of cyclosporine-induced interstitial fibrosis.' *Transplant Proc.*, 1994, **26**, 2588–2589.

67. T. Yamamoto, N. A. Noble, D. E. Miller *et al.*, 'Sustained expression of TGF-beta1 underlies development of progressive kidney fibrosis.' *Kidney Int.*, 1994, **45**, 916–927.

68. E. A. Burdmann, S. Rosen, J. Lindsley *et al.*, 'Production of less chronic nephrotoxicity by cyclosporine G than cyclosporine A in a low-salt rat model.' *Transplantation*, 1993, **55**, 963–966.

69. L. D. Truong, A. Farhood, J. Tasby *et al.*, 'Experimental chronic renal ischemia: morphologic and immunologic studies.' *Kidney Int.*, 1992, **41**, 1676–1689.

70. R. J. Johnson, C. E. Alpers, A. Yoshimura *et al.*, 'Renal injury from angiotensin II mediated hypertension.' *Hypertension*, 1992, **19**, 464–474.

71. B. Ryffel, 'Cyclosporin binding proteins. Identification, distribution, function and relation to FK binding proteins.' *Biochem. Pharmacol.*, 1993, **46**, 1–12.

72. M. J. Gething and J. Sambrook, 'Protein folding in the cell.' *Nature*, 1992, **355**, 33–45.

73. J. Mason, 'The pathophysiology of Sandimmune (cyclosporine) in man and animals.' *Pediatr. Nephrol.*, 1990, **4**, 686–704.

74. S. L. Schreiber and G. R. Crabtree, 'The mechanism of action of cyclosporin A and FK506.' *Immunol. Today*, 1992, **13**, 136–142.

75. J. A. Tumlin, J. T. Someren, C. E. Swanson *et al.*, 'Expression of calcineurin activity and alpha-subunit isoforms in specific segments of the rat nephron.' *Am. J. Physiol.*, 1995, **269**, F558–F563.

76. J. P. Lea, J. M. Sands, S. J. McMahon *et al.*, 'Evidence that the inhibition of Na$^+$/K$^+$-ATPase activity by FK506 involves calcineurin.' *Kidney Int.*, 1994, **46**, 647–652.

77. B. J. Ballermann, M. L. Zeidel, M. E. Gunning *et al.*, in 'The Kidney,' 4th edn., eds. B. M. Brenner and F. C. Rector, Jr., W.B. Saunders Co., Philadelphia, PA, 1991, p. 514.

78. B. E. Bierer, G. Hollaender, D. Fruman *et al.*, 'Cyclosporin A and FK506: molecular mechanisms of immunosuppression and probes for transplantation biology.' *Curr. Opin. Immunol.*, 1993, **5**, 763–773.

79. R. Morris, 'Modes of action of FK506, cyclosporin A, and rapamycin.' *Transplant. Proc.*, 1994, **26**, 3272–3275.

80. G. Wiederrecht, E. Lam, S. Hung *et al.*, 'The mechanism of action of FK506 and cyclosporin A.' *Ann. NY Acad. Sci.*, 1993, **696**, 9–19.

81. F. Vincenti, D. A. Laskow, J. F. Neylan *et al.*, 'One-year follow-up of an open-label trial of FK506 for primary kidney transplantation. A report of the U.S. Multicenter Kidney Transplant Group.' *Transplantation*, 1996, **6**, 1576–1581.

82. D. A. Laskow, F. Vincenti, J. F. Neylan *et al.*, 'Phase II FK506 multicenter concentration control study: one year follow-up study.' *Transplant. Proc.*, 1995, 809–811.

83. P. S. Randhawa, R. Shapiro, M. L. Jordan *et al.*, 'The histopathological changes associated with allograft rejection and drug toxicity in renal transplant recipients maintained on FK506. Clinical significance and comparison with cyclosporine.' *Am. J. Surg. Pathol.*, 1993, **17**, 60–68.

84. J. F. DiJoseph, M. J. Mihatsch and S. N. Sehgal, 'Influence of rat strain on rapamycin's kidney effects.' *Transplant. Proc.*, 1993, **25**, 714–715.

85. J. F. DiJoseph, R. N. Sharma and J. Y. Chang, 'The effect of rapamycin on kidney function in the Sprague-Dawley rat.' *Transplantation*, 1992, **53**, 507–513.

86. P. H. Whiting, J. Woo, B. J. Adam *et al.*, 'Toxicity of rapamycin—a comparative and combination study with cyclosporine at immunotherapeutic dosage in the rat.' *Transplantation*, 1991, **52**, 203–208.

87. D. Casellas, M. Dupont, F. J. Kaskel *et al.*, 'Direct visualization of renin-cell distribution in preglomerular vascular trees dissected from rat kidney.' *Am. J. Physiol.*, 1993, **265**, F151–F156.

7.30
Analgesics and Nonsteroidal Anti-inflammatory Drugs

JOAN B. TARLOFF
Philadelphia College of Pharmacy and Science, PA, USA

7.30.1   INTRODUCTION

Nonsteroidal anti-inflammatory drugs (NSAIDs) such as aspirin and ibuprofen inhibit the cyclo-oxygenase activity of prostaglandin H synthase (PHS), the enzyme catalyzing the initial step in prostaglandin synthesis (formation of PGG_2 and PGH_2 from arachidonic acid; Figure 1). Since prostaglandins are involved in inflammation, interrupting the synthesis of prostaglandins acts to relieve inflammation that accompanies many diseases, including rheumatoid arthritis, osteoarthritis, and systemic lupus erythematosus. In addition, prostaglandins may be involved in body temperature elevation and pain production, accounting for the use of NSAIDs for fever reduction and analgesia. In general, NSAID therapy is safe and effective, and the most significant adverse effects involve gastrointestinal irritation and ulceration.

Acute inhibition of prostaglandin synthesis in the kidney, while of minimal safety concern in normal healthy individuals, may lead to serious deterioration of renal function in patients with renal blood flow that is dependent on prostaglandins (e.g., congestive heart failure, nephrotic syndrome, cirrhosis, and salt depletion).[1] In addition, NSAIDs exhibit other potentially

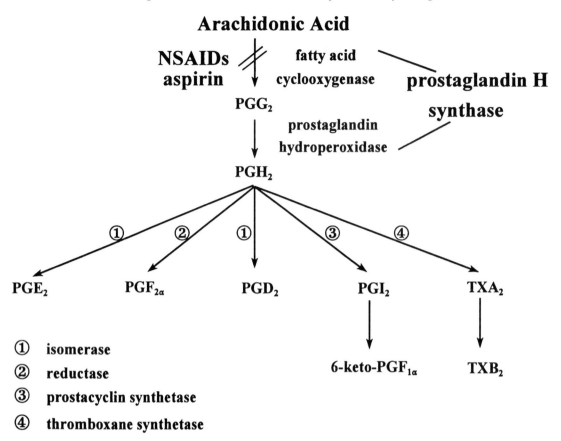

Figure 1 Prostaglandin H synthase-catalyzed formation of prostacyclins (PGG$_2$, PGH$_2$, and PGI$_2$), prostaglandins (PGE$_2$, PGF$_{2\alpha}$, and PGD$_2$), and thromboxanes (TXA$_2$, TXB$_2$) from arachidonic acid. The fatty acid cyclooxygenase component (COX1) of prostaglandin H synthase is reversibly inhibited by NSAIDs and irreversibly inhibited by aspirin, accounting for the ability of these drugs to interrupt the synthesis of prostacyclins, prostaglandins, and thromboxanes.

serious renal toxicities, including acute tubular necrosis, interstitial nephritis with nephrotic syndrome, and papillary necrosis, that may or may not be related to inhibition of prostaglandin synthesis.

NSAIDs are among the most widely used drugs and are available both by prescription and over-the-counter. In the United States, numerous NSAIDs are available to consumers who receive more than 70 million prescriptions annually at a cost of over $1 billion.[2] In 1987, about 74 million prescriptions for NSAIDs were dispensed by community pharmacists, representing 4.5% of all prescription orders.[3] At least 50 million Americans probably use NSAIDs intermittently or routinely and one in seven Americans is likely to be treated with an NSAID for a chronic rheumatologic disorder.[4] Whelton and Hamilton estimate that at least 500 000 of those patients using NSAIDs will be expected to develop some degree of renal functional abnormality, giving a 1% incidence of adverse renal reactions due to NSAID therapy.[5]

7.30.2 ARACHIDONIC ACID METABOLISM

There are three principal pathways involved in the metabolism of arachidonic acid. In one pathway, prostacyclins, prostaglandins, and thromboxanes are formed from arachidonic acid by the action of PHS (Figure 1). This membrane-bound enzyme contains both cyclo-oxygenase and peroxidase activities. There are at least two forms of cyclo-oxygenase, COX1 and COX2. COX1 is a constitutive enzyme found in high activities in numerous tissues

Table 1 Comparison of PHS-1 IC_{50} potency and therapeutic plasma concentrations of some nonsteroidal anti-inflammatory agents.

Compound	IC_{50} (μM)[a]			Pharmacokinetic profile	
	Murine COX1	Human COX1	Sheep COX1	Dose (mg)	C_{max} (μM)[b]
Salicylates					
Aspirin	1.67 ± 1.11[8]		83[10]	650	345
Sodium salicylate	219 ± 69[8]	>1000[9]		650	340
Aminophenols					
Acetaminophen[c]	17.9 ± 13.2[8]		>100	325	250
Enolic acids					
Phenylbutazone		16.0[9]	12.6[10]	100	27
Piroxicam	9.0–24[6]	17.7 ± 3.3[9]		10	3.6
Carboxylic acids					
Fenamic acids					
Mefenamic acid			2.1[10]	250	98.7
Meclofenamic acid	2.0–2.5[6]	1.5 ± 0.6[9]	0.19 ± 0.13[11]	50	14
Acetic acids					
Indomethacin	4.9–8.1[6]	13.5 ± 3.5[9]	0.67 ± 0.09[11]	25	8.3
Sulindac sulfide	0.3–0.5[6]	1.3 ± 1.0[9]		150	50
Diclofenac	1.6 ± 0.6[8]	2.7 ± 1.0[9]		50	18.7
Propionic acids					
Ibuprofen	8.9–14[6]	4.0 ± 1.0[9]	1.5[10]	400	230
Naproxen	9.6 ± 4.5[8]	4.8 ± 1.8[9]	6.1[10]	250	130
Ketoprofen		31.5 ± 9.2[9]		50	25.5
Flurbiprofen	0.46–0.50[6]	0.5 ± 0.1[9]	0.30 ± 0.03[11]	100	5.6

[a]Concentration *in vitro* required for 50% inhibition of oxygen consumption or PGE_2 synthesis. [b]Calculated as dose/V_d. Volumes of distribution obtained from Verbeeck *et al.*[12] and C_{max} calculated assuming total body weight of 70 kg. [c]IC_{30} because 50% inhibition was not achieved at concentrations up to $1\,mg\,mL^{-1}$.

including kidney[6] and is effectively inhibited by NSAIDs (Table 1). COX2 is highly inducible and is also inhibited by NSAIDs. In addition, COX2 is inhibited by glucocorticoids such as dexamethasone.[6] As illustrated in Figure 2, PHS-associated cyclo-oxygenase activities are present throughout the renal vasculature (afferent and efferent arterioles, glomerular and peritubular capillaries), in epithelial cells of the collecting tubule, and in interstitial cells adjacent to the thick ascending limb of the loop of Henle and in the medulla.[2,7]

In the second pathway, leukotrienes are formed from arachidonic acid by the action of lipoxygenase (Figure 3). In the third pathway, arachidonic acid is converted to epoxyeicosatrienoic acids (EET) and hydroxyeicosatetraenoic acids (HETE) by cytochrome P450-dependent monooxygenases (Figure 4). In the kidney, cytochrome P450 monooxygenase activities are present in the proximal tubule and the medullary thick ascending limb of the loop of Henle (Figure 2).

7.30.2.1 Prostaglandin H Synthase-dependent Arachidonic Acid Metabolism

Cyclooxygenase catalyzes the first step in the metabolism of arachidonic acid, converting this fatty acid to prostaglandin G_2 (PGG_2), a prostacyclin (Figure 1). COX1 is inhibited reversibly by NSAIDs and irreversibly by aspirin (Table 1). For some NSAIDs, such as ibuprofen and naproxen, a single therapeutic dose may achieve plasma drug concentrations in marked excess of the IC_{50} and thus may inhibit COX1 (Table 1). The second enzyme activity associated with PHS is that of a peroxidase, specifically prostaglandin hydroperoxidase, that converts PGG_2 to a second prostacyclin, prostaglandin H_2 (PGH_2). There are no known specific inhibitors for PHS-catalyzed peroxidation.[13] PGH_2 is further metabolized by prostacyclin synthetase (Figure 1), to generate prostaglandin I_2 (PGI_2). All three prostacyclins, PGG_2, PGH_2, and PGI_2, act as vasodilators or vasoconstrictors, depending on the vascular bed. In the kidney, prostacyclins

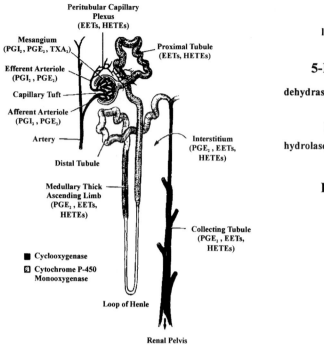

Arachidonic Acid

lipoxygenase

5-HPETE → hydroperoxidase → **5-HETE**

dehydrase

glutathione

LTA$_4$

hydrolase   glutathione *S*-transferase

LTB$_4$ **LTC$_4$**

γ-glutamyl transpeptidase

glutamate ←

LTD$_4$

dipeptidase

glycine ←

LTE$_4$

glutamate ↘ γ-glutamyl transpeptidase

LTF$_4$

Figure 2 Sites of cyclooxygenase and cytochrome P450 activities along the nephron. Different products may be formed from arachidonic acid along the nephron because cyclooxygenase and cytochrome P450 activities are distributed differently along the nephron. For example, glomeruli and renal blood vessels contain cyclooxygenase and produce the metabolites illustrated in Figure 1. In contrast, tubular epithelial cells contain primarily cytochrome P450 and produce 20-OH-eicosatetraenoic and eicosatetraen-1,20-dioic acid (EETs) and 19-OH- and 19-oxo-eicosatetraenoic acids (HETE) (reproduced, with permission, from the Annual Review of Pharmacology and Toxicology, Volume 32, © 1993, by Annual Reviews Inc.).

Figure 3 Lipoxygenase-catalyzed arachidonic acid metabolism to form leukotrienes. Arachidonic acid is metabolized by lipoxygenase to 5-hydroperoxyeicosatetranoic acid (5-HPETE). Leukotrienes (LTA$_4$, LTB$_4$, LTC$_4$, LTD$_4$, LTE$_4$, and LTF$_4$) are formed from 5-HPETE by various enzymatic reactions. Alternately, 5-HPETE may be converted to 5-monohydroxyeicosatetranoic acid (5-HETE) by the action of a hydroperoxidase enzyme.

promote diuresis, natriuresis, and kaliuresis, suggesting that these substances promote vasodilation of renal blood vessels (Table 2). Prostacyclin production in the kidney may be monitored by measuring the urinary excretion of 6-keto-prostaglandin F$_{1\alpha}$ (6-keto-PGF$_{1\alpha}$), a stable product of PGI$_2$ metabolism (Figure 1).

PGH$_2$ may be converted either enzymatically or nonenzymatically to several prostaglandins, including prostaglandin E$_2$ (PGE$_2$), prostaglandin F$_{2\alpha}$ (PGF$_{2\alpha}$), and prostaglandin D$_2$ (PGD$_2$), illustrated in pathways 1 and 2 in Figure 1. These prostaglandins, in particular PGE$_2$, have numerous actions in the kidney, including vasodilation of renal vessels which causes an increase in renal blood flow (RBF) and glomerular filtration rate (GFR); stimulation of renin release; and inhibition of sodium reabsorption in the

distal tubule, antidiuretic hormone (ADH)-dependent water reabsorption in the collecting duct, and chloride reabsorption in the thick ascending limb of the loop of Henle (Table 2) resulting in diuresis, natriuresis, and kaliuresis.

PGH$_2$ also may be metabolized by thromboxane synthetase (Figure 1) to thromboxane A$_2$ (TXA$_2$), which in turn spontaneously rearranges to yield thromboxane B$_2$ (TXB$_2$). Thromboxanes are potent vasoconstrictors and may antagonize the 'vasodilating effects of PGE$_2$ in the kidney (Table 2).

7.30.2.2 Lipoxygenase-dependent Arachidonic Acid Metabolism

The enzyme lipoxygenase catalyzes the formation of 5-hydroperoxyeicosatetranoic acid

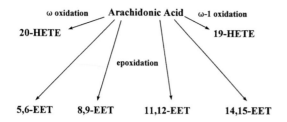

Figure 4 Cytochrome P450-catalyzed arachidonic acid metabolism. Arachidonic acid undergoes ω oxidation to form 20-hydroxyeicosatetranoic acid (20-HETE) and ω − 1 oxidation to form 19-hydroxyeicosatetraenoic acid (19-HETE). Additionally, arachidonic acid may undergo cytochrome P450-catalyzed epoxidation to form various epoxyeicosatrienoic acids (EETs).

(5-HPETE) from arachidonic acid. 5-HPETE may be converted to 5-monohydroxyeicosatetranoic acid (5-HETE) by the action of a hydroperoxidase enzyme or to leukotriene A_4 (LTA_4) by the action of a dehydrase enzyme.

LTA_4 may be converted to leukotriene B_4 (LTB_4) by the action of a hydrolase enzyme or to leukotriene C_4 (LTC_4) following conjugation of LTA_4 with reduced glutathione (GSH) catalyzed by glutathione S-transferase (GST) (Figure 3). LTC_4 undergoes a series of reactions involving enzymes of the γ-glutamyl cycle. LTC_4 is converted to leukotriene D_4 (LTD_4) following cleavage of the γ-glutamyl group of GSH catalyzed by γ-glutamyl transpeptidase (GGT). LTD_4 is further metabolized by a dipeptidase enzyme, resulting in the loss of the glycine residue of GSH and formation of leukotriene E_4 (LTE_4). LTE_4 is further metabolized by GGT resulting in the addition of a γ-glutamyl group from glutamate and the formation of leukotriene F_4 (LTF_4) (Figure 3). The kidney contains abundant amounts of lipoxygenase, GGT, and dipeptidases as well as GST and GSH, so that renal synthesis of all the leukotrienes undoubtedly occurs. In the kidney, leukotrienes cause a transient increase in renal perfusion pressure followed by a sustained hypotension (Table 2). These vascular effects would cause an initial increase followed by a sustained decrease in RBF and GFR.

Table 2 Effects of prostaglandins and leukotrienes on the kidney.

	Vascular	*Tubular*
Prostacyclins $PGG_2/PGH_2/PGI_2$	Vasodilation ↑ Renin release Maintain glomerular filtration rate Mesangial cell relaxation	Diuresis, natriuresis, kaliuresis
Prostaglandins PGE_2	Vasodilation ↑ Renal blood flow ↑ Renin release Maintain glomerular filtration rate Mesangial cell relaxation	Diuresis, natriuresis, kaliuresis ↓ ADH-dependent water reabsorption ↓ Cl^- reabsorption in TAL ↑ Erythropoietin release
$PGF_{2\alpha}$ PGD_2	Vasodilation (humans) ↑ Renin release Vasodilation in resistance vessels	Diuresis, natriuresis
PGA_2	Vasodilation ↑ Renal blood flow	↑ Erythropoietin release
Thromboxanes (TXA_2/TXB_2)	Vasoconstriction Decrease glomerular filtration rate Mesangial cell contraction	
Leukotrienes (LTC_4/LTD_4)	Hypertension followed by prolonged hypotension	
EETs/HETEs	↑ Renin release	Inhibition of $Na^+K^+ATPase$ in PT, TAL

TAL = thick ascending limb of the loop of Henle. PT = proximal tubule.

7.30.2.3 Cytochrome P450-dependent Arachidonic Acid Metabolism

Cytochromes P450 form 20-hydroxyeicosa-tetraenoic acid by ω oxidation and 19-hydroxy-eicosatetraenoic acid by ω − 1 oxidation (19-HETE and 20-HETE in Figure 4) and also catalyze epoxidation of arachidonic acid to form various epoxyeicosatrienoic acids (EET in Figure 4).[14] Cytochrome P450 isozymes that metabolize arachidonic acid have been localized in the proximal tubule and thick ascending limb of the loop of Henle (Figure 2). HETE and EET act at their sites of production in the proximal tubule and medullary thick ascending limb of the loop of Henle to inhibit $Na^+K^+ATPase$, resulting in natriuresis and diuresis (Table 2).[2]

7.30.3 EFFECTS OF ANALGESICS AND NSAIDS ON RENAL FUNCTION

Prostaglandins operate in conjunction with a variety of other mediators, including ADH, angiotensin II, norepinephrine, and kinins (Figure 5). However, under normal conditions, prostaglandins exert little or no important control over RBF.[15] Rather, prostaglandins are important in sustaining renal function under conditions where the vasoconstrictor system is activated, as by angiotensin II, norepinephrine, vasopressin, and endothelin.[7] During activation of vasoconstrictor systems, such as in circulatory collapse or stress, unopposed vasoconstriction would lead to decreased RBF, GFR, and renal excretory capacity. To maintain normal renal function, prostaglandins are synthesized locally and attenuate these vasoconstrictor effects.[7] In addition, some of the vasoconstrictor hormones, such as angiotensin II and endothelin, stimulate local production of vasodilator prostaglandins, including PGE_2 and $PGF_{2\alpha}$, resulting in a negative feedback loop between vasoconstrictors and prostaglandins.[7] Interrupting prostaglandin synthesis by NSAID administration under conditions in which RBF is dependent on prostaglandins disturbs the normal balance between vasodilators and vasoconstrictors, allowing vasoconstriction to predominate.

In general, NSAID-induced adverse effects on the kidneys are relatively uncommon. However, a high incidence of acute renal failure has been reported in patients with hypovolemia or reduced circulatory volume due to congestive heart failure, cirrhosis, or intensive diuretic therapy. Also at high risk are patients with nephrotic syndrome or preexisting renal impairment, including elderly patients and patients with diabetes.[2,3,7] In all of these conditions, RBF is at least partially dependent on prostaglandin production so that NSAID administration would be expected to produce precipitous declines in RBF and GFR, resulting in acute renal failure.[2]

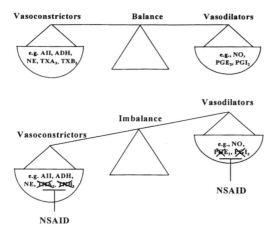

Figure 5 Schematic representation of the balance between vasoconstrictors and vasodilators involved in renal blood flow and glomerular filtration. Vasoconstrictors include angiotensin II (AII), anti-diuretic hormone (ADH), norepinephrine (NE), and endothelin. Vasodilators include nitric oxide (NO), PGE_2, and possibly PGI_2. NSAID therapy disturbs this balance by inhibiting the synthesis of PGE_2 and PGI_2. Under these circumstances, vasoconstrictor influences predominate and renal blood flow and glomerular filtration may be decreased (reproduced by permission of Blackwell Science from *Kidney Int.*, 1993, **44**, 643–653).

7.30.3.1 Sodium and Water Retention

Under normal conditions, natriuresis is mediated by two prostaglandin-dependent effects: (i) vasodilation resulting in increased RBF and reduced proximal tubular reabsorption of solutes and water,[16] and (ii) direct inhibition of sodium reabsorption at the thick ascending limb of the loop of Henle[17] (Table 2). Thus, NSAIDs would be expected to (and in fact do) cause a transient degree of sodium and water retention, as well as mild hypertension, even in clinically normal patients.[1,18] NSAID-induced antidiuresis and antinatriuresis occur in the absence of measurable changes in RBF or GFR,[17] suggesting that NSAIDs act at least in part by directly blocking the inhibitory effects of PGs on tubular reabsorption of sodium. In individuals with preexisting renal dysfunction, clinically significant edema may develop during NSAID therapy because these patients are unable to excrete salt and water loads normally.

In normal individuals, NSAID-induced fluid retention is minor, reversible upon discontinuation of NSAID therapy, and easily managed in patients who require continued NSAID therapy.[5]

Transient changes in renal function have also been observed in laboratory animals following acute administration of NSAIDs. For example, sodium excretion is significantly reduced while urine volume is unaltered following acute administration of sodium meclofenamate to anesthetized, sodium-replete dogs.[19] In these normal dogs, NSAID administration has no effect on RBF or GFR. In laboratory animals as well as humans, salt and water retention following NSAID administration are transient and easily reversible when NSAID therapy is discontinued.

7.30.3.2 Hyperkalemia

Hyperkalemia is an unusual complication of NSAID therapy (Table 3), although severe hyperkalemia has been reported in patients with mild renal insufficiency who received indomethacin for gout.[4] Other predisposing factors for the development of NSAID-induced hyperkalemia include cardiac failure, diabetes, multiple myeloma, concurrent administration of potassium supplements or potassium-sparing diuretics, or administration of angiotensin converting enzyme inhibitors.[5] Prostaglandins stimulate renin release (Table 2) so that NSAID therapy would be expected to reduce renin release and angiotensin II formation.[15] Since angiotensin II is a prime stimulus for aldosterone secretion, NSAID therapy would be expected to secondarily decrease plasma aldosterone concentrations, as outlined in Figure 6. Aldosterone is the primary factor regulating plasma potassium concentration under normal circumstances. In the kidney, aldosterone stimulates potassium secretion, at least in part in exchange with sodium reabsorption, thereby promoting potassium excretion and maintaining potassium balance. With NSAID-induced hypoaldosteronism, potassium secretion would be decreased allowing the possibility for hyperkalemia to occur (Figure 6).[4,5] In addition, NSAIDs (or at least indomethacin) may exert a direct effect on renal tubular cells to reduce potassium uptake.[4,5]

Table 3 Renal effects of nonsteroidal anti-inflammatory drugs.[a]

Drug class	Generic name	Edema	↑ K	ARF	NS	PN
Salicylates						
Acetylated	Aspirin	Cl		Cl		Cl
Nonacetylated	Diflunisal	Cl		Cl	Cl	An
Propionic acids	Ibuprofen	Cl	Cl	Cl	Cl	Cl
	Naproxen	Cl		Cl	Cl	An
	Fenoprofen calcium	Cl		Cl	Cl	Cl
	Ketoprofen	Cl		Cl	Cl	
	Fluriboprofen	Cl	Cl	Cl	Cl[b]	An
Indolacetic acids	Indomethacin	Cl	Cl	Cl	Cl	Cl[c]
	Sulindac	Cl	Cl	Cl	Cl	An
	Tolmetin	Cl		Cl	Cl	An
	Zomepirac	Cl		Cl	Cl	An
	Diclofenac	Cl		Cl	Cl	Cl
Anthranilic acids	Meclofenamate sodium	Cl		Cl	Cl	An
	Mefenamic acid	Cl		Cl	Cl	Cl
Pyrazolones	Phenylbutazone	Cl		Cl	Cl	Cl
Oxicams	Piroxicam	Cl	Cl	Cl	Cl	Cl

Source: Clive and Stoff.[4]

[a]ARF = acute renal failure; NS = interstitial nephritis and nephrotic syndrome; PN = papillary necrosis; K = hyperkalemia; Cl = reported in clinical studies; An = described in studies in animals but not in humans. [b]Causes interstitial nephritis without nephrotic syndrome. [c]Reported in combination with phenylbutazone.

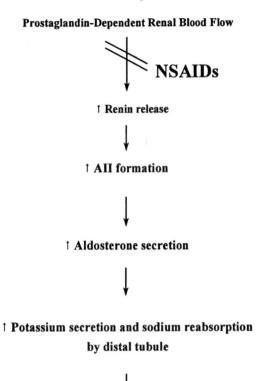

Prostaglandin-Dependent Renal Blood Flow

NSAIDs

↑ **Renin release**

↓

↑ **AII formation**

↓

↑ **Aldosterone secretion**

↓

↑ **Potassium secretion and sodium reabsorption by distal tubule**

↓

Normal potassium balance

Figure 6 Mechanism of NSAID-induced hyperkalemia. Under normal circumstances, aldosterone secretion is stimulated by angiotensin II. Aldosterone promotes potassium secretion and sodium reabsorption in the distal tubule of the nephron. Angiotensin II formation is dependent on renin secretion to initiate the reactions that convert angiotensinogen to angiotensin II. Prostaglandins are among many stimuli for renin secretion and when prostaglandin synthesis is interrupted, as by NSAID therapy, renin secretion is decreased. Through the pathway outlined, inhibition of renin secretion may cause secondary hypoaldosteronism and hyperkalemia.

7.30.3.3 Acute Renal Failure

In patients with preexisting renal insufficiency, NSAID therapy may cause abrupt declines in RBF and GFR due to the vascular effects of PGE_2 withdrawal (Figure 7).[2] Abrupt declines in renal function have been reported to occur with numerous NSAIDs, as indicated in Table 3. Any condition that causes a decrease in effective circulating blood volume make RBF at least partially dependent on prostaglandin production. Such conditions include congestive heart failure, dehydration, hemorrhage, cirrhosis with ascites, and excessive diuretic therapy, as outlined in Figure 7. In all these conditions, the decrease in effective circulating blood volume acts as a stimulus to produce vasoconstrictors such as angiotensin II, norepinephrine, and endothelin. In this setting, the vasodilator renal prostaglandins are necessary to counteract the vasoconstrictors and maintain RBF and GFR. When NSAID therapy is introduced, renal prostaglandin production is abruptly curtailed and RBF and GFR decline precipitously (Figure 7).[2] This NSAID-induced acute renal failure is the most common renal side effect of NSAID therapy, occurring in 0.5–1% of patients taking NSAIDs on a chronic basis (Table 3).[5] NSAID-induced acute renal failure is reversible and no structural damage is observed during or following the renal dysfunction.

In patients with renal insufficiency but not in normal individuals, NSAIDs cause about 30% decreases in RBF (measured as *para*-aminohippurate clearance) and GFR (measured as inulin clearance).[1,2,20] Thus, in normal patients, NSAID administration does not alter renal hemodynamics, probably because RBF is not dependent on continuing prostaglandin production. In patients with renal insufficiency, however, in whom prostaglandin production partially counteracts the effects of vasoconstrictor agents such as angiotensin II and norepinephrine, NSAID administration causes declines in RBF and GFR.[7,15] In salt-deprived patients, aspirin administration produces 12–15% reductions in creatinine and inulin clearances,[21] demonstrating that renal function may be depressed when effective circulatory volume is depleted, as by salt restriction. However, the hemodynamic and urinary changes caused by NSAID administration in both normal patients and patients with renal insufficiency are fully reversible within a few hours of NSAID administration.[1,2] Further, chronic therapy with NSAIDs for as long as 1 month causes no sustained decrement in renal function,[2] suggesting again that the effects of NSAIDs on renal hemodynamics and glomerular filtration are fully reversible.

Similarly, in conscious, unstressed, sodium replete laboratory dogs, administration of NSAIDs produces no significant changes in RBF or GFR, indicating that, under basal conditions, RBF is not dependent on prostaglandin synthesis.[19,22,23] However, administration of NSAIDs to conscious or anesthetized sodium-deprived dogs[24] or the anesthetized surgically operated animals produces significant decreases in RBF and increases in renal vascular resistance without changing GFR.[15]

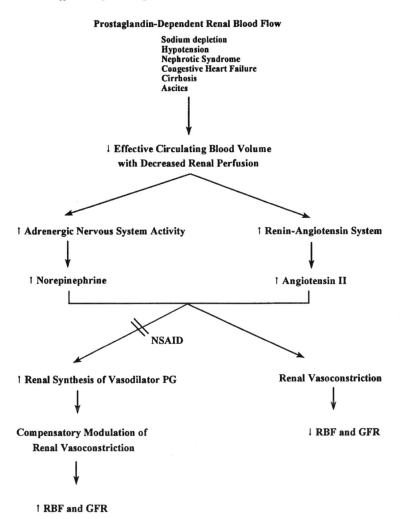

Figure 7 Mechanism of NSAID-induced decreases in renal blood flow and glomerular filtration. When renal blood flow is dependent on prostaglandin production, interruption of prostaglandin synthesis allows vaso-constriction to predominate. When renal blood vessels are constricted, renal blood flow and glomerular filtration rate are reduced (reproduced by permission of Cahners from *Am. J. Med.*, Suppl. 1986, **80**, 12–21.)

The renal vasoconstrictor response in anesthe-tized, surgically operated animals is at least partially dependent on prostaglandin synthesis as the response to NSAID therapy is blunted or abolished by exogenous administration of PGE_2.[25] In addition, NSAID-induced declines in RBF are at least partially dependent on activation of the renin-angiotensin system, as administration of saralasin blunts the effects of meclofenamate in sodium-depleted dogs.[19]

7.30.3.4   Acute Tubular Necrosis

Oliguria and nonoliguric renal failure con-sistent with acute tubular necrosis has been reported following clinical overdoses of aspirin, acetaminophen, and other NSAIDs (indomethacin, phenylbutazone, fenoprofen, ibuprofen, and mefenamic acid) (Table 3).[26] Histologic lesions ranging from minor tubular changes to frank tubular necrosis have been observed experimentally with acetaminophen, salicylates, phenylbutazone, ibuprofen, and mefenamic acid.[26] Histologic damage with NSAIDs is largely confined to the proximal tubule of the nephron.

7.30.3.4.1   Salicylates

Sodium salicylate causes increases in blood urea nitrogen concentration and excretion of glucose, protein, blood, and tubular epithelial cells in urine of humans as well as experimental animals.[27,28] The effects of sodium salicylate are transient and signs of regeneration are evident within 24 h of administration to laboratory rats.[29] In rats, administration of 500 mg sodium salicylate kg^{-1} (i.p.) is associated with morphological changes including dilation and vacuolization of proximal tubular epithelial cells within 6 h of administration.[28] These morphological changes are transient and renal morphology in salicylate-treated rats is not different from control animals within 12 h of administration.[28] Similarly, blood urea nitrogen concentrations are elevated from 2 h through 12 h after salicylate administration but return to control values by 24 h.[28] Salicylate-induced proximal tubular damage is more severe and more persistent in 12-month old rats as compared to 3-month old rats.[28] For example, histological changes in 12-month old rat kidneys include areas of focal proximal tubular necrosis and interstitial edema and these changes persist through 12 h following salicylate administration.[28] Regenerative changes are not observed in older rats until 24 h after salicylate administration. Similarly, blood urea nitrogen concentration increases to a greater extent and remains elevated for a longer period of time (through 24 h) in 12-month old rats as compared to 3-month old rats receiving the same dosage of salicylate.[28] Although the precise mechanisms involved in salicylate nephrotoxicity remain undefined, covalent binding may play a role in the development of proximal tubular necrosis in rats. Specifically, radiolabel from salicylate becomes covalently bound to macromolecules in rat renal cortex but not medulla. In the cortex, greater than 50% of total covalently bound radioactivity in the renal cortex is associated with the mitochondrial fraction.[28] Thus, salicylate-induced proximal tubular necrosis may involve bioactivation of salicylate to a reactive intermediate that undergoes covalent binding to mitochondrial proteins, thereby interrupting normal mitochondrial function.[30]

Aspirin administration in rats produces mild proximal tubular damage, as indicated by enzymuria (GGT and N-acetyl-β-D-glucosaminidase), glucosuria, proteinuria, and elevations of blood urea nitrogen concentrations within 24 h of aspirin administration to rats.[31,32] Massive proximal tubular necrosis is evident as early as 2 h following oral administration of aspirin (900 mg kg^{-1}) to rats. Tubular necrosis is maximal at 12 h and regeneration is observed 24–48 h following aspirin administration.[27] Aspirin produces tubular necrosis in both males and females and cortical necrosis develops in some rats after dosages of aspirin as low as 100 mg kg^{-1}.[22] Aspirin-induced tubular necrosis may be due to a reactive metabolite generated by cytochromes P450.[30] Alternately, since aspirin undergoes rapid deacetylation to form salicylate, the nephrotoxicity of aspirin may be due to salicylate rather than aspirin

itself. Either aspirin or a metabolite undergoes covalent binding with cellular macromolecules,[33] significantly depletes renal glutathione content,[34] and inhibits PHS activity in renal cortex and medulla.[33] In addition, aspirin inhibits the hexose-monophosphate shunt metabolic pathway, which may contribute to further decreases in renal glutathione concentrations.[26] Aspirin uncouples oxidative phosphorylation leading to decreases in intracellular ATP and increases in AMP in renal cortex but not medulla.[35,36] The lack of effect on renal medulla may relate to the fact that medullary metabolism is primarily glycolytic rather than aerobic[37] so that uncoupling of oxidative phosphorylation may have fewer dramatic effects on medullary metabolism.

7.30.3.4.2 Acetaminophen

Massive dosages of acetaminophen (500 mg kg^{-1} or greater) cause acute proximal tubular necrosis in humans, rats, and mice.[38–40] In both rats and mice, acetaminophen-induced proximal tubular necrosis is accompanied by marked elevations of blood urea nitrogen concentrations and kidney weights within 24 h of administration.[39] In addition, renal glutathione is transiently depleted and radiolabel from acetaminophen becomes covalently bound to renal proteins within a few hours of acetaminophen administration.[40–41] Covalently bound material in both liver and kidneys of acetaminophen-treated mice reacts immunochemically with an antibody directed against the carbonyl group of acetaminophen.[40–42] Covalent binding of radiolabel to renal proteins is not a random event but rather, certain proteins are selective targets for acetaminophen. Specifically, proteins with molecular weights of 27, 33, and 56–58 kDa are arylated by acetaminophen.[40] The 56–58 kDa protein arylated by acetaminophen in kidney is also a target for acetaminophen in liver and has been identified as a 56 kDa selenium binding protein in mouse liver.[43] The proteins of 27 and 33 kDa arylated by acetaminophen in kidney have not been purified, sequenced, or identified as yet.

The mechanisms leading to acetaminophen-induced nephrotoxicity are different in rats and mice. In mice, acetaminophen nephrotoxicity involves cytochrome P450-catalyzed oxidation of acetaminophen to N-acetyl-benzoquinoneimine, similar to the pathway involved in acetaminophen hepatotoxicity. As a reactive intermediate, N-acetyl-benzoquinoneimine may covalently bind to renal macromolecules and/or induce oxidative stress. In mice, acetaminophen nephrotoxicity is highly correlated with cytochrome P450 2E1 content.[44] Male mice have a higher content of cytochrome P450 2E1 and are more susceptible to acetaminophen nephrotoxicity than are female mice.[44] Inhibition of cytochrome P450 activity with piperonyl butoxide protects mice against acetaminophen-induced nephrotoxicity whereas inhibition of acetaminophen deacetylation with tri-*ortho*-tolyl phosphate, a carboxyesterase inhibitor, affords no protection from toxicity.[42] These data suggest that in mice, acetaminophen nephrotoxicity involves a cytochrome P450-dependent pathway similar to the mechanism involved in acetaminophen-induced hepatotoxicity.

In rats, however, a different pathway involving deacetylation of acetaminophen to yield *para*-aminophenol seems to be involved in acetaminophen nephrotoxicity.[45] Pretreating rats with inhibitors and inducers of cytochrome P450 produces inconsistent effects on the severity of acetaminophen nephrotoxicity.[46,47] Pretreating rats with bis-(*para*-nitrophenyl) phosphate, a carboxyesterase inhibitor, protects rats from acetaminophen-induced nephrotoxicity,[48] suggesting that deacetylation to *para*-aminophenol is an obligatory step. Covalent binding of radiolabel derived from acetaminophen to renal macromolecules is greater when the radioactive tag is located on the aromatic ring rather than on the carbonyl moiety.[47] Using an antibody directed against the carbonyl group of acetaminophen, immunochemical binding is detectable in liver but not kidneys of acetaminophen-treated rats,[49] supporting the hypothesis that a deacetylated metabolite of acetaminophen, probably *para*-aminophenol, is responsible for covalent binding and nephrotoxicity in rats.

Para-aminophenol produces acute proximal tubular necrosis in rats[39,45] and small amounts of *para*-aminophenol are excreted in urine from rats or hamsters treated with acetaminophen.[41,50] Thus it is possible that acetaminophen-induced nephrotoxicity may involve deacetylation to *para*-aminophenol, which ultimately causes renal damage due to enzymatic or nonenzymatic oxidation.[51–53] However, both liver and kidney deacetylases catalyze formation of *para*-aminophenol from acetaminophen[54] while *para*-aminophenol produces nephrotoxicity but not hepatotoxicity.[45,55] If autoxidation is responsible for the toxicity of *para*-aminophenol, then both liver and kidney should be damaged following administration of acetaminophen or *para*-aminophenol.

Evidence suggests that *para*-aminophenol undergoes sequential metabolism by liver and kidney to form nephrotoxic metabolites. For example, cannulation of the common bile duct partially prevents *para*-aminophenol-induced nephrotoxicity in Fischer-344 rats,[56] suggesting that one or more hepatic metabolites of *para*-aminophenol gain access to the kidney and contribute to nephrotoxicity following biliary excretion. These putative toxic metabolites may be glutathione conjugates of *para*-aminophenol since pretreatment of Fischer-344 rats with buthionine sulfoximine to deplete renal and hepatic glutathione completely prevents nephrotoxicity following administration of *para*-aminophenol.[56] Further, synthetic glutathione conjugates of *para*-aminophenol produce nephrotoxicity *in vivo*[57,58] and *in vitro*.[59] Nephrotoxicity of *para*-aminophenol-glutathione conjugates is prevented by pretreating rats with ascorbic acid, while pretreatment with probenecid (an inhibitor of organic anion transport in the kidney), acivicin (an inhibitor of γ-glutamyl transpeptidase), or aminooxyacetic acid (an inhibitor of cysteine conjugate β-lyase) is without effect.[58,60,61] When administered to male Fischer-344 rats, *para*-aminophenol-glutathione conjugates are more toxic than *para*-aminophenol.[57] In contrast, *para*-aminophenol and *para*-aminophenol-glutathione conjugates are approximately equitoxic in renal epithelial cells obtained from male Wistar rats.[59] It is clear that *para*-aminophenol undergoes glutathione conjugation in the liver and that these conjugates are capable of injuring the kidney. The mechanism involved in conjugate formation and the contribution that these conjugates make to the nephrotoxicity of *para*-aminophenol has yet to be defined.

The precise mechanism by which *para*-aminophenol or its metabolite produce renal toxicity is not entirely clear. Radiolabel from *para*-aminophenol is preferentially incorporated into proteins in kidney as compared with liver.[60,62] Covalently bound material derived from radiolabeled *para*-aminophenol is distributed in mitochondria, microsomes, and cytosol and associated with protein, DNA, mitochondrial enzymes, and glucose-6-phosphatase.[62] In Wistar rats treated with *para*-aminophenol, glucose synthesis from lactate or pyruvate is inhibited in kidney but not liver within 30–60 min of treatment.[63] In addition, renal mitochondria obtained from Sprague–Dawley rats treated with *para*-aminophenol display decreased state 3 respiration and respiratory control ratio within 60 min of treatment.[64] Thus, *para*-aminophenol or its metabolites may injure proximal tubular cells by interfering with mitochondrial function, intermediary metabolism, or a combination of events.

7.30.3.4.3 Other NSAIDs

Although acute tubular necrosis has been reported following overdoses of phenylbutazone, ibuprofen, and mefenamic acid,[26] nephrotoxicity due to these agents has not been investigated as thoroughly as with aspirin, salicylate, or acetaminophen. The time course, functional characteristics, and mechanisms leading to acute tubular necrosis with other NSAIDs are unclear, and the relevance to human consumption has not been established.

7.30.3.5 Interstitial Nephritis and Nephrotic Syndrome

Interstitial nephritis is a rare, idiosyncratic reaction to therapy with numerous NSAIDs (Table 3).[3] Interstitial nephritis is estimated to occur in one of every 5000 to 10 000 patients receiving NSAID therapy and differs from acute ischemic renal insufficiency in onset, severity, and duration.[3] Interstitial nephritis can occur within 1 week of NSAID administration but more often occurs several months to a year after the start of NSAID therapy.[2] Whereas acute renal insufficiency causes minimal decrements in GFR, patients with interstitial nephritis may present with elevated serum creatinine concentrations ($>6\,mg\,dL^{-1}$ compared with the normal of 0.8–$2.0\,mg\,dL^{-1}$).[2] In NSAID-induced acute renal insufficiency, withdrawal of the causative agent usually restores renal function back to normal within a day or two. In contrast, NSAID-induced interstitial nephritis may persist for 1–3 months following discontinuation of NSAID therapy.[2] Clinical symptoms of NSAID-induced interstitial nephritis also include edema, oliguria, and/or proteinuria. Systemic symptoms of allergic interstitial nephritis, such as fever, drug rash, eosinophilia, and eosinophiluria, are absent with NSAID-induced interstitial nephritis.[5] Histological findings included diffuse interstitial edema with evidence of mild to moderate inflammation, and cellular infiltration of cytotoxic T cells and smaller numbers of B cells and eosinophils.[65,66] Glomerular membrane involvement, such as fusion of epithelial foot processes, is observed in NSAID-induced acute interstitial nephritis.[67] Microscopically, glomeruli resemble those seen in minimal-change nephrotic syndrome with effacement of foot processes of epithelial cells visible by electron microscopy but little or no abnormality visible by light microscopy.[68] Characteristically, NSAID-induced interstitial nephritis consists of minimal change glomerulonephritis with interstitial nephritis, accounting for albuminuria in the nephrotic syndrome

range.[5] The most marked histologic changes occur in the interstitium surrounding renal tubules and consist of focal diffuse inflammatory infiltrate containing cytotoxic T lymphocytes, B cells, other T cells, and plasma cells.[65]

Interstitial nephritis is not related to the dose of NSAID used and does not appear to be directly related to inhibition of prostaglandin synthesis *per se*.[1,68] Rather, by inhibiting cyclooxygenase, NSAIDs may make more arachidonic acid available for lipoxygenase-catalyzed leukotriene formation.[5] Since leukotrienes are chemotactic agents, leukotrienes may participate in the development of interstitial nephritis and proteinuria by attracting T lymphocytes, promoting inflammation, and increasing vascular permeability.[5,65] Fortunately, interstitial nephritis due to NSAID therapy is reversible when the drug is discontinued.

Risk factors for NSAID-induced interstitial nephritis are not entirely clear. Preexisting renal insufficiency does not appear to predispose a patient to develop interstitial nephritis.[5] Old age may be a risk factor but the increased incidence of interstitial nephritis in the elderly may be a reflection of the tendency for older patients to be the more likely candidates for chronic NSAID therapy.[5]

7.30.3.6 Papillary Necrosis

Chronic administration of nearly all NSAIDs produces papillary necrosis in laboratory animals and clinical case reports of papillary necrosis may be found in the medical literature (Table 3).[5] Eknoyan estimates that 15–20% of clinical cases of renal papillary necrosis are associated with chronic analgesic consumption.[69] In some patients, analgesic use in the therapeutic range may cause renal impairment, such as an inability to excrete a sodium or water load, whereas cumulative ingestion of more than 3 kg of an analgesic agent have been associated with renal failure, such as papillary necrosis or oliguria.[70] While the cumulative dose of 3 kg may seem quite high, patients with arthritis may easily consume 10 aspirin tablets per day (3.25 g) or about 1.2 kg aspirin per year. Estimates of the prevalence of chronic analgesic consumption range from as low as 2% to as high as 30%, depending on the type and geographic location of the population surveyed.[71] As indicated in Table 4, epidemiological surveys indicate a wide geographic variation in the incidence of analgesic-associated papillary necrosis. Phenacetin was initially implicated as the causative agent of papillary necrosis and was removed from nonprescription analgesics as early as 1961 in Sweden and

Table 4   Autopsy incidence of renal papillary necrosis (PN) or chronic interstitial nephritis (CIN) in different countries.

Country	Years	Total no. of autopsies	No. with PN or CIN (%)
Australia			
Brisbane	1964	457	32 (7.0)
	1975–1976	628	54 (8.6)
Sydney	1959–1962	1 530	50 (3.7)
Denmark	1957–1960	3 969	90 (3.4)
	1960–1967	10 241	288 (2.8)
	1961–1975	25 358	224 (0.9)
Switzerland	1948–1957	14 942	192 (1.3)
	1970–1972	2 819	50 (1.75)
New Zealand	1971	184	3 (1.6)
Scotland	1957–1966	4 982	27 (0.54)
	1959–1962	1 987	12 (0.60)
England	1970–1975	20 229	83 (0.41)
	1961–1967	18 866	31 (0.16)
United States	1947–1961	11 000	25 (0.23)

Source: Prescott.[38]

Denmark.[38] However, it is now apparent that papillary necrosis may be associated with a variety of analgesic agents in the absence of phenacetin ingestion.[2] Further, the incidence of papillary necrosis has not declined despite removal of phenacetin from public consumption, suggesting that other agents or combinations of agents are responsible for NSAID-induced papillary necrosis.[38] In humans, papillary necrosis has been reported following administration of aspirin (alone or in combination with other analgesics but not phenacetin), ibuprofen, fenoprofen, indomethacin, diclofenac, mefenamic acid, phenylbutazone, and piroxicam (Table 3).[38,68]

Clinical manifestations of papillary necrosis include loin pain, hematuria, ureteral obstruction and/or uremia. Functional changes include salt wastage, hyperkalemia, and metabolic acidosis. Patients often complain of polyuria and exhibit a defect in urinary concentrating ability that is not restored by ADH administration.[26,72] A decline in GFR may be a late reversible or irreversible clinical feature of analgesic nephropathy.[70] When analgesic preparations contained phenacetin, papillae in patients with papillary necrosis contained a black pigment, possibly due to a breakdown product of phenacetin.[68] However, papillary necrosis associated with aspirin or other NSAIDs may show no pigmentation.[68] Urinary tract infection and hypertension are common secondary symptoms.[5,38] In addition, patients either present with or develop hypertension during the course of analgesic nephropathy.[26] In severe cases, renal function may deteriorate rapidly despite vigorous dialysis and septicemia and death may ensue.[72]

In more moderate cases where only some of the papillae are involved, a characteristic "ring sign" will be seen on intravenous pyelography and renal function may stabilize if the causative agent is removed.[72] Initially, the lesion is localized in the papilla and may ascend into the cortex. The lesion is usually bilateral.[73] In humans, there are three stages of development identified by the severity of the lesion.[73,74] Early changes are confined to the inner medulla and papilla while the outer medulla and cortex are normal. In the papilla there is necrosis of interstitial cells, medullary loops of Henle, and medullary capillaries. Collecting ducts running through the areas of necrosis show little change although there may be slight dysplasia and some basement membrane thickening. The intermediate stage is characterized by evidence of necrosis throughout the papillae reaching the junction of inner and outer medulla. Cortex and outer medulla remain normal macroscopically and microscopically. Toward the tip of the papilla, all elements are necrotic except for collecting ducts and occasional vasa recta. The advanced stage involves total destruction of the papilla and the papilla may become separated from the renal parenchyma or be entirely absent.[73,74] A border of calcification may be present proximal to the advancing edge of necrosis. In the advanced stage, the renal cortex may be fibrotic and contain infiltrates of a variety of inflammatory cells. Cortical involvement has previously been ascribed to "chronic interstitial nephritis" but is now believed to be a sequelae of papillary necrosis.[38] Tubular atrophy may be present and the kidneys appear shrunken with irregular borders upon radiological examination.[75]

In severe papillary necrosis, areas of the renal medulla may become damaged. In addition to containing medullary portions of the loop of Henle and collecting duct, the medulla contains three types of interstitial cells.[76] Type 1 interstitial cells, which are the only interstitial cells in the inner medulla and papilla, are stellate in appearance with long slender processes containing numerous lipid droplets.[76] These lipid droplets include triglycerides, cholesterol esters, and phospholipids.[77] The triglycerides include long-chain fatty acids such as homo-γ-linolenic acid and arachidonic acid, leading to the hypothesis that renomedullary interstitial cells are involved in the production and storage of medullary prostaglandins. The interstitial cells are located close to loops of Henle and capillaries but rarely close to collecting tubules,[76]

suggesting that prostaglandins formed by these interstitial cells would exert actions in the loop of Henle and on medullary blood flow. Renomedullary interstitial cells may also function to provide mechanical support, in the synthesis of extracellular matrix, and assist in urinary concentration and blood pressure control.[76,78,79] In particular, renomedullary interstitial cells synthesize and secrete an antihypertensive substance (medullipin I) in response to increases in renal perfusion pressure.[80] Loss of renomedullary interstitial cells, as by chemical papillectomy, bilateral hydronephrosis, and papillary ablation, makes rats hypertensive following salt loading.[78,80] Administration of medullipin I or transplantation of renomedullary interstitial cells restores blood pressure toward normal following papillectomy, demonstrating the antihypertensive properties of renomedullary interstitial cells.[80] Thus, it is possible that loss of renomedullary interstitial cells due to NSAID-induced papillary necrosis may predispose individuals to the development of hypertension due to loss of the antihypertensive properties of these cells.

The pathology of analgesic-associated papillary necrosis has been established by studying the lesion in laboratory animals. However,

Table 5 Dosages of nonnarcotic analgesics shown to produce experimental renal papillary necrosis.[a]

Preparation	Oral daily dose (mg kg^{-1})	Duration (weeks)	Renal papillary necrosis (% incidence)
Combination analgesics			
A + P + C/A + P	280–900	8–72	37.5–100
A + C/A + APAP	500–900	12–72	56.3–100
A + salicylamide + C	840	12–72	53.9
Antipyrine + P + C	840	12–72	36.4
Single analgesics			
Aspirin	200–500	8–66	33.3–75
Phenacetin	200–3000	4–48	37.5–80
Acetaminophen	625–3000	8–20	42.9–60
Phenylbutazone	10	8–20	11.1
	400 (i.v.)	2–54 (hours)	19
Indomethacin	12	12–30	28.6
	75 (i.v.)	2–54 (hours)	21
Antipyrine	1000	12–30	12.5
Amidopyrine	500–1200	5	83.3
Mefenamic acid	100	8–20	66.7
Flufenamic acid	50–100	–	+
Meclofenamic acid	–	–	+
Fenoprofen	40–305	72	+
Naproxen	–	–	+
Sudoxicam	2	–	+

Source: Nunra.[26]
[a]A = aspirin, P = phenacetin, C = caffeine, APAP = acetaminophen, i.v. = intravenously, – = not reported, + = present but not quantitated for incidence.

commonly used laboratory animals (dog, rat, and rabbit) are not reliably susceptible to salicylate-, phenacetin-, and NSAID-induced papillary necrosis so that these compounds must be given in large dosages over long periods of time (Table 5).[81] For example, papillary necrosis occurred in only 50% of rats given acetylsalicylic acid at a dose of 500 mg kg^{-1}day^{-1} after 10–60 weeks.[82] Renal papillary necrosis has been produced experimentally with phenacetin, aspirin, acetaminophen, antipyrine, amino-pyrine, and a number of NSAIDs including phenylbutazone, indomethacin, mefenamic acid, flufenamic acid, meclofenamic acid, feno-profen, naproxen, and sudoxicam (Table 5).[26,81] When papillary necrosis is experimentally induced in laboratory animals (e.g., with 2-bromoethylamine), morphological changes similar to those observed in human kidneys are seen.[72] Functional changes observed during the course of analgesic nephropathy in rats include impaired concentrating capacity, hypertension, reduced total body sodium, reduced renal vein plasma PGE_2, and increased renal vein plasma renin activity.[26]

The mechanisms involved in papillary necro-sis have received considerable attention but remain unresolved. Duggin suggests that at least one of two factors must be present for the development of papillary necrosis.[34] First, the tissue must have a metabolic or functional predisposition such as susceptibility to ischemia from reduced RBF. It has been postulated that the mechanism of NSAID-induced papillary necrosis is the result of prostaglandin-mediated acute ischemic insult. Analgesic-associated nephropathy in rats occurs with a variety of structurally unrelated NSAIDs, indirectly sup-porting an ischemic mechanism.[2] NSAIDs inhibit prostaglandin synthesis and with pro-longed NSAID therapy, papillary ischemia followed by necrosis may occur.[2]

The second requirement proposed by Duggin is that the causative agent must selectively concentrate within the renal papilla.[34] Both aspirin and acetaminophen, agents associated with papillary necrosis, selectively concentrate within the papilla.[26,68] Although acetamino-phen is reported to be an ineffective COX1 inhibitor outside of the central nervous system,[83] acetaminophen effectively lowers PGE_2 and $PGF_{2\alpha}$ production in renal medul-lary slices from rat kidney.[84] The IC_{50} for acetaminophen-induced inhibition of prosta-glandin synthesis is 100–200 μM and the effect of acetaminophen is reversible.[84] By compar-ison, 1 mM aspirin and 0.28 mM indomethacin reduce PGE_2 synthesis to the same extent as did 1 mM acetaminophen.[85] Rats fed a combina-tion of aspirin, phenacetin, and caffeine for

8–20 weeks have a 38.8% incidence of papillary necrosis when deprived of water for 16 h per day whereas rats maintained in a constant diuretic state (by substituting 5% glucose for drinking water) have a 16.7% incidence of papillary necrosis.[82]

Biochemical mechanisms that may contri-bute to analgesic nephropathy include deple-tion of cellular glutathione by salicylates and metabolism of acetaminophen to a highly reactive intermediate that binds covalently to tissue proteins when glutathione concentrations are low.[54,68] As previously mentioned, PHS, which is highly active in inner medulla and papilla, has a peroxidase activity that may participate in the formation of reactive inter-mediates.[85] These reactive intermediates can arylate tissue nucleophiles such as proteins, RNA, and DNA, to initiate cellular damage.[85] It is possible that acetaminophen undergoes peroxidase-catalyzed co-oxidation that may contribute to acetaminophen-induced papillary necrosis.[86] However, acetaminophen inhibits PHS-associated cyclooxygenase activity with an IC_{50} of 0.1 mM acetaminophen for inhibi-tion of PGE_2 formation by rabbit renal medullary microsomes.[87] If acetaminophen selectively concentrates in the renal medulla and papilla, it is possible that regional concen-trations could approach or exceed 0.1 mM, leading to inactivation of cyclooxygenase activ-ity and, thereby, actually protect the renal medulla and papilla from damage caused by reactive acetaminophen metabolites. In addi-tion, the hypothesis of PHS-catalyzed bioacti-vation of aspirin is untenable because aspirin irreversibly inhibits PHS-associated cyclooxy-genase activity (Table 1),[87] so that aspirin would be expected to inhibit its own PHS-associated cooxidation.

Nonnarcotic analgesics have the potential to cause direct cell injury. For example, salicylates inhibit the hexose monophosphate shunt in medullary tissue whereas acetaminophen does not.[88] In dog medullary slices, only the combi-nation of salicylic acid and acetaminophen alters metabolism.[88] Salicylates depress protein synthesis, probably due to inhibition of amino acyl t-RNA synthetases, and salicylates deplete cells of ATP.[36] NSAIDs uncouple oxidative phosphorylation and deplete ATP in rat liver mitochondria *in vitro*, and the rank order of potency for these effects is roughly: aspirin 1, ibuprofen 10, dinitrophenol 40–50, flufenamic acid 200.[89]

Papillary necrosis occurs with structurally dissimilar compounds that share the ability to inhibit prostaglandin synthesis, suggesting that redistribution of prostaglandin-dependent medullary blood flow may contribute to the development of papillary necrosis. With prolonged administration, NSAIDs may cause sustained papillary and medullary ischemia that ultimately may culminate in papillary necrosis if the causative agent is not discontinued.

7.30.4 CONCLUSIONS

NSAIDs are indispensable therapeutic agents for a variety of disorders. In general, occasional use of NSAIDs for analgesia presents little or no risk for adverse renal effects. However, chronic use of NSAIDs, as for various types of arthritis, may predispose patients to potentially serious side effects. In particular, patients with pre-existing risk factors such as congestive heart failure, cirrhosis, or nephrotic syndrome may be more vulnerable to serious toxicities associated with NSAIDs than the general population. In addition, the elderly may be more susceptible to NSAID toxicities than the general population. However, the potential benefit of NSAID therapy far outweighs the risks associated with NSAID use and these agents remain widely used in clinical medicine.

7.30.5 REFERENCES

1. R. D. Toto, 'The role of prostaglandins in NSAID induced renal dysfunction.' *J. Rheumatol.*, 1991, **28** (Suppl.) , 22–25.
2. M. D. Murray and D. C. Brater, 'Renal toxicity of the nonsteroidal anti-inflammatory drugs.' *Annu. Rev. Pharmacol. Toxicol.*, 1993, **32**, 435–465.
3. L. C. Knodel, 'NSAID adverse effects and interactions: who is at risk?' *Am. Pharm.*, 1992, **NS32**, 39–47.
4. D. M. Clive and J. S. Stoff, 'Renal syndromes associated with nonsteroidal anti-inflammatory drugs.' *N. Engl. J. Med.*, 1984, **9**, 563–572.
5. A. Whelton and C. W. Hamilton, 'Nonsteroidal anti-inflammatory drugs: effects on kidney function.' *J. Clin. Pharmacol.*, 1991, **31**, 588–598.
6. E. A. Meade, W. L. Smith and D. L. DeWitt, 'Differential inhibition of prostaglandin endoperoxide synthase (cyclooxygenase) isozymes by aspirin and other non-steroidal anti-inflammatory drugs.' *J. Biol. Chem.*, 1993, **268**, 6610–6614.
7. D. Schlondorff, 'Renal complications of nonsteroidal anti-inflammatory drugs.' *Kidney Int.*, 1993, **44**, 643–653.
8. J. A. Mitchell, P. Akarasereenont, C. Thiemermann *et al.*, 'Selectivity of nonsteroidal antiinflammatory drugs as inhibitors of constitutive and inducible cyclooxygenase.' *Proc. Natl. Acad. Sci. USA*, 1993, **90**, 11693–11697.
9. O. Laneuville, D. K. Breuer, D. L. Dewitt *et al.*, 'Differential inhibition of human prostaglandin endoperoxide H synthases-1 and -2 by nonsteroidal anti-inflammatory drugs.' *J. Pharmacol. Exp. Ther.*, 1994, **271**, 927–934.
10. R. J. Flower, 'Drugs which inhibit prostaglandin biosynthesis.' *Pharmacol. Rev.*, 1974, **26**, 33–67.
11. R. J. Kulmacz and W. E. M. Lands, 'Stoichiometry and kinetics of the interaction of prostaglandin H synthase with anti-inflammatory agents.' *J. Biol. Chem.*, 1985, **260**, 12572–12578.
12. R. K. Verbeeck, J. L. Blackburn and G. R. Loewen, 'Clinical pharmacokinetics of non-steroidal anti-inflammatory drugs.' *Clin. Pharmacokinet.*, 1983, **8**, 297–331.
13. T. E. Eling and J. F. Curtis, 'Xenobiotic metabolism by prostaglandin H synthase.' *Pharmacol. Ther.*, 1992, **53**, 261–273.
14. A. R. Morrison, 'Biochemistry and pharmacology of renal arachidonic acid metabolism.' *Am. J. Med.*, 1986, **80**, (1A), 3–11.
15. M. J. Dunn and E. J. Zambraski, 'Renal effects of drugs that inhibit prostaglandin synthesis.' *Kidney Int.*, 1980, **18**, 609–622.
16. I. Ichikawa and B. M. Brenner, 'Importance of efferent arteriolar vascular tone in regulation of proximal tubule fluid reabsorption and glomerulotubular balance in the rat.' *J. Clin. Invest.*, 1980, **65**, 1192–1201.
17. S. Kaojarern, P. Chennavasin, S. Anderson *et al.*, 'Nephron site of effect of nonsteroidal anti-inflammatory drugs on solute excretion in humans.' *Am. J. Physiol.*, 1983, **244**, F134–F139.
18. K. H. Raymond and M. D. Lifschitz, 'Effect of prostaglandins on renal salt and water excretion.' *Am. J. Med.*, 1986, **80**, 22–33.
19. M. C. Blasingham and A. Nasjletti, 'Differential renal effects of cyclooxygenase inhibition in sodium-replete and sodium-deprived dog.' *Am. J. Physiol.*, 1980, **239**, F360–F365.
20. D. C. Brater, S. A. Anderson, D. Brown-Cartwright *et al.*, 'Effect of etodolac in patients with moderate renal impairment compared with normal subjects.' *Clin. Pharmacol. Ther.*, 1985, **38**, 674–679.
21. R. S. Muther, D. M. Potter and W. M. Bennett, 'Aspirin-induced depression of glomerular filtration rate in normal humans: role of sodium balance.' *Ann. Intern. Med.*, 1981, **94**, 317–321.
22. G. F. DiBona, 'Prostaglandins and nonsteroidal anti-inflammatory drugs. Effects on renal hemodynamics.' *Am. J. Med.*, 1986, **80**, (1A), 12–21.
23. N. A. Terragno, D. A. Terragno and J. C. McGiff, 'Contribution of prostaglandins to the renal circulation in conscious, anesthetized, and laparotomized dogs.' *Circ. Res.*, 1977, **40**, 590–595.
24. M. C. Blasingham, R. E. Shade, L. Share *et al.*, 'The effect of meclofenamate on renal blood flow in the unanesthetized dog: relation to renal prostaglandins and sodium balance.' *J. Pharmacol. Exp. Ther.*, 1980, **214**, 1–4.
25. S. G. Chrysant, 'Renal functional changes induced by prostaglandin E_1 and indomethacin in the anesthetized dog.' *Arch. Int. Pharmacodyn. Ther.*, 1978, **234**, 156–163.
26. R. S. Nanra, 'Renal effects of antipyretic analgesics.' *Am. J. Med.*, 1983, **75** (5A), 70–81.
27. L. Arnold, C. Collins and G. A. Starmer, 'The short-term effects of analgesics on the kidney with special reference to acetylsalicylic acid.' *Pathology*, 1973, **5**, 123–134.
28. M. E. Kyle and J. J. Kocsis, 'The effect of age on salicylate-induced nephrotoxicity in male rats.' *Toxicol. Appl. Pharmacol.*, 1985, **81**, 337–347.
29. M. J. Robinson, E. A. Nichols and L. Taitz, 'Nephrotoxic effect of acute sodium salicylate intoxication in the rat.' *Arch. Pathol.*, 1967, **84**, 224–226.
30. J. R. Mitchell, R. J. McMurtry, C. N. Statham *et al.*, 'Molecular basis for several drug-induced nephropathies.' *Am. J. Med.*, 1977, **62**, 518–526.
31. R. A. Owen and R. Heywood, 'Age-related suscept-

ibility to aspirin-induced nephrotoxicity in female rats.' *Toxicol. Lett.*, 1983, **18**, 167–170.

32. S. Chakrabarti and S. Yamaguchi, 'Ethanol-induced protection against acute nephrotoxicity due to aspirin.' *Res. Commun. Subst. Abuse*, 1986, **7**, 153–160.

33. R. J. Caterson, G. G. Duggin, J. Horvath *et al.*, 'Aspirin, protein transacetylation and inhibition of prostaglandin synthetase in the kidney.' *Br. J. Pharmacol.*, 1978, **207**, 353–358.

34. G. G. Duggin, 'Mechanisms in the development of analgesic nephropathy.' *Kidney Int.*, 1980, **18**, 553–561.

35. A. Quintanilla and R. H. Kessler, 'Direct effect of salicylates on renal function in the dog.' *J. Clin. Invest.*, 1973, **52**, 3143–3153.

36. A. G. Dawson, 'Effects of acetylsalicylate on gluconeogenesis in isolated rat kidney tubules.' *Biochem. Pharmacol.*, 1975, **24**, 1407–1411.

37. P. Silva, 'Renal fuel utilization, energy requirements, and function.' *Kidney Int.*, Suppl. 1987, **22**, S9–S14.

38. L. F. Prescott, 'Analgesic nephropathy: a reassessment of the role of phenacetin and other analgesics.' *Drugs*, 1982, **23**, 75–149.

39. J. B. Tarloff, R. S. Goldstein, D. G. Morgan *et al.*, 'Acetaminophen and *p*-aminophenol nephrotoxicity in aging male Sprague–Dawley and Fischer 344 rats.' *Fundam. Appl. Toxicol.*, 1989, **12**, 78–91.

40. S. G. Emeigh Hart, W. P. Beierschmitt, D. S. Wyand *et al.*, 'Acetaminophen nephrotoxicity in CD-1 mice. I. Evidence of a role for *in situ* activation in selective covalent binding and toxicity.' *Toxicol. Appl. Pharmacol.*, 1994, **126**, 267–275.

41. J. F. Newton, M. Yoshimoto, J. Bernstein *et al.*, 'Acetaminophen nephrotoxicity in the rat. I. Strain differences in nephrotoxicity and metabolism.' *Toxicol. Appl. Pharmacol.*, 1983, **69**, 291–306.

42. S. G. Emeigh Hart, W. P. Beierschmitt, J. B. Bartolone *et al.*, 'Evidence against deacetylation and for cytochrome P450-mediated activation in acetaminophen-induced nephrotoxicity in the CD-1 mouse.' *Toxicol. Appl. Pharmacol.*, 1991, **107**, 1–15.

43. J. B. Bartolone, R. B. Birge, S. J. Bulera *et al.*, 'Purification, antibody production, and partial amino acid sequence of the 58-kDa acetaminophen-binding liver proteins.' *Toxicol. Appl. Pharmacol*, 1992, **113**, 19–29.

44. J. J. Hu, M. J. Lee, M. Vapiwala *et al.*, 'Sex-related differences in mouse renal metabolism and toxicity of acetaminophen.' *Toxicol. Appl. Pharmacol.*, 1993, **122**, 16–26.

45. J. F. Newton, C. H. Kuo, M. W. Gemborys *et al.*, 'Nephrotoxicity of *p*-aminophenol, a metabolite of acetaminophen, in the Fischer 344 rat.' *Toxicol. Appl. Pharmacol.*, 1982, **65**, 336–344.

46. R. J. McMurtry, W. R. Snodgrass and J. R. Mitchell, 'Renal necrosis, glutathione depletion, and covalent binding after acetaminophen.' *Toxicol. Appl. Pharmacol.*, 1978, **46**, 87–100.

47. J. F. Newton, D. A. Pasino and J. B. Hook, 'Acetaminophen nephrotoxicity in the rat: quantitation of renal metabolic activation *in vivo*.' *Toxicol. Appl. Pharmacol.*, 1985, **78**, 39–46.

48. J. F. Newton, C. H. Kuo, G. M. DeShone *et al.*, 'The role of *p*-aminophenol in acetaminophen-induced nephrotoxicity: effect of bis(*p*-nitrophenyl)phosphate on acetaminophen and *p*-aminophenol nephrotoxicity and metabolism in Fischer 344 rats.' *Toxicol. Appl. Pharmacol.*, 1985, **81**, 416–430.

49. J. B. Tarloff, E. A. Khairallah, S. D. Cohen *et al.*, 'Sex- and age-dependent acetaminophen hepato- and nephrotoxicity in Sprague–Dawley rats: role of tissue accumulation, nonprotein sulfhydryl depletion, and covalent binding.' *Fundam. Appl. Toxiol.*, 1996, 13–22.

50. M. W. Gemborys and G. H. Mudge, 'Formation and disposition of the minor metabolites of acetaminophen in the hamster.' *Drug Metab. Dispos.*, 1981, **9**, 340–351.

51. I. C. Calder, A. C. Yong, R. A. Woods *et al.*, 'The nephrotoxicity of *p*-aminophenol. II. The effect of metabolic inhibitors and inducers.' *Chem. Biol. Interact.*, 1979, **27**, 245–254.

52. P. D. Josephy, T. E. Eling and R. P. Mason, 'Oxidation of *p*-aminophenol catalyzed by horseradish peroxidase and prostaglandin synthase.' *Mol. Pharmacol.*, 1983, **23**, 461–466.

53. P. J. Boogaard, J. F. Nagelkerke and G. J. Mulder, 'Renal proximal tubular cells in suspension or in primary culture as *in vitro* models to study nephrotoxicity.' *Chem. Biol. Interact.*, 1990, **76**, 251–291.

54. C. A. Mugford and J. B. Tarloff, 'Contribution of oxidation and deacetylation to the bioactivation of acetaminophen *in vitro* in liver and kidney from male and female Sprague-Dawley rats.' *Drug Metab. Dispos.*, 1995, **23**, 290–295.

55. C. M. Burnett, T. A. Re, S. Rodriguez *et al.*, 'The toxicity of *p*-aminophenol in the Sprague–Dawley rat: effects on growth, reproduction and foetal development.' *Food Chem. Toxicol.*, 1989, **27**, 691–698.

56. K. P. Gartland, C. T. Eason, F. W. Bonner *et al.*, 'Effects of biliary cannulation and buthionine sulphoximine pretreatment on the nephrotoxicity of para-aminophenol in the Fischer 344 rat.' *Arch. Toxicol.*, 1990, **64**, 14–25.

57. L. M. Fowler, R. B. Moore, J. R. Foster *et al.*, 'Nephrotoxicity of 4-aminophenol glutathione conjugate.' *Hum. Exp. Toxicol.*, 1991, **10**, 451–459.

58. L. M. Fowler, J. R. Foster and E. A. Lock, 'Nephrotoxicity of 4-amino-3-*S*-glutathionylphenol and its modulation by metabolism or transport inhibitors.' *Arch. Toxicol.*, 1994, **68**, 15–23.

59. C. Klos, M. Koob, C. Kramer *et al.*, '*p*-Aminophenol nephrotoxicity: biosynthesis of toxic glutathione conjugates.' *Toxicol. Appl. Pharmacol.*, 1992, **115**, 98–106.

60. L. M. Fowler, J. R. Foster and E. A. Lock, 'Effect of ascorbic acid, acivicin and probenecid on the nephrotoxicity of 4-aminophenol in the Fischer 344 rat.' *Arch. Toxicol.*, 1993, **67**, 613–621.

61. M. L. Anthony, C. R. Beddell, J. C. Lindon *et al.*, 'Studies on the effects of L(αS,5S)-α-amino-3-chloro-4,5-dihydro-5-isoxazoleacetic acid (AT-125) on 4-aminophenol-induced nephrotoxicity in the Fischer 344 rat.' *Arch. Toxicol.*, 1993, **67**, 696–705.

62. C. A. Crowe, A. C. Yong, I. C. Calder *et al.*, 'The nephrotoxicity of *p*-aminophenol. I. The effect on microsomal cytochromes, glutathione and covalent binding in kidney and liver.' *Chem. Biol. Interact.*, 1979, **27**, 235–243.

63. J. D. Tange, B. D. Ross and J. G. G. Ledingham, 'Effects of analgesics and related compounds on renal metabolism in rats.' *Clin. Sci. Mol. Med.*, 1977, **53**, 485–492.

64. C. A. Crowe, I. C. Calder, N. P. Madsen *et al.*, 'An experimental model of analgesic-induced renal damage—some effects of *p*-aminophenol on rat kidney mitochondria.' *Xenobiotica*, 1977, **7**, 345–356.

65. W. L. Bender, A. Whelton, W. E. Beschorner *et al.*, 'Interstitial nephritis, proteinuria, and renal failure caused by nonsteroidal anti-inflammatory drugs. Immunological characterization of the inflammatory infiltrate.' *Am. J. Med.*, 1984, **76**, 1006–1012.

66. A. B. Abt and J. A. Gordon, 'Drug-induced interstitial nephritis. Co-existence with glomerular disease.' *Arch. Intern. Med.*, 1985, **145**, 1063–1067.

67. C. L. Pirani, A. Valeri, V. D'Agati *et al.*, 'Renal toxicity of nonsteroidal anti-inflammatory drugs.' *Contrib. Nephrol.*, 1987, **55**, 159–175.

68. P. Kincaid-Smith, 'Effects of non-narcotic analgesics on the kidney.' *Drugs*, 1986, **32** Suppl. 4, 109–128.

69. G. Eknoyan, in 'Primer on Kidney Diseases,' ed. A. Greenberg, The National Kidney Foundation, 1994, pp. 182–184.

70. G. E. Schreiner, J. F. McAnally and J. F. Winchester, 'Clinical analgesic nephropathy.' *Arch. Intern. Med.*, 1981, **141**, 349–357.

71. V. M. Buckalew, in 'Primer on Kidney Diseases,' ed. A. Greenberg, National Kidney Foundation, Academic Press, New York, 1994, pp. 192–196.

72. S. Sabatini, 'Analgesic-induced papillary necrosis.' *Semin. Nephrol.*, 1988, **8**, 41–54.

73. F. J. Gloor, 'Changing concepts in pathogenesis and morphology of analgesic nephropathy as seen in Europe.' *Kidney Int.*, 1978, **13**, 27–33.

74. A. Burry, 'Pathology of analgesic nephropathy: Australian experience.' *Kidney Int.*, 1978, **13**, 34–40.

75. N. Lindvall, 'Radiological changes of renal papillary necrosis.' *Kidney Int.*, 1978, **13**, 93–106.

76. S. O. Bohman, The ultrastructure of the rat renal medulla as observed after improved fixation methods.' *J. Ultrastruct. Res.*, 1974, **47**, 329–360.

77. E. Änggård, S. O. Bohman, J. E. Griffin 3rd *et al.*, 'Subcellular localization of the prostaglandin system in the rabbit renal papilla.' *Acta Physiol. Scand.*, 1972, **84**, 231–246.

78. W. H. Sternberg, E. Farber and C. E. Dunlap, 'Histochemical localization of specific oxidative enzymes: II. Localization of diphosphopyridine nucleotide and triphosphopyridine nucleotide diaphorases and the succindehydrogenase system in the kidney.' *J. Histochem. Cytochem.*, 1956, **4**, 266–283.

79. E. E. Muirhead, 'Antihypertensive functions of the kidney: Arthur C. Corcoran Memorial Lecture' *Hypertension*, 1980, **2**, 444–464.

80. E. E. Muirhead, 'The medullipin system of blood pressure control.' *Am. J. Hypertens.*, 1991, **4**, 556S–568S.

81. E. H. Wiseman and H. Reinert, 'Anti-inflammatory drugs and renal papillary necrosis.' *Agents Actions*, 1975, **5**, 322–325.

82. R. S. Nanra and P. Kincaid-Smith, 'Papillary necrosis in rats caused by aspirin and aspirin-containing mixtures.' *Br. Med. J.*, 1970, **3**, 559–561.

83. R. J. Flower and J. R. Vane, 'Inhibition of prostaglandin synthetase in brain explains the anti-pyretic activity of paracetamol (4-acetamidophenol).' *Nature*, 1972, **240**, 410–411.

84. T. V. Zenser, M. B. Mattammal, C. A. Herman *et al.*, 'Effect of acetaminophen on prostaglandin E_2 and prostaglandin $F_{2\alpha}$ synthesis in the renal inner medulla of rat.' *Biochim. Biophys. Acta*, 1978, **542**, 486–495.

85. B. B. Davis, M. B. Mattammal and T. V. Zenser, 'Renal metabolism of drugs and xenobiotics.' *Nephron*, 1981, **27**, 187–196.

86. J. Mohandas, G. G. Duggin, J. S. Horvath *et al.*, 'Metabolic oxidation of acetaminophen (paracetamol) mediated by cytochrome P-450 mixed-function oxidase and prostaglandin endoperoxide synthetase in rabbit kidney.' *Toxicol. Appl. Pharmacol.*, 1981, **61**, 252–259.

87. M. B. Mattammal, T. V. Zenser, W. W. Brown *et al.*, 'Mechanism of inhibition of renal prostaglandin production by acetaminophen.' *J. Pharmacol. Exp. Ther.*, 1979, **210**, 405–409.

88. W. Davidson, L. Bassist and W. Shippey, '*In vitro* effects of aspirin and phenacetin metabolites on the metabolism of dog renal medulla.' *Clin. Res.*, 1973, **21**, 227.

89. Y. Tokumitsu, S. Lee and M. Ui, '*In vitro* effects of nonsteroidal anti-inflammatory drugs on oxidative phosphorylation in rats liver mitochondria.' *Biochem. Pharmacol.*, 1977, **26**, 2101–2106.

7.31
Halogenated Hydrocarbons

ADNAN A. ELFARRA
University of Wisconsin–Madison, WI, USA

7.31.1 INTRODUCTION

Halogenated hydrocarbons represent a large group of aromatic and aliphatic compounds with diverse industrial, agricultural, medical, and public health applications. Only the renal toxicology of selected aliphatic halogenated hydrocarbons will be discussed in this chapter. The toxicology of bromobenzene, an aromatic halogenated hydrocarbon, and related chemicals, are covered in Chapter 32, this volume.

Aliphatic halogenated hydrocarbons comprise many haloalkanes, haloalkenes, and haloalkynes. These chemicals are used as industrial solvents, pesticides, and starting materials in the manufacture of other chemicals. The major routes of human exposure to these chemicals at industrial sites are via inhalation and dermal contact. The general public may also be exposed to these chemicals by the oral or inhalation route as many aliphatic halogenated hydrocarbons are released into the environment or detected in drinking water. Because aliphatic halogenated hydrocarbons can induce toxicity to multiple organs in experimental animals, there is concern about the potential human health effects of these chemicals after acute, subchronic, or chronic exposure.

Most aliphatic halogenated hydrocarbons are chemically stable. Thus, they must undergo biotransformation to yield the ultimate toxic metabolites. The objective of this chapter is to review knowledge on the renal toxicology of representative haloalkanes and haloalkenes, the biochemical basis for their kidney selectivity, and, when possible, to correlate species- and sex-related differences in metabolism and nephrotoxicity. Additional reviews on nephrotoxicity and bioactivation mechanisms of aliphatic halogenated hydrocarbons are available.[1–10]

7.31.2 HALOALKANES

Haloalkanes vary immensely from each other in terms of their metabolism, bioactivation, and the chemical nature of the reactive intermediates formed, all of which depend in part upon the nature of the halogen substituent and the number of carbon atoms in the molecule. Thus, these chemicals cannot be considered uniform in their toxicological properties. The renal toxicology of chloroform, carbon tetrachloride, and 1,2-dichloroethane will be discussed below. Other nephrotoxic haloalkanes (e.g., dichloromethane, 1,2-dibromo-3-chloropropane and 1,2-dibromoethane) have similar bioactivation mechanisms.[11–17]

7.31.2.1 Chloroform (Trichloromethane)

Chloroform is used as an industrial solvent and as an intermediate in the manufacture of polymeric materials. Low levels (0.1–300 $\mu g\,L^{-1}$) of chloroform have been detected in drinking water subsequent to various chlorination processes.[18] Reports from several laboratories have demonstrated that the acute nephrotoxicity of chloroform is species, strain, and sex dependent;[19–26] male mice are more susceptible than rats, rabbits, or dogs, whereas female mice are resistant. Tubular swelling, necrosis, and casts, localized primarily in the proximal tubules, are the major histopathological changes in the kidney after exposure of experimental animals to chloroform. Chloroform-induced nephrotoxicity is also associated with elevated blood urea nitrogen concentrations, proteinuria, and glucosuria. *In vitro* uptake of organic anions and cations by renal cortical slices is also inhibited by *in vivo* treatment with chloroform.[27] While human exposure to chloroform has been associated with oliguria, proteinuria, increase in blood urea nitrogen, and renal tubular necrosis,[28–30] the threshold dose for acute chloroform kidney toxicity in humans is unknown. The localization of the human kidney lesion to the proximal tubules suggests a common mechanism of chloroform nephrotoxicity in most mammalian species.

Laboratory animals and humans metabolize chloroform to CO_2 (Figure 1); mouse, rat, and human conversion rates of an orally administered dose were 80%, 60%, and 50%, respectively.[3] Cytochrome P450s have been implicated in this biotransformation reaction; cytochrome P450-dependent oxidation of chloroform yields trichloromethanol which nonenzymatically undergoes conversion to phosgene, an electrophilic metabolite. Phosgene can then be hydrolyzed to yield CO_2. Alternatively, phosgene can react with glutathione to yield diglutathione carbonate; once cellular glutathione concentrations are depleted, phosgene can react with cellular macromolecules to form covalent adducts that

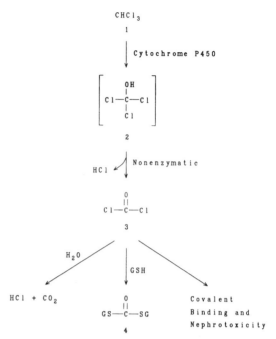

Figure 1 Oxidative metabolism of chloroform (1) to phosgene (3) by cytochrome P450s; (2), trichloromethanol; (4) diglutathione carbonate.

have been implicated in chloroform-induced nephrotoxicity. Further support for the involvement of P450s in chloroform-induced nephrotoxicity has been provided by the finding that deuterated chloroform, which is metabolized to phosgene more slowly than chloroform, is less nephrotoxic to mice than chloroform.[31,32] Covalent binding of chloroform-derived metabolites to kidney macromolecules is greater in male mice than in female mice.[31]

Although chloroform bioactivation to nephrotoxic metabolites could potentially occur in liver as well as in kidney, several studies have shown that chloroform-induced hepatotoxicity and nephrotoxicity can be modulated differently by various drug, chemical, or hormonal treatments, suggesting that chloroform is bioactivated by independent mechanisms in the liver and kidney.[1,31] The renal metabolism of chloroform by P450 enzymes correlates well with chloroform-induced nephrotoxicity.[21,24,33] The ability of human cytochrome P450 2E1 to metabolize chloroform *in vitro* has been demonstrated.[34] Thus, the findings that the level of this enzyme in the male mouse kidney is significantly higher than in the female mouse kidney and that treatment of female mice with testosterone, an agent that potentiates chloroform nephrotoxicity in female mice, significantly increases the level of this enzyme in the female mouse kidney,[35] suggest a role for renal P450 2E1 in chloroform-induced nephrotoxicity. The extent of P450 2E1 expression in human kidney and its regulation by various genetic, nutritional, and environmental factors remain to be determined. P450 enzymes, other than P450 2E1, may also metabolize chloroform. The availability of several cDNA-expressed human P450s should make it possible to identify additional P450 isoforms that may be involved in chloroform bioactivation. These studies may help determine which animal species might be a suitable model to assess risk to humans. In addition, because macromolecules are targets of phosgene alkylation, identification of critical targets may allow a better understanding of how covalent modification of renal macromolecules by phosgene may lead to cell necrosis.

7.31.2.2 Carbon Tetrachloride (Tetrachloromethane)

Carbon tetrachloride is commonly used as a chemical intermediate, solvent, and dry cleaning fluid. Human exposure to carbon tetrachloride from occupational or environmental sources is low and unlikely to produce acute kidney toxicity. However, a few cases of acute nephrotoxicity, especially when exposure occurs by inhalation, have been reported.[36-40] Based on these cases and the relative resistance of Sprague-Dawley or Wistar rats to carbon tetrachloride nephrotoxicity,[1] humans appear to be more susceptible to carbon tetrachloride-induced nephrotoxicity than most common strains of laboratory rats. In contrast to the early onset of carbon tetrachloride-induced hepatotoxicity (within 24 h of exposure), renal injury in humans and nonrodent species (domestic cats and rabbits) is typically encountered several days after carbon tetrachloride

exposure.[1,36-40] Nephrotoxicity of carbon tetrachloride is indicated by proteinuria, anuria, or oliguria and elevated concentrations of blood urea nitrogen and serum creatinine. In addition, these biochemical indicators of nephrotoxicity are usually associated with subtle morphological changes in the proximal tubular region of the nephron.[1,36-40]

Carbon tetrachloride is known to be metabolized by mouse and rat liver and kidney microsomes, presumably by cytochrome P450s, to yield a trichloromethyl radical which may be oxygenated to yield phosgene (Figure 2). Hydrolysis of phosgene would then result in the formation of CO_2, a major *in vivo* and *in vitro* metabolite of carbon tetrachloride. Phosgene may also react with glutathione to yield diglutathione carbonate. Alternatively, the trichloromethyl radical may abstract a proton from unsaturated fatty acids to yield chloroform and a fatty acid radical, which can lead to lipid peroxidation. The trichloromethyl radical may also bind covalently to proteins and lipids.[41-44] However, the role of metabolism in carbon tetrachloride-induced nephrotoxicity remains unclear. Furthermore, without a mechanistic-based understanding of carbon tetrachloride nephrotoxicity in experimental animals, it is difficult to make extrapolations between the responses of man and animals.

7.31.2.3 1,2-Dichloroethane (Ethylene Dichloride)

1,2-Dichloroethane is one of the most abundant synthetic chemicals in the United States and is used mainly as an intermediate in the production of vinyl chloride. Human exposure to 1,2-dichloroethane occurs mostly via inhalation

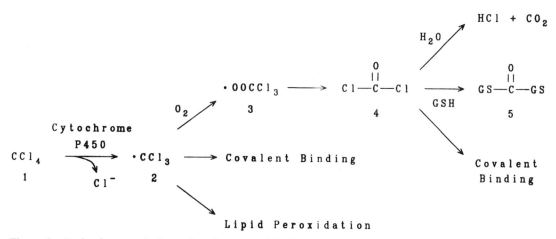

Figure 2 Reductive metabolism of carbon tetrachloride (1) to trichloromethyl free radical (2) by cytochrome P450s, and the subsequent reactions of the trichloromethyl free radical to yield trichloromethylperoxy free radical (3), phosgene (4), and diglutathione carbonate (5).

of vapors at industrial sites. However, drinking water may also be a source of human exposure.[45] 1,2-Dichloroethane intoxication in humans can produce acute renal failure that is associated with oliguria, albuminuria, and increased blood urea nitrogen concentrations.[46] The renal lesion, characterized by accumulation of fat and degeneration of tubular epithelium, appears to be selectively localized to the proximal tubular epithelium.[46] Single and repeated inhalation exposure of rats, guinea pigs, and monkeys to 1,2-dichloroethane at concentrations of 400–2400 ppm resulted in fatty degeneration, cast formation, and proliferation of the tubular epithelium.[46–49] However, when 1,2-dichloroethane was given in drinking water at 8000 ppm for 13 weeks to both sexes of F344/N rats, Sprague-Dawley rats, and Osborne–Mendel rats, only female F344/N rats exhibited renal lesions.[45] These results suggest that animals may be more susceptible to 1,2-dichloroethane-induced nephrotoxicity after inhalation exposure as compared with exposure by drinking contaminated water. While the above results also suggest that female F344/N rats are unusually sensitive to 1,2-dichloroethane nephrotoxicity when administered orally, the biochemical basis for this sensitivity remains unclear. Whereas *in vitro* studies with 1,2-dichloropropane, a nephrotoxic analogue of 1,2-dichloroethane, show that renal cortical slices from male Wistar rats are more sensitive to 1,2-dichloropropane-induced toxicity than females, possibly because of higher levels of expression of P450 2E1 in the male rat kidney,[50] the effect of gender on 1,2-dichloropropane *in vivo* nephrotoxicity has not been investigated.

1,2-Dichloroethane bioactivation can occur via two major metabolic pathways catalyzed by cytochrome P450s and glutathione *S*-transferases (Figures 3 and 4).[51–59] 1,2-Dichloroethane is a known substrate for human P450 2E1.[34] Cytochrome P450 oxidation at one of the carbon atoms forms a highly unstable *gem*-chlorohydrin which spontaneously decomposes to yield the electrophilic metabolite, 2-chloroacetaldehyde. 2-Chloroacetaldehyde may react with glutathione to yield 2-(*S*-glutathionyl)acetaldehyde, which can then be reduced by alcohol dehydrogenase to yield *S*-(2-hydroxyethyl)glutathione. 2-Chloroacetaldehyde can also be oxidized by aldehyde dehydrogenase to chloroacetic acid, a major urinary metabolite of 1,2-dichloroethane. The alternative metabolic pathway involves glutathione conjugation by cytosolic glutathione *S*-transferases to yield *S*-(2-chloroethyl)glutathione, a half-sulfur mustard, which may rearrange to an electrophilic episulfonium ion. This episulfonium ion

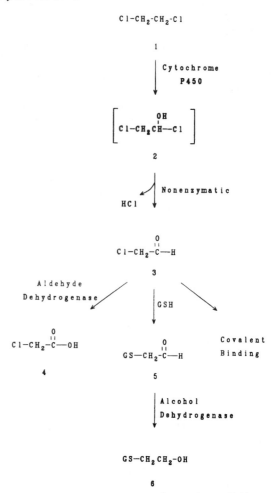

Figure 3 Oxidative metabolism of 1,2-dichloroethane (1) to 1,2-dichloroethanol (2), 2-chloroacetaldehyde (3), and 2-chloroacetic acid (4) by cytochrome P450s; (5) 2-(*S*-glutathionyl)acetaldehyde; (6) 2-hydroxyethylglutathione.

can either be hydrolyzed to yield *S*-(2-hydroxyethyl)glutathione, or can react with glutathione to yield *S*,*S'*-1,2-ethanediylbisglutathione or ethylene. An *in vitro* study in rat hepatocytes suggests glutathione *S*-transferase-catalyzed reactions are responsible for the majority of glutathione depletion induced by 1,2-dichloroethane, 1,2-dibromoethane, and 1-bromo-2-chloroethane.[58] Because chloroacetaldehyde and *S*-(2-chloroethyl)glutathione, and its putative metabolite *S*-(2-chloroethyl)-cysteine, have been shown to cause nephrotoxicity *in vivo*, cytotoxicity *in vitro*, and/or can act as alkylating agents,[59–65] the formation of these metabolites is implicated in 1,2-dichloroethane-induced nephrotoxicity.

1,2-Dichloroethane metabolism by glutathione *S*-transferases and cytochrome P450s to yield nephrotoxic metabolites may occur in both liver and kidney. The relative roles of hepatic and renal metabolism by either pathway is likely

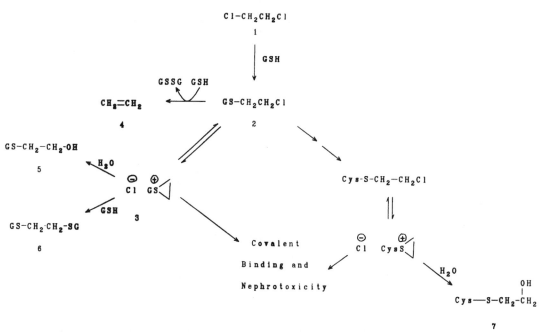

Figure 4 Glutathione *S*-transferase-dependent metabolism of 1,2-dichloroethane (1) to *S*-(2-chloroethyl)-glutathione (2); 3, episulfonium ion derived from *S*-(2-chloroethyl glutathione; (4) ethylene; (5) *S*-(2-hydroxy-ethyl)glutathione; (6) *S*,*S*-1,2-ethanediyl-bisglutathione; (7) *S*-(2-hydroxyethyl)-L-cysteine.

to be species, sex, and age dependent.[34–35,66–69] For *S*-(2-chloroethyl)glutathione and *S*-(2-chloroethyl)cysteine, renal transport mechanisms may play a major role in determining the target organ of toxicity caused by these *S*-conjugates. Thus, conjugates formed in liver are likely to be translocated via the circulation and/or the enterohepatic circulation to the kidney, where they can be actively transported and concentrated within the tubular epithelium.[10,59–61,70–73]

7.31.3 HALOALKENES

Several haloalkenes, including 1,1-dichloro-ethylene (vinylidene chloride), vinyl chloride, allyl chloride, 1,2-dichloroethylene, trichloroethylene, tetrachloroethylene, chlorotrifluoroethylene, tetrafluoroethylene, hexachlorobutadiene, and hexafluoropropene, are known to induce renal damage.[1,2,74–101] The nephrotoxicity and bio-activation mechanisms of two representative haloalkenes, namely, 1,1-dichloroethylene and trichloroethylene, will be discussed below. Other nephrotoxic haloalkenes share similar bioactivation mechanisms.

7.31.3.1 1,1-Dichloroethylene (Vinylidene Chloride)

1,1-Dichloroethylene is widely used as a monomer in the industrial manufacture of

plastics. Long-term or acute animal exposure to 1,1-dichloroethylene has been associated with nephrotoxicity.[75–78,102] In the rat, 1,1-dichloro-ethylene nephrotoxicity is characterized by severe cortical tubular necrosis and increased blood urea nitrogen and serum creatinine concentrations. Treatment of mice with carbon disulfide, an inhibitor of hepatic and renal cytochrome P450s, protects against the renal toxicity of 1,1-dichloroethylene.[103] In rats, however, pretreatment with phenobarbital or poly-chlorinated biphenyls, which induced or had no effect on renal P450s, protected against 1,1-dichloroethylene-induced nephrotoxicity.[77] While mouse kidney microsomes could not bioactivate 1,1-dichloroethylene to electrophi-lic metabolites *in vitro*, mice given [^{14}C]-1,1-dichloroethylene exhibited the highest covalent binding in the kidney.[104,105] Mice pretreated with buthionine sulfoximine, an irreversible in-hibitor of glutathione biosynthesis, exhibited an increase in renal covalent binding and nephro-toxicity of 1,1-dichloroethylene.[102] While the above results provide evidence for the role of reactive metabolites in 1,1-dichloroethylene covalent binding and nephrotoxicity, the roles of hepatic and renal P450s and glutathione *S*-transferases in 1,1-dichloroethylene-induced nephrotoxicity remain unclear.

Rat liver P450-dependent metabolism of 1,1-dichloroethylene yields three electrophilic meta-bolites, namely, 1,1-dichloroethylene oxide, 2-chloroacetyl chloride, and 2,2-dichloroacet-

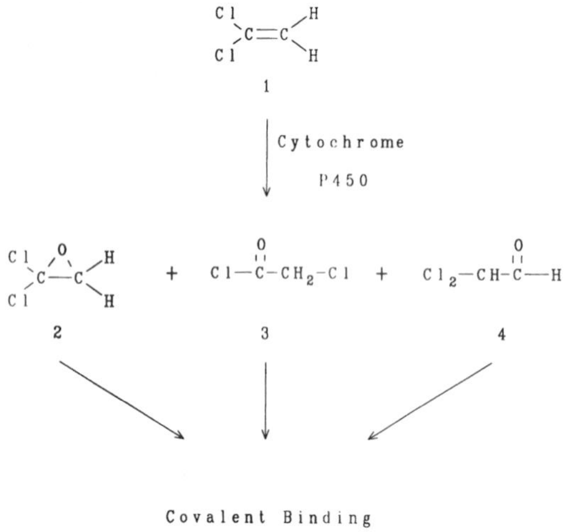

Figure 5 Oxidative metabolism of 1,1-dichloro-ethylene (1) to 1,1-dichloroethylene oxide (2), 2-chloroacetyl chloride (3) and 2,2-dichloroacetaldehyde (4) by cytochrome P450s.

1,1-dichloroethylene to yield chloroacetic acid, a hydrolysis product of 2-chloroacetyl chloride.[107] *In vitro* studies with mouse liver and kidney microsomes suggest the involvement of P450 2E1 in these metabolic reactions and show that 1,1-dichloroethylene oxidation rates in mouse liver microsomes are higher than in rat liver microsomes.[107–109] However, the role of 1,1-dichloroethylene oxidative metabolites in nephrotoxicity remains unclear. In addition, while the reaction of glutathione with these oxidative metabolites is known to yield electrophilic glutathione *S*-conjugates[109,110], the role of *S*-conjugates in 1,1-dichloroethylene-induced nephrotoxicity is unknown.

7.31.3.2 Trichloroethylene (Trichloroethene)

Trichloroethylene, a common industrial solvent used for degreasing metals, is produced in the United States at about 130 000 metric tons per year. In addition to being present at industrial sites, trichloroethylene is a contaminant at many chemical waste sites. It has also

aldehyde (Figure 5); 1,1-dichloroethylene oxide is not an obligate precursor to 2-chloroacetyl chloride or 2,2-dichloroacetaldehyde.[106] Microsomes from male mouse kidney biotransform

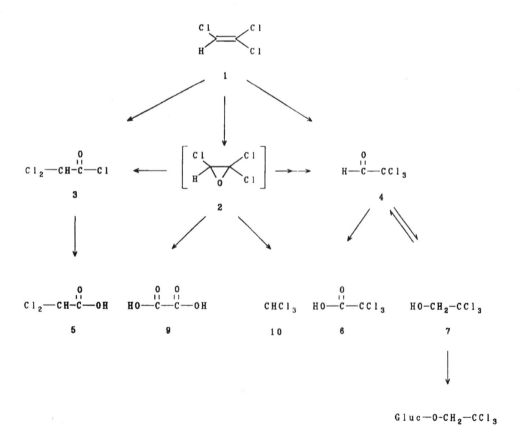

Figure 6 Oxidative metabolism of trichloroethylene (1) to trichloroethylene epoxide (2), 2,2-dichloroacetyl chloride (3), and chloral (4) by cytochrome P450s; (5) dichloroacetic acid; (6) trichloroacetic acid; (7) trichloroethanol; (8) trichloroethanol glucuronide; (9) oxalic acid; (10) chloroform.

been detected in natural water, municipal drinking water, air, and the food chain. The production, use, metabolism, pharmacokinetics, and toxicity of trichloroethylene have been reviewed.[111–113] Epidemiological studies suggest a correlation between human chronic exposure to low levels of trichloroethylene and the development of kidney tumors, whereas acute or chronic exposure to high levels may induce tubular necrosis.[111–117] In experimental animals, chronic exposure of male and female rats, but not mice, to trichloroethylene causes nephrotoxicity.[111–112]

Oxidative metabolism by cytochrome P450 and glutathione conjugation by cytosolic and microsomal glutathione S-transferases are the major metabolic pathways of trichloroethylene metabolism[111–113,118–120] (Figures 6 and 7). Major end products of oxidative metabolism include trichloroacetic acid, trichloroethanol, and trichloroethanol glucuronide. Chloral, dichloroacetic acid, chloroform, and oxalic acid are other detectable oxidative metabolites which have been implicated in hepatotoxicity and hepatocarcinogenicity in mice.

The glutathione conjugation pathway yields S-(1,2-dichlorovinyl)glutathione and S-(1,2-dichlorovinyl)-L-cysteine. Some studies suggest that although the glutathione conjugation pathway is not the predominant metabolic pathway of trichloroethylene, it may be the most relevant to the development of nephrotoxicity. Glutathione conjugation of trichloroethylene has been demonstrated *in vivo* in rats and humans exposed to trichloroethylene by the detection of two mercapturic acids of trichloroethylene in urine.[80,81,121,122] That these mercapturic acids and/or the corresponding glutathione- and cysteine S-conjugates are potent nephrotoxicants *in vivo* and cytotoxicants

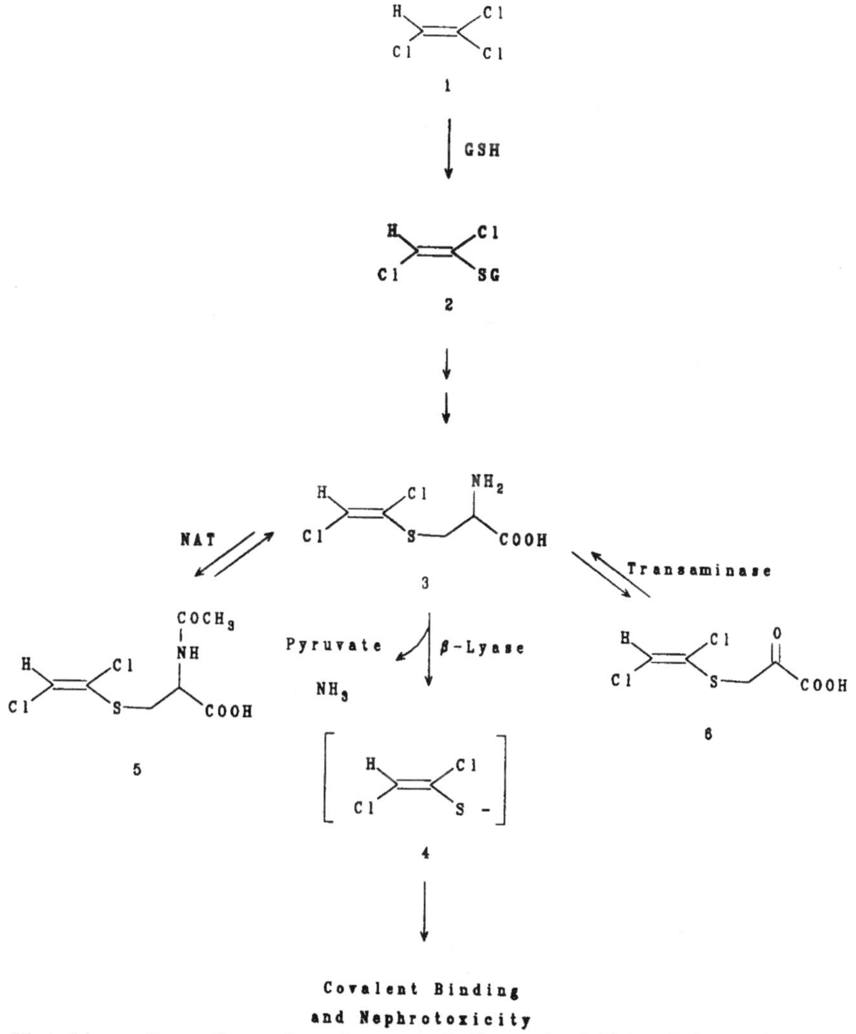

Figure 7 Glutathione S-transferase-dependent metabolism of trichloroethylene (1) to S-(1,2-dichlorovinyl)glutathione (2) and S-(1,2-dichlorovinyl)-L-cysteine (3); (4) 1,2-dichlorovinylthiol; (5) S-(1,2-dichlorovinyl)-N-acetyl-L-cysteine; (6) the α-keto acid metabolite of S-(1,2-dichlorovinyl)-L-cysteine.

in vitro[123-135] is evidence that glutathione conjugation may be important to the development of the nephrotoxicity of trichloroethylene.

7.31.3.3   *S*-(1,2-Dichlorovinyl)glutathione and *S*-(1,2-Dichlorovinyl)-L-cysteine

The mechanisms of *S*-(1,2-dichlorovinyl)glutathione- and *S*-(1,2-dichlorovinyl)-L-cysteine-induced nephrotoxicity have been investigated in several laboratories. The finding that L-(αS,5S)-α-amino-3-chloro-4,5-dihydro-5-isoxazoleacetic acid (AT-125), an inhibitor of γ-glutamyltranspeptidase, which is minimally present in liver and predominantly present in kidney, protects against *S*-(1,2-dichlorovinyl)-glutathione-induced nephrotoxicity suggests a role for this enzyme in trichloroethylene-induced nephrotoxicity.[124,128,129] These results also show that the differential localization of γ-glutamyltranspeptidase may play a major role in the target organ selectivity of glutathione *S*-conjugates. Further evidence for this hypothesis is provided by the finding that addition of γ-glutamyltranspeptidase increased the cytotoxicity of *S*-(1,2-dichlorovinyl)glutathione in primary cultures of human proximal tubular cells.[133] *S*-(1,2-Dichlorovinyl)-L-cysteinylglycine, the expected metabolite of *S*-(1,2-dichlorovinyl)glutathione, is also cytotoxic and cell death can be prevented by the addition of 1,10-phenanthroline or phenylalanylglycine, inhibitors of aminopeptidase M and cysteinylglycine dipeptidase, respectively.[128] These results suggest a role for *S*-(1,2-dichlorovinyl)-L-cysteine in *S*-(1,2-dichlorovinyl)glutathione-induced nephrotoxicity. The findings that the nephrotoxicity of the cysteine *S*-conjugate can be blocked by aminooxyacetic acid and probenecid provide evidence for the roles of cysteine conjugate β-lyase and the organic anion transport system, respectively, in *S*-(1,2-dichlorovinyl)-L-cysteine-induced nephro-toxicity.[124,128,129] Metabolism of *S*-(1,2-dichlorovinyl)-L-cysteine by cysteine conjugate β-lyase yields pyruvate, ammonia, and a reactive sulfur intermediate which covalently binds to cellular macromolecules, leading to cytotoxicity and nephrotoxicity.[126,129,130-140] In this regard, it is of interest to note that multiple cysteine conjugate β-lyase activities exist in rat kidney cytosol and mitochondria.[141-145] A cysteine conjugate β-lyase activity has also been isolated and characterized from human kidney cytosol.[146] The cDNA for human glutamine transaminase K, an enzyme that has cysteine conjugate β-lyase activity, has been cloned and expressed in Cos-1 cells.[147]

The involvement of cysteine conjugate β-lyase in *S*-(1,2-dichlorovinyl)-L-cysteine nephrotoxicity is also supported by the findings that *S*-(1,2-dichlorovinyl)-DL-α-methylcysteine, which cannot be cleaved by β-lyase, is not nephrotoxic,[124] and that addition of α-keto acids, which increase β-lyase activity, potentiated the cytotoxicity of *S*-(1,2-dichlorovinyl)-L-cysteine.[129,136] In addition, treatment of rats with aminooxyacetic acid increased the fraction of the *S*-(1,2-dichlorovinyl)-L-cysteine dose excreted in urine as the corresponding mercapturic acid.[137] Thus, activities of the renal organic acid transport system and cysteine conjugate β-lyase contribute to predisposing the kidney to trichloroethylene toxicity. Furthermore, formation of the mercapturic acid, which cannot be cleaved by β-lyase, is likely to decrease toxicity by enhancing elimination of the toxicant into urine.

Because species differences in *S*-(1,2-dichlorovinyl)-L-cysteine-induced nephrotoxicity have been reported,[2,123,131,138] *in vivo* studies with the model cysteine *S*-conjugates, *S*-(2-benzothiazolyl)-L-cysteine, *S*-(purin-6-yl)-L-cysteine and *S*-(guanin-6-yl)-L-cysteine (Figure 8), which can be metabolized by cysteine conjugate β-lyase to stable thiols, were carried out in rats and in guinea pigs to correlate β-lyase activity with *S*-(1,2-dichlorovinyl)-L-cysteine-induced nephrotoxicity.[148-152] The extent of metabolism of *S*-(purin-6-yl)-L-cysteine or *S*-(guanin-6-yl)-L-cysteine in the rat was lower than that detected with *S*-(2-benzothiazolyl)-L-cysteine. The percentage of the dose recovered as 2-mercaptobenzothiazole and 2-mercaptobenzothiazole *S*-glucuronic acid in the urine of guinea pigs given *S*-(2-benzothiazolyl)-L-cysteine was approximately three times higher than that excreted in rat urine (60% vs. 20%), even though the two species have similar *in vitro* β-lyase activities. The higher net *in vivo* β-lyase-dependent metabolism in the guinea pig is likely to be a consequence of the poor ability of the guinea pig to metabolize cysteine *S*-conjugates to mercapturic acids which cannot be cleaved by cysteine conjugate β-lyase. Species differences in the net *in vivo* cysteine conjugate β-lyase-dependent metabolism correlate well with the higher sensitivity of guinea pigs to *S*-(1,2-dichlorovinyl)-L-cysteine nephrotoxicity in comparison with rats. Thus, species differences in cysteine *S*-conjugate-induced nephrotoxicity may be attributable to species differences in detoxication by cysteine conjugate *N*-acetyltransferase(s).[153-155] Because nephrotoxic cysteine *S*-conjugates may also be detoxified by transamination catalyzed by transaminase enzymes, and/or by oxidation reactions catalyzed by L-α-hydroxy acid oxidase,[156] species

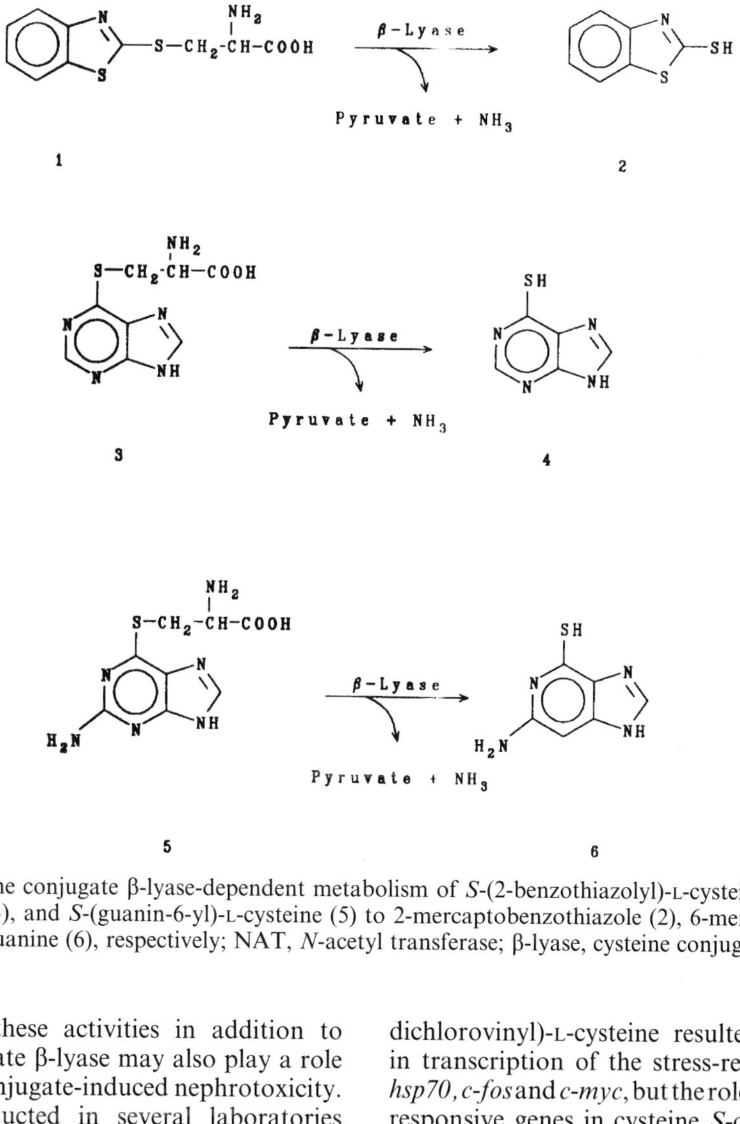

Figure 8 Cysteine conjugate β-lyase-dependent metabolism of *S*-(2-benzothiazolyl)-L-cysteine (1), *S*-(purin-6-yl)-L-cysteine (3), and *S*-(guanin-6-yl)-L-cysteine (5) to 2-mercaptobenzothiazole (2), 6-mercaptopurine (4), and 6-thioguanine (6), respectively; NAT, *N*-acetyl transferase; β-lyase, cysteine conjugate β-lyase.

differences in these activities in addition to cysteine conjugate β-lyase may also play a role in cysteine *S*-conjugate-induced nephrotoxicity.

Studies conducted in several laboratories indicate that the cytotoxicity of *S*-(1,2-dichlorovinyl)-L-cysteine is mediated at the mitochondrial level (Figure 9).[128,157-166] Depletion of mitochondrial ATP concentrations, initiation of mitochondrial lipid peroxidation and GSSG formation, inhibition of mitochondrial lipoyl dehydrogenase activity, release of Ca^{2+} from the mitochondria, and inhibition of mitochondrial membrane potential have been observed prior to renal cell death and correlated well with cytotoxicity. In LLC-PK1 cells, cytotoxicity is preceded by an elevation in the cytosolic Ca^{2+} concentration followed by an increase in DNA-double strand breaks and an increase in poly(ADP-ribosylation) of nuclear proteins.[161] Studies in the Fischer 344 rat indicate that mitochondrial heat shock proteins (P1 protein and mortalin) are major targets for covalent modification by reactive metabolites derived from nephrotoxic cysteine *S*-conjugates.[167] Treatment of LLC-PK1 cells with *S*-(1,2-

dichlorovinyl)-L-cysteine resulted in increases in transcription of the stress-responsive genes *hsp70, c-fos* and *c-myc*, but the roles of these stress responsive genes in cysteine *S*-conjugate cytotoxicity remain unclear.[167-171] Elevation of intra-cellular calcium in the LLC-PK1 line of renal epithelial cells, which can be blocked by aminooxyacetic acid or by chelators such as Quin-2AM and EGTA-AM, is not blocked by antioxidants,[166] such as dithiothreitol, butylated hydroxytoluene, or *N,N'*-diphenyl-*p*-phenylene-diamine, which inhibited the *in vitro* cytotoxicity of *S*-(1,2-dichlorovinyl)-L-cysteine.[164-166] These results suggest that subsequent to increased intracellular Ca^{2+} concentration, mitochondrial and/or cellular lipid peroxidation may occur leading to cell death.

Evidence for the role of lipid peroxidation in *S*-(1,2-dichlorovinyl)-L-cysteine-induced *in vivo* nephrotoxicity is indicated by the findings that mice given *S*-(1,2-dichlorovinyl)-L-cysteine exhaled ethane, had depleted renal glutathione concentrations, and had increased malondial-dehyde formation in the renal cortex.[172] Rats given the antioxidant methimazole were

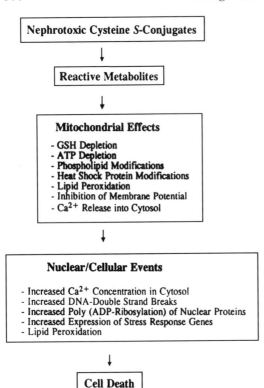

Figure 9 Proposed scheme for the *S*-(1,2-dichloro-vinyl)-L-cysteine-induced events that precede cyto-toxicity in renal cells.

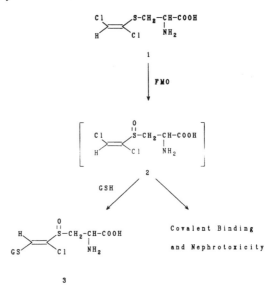

Figure 10 Proposed scheme for the bioactivation of *S*-(1,2-dichlorovinyl)-L-cysteine (1) by flavin-containing monooxygenases (FMO); 2;, *S*-(1,2-dichloro-vinyl)-L-cysteine sulfoxide; 3, *S*-[1-chloro-2-(*S*-glutathionyl)vinyl]-L-cysteine sulfoxide.

protected from the nephrotoxicity of *S*-(1,2-dichlorovinyl)-L-cysteine.[173–176] Other agents that have been shown to protect renal proximal tubules against *S*-(1,2-dichlorovinyl)-L-cysteine-induced cytotoxicity *in vitro* are strychnine and glycine.[177,178] While these agents also act in the late phase of necrotic cell injury, their mechanism of action is different from that of antioxidants. Strychnine and glycine may block cytotoxicity by inhibiting a receptor, related to the neuronal strychnine-sensitive glycine receptor, which mediates chloride influx leading to cell swelling and lysis. *S*-(1,2-Dichlorovinyl)-L-cysteine nephrotoxicity in the rat is usually followed by a nephrogenic repair response, which is characterized by an early stage of increase in proliferation of cells at the wound site by 24 h after exposure, loss of differentiated character in the regenerative epithelium, and cessation of cell growth and redifferentiation between days 5 and 13.[179]

S-(1,2-Dichlorovinyl)-L-cysteine sulfoxide (Figure 10), a reactive metabolite of *S*-(1,2-dichlorovinyl)-L-cysteine, has been shown to be much more nephrotoxic than *S*-(1,2-dichloro-vinyl)-L-cysteine *in vivo* and *in vitro*.[180–182] Pretreatment of rats with aminooxyacetic acid did not protect against *S*-(1,2-dichloro-vinyl)-L-cysteine sulfoxide-induced nephrotoxi-

city, whereas partial protection was observed in rats given *S*-(1,2-dichlorovinyl)-L-cysteine. These results suggest that in addition to bioactivation by cysteine conjugate β-lyase, sulfoxidation may play a role in *S*-(1,2-dichlorovinyl)-L-cysteine-induced nephrotoxicity. Further evidence for this hypothesis is provided by correlating *in vitro* cytotoxicity in proximal and distal tubular cells with the cellular activities of cysteine conjugate β-lyase and cysteine conjugate *S*-oxidase.[181] The renal cysteine conjugate *S*-oxidase activity has been purified from rat kidney microsomes and characterized as a flavin-containing mono-oxygenase.[180,183] The findings that *S*-(1,2-dichlorovinyl)-L-cysteine sulfoxide can act as a Michael acceptor in its reaction with glutathione both *in vitro* and *in vivo* (Figure 10) suggest that alkylation of thiol groups on critical renal macromolecules may play a role in the mechanism of nephrotoxicity of *S*-(1,2-dichlorovinyl)-L-cysteine sulfoxide.[184] Similarly, P450 3A can biotransform the mercapturic acid of *S*-(1,2-dichlorovinyl)-L-cysteine into the corresponding sulfoxide, which may act as a direct-acting electrophile.[185] While the sulfoxides of *S*-(1,2-dichlorovinyl)-L-cysteine and its mercapturic acid have not been identified as *in vivo* metabolites of *S*-(1,2-dichlorovinyl)-L-cysteine, inorganic sulfate, a potential metabolic and/or decomposition product of these sulfoxides, is the major plasma metabolite detected after rats are given [^{35}S]-(1,2-dichloro-vinyl)-L-cysteine.[137]

7.31.4 SUMMARY

Since aliphatic halogenated hydrocarbons are major environmental contaminants, studies aimed at determining the factors that predispose the kidney to halogenated hydrocarbon-induced toxicity, such as age, sex, and coexposure to other drugs or chemicals, are warranted. Furthermore, while cytochrome P450 enzymes and/or glutathione *S*-transferases are implicated in the initial bioactivation reactions of many haloalkanes and haloalkenes, the existence of a large number of these enzymes[34,67,69,186–188] makes it difficult to assess their relative roles in nephrotoxicity. This may be important because of the genetic polymorphism of humans for several of these isoforms. In addition, various isoforms can be induced or inhibited after consumption of certain foods, alcohol, or medications. Thus, for a better understanding of the roles of these enzymes in nephrotoxicity and to extrapolate human risk, the roles of different isoforms in haloalkane and haloalkene bioactivation need to be determined. The availability of human tissues and cDNA-expressed human enzymes for biomedical research will make it possible to address this question. Future studies should also provide detailed information on the relative roles of cysteine conjugate β-lyases and cysteine conjugate *S*-oxidases in *S*-conjugate-induced nephrotoxicity.

ACKNOWLEDGMENTS

The author thanks Renee Krause and Sharon Ripp for their help, comments, and suggestions during the preparation of this chapter. Studies conducted in the author's laboratory were supported by grants from NIH (R01 DK44295 and T32 ES07015).

7.31.5 REFERENCES

1. W. M. Kluwe, in 'Toxicology of the Kidney,' ed. J. B. Hook, Raven Press, New York, 1981, pp. 179–226.
2. A. A. Elfarra, in 'Toxicology of the Kidney,' 2nd edn., eds. J. B. Hook and R. S. Goldstein, Raven Press, New York, 1993, pp. 387–413.
3. R. J. Laib, in 'Reviews on Drug Metabolism and Drug Interactions,' eds. A. H. Beckett and J. W. Gottod, Freund, London, 1982, vol. IV, no. 1, pp. 1–48.
4. T. J. Monks, M. W. Anders, W. Dekant *et al.*, 'Glutathione conjugate mediated toxicities.' *Toxicol. Appl. Pharmacol.*, 1990, **106**, 1–19.
5. L. H. Lash, 'Role of renal metabolism in risk to toxic chemicals.' *Environ. Health Perspect.*, 1994, **102**, 75–79.
6. M. W. Anders, W. Dekant and S. Vamvakas, 'Glutathione-dependent toxicity.' *Xenobiotica*, 1992, **22**, 1135–1145.
7. J. F. Nagelkerke and P. J. Boogaard, 'Nephrotoxicity of halogenated alkenyl cysteine-*S*-conjugates.' *Life Sci.*, 1991, **49**, 1769–1776.
8. E. A. Lock, 'Mechanism of nephrotoxic action due to organohalogenated compounds.' *Toxicol. Lett.*, 1989, **46**, 93–106.
9. W. Dekant, S. Vamvakas and M. W. Anders, 'Bioactivation of nephrotoxic haloalkenes by glutathione conjugation: formation of toxic and mutagenic intermediates by cysteine conjugate β-lyase.' *Drug Metab. Rev.*, 1989, **20**, 43–83.
10. T. J. Monks and S. S. Lau, 'Commentary: renal transport processes and glutathione conjugate-mediated nephrotoxicity.' *Drug Metab. Dispos.*, 1987, **15**, 437–441.
11. E. J. Soderlund, M. Låg, J. A. Holme *et al.*, 'Species differences in kidney necrosis and DNA damage, distribution and glutathione-dependent metabolism of 1,2-dibromo-3-chloropropane (DBCP).' *Pharmacol. Toxicol.*, 1990, **66**, 287–293.
12. P. G. Pearson, E. J. Soderlund, E. Dybing *et al.*, 'Metabolic activation of 1,2-dibromo-3-chloropropane: evidence for the formation of reactive episulfonium ion intermediates.' *Biochemistry*, 1990, **29**, 4971–4981.
13. M. Låg, J. G. Omichinski, E. J. Soderlund *et al.*, 'Role of P-450 activity and glutathione levels in 1,2-dibromo-3-chloropropane tissue distribution, renal necrosis, and *in vivo* DNA damage.' *Toxicology*, 1989, **56**, 273–288.
14. W. M. Kluwe, F. W. Harrington and S. E. Cooper, 'Toxic effects of organohalide compounds on renal tubular cells *in vivo* and *in vitro*.' *J. Pharmacol. Exp. Ther.*, 1982, **220**, 597–603.
15. W. M. Kluwe, R. McNish and J. B. Hook, 'Acute nephrotoxicities and hepatotoxicities of 1,2-dibromo-3-chloropropane and 1,2-dibromoethane in male and female F344 rats.' *Toxicol. Lett.*, 1981, **8**, 317–321.
16. F. A. Blocki, M. S. P. Logan, C. Baoli *et al.*, 'Reaction of rat liver glutathione *S*-transferases and bacterial dichloromethane dehalogenase with dihalomethanes.' *J. Biol. Chem.*, 1994, **269**, 8826–8830.
17. R. Thier, J. B. Taylor, S. E. Pemble *et al.*, 'Expression of mammalian glutathione *S*-transferase 5-5 in *Salmonella typhimurium* TA1535 leads to base-pair mutations upon exposure to dihalomethanes.' *Proc. Natl. Acad. Sci. USA*, 1993, **90**, 8575–8580.
18. J. M. Symons, T. A. Bellar, J. K. Carswell *et al.*, 'National organics reconnaissance survey for halogenated organics.' *J. Am. Water Works Assoc.*, 1975, **67**, 634–647.
19. A. B. Eschenbrenner and E. Miller, 'Sex differences in kidney morphology and chloroform necrosis.' *Science*, 1945, **102**, 302–303.
20. T. R. Torkelson, F. Oyen and V. K. Rowe, 'The toxicology of chloroform as determined by single and repeated exposure of laboratory animals.' *Am. Ind. Hyg. Assoc. J.*, 1976, **37**, 697–705.
21. J. H. Smith, K. Maita, S. D. Sleight *et al.*, 'Mechanism of chloroform nephrotoxicity. I. Time course of chloroform toxicity in male and female mice.' *Toxicol. Appl. Pharmacol.*, 1983, **70**, 467–479.
22. J. L. Larson, D. C. Wolf and B. E. Butterworth, 'Acute and nephrotoxic effects of chloroform in male F-344 rats and female B6C3F1 mice.' *Fundam. Appl. Toxicol.*, 1993, **20**, 302–315.
23. J. L. Larson, D. C. Wolf, K. T. Morgan *et al.*, 'The toxicity of 1-week exposures to inhaled chloroform in female B6C3F$_1$ mice and male F-344 rats.' *Fundam. Appl. Toxicol.*, 1994, **22**, 431–446.
24. L. R. Pohl, J. W. George and H. Satoh, 'Strain and sex differences in chloroform-induced nephrotoxicity. Different rates of metabolism of chloroform to

phosgene by the mouse kidney.' *Drug Metab. Dispos.*, 1984, **12**, 304–308.

25. J. H. Smith, K. Maita, S. D. Sleight *et al.*, 'Effect of sex hormone status on chloroform nephrotoxicity and renal mixed function oxidases in mice.' *Toxicology*, 1984, **30**, 305–316.

26. R. N. Hill, T. L. Clemens, D. K. Liu *et al.*, 'Genetic control of chloroform toxicity in mice.' *Science*, 1975, **190**, 159–161.

27. W. M. Kluwe and J. B. Hook, 'Polybrominated biphenyl-induced potentiation of chloroform toxicity.' *Toxicol. Appl. Pharmacol.*, 1978, **45**, 861–869.

28. G. F. Gibberd, 'Delayed chloroform poisoning in obstetric practice—a clinical study with reports on three cases.' *Guys Hosp. Bull.*, 1935, **85**, 142–160.

29. R. L. Lunt, 'Delayed chloroform poisoning in obstetric practice.' *Br. Med. J.*, 1953, **1**, 489–490.

30. W. F. Von Oettingen, 'The Halogenated Hydrocarbons of Industrial and Toxicological Importance,' Elsevier, New York, 1964.

31. M. B. Bailie, J. H. Smith. J. F. Newton *et al.*, 'Mechanism of chloroform nephrotoxicity. IV. Phenobarbital potentiation of *in vitro* chloroform metabolism and toxicity in rabbit kidneys.' *Toxicol. Appl. Pharmacol.*, 1984, **74**, 285–292.

32. G. F. Rush, J. H. Smith, J. F. Newton *et al.*, 'Chemically induced Nnephrotoxicity: role of metabolic activation.' *CRC Crit. Rev. Toxicol.*, 1984, **13**, 99–160.

33. M. Ahmadizadeh, C. H. Kuo and J. B. Hook, 'Nephrotoxicity and hepatotoxicity of chloroform in mice: effect of deuterium substitution.' *J. Toxicol. Environ. Health*, 1981, **8**, 105–111.

34. F. J. Gonzalez and H. V. Gelboin, 'Role of human cytochromes P450 in the metabolic activation of chemical carcinogens and toxins.' *Drug Metab. Rev.*, 1994, **26**, 165–183.

35. J. J. Hu, M. J. Lee, M. Vapiwala *et al.*, 'Sex-related differences in mouse renal metabolism and toxicity of acetaminophen.' *Toxicol. Appl. Pharmacol.*, 1993, **122**, 16–26.

36. W. R. Guild, J. V. Young and J. P. Merril, 'Anuria due to carbon tetrachloride intoxication.' *Am. J. Pathol.*, 1958, **48**, 1221–1227.

37. H. D. Moon, 'The pathology of fatal carbon tetrachloride poisoning with special reference to the histogenesis of the hepatic and renal lesions.' *Am. J. Pathol.*, 1950, **26**, 1041–1057.

38. J. Sirota, 'Carbon tetrachloride poisoning in man. I. The mechanisms of renal failure and recovery.' *J. Clin. Invest.*, 1949, **28**, 1412–1422.

39. H. Smetana, 'Nephrosis due to carbon tetrachloride.' *Arch. Intern. Med.*, 1939, **63**, 760–777.

40. R. D. Stewart, E. A. Boettner, R. R. Southworth *et al.*, 'Acute carbon tetrachloride intoxication.' *J. Am. Med. Assoc.*, 1963, **183**, 994–997.

41. M. D. Villarruel, E. G. de Toranzo and J. A. Castro, 'Carbon tetrachloride activation, lipid peroxidation and the mixed function oxygenase activity of various rat tissues.' *Toxicol. Appl. Pharmacol.*, 1977, **41**, 337–344.

42. T. C. Butler, 'Reduction of carbon tetrachloride *in vivo* and reduction of carbon tetrachloride and chloroform *in vitro* by tissue and tissue constituents,' *J. Pharmacol. Exp. Ther.*, 1961, **134**, 311–319.

43. B. B. Paul and D. Rubenstein, 'Metabolism of carbon tetrachloride and chloroform by the rat.' *J. Pharmacol. Exp. Ther.*, 1963, **141**, 141–148.

44. R. Reiter and R. F. Burk, 'Formation of glutathione adducts of carbon tetrachloride metabolites in a rat liver microsomal incubation system.' *Biochem. Pharmacol.*, 1988, **37**, 327–331.

45. D. L. Morgan, J. R. Bucher and M. R. Elwell, 'Comparative toxicity of ethylene dichloride in F344/N, Sprague-Dawley and Osborne–Mendel rats.' *Food Chem. Toxicol.*, 1990, **28**, 839–845.

46. R. E. Yodaiken and J. R. Babcock, '1,2-Dichloroethane poisoning.' *Arch. Environ. Health*, 1973, **26**, 281–284.

47. L. A. Heppel, P. A. Neal, T. L. Perrin *et al.*, 'The toxicology of 1,2-dichloroethane (ethylene dichloride). V. The effects of daily inhalation.' *J. Ind. Hyg. Toxicol.*, 1948, **28**, 113–120.

48. D. D. McCollister, R. L. Hollingsworth, F. Oyen *et al.*, 'Comparative inhalation toxicology of fumigant mixtures.' *AMA Arch. Ind. Med.*, 1955, **13**, 1–7.

49. H. C. Spencer, V. K. Rowe, E. M. Adams *et al.*, 'Vapor toxicity of ethylene dichloride determined by experiments on laboratory animals.' *Arch. Ind. Hyg. Occup. Med.*, 1951, **4**, 482–493.

50. A. Odinecs, S. Maso, G. Nicoletto *et al.*, 'Mechanism of sex-related differences in nephrotoxicity of 1,2-dichloropropane in rats.' *Renal Fail.*, 1995, **17**, 517–524.

51. F. P. Guengerich, W. M. Crawford, Jr., J. Y. Domoradski *et al.*, '*In vitro* activation of 1,2-dichloroethane by microsomal and cytosolic enzymes.' *Toxicol. Appl. Pharmacol.*, 1980, **55**, 303–317.

52. M. W. Anders and J. C. Livesey, 'Metabolism of 1,2-Dichloroethanes, Banbury Report 5: Ethylene Dichloride: a Potential Health Risk,' Cold Spring Harbor Laboratory Press, Cold Spring Harbor, NY, 1980, pp. 331–341.

53. D. H. Marchand and D. J. Reed, 'Identification of the reactive glutathione conjugate S-(2-chloroethyl)glutathione in the bile of 1-bromo-2-chloroethane-treated rats by high-pressure liquid chromatography and precolumn derivatization with o-phthaldehyde.' *Chem. Res. Toxicol.*, 1989, **2**, 449–454.

54. W. W. Webb, A. A. Elfarra, K. D. Webster *et al.*, 'Role for an episulfonium ion in S-(2-chloroethyl)-DL-cysteine-induced cytotoxicity and its reaction with glutathione.' *Biochemistry*, 1987, **26**, 3017–3023.

55. C. S. Schasteen and D. J. Reed, 'The hydrolysis and alkylation activities of S-(2-haloethyl)-L-cysteine analogs—evidence for extended half-life.' *Toxicol. Appl. Pharmacol.*, 1983, **70**, 423–432.

56. D. J. Reed and G. L. Foureman, 'A comparison of the alkylating capabilities of the cysteinyl and glutathionyl conjugates of 1,2-dichloroethane.' *Adv. Exp. Med. Biol.*, 1986, **197**, 469–475.

57. L. A. Peterson, T. M. Harris and F. P. Guengerich, 'Evidence for an episulfonium ion in the formation of S-[2-N^7-guanyl)ethyl]glutathione in DNA.' *J. Am. Chem. Soc.*, 1988, **110**, 3284–3291.

58. P. A. Jean and D. J. Reed, 'Utilization of glutathione during 1,2-dihaloethane metabolism in rat hepatocytes.' *Chem. Res. Toxicol.*, 1992, **5**, 386–391.

59. A. A. Elfarra, R. B. Baggs and M. W. Anders, 'Structure–nephrotoxicity relationships of S-(2-chloroethyl)-DL-cysteine and analogs: role for an episulfonium ion.' *J. Pharmacol. Exp. Ther.*, 1985, **233**, 512–516.

60. R. A. Kramer, G. Foureman, K. E. Greene *et al.*, 'Nephrotoxicity of S-(2-chloroethyl)glutathione in the Fischer rat: evidence for γ-glutamyltranspeptidase-independent uptake by the kidney.' *J. Pharmacol. Exp. Ther.*, 1987, **242**, 741–748.

61. W. X. Guo, S. Chakrabarti, M. A. Malick *et al.*, 'Effect of S-(2-chloroethyl)-DL-cysteine on the transport of p-aminohippurate ion in renal plasma membrane vesicles.' *Arch. Biochem. Biophys.*, 1990, **283**, 206–209.

62. M. J. Zamlauski-Tucker, M. E. Morris and J. E. Springate, 'Ifosfamide metabolite chloroacetaldehyde causes Fanconi syndrome in the perfused rat kidney.' *Toxicol. Appl. Pharmacol.*, 1994, **129**, 170–175.

63. C. Sood and P. J. O'Brien, 'Chloroacetaldehyde-induced hepatocyte cytotoxicity. Mechanisms for cytoprotection.' *Biochem. Pharmacol.*, 1994, **48**, 1025–1032.

64. M. Meyer, O. N. Jensen, E. Barofsky *et al.*, 'Thioredoxin alkylation by a dihaloethane–glutathione conjugate.' *Chem. Res. Toxicol.*, 1994, **7**, 659–665.

65. J. C. L. Erve, M. L. Deinzer and D. J. Reed, 'Alkylation of oxytocin by *S*-(2-chloroethyl)glutathione and characterization of adducts by tandem mass spectrometry and Edman degradation.' *Chem. Res. Toxicol.*, 1995, **8**, 414–421.

66. B. Rozell, H. A. Hansson, C. Guthenberg *et al.*, 'Glutathione transferases of classes α, μ and π show selective expression in different regions of rat kidney.' *Xenobiotica*, 1993, **23**, 835–849.

67. S. S. Singhal, M. Saxena, H. Ahmad *et al.*, 'Glutathione *S*-transferases of mouse liver: sex-related differences in the expression of various isozymes.' *Biochim. Biophys. Acta*, 1992, **1116**, 137–146.

68. C. A. Hinchman, H. Matsumoto, T. W. Simmons *et al.*, 'Intrahepatic conversion of a glutathione conjugate to its mercapturic acid: metabolism of 1-chloro-2,4-dinitrobenzene in isolated perfused rat and guinea pig livers.' *J. Biol. Chem.*, 1991, **266**, 22179–22185.

69. R. M. E. Vos and P. J. Van Bladeren, 'Glutathione *S*-transferases in relation to their role in the biotransformation of xenobiotics.' *Chem. Biol. Interact.*, 1990, **75**, 241–265.

70. W. H. Dantzler, K. K. Evans and S. H. Wright, 'Kinetics of interactions of *para*-aminohippurate, probenecid, cysteine conjugates and *N*-acetyl cysteine conjugates with basolateral organic anion transporter in isolated rabbit proximal renal tubules.' *J. Pharmacol. Exp. Ther.*, 1995, **272**, 663–672.

71. S. Silbernagl and A. Heuner, 'Renal transport and metabolism of mercapturic acids and their precursors.' *Toxicol. Lett.*, 1990, **53**, 45–51.

72. V. H. Schaeffer and J. L. Stevens, 'The transport of *S*-cysteine conjugates in LLC-PK$_1$ cells and its role in toxicity.' *Mol. Pharmacol.*, 1987, **31**, 506–512.

73. L. H. Lash and M. W. Anders, 'Uptake of nephrotoxic *S*-conjugates by isolated rat renal proximal tubular cells.' *J. Pharmacol. Exp. Ther.*, 1989, **248**, 531–537.

74. G. L. Plaa and R. E. Larson, 'Relative nephrotoxic properties of chlorinated methane, ethane, and ethylene derivatives in mice.' *Toxicol. Appl. Pharmacol.*, 1965, **7**, 37–44.

75. J. A. Pendergast, R. A. Jones, L. J. Jenkins, Jr. *et al.*, 'Effects on experimental animals of long-term inhalation of trichloroethylene, carbon tetrachloride, 1,1,1-trichloroethane, dichlorodifluoromethane and 1,1-dichloroethylene.' *Toxicol. Appl. Pharmacol.*, 1967, **10**, 270–289.

76. L. J. Jenkins, Jr. and M. E. Andersen, '1,1-Dichloroethylene nephrotoxicity in the rat.' *Toxicol. Appl. Pharmacol.*, 1978, **46**, 131–141.

77. N. M. Jackson and R. B. Conolly, 'Acute nephrotoxicity of 1,1-dichloroethylene in the rat after inhalation exposure.' *Toxicol. Lett.*, 1985, **29**, 191–199.

78. Environmental Protection Agency, 'Vinylidene Chloride: Health and Environmental Impacts,' EPA Office of Toxic Substances, Washington, DC, 1976, 560/6–76–023.

79. National Toxicology Program, 'Toxicology and Carcinogenesis Studies of Trichloroethylene in F344/N Rats and B6C3F1 Mice (Gavage).' Pub. No. (NH), 1982, pp. 83–1799.

80. W. Dekant, M. Metzler and D. Henschler, 'Identification of *S*-1,2-dichlorovinyl-*N*-acetyl cysteine as a urinary metabolite of trichloroethylene: a possible explanation for its nephrocarcinogenicity in male Rats.' *Biochem. Pharmacol.*, 1986, **35**, 2455–2458.

81. W. Dekant, M. Koob and D. Henschler, 'Metabolism of trichloroethene—*in vivo* and *in vitro* evidence for activation by glutathione conjugation.' *Chem. Biol. Interact.*, 1990, **73**, 89–101.

82. Current Intelligence Bulletin, No. 20, 'Tetrachloroethylene (perchloroethylene).' US Department of Health, Education, and Welfare, PHS, NIOSH, Washington, DC, January, 1978.

83. National Toxicology Program NTP TR 311. US Department of Health, Education and Welfare, US Public Health Service, Washington, DC, 1986.

84. J. Odum and T. Green, 'The metabolism and nephrotoxicity of tetrafluoroethylene in the rat.' *Toxicol. Appl. Pharmacol.*, 1984, **76**, 306–318.

85. C. L. Potter, A. J. Gandolfi, R. Nagle *et al.*, 'Effects of inhaled chlorotrifluoroethylene and hexafluoropropene on the rat kidney.' *Toxicol. Appl. Pharmacol.*, 1981, **59**, 431–440.

86. L. A. Buckley, J. W. Clayton, R. B. Nagle *et al.*, 'Chlorotrifluoroethylene nephrotoxicity in rats: a subacute study.' *Fundam. Appl. Toxicol.*, 1982, **2**, 181–186.

87. J. N. M. Commandeur, R. A. J. Oosterdorp, P. R. Schoofs *et al.*, 'Nephrotoxicity and hepatotoxicity of 1,1-dichloro-2,2-difluoroethylene in the rat. Implications for the differential mechanisms of bioactivation.' *Biochem. Pharmacol.*, 1987, **36**, 4229–4237.

88. J. N. M. Commandeur, J. P. G. Brakenhoff, F. J. J. De Kanter *et al.*, 'Nephrotoxicity of mercapturic acids of three structurally related 2,2-difluoroethylenes in the rat. Implications for different bioactivation mechanisms.' *Biochem. Pharmacol.*, 1988, **37**, 4495–4504.

89. T. R. Torkelson and F. Oyen, 'The toxicity of 1,3-dichloropropene as determined by repeated exposure of laboratory animals.' *Am. Ind. Hyg. Assoc. J.*, 1977, **38**, 217–223.

90. J. Osterloh and X.-W. He, 'Effects of 1,3-dichloropropene on the kidney of Fischer 344 rats after pretreatment with diethymaleate, buthionine sulfoximine, and aminooxyacetic acid.' *J. Toxicol. Environ. Health*, 1990, **29**, 247–255.

91. S. Vamvakas, E. Kremling and W. Dekant, 'Metabolic activation of the nephrotoxic haloalkene 1,1,2-trichloro-3,3,3-trifluoro-1-propene by glutathione conjugation.' *Biochem. Pharmacol.*, 1989, **38**, 2297–2304.

92. W. O. Berndt and H. M. Mehendale, 'Effects of hexachlorobutadiene (HCBD) on renal function and renal organic ion transport in the rat.' *Toxicology*, 1979, **14**, 55–65.

93. E. A. Lock and J. Ishmael, 'The acute toxic effects of hexachloro-1,3-butadiene on rat kidney.' *Arch. Toxicol.*, 1979, **43**, 47–57.

94. J. B. Hook, J. Ishmael and E. A. Lock, 'Nephrotoxicity of hexachloro-1,3-butadiene in the rat: the effect of age, sex and strain.' *Toxicol. Appl. Pharmacol.*, 1983, **67**, 122–131.

95. J. A. Nash, L. J. King, E. A. Lock *et al.*, 'The metabolism and disposition of hexachloro-1,3-butadiene in the rat and its relevance to nephrotoxicity.' *Toxicol. Appl. Pharmacol.*, 1984, **73**, 124–137.

96. E. A. Lock, J. Ishmael and J. B. Hook, 'Nephrotoxicity of hexachloro-1,3-butadiene in the mouse: the effect of age, sex, strain, monooxygenase modifiers, and the role of glutathione.' *Toxicol. Appl. Pharmacol.*, 1984, **72**, 484–494.

97. E. M. Adams, H. C. Spencer and D. D. Irish, 'The acute vapor toxicity of allyl chloride.' *J. Ind. Hyg. Toxicol.*, 1940, **22**, 79–86.

98. T. R. Torkelson, M. A. Wolf, F. Oyen *et al.*, 'Vapor toxicity of allyl chloride as determined on

laboratory animals.' *Am. Ind. Hyg. Assoc. J.*, 1959, **20**, 212–223.

99. T. R. Torkelson, F. Oyen and V. K. Rowe, 'The toxicity of vinyl chloride as determined by repeated exposure of laboratory animals.' *Am. Ind. Hyg. Assoc. J.*, 1961, **22**, 354–361.

100. G. Birner, M. Werner, M. M. Ott *et al.*, 'Sex differences in hexachlorobutadiene biotransformation and nephrotoxicity.' *Toxicol. Appl. Pharmacol.*, 1995, **132**, 203–212.

101. A. A. Elfarra and M. W. Anders, 'Renal processing of glutathione conjugates. Role in nephrotoxicity.' *Biochem. Pharmacol.*, 1984, **33**, 3729–3732.

102. E. B. Brittebo, P. O. Darnerud, C. Eriksson *et al.*, 'Nephrotoxicity and covalent binding of 1,1-dichloroethylene in buthionine sulphoximine-treated mice.' *Arch. Toxicol.*, 1993, **67**, 605–612.

103. Y. Masuda and N. Nakayama, 'Protective action of diethyldithiocarbamate and carbon disulfide against acute toxicities induced by 1,1-dichloroethylene in mice.' *Toxicol. Appl. Pharmacol.*, 1983, **71**, 42–53.

104. L. K. Okine, J. M. Goochee and T. E. Gram, 'Studies on the distribution and covalent binding of 1,1-dichloroethylene in the mouse. Effect of various pretreatments on covalent binding *in vivo*.' *Biochem. Pharmacol.*, 1985, **34**, 4051–4057.

105. L. K. Okine and T. E. Gram, '*In vitro* studies on the metabolism and covalent binding of [^{14}C]1,1-dichloroethylene by mouse liver, kidney, and lung.' *Biochem. Pharmacol.*, 1986, **35**, 2789–2795.

106. D. C. Liebler and F. P. Guengerich, 'Olefin oxidation by cytochrome P-450: evidence for group migration in catalytic intermediates formed with vinylidene chloride and *trans*-1-phenyl-1-butene.' *Biochemistry*, 1983, **22**, 5482–5489.

107. P. Speerschneider and W. Dekant, 'Renal tumorigenicity of 1,1-dichloroethene in mice: the role of male-specific expression of cytochrome P450 2E1 in the renal bioactivation of 1,1-dichloroethene.' *Toxicol. Appl. Pharmacol.*, 1995, **130**, 48–56.

108. R. P. Lee and P. G. Forkert, '*In vitro* biotransformation of 1,1-dichloroethylene by hepatic cytochrome P-450 2E1 in mice.' *J. Pharmacol. Exp. Ther.*, 1994, **270**, 371–376.

109. T. F. Dowsley, P. G. Forkert, L. A. Benesch *et al.*, 'Reaction of glutathione with the electrophilic metabolites of 1,1-dichloroethylene.' *Chem. Biol. Interact.*, 1995, **95**, 227–244.

110. D. C. Liebler, D. G. Latwesen and T. C. Reeder, '*S*-(2-Chloroacetyl)glutathione, a reactive glutathione thiol ester and a putative metabolite of 1,1-dichloroethylene.' *Biochemistry*, 1988, **27**, 3652–3657.

111. I. W. F. Davidson and R. P. Beliles, 'Consideration of the target organ toxicity of trichloroethylene in terms of metabolite toxicity and pharmacokinetics.' *Drug Metab. Rev.*, 1991, **23**, 493–599.

112. A. R. Goeptar, J. N. M. Commandeur, B. van Ommen *et al.*, 'Metabolism and kinetics of trichloroethylene in relation to toxicity and carcinogenicity. relevance of the mercapturic acid pathway.' *Chem. Res. Toxicol.*, 1995, **8**, 3–21.

113. T. Green, 'Chloroethylenes: a mechanistic approach to human risk evaluation.' *Annu. Rev. Pharmacol. Toxicol.*, 1990, **30**, 73–89.

114. D. Henschler, S. Vamvakas, M. Lammert *et al.*, 'Increased incidence of renal cell tumors in a cohort of cardboard workers exposed to trichloroethene.' *Arch. Toxicol.*, 1995, **69**, 291–299.

115. N. J. David, R. Wolman, F. J. Milne *et al.*, 'Acute renal failure due to trichloroethylene poisoning.' *Br. J. Ind. Med.*, 1989, **46**, 347–349.

116. C. F. Gutch, W. G. Tomhave and S. C. Stevens, 'Acute renal failure due to inhalation of trichloroethylene.' *Ann. Intern. Med.*, 1965, **63**, 128–134.

117. A. Seldén, B. Hultberg, A. Ulander *et al.*, 'Trichloroethylene exposure in vapour degreasing and the urinary excretion of *N*-Acetyl-β-D-glucosaminidase.' *Arch. Toxicol.*, 1993, **67**, 224–226.

118. B. Soucek and D. Vlachova, 'Excretion of trichloroethylene metabolites in human urine.' *Br. J. Ind. Med.*, 1960, **17**, 60–64.

119. R. E. Miller and F. P. Guengerich, 'Oxidation of trichloroethylene by liver microsomal cytochrome P-450: evidence for chlorine migration in a transition state not involving trichloroethylene oxide.' *Biochemistry*, 1982, **21**, 1090–1097.

120. L. H. Lash, Y. Xu, A. A. Elfarra *et al.*, 'Glutathione-dependent metabolism of trichloroethylene in isolated liver and kidney cells of rats and its role in mitochondrial and cellular toxicity.' *Drug Metab. Dispos.*, 1995, **23**, 846–853.

121. J. N. M. Commandeur and N. P. E. Vermeulen, 'Identification of *N*-acetyl(2,2-dichlorovinyl)- and *N*-acetyl-(1,2-dichlorovinyl)-L-cysteine as two regioisomeric mercapturic acids of trichloroethylene in the rat.' *Chem. Res. Toxicol.*, 1990, **3**, 212–218.

122. G. Birner, S. Vamvakas, W. Dekant *et al.*, 'Nephrotoxic and genotoxic *N*-acetyl-*S*-dichlorovinyl-L-cysteine is a urinary metabolite after occupational 1,1,2-trichloroethene exposure in humans: implications for the risk of trichloroethene exposure.' *Environ. Health Perspect.*, 1993, **99**, 281–284.

123. B. Terracini and V. H. Parker, 'A pathological study on the toxicity of *S*-dichlorovinyl-L-cysteine.' *Food Cosmet. Toxicol.*, 1965, **3**, 67–74.

124. A. Elfarra, I. Jakobson and M. W. Anders, 'Mechanism of *S*-(1,2-dichlorovinyl)glutathione-induced nephrotoxicity.' *Biochem. Pharmacol.*, 1986, **35**, 283–288.

125. A. A. Elfarra, L. H. Lash and M. W. Anders, 'Metabolic activation and detoxication of nephrotoxic cysteine and homocysteine *S*-conjugates.' *Proc. Natl. Acad. Sci. USA*, 1986, **83**, 2667–2671.

126. P. O. Darnerud, I. Brandt, V. J. Feil *et al.*, '*S*-(1,2-Dichloro-[^{14}C]vinyl)-L-cysteine (DCVC) in the mouse kidney: correlation between tissue-binding and toxicity.' *Toxicol. Appl. Pharmacol.*, 1988, **95**, 423–434.

127. J. N. M. Commandeur, P. J. Boogaard, G. J. Mulder *et al.*, 'Mutagenicity and cytotoxicity of two regioisomeric mercapturic acids and cysteine *S*-conjugates of trichloroethylene.' *Arch. Toxicol.*, 1991, **65**, 373–380.

128. L. H. Lash and M. W. Anders, 'Cytotoxicity of *S*-(1,2-dichlorovinyl)glutathione and *S*-(1,2-dichlorovinyl)-L-cysteine in isolated rat kidney cells.' *J. Biol. Chem.*, 1986, **261**, 13076–13081.

129. J. Stevens, P. Hayden and G. Taylor, 'The role of glutathione conjugate metabolism and cysteine conjugate β-lyase in the mechanism of *S*-cysteine conjugate toxicity in LLC-PK1 cells.' *J. Biol. Chem.*, 1986, **261**, 3325–3332.

130. G. H. I. Wolfgang, A. J. Gandolfi, J. L. Stevens *et al.*, '*N*-Acetyl-*S*-(1,2-dichlorovinyl)-L-cysteine produces a similar toxicity to *S*-(1,2-dichlorovinyl)-L-cysteine in rabbit renal slices: differential transport and metabolism.' *Toxicol. Appl. Pharmacol.*, 1989, **101**, 205–219.

131. D. A. Koechel, M. E. Krejci and R. E. Ridgewell, 'The acute effects of *S*-(1,2-dichlorovinyl)-L-cysteine and related chemicals on renal function and ultrastructure in the pentobarbital-anesthetized dog: structure–activity relationships, biotransformation, and unique site-specific nephrotoxicity.' *Fundam. Appl. Toxicol.*, 1991, **17**, 17–33.

132. M. L. Anthony, C. R. Beddell, J. C. Lindon *et al.*, 'Studies on the comparative toxicity of *S*-(1,2-dichlorovinyl)-L-cysteine, *S*-(1,2-dichlorovinyl)-L-homocys-

teine and 1,1,2-trichloro-3,3,3-trifluoro-1-propene in the Fischer 344 rat.' *Arch. Toxicol.*, 1994, **69**, 99–110.

133. J. C. Chen, J. L. Stevens, A. L. Trifillis *et al.*, 'Renal cysteine conjugate β-lyase-mediated toxicity studied with primary cultures of human proximal tubular cells.' *Toxicol. Appl. Pharmacol.*, 1990, **103**, 463–473.

134. G. H. Wolfgang, A. J. Gandolfi, R. B. Nagle *et al.*, 'Assessment of S-(1,2-dichlorovinyl)-L-cysteine induced toxic events in rabbit renal cortical slices. biochemical and histological evaluation of uptake, covalent binding, and toxicity.' *Chem. Biol. Interact.*, 1990, **75**, 153–170.

135. G. J. Stijntjes, J. N. M. Commandeur, J. M. te Koppele *et al.*, 'Examination of the structure–toxicity relationships of L-cysteine-S-conjugates of halogenated alkenes and their corresponding mercapturic acids in rat renal tissue slices.' *Toxicology*, 1993, **79**, 67–79.

136. A. A. Elfarra, L. H. Lash and M. W. Anders, 'α-Keto acids stimulate rat renal cysteine conjugate β-lyase activity and potentiate the cytotoxicity of S-(1,2-dichlorovinyl)-L-cysteine.' *Mol. Pharmacol.*, 1987, **31**, 208–212.

137. M. B. Finkelstein, N. J. Patel and M. W. Anders, 'Metabolism of [^{14}C]- and [^{35}S]S-(1,2-dichlorovinyl)-L-cysteine in the male Fischer 344 rat.' *Drug Metab. Dispos.*, 1995, **23**, 124–128.

138. R. K. Bhattacharya and M. O. Schultze, 'Enzymes from bovine and turkey kidneys which cleave S-(1,2-dichlorovinyl)-L-cysteine.' *Comp. Biochem. Physiol.*, 1967, **22**, 723–735.

139. W. Dekant, G. Urban, C. Görsmann *et al.*, 'Thioketene formation from α-haloalkenyl 2-nitrophenyl disulfide: models for biological reactive intermediates of cytotoxic S-conjugates.' *J. Am. Chem. Soc.*, 1991, **113**, 5120–5122.

140. J. L. Stevens and W. B. Jakoby, 'Cysteine conjugate β-lyase.' *Methods Enzymol.*, 1985, **113**, 510–515.

141. L. H. Lash, A. A. Elfarra and M. W. Anders, 'Renal cysteine conjugate β-lyase. Bioactivation of nephrotoxic cysteine S-conjugates in mitochondrial outer membrane.' *J. Biol. Chem.*, 1986, **261**, 5930–5935.

142. J. L. Stevens, J. D. Robbins and R. A. Byrd, 'A purified cysteine conjugate β-lyase from rat kidney cytosol. Requirement for an α-keto acid or an amino acid oxidase for activity and identity with soluble glutamine transaminase K.' *J. Biol. Chem.*, 1986, **261**, 15529–15537.

143. T. Jones, C. Qin, V. H. Schaeffer *et al.*, 'Immunohistochemical localization of glutamine transaminase K, a rat kidney cysteine conjugate β- lyase, and the relationship to the segment specificity of cysteine conjugate nephrotoxicity.' *Mol. Pharmacol.*, 1988, **34**, 621–627.

144. D. G. Abraham, P. P. Patel and A. J. L. Cooper, 'Isolation from rat kidney of a cytosolic high molecular weight cysteine-S-conjugate β-lyase with activity toward leukotriene E$_4$.' *J. Biol. Chem.*, 1995, **270**, 180–188.

145. J. L. Stevens, 'Cysteine conjugate β-lyase activities in rat kidney cortex: subcellular localization and relationship to the hepatic enzyme.' *Biochem. Biophys. Res. Commun.*, 1985, **129**, 499–504.

146. L. H. Lash, R. M. Nelson, R. A. Van Dyke *et al.*, 'Purification and characterization of human kidney cytosolic cysteine conjugate β-lyase activity.' *Drug Metab. Dispos.*, 1990, **18**, 50–54.

147. S. Perry, H. Harries, C. Scholfield *et al.*, 'Molecular cloning and expression of a cDNA for human kidney cysteine conjugate β-lyase.' *FEBS Lett.*, 1995, **360**, 277–280.

148. A. A. Elfarra and I. Y. Hwang, '*In vivo* metabolites of S-(2-benzothiazolyl)-L-cysteine as markers of *in vivo* cysteine conjugate β-lyase and thiol glucuronosyl transferase activities.' *Drug Metab. Dispos.*, 1990, **18**, 917–922.

149. I. Y. Hwang and A. A. Elfarra, 'Cysteine S-conjugates may act as kidney-selective prodrugs: formation of 6-mercaptopurine by the renal metabolism of S-(6-purinyl)-L-cysteine.' *J. Pharmacol. Exp. Ther.*, 1989, **251**, 448–454.

150. A. A. Elfarra and I. Y. Hwang, 'Targeting of 6-mercaptopurine to the kidneys. Metabolism and kidney-selectivity of S-(6-purinyl)-L-cysteine analogs in rats.' *Drug Metab. Dispos.*, 1993, **21**, 841–845.

151. I. Y. Hwang and A. A. Elfarra, 'Kidney-selective prodrugs of 6-mercaptopurine: biochemical basis of the kidney selectivity of S-(6-purinyl)-L-cysteine and metabolism of new analogs in rats.' *J. Pharmacol. Exp. Ther.*, 1991, **258**, 171–177.

152. A. A. Elfarra, R. J. Duescher, I. Y. Hwang *et al.*, 'Targeting 6-thioguanine to the kidney with S-(guanin-6-yl)-L-cysteine.' *J. Pharmacol. Exp. Ther.*, 1995, **274**, 1298–1304.

153. M. W. Duffel and W. B. Jakoby, 'Cysteine S-conjugate N-acetyltransferase from rat kidney microsomes.' *Mol. Pharmacol.*, 1982, **21**, 444–448.

154. J. N. M. Commandeur, F. J. J. De Kanter and N. P. E. Vermeulen, 'Bioactivation of the cysteine-S-conjugate and mercapturic acid of tetrafluoroethylene to acylating reactive intermediates in the rat: dependence of activation and deactivation activities on acetyl coenzyme A availability.' *Mol. Pharmacol.*, 1989, **36**, 654–663.

155. J. N. M. Commandeur, G. J. Stijntjes, J. Wijngaard *et al.*, 'Metabolism of L-cysteine S-conjugates and N-(trideuteroacetyl)-L-cysteine S-conjugates of four fluoroethylenes in the rat. Role of balance of deacetylation and acetylation in relation to the nephrotoxicity of mercapturic acids.' *Biochem. Pharmacol.*, 1991, **42**, 31–38.

156. J. L. Stevens, P. B. Hatzinger and P. J. Hayden, 'Quantitation of multiple pathways for the metabolism of nephrotoxic cysteine conjugates using selective inhibitors of L-α-hydroxy acid oxidase (L-amino acid oxidase) and cysteine conjugate β-lyase.' *Drug Metab. Dispos.*, 1989, **17**, 297–303.

157. V. H. Parker, 'A biochemical study of the toxicity of S-dichlorovinyl-L-cysteine.' *Food Cosmet. Toxicol.*, 1965, **3**, 75–84.

158. J. L. Stevens, N. Ayoubi and J. D. Robbins, 'The role of mitochondrial matrix enzymes in the metabolism and toxicity of cysteine conjugates.' *J. Biol. Chem.*, 1988, **263**, 3395–3401.

159. L. H. Lash and M. W. Anders, 'Mechanism of S-(1,2-dichlorovinyl)-L-cysteine- and S-(1,2-dichlorovinyl)-L-homocysteine-induced renal mitochondrial toxicity.' *Mol. Pharmacol.*, 1987, **32**, 549–556.

160. B. van de Water, J. P. Zoetewey, H. J. G. M. de Bont *et al.*, 'The relationship between intracellular Ca^{2+} and the mitochondrial membrane potential in isolated proximal tubular cells from rat kidney exposed to the nephrotoxin 1,2-dichlorovinyl-cysteine.' *Biochem. Pharmacol.*, 1993, **45**, 2259–2267.

161. S. Vamvakas, D. Bittner, W. Dekant *et al.*, 'Events that precede and that follow S-(1,2-dichlorovinyl)-L-cysteine-induced release of mitochondrial Ca^{2+} and their association with cytotoxicity to renal cells.' *Biochem. Pharmacol.*, 1992, **44**, 1131–1138.

162. E. A. Lock and R. G. Schnellmann, 'The effect of haloalkene cysteine conjugates on rat renal glutathione reductase and lipid dehydrogenase activities.' *Toxicol. Appl. Pharmacol.*, 1990, **104**, 180–190.

163. P. J. Hayden and J. L. Stevens, 'Cysteine conjugate

toxicity, metabolism, and binding to macromolecules in isolated rat kidney mitochondria.' *Mol. Pharmacol.*, 1990, **37**, 468–476.

164. Q. Chen, T. W. Jones, P. C. Brown *et al.*, 'The mechanism of cysteine conjugate cytotoxicity in renal epithelial cells. Covalent binding leads to thiol depletion and lipid peroxidation.' *J. Biol. Chem.*, 1990, **265**, 21603–21611.

165. C. E. Groves, E. A. Lock and R. G. Schnellmann, 'Role of lipid peroxidation in renal proximal tubule cell death induced by haloalkene cysteine conjugates.' *Toxicol. Appl. Pharmacol.*, 1991, **107**, 54–62.

166. Q. Chen, T. W. Jones and J. L. Stevens, 'Early cellular events couple covalent binding of reactive metabolites to cell killing by nephrotoxic cysteine conjugates.' *J. Cell. Physiol.*, 1994, **161**, 293–302.

167. S. A. Bruschi, K. A. West, J. W. Crabb *et al.*, 'Mitochondrial HSP60 (P1 protein) and a HSP70-like protein (mortalin) are major targets for modification during *S*-(1,1,2,2-tetrafluoroethyl)-L-cysteine-induced nephrotoxicity.' *J. Biol. Chem.*, 1993, **268**, 23157–23161.

168. Q. Chen, K. Yu and J. L. Stevens, 'Regulation of the cellular stress response by reactive electrophiles. The role of covalent binding and cellular thiols in transcriptional activities of the 70-kilodalton heat shock protein gene by nephrotoxic cysteine conjugates.' *J. Biol. Chem.*, 1992, **267**, 24322–24327.

169. S. Vamvakas, D. Bittner and U. Koster, 'Enhanced expression of the protooncogenes *c-myc* and *c-fos* in normal and malignant renal growth.' *Toxicol. Lett.*, 1993, **67**, 161–172.

170. S. Vamvakas and U. Koster, 'The nephrotoxin dichlorovinylcysteine induces expression of the protooncogenes *c-fos* and *c-myc* in LLC-PK$_1$ cells—a comparative investigation with growth factors and 12-*O*-tetradecanoylphorbolacetate.' *Cell Biol. Toxicol.*, 1993, **9**, 1–13.

171. K. Yu, Q. Chen, H. Liu *et al.*, 'Signalling the molecular stress response to nephrotoxic and mutagenic cysteine conjugates: differential roles for protein synthesis and calcium in the induction of *c-fos* and *c-myc* mRNA in LLC-PK1 cells.' *J. Cell. Physiol.*, 1994, **161**, 303–311.

172. W. Beuter, C. Cojocel, W. Müller *et al.*, 'Peroxidative damage and nephrotoxicity of dichlorovinylcysteine in mice.' *J. Appl. Toxicol.*, 1989, **9**, 181–186.

173. P. J. Sausen, A. A. Elfarra and A. J. Cooley, 'Methimazole protection of rats against chemically induced kidney damage *in vivo*.' *J. Pharmacol. Exp. Ther.*, 1992, **260**, 393–401.

174. D. M. Vail, A. A. Elfarra, A. J. Cooley *et al.*, 'Methimazole as a protectant against cisplatin-induced nephrotoxicity using the dog as a model.' *Cancer Chemother. Pharmacol.*, 1993, **33**, 25–30.

175. A. A. Elfarra, R. J. Duescher, P. J. Sausen *et al.*, 'Methimazole protection of rats against gentamicin-induced nephrotoxicity.' *Can. J. Physiol. Pharmacol.*, 1994, **72**, 1238–1244.

176. D. M. Vail, A. A. Elfarra, D. L. Panciera *et al.*, 'Pharmacokinetics and short-term clinicopathologic changes after intravenous administration of a high dose of methimazole in dogs.' *Am. J. Vet. Res.*, 1994, **55**, 1597–1601.

177. G. W. Miller, E. A. Lock and R. G. Schnellmann, 'Strychnine and glycine protect renal proximal tubules from various nephrotoxicants and act in the late phase of necrotic cell injury.' *Toxicol. Appl. Pharmacol.*, 1994, **125**, 192–197.

178. G. W. Miller and R. G. Schnellmann, 'Inhibitors of renal chloride transport do not block toxicant-induced chloride influx in the proximal tubule.' *Toxicol. Lett.*, 1995, **76**, 179–184.

179. A. Wallin, G. Zhang, T. W. Jones *et al.*, 'Mechanism of the nephrogenic repair response. Studies on proliferation and vimentin expression after ^{35}S-1,2-dichlorovinyl-L-cysteine nephrotoxicity *in vivo* and in cultured proximal tubule epithelial cells.' *Lab. Invest.*, 1992, **66**, 472–484.

180. P. J. Sausen and A. A. Elfarra, 'Cysteine conjugate *S*-oxidase. Characterization of a novel enzymatic activity in rat hepatic and renal microsomes.' *J. Biol. Chem.*, 1990, **265**, 6139–6145.

181. L. H. Lash, P. J. Sausen, R. J. Duescher *et al.*, 'Roles of cysteine conjugate β-lyase and *S*-oxidase in nephrotoxicity: studies with *S*-(1,2-dichlorovinyl)-L-cysteine and *S*-(1,2-dichlorovinyl)-L-cysteine sulfoxide.' *J. Pharmacol. Exp. Ther.*, 1994, **269**, 374–383.

182. S. Ripp, R. Philpot and A. A. Elfarra, 'Cysteine *S*-conjugate oxidation by flavin-containing monooxygenases.' *Fundam. Appl. Toxicol.*, 1996, **30**, 214 (Abstract).

183. P. J. Sausen, R. J. Duescher and A. A. Elfarra, 'Further characterization and purification of the flavin-dependent *S*-benzyl-L-cysteine *S*-oxidase activities of rat liver and kidney microsomes.' *Mol. Pharmacol.*, 1993, **43**, 388–396.

184. P. J. Sausen and A. A. Elfarra, 'Reactivity of cysteine *S*-conjugate sulfoxides: formation of *S*-[1-chloro-2-(*S*-glutathionyl)vinyl]-L-cysteine sulfoxide by the reaction of *S*-(1,2-dichlorovinyl)-L-cysteine sulfoxide with glutathione.' *Chem. Res. Toxicol.*, 1991, **4**, 655–660.

185. M. Werner, G. Birner and W. Dekant, 'Sulfoxidation of mercapturic acids derived from tri- and tetrachloroethene by cytochromes P450 3A: a bioactivation reaction in addition to deacetylation and cysteine conjugate β-lyase mediated cleavage.' *Chem. Res. Toxicol.*, 1996, **9**, 41–49.

186. R. M. E. Vos, R. T. H. van Weile, W. H. M. Peters *et al.*, 'Genetic deficiency of human class mu glutathione *S*-transferase isoenzymes in relation to the urinary excretion of the mercapturic acids of *Z*- and *E*-1,3-dichloropropene.' *Arch. Toxicol.*, 1991, **65**, 95–99.

187. B. Ketterer, J. M. Harris, G. Talaska *et al.*, 'The human glutathione *S*-transferase supergene family, its polymorphism, and its effects on susceptibility to lung cancer.' *Environ. Health Perspect.*, 1992, **98**, 87–94.

188. S. S. Singhal, P. Zimniak, S. Awasthi *et al.*, 'Several closely related glutathione *S*-transferase isozymes catalyzing conjugation of 4-hydroxynonenal are differentially expressed in human tissues.' *Arch. Biochem. Biophys.*, 1994, **311**, 242–250.

7.32

Bromobenzene Nephrotoxicity: A Model of Metabolism-dependent Toxicity

SERRINE S. LAU and TERRENCE J. MONKS
University of Texas at Austin, TX, USA

7.32.1 INTRODUCTION: METABOLISM AS A DETERMINANT OF TOXICITY

Halogenated hydrocarbons are a diverse class of compounds that are used widely as solvents, insecticides and in industrial chemical syntheses. Human exposure to this class of compounds is therefore widespread and the diverse toxicity of halogenated hydrocarbons is well documented. Because the halogenated hydrocarbons constitute such a large number of compounds, selected chemicals from this group have been used as surrogates to determine the basis of their adverse effects. Thus, bromobenzene has been used as a model

617

halogenated hydrocarbon primarily because bromobenzene administration to rats and mice causes severe hepatic centrilobular necrosis, pulmonary necrosis, and renal proximal tubular necrosis, providing opportunities to study the molecular and cellular basis of the target organ selectivity of these compounds. Reid[1] reported the development of extensive coagulation necrosis of the proximal convoluted renal tubules following a single dose of bromobenzene to mice (4.85 mmol kg^{-1}) and rats (9.3 mmol kg^{-1}). Histopathological changes began to appear 18–24 h after dosing, and were maximal after 36–48 h. Microscopic examination of the lesions revealed dilated proximal tubules in various stages of degeneration. Some dilated proximal tubules were filled with acidophillic casts. In contrast, the glomeruli and most of the distal convoluted tubules appeared normal. These early studies suggested that the nephrotoxicity was probably caused by a metabolite formed in the liver and subsequently transported to the kidney.[1] The nature of this liver-derived circulating metabolite was unclear, but it seemed possible that bromobenzene-3,4-oxide, the putative hepatotoxic metabolite, might be sufficiently stable to escape the liver and produce toxicity in the kidney. Although the 3,4-oxide is capable of diffusing out of hepatocytes[2,3] and can be found in retro-orbital sinus blood of rats treated with bromobenzene,[4] the inability of the kidney to catalyze the formation of the 3,4-oxide, combined with the relatively high renal activity of the glutathione (GSH) *S*-transferase required for detoxication of the 3,4-oxide[5] suggests that any of the epoxide reaching the kidney from the liver is efficiently detoxified there. Subsequently, Rush *et al.*[6] demonstrated that *ortho*-bromophenol, *meta*-bromophenol, *para*-bromophenol and 4-bromocatechol, phenolic metabolites of bromo-benzene, are each capable of producing renal damage in ICR mice (0.56 mmol kg^{-1}, i.v.). 4-Bromocatechol is the most potent of these metabolites, and the renal damage it produces cannot be distinguished, morphologically, from that produced by bromobenzene (9.3 mmol kg^{-1}, i.p.). However, unlike bromobenzene, 4-bromocatechol (2.55 mmol kg^{-1}, i.p.) causes no apparent adverse effects to the kidney when administered to male Sprague-Dawley rats.[7] Although both rats and mice are also sensitive to the nephrotoxic effects of bromobenzene[1] and species differences exist in bromobenzene metabolism[8] the basis for the differences seen in the susceptibility of mice and rats to 4-bromocatechol-mediated nephrotoxicity, and the basis for the tissue selectivity of 4-bromocatechol in mice, remain to be determined.

7.32.1.1 Hepatic Generation of a Potent Nephrotoxicant, 2-Bromo-bis(glutathion-*S*-yl)hydroquinone

In contrast to 4-bromocatechol, administration of *ortho*-bromophenol (1.8 mmol kg^{-1}, i.p.) to rats produces nephrotoxicity similar to that caused by bromobenzene. Moreover, the toxicity is caused by a metabolite of *ortho*-bromophenol, formed in the liver and transported by the circulation to the kidney.[9] Subsequently, 2-bromohydroquinone (2-BrHQ) was identified as a major metabolite of both bromobenzene and *ortho*-bromophenol in male Sprague-Dawley rats.[10] Administration of 2-BrHQ (0.8 mmol kg^{-1}; i.p.) to rats results in severe renal proximal tubular necrosis and elevations in blood urea nitrogen (BUN) concentrations. Interestingly, 2-BrHQ formation from *ortho*-bromophenol is 50-fold higher in liver microsomes than in kidney microsomes, yet neither *ortho*-bromophenol nor 2-BrHQ have any adverse effects on liver morphology *in vivo*.[10] This contrasts with the ability of *ortho*-bromophenol to severely deplete GSH in isolated hepatocytes, and to cause cytotoxicity. Thus, *in vitro* screening would predict this compound to be a hepatotoxicant, yet the target organ *in vivo* is the kidney, not the liver. These are important observations because they underscore the importance of understanding both metabolism and interorgan cooperativity in predicting target organ toxicity. Mechanisms of chemical-induced toxicity that are dependent upon the involvement of more than one tissue (or even cell type) are difficult to model *in vitro*. Data from *in vivo* studies should therefore always drive *in vitro* mechanistic studies, and not vice versa.

2-BrHQ is readily oxidized to the corresponding 1,4-benzoquinone, and in the presence of GSH gives rise to 2-bromo-3-(glutathion-*S*-yl)-hydroquinone, 2-bromo-5-(glutathion-*S*-yl)-hydroquinone, 2-bromo-6-(glutathion-*S*-yl)-hydroquinone, and 2-bromo-bis(glutathion-*S*-yl)hydroquinone[11] (Figure 1). Each of these conjugates has been identified as an *in vivo* metabolite of 2-BrHQ in rat bile.[12] 2-Bromo-bis(glutathion-*S*-yl)HQ is a very potent nephrotoxicant in male rats, producing elevations in BUN following a single intravenous dose of 15–30 µmol kg^{-1} [11] and proteinuria, enzymuria, and glucosuria at 10 µmol kg^{-1}.[13] In addition, the histopathological alterations in the kidney are indistinguishable from those observed following either bromobenzene (9.3 mmol kg^{-1}), *ortho*-bromophenol (1.8 mmol kg^{-1}) or 2-BrHQ (0.8 mmol kg^{-1}) administration. The histological changes are essentially localized to the cortico-medullary junction, where extensive

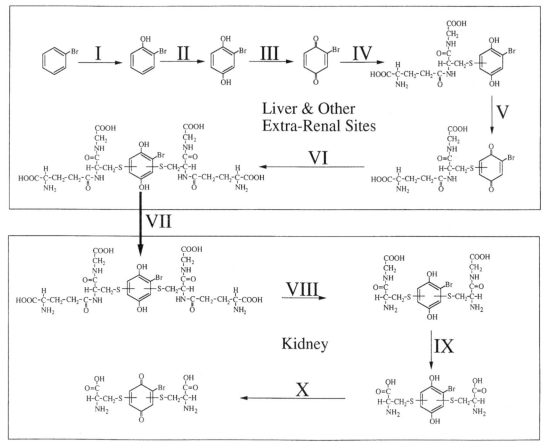

Figure 1 Interorgan cooperation in the metabolism-dependent nephrotoxicity of bromobenzene. Bromobenzene is metabolized in successive cytochrome P450-mediated oxidations to 2-bromohydroquinone (I and II). 2-Bromohydroquinone is oxidized by cytochrome P450, and perhaps other (per)oxidases (III) to 2-bromo-1,4-benzoquinone, which readily reacts with glutathione (IV) to form either 2-bromo-3-(glutathion-S-yl)HQ, 2-bromo-5-(glutathion-S-yl)HQ, and 2-bromo-6-(glutathion-S-yl)HQ. The glutathione conjugates may also undergo oxidation (V) coupled to the subsequent addition of a second molecule of glutathione (VI) to form 2-bromo-bis(glutathion-S-yl)HQ, which is then transported to the kidney (VII). Metabolism of 2-bromo-bis(glutathion-S-yl)HQ by renal brush border γ-glutamyl transpeptidase (VIII) and dipeptidases (IX) yields 2-bromo-bis(cystein-S-yl)hydroquinone, a highly reactive metabolite that is readily oxidized to the corresponding 2-bromo-bis(cystein-S-yl)-1,4-benzoquinone (X). The fate of 2-bromo-bis(cystein-S-yl)HQ is depicted in Figure 2.

coagulative necrosis of the proximal tubules is observed. The most severe damage appears in the S3 segments of the proximal tubules, which contain esosinophilic cells with pyknotic nuclei. The collecting ducts and tubules show occasional hyaline casts. Some of the tubules in the midcortex and outer cortex are markedly dilated, and contain proteinaceous fluid. The glomeruli appear unchanged. Thus, the dose of 2-bromo-bis(glutathion-S-yl)HQ required to produce renal proximal tubular necrosis in the rat is approximately one-thousandth of that of bromobenzene. 2-bromo-bis(glutathion-S-yl)HQ is therefore a relatively potent and specific nephrotoxicant. Each of the 2-bromo-mono(glutathion-S-yl)HQs also cause increases in BUN when administered to male

rats (50 µmol kg⁻¹) but the magnitude of this elevation is substantially less than that caused by 2-bromo-bis(glutathion-S-yl)HQ.[14]

7.32.2 ROLE OF RENAL γ-GLUTAMYL TRANSPEPTIDASE

The first step in the metabolism of either GSH or its S-conjugates involves either hydrolysis or transpeptidation by γ-glutamyl transpeptidase (γ-GT) and transfer of the γ-glutamyl group to an appropriate acceptor (water for hydrolysis, an amino acid for transpeptidation). γ-GT is an ubiquitous enzyme, it is membrane bound, and its active site is oriented on the outer surface of cell membranes.[15] In the kidney, γ-GT is found in high concentrations within the brush border membrane of renal proximal tubular cells (the active site being exposed to the tubular lumen) and at lower concentrations in the basolateral membrane. Evidence implicating a role for γ-GT in the nephrotoxicity of 2-BrHQ and 2-bromo-bis(glutathion-*S*-yl)HQ was obtained from studies using L-(α-5*S*)-α-amino-*S*-chloro-4,5-dihydro-5-isoxazoleacetic acid (AT-125; Acivicin) an inhibitor of γ-GT.[16] Pretreatment of rats with AT-125 (10 mg kg⁻¹, i.p.) significantly inhibits renal γ-GT activity, decreases the γ-GT-mediated proximal tubular accumulation of 2-bromo-(glutathion-*S*-yl)HQ conjugates, as evidenced by an increase in their urinary excretion, decreases the amount of reactive electrophilic metabolites covalently bound to renal protein, and protects animals from 2-BrHQ (0.8 mmol kg⁻¹) mediated nephrotoxicity.[12]

In contrast to AT-125, aminooxyacetic acid, which inhibits cysteine conjugate β-lyase[17] does not protect against either 2-bromo-bis(glutathion-*S*-yl)HQ,[13] or 2-bromo-bis(cystein-*S*-yl)-HQ[18]-mediated renal necrosis. In addition, although 6-bromo-2,5-dihydroxythiophenol (0.27 mmol kg⁻¹, i.p.) a putative β-lyase catalyzed metabolite of 2-bromo-3-(glutathion-*S*-yl)HQ[19] causes elevations in BUN and renal proximal tubular necrosis, administration of a series of isomeric bromothiophenols (0.2–0.8 mmol kg⁻¹, i.p.) causes no apparent histological alterations in the kidney, indicating that the quinone function of 6-bromo-2,5-dihydroxythiophenol, rather than the thiol function, is necessary for the expression of toxicity. Thus, in contrast to the nephrotoxicity of the haloalkanes and haloalkenes, the nephrotoxicity of 2-bromo-bis(glutathion-*S*-yl)HQ does not appear to involve the β-lyase pathway.

2-Bromo-mono(glutathion-*S*-yl)hydroquinones are substantially less toxic than 2-bromo-bis(glutathion-*S*-yl)HQ, and differences in the position of GSH addition also affect the relative toxicity of the isomeric 2-bromo-mono(glutathion-*S*-yl)HQ's.[11,14] 2,3,5-tris(glutathion-*S*-yl)hydroquinone is also a more potent nephrotoxicant than either 2,3-bis(glutathion-*S*-yl)hydroquinone, 2,5-bis(glutathion-*S*-yl)-hydroquinone, 2,6-bis(glutathion-*S*-yl)hydroquinone or 2-(glutathion-*S*-yl)-hydroquinone[20] and 2,3-bis(glutathion-*S*-yl)-1,4-naphthoquinone is more toxic than 2-(glutathion-*S*-yl)-1,4-naphthoquinone,[21] although the reasons for such differences in toxicity are unclear. Inhibition of renal γ-GT by pretreatment of animals with AT-125 completely protects against the nephrotoxicity caused by both 2-bromo-bis(glutathion-*S*-yl)HQ[13] and 2,3,5-trisglutathion-*S*-yl)-hydroquinone,[20] indicating that metabolism by γ-GT is required for toxicity. However, the

monosubstituted conjugates are better substrates for renal γ-GT than the bis-substituted conjugates.[22,23] It is therefore unlikely that differences in the rate of metabolism of the GSH conjugates by γ-GT contribute to their differential toxicity. In support of this view, the capacity of total renal γ-GT (~4×10⁵ units kidney⁻¹) to metabolize 2-bromo-bis(glutathion-*S*-yl)HQ exceeds the dose required to elicit toxicity (10–30 µmol kg⁻¹ or 2–6 µmol 200 g⁻¹ rat). The V_{max} for the hydrolysis and transpeptidation of 2-bromo-bis(glutathion-*S*-yl)HQ are almost identical, 9 and 11 pmol unit⁻¹ min⁻¹, respectively.[23] A 200 g rat therefore has the capacity to metabolize 7–9 µmol 2-bromo-bis(glutathion-*S*-yl)HQ min⁻¹. The corresponding values for the 2-bromo-mono(glutathion-*S*-yl)HQ conjugates range between 11 µmol min⁻¹ for the hydrolysis reaction, and 102–222 µmol min⁻¹ for transpeptidation.[23] Thus, the activity of renal γ-GT is unlikely to be either the rate-limiting step in the metabolism of the GSH conjugates, nor a major factor governing differences in toxicity between 2-bromo-bis(glutathion-*S*-yl)HQ and the 2-bromo-mono(glutathion-*S*-yl)HQs. Furthermore, species differences in susceptibility to 2-bromo-bis(glutathion-*S*-yl)-HQ-mediated nephrotoxicity do not correlate with differences in renal γ-GT activity (see below). Therefore, although γ-GT is clearly important for the onset and development of nephrotoxicity, factors other than, or in addition to, renal γ-GT are probably responsible for the differences in toxicity observed between mono- and multi-*S*-substituted quinones.

7.32.3   NEPHROTOXICITY OF 2-BROMO-(CYSTEIN-*S*-YL)HYDROQUINONES

γ-GT catalyzes the first step in a series of reactions that usually results in mercapturic acid formation. Mercapturates, the *S*-conjugates of *N*-acetylcysteine, are formed via the *N*-acetylation of cysteine thioethers by a requisite cysteine S-conjugate *N*-acetyltransferase.[24] Although neither cystein-*S*-yl nor *N*-acetylcystein-*S*-yl conjugates are excreted as major urinary metabolites of 2-BrHQ[12] their potential reactivity does not preclude the possibility that they contribute to nephrotoxicity following 2-bromo-bis(glutathion-*S*-yl)HQ administration. Consistent with the relative toxicity of the 2-bromo-(glutathion-*S*-yl)HQ conjugates, 2-bromo-bis(cystein-*S*-yl)HQ is a more potent nephrotoxicant than the 2-bromo-mono-(cystein-*S*-yl)HQ conjugates.[18] In addition, the cysteine conjugates are more potent nephrotoxicants than the *N*-acetylcysteine conjugates (see below). Systemic administration of 2-bromo-bis(cystein-S-yl)HQ, or its *in situ* formation via metabolism of 2-bromo-bis(glutathion-S-yl)-HQ by renal γ-GT, results in its rapid delivery to renal proximal tubule cells, where several competing reactions may occur (Figure 2). The cysteine conjugate may either be *N*-acetylated (I) followed by either regeneration of the cysteine conjugate by deacetylases (II), or excreted into urine (III). Alternatively, 2-bromo-bis(cystein-S-yl)HQ is readily oxidized, followed by either the generation of reactive oxygen species (IV), arylation of tissue macromolecules (V) or intramolecular cyclization (VI) and polymerization (VII). Whether the protein bound adduct retains the ability to catalyze the generation of reactive oxygen species (VIII) is not known. Oxidation of the cysteine conjugates (and the consequences thereof) probably occurs much faster than either *N*-acetylation (or β-elimination). In support of this view, the efficiency of the *N*-acetylation of various bromo-(cystein-S-yl)HQs, by rat kidney microsomes, in the presence of acetyl-coenzyme A, is substantially increased in the presence of ascorbic acid.[23] The omission of ascorbic acid

has a dramatic effect on the *N*-acetylation of 2-bromo-bis(cystein-S-yl)HQ, whereas the effects of ascorbic acid on the *N*-acetylation of 2-bromo-mono(cystein-S-yl)HQs are less pronounced. It appears that a greater fraction of 2-bromo-mono(cystein-S-yl)HQs is available as a potential substrate for β-lyase than 2-bromo-bis(cystein-S-yl)HQ, inasmuch as 2-bromo-bis(cystein-S-yl)HQ has a lower oxidation potential than the 2-bromo-mono(cystein-S-yl)HQs and is more readily oxidized to the reactive quinone. In addition, intramolecular cyclization and 1,4-benzothiazine formation (see below) is probably a minor pathway of 2-bromo-bis(cystein-S-yl)-1,4-benzoquinone disposition, since this reaction eliminates the reactive quinone function from the molecule and would therefore be a pathway of detoxication. Inhibition of cysteine conjugate β-lyase with aminooxyacetic acid does not prevent 2-bromo-bis(cystein-S-yl)HQ mediated nephrotoxicity,[18] which is in accord with the inability of aminooxyacetic acid to inhibit the nephrotoxicity of 2-bromo-bis(glutathion-S-yl)HQ. Aminooxyacetic acid did partially protect against 2-bromo-mono(cystein-S-yl)HQ nephrotoxicity[18] an effect which may be related to their conversion to the corresponding mercapturic acid.

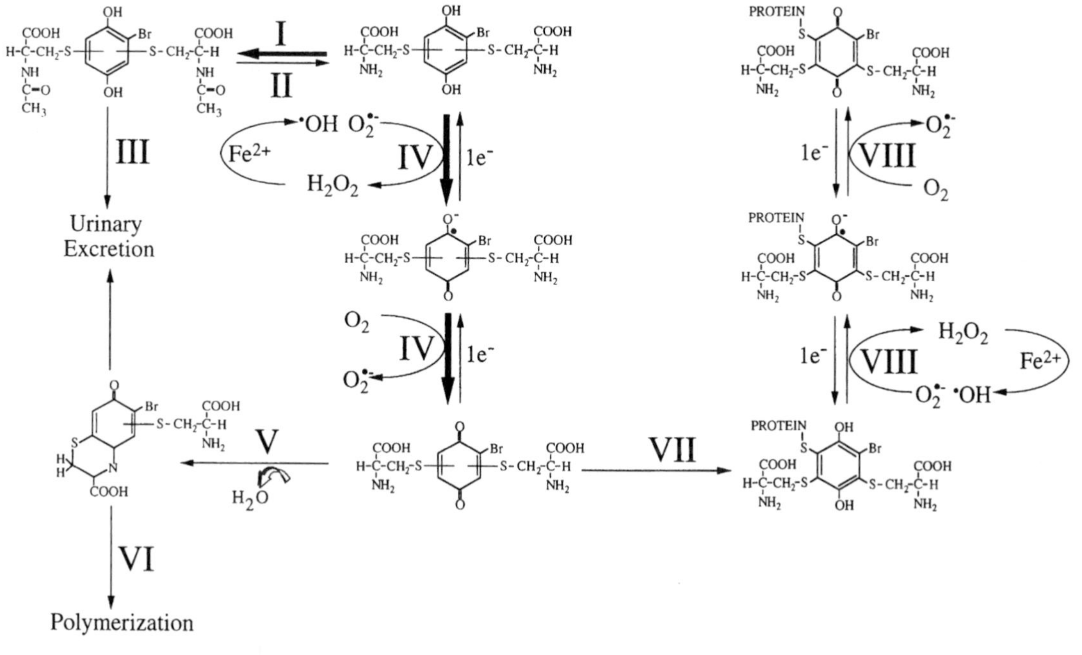

Figure 2 The disposition of 2-bromo-bis(cystein-S-yl)HQ. The cysteine conjugate may undergo *N*-acetylation (I) followed by either regeneration of the cysteine conjugate by deacetylases (II) or excretion of the mercapturic acid into urine (III). Alternatively 2-bromo-bis(cystein-S-yl)HQ is readily oxidized, accompanied by either the generation of reactive oxygen species (IV), intramolecular cyclization (V), polymerization (VI), or arylation of tissue macromolecules (VII). Whether the protein bound adduct retains the ability to catalyze the generation of reactive oxygen species (VIII) is not known.

7.32.3.1   2-Bromo-bis(*N*-acetylcystein-*S*-yl) hydroquinone. A Red Herring?

Although intravenous administration of 2-bromo-bis(*N*-acetylcystein-*S*-yl)HQ causes nephrotoxicity, its contribution to the nephrotoxicity of 2-bromo-bis(glutathion-*S*-yl)HQ *in vivo* is probably minor. In support of this view, both aminooxyacetic acid and probenecid almost completely inhibit the nephrotoxicity of systemically administered 2-bromo-bis-(*N*-acetylcystein-*S*-yl)HQ[18] but not that of 2-bromo-bis(glutathion-*S*-yl)HQ[13] or 2-bromo-bis(cystein-*S*-yl)HQ.[18] Therefore, either 2-bromo-bis(*N*-acetylcystein-*S*-yl)HQ is not formed as a major metabolite of 2-bromo-bis(glutathion-*S*-yl)HQ, or the disposition of the mercapturate, when administered systemically, is different from that when it is formed *in situ* from 2-bromo-bis(cystein-*S*-yl)HQ. Such differences may be related to differences in the site of delivery of the conjugates to renal proximal tubule cells. Thus, 2-bromo-bis-(*N*-acetylcystein-*S*-yl)HQ is probably transported into renal proximal tubule cells via the probenecid sensitive organic anion carrier[18] located on the basolateral membrane, whereas 2-bromo-bis(cystein-*S*-yl)HQ is probably transported into renal cells via amino acid carriers located on both the apical and basolateral membranes. The ability of aminooxyacetic acid to inhibit the toxicity of 2-bromo-bis(*N*-acetylcystein-*S*-yl)HQ may also imply that deacetylation is a prerequisite for the toxicity of the mercapturate, since *N*-acetylcysteine conjugates are not substrates for β-lyase. Alternatively, aminooxyacetic acid may affect the uptake of the mercapturates into renal tubular cells.

Additional studies support the contention that the disposition of the metabolites arising from the γ-GT catalysed metabolism of the GSH conjugates *in vivo* is different from their disposition when administered systemically. For example, aminooxyacetic acid almost completely inhibits the nephrotoxicity of 2-bromo-6-(*N*-acetylcystein-*S*-yl)HQ and 2-bromo-bis(*N*-acetylcystein-*S*-yl)HQ, yet only partially inhibits that of 2-bromo-5-, and 2-bromo-6-(cystein-*S*-yl)HQ, and has no effect on 2-bromo-bis(cystein-*S*-yl)HQ nephrotoxicity.[18] Thus, when the cysteine conjugates are generated *in vivo* via deacetylation of the *N*-acetylcysteine conjugates, aminooxyacetic acid effectively prevents cytotoxicity. In contrast, when the cysteine conjugate is delivered to renal proximal tubular cells extracellularly, either via the activity of γ-GT and dipeptidases on the GSH conjugates, or via systemic administration of the cysteine conjugates, amino-oxyacetic acid is less effective at reducing the ensuing toxicity.

7.32.3.2   *N*-Deacetylation/*N*-Acetylation as a Determinant of 2-Bromo-(*N*-acetylcystein-*S*-yl)hydro-quinone-mediated Nephrotoxicity

Substantial differences exist in the relative toxicity of the 2-bromo-mono(*N*-acetylcystein-*S*-yl)HQ conjugates,[18] the position of thiol addition significantly influences nephrotoxicity; 2-bromo-6-(*N*-acetylcystein-*S*-yl)HQ is a more potent nephrotoxicant than 2-bromo-5-(*N*-acetylcystein-*S*-yl)HQ whereas 2-bromo-3-(*N*-acetylcystein-*S*-yl)HQ (200 μmol kg^{-1}, i.v.) is essentially without effect. Because the cystein-*S*-yl conjugates are easier to oxidize than the mercapturates,[23,25] deacetylation of the mercapturic acids may be necessary for the oxidation of the hydroquinone moiety and for the generation of reactive quinones. Indeed, the relative nephrotoxic potency of the mercapturates correlates with the rate at which the conjugates undergo *N*-deacetylation/*N*-acetylation cycling. Increasing the rate of *N*-deacetylation with respect to the reverse *N*-acetylation reaction increases the toxicity. For example, 2-bromo-5-(*N*-acetylcystein-*S*-yl)HQ and 2-bromo-6-(*N*-acetylcystein-*S*-yl)HQ, which are the most toxic of the 2-bromo-mono(*N*-acetylcystein-*S*-yl)HQs,[18] are the most rapidly deacetylated.[23] In contrast 2-bromo-3-(*N*-acetylcystein-*S*-yl)-HQ, which exhibits no adverse effects following *in vivo* administration, is not a substrate for the *N*-deacetylase.[18] 2-bromo-bis(*N*-acetylcystein-*S*-yl)HQ, the most potent of the mercapturates, is also effectively *N*-deacetylated,[18] and 2-bromo-bis(cystein-*S*-yl)HQ is a relatively poor substrate for the *N*-acetyl-transferase,[18] consistent with the view that the efficiency of the *N*-acetylation reaction is dependant upon the lipophilicity of the substrate.[24] Consequently, the ratio of *N*-deacetylation to *N*-acetylation is highest for the 2-bromo-bis(*N*-acetylcystein-*S*-yl)HQ/2-bromo-bis(cystein-*S*-yl)HQ pair. The results indicate that the *N*-deacetylation/*N*-acetylation ratio plays an important role in modulating the toxicity of quinone-thioethers.[18]

Although 2-bromo-3-(*N*-acetylcystein-*S*-yl)-HQ is essentially inactive as a renal toxicant *in vivo* (see above) concentrations of 25–50 μM do cause cytotoxicity when incubated with rat renal tubular epithelial cells *in vitro*.[26] Moreover, the cytotoxicity is not prevented by aminooxyacetic acid but is inhibited by 1 mM probenecid.[26] The reasons for the discrepancy between the *in vivo* and *in vitro* data are unclear but again highlight the problems involved when extrapolating *in vitro* data to the more complex system that exists *in vivo*. Because thiol addition to polyphenols does not hinder their ability to undergo oxidation, the resulting quinones may interact with nucleophilic targets close to their site of formation, such as the brush border membrane of renal proximal tubular cells. Interestingly, the membrane fraction of kidneys obtained from rats treated with 2-bromo-[^{14}C]HQ contain [^{14}C]-radiolabeled proteins, and the excretion of γ-GT into urine of animals treated with 2-bromo-bis(glutathion-*S*-yl)HQ, and other quinone-thioethers, is an extremely sensitive indicator of the damage to proximal tubular cells[27] (see below).

The relative nephrotoxicity of 2-BrHQ-thioether conjugates is therefore determined by a combination of enzymatic (*N*-acetylation/*N*-deacetylation) and redox factors. Differences exhibited by the mono- and bis-substituted thioethers correlate with the relative rate at which they undergo *N*-deacetylation/*N*-acetylation cycling and by the ease with which they undergo oxidation to the corresponding quinone-thioether. Differences in the balance of *N*-deacetylation and *N*-acetylation may also confer species susceptibility to 2-bromo-bis-(glutathion-*S*-yl)HQ-mediated nephrotoxicity (see below).

7.32.4 OXIDATIVE CYCLIZATION AND 1,4-BENZOTHIAZINE FORMATION

7.32.4.1 The Yin and Yang of γ-Glutamuyl Transpeptidase

Since further metabolism of the GSH conjugates by γ-GT is required for the expression of toxicity, the fate of 2-bromo-3-(glutathion-*S*-yl)HQ was examined in more detail.[28] Initial attempts to synthesize the cystein-*S*-yl conjugates of 2-BrHQ resulted in the formation and deposition of an insoluble black precipitate. The same insoluble material is formed as the end product of the reaction of γ-GT with 2-bromo-3-(glutathion-*S*-yl)HQ. When the precipitate is dissolved in strong acid or base it gives rise to either a violet or red solution, respectively. The product(s) therefore exhibit properties of a pH indicator. When the amount of γ-GT used to catalyse this reaction is substantially decreased, and the reaction terminated prior to the deposition of the insoluble pigment, intermediate products of the reaction can be isolated. The first major intermediate of the reaction identified by NMR and mass spectroscopy is 2*H*-(3-glycyl)-7-hydroxy-

8-bromo-1,4-benzothiazine.[28] This product arises via the oxidative cyclization of 2-bromo-3-(cystein-*S*-yl-glycyl)-HQ. Removal of the γ-glutamate moiety releases the cysteinyl amino group from the peptide bond, exposing the free α-amino group. Oxidation of the hydroquinone function in the conjugate then permits condensation of the free α-amino group with the quinone carbonyl, followed by elimination of water, and 1,4-benzothiazine formation. The subsequent acid catalyzed elimination of glycine initiates the formation of dimeric and possibly polymeric products that eventually constitute the final insoluble product of these reactions. The pH indicator properties of the pigment(s) arising from these reactions are similar to those of the trichochrome polymers formed during phaeomelanin biosynthesis from *S*-(3,4-dihydroxyphenylalanine)-L-cysteine.[29] The elucidation of this pathway also offers a possible explanation for the inability to identify either cystein-*S*-yl or *N*-acetylcystein-*S*-yl conjugates as major metabolites in the urine of rats treated with 2-Br-[^{14}C]-HQ.[12] Oxidative cyclization of the cystein-*S*-ylglycine and cystein-*S*-yl conjugates diverts the products of the γ-GT catalyzed reaction away from the classic mercapturic acid pathway enzymes.

Intramolecular cyclization and 1,4-benzothiazine formation probably constitutes a detoxication reaction. In support of this view, administration of 2-bromo-(homocystein-*S*-yl)-HQ conjugates to rats (100 mg kg^{-1}) causes significant elevations in BUN and histopathological alterations to the kidney[14] consistent with those seen following treatment of rats with 2-bromo-bis(glutathion-*S*-yl)HQ. The additional methylene group in 2-bromo-(homocystein-*S*-yl)HQ should hinder the cyclization reaction. Consequently, if polymerization is necessary for the toxicity of the 2-bromo-(glutathion-*S*-yl)HQs, then the 2-bromo-(homocystein-*S*-yl)HQ conjugates should exhibit little if any toxicity. Furthermore, only the *N*-acetylcysteine conjugate of menadione (2-methyl-1,4-naphthoquinone), and not the GSH conjugate, causes nephrotoxicity when administered to rats[22] and only the mercapturate is cytotoxic when incubated with isolated rat kidney cortical epithelial cells.[30] Moreover, treatment of renal epithelial cells with the glutathione disulfide (GSSG) reductase inhibitor, *N*,*N*-bis(2-chloro-ethyl)-*N*-nitroso-urea, potentiates the toxicity of 2-methyl-3-(*N*-acetylcystein-*S*-yl)-1,4-naphtho-quinone.[30] Since both 2-methyl-3-(glutathion-*S*-yl)-1,4-naphthoquinone[31] and 2-methyl-3-(*N*-acetylcystein-*S*-yl)-1,4-naphtho-quinone[30] redox cycle with the concomitant generation of reactive oxygen species, differences between the mercapturate and GSH conjugate of menadione are probably due to the ability of the latter to undergo γ-GT catalyzed oxidative cyclization and 1,4-benzothiazine formation.[28,30] This reaction eliminates the reactive quinone function from the molecule and effectively prevents redox cycling of the thioether. In contrast, the presence of the *N*-acetyl group in the mercapturate prohibits condensation of the cysteine amino group with the quinone carbonyl group. Consequently, 2-methyl-3-(*N*-acetylcystein-*S*-yl)-1,4-naphthoquinone retaines the ability to redox cycle with the concomitant formation of reactive oxygen species. Consistent with this view, basolateral exposure of rat renal proximal tubular cells to 2-methyl-3-(*N*-acetylcystein-*S*-yl)-1,4-naphthoquinone, in the presence of probenecid, an inhibitor of renal organic anion transport, potentiates cytotoxicity.[32] In addition, inhibition of the intracellular *N*-deacetylation of the mercapturate with paraoxon also potentiates cytotoxicity.[32] These two experiments indicate that (i) interventions which prevent removal of the *N*-acetyl group, and hence prevent condensation of the free cysteinyl amino group with the quinone carbonyl, preserve the redox properties of the quinone-thioether and (ii) intracellular *N*-deacetylation of 2-methyl-3-(*N*-acetylcystein-*S*-yl)-1,4-naphthoquinone represents a detoxication reaction by facilitating cyclization

The GSH conjugate of menadione is only toxic to renal epithelial cells following basolateral exposure, indicating apical detoxication by brush border γ-GT. In accord with this view, 2-methyl-3-(glutathion-*S*-yl)-1,4-naphthoquinone is only toxic to proximal tubular cells following apical exposure when γ-GT was inhibited.[32] In contrast, although 2-methyl-3-(glutathion-*S*-yl)-1,4-naphthoquinone (600 μM) produces toxicity in the isolated perfused rat kidney[33] it is not nephrotoxic when administered *in vivo* (200 μmol kg^{-1}, i.v.)[22] nor when incubated with isolated rat renal epithelial cells (250 μM).[33] A review of the implications of this reaction and how it may modulate the toxicity of quinone thioethers, has been published.[34]

7.32.5   CELLULAR TARGETS OF 2-BROMO-BIS(GLUTATHION-*S*-YL)HYDROQUINONE

Early subcellular targets of 2-bromo-bis(glutathion-*S*-yl)HQ-mediated nephrotoxicity in rats were determined by morphological and biochemical criteria. Ultrastructural changes to the proximal tubular brush border, nuclei,

and endoplasmic reticulum are observed within 30 min of 2-bromo-bis(glutathion-*S*-yl)HQ administration (30 μmol kg^{-1}).[27] These changes consist of loss of the brush border membrane, margination of heterochromatin, and reorganization of the endoplasmic reticulum into discrete aggregates. The desquamation of the brush border membrane into the tubular lumen corresponds with the excretion of γ-GT and alkaline phosphatase in urine. Indeed, the majority of the γ-GT excreted following 2-bromo-bis(glutathion-*S*-yl)HQ occurs in the first or second voiding of the bladder (4–6 h after treatment), prior to any sign of overt renal damage, consistent with the finding that significant elevations in BUN concentrations only occur eight hours after treatment. Although the rapid loss of the brush border membrane induced by 2-bromo-bis(glutathion-*S*-yl)HQ may be a critical early event in proximal tubular cell death, it is a common feature of chemical-induced renal tubular injury. However, the significance of the loss of the brush border membrane in response to structurally diverse chemicals or hypoxic stress[35] is unclear. Since loss of the apical membrane occurs in response to diverse stimuli, perhaps a common signaling pathway (stress response?) is involved. Alternatively, since the activity of γ-GT is necessary for 2-bromo-bis(glutathion-*S*-yl)HQ-mediated nephrotoxicity, either by facilitating the uptake of the corresponding cystein-*S*-yl-glycine or cystein-*S*-yl conjugate into renal cells, and/or the oxidation of the quinol moiety[23,26] it is possible that a direct interaction occurs between reactive electrophilic metabolites of 2-bromo-bis(glutathion-*S*-yl)HQ and the brush border membrane. In support of this view, covalently bound radiolabel is associated with the brush border/plasma membrane fraction of renal tissue obtained 0.5–4 h after treatment of rats with 2-Br-[^{14}C]HQ (unpublished data).

As the injury in response to 2-bromo-bis-(glutathion-*S*-yl)HQ develops, cell swelling with loss of cytosolic density, and loss of chromatin staining is observed.[27] Between 2 h and and 4 h the nuclei undergo extensive karyorrhexis and karyolysis. DNA isolated from the corticomedullary junction four hours after compound administration, exhibits extensive and random fragmentation. Mitochondria are not initial targets of 2-bromo-bis(glutathion-*S*-yl)HQ.[27] Mitochondrial respiratory function is not altered 30 min after 2-bromo-bis(glutathion-*S*-yl)HQ administration, but by 2 h exhibits a significant decrease in respiratory control ratios, a consequence of an increase in state 4 respiration.[27] Similar increases in state 4 respiration after 2,3,5-tris(glutathion-*S*-yl)hydroquinone administration are also unrelated to the development of cytotoxicity.[36] At later time points (8 h), state 4 respiration returns to control values, but the respiratory control ratio remains significantly depressed due to decreases in state 3 respiration. At this time BUN concentrations are significantly elevated.

In support of the biochemical evaluations of mitochondrial respiratory function, morphological studies (electron microscopic) indicate the presence of intact mitochondria 30 min after 2-bromo-bis(glutathion-S-yl)HQ administration. At this time the mitochondria appear normal or slightly rounded, but otherwise unaltered. The morphology of mitochondria between two and four hours are also consistent with maintenance of mitochondrial function. Condensation of the mitochondrial matrix and intercristal swelling is observed 2 h after 2-bromo-bis(glutathion-S-yl)HQ, but this represents a reversible alteration, associated with an active state 3 respiration in the presence of an increased ADP to ATP ratio.[37–39] In addition, studies with the *in situ* perfused rat kidney[40] demonstrate that neither 2-bromo-bis(glutathion-S-yl)HQ nor 2-bromo-6-(glutathion-S-yl)HQ have any adverse effects on mitochondrial respiratory function, or succinate dehydrogenase activity, at concentrations that cause disruption of the brush border membrane (evidenced by significant elevations in the urinary excretion of γ-GT). The finding that mitochondrial condensation is not followed by secondary high amplitude swelling prior to cell death and necrosis in 2-bromo-bis(glutathion-S-yl)HQ-induced nephrotoxicity represents a finding that has not been described with any other nephrotoxic agent. In the general sequence of events leading to necrotic cell death, mitochondrial swelling usually preceeds effects on the nuclei.[41] Studies with 2-bromo-bis(glutathion-S-yl)HQ are in contrast to this scenario, in that 2-bromo-bis-(glutathion-S-yl)HQ induces severe karyolysis prior to any apparent adverse effects on mitochondria. Early changes in nuclear structure, in the presence of normal mitochondria, do occur during apoptotic cell death[41–43] but the decrease in cytosolic density caused by 2-bromo-bis(glutathion-S-yl)HQ contrasts with the increase in cytosolic density that occurs during apoptotic cell death. Thus, 2-bromo-bis-(glutathion-S-yl)HQ-mediated cell death exhibits some distinctive morphological features. Collectively, the data suggest that the brush border membrane and the nucleus are early targets of 2-bromo-bis(GSyl)HQ-induced cytotoxicity, and that alterations in mitochondrial structure and respiratory function occur subsequent to the initial injury, concomitant with the onset of cell death.

7.32.5.1 Arylation and Redox Cycling Between Scylla and Charybdis

The ability of quinones to either arylate tissue macromolecules and/or undergo redox cycling, with the concomitant generation of reactive oxygen species, suggests that 2-bromo-bis(glutathion-S-yl)HQ mediated nephrotoxicity could be initiated by either, or both of these reactions. The extent of the *in vivo* covalent binding of 2-bromo-[^{14}C]-HQ to renal tissue correlates with the degree of nephrotoxicity (as monitored by increases in BUN).[12] The covalent binding of reactive electrophilic metabolites of 2-bromo-(glutathion-[^{35}S]-yl)HQs to rat kidney homogenate is inhibited by ascorbic acid,[13] implying that oxidation to the corresponding 1,4-benzoquinones plays a role in bioactivation. Binding of 2-bromo-bis(glutathion-[^{35}S]-yl)HQ, and/or its metabolites, to rat kidney homogenates is two- to threefold lower than that of the 2-bromo-mono(glutathion-[^{35}S]-yl)HQ conjugates, and ascorbic acid inhibits binding of the 2-bromo-mono(glutathion-[^{35}S]-yl)HQ conjugates by 62–87%, but that of 2-bromo-bis(glutathion-[^{35}S]-yl)HQ by only 28%.[13] The differential effects of ascorbic acid on the covalent binding of 2-bromo-(glutathion-[^{35}S]-yl)HQs may be related to the relative ease of oxidation of the conjugates. Thus, at pH 7.4, 2-bromo-bis(glutathion-S-yl)HQ is more stable to oxidation than the 2-bromo-mono(glutathion-S-yl)HQs,[23] and the relative binding of 2-bromo-bis(glutathion-[^{35}S]-yl)HQ to kidney homogenates is correspondingly less than that of the 2-bromo-mono(glutathion-[^{35}S]-yl)HQ conjugates. The ability of ascorbic acid to inhibit the covalent binding of the conjugates correlates with their oxidation potentials, as determined by cyclic voltammetry at pH 4.0.[13] In addition, although differences in the extent of [^{14}C]-radiolabel covalently bound to renal macromolecules following treatment of rats with 2-bromo-[^{14}C]HQ correlate with the degree of nephrotoxicity,[25] a causal relationship between the two observations remains to be established. Moreover, because the toxicity of 2-bromo-bis(glutathion-S-yl)HQ in renal epithelial cell culture can be inhibited by agents that prevent the generation of hydroxyl radicals (see Section 7.32.6), the relative importance of alkylation and oxidation (Scylla and Charybdis) in 2-bromo-bis(glutathion-S-yl)HQ-mediated renal toxicity remains to be established. However, it is our own view that these two processes are interdependent, and that the consequences of both processes combine to produce cell death.

Although 2-bromo-bis(glutathion-S-yl)HQ is more difficult to oxidize than either of the 2-bromo-mono(glutathion-*S*-yl)HQ conjugates, and although oxidation to the quinone appears necessary for covalent binding, 2-bromo-bis(glutathion-*S*-yl)HQ is a more potent nephrotoxicant than the 2-bromo-mono (glutathion-*S*-yl)HQ conjugates. This apparent paradox was resolved when the effects of γ-GT, and metabolism through the mercapturic acid pathway, on the ease of oxidation of the various intermediates was investigated.[27] Interestingly, although 2-bromo-bis(glutathion-*S*-yl)HQ is more difficult to oxidize than the 2-bromo-mono(glutathion-*S*-yl)HQ conjugates, 2-bromo-bis(cystein-*S*-yl)HQ has a far lower oxidation potential than 2-bromo-bis-(glutathion-*S*-yl)HQ and is more readily oxidized than the 2-bromo-mono(cystein-*S*-yl)HQ conjugates. Thus it appears that γ-GT serves at least two functions in the nephrotoxicity of 2-bromo-bis(glutathion-*S*-yl)HQ. First it serves a transport function, in that metabolism to the cysteinylglycine and/or cysteine conjugate may be a prerequisite for transport across the apical (luminal) and/or basolateral membranes. Second, it serves to convert the relatively stable 2-bromo-bis(glutathion-*S*-yl)HQ into a readily oxidizable, and therefore potentially more reactive metabolite, namely 2-bromo-bis(cystein-*S*-yl)HQ.

7.32.6 2-BROMO-(GLUTATHION-*S*-YL)-HYDROQUINONE-MEDIATED CYTOTOXICITY IN RENAL EPITHELIAL CELLS. USE OF AN *IN VITRO* MODEL TO PROBE MECHANISM OF ACTION

The LLC-PK$_1$ renal epithelial cell line exhibits characteristics indicative of a proximal tubular origin.[44] For example, γ-GT activity is predominantly expressed on the apical (brush border) membrane. The usefulness of this cell line as a model for investigating the mechanism of toxicity of a variety of aliphatic GSH conjugates, which require metabolism by γ-GT to produce toxicity, has been established.[45–47] 2-Bromo-bis(glutathion-*S*-yl)HQ, 2-bromo-6-(glutathion-*S*-yl)HQ, and 2-bromo-3-(glutathion-*S*-yl)HQ cause the formation of single strand breaks in DNA in LLC-PK$_1$ cells within 15 min of exposure.[48] When neutral red accumulation, 3-[4,5-dimethylthiazol-2-yl]-2,5-diphenyltetrazolium bromide (MTT)-formazan formation, and intracellular lactate dehydrogenase (LDH) activity are used as indicators of lysosomal and mitochondrial function,

and plasma membrane integrity, respectively, decreases in lysosomal neutral red accumulation, indicative of H$^+$ loss from lysosomes, is the most sensitive indicator of changes in cellular homeostasis. Significant decreases in neutral red accumulation occur after 30 min exposure to 2-bromo-6-(glutathion-*S*-yl)HQ (0.5 mM) and 60 min after exposure to 2-bromo-bis(glutathion-*S*-yl)HQ (0.5 mM).[48] Consistent with these findings, intracellular acidification occurs rapidly (<30 min) in LLC-PK$_1$ cells exposed to 2-bromo-bis(glutathion-*S*-yl)HQ and 2-bromo-6-(glutathion-*S*-yl)HQ.[49] Cellular viability determined by either MTT-formazan accumulation or intracellular LDH activity, is essentially identical in 2-bromo-bis(glutathion-*S*-yl)HQ treated cultures, suggesting that damage to mitochondria occurs concomitant with the loss of plasma membrane integrity. Thus, the LLC-PK$_1$ cell model appears to reflect the temporal changes in renal proximal tubules caused by 2-bromo-bis(glutathion-*S*-yl)HQ *in vivo*. The finding that the nucleus and DNA are sensitive targets of 2-bromo-bis(glutathion-*S*-yl)HQ is further supported by the observation that *gadd*-153, a gene induced by growth arrest and DNA damage is activated by relatively low concentrations (50 µM) of 2-bromo-bis(glutathion-*S*-yl)HQ.[50]

DNA damage induces a variety of cellular responses which, in a coordinated fashion, are designed to repair damage prior to cell division. One such repair response involves the activation of poly(ADP-ribose)polymerase, which catalyzes the conversion of NAD to nicotinamide and protein-bound poly(ADP-ribose).[51,52] When poly(ADP-ribose)polymerase is inhibited by 3-aminobenzamide,[53] the repair of DNA strand breaks is repressed,[54–56] and this can result in either increased[56–59] or decreased[60,61] toxicity, depending on the severity of the DNA-damage. Because incubation of LLC-PK$_1$ cells with 3-aminobenzamide following treatment with 2-bromo-6-(glutathion-*S*-yl)HQ or 2-bromo-bis(glutathion-*S*-yl)HQ partially decreases cytotoxicity,[48] the recruitment of factors (NAD$^+$; and ATP) required for efficient DNA repair, may adversely affect other cellular processes and exacerbate toxicity. Consistent with an increased demand for ATP to effect DNA repair, mitochondrial dehydrogenases are activated within 30 min after exposure to 2-bromo-bis(glutathion-*S*-yl)HQ (400 µM), whereas extensive DNA fragmentation occurs within 15 min. Furthermore, subsequent to the initial DNA fragmentation, a decrease in the severity of DNA single strand breaks occur, consistent with the activation of DNA repair processes. The pattern of DNA repair seen in LLC-PK$_1$ cells treated with 2-bromo-bis(glutathion-S-yl)HQ may also reflect the formation of quinone-thioether derived hydroxyl radicals (see below) which can catalyze the formation of several distinct modified DNA bases, and which are likely to be repaired at different rates. 2-Bromo-bis-(glutathion-S-yl)HQ-mediated cytotoxicity and DNA fragmentation also appear to be independent of endonuclease activation. When cells are pretreated with aurintricarboxylic acid, an inhibitor of protein-nucleic acid interactions[62,63] and of endonucleases,[64] no effect on 2-bromo-bis(glutathion-S-yl)HQ-mediated cytotoxicity or DNA fragmentation is observed.

The 2-bromo-bis(glutathion-S-yl)HQ catalyzed formation of DNA single strand breaks was determined by alkaline elution, a technique which can also provide information on the nature of strand breakage. Following the initial unwinding of the DNA produced at alkaline pH, first-order elution kinetics are expected.[65] The elution profiles obtained with 2-bromo-bis-(glutathion-S-yl)HQ and the 2-bromo-mono-(glutathion-S-yl)HQs are nonlinear.[48] Similar profiles are observed in studies with H_2O_2[66] and the phorbol ester tumor promoters.[67] It is possible that such nonlinear elution profiles are a consequence of the nonrandom induction of single strand breaks within the genome and/or the presence of both frank and alkaline-labile lesions.

A role for reactive oxygen species in 2-bromo-bis(glutathion-S-yl)HQ and 2-bromo-6-(glutathion-S-yl)HQ mediated cytotoxicity is supported by the finding that treatment of cells with catalase, or pretreatment with deferoxamine mesylate, protects against 2-bromo-6-(glutathion-S-yl)HQ- and 2-bromo-bis(glutathion-S-yl)HQ-mediated DNA single strand breaks[68] and cytotoxicity.[48] Toxicity therefore involves the iron-catalyzed Haber–Weiss reaction, in which $O_2^{\cdot-}$ undergoes dismutation to form H_2O_2, and reduces Fe^{3+} to Fe^{2+}. The H_2O_2 then reacts with Fe^{2+} to generate the hydroxyl radical, which is probably the reactive species responsible for the DNA damage. Removal of deferoxamine prior to exposure of the cells to the conjugates ensures that cytoprotection is unlikely to be a consequence of its radical-scavenging properties. In support of a role for H_2O_2 in 2-bromo-bis(glutathion-S-yl)HQ and 2-bromo-6-(glutathion-S-yl)HQ mediated cytotoxicity, exposure of bacterial and mammalian cells to H_2O_2 or hydroquinone causes extensive DNA damage[51,69–71] and catalase and deferoxamine mesylate protect against toxicity mediated by H_2O_2 directly,[72] or via its generation from hypoxanthine–xanthine oxidase[72,73] and glucose–glucose oxidase.[74] Phosphate within the DNA backbone has a high affinity for Fe^{3+}, thus catalyzing rapid Fe^{2+} oxidation[75] and concomitant reduction of H_2O_2 into water and hydroxyl radicals. In particular, the hydroxyl radical initiates DNA fragmentation via base modifications[76] and/or via attack on the sugar phosphate backbone,[77] eventually resulting in cell death.[70,72,78] The cellular source of the Fe^{3+}/Fe^{2+} redox couple, and the site of H_2O_2 formation in 2-bromo-bis(glutathion-S-yl)HQ-mediated cytotoxicity are not known.

In contrast to deferoxamine, bathocuproine does not protect against 2-bromo-6-(glutathion-S-yl)HQ or 2-bromo-bis(glutathion-S-yl)HQ mediated cytotoxicity[48] indicating that Cu^{2+} ions are not required for the generation of reactive oxygen species. In contrast, macromolecular associated copper may play an important role in hydroquinone-mediated DNA damage since it reduces hydroquinone mediated GSH depletion in mouse bone marrow stromal cells,[79] and prevents hydroquinone/Cu^{2+} induced DNA strand breaks in a cell free system.[80] Consistent with this finding, CuZnSOD accelerates the cell-free oxidation of hydroquinone.[81,82] Interestingly, superoxide dismutase (SOD) potentiates 2-bromo-bis(glutathion-S-yl)HQ-mediated cytotoxicity but has no effect on 2-bromo-6-(glutathion-S-yl)HQ-induced cytotoxicity.[48] Moreover, catalase is more effective at protecting cells from 2-bromo-bis(glutathion-S-yl)HQ toxicity than from 2-bromo-6-(glutathion-S-yl)HQ.[48] The reason for such differences are not known, but may be due to differences in the mechanisms and kinetics of oxidation, similar to that described for ferri-hemoglobin formation from 4-(N-dimethyl)-aminophenol and its bis(glutathionyl) conjugate.[83] Thus, $O_2^{\cdot-}$ produced in the reaction between 4-di-methyl-p-amino-2,6-bis(-glutathion-S-yl)-phenol and hemoglobin, or in the autooxidation of 4-dimethyl-p-amino-2,6-bis(glutathion-S-yl)-phenol, increases the rate of ferrihemoglobin formation by 4-dimethyl-p-amino-2,6-bis(glutathion-S-yl)-phenol, a portion of which is attributable to H_2O_2 produced via dismutation of $O_2^{\cdot-}$. In contrast, $O_2^{\cdot-}$ and H_2O_2 are not detected in the reaction of 4-(N-dimethyl)-aminophenol with hemoglobin.[83]

The cytotoxicity of 2-Br-mono(glutathion-S-yl)HQs and 2-bromo-bis(glutathion-S-yl)HQ in cultured renal epithelial cells is therefore probably due to the generation of hydrogen peroxide and the subsequent iron-mediated formation of hydroxyl radicals. A major target of these reactions is the nucleus, since DNA single strand breaks occur rapidly at relatively low concentrations.

7.32.6.1 Limitations of the *In Vitro* Model

The relative potency of the conjugates *in vitro* is different from that seen *in vivo*. 2-Bromo-bis(glutathion-*S*-yl)HQ is the most nephrotoxic *in vivo* and the least toxic *in vitro*. This difference is probably a kinetic phenomenon. Significant differences exist in the metabolism of 2-bromo-(glutathion-*S*-yl)HQs by γ-GT. Thus, the V_{max}/K_m values for the 2-bromo-mono-(glutathion-*S*-yl)HQs are sevenfold higher (hydrolysis) and 33-fold higher (transpeptidation) than for 2-bromo-bis(glutathion-*S*-yl)HQ.[23] These differences are unlikely to be rate-limiting *in vivo*, since each kidney contains roughly 4×10^5 units γ-GT (see above). Thus differences in the *in vivo* toxicity of 2-bromo-mono(glutathion-*S*-yl)HQs and 2-bromo-bis-(glutathion-*S*-yl)HQ probably reflect inherent differences in the reactivity of 2-bromo-mono-(cystein-*S*-yl)HQs and 2-bromo-bis(cystein-*S*-yl)HQ. In contrast, γ-GT may be the rate limiting step *in vitro* because the amount of the enzyme in cell monolayers is likely to be orders of magnitude lower than *in vivo*.

The *in vitro* data with 2-bromo-(glutathion-*S*-yl)HQs suggest a mechanism whereby the conjugates generate H_2O_2, and the subsequent iron-catalyzed generation of hydroxyl radicals causes DNA fragmentation and cytotoxicity. Because the toxicity of 2-methyl-3-(*N*-acetylcystein-*S*-yl)-1,4-naphthoquinone in isolated rat renal epithelial cells is potentiated by inhibition of GSSG reductase with *N,N*-bis(2-chloroethyl)-*N*-nitrosourea,[32] reactive oxygen species also play a role in the cytotoxicity of 2-methyl-3-(*N*-acetylcystein-*S*-yl)-1,4-naphthoquinone.[32] In contrast, the toxicity of 2-BrHQ to rabbit renal proximal tubule suspensions is suggested to be a consequence of covalent binding[84] rather than of oxidative stress.[85] Thus, 2-BrHQ neither increases oxidized GSH (GSSG) content nor initiates lipid peroxidation; the antioxidant butylated hydroxytoluene does not protect against 2-BrHQ induced mitochondrial dysfunction; and inhibition of GSSG reductase neither increases GSSG content nor potentiates 2-BrHQ induced mitochondrial dysfunction and cell death.[85] It therefore appears that the mechanisms of 2-BrHQ and 2-bromo-(gluta-thion-*S*-yl)HQ mediated cytotoxicity *in vitro* differ. 2-Bromo-bis(glutathion-*S*-yl)HQ is a more potent *in vivo* nephrotoxicant than 2-BrHQ, whereas the reverse occurs *in vitro* (see above). Because 2-BrHQ is more lipophilic than 2-bromo-bis(glutathion-*S*-yl)HQ, 2-BrHQ will cross the cell membrane and enter the cell faster than 2-bromo-bis(gluta-thion-*S*-yl)HQ. Moreover, because 2-bromo-bis(glutathion-*S*-yl)HQ requires metabolism by γ-GT in order for 2-bromo-bis(cystein-*S*-yl)HQ to be transported into cells, and because this enzyme is rate-limiting *in vitro*, the kinetics of accumulation of 2-BrHQ and 2-bromo-bis(glutathion-*S*-yl)HQ into LLC-PK$_1$ cells will differ substantially, and favor uptake of the former.

7.32.7 SPECIES DIFFERENCES IN SUSCEPTIBILITY TO 2-BROMO-BIS(GLUTATHION-*S*-YL)-HYDROQUINONE-MEDIATED NEPHROTOXICITY

Although Sprague-Dawley and Fischer 344 rats exhibit the highest level of renal γ-GT and are the most susceptible rodent species to 2-bromo-bis(glutathion-*S*-yl)HQ (20 μmol kg^{-1}, i.v.) induced nephrotoxicity, species differences in renal γ-GT do not correlate to susceptibility in the other species examined.[86] The guinea pig, which exhibits relatively low renal γ-GT activity, is the only other species found to be susceptible to 2-bromo-bis(glutathion-*S*-yl)HQ (200 μmol kg^{-1}, intracardiac) although at a dose 10 times higher than that required to produce toxicity in rats. Although the activity of renal γ-GT in the guinea pig is lower than that in several other species, it may still be sufficient to allow considerable processing of GSH conjugates (see above). However, factors in addition to γ-GT must play an important role in modulating species susceptibility to 2-bromo-bis(glutathion-*S*-yl)HQ. Guinea pigs are known to excrete low levels of mercapturic acid, and this inability to excrete mercapturic acids is considered to be a consequence of an inability to *N*-acetylate cysteine conjugates.[87] However, the guinea pig kidney can catalyze the *N*-acetylation of 2-bromo-(cystein-*S*-yl)-HQs, albeit at comparatively low rates.[88] In contrast, the activity of the corresponding *N*-deacetylase in the guinea pig, compared to other species, is very high[88] and this might provide a basis upon which to explain the susceptibility of this species to 2-bromo-bis-(glutathion-*S*-yl)HQ nephrotoxicity. Thus, the concentration of the cysteine conjugate generated by each species depends not only upon the activity of γ-GT, but also upon the relative activities of the *N*-acetyl transferase and *N*-deacetylase enzymes (differences in the rate of transport of the cysteine and *N*-acetyl-cysteine conjugates both into and out of renal proximal tubular cells will also be important). Since the cystein-*S*-yl conjugates are easier to oxidize than the *N*-acetylcystein-*S*-yl conjugates, conditions favoring formation of the former will most likely increase the chances of initiating toxicity. In the guinea pig, the balance of these activities favors the formation of high cysteine conjugate concentrations.

Finally, since oxidation is important in the toxicity of these conjugates, species differences in the balance between those factors that contribute to quinol-thioether oxidation and quinone-thioether reduction, will also influence species susceptiblity to nephrotoxicity. Such factors might include the availability of reducing equivalents such as NADPH and GSH and enzymatic determinants such as the specific activity of NADPH quinone oxidoreductase. Quinone–GSH conjugates are substrates for this enzyme.[89] However, inhibition of rat renal NADPH:quinone oxidoreductase with dicumarol does not significantly affect the nephrotoxicity of either 2-bromo-mono(glutathion-S-yl)HQ conjugates ($100 \mu mol \, kg^{-1}$, i.v.) or of 2-bromo-bis(glutathion-S-yl)HQ ($10 \mu mol \, kg^{-1}$, i.v.).[14] Although the mechanism of 2-BrHQ-thioether oxidation remains unclear, it seems logical that differences in the ability of each species to catalyze this oxidation will also determine susceptibility to nephrotoxicity.

7.32.8 CONCLUSION

Bromobenzene-mediated nephrotoxicity has been used as a model of metabolism-dependant target-organ toxicity. However, the information obtained from this model is applicable to drug metabolism and toxicology beyond the elucidation of the mechanisms of bromobenzene-mediated nephrotoxicity. 2-Bromo-bis(glutathion-S-yl)HQ is a potent nephrotoxicant, producing changes in renal structure and function at a dose of $10 \mu mol \, kg^{-1}$, approximately one-thousandth of the dose of bromobenzene required to produce a similar toxicity. This potency challenges our definition of a "minor" metabolite. Thus, a quantitatively "minor" metabolite may be a major metabolite in terms of its ability to produce a biological response. Theoretically, we may be able to account for >99% of an administered dose, and still fail to identify a proximate (re)active metabolite. The rapid evolution of modern analytical technology, particularly HPLC coupled to electrospray mass spectrometry, and capillary gel electrophoresis, should go some way to alleviating some of these concerns.

The renal specific toxicity of 2-bromo-bis-(glutathion-S-yl)HQ is in great part mediated by γ-GT. However, cells other than the epithelial cells of renal proximal tubules, such as pancreatic acinar and ductile epithelial cells and the epithelial cells of the jejunum, bile duct, epididymis, seminal vesicles, bronchioles, thyroid follicles, choroid plexus, ciliary body, and retinal pigment epithelium, also express relatively high γ-GT activity. These cells may therefore be exposed to higher concentrations of potentially toxic quinone-thioethers than other tissues. There is also increasing evidence indicating that quinone-thioethers exhibit a variety of toxicological activity.[89] However, we are only just beginning to unravel the mechanisms by which these compounds exert their toxicological actions, and identifying factors that might regulate this activity. Quinone-thioether toxicology should therefore prove a fruitful area for future research.

7.32.9 REFERENCES

1. W. D. Reid, 'Mechanism of renal necrosis induced by bromobenzene or chlorobenzene.' *Exp. Mol. Pathol.*, 1973, **19**, 197–214.
2. T. J. Monks, S. S. Lau and J. R. Gillette, 'Diffusion of reactive metabolites out of hepatocytes: studies with bromobenzene.' *J. Pharmacol. Exp. Ther.*, 1984; **228**, 393–399.
3. J. R. Gillette, S. S. Lau and T. J. Monks, 'Intra- and extra-cellular formation of metabolites from chemically reactive species.' *Biochem. Soc. Trans.*, 1984, **12**, 4–7.
4. S. S. Lau, T. J. Monks, K. E. Greene *et al.*, 'Detection and half-life of bromobenzene-3,4-oxide in blood.' *Xenobiotica*, 1984, **14**, 539–543.
5. T. J. Monks and S. S. Lau, 'Activation and detoxification of bromobenzene in extrahepatic tissues.' *Life Sci.*, 1984, **35**, 561–568.
6. G. F. Rush, J. F. Newton, K. Maita *et al.*, 'Nephrotoxicity of phenolic bromobenzene metabolites in the mouse.' *Toxicology*, 1984, **30**, 259–272.
7. T. J. Monks, S. S. Lau and R. J. Highet, 'Formation of nontoxic reactive metabolites of *p*-bromophenol. Identification of a new glutathione conjugate.' *Drug Metab. Dispos.*, 1984, **12**, 432–437.
8. J. R. Mitchell, W. D. Reid, B. Christie *et al.*, 'Bromobenzene-induced hepatic necrosis: species differences and protection by SKF 525-A.' *Res. Commun. Chem. Pathol. Pharmacol.*, 1971, **2**, 877–888.
9. S. S. Lau, T. J. Monks, K. E. Greene *et al.*, 'The role of ortho-bromophenol in the nephrotoxicity of bromobenzene in rats.' *Toxicol. Appl. Pharm.*, 1984, **72**, 539–549.
10. S. S. Lau, T. J. Monks and J. R. Gillette, 'Identification of 2-bromohydroquinone as a metabolite of bromobenzene and *o*-bromophenol: implications for bromobenzene-induced nephrotoxicity.' *J. Pharmacol. Exp. Ther.*, 1984, **230**, 360–366.
11. T. J. Monks, S. S. Lau, R. J. Highet *et al.*, 'Glutathione conjugates of 2-bromohydroquinone are nephrotoxic.' *Drug Metab. Dispos.*, 1985, **13**, 553–559.
12. S. S. Lau and T. J. Monks, 'The *in vivo* disposition of 2-bromo-[^{14}C] hydroquinone and the effect of γ-glutamyl transpeptidase inhibition.' *Toxicol. Appl. Pharmacol.*, 1990, **103**, 121–132.
13. T. J. Monks, R. J. Highet and S. S. Lau, '2-Bromo-(diglutathion-S-yl) hydroquinone nephrotoxicity:

physiological, biochemical, and electrochemical determinants.' *Mol. Pharmacol.*, 1988, **34**, 492–500.

14. S. S. Lau and T. J. Monks, 'Glutathione conjugation as a mechanism of targeting latent quinones to the kidney.' *Adv. Exp. Biol. Med.*, 1991, **283**, 457–464.

15. S. S. Tate, in 'Enzymatic Basis of Detoxication,' ed. W. B. Jakoby, Academic Press, New York, 1980, vol. 2, pp. 95–120.

16. G. Weber, 'Biochemical strategy of cancer cells and the design of chemotherapy: G. H. A. Clowes Memorial Lecture.' *Cancer Res.*, 1983, **43**, 3466–3492.

17. W. Dekant, S. Vamvakas and M. W. Anders, in 'Conjugation-Dependent Carcinogenicity and Toxicity of Foreign Compounds,' eds. M. W. Anders and W. Dekant, Academic Press, New York, 1994, pp. 115–162.

18. T. J. Monks, T. W. Jones, B. A. Hill *et al.*, 'Nephrotoxicity of 2-bromo-(cystein-*S*-yl) hydroquinone and 2-bromo-(*N*-acetyl-L-cystein-*S*-yl) hydroquinone thioethers.' *Toxicol. Appl. Pharmacol.*, 1991, **111**, 279–298.

19. T. J. Monks, R. J. Highet, P. S. Chu *et al.*, 'Synthesis and nephrotoxicity of 6-bromo-2,5-dihydroxythiophenol.' *Mol. Pharmacol.*, 1988, **34**, 15–22.

20. S. S. Lau, B. A. Hill, R. J. Highet *et al.*, 'Sequential oxidation and glutathione addition to 1,4-benzoquinone: correlation of toxicity with increased glutathione substitution.' *Mol. Pharmacol.*, 1988, **34**, 829–836.

21. S. S. Lau, T. W. Jones, R. J. Highet *et al.*, 'Differences in the localization and extent of the renal proximal tubular necrosis caused by mercapturic acid and glutathione conjugates of 1,4-naphthoquinone and menadione.' *Toxicol. Appl. Pharmacol.*, 1990, **104**, 334–350.

22. B. A. Hill, H-H. Lo, T. J. Monks *et al.*, 'The role of γ-glutamyl transpeptidase in hydroquinone-glutathione conjugate mediated nephrotoxicity.' *Adv. Exp. Med. Biol*, 1991, **283**, 749–751.

23. T. J. Monks, H. H. Lo and S. S. Lau, 'Oxidation and acetylation as determinants of 2-bromocystein-*S*-yl hydroquinone-mediated nephrotoxicity.' *Chem. Res. Toxicol.*, 1994, **7**, 495–502.

24. M. W. Duffel and W. B. Jakoby, 'Cysteine *S*-conjugate *N*-acetyltransferase from rat kidney microsomes.' *Mol. Pharmacol.*, 1982, **21**, 444–448.

25. T. J. Monks and S. S. Lau, 'Glutathione, γ-glutamyl transpeptidase, and the mercapturic acid pathway as modulators of 2-bromohydroquinone oxidation.' *Toxicol. Appl. Pharmacol.*, 1990, **103**, 557–563.

26. S. Vamvakas, D. Bittner, M. Koob *et al.*, 'Glutathione depletion, lipid peroxidation, DNA double-strand breaks and the cytotoxicity of 2-bromo-3-(*N*-acetylcystein *S*-yl) hydroquinone in rat renal cortical cells.' *Chem. Biol. Interact.*, 1992, **83**, 183–199.

27. M. I. Rivera, T. W. Jones, S. S. Lau *et al.*, 'Early morphological and biochemical changes during 2-Br-(diglutathion-*S*-yl) hydroquinone-induced nephrotoxicity.' *Toxicol. Appl. Pharm.*, 1994, **128**, 239–250.

28. T. J. Monks, R. J. Highet and S. S. Lau, 'Oxidative cyclization, 1,4-benzothiazine formation and dimerization of 2-bromo-3-(glutathion-*S*-yl) hydroquinone.' *Mol. Pharmacol.*, 1990, **38**, 121–127.

29. G. Prota, 'Progress in the chemistry of melanins and related metabolites.' *Med. Res. Rev.*, 1988, **87**, 525–556.

30. P. C. Brown, D. M. Dulik and T. W. Jones, 'The toxicity of menadione (2-methyl-1,4-naphthoquinone) and two thioether conjugates studied with isolated renal epithelial cells.' *Arch. Biochem. Biophys.*, 1991, **285**, 187–196.

31. H. Wefers and H. Sies, 'Hepatic low-level chemiluminescence during redox cycling of menadione and the menadione-glutathione conjugate: relation to glutathione and NAD(P)H: quinone reductase (DT-diaphorase) activity.' *Arch Biochem Biophys.*, 1983, **224**, 568–578.

32. H. E. Haenen, P. Rogmans, J. H. Temmink *et al.*, 'Differential detoxification of two thioether conjugates of menadione in confluent monolayers of rat renal proximal cells.' *Toxicol. In Vitro*, 1994, **8**, 207–214.

33. F. A. Redegeld, G. A. Hofman, P. G. van der Loo *et al.*, 'Nephrotoxicity of the glutathione conjugate of menadione (2-methyl-1,4-naphthoquinone) in the isolated perfused rat kidney. Role of metabolism by γ-glutamyl transpeptidase and probenecid-sensitive transport.' *J. Pharmacol. Exp. Ther.*, 1991, **256**, 665–669.

34. T. J. Monks, 'Modulation of quinol/quinine-thioether toxicity by intramolecular detoxication.' *Drug Metab. Rev.*, 1995, **27**, 93–106.

35. K. A. Reimer, C. E. Ganote and R. B. Jennings, 'Alterations in renal cortex following ischemic injury. 3. Ultrastructure of proximal tubules after ischemia or autolysis.' *Lab. Invest.*, 1972, **26**, 347–363.

36. B. A. Hill, T. J. Monks and S. S. Lau, 'The effects of 2,3,5-(triglutathion-*S*-yl) hydroquinone on renal mitochondrial respiratory function *in vivo* and *in vitro*: possible role in cytotoxicity.' *Toxicol. Appl. Pharmacol.*, 1992, **117**, 165–171.

37. B. F. Trump, P. J. Goldblatt and R. E. Stowell, 'Studies on necrosis of mouse liver *in vitro*. Ultrastructural alterations in the mitochondria of hepatic parenchymal cells.' *Lab. Invest.*, 1965, **14**, 343–371.

38. C. R. Hackenbrock, 'Ultrastructural bases for metabolically linked mechanical activity in mitochondria. I. Reversible ultrastructural changes with change in metabolic steady state in isolated liver mitochondria.' *Cell Biol.*, 1966, **30**, 269–297.

39. A. Tzagoloff, 'Mitochondria Strucure and Compartmentalization,' Plenum Press, New York, 1982, pp. 15–38.

40. M. I. Rivera, L. M. Hinojosa, B. A. Hill *et al.*, 'Metabolism and toxicity of 2-bromo-(diglutathion-*S*-yl)-hydroquinone and 2-bromo-3-(glutathion-*S*-yl)-hydroquinone in the *in situ* perfused rat kidney.' *Drug Metab. Disp.*, 1994, **22**, 503–510.

41. A. H. Wyllie, J. F. Kerr and A. R. Currie, 'Cell death: the significance of apoptosis.' *Int. Rev. Cytol.*, 1980, 68, 251–306.

42. A. R. Boobis, D. J. Fawthrop and D. S. Davies, 'Mechanisms of cell death.' *Trends Pharmacol. Sci.*, 1989, **10**, 275–280.

43. W. Bursch, F. Oberhammer and R. Schulte-Hermann, 'Cell death by apoptosis and its protective role against disease.' *Trends Pharmacol. Sci.*, 1992, **13**, 245–251.

44. R. N. Hull, W. R. Cherry and G. W. Weaver, 'The origin and characteristics of a pig kidney cell strain, LLC-PK.' *In Vitro*, 1976, **12**, 670–677.

45. J. Stevens, P. Hayden and G. Taylor, 'The role of glutathione conjugate metabolism and cysteine conjugate β-lyase in the mechanism of *S*-cysteine conjugate toxicity in LLC-PK1 cells.' *J. Biol. Chem.*, 1986, **261**, 3325–3332.

46. J. J. Mertens, J. G. Weijnen, W. J. van Doorn *et al.*, 'Differential toxicity as a result of apical and basolateral treatment of LLC-PK1 monolayers with *S*-(1,2,3,4,4-pentachlorobutadienyl) glutathione and *N*-acetyl-*S*-(1,2,3,4,4-pentachlorobutadienyl)-L-cysteine.' *Chem. Biol. Interact.*, 1988, **65**, 283–293.

47. S. Vamavakas, V. K. Sharma, S. S. Sheu *et al.*, 'Perturbations of intracellular calcium distribution in kidney cells by nephrotoxic haloalkenyl cysteine *S*-conjugates.' *Mol. Pharmacol.*, 1990, **38**, 455–461.

48. J. J. Mertens, N. W. Gibson, S. S. Lau *et al.*, 'Reactive oxygen species and DNA damage in 2-bromo-(glutathion-*S*-yl) hydroquinone-mediated cytotoxicity.' *Arch. Biochem. Biophys.*, 1995, **320**, 51–58.

49. S. J. Albuquerque, T. J. Monks and S. S. Lau, 'Modulation of intracellular pH by 2-bromo-6-(glutathion-*S*-yl)hydroquinone.' *Toxicologist*, 1995, **15**, 305 (Abstract).

50. J. K. Jeong, J. L. Stevens, S. S. Lau *et al.*, 'Quinone-thioether-mediated DNA damage, growth arrest, and *gadd153* expression in renal proximal tubular epithelial cells.' *Mol. Pharmacol.*, 1996, **50**, 592–598.

51. I. U. Schraufstatter, D. B. Hinshaw, P. A. Hyslop *et al.*, 'Oxidant injury of cells. DNA strand-breaks activate polyadenosine diphosphate-ribose polymerase and lead to depletion of nicotinamide adenine dinucleotide.' *J. Clin. Invest.*, 1986, **77**, 1312–1320.

52. K. Ueda and O. Hayaishi, 'ADP-ribosylation.' *Annu. Rev. Biochem.*, 1985, **54**, 73–100.

53. M. Purnell and W. J. Whish, 'Novel inhibitors of poly (ADP-ribose) synthetase.' *Biochem. J.*, 1980, **185**, 775–777.

54. M. R. James and A. R. Lehman, 'Role of poly-(adenosine diphosphate ribose) in deoxyribonucleic acid repair in human fibroblasts.' *Biochemistry*, 1982, **21**, 4007–4013.

55. L. A. Zwelling, D. Kerrigan and Y. Pommier, 'Inhibitors of poly-(adenosine diphosphoribose) synthesis slow the resealing rate of x-ray-induced DNA strand breaks.' *Biochem. Biophys. Res. Commun.*, 1982, **104**, 897–902.

56. G. Ahnström and M. Ljungman, 'Effects of 3-aminobenzamide on the rejoining of DNA-strand breaks in mammalian cells exposed to methyl methanesulphonate; role of poly (ADP-ribose) polymerase.' *Mutat. Res.*, 1988, **194**, 17–22.

57. N. Nduka, C. J. Skidmore and S. Shall, 'The enhancement of cytotoxicity of *N*-methyl-*N*-nitrosourea and of gamma-radiation by inhibitors of poly-(ADP-ribose) polymerase.' *Eur. J. Biochem.*, 1980, **105**, 525–530.

58. J. E. Cleaver and W. F. Morgan, 'Poly (ADP-ribose) synthesis is involved in the toxic effects of alkylating agents but does not regulate DNA repair.' *Mutat. Res.*, 1985, **150**, 69–76.

59. W. Shen, M. Kamendulis, S. D. Ray *et al.*, 'Acetaminophen-induced cytotoxicity in cultured mouse hepatocytes: effects of Ca^{2+}-endonuclease, DNA repair, and glutathione depletion inhibitors on DNA fragmentation and cell death.' *Toxicol. Appl. Pharmacol.*, 1992, **112**, 32–40.

60. A. Tanizawa, M. Kubota, T. Takimoto *et al.*, 'Prevention of adriamycin-induced interphase death by 3-aminobenzamide and nicotinamide in a human promyelocytic leukemia cell line.' *Biochem. Biophys. Res. Commun.*, 1987, **144**, 1031–1036.

61. S. Seto, C. J. Carrera, M. Kubota *et al.*, 'Mechanism of deoxyadenosine and 2-chlorodeoxyadenosine toxicity to nondividing human lymphocytes.' *J. Clin. Invest.*, 1985, **75**, 377–383.

62. R. G. González, R. S. Haxo and T. Schleich, 'Mechanism of action of polymeric aurintricarboxylic acid, a potent inhibitor of protein–nucleic acid interactions.' *Biochemistry*, 1980, **19**, 4299–4303.

63. T. Blumenthal and T. A. Landers, 'The inhibition of nucleic acid-binding proteins by aurintricarboxylic acid.' *Biochem. Biophys. Res. Commun.*, 1973, **55**, 680–688.

64. R. B. Hallick, B. K. Chelm, P. W. Gray *et al.*, 'Use of aurintricarboxylic acid as an inhibitor of nucleases during nucleic acid isolation.' *Nucleic Acid Res.*, 1977, **4**, 3055–3064.

65. K. W. Kohn, R. A. Ewig, L. C. Erickson *et al.*, in 'DNA Repair. A Laboratory Manual of Research Procedures,' eds. E. C. Friedberg and P. C. Hanawalt, Dekker, New York, 1981, vol. 1, part B, pp. 379–401.

66. Z. Djuric, C. K. Everett and D. A. Luongo, 'Toxicity, single-strand breaks, and 5-hydroxymethyl-2′-deoxy-uridine formation in human breast epithelial cells treated with hydrogen peroxide.' *Free Radic. Biol. Med.*, 1993, **14**, 541–547.

67. Y. Sun, Y. Pommier and N. H. Colburn, 'Acquisition of a growth-inhibitory response to phorbol ester involves DNA damage.' *Cancer Res.*, 1992, **52**, 1907–1915.

68. J. K. Jeong, E. Dybing, E. Soderlund *et al.*, 'DNA damage, *gadd-153* expression, and cytotoxicity in renal proximal tubular epithelial cells.' *Fundam. Appl. Toxicol.*, 1996, **30**, 328.

69. A. C. Mello Filho and R. Meneghini, 'In vivo formation of single-strand breaks in DNA by hydrogen peroxide is mediated by the Haber–Weiss reaction.' *Biochim. Biophys. Acta*, 1984, **781**, 56–63.

70. H. C. Birnboim, 'DNA strand breaks in human leukocytes induced by superoxide anion, hydrogen peroxide and tumor promotors are repaired slowly compared to breaks induced by ionizing radiation.' *Carcinogenesis*, 1986, 7, 1511–1517.

71. P. Leanderson and C. Tagesson, 'Cigarette smoke-induced DNA damage in cultured human lung cell: role of hydroxyl radicals and endonuclease activation.' *Chem. Biol. Interact.*, 1992, **81**, 197–208.

72. P. R. Kvietys, W. Inauen, B. R. Bacon *et al.*, 'Xanthine oxidase-induced injury to endothelium: role of intracellular iron and hydroxyl radical.' *Am. J. Physiol.*, 1989, **257**, H1640–H1646.

73. S. P. Andreoli and J. A. McAteer, 'Reactive oxygen molecule-mediated injury in endothelial and renal tubular epithelial cells *in vitro*.' *Kidney Int.*, 1990, **38**, 578–594.

74. P. D. Walker and S. V. Shah, 'Hydrogen peroxide cytotoxicity in LLC-PK1 cells: a role for iron.' *Kidney Int.*, 1991, **40**, 891–898.

75. D. M. Miller, G. R. Buettner and S. D. Aust, 'Transition metals as catalysts of "autoxidation" reactions.' *Free Radic. Biol. Med.*, 1990, **8**, 95–108.

76. M. Dizdaroglu, Z. Nackerdien, B. C. Chao *et al.*, 'Chemical nature of *in vivo* DNA base damage in hydrogen peroxide-treated mammalian cells.' *Arch. Biochem. Biophys.*, 1991, **285**, 388–390.

77. J. A. Imlay and S. Linn, 'DNA damage and oxygen radical toxicity.' *Science*, 1988, **240**, 1302–1309.

78. A. C. Mello Filho, M. E. Hoffmann and R. Meneghini, 'Cell killing and DNA damage by hydrogen peroxide are mediated by intracellular iron.' *Biochem. J.*, 1984, **218**, 273–275.

79. Y. Li and M. A. Trush, 'Oxidation of hydroquinone by copper: chemical mechanism and biological effects.' *Arch. Biochem. Biophys.*, 1993, **300**, 346–355.

80. Y. Li and M. A. Trush, 'DNA damage resulting from the oxidation of hydroquinone by copper: role for a Cu(II)/Cu(I) redox cycle and reactive oxygen generation.' *Carcinogenesis*, 1993, **14**, 1303–1311.

81. P. Eyer, 'Effects of superoxide dismutase on the autoxidation of 1,4-hydroquinone.' *Chem. Biol. Interact.*, 1991, **80**, 159–176.

82. M. A. Trush, P. Kuppusamy, J. L. Zweier *et al.*, 'Chemical mechanism and biological effects of the CuZnSOD-accelerated oxidation of 1,4-hydroquinone.' *Toxicologist*, 1994, **14**, 167 (abstract).

83. P. Eyer and M. Kiese, 'Biotransformation of 4-dimethylaminophenol: reaction with glutathione, and some properties of the reaction products.' *Chem. Biol. Interact.*, 1976, **14**, 165–178.

84. R. G. Schnellmann, T. J. Monks, L. J. Mandel *et al.*, '2-Bromohydroquinone-induced toxicity to rabbit

renal proximal tubules: the role of biotransformation, glutathione, and covalent binding.' *Toxicol. Appl. Pharmacol.*, 1989, **99**, 19–27.

85. R. G. Schnellmann, '2-Bromohydroquinone-induced toxicity to rabbit renal proximal tubules: evidence against oxidative stress.' *Toxicol. Appl. Pharmacol.*, 1989, **99**, 11–18

86. S. S. Lau, T. W. Jones, R. Sioco *et al.*, 'Species differences in renal γ-glutamyl transpeptidase activity do not correlate with susceptibility to 2-bromo-(diglutathion-S-yl)-hydroquinone nephrotoxicity.' *Toxicology*, 1990, **64**, 291–311.

87. H. G. Bray, T. J. Franklin and S. P. James, 'The formation of mercapturic acids. 3. *N*-acetylation of *S*-substituted cysteines in the rabbit, rat and guinea pig.' *Biochem. J.*, 1959, **73**, 465–473.

88. S. S. Lau, H. E. Kleiner and T. J. Monks, 'Metabolism as a determinant of species susceptibility to 3,5,9-tri(glutathion-S-yl) hydroquinone-mediated nephrotoxicity. The Role of *N*-acetylation and *N*-deacetylation.' *Drug Metab. Dispos.*, 1995, **23**, 1136–1142.

89. T. J. Monks and S. S. Lau, 'Toxicology of quinone-thioethers.' *Crit. Rev. Toxicol.*, 1992, **22**, 243–270.

7.33
Renal To...

RUDOLFS K. ZA...
Mercer Universit...   USA

7.33.1   INTRODUCTION   633

7.33.2   RENAL DISPOSITION OF MERCURY   634
   7.33.2.1   Intrarenal Distribution and Localization of Mercury   634
   7.33.2.2   Intracellular Distribution of Mercury   635

7.33.3   URINARY EXCRETION OF MERCURY   636

7.33.4   RENAL TOXICITY OF MERCURY   637
   7.33.4.1   Site of Tubular Injury   638
   7.33.4.2   Markers of Renal Cellular Injury and Impaired Renal Function   638
   7.33.4.3   Renal Autoimmunity Induced by Mercury   639

7.33.5   RENAL CELLULAR EFFECTS INDUCED BY MERCURY   639
   7.33.5.1   Effects of Mercury on Intracellular Thiol Metabolism and Status   639
   7.33.5.2   Role of Lipid Peroxidation and Oxidative Stress in Mercury-induced Renal Cellular Injury   640
   7.33.5.3   Effects of Mercury on Renal Mitochondrial Function   641
   7.33.5.4   Effects of Mercury on Intracellular Distribution of Calcium Ions   642
   7.33.5.5   Alterations in Plasma Membrane ($Na^+ + K^+$)-stimulated ATPase Induced by Mercury   642
   7.33.5.6   Influence of Mercury on Heme Metabolism   643
   7.33.5.7   Expression of Stress Proteins After Exposure to Mercury   643

7.33.6   FACTORS THAT MODIFY THE RENAL TOXICITY OF MERCURY   644
   7.33.6.1   Influence of Intracellular Thiols on the Renal Accumulation and Toxicity of Mercury   644
   7.33.6.2   Modulation of Renal Disposition and Toxicity of Mercury by Extracellular Thiols   644
   7.33.6.3   Effects of Reduced Nephron Number and Compensatory Renal Growth on the Renal Disposition
              and Toxicity of Mercury   647

7.33.7   CONCLUSION   647

7.33.8   REFERENCES   647

7.33.1   INTRODUCTION

Mercury is a unique element found naturally in the environment in several forms. Metallic mercury, mercuric sulfide, mercuric chloride, and methylmercury are the most common forms of mercury that are found in nature.[1] At room temperature, elemental (or metallic) mercury exists as a liquid. Because of the high vapor pressure of metallic mercury, mercury vapor is one form of mercury that is released into the environment as a result of natural processes. Mercury can also exist as a cation having an oxidation state of +1; (mercurous) or +2 (mercuric). The most common cationic form of mercury encountered in occupational and environmental settings is the mercuric form, which can have a valence of +1 or +2, depending on whether the mercuric ion is covalently bound to a carbon of an organic side group, such as an alkyl group. Of the organic mercuric compounds, methylmercury is perhaps the most widely found in the environment, mainly as the result of conversion of inorganic forms of mercury to methylmercury by microorganisms in soil and water.

Since mercury is found in so many sources in the environment, it is nearly impossible for most humans to avoid exposure to low levels of some form(s) of mercury on a regular basis. For example, many individuals are frequently exposed to elemental mercury vapor from dental amalgams (which are composed of 50% metallic mercury) during the process of mastication.[2] The risk of exposure to elevated levels of the various forms of mercury is of considerable concern due to the increasing deposition of mercury in the environment.[3] Exposure to elevated levels of mercury that have well documented effects on human health generally occurs in industrial settings, either as a result of exposure to mercury in chemical, manufacturing or medical facilities, or exposure to air, soil, and/or water contaminated with mercury in and around these facilities, or at hazardous waste disposal sites. However, there are a number of other situations and ways in which the general public can be exposed to elevated or toxic levels of mercury. One way is by breathing in air containing mercury vapor from spills of metallic mercury (such as that from thermometers) or from around incinerators that burn fossils fuels having a high content of mercury (such as coal). Ingestion of fish that are highly contaminated with methylmercury is another way that the general public can be exposed to elevated or toxic levels of mercury. Careless use of certain antiseptics, disinfectants, and other medicinal products containing inorganic forms of mercury is also a means by which one can be exposed to greater levels of mercury.

Elemental mercury and inorganic and organic mercury-containing compounds cause toxic effects in a number of systems and organs, depending on the chemical form of mercury, the level of exposure, the duration of exposure, and the route of exposure. One organ that is particularly vulnerable to the toxic effects of mercury, especially the mercurous and mercuric forms of this metal, is the kidney. In fact, all forms of mercury are nephrotoxic, depending on the conditions of exposure.

A considerable body of scientific study has been carried out since the 1960s on the disposition and toxicity of mercury in the kidneys of humans and other mammals (for review, see Zalups and Lash[4]). It is the aim of this chapter to summarize some of the scientific findings regarding the renal disposition and toxicity of the various forms of mercury.

7.33.2 RENAL DISPOSITION OF MERCURY

The kidneys are the primary target organs where mercury accumulates in mammals exposed to elemental mercury or inorganic forms of mercury.[4] Renal uptake of mercury *in vivo* is very rapid. Findings indicate that as much as 50% of a low-dose of inorganic mercury is present in the kidneys of rats within a few hours after exposure.[5] A significant amount of mercury also accumulates in the kidneys after exposure to organic forms of mercury, but to a lesser extent than that which occurs after exposure to inorganic or elemental forms of mercury.

7.33.2.1 Intrarenal Distribution and Localization of Mercury

The renal cortex and outer stripe of the outer medulla are the main zones where both inorganic and organic forms of mercury accumulate.[4] On the basis of histochemical and autoradiographic studies in mice and rats,[6-11] and tubular microdissection studies in rats and rabbits,[12,13] it appears that the accumulation of inorganic mercury in the renal cortex and outer stripe of the outer medulla occurs mainly along the convoluted and straight segments of the proximal tubule. However, one cannot exclude, at present, the possibility that some uptake and transport of mercury occurs along other segments of the nephron and collecting duct.

It is interesting that deposits of presumed inorganic mercury have also been found along segments of proximal tubules in the kidneys of rats and mice treated with organic forms of mercury.[8,9] Other data indicate that a significant fraction of the mercury in the kidneys of animals exposed to methylmercury is in the inorganic form,[14-18] suggesting that some organic mercury must be oxidized to inorganic mercury prior to and/or after it enters the renal tubular epithelial cells. There is evidence that intracellular conversion of methylmercury to inorganic mercury can occur.[19] However, the mechanism for this conversion is unknown.

Although the precise mechanisms involved in the proximal tubular uptake of mercury have not yet been defined, experimental evidence from mice and rats tends to implicate two primary mechanisms. One of these mechanisms appears to involve the activity of the brush border enzyme γ-glutamyltranspeptidase (GGT),[20-26] and the other appears to involve the activity of the organic anion transport system.[25-28] *In vivo* findings from rats indicate that treatment with acivicin to inhibit the activity of renal GGT or treatment with *p*-aminohippurate (PAH) to inhibit the organic anion transport system, causes a significant reduction in the renal uptake of inorganic mercury.[25] Moreover, these findings show that an additive effect occurs when treatment with acivicin and PAH are combined (see Figure 1 for details). In fact, the additive effect causes an approximate 90% reduction in the renal uptake of inorganic mercury. Findings from another study provide additional evidence that the mechanism involving the GGT is localized on the luminal membrane and the mechanism that can be blocked by PAH is localized on the basolateral membrane.[26] A scheme summarizing potential mechanisms involved in the proximal tubular uptake of mercury is provided in Figure 2. Further studies are underway to better define the mechanisms involved in the uptake of mercury along the segments of the proximal tubule.

7.33.2.2 Intracellular Distribution of Mercury

Inorganic mercury appears to distribute throughout all intracellular pools once it has been taken up by the proximal tubular epithelium.[17,29–31] Cellular fractionation studies using the renal cortex of rats treated with mercuric chloride acutely and chronically have revealed that mercury distributes between nuclear, lysosomal, mitochondrial, brush border, and supernatant fractions, with the nuclear fraction containing the greatest amount of mercury among all organelle fractions.[31,32] Comparable findings have also been obtained in other studies using the renal cortex of normal and uninephrectomized rats treated with mercuric chloride.[29,30] In these studies, the cytosolic fraction was found to contain the greatest content of mercury. Interestingly, the relative specific content of mercury increased with the

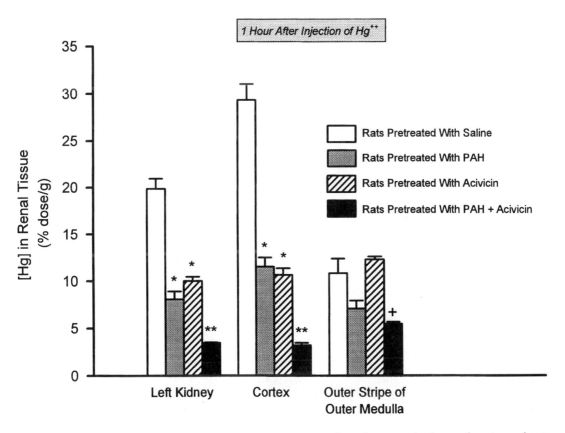

Figure 1 Content of mercury in the left kidney and concentration of mercury in the renal cortex and outer stripe of the outer medulla of rats ($n = 4$ per group) pretreated with saline, *p*-aminohippurate (PAH), acivicin, or PAH plus acivicin 1 h after the i.v. administration of a $0.5\,\mu\text{mol}\,\text{kg}^{-1}$ nontoxic dose of inorganic mercury. The rats pretreated with PAH received a $10\,\text{mmol}\,\text{kg}^{-1}$ dose (i.v.) of the compound 5 min prior to injection of inorganic mercury. The rats pretreated with acivicin received a $10\,\text{mg}\,\text{kg}^{-1}$ dose (i.p) of the compound 150 min and 60 min prior to injection of mercury. For more details of the experimental protocol, refer to Zalups.[25] All values represent mean SE. * = significantly different ($P < 0.05$) from the corresponding mean for the group pretreated with only saline. ** = signficantly different ($P < 0.05$) from the corresponding means for the remaining three groups. + = significantly different ($P < 0.05$) from the corresponding mean for the group pretreated with acivicin and the group pretreated with saline.

greatest amount being in the lysosomal fraction when rats were made proteinuric with an aminoglycoside,[31] or when rats were treated chronically with mercuric chloride.[32] These increases in the lysosomal content of mercury may reflect the fusion of primary lysosomes with endocytotic or cytosolic vesicles containing complexes of inorganic mercury bound to proteins.

7.33.3 URINARY EXCRETION OF MERCURY

Both urinary and the fecal excretion of mercury are the principal means by which humans and other mammals eliminate mercury from the body after exposure to elemental, organic or inorganic forms of mercury. Generally, more mercury is excreted in the feces than in the urine.[2,4,12,18,33–35] In rats, it has been shown that more than twice as much inorganic mercury is excreted in the feces than in the urine during the initial few days after exposure to a nontoxic dose of mercuric chloride.[4,12,34,35] Generally, less than 10% of the dose is excreted

in the urine during this time.[12,34,35] In one study, rats injected intravenously (i.v.) with a non-toxic dose of mercuric chloride excreted about 20% of the dose of mercury in the urine and 30% of the dose in the feces by the end of 54 d following injection.[34] Even less mercury is excreted in the urine after exposure to organic forms of mercury. In a recent study, both normal and uninephrectomized rats excreted only about 3% of the dose of mercury in the urine by the end of the initial 7 d following the i.v. injection of a low-dose of methylmercury.[18] By contrast, greater than 15% of the administered dose was excreted in the feces by these animals during the same period of time. In a study where 7 male adult human subjects were administered a tracer amount of mercury-203 labeled methylmercury, the cumulative fecal excretion of mercury over 70 d was much greater than the cumulative urinary excretion of mercury.[36] In fact, about 30% of the dose was excreted in the feces in 70 d while only about 4% of the dose was excreted in the urine. Very little is known about the mechanisms involved in the urinary excretion of mercury. It is clear that inorganic and organic forms of mercury are taken up very avidly by proximal

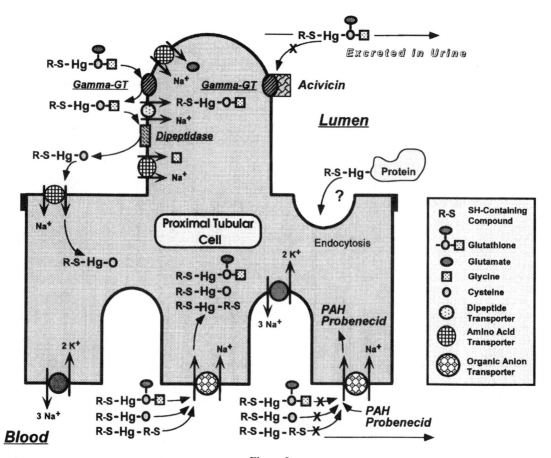

Figure 2

tubular epithelial cells. Whether this uptake is mediated through reabsorptive and/or secretory pathways or whether there is addition of mercury to the luminal fluid through a secretory mechanism is not clear. More studies are needed to better define the mechanisms involved in the urinary excretion of mercury-containing compounds.

7.33.4 RENAL TOXICITY OF MERCURY

As mentioned earlier, all forms of mercury are nephrotoxic *in vivo*,[37] although the inorganic forms of mercury are the most nephrotoxic.[37] With organic mercuric compounds, multiple exposures to relatively large amounts of these compounds are generally required to induce renal injury.[8,38,39] By contrast, renal injury induced by inorganic mercury is generally fully expressed during the initial 24 h following exposure, and can be induced in rats with a single dose of as little as 1.5 μmol

mercury per kilogram (Hg kg^{-1}).[40,41] It should be pointed out, however, that it has been found that rats tend to be more vulnerable to the nephrotoxic effects of inorganic mercury than New Zealand white rabbits or several strains of mice (Zalups, unpublished findings). Additionally, some strain-differences in the severity of the nephropathy induced by inorganic mercury in rats have been found (Zalups, unpublished findings).

In rats, the oral LD$_{50}$ for inorganic mercury, in the form of mercuric chloride, has been reported to be in the range of 25.9–77.7 mg kg^{-1}.[42] A slightly lower range of doses of inorganic mercury (10–42 mg kg^{-1}), in the form of mercuric chloride, has been estimated to be fatal in humans.[43] In one study of human poisoning with mercuric chloride, 9 patients died after ingesting a single dose of inorganic mercury ranging from 29 to more than 50 mg kg^{-1}.[2,44] It should be pointed out that death due to ingestion of a single dose of inorganic mercury is generally due to multiple

Figure 2 Scheme of potential mechanisms involved in the proximal tubular uptake of inorganic mercury. This scheme shows three potential mechanisms by which mercuric ions can be taken up by epithelial cells lining the proximal tubule. Findings provide strong support for the existence of two these mechanisms, One of them involves the action of the γ-glutamyltranspeptidase (γ-GT) on the luminal membrane, and the other involves the sodium-dependent, probenecid and *p*-aminohippurate (PAH)-sensitive organic anion transport system on the basolateral membrane. Findings from a study[25] show that near complete inhibition of the renal γ-glutamyltranspeptidase (with acivicin) and the organic anion transport system (with PAH) causes close to a 90% reduction in the renal cortical uptake of an administered nontoxic dose of inorganic mercury. When acivicin is used to inhibit the renal γ-glutamyltranspeptidase, there is enhanced urinary excretion of mercury and glutathione (GSH). It is on the basis of these findings and the findings of other investigators [22–29] that the following scheme was generated. According to the presented scheme, the mechanism of action of the γ-glutamyltranspeptidase in the proximal tubular uptake of mercury probably involves the catalytic cleavage of the γ-glutamylcysteine bond on molecules of GSH bound to inorganic mercury (that could be bound to another molecule of GSH or some other nonprotein thiol like cysteine). After the action of the γ-glutamyltranspeptidase, the resulting cysteinylglycine remains bound to inorganic mercury and can potentially enter two pathways. One pathway involves cysteinylglycine cotransporting inorganic mercury into the proximal tubular cell by the sodium-dependent dipeptide transport system, and the second pathway involves the catalytic cleavage of the cysteinylglycinyl bond by a dipeptidase located on the luminal membrane, leaving cysteine bound to inorganic mercury. The resulting cysteine then could potentially cotransport inorganic mercury into the proximal tubular epithelial cells by the sodium-dependent neutral amino acid transporter. The later possibility is more likely because of the abundance of dipeptidase activity on the luminal membrane. It should be pointed out that sequential enzymatic degradation of GSH to cysteine while the sulfhydryl of cysteine is bound to a mercuric ion has been demonstrated *in vitro*.[22] The scheme also shows that GSH, cysteine, or other *S*-conjugates of inorganic mercury can potentially be transported across the basolateral membrane by the organic anion transport system, which can be inhibited by the organic anions PAH or probenecid. Although it appears that a significant fraction of mercury enters into proximal tubular cells via the organic transport system located on the basolateral membrane, it is not clear as to which ligand(s) mercury is bound to prior to being transported into the epithelial cells. GSH and cysteine may be likely candidates, since they are both found in the blood at in low micromolar concentrations, and *S*-conjugates of both cysteine and GSH have been demonstrated to be taken up by proximal tubular epithelial cells by the organic anion transport system. The third mechanism presented in the scheme involves the endocytosis of inorganic mercury bound to a filtered protein, such as albumin. There is some indirect evidence from a few studies suggesting that this may occur. If this mechanism does exist, it would seem at present that it plays a minor role relative to the other two mechanisms presented in the scheme. The scheme presented in this figure is meant to summarize some of the current lines of thinking regarding the mechanisms involved in the proximal tubular uptake of inorganic mercury, and it is in no way meant to exclude other possibilities, for they may very well exist. It is also possible that the same mechanisms involved in the proximal tubular uptake of inorganic mercury are also involved in the uptake of organic forms of mercury.

effects. In addition to acute renal failure, cardiovascular collapse, shock, and severe gastrointestinal damage and bleeding are contributing causes of death.[1]

7.33.4.1 Site of Tubular Injury

The pars recta (straight segment) of the proximal tubule (particularly the portion at the junction of the cortex and outer medulla) is the segment of the nephron that is most vulnerable to the toxic effects of both inorganic and organic forms of mercury.[11,40-41,45-52] Depending on the severity of the nephropathy induced by mercury, cellular injury and necrosis can occur along the entire length of the pars recta, from just underneath the capsule to the junction of the outer and inner stripes of the outer medulla. Convoluted portions of proximal tubules and sometimes distal segments of the nephron can be involved when the nephropathy is very severe. There is no direct evidence, however, indicating that mercury has direct *in vivo* toxic effects on segments distal to the proximal tubule and, thus, the involvement of these segments may be due to secondary effects elicited by severe damage to the pars recta of proximal tubules. Further research is clearly needed to clarify this issue.

It is interesting, that in contrast to the effects of mercury *in vivo*, all three segments (S1, S2, and S3) of the proximal tubule (of the rabbit) become intoxicated with either inorganic mercury or methylmercury when the mercury-containing compounds are perfused through the lumen of each of these segments *in vitro*.[53-55] The differences between the *in vivo* and *in vitro* findings are somewhat perplexing since all segments of the proximal tubule accumulate mercury under both experimental conditions.[12,13,53,55] Another interesting difference between the *in vivo* and *in vitro* situation is that, *in vitro*, organic mercury (specifically methylmercury) is more toxic to proximal tubular epithelial cells than inorganic mercury. This has been demonstrated in primary cultures of proximal tubular epithelial cells[56] and in isolated perfused segments of the proximal tubule.[54]

7.33.4.2 Markers of Renal Cellular Injury and Impaired Renal Function

A number of methods have been used to detect renal tubular injury induced by mercury. One noninvasive method that has been employed frequently is measuring the urinary excretion of a number of cellular enzymes.[1,2,57-64] The rationale for using the urinary excretion of cellular enzymes as an indicator of renal cellular injury is based on the close association between renal cellular injury (and necrosis) and enzymuria.[40,62] After renal epithelial cells have undergone cellular necrosis, most, if not all, of the contents of the necrotic epithelial cells, including numerous cellular enzymes, are released into the tubular lumen and are excreted in the urine. The usefulness of any particular cellular enzyme as a marker of renal cellular injury or necrosis depends on the stability of the enzyme in urine, whether the enzyme or the activity of the enzyme is greatly influenced by the toxicant that is being studied and the subcellular localization of the enzyme relative to the subcellular site of injury.

During the early stages of the nephropathy induced by mercury, prior to tubular necrosis, cells along the proximal tubule undergo a number of degenerative changes and begin to lose some of their luminal (brush-border) membrane.[51] Evidence from several studies shows that the urinary excretion of the brush-border enzymes, alkaline phosphatase (AP) and GGT, increases during the nephropathy induced by mercury-containing compounds.[41,59,62,65] When tubular injury becomes severe and necrosis of tubular epithelial cells is apparent, then the urinary excretion of a number of other intracellular enzymes, such as lactate dehydrogenase (LDH), aspartate aminotransferase (AST), alanine aminotransferase (ALT), and N-acetyl-β-D-glucosaminidase (NAG), increases.[1,2,40,41,62,65]

Once a significant number of proximal tubules have become functionally compromised by the toxic effects of mercury, the capacity for the reabsorption of filtered plasma solutes and water is greatly diminished. As a consequence of this diminished reabsorptive capacity, there is an increase in the urinary excretion of both water and a number of plasma solutes, such as glucose, amino acids, albumin, and other plasma proteins.[40,41,65,66] In a study of workers exposed to mercury vapor, it was demonstrated that increased urinary excretion of Tamm–Horsfall glycoprotein and tubular antigens, and decreased urinary excretion of prostaglandin E2 and F2α and thromboxane B2, can also be used as indices of renal dysfunction induced by mercury.[67]

In a couple of studies, the urinary excretion of mercury (factored by the total renal mass) was demonstrated (in normal and uninephrectomized rats) to correlate very closely with the level of injury to pars recta segments of proximal tubules during the acute nephropathy induced by low toxic doses of inorganic mercury.[40,41] Urinary excretion of mercury was shown to correlate with the histopathological scoring of injury to the pars recta of proximal tubules and with the increase in the urinary excretion of albumin and total protein and the urinary excretion of the cellular enzymes LDH, GGT, and NAG.[40] Data from this study indicate that as the level of renal injury increases, there was a corresponding increase in the urinary excretion of mercury. Other nephrotoxic agents have also been shown to decrease the retention of mercury in the kidney and to increase the excretion of mercury in the urine,[68–70] presumably by causing the release of mercury from renal epithelial cells undergoing necrosis and decreased uptake. Although the urinary excretion of mercury appears to correlate well with the level of acute renal injury induced by mercuric chloride, there does not appear to be a close correlation between the severity of renal injury and the renal concentration or content of mercury.[40,41]

When renal tubular injury becomes severe during the nephropathy induced by mercury, the concentration of creatinine in plasma increases due to a decrease in glomerular filtration rate (GFR),[50,65,71] although the mechanisms responsible for the decreased GFR are not well defined. In addition to decreases in GFR, the fractional excretion of sodium and potassium increase.[50] These functional changes reflect a significant decrease in the number of functioning nephrons since similar changes occur in rats and mice when their total renal mass has been reduced surgically by about 75%.[72–74] Blood urea nitrogen (BUN) also increases when glomerular filtration rate decreases significantly and, thus, may be used as an indicator of impaired renal function.[50] However, it is preferable to use the clearance of creatinine or inulin over measuring BUN as an index of renal function, since BUN can be elevated due to nonrenal causes and is not as sensitive an indicator of renal function. After exposure to high doses of mercury, an oliguric or anuric acute renal failure ensues.

7.33.4.3 Renal Autoimmunity Induced by Mercury

Evidence from studies with rabbits, inbred Brown–Norway rats, and a cross between Brown–Norway and Lewis rats indicates that multiple exposures to inorganic mercury can lead to the production of antibodies against the glomerular basement membrane and results in an immunologically mediated membranous glomerular nephritis.[75,76] This type of nephropathy is characterized by the binding of antibodies to the glomerular basement membrane, followed by the deposition of immune complexes in the glomerulus.[77–79] There is also evidence from studies with several strains of both mice and rats that repeated exposures to inorganic mercury can lead to the deposition of immune complexes in the mesangium and along the glomerular basement membrane, which leads to an immune-complex glomerulonephritis.[75,80,81] Whether mercury can induce an autoimmune glomerulonephritis in humans is not clear. It should be pointed out that a great majority of the cases of glomerulonephritis in humans are classified as idiopathic. It is possible that some of the idiopathic forms of glomerulonephritis could be induced by exposure to mercury or other environmental or occupational toxicants.

7.33.5 RENAL CELLULAR EFFECTS INDUCED BY MERCURY

7.33.5.1 Effects of Mercury on Intracellular Thiol Metabolism and Status

A prominent effect of mercury on intracellular thiol status is induction of metallothionein.[82] Metallothioneins are a group of small proteins, having an approximate molecular weight of 6–7 kDa, that contain multiple sulfhydryl groups and have the capacity to bind numerous metal ligands, including inorganic mercury, cadmium, zinc, copper, silver, and platinum. Administration of a single, daily, nontoxic dose of mercuric chloride ($0.25\,\mu mol\,kg^{-1}$) over several days has been shown to cause almost a doubling in the concentration of metallothionein in the renal cortex or outer stripe of the outer medulla in rats.[83] Exposure of rats to elemental mercury vapor over the course of several days has also been shown to cause induction of synthesis of metallothionein in kidney.[84] The increase appears to be tissue-selective, as changes in hepatic metallothionein synthesis have not been demonstrated. Elemental mercury vapor is converted into inorganic mercury, which is predominantly (>98%) recovered in the kidneys, suggesting that the induction of metallothionein due to elemental mercury may be mediated by inorganic mercury.

Both inorganic and organic forms of mercury also cause changes in intracellular GSH status and metabolism. These effects are observed acutely after short-term, single treatments, and are concentration-dependent. Several reports, using both *in vivo* and *in vitro* systems, have demonstrated increases in intracellular contents of GSH in renal tubular epithelial cells after administration of relatively low, toxic or subtoxic doses of either methylmercury[85] or inorganic mercury.[86–90] At higher doses of inorganic mercury, however, decreases in renal content of GSH, which are often substantial, are observed.[52,91–95]

The dose-dependent effect of inorganic mercury on renal GSH was demonstrated in male Sprague-Dawley rats that received one of several nontoxic or nephrotoxic intravenous doses of mercuric chloride 24 h before renal homogenates were fractionated and analyzed for the content of GSH.[52] At the level of the whole kidney, or in samples derived from the cortex or the outer stripe of the outer medulla, the subtoxic or the moderately nephrotoxic dose of mercuric chloride produced significant increases in the renal concentration of GSH. This effect was most marked in the outer stripe of the outer medulla, where the concentration of GSH increased by as much as 85%. There is toxicological significance in this finding, since the outer stripe of the outer medulla is where tubular injury induced by mercury is mainly localized. At the highest nephrotoxic dose of mercuric chloride, the concentrations of GSH in the renal cortex and outer stripe of the outer medulla were similar to those in controls.

Since the cellular content of GSH is under feedback control, the large increases observed after treatment with inorganic mercury suggest that subtoxic or moderately toxic doses of inorganic mercury induce synthesis of γ-gluta-mylcysteine synthetase (GCS), which is the rate-limiting enzyme involved in the biosynthesis of GSH. To test this hypothesis, Lash and Zalups[96] measured the activity of GCS in freshly isolated renal proximal tubular and distal tubular cells from control rats or from rats treated with a subtoxic dose of mercuric chloride 24 h before isolation of cells. They found that the activity of GCS was increased in the renal epithelial cells isolated from rats treated with inorganic mercury. Further support for this hypothesis comes from a study by Woods *et al.*,[85] who showed that the mRNA for GCS increased (by 4.4-fold) in kidneys of male Fischer-344 rats treated with methylmercury hydroxide for 3 wk. Therefore, subtoxic doses of both inorganic and organic forms of mercury appear to induce synthesis of GSH via the activity of GCS.

In addition to causing upregulation of GCS, inorganic mercury also alters, in a dose-depedent manner, the activity of other GSH-dependent enzymes. The effects of inorganic mercury on these enzymes differ depending on whether a subtoxic, a moderately toxic, or a highly toxic dose is administered. Subtoxic doses of mercuric chloride apparently cause increases in activities of GSSG reductase and GSH peroxidase in isolated epithelial cells from

both proximal tubular and distal tubular regions of the rat nephron.[96] In contrast, marked decreases in the activities of renal GSSG reductase and GSH peroxidase have been observed in male rats treated chronically with a relatively high-dose of mercuric chloride by one group of investigators.[91] They also found apparent adaptive increases in catalase activity.[91] Similarly, others have found significant decreases in the activity of GSSG reductase after the administration of highly nephrotoxic doses of mercuric chloride.[87,92,94] While two groups of investigators found small (20–35%), but statistically significant, decreases in the activity of GSH peroxidase,[87,94] another group did not see any change in the activity of this enzyme.[92]

7.33.5.2　Role of Lipid Peroxidation and Oxidative Stress in Mercury-induced Renal Cellular Injury

Findings from several studies suggest that an important mechanism involved in renal cellular injury induced by either *in vivo* or *in vitro* exposure to inorganic or organic forms of mercury involves production of oxidative stress. The high affinity of mercuric ions for binding to thiols naturally suggests that the ensuing depletion of intracellular thiols either directly causes or predisposes proximal tubular cells to oxidative stress. Furthermore, other cellular antioxidants, including ascorbic acid and vitamin E, have been reported to be depleted in the kidneys of rats treated with mercuric chloride.[92] Activity of several antioxidant enzymes also appears to be markedly diminished after *in vivo* exposure of rats to nephrotoxic doses of mercuric chloride. For example, it has been reported that administration of mercuric chloride to male Sprague-Dawley rats caused marked decreases in activity of superoxide dismutase, catalase, GSH peroxidase, and GSSG reductase in the renal cortex.[94]

Decreases in activities of these protective enzymes would be expected to increase the susceptibility of renal epithelial cells to oxidative injury. There has been some disagreement as to whether mercury itself causes oxidative injury or whether it merely makes the renal epithelial cells more sensitive to agents that produce oxidative stress. Fukino *et al.*[92] found that thiobarbiturate reactants, which indicate occurrence of lipid peroxidation, were markedly increased in renal cortical homogenates from rats 12 h after a subcutaneous injection of a nephrotoxic dose of mercuric chloride. Gstraunthaler *et al.*[94] observed increases in formation of malondialdehyde in renal cortical homogenates obtained from mercuric chloride-treated rats as compared with control rats treated only with cumene hydroperoxide. Thus, mercuric chloride enhanced the ability of other agents to produce lipid peroxidation. Since the two groups administered equivalent doses of mercuric chloride, concentration dependence cannot be invoked to explain the difference in observed responses.

The strong dependence of renal function and maintenance of cellular content of GSH and redox status on mitochondrial generation of ATP led Lund *et al.*[97] to investigate the role of mercury-induced oxidative stress localized to the mitochondrion as a mechanism of mercury-induced injury. They investigated the effects of inorganic mercury on hydrogen peroxide production by isolated renal cortical mitochondria from rats. Depending on the supply and coupling site specificity of respiratory substrates, variable increases in the formation of hydrogen peroxide were observed; incubation of isolated mitochondria with 30 nmol mercuric chloride per mg protein increased formation of hydrogen peroxide fourfold at the ubiquinone–cytochrome b region and twofold at the NADH dehydrogenase region. Additionally, iron-dependent lipid peroxidation was increased 3.5-fold at the NADH dehydrogenase region and by 25% at the ubiquinone–cytochrome b region. Intramitochondrial GSH was decreased in a time- and concentration-dependent manner by mercuric chloride. In fact, a 12 nmol mg^{-1} protein concentration of mercuric chloride completely depleted mitochondrial GSH within 30 min, suggesting that targeting of mitochondrial GSH by mercury may be responsible for the intramitochondrial oxidative stress. Lund *et al.*[98] have demonstrated that production of hydrogen peroxide, depletion of GSH, and lipid peroxidation increased in mitochondria isolated from renal cortical homogenates of rats treated *in vivo* with mercuric chloride ($HgCl_2$) after addition of appropriate respiratory substrate. These findings support *in vitro* data and lead one to suggest that mercury-induced oxidative stress within mitochondria is an important mechanism involved in renal tubular injury induced by mercury.

7.33.5.3 Effects of Mercury on Renal Mitochondrial Function

As described above, Lund *et al.*[97,98] have demonstrated that inorganic mercury interferes with respiratory function, leading to enhanced production of hydrogen peroxide, particularly at coupling site II of the electron transport chain. Furthermore, the demonstration of oxidative stress in mitochondria isolated from kidneys after *in vivo* treatment of rats with $HgCl_2$ indicate that an oxidative stress localized to the mitochondria may be responsible for mercury-induced inhibition of various energy-dependent processes in the kidneys.

In an earlier series of studies, Weinberg *et al.*[99,100] compared the effects of mercuric chloride on mitochondrial function *in vitro* after either *in vivo* or *in vitro* treatment with mercuric chloride. If mitochondria were isolated from male Sprague-Dawley rats and then treated *in vitro* with inorganic mercury,[99] a marked uncoupling of respiration (i.e., increase in state 4 rate of oxygen consumption) and a significant decrease in the rate of substrate-stimulated respiration (i.e., state 3 respiration) were observed. Additionally, atractyloside-insensitive ADP uptake and both basal- and magnesium +2-activated oligomycin-sensitive ATPase activities were markedly increased by inorganic mercury. These *in vitro* effects occurred with a threshold concentration of mercuric chloride of 2 nmol per mg protein. Similarly, if renal cortical mitochondria were isolated from rats treated with mercuric chloride *in vivo* (5 mg kg^{-1}, s.c.), the most prominent effects were inhibition of ADP uptake and decreases in the rates of state 3 and uncoupler-stimulated respiration.[100] These effects were not attributed to interaction of mercury with mitochondria during the isolation procedure. However, with both *in vivo* and *in vitro* treatment, inorganic mercury was not readily washed out of mitochondria, suggesting complex formation between mercuric ions and mitochondrial thiols, possibly protein sulfhydryl groups.

Chavez and Holguin[101] and Chavez *et al.*[102] also found uncoupling of mitochondrial respiration after either *in vivo* or *in vitro* treatment of male Wistar rats with mercuric chloride. Consistent with this finding, they found that inorganic mercury induced calcium efflux from mitochondria, oxidation of pyridine nucleotides, and a collapse of the membrane potential. In support for the mechanism of mercury-induced mitochondrial injury involving complex formation between mercuric ions and mitochondrial sulfhydryl groups, Chavez and Holguin[101] reported that inorganic mercury bound to mitochondrial protein in a concentration-dependent pattern, with saturation at approximately 9 nmol mercuric chloride per mg protein.

Jung *et al.*[103] employed ATP depletion by different chemical agents in microdissected nephron segments to localize the nephron site specificity of injury. They found that 1 μM mercuric chloride produced a significant depletion of intracellular ATP content only in S2 segments; nephron segments derived from the other zones of the proximal tubule (i.e., S1 or S3) or other regions of the nephron, such as the distal convoluted tubule or the medullary thick ascending limb of the loop of Henle, were not as sensitive to ATP depletion after incubation with inorganic mercury. This pattern agrees with histopathological data, which demonstrates that the pars recta of the proximal tubule is the primary target of inorganic mercury. These data also support the conclusion that renal mitochondria are early intracellular targets of inorganic mercury. This is logical considering the extremely high content of sulfhydryl-containing proteins in both mitochondrial matrix and in mitochondrial inner membrane.

Zalups et al.[104] studied the accumulation and toxicity of inorganic mercury in suspensions of isolated segments of renal proximal tubules from the rabbit. Incubation of isolated proximal tubules with mercuric chloride, in the absence of extracellular thiols, caused a marked time- and concentration-dependent inhibition of nystatin-stimulated oxygen consumption, demonstrating mitochondrial toxicity in an intact *in vitro* renal cellular model. Furthermore, inhibition of oxygen consumption by mercuric chloride preceded the development of irreversible cellular injury, as assessed by release of LDH from the tubular segments, suggesting that inhibition of cellular energetics is a critical component of the nephrotoxic response to inorganic mercury.

7.33.5.4   Effects of Mercury on Intracellular Distribution of Calcium Ions

Inorganic mercury induces efflux of calcium ions from renal mitochondria of rats both *in vivo* and *in vitro*.[101,102] The importance of maintaining appropriate intracellular concentrations of calcium for proper cellular function is well-documented, suggesting that the prominent effects of mercury on mitochondrial calcium status may play an important part in the acute nephropathy induced by mercury.

Smith et al.[105] used primary cultures of renal tubular cells from rabbits that were mostly of proximal tubular origin as an *in vitro* model system to study the effects of inorganic mercury on the intracellular distribution of ionic calcium. They employed the fluorescent dye Fura 2 to quantitate the cytosolic content of free

ionized calcium. Treatment of cells with low (2.5–10 μM) concentrations of mercuric chloride produced 2–10-fold increases in the intracellular content of calcium. In contrast to this, exposure of cells to higher (25–100 μM) concentrations of mercuric chloride produced an initial, rapid 10–12-fold increase in intracellular calcium, which quickly returned to levels that were about twice those in control cells. This was subsequently followed by a second, more gradual, increase in the intracellular content of calcium that was dependent on the presence of extracellular calcium. Cytotoxicity was also associated with this phase of increase in the intracellular content of calcium and was similarly dependent on the presence of extracellular calcium. The increases in the cytosolic content of calcium that were independent of extracellular concentrations of calcium were primarily due to release of intracellular calcium ions from nonmitochondrial intracellular stores, presumably derived from the endoplasmic reticulum. The subsequent decrease in intracellular calcium content may be due to buffering processes, such as uptake though the microsomal calcium plus magnesium ($Ca^{2+} + Mg^{2+}$)-ATPase, or through the mitochondrial uniporter. The dependence of the slow, late phase of increase in cytosolic calcium content on extracellular calcium with higher concentrations of inorganic mercury, suggests that nonlethal effects of inorganic mercury in renal cells are associated with redistribution of intracellular stores of calcium, but that toxic effects are associated with mercury-induced changes in permeability of the plasma membrane.

7.33.5.5   Alterations in Plasma Membrane ($Na^+ + K^+$)-stimulated ATPase Induced by Mercury

Cellular plasma membranes contain a large number of proteins containing sulfhydryl groups that are critical for enzymatic activity and membrane structure.[106] Among them is the sodium plus potassium ($Na^+ + K^+$)-stimulated ATPase located on the basolateral membrane of epithelial cells in both the proximal and distal regions of the nephron, which is markedly inhibited by alkylation or oxidation of its sulfhydryl group. Anner and colleagues[107–109] have conducted a detailed series of studies on the interaction between mercury-containing compounds and purified and reconstituted ATPase protein from the renal outer medulla of the rat, rabbit, and sheep. To determine the molecular details of the interaction between mercury-containing compounds and the

$(Na^+ + K^+)$-stimulated ATPase, studies had to be done with purified and reconstituted enzyme rather than intact renal epithelial cells or renal tubules.

Anner et al.[108] showed that a number of mercury-containing compounds, including mercuric chloride, mersalyl, and p-mercuribenzenesulfonic acid, potently inhibited the activity of the ATPase by binding to a site distinct from that at which the cardiac glycosides (e.g., digoxin and ouabain) bind. The binding of inorganic mercury was concentration dependent and was modulated by addition of mercury chelators, such as EDTA or 2,3-dimercapto-1-propanesulfonic acid, indicating that inorganic mercury binding to the enzyme is reversible.

Imesch et al.[109] showed that the inactivation of the $(Na^+ + K^+)$-stimulated ATPase by mercuric chloride (0.1–100 μM) apparently loosens the interaction between the α- and β-subunits of the ATPase molecule, thereby altering the sensitivity of the enzyme to extracellular drugs, hormones, and antibodies.

Moreover, Anner and Moosmayer[107] showed that the binding of inorganic mercury to the $(Na^+ + K^+)$-stimulated ATPase molecule occurs primarily at the cytosolic surface. Binding of mercury was closely correlated with inhibition of uptake of rubidium-86, indicating that the metal-binding site is critical to the active transport function of the ATPase.

An important extension of these studies will be to design experiments to investigate effects of mercury on $(Na^+ + K^+)$-stimulated ATPase function in more intact renal systems, such as isolated perfused tubule segments or isolated cells. Considering the potency of the interaction and the fact that the plasma membrane is a very early target site for mercury, it is likely that this interaction will be important in the mechanism of mercury-induced renal cellular injury. It is also likely that sulfhydryl groups on other membrane proteins, particularly those in the epithelial cells lining the proximal tubule, interact with mercury and may play a role in the nephropathy induced by mercury.

7.33.5.6 Influence of Mercury on Heme Metabolism

A prominent effect of mercury intoxication *in vivo* is porphyrinuria.[110] The porphyrinogenic properties of mercury-containing compounds were initially attributed to metal-induced alterations in the regulation of enzymes involved in heme biosynthesis or degradation in target tissues. However, since the magnitude of porphyrin excretion during prolonged exposure to either methylmercury or inorganic mercury is greater than can be accounted for by changes in heme metabolism alone, Woods and colleagues[110,111] needed to invoke alternative biochemical mechanisms to explain their findings. They showed that mercuric ions promoted free radical-mediated oxidation of reduced porphyrins. The mechanism involved depletion or interference of normal antioxidants in renal epithelial cells, such as endogenous thiols like GSH. Furthermore, the ability of inorganic mercury and GSH to react with endogenously produced reactive oxygen metabolites from both hepatic and renal mitochondria from rats was correlated with porphyrinogen oxidation.

An important clinical application of this effect of mercury is illustrated in a study by Bowers et al.,[112] who evaluated patterns of urinary excretion of porphyrin in male Fischer-344 rats as a diagnostic tool to assess exposure to inorganic mercury or methyl mercury. Evaluation of the urinary excretion of porphyrins is a noninvasive method that can be applied to human populations suspected of being exposed to mercury-containing compounds.[113]

7.33.5.7 Expression of Stress Proteins After Exposure to Mercury

Various environmental stimuli, including toxic chemicals, increase the synthesis of a class of proteins known as "stress proteins." Goering et al.[114] evaluated the effect of a nephrotoxic dose of mercuric chloride (1 mg kg^{-1}) on patterns of protein synthesis in the kidneys of male Sprague-Dawley rats. Enhanced *de novo* synthesis of 70 kDa and 90 kDa molecular mass proteins were detected as early as 2 h after exposure to inorganic mercury, and maximal increases in protein levels were observed at 4–8 h posttreatment. By 16 h postinjection, rates of synthesis of the stress proteins decreased back towards basal levels. Changes in protein expression also occurred in liver, but were of smaller magnitudes and were not observed until 16–24 h postinjection.

Goering et al.[114] concluded that alterations in expression of stress proteins precede overt renal injury and are target organ-specific, suggesting that they may serve as biomarkers of renal injury. Furthermore, once the biological function(s) of these proteins are identified, a more complete understanding of the early effects of mercury can be obtained.

**7.33.6 FACTORS THAT MODIFY THE
RENAL TOXICITY OF MERCURY**

**7.33.6.1 Influence of Intracellular Thiols on
the Renal Accumulation and Toxicity
of Mercury**

Two major intracellular thiols, GSH and metallothionein, appear to be important in regulating the renal accumulation of mercury and ultimately, the renal susceptibility to mercury-induced renal cellular injury. It is not clear whether other molecules within cells, including the large supply of nonmetallothionein protein sulfhydryls, can also be manipulated to alter renal cellular accumulation and toxicity of mercury.

Intracellular contents of GSH can be readily manipulated within a relatively brief time. Several investigators have employed diethyl maleate to conjugate GSH, thereby lowering the amount of intracellular GSH available to interact with mercury compounds. In two studies Berndt and colleagues,[20,115] and in another study Johnson,[116] showed that depletion of intracellular GSH or nonprotein thiols was accompanied by decreases in the renal accumulation of inorganic mercury in animals treated with mercuric chloride. In the studies by Berndt and colleagues,[20,115] depletion of intracellular GSH also resulted in increased severity of renal injury induced by treatment with mercuric chloride. Zalups and Lash[52] also found a close correlation between intrarenal concentrations of GSH and accumulation of inorganic mercury.

By contrast, Tanaka et al.[23] lowered intracellular content of GSH in the kidneys of mice by administration of buthionine sulfoximine, which is a potent inhibitor of the intracellular synthesis of GSH, and then inhibited degradation of GSH with the potent inhibitor of GGT, acivicin, and observed no changes in accumulation of either inorganic mercury or methylmercury, when compared with control animals. However, Girardi and Elias[93] reported increases in intrarenal accumulation of inorganic mercury in mice treated with diethyl maleate. A possible explanation for the discrepancy in results may involve patterns of urinary excretion of GSH. In studies where acivicin was used to inhibit GGT, Berndt *et al.*[20] showed in rats, and Tanaka *et al.*[23] showed in mice, that the urinary excretion of both GSH and inorganic mercury increased after inhibition of GSH-degradation. Tanaka-Kagawa *et al.*[24] also found that the urinary excretion of inorganic mercury increased while the renal accumulation of either inorganic mercury and methylmercury decreased.

In the study by Tanaka et al.,[23] some mice were also pretreated with 1,2-dichloro-4-nitrobenzene, which specifically depletes hepatic GSH, prior to injection of mercuric chloride. They found that prior depletion of hepatic GSH resulted in a marked reduction in the renal accumulation of mercury and a significant decrease in the level of renal cellular injury induced by inorganic mercury. These findings tend to suggest that hepatically synthesized GSH and the activity of GGT are involved in the renal uptake of mercury.

Increases in the intracellular contents of GSH and other nonprotein thiols can also be achieved by several means. Girardi and Elias[93,117] treated mice with *N*-acetylcysteine and found that intracellular accumulation of inorganic mercury was lower in both kidney and liver as compared with animals possessing normal intracellular thiol status. Inasmuch as transport of inorganic mercury with GSH has been established in liver[118] and higher intracellular contents of GSH would be expected to provide increased numbers of ligands for binding to inorganic mercury, the seemingly paradoxical results of Girardi and Elias[93] and the discrepancies described above suggest that the intrarenal and intracellular disposition of mercury-containing compounds must be regulated by a more complex array of factors than the availability of thiol ligands on GSH.

Additional experiments by Tanaka-Kagawa et al.,[24] involving changes in intracellular metallothionein levels, may provide some clarification of the contradictory reports of GSH-depletion and the accumulation of mercury. These investigators reported that induction of renal metallothionein with $Bi(NO_3)_3$ diminished the ability of acivicin to decrease intrarenal accumulation of either inorganic mercury or methylmercury. They interpreted this as indicating that inorganic mercury or methylmercury that is bound to ligands, other than metallothionein, in renal cells can be readily secreted into the tubular lumen with intracellular GSH. Other studies[83,88,92,119] have documented increases in intrarenal mercury accumulation and decreases in nephrotoxicity induced by either organic or inorganic mercury after induction of renal metallothionein. Thus, it appears that there is a complex interplay between protein and nonprotein thiols that ultimately determines the renal disposition to mercury.

**7.33.6.2 Modulation of Renal Disposition
and Toxicity of Mercury by
Extracellular Thiols**

While manipulation of intracellular thiols is sometimes employed therapeutically to alter the accumulation of mercury and to modulate effects of mercury once it enters target sites, administration of thiol-containing compounds can be applied prior to, or simultaneously with, mercury-containing compounds to alter the pharmacokinetics and pharmacodynamics of mercury. Both 2,3-dimercaptopropane-1-sulfonate (DMPS) and *meso*-2,3-dimercaptosuccinic acid (DMSA) are becoming two of the more commonly used metal chelators employed as antidotes for mercury poisoning, and their chemical and pharmacological properties have been reviewed.[120,121] Examples of some of their most distinguishing features are that in contrast to the earlier chelator dimercaprol (also known as British AntiLewisite or BAL), DMPS and DMSA are fairly nontoxic, are very water soluble, are not very lipid soluble, and are effective if administered orally. The two chemicals are quite versatile, being capable of chelating arsenic, lead, cadmium, and mercury. However, they differ in potency and specificity; DMSA is generally the more effective of the two in chelating organic mercury, whereas DMPS is the more effective of the two in chelating inorganic mercury.[120–122] Additional extracellular thiol reagents that have been employed clinically for removal of methylmercury are D-penicillamine and *N*-acetyl-DL-penicillamine.[120] Some of the reported variabilities in effectiveness and potency of the various chelators of mercury may be attributed to species differences, routes of administration, and doses of chelators given.

Zalups *et al.*[65] demonstrated dose-dependent protection in rats from the nephropathy induced by inorganic mercury. Their data suggest that the protective effects of DMPS may be attributed to decreases in the renal mercury burden and increases in urinary excretion of mercury. Furthermore, Maiorino *et al.*[123] demonstrated a high correlation between effectiveness of DMPS and urinary excretion of both inorganic mercury and DMPS in humans. In a study[122] the same dose of DMPS or DMSA, administered to rats 24 h after the animals were given an intravenous nontoxic dose of mercuric chloride, was shown to reduce the renal burden of mercury significantly during the subsequent 24 h after treatment with the respective chelator. Treatment with DMPS caused a reduction in the renal burden of mercury by more than 80%, while DMSA caused a reduction in the renal burden of mercury by about 50%. These findings indicate that DMPS is more effective (on a per mole basis) in reducing the renal burden of mercury after exposure to inorganic mercury. The kinetics involved in the rapid reduction of the renal burden of mercury after treatment with

DMPS or DMSA appear to indicate that transport of both of these chelating agents, by the epithelial cells along the proximal tubule, is involved in the reduction of the renal tubular burden of mercury. It is well established that both organic anions, such as sulfonates, and dicarboxylic acids, such as succinic acid, are transported by proximal tubular epithelial cells.

Additional support for the hypothesis that transport of DMPS and intracellular chelation of mercury occur along segments of the proximal tubule after treatment with DMPS comes from the study by Klotzbach and Diamond.[124] With isolated perfused kidneys from male Long-Evans rats, they showed that DMPS undergoes net tubular secretion by a kinetically saturable process that is inhibited by *p*-aminohippurate and probenecid. They also found that DMPS produced a dose-dependent decrease in the retention of inorganic mercury and an increase in urinary excretion of inorganic mercury. Furthermore, both effects were blocked by probenecid, suggesting that the mechanism of protection by DMPS is by chelation of inorganic mercury within the proximal tubular cell. Many investigators have observed that DMPS is readily oxidized in perfusates or in plasma to the disulfide form. To enable interaction with metals, DMPS is reduced back to the dithiol form within proximal tubular cells by a GSH-dependent thiol–disulfide exchange reaction.[124,125]

Other low molecular weight thiols have been employed experimentally to modulate the nephrotoxicity of mercury. Because of its prominence as the primary intracellular, non-protein thiol, exogenous GSH is a logical choice to try as a modulatory agent. Work by Jones and colleagues[126] has demonstrated that oral administration of GSH can significantly increase the content of GSH in the lung, kidney, heart, brain, small intestine, and skin, but not in liver, under conditions where GSH is depleted. This suggests that GSH taken orally may supplement cellular GSH in some tissues under certain toxicological or pathological conditions. A large body of data have been accumulated both *in vivo* and in *in vitro* systems showing that exogenous GSH can protect against mercury-induced renal injury. Zalups *et al.*[55] perfused isolated rabbit proximal tubules with 18.4 μM mercuric chloride and various thiols, including GSH or cysteine. Both thiols, present in the perfusate at fourfold higher concentrations than inorganic mercury, either prevented or significantly decreased the extent of tubular injury. An ultrafiltrate of rabbit plasma was similarly protective. The mechanism of protection by GSH, cysteine, or plasma ultrafiltrate appeared to involve decreased movement of inorganic mercury across the luminal membrane, as measured by the disappearance flux of inorganic mercury from the lumen, was significantly decreased in the presence of these thiols or thiol-containing compounds. Houser et al.[30] administered GSH monoethyl ester to rats and found that both renal cortical accumulation of inorganic mercury and the severity of mercury-induced renal injury were diminished.

The protective effects of exogenous GSH and DMPS have also been demonstrated in suspensions of isolated proximal tubular cells from rats.[95] They incubated proximal tubular cells for 15 min in an extracellular buffer containing bovine serum albumin and various concentrations of GSH or DMPS. Cells were then incubated for an additional 1 h in the presence of 250 µM mercuric chloride, which was found to be the threshold concentration of inorganic mercury that produced cellular injury under these incubation conditions. GSH provided concentration-dependent protection from mercury-induced cytotoxicity, as assessed by decreases in total cellular lactate dehydrogenase activity. A GSH concentration of 500 µM, or twice that of inorganic mercury, was required to completely protect the proximal tubular cells. DMPS, in contrast, provided complete protection against 250 µM mercuric chloride at a concentration (175 µM) that was less than that of inorganic mercury. This difference in behavior between GSH and DMPS most likely arises from differences in chemistry and renal handling of the two compounds.

In contrast to the results described above, Tanaka et al.[23] found that coadministration of GSH and mercuric chloride to mice caused the renal content of mercury to increase relative to that in mice that were given mercuric chloride alone. These investigators concluded that transport of inorganic mercury to the kidney may occur as a mercury–GSH complex and that the simultaneous presence of GSH enhances uptake of mercury. Zalups and Barfuss[127,128] have observed similar effects in rats coadministered a nontoxic dose of inorganic mercury with GSH or cysteine. Consistent with these findings, Miller and Woods[129] showed that complexes of GSH and mercury +2 or GSSG and Hg$^+$ promoted uroporphyrinogen oxidation and catalyzed decomposition of hydrogen peroxide, indicating that mercury–GSH (or other thiol) complexes likely contribute to mercury-induced toxicity. Some of these results have also been confirmed in the laboratory of Zalups et al. (unpublished data). There is data for rats indicating that when a toxic 2.0 µmol kg^{-1} dose of mercuric chloride is coadministered with a 2:1 mole ratio of GSH or cysteine, the nephropathy induced by the inorganic mercury is made more severe. Resolution of the marked contrast between these findings and those described above will require a detailed mechanistic description of the renal transport of inorganic mercury. Although advances have been made in the understanding of mechanisms of renal transport of mercury, the role of thiols in the renal cellular uptake of mercury is still unclear.

Although DMPS and DMSA are dithiols and are highly effective protective agents against mercury-induced injury, less definitive results have been obtained with two other dithiols, dithioerythritol and dithiothreitol. Barnes et al.[130] observed, in rats, evidence of protection against morphologic lesions and losses of activities of key marker enzymes for plasma membrane and mitochondria induced by mercury with dithiothreitol. Weinberg et al.[99] provided evidence of protection for isolated renal mitochondria from mercuric chloride-induced dysfunction by dithioerythritol, but only if the dithiol was added *in vitro* simultaneously with mercuric chloride; when the dithiol agent was added *in vitro* after the rats had been treated with mercuric chloride *in vivo*, no protection or reversal of toxicity was observed. To complicate further the understanding of how dithiols interact with mercury-containing compounds in biological systems, Chavez and Holguin[101] reported that addition of dithiothreitol to renal mitochondria isolated from the rat that had been treated with inorganic mercury actually increased the degree of mitochondrial injury induced by mercury. They suggested that the dithiol made additional sulfhydryl-sensitive sites available for interaction with mercury, thereby enhancing the toxic response. In the same study,[101] the investigators also reported that the monothiol 2-mercaptoethanol also enhanced mercuric chloride-induced mitochondrial injury, although higher concentrations than those of the dithiol were required to reproduce the effect.

Chavez et al.[102] have also reported that the angiotensin converting enzyme inhibitor captopril (1-(3-mercapto-2-methyl-1-oxopropyl)-1-proline) was an effective protective agent both *in vivo* and *in vitro* against mercuric chloride-induced mitochondrial injury and morphological damage.

Besides low-molecular weight thiols, administration of inorganic mercury complexed to the small sulfhydryl-containing protein metallothionein, although not providing protection against the toxicity induced by inorganic mercury, altered the renal site of injury.[131] Whereas the primary target of renal injury due to mercuric chloride are the pars recta (S2, S3) segments of the proximal tubule, the primary target of renal injury due to mercury–metallothionein appears to be the pars convoluta and early pars recta (S1, S2) segments of the proximal tubule. Intrarenal accumulation and urinary excretion of inorganic mercury in rats was also greater when mercury was administered with metallothionein than when mercury was administered by itself.[132]

7.33.6.3 Effects of Reduced Nephron Number and Compensatory Renal Growth on the Renal Disposition and Toxicity of Mercury

Reduction in the number of functioning nephrons, which can occur as a consequence of aging, renal disease, or surgical removal of renal tissue, has profound effects on renal cellular function and, consequently, on the renal handling of exogenous chemicals and on the susceptibility of renal tissue to chemically-induced injury.[133] The remnant renal tissue undergoes compensatory growth, which is predominantly (i.e., >85%) due to cellular hypertrophy, particularly in segments of the proximal tubule. One of the more prominent changes in renal function that occurs as a result of compensatory hypertrophy includes marked increases in mitochondrial metabolism, which may lead to an enhanced susceptibility of renal tissue to oxidative stress.[134]

Numerous animal studies have shown that rats that have undergone a reduction in renal mass, such as unilateral nephrectomy, are more susceptible to the nephropathy induced by inorganic mercury.[30,41,52,135] The biochemical changes that occur as a consequence of reduced renal mass and compensatory hypertrophy are retained *in vitro* when proximal tubular cells are isolated from rats.[95,96] Furthermore, the enhanced susceptibility of rats to the nephro-toxic effects of inorganic mercury is also retained *in vitro*. In the absence of exogenous thiols in the extracellular incubation medium, isolated proximal tubular cells from unilaterally nephrectomized rats in which compensatory renal growth has occurred (NPX) exhibit irreversible cellular injury at significantly lower concentrations of mercuric chloride than isolated proximal tubular cells from sham-operated (SHAM) rats.

Although the mechanism(s) for the enhanced susceptibility of proximal tubular cells from NPX rats to injury induced by mercury are not well characterized, it appears that enhanced accumulation of mercury is a contributing factor. Findings from studies with both mercuric chloride[11,35,40,52,117,136] and methylmercuric chloride[18] indicate that greater amounts of mercury, on a per gram tissue basis, accumulate in the remnant kidney of NPX rats than in the kidneys of SHAM or control rats. Moreover, the findings indicate that greatest increase in the accumulation of mercury occurs in the outer stripe of the outer medulla, and specifically in pars recta segments of proximal tubules,[11] which coincides with the site where the toxicity of mercury is expressed in the kidney. Other factors, such as changes in intrarenal handling of mercury, are also probably involved in changing the cellular response to mercury exposure. Some of the altered accumulation of mercury that occurs in the remnant kidney is probably related to alteration in the renal concentrations of intracellular thiols. Other findings show that both the intracellular metabolism of GSH[52] and metallothionein[83,119] are altered significantly after renal mass is reduced following unilateral nephrectomy and compensatory renal growth.

7.33.7 CONCLUSION

Despite the fact that a great deal of information has been gathered regarding the renal tubular handling and toxicity of the various forms of mercury, much more experimental research is needed to gain a better understanding of the precise mechanisms involved in the transport of mercury along the nephron, and the specific mechanisms involved in the nephropathy induced by the different forms of mercury.

7.33.8 REFERENCES

1. ATSDR, 'Toxicological Profile for Mercury.' US Department of Health and Human Services, Public Health Service, Agency for Toxic Substance and Disease Registry, 1994, TP-93/10.
2. WHO, 'Environmental Health Criteria 118: Inorganic Mercury.' World Health Organization, Geneva, 1991.
3. W. F. Fitzgerald and T. W. Clarkson, 'Mercury and monomethylmercury: present and future concerns.' *Environ. Health Perspect.*, 1991, **96**, 159–166.
4. R. K. Zalups and L. H. Lash, 'Advances in understanding the renal transport and toxicity of mercury.' *J. Toxicol. Environ. Health*, 1994, **42**, 1–44.
5. R. K. Zalups, 'Early aspects of the intrarenal distribution of mercury after the intravenous administration of mercuric chloride.' *Toxicology*, 1993, **79**, 215–228.
6. P. Hultman and S. Enestrom, 'Localization of mercury in the kidney during experimental acute tubular necrosis studied by the cytochemical Silver Amplification method.' *Br. J. Exp. Pathol.*, 1986, **67**, 493–503.
7. P. Hultman and S. Enestrom, 'Dose-response studies in murine mercury-induced autoimmunity and immune-complex disease.' *Toxicol. Appl. Pharmacol.*, 1992, **113**, 199–208.

8. L. Magos, A. W. Brown, S. Sparrow *et al.*, 'The comparative toxicology of ethyl- and methyl-mercury.' *Arch. Toxicol.*, 1985, **57**, 260–267.

9. P. M. Rodier, B. Kates and R. Simons, 'Mercury localization in mouse kidney over time: autoradiography vs. silver staining.' *Toxicol. Appl. Pharmacol.*, 1988, **92**, 235–245.

10. R. Taugner, K. Winkel and J. Iravani, 'Lokalization der sublimatanreicerung in der rattenneire.' *Virchows Arch. Pathol. Anat. Physiol.*, 1966, **340**, 369–383.

11. R. K. Zalups, 'Autometallographic localization of inorganic mercury in the kidneys of rats: Effect of unilateral nephrectomy and compensatory renal growth.' *Exp. Mol. Pathol.*, 1991, **54**, 10–21.

12. R. K. Zalups, 'Method for studying the *in vivo* accumulation of inorganic mercury in segments of the nephron in the kidneys of rats treated with mercuric chloride.' *J. Pharmacol. Meth.*, 1991, **26**, 89–104.

13. R. K. Zalups and D. W. Barfuss, 'Accumulation of inorganic mercury along the renal proximal tubule of the rabbit.' *Toxicol. Appl. Pharmacol.*, 1990, **106**, 245–253.

14. J. C. Gage, 'Distribution and excretion of methyl and phenylmercury salts.' *Br. J. Ind. Med.*, 1964, **21**, 197–202.

15. T. Norseth and T. W. Clarkson, 'Studies on the biotransformation of ^{203}Hg-labeled methylmercury chloride in rats.' *Arch. Environ. Health*, 1970, **21**, 717–727.

16. T. Norseth and T. W. Clarkson, 'Biotransformation of methylmercury salts in the rat studied by specific determination of inorganic mercury.' *Biochem. Pharmacol.*, 1970, **19**, 2775–2783.

17. S. Omata, M. Sato, K. Sakimura *et al.*, 'Time-dependent accumulation of inorganic mercury in subcellular fractions of kidney, liver, and brain or rats exposed to methylmercury.' *Arch. Toxicol.*, 1980, **44**, 231–241.

18. R. K. Zalups, D. W. Barfuss and P. J. Kostyniak, 'Altered intrarenal accumulation of mercury in uninephrectomized rats treated with methylmercury chloride.' *Toxicol. Appl. Pharmacol.*, 1992, **115**, 174–182.

19. J. D. Dunn and T. W. Clarkson, 'Does mercury exhalation signal demethylation of methylmercury?' *Health Phys.*, 1980, **38**, 411–414.

20. W. O. Berndt, J. M. Baggett, A. Blacker *et al.*, 'Renal gluthathione and mercury uptake by kidney.' *Fundam. Appl. Toxicol.*, 1985, **5**, 832–839.

21. J. de Ceaurriz, J. Payan, G. Morel *et al.*, 'Role of extracellular glutathione and γ-glutamyltranspeptidase in the dispostion and kidney toxicity of inorganic mercury in rats.' *J. Appl. Toxicol.*, 1994, **14**, 201–206.

22. A. Naganuma, N. Oda-Urano, T. Tanaka *et al.*, 'Possible role of hepatic glutathione in transport of methylmercury into mouse kidney.' *Biochem. Pharmacol.*, 1988, **37**, 291–296.

23. T. Tanaka, A. Naganuma and N. Imura, 'Role of γ-glutamyltranspeptidase in renal uptake and toxicity of inorganic mercury in mice.' *Toxicology*, 1990, **60**, 187–198.

24. T. Tanaka-Kagawa, A. Naganuma and N. Imura, 'Tubular secretion and reabsorption of mercury compounds in mouse kidney.' *J. Pharmacol. Exp. Ther.*, 1993, **264**, 776–782.

25. R. K. Zalups, 'Organic anion transport and action of γ-glutamyltranspeptidase in kidney linked mechanistically to renal tubular uptake of inorganic mercury.' *Toxicol. Appl. Pharmacol.*, 1995, **132**, 289–298.

26. R. K. Zalups and K. H. Minor, 'Luminal and basolateral mechanisms involved in the renal tubular uptake of inorganic mercury.' *J. Toxicol. Environ Health*, 1995, **46**, 73–100.

27. T. Tanaka, A. Naganuma and N. Imura, 'Routes for renal transport of methylmercury in mice.' *Eur. J. Pharmacol.*, 1992, **228**, 9–14.

28. R. K. Zalups, and D. W. Barfuss, 'Pretreatment with *p*-aminohippurate inhibits the renal uptake and accumulation of injected inorganic mercury in the rat.' *Toxicology*, 1995, **103**, 23–35.

29. J. M. Baggett and W. O. Berndt, 'The effect of potassium dichromate and mercuric chloride on urinary excretion and organ and subcellular distribution of (^{203}Hg) mercuric chloride in rats.' *Toxicol. Lett.*, 1985, **29**, 115–121.

30. M. T. Houser and W. O. Berndt, 'Unilateral nephrectomy in the rat: effects on mercury handling and renal cortical subcellular distribution.' *Toxicol. Appl. Pharmacol.*, 1988. **93**, 187–194.

31. K. M. Madsen, 'Mercury accumulation in kidney lysosomes on proteinuric rats.' *Kidney Int.*, 1980, **18**, 445–453.

32. K. M. Madsen and J. C. Hansen, 'Subcellular distribution of mercury in the rat kidney cortex after exposure to mercuric chloride.' *Toxicol. Appl. Pharmacol.*, 1980, **54**, 443–453

33. N. Ballatoriand and T. W. Clarkson, 'Billiary secretion of glutathione and of glutathione–metal complexes.' *Fundam. Appl. Toxicol.*, 1985, **5**, 816–831.

34. A. Rothstein and A. D. Hayes, 'The metabolism of mercury in the rat studied by isotope techniques.' *J. Pharmacol. Exp. Ther.*, 1960, **130**, 166–176.

35. R. K. Zalups, J. M. Klotzbach and G. L. Diamond, 'Enhanced accumulation of inorganic mercury in renal outer medullar after unilateral nephrectomy.' *Toxicol. Appl. Pharmacol.*, 1987, **89**, 226–236.

36. J. C. Smith, P. V. Allen, M. D. Turner *et al.*, 'The kinetics of intravenously administered methylmercry in man.' *Toxicol. Appl. Pharmacol.*, 1994, **128**, 251–256.

37. L. Magos and T. W. Clarkson, in 'Handbook of Physiology,' Section 9, Renal Physiology, ed. D. H. K. Lee, American Physiological Society, Bethesda, MD, 1977, pp. 503–512.

38. L. W. Chang, R. A. Ware and P. A. Desnoyers, 'A histochemical study on some enzymes changes in the kidney, liver and brain after chronic mercury intoxication in the rat.' *Food Cosmet. Toxicol.*, 1973, **11**, 283–286

39. S. I. McNeil, M. K. Bhatnagar and C. J. Turner, 'Combined toxicity of ethanol and methylmercury in rat.' *Toxicology*, 1988, **53**, 345–363.

40. R. K. Zalups and G. L. Diamond, 'Mercuric chloride-induced nephrotoxicity in the rat following unilateral nephrectomy and compensatory renal growth.' *Virchows Arch. B Cell Pathol. Incl. Mol. Pathol.*, 1987, **53**, 336–346.

41. R. K. Zalups, C. Cox and G. L. Diamond, in 'Biological Monitoring of Toxic Metals,' eds. T. W. Clarkson, L. Friberg, G. F. Nordberg *et al.*, Plenum, New York, 1988, pp. 531–545.

42. K. Kostial, D. Kello, S. Jugo *et al.*, 'Influence of age on metal metabolism and toxicity.' *Environ. Health Perspect.*, 1978, **25**, 81–86.

43. M. N. Gleason, R. E. Gosselin and D. C. Hodge, 'Clinical Toxicology of Commercial Products,' Williams and Wilkins, Baltimore, 1957, p. 154.

44. P. Troen, S. A. Kaufman and K. H. Katz, 'Mercuric bichloride poisoning.' *New Engl. J. Med.*, 1951, **244**, 459–463.

45. F. E. Cuppage and A. Tate, 'Repair of the nephron following injury with mercuric chloride.' *Am. J. Pathol.*, 1967, **51**, 405–429.

46. B. A. Fowler, 'The morphological effects of dieldrin and methylmercuric chloride on pars recta segments of the rat kidney proximal tubules.' *Am. J. Pathol.*, 1972, **69**, 163–178.

47. C. E. Ganote, K. A. Reimer and R. B. Jennings, 'Acute mercuric chloride nephrotoxicity. An electron microscopic and metabolic study.' *Lab. Invest.*, 1974, **31**, 633–647.

48. T. L. Gritzka, and B. F. Trump, 'Renal tubular lesions caused by mercuric chloride. Electron microscopic observations: degeneration of the pars recta.' *Am. J. Pathol.*, 1968, **52**, 1225–1277.

49. R. Klein, S. P. Herman, B. C. Bullock *et al.*, 'Methylmercury intoxication in rat kidneys. Functional acid pathological changes.' *Arch. Pathol.*, 1973, **96**, 83–90.

50. E. M. McDowell, R. B. Nagle, R. C. Zalme *et al.*, 'Studies on the pathophysiology of acute renal failure. I. Correlation of ultrastructure and function in the proximal tubule of the rat following administration of mercuric chloride.' *Virchows Arch. B. Cell Pathol.*, 1976, **22**, 173–196.

51. R. C. Zalme, E. M. McDowell, R. B. Nagle *et al.*, 'Studies on the pathophysiology of acute renal failure. II. A histochemical study of the proximal tubule of the rat following administration of mercuric chloride.' *Virchows Arch. B. Cell Pathol.*, 1976, **22**, 197–216.

52. R. K. Zalups and L. H. Lash, 'Effects of uninephrectomy and mercuric chloride on renal glutathione homeostasis.' *J. Pharmacol. Exp. Ther.*, 1990, **254**, 962–970.

53. D. W. Barfuss, M. K. Robinson and R. K. Zalups, 'Inorganic mercury transport in the proximal tubule of the rabbit.' *J. Am. Soc. Nephrol.*, 1990, **1**, 910–917.

54. R. K. Zalups and D. W. Barfuss, 'Transport and toxicity of methylmercury along the proximal tubule of the rabbit.' *Toxicol. Appl. Pharmacol.*, 1993, **121**, 176–185.

55. R. K. Zalups, M. K. Robinson and D. W. Barfuss, 'Factors affecting inorganic mercury transport and toxicity in the isolated perfused proximal tubule.' *J. Am. Soc. Nephrol.*, 1991, **2**, 866–878.

56. M. D. Aleo, M. L. Taub and P. J. Kostyniak, 'Primary cultures of rabbit renal proximal tubule cells. III. Comparative cytotoxicity of inorganic and organic mercury.' *Toxicol. Appl. Pharmacol.*, 1992, **112**, 310–317.

57. J. P. Buchet, H. Roels, A. Bernard *et al.*, 'Assessment of renal function of workers exposed to inorganic lead, cadmium or mercury vapor.' *J. Occup. Med.*, 1980, **22**, 741–750.

58. B. G. Ellis, R. G. Price and J. C. Topham, 'The effect of tubular damage by mercuric chloride on kidney function and some urinary enzymes in the dog.' *Chem. Biol. Interact.*, 1973, **7**, 101–113.

59. C. A. Gottelli, E. Astolfi, C. Cox *et al.*, 'Early biochemical effects of an organic mercury fungicide on infants: dose makes the poison.' *Science*, 1985, **227**, 638–640.

60. B. B. Kirschbaum, 'Alanine aminopeptidase excretion after mercuric chloride renal failure.' *Biochem. Med.*, 1979, **21**, 220–225.

61. F. Planas-Bohne, 'The effect of mercuric chloride on the excretion of two urinary enzymes in the rat.' *Arch. Toxicol.*, 1977, **37**, 219–225.

62. R. G. Price, 'Urinary enzymes, nephrotoxicity and renal disease.' *Toxicology*, 1982, **23**, 99–134.

63. M. D. Stonard, B. V. Chater, D. P. Duffield *et al.*, 'An evaluation of renal function in workers occupationally exposed to mercury vapour.' *Int. Arch. Occup. Environ. Health*, 1983, **52**, 177–189.

64. W. E. Stroo and J. B. Hook, 'Enzymes of renal origin in urine as indicators of nephrotoxicity.' *Toxicol. Appl. Pharmacol.*, 1977, **39**, 423–434.

65. R. K. Zalups, R. M. Gelein and E. Cernichiari, 'DMPS as a rescue agent for the nepphropathy induced by mercuric chloride.' *J. Pharmacol. Exp. Ther.*, 1991, **256**, 1–10.

66. G. L. Diamond, in 'Biological Monitoring of Toxic Metals,' eds. T. W. Clarkson, L. Friberg, G. F. Nordberg *et al.*, Plenum, New York, 1988, pp. 515–529.

67. A. Cardenas, H. Roels, A. M. Bernard *et al.*, 'Markers of early renal changes induced by industrial pollutants. I. Application to workers exposed to mercury vapour.' *Br. J. Ind. Med.*, 1993, **50**, 17–27.

68. T. W. Clarkson and L. Magos, 'The effect of sodium maleate on the renal disposition and excretion of mercury.' *Br. J. Pharmacol. Chemother.*, 1967, **31**, 560–567.

69. L. Magos and T. Stoychev, 'Combined effect of sodium maleate and some thiol compounds on mercury excretion and redistribution in rats.' *Br. J. Pharmacol.*, 1969, **35**, 121–126.

70. B. Trojanowska, J. K. Piotrowski and S. Szendzikowski, 'The influence of thioacetamide on the excretion of mercury in rats.' *Toxicol. Appl. Pharmacol.*, 1971, **18**, 374–386.

71. R. L. Barenberg, S. Solomon, S. Papper *et al.*, 'Clearance and micropuncture study of renal function in mercuric chloride treated rats.' *J. Lab. Clin. Med.*, 1968, **72**, 473–484.

72. R. K. Zalups, 'Effect of dietary K$^+$ and 75% nephrectomy on the morphology of principal cells in CCDs.' *Am. J. Physiol.*, 1989, **257**, F387–F396.

73. R. K. Zalups and D. A. Henderson, 'Cellular morphology in outer medullary collecting duct: effect of 75% nephrectomy and K$^+$ deplation.' *Am. J. Physiol.*, 1992, **263**, F1119–F1127.

74. R. K. Zalups, B. A. Stanton, J. B. Wade *et al.*, 'Structural adaptation in initial collecting tubule following reduction in renal mass.' *Kidney Int.*, 1985, **27**, 636–642.

75. P. E. Bigazzi, 'Autoimmunity induced by chemicals.' *J. Toxicol. Clin. Toxicol.*, 1988, **26**, 125–156.

76. P. E. Bigazzi, 'Lessons from animal models: the scope of mercury-induced autoimmunity.' *Clin. Immunol. Immunopathol.*, 1992, **65**, 81–84.

77. P. Druet, E. Druet, F. Potdevin *et al.*, 'Immune type glomerulonephritis induced by HgCl$_2$ in the Brown-Norway rat.' *Ann. Immunol. (Paris)*, 1978, **129C**, 777–792.

78. A. A. Roman-Franco, M. Turiello, B. Albini, *et al.*, 'Antibasement membrane antibodies and antigen–antibody complexes in rabbits injected with mercuric chloride.' *Clin. Immunol. Immunopathol.*, 1978, **9**, 464–481.

79. C. Sapin, E. Dreut and P. Dreut, 'Induction of anti-glomerular basement membrane antibodies in the brown Norway rat by mercuric chloride.' *Clin. Exp. Immunol.*, 1977, **28**, 173–179.

80. S. Enestrom and P. Hultman, 'Immune-mediated glomerulonephritis induced by mercuric chloride in mice.' *Experientia*, 1984, **40**, 1234–1240.

81. P. Hultman, S. Enestrom and H. von Schenck, 'Renal handling of inorganic mercury in mice. The early excretion phase following a single intravenous injection of mercuric chloride studied by the Silver Amplification method.' *Virchows Arch. B. Cell Pathol. Incl. Mol. Pathol.*, 1985, **49**, 209–224.

82. J. K. Piotrowski, B. Trojanowska, J. M. Wisniewska-Knypl *et al.*, 'Mercury binding in the kidney and liver of rats repeatedly exposed to mercuric chloride: induction of metallothionein by mercury and cadmium.' *Toxicol. Appl. Pharmacol.*, 1974, **27**, 11–19.

83. R. K. Zalups and M. G. Cherian, 'Renal metallothionein metabolism after a reduction of renal mass. I. Effect of unilateral nephrectomy and compensatory growth on basal and metal-induced renal metallothionein metabolism.' *Toxicology*, 1992, **71**, 83–102.

84. M. G. Cherian and T. W. Clarkson, 'Biochemical changes in rat kidney on exposure to elemental mercury vapor: effect on biosynthesis of metallothionein.' *Chem. Biol. Interact.*, 1976, **12**, 109–120.

85. J. S. Woods, H. A. Davis and R. P. Baer, 'Enhancement of γ-glutamylcysteine synthetase mRNA in rat kidney and methylmercury.' *Arch. Biochem. Biophys.*, 1992, **296**, 350–353.

86. M. D. Aleo, M. L. Taub, J. R. Olson *et al.*, in '*In Vitro* Toxicology: Approaches to Validation,' ed. A. M. Goldberg, Mary Ann Liebert, New York, 1987, pp. 211–226.

87. A. S. Chung, M. D. Maines and W. A. Reynolds, 'Inhibition of the enzymes of glutathione metabolism by mercuric chloride in the rat kidney: reversal by selenium.' *Biochem. Pharmacol.*, 1982, **31**, 3093–3100.

88. H. Fukino, M. Hirai, Y. M. Hsueh *et al.*, 'Mechanismn of of protection by zinc against mercuric chloride toxicity in rats: effects of zinc and mercury on glutathione metabolism.' *J. Toxicol. Environ. Health*, 1986, **19**, 75–89.

89. C. P. Siegers, M. Schenke and M. Younes, 'Influence of cadmium chloride, mercuric chloride, and sodium vanadate on the glutathione-conjugating enzyme system in liver, kidney, and brain of mice.' *J. Toxicol. Environ. Health*, 1987, **22**, 141–148.

90. R. K. Zalups and J. C. Veltman, 'Renal glutathione homeostasis in compensatory renal growth.' *Life Sci.*, 1988, **42**, 2171–2176.

91. S. Addya, K. Chakravarti, A. Basu *et al.*, 'Effects of mercuric chloride on several scavenging enzymes in rat kidney and influence of vitamin E supplementation.' *Acta Vitaminol. Enzymol.*, 1984, **6**, 103–107.

92. H. Fukino, M. Hirai, Y. M. Hsueh *et al.*, 'Effect of zinc pretreatment on mercuric chloride-induced lipid peroxidation in the rat kidney.' *Toxicol. Appl. Pharmacol.*, 1984, **73**, 395–401.

93. G. Girardi and M. M. Elias, 'Effectiveness of *N*-acetylcysteine in protecting against mercuric chloride-induced nephrotoxicity.' *Toxicology*, 1991, **67**, 155–164.

94. G. Gstraunthaler, W. Pfaller and P. Kotanko, 'Glutathione depletion and *in vitro* lipid peroxidation in mercury or maleate-induced acute renal failure.' *Biochem. Pharmacol.*, 1983, **32**, 2969–2972.

95. L. H. Lash and R. K. Zalups, 'Mercuric chloride-induced cytotoxicity and compensatory hypertrophy in rat kidney proximal tubular cells.' *J. Pharmacol. Exp. Ther.*, 1992, **261**, 819–829.

96. L. H. Lash and R. K. Zalups, 'Activities of enzymes involved in renal cellular glutathione metabolism after uninephrectomy in the rat.' *Arch. Biochem. Biophys.*, 1994, **309**, 129–138.

97. B. O. Lund, D. M. Miller and J. S. Woods, 'Mercury-induced H_2O_2 production and lipid peroxidation *in vitro* in rat kidney mitochondria.' *Biochem. Pharmacol.*, 1991, **42**, Suppl., S181–S187.

98. B. O. Lund, D. M.Miller and J. S. Woods, 'Studies on Hg(II)-induced H_2O_2 formation and oxidative stress *in vivo* and *in vitro* in rat kidney mitochondria.' *Biochem. Pharmacol.*, 1993, **45**, 2017–2024.

99. J. M. Weinberg, P. G. Harding and H. D. Humes, 'Mitochondrial bioenergetics during the initiation of mercuric chloride-induced renal injury. I. Direct effects of *in vitro* mercuric chloride on renal mitochondrial function.' *J. Biol. Chem.*, 1982, **257**, 60–67.

100. J. M. Weinberg, P. G. Harding and H. D. Humes, 'Mitochondrial bioenergetics during the initiation of mercuric chloride-induced renal injury. II. Functional alterations of renal cortical mitochondria isolated after mercuric chloride treatment.' *J. Biol. Chem.*, 1982, **257**, 68–74.

101. E. Chavez and J. A. Holguin, 'Mitochondrial calcium release induced by Hg^{2+}.' *J. Biol. Chem.*, 1988, **263**, 3582–3587.

102. E. Chavez, C. Zazueta, A. Osornio *et al.*, 'Proetective behavior of captopril on Hg^{2+}-induced toxicity on kidney mitochondria. *In vivo* and *in vitro* experiments.' *J. Pharmacol. Exp. Ther.*, 1991, **256**, 385–390.

103. K. Y. Jung, S. Uchida and H. Endou, 'Nephrotoxicity assessment by measuring cellular ATP content. I. Substrate specifiicities in the maintenance of ATP content in isolated rat nephron segments.' *Toxicol. Appl. Pharmacol.*, 1989, **100**, 369–382.

104. R. K. Zalups, K. L. Knutson and R. G. Schnellmann, '*In vitro* analysis of the accumulation and toxicity of inorganic mercury in segments of the promal tubule isolated from the rabbit kidney.' *Toxicol. Appl. Pharmacol.*, 1993, **119**, 221–227.

105. M. W. Smith, I. S. Ambudkar, P. C. Phelps *et al.*, '$HgCl_2$-induced changes in cytosolic Ca^{2+} of cultured rabbit renal tubular cells.' *Biochim. Biophys. Acta*, 1987, **931**, 130–142.

106. A. Rothstein, 'Sulfhydryl groups in membrane stucture and function.' *Curr. Top. Membr. Transport*, 1970, **1**, 135–176.

107. B. M. Anner and M. Moosmayer, 'Mercury inhibits Na-K-ATPase primarily at the cytosolic side.' *Am. J. Physiol.*, 1992, **262**, F843–F842.

108. B. M. Anner M. Moosmayer and E. Imesch, 'Mercury blocks Na-K-ATPase by a ligand-dependent and reversible mechanism.' *Am. J. Physiol.*, 1992, **262**, F830–F836.

109. E. Imesch, M. Moosmayer and B. M. Anner, 'Mercury weakens membrane anchoring of Na-K-ATPase.' *Am. J. Physiol.*, 1992, **262**, F837–F842.

110. J. S. Woods, C. A. Calas, L. D. Aicher *et al.*, 'Stimulation of porphyrinogen oxidation by mercuric ion. I. Evidence of free radical formation in the presence of thiols and hydrogen peroxide.' *Mol. Pharmacol.*, 1990, **38**, 253–260.

111. J. S. Woods, C. A. Calas, L. D. Aicher, 'Stimulation of porphyrinogen oxidation by mercuric ion. II. Promotion of oxidation from the interaction of mercuric ion, glutathione, and mitochondria-generated hydrogen peroxide.' *Mol. Pharmacol.*, 1990, **38**, 261–266.

112. M. A. Bowers, L. D. Aicher, H. A. Davis *et al.*, 'Quantirtative determination of porphyrins in rat and human urine and evaluation of urinary porphyrin profiles during mercury and lead exposures.' *J. Lab. Clin. Med.*, 1992, **120**, 272–281.

113. J. S. Woods, M. D. Martin, C. A. Naleway *et al.*, 'Urinary porphyrin profiles as a biomarker of mercury exposure: studies on dentists with occupational exposure to mercury vapor.' *J. Toxicol. Environ. Health*, 1993, **40**, 239–250.

114. P. L. Goering, B. R. Fisher, P. P. Chaudhary *et al.*, 'Relationship between stress protein induction in rat kidney by mercuric chloride and nephrotoxicity.' *Toxicol. Appl. Pharmacol.*, 1992, **113**, 184–191.

115. J. M. Baggett and W. O. Berndt, 'The effect of depletion of nonprotein sulfhydryls by diethyl maleate plus buthionine sulfoximine on renal uptake of mercury in the rat.' *Toxicol. Appl. Pharmacol.*, 1986, **83**, 556–562.

116. D. R. Johnson, 'Role of renal cortical sulfhydryl groups in development of mercury-induced renal toxicity.' *J. Toxicol. Environ. Health*, 1982, **9**, 119–126.

117. G. Girardi and M. M. Elias, 'Effect of different renal glutathione levels on renal mercury disposition and excretion in the rat.' *Toxicology*, 1993, **81**, 57–67.

118. N. Ballatori, 'Mechanisms of metal transport across liver cell plasma membranes.' *Drug Metab. Rev.*, 1991, **23**, 83–132.

119. R. K. Zalups and M. G. Cherian, 'Renal metallothionein metabolism after a reduction of renal mass. II. Effect of zinc pretreatment on the renal toxicity and intrarenal accumulation of inorganic mercury.' *Toxicology*, 1992, **71**, 103–117.

120. H. V. Aposhian, 'DMSA and DMPS-water soluble antidotes for heavy metal poisoning.' *Annu. Rev. Pharmacol. Toxicol.*, 1983, **23**, 193–215.

121. H. V. Aposhian and M. M. Aposhian, '*meso*-2,3-Dimercaptosuccinic acid: chemical and toxicoological properties of an orally effective metal chelating agent.' *Annu. Rev. Pharmacol. Toxicol.*, 1990, **30**, 279–306.

122. R. K. Zalups, 'Influence of 2,3-dimercaptopropane-1-sulfonate (DMPS) and *meso*-2,3-dimercaptosuccinic acid (DMSA) on the renal disposition of mercury in normal and uninephrectomized rats exposed to inorganic mercury.' *J. Pharmacol. Exp. Ther.*, 1993, **267**, 791–800.

123. R. M. Maiorino, R. C. Dart, D. E. Carter *et al.*, 'Determination and metabolism of dithiol chelating agents. XII. Metabolism and pharmacokinetics of sodium 2,3-dimercaptopropane-1-sulfonate in humans.' *J. Pharmacol. Exp. Ther.*, 1991, **259**, 808–814.

124. J. M. Klotzbach and G. L. Diamond, 'Complexing activity and excretion of 2,3-dimercapto-1-propane sulfonate in rat kidney.' *Am. J. Physiol.*, 1988, **254**, F871–F878.

125. J. R. Stewart and G. L. Diamond, '*In vivo* tubular secretion and metabolism of the disulfide of 2,3-dimercaptopropane-1-sulfonate.' *Drug Metab. Dispos.*, 1988, **16**, 189–195.

126. T. Y. Aw, G. Wierzbicka and D. P. Jones, 'Oral glutathione increases tissue glutathione *in vivo*.' *Chem. Biol. Interact.*, 1991, **80**, 89–97.

127. R. K. Zalups and D. W. Barfuss, 'Accumulation and handling of inorganic mercury in the kidney after co-administration with glutathione.' *J. Toxicol. Environ. Health*, 1995, **44**, 385–399.

128. R. K. Zalups and D. W. Barfuss, 'Renal disposition of mercury in rats after intravenous injection of inorganic mercury and cysteine.' *J. Toxicol. Environ. Health*, 1995, **44**, 401–413.

129. D. W. Miller and J. S. Woods, 'Reox activities of mercury-thiol complexes: implications for mercury-induced porphyria and toxicity.' *Chem. Biol. Interact.*, 1993, **88**, 23–35.

130. J. L. Barnes, E. M. McDowell, J. S. McNeil *et al.*, 'Studies on the pathophysiology of acute renal failure. IV. Protective effect of dithiothreitol following administration of mercuric chloride in the rat.' *Virchows Arch. B Cell Pathol. Incl. Mol. Pathol.*, 1980, **32**, 201–232.

131. H. M. Chan, M. Satoh, R. K. Zalups *et al.*, 'Exogenous metallothionein and renal toxicity of cadmium and mercury in rats.' *Toxicology*, 1992, **76**, 15–26.

132. R. K. Zalups, M. G. Cherian and D. W. Barfuss, 'Mercury–metallothionein and the renal accumulation and handling of mercury.' *Toxicology*, 1993, **83**, 61–78.

133. T. W. Meyer, J. V.Scholey and B. M. Brenner, in 'The Kidney,' eds. B. M. Brenner and F. C. Rector, Jr., 4th edn., W. B. Saunders, Philadelphia, PA, 1991, chap. 39, pp. 1871–1908.

134. K. A. Nath, A. J. Croatt and T. H. Hostetter, 'Oxygen consumption and oxidant stress in surviving nephrons.' *Am. J. Physiol.*, 1990, **258**, F1354–F1362.

135. M. T. Houser and W. O. Berndt, 'The effect of unilateral nephrectomy on the nephrotoxicity of mercuric chloride in the rat.' *Toxicol. Appl. Pharmacol.*, 1986, **83**, 506–515.

136. R. K. Zalups, 'Renal accumulation and intrarenal distribution of inorganic mercury in the rabbit: effect of unilateral nephrectomy and dose of mercuric chloride.' *J. Toxicol. Environ. Health*, 1991, **33**, 213–228.

7.34
Other Metals

BRUCE A. FOWLER and KATHERINE S. SQUIBB
University of Maryland Baltimore County, MD, USA

7.34.1 INTRODUCTION

The nephrotoxic effects of metals and metalloids such as arsenic have been known for many years as a result of environmental, occupational, or therapeutic use of these agents. The mechanisms by which metals or metalloids produce cellular injury in the kidney vary greatly as a function of dose, duration of exposure, chemical species, and intracellular target organelle. The roles of several metal-binding proteins in mediating cellular injury from a number metals such as cadmium, mercury and lead have also been more completely understood at the cellular and molecular levels of biological organization. Interactions between metals in the kidney have also been documented for a number of metals with regard to alterations in intracellular binding patterns, histopathological effects and biomarkers of renal cell injury. The kidney is a multifunctional organ system which is comprised of a variety of different cell types that also vary greatly in their susceptibility to toxic metals and metalloids. In general, cells of the proximal tubule are a major target cell population for metal-induced nephrotoxicity due in part to their reabsorptive functions and high metabolic activity. The actions of various chemical species of metals/metalloids on the proximal tubule vary along the three distinct anatomical segments. The relatively specific nephrotoxic effects of mercuric ion the third segment of the proximal tubule have been known for many years while the preferential toxic effects of the cadmium metallothioneincomplex on the first and second segments of the the proximal tubule have only been appreciated in the last 20 years. The expanding role of biomarkers for assessing early signs of renal toxicity from metals is an important new area of intense research.

This chapter will attempt to provide an up-to-date review of current knowledge regarding mechanisms of metal/metalloid toxicity in the kidney and discuss the current and possible future roles of biomarkers in the early detection of kidney cell injury from this class of toxic agents. The nephrotoxic elements discussed in this chapter will be taken in alphabetical order.

7.34.2 ARSENIC

The metalloid arsenic may exist in three oxidation states ($+/-3, +5$) and is metabolized into several methylated species *in vivo*. The nephrotoxicity of arsenicals has received much less attention than that of toxic elements despite the fact that urinary elimination is the major route for arsenical excretion from the body and the kidney actively accumulates arsenicals such as As^{5+} during this process.[1,2] Inorganic arsenic exposure has been reported to cause renal failure in humans.[3]

The proximal tubule cell mitochondria appear to be the major intracellular target for arsenicals. Ultrastructural/biochemical studies by Brown *et al.*,[4] demonstrated marked *in situ* swelling following prolonged oral exposure to As^{5+} in the drinking water.This ultrastructural effect was associated with decreases in respiratory control and state 3 respiration. This is consistent with other studies which have demonstrated the marked uncoupling effects of arsenicals on mitochondria.[5] *In vitro* studies by Aoki *et al.*[6] using rat proximal tubule cell primary cultures, have shown that relatively low concentrations of As^{3+} also induces the expression of the major classes of stress proteins indicating that arsenicals also produce alterations in gene expression in this kidney cell type.

7.34.3 BISMUTH

Bismuth nephrotoxicity is also primarily directed at the renal proximal tubule cells with necrosis and formation of distinctive bismuth containing[7,8] intranuclear inclusion bodies that are morphologically different from those seen with lead exposure. Unlike lead, bismuth has also been shown to induce and bind to metallothionein[9] indicating that its intracellular bioavailability may be modulated by this protein indicating that the intracellular handling of this element is different from that of lead as described below. The biochemical mechanisms of bismuth are less studied than those of lead but bismuth has been shown to interfere with enzymes in heme biosynthetic pathway[10] indicating at least some similarities with this element.

7.34.4 CADMIUM

Cadmium is a group IIB transition metal that exists in a single valence state (cadmium^{+2}). Cadmium has one of the longest biological half-lives of all of the nephrotoxic metals in mammalian organisms ($t_{1/2} > 30$ years), due to its accumulation and long term retention in liver and kidney tissue.[11] The primary reason for the long tissue half-life of this metal is its strong binding to a metal sequestering protein called metallothionein (MT). MT is a small (6000 Da), soluble, intracellular protein that contains a high percentage (30%) of cysteine residues that form two metal binding sites within the protein molecule. These two sites bind a number of metals, including cadmium, zinc, copper, and mercury, through sufhydryl linkages.[12] The binding constants for the metals vary, however cadmium is one of the most tightly bound metals[13,14] with an equilibrium binding constant of 2×10^{-16}. This strong binding of cadmium to MT accounts for the long retention of cadmium by cells within the liver and kidney. As the protein molecule is turned over, cadmium rebinds to newly synthesized MT rather than being excreted by the cell. MT synthesis is induced by cadmium and this metal squestering protein provides a mechanism by which cadmium can be retained within the cell in a "nontoxic" form. Binding of cadmium by MT decreases the interaction of cadmium with other −SH groups and other sensitive metal binding ligands within the cell, thereby preventing the ion from inhibiting enzymes and regulatory proteins.[15]

Although MT serves as a relatively high-capacity "detoxification" system for metal ions, this system can be overwhelmed either by acute, high doses of cadmium or long-term cumulative exposure to the metal. During chronic exposures, cadmium accumulates in both the liver and kidney with relatively little damage to either organ until cadmium concentrations reach about $60 \,\mu g \, g^{-1}$ in the liver and about $200–300 \,\mu g \, g^{-1}$ in the kidney.[16–19] At this time, damage to the renal proximal tubule cells becomes evident as cadmium and MT begins to be excreted in the urine.[11]

The nephrotoxicity caused by cadmium exposure is primarily proximal tubule damage as evidenced by the development of low molecular weight proteinuria (β-2-microglobulin and retinol binding protein), glucosuria, amionoaciduria, calciuria, and hyperphosphaturia.[11,20–32] Mixed proteinuria (high and low molecular weight proteinuria) has been reported in some individuals with occupational exposure to cadmium, suggesting that glomerular effects of cadmium may sometimes occur as well.[30] In rats treated with cadmium, chronic exposure causes renal proximal tubule cell necrosis consistent with effects observed in humans.[33–35]

The exact mechanism by which cadmium causes proximal tubule cell necrosis is still unknown. There is evidence that CdMT released from the liver is filtered through the glomerulus and is absorbed by renal proximal tubule cells normally involved in low molecular weight protein reabsorption[36–53] as presented in Figure 1. Injections of CdMT in rats have shown that very low doses of cadmium in this form can cause proximal tubule cell damage,[43–45,50,54,55] suggesting that

relatively high doses of cadmium are delivered to a limited number of specific cells within the kidney by this mechanism. The S1 and S2 segments of the proximal tubule have been shown to be the target cells for cadmium in chronic cadmium exposure studies in rats.[56] Whether damage occurs from an interaction of CdMT with the brush border membrane as the protein is reabsorbed from the filtrate[50] or whether the protein is internalized by the proximal tubule cells and degraded by the lysosomal system with a release of relatively high concentrations of the toxic cadmium ion within the cell[43–45] is unknown. Ultrastructural studies have shown that injections of CdMT cause vessiculation of proximal tubule cells and an accumulation of small, dense lysosomes that contain cadmium[47,57] suggesting that the lysosomal fusion process may be disrupted. It has been suggested that cadmium ions may interact with calmodulin within affected tubule cells,[58–61] thereby damaging the cytoskeletal system that plays a role in mediating lysosomal biogenesis.[62] Other studies suggest that cadmium induces apoptosis in renal proximal tubule cells following chronic cadmium

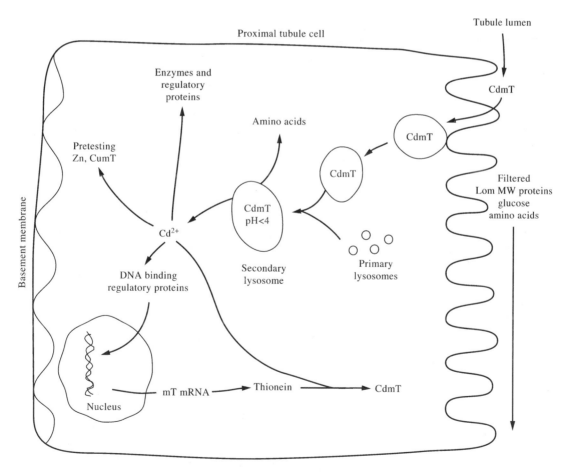

Figure 1 Proposed mechanism of proximal tubule cell necrosis caused by cadmium.

exposure.[63] *In vitro* studies using primary cultured renal tubular cells have shown that cadmium can be detected in the nucleus of cells within 60–90 min after exposure.[64]

Because cellular levels of MT play an important role in regulating cadmium toxicity, factors that influence MT synthesis will influence cadmium toxicity as well. The essential metal, zinc, also induces MT in the kidney and binds to this protein. Because cadmium has a stronger binding affinity than zinc for MT,[13] cadmium can replace zinc in cellular MT pools.[14,65] *In vivo* and *in vitro* animal studies have shown that pretreatment with zinc decreases the renal toxicity of cadmium.[66,67]

7.34.5 CHROMIUM

Chromium is a group 6 transition element that can exist in a range of oxidation states from −2 to +6, although Cr^{3+} and Cr^{6+} are the most common natural forms. The toxicity of chromium in mammalian systems is dependent upon its oxidation state, which greatly influences its bioavailability, tissue distribution and excretion.[68] As a cation, Cr^{3+} does not readily cross cell membranes, while Cr^{6+} exists as the oxyanion in aqueous solutions and readily enters cells. The primary target organs of chromium are the respiratory tract, skin, liver and kidney.[69,70]

Renal effects of chromium have been studied in both humans and experimental animals. Tubular necrosis has been reported in humans accidentally ingesting chromate salts and in individuals exposed dermally to solutions containing Cr^{6+}.[71-74] There is some evidence that occupational exposure to chromium can cause alterations in renal function. Lingberg and Vesterberg[75] reported that urinary β-2 microglobulin was higher in chromium platers than in controls although urinary albumin levels were not elevated, suggesting the presence of renal tubular damage. This finding was not supported by Saner and co-workers[76] in a study of tannery workers, however. Other studies in occupationally exposed humans suggest that chromium causes glomerular alterations[77-79] in the form of lipid degeneration and cell proliferation of the Bowman's capsule.[80]

Animal studies support the findings in humans that exposure to high doses of Cr^{6+} induce tubular necrosis. Ultrastructural studies have shown that disruption of the brush border membrane of cells in the S1 and S2 segments of the proximal tubule occurs rapidly after chromium injection. Associated functional alterations include a decreased reabsorptive capacity,

evidenced by glucosura, proteinuria, and enzymuria, followed at high doses by increased BUN and creatinine.[73,81-86] There is no reported evidence of glomerular damage due to chromium exposure in animal studies; most, however, have concentrated on tubular effects of this element. Due to the more rapid metabolism and excretion of chromium than most nephrotoxic metals, many chronic low dose exposures to Cr^{6+} have failed to detect signs of renal damage.[87-91] Thomann *et al.*[92] however have shown that chromium continues to accumulate in the kidney with time when rats are exposed to 100 ppm Cr^{6+} in their drinking water for up to 3 months suggesting that long term exposure to low levels of chromium could eventually result in kidney damage. There is considerable work that needs to be done to fully understand chromium effects under long term exposure conditions.

7.34.6 GALLIUM

The nephrotoxicity of gallium has been studied in relation to the use of this element as an antitumor agent.[93] It has been shown to produce necrosis of the renal proximal tubules following acute injection. Injection of hamsters with the III–V semiconductor gallium arsenide has been shown to inhibit renal delta aminolevulinic acid dehydratase (ALAD)[94,95] and this effect was found to be principally due to the gallium moiety. Other *in vitro* studies using rat kidney proximal tubular epithelial cell cultures[6] showed little effect on protein synthesis patterns presumably due to minimal uptake of this element under these short-term exposure conditions. *In vivo* exposures, however, indicated that renal proximal tubule cell cultures from gallium arsenide treated animals showed highly specific alterations in gene expression patterns[96-98] indicating that prolonged exposures to this element do result in alterations in gene expression.

7.34.7 INDIUM

The element indium has been used for radiocontrast studies and the nephrotoxicity of this element has been evaluated for this purpose.[99] Like gallium, it produces toxicity to the renal proximal tubules.[99-100] It has also been shown to markedly inhibit the activity of renal ALAD[101] and other studies using *in vivo* exposure to the III–V semiconductor indium arsenide have reported similar findings and the development of a characteristic porphyrinuria pattern.[98] Other studies have shown marked alterations in protein synthetic patterns in renal proximal tubule cells that were different than those observed with gallium arsenide. In general, the element indium inhibited the stress protein response in renal proximal tubule cells with increased duration of exposure.[96-98]

7.34.8 LEAD

Human lead exposure occurs via air, food, and water.[102] The kidney is a major target organ for lead with well documented effects on renal control of both blood pressure and the proximal tubules. It is clear that only a small fraction of the lead in the kidney is biologically available. The pathognomonic lead inclusion bodies which are found in the nuclei of the proximal tubule cells constitute a large fraction of total renal lead burden once their formation is initiated. It has been discovered that renal lead binding proteins in rats,[103] monkeys,[104] and humans[105,106] appear to play major roles in the intracellular handling of lead at lower dose levels in this organ system. The fact that chemically similar proteins exist in these three species suggest similar biochemical mechanisms for lead action in kidneys of these three species. If this is true, then this may have implications for risk assessments of lead nephrotoxicity and renal carcinogenicity. The discussion given below is focused on examining mechanisms of lead toxicity in the kidney from the perspective of the intracellular bioavailability of lead to sensitive target processes.

7.34.8.1 Lead Nephropathy

Prolonged high-dose lead exposure of humans is known to produce a Fanconi syndrome characterized by glucosuria, phosphaturia and amino aciduria.[107] Proteinuria has also been variably reported in some studies.[108] Morphological studies at the light and electron microscopic levels[109-116] have demonstrated mitochondrial and cellular swelling, karyomegally, increased cellular division, and the formation of lead-containing intranuclear inclusion bodies as common cellular alterations. The morphological changes in the mitochondrial and nuclear compartments have been closely linked with biochemical disturbances in mitochondrial respiration and alterations in gene expression indicating a close relationship between structural alterations in these organelle compartments and biochemical functionality.[110] Studies from a number of laboratories have shown that lead in the

circulation is bound to both the erythrocytes and components of the plasma. The biologically available fraction of lead which is taken up by cells in the kidney appears to reside in the plasma fraction rather than with the erythrocytes. Studies by Victery *et al.*[117] using an *in vitro* membrane vesicle system showed extensive Pb^{2+} membrane binding but failed to demonstrate the existence of an active Pb^{2+} membrane transport system. Data from these studies suggest that the renal uptake of any Pb^{2+} in blood most likely occurs via endocytosis of lead bound to the cell membrane. Passive diffusion of lead across the membrane is also not excluded by these studies. It is more likely that the biologically active fraction of lead is associated with diffusible proteins that are taken up by the proximal tubule cells along with other filtered proteins from the circulation (Figure 2).

Studies in rats[118] have shown that lead is bound to the protein alpha-2-U-globulin which is synthesized in the liver and transported to the kidney[119] where it is taken up by renal proximal tubule cells. This phenomenon would provide a molecular mechanism by which lead would be rapidly transported from the liver to the kidney. The kidney lead-binding proteins in humans recently identified as diazapine-binding inhibitor (DBI) and thymosin beta-4[105,106] have similar chemical properties to alpha-2-U-globulin but are not members of the same gene family. The hypothesized pivotal roles of these chemically similar proteins in mediating the cellular toxicity and perhaps carcinogenicity of lead in the kidney across species are discussed in the following sections.

7.34.8.2 Biochemical Characteristics of Kidney Lead Binding Proteins

Studies by Oskarsson *et al.*[103] using ^{203}Pb demonstrated two soluble protein fractions in rat kidney with estimated molecular masses of 11 500 Da and 63 000 Da by Sephadex column chromatography. Injection of ^{203}Pb injection into rats demonstrated an apparent decrease in binding from the 11 500 Da to the 63 000 Da fraction. Subsequent studies by DuVal and Fowler,[120] demonstrated that purified lead binding proteins (PbBP) did demonstrate aggregation on SDS gels as a function of metal binding. The binding of ^{203}Pb to the 63 000 Da entity also appeared to be quite stable as evidenced by ^{203}Pb autoradiograms of this fraction on SDS polyacrylamide gels. Subsequent characterization studies[121] showed that the low molecular mass PbBP was alpha-2-U-globulin by N-terminal sequence analysis.

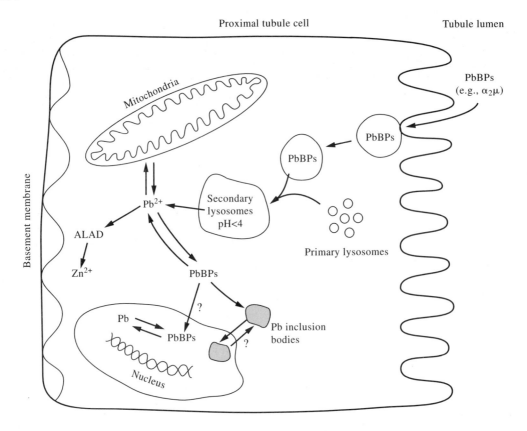

Figure 2 Proposed mechanism for the movement of lead in the proximal tubule cells.

Other researchers[122,123] have shown the increased presence of an acidic nuclear protein with an approximate molecular mass of 32 000 Da that is associated with the lead intranuclear inclusion body fraction which is similar to the 28 000 Da inclusion body protein described by Moore *et al.*[124] The possbile relationships of these various molecular size classes to one another awaits futher study. Other studies on PbBPs from human kidney[105,106] have shown the soluble lead binding protein to be diazapine binding inhibitor (DBI). The exact relationships between lead-binding to the monomeric forms of these proteins, initiation of aggregation[120] and formation of inclusion bodies is unclear. It is worth noting, however, that all of these proteins are acidic in nature and possess a high content of glutamic and aspartic amino acids and K_d values for lead on the order of 10^{-8} M suggesting that they behave in a similar manner with regard to the intracellular bioavailability to sensitive biochemical processes in both species as discussed below.

The PbBPs appear to be capable of mediating the bioavailability to lead to sensitive enzymatic processes such as the enzyme ALAD. Fowler

et al.[125] reported that rat renal ALAD was relatively insensitive to lead inhibition following *in vivo* exposure for up to 9 months. *In vitro* studies[126,127] showed that the PbBP was able to attenuate lead inhibition of this enzyme via both lead chelation and zinc donation to this zinc-activated enzyme. It is also possible that some of the resistance of rat renal ALAD to lead inhibition could also be due to metallothionein zinc donation to this enzyme from the endogenous renal Zn,Cu metallothionein pool.[128,129]

Mistry and co-workers[130,131] using partially purified PbBPs from the previously indentified 11 500 Da and 63 000 Da fractions, reported a K_d value on the order of 10^{-8} M for lead binding to both fractions as well as the total cytosol. These data were subsequently confirmed by DuVal and Fowler[120] using HPLC-purified fractions of the rat renal PbBP. Studies by Smith *et al.*[106] obtained similar K_d values for the human kidney PbBP again indicating similar chemical properties for these molecules between the two species. Studies by Mistry and co-workers[130,131] also demonstrated that the rat renal PbBP facilitated the cell-free nuclear translocation of lead. The lowest molecular mass PbBP demonstrated an apparent sedimentation coefficient via sucrose density gradient studies of 2S while the larger PbBP fraction showed a sedimentation coefficient of 4.6 S. A third component of 7–8 S was also indentified on the gradient. Lead-203 peaks with sedimentation coefficients of 2 S and 4 S were recovered from KCl nuclear extracts, following cell-free nuclear translocation studies using purified rat kidney nuclei, demonstrating an apparent *in vitro* association of these fractions with renal nuclear chromatin. The 4 S peak may represent a modified form of the 4.6 S cytosolic peak that underwent some undefined structural alteration as a result binding to chromatin. Metal displacement studies conducted on these protein peaks showed surprisingly that while calcium did not effect [203]Pb binding from the peaks at any concentration, cadmium and zinc were found to be highly effective competitors for displacing lead from these peaks. Zinc was also found to increase the cell free nuclear translocation of radio-labeled lead suggesting that these molecules probably normally bind zinc and that lead is serendipitously accepted into zinc binding sites. These data support the hypothesis[131,132] that the renal PbBPs mediate the effects of lead on gene expression in the kidney and may play a role in development of renal cancer following exposure to this element as discussed below. This hypothesis may be conceptually important given the existence of chemically similar PbBPs in humans and the apparent increased rate of renal adenocarcinoma in lead exposed smelter workers.[133]

The ability of lead to induce renal adenocarcinomas in rats and mice has been extensively studied but the mechanisms underlying these processes are not well understood at the molecular level. Choie and Richter[111–114] showed marked increases in DNA, RNA and protein synthesis in renal proximal tubule cells of rats and mice. Further studies by others[134] showed that these alterations occurred at lead dose levels below those required to initiate cell turnover. It would appear that lead exerts some highly specific effects on the nuclear machinery regulating cellular replication in the kidney which is not dependent upon the actually killing the cells. Such low dose effects are again consistent with the hypothesized role of the renal PbBPs in mediating lead-induced renal carcinogenesis.

7.34.9 URANIUM

Uranium is a radioactive metal that consists naturally as three primary isotopes ([234]U, [235]U,

and [238]U). Uranium-238 is the most abundant, making up more than 99% of the uranium found naturally in rocks and soils. A member of the actinide series of metals, uranium is the heaviest naturally existing element (atomic weight 238.03).

Occupational exposure to uranium has been a major concern since the early 1940s when exposure to this element increased dramatically due to its use by the nuclear industry. Hazards associated with uranium exposure are based not only on its radioactive properties, but also on its chemical nature. Studies in animals and humans have shown that the kidney is the most sensitive organ in uranium toxicity[135–141] due to the interaction of uranium ions with renal proximal tubule epithelial cells. Since 1853, it has been known that acute nephritis can be induced by uranium exposure.[142]

Uranium has two valence states, U^{2+} and U^{6+}. The uranyl ion, UO_2^{2+}, which is the most stable form of uranium in solution, is thought to be the most prevalent form of uranium in body fluids.[143,144] In circulation, approximately 40% of the UO_2^{2+} is loosely bound to the protein, transferrin, while the remainder is in low molecular weight anionic complexes, mainly with the carbonate ion.[145,147] Uranium-carbonate filters through the glomerulus, thereby reaching the lumen of the kidney tubule. As the acidity of the filtrate increases, the uranium ion is released from the bicarbonate complex and is free to bind to cells along the proximal tubule.

Effects of uranium exposure include alterations in the reabsorption of glucose, sodium, amino acids, protein, water, and other substances by kidney cells. Mechanistic studies suggest that UO_2^{2+} inhibits absorptive functions of proximal tubule cells either through an interaction with high affinity anionic binding sites on the brush border membranes or by inhibiting lysosomal or mitochondrial function following uptake of the uranyl ion into the cell.[137,148] These studies have lead to an established "critical concentration" of uranium in the kidney of $3 \mu g \, g^{-1}$ of kidney mass.[149,150]

A shortcoming in our knowledge of uranium toxicity is that most mechanistic studies have been conducted using acute, relatively high doses of uranium, which do not mimic the type of exposure to this metal that is of greatest concern today. Thun *et al.*[151] conducted a study of uranium mill workers and found evidence of β-2-microglobulinuria and aminoaciduria. In addition to workers in the nuclear industry, DOE workers in depleted uranium (DU) weapons production plants and exposure of armed services personnel to uranium dusts created by the use of depleted uranium in weapons is of growing concern. As with other nephrotoxic metals, a better understanding of the effects of chronic, low doses of uranium is needed.[137]

7.34.10 METAL INTERACTIONS IN THE KIDNEY

Interaction studies using rats exposed to lead, cadmium and arsenic *in vivo* by Mahaffey and *et al.*[152,153] have shown that concomitant exposure to cadmium reduces deposition of lead in the kidney and ablates the formation of lead intranuclear inclusion bodies. These data are consistent with the *in vitro* studies by Mistry *et al.*[130] which demonstrated that cadmium was a highly effective competitor for displacing lead binding to the rat renal PbBP as discussed above and support the hypothesis that these proteins are involved in the formation of the lead intranuclear inclusion bodies.[154]

7.34.11 REFERENCES

1. J. M. Ginsberg and W. W. Lotspeich, *Am. J. Physiol.*, 1963, **205**, 707.
2. J. M. Ginsberg, *Am. J. Physiol.*, 1965, **208**, 832.
3. J. P. Frejaville, J. Bescol, J. P. LeClere *et al.*, 'Intoxication aigue par les derives arsenicaux; (a propos de 4 observations personnelles); troubles de l'hemostase; etude ultramicroscopique du foie et du rein. [Acute poisoning with arsenic derivatives; (apropos of 4 personal cases); hemostasis disorders; ultramicroscopic study of the liver and kidney.]' *Ann. Med. Interne (Paris)*, 1972, **123**, 713.
4. M. M. Brown, B. C. Rhyne, R. A. Goyer *et al.*, 'Intracellular effects of chronic arsenic administration on renal proximal tubule cells.' *J. Toxicol. Environ. Health*, 1976, **1**, 505–514.
5. A. L. Fluharty and D. R. Sanadi, *J. Biol. Chem.*, 1961, **236**, 2772.
6. Y. Aoki, M. M. Lipsky and B. A. Fowler, 'Alteration in protein synthesis in primary cultures of rat kidney proximal tubule epithelial cells by exposure to gallium, indium, and arsenite.' *Toxicol. Appl. Pharmacol.*, 1990, **106**, 462–468.
7. D. L. Beaver and R. E. Burr, *Am. J. Pathol.*, 1963, **42**, 609.
8. B. A. Fowler and R. A. Goyer, 'Bismuth localization within nuclear inclusions by x-ray microanalysis. Effects of accelerating voltage.' *J. Histochem. Cytochem.*, 1975, **23**, 722–726.
9. J. A. Szymanska and A. J. Zelasowski, 'Effect of cadmium, mercury, and bismuth on the copper content in rat tissues.' *Environ. Res.*, 1979, **19**, 121–126.
10. J. S. Woods and B. A. Fowler, 'Alteration of mitochondrial structure and heme biosynthetic parameters in liver and kidney cells by bismuth.' *Toxicol. Appl. Pharmcaol.*, 1987, **90**, 274–283.
11. L. Friberg, C.-F. Elinder, T. Kjellstrom *et al.*, 'Cadmium and Health: a Toxicological and Epidemiological Appraisal. Vol. 2. Effects and Response,' CRC Press, Boca Raton, FL, 1986.
12. J. H. R. Kagi, in 'Metallothionein III,' eds. K. T. Suzuki, N. Imura and M. Kinura, Birkhauser, Basel, 1993, pp. 29–55.
13. J. H. R. Kagi and A. Schaffer, 'Biochemistry of metallothionein.' *Biochemistry*, 1988, **27**, 8509–8515.
14. B. L. Vallee and W. Maret, in 'Metallothionein III,' eds. K. T. Suzuki, N. Imura and M. Kinura, Birkhauser, Basel, 1993, pp. 1–27.
15. K. S. Squibb, in 'Toxicology of Metals,' ed. C. L. Chang, CRC Press, Boca Raton, FL, 1996, vol. I, pp. 727–732.
16. H. Roels, R. Lauwerys and A. N. Dardenne, 'The critical level of cadmium in human renal cortex: a reevaluation.' *Toxicol. Lett.*, 1983, **15**, 357–360.
17. S. Skerfving, J.-O. Christoffersson, A. Schutz *et al.*, *Biol. Trace Elem. Res.*, 1987, **13**, 241.
18. K. J. Ellis, K. Yuen, S. Yasamura *et al.*, 'Dose–response analysis of cadmium in man: body burden vs. kidney dysfunction.' *Environ. Res.*, 1984, **33**, 216–226.
19. K. J. Ellis, S. H. Cohn and T. J. Smith, 'Cadmium inhalation exposure estimates: their significance with respect to kidney and liver cadmium burden.' *J. Toxicol. Environ. Health*, 1985, **15**, 173–187.
20. R. G. Adams, J. F. Harrison and P. Scott, 'The development of cadmium-induced proteinuria, impaired renal function, and osteomalacia in alkaline battery workers.' *Q. J. Med.*, 1969, **38**, 425–443.
21. T. Kjellstrom, in 'Cadmium and Health: A Toxicological and Epidemiological Appraisal,' eds. L. Friberg, C.-G. Erlinder and T. Kjellstrom, CRC Press, Boca Raton, FL, 1986, vol. 2, pp. 47–86.
22. G. F. Nordberg, T. Kjellstrom and M. Nordberg, in 'Cadmium and Health: a Toxicological and Epidemiological Appraisal,' eds. L. Friberg, C.-G. Erlinder and T. Kjellstrom, CRC Press, Boca Raton, FL, vol. 2, 1986.
23. G. Kanzanzis, *Environ. Health Perspect.*, 1979, **28**, 155.
24. B. A. Fowler, P. L. Goering and K. S. Squibb, in 'Metallothionein,' eds. J. H. R. Kagi and Y. Kojima, Birkhauser, Basel, 1987, pp. 66–68.
25. T. Y. Jin, P. Leffler and G. F. Nordberg, 'Cadmium-metallothionein nephrotoxicity in the rat: transient calcuria proteinuria.' *Toxicology*, 1987, **45**, 307–317.
26. P. Leffler, T. Y. Jin and G. F. Nordberg, 'Cadmium-metallothionein-induced kidney dysfunction increases magnesium excretion in the rat.' *Toxicol. Appl. Pharmacol.*, 1990, **103**,180–184.
27. B. A. Fowler and M. Akkerman, in 'Proceedings of the International Symposium on Cadmium in the Human Environment: Toxicity and Carcinogenicity. Lyon, France,' IARC, 1991.
28. B. A. Fowler, R. E. Gandley, M. M. Akkerman *et al.*, in 'US Japan Conference on Metallothionein in Biology and Medicine,' eds. C. D. Klaassen and K. T. Suzuki, CRC Press, Boca Raton, FL, 1991, pp. 311–322.
29. R. A. Goyer and M. G. Cherian, in 'Metal Toxicology,' eds. R. A. Goyer, C. D. Klaassen and M. P. Waalkes, Academic Press, New York, 1995, pp. 389–412.
30. A. Bernard, J. P. Buchet, H. Roels *et al.*, 'Renal excretion of proteins and enzymes in workers exposed to cadmium.' *Eur. J. Clin. Invest.*, 1979, **9**, 11–22.
31. H. A. Roels, R. R. Lauwerys, J. P. Buchet *et al.*, '*In vivo* measurement of liver and kidney cadmium in workers exposed to this metal: its significance with respect to cadmium in blood and urine.' *Environ. Res.*, 1981, **26**, 217–240.
32. C. J. Guthrie, D. R. Chettle, D. M. Franklin *et al.*, 'The use of multiple parameters to characterize cadmium-induced renal dysfunction resulting from occupational exposure.' *Environ. Res.*, 1994, **65**, 22–41.
33. R. E. Dudley, L. M. Gammal and C. D. Klaassen, 'Cadmium-induced hepatic and renal injury in chronically exposed rats: likely role of hepatic cadmium-metallothionein in nephrotoxicity.' *Toxicol. Appl. Pharmacol.*, 1985, **77**, 414–426.

34. K. Kobota, R. Matsui, Y. Fukuyama *et al.*, *Igaku To Seibutsugaku*, 1970, **80**, 315.

35. R. A. Goyer, C. R. Miller, Z. Y. Zhu *et al.*, 'Non-metallothionein-bound cadmium in the pathogenesis of cadmium nephrotoxicity in the rat.' *Toxicol. Appl. Pharmacol.*, 1989, **101**, 232–244.

36. J. S. Garvey and C. C. Chang, 'Detection of circulating metallothionein in rats injected with zinc or cadmium.' *Science*, 1981, **214**, 805–807.

37. C. C. Chang, R. J. Vander Mallie and J. S. Garvey, 'A radioimmunoassay for human metallothionein.' *Toxicol. Appl. Pharmacol.*, 1980, **55**, 94–102.

38. Z. A. Shaikh and K. Hirayama, 'Metallothionein in the extracellular fluids as an index of cadmium toxicity.' *Environ. Health Perspect.*, 1979, **28**, 267–271.

39. M. Sato, P. K. Mehra and I. Bremner, 'Measurement of plasma metallothionein-I , in the assessment of the zinc status of zinc-deficient and stressed rats.' *J. Nutr.*, 1984, **114**, 1683–1689.

40. R. J. Vander Mallie and J. S. Garvey, 'Radioimmunoassay of metallothioneins.' *J. Biol. Chem.*, 1979, **254**, 8416–8421.

41. I. Bremner, P. K. Mehra, J. N. Morrison *et al.*, 'Effects of dietary copper supplementation of rats on the occurrence of metallothionein-I in liver and its secretion into blood, bile and urine.' *Biochem. J.*, 1986, **235**, 735–739.

42. R. K. Mehra and I. Bremner, 'Development of a radioimmunoassay for rat liver metallothionein-I and its application to the analysis of rat plasma and kidneys.' *Biochem. J.*, 1983, **213**, 459–465.

43. K. S. Squibb, J. W. Ridlington, N. G. Carmicheal *et al.*, 'Early cellular effects of circulating cadmium-thionein on kidney proximal tubules.' *Environ. Health Perspect.*, 1979, **28**, 287–296.

44. K. S. Squibb, J. B. Pritchard and B. A. Fowler, 'Cadmium-metallothionein nephropathy: relationships between ultrastructural/biochemical alterations and intracellular cadmium binding.' *J. Pharmacol. Exp. Ther.*, 1984, **229**, 311–321.

45. K. S. Squibb and B. A. Fowler, 'Intracellular metabolism and effects of circulating cadmium-metallothionein in the kidney.' *Environ. Health Perspect.*, 1984, **54**, 31–35.

46. E. C. Foulkes, 'Renal tubular transport of cadmium-metallothionein.' *Toxicol. Appl. Pharmacol.*, 1978, **45**, 505–512.

47. K. S. Squibb, J. B. Pritchard and B. A. Fowler, in 'Biological Roles of Metallothionein,' ed. E. C. Foulkes, Elsevier/North-Holland, New York, 1982, pp. 181–192.

48. M. Webb and A. T. Etienne, 'Studies on the toxicity and metabolism of cadmium-thionein.' *Biochem. Pharmacol.*, 1977, **26**, 25–30.

49. B. A. Fowler and G. F. Nordberg, 'The renal toxicity of cadmium metallothionein: morphometric and x-ray microanalytical studies.' *Toxicol. Appl. Pharmacol.*, 1978, **46**, 609–623.

50. M. G. Cherian, R. A. Goyer and L. Delaquerriere-Richardson, 'Cadmium-metallothionein-induced nephropathy.' *Toxicol. Appl. Pharmacol.*, 1976, **38**, 399–408.

51. M. G. Cherian and Z. A. Shaikh, 'Metabolism of intravenously injected cadmium-binding protein.' *Biochem. Biophys. Res. Commun.*, 1975, **65**, 863–869.

52. Z. A. Shaikh and J. C. Smith, 'The biosynthesis of metallothionein rat liver and kidney after administration of cadmium.' *Chem. Biol. Interact.*, 1976, **15**, 327–336.

53. K. Cain and D. E. Holt, 'Studies of cadmium-thionein induced nephropathy: time course of cadmium-thionein uptake and degradation.' *Chem. Biol. Interact.*, 1983, **43**, 223–237.

54. R. A. Goyer and B. A. Fowler, *Environ. Health Perspect.*, 1984, **54**, 1.

55. X. P. Wang, H. M. Chan, R. A. Goyer *et al.*, 'Nephrotoxicity of repeated infections of cadmium-metallothionein in rats.' *Toxicol. Appl. Pharmacol.*, 1993, **119**, 11–16.

56. C. Dorian, V. H. Gattone, II and C. D. Klaassen, 'Renal cadmium deposition and injury as a result of accumulation of cadmium-metallothionein (CdMT) by the proximal convoluted tubules—a light microscopic autoradiography study with 109CdMT.' *Toxicol. Appl. Pharmacol.*, 1992, **114**, 173–181.

57. K. S. Squibb, J. B. Pritchard and B. A. Fowler, 'Cadmium-metallothionein nephropathy: relationships between ultrastructural/biochemical alterations and intracellular cadmium binding.' *J. Pharm. Exp. Ther.*, 1984, **229**, 311–321.

58. E. A. Conner and B. F. Fowler, in 'Toxicology of the Kidney,' eds. J. B. Hook and R. S. Goldstein, Raven Press, New York, 1993, pp. 437–457.

59. S. H. Chao, Y. Suzuki, J. R. Zysk *et al.*, 'Activation of calmodulin by various metal cations as a function of ionic radius.' *Mol. Pharmacol.*, 1984, **26**, 75–82.

60. J. S. Mills and J. D. Johnson, *Arch. Toxicol.*, 1985, **57**, 205.

61. Y. Suzuki, S. H. Charo, J. R. Zysk *et al.*, 'Stimulation of calmodulin by cadmium ion.' *Arch. Toxicol.*, 1985, **57**, 205–211.

62. D. A. Wall and T. Maack, 'Endocytic uptake, transport, and catabolism of proteins by epithelial cells.' *Am. J. Physiol.*, 1985, **248**, C12–C20.

63. T. Hamada, S. Nakano, S. Iwai *et al.*, 'Pathological study on beagles after long-term oral administration of cadmium.' *Toxicol. Pathol.*, 1991, **19**, 138–147.

64. T. Hamada, A. Tanimoto, S. Iwai *et al.*, 'Cytopathological changes induced by cadmium-exposure in canine proximal tubular cells: a cytochemical and ultrastructural study.' *Nephron*, 1994, **68**, 104–111.

65. A. H. Robbins, D. E. McRee, M. Williamson *et al.*, 'Refined crystal structure of Cd, Zn metallothionein at 2.0 A resolution.' *J. Mol. Biol.*, 1991, **221**, 1269–1293.

66. X. Y. Liu, T. Y. Jin, G. F. Nordberg *et al.*, 'A multivariate study of protective effects of Zn and Cu against nephrotoxicity induced by cadmium metallothionein in rats.' *Toxicol. Appl. Pharm.*, 1992, **114**, 239–245.

67. J. Liu, K. S. Squibb, M. Akkerman *et al.*, *Renal failure*, in press.

68. K. S. Squibb and E. T. Snow, in 'Handbook of Hazardous Materials,' ed. M. Corn, Academic Press, New York, 1993, pp. 127–144.

69. S. K. Tandon, *Top. Environ. Health*, 1982, **5**, 209.

70. J. O. Nriagu and E. Nieboer, 'Chromium in the Natural and Human Environments,' Wiley, New York, 1988.

71. S. Langard and T. Norseth, in 'Handbook on the Toxicology of Metals,' ed. L. Friberg, Elsevier Science, Amsterdam, 1986, vol. II, pp. 185–210.

72. U. Korallus, C. Harzdorf and J. Lewalter, 'Experimental bases for ascorbic acid therapy of poisoning by hexavalent chromium compounds.' *Int. Arch. Occup. Environ. Health.*, 1984, **53**, 247–256.

73. I. Franchini, A. Mutti, A. Cavatorta *et al.*, 'Nephrotoxicity of chromium. Remarks on an experimental and epidemiological investigation.' *Contrib. Nephrol.*, 1978, **10**, 98–110.

74. W. Jao, J. R. Manaligod, L. T. Gerardo *et al.*, 'Myloid bodies in drug-induced acute tubular necrosis.' *J. Pathol.*, 1983, **139**, 33–40.

75. E. Linberg and O. Vesterberg, 'Urinary excretion of proteins in chromeplaters, exchromeplaters and referents.' *Scand. J. Work Environ. Health*, 1983, **9**,

505–510.

76. G. Saner, V. Yuzbasiyan and S. Cigdem, 'Hair chromium concentration and chromium excretion in tannery workers.' *Br. J. Ind. Med.*, 1984, **41**, 263–266.

77. M. A. Verschoor, P. C. Bragt, R. F. Herber *et al.*, 'Renal function of chrome-plating workers and welders.' *Int. Arch. Occup. Environ. Health*, 1988, **60**, 67–70.

78. M. Hagberg, B. Lindqvist and S. Wall, 'Exposure to welding fumes and chronic renal diseases, a negative case-referent study.' *Int. Arch. Occup. Environ. Health*, 1986, **58**, 191–195.

79. B. Lindqvist, *IRCS Med. Sci. Pathol. Soc. Occup. Med.*, 1983, **11**, 99.

80. T. Sano and I. Ibihara, *J. Sci. Lab.*, 1979, **55**, 21.

81. A. P. Evan and W. G. Dail, Jr., 'The effects of sodium chromate on the proximal tubules of the rat kidney. Fine structural damage and lysozymuria.' *Lab. Invest.*, 1974, **30**, 704–715.

82. W. O. Berndt, 'Renal chromium accumulation and its relationship to chromium-induced nephrotoxicity.' *J. Toxicol. Environ. Health*, 1976, **1**, 449–459.

83. M. W. Kluwe, 'Developed resistance to mercuric chloride nephrotoxicity: failure to protect against other nephrotoxicants.' *Toxicol. Lett.*, 1982, **12**, 19–25.

84. H. Miyajima, W. R. Hewitt, M. G. Cote *et al.*, 'Relationships between histological and functional indices of acute chemically induced nephrotoxicity.' *Fundam. Appl. Toxicol.*, 1983, **3**, 543–551.

85. A. Kumar and S. V. Rana, 'Enzymological effects of hexavalent chromium in the rat kidney.' *Int. J. Tissue React.*, 1984, **6**, 135–139.

86. J. M. Baggette and W. O. Berndt, 'Interaction of potassium dichromate with the nephrotoxins, mercuric chloride and citrinin.' *Toxicology*, 1984, **33**, 157–169.

87. R. D. MacKensie, R. U. Byerrum, C. F. Decker *et al.*, *AMA Arch. Ind. Health*, 1958, **18**, 232.

88. U. Glaser, D. Hochrainer, H. Kloppel, 'Low level chromium (VI) inhalation effects on alveolar macrophages and immune functions in Wistar rats.' *Arch. Toxicol.*, 1985, **57**, 250–256.

89. U. Glaser, D. Hochrainer, H. Kloppel *et al.*, 'Carcinogenicity of sodium dichromate and chromium (VI/III) oxide aerosols inhaled by male Wistar rats.' *Toxicology*, 1986, **42**, 219–232.

90. U. Glaser, D. Hochrainer and H. Oldiges, *Env. Hyg.*, 1988, **1**, 111.

91. J. Diaz-Mayans, R. Laborda and A. Nunez, 'Hexavalent chromium effects on motor activity and some metabolic aspect of Wistar albino rats.' *Comp. Biochem. Physiol. [C]*, 1986, **83**, 191–195.

92. R. V. Thomann, C. A. Snyder and K. S. Squibb, 'Development of a pharmacokinetic model for chromium in the rat following subchronic exposure. I. The importance of incorporating long-term storage compartment.' *Toxicol. Appl. Pharmacol.*, 1994, **128**, 189–198.

93. R. H. Adamson, G. P. Canellos and S. M. Siever, 'Studies on the antitumor activity of gallium nitrate (NSC-15200) and other group IIIa metal salts.' *Cancer Chemother. Rep.*, 1975, **59**, 599–610.

94. P. L. Goering and B. A. Fowler, *Pharmacol.*, 1986, **28**, 229.

95. P. L. Goering, R. R. Maronpot and B. A. Fowler, 'Effect of intratracheal gallium arsenide administration on delta-aminolevulinic acid dehydratase in rats: relationship to urinary exeretion of aminolevulinic acid.' *Toxicol. Appl. Pharmacol.*, 1988, **92**, 179–193.

96. B. A. Fowler, H. Yamauchi and M. Akkerman, *The Toxicologist*, 1993, **3**, 38.

97. E. A. Conner, H. Yamauchi, M. Akkerman *et al.*, *The Toxicologist*, 1994, **14**, 266.

98. E. A. Conner, H. Yamauchi and B. A. Fowler, 'Alterations in the heme biosynthetic pathway from the III-V semiconductor metal, indium arsenide (InAs).' *Chem. Biol. Interact.*, 1995, **96**, 273–285.

99. F. P. Castronovo, 'Factors affecting the toxicity of the element indium.' Ph.D. Thesis, 1970, pp. 73.

100. F. P. Castronovo and H. N. Wagner, 'Factors affecting the toxicity of the element indium.' *Br. J. Exp. Pathol.*, 1971, **52**, 543–559.

101. J. S. Woods and B. A. Fowler, 'Selective inhibition of delta-aminolevulinic acid dehydratase by indium chloride in rat kidney: biochemical and ultrastructural studies.' *Exp. Mol. Pathol.*, 1982, **36**, 306–315.

102. NAS (National Academy Science), 'Report of the Committee on Measuring Lead Exposure in Infants, Children and Other Sensitive Populations,' NAS/NRC Press, Washington, DC, 1993, p. 337.

103. A. Oskarsson, K. S. Squibb and B. A. Fowler, 'Intracellular binding of lead in the kidney: the partial isolation and characterization of postmitochondrial lead binding components.' *Biochem. Biophys. Res. Commun.*, 1982, **104**, 290–298.

104. E. A. Connor, H. Yomauchi, B. A. Fowler *et al.*, 'Biological indicators for monitoring exposure/toxicity from III-V semiconductors.' *J. Expos. Anal. Environ. Epidemiol.*, 1993, **3**, 431–440.

105. D. R. Smith, M. W. Kahng, E. A. Conner *et al.*, *The Toxicologist*, 1993, **13**, 443.

106. D. R. Smith, M. W. Kahng, B. Quintanilla-Vega *et al.*, *The Toxicologist*, 1994, **14**, 84.

107. R. A. Goyer and B. C. Rhyne, in 'International Review of Experimental Pathology,' eds. G. W. Richter and M. A. Epstein, Academic Press, New York, 1973, vol. 12, p. 1.

108. C. V. Vacca, J. D. Hines and P. W. Hall 3rd, 'The proteinuria of industrial lead intoxication.' *Environ. Res.*, 1986, **41**, 440–446.

109. R. A. Goyer, 'The renal tubule in lead poisoning. I. Mitochondrial swelling and aminoacidura.' *Lab Invest.*, 1968, **19**, 71–77.

110. R. A. Goyer and R. Krall, 'Ultrastructural transformation in mitochondria isolated from kidneys of normal and lead-intoxicated rats.' *J. Cell Biol.*, 1969, **41**, 393–400.

111. D. D. Choie and G. W. Richter, 'Cell proliferation in rat kidneys after prolonged treatment with lead.' *Am. J. Pathol.*, 1972, **68**, 359–370.

112. D. D. Choie, G. W. Richter and L. B. Young, 'Biogenesis of intranuclear lead-protein inclusions in mouse kidney.' *Beitr Pathol.*, 1975, **155**, 197–203.

113. D. D. Choie and G. W. Richter, 'Cell proliferation in mouse kidney induced by lead. I. Synthesis of deoxyribonucleic acid.' *Lab. Invest.*, 1974, **30**, 647–651.

114. D. D. Choie and G. W. Richter, 'Cell proliferation in mouse kidney induced by lead. II. Synthesis of ribonucleic acid and protein.' *Lab. Invest.*, 1974, **30**, 652–656.

115. G. W. Richter, 'Evolution of cytoplasmic fibrillar bodies induced by lead in rat and mouse kidneys.' *Am. J. Pathol.*, 1976, **83**, 135–148.

116. M. Murakami, R. Kawamura, S. Nishii *et al.*, 'Early appearance and localization of intranuclear inclusions in the segments of renal proximal tubules of rats following ingestion of lead.' *Br. J. Exp. Pathol.*, 1983, **64**, 144–155.

117. W. Victery C. R. Miller and B. A. Fowler, 'Lead accumulation by rat renal brush border membrane vesicles.' *J. Pharmacol. Exp. Ther.*, 1984, **231**, 589–596.

118. B. A. Fowler and G. DuVal, 'Effects of lead on the kidney: roles of high-affinity lead-binding proteins.'

Environ. Health Perspect., 1991, **91**, 77–80.

119. J. A. Swenberg, B. Short, S. Borghoff *et al.*, 'The comparative pathobiology of alpha 2u-globulin nephropathy. [Review].' *Toxicol. Appl. Pharmacol.*, 1989, **97**, 35–46.

120. G. DuVal and B. A. Fowler, *The Toxicologist*, 1990, **10**, 160.

121. B. A. Fowler and G. DuVal, 'Effects of lead on the kidney: roles of high-affinity lead-binding proteins.' *Environ. Health. Perspect.*, 1991, **91**, 77–80.

122. K. R. Shelton and P. M. Egle, 'The proteins of lead-induced intranuclear inclusion bodies.' *J. Biol. Chem.*, 1982, **257**, 11802–11807.

123. K. R. Shelton, J. M. Todd and P. M. Egle, *J. Biol. Chem.*, 1986, **261**, 1935.

124. J. F. Moore, R. A. Goyer and M. Wilson, 'Lead-induced inclusion bodies. Solubility, amino acid content, and relationship to residual acidic nuclear proteins.' *Lab. Invest.*, 1973, **29**, 488–494.

125. B. A. Fowler, C. A. Kimmel, J. S. Woods *et al.*, 'Chronic low-level lead toxicity in the rat. III. An integrated assessment of long-term toxicity with special reference to the kidney.' *Toxicol. Appl. Pharmacol.*, 1980, **56**, 59–77.

126. P. L. Goering and B. A. Fowler, 'Regulation of lead inhibition of delta-aminolevulinic acid dehydratase by a low molecular weight, high affinity renal lead-binding protein.' *J. Pharmacol. Exp. Ther.*, 1984, **231**, 66–71.

127. P. L. Goering and B. A. Fowler, 'Mechanism of renal lead-binding protein reversal of delta-aminolevulinic acid dehydratase inhibition by lead.' *J. Pharamacol. Exp. Ther.*, 1985, **234**, 365–371.

128. P. L. Goering and B. A. Fowler, 'Kidney zinc-thionein regulation of delta-aminolevulinic acid dehydratase inhibition by lead.' *Arch. Biochem. Biophys.*, 1987, **253**, 48–55.

129. P. L. Goering and B. A. Fowler, 'Metal constitution of metallothionein influences inhibition of delta-aminolaevulinic acid dehydratase (porphobilinogen synthase) by lead.' *Biochem. J.*, 1987, **245**, 339–345.

130. P. Mistry, C. Mastri and B. A. Fowler, 'Influence of metal ions on renal cytosolic lead-binding proteins and nuclear uptake of lead in the kidney.' *Biochem. Pharmacol.*, 1986, **35**, 711–713.

131. B. A. Fowler, *Comm. Toxicol.*, 1989, **3**, 27.

132. B. A. Fowler, M. W. Kahng and D. R. Smith, 'Role of lead-binding proteins in renal c:ancer.' *Environ. Health Perspect.*, 1994, **102**, 115–116.

133. K. Steenland, S. Selevan and P. Landrigan, 'The mortality of lead smelter workers: an update [published erratum appears in *Am. J. Public Health* 1993, **83(1)**, 60].' *Am. J. Public Health*, 1992, **82**, 1641–1644.

134. P. Mistry, C. Mastri and B. A. Fowler, *Toxicology*, 1987, **7**, 78.

135. ATSDR (Agency for Toxic Disease Registry), in 'Toxicological Profile for Uranium,' USPHS, TP90-29, 1990, p. 201.

136. H. C. Hodge, in 'Uranium, Plutonium, Transplutonic Elements: Handbook of Experimental Pharmacology,' eds. H. C. Hodge, J. N. Stannard and J. B. Hursh, Springer-Verlag, New York, 1973, vol. 36, pp. 5–68.

137. R. W. Leggett, 'The behavior and chemical toxicity of U in the kidney: a reassessment.' *Health Phys.*, 1989, **57**, 365–383.

138. J. L. Domingo, J. M. Llobet, J. M. Tomas *et al.*, 'Acute toxicity of uranium in rats and mice.' *Bull. Environ. Contam. Toxicol.*, 1987, **39**, 168–174.

139. R. L. Kathren, J. F. McInney, R. H. Moore *et al.*, 'Uranium in the tissues of an occupationally exposed individual.' *Health Phys.*, 1989, **57**, 17–21.

140. H. E. Stokinger, in 'Patty's Industrial Hygiene and Toxicology,' eds. G. D. Clayton and F. E. Clayton, Wiley, New York, vol. 2A, 1981.

141. M. E. Wrenn, P. W. Durbin, B. Howard *et al.*, 'Metabolism of ingested U and Ra.' *Health Phys.*, 1985, **48**, 601–633.

142. C. Le Conte, *C. R. Soc. Biol. (Paris)*, 1853, **5**, 171 as cited in D. R. Fisher, in 'Handbook on Toxicity of Inorganic Compounds,' eds. H. G. Seiler, H. Sigel and A. Sigel. Dekker, New York, 1988, pp. 739–748.

143. P. W. Durbin, in 'Biokinetics and Analysis of Uranium in Man,' ed. R. H. Moore, United States Uranium Registry; USUR-05, HEHF-47, F1-F65, 1984.

144. J. E. Gindler, in 'Uranium, Plutonium, Transplutonic Elements: Handbook of Experimental Pharmacology,' eds. H. C. Hodge, J. N. Stannard and J. B. Hursh, Springer-Verlag, New York, 1973, vol. 36, pp. 69–164.

145. R. L. Moore, in 'Biokinetics and Analysis of Uranium in Man,' USUR-05, HEHF-47, Hanford Environmental Health Foundation, Richland, Washington, 1984.

146. W. L. Stevens, F. W. Bruenger, D. R. Atherton *et al.*, 'The distribution and retention of hexavalent 233U in the beagle.' *Radiat. Res.*, 1980, **83**, 109–126.

147. M. E. Wrenn, *et al.*, *J. Amer. Water Works Assoc.*, 1987, April, 177.

148. F. N. Ghadially, J. M. Lalonde and S. Yang-Steppuhn, *Virchows Arch.*, 1982, **B39**, 21.

149. G. Diamond, *Radiation Protection Dosimetry*, 1989, **26**, 23.

150. ICRP (International Commission on Radiation Protection), *Health Phys.*, 1960, **3**, 1.

151. M. J. Thun, D. B. Baker, K. Steenland *et al.*, 'Renal toxicity in uranium mill workers.' *Scand. J. Work Environ. Health.*, 1985, **11**, 83–90.

152. K. R. Mahaffey and B. A. Fowler, 'Effects of concurrent administration of lead, cadmium, and arsenic in the rat.' *Environ. Health Perspect.*, 1977, **19**, 165–167.

153. K. R. Mahaffey, S. G. Capar, B. C. Gladen *et al.*, 'Concurrent exposure to lead, cadmium, and arsenlc. Effects on toxicity and tissue metal concentrations in the rat.' *J. Lab. Clin. Med.*, 1981, **98**, 463–468.

154. B. A. Fowler, 'Mechanisms of kidney cell injury from metals.' *Environ. Health. Perspect.*, 1992, **100**, 57–63.

7.35
Succinimides

GARY O. RANKIN

Marshall University School of Medicine, Huntington, WV, USA

7.35.1 INTRODUCTION

7.35.1.1 Chemistry

The parent compound for the nephrotoxicant succinimides is succinimide (1H-2,5-pyrrolidinedione, Figure 1). Although succinimide is not a nephrotoxicant, substitution of certain aryl groups on the nitrogen atom of the imide group (*N*-arylsuccinimides) or on the ethylene bridge (*C*-arylsuccinimides) leads to succinimide derivatives capable of inducing acute and/or chronic nephrotoxicity.

The nephrotoxicant succinimides identified to date are all synthetic chemicals prepared from the appropriate succinic acid or succinic anhydride derivative via reaction with a primary amine or ammonia. Several variations of these synthetic pathways have been attempted and are described in more detail in a review.[1]

While laboratory syntheses of *C*- or *N*-arylsuccinimides are the main routes to the formation of the agents, biosynthesis of *N*-(chlorophenyl)succinimides from chloroanilines by fungi is also possible.[2]

7.35.1.2 Uses of Succinimides

Succinimide derivatives have been evaluated for an extensive range of agricultural, industrial, and pharmaceutical applications. Many of the *N*-arylsuccinimides were initially prepared in Japan in order to find newer, more efficacious agricultural antimicrobial agents.[3–5] Although these compounds were effective against a wide range of micro-organisms, the *N*-arylsuccinimides were most widely evaluated as agricultural fungicides.[3–6] Fungicidal activity was enhanced by the addition of halogen atoms to

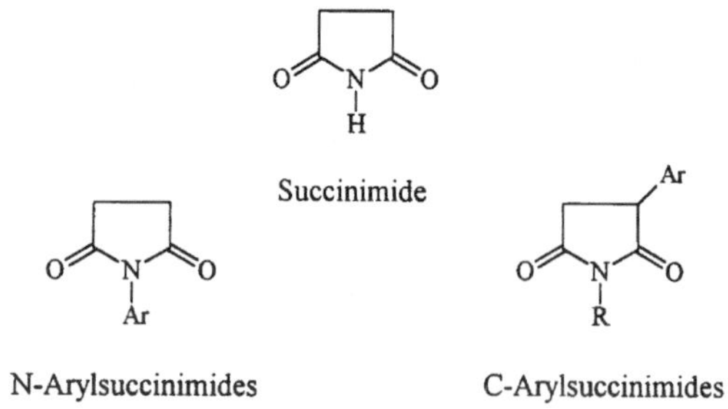

Figure 1 Structure of succinimide and its aryl derivatives.

the phenyl ring in the *N*-phenylsuccinimides,[3,4] and this observation eventually led to the marketing and field testing of the highly efficacious fungicide *N*-(3,5-dichlorophenyl)-succinimide (NDPS, dimetachlone; Figure 2) as Ohric.[7,8]

The *C*-arylsuccinimides have been used clinically as antiepileptic drugs to treat absence seizures.[9] Phensuximide (Milontin) and methsuximide (Celontin) (Figure 2) have been available since the early 1950s,[10–12] but they have largely been replaced by a less toxic succinimide, ethosuximide (2-ethyl-2-methyl-succinimide, Zarontin), and non-succinimide antiepileptics. However, derivatives and analogs of phensuximide and methsuximide continue to be synthesized and evaluated as antiepileptic agents in an attempt to discover safer and more efficacious drugs to treat absence seizures.[13–15]

7.35.2 *N*-ARYLSUCCINIMIDES

7.35.2.1 Introduction

The promising agricultural antimicrobial activity exhibited by *N*-(3,5-dichlorophenyl)-succinimide (NDPS, Figure 2) led to extensive field testing of this efficacious fungicide and to studies of the toxic potential of NDPS in

mammalian systems. Ito *et al.*[16] reported that NDPS administered at 5000 ppm in the diet for eight weeks to male Wistar rats followed by commercial stock diet for a total of 20–34 weeks did not induce renal carcinoma but induced severe interstitial nephritis and promoted the nephrocarcinogenicity of dimethyl-nitrosamine. Subsequently, Sugihara *et al.*[17] demonstrated that 5000 ppm NDPS in the diet of rats for 4–24 weeks induced only nephrotoxicity (elevated blood urea nitrogen (BUN) concentration and kidney weight, decreased proximal tubular enzyme activities, histological changes), with proximal convoluted tubular effects appearing within one week. Administration of 5000 ppm NDPS in the diet for 12 or more weeks resulted in interstitial nephritis. The nephrotoxicity induced by NDPS was dose dependent, as 2500 ppm NDPS in the diet had little effect on renal function or morphology.[17]

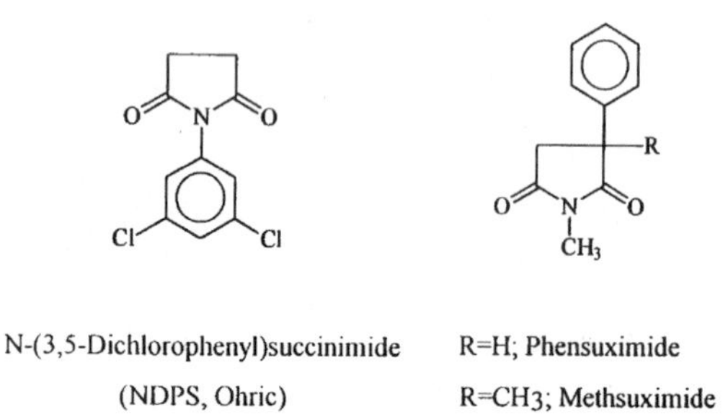

NDPS also induces marked nephrotoxicity following intraperitoneal (i.p.) administration of a dose of 0.4 mmol kg^{-1} or greater to Sprague–Dawley[18] or Fischer 344[19] rats (Table 1). Acute NDPS nephrotoxicity is characterized by diuresis, increased proteinuria, glucosuria, hematuria, elevated BUN concentration, and kidney weight, a decreased accumulation of organic anions and cations by renal cortical slices, and proximal tubular necrosis.[18–20] The initial renal lesion in male rats is observed in the S1 and S2 segments of the proximal tubule, but damage is also seen in some S3 segments at higher doses.[18–21] Interestingly, female Fischer 344 rats appear more susceptible to NDPS than males with marked nephrotoxicity apparent after an i.p. administration of 0.2 mmol kg^{-1} NDPS in females.[21] In addition, the primary site of the proximal tubular lesion in females is the S3 segment which indicates that the nature of acute NDPS nephrotoxicity is gender dependent.

Acute NDPS nephrotoxicity is rapid in onset and requires up to seven days for recovery of most renal functional parameters following the administration of 0.4 or 0.8 mmol kg^{-1} NDPS.[20] Organic anion (*p*-aminohippurate, PAH) accumulation by renal cortical slices is the most sensitive indicator of NDPS nephrotoxicity. PAH accumulation is decreased by 1 h post-NDPS (0.8 mmol kg^{-1}) and remains depressed after seven days. Diuresis and morphological changes are evident by three hours, while BUN concentration is increased at 24 hours. Both altered urine volume and BUN concentration return to normal by seven days post-treatment, but kidney weight remains elevated.

Based on these observations, coupled with

N-(3,5-Dichlorophenyl)succinimide
(NDPS, Ohric)

R=H; Phensuximide

R=CH$_3$; Methsuximide

Figure 2 Prototypic *N*- and *C*-arylsuccinimide nephrotoxicants.

Table 1   Characteristics of NDPS nephrotoxicity in male rats.[18–20,32]

Parameter	Post-treatment day[a]	
	Day 1	*Day 2*
Urine volume	↑↑	↑↑
Proteinuria	↑↑	↑
Glucosuria	Trace	Negative
Hematuria	Trace	Negative
Kidney weight	↑↑	↑↑
PAH accumulation[b]	↓	↓
TEA accumulation[b]	↓	↓
BUN concentration	↑↑	↑↑
Renal morphology	Severe proximal tubular necrosis	Severe proximal tubular necrosis

Source: Rankin[18]; Rankin *et al.*[19,20,32]
[a]Rats received a single i.p. dose of 0.4 mmol kg^{-1}. The arrows refer to effects compared to controls.   [b]PAH and TEA accumulation were measured *ex vivo* using renal cortical slices.

extensive structure–nephrotoxicity relationship studies, NDPS has been used as a prototypic *N*-arylsuccinimide for evaluating the acute nephrotoxicity induced by this class of chemicals. It has also been proposed that NDPS is an excellent model compound for inducing chronic renal interstitial fibrosis in rats.[22]

7.35.2.2   Structure–Nephrotoxicity Relationships

It is clear from numerous studies that the acute nephrotoxicity induced by the *N*-arylsuccinimides is very closely linked to substitutions on the *N*-aryl group or modifications to the succinimide ring (Table 2). However, it is unclear whether these substitutions alter the intrinsic nephrotoxic potential of NDPS or whether the renal exposure/accumulation of these substituted compounds is different from that of NDPS in rats.

The parent compound for the *N*-arylsuccinimides, *N*-phenylsuccinimide (NPS), does not induce nephrotoxicity at i.p. doses up to 1.0 mmol kg^{-1} in male rats.[23] However, certain halogen substitutions can produce NPS derivatives capable of inducing marked nephrotoxicity. Among the monochlorophenyl derivatives of NPS, 1.0 mmol kg^{-1} *N*-(3- chlorophenyl)succinimide induces weak diuresis, decreases in PAH accumulation, and a small increase in BUN concentration,[23,24] while the 2-chloro and 4-chlorophenyl isomers at equimolar doses are

Table 2   Correlation of phenyl ring substitution among the *N*-phenylsuccinimides and nephrotoxicity in rats.[a,b,c]

No nephrotoxicity		Mild nephrotoxicity	Marked nephrotoxicity
H	2,5-Cl$_2$	3-Cl	2,4-Cl$_2$
2-Cl	2,6-Cl$_2$	3-CH$_3$	3,5-Cl$_2$
3-F	3,5-F$_2$	4-NO$_2$	3,5-I$_2$
3-Br	3,5-(O$_2$ CCH$_3$)$_2$	4-COCH$_3$	3,4,5-Cl$_3$
3-I	3,5-(CH$_3$)$_2$	2,3-Cl$_2$	
4-Cl	3,5-(OCH$_3$)$_2$	3,4-Cl$_2$	
4-CH$_3$		3,5-Br$_2$	
4-C(CH$_3$)$_3$		3,5-(NO$_2$)$_2$	
4-OCH$_3$			

[a]Compounds were administered i.p. at 1.0 mmol kg^{-1}, except for 3,5-I$_2$, which was administered at 0.8 mmol kg^{-1}.   [b]Degree of nephrotoxicity was determined as: no nephrotoxicity = no increase in BUN concentration at 48 h post-treatment; mild nephrotoxicity = BUN concentration > control levels but < 50 mg%; marked nephrotoxicity = BUN concentration > 100 mg%.   [c]Data are from Refs. 23–29. Substitutions are listed by their position on the phenyl ring (e.g. 2-) and functional group type (e.g. -Cl).

non-nephrotoxicants.[23] Substitution of other halogens (fluoride, bromide, or iodide) for chloride at the 3-position does not produce a derivative which induces greater nephrotoxicity when administered at a dose of $1.0 \, \text{mmol kg}^{-1}$ than the 3-chloro isomer.[24] However, addition of a second chloride group to the 5-position of N-(3-chlorophenyl)succinimide produces NDPS, the most potent nephrotoxicant among the three monochlorophenyl and six dichlorophenyl isomers of NPS.[25] Increasing the number of chloride groups to three does not further enhance nephrotoxicity,[26] while iodide substitution for chloride in NDPS produces a compound which also induces marked nephrotoxicity at an i.p. dose of $0.8 \, \text{mmol kg}^{-1}$ in male Fischer 344 rats.[27] In contrast, at equimolar doses to NDPS, N-(3,5-difluorophenyl)succinimide is a non-nephrotoxicant.[27] Thus, at least two non-fluoride halogen groups in the 3- and 5-positions of the phenyl ring will enhance nephrotoxic potential *in vivo*.

The role of the halogen atoms for N-arylsuccinimide nephrotoxicity appears to be extremely important. Replacement of halogen atoms by various electron-donating or -withdrawing groups produces compounds which have little effect on renal function.[24,28,29] However, the exact role that these 3,5-dihalide groups play in N-arylsuccinimide nephrotoxicity is unclear.

Modifications to the succinimide ring can also markedly influence the nephrotoxicity of N-arylsuccinimides. Using NDPS as the model nephrotoxicant, Yang *et al.*[30] found that addition of methyl groups, removal of one or both carbonyl groups, splitting the ethylene bridge, or hydrolysis of the succinimide ring produced NDPS derivatives or analogues with little, if any, ability to induce nephrotoxicity. Kellner-Weibel and Harvison[31] noted that the glutarimide derivative of NDPS was also a non-nephrotoxicant. The only, single modification to the succinimide ring of NDPS which has enhanced nephrotoxicity is the addition of a hydroxyl group to yield the NDPS metabolite N-(3,5-dichlorophenyl)-2-hydroxysuccinimide (NDHS) (see Figure 3).[32]

Thus, the succinimide ring appears to be of critical importance in NDPS nephrotoxicity

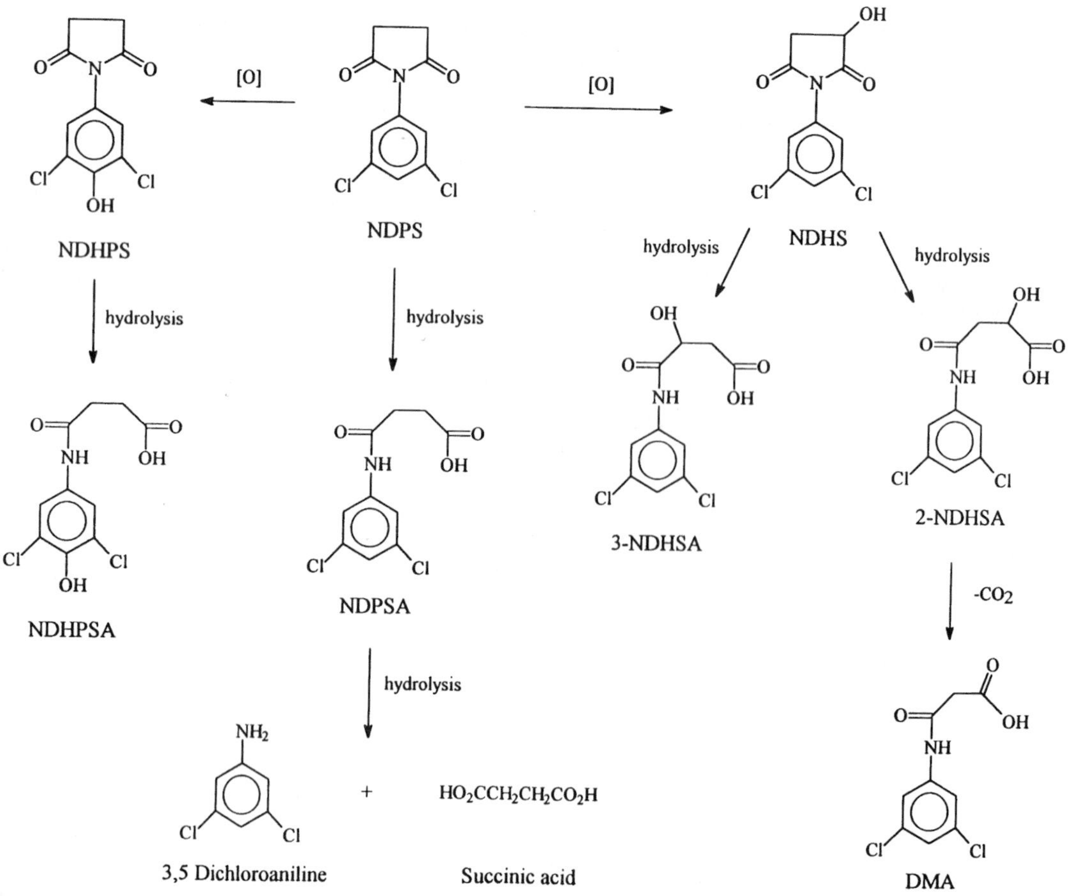

Figure 3 Biotransformation pathways for NDPS.

and the nephrotoxic potential of *N*-arylsucci-nimides can be further enhanced by the addi-tion of two nonfluoride halogens to the 3- and 5-positions of the phenyl ring and the addition of a hydroxy group to the ethylene bridge of the succinimide ring.

7.35.2.3 Role of Biotransformation in Acute Nephrotoxicity

The role of biotransformation in acute *N*-arylsuccinimide nephrotoxicity has been exam-ined in detail only for NDPS. Ohkawa *et al.*[33] demonstrated that NDPS was initially bio-transformed by rat and dog via at least two pathways: (i) hydrolysis of the succinimide ring to form *N*-(3,5-dichlorophenyl)succinamic acid (NDPSA) and (ii) oxidation of the ethylene bridge by microsomal enzymes to form NDHS (Figure 3). NDHS can hydrolyze (enzymatic or nonenzymatic hydrolysis) to form *N*-(3,5-dichlorophenyl)-2-hydroxysuccinamic acid (2-NDHSA), which decarboxylates to form *N*-(3,5-dichlorophenyl)malonamic acid (DMA). Ohkawa *et al.*[33] also noted that liver and kidney homogenates could degrade NDPS to NDPSA and succinic acid. Nyarko and Harvison[34] found that hepatocytes from male Fischer 344 rats were capable of biotransform-ing NDPS by these same pathways and also identified several new NDPS metabolites including 3-NDHSA, 3,5-dichloroaniline, and metabolites arising from *p*-hydroxylation of NDPS (*N*-(3,5-di-chloro-4-hydroxyphenyl)-succinimide (NDHPS) and *N*-(3,5-dichloro-4-hydroxyphenyl)succinamic acid (NDHPSA)) (Figure 3).

The relative contribution of each of these biotransformation pathways to NDPS nephro-toxicity has been examined. Metabolites arising from *p*-hydroxylation of NDPS (NDHPS, NDHPSA) are not nephrotoxicants *in vivo* when administered to rats at doses 2.5 times higher than the minimal nephrotoxic dose of NDPS.[35] Also, 3-methylcholanthrene, an indu-cer of aromatic ring hydroxylation, attenuates some aspects of NDPS nephrotoxicity,[36] which supports aromatic hydroxylation being a detox-ification pathway. Similarly, hydrolysis of NDPS to form NDPSA and/or 3,5-dichloroani-line also seems to contribute little to NDPS nephrotoxicity, since NDPSA[30] and 3,5-dichloroaniline[37] are weak nephrotoxicants compared with NDPS.

Although *p*-hydroxylation and hydrolysis of NDPS are detoxification pathways for NDPS, oxidation of the succinimide ring to form NDHS is a critical bioactivation step. Pretreat-ment of rats with microsomal enzyme inhibitors (e.g., cobalt chloride, piperonyl butoxide) markedly attenuates NDPS nephrotoxicity, while the microsomal enzyme inducer pheno-barbital potentiates NDPS-induced renal effects.[36] NDPS nephrotoxicity is also reduced by deuterium labeling of the succinimide ring of NDPS, a measure which would slow the rate of hydroxylation at the ethylene bridge.[38] It is also likely that formation of NDHS occurs primar-ily in the liver, as phenobarbital induces hepatic but not renal microsomal enzymes in rat.[39] Subsequent studies have found that NDHS and 2-NDHSA are potent nephrotoxicants when administered at an i.p. dose of $0.1 \, \text{mmol kg}^{-1}$, a dose four times lower than the minimal nephrotoxic dose of NDPS,[32,40] while DMA is a non-nephrotoxicant at doses up to $1.0 \, \text{mmol kg}^{-1}$.[32] Initial *in vivo* studies sug-gested that 3-NDHSA induced nephrotoxicity which was of the same magnitude as NDHS and 2-NDHSA at eqimolar i.p. doses.[41] How-ever, subsequent studies have found that warming the injection solution facilitated cycli-zation of 3-NDHSA to NDHS and when administered i.p. in 25% DMSO in sesame oil without heating, 3-NDHSA induced nephro-toxicity at doses of $0.4 \, \text{mmol kg}^{-1}$ or greater (unpublished data).

These observations suggest that the ultimate nephrotoxicant species is formed following biotransformation of NDPS to NDHS or an NDHSA isomer. However, determination of the exact chemical nature of the ultimate nephrotox-icant species has been complicated by findings that interconversion between NDHS and the two NDHSA isomers can occur *in vivo* and *in vitro* (unpublished observations). In several *in vitro* models (renal cortical slices, isolated renal proximal tubules, isolated renal mitochondria), NDPS and its known metabolites have failed to induce significant nephrotoxicity, which has further complicated identification of the ultimate nephrotoxicant species.[32,42]

There is evidence that NDPS induces nephro-toxicity via a reactive metabolite of extrarenal origin. Ohkawa *et al.*[33] noted that labile NDPS metabolites were present in the urine of rats treated orally with $100 \, \text{mg kg}^{-1}$ NDPS. It was believed that these metabolites were conjugates of 2-NDHSA, although their identity was not fully determined. Shantz and Harvison[43] found that in rats pretreated with phenobarbital, covalent binding in kidney was increased greater than threefold following [^{14}C]-NDPS $0.2 \, \text{mmol kg}^{-1}$ administration as compared with renal covalent binding in non-phenobar-bital pretreated rats. Shantz and Harvison[43] also noted an increase in urinary excretion of 2-NDHSA and a decrease in NDPSA excretion in the phenobarbital pretreated group. These observations are consistent with the extrarenal production of a potentially reactive NDPS metabolite, since (i) phenobarbital increases hepatic, but not renal, microsomal enzyme activities in rat,[39] (ii) the kidney does not appear to metabolize NDPS or any of the known NDPS metabolites produced by the liver,[44] and (iii) NDPS and its known metabolites are not directly cytotoxic to renal tissue in any *in vitro* model examined.[32,42,44]

The exact nature of the reactive NDPS metabolite(s) is unclear. It has been suggested that in the liver, NDHS is converted into a conjugate (e.g., a glucuronide) which is then transported to the kidney and into proximal tubule cells.[42] The metabolite could then eliminate the conjugating group (e.g., glucuronic acid) to form *N*-(3,5-dichlorophenyl)maleimide (NDPM)(Figure 4), a highly reactive and cytotoxic compound. Aleo *et al.*[42] found that NDPM, but not its hydrolysis product *N*-(3,5-dichlorophenyl)maleamic acid (NDPMA), was cytotoxic to isolated rat renal proximal tubules as indicated by a marked decrease in basal and nystatin-stimulated oxygen consumption and a marked increase in lactate dehydrogenase (LDH) release.

NDPM was also markedly toxic to isolated rat renal mitochondria.[42] Following exposure to NDPM for 3.5 min, state 3 and state 4 respiration were decreased in a concentration-dependent manner, with an IC_{50} for inhibition of state 3 respiration of ~25 μM and ~50 μM for state 4 respiration. Additional studies indicated that site 1 was more sensitive to inhibition by NDPM than site 2, while cytochrome *c*-cytochrome oxidase and the electron transport chain were not inhibited. In contrast, NDPMA only slightly altered mitochondrial function at bath concentrations as high as 100 μM.[42]

A similar reaction sequence to that proposed for the formation of NDPM has been reported for glutethimide, which is hydroxylated on the

carbon atom adjacent to a carbonyl group followed by dehydration to form an α,β-unsaturated carbonyl moiety.[45] In addition, phenobarbital potentiates NDHS nephrotoxicity via a mechanism which does not involve further oxidative metabolism of NDHS[46] and could be related to the induction of phase II enzymes (e.g., UDP-glucuronyltransferase, glutathione *S*-transferase).[47,48] However, no evidence for the formation of NDHS conjugates or NDPM has been obtained in previous biotransformation studies,[33,34,44] and the nature of the reactive NDPS metabolite remains to be discovered.

7.35.2.4 Mechanism of Acute Nephrotoxicity

Although numerous studies have examined the structure–nephrotoxicity relationships for *N*-arylsuccinimides and the role of metabolites in NDPS nephrotoxicity, very little is known about the mechanism by which NDPS or other *N*-arylsuccinimides induce acute nephrotoxicity. However, one important aspect of the mechanism for NDPS-induced nephrotoxicity which has been examined is the role of altered renal hemodynamics.

Creatinine clearance is reduced by over 50% within 3 h following administration of 1.0 mmol kg^{-1} NDPS to rats and is reduced by 75% at 6 h post-treatment.[49] This observation suggests that glomerular filtration rate (GFR) is markedly reduced very soon after exposure to NDPS. Beers and Rankin[49] found that in the anesthetized rat 0.4 mmol kg^{-1} NDHS-induced decreases in GFR were not accompanied by similar decreases in renal blood flow, suggesting that the observed decreases in GFR were due to mechanisms other than decreased renal hemodynamics.

Reduced glutathione (GSH) also plays a role in NDPS nephrotoxicity, although the mechanism is unclear. Yang *et al.*[50] reported that GSH

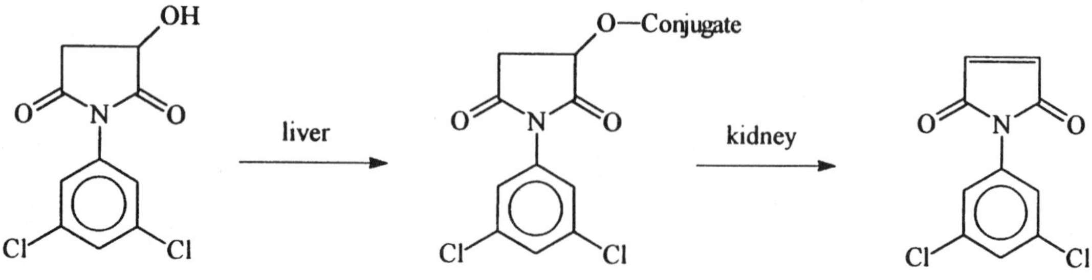

Figure 4 Postulated route of formation of NDPM.

depletion with diethyl maleate pretreatment attenuated *in vivo* NDPS nephrotoxicity in rats. It was subsequently shown that pretreatment with the GSH synthesis inhibitor buthionine sulfoximine also attenuatedNDPS nephrotoxicity[51] as well as NDHS and 2-NDHSA nephrotoxicity.[40] These findings indicate that GSH plays a central role in the mechanism of NDPS nephrotoxicity. Attenuation of NDPS and NDPS metabolite nephrotoxicity following GSH depletion does not appear to be due to the inhibition of formation of GSH or cysteine conjugates of NDPS metabolites.[52] No GSH conjugate or GSH conjugate metabolite of NDPS has been identified in previous biotransformation studies.[33,34,44] GSH depletion also attenuates the toxicity induced by methyl chloride,[53] cisplatin,[54] galactosamine/endotoxin,[55] and α-naphthyl isothiocyanate[56] through mechanisms which do not appear to involve the formation of GSH or cysteine conjugates from the toxicants, and could indicate that a cellular function of GSH other than xenobiotic conjugation plays a role in NDPS nephrotoxicity.

7.35.2.5 Chronic Nephrotoxicity

Chronic exposure to NDPS results in the development of severe interstitial nephritis in rats. Sugihara *et al.*[17] reported that dietary concentrations of 5000 ppm NDPS in the food for 12–24 weeks induced marked interstitial nephritis. These findings were similar to the observations of Barrett *et al.*,[22] who administered NDPS (75 mg day^{-1}) orally via a gastric tube and observed moderate tubulo-interstitial nephritis by 28 days and severe interstitial nephritis by 108 days.

In both of these studies, proximal tubular damage preceded the development of the nephritis and extensive fibrosis. Barrett *et al.*[22] noted that tubular damage was present within three days of beginning NDPS administration. At this time, protocollagen proline hydroxylase was also elevated about 50% and continued to increase throughout the 108 days of treatment. By nine days, renal collagen levels also increased and continued to increase throughout the remainder of the study. It was also noted that the effects of NDPS on proline hydroxylase activity were reversible, since discontinuation of NDPS administration after 54 days resulted in a return of proline hydroxylase activity to control levels 54 days later. These observations led Barrett *et al.*[22] to propose that NDPS would be an excellent model compound to study chemically induced interstitial nephritis.

Although NDPS is not a rodent carcinogen, continued administration of NDPS in food for eight weeks can enhance the nephrocarcinogenic properties of a number of carcinogens including dimethylnitrosamine,[16,57] citrinin,[57] streptozotocin,[58] and 2-(ethylnitrosamine)ethanol.[59] The mechanism of this enhancing effect of NDPS on renal carcinogens is unclear, but could be due to the nephrotoxic properties of NDPS, the ability of NDPS to induce microsomal enzyme activities,[60] or a combination of these effects.

7.35.3 C-ARYLSUCCINIMIDES

7.35.3.1 Introduction

The *C*-arylsuccinimides have been developed primarily to treat the absence seizures which characterize petit mal epilepsy. The three succinimide drugs which have been used most frequently are phensuximide, methsuximide, and ethosuximide.[9] Although ethosuximide (2-ethyl-2-methylsuccinimide) is still a drug of choice for treating absence seizures, the *C*-arylsuccinimides, phensuximide and methsuximide (Figure 2), have largely been replaced clinically by less toxic drugs. However, the potential for the discovery of safer and more efficacious antiepileptic drugs based on the *C*-arylsuccinimide structure has led to the continued synthesis and evaluation of phensuximide and methsuximide derivatives and analogues.[13–15,61–63]

Ethosuximide does not induce nephrotoxicity in humans[9] or laboratory animals[64,65] at or slightly above therapeutic doses. However, clinical studies with phensuximide and methsuximide revealed the potential of both drugs to induce nephrotoxicity,[9,10] and the few studies examining the nephrotoxic potential of the *C*-arylsuccinimides have focused on phensuximide, methsuximide, and their derivatives.

7.35.3.2 Nephrotoxicity in Humans

In humans, the common side effects induced by phensuximide and methsuximide are gastrointestinal distress and central nervous system effects, but rashes, hematotoxicity, fever, hepatotoxicity, and nephrotoxicity have also been reported.[9,66] Nephrotoxicity is most frequently seen as proteinuria and microscopic hematuria. In some cases, nephrotoxicity is severe enough to cause discontinuation of the drug, while in other patients the nephrotoxicity reverses in spite of continued administration of the succinimide.

Millichap[10] reported the efficacy and toxicity of phensuximide in 21 patients with petit mal epilepsy. Thirteen patients exhibited signs of toxicity with abnormal urinary findings being seen in 10 patients. Nephrotoxicity was seen as proteinuria, microscopic hematuria and the presence of granular casts in the urine. These effects were transient in seven patients, but persisted in three patients, and medication had to be discontinued in a seven year old child due to phensuximide-induced nephrotoxicity.

The parent drug (methsuximide or phensuximide) appears to be responsible for the effects on the central nervous system and the gastrointestinal tract. However, methsuximide and phensuximide metabolites may play a role in the organ-directed toxicity induced by these antiepileptics.[66] No studies in humans have addressed the nature of nephrotoxicant metabolites being produced from phensuximide or methsuximide and, with the replacement of these antiepileptic agents by less toxic drugs, it is unlikely that such studies will be forthcoming.

7.35.3.3 Nephrotoxicity in Laboratory Animals

Although phensuximide and methosuximide induce occasional nephrotoxicity in humans, early chronic studies in mice, rats, dogs, and rhesus monkeys failed to demonstrate any evidence of nephrotoxicity.[64] Rankin et al.[65,67] also determined that the three antiepileptic succinimides are not acute nephrotoxicants in rats at several times their therapeutic dose. However, daily administration of phensuximide (0.3 or 0.6 mmol kg^{-1} day^{-1}, i.p.) to male Fischer 344 rats did induce transient proteinuria and hematuria.[67] Proteinuria was increased after four days of treatment, while hematuria appeared between three and four days of phensuximide administration. Proximal tubular cells were the primary renal targets for phensuximide. Proximal tubular changes were characterized by a distention of the basal infoldings, the appearance of large cytoplasmic vacuoles, protrusion of nuclei into the lumen, and the occlusion of many proximal tubular lumina. The bladder also exhibited morphological changes with abnormal accumulations of erythrocytes noted within the wall of the urinary bladder. These observations could indicate that phensuximide-induced hematuria results from damage to the bladder as well as to the kidney. While this possibility was not explored in clinical studies, the bladder is clearly a target for phensuximide toxicity in Fischer 344 rats.

The renal morphological changes and bladder toxicity seen with multiple phensuximide dosing are also seen within 24–48 h in phenobarbital-pretreated rats administered a single dose of phensuximide, but not methsuximide or ethosuximide.[65] Repeated administration of phensuximide, like phenobarbital, results in hepatic microsomal induction[68] and could indicate that a phensuximide metabolite is responsible for or contributes to phensuximide nephrotoxicity.

The nature of a nephrotoxic metabolite from phensuximide has not been established. However, the only difference in the chemical structures of phensuximide and methsuximide is the presence of a methyl group on the succinimide ring of methsuximide (Figure 5). This minor structural difference between the two succinimides could indicate that biotransformation at the benzylic position of phensuximide may be important for the bioactivation of phensuximide. It is also possible that the presence of the methyl group on the benzylic position of methsuximide retards biotransformation at sites on methsuximide other than the benzylic sites which are accessible in phensuximide. However, studies have not been conducted to explore these two possibilities.

One toxic metabolite that might be formed from either methsuximide or phensuximide is the arene oxide formed during hydroxylation of the phenyl ring. However, the addition of various substituents to the *meta* or *para* positions of the phenyl ring in phensuximide can produce phensuximide derivatives with enhanced nephrotoxic potential[69–71] and minimizes the likelihood that arene oxide metabolites contribute to phensuximide nephrotoxicity in animal models.

Thus, it appears that succinimide ring oxidation probably contributes to the nephrotoxic mechanism of the *C*-arylsuccinimides, but more definitive studies are needed to clearly determine the nature of nephrotoxicant metabolites produced from the *C*-arylsuccinimides and the mechanism of toxicity induced by this class of compounds.

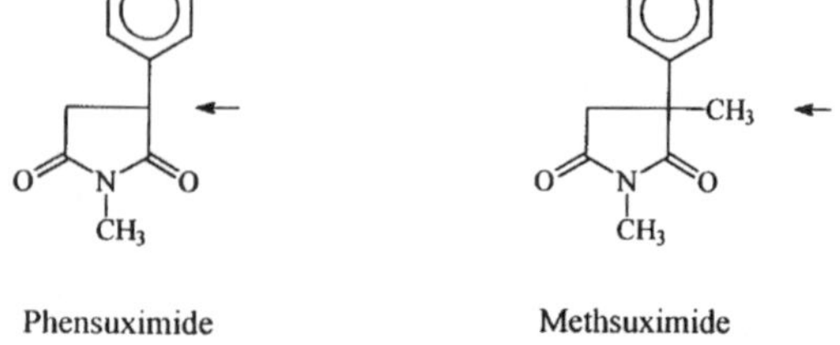

Phensuximide   Methsuximide

Figure 5 Chemical structural differences between phensuximide and methsuximide. Arrows denote unique sites for biotransformation in each drug.

7.35.4 SUMMARY

Succinimides have been widely evaluated as agricultural, industrial, and pharmaceutical agents. However, nephrotoxicity is an important toxicity associated with exposure to either *N*-arylsuccinimide or *C*-arylsuccinimide derivatives. The prototypic *N*-arylsuccinimide, NDPS, induces polyuric renal failure and proximal tubular necrosis following administration of a single i.p. injection of NDPS at a dose as low as 0.4 mmol kg^{-1}. The ultimate nephrotoxicant species derived from NDPS is not known with certainty, but oxidation of the succinimide ring and, possibly, extrarenal conjugation (e.g., glucuronidation) of an oxidative metabolite (NDHS and/or 2-NDHSA) appear to be important bioactivation steps for NDPS. Female Fischer 344 rats are more sensitive than males to NDPS nephrotoxicity, although the mechanism of this difference is not known. Among the *C*-arylsuccinimide antiepileptic agents, phensuximide and methsuximide can induce nephrotoxicity (proteinuria, microscopic hematuria, and urinary casts) in humans. Phensuximide, but not methsuximide, induces similar renal effects in rats, with renal proximal tubular cells and the urinary bladder primarily affected. Oxidation of the succinimide ring of phensuximide appears to be an important bioactivation step, but the nature of the ultimate urotoxicant species derived from phensuximide remains to be determined. However, with the continued development of new compounds based on the succinimide group, it is important to understand how these compounds induce nephrotoxicity.

7.35.5 REFERENCES

1. G. O. Rankin, D. J. Yang, H. C. Shih *et al.*, 'Synthesis and toxicity of *C*- and *N*-arylsuccinimides.' *Trends Heterocycl. Chem.*, 1993, **3**, 83–93.
2. M. Arjmand and H. Sandermann, Jr., '*N*-(Chlorophenyl)succinimides: a novel metabolite class isolated from *Phanerochaete chrysosporium*.' *Pestic. Biochem. Physiol.*, 1987, **27**, 173–181.
3. A. Fujinami, T. Ozaki, K. Nodera *et al.*, 'Studies on biological activity of cyclic imide compounds. II. Antimicrobial activity of 1-phenylpyrrolidine-2,5-diones and related compounds'. *Agric. Biol. Chem.*, 1972, **36**, 318–323.
4. C. Takayama and A. Fujinami, 'Quantitative structure activity relationships of antifungal *N*-phenylsuccinimides and *N*-phenyl-1,2-dimethylcyclopropanedicarboximides.' *Pestic. Biochem. Physiol.*, 1979, **12**, 163–171.
5. W. Kramer, in 'Chemistry of Pesticides,' ed. K. H. Buchel, Wiley-Interscience, New York, 1983, pp. 227–321.
6. C. Takayama, A. Fujinami, O. Kirino *et al.*, 'Biological activity of cyclic imide compounds. Part V. Quantitative structure–activity relationships of antifungal 1-(3,5-dichlorophenyl)-2,5-pyrrolidinediones and 3-(3,5-dichlorophenyl)-2,4-oxazolidinediones.' *Agric. Biol. Chem.*, 1982, **46**, 2755–2758.
7. C. C. Chein, Y. C. Hung and C. L. Chu, 'Greenhouse study and field experimentation on the control of sheath blight of rice with systemic fungicides.' *Nung Yeh Yen Chiu*, 1972, **21**, 191–202.
8. J. B. Heaton, 'Control of brown rot and transit rot of peaches with post-harvest fungicidal dips.' *Queensl. J. Agric. Anim. Sci.*, 1980, **37**, 155–159.
9. T. W. Rall and L. S. Schleifer, in 'The Pharmacological Basis of Therapeutics,' 8th edn., eds. A. G. Gilman, T. W. Rall, A. S. Nies *et al.*, Pergamon, New York, 1990, pp. 436–462.
10. J. G. Millichap, 'Milontin: a new drug in the treatment of petit mal.' *Lancet*, 1952, **2**, 907–910.
11. C. H. Carter, 'Use of Milontin in the control of petit mal epilepsy.' *Neurology*, 1954, **4**, 935–937.
12. F. T. Zimmerman, 'Evaluation of *N*-methyl-A,A-methylphenyl succinimide in the treatment of petit mal epilepsy.' *NY State J. Med.*, 1956, 1460–1465.
13. J. Lange, S. Rump, I. Ilczuk *et al.*, 'Synthesis and properties of cyclic derivatives of succinic acid with anticonvulsant activity. Part 3.' *Pharmazie*, 1979, **34**, 794–795.
14. A. Zejc, J. Obniska, E. Chojnacka-Wojcik *et al.*, 'Synthesis and anticonvulsant properties of *N*-methylpyridyl derivatives of *m*- and *p*-bromophenylsuccinimides.' *Pol. J. Pharmacol. Pharm.*, 1987, **39**, 91–95.
15. J. Lange, W. Kazmierski and J. Daroszewski, 'A quantitative structure–activity relationship (QSAR) analysis of the effects of aromatic substitution on the anticonvulsant activity and toxicity of arylsuccinimides.' *Pol. J. Pharmacol. Pharm.*, 1991, **43**, 71–77.
16. N. Ito, S. Sugihara, S. Makiura *et al.*, 'Effect of *N*-(3,5-dichlorophenyl)succinimide on the histological pattern and incidence of kidney tumors in rats induced by dimethylnitrosoamine.' *Gann*, 1974, **65**, 131–138.
17. S. Sugihara, Y. Shinohara, Y. Miyata *et al.*, 'Pathologic analysis of chemical nephritis in rats induced by *N*-(3,5-dichlorophenyl)succinimide.' *Lab. Invest.*, 1975, **33**, 219–230.
18. G. O. Rankin, 'Nephrotoxicity following acute administration of *N*-(3,5-dichlorophenyl)succinimide in rats.' *Toxicology*, 1982, **23**, 21–31.
19. G. O. Rankin, D. J. Yang, K. Cressey-Veneziano *et al.*, '*N*-(3,5-Dichlorophenyl)succinimide nephrotoxicity in the Fischer-344 rat.' *Toxicol. Lett.*, 1985, **24**, 99–105.
20. G. O. Rankin, K. Cressey-Veneziano and P. I. Brown, 'Onset of and recovery from acute *N*-(3,5-dichlorophenyl)succinimide-induced nephrotoxicity in Sprague–Dawley rats.' *Toxicology*, 1984, **30**, 205–216.
21. G. O. Rankin, K. W. Beers, V. J. Teets *et al.*, 'Acute *N*-(3,5-dichlorophenyl)succinimide nephrotoxicity in female Fischer 344 rats.' *Toxicology*, 1994, **88**, 151–164.
22. M. C. Barrett, S. J. Cashman and J. Moss, 'Experimental interstitial renal fibrosis in rats: nephritis induced by *N*-(3,5-dichlorophenyl)succinimide.' *Br. J. Exp. Pathol.*, 1983, **64**, 425–436.
23. G. O. Rankin, D. J. Yang, E. P. Lahoda *et al.*, 'Acute nephrotoxicity of *N*-phenyl and *N*-(monochlorophenyl)succinimides in Fischer 344 and Sprague–Dawley rats.' *Toxicology*, 1985, **34**, 299–308.
24. D. J. Yang, P. I. Brown, H. H. Lo *et al.*, 'Structure-nephrotoxicity relationships for *meta*-substituted *N*-phenylsuccinimides.' *J. Appl. Toxicol.*, 1987, **7**, 153–160.
25. D. J. Yang, E. P. Lahoda, P. I. Brown *et al.*, 'Acute nephrotoxicity of isomeric *N*-(dichlorophenyl)succinimides in Sprague–Dawley and Fischer 344 rats.' *Fundam. Appl. Toxicol.*, 1985, **5**, 1119–1127.

26. D. J. Yang, E. P. Lahoda, P. I. Brown *et al.*, 'Acute *N*-(3,4,5-trichlorophenyl)succinimide-induced nephrotoxicity in Sprague–Dawley and Fischer-344 rats.' *Toxicol. Lett.*, 1986, **31**, 219–228.

27. D. J. Yang, H. H. Lo, V. J. Teets *et al.*, 'Nephrotoxicity of *N*-(3,5-dihalophenyl) succinimides in Fischer 344 rats.' *J. Toxicol. Environ. Health*, 1987, **20**, 333–346.

28. G. O. Rankin, V. J. Teets, H. C. Shih *et al.*, 'Renal effects of *N*-(3,5-disubstituted phenyl)succinimides in the Fischer 344 rat.' *J. Appl. Toxicol.*, 1992, **12**, 211–216.

29. D. J. Yang, E. P. Lahoda, P. I. Brown *et al.*, 'Structure–nephrotoxicity relationships for *para*-substituted *N*-phenylsuccinimides in Sprague–Dawley and Fischer 344 rats.' *Toxicology*, 1985, **36**, 23–35.

30. D. J. Yang, C. D. Richmond, V. J. Teets *et al.*, 'Effect of succinimide ring modification on *N*-(3,5-dichlorophenyl)succinimide-induced nephrotoxicity in Sprague–Dawley and Fischer 344 rats.' *Toxicology*, 1985, **37**, 65–77.

31. G. L. Kellner-Weibel, R. Tchao and P. J. Harvison, 'Nephrotoxic potential of *N*-(3,5-dichlorophenyl)glutarimide and *N*-(3,5-dichlorophenyl)glutaramic acid in Fischer 344 rats.' *Toxicol. Lett.*, 1995, **80**, 123–129.

32. G. O. Rankin, H. C. Shih, D. J. Yang *et al.*, 'Nephrotoxicity of *N*-(3,5-dichlorophenyl)succinimide metabolites *in vivo* and *in vitro*.' *Toxicol. Appl. Pharmacol.*, 1988, **96**, 405–416.

33. H. Ohkawa, Y. Hisada, N. Fujiwara *et al.*, 'Metabolism of *N*-(3′,5′-dichlorophenyl)succinimide in rats and dogs.' *Agric. Biol. Chem.*, 1974, **38**, 1359–1369.

34. A. K. Nyarko and P. J. Harvison, 'Metabolism of the nephrotoxicant *N*-(3,5-dichlorophenyl)succinimide by isolated rat hepatocytes.' *Drug Metab. Dispos.*, 1995, **23**, 107–112.

35. P. J. Harvison, R. J. Griffin, V. J. Teets *et al.*, 'Nephrotoxic potential of *N*-(3,5-dichloro-4-hydroxyphenyl)succinimide and *N*-(3,5-dichloro-4-hydroxyphenyl)succinamic acid in Fischer-344 rats.' *Toxicol. Lett.*, 1992, **60**, 221–226.

36. G. O. Rankin, D. J. Yang, C. D. Richmond *et al.*, 'Effect of microsomal enzyme activity modulation on *N*-(3,5-dichlorophenyl)succinimide-induced nephrotoxicity.' *Toxicology*, 1987, **45**, 269–289.

37. G. O. Rankin, D. J. Yang, V. J. Teets *et al.*, '3,5-Dichloroaniline-induced nephrotoxicity in the Sprague–Dawley rat.' *Toxicol. Lett.*, 1986, **30**, 173–179.

38. G. O. Rankin, D. J. Yang, V. J. Teets *et al.*, 'Deuterium isotope effect in acute *N*-(3,5-dichlorophenyl)succinimide-induced nephrotoxicity.' *Life Sci.*, 1986, **39**, 1291–1299.

39. M. W. Anders, 'Metabolism of drugs by the kidney.' *Kidney Int.*, 1980, **18**, 636–647.

40. G. O. Rankin, V. J. Teets, D. W. Nicoll *et al.*, 'Effect of buthionine sulfoximine on *N*-(3,5-dichlorophenyl)-2-hydroxysuccinimide and *N*-(3,5-dichlorophenyl)-2-hydroxysuccinamic acid nephrotoxicity.' *Toxicol. Lett.*, 1991, **57**, 297–308.

41. G. O. Rankin, H. C. Shih, V. J. Teets *et al.*, 'Acute nephrotoxicity induced by *N*-(3,5-dichlorophenyl)-3-hydroxysuccinamic acid in Fischer 344 rats.' *Toxicol. Lett.*, 1989, **48**, 217–223.

42. M. D. Aleo, G. O. Rankin, T. J. Cross *et al.*, 'Toxicity of *N*-(3,5-dichlorophenyl)succinimide and metabolites to rat renal proximal tubules and mitochondria.' *Chem. Biol. Interact.*, 1991, **78**, 109–121.

43. C. M. Shantz and P. J. Harvison, 'Phenobarbital modulation of the *in vivo* metabolism and distribution of *N*-(3,5-dichlorophenyl)succinimide.' *Toxicologist*, 1995, **15**, 298.

44. C. M. Shantz and P. J. Harvison, 'Potential metabolism and cytotoxicity of *N*-(3,5-dichlorophenyl)succinimide and its hepatic metabolites in isolated rat renal cortical tubule cells.' *Toxicology*, in press.

45. K. A. Kennedy and L. J. Fischer, 'Quantitative and stereochemical aspects of glutethimide metabolism in humans.' *Drug Metab. Dispos.*, 1979, **7**, 319–324.

46. K. W. Beers, D. W. Nicoll, D. K. Anestis *et al.*, 'Effect of microsomal enzyme modulators on *N*-(3,5-dichlorophenyl)-2-hydroxysuccinimide (NDHS)-induced nephrotoxicity in the Fischer 344 rat.' *Toxicology*, 1993, **84**, 141–155.

47. R. M. E. Vos and P. J. Van Bladeren, 'Glutathione *S*-transferases in relation to their role in the biotransformation of xenobiotics.' *Chem. Biol. Interact.*, 1990, **75**, 241–265.

48. D. Goon and C. D. Klaassen, 'Effects of microsomal enzyme inducers upon UDP-glucuronic acid concentration and UDP-glucuronyltransferase activity in rat intestine and liver.' *Toxicol. Appl. Pharmacol.*, 1992, **115**, 253–260.

49. K. W. Beers and G. O. Rankin, 'Effect of *N*-(3,5-dichlorophenyl)-2-hydroxysuccinimide on renal function and hemodynamics in the anesthetized rat.' *Toxicology*, 1993, **79**, 139–148.

50. D. J. Yang, V. J. Teets, B. Bolton *et al.*, 'Role of glutathione in acute *N*-(3,5-dichlorophenyl)succinimide-induced nephrotoxicity in Sprague–Dawley and Fischer 344 rats.' *Toxicology*, 1987, **45**, 25–44.

51. G. O. Rankin, V. J. Teets, D. W. Nicoll *et al.*, 'Effect of buthionine sulfoximine on acute *N*-(3,5-dichlorophenyl)succinimide-induced nephrotoxicity in Fischer 344 rats.' *Toxicol. Lett.*, 1990, **52**, 91–100.

52. G. O. Rankin, H. C. Shih, V. J. Teets *et al.*, '*N*-(3,5-Dichlorophenyl)succinimide nephrotoxicity: evidence against the formation of nephrotoxic glutathione or cysteine conjugates.' *Toxicology*, 1991, **68**, 307–325.

53. G. J. Chellman, R. D. White, R. M. Norton *et al.*, 'Inhibition of acute toxicity of methyl chloride in maleB6C3F1 mice by glutathione depletion.' *Toxicol. Appl. Pharmacol.*, 1986, **86**, 93–104.

54. R. D. Mayer, K. E. Lee and A. T. K. Crockett, 'Inhibition of cisplatin-induced nephrotoxicity in rats by buthionine sulfoximine, a glutathione synthesis inhibitor.' *Cancer Chemother. Pharmacol.*, 1987, **20**, 207–210.

55. G. Tiegs and A. Wendel, 'Leukotriene-mediated liver injury.' *Biochem. Pharmacol.*, 1988, **37**, 2569–2573.

56. L. J. Dahm and R. A. Roth, 'Protection against α-naphthyl isothiocyanate-induced liver injury by decreased hepatic non-protein sulfhydryl content.' *Biochem. Pharmacol.*, 1991, **42**, 1181–1188.

57. Y. Shinohara, M. Aria, K. Hirao *et al.*, 'Combination effect of citrinin and other chemicals on rat kidney tumorigenesis.' *Gann*, 1976, **67**, 147–155.

58. Y. Shinohara, Y. Miyata, G. Murasaki *et al.*, 'Effect of *N*-(3,5-dichlorophenyl)succinimide on the histological pattern and incidence of kidney tumors induced by streptozoticin in rats.' *Gann*, 1977, **68**, 397–404.

59. T. Shirai, M. Ohshima, A. Masuda *et al.*, 'Promotion of 2-(ethylnitrosamino)ethanol-induced renal carcinogenesis in rats by nephrotoxic compounds: positive responses with folic acid, basic lead acetate, and *N*-(3,5-dichlorophenyl)succinimide but not with 2,3-dibromo-1-propanol phosphate.' *J. Natl. Cancer Inst.*, 1984, **72**, 477–482.

60. M. Valentovic, C. Elliott, V. J. Teets *et al.*, 'Enzyme induction produced by *N*-(3,5-dichlorophenyl)succinimide (NDPS) in rats.' *Biochem. Pharmacol.*, 1988, **37**, 768–770.

61. R. R. Goehring, T. D. Greenwood, J. S. Pisipati *et al.*, 'Synthesis and anticonvulsant evaluation of some new 2-benzylsuccinimides.' *J. Pharm. Sci.*, 1991, **80**, 790–792.

62. J. Lange, J. Lapszewicz, B. Migaj *et al.*, 'Synthesis and properties of cyclic derivatives of phenylsuccinic acid. IX. New phenylsuccinimide derivatives with anticonvulsant properties.' *Acta Pol. Pharm.*, 1987, **44**, 249–254.

63. M. Amir and E. Singh, 'Some new *N*-substituted α-aryl/alkyl succinimides as possible anticonvulsants.' *Pharmazie*, 1991, **46**, 705–707.

64. G. Chen, J. K. Weston and A. C. Bratton, Jr., 'Anticonvulsant activity and toxicity of phensuximide, methsuximide and ethosuximide'. *Epilepsia*, 1963, **4**, 66–76.

65. G. O. Rankin, V. J. Teets, D. W. Nicoll *et al.*, 'Acute effects of the antiepileptic succinimides on the urinary tract and potentiation of phensuximide-induced urotoxicity by phenobarbital.' *J. Appl. Toxicol.*, 1990, **10**, 203–209.

66. M. J. Ellenhorn and D. G. Barceloux, 'Medical Toxicology: Diagnosis and Treatment of Human Poisoning,' New York, 1988, pp. 229–266.

67. G. O. Rankin, K. Cressey-Veneziano, R. T. Wang *et al.*, 'Urinary tract effects of phensuximide in the Sprague–Dawley and Fischer 344 rat.' *J. Appl. Toxicol.*, 1986, **6**, 349–356.

68. T. C. Orton and P. J. Nicholls, 'Effects in rats of subacute administration of ethosuximide, methsuximide and phensuximide on hepatic microsomal enzymes and porphyrin turnover.' *Biochem. Pharmacol.*, 1972, **21**, 2253–2261.

69. G. O. Rankin, H. C. Shih, V. J. Teets *et al.*, 'Acute toxicity induced by 2-aryl-*N*-methylsuccinimides.' *J. Appl. Toxicol.*, 1990, **10**, 143–152.

70. G. O. Rankin, K. W. Beers, D. W. Nicoll *et al.*, 'Role of *para*-hydroxylation in phensuximide-induced urotoxicity in the Fischer 344 rat.' *Toxicology*, 1992, **74**, 77–88.

71. A. Jaklinski and R. Bryc, 'Microscopic examination of rat organs after administration of *m*-bromophenylsuccinimide for 6 and 14 weeks.' *Acta Polon. Pharm.*, 1975, **32**, 379–385.

7.36
α2u-Globulin Nephropathy

LOIS D. LEHMAN-MCKEEMAN
Procter & Gamble, Cincinnati, OH, USA

7.36.1 INTRODUCTION

α2u-Globulin nephropathy is a renal syndrome that occurs exclusively in male rats. It is produced by a diverse group of chemicals and is manifested acutely as the accumulation of protein in phagolysosomes of renal proximal tubule cells. The syndrome was named to reflect the experimental observation that the male rat-specific protein, α2u-globulin, was the only protein accumulating in the histologically-evident droplets. With chronic exposure, renal toxicity progresses from protein overload to renal cell injury, compensatory cell proliferation and ultimately a low, but significant incidence of renal tubular tumors. The hallmark of the syndrome is that the toxic and carcinogenic responses are observed exclusively in male rats. The species specificity of this nephropathy is determined by the fact that α2u-globulin is synthesized only by adult male rats, and the rate-limiting step in the development of the syndrome is the ability of a chemical (or its metabolites) to bind reversibly, but specifically to α2u-globulin. Furthermore, although other species, including humans, synthesize proteins that are similar to α2u-globulin, differences in ligand-binding properties and renal handling of these homologues preclude their involvement in this protein droplet nephropathy. A detailed understanding of the biochemical and molecular mechanisms underlying the development of this syndrome has led to the conclusion that chemicals which cause α2u-globulin nephropathy do not present a risk for similar toxicity or carcinogenic response in humans.

7.36.2 CHEMICAL INDUCERS OF α2u-GLOBULIN NEPHROPATHY

A diverse and ever-growing group of chemicals has been shown to cause α2u-globulin nephropathy (Table 1). Historically, the

Table 1 Chemicals which produce α2u-globulin nephropathy.

Chemical	Ref.
Decalin	6
Tricyclodecane	7
1-Decalone	8
2-Decalone	8
Naphthas	9
60 Solvent, 70 Solvent	10
C10–12 Isoparaffinic Solvent	11
Stoddard Solvent	10
Tetralin	12
2,4,4-Trimethylpentane	13
Tridecyl acetate	14
t-Butylcyclohexane	15
Isopropylcyclohexane	16
Diisobutylketone	17
Methylisobutylketone	18
3,5,5-Trimethyl-hexanoyloxybenzenesulfonate	19
BW540C, BW58C	20
Levamisole	20
Gabapentin	5
Sodium barbital	21
Diethylacetyl urea	21
Lindane	22
1,3,6-Tricyanohexane	23
4-Chloro-α,α,α-trifluorotoluene	24

Table 2 Chemicals which produce α2u-globulin nephropathy and renal tubular tumors in male rats.

Chemical	Ref.
Unleaded gasoline	10,25
Jet fuels (JP-5, JP-4, JP-10, RJ-5)	26,27,28
Isophorone	29
Pentachloroethane	30,31
Perchloroethylene	30,32,33
1,4-Dichlorobenze	34,35
d-Limonene	36,37,38
Dimethyl methylphosphonate	39

nephropathy was characterized following exposure to a variety of small, branched-chain petroleum hydrocarbons,[1] but, as noted in Table 1, both aliphatic and aromatic compounds representing a variety of solvents, fuels, pesticides and pharmaceutical agents have since been shown to produce this toxicity. Structure–activity relationships are determined primarily by the hydrophobicity and molecular volume of the binding cavity in α2u-globulin.[2-4] Whereas a variety of chemicals has been shown to cause the acute syndrome of renal protein overload, only a small subset of chemicals has been tested in chronic bioassays. However, all chemicals shown to produce renal tubular tumors exclusively in male rats (Table 2) have also produced the acute nephropathy, suggesting that there is a link between the nephropathy and the carcinogenic outcome. The only exception to this generalization is gabapentin,[5] which produced the nephropathy but did not produce renal tumors.

Most of the research conducted to determine the mechanisms underlying the development of this syndrome has been done with unleaded gasoline, 2,4-4-trimethylpentane (the active component in unleaded gasoline) or *d*-limonene as model compounds. For this chapter, data obtained with *d*-limonene is primarily used to summarize the biochemical and pathophysiological mechanisms underlying the renal protein overload and development of renal tubular tumors seen with α2u-globulin nephropathy. *d*-Limonene is a naturally occurring monoterpene hydrocarbon found in a variety of citrus fruits and used widely as a flavor, fragrance and industrial solvent. Despite its ability to produce renal tubular tumors in male rats,[38] it has been considered for therapeutic use as an anticancer drug.[40]

7.36.3 HISTOPATHOLOGICAL CHARACTERIZATION OF α2u-GLOBULIN NEPHROPATHY

The acute, pathognomonic feature of α2u-globulin nephropathy is the rapid accumulation of protein (or hyaline) droplets in the proximal tubule cells following chemical treatment. Male rats are unique among all species in that they present with a background of spontaneous protein droplets in the proximal tubule, particularly the cells of the P2 segment. (also referred to as the S2 segment). These spontaneous droplets have been attributed to the large amount of filtered protein that is reabsorbed and degraded in the lysosomal compartment of these cells.[41,42] By light microscopy, these droplets are somewhat difficult to detect with routine hematoxylin and eosin staining, but can be readily visualized with special stains including the Mallory–Heidenhain stain[43] or Lee's methylene basic blue fuschin stain.[13] Figure 1 (top panel) shows the typical appearance of an untreated male rat kidney prepared with Mallory–Heidenhain staining, in which a few, small protein droplets are noted. In contrast, following *d*-limonene treatment (Figure 1, bottom panel), there is an obvious and significant increase in the number, size and staining intensity of the protein droplets. At the ultrastructural level, these large droplets represent

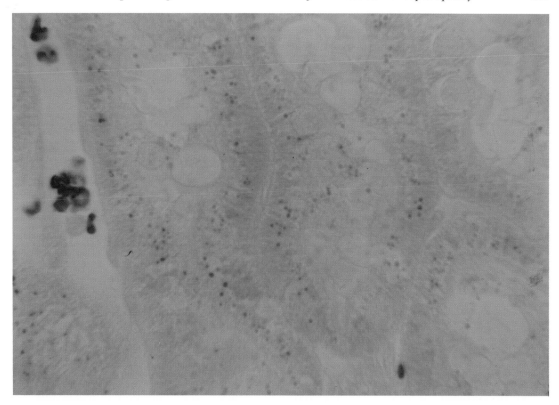

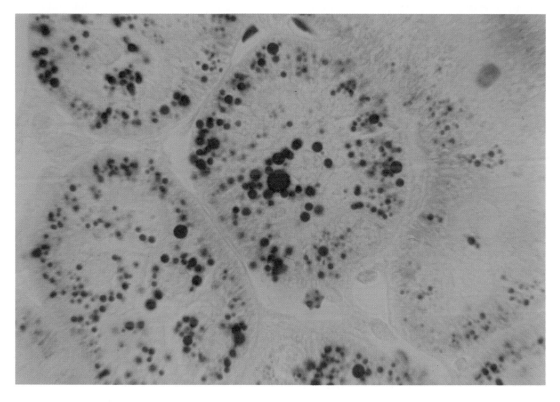

Figure 1 Histopathological characterization of the spontaneous and exacerbated formation of hyaline droplets, the hallmark of α2u-globulin nephropathy, in male rat kidneys (×320). When stained with Mallory–Heidenhain stain, small, lightly stained eosinophilic droplets are observed in the proximal tubular epithelium of untreated male rats (top panel). No similar spontaneous droplet formation is observed in the kidneys of any other species. Following a single oral dosage of *d*-limonene to male rats (150 mg kg^{-1}), the number, size and staining intensity of the droplets is markedly increased within 24 h (lower panel).

phagolysosomes that can be so engorged with protein, that they become irregular and polyangular in shape and contain crystalloid inclusions.[44,45]

Whereas the accumulation of protein droplets occurs rapidly, continued chemical treatment results in additional histological changes in the kidney. Specifically, with 3–4 weeks of dosing, progressive renal injury, characterized by single cell degeneration and necrosis in the renal proximal tubule, is noted. Dead cells are sloughed into the lumen of the nephron, and while moving through the nephron contribute to the development of granular casts at the cortico-medullary junction. There is also an increase in mitotic figures in the P2 segment.[13,43,46–48] If treatment is stopped at this time, recovery occurs, and normal renal architecture is restored.[49] After several months of treatment, linear papillary mineralization, caused by the accumulation of calcium hydroxyapatite in the thin limbs of Henle, is noted, and there is an accelerated onset of the cortical changes typical of chronic progressive nephropathy typically seen in older male rats.[13,43,48] Renal cell death and degeneration cause compensatory cell proliferation in the cortex, and sporadic foci of atypical hyperplasia, defined as the aggregation of morphologically abnormal cells, can be observed in the proximal tubules. Again, these changes are observed only in male rats. No renal changes have been observed in female rats or in any other species including mice, dogs and monkeys.[29,31,33,35,38,39,50] As discussed below, the increased cell turnover and cell proliferation are prerequisite to and ultimately involved in the tumor development associated with this syndrome.

7.36.4 α2u-GLOBULIN

The histological characteristics of this syndrome were determined well before the role of α2u-globulin in renal tumorigenesis was understood. Given that the droplets observed histologically appeared to contain protein, initial biochemical experiments focused on identifying what protein or proteins had accumulated in the droplets following exposure. Using two-dimensional electrophoresis, Alden et al.[43] demonstrated that only one protein had accumulated in the kidney after chemical treatment, and this protein was subsequently identified as α2u-globulin. Moreover, immunohistochemical studies have localized the accumulation of α2u-globulin to the protein droplets.[51,52]

The identification of α2u-globulin immediately provided some perspective concerning the male rat specificity of this syndrome. Specifically, α2u-globulin is an unusual protein synthesized exclusively by adult male rats (Table 3). It was originally isolated from male rat urine by Roy et al.[53] It is synthesized predominantly in liver,[54,55] but extra-hepatic sites of synthesis, including the salivary, lachrymal, preputial, meibomian, and perianal gland, have been identified.[56–58] However, α2u-globulin purified from these accessory glands is electrophoretically-distinct from the hepatic form of the protein.

α2u-Globulin is encoded by a multigene family clustered on chromosome 5,[55] and its expression is regulated by a complex interaction of testosterone, glucocorticoids, insulin, thyroid hormone, and growth hormone.[59–64] It is generally recognized that gene expression is maximal in a hormonally intact, sexually mature male rat, and expression cannot be increased further.[59] α2u-Globulin represents the second most abundant mRNA species in the male rat liver, with albumin the most abundant.[59,65] Although female rats possess the entire complement of hepatic α2u-globulin genes, estrogen is a very effective repressor of the expression of the genes.[66] Masculinization of female rats will increase expression of α2u-globulin, but not to the levels seen in males.[67,68] In extrahepatic sites, the hormonal regulation of α2u-globulin gene expression is not sex-specific as the protein has been detected in the salivary, lachrymal, and preputial gland of female rats.[56,57]

The α2u-globulin detected in the kidney is considered to be hepatic in origin.[54,56] Specifically, once synthesized, the protein is rapidly secreted from the liver into the general circulation, and only very low hepatic levels are detected at steady state.[53] The molecular weight of α2u-globulin is approximately 18.5 kDa, enabling it to be freely filtered across the

Table 3 Characteristics of α2u-globulin.

1. Synthesized primarily in the liver of adult male rats
2. Not synthesized by any other species
3. Hormonal regulation of hepatic gene expression
   Required for expression: testosterone, glucocorticoids insulin, thyroid hormone, growth hormone
   Inhibits expression: estrogen
4. Low molecular weight (18.5 kDa) allows for glomerular filtration
5. Major urinary protein excreted by adult male rats
6. Rapidly accumulates in protein droplets following chemical exposure
7. Unknown function

glomerulus.[53] In fact, renal clearance of the protein is so effective that plasma concentrations of α2u-globulin are generally low, typically no higher than 3 mg%.[54] It is estimated that adult male rats synthesize about 50 mg α2u-globulin day[-1], and essentially all of this protein is filtered by the kidney.[53,69,70] The renal handling of the protein is unlike most other proteins, however, as it is only partially reabsorbed. Specifically, about 60% of the filtered α2u-globulin (or approximately 30 mg day[-1]) is reabsorbed,[69,71] which is in direct contrast to most other proteins which are effectively entirely reabsorbed by the proximal tubule cells.[72] The large amount of α2u-globulin which is reabsorbed is thought to contribute to the spontaneous formation of protein droplets in the male rat kidney. The remaining, nonabsorbed protein (15–20 mg day[-1]) is excreted in the urine, where α2u-globulin is the most abundant protein species, accounting for about 35% of the total urinary protein.[53,60,70,73]

Although male rats synthesize a large quantity of α2u-globulin, the function of this protein remains unknown. Advances in protein biochemistry have indicated that proteins which are structurally similar to α2u-globulin serve to bind and transport a variety of hydrophobic ligands.[74–76] In this regard, the detection of α2u-globulin in pheromone producing glands (preputial, meibomian, and perianal glands) may be suggestive of a function relating to the transport of pheromones to the body surface or out into the urine. However, no endogenous ligand for α2u-globulin has ever been identified.[53,77] Other functions suggested for the protein include a role in renal fatty acid binding[78] or in the regulation or modulation of spermatogenesis.[79]

Whereas a physiological function for α2u-globulin is not known, its role in the pathophysiology of male rat specific protein droplet nephropathy and renal carcinogenesis is unequivocal. Only α2u-globulin accumulates in the protein droplets, only male rats synthesize α2u-globulin, and only male rats develop this syndrome. The nephropathy is not seen in female rats or any other species.[43,47,80–82] Furthermore, the nephropathy does not develop in juvenile male rats[43] since synthesis of the protein is not detected until puberty,[83] nor is it observed in the male NCI Black Reiter rat,[84] an unusual strain which does not synthesize α2u-globulin.[85] Finally, although mice are refractory to this toxicity the nephropathy can be produced in a transgenic mouse engineered to express α2u-globulin.[86]

7.36.5 BIOCHEMICAL MECHANISMS OF α2u-GLOBULIN NEPHROPATHY

The common characteristic of all chemicals that produce α2u-globulin nephropathy is the ability to bind to the protein. Indeed, the ability to bind to α2u-globulin is the rate-limiting step in the development of the nephropathy.[47,80] In the kidney there is a sex-dependent retention of chemical, with more compound distributing to and being retained by the male rat kidney relative to female rats,[19,37,87] and in the male rat kidney, between 20% and 40% of the chemical is bound specifically to α2u-globulin.[19,37,88,89] This binding is reversible, but the ligand–protein complex will survive tissue homogenization and centrifugation. Results from *in vitro* equilibrium saturation binding studies have indicated that the dissociation constant for xenobiotic binding to α2u-globulin is approximately 10^{-7} M.[2,69,77]

For several chemicals, the moieties that bind to α2u-globulin have been identified (Table 4). These ligands are as structurally diverse as the class of compounds which produce the syndrome. In some cases, parent compound will bind directly to the protein (isophorone),

Table 4 Affinity of chemicals that bind to α2u-globulin.

Ligand	K_d or K_i	Ref.
2,4,4-Trimethyl-2-pentanol	6.4×10^{-7} M	2,8,77
d-Limonene-1,2-epoxide	5.6×10^{-7} M	2,8,69,77
Isophorone	7.7×10^{-6} M	2
1,4-Dichlorobenzene	5.2×10^{-4} M	2
2,5-Dichlorophenol	3.9×10^{-4} M	2
d-Limonene	1.0×10^{-4} M	2,8
3,5,5-Trimethylhexanoic acid	ND[a]	19

[a]Not determined

whereas oxidative metabolites, including epoxides (*d*-limonene-1,2-epoxide), ketones (γ-lactone of 3,5,5-trimethylhexanoic acid) and hydroxylated metabolites of aliphatic (2,4,4-trimethyl-2-pentanol) or aromatic compounds (2,5-dichorophenol) have also been isolated from α2u-globulin. It is particularly interesting to note that the limonene epoxide intermediate binds reversibly to this protein.[37,77] X-ray crystallographic results indicate that the epoxide moiety is likely oriented in the α2u-globulin binding pocket in the vicinity of a methionine residue, but it is not clear whether the methionine residue is essential for the binding of this ligand.[4] More information with respect to the nature of the binding site in α2u-globulin and the orientation of the chemical in the binding site will become available as the x-ray crystal structure of the protein is solved (see below). However, it is clear that ligands must be hydrophobic in nature and have a molecular volume of no greater than 100 Å^3 to fit into the protein binding site.[2,4,90]

Although the ligand–protein complex can be isolated from the kidney *in vivo*, it is not known where the interaction between the xenobiotic and α2u-globulin occurs. However, given that the complex appears to be localized in the phagolysosomal compartment, it is likely that the complex does not form in the kidney. Rather, the interaction probably occurs outside the kidney (in liver or blood), and following glomerular filtration, the complex enters the proximal tubule cells by endocytosis and accumulates in the phagolysosomes after lysosomal fusion.

Following binding of a ligand to α2u-globulin, the lysosomal degradation of the protein is altered relative to the native protein. Specifically, in *in vitro* experiments with renal cortical lysosomal lysates as enzyme source, the rate of degradation of α2u-globulin to which *d*-limonene-1,2-epoxide, isophorone or 2,5-dichlorophenol was bound decreased by about 30%.[91] In contrast, however, the lysosomal degradation of other proteins such as albumin was not altered by the presence of any of these chemicals, nor was there any change in lysosomal cathepsin activity towards model substrates in the presence of the chemicals or the α2u-globulin complexes.[91] Using immunoblotting techniques, it has been found that α2u-globulin exists in two forms in the kidney.[2,92] One form is the native protein (18.5 kDa), whereas the second form is a smaller form (~16 kDa) representing the native protein lacking the first nine amino acids. In general, both forms of the protein increase after chemical treatment.[2,92] Thus, the hallmark exacerbation of protein droplets seen following

chemical treatment occurs because the rate of degradation of α2u-globulin is reduced by ligand binding, and with chemical binding, protein begins to accumulate in the phagolysosomes.

As chemical treatment continues, the lysosomes become enlarged, engorged with protein and polyangular.[44,45] Although the mechanisms of cell death are not understood for this syndrome, there is cytotoxicity evidenced by single cell necrosis in the P2 segment of the proximal tubule[13,80] and a dose- and time-dependent loss in renal function. Renal functional changes observed include reduced uptake of organic anions, cations, and amino acids and

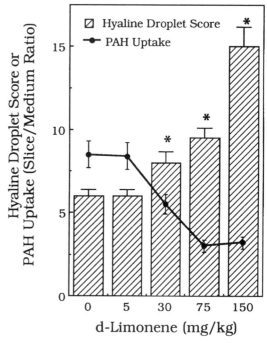

Figure 2 Dose–response relationship for *d*-limonene-induced acute (single exposure) exacerbation of hyaline droplet formation and subchronic changes (91 days of repeat dosing) in renal function. Exacerbation of hyaline droplets was evaluated histologically 24 h after a single oral dosage of *d*-limonene (150 mg kg^{-1}). Droplets were graded on a scale from 0 to 16 (with a maximum score of 16) after staining with Mallory–Heidenhain stain. In this example, renal function was assessed as the *in vitro* uptake of *p*-aminohippurate (PAH) in cortical slices prepared from rats dosed daily with *d*-limonene for 91 days. Asterisks indicate dosages for which both hyaline droplet scores and PAH uptake were statistically different from control ($P < 0.05$). A dosage of 5 mg kg^{-1} did not exacerbate hyaline droplet formation and did not change PAH uptake. At dosages of 30 mg kg^{-1} and higher, dose-dependent increases in hyaline droplet severity were observed along with a dose-dependent decrease in PAH uptake.

a mild proteinuria resulting from a large increase in the amount of α2u-globulin excreted in urine.[93,94] These functional changes occur only at dosages which exacerbate the protein droplet formation (Figure 2). In response to the cell death and functional changes, there is cell proliferation in the kidney, most notably in the P2 segment of the proximal tubules, the site of the protein accumulation.[94–96] With continued treatment, the cell proliferation continues, but does not restore renal function. As shown in Figure 3, loss of renal function is both dose- and time-dependent, and reflects the severity of the acute nephropathy.[94,95] The increase in renal cell proliferation is believed to exert a promotional influence on the kidney, such that sustained cell turnover is mechanistically-linked to the development of renal tubular tumors.[94–98] Thus, α2u-globulin nephropathy begins acutely as protein accumulation, but represents a continuum of changes that ultimately progress to renal tumors.[80–82,99,100] The major events in this continuum are summarized in Figure 4 with binding of a chemical to α2u-globulin as the rate-limiting step in the development of this syndrome.

7.36.6 RENAL TUBULAR TUMOR FORMATION IN α2u-GLOBULIN NEPHROPATHY

Renal tubular tumors are uncommon in rats, occurring spontaneously at incidences of 0.35% and 0.17% for male and female F344 rats, respectively.[101] As noted in Table 2, eight chemicals causing renal tubular tumors exclusively in male rats have been identified, and the highest tumor rate (seen with *d*-limonene) was approximately 25%.[38] Thus, the chemicals listed in Table 2 are distinguished from classical

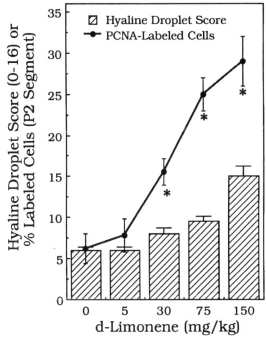

Figure 3 Dose–response relationship for *d*-limonene-induced acute exacerbation of hyaline droplet formation and subchronic changes in proximal tubule cell proliferation. Hyaline droplets were graded histologically on a scale from 0 to 16 (with a maximum score of 16; as described in Figure 2) after staining with Mallory–Heidenhain stain, and cell proliferation was evaluated by histological detection of proliferating cell nuclear antigen (PCNA) after 91 days of dosing with *d*-limonene. Asterisks indicate dosages for which both hyaline droplet scores and PCNA-labeled cells were statistically different from control ($P < 0.05$). There was no exacerbation of hyaline droplet formation and no increase in PCNA staining at a dosage of 5 mg kg^{-1}. However, cell proliferation, localized to the P2 segment of the proximal tubule, increased in a dosage-dependent manner at 30 mg kg^{-1} and higher, and these changes were consistent with the dosage-dependent increase in hyaline droplet severity.

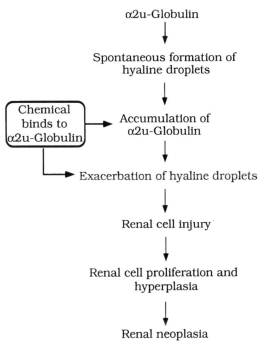

Figure 4 Schematic representation of the continuum of changes initiated by the binding of a xenobiotic to α2u-globulin. Small droplets form spontaneously in male rat kidneys due to the large amount of α2u-globulin in the kidney. When a chemical binds to α2u-globulin, the acute, pathognomonic response is the rapid exacerbation of hyaline droplet formation, and with chronic exposure, the syndrome progresses from protein overload to renal cell injury and compensatory hyperplasia prior to the development of renal tubular tumors.

renal carcinogens such as the nitrosamines where tumor incidences approach 100%.[102,103] Furthermore, the renal tumors associated with α2u-globulin nephropathy are distinguished from classical renal carcinogens, in that they show a much longer latency period, requiring at least 18 months of continued dosing, and the chemicals that cause the male rat-specific renal tumors are not highly mutagenic.[102–104] For the chemicals listed in Table 2, there is generally no evidence of mutagenicity in Salmonella (with or without metabolic activation) or mouse lymphoma cells, or clastogenicity in Chinese hamster ovary cells.[29,31,33,35,38,39] In this regard, the identification of a potentially reactive epoxide intermediate of *d*-limonene that binds to α2u-globulin might raise some concern that the epoxide was directly involved in the carcinogenicity of *d*-limonene. However, this epoxide is an unusually stable chemical, which is also negative in a battery of mutagenicity tests.[105–108]

The lack of genotoxicity among the chemicals causing α2u-globulin nephropathy and male rat specific renal tumors supports the concept that nongenotoxic mechanisms are involved in the carcinogenic response. The cell proliferation seen with subchronic and chronic exposure to these chemicals is likely to be involved in the tumor formation associated with this syndrome.[80,95,96] Specifically, sustained cell proliferation is believed to be a primary mechanism underlying the tumor response to many nongenotoxic chemicals, as accelerated cell turnover can promote clonal expansion of spontaneously-initiated cells.[109,110] As shown in Figure 3, the dose-related increase in renal cell proliferation is parallel to the dose–response curve for the development of the hyaline droplets. Although this does not prove a causal relationship, it suggests that sustained cell proliferation is the pivotal link between the acute nephropathy and the chronic renal tumor formation. Furthermore, in initiation–promotion protocols, both unleaded gasoline[97] and *d*-limonene[98] have been shown to act as promoters of *N*-ethyl-*N*-hydroxyethylnitrosamine (EHEN) initiated renal tumors. However, when tested in a similar regimen, *d*-limonene does not promote tumor formation in the α2u-globulin-deficient, NCI Black Reiter rat,[98] again supporting an association between α2u-globulin and renal tumors. Finally, sodium barbital, used historically as a renal tumor promoter[111] has also been shown to cause α2u-globulin nephropathy in male rats,[21] again supporting a role for sustained cell proliferation in linking the nephropathy to the development of renal tumors as presented in Figure 4.

7.36.7 HUMAN RELEVANCE OF α2u-GLOBULIN NEPHROPATHY

Given the biochemical evidence that α2u-globulin is necessary for the development of this syndrome and that the protein is synthesized exclusively by adult male rats, it seems reasonable to conclude that α2u-globulin nephropathy is unique to the male rat. However, α2u-globulin is a member of the lipocalin protein superfamily,[75] a large and diverse group of proteins that have relatively low amino acid sequence homology, but which share a very unusual tertiary structure. The proteins in this family, originally referred to as the α2u-globulin protein superfamily,[112] are synthesized across many species, including humans. Moreover, the primary function of the lipocalins is to bind and transport small hydrophobic molecules, and for many of them, endogenous ligands have been identified (Table 5). The proteins in this superfamily are distinguished by a three-dimensional structure that features a β-barrel with eight antiparallel β-strands that fold into an orthogonal calyx, or cup-like structure. Ligands are bound within the β-barrel cavity which is lined almost exclusively by hydrophobic amino acids that contribute to ligand specificity.[25,113,114]

The superfamily protein that is most similar to α2u-globulin is mouse urinary protein (MUP). This protein is essentially the mouse homologue of α2u-globulin, as it shares approximately 90% sequence homology and is hormonally regulated in a manner similar to that described previously for α2u-globulin.[115,139] Like α2u-globulin, MUP is the major urinary protein excreted in mice,[69,140,141] and represents the most abundant mRNA species in the mouse liver.[139] It is estimated that male mice excrete about 15 mg of MUP daily.[69,140] One distinction between MUP and α2u-globulin is that MUP expression is not completely repressed by estrogen, so that female mice synthesize and excrete MUP, albeit at levels that are significantly less than male mice.[69,139,142,143]

If the similarities among the lipocalins are sufficient for these proteins to bind to chemicals that cause α2u-globulin nephropathy, then, because of the very high homology between α2u-globulin and MUP, mice should be sensitive to developing a similar renal syndrome. However, studies conducted in mice have indicated that there are no renal changes, including renal tumors, observed following exposure to any of the chemicals noted in Table 2 nor any evidence of an acute nephropathy.[19,29,31,33,35,38,39,69] Furthermore, in α2u-globulin transgenic mice, *d*-limonene treatment

Table 5 Members of the lipocalin or α2u-globulin superfamily.

Protein	Endogenous ligand	Ref.
α2u-Globulin	Unknown	4,112
Mouse urinary protein	2-(s-butyl)-4,5-dihydrothiazole	90,115
Retinol binding protein	Retinol	113,116
α₁-Acid glycoprotein	Drugs, steroids	117,118
β-Lactoglobulin	Retinol	117,119,120
Odorant-binding protein	Odorants	121,122
Apolipoprotein D	Cholesterol	123
Von Ebner's gland protein	Unknown	124
Protein-HC	Unknown	125
Pregnancy-associated endometrial α2-globulin	Unknown	126
Androgen-dependent secretory protein	Retinoic acid	127,128
Aphrodesin	Unknown	129
Probasin	Unknown	130
Complement C8	Retinol	131
Mouse 24p3 protein	Unknown	132
Prostaglandin D synthase	PGH₂	133
Tear prealbumin (LNCI)	Retinol	134
Neutrophil gelatinase-associated lipocalin	Unknown	135
Quiescence specific protein	Unknown	136
Purpurin	Retinol, heparin	137
Insecticyanin	Biliverdin 1X	138

causes α2u-globulin to accumulate while not affecting the renal handling of MUP in any way.[86] The lack of toxicity in mice treated with hyaline droplet inducing agents is attributed to several major differences between MUP and α2u-globulin. Specifically, MUP does not bind any of the chemicals shown to bind to α2u-globulin,[69,77] and MUP is not reabsorbed into the kidney to any significant extent.[69,144]

There are two other major differences between α2u-globulin and the lipocalin superfamily proteins that distinguish α2u-globulin from all of these proteins. The first difference is that for most of the proteins in the superfamily, endogenous ligands have been identified, whereas no endogenous ligand has been detected for α2u-globulin (Table 5). For example, it is well-established that retinol-binding protein (RBP) binds and transports retinol,[116] whereas apoliprotein D transports cholesterol in the circulation.[123] Similarly, a variety of ligands have been identified for odorant binding protein[122] and α1-acid glycoprotein.[118] For MUP, a dihydrothiazole compound has been isolated from the protein excreted in urine.[90,145] Thus, the presence of these endogenous ligands is likely to prevent the binding of hyaline droplet inducing agents to other superfamily proteins, whereas the absence of an endogenous ligand for α2u-globulin allows for chemicals like d-limonene-1,2-epoxide to bind to the protein. In this regard, it also appears that the binding of ligands to other superfamily proteins

does not affect the renal handling of that protein, whereas for α2u-globulin, binding of a chemical clearly alters the renal degradation of the protein.[91,116,117] In this manner, although the other superfamily proteins are reabsorbed

Table 6 Amino acid residues that line the binding pockets of α2u-globulin and MUP.

α2u-Globulin[4]	MUP[90]
	Ile19
	Leu28
Met38	Phe38
Val40	Leu40
Met42	Leu42
	Ile45
	Leu52
Phe54[a]	Leu54[a]
Phe56	Phe56
Leu69	Met69
Val82	Val82
Tyr84	Try84
Ans88	Ans88
Phe90	Phe90
	Ile96
	Ile101
Phe103[a]	Ala103[a]
Leu105	Leu105
Leu116	
Val118	
Tyr120	Tyr120

[a]Affect shape of binding pocket.

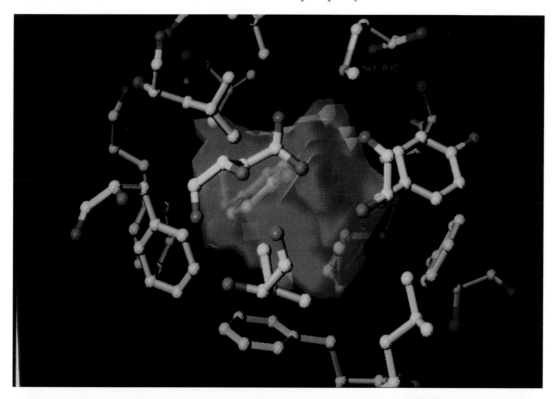

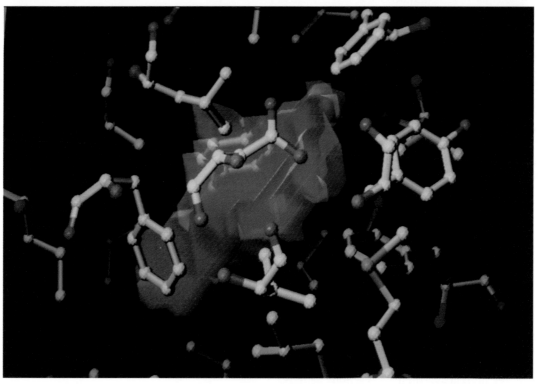

Figure 5 Depiction of the ligand-binding cavities in α2u-globulin (top panel) and MUP (bottom panel) as determined by x-ray crystallography. For α2u-globulin, the blue cloud indicates the location and shape of the binding cavity which is essentially spherical, resulting from the orientation of two phenylalanine residues which close off the back of the cavity. The two phenylalanine residues in α2u-globulin are replaced by leucine and alanine, respectively, in MUP, resulting in a binding cavity that is elongated and flattened (red cloud) relative to the binding pocket in α2u-globulin.

by the proximal tubule, they do not accumulate in the kidney either spontaneously or following chemical treatment.[43,77]

The last and most important difference between α2u-globulin and the other members of the lipocalin superfamily concerns the nature of the binding cavity in α2u-globulin. Based on x-ray crystallographic data,[24] the binding pocket in α2u-globulin is characterized as: (i) closed off to water; (ii) lined by extremely hydrophobic amino acid residues; (iii) having a probe accessible volume of $84.0 \pm 0.9 \text{Å}^3$, and (iv) very similar to mouse urinary protein (MUP) in both hydrophobicity and volume (75Å^3).[90,146] However, subtle differences in the amino acids that line the binding pocket affect the shape of the cavities (Table 6) so that the overlap between the cavities in these homologous proteins is only about 42Å^3. Specifically the location of two phenylalanine residues in α2u-globulin (Phe-54 and Phe-103) serve to close off the back of the binding cavity in α2u-globulin, creating a pocket that is essentially spherical in shape (Figure 5, top panel). In contrast, the amino acid residues at position 54 and 103 in MUP are leucine and alanine, respectively. These amino acids do not close off the pocket, yielding a cavity which is flattened and elongated (Figure 5, bottom panel). In this manner, the MUP cavity cannot accommodate chemicals such as *d*-limonene-1,2-epoxide.[4,69]

Other lipocalins, such as retinol-binding protein and bilin-binding protein also have flattened binding cavities and have much less overlap (about 20Å^3) with α2u-globulin.[4,113] Thus, the specific binding of chemicals to α2u-globulin, is ultimately attributed to the unique nature and shape of its ligand binding site.

7.36.8 RISK ASSESSMENT FOR CHEMICALS CAUSING α2u-GLOBULIN NEPHROPATHY

The histopathological, biochemical and molecular data provide compelling evidence that α2u-globulin nephropathy is unique to the male rat and is unlikely to occur in any other species, including humans. Furthermore, those chemicals shown to produce clear evidence of renal cancer in male rats are also not likely to be carcinogenic in humans.[99,100] Following extensive review, the USEPA concluded that renal toxicity or tumors in male rats arising from a process involving α2u-globulin should not be used to estimate either nephrotoxic hazard or cancer risk for humans, and established criteria to be used to categorize chemicals as belonging

Table 7 Regulatory criteria for establishing the role of α2u-globulin nepropathy in male rat renal carcinogenesis.

Essential evidence
- Renal tumors occur only in male rats
- Acute exposure exacerbates hyaline droplet formation
- α2u-Globulin accumulates in hyaline droplets
- Subchronic histopathological changes including granular cast formation and linear papillary mineralization
- Absence of hyaline droplets and characteristic histopathological changes in female rats and mice
- Negative for genotoxicity in a battery of tests

Additional supportive evidence
- Reversible binding of chemical (or metabolites) to α2u-globulin
- Increased and sustained cell proliferation in P2 segment of proximal tubules in male rat kidneys
- Dose–response relationship between hyaline droplet severity and renal tumor incidence

Source: USEPA.[147]

to this class of male rat specific renal carcinogens (Table 7).[82,147] Given these conclusions, the syndrome of α2u-globulin nephropathy has helped to demonstrate the utility of biochemical and molecular information for understanding toxic mechanisms and has provided an excellent opportunity to incorporate mechanistic data into the human risk assessment process.

ACKNOWLEDGMENT

The author gratefully acknowledges the technical contributions of Douglas Caudill in completing much of the work described in this chapter.

7.36.9 REFERENCES

1. C. P. Carpenter, E. R. Kinkead, K. L. Geary, Jr. *et al.*, 'Petroleum hydrocarbon toxicity studies. II. Animal and human response to vapors of varnish makers' and painters' naphtha.' *Toxicol. Appl. Pharmacol.*, 1975, **32**, 263–281.
2. S. J. Borghoff, A. B. Miller, J. P. Bowen *et al.*, 'Characteristics of chemical binding to α2u-globulin *in vitro*-evaluating structure-activity relationships.' *Toxicol. Appl. Pharmacol.*, 1991, **107**, 228–238.
3. M. D. Barratt, 'A quantitative structure-activity relationship (QSAR) for prediction of α2u-globulin nephropathy.' *Toxic. In Vitro*, 1994, **8**, 885–887.
4. L. D. Lehman-McKeeman, G. J. Kleywegt, J. D. Oliver *et al.*, 'Characterization of the ligand binding site in α2u-globulin by x-ray crystallography.' *Toxicologist*, 1995, **15**, 300.

5. M. A. Dominick, D. G. Robertson, M. R. Bleavins *et al.*, 'α2u-Globulin without nephrocarcinogenesis in male Wistar rats administered 1-(aminomethyl)cyclohexaneacetic acid.' *Toxicol. Appl. Pharmacol.*, 1991, **111**, 375–387.

6. R. L. Kanerva, M. S. McCracken, C. L. Alden *et al.*, 'Morphogenesis of decalin-induced renal alterations in the male rat.' *Food Chem. Toxicol.*, 1987, **25**, 53–61.

7. E. Bomhard, M. Marsmann, C. Ruhl-Fehlert *et al.*, 'Relationships between structure and induction of hyaline droplet accumulation in the renal cortex of male rats by alipatic and alicyclic hydrocarbons.' *Arch. Toxicol.*, 1990, **64**, 530–538.

8. S. K. Wolf, D. Caudill and L. D. Lehman-McKeeman, 'Development of an *in vitro* assay to predict hyaline droplet nephropathy.' *Toxicologist*, 1993, **13**, 453.

9. C. A. Halder, T. M. Warne and N. S. Hatoum, in 'Renal Effects of Petroleum Hydrocarbons. Advances in Modern Environmental Toxicology,' eds. M. A. Mehlman, C. P. Hemstreet, III, J. J. Thorpe *et al.*, Princeton Scientific Publishers, NJ, 1984, pp. 73–88.

10. R. D. Phillips and B. Y. Cockrell, in 'Renal Effects of Petroleum Hydrocarbons,' eds. M. A. Mehlman, C. P. Hemstreet, III, J. J. Thorpe *et al.*, Princeton Scientific Publishers, NJ, 1984, pp. 89–105.

11. C. Viau, A. Bernard, F. Gueret *et al.*, 'Isoparaffinic solvent-induced nephrotoxicity in the rat.' *Toxicology*, 1986, **38**, 227–240.

12. M. P. Servé, B. M. Llewelyn, K. O. Yu *et al.*, 'Metabolism and nephrotoxicity of tetralin in male Fischer 344 rats.' *J. Toxicol. Environ. Health*, 1989, **26**, 267–275.

13. B. G. Short, V. L. Burnett, M. G. Cox *et al.*, ' Site-specific renal cytotoxicity and cell proliferation in male rats exposed to petroleum hydrocarbons.' *Lab. Invest.*, 1987, **57**, 564–577.

14. W. C. Daughtrey, J. H. Smith, J. P. Hinz *et al.*, 'Subchronic toxicity evaluation of tridecyl acetate in rats.' *Fundam. Appl. Toxicol.*, 1990, **14**, 104–112.

15. G. M. Hennigsen, K. O. Yu, R. A. Salomon *et al.*, 'The metabolism of *t*-butylcyclohexane in Fischer-344 male rats with hyaline droplet nephropathy.' *Toxicol. Lett.*, 1987, **39**, 313–318.

16. G. M. Henningsen, R. A. Salomon, K. O. Yu *et al.*, 'Metabolism of nephrotoxic isopropylcyclohexane in male Fischer 344 rats.' *J. Toxicol. Environ. Health*, 1988, **24**, 19–25.

17. D. E. Dodd, P. E. Losco, C. M. Troup *et al.*, 'Hyaline droplet nephrosis in male Fischer-344 rats following inhalation of diisobutyl ketone.' *Toxicol. Ind. Health*, 1987, **3**, 443–457.

18. R. D. Phillips, E. J. Moran, D. E. Dodd *et al.*, 'A 14-week vapor inhalation toxicity study of methyl isobutyl ketone.' *Fundam. Appl. Toxicol.*, 1987, **9**, 380–388.

19. L. D. Lehman-McKeeman, P. A. Rodriguez, D. Caudill *et al.*, 'Hyaline droplet nephropathy resulting from exposure to 3,5,5-trimethyl-hexanoyloxybenzene sulfonate.' *Toxicol. Appl. Pharmacol.*, 1991, **107**, 429–438.

20. N. G. Read, P. J. Astbury, R. J. I. Morgan *et al.*, 'Induction and exacerbation of hyaline droplet formation in the proximal tubular cells of the kidneys from male rats receiving a variety of pharmacological agents.' *Toxicology*, 1988, **52**, 81–101.

21. Y. Kurata, B. A. Diwan, L. D. Lehman-McKeeman, *et al.*, 'Comparative hyaline droplet nephropathy in male F344/NCr rats induced by sodium barbital and diacetylurea, a breakdown product of sodium barbital.' *Toxicol. Appl. Pharmacol.*, 1994, **126**, 224–232.

22. D. R. Dietrich and J. A. Swenberg, 'Lindane induces nephropathy and renal accumulation of α2u-globulin in male but not in female Fischer 344 rats or male NBR rats.' *Toxicol. Lett.*, 1990, **53**, 179–181.

23. D. E. Johnson, in 'The Pharmacology and Toxicology of Proteins', eds. J. S. Holdenberg and J. L. Winkelhake, Alan R. Liss, New York, pp. 165–171.

24. A. Macri, C. Ricciardi, A. V. Stazi *et al.*, 'Subchronic oral toxicity of 4-chloro-α,α,α-trifluoro-toluene in Sprague–Dawley rats.' *Food Chem. Toxicol.*, 1987, **25**, 781–786.

25. D. N. Kitchen, in 'Renal Effects of Petroleum Hydrocarbons,' eds. M. A. Mehlman, C. P. Hemstreet III, J. J. Thorpe *et al.*, Princeton Scientific Publishers, NJ, 1984, pp. 65–71.

26. R. H. Bruner, in 'Renal Effects of Petroleum Hydrocarbons,' eds. M. A. Mehlman, C. P. Hemstreet III, J. J. Thorpe *et al.*, Princeton Scientific Publishers, NJ, 1984, pp. 133–140.

27. R. H. Bruner, E. R. Kinkead, T. P. O'Neill *et al.*, 'The toxicologic and oncogenic potential of JP-4 jet fuel vapors in rats and mice: 12-month intermittent inhalation exposures.' *Fundam. Appl. Toxicol.*, 1993, **20**, 97–110.

28. C. L. Gaworski, J. D. MacEwen, E. H. Vernot *et al.*, in 'Applied Toxicology of Petroleum Hydrocarbons. Advances in Modern Environmental Toxicology,' eds. H. N. MacFarland, C. E. Holdsworth, J. A. MacGregor *et al.*, Princeton Scientific Publishers, NJ, 1984, pp. 33–47.

29. National Toxicology Program, 'Toxicology and carcinogenesis studies of isophorone in F344/N rats and B6C3F1 mice (gavage study).' NTP TR 291, National Toxicology Program, Research Triangle Park, NC, 1986.

30. T. L. Goldsworthy, O. Lyght, V. L. Burnett *et al.*, 'Potential role of α2u-globulin, protein droplet accumulation, and cell replication in the renal carcinogenicity of rats exposed to trichloroethylene, perchloroethylene and pentachloroethane.' *Toxicol. Appl. Pharmacol.*, 1988, **96**, 367–379.

31. National Toxicology Program, 'Carcinogenesis bioassay of pentachloroethane in F-344/N rats and B63F1 mice (gavage study).' NTPTR 232, US Dept. Health and Human Services, Public Health Service, National Institutes of Health, 1983.

32. T. Green, J. Odum, J. A. Nash *et al.*, 'Perchloroethylene-induced rat kidney tumors: an investigation of the mechanisms involved and their relevance to humans.' *Toxicol. Appl. Pharmacol.*, 1990, **103**, 77–89.

33. National Toxicology Program, 'Carcinogenesis bioassay of tetrachloroethylene (perchloroethylene) in F-344/N rats and B6C3F1 mice (inhalation study).' NTP TR 311, US Dept. Health and Human Services, Public Health Service, National Institutes of Health, 1986.

34. E. Bomhard, G. Luckhaus, W. H. Voight *et al.*, 'Induction of light hydrocarbon nephropathy by *p*-dichlorobenzene.' *Arch. Toxicol.*, 1988, **61**, 433–439.

35. National Toxicology Program, 'Toxicology and carcinogenesis studies of 1,4-dichlorobenzene in F344/N rats and B6C3F1 mice.' NTPTR 319, US Dept. Health and Human Services, Public Health Service, National Institutes of Health, 1987.

36. D. R. Webb, G. M. Ridder and C. L. Alden, 'Acute and subchronic nephrotoxicity of *d*-limonene in Fischer 344 rats.' *Food. Chem. Toxicol.*, 1989, **27**, 639–649.

37. L. D. Lehman-McKeeman, P. A. Rodriguez, R. Takigiku *et al.*, '*d*-Limonene-induced male rat-specific nephrotoxicity: evaluation of the association between *d*-Limonene and α2u-globulin.' *Toxicol. Appl. Pharmacol.*, 1989, **99**, 250–259.

38. National Toxicology Program, 'Toxicology and carcinogenesis studies of *d*-limonene in F344/N rats and B6C3F1 mice (gavage studies).' NTPTR 347, US

Dept. Health and Human Services, Public Health Service, National Institutes of Health, 1990.

39. National Toxicology Program, 'Carcinogenesis studies of dimethyl methylphosphonate in F-344/N rats and B6C3F1 mice (gavage study).' NTP TR 323, US Dept. Health and Human Services, Public Health Service, National Institutes of Health, 1987.

40. P. L. Crowell and M. N. Gould, 'Chemoprevention and therapy of cancer by *d*-limonene.' *Crit. Rev. Oncog.*, 1994, **5**, 1–22.

41. A. B. Maunsbach, 'Electron microscopic observations of cytoplasmic bodies with crystalline patterns in rat kidney proximal tubule cells.' *J. Ultrastuct. Res.*, 1966, **14**, 167.

42. N. Kretchmer and J. Bernstein, 'The dynamic morphology of the nephron: morphogenesis of the "protein droplet".' *Kidney Int.*, 1974, **5**, 96–105.

43. C. L. Alden, R. L. Kanerva, G. Ridder *et al.*, in 'Renal Effects of Petroleum Hydrocarbons. Advances in Modern Environmental Toxicology,' eds. M. A. Mehlman, C. P. Hemstreet, III, J. J. Thorpe *et al.*, Princeton Scientific Publishers, NJ, 1984.

44. L. C. Stone, R. L. Kanerva, J. L. Burns *et al.*, 'Decalin-induced nephrotoxicity: light and electron microscopic examination of the effects of oral dosing on the development of kidney lesions in the rat.' *Food Chem. Toxicol.*, 1987, **25**, 43–52.

45. B. D. Garg, M. J. Olson, L. C. Li *et al.*, 'Phagolysosomal alterations induced by unleaded gasoline in epithelial cells of the proximal convoluted tubules of male rats: effect of dose and treatment duration.' *J. Toxicol. Environ. Health.*, 1989, **26**, 101–118.

46. B. G. Short, V. L. Burnett and J. A. Swenberg, 'Histopathology and cell proliferation induced by 2,2,4-trimethylpentane in the male rat kidney.' *Toxicol. Pathol.*, 1986, **14**, 194–203.

47. S. J. Borghoff, B. G. Short and J. A. Swenberg, 'Biochemical mechanisms and pathobiology of α2u-globulin nephropathy.' *Annu. Rev. Pharmacol. Toxicol.*, 1990, **30**, 349–367.

48. B. F. Trump, M. M. Lipsky, T. W. Jones *et al.*, in 'Renal Effects of Petroleum Hydrocarbons. Advances in Modern Environmental Toxicology,' eds. M. A. Mehlman, C. P. Hemstreet, III, J. J. Thorpe *et al.*, Princeton Scientific Publishers, NJ, 1984, pp. 273–288.

49. D. R. Mattie, C. L. Alden, T. K. Newell *et al.*, 'A 90-day continuous vapor inhalation toxicity study of JP-8 jet fuel followed by 20 or 21 months of recovery in Fischer 344 rats and C57BL/6 mice.' *Toxicol. Pathol.*, 1991, **19**, 77–87.

50. D. R. Webb, R. L. Kanerva, D. K. Hysell *et al.*, 'Assessment of the subchronic oral toxicity of *d*-limonene in dogs.' *Food Chem. Toxicol.*, 1990, **28**, 669–675.

51. V. L. Burnett, B. G. Short and J. A. Swenberg, 'Localization of α2u-globulin within protein droplets of male rat kidney: immunohistochemistry using perfusion-fixed, GMA-embedded tissue sections.' *J. Histochem. Cytochem.*, 1989, **37**, 813–818.

52. D. R. Dietrich and J. A. Swenberg, 'NCI-Black-Reiter (NBR) male rats fail to develop renal disease following exposure to agents that induce α2u-globulin (α2u) nephropathy.' *Fundam. Appl. Toxicol.*, 1991, **16**, 749–762.

53. A. K. Roy, O. W. Neuhaus and C. R. Harmison, 'Preparation and characterization of a sex-dependent rat urinary protein.' *Biochim. Biophys. Acta*, 1966, **127**, 72–81.

54. A. K. Roy and O. W. Neuhaus, 'Proof of the hepatic synthesis of a sex-dependent protein in the rat.' *Biochim. Biophys. Acta*, 1966, **127**, 82–87.

55. D. T. Kurtz, 'Rat α2u globulin is encoded by a multigene family.' *J. Mol. Appl. Genet.*, 1981, **1**, 29–38.

56. J. I. MacInnes, E. S. Nozik and D. T. Kurtz, 'Tissue-specific expression of the rat alpha 2u globulin gene family.' *Mol. Cell. Biol.*, 1986, **6**, 3563–3567.

57. C. V. Murty, F. H. Sarkar, M. A. Mancini *et al.*, 'Sex-independent synthesis of α2u-globulin and its messenger ribonucleic acid in the rat preputial gland: biochemical and immunocytochemical analyses.' *Endocrinology*, 1987, **121**, 1000–1005.

58. M. A. Mancini, D. Majumdar, B. Chartterjee *et al.*, 'α2u-Globulin in modified sebaceous glands with pheromonal functions: localization of the protein and its mRNA in preputial, meibomian and perianal glands.' *J. Histochem. Cytochem.*, 1989, **37**, 149–157.

59. D. T. Kurtz and P. Feigelson, in 'Biochemical Actions of Hormones,' ed. G. Litwack, Academic Press, NY, 1978, vol. V, pp. 433–455.

60. D. T. Kurtz, A. E. Sippel and P. Feigelson, 'Effect of thyroid hormones on the level of the hepatic mRNA for α2u-globulin.' *Biochemistry*, 1976, **15**, 1031–1036.

61. D. T. Kurtz, A. E. Sippel, R. Ansah-Yiadom *et al.*, 'Effects of sex hormones on the level of messenger RNA for the rat hepatic protein α2u-globulin.' *J. Biol. Chem.*, 1976, **251**, 3594–3598.

62. D. T. Kurtz, K. M. Chan and P. Feigelson, 'Translational control of hepatic α2u-globulin synthesis by growth hormone.' *Cell*, 1978, **15**, 743–750.

63. D. T. Kurtz, K. M. Chan and P. Fiegelson, 'Glucocorticoid induction of hepatic α2u-globulin synthesis and messenger RNA level in castrated male rats *in vivo*.' *J. Biol. Chem.*, 1978, **253**, 7886–7890.

64. A. K. Roy, B. Chatterjee, M. S. Prasad *et al.*, 'Role of insulin in the regulation of the hepatic messenger RNA for α2u-globulin in diabetic rats.' *J. Biol. Chem.*, 1980, **255**, 11614–11618.

65. A. E. Sippel, D. T. Kurtz, H. P. Morris *et al.*, 'Comparison of *in vivo* translational rates and messenger RNA levels of α2u-globulin in rat liver and Morris hepatoma 5123D.' *Cancer Res.*, 1976, **36**, 3588–3593.

66. A. K. Roy, D. M. McMinn and N. M. Biswas, 'Estrogenic inhibition of the hepatic synthesis of α2u globulin in the rat.' *Endocrinology*, 1975, **97**, 1501–1508.

67. A. K. Roy and O. W. Neuhaus, 'Androgenic control of a sex-dependent protein in the rat.' *Nature*, 1967, **214**, 618–618.

68. B. Chatterjee, W. F. Demyan, C. S. Song *et al.*, 'Loss of androgenic induction of α2u-globulin gene family in the liver of NIH black rats.' *Endocrinology*, 1989, **125**, 1385–1388.

69. L. D. Lehman-McKeeman and D. Caudill. 'Biochemical basis for mouse resistance to hyaline droplet nephropathy: lack of relevance of the α2u-globulin protein superfamily in this male rat-specific syndrome.' *Toxicol. Appl. Pharmacol.*, 1992, **112**, 214–221.

70. O. W. Neuhaus, W. Flory, N. Biswas *et al.*, 'Urinary excretion of α2u-globulin by adult male rats following treatment with nephrotoxic agents.' *Nephron*, 1981, **28**, 133–140.

71. O. W. Neuhaus, 'Renal reabsorption of low molecular weight proteins in adult male rats: α2u-globulin.' *Proc. Soc. Exp. Biol. Med.*, 1986, **182**, 531–539.

72. T. Maack, V. Johnson, S. T. Kau, *et al.*, 'Renal filtration, transport and metabolism of low-molecular-weight proteins: a review.' *Kidney Int.*, 1979, **16**, 251–270.

73. L. D. Lehman-McKeeman, and D. Caudill, 'Quantitation of α2u-globulin and albumin by reverse-phase

high performance liquid chromatography.' *J. Pharmacol. Methods*, 1991, **26**, 239–247.

74. A. Sivaprasadarao, M. Boujelal and J. B. C. Findlay, 'Lipocalin structure and function.' *Biochem. Soc. Trans.*, 1993, **21**, 619–622.

75. D. R. Flower, A. C. North and T. K. Attwood, 'Structure and sequence relationships in the lipocalins and related proteins.' *Protein Sci.*, 1993, **2**, 753–761.

76. J. Godovac-Zimmermann, 'The structural motif of β-lactoglobulin and retinol-binding protein: a basic framework for binding and transport of small hydrophobic molecules?' *Trends Biochem. Sci.*, 1988, **13**, 64–66.

77. L. D. Lehman-McKeeman and D. Caudill, 'α2u-Globulin is the only member of the lipocalin protein superfamily that binds to hyaline droplet inducing agents.' *Toxicol. Appl. Pharmacol.*, 1992, **116**, 170–176.

78. H. Kimura, S. Odani, J. I. Suzuki *et al.*, 'Kidney fatty acid-binding protein: identification as α2u-globulin.' *FEBS Lett.*, 1989, **246**, 101–104.

79. A. K. Roy, J. G. Byrd, N. M. Biswas *et al.*, 'Protection of spermatogenesis by α2u-globulin in rats treated with oestrogen.' *Nature*, 1976, **260**, 719–721.

80. J. A. Swenberg, B. Short, S. Borghoff *et al.*, 'The comparative pathobiology of α2u-globulin nephropathy.' *Toxicol. Appl. Pharmacol.*, 1989, **97**, 35–46.

81. L. D. Lehman-McKeeman, in 'Toxicology of the Kidney,' Raven Press, New York, 1993, pp. 477–494.

82. G. C. Hard, I. S. Rodgers, K. P. Baetcke *et al.*, 'Hazard evaluation of chemicals that cause α2u-globulin, hyaline droplet nephropathy, and tubule neoplasia in the kidneys of male rats.' *Environ. Health. Perspect.*, 1993, **99**, 313–349.

83. A. K. Roy, T. S. Nath, N. M. Motwani *et al.*, 'Age-dependent regulation of the polymorphic forms of α2u-globulin.' *J. Biol. Chem.*, 1983, **258**, 10123–10127.

84. D. R. Dietrich and J. A. Swenberg, 'NCI-Black Reiter (NBR) male rats fail to develop renal disease following exposure to agents that induce α2u-globulin (α2u) nephropathy.' *Fundam. Appl. Toxicol.*, 1991, **16**, 749–762.

85. B. Chatterjee, W. F. Demyan, C. S. Song *et al.*, 'Loss of androgenic induction of α2u-globulin gene family in the liver of NIH black rats.' *J. Endocrinology*, 1989, **125**, 1385–1388.

86. L. D. Lehman-McKeeman and D. Caudill, '*d*-Limonene-induced hyaline droplet nephropathy in α2u-globulin transgenic mice.' *Fundam. Appl. Toxicol.*, 1994, **23**, 562–568.

87. M. Charbonneau, E. A. Lock, J. Strasser *et al.*, '2,2,4-Trimethylpentane-induced nephrotoxicity. I. Metabolic disposition of TMP in male and female Fischer 344 rats.' *Toxicol. Appl. Pharmacol.*, 1987, **91**, 171–181.

88. E. A. Lock, M. Charbonneau, J. Strasser *et al.*, '2,2,4-Trimethylpentane-induced nephrotoxicity. II. The reversible binding of a TMP metabolite to a renal protein fraction containing α2u-globulin.' *Toxicol. Appl. Pharmacol.*, 1987, **91**, 182–192.

89. M. Charbonneau, J. Strasser, E. A. Lock *et al.*, 'Involvement of reversible binding for α2u-globulin in 1,4-dichlorobenzene-induced nephrotoxicity.' *Toxicol. Appl. Pharmacol.*, 1989, **99**, 122–132.

90. Z. Bocskei, C. R. Groom, D. R. Flower *et al.*, 'Pheromone binding to two rodent urinary proteins revealed by x-ray crystallography.' *Nature*, 1992, **360**, 186–188.

91. L. D. Lehman-McKeeman, M. I. Rivera-Torres and D. Caudill, 'Lysosomal degradation of α2u-globulin and α2u-globulin-xenobiotic conjugates.' *Toxicol. Appl. Pharmacol*, 1990, **103**, 539–548.

92. K. Saito, S. Uwagawa, H. Kaneko *et al.*, 'Behavior of α2u-globulin accumulating in kidneys of male rats treated with *d*-limonene: kidney-type α2u-globulin in the urine as a marker of *d*-limonene nephropathy.' *Toxicology*, 1991, **79**, 173–183.

93. L. D. Lehman-McKeeman, G. P. Daston, D. Caudill *et al.*, 'Renal functional changes associated with *d*-limonene-induced hyaline droplet nephropathy.' *Toxicologist*, 1991, **11**, 137.

94. L. D. Lehman-McKeeman, in 'Dose–response relationships for male rat specific α2u-globulin nephropathy and renal carcinogenesis. Monograph of the Cancer Dose Response Working Group', ILSI Press, 1995, pp. 175–183.

95. B. G. Short, V. L. Burnett and J. A. Swenberg, 'Elevated proliferation of proximal tubule cells and localization of accumulated α2u-globulin in F344 rats during chronic exposure to unleaded gasoline or 2,2,4-trimethylpentane.' *Toxicol. Appl. Pharmacol.*, 1989, **101**, 414–431.

96. T. Umemura, K. Tokumo and G. M. Williams, 'Cell proliferation induced in the kidneys and livers of rats and mice by short term exposure to the carcinogen *p*-dichlorobenzene.' *Arch. Toxicol.*, 1992, **66**, 503–507.

97. B. G. Short, W. H. Steinhagen and J. A. Swenberg, 'Promoting effects of unleaded gasoline and 2,4,4-trimethylpentane on the development of atypical cell foci and renal tubular cell tumors in rats exposed to *N*-ethyl-*N*-hydroexyethylnitrosamine.' *Cancer Res.*, 1989, **49**, 6369–6378.

98. D. R. Dietrich and J. A. Swenberg, 'The presence of α2u-globulin is necessary for a *d*-limonene promotion of male rat kidney tumors.' *Cancer Res.*, 1991, **51**, 3512–3521.

99. W. G. Flamm and L. D. Lehman-McKeeman, 'The human relevance of the renal turmor-inducing potential of *d*-limonene in male rats: implications for risk assessment.' *Regul. Toxicol. Pharmacol.*, 1991, **13**, 70–86.

100. G. C. Hard and J. Whysner, 'Risk assessment of *d*-limonene: an example of male rat-specific renal tumorigens.' *Crit. Rev. Toxicol.*, 1994, **24**, 231–254.

101. H. A. Solleveld, J. K. Haseman and E. E. McConnell, 'Natural history of body weight gain, survival, and neoplasia in the F344 rats.' *J. Natl. Cancer Inst.*, 1984, **72**, 929–940.

102. G. C. Hard, in 'Nephrotoxicity in the Experimental and Clinical Situation. Part 1,' eds. P. H. Bach and E. A. Lock, Martinus Nighoff, Boston, MA, 1987, pp. 211–250.

103. D. R. Dietrich and J. A. Swenberg, in 'Toxicology of the Kidney,' Raven Press, New York, 1993, pp. 495–537.

104. G. C. Hard, 'High frequency, single-dose model of renal adenoma/carcinoma induction using dimethylnitrosamine in Crl: (W)BR rats.' *Carcinogenesis*, 1984, **5**, 1047–1050.

105. T. Watabe, A. Hiratsuka, M. Isobe *et al.*, 'Metabolism of *d*-limonene by hepatic microsomes to non-mutagenic epoxides toward *Salmonella typhimurium*.' *Biochem. Pharmacol.*, 1980, **29**, 1068–1071.

106. T. Watabe, A. Hiratsuka, N. Ozawa *et al.*, 'A comparative study on the metabolism of d-limonene and 4-vinylcyclohex-1-ene by hepatic microsomes.' *Xenobiotica*, 1981, **11**, 333–344.

107. A. Busler, W. von der Hude and A. Seelbach, 'Genotoxicity of epoxides. I. Investigations with the SOS Chromotest and the Salmonella/mammalian microsome test.' *Mutagenesis*, 1989, **4**, 313–314.

108. W. von der Hude, A. Mateblowski and G. Obe, 'Genotoxicity of epoxides. II. *In vitro* investigations with the sister-chromatid exchange (SCE) test and the unscheduled DNA synthesis (UDS) test.' *Mutagenesis*, 1989, **4**, 323–324.

109. P. Grasso, 'Persistent organ damage and cancer production in rats and mice.' *Arch. Toxicol., Suppl.*, 1987, **11**, 75–83.

110. S. M. Cohen and L. B. Ellwein, 'Cell proliferation in carcinogenesis.' *Science*, 1990, **249**, 1007–1011.

111. B. A. Diwan, J. M. Rice, M. Ohshima *et al.*, 'Comparative tumor-promoting activities of phenobarbital, amobarbital, barbital sodium, and barbituric acid on livers and other organs of male F344/NCr rats following initiation with *N*-nitrosodiethylamine.' *J. Natl. Cancer Inst.*, 1985, **74**, 509–516.

112. D. E. Brooks, 'The major androgen-regulated secretory proteins of the rat epididymus bear sequence homology with members of the α2u-globulin superfamily.' *Biochem Int.*, 1987, **14**, 235–240.

113. S. A. Cowan, M. E. Newcomer and T. A. Jones, 'Crystallographic refinement of human serum retinol binding protein at 2 Å resolution.' *Proteins*, 1990, **8**, 44–61.

114. L. Sawyer, 'Protein structure. One fold among many.' *Nature*, 1987, **327**, 659.

115. P. H. Shaw, W. A. Held and N. D. Hastie, 'The gene family for major urinary proteins: expresson in several secretory tissues of the mouse.' *Cell*, 1983, **32**, 755–761.

116. W. S. Blaner, 'Retinol-binding protein: the serum transport protein for vitamin A.' *Endocr. Rev.*, 1989, **10**, 308–316.

117. M. Ganguly, R. H. Carnigham and U. Westphal, 'Steroid–protein interactions. XIV. Interaction between human α1-acid glycoprotein and progesterone.' *Biochemistry*, 1967, **6**, 2803–2814.

118. S. Urien, E. Albengres, R. Zini *et al.*, 'Evidence for binding of certain acidic drugs to α1-acid glycoprotein.' *Biochem. Pharmacol.*, 1982, **31**, 3687–3689.

119. A. Wishnia and T. W. Pinder, Jr., 'Hydrophobic interactions in proteins. The alkane binding site of β-lactoglobulins A and B.' *Biochemistry*, 1966, **5**, 1534–1542.

120. M. Z. Papiz, L. Sawyer, E. E. Eliopoulos *et al.*, 'The structure of β-lactoglobulin and its similarity to plasma retinol-binding protein.' *Nature*, 1986, **324**, 383–385.

121. S. H. Snyder, P. B. Sklar and J. Pevsner, 'Molecular mechanisms of olfaction.' *J. Biol. Chem.*, 1988, **263**, 13971–13974.

122. J. Pevsner, V. Hou, A. M. Snowman *et al.*, 'Odorant-binding protein. Characterization and ligand binding.' *J. Biol. Chem.*, 1990, **265**, 6118–6125.

123. D. T. Drayna, J. W. McLean, K. L. Wion *et al.*, 'Human apolipoportein D gene: gene sequence, chromosome localization, and homology to the α2u-globulin superfamily.' *DNA*, 1987, **6**, 199–204.

124. H. Schmale, H. Holtgreve-Grez and H. Christiansen, 'Possible role for salivary gland protein in taste reception indicated by homology to lipophilic-ligand carrier proteins.' *Nature*, 1990, **343**, 366–369.

125. C. Lopez Otin, A. O. Grubb and E. Mendez, 'The complete amino acid sequence of human complex-forming glycoprotein heterogeneous in charge (protein HC) from one individual.' *Arch. Biochem. Biophys.*, 1984, **228**, 544–554.

126. M. Julkunen, M. Seppala and O. A. Lanne, 'Complete amino acid sequence of human placental protein 14: a progesterone-regulated urine protein homologous to β-lactoglobulins.' *Proc. Natl. Acad. Sci. USA*, 1988, **85**, 8845–8849.

127. D. E. Brooks, A. R. Means, E. J. Wright *et al.*, 'Molecular cloning of the cDNA for two major androgen-dependent secretory proteins of 18.5 kilodaltons synthesized by the rat epididymus.' *J. Biol. Chem.*, 1986, **261**, 4956–4961.

128. D. R. Flower, 'The lipocalin protein family: a role in cell regulation.' *FEBS Lett.*, 1994, **354**, 7–11.

129. W. J. Henzel, H. Rodriguez, A. G. Singer *et al.*, 'The primary structure of aphrodisin.' *J. Biol. Chem.*, 1988, **263**, 16682–16687.

130. A. M. Spence, P. C. Sheppard, J. R. Davie *et al.*, 'Regulation of a bifunctional mRNA results in synthesis of secreted and nuclear probasin.' *Proc. Natl. Acad. Sci. USA*, 1989, **86**, 7843–7847.

131. J. A. Haefliger, M. C. Peitsch, D. E. Jenne *et al.*, 'Structural and functional characterization of complement C8γ, a member of the lipocalin protein family.' *Mol. Immunol.*, 1991, **28**, 123–131.

132. D. R. Flower, A. C. T. North and T. K. Attwood, 'Mouse oncogene protein 24p3 is a member of the lipocalin protein family.' *Biochem. Biophys. Res. Commun.*, 1991, **180**, 69–74.

133. A. Nagata, Y. Suzuki, M. Igarashi *et al.*, 'Human brain prostaglandin D synthase has been evolutionarily differentiated from lipophilic-ligand carrier proteins.' *Proc. Natl. Acad. Sci. USA*, 1991, **88**, 4020–4024.

134. B. Redl, P. Holzfeind and F. Lottspeich, 'cDNA cloning and sequencing reveals human tear pre-albumin to be a member of the lipocalin-ligand carrier protein superfamily.' *J. Biol. Chem.*, 1992, **267**, 20282–20287.

135. L. Kjeldsen, A. H. Johnsen, H. Sengelov *et al.*, 'Isolation and primary structure of NGAL, a novel protein associated with human neurophil gelatinase.' *J. Biol. Chem.*, 1993, **268**, 10425–10432.

136. P. Berman, P. Gray, E. Chen *et al.*, 'Sequence analysis, cellular localization and expression of a neuroretina adhesion and cell survival molecule.' *Cell*, 1987, **51**, 135–142.

137. F. D. Cancedda, B. Dozin, F. Rossi *et al.*, 'The Ch21 protein, developmentally regulated in chick embryo belongs to the superfamily of lipophilic molecule carrier proteins.' *J. Biol. Chem.*, 1990, **265**, 19060–19064.

138. H. M. Holden, W. R. Rypniewski, J. H. Law *et al.*, 'The molecular structure of insecticyanin from the tobacco hornworm *Manduca sexta* L. at 2.6 Å resolution.' *EMBO J.*, 1987, **6**, 1565–1570.

139. N. D. Hastie, W. A. Held and J. J. Toole, 'Multiple genes coding for the androgen-regulated major urinary proteins of the mouse.' *Cell*, 1979, **17**, 449–457.

140. L. D. Lehman-McKeeman, D. Caudill, R. Takigiku *et al.*, 'Comparative disposition of *d*-limonene in rats and mice: relevance to male-rat-specific nephrotoxicity.' *Toxicol. Lett.*, 1990, **53**, 193–195.

141. J. S. Finlayson, M. Potter and C. R. Runner, 'Electrophoretic variation and sex dimorphism of the major urinary protein complex in inbred mice: a new genetic marker.' *J. Natl. Cancer Inst.*, 1963, **31**, 91–107.

142. J. L. Knopf, J. F. Gallagher and W. A. Held, 'Differential, multihormonal regulation of the mouse major urinary protein gene family in the liver.' *Mol. Cell Biol.*, 1983, **3**, 2232–2240.

143. P. H. Rumke and P. J. Thung, 'Immunological studies on the sex-dependent prealbumin in mouse urine and its occurrence in the serum.' *Acta Endocrinol.*, 1964, **47**, 156–164.

144. G. L. Larsen, A. Bergman and E. Klasson-Wehler, 'A methylsulphonyl metabolite of a polychlorinated biphenyl can serve as a ligand for α2u-globulin in rat and major-urinary-protein in mice.' *Xenobiotica*, 1990, **20**, 1343–1352.

145. H. M. Liebich, A. Zlatkis, W. Bertsch *et al.*, 'Identification of dihydrothiazoles in urine of male mice.' *Biomed. Mass Spectrom.*, 1977, **4**, 69–72.

146. Z. Bocskei, J. B. Findlay, A. C. North *et al.*, 'Crystallization and preliminary x-ray data for the mouse major-urinary-protein and rat α2u-globulin.' *J. Mol. Biol.*, 1991, **218**, 699–701.

147. USEPA, 'α2u-Globulin: association with chemically induced renal toxicity and neoplasia in the male rat.' *Risk Assessment Forum.* 1991. EPA/625/3-91/019F.

Subject Index

Every effort has been made to index as comprehensively as possible, and to standardize the terms used in the index in line with the following standards:

EMTREE Thesaurus as a general guide to the selection of preferred terms.

IUPAC Recommendations for the nomenclature of chemical terms, with trivial names being employed where normal usage dictates.

In view of the diverse nature of the terminology employed by the different authors, the reader is advised to search for related entries under the appropriate headings.

The index entries are presented in letter-by-letter alphabetical sequence. Chemical terms are filed under substituent prefixes, where appropriate, rather than under the parent compound name; this is in line with the presentation given in the EMTREE Thesaurus.

The index is arranged in set-out style, with a maximum of three levels of heading. Location references refer to page number; major coverage of a subject is indicated by bold, italicized, elided page numbers; for example,

Risk assessment (RA), pesticides *1234–55*
    toxicological data 345

See cross-references direct the user to the preferred term; for example,

Vitamin A *See* Retinol

See also cross-references provide the user with guideposts to terms of related interest, from the broader term to the narrower term, and appear at the end of the main heading to which they refer; for example

Smoking
    See also Cigarette smoking; Pipe smoking